T0180346

Advanced Dynamics of Mechanical Systems

Federico Cheli · Giorgio Diana

Advanced Dynamics of Mechanical Systems

Springer

Federico Cheli
Department of Mechanical Engineering
Polytechnic University of Milan
Milan
Italy

Giorgio Diana
Department of Mechanical Engineering
Polytechnic University of Milan
Milan
Italy

ISBN 978-3-319-36761-3 ISBN 978-3-319-18200-1 (eBook)
DOI 10.1007/978-3-319-18200-1

Springer Cham Heidelberg New York Dordrecht London

Softcover reprint of the hardcover 1st edition 2015

Printed on acid-free paper

Springer International Publishing AG Switzerland is part of Springer Science+Business Media
(www.springer.com)

Preface

This book represents the natural evolution of the lecture notes of the course "Dynamics and Vibrations of Machines" held at the Politecnico di Milano in the academic years 1981–1992 and of a book already published for the course "Simulation and modelling of mechanical systems" (academic years 1993–2014).

These collected works can be considered as a natural extension of the didactic work carried out in this area, initially by Prof. O. Sesini and later by Profs. A. Capello, E. Massa and G. Bianchi. The contents of this book also sum up decades of experience gained by the research group which is part of the Section of Mechanical Systems at the Politecnico di Milano (former Institute of Applied Mechanics). It also draws upon the research topics developed by a research group of the Department of Mechanics, to which the authors belong. Said research was generally based on problems encountered in the industrial world, performed in collaboration with organizations and research centres including ABB, ENEL (Italian General Electricity Board), FS (Italian Railways), Bombardier, Alstom, Ansaldo, ENEL-CRIS, ISMES, Fiat, Ferrari and Pirelli, as well as countless others. In this context, the following research topics were considered of prime importance:

- analytical and experimental investigations on the vibration of power lines;
- slender structures—wind interaction;
- aeroelastic behaviour of suspension bridges;
- dynamic behaviour of structures subjected to road and rail traffic;
- rail vehicle dynamics, pantograph—catenary interaction, train—railway infrastructure interaction, etc.;
- ground vehicle dynamics; and
- rotor dynamics.

These themes impacted significantly on the development of this book.

The educational content of this volume is primarily addressed to students of engineering taking courses in mechanics, aerospace, automation and energy, disciplines introduced recently by the Italian Ministry of Education in compliance with the New Italian University Order. However, given its organic structure and the comprehensive overview of the subjects dealt with, the book could also serve as a useful tool to professionals in the industry.

In this book, an engineering approach for the schematization of a generic mechanical system, applicable both to rigid and deformable bodies, is introduced. Such an approach is necessary to identify the behaviour of a mechanical system subject to different excitation sources. In addition to the traditional aspects associated with the dynamics and vibrations of mechanical systems, the engineering approach illustrated herein allows us to reproduce the interaction of mechanical systems with different force fields acting on its various components (e.g. action of fluids and contact forces), i.e. forces dependent on system motion, and, consequently, its state.

This concept, dealing with the interaction of force fields and mechanical systems, gives rise to a new system on which the dynamic behaviour is considered, focusing, in particular, on the analysis of motion stability.

Controlled systems, in which the action of the actuator, controlled in a closed loop, defines forces as a function of the state of the system, can also be assimilated to systems interacting with force fields and, for this reason, dealt with in a similar way.

Traditionally, however, there are typical approaches in this area that cannot be ignored and, for this reason, controlled systems are treated in a separate text.[1]

In this text, however, an effort has been made to reference the symbols and main techniques used in control engineering, in order to create an easy interface for mechanical engineers dealing with electronic control.

More specifically, in the first part of this book, we will analyse mechanical systems with 1 or more degrees of freedom (d.o.f.), generally in large motion and, subsequently, the small motion of systems in the neighbourhood of either the steady-state motion or the static equilibrium position. In this phase, we will analyse both discrete and continuous systems, for which certain discretization procedures will be discussed (modal approach, finite elements).

Conversely, the second part of this text deals with the study of mechanical systems subjected to force fields, with many examples such as fluid–elastic interaction, train and railway interaction, rotor dynamics, experimental techniques related to parameter identification and random excitations.

The first part of the text can be a useful tool for undergraduate courses to approach the dynamics and the vibration problems in the mechanical systems.

The second part is more suitable for graduate and Ph.D. students to analyse many real problems due to the interaction of mechanical systems with different surrounding fields of forces. The main problems related to the behaviour and stability of these systems are fully described in the last part of the book and will be very useful for the students.

---

[1]Diana and Resta [1].

We would like to extend our special thanks to all the lecturers and researchers of the Section of System Mechanics of the Department of Mechanics at the Politecnico di Milano for all their help and input provided during the drafting of this book.

The authors would also like to express their gratitude especially to Professor Bruno Pizzigoni for his hard and excellent work for the audit and the check of the English text. It goes without saying that, as always, there are likely to be omissions and errors for which we hope you will forgive us.

Federico Cheli
Giorgio Diana

Reference

1. Diana G, Resta F (2006) Controllo dei sistemi meccanici. Polipress, Milano

Contents

Introduction

Dynamic analysis is a necessary step to design, verify, edit and then operate, diagnose and monitor a generic mechanical system. The term "dynamic analysis" refers to the study aimed at identifying the dynamic behaviour (displacements, velocities, accelerations, strains and stresses) of the different components of a mechanical system subject to the different forces (excitation causes) occurring under normal operating conditions. Mechanical systems are generally constituted by mutually interconnected bodies subjected to forces that can be explicit functions of time (i.e. independent from the dynamics of the system itself) or forces that, generically, depend on time and possibly on the motion of the system itself. In the latter case, we speak generically of "force fields". Figure 1 shows some examples of mechanical systems in the presence of force fields.

Generically speaking, mechanical systems can perform large motions or small motions about a position of static equilibrium (rest): this allows for the immediate classification of mechanical systems into two different categories:

- "structures", i.e. systems that admit a static equilibrium position (rest position) about which "small motions" are studied;
- "machines", i.e. systems in which this type of rest position is not always present, and which, therefore, either have generic large motions or are in "steady-state" motion.

For the structures, motion is permitted by the deformability of the various components: in this case, the dynamic analysis will concern the vibratory motion about the equilibrium position.

Conversely, even without considering the deformability of their components, machines possess motion: typical examples are a shaft rotating around its own axis, a road or rail vehicle (Fig. 1), the slider–crank mechanism of an internal combustion engine or any machine element in motion (articulated systems, robots, etc.).

To evaluate the behaviour of a generic mechanical system at design level or during tuning operation, it is necessary to realize a mathematical model that attempts to simulate its dynamic behaviour.

Fig. 1 Some examples of mechanical systems subject to force fields: high speed trains, gas turbines and new skyscrapers in Milan

The schematizations that can be adopted in the creation of mathematical models of mechanical systems can be more or less complex, depending on the type of problem that needs to be solved. Mathematical models may be more or less accurate:

- rigid body models [with one or more degree of freedom (d.o.f.)];
- deformable bodies (i.e. with infinite degrees of freedom);
- linear models or models linearized about an equilibrium condition (rest or steady state) and;
- nonlinear models to simulate generically large motions.

In some cases, a scheme of interconnected rigid bodies can be sufficiently accurate; conversely, in others, it is necessary to consider the distributed deformability of the various elements. As an example, the analysis of the dynamics of motion of a rotor subjected to torques and resistant moments can be carried out by considering the rotor as rigid while for the study of bending and torsional motions around the steady-state speed of rotation, it is necessary to keep account of the deformability of the rotor itself.

In addition to the mechanical system, it is also necessary to model the forces that are applied to it. In this case too, different schematization "levels" of these forces exist:

- models in which constraints are considered as ideal;
- models where real constraints are considered, in which the contact forces due to the same constraints can be attributed to force fields;
- models in which the interactions between any pair of bodies or between body and fluid are reproduced through the definition of force fields; and
- models in which a control action is taken into account and where both the control logic and the actuators, used to impose the necessary forces, have to be described.

Therefore, in the analysis of a dynamic system, an in-depth knowledge of the different force fields acting on the various elements is required (see Fig. 1, fluid action, contact forces between bodies, action of electromagnetic fields, etc.), since such fields can significantly affect the behaviour of the system itself.

An approach of this kind leads to an accurate, and, as such, complex modelling. Let us consider, as an example, the slider–crank mechanism of internal combustion engines: despite considering the crank, connecting rod and piston as rigid bodies, each of them should be allocated (though only considering plane motion for the sake of simplicity) three degrees of freedom, allowed by the deformability of the lubricant films present in the various pairs. Due to the presence of fluid in the various mechanism components (lubricant in the couplings, fluid contained in the combustion chamber subject to thermodynamic transformation), force fields arise as a function of their positions and relative speeds. To write the equations of motion of the system, the definition of these force fields is essential. Depending on the particular nature of the problem considered and the aim of the analysis, more simple models can be obtained. If, in the previous example, the law of motion of the piston has to be calculated, it seems sufficiently accurate to neglect the relative motions due to the deformability of the lubricant fluid films in the mechanism pairs. In this case, the mechanism is reduced to a one degree-of-freedom system. If the objective of the analysis is, for example, to define the pressure and temperature distributions on the piston skirt, as previously mentioned, system modelling must be carried out.

Once the overall mathematical model of the mechanical system has been defined (i.e. mechanical system model and applied forces), the equations of motion must be written and solved to evaluate the dynamic response of same.

To write the equations of motions, various methods can be applied. Methods that best lend themselves to a systematic analysis of various problems are definitely those of dynamic equilibrium and Lagrange's equations:

- dynamic equilibrium can definitely be used for simple rigid body schematizations;
- in systems subject to ideal constraints, by using Lagrange's method, it is not necessary to introduce constraint reactions;
- Lagrange's equations must be used if body deformability is considered, e.g. by using a discretization method of the finite element type.

The differential equations that describe systems subject to large motions, obtained using the methods described above, appear to be nonlinear. Small motions about an equilibrium position are described by linear differential equations.

When analysing machines, it may often be useful to linearize the equations of motion about a static or dynamic equilibrium condition (i.e. about a rest or steady-state condition), like, for example, when calculating natural frequencies or analysing incipient instability conditions.

At this point, the choice of the algorithm used to solve the equations of motion becomes important: in the case of linear or linearized systems, the analysis often provides solutions to various problems in a closed form while, for nonlinear systems, a solution has to be found by using numerical techniques.

As regards the analysis of a generic mechanical system, a fundamental aspect that will be taken into account in this book is the definition of a systematic approach to the schematization of the system itself. The study of the dynamics of systems with rigid or deformable interconnected bodies has had recent developments, based on the use of coordinate transformation matrices, thus giving rise to a method, now known as the "Multi-Body System Method", particularly suited to computer implementation in order to represent the kinematics of the various elements.[2]

The study of a dynamics problem consists in writing the equations of motion—generally nonlinear—in finding an existing equilibrium condition and in linearizing the equations of motion around this condition, for stability analysis.

Dynamic analysis is also aimed at studying the transient motion performed to possibly attain a steady-state situation or limit cycle. Stability analysis is usually

[2]Cheli and Pennestrì [1], Shabana [2].

carried out by using a linear approach; conversely, the study of the transient motion or the determination of any limit cycles requires the solution of nonlinear problems.

For example, when wishing to tackle problems related to the dynamics of a vehicle in motion on a straight road or on a curve, once the equations of motion have been obtained, it is first necessary to find a steady-state solution (i.e. a vehicle moving at a constant speed on a straight line or on a curve), subsequently going on to analyse the perturbed motion around the main trajectory. This analysis may either be performed by means of equations linearized around steady-state conditions, or by integrating nonlinear equations with assigned initial conditions: in this way, in addition to the analysis of the stability, any "large" motion of the vehicle can be defined. Although it is not easy to make a clear distinction of subjects, it can be stated that the analysis of large motion is a subject of "Applied Mechanics",[3] although in Applied Mechanics courses simplified rigid body schematizations, typically either with one or a few degree of freedom, are usually presented.

Conversely, in this book, the foundations for an engineering approach to the general problem are given, mainly by investigating the aspects associated with the analysis of the stability and vibrations of mechanical systems. In the first part of the text, we will consider the small motions of systems arising from perturbations of stable static equilibrium conditions, which represent the classic study of vibrating systems. By considering an energy approach (Lagrange's equations), the forms of energy involved in the vibratory phenomenon are kinetic energy due to the mass of the system, the elastic potential energy, possibly gravity and, finally, energy dissipation either due to the imperfect elasticity of the materials or comparable to viscous effects. Mechanical systems falling into this category are termed dissipative. For this class of systems, we will consider motions arising from small perturbations of static equilibrium conditions, perturbed free motions and the forced motions due to external excitation forces, generally functions of time.

These systems will be analysed in order of complexity, starting gradually from systems with one degree of freedom, systems with 2-n degree of freedom (in this context, multi-body methodologies for writing equations of motion will be mentioned, see Chaps. 1 and 2), finally extending the discussion to systems with infinite degree of freedom (continuous systems, Chap. 3).

As far as the latter are concerned, procedures pertaining to their discretization will be described, with particular reference to the modal approach (Chap. 3), the finite element technique (Chap. 4) and, finally, the identification techniques of modal parameters from experimental tests (Chap. 8).

The second part of this book will be devoted to the study of mechanical systems subjected to force fields (Chap. 5): In addition to the possible steady-state solution, the perturbed motion about same will be also studied. To achieve this, it is

[3]Bachschmid et al. [3].

necessary to define the motion of the system in a more complete and complex form than that used in the classical treatises of Applied Mechanics, i.e. by also considering the deformability of the system, in the event of this being deemed important to define its dynamic behaviour. The approach is of the general type: after writing the equations of motion, a steady-state solution is looked for. If such a solution is found, the subsequent dynamic analysis can be performed by linearizing the system about this steady-state solution in order to check stability. The linearized equations may present constant coefficients or coefficients as functions of time. The methodology of analysis depends on the structure of these equations: more in particular, for the first category of equations, a systematic study is possible, while, with coefficients as functions of time, analysis procedures are more complex. The dynamic behaviour of the generic system can also be analysed by using a nonlinear approach consisting of a numerical integration of the equations of motion or, for some kind of nonlinear problems, of the use of approximate analytical methods. To write the equations that govern the mechanical system in its most general form, it is necessary to consider not only the kinetic energy, potential energy and dissipative function, but also the presence of force fields surrounding the system: in addition to the elastic one, force fields due to the action of a fluid, contact forces between bodies or electromagnetic fields may be present. The forces that the various elements of a mechanical system exchange, not only with each other but also with the surrounding environment, are generally functions of the independent variables that define the motion of the system itself and their derivatives. In a broad sense, these interactions are considered to be due to "force fields". Generally speaking, these force fields (Chap. 5) are nonlinear and characterize system behaviour significantly.

If the forces are solely functions of the configuration of the system, they will be referred to as "positional" force field; on the contrary, if they depend on the velocity of the system, they will be termed "velocity" force field (force fields that have not been dealt with in other courses will be described briefly in the same chapter).

Not only does the presence of force fields condition steady state but also free motion, influencing system stability: for example, owing to the fact that positional non-conservative force fields are able to introduce energy into the system, they can generate forms of instability. Velocity force fields are non-conservative by definition, meaning that they are also capable of modifying the stability of a system. The development of the study of mechanical systems subjected to force fields, in addition to those regarding elastic and dissipative ones, is common to many disciplines and falls within the analysis of mechanical systems in a more general sense.

Controlled systems can also be classified in this category, in the sense that a control system applies forces proportional to the values of the independent variables or to the difference between a reference quantity value and the actual value of same (function of the independent variables). From this point of view, control systems can also be considered as force fields. However, although this course is not aimed at the systematic analysis of control problems, it definitely lays the foundations for

[3] Bachschmid et al. [3].

control engineering, with which it shares innumerable problems and solution methods, at least as far as the study of stability is concerned. Therefore, where possible, an effort will be made to approach the structure of equations and the symbolism adopted in the field of control engineering, so as to deal with the study of mechanical systems in a similar way to that of controlled systems. The equations of motion of a mechanical system are generally written in the form:

$$[M]\ddot{\underline{x}} + [R]\dot{\underline{x}} + [K]\underline{x} = \underline{F}(\underline{x}, \dot{\underline{x}}, \ddot{\underline{x}}, t) \tag{1}$$

where $\underline{x}$ is the vector of independent variables, $\underline{F}$ the applied external forces (due to force fields or known functions of time or due to the action of a controller), $[M]$, $[R]$ and $[K]$ the mass, damping and stiffness matrices. By linearizing these equations about an equilibrium or steady-state position, in addition to the structural matrices, terms arising from the linearization of the force fields also appear. These terms are called the Jacobians of the force fields: the equation of motion in this case can be rewritten in the form:

$$[[M] + [M_F]]\ddot{\underline{x}} + [[R] + [R_F]]\dot{\underline{x}} + [[K] + [K_F]]\underline{x} = \underline{F}(t) \tag{2}$$

where $[M_F]$, $[R_F]$ and $[K_F]$ are the matrices arising from the linearization of the force fields while, on the right hand side, only the external forces, explicit functions of time, remain. The analysis of the overall structure of the matrices (whether symmetric, definite positive or not) allows us to check system stability. An approach of this type (see Eq. 2) is typical of mechanics; however, the same equations can be rewritten in the equivalent form:

$$\dot{\underline{z}} = [A]\underline{z} + \underline{u}(t) \tag{3}$$

in which $[A]$ is formed by the matrices $[M]$, $[R]$ and $[K]$: in turn, this type of matrix can be a function of time if the equations of motion arising from the linearization do not present constant coefficients. In Eq. (3), the vector of so-called state variables is indicated:

$$\underline{z} = \begin{bmatrix} \dot{\underline{x}} \\ \underline{x} \end{bmatrix} \tag{4}$$

while $\underline{u}(t)$ is the vector of known terms, solely a function of time, owing to the fact that the state-dependent control forces are already included in a linear way in $[M_F]$, $[R_F]$ and $[K_F]$. The two approaches (1) and (3) only differ in terms of a symbolic aspect, even though matrix $[A]$ loses the physical meanings of the problem with respect to matrices $[M]$, $[R]$ and $[K]$.

More in detail, the second part (Chap. 5) is structured as follows:

- a general discussion of the problem of systems subjected to force fields;

- an analysis of systems subjected to positional force fields, differing from gravitational and elastic ones, and a discussion of stability for systems with one and two degrees of freedom;
- a steady-state solution and linearization of the force field;
- analysis of systems with one or two degrees of freedom subjected to positional and velocity force fields; and
- the development of several examples including the analysis of a 2D airfoil flow, motion of a journal inside a bearing with hydrodynamic lubrication, motion of a train axle and of a road vehicle, or extensions to continuous systems (finite element models).

In dedicated chapters, the course will also deal with problems related to:

- rotor dynamics (balancing, oil film instability, interaction with the foundation, etc., Chap. 6);
- the definition of different types of random excitation forces including those due to turbulent wind, waves and earthquakes (Chap. 7): in this chapter, the problem of vortex shedding will be illustrated, referencing the more general problem of vibrations induced by fluids, outlined in previous chapters; and
- the experimental identification of parameters of a real system (Chap. 8): this aspect is fundamental in modelling.

The creation of mathematical models, targeted at defining the dynamic behaviour of mechanical systems in the terms mentioned above, is a well-established discipline. Algorithms developed to simulate, as accurately as possible, the behaviour of structures and machines subjected to different excitation causes have become essential tools for the design and operation of such systems.

References

1. Cheli F, Pennestrì E (2006) Cinematica e dinamica dei sistemi multibody. Casa Editrice Ambrosiana, Milano
2. Shabana AA (2005) Dynamics of multibody systems, 3rd edn. Cambridge University Press
3. Bachschmid N, Bruni S, Collina A, Pizzigoni B, Resta F (2003) Fondamenti di meccanica teorica ed applicata. McGraw-Hill

Chapter 1
Nonlinear Systems with 1-n Degrees of Freedom

1.1 Introduction

Our aim is to simulate the behaviour of systems with 1 or n degrees of freedom, which undergo large motion or rather in which the displacements of the single bodies composing the system itself are not small compared with the body dimensions: under these conditions it is not possible to linearize the system. The aim of this chapter is not only to provide the necessary instruments for the analysis of motion in the large which, as will be seen, always gives rise to nonlinear differential equations but also to show how the linearization of these equations can be achieved within about an equilibrium or steady-state configuration, in case of this existing. Conversely, in the following chapter we will deal with linear or linearized systems within the neighbourhood of an equilibrium position (rest or steady-state).

1.2 Cartesian Coordinates, Degrees of Freedom, Independent Coordinates

The configuration of a generic multi-body system with rigid bodies (such as the one shown in Fig. 1.1) can be described by the displacements (translations and rotations), velocities and accelerations of the single bodies defined with respect to a reference system: generally speaking, two types of reference systems [1] are necessary for a multi-body system:

- A global or inertial reference system $(O\text{-}X\text{-}Y\text{-}Z)$;
- A local reference system $(O_c\text{-}X_c\text{-}Y_c\text{-}Z_c)$, that is connected to each body, so that this reference system translates and rotates with the body itself.

Without loss of generality, it is possible to assume the origin of the local reference system O_c coinciding with centre of mass G_c of the generic body and the local axes parallel to the main axes of inertia [2–4], Fig. 1.2.

F. Cheli and G. Diana, *Advanced Dynamics of Mechanical Systems*,
DOI 10.1007/978-3-319-18200-1_1

Fig. 1.1 A typical multi-body system

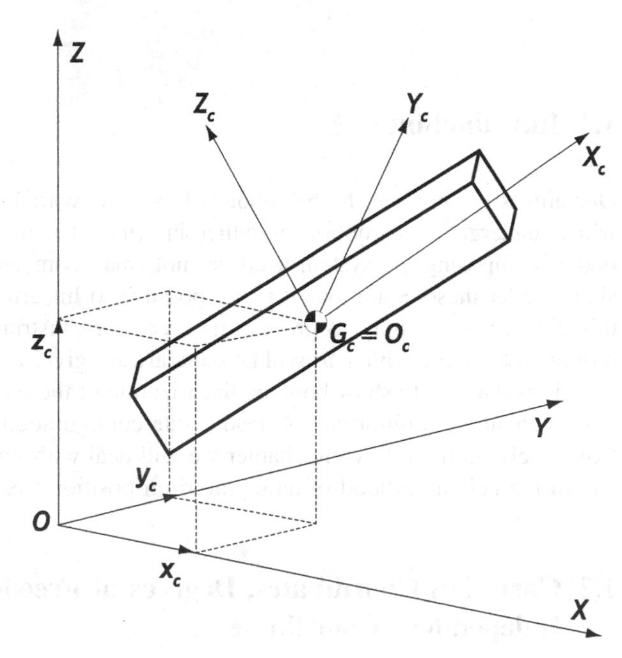

Fig. 1.2 Global (O-X-Y-Z) and local (O_c-X_c-Y_c-Z_c) reference systems

The configuration assumed by each rigid body in space can be identified by 6 Cartesian coordinates $\underline{q}_c$:

$$\underline{q}_c = \begin{Bmatrix} x_c \\ y_c \\ z_c \\ \rho_c \\ \beta_c \\ \sigma_c \end{Bmatrix} \tag{1.1}$$

three of which (x_c, y_c, z_c) describe the translation of the body's centre of mass (i.e. the origin of the moving reference frame) with respect to the absolute reference system (O-X-Y-Z), while the other three coordinates $(\rho_c, \beta_c, \sigma_c)$ define the orientation in space of the body, i.e. the orientation of the moving reference frame with respect to the inertial reference frame. It is possible to assume Euler angles, Cardan angles, Rodriguez parameters and quaternions (as will be described in detail further on)[1] as rotation coordinates [1, 2, 5]. The complete system, constituted by n_c rigid bodies can thus be described by n_t Cartesian coordinates with:

$$n_t = 6 \cdot n_c \tag{1.2}$$

formally defining the vector $\underline{q}_t$ of the total Cartesian coordinates:

$$\underline{q}_t = \begin{Bmatrix} \underline{q}_1 \\ \underline{q}_2 \\ \cdots \\ \underline{q}_c \\ \cdots \\ \underline{q}_{nc} \end{Bmatrix} \tag{1.3}$$

Considering the single bodies as separate from each other and not constrained to the ground, the equations of motion of a generic mechanical system can be obtained by means of Lagrange's equations[2] [3–5]:

$$\left\{\frac{d}{dt}\left(\frac{\partial E_c}{\partial \underline{\dot{q}}_t}\right)\right\}^T - \left\{\frac{\partial E_c}{\partial \underline{q}_t}\right\}^T + \left\{\frac{\partial D}{\partial \underline{\dot{q}}_t}\right\}^T + \left\{\frac{\partial V}{\partial \underline{q}_t}\right\}^T = \underline{Q}_t \tag{1.4}$$

where E_c, D and V respectively represent the kinetic energy, the dissipation function and the potential energy of the system and, finally, $\underline{Q}_t$ the vector of the nonconservative forces obtained from the virtual work $\delta^* L$ performed by all the forces not considered in the various forms of energy in (1.4)[3]:

---

[1]When writing the equations of motion of a multi-body system, it is also possible to assume, as independent variables, those associated with the relative motion of one reference frame with respect to another.

[2]The equations of motion can obviously also be obtained by means of other methodologies [1, 2, 5], such as, for example, the cardinal equations of dynamics.

[3]The virtual work, in scalar form defined as:

$$\delta^* L = \sum_{i=1}^{nt} Q_i \delta * q_i \tag{1.3.1}$$

Having organized both the independent coordinates q_i $(i = 1, nt)$ and the Lagrangian components Q_i $(i = 1, nt)$ in column matrices:

$$\underline{q}_t^T = \{q_1 \quad \ldots \quad q_i \quad \ldots \quad q_{nt}\}; \quad \underline{Q}_t^T = \{Q_1 \quad \ldots \quad Q_i \quad \ldots \quad Q_{nt}\} \tag{1.3.2}$$

can be rewritten in matrix form as:

$$\delta^* L = \underline{Q}_t^T \delta^* \underline{q}_t \tag{1.3.3}$$

$$\delta^* L = \underline{Q}_t^T \, \delta^* \, \underline{q}_t \tag{1.5}$$

where $\delta^* \, \underline{q}_t$ is the virtual displacement of the Cartesian coordinates.[4]

The constraints to which the mechanical system is subjected reduce its possibilities of motion and, as a consequence, reduce its degrees of freedom. The kinematic constraint conditions can be described in terms of n_v nonlinear algebraic equations of the type[5, 6]:

$$\underline{f}_v(\underline{q}_t) = \underline{0} \tag{1.6}$$

which are able to represent the fact that the generic body comprising the system as a whole is constrained to the ground in one or two points or that relative constraints exist between different bodies.

In this case, the total Cartesian coordinates are no longer all independent one from the other: the number of free coordinates of the system (corresponding to the

[4]In the matrix formulation adopted, we used the convention whereby the derivative of a generic scalar s with respect to a vector $\underline{v}$ (column matrix of n terms) is defined by means of a vector line consisting of n terms:

$$\frac{\partial s}{\partial \underline{v}} = \left\{ \frac{\partial s}{\partial v_1} \quad \frac{\partial s}{\partial v_2} \quad \cdots \quad \frac{\partial s}{\partial v_{n-1}} \quad \frac{\partial s}{\partial v_n} \right\} = \underline{v}_s^T \tag{1.4.1}$$

Consequently, the derivative of a generic vector $\underline{v}$ (consisting of n terms) with respect to another just as generic vector $\underline{w}$ (consisting of m terms) is a matrix (n m) defined as:

$$\frac{\partial \underline{v}}{\partial \underline{w}} = \begin{bmatrix} \frac{\partial v_1}{\partial w_1} & \frac{\partial v_1}{\partial w_2} & \cdots & \cdots & \frac{\partial v_1}{\partial w_{m-1}} & \frac{\partial v_1}{\partial w_m} \\ \frac{\partial v_2}{\partial w_1} & \frac{\partial v_2}{\partial w_2} & \cdots & \cdots & \frac{\partial v_2}{\partial w_{m-1}} & \frac{\partial v_2}{\partial w_m} \\ \cdots & \cdots & \cdots & \cdots & \cdots & \cdots \\ \cdots & \cdots & \cdots & \cdots & \cdots & \cdots \\ \frac{\partial v_{n-1}}{\partial w_1} & \frac{\partial v_{n-1}}{\partial w_2} & \cdots & \cdots & \frac{\partial v_{n-1}}{\partial w_{m-1}} & \frac{\partial v_{n-1}}{\partial w_m} \\ \frac{\partial v_n}{\partial w_1} & \frac{\partial v_n}{\partial w_2} & \cdots & \cdots & \frac{\partial v_n}{\partial w_{m-1}} & \frac{\partial v_n}{\partial w_m} \end{bmatrix} = [v_w] \tag{1.4.2}$$

[5]For example, a spherical hinge on the ground means that, in the point where the hinge is positioned, the three components of absolute displacement of the body are null and this condition is analytically introduced by means of three constraint equations.

[6]Let us remember that, generally speaking, the equations of kinematic constraint can also be a function of time, i.e.:

$$\underline{f}_v(\underline{q}_t, t) = \underline{0} \tag{1.6.1}$$

In this particular text, we will not consider constraints whose characteristics change over time, or rather, we will only analyze the condition of holonomic constraints.

actual degrees of freedom), which we will subsequently indicate by means of $\underline{q}_l$, is defined by the following relation[7]:

$$n_l = n_t - n_v = 6 \cdot n_c - n_v \tag{1.7}$$

By using this approach, vector $\underline{q}_t$ of all the Cartesian coordinates (Eq. (1.3)) can be divided into two sub-vectors respectively containing the free coordinates $\underline{q}_l$ and the dependent coordinates $\underline{q}_d$:

$$\underline{q}_t = \begin{Bmatrix} \underline{q}_l \\ \underline{q}_d \end{Bmatrix} \tag{1.8}$$

Imposing constraints (1.5) on the system involves a relationship between the dependent $\underline{q}_d$ and independent variables $\underline{q}_l$ formally of the type:

$$\underline{q}_d = \underline{q}_d(\underline{q}_l) \tag{1.9}$$

This relationship can be expressed in terms of velocity, by directly deriving (1.9) with respect to time:

$$\underline{\dot{q}}_d = \left[\frac{\partial \underline{q}_d}{\partial \underline{q}_l}\right] \underline{\dot{q}}_l \tag{1.10}$$

thus obtaining a linear relationship (defined by means of the Jacobian $\left[\frac{\partial \underline{q}_d}{\partial \underline{q}_l}\right]$) between the velocities of the free coordinates and the dependent ones.

The equations of motion of a generic mechanical system in which constraints are considered prove to be a system of hybrid equations, partially non-linear differential (1.4) and partially non-linear algebraic (1.6):

$$\begin{cases} \left\{\frac{d}{dt}\left(\frac{\partial E_c}{\partial \underline{\dot{q}}_t}\right)\right\}^T - \left\{\frac{\partial E_c}{\partial \underline{q}_t}\right\}^T + \left\{\frac{\partial D}{\partial \underline{\dot{q}}_t}\right\}^T + \left\{\frac{\partial V}{\partial \underline{q}_t}\right\}^T = \underline{Q}_t \\ \underline{f}_v(\underline{q}_t) = \underline{0} \end{cases} \tag{1.11}$$

[7]The number of Cartesian coordinates and free coordinates and, therefore, the number of degrees of freedom, of a multi-body mechanical system depends on the schematization with which we wish to simulate the dynamics of the system which, in turn, depends on the complexity of the problem that we wish to solve and on the level of schematization adopted for the model, as well as on the constraints to which this is subjected. This aspect of the problem will be dealt with in the chapter on fields of force (Chap. 5).

In this text, as regards the various techniques available in literature [1, 2, 6, 7] for the numerical solution of the hybrid system (1.11) including the differential equations of motion and the nonlinear algebraic equations, reference will be made to the three most widely used methods:

- the *method of Lagrange multipliers*, in which all the $\underline{q}_t$ are considered as independent coordinates $\underline{q}$

$$\underline{q} = \underline{q}_t \tag{1.12}$$

By using this approach, the constraints are removed and the work of the constraint forces $\underline{R}$, (in a number equal to n_v), is added into the equations. By using this approach, the n_v constraint equations are added to the system equations of motion ($6 \cdot n_c$ differential equations) thus obtaining $n_t + n_v$ equations, in the unknowns $\underline{q}_t + \underline{R}$. As will be seen further on, by deriving the constraint equations with respect to time, it is possible to obtain a single system of differential equations.

- the *minimal set* method in which the constraint equations (1.6) are used to directly reduce the degrees of freedom of the system: assuming only the free coordinates as independent coordinates:

$$\underline{q} = \underline{q}_l \tag{1.13}$$

By using this approach, the total number of differential equations to be solved is equal to the number of degrees of freedom of system n_l and the reaction forces due to constraints do not appear. Using this method, in order to apply Lagrange's equations:

$$\left\{\frac{d}{dt}\left(\frac{\partial E_c}{\partial \underline{\dot{q}}_l}\right)\right\}^T - \left\{\frac{\partial E_c}{\partial \underline{q}_l}\right\}^T + \left\{\frac{\partial D}{\partial \underline{\dot{q}}_l}\right\}^T + \left\{\frac{\partial V}{\partial \underline{q}_l}\right\}^T = \underline{Q}_l \tag{1.14}$$

it is necessary to introduce the relationship between dependent and free coordinates (1.9).

- *method involving the introduction of real constraints by means of fields of forces*: this approach is used in more refined schematization, in which ideal constraints no longer exist and the external and internal constraint reactions exchanged between the bodies are defined, as specified further on, as fields of forces which are functions of the independent variables themselves (Chap. 5).

Before attempting a more in-depth description of the methodologies necessary to write the equations of motion of a generic mechanical system, it might be useful to use an example to clarify the concept of Cartesian coordinates and independent coordinates.

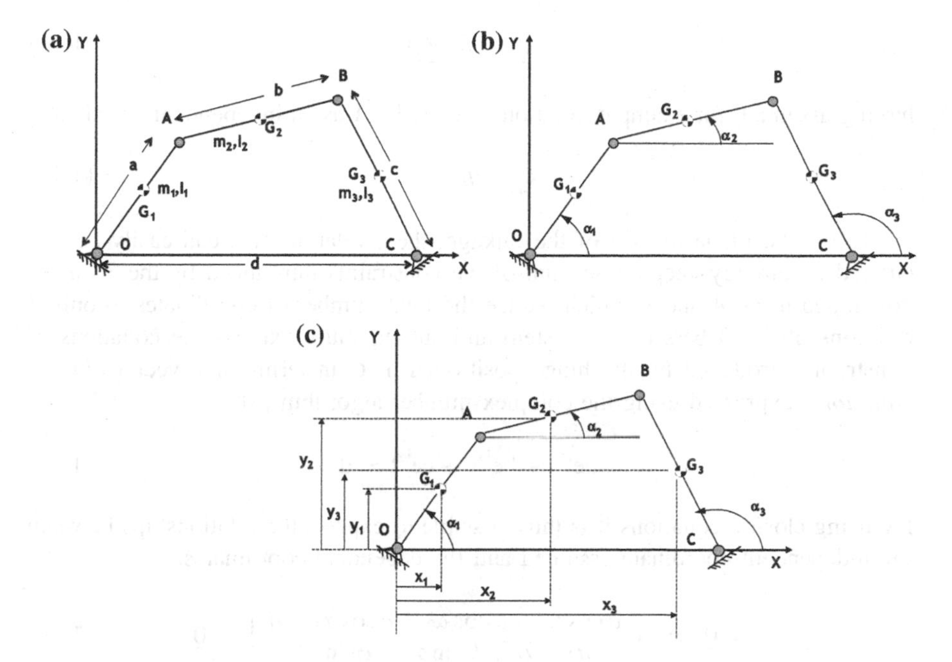

Fig. 1.3 Articulated quadrilateral: kinematic analysis

For this purpose, for reasons of simplicity, let us consider a flat system constituted by a plane four bar linkage, as shown in Fig. 1.3a. Members of the four bar linkage will be termed a, b and c while the corresponding grounded bar will be termed d: the centres of gravity of the moving bars, each having a mass, $m_i(i = 1, 2, 3)$ and a mass moment of inertia $I_i(i = 1, 2, 3)$ with respect to the centre of mass, will be termed G_1, G_2 and G_3.

Considering as independent coordinates of each body the two Cartesian components of the displacement of its centre of gravity and its rotation (see Figs. 1.3b and c), the total number of degrees of freedom is equal to $n_t = 3n_c = 9$:

$$\underline{q}_t^T = \{\alpha_1 \quad \alpha_2 \quad \alpha_3 \quad x_1 \quad y_1 \quad x_2 \quad y_2 \quad x_3 \quad y_3\} \tag{1.15}$$

The number of constraint degrees is $n_v = 8$, since 2 degrees of freedom have been eliminated from each hinge (in correspondence to which, absolute or relative displacements are prevented).

Thus, the system analysed has only one degree of freedom $n_l = 1$: in order to describe motion, using the minimal set method, it is sufficient to assume only one independent variable (e.g. rotation α_1 of bar 1) while the other dependent coordinates need to be expressed as a function of the only independent variable q_l:

$$\underline{q}_d = \underline{q}_d(\underline{q}_l) \tag{1.16}$$

having assumed, for example, rotation α_1 of rod $\overline{OA}$ as an independent coordinate:

$$\underline{q}_l = q_l = \alpha_1 \tag{1.17}$$

Due to the plane motion of the linkage, these relationships can easily be calculated by directly keeping account of the constraints introduced by the 3 hinges positioned in O, A and B which reduce the total number of coordinates to only 3 rotations of the 3 bars of the system and subsequently express the equations of constraint introduced by the hinge positioned in C in terms of a vector *closure equations*, expressed using the complex number algorithm [8][8]:

$$ae^{i\alpha_1} + be^{i\alpha_2} + ce^{i\alpha_3} = d \tag{1.18}$$

By using closure equations it is thus possible to express the relationships between the independent coordinate assumed and the dependent coordinates:

$$\underline{f}_v(\underline{q}_t) = \underline{0} \Rightarrow \begin{Bmatrix} a\cos\alpha_1 + b\cos\alpha_2 + c\cos\alpha_3 - d \\ a\sin\alpha_1 + b\sin\alpha_2 - c\sin\alpha_3 \end{Bmatrix} = \underline{0} \tag{1.19}$$

From which, having indicated with:

$$\underline{q}_d^T = \{\alpha_2 \quad \alpha_3\} \tag{1.20}$$

it is possible to obtain:

$$\underline{q}_d = \underline{q}_d(q_l) \Rightarrow \begin{Bmatrix} \alpha_2 \\ \alpha_3 \end{Bmatrix} = \begin{Bmatrix} \alpha_2(\alpha_1) \\ \alpha_3(\alpha_1) \end{Bmatrix} \tag{1.21}$$

If Eq. (1.21) is derived with respect to time, it is possible to express the velocities of the dependent variables as a function of the independent variable

$$\underline{\dot{q}}_d = \left[\frac{\partial \underline{q}_d}{\partial \underline{q}_l}\right] \underline{\dot{q}}_l \Rightarrow \begin{Bmatrix} \dot{\alpha}_2 \\ \dot{\alpha}_3 \end{Bmatrix} = \begin{bmatrix} \partial\alpha_2/\partial\alpha_1 \\ \partial\alpha_3/\partial\alpha_1 \end{bmatrix} \dot{\alpha}_1 \tag{1.22}$$

where it is possible to observe a linear relationship between these derivatives. As previously mentioned, in order to obtain the Jacobian $\left[\frac{\partial \underline{q}_d}{\partial \underline{q}_l}\right]$, it is possible to derive Eq. (1.19) with respect to time, thus obtaining:

[8]More generally speaking, by analysing a mechanical system in space (see Sect. 1.6) it is more practical to use the natural algorithm of the vectors, using a matrix type approach and reasoning in terms of components.

$$\begin{aligned} -a\,\dot{\alpha}_1 \sin\alpha_1 - b\,\dot{\alpha}_2 \sin\alpha_2 + c\,\dot{\alpha}_3 \sin\alpha_3 &= 0 \\ a\,\dot{\alpha}_1 \cos\alpha_1 + b\,\dot{\alpha}_2 \cos\alpha_2 - c\,\dot{\alpha}_3 \cos\alpha_3 &= 0 \end{aligned} \tag{1.23}$$

or rather, in matrix form:

$$\left\{\begin{matrix}\dot{\alpha}_2\\ \dot{\alpha}_3\end{matrix}\right\} = \left\{\begin{matrix}\frac{\partial\alpha_2}{\partial\alpha_1}\\ \frac{\partial\alpha_3}{\partial\alpha_1}\end{matrix}\right\}\dot{\alpha}_1 = \begin{bmatrix}-b\sin\alpha_2 & c\sin\alpha_3\\ b\cos\alpha_2 & -c\cos\alpha_3\end{bmatrix}^{-1}\left\{\begin{matrix}a\sin\alpha_1\\ -a\cos\alpha_1\end{matrix}\right\}\dot{\alpha}_1 = [J(\alpha_1)]\,\dot{\alpha}_1 \tag{1.24}$$

having indicated the Jacobian matrix with $[J(\alpha_1)]$:

$$[J(\alpha_1)] = \left\{\begin{matrix}J_2(\alpha_1)\\ J_3(\alpha_1)\end{matrix}\right\} = \begin{bmatrix}-b\sin\alpha_2 & c\sin\alpha_3\\ b\cos\alpha_2 & -c\cos\alpha_3\end{bmatrix}^{-1}\left\{\begin{matrix}a\sin\alpha_1\\ -a\cos\alpha_1\end{matrix}\right\} \tag{1.25}$$

After expressing the relationship between the dependent variables $\underline{q}_d$ and the independent variables $\underline{q}_I$ (1.19), (1.24), it is now possible to introduce this relationship into the various forms of energy in order to obtain the equations of motion using Lagrange's equation (1.14).

Conversely, if these equations are obtained using the cardinal equations of dynamics [1, 8, 9], i.e. by directly introducing the inertia forces, it is necessary to define the accelerations relative to the dependent variables in terms of the independent variables deriving from (1.23):

$$\begin{aligned} -a\,\ddot{\alpha}_1 \sin\alpha_1 - a\,\dot{\alpha}_1^2 \cos\alpha_1 - b\,\ddot{\alpha}_2 \sin\alpha_2 - b\,\dot{\alpha}_2^2 \cos\alpha_2 + c\,\ddot{\alpha}_3 \sin\alpha_3 + c\,\dot{\alpha}_3^2 \cos\alpha_3 &= 0 \\ a\,\ddot{\alpha}_1 \cos\alpha_1 - a\,\dot{\alpha}_1^2 \sin\alpha_1 + b\,\ddot{\alpha}_2 \cos\alpha_2 - b\,\dot{\alpha}_2^2 \sin\alpha_2 - c\,\ddot{\alpha}_3 \cos\alpha_3 + c\,\dot{\alpha}_3^2 \sin\alpha_3 &= 0 \end{aligned} \tag{1.26}$$

i.e. in matrix form, keeping account of (1.25):

$$\begin{aligned} &\begin{bmatrix}-b\sin\alpha_2 & c\sin\alpha_3\\ b\cos\alpha_2 & -c\cos\alpha_3\end{bmatrix}\left\{\begin{matrix}\ddot{\alpha}_2\\ \ddot{\alpha}_3\end{matrix}\right\} = \left\{\begin{matrix}a\sin\alpha_1\\ -a\cos\alpha_1\end{matrix}\right\}\ddot{\alpha}_1 + \left\{\begin{matrix}a\cos\alpha_1\\ a\sin\alpha_1\end{matrix}\right\}\dot{\alpha}_1^2 + \left\{\begin{matrix}b\cos\alpha_2\\ b\sin\alpha_2\end{matrix}\right\}\dot{\alpha}_2^2 - \left\{\begin{matrix}c\cos\alpha_3\\ c\sin\alpha_3\end{matrix}\right\}\dot{\alpha}_3^2 \\ &= \left\{\begin{matrix}a\sin\alpha_1\\ -a\cos\alpha_1\end{matrix}\right\}\ddot{\alpha}_1 + \left\{\left\{\begin{matrix}a\cos\alpha_1\\ a\sin\alpha_1\end{matrix}\right\} + \left\{\begin{matrix}b\cos\alpha_2\\ b\sin\alpha_2\end{matrix}\right\}J_2^2(\alpha_1) - \left\{\begin{matrix}c\cos\alpha_3\\ c\sin\alpha_3\end{matrix}\right\}J_3^2(\alpha_1)\right\}\dot{\alpha}_1^2 \end{aligned} \tag{1.27}$$

which can be rewritten as, see Eq. (1.25):

$$\left\{\begin{matrix}\ddot{\alpha}_2\\ \ddot{\alpha}_3\end{matrix}\right\} = \left\{\begin{matrix}\frac{\partial^2\alpha_2}{\partial^2\alpha_1}\\ \frac{\partial^2\alpha_3}{\partial^2\alpha_1}\end{matrix}\right\}\dot{\alpha}_1^2 + \left\{\begin{matrix}\frac{\partial\alpha_2}{\partial\alpha_1}\\ \frac{\partial\alpha_3}{\partial\alpha_1}\end{matrix}\right\}\ddot{\alpha}_1 = [H(\alpha_1)]\,\dot{\alpha}_1^2 + [J(\alpha_1)]\,\ddot{\alpha}_1 \tag{1.28}$$

having indicated the Hessian matrix by means of $[H(\alpha_1)]$:

$$[H(\alpha_1)] = \begin{bmatrix} -b\sin\alpha_2 & c\sin\alpha_3 \\ b\cos\alpha_2 & -c\cos\alpha_3 \end{bmatrix}^{-1} \left\{ \begin{Bmatrix} a\cos\alpha_1 \\ -a\sin\alpha_1 \end{Bmatrix} + \begin{Bmatrix} b\cos\alpha_2 \\ b\sin\alpha_2 \end{Bmatrix} J_2^2(\alpha_1) + \begin{Bmatrix} -c\cos\alpha_3 \\ -c\sin\alpha_3 \end{Bmatrix} J_3^2(\alpha_1) \right\} \dot{\alpha}_1^2 \tag{1.29}$$

Equations (1.1), (1.24) and (1.28) represent the relationship between the physical variables, i.e. rotations of members of the quadrilateral (Fig. 1.3) and the only independent variable assumed α_1.

Even the relationship between physical variables, components of the displacement of the centers of mass of the single rod can be obtained by means of the same procedure, where:

$$\begin{cases} x_1 = \dfrac{a}{2}\cos\alpha_1 \\ y_1 = \dfrac{a}{2}\sin\alpha_1 \end{cases} \quad \begin{cases} x_2 = a\cos\alpha_1 + \dfrac{b}{2}\cos\alpha_2 \\ y_2 = a\sin\alpha_1 + \dfrac{b}{2}\sin\alpha_2 \end{cases} \quad \begin{cases} x_3 = a\cos\alpha_1 + b\cos\alpha_2 + \dfrac{c}{2}\cos\alpha_3 \\ y_3 = a\sin\alpha_1 + b\sin\alpha_2 - \dfrac{c}{2}\sin\alpha_3 \end{cases} \tag{1.30}$$

By deriving these relations with respect to time, it is possible to obtain the relationship between the physical variables and the only independent variable used to describe the motion of the crank mechanism (Fig. 1.3) in terms of velocity:

$$\begin{cases} \dot{x}_1 = \left(-\dfrac{a}{2}\sin\alpha_1\right)\dot{\alpha}_1 = \left(\dfrac{\partial x_1}{\partial\alpha_1}\right)\dot{\alpha}_1 \\ \dot{y}_1 = \left(\dfrac{a}{2}\cos\alpha_1\right)\dot{\alpha}_1 = \left(\dfrac{\partial y_1}{\partial\alpha_1}\right)\dot{\alpha}_1 \end{cases}$$
$$\begin{cases} \dot{x}_2 = -\dot{\alpha}_1 a\sin\alpha_1 - \dot{\alpha}_2\dfrac{b}{2}\sin\alpha_2 = \left(-a\sin\alpha_1 - J_2(\alpha_1)\dfrac{b}{2}\sin\alpha_2\right)\dot{\alpha}_1 = \left(\dfrac{\partial x_2}{\partial\alpha_1}\right)\dot{\alpha}_1 \\ \dot{y}_2 = \dot{\alpha}_1 a\cos\alpha_1 + \dot{\alpha}_2\dfrac{b}{2}\cos\alpha_2 = \left(a\cos\alpha_1 + J_2(\alpha_1)\dfrac{b}{2}\cos\alpha_2\right)\dot{\alpha}_1 = \left(\dfrac{\partial y_2}{\partial\alpha_1}\right)\dot{\alpha}_1 \end{cases}$$
$$\begin{cases} \dot{x}_3 = -\dot{\alpha}_1 a\sin\alpha_1 - \dot{\alpha}_2 b\sin\alpha_2 - \dot{\alpha}_3\dfrac{c}{2}\sin\alpha_3 = \left(-a\sin\alpha_1 - J_2(\alpha_1)b\sin\alpha_2 - J_3(\alpha_1)\dfrac{c}{2}\sin\alpha_3\right)\dot{\alpha}_1 = \left(\dfrac{\partial x_3}{\partial\alpha_1}\right)\dot{\alpha}_1 \\ \dot{y}_3 = \dot{\alpha}_1 a\cos\alpha_1 + \dot{\alpha}_2 b\cos\alpha_2 - \dot{\alpha}_3\dfrac{c}{2}\cos\alpha_3 = \left(a\cos\alpha_1 + J_2(\alpha_1)b\cos\alpha_2 - J_3(\alpha_1)\dfrac{c}{2}\cos\alpha_3\right)\dot{\alpha}_1 = \left(\dfrac{\partial y_3}{\partial\alpha_1}\right)\dot{\alpha}_1 \end{cases} \tag{1.31}$$

and accelerations:

$$
\begin{cases}
\ddot{x}_1 = \left(-\frac{a}{2}\sin\alpha_1\right)\ddot{\alpha}_1 + \left(-\frac{a}{2}\cos\alpha_1\right)\dot{\alpha}_1^2 = \left(\frac{\partial x_1}{\partial \alpha_1}\right)\ddot{\alpha}_1 + \left(\frac{\partial^2 x_1}{\partial \alpha_1^2}\right)\dot{\alpha}_1^2 \\
\ddot{y}_1 = \left(\frac{a}{2}\cos\alpha_1\right)\ddot{\alpha}_1 + \left(-\frac{a}{2}\sin\alpha_1\right)\dot{\alpha}_1^2 = \left(\frac{\partial y_1}{\partial \alpha_1}\right)\ddot{\alpha}_1 + \left(\frac{\partial^2 y_1}{\partial \alpha_1^2}\right)\dot{\alpha}_1^2
\end{cases}
$$

$$
\begin{cases}
\ddot{x}_2 = \left(-a\sin\alpha_1 - J_2(\alpha_1)\frac{b}{2}\sin\alpha_2\right)\ddot{\alpha}_1 + \left(-a\cos\alpha_1 - \frac{\partial J_2(\alpha_1)}{\partial \alpha_1}\frac{b}{2}\sin\alpha_2 - J_2^2(\alpha_1)\frac{b}{2}\cos\alpha_2\right)\dot{\alpha}_1^2 \\
\quad = \left(\frac{\partial x_2}{\partial \alpha_1}\right)\ddot{\alpha}_1 + \left(\frac{\partial^2 x_2}{\partial \alpha_1^2}\right)\dot{\alpha}_1^2 \\
\ddot{y}_2 = \left(a\cos\alpha_1 + J_2(\alpha_1)\frac{b}{2}\cos\alpha_2\right)\ddot{\alpha}_1 + \left(-a\sin\alpha_1 + \frac{\partial J_2(\alpha_1)}{\partial \alpha_1}\frac{b}{2}\cos\alpha_2 - J_2^2(\alpha_1)\frac{b}{2}\sin\alpha_2\right)\dot{\alpha}_1^2 \\
\quad = \left(\frac{\partial y_2}{\partial \alpha_1}\right)\ddot{\alpha}_1 + \left(\frac{\partial^2 y_2}{\partial \alpha_1^2}\right)\dot{\alpha}_1^2
\end{cases}
$$

$$
\begin{cases}
\ddot{x}_3 = \cdots = \left(\frac{\partial x_3}{\partial \alpha_1}\right)\ddot{\alpha}_1 + \left(\frac{\partial^2 x_3}{\partial \alpha_1^2}\right)\dot{\alpha}_1^2 \\
\ddot{y}_3 = \cdots = \left(\frac{\partial y_3}{\partial \alpha_1}\right)\ddot{\alpha}_1 + \left(\frac{\partial^2 y_3}{\partial \alpha_1^2}\right)\dot{\alpha}_1^2
\end{cases}
\tag{1.32}
$$

Having introduced the general concept that enables us to describe the motion of a generic mechanical system, we will now go on to give a more accurate description of the methodology that can be used to write the equations of motion of same.

1.3 Writing Equations of Motion

As is known, a useful, widely-used approach to write the equations of motion of a generic mechanical system [1] is using Lagrange's equations [1, 2, 8]. For this purpose, let us consider a generic mechanical system composed by n_c rigid bodies and let us indicate the vector of the variables chosen to describe motion by $\underline{q}$: as is known the equations can thus be defined by the following relation:

$$
\left\{\frac{d}{dt}\left(\frac{\partial E_c}{\partial \underline{\dot{q}}}\right)\right\}^T - \left\{\frac{\partial E_c}{\partial \underline{q}}\right\}^T + \left\{\frac{\partial D}{\partial \underline{\dot{q}}}\right\}^T + \left\{\frac{\partial V}{\partial \underline{q}}\right\}^T = \underline{Q} \tag{1.33}
$$

where E_c, D and V respectively represent kinetic energy, the dissipation function and the potential energy with which the system is equipped and finally $\underline{Q}$ is the vector of nonconservative forces acting on same obtained from the relation:

$$\delta^* L = \underline{Q}^T \, \delta^* \underline{q} \tag{1.34}$$

where $\delta^* \underline{q}$ is the virtual displacement of the same variables. According to the theory outlined in this paragraph, the vector $\underline{q}$ can be represented by the independent variables (minimal set method) or by the total number of variables ($\underline{q}_t = n_c * 6$ or $\underline{q}_t = n_c * 3$, method of Lagrange multipliers): the only difference lies in the fact that, with the first approach, the constraint reactions $\underline{R}$ do not appear in the equations of motions if, obviously, they do not perform any work. Following the second method, in the virtual work of the external forces it is necessary to introduce the work performed by these reactions, since, having freed the constraints, the same $\underline{R}$ are external forces, performing work. In order to apply (1.33) it is thus necessary to clearly express the various forms of energy and the virtual work of the external forces (1.34) as a function of the variables $\underline{q}$ and of their derivatives.

1.3.1 Definition of the Various Forms of Energy as a Function of the Physical Variables

In order to easily define the expressions of the energy forms, it is useful to refer to *physical variables*, by choosing the most convenient ones describing the system.

The generic kth physical variable Y_{fk} is always correlated to independent variables $\underline{q}$ according to relations (generally non-linear) of the type:

$$Y_{fk} = Y_{fk}(\underline{q}) \qquad (k = 1, 2, \ldots, m) \tag{1.35}$$

By introducing this relationship into the expressions of the various forms of energy (1.33), by applying Lagrange's equations it is possible to obtain the equations of motion of the mechanical system analysed.

Let us remember that [1–3] the kinetic energy E_{cj} associated with a standard jth rigid body constituting the entire system is given by:

$$E_{cj} = \frac{1}{2} m_j \dot{x}_j^2 + \frac{1}{2} m_j \dot{y}_j^2 + \frac{1}{2} m_j \dot{z}_j^2 + \frac{1}{2} J_{xj} \omega_{xj}^2 + \frac{1}{2} J_{yj} \omega_{yj}^2 + \frac{1}{2} J_{zj} \omega_{zj}^2 \tag{1.36}$$

which can be expressed in matrix form as:

$$E_{cj} = \frac{1}{2} \dot{\underline{y}}_{mj}^T \left[M_{fj} \right] \dot{\underline{y}}_{mj} \tag{1.37}$$

having used:

$$\dot{\underline{y}}_{mj} = \begin{Bmatrix} \dot{x}_j \\ \dot{y}_j \\ \dot{z}_j \\ \omega_{xj} \\ \omega_{yj} \\ \omega_{zj} \end{Bmatrix} \tag{1.38}$$

to indicate the vector of the *physical variables*, which are convenient to define the kinetic energy of the generic rigid body constituting a part of the mechanical system and

$$[M_{fj}] = \begin{bmatrix} m_j & 0 & 0 & 0 & 0 & 0 \\ 0 & m_j & 0 & 0 & 0 & 0 \\ 0 & 0 & m_j & 0 & 0 & 0 \\ 0 & 0 & 0 & J_{xj} & 0 & 0 \\ 0 & 0 & 0 & 0 & J_{yj} & 0 \\ 0 & 0 & 0 & 0 & 0 & J_{zj} \end{bmatrix} \tag{1.39}$$

to indicate the mass matrix, as a function of the same physical variables (1.38).

In vector $\dot{\underline{y}}_{mj}$, the first three terms keep account of the contribution to the translation in terms of the components $\dot{x}_j, \dot{y}_j$ and $\underline{z}_j$ of vector $\vec{V}_j$ (Fig. 1.4) which defines the absolute velocity of the centre of gravity in the absolute reference system, while the next three terms keep account of the contribution to rotation by

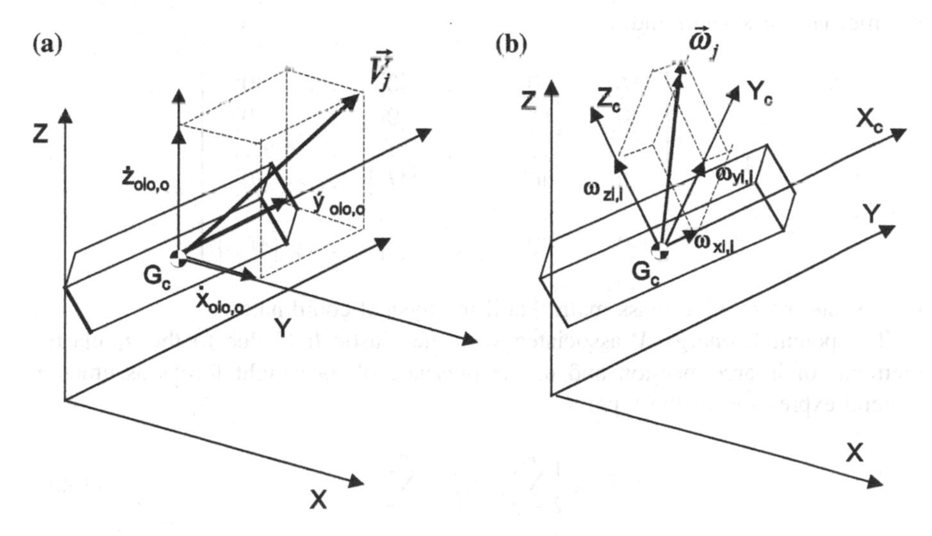

Fig. 1.4 Definition of the velocity components $\vec{V}_j$ ed $\vec{\omega}_j$ for a generic rigid body

expressing the absolute angular velocity vector $\vec{\omega}_j$ of the body as a function of the three components ω_{xj}, ω_{yj} and ω_{zj}, expressed in a Cartesian reference system $\left(G_j - \xi_{j1} - \xi_{j2} - \xi_{j3}\right)$. This reference system is connected to the body itself, with origin O_j in its centre of gravity G_j and axes parallel to the principal axes of inertia (in this way and in this case, the inertia tensor is diagonal [3, 4, 10].

For a generic vibrating system constituted by n_c rigid bodies each with 6 d.o.f., the total kinetic energy E_c will be given by the sum of the kinetic energies associated with the single bodies constituting the vibrating system itself:

$$E_c = \sum_{j=1}^{n_c} E_{cj} \tag{1.40}$$

Equation (1.40) can be expressed in matrix forma as:

$$E_c = \frac{1}{2} \underline{\dot{y}}_m^T \left[M_f\right] \underline{\dot{y}}_m \tag{1.41}$$

and having used:

$$\underline{\dot{y}}_m = \begin{Bmatrix} \underline{\dot{y}}_{m1} \\ \underline{\dot{y}}_{m2} \\ \cdots \\ \underline{\dot{y}}_{mj} \\ \cdots \\ \underline{\dot{y}}_{mnc} \end{Bmatrix} \tag{1.42}$$

to indicate the vector of the physical variables relative to all the bodies composing the mechanical system and:

$$\left[M_f\right] = \begin{bmatrix} \left[M_{f1}\right] & [0] & \cdots & [0] & \cdots & [0] \\ [0] & \left[M_{f2}\right] & \cdots & [0] & \cdots & [0] \\ \cdots & \cdots & \cdots & \cdots & \cdots & \cdots \\ [0] & [0] & \cdots & \left[M_{fj}\right] & \cdots & [0] \\ \cdots & \cdots & \cdots & \cdots & \cdots & \cdots \\ [0] & [0] & \cdots & [0] & \cdots & \left[M_{fnc}\right] \end{bmatrix} \tag{1.43}$$

to indicate the relative mass matrix, still in physical coordinates.

The potential energy V associated with the elastic field due to the n_k elastic elements of interconnection and to the presence of n_P weight forces assumes a general expression of the type:

$$V = \frac{1}{2} \sum_{j=1}^{n_k} k_j \Delta l_j^2 + \sum_{j=1}^{n_p} p_j h_j \tag{1.44}$$

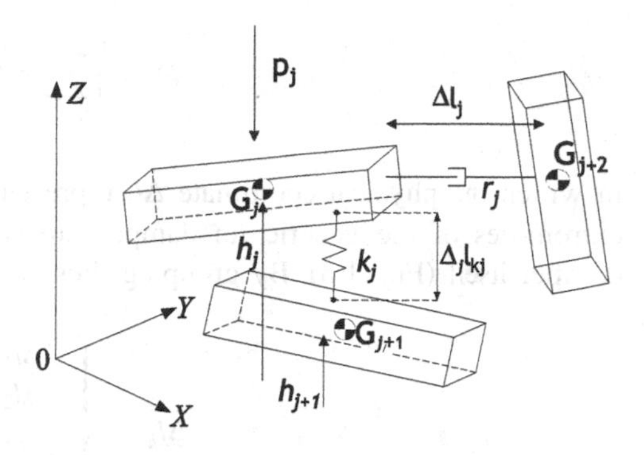

Fig. 1.5 Definition of the terms connected to the potential energy

in which k_j represents the stiffness of the generic jth elastic element and p_j the weight force applied to the centre of gravity G_j of the generic jth body. In (1.44) the physical coordinates Δl_j and h_j respectively represent (Fig. 1.5) the relative displacement of the extremities of the generic spring, in the direction of the spring itself and the elevation of the centre of gravity of the generic body which, on account of being organized in matrix form:

$$\underline{\Delta l}_k = \begin{Bmatrix} \Delta l_1 \\ \Delta l_2 \\ \dots \\ \Delta l_{nk-1} \\ \Delta l_{nk} \end{Bmatrix}; \qquad \underline{h}_p = \begin{Bmatrix} h_1 \\ h_2 \\ \dots \\ h_{np-1} \\ h_{np} \end{Bmatrix} \tag{1.45}$$

allow us to rewrite the expression of the potential energy V (1.44), as a function of the assumed physical variables, as:

$$V = \frac{1}{2}\underline{\Delta l}_k^T [K_{\Delta l}]\, \underline{\Delta l}_k + \underline{p}^T\, \underline{h}_p \tag{1.46}$$

having organized the stiffness matrix in physical coordinates $[K_{\Delta l}]$ and the vector of the weight forces $\underline{p}$ as follows:

$$[K_{\Delta l}] = \begin{bmatrix} k_1 & 0 & \dots & 0 & 0 \\ 0 & k_2 & \dots & 0 & 0 \\ \dots & \dots & \dots & \dots & \dots \\ 0 & 0 & \dots & k_{nk-1} & 0 \\ 0 & 0 & \dots & 0 & k_{nk} \end{bmatrix}; \quad \underline{p} = \begin{Bmatrix} p_1 \\ p_2 \\ \dots \\ p_{np-1} \\ p_{np} \end{Bmatrix} \tag{1.47}$$

The dissipation energy, due to the presence of n_s viscous dampers of constant r_j can be defined as:

$$D = \frac{1}{2}\sum_{j=1}^{n_s} r_j \dot{\Delta l}_j^2 \quad (1.48)$$

in which the physical coordinate $\dot{\Delta l}_j$ represents the relative velocity to which the extremities of the generic jth damper are subjected, along the direction of the damper itself (Fig. 1.5). By grouping these components in only one vector:

$$\underline{\dot{\Delta l}}_k = \begin{Bmatrix} \dot{\Delta l}_1 \\ \dot{\Delta l}_2 \\ \cdots \\ \dot{\Delta l}_{ns-1} \\ \dot{\Delta l}_{ns} \end{Bmatrix} \quad (1.49)$$

it is possible to rewrite (1.48) in matrix form as:

$$D = \frac{1}{2}\underline{\dot{\Delta l}}_r^T [R_{\Delta l}] \underline{\dot{\Delta l}}_r \quad (1.50)$$

having organized the damping matrix in physical coordinates $[R_{\Delta l}]$ as:

$$[R_{\Delta l}] = \begin{bmatrix} r_1 & 0 & \cdots & 0 & 0 \\ 0 & r_2 & \cdots & 0 & 0 \\ \cdots & \cdots & \cdots & \cdots & \cdots \\ 0 & 0 & \cdots & r_{ns-1} & 0 \\ 0 & 0 & \cdots & 0 & r_{ns} \end{bmatrix} \quad (1.51)$$

Finally, let us consider the virtual work $\delta^* L$ performed by the n_f forces $\vec{F}_j$ for a virtual displacement $\delta^* \vec{y}_{fj}$: for sake of simplicity, we will use $\delta^* y_{fj}$ to indicate the displacement component in the direction of the generic force $\vec{F}_j$ so that:

$$\delta^* L = \sum_{j=1}^{n_f} \vec{F}_j \times \delta^* \vec{y}_{fj} = \sum_{j=1}^{n_f} F_j \delta^* y_{fj} \quad (1.52)$$

By assembling the forces and the virtual displacements in two vectors:

$$\underline{F} = \begin{Bmatrix} F_1 \\ F_2 \\ \cdots \\ F_{nf-1} \\ F_{nf} \end{Bmatrix}; \quad \underline{y}_f = \begin{Bmatrix} y_{f1} \\ y_{f2} \\ \cdots \\ y_{nf-1} \\ y_{nf} \end{Bmatrix} \quad (1.53)$$

it is possible to rewrite (1.52) in matrix form as:

$$\delta^* L = \underline{F}^T \, \delta^* \, \underline{y}_f \tag{1.54}$$

The forces can either depend explicitly on time $\underline{F} = \underline{F}_e(t)$ or, more generally speaking, on the state of the system $\underline{q}$ and $\underline{\dot{q}}$: this is applicable to the field forces $\underline{F} = \underline{F}_c(\underline{q}, \underline{\dot{q}})$ (dealt with in-depth in Chap. 5) or to the control forces (see bibliography [11]), thus obtaining

$$\underline{F} = \underline{F}_e(t) + \underline{F}_c(\underline{q}, \underline{\dot{q}}) \tag{1.55}$$

All the forces that do not come from of the kinetic (inertia forces), potential and dissipation energy, such as, for example, the non-linear elastic and dissipation ones, are finally established and represented in the $\underline{F}$ vector. All the conservative forces (both in terms of module and direction) (1.46) and (1.47) have been assembled in vector $\underline{p}$ of (1.45).

1.3.2 Definition of the Various Forms of Energy as a Function of the Independent Variables

Now that we have described all the various forms of energy, it is necessary to express the physical variables $\underline{\dot{y}}_m$(1.41), $\underline{h}_p$ and $\underline{\Delta l}_k$(1.46), $\underline{\dot{\Delta l}}_r$(1.50), $\delta^* \underline{y}_f$ (1.54) as a function of the independent variables $\underline{q}$, respectively as:

$$\begin{aligned}
\underline{y}_m &= \underline{y}_m(\underline{q}) \Rightarrow \underline{\dot{y}}_m = \left[\frac{\partial \underline{y}_m}{\partial \underline{q}}\right] \underline{\dot{q}} \\
\underline{\Delta l}_k &= \underline{\Delta l}_k(\underline{q}) \\
\underline{h}_p &= \underline{h}_p(\underline{q}) \\
\underline{\Delta l}_r &= \underline{\Delta l}_r(\underline{q}) \Rightarrow \underline{\dot{\Delta l}}_r = \left[\frac{\partial \underline{\Delta l}_r}{\partial \underline{q}}\right] \underline{\dot{q}} \\
\underline{y}_f &= \underline{y}_f(\underline{q}) \Rightarrow \delta^* \, \underline{y}_f = \left[\frac{\partial \underline{y}_f}{\partial \underline{q}}\right] \delta^* \underline{q}
\end{aligned} \tag{1.56}$$

By keeping account of these relations, it is thus possible to express the various forms of energy as a function of the independent coordinates assumed to describe system motion

By keeping account of coordinate transformations (1.56), the kinetic energy of the entire system (1.41) becomes:

$$E_c = \frac{1}{2}\dot{\underline{y}}_m^T [M_y]\, \dot{\underline{y}}_m = \frac{1}{2}\dot{\underline{q}}^T \left[\frac{\partial \underline{y}_m}{\partial \underline{q}}\right]^T [M_f] \left[\frac{\partial \underline{y}_m}{\partial \underline{q}}\right] \dot{\underline{q}} = \frac{1}{2}\dot{\underline{q}}^T [M] \dot{\underline{q}} \tag{1.57}$$

having used:

$$[M] = \left[\frac{\partial \underline{y}_m}{\partial \underline{q}}\right]^T [M_f] \left[\frac{\partial \underline{y}_m}{\partial \underline{q}}\right] = \left[M\left(\underline{q}\right)\right] \tag{1.58}$$

to indicate the mass matrix expressed as a function of the independent coordinates.

By keeping account of Eq. (1.56), potential energy (1.146) can be expressed as:

$$V = \frac{1}{2}\Delta\underline{l}_k^T(\underline{q})[K_{\Delta l}]\, \Delta\underline{l}_k(\underline{q}) + \underline{p}^T\, \underline{h}_p(\underline{q}) \tag{1.59}$$

By keeping account of Eq. (1.56), the dissipation function (1.50) can be rewritten in the independent variables as:

$$D = \frac{1}{2}\Delta\dot{\underline{l}}_r^T [R_y]\, \Delta\dot{\underline{l}}_r = \frac{1}{2}\dot{\underline{q}}^T \left[\frac{\partial \Delta\underline{l}_r}{\partial \underline{q}}\right]^T [R_y] \left[\frac{\partial \Delta\underline{l}_r}{\partial \underline{q}}\right] \dot{\underline{q}} = \frac{1}{2}\dot{\underline{q}}^T [R] \dot{\underline{q}} \tag{1.60}$$

having used:

$$[R] = \left[\frac{\partial \Delta\underline{l}_r}{\partial \underline{q}}\right]^T [R_y] \left[\frac{\partial \Delta\underline{l}_r}{\partial \underline{q}}\right] = \left[R\left(\underline{q}\right)\right] \tag{1.61}$$

to indicate the damping matrix in independent coordinates.

Finally, by introducing the transformation of variables (1.56), the virtual work done by external forces (1.54), becomes:

$$\delta^* L = \underline{F}^T\, \delta^* \underline{y}_f = \underline{F}^T \left[\frac{\partial \underline{y}_f}{\partial \underline{q}}\right] \delta^* \underline{q} = \underline{Q}^T\, \delta^* \underline{q} \tag{1.62}$$

having used:

$$\underline{Q} = \left[\frac{\partial \underline{y}_f}{\partial \underline{q}}\right]^T \underline{F} = \underline{Q}(\underline{q}) \tag{1.63}$$

to indicate the vector of the generalized forces of the system.

1.3.3 Application of Lagrange's Equations

Having expressed the various forms of energy as a function of the independent variables, Eqs. (1.57) ÷ (1.64), it is now possible to write the equations of motion of the system, by applying the Lagrange's Equation (1.33).

The first inertia term present in Lagrange's Equation (1.33) can be obtained by keeping account of expressions (1.57) and relationships (1.56) that lead to:

$$\left\{\frac{\partial E_c}{\partial \dot{\underline{q}}}\right\}^T = [M]\dot{\underline{q}} \Rightarrow \left\{\frac{d}{dt}\left(\frac{\partial E_c}{\partial \dot{\underline{q}}}\right)\right\}^T = [M]\ddot{\underline{q}} + [\dot{M}]\dot{\underline{q}} \tag{1.64}$$

where

$$[\dot{M}] = \frac{d[M]}{dt} \tag{1.65}$$

while the second term can be expressed, in matrix form as:

$$\left\{\frac{\partial E_c}{\partial \underline{q}}\right\}^T = \underline{Q}_c = \begin{Bmatrix} \cdots \\ \cdots \\ \frac{1}{2}\dot{\underline{q}}^T\left[\frac{\partial [M]}{\partial q_j}\right]\dot{\underline{q}} \\ \cdots \\ \cdots \end{Bmatrix} \tag{1.66}$$

The Lagrangian component of conservative forces, i.e. the derivative with respect to the independent variables of potential energy (1.59), can formally be obtained as follows:

$$\begin{aligned}\left\{\frac{\partial V}{\partial \underline{q}}\right\}^T &= \left\{\left\{\frac{\partial V}{\partial \underline{\Delta l}_k}\right\}\left[\frac{\partial \underline{\Delta l}_k}{\partial \underline{q}}\right]\right\}^T + \left\{\left\{\frac{\partial V}{\partial \underline{h}_p}\right\}\left[\frac{\partial \underline{h}_p}{\partial \underline{q}}\right]\right\}^T \\ &= \left[\frac{\partial \underline{\Delta l}_k}{\partial \underline{q}}\right]^T [K_{\Delta l}]\,\underline{\Delta l}_k + \left[\frac{\partial \underline{h}_p}{\partial \underline{q}}\right]^T \underline{p} = \underline{Q}_p(\underline{q})\end{aligned} \tag{1.67}$$

and thus results as a nonlinear function of $\underline{q}$, since V (1.59) is a non-quadratic function of the independent variables.

The forcing terms associated with the dissipation function (1.60) in turn become:

$$\left\{\frac{\partial D}{\partial \dot{\underline{q}}}\right\}^T = [R(\underline{q})]\dot{\underline{q}} \tag{1.68}$$

Since dissipation energy D (1.60) is generally a non-quadratic form, the Jacobians $\left[\frac{\partial \Delta l_r}{\partial \underline{q}}\right]$ which appear in Eq. (1.56) are themselves, given the nonlinear relationship between the physical variables and the independent variables, functions of $\underline{q}$ and, therefore, as previously mentioned, the damping matrix:

$$[R] = \left[\frac{\partial \Delta l_r}{\partial \underline{q}}\right]^T [R_y] \left[\frac{\partial \Delta l_r}{\partial \underline{q}}\right] = \left[R\left(\underline{q}\right)\right] \tag{1.69}$$

is also generally a function of $\underline{q}$.

Application of the Lagrange's equations thus leads to the definition of the Lagrangian component of the active external non-conservative forces by keeping account of Eq. (1.55):

$$\underline{Q} = \left[\frac{\partial \underline{y}_f}{\partial \underline{q}}\right]^T \underline{F}(\underline{q}, \underline{\dot{q}}, t) = \underline{Q}(\underline{q}, \underline{\dot{q}}, t) \tag{1.70}$$

which will be a function of time t due to the presence of the forces which are explicit functions of both time and the free coordinates, due to their presence in $\underline{F}(\underline{q}, \underline{\dot{q}}, t)$ and in the Jacobian $\left[\frac{\partial \underline{y}_f}{\partial \underline{q}}\right]$. This Jacobian is constant only in the case of a linear relationship between $\underline{y}_f$ and $\underline{q}$.

By keeping account of Eqs. (1.64) ÷ (1.70), the equations of motion of the system can thus generally be expressed by a set of second order, nonlinear, differential equations which can be expressed as a function of the independent coordinates $\underline{q}$:

$$[M(\underline{q})]\underline{\ddot{q}} = -[\dot{M}(\underline{q}, \underline{\dot{q}})]\underline{\dot{q}} + \underline{Q}_c(\underline{q}, \underline{\dot{q}}) - \underline{Q}_p(\underline{q}) - [R(\underline{q})]\underline{\dot{q}} + \underline{Q}(\underline{q}, \underline{\dot{q}}, t) \tag{1.71}$$

where, depending on the position assumed by same, the system's mass matrix is expressed by $[M(\underline{q})]$, the terms $-[\dot{M}(\underline{q}, \underline{\dot{q}})]\underline{\dot{q}} + \underline{Q}_c(\underline{q}, \underline{\dot{q}})$ contain the gyroscopic effects and the Coriolis components, $[R(\underline{q})]\underline{\dot{q}}$ represent the contribution of the dissipation elements and $\underline{Q}_p(\underline{q})$ contains the elastic terms.

1.3.4 Lagrange Multiplier Method

As previously mentioned, an alternative approach is that of the Lagrange multipliers. This consists in considering all the $\underline{q}_t$ as independent variables and in freeing the constraints by also introducing the work of the n_v constraining reactions $\underline{R}$ into the equations. By using this approach, the virtual work of the forces also includes the constraining reactions, external and internal to system $\underline{R}$:

$$\delta^* L = \underline{F}^T \delta^* y_f + \underline{R}^T \delta^* \underline{y}_R = \left\{ \underline{F}^T \left[\frac{\partial \underline{y}_f}{\partial \underline{q}_t} \right] + \underline{R}^T \left[\frac{\partial \underline{y}_R}{\partial \underline{q}_t} \right] \right\} \delta^* \underline{q}_t = \underline{Q}_t^T \delta^* \underline{q}_t \quad (1.72)$$

and vector $\underline{Q}_t$ of the generalized forces can be expressed using this approach as:

$$\underline{Q}_t = \underline{Q}_{tF} + \underline{Q}_{tR} = \left[\frac{\partial \underline{y}_f}{\partial \underline{q}_t} \right]^T \underline{F} + \left[\frac{\partial \underline{y}_R}{\partial \underline{q}_t} \right]^T \underline{R} = [C_{tF}]^T \underline{F} + [C_{tR}]^T \underline{R} \quad (1.73)$$

where $[C_{tR}]$ is the Jacobian matrix associated with the constraints, (1.6):

$$[C_{tR}] = \left[\frac{\partial \underline{y}_R}{\partial \underline{q}_t} \right] \quad (1.74)$$

By using the Lagrange multiplier method, we thus consider both the equations of motion and the constraint Equation (1.11), which, by keeping account of (1.72) and (1.73) generally become:

$$\begin{cases} [M_t(\underline{q}_t)]\underline{\ddot{q}}_t = -[\dot{M}_t(\underline{q}_t, \underline{\dot{q}}_t)]\underline{\dot{q}}_t + \underline{Q}_{tc}(\underline{q}_t, \underline{\dot{q}}_t) - \underline{Q}_{tp}(\underline{q}_t) - [R_t(\underline{q}_t)]\underline{\dot{q}}_t + \underline{Q}_{tF} + [C_{tR}]^T \underline{R} \\ \underline{f}_v(\underline{q}_t) = 0 \end{cases} \quad (1.75)$$

By using this approach, the dynamics of the multi-body systems is therefore described by a hybrid system of differential and algebraic Equation (1.74) coupled with each other (called in literature *differential algebraic equations*, see [6, 9, 7, 12, 13]. In order to transform the system of differential algebraic equations into a system of ordinary differential equations, it is customary to suitably manipulate the algebraic equations relative to the constraint conditions [2], in order to give them a better formulation. For this purpose, the constraint equations[9] are derived with respect to time:

$$\underline{f}_v(\underline{q}_t) = \underline{0} \Rightarrow \left[\frac{\partial \underline{f}_v}{\partial \underline{q}_t} \right] \underline{\dot{q}}_t = [C_{tR}]\underline{\dot{q}}_t = \underline{0} \Rightarrow [C_{tR}]\underline{\ddot{q}}_t + \frac{\partial \left\{ [C_{tR}]\underline{\dot{q}}_t \right\}}{\partial \underline{q}_t} \underline{\dot{q}}_t = 0 \quad (1.76)$$

thus obtaining:

[9]It is easy to demonstrate [1, 2] that the Jacobian $[C_{tR}]$ di (1.74), associated with the constraint equations, is the same Jacobian that appears in the Lagrangian component of the constraining reactions (1.76).

$$\begin{cases} [M_t(\underline{q}_t)]\underline{\ddot{q}}_t = -[\dot{M}_t(\underline{q}_t, \underline{\dot{q}}_t)]\underline{\dot{q}}_t + \underline{Q}_{tc}(\underline{q}_t, \underline{\dot{q}}_t) - \underline{Q}_{tp}(\underline{q}_t) - [R_t(\underline{q}_t)]\underline{\dot{q}}_t + \underline{Q}_{tF} + [C_{tR}]^T \underline{R} \\ [C_{tR}]\underline{\ddot{q}}_t = -\dfrac{\partial\{[C_{tR}]\underline{\dot{q}}_t\}}{\partial \underline{q}_t} \underline{\dot{q}}_t \end{cases} \tag{1.77}$$

The same equations written in the following compact form:

$$\begin{bmatrix} [M_t(\underline{q}_t)] & -[C_{tR}(\underline{q}_t)]^T \\ [C_{tR}(\underline{q}_t)] & [0] \end{bmatrix} \begin{Bmatrix} \underline{\ddot{q}}_t \\ \underline{R} \end{Bmatrix} = \begin{Bmatrix} \underline{Q}_{tt}(t, \underline{q}_t, \underline{\dot{q}}_t) \\ \underline{Q}_{tv}(\underline{q}_t, \underline{\dot{q}}_t) \end{Bmatrix} \tag{1.78}$$

Having indicated the above by

$$\begin{aligned} \underline{Q}_{tt}(t, \underline{q}_t, \underline{\dot{q}}_t) &= -[\dot{M}_t(\underline{q}_t, \underline{\dot{q}}_t)]\,\underline{\dot{q}}_t + \underline{Q}_{tc}(\underline{q}_t, \underline{\dot{q}}_t) - \underline{Q}_{tp}(\underline{q}_t) - [R_t(\underline{q}_t)]\,\underline{\dot{q}}_t + \underline{Q}_{tF} \\ \underline{Q}_{tv}(\underline{q}_t, \underline{\dot{q}}_t) &= -\left[\frac{\partial\{[C_{tR}]\,\underline{\dot{q}}_t\}}{\partial \underline{q}_t}\right] \underline{\dot{q}}_t \end{aligned} \tag{1.79}$$

Minimum Set of Variables Method

As previously mentioned, in (1.70) $\underline{Q}(t)$ is the vector of the generalized external forces applied to the system. By adopting the minimum set of variables method, this term is defined by all the forces external to system $\underline{F}$ (in this case, the constraining reactions $\underline{R}$ do not appear owing to the fact that these, by keeping account of the external forces of the constraint conditions do not by definition, perform any work):

$$\delta^* L = \underline{F}^T \delta^* \underline{y}_f = \underline{F}^T \left[\frac{\partial \underline{y}_f}{\partial \underline{q}_l}\right] \delta^* \underline{q}_l = \underline{Q}_{lF}^T \delta^* \underline{q}_l \tag{1.80}$$

from which:

$$\underline{Q}_{lF} = \left[\frac{\partial \underline{y}_f}{\partial \underline{q}_l}\right]^T \underline{F} = [C_{lF}]^T \underline{F} \tag{1.81}$$

and the final solving system consists of a set of n_l (equal to degrees of freedom of the system) nonlinear, second order differential equations, of the type:

$$[M_l(\underline{q}_l)]\underline{\ddot{q}}_l = -[\dot{M}_l(\underline{q}_l, \underline{\dot{q}}_l)]\underline{\dot{q}}_l + \underline{Q}_{lc}(\underline{q}_l, \underline{\dot{q}}_l) - \underline{Q}_{lp}(\underline{q}_l) - [R_l(\underline{q}_l)]\underline{\dot{q}}_l + \underline{Q}_{lF} \tag{1.82}$$

1.3.5 Method of Introducing Real Constraints Using Force Fields

Both minimum set of variables (Sect. 1.3.5) and Lagrange multiplier (Sect 1.3.4) methods presuppose the presence of ideal non-deformable constraints that do not dissipate energy. In reality, the contact actions exercised by the constraints are, in a more realistic schematization of the problem, attributable to *field forces*, functions of the independent coordinates themselves (see Chap. 5).

To simplify matters, in the text that follows we will first introduce and then consider the ideal constraint condition. Then, at a later stage, we will keep account of the real field forces: by using this approach we will eliminate the constraints and replace them with a more accurate schematization, by means of field forces.

By using this approach:

- ideal constraints no longer exist;
- all the 6*N (3*N for in-plane motion) physical coordinates relative to all the N bodies are considered as independent variables;
- the internal and external actions $\underline{R}_{tc}$ exchanged between the bodies are defined as *force fields*, i.e. functions of the independent variables themselves:

$$\underline{R}_{tc} = \underline{R}_{tc}(\underline{q}_t, \underline{\dot{q}}_t) \tag{1.83}$$

By using this approach, the equations of motion (1.71) become:

$$\begin{aligned}[M_t(\underline{q}_t)]\underline{\ddot{q}}_t = &-[\dot{M}_t(\underline{q}_t, \underline{\dot{q}}_t)]\underline{\dot{q}}_t + \underline{Q}_{tc}(\underline{q}_t, \underline{\dot{q}}_t) - \underline{Q}_{tp}(\underline{q}_t) - [R_t(\underline{q}_t)]\underline{\dot{q}}_t + \underline{Q}_{tF} \\ &+ [C_{tR}]^T \underline{R}_{tc}(\underline{q}_t, \underline{\dot{q}}_t)\end{aligned} \tag{1.84}$$

i.e. $n_t = 6 \cdot n_c$ nonlinear differential equations in the n_t unknown $\underline{q}_t$.

Examples of this approach will be described in Chaps. 5 and 6 to simulate the behaviour of the force fields applied to the mechanical system analysed, as, for example, in the case of hydrodynamic lubricated bearings (see Chap. 6) or during contact between the wheel and the rail or the tyre and the road (see Chap. 5).

1.4 Nonlinear Systems with One Degree of Freedom

We will now apply the approaches previously described to a system with one degree of freedom.

As previously mentioned, the assumption of considering a system to have one degree of freedom obviously depends on the schematization adopted and, in general, on the assumption that the constraints of the mechanical system analysed are ideal.

1.4.1 Writing Equations of Motion "in the Large"

In the event of using the minimum set of variables method, the mechanical system analysed can have $n_l = n_t - n_v = 1$, in other words it can only have one degree of freedom (1 d.o.f.): in this case, the motion of the system can be described by only one independent variable:

$$\underline{q} = \underline{q}_l = q \tag{1.85}$$

and the various forms of energy, see Eqs. (1.57) ÷ (1.63), thus become:

$$\begin{aligned}
E_c &= \frac{1}{2}\dot{\underline{y}}_m^T [M_f]\,\dot{\underline{y}}_m = \frac{1}{2}\dot{q}^T \left[\frac{\partial \underline{y}_m}{\partial q}\right]^T [M_f] \left[\frac{\partial \underline{y}_m}{\partial q}\right] \dot{q} = \frac{1}{2} m(q)\,\dot{q}^2 \\
V &= \frac{1}{2}\underline{\Delta l}_k^T(q)[K_{\Delta l}]\,\underline{\Delta l}_k(q) + \underline{p}^T\,\underline{h}_p(q) = V(q) = \frac{1}{2}\sum_{j=1}^{nk} K_j \Delta l_j^2(q) + \sum_{j=1}^{np} p_j h_j(q) \\
D &= \frac{1}{2}\dot{\underline{\Delta l}}_r^T [R_{\Delta l}]\,\dot{\underline{\Delta l}}_r = \frac{1}{2}\dot{q}^T \left[\frac{\partial \underline{\Delta l}_r}{\partial q}\right]^T [R_{\Delta l}] \left[\frac{\partial \underline{\Delta l}_r}{\partial q}\right] \dot{q} = \frac{1}{2} r(q)\,\dot{q}^2 \\
\delta^* L &= \underline{F}^T\,\delta^* \underline{y}_f = \underline{F}^T \left[\frac{\partial \underline{y}_f}{\partial q}\right] \delta^* q = \sum_{j=1}^{nf} F_j \left(\frac{\partial y_{fj}}{\partial q}\right) \delta^* q \; = Q^T \delta^* q
\end{aligned} \tag{1.86}$$

By applying Lagrange's equations (in scalar form, since in this case q is not a vector now)

$$\frac{d}{dt}\left(\frac{\partial E_c}{\partial \dot{q}}\right) - \frac{\partial E_c}{\partial q} + \frac{\partial D}{\partial \dot{q}} + \frac{\partial V}{\partial q} = Q, \tag{1.87}$$

we obtain:

$$\begin{aligned}
\left(\frac{\partial E_c}{\partial \dot{q}}\right) &= m(q)\dot{q} \\
\frac{d}{dt}\left(\frac{\partial E_c}{\partial \dot{q}}\right) &= m(q)\,\ddot{q} + \frac{\partial m(q)}{\partial q}\dot{q}^2 \\
\frac{\partial E_c}{\partial q} &= \frac{1}{2}\frac{\partial m(q)}{\partial q}\dot{q}^2 \\
\frac{\partial V}{\partial q} &= \sum_{j=1}^{nk} K_j \Delta l_j \left(\frac{\partial \Delta l_j}{\partial q}\right) + \sum_{j=1}^{np} p_j \left(\frac{\partial h_j}{\partial q}\right) \\
\frac{\partial D}{\partial \dot{q}} &= r(q)\dot{q} \\
Q &= \sum_{j=1}^{nf} \left(\frac{\partial y_{fi}}{\partial q}\right) F_j
\end{aligned} \tag{1.88}$$

and therefore:

$$m(q)\,\ddot{q}+\frac{1}{2}\frac{\partial m(q)}{\partial q}\dot{q}^2+r(q)\dot{q}+\sum_{j=1}^{nk}K_j\Delta l_j\left(\frac{\partial \Delta l_j}{\partial q}\right)+\sum_{j=1}^{np}p_j\left(\frac{\partial h_j}{\partial q}\right)=\sum_{j=1}^{nf}\left(\frac{\partial y_{fj}}{\partial q}\right)F_j \quad (1.89)$$

where

$$\begin{aligned} m(q) &= \left\{\frac{\partial \underline{y}_m}{\partial q}\right\}^T [M_f]\left\{\frac{\partial \underline{y}_m}{\partial q}\right\} = \left\{\frac{\partial \underline{y}_m}{\partial q}\right\}^T diag[m_i]\left\{\frac{\partial \underline{y}_m}{\partial q}\right\} \\ r(q) &= \left\{\frac{\partial \Delta \underline{l}_r}{\partial q}\right\}^T [R_f]\left\{\frac{\partial \Delta \underline{l}_r}{\partial q}\right\} = \left\{\frac{\partial \Delta \underline{l}_r}{\partial q}\right\}^T diag[r_i]\left\{\frac{\partial \Delta \underline{l}_r}{\partial q}\right\} \end{aligned} \quad (1.90)$$

The problem involved in writing this equation basically lies in solving the kinematic problem, i.e. in the definition of the relationships between the physical variables and independent variables and, consequently, in the evaluation of the terms $\left[\frac{\partial y_m}{\partial q}\right]$, $\left[\frac{\partial \Delta l_k}{\partial q}\right]$, $\left[\frac{\partial \Delta l_r}{\partial q}\right]$, $\left[\frac{\partial y_f}{\partial q}\right]$.

As we proceed, we will try to outline the methodologies used for this purpose, methodologies which, for in-plane kinematical problem, use the definition of vectors through complex numbers (see [8] and Sect. 1.2). For 3D problems, a real vectorial algorithm, developed by means of a matrix-type algorithm (see [1, 2, 5] and Sects. 1.6 and 1.7) is referred to.

The equation of motion obtained in this way (1.89) is neither linear nor generally can it be integrated analytically. In the event of our being interested in studying the problem of large oscillations, it is essential to keep account of the intrinsic nonlinearity of the system by directly integrating expressions (1.89): in order to define the solution, it is possible to resort either to numerical integration (see Chap. 4, Sect. 4.8 and bibliography [1, 14, 15]) or to approximate analytical methods that are generally valid for specific classes of equations of motion.

Let us now apply the methodology described in a general form for 1 d.o.f. systems to a standard slider-crank mechanism shown in Fig. 1.6, subjected to in-plane motion (in the horizontal plane) consisting of a connecting rod, its relative slider and a crank. We assume ideal constraints in the crank, crank-connecting rod and connecting rod-piston journal bearings and in the prismatic piston-cylinder coupling. A more suitable schematization should keep account of the relative

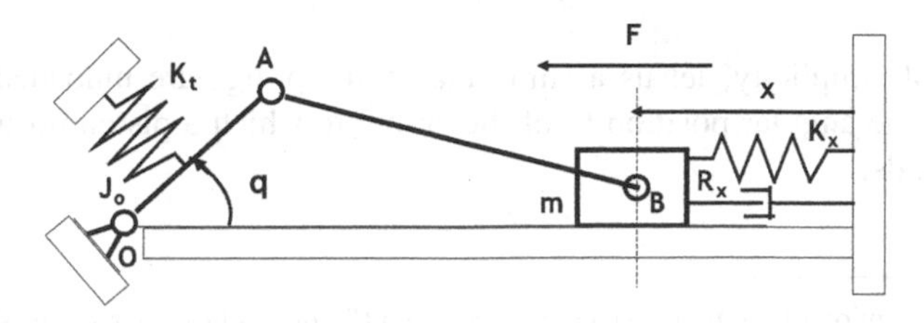

Fig. 1.6 Standard centered crank gear

motion of the journals in their housings, the ideal constraints should be released and the forces exerted by the lubricant films should be introduced instead, as functions of the variables governing the relative motion.

Let us indicate the mass of the piston by m and the mass moment of inertia of the crank with respect to its rotation axis by J_o (this will result from a *reduced mass* model of the connecting rod [8], so that this component is devoid of mass: part of it is added to the piston, which becomes of mass m, and the remaining part to the crank, whose resulting mass moment of inertia becomes J_o). The piston (see Fig. 1.6) is connected to the ground by means of a spring of stiffness K_X and a viscous damper of constant R_X: actually, these terms can represent the damping and equivalent linearized elastic effects of the fluid contained in the cylinder.[10] Furthermore, we also assume that a force, which is a function of time $F(t)$ (which allows us, for example, to represent the part of the forces transmitted by the fluid, an explicit function of time) acts on the piston while the crank, connected to the ground by means of a spring of flexural stiffness K_T, is subject to a constant torque M_o.

The overall number of degrees of freedom of the system (in plane motion) is equal to $n_t = 3 \times n_c = 9$, the number of constraints is $n_v = 8$ (i.e.: 2 for the hinge on the ground, 2 for the relative displacements for the other two hinges and finally 2 for the piston, for which vertical displacement and rotation are inhibited): the system thus only has one degree of freedom $n_l = 1$.

To consider this system as having only one degree of freedom is, as previously mentioned, a simplification, associated with the hypothesis of ideal constraints, i.e. by assuming that the relative displacements in the revolute and prismatic joints, allowed by the lubricant film, are negligible, neglecting the internal frictions and finally considering as rigid all the bodies of the system itself: thanks to this hypothesis, in order to define the motion of the system in question it is not necessary to evaluate the constraint reactions on the ground or the internal reactions in the mechanism.

A general methodology used to deal with this problem is based on the consideration that the entire system is composed of several rigid bodies, interconnected by ideal constraints and (since this is an in- plane problem) on the representation of the displacement vectors through complex numbers in exponential form (Fig. 1.7, [8]).

The horizontal displacement x of piston, measured from the attachment point of spring K_x to the ground, and rotation θ of the crank from the horizontal axis y (Fig. 1.7) are chosen as physical coordinates. The rotation of the crank is chosen as a free coordinate q (i.e. as an independent coordinate):

$$\underline{q}_l = q = \theta \tag{1.91}$$

For reasons of simplicity, let us assume that both springs are unloaded in correspondence of the angular position θ_s of the crank, to which a piston position $x = l_s$ also corresponds:

[10] A detailed description of the fluid-dynamic phenomena (fluid-structure interaction) will be given in Chap. 5.

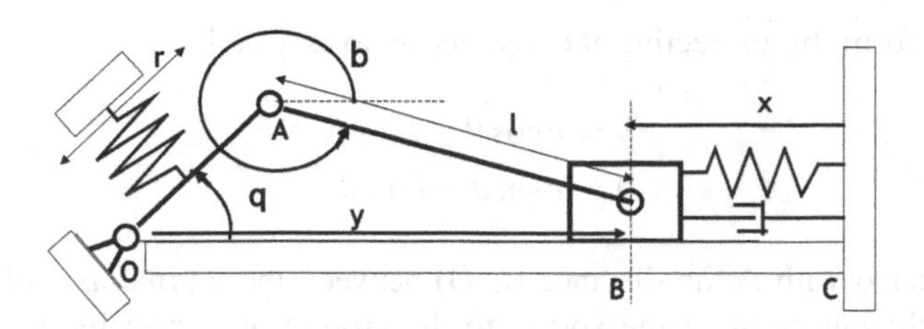

Fig. 1.7 A general approach using complex numbers

$$\begin{aligned} x &= l_s \\ \theta &= \theta_s \Rightarrow q = q_s \end{aligned} \tag{1.92}$$

We will use Lagrange's equations to write the equations of motion of the system: this choice is convenient, first and foremost, to write the various forms of energy and virtual work (1.54) as a function of the physical coordinates. As regards the system in question, the various forms of energy as a function of the physical variables chosen are:

$$\begin{aligned} E_c &= \frac{1}{2} m\dot{x}^2 + \frac{1}{2} J_o \dot{\theta}^2 \\ V &= \frac{1}{2} K_T (\theta - \theta_s)^2 + \frac{1}{2} K_X (x - l_s)^2 - M_o \theta \\ D &= \frac{1}{2} R_X \dot{x}^2 \\ \delta^* L &= F(t) \delta^* x \end{aligned} \tag{1.93}$$

The next step is to impose the relationship between the physical coordinates $\underline{y}_f$ and the independent coordinate assumed:

$$\underline{y}_f = \underline{y}_f(q) \tag{1.94}$$

In order to do this, it is possible to write the equations of constraint imposed by the kinematics of the system (see Fig. 1.7), which, in vector form, becomes:

$$(B - O) = (B - A) + (A - O) \tag{1.95}$$

This constraint (*equation of closure* of the analysed mechanism [8]) can be described in terms of complex numbers by means of the relation:

$$y = re^{i\theta} + le^{i\beta} \tag{1.96}$$

i.e. in scalar form, by projecting the vectors on axes y and z:

$$\begin{aligned} y &= r\cos\theta + l\cos\beta \\ 0 &= r\sin\theta + l\sin\beta \end{aligned} \tag{1.97}$$

Having indicated with Δ the distance (C-O) between the journal axis of the crank and the attachment point of the spring to the ground, it is possible to obtain the dependence of coordinate x from independent variable θ by means of the relation (see Fig. 1.7):

$$x = (B - C) = \Delta - re^{i\theta} - le^{i\beta} \tag{1.98}$$

Which, in scalar form, can be rewritten as:

$$\begin{cases} x = \Delta - r\cos\theta - l\cos\beta \\ 0 = -r\sin\theta - l\sin\beta \end{cases} \tag{1.99}$$

By substituting the second relation in the first, we obtain the relationship between the physical variable x and the coordinate θ, assumed as independent:

$$\begin{cases} \sin\beta = -\dfrac{r}{l}\sin\theta \\ x = \Delta - r\cos\theta - l\sqrt{1 - \sin^2\beta -} = \Delta - r\cos\theta - l\sqrt{1 - \left(\dfrac{r}{l}\right)^2 \sin^2\theta} \end{cases} \tag{1.100}$$

In order to simplify Eq. (1.100), by considering the connecting rod as infinitely long [8, 16, 17] and keeping account of (1.91), relation (1.100) becomes:

$$\begin{aligned} x &= \Delta - l - r\cos\theta = \Delta - l - r\cos q \\ \dot{x} &= r\sin q\dot{q} \end{aligned} \tag{1.101}$$

By now introducing Eq. (1.101) into the expressions of the various forms of energy (1.93), it is possible to express them as a function of the only independent variable q:

$$\begin{aligned} E_c &= \frac{1}{2}(J_o + mr^2\sin^2 q)\,\dot{q}^2 = \frac{1}{2}m(q)\,\dot{q}^2 \\ V &= \frac{1}{2}K_T(q - q_s)^2 + \frac{1}{2}K_X r^2(\cos q_s - \cos q)^2 - M_o q = V(q) \\ D &= \frac{1}{2}R_X\,\dot{x}^2 = \frac{1}{2}(R_X r^2\sin^2 q)\,\dot{q}^2 = \frac{1}{2}r*(q)\,\dot{q}^2 \\ \delta * L &= F(t)\delta^* x = F(t)\left(\frac{\partial x}{\partial q}\right)\delta^* q = (F(t)r\sin q)\delta^* q = Q(t,q)\delta^* q \end{aligned} \tag{1.102}$$

where:

$$m*(q) = J_o + mr^2 \sin^2 q \tag{1.103}$$

is used to indicate the generalized mass of the system:

$$r*(q) = R_X r^2 \sin^2 q \tag{1.104}$$

the generalized relative damping and

$$Q(t,q) = F(t) r \sin q \tag{1.105}$$

the Lagrangian component of the external force.

Being x a nonlinear function of q (1.101), both the kinetic energy E and the dissipation function D are a non-quadratic form, resulting in the generalized mass $m*(q)$ and generalized viscous damping $r*(q)$. It is at this point that the Lagrange's equations can be applied, deriving the various forms of energy (1.102) with respect to thc independent variable assumed.

By deriving the kinetic energy we obtain:

$$\begin{aligned}
\frac{\partial E_c}{\partial \dot{q}} &= \mathrm{m}^*(q)\dot{q} = (J_o + m r^2 \sin^2 q)\dot{q} \\
\frac{d}{dt}\left(\frac{\partial E_c}{\partial \dot{q}}\right) &= m*(q)\ddot{q} + \left(\frac{\partial m(q)}{\partial q}\right)\dot{q}^2 = (J_o + m r^2 \sin^2 q)\ddot{q} + 2m r^2 \sin q \cos q \dot{q}^2 \\
\left(\frac{\partial E_c}{\partial q}\right) &= \frac{1}{2}\left(\frac{\partial m*(q)}{\partial q}\right)\dot{q}^2 = m r^2 \sin q \cos q \dot{q}^2 \\
\frac{d}{dt}\left(\frac{\partial E_c}{\partial \dot{q}}\right) - \left(\frac{\partial E_c}{\partial q}\right) &= (J_o + m r^2 \sin^2 q)\ddot{q} + m r^2 \sin q \cos q \dot{q}^2
\end{aligned} \tag{1.106}$$

in which the presence of non-linear terms in q and $\dot{q}^2$ can be noted.

Since this is not a quadratic form in q, the derivative of potential energy provides a nonlinear term:

$$\frac{\partial V}{\partial q} = K_T q - K_T q_s + K_x r^2 \cos q_s \sin q - K_x r^2 \cos q \sin q - M_o \tag{1.107}$$

and, finally, the derivative of dissipation function D becomes:

$$\frac{\partial D}{\partial \dot{q}} = (R_x r^2 \sin^2 q)\dot{q} \tag{1.108}$$

The equation of motion in the large of the system analysed (Fig. 1.6) is thus a differential equation with total derivatives:

$$\begin{aligned}&(J_o + m r^2 \sin^2 q)\, \ddot{q} + m r^2 \sin q \cos q\, \dot{q}^2 + (R_x r^2 \sin^2 q)\, \dot{q}\\&+ K_T q - K_T q_s + K_x r^2 \cos q_s \sin q - K_x r^2 \cos q \sin q = M_o + F(t)\, r \sin q\end{aligned} \quad (1.109)$$

As can be seen, this equation is strongly nonlinear: in the event of our wanting to integrate (1.109), it is necessary to adopt the step-by-step numerical integration procedures described in Chap. 4, Sect. 4.7 (see also [14, 15, 18]).

1.4.2 Writing Linearized Equations

Conversely, limiting our study to small oscillation about a static equilibrium position, it is possible to linearize the same equations around the equilibrium position itself (rest or steady-state) in order to obtain a linear equation with constant coefficients. As opposed to the numeric integration of nonlinear equations, the linearization of the equations of motion permits the use of standard algorithms for the analysis of linear systems (this system will be analysed in detail in Chap. 2). In this case it is necessary to find the position, if this exists, of either static equilibrium (rest) or steady state motion, by solving generally a nonlinear equation and subsequently linearizing the equation of motion itself.

To simplify the issues and due to educational nature of this text, as we proceed we will make reference to:

- Systems with one degree of freedom, with reference, in particular, to the minimum set of variables method;
- Excitation forces applied to the system depending both on time alone and on the state of the system:

$$\underline{F} = \underline{F}(t) + \underline{F}_c(q, \dot{q}) \quad (1.110)$$

- the linearization of the same equations about a static equilibrium position (rest).

The static equilibrium position q_o is defined by solving the nonlinear equation in the variable q representing the static equilibrium condition, obtained as a particular case of the Lagrange's equation by keeping account of only the constant terms of the external forces:

$$\left(\frac{\partial V}{\partial q}\right)_o = 0 \quad (1.111)$$

In this case in question, this equation becomes:

$$K_T q_o - K_T q_s + K_x r^2 \cos q_s \sin q_o - K_x r^2 \cos q_o \sin q_o - M_o = 0 \tag{1.112}$$

The solution, i.e. the static position q_o, if this exists, is generally obtained using suitable numerical methods of the Newton-Raphson type [18].

In the neighbourhood of this position, the equation of motion is linearized by means of suitable series developments of the various functions that appear in the Lagrange's equation by introducing, as an independent variable $\bar{q}$, which describes the perturbed motion of the system in the neighbourhood of the same static equilibrium position:

$$q = q_o + \bar{q} \Rightarrow \bar{q} = q - q_o + \Rightarrow \dot{\bar{q}} = \dot{q} \Longrightarrow \ddot{\bar{q}} = \ddot{q} \tag{1.113}$$

To obtain a linear equation of motion of the vibrating system, kinetic energy E_c, dissipation function D and potential energy V (1.102) must assume a quadratic form in the independent variables.

Let us first consider kinetic energy E_c [Eqs. (1.57), (1.86)]: in the case in question, i.e. a system with only one degree of freedom, in order to obtain a quadratic form the generalized mass is obtained by a series expansion in the neighbourhood of the static equilibrium position $q = q_o$, only keeping account of the constant term, calculated in correspondence to position q_o, thus obtaining:

$$[M(q)] = \left[\frac{\partial \underline{y}_m}{\partial q}\right]^T [M_y] \left[\frac{\partial \underline{y}_m}{\partial q}\right] \approx \left[\frac{\partial \underline{y}_m}{\partial q}\right]_o^T [M_y] \left[\frac{\partial \underline{y}_m}{\partial q}\right]_o = [m(q_o)] = m_o \tag{1.114}$$

In the case in question (Fig. 1.6), by substituting (1.113) in the expression of kinetic energy (1.102) we obtain the expression, in quadratic form, of E_c as a function of the independent variable $\dot{\bar{q}}$:

$$E_{cL} = \frac{1}{2}\dot{\bar{q}}^T (J_o + mr^2 \sin^2 q_o)\dot{\bar{q}} = \frac{1}{2}\dot{\bar{q}}^T m_o \dot{\bar{q}} = \frac{1}{2} m_o \dot{\bar{q}}^2 \tag{1.115}$$

By now applying the Lagrange's equations to expression E_{cL}, (1.115), expressed in quadratic form, the linearized inertia terms in the neighbourhood of the static equilibrium position are obtained:

$$\left\{\frac{d}{dt}\left(\frac{\partial E_{cL}}{\partial \dot{\bar{q}}}\right)\right\} - \left\{\frac{\partial E_{cL}}{\partial \bar{q}}\right\} = m_o \ddot{\bar{q}} \tag{1.116}$$

As with the kinetic energy, dissipation function D (1.60), (1.86) must also be expressed in quadratic form, making reference to Taylor series expansion of the damping matrix $[R]$ up to the constant term:

$$[R(q)] = \left[\frac{\partial \Delta l_r}{\partial q}\right]^T [R_y] \left[\frac{\partial \Delta l_r}{\partial q}\right] \approx \left[\frac{\partial \Delta l_r}{\partial q}\right]_o^T [R_y] \left[\frac{\partial \Delta l_r}{\partial q}\right]_o = [R(q_o)] = r_o \quad (1.117)$$

In the case of the slider-crank mechanism analysed (1.102), it is possible to make the dissipation function quadratic:

$$D_L = \frac{1}{2}\dot{\overline{q}}^T \left(R_x r^2 \sin^2 q_o\right)\dot{\overline{q}} = \frac{1}{2}\dot{\overline{q}}^T r_o \dot{\overline{q}} = \frac{1}{2} r_o \dot{\overline{q}}^2 \quad (1.118)$$

from which:

$$\left\{\frac{\partial D_L}{\partial \dot{\overline{q}}}\right\} = r_o \dot{\overline{q}} \quad (1.119)$$

In order to linearize the equations of motion of the general nonlinear vibrating system, potential energy V must also be made quadratic: for this purpose, we proceed with the Taylor series expansion of function $V = V(q)$ (1.59) and (1.86) about the static equilibrium position, defined by the relation $q = q_o$, up to the quadratic term:

$$V_L = V(q_o) + \left\{\frac{\partial V}{\partial q}\right\}_o \overline{q} + \frac{1}{2}\left[\frac{\partial^2 V}{\partial q^2}\right]_o \overline{q}^2 = V(q_o) + \left\{\frac{\partial V}{\partial q}\right\}_o \overline{q} + \frac{1}{2} k_o \overline{q}^2 \quad (1.120)$$

after using, as mentioned in (1.113), variable $\overline{q}$ to define the perturbed motion in the neighbourhood of the static equilibrium position and having indicated generalized stiffness with:

$$k_o = \left[\frac{\partial^2 V}{\partial q^2}\right]_o \quad (1.121)$$

By remembering the equation that permitted us to define the static equilibrium position $\left(\frac{\partial V}{\partial q}\right)_o = 0$ (1.111) and by recalling that in Lagrange's equations it is necessary to derive the potential energy with respect to the independent variable $\overline{q}$ (to whose derivative the constant terms and, in particular, term $V(q_o)$, do not give any contribution), the expression of the linearized potential energy itself can always be rewritten as:

$$V_L = \frac{1}{2} k_o \overline{q}^2 \quad (1.122)$$

from which:

$$\frac{\partial V_L}{\partial q} = k_o \overline{q} \quad (1.123)$$

By analysing a generic system with one degree of freedom, the expression of potential energy V (1.86) in scalar form is:

$$V = \frac{1}{2}\sum_{j=1}^{nk} k_j \Delta l_j^2 + \sum_{j=1}^{np} p_j h_j \tag{1.124}$$

we obtain:

$$\frac{\partial V}{\partial q} = \frac{1}{2}\sum_{j=1}^{nk} \frac{\partial\left(k_j \Delta l_j^2\right)}{\partial \Delta l_j}\frac{\partial \Delta l_j}{\partial q} + \sum_{j=1}^{np} p_j \frac{\partial h_j}{\partial q} = \sum_{j=1}^{nk} k_j \Delta l_j \left(\frac{\partial \Delta l_j}{\partial q}\right) + \sum_{j=1}^{np} p_j \left(\frac{\partial h_j}{\partial q}\right) \tag{1.125}$$

and similarly the second derivative of potential energy can be expressed as:

$$\frac{\partial^2 V}{\partial q^2} = \sum_{j=1}^{nk} k_j \Delta l_j \left(\frac{\partial^2 \Delta l_j}{\partial q^2}\right) + \sum_{j=1}^{nk} k_j \left(\frac{\partial \Delta l_j}{\partial q}\right)^2 + \sum_{j=1}^{np} p_j \left(\frac{\partial^2 h_j}{\partial q^2}\right) \tag{1.126}$$

The stiffness of the linearized system (1.121), (1.126), thus becomes:

$$\begin{aligned} \left.\frac{\partial^2 V}{\partial q^2}\right|_0 &= \sum_{j=1}^{nk} k_j \Delta l_{jo} \left(\frac{\partial^2 \Delta l_j}{\partial q^2}\right)_o + \sum_{j=1}^{nk} k_j \left(\frac{\partial \Delta l_j}{\partial q}\right)_o^2 + \sum_{j=1}^{np} p_j \left(\frac{\partial^2 h_j}{\partial q^2}\right)_o \\ &= k'_o + k'_o + k'''_o = k_o \end{aligned} \tag{1.127}$$

In particular, in (1.127), the single terms assume the following meaning. The term:

$$k'_o = \sum_{j=1}^{nk} k_j \Delta l_{jo} \left(\frac{\partial^2 \Delta l_j}{\partial q^2}\right)_o = \sum_{j=1}^{nk} P_{recjo} \left(\frac{\partial^2 \Delta l_j}{\partial q^2}\right)_o \tag{1.128}$$

represents the elastic restoring force due to the static pre-load $P_{recjo} = k_j \Delta l_{jo}$ of the springs, a non-null term if the second derivative of elongation is different from zero, e.g. if the link between the elongation of the single elastic elements Δl_j depends in linear form on the independent variable q. The term:

$$k''_o = \sum_{j=1}^{nk} k_j \left(\frac{\partial \Delta l_j}{\partial q}\right)_o^2 \tag{1.129}$$

keeps account of the stiffness of the generic jth spring, according to the free coordinate $\overline{q}$ of the system: this term is what is usually identified as "system stiffness", normally different from zero. Finally, in (1.127):

$$k'''_o = \sum_{j=1}^{np} p_j \left(\frac{\partial^2 h_j}{\partial q^2}\right)_o \tag{1.130}$$

is the term due to the constant loads p_j which are non- null if the second derivative $\left(\frac{\partial^2 h_j}{\partial q^2}\right)_o$ is different from zero: this term is conceptually similar to that of (1.128), even though, in this case, it is a question of a constant force that has been specifically applied and which does not derive from the preload of a spring.

The constant forces (whether external or due to preloaded springs) can thus give a non-null contribution to the system stiffness in the equations that define the perturbed motion of same: this occurs if the displacement of their application point is a nonlinear function of the independent variable q, with a non-null second derivative in correspondence to the static solution q_o.

Every time that constant forces (such as e.g. weight) or preloads, due to springs, act on the system, these will not appear in the linearized equation of motion around the static equilibrium position if the relationship between the physical variables (components of the displacements of the application points of the forces or elongations of the springs) are linear functions of the free coordinate q. In fact, in the case of a linear relation, the second derivative of functions v and $h_j(q)$ is null and thus the terms (1.128) and (1.130) are null. Viceversa, in the case of a nonlinear relationship, the constant forces (or the preloads of the springs, deriving from the static solution), multiplied by terms $\left(\frac{\partial^2 \Delta l_j}{\partial q^2}\right)_o$ o $\left(\frac{\partial^2 h_j}{\partial q^2}\right)_o$, calculated in correspondence to the static equilibrium position, will generally appear in (1.127).

In the case of the standard slider-crank mechanism analysed (Fig. 1.6), the various terms of (1.120), keeping account of (1.125) and (1.126), starting from expression (1.102) and (1.107), become:

$$
\begin{aligned}
V(q_o) &= \frac{1}{2}K_T(q_o - q_s)^2 + \frac{1}{2}K_X r^2(\cos q_s - \cos q_o)^2 - M_o q_o = V_o \\
\left(\frac{\partial V}{\partial q}\right) &= K_T q - K_T q_s + K_x r^2 \cos q_s \sin q - K_x r^2 \cos q \sin q - M_o \\
\left(\frac{\partial V}{\partial q}\right)_o &= K_T q_o - K_T q_s + K_x r^2 \cos q_s \sin q_o - K_x r^2 \cos q_o \sin q_o - M_o = 0 \\
\left[\frac{\partial^2 V}{\partial q^2}\right] &= K_T + K_x r^2 \cos q_s \cos q + K_x r^2 \sin^2 q - K_x r^2 \cos^2 q \\
&= K_T + K_x r^2 \cos q_s \cos q - K_x r^2 \cos 2q \\
\left[\frac{\partial^2 V}{\partial q^2}\right]_o &= K_T + K_x r^2 \cos q_s \cos q_o - K_x r^2 \cos 2q_o = k_o
\end{aligned}
\tag{1.131}
$$

In the case in question, the expression of potential energy that has been rendered quadratic (1.120), thus becomes:

$$
V_L = \frac{1}{2}\left[\frac{\partial^2 V}{\partial q^2}\right]_o \bar{q}^2 = \frac{1}{2}k_o \bar{q}^2 = \frac{1}{2}\left(K_T + K_x r^2 \cos q_s \cos q_o - K_x r^2 \cos 2q_o\right)\bar{q}^2 \tag{1.132}
$$

By using Lagrange's equation to derive the quadratic form of potential energy V_L, (1.132), we obtain:

$$\left(\frac{\partial V_L}{\partial q}\right) = (K_T + K_x r^2 \cos q_s \cos q_o - K_x r^2 \cos 2q_o)\overline{q} = k_o \overline{q} \quad (1.133)$$

As a last step, we consider the virtual work $\delta^* L$ of the non-conservative forces: these forces can be rewritten, always for a system with one degree of freedom, as:

$$\underline{F}(t) + \underline{F}_c(q, \dot{q}) = \overline{\underline{F}}(t) + \overline{\underline{F}}_c(\overline{q}, \dot{\overline{q}}) \quad (1.134)$$

By expanding the Lagrangian component of these forces around the static equilibrium position (1.111) and by keeping account of expressions (1.134), this becomes:

$$\begin{aligned} Q &= \left\{\frac{\partial \underline{y}_f}{\partial q}\right\}^T \{\overline{\underline{F}}(t) + \overline{\underline{F}}_c(\overline{q}, \dot{\overline{q}})\} \\ &\approx \left\{\left\{\frac{\partial \underline{y}_f}{\partial q}\right\}_o^T + \left\{\frac{\partial^2 \underline{y}_f}{\partial q^2}\right\}_o^T \overline{q}\right\}\left\{\overline{\underline{F}}(t) + \left\{\frac{\partial \overline{\underline{F}}_c}{\partial q}\right\}_o \overline{q} + \left\{\frac{\partial \overline{\underline{F}}_c}{\partial \dot{q}}\right\}_o \dot{\overline{q}}\right\} \end{aligned} \quad (1.135)$$

i.e., by reorganizing the relation and eliminating the higher order terms:

$$\begin{aligned} Q &\approx \left\{\frac{\partial \underline{y}_f}{\partial q}\right\}_o^T \overline{\underline{F}}(t) + \left\{\frac{\partial^2 \underline{y}_f}{\partial q^2}\right\}_o^T \overline{\underline{F}}(t)\overline{q} + \left\{\frac{\partial \underline{y}_f}{\partial q}\right\}_o^T \left\{\frac{\partial \overline{\underline{F}}_c}{\partial q}\right\}_o \overline{q} + \left\{\frac{\partial \underline{y}_f}{\partial q}\right\}_o^T \left\{\frac{\partial \overline{\underline{F}}_c}{\partial \dot{q}}\right\}_o \dot{\overline{q}} \\ &= \overline{Q}(t) + k'_{Fqo}(t)\overline{q} + k_{Fqo}\overline{q} + r_{Fqo}\dot{\overline{q}} \end{aligned} \quad (1.136)$$

being:

$$\begin{aligned} \overline{Q}(t) &= \left\{\frac{\partial \underline{y}_f}{\partial q}\right\}_o^T \overline{\underline{F}}(t) \\ k'_{Fo}(t) &= \left\{\frac{\partial^2 \underline{y}_f}{\partial q^2}\right\}_o^T \overline{\underline{F}}(t) \\ k_{Fqo} &= \left\{\frac{\partial \underline{y}_f}{\partial q}\right\}_o^T \left\{\frac{\partial \overline{\underline{F}}_c}{\partial q}\right\}_o \\ r_{Fqo} &= \left\{\frac{\partial \underline{y}_f}{\partial q}\right\}_o^T \left\{\frac{\partial \overline{\underline{F}}_c}{\partial \dot{q}}\right\}_o \end{aligned} \quad (1.137)$$

By keeping account of (1.114), (1.119), (1.127) and (1.136), the equations of motion of the system with 1 degree of freedom linearized in the neighbourhood of the static equilibrium position thus become:

$$\begin{aligned} m_o\ddot{\overline{q}} + r_o\dot{\overline{q}} + k_o\overline{q} &= \overline{Q}(t) + k'_{Fo}(t)\,\overline{q} + k_{Fqo}\overline{q} + r_{Fqo}\dot{\overline{q}} \\ m_o\ddot{\overline{q}} + \left[r_o - r_{Fqo}\right]\dot{\overline{q}} + \left[k_o - k'_{Fo}(t) - k_{Fqo}\right]\overline{q} &= \overline{Q}(t) \end{aligned} \tag{1.138}$$

We notice that also keeping account of the second term of expansion (1.136) once again a linear equation in $\overline{q}$ is obtained but with coefficients that are functions of time $k'_{Fo}(t)$; this term gives rise to a variable stiffness in time k_{tot} which makes the system itself parametric:

$$m_o\ddot{\overline{q}} + r_{toto}\dot{\overline{q}} + k_{tot}(t)\,\overline{q} = \overline{Q}(t) \tag{1.139}$$

In the event of our wishing to obtain an equation with constant coefficients, it is not necessary to consider the linear term of the series expansion for the forces that are functions of time. In this case, the writing of the linearized equation with constant coefficients is as follows:

$$m_o\ddot{\overline{q}} + r_{toto}\dot{\overline{q}} + k_{toto}\overline{q} = \overline{Q}(t) \tag{1.140}$$

where:

$$k_{toto} = k_o - k_{Fqo} \tag{1.141}$$

As can be seen (1.141), in the equilibrium equations linearized in the neighbourhood of the static equilibrium position (rest), only the Lagrangian component of the dynamic part (i.e. variable in time) of the non-conservative forces $\overline{Q}(t)$ is present. It has been obtained by multiplying the physical forces by the Jacobian of the application points of the excitation forces, evaluated in the same static equilibrium position: for this reason, this term proves to be a function of time only.

By returning to the slider-crank example (Fig. 1.6), the virtual work (1.102) is defined by linearizing the equations of motions as:

$$\begin{aligned} \delta^* L &= (F(t)\, r\sin q)\delta * q \cong (F(t)\, r\sin q_o - (F(t)\, r\cos q_o)\,\overline{q})\delta * q \\ &= \left(\overline{F}(t) - k'_{Fo}(t)\,\overline{q}\right)\delta * q \end{aligned} \tag{1.142}$$

from which the Lagrangian component becomes:

$$\overline{Q}(t) = F(t) r\sin q_o - (F(t) r\cos q_o)\overline{q} = \overline{F}(t) - k'_{Fo}(t)\overline{q} \tag{1.143}$$

By keeping account of (1.115), (1.118), (1.133) and (1.143), the linearized equation of motion of the slider-crank in Fig. 1.6 (see 1.139), is the following:

$$\begin{aligned} &(J_o + mr^2\sin^2 q_o)\,\ddot{\overline{q}} + (R_x r^2\sin^2 q_o)\,\dot{\overline{q}} + (K_T + K_x r^2\cos q_s\cos q_o - K_x r^2\cos 2q_o)\overline{q} \\ &= F(t)\, r\sin q_o - (F(t)\, r\cos q_o)\,\overline{q}(J_o + mr^2\sin^2 q_o)\,\ddot{\overline{q}} + (R_x r^2\sin^2 q_o)\,\dot{\overline{q}} \\ &\quad + \left(K_T + K_x r^2\cos q_s\cos q_o - K_x r^2\cos 2q_o + (F(t)\, r\cos q_o)\right)\overline{q} = F(t)\, r\sin q_o \end{aligned} \tag{1.144}$$

which, by neglecting the parametric stiffness term, which is a function of time, is reduced to:

$$\begin{aligned}&\left(J_o + mr^2 \sin^2 q_o\right)\ddot{\overline{q}} + \left(R_x r^2 \sin^2 q_o\right)\dot{\overline{q}} \\ &+ \left(K_T + K_x r^2 \cos q_s \cos q_o - K_x r^2 \cos 2q_o\right)\overline{q} = F(t)\, r \sin q_o\end{aligned} \tag{1.145}$$

that is:

$$m_o \ddot{\overline{q}} + r_o \dot{\overline{q}} + k_o \overline{q} = \overline{Q}(t) \tag{1.146}$$

i.e. a differential equation in $\overline{q}$ with constant coefficients.

1.5 Nonlinear Systems with 2 Degrees of Freedom

As an example of a nonlinear system with 2 degrees of freedom, let us consider the system in Fig. 1.8 consisting of a mass m (weight $P = mg$) with 2 degrees of freedom, associated with the horizontal and vertical translation. Let us now define the horizontal and vertical displacements x and y of the centre of gravity of the body as independent variables, whereby the position of the unloaded vertical and horizontal spring corresponds to $x = 0, y = 0$.

The equations of motion of the system can be obtained by using Lagrange's equations: the relative forms of energy of the system being examined are:

$$\begin{aligned}&E_c = \frac{1}{2} m\dot{x}^2 + \frac{1}{2} m\dot{y}^2 \\ &V = \frac{1}{2} K_h \Delta l_h^2 + \frac{1}{2}\frac{K_v}{2}\Delta l_v^2 + \frac{1}{2}\frac{K_v}{2}\Delta l_v^2 + Ph = \frac{1}{2} K_h \Delta l_h^2 + \frac{1}{2} K_v \Delta l_v^2 + Ph \\ &D = \frac{1}{2} R_h \Delta\dot{l}_h^2 + \frac{1}{2}\frac{R_v}{2}\Delta\dot{l}_v^2 + \frac{1}{2}\frac{R_v}{2}\Delta\dot{l}_v^2 \\ &\delta * L = 0\end{aligned} \tag{1.147}$$

where Δl_h and Δl_v are respectively the elongations of the horizontal and vertical spring.

As is customary, for reasons of convenience, these energy forms are expressed as a function of the physical variables by means of which it is possible to define the expressions themselves. These physical variables (elongation of the vertical Δl_v and lateral Δl_h elastic elements and elevation of the centre of gravity of body h) are functions of the independent variables x and y assumed to describe the motion of the system by means of the following expressions (also see Fig. 1.9)[11]:

[11]Spring K_h is constrained to the ground by a sliding block meaning that its elongation coincides with the value associated with coordinate x.

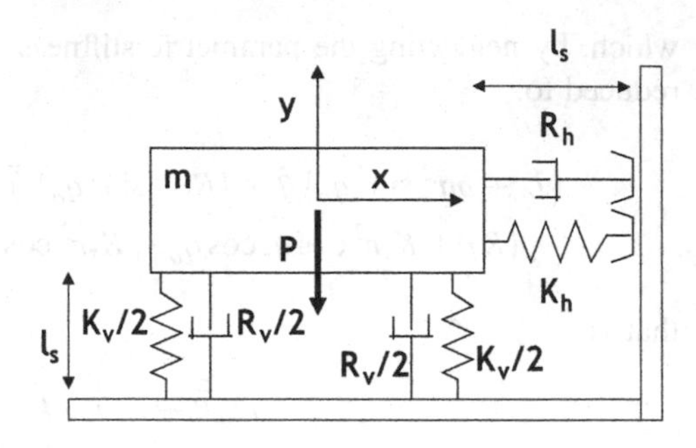

Fig. 1.8 Vibrating system with two d.o.f

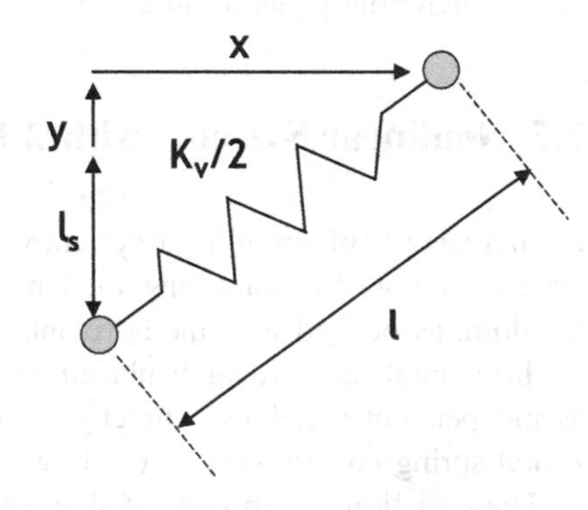

Fig. 1.9 Elongation of the vertical elastic element

$$
\begin{aligned}
\Delta l_h &= x \\
\Delta l_v &= \sqrt{x^2 + (l_s + y)^2} - l_s = l - l_s \\
h &= y \\
\dot{\Delta l}_h &= \dot{x} \\
\dot{\Delta l}_v &= \frac{\partial \Delta l_v}{\partial x}\dot{x} + \frac{\partial \Delta l_v}{\partial y}\dot{y}
\end{aligned} \tag{1.148}
$$

in which the length of the vertical springs in undeformed configuration has been indicated by l_s and the corresponding Jacobians by $\frac{\partial \Delta l_v}{\partial x}$ and $\frac{\partial \Delta l_v}{\partial y}$:

$$
\begin{aligned}
\frac{\partial \Delta l_v}{\partial x} &= \frac{1}{\sqrt{x^2 + (l_s + y)^2}} x = \frac{x}{l} \\
\frac{\partial \Delta l_v}{\partial y} &= \frac{1}{\sqrt{x^2 + (l_s + y)^2}} (l_s + y) = \frac{(l_s + y)}{l}
\end{aligned} \tag{1.149}
$$

By keeping account of expressions (1.147) the various forms of energy (1.146) become:

$$\begin{aligned}
E_c &= \frac{1}{2}m\dot{x}^2 + \frac{1}{2}m\dot{y}^2 \\
V &= \frac{1}{2}K_h x^2 + \frac{1}{2}K_v\left(\sqrt{x^2 + (l_s + y)^2} - l_s\right)^2 + Py \\
D &= \frac{1}{2}R_h\dot{x}^2 + \frac{1}{2}R_v\left(\left(\frac{\partial \Delta l_v}{\partial x}\right)\dot{x} + \left(\frac{\partial \Delta l_v}{\partial y}\right)\dot{y}\right)^2 \\
\delta * L &= 0
\end{aligned} \tag{1.150}$$

By keeping account of (1.150) and (1.149), by means of equation:

$$\frac{d}{dt}\left(\frac{\partial E_c}{\partial \dot{q}_i}\right) - \frac{\partial E_c}{\partial q_i} + \frac{\partial D}{\partial \dot{q}_i} + \frac{\partial V}{\partial q_i} = Q_i \quad (i = 1, 2) \tag{1.151}$$

where $(q_i = x, y)$, it is possible to obtain the nonlinear equations of motion of the system:

$$\begin{cases}
m\ddot{x} + R_h\dot{x} + R_v\left(\frac{\partial \Delta l_v}{\partial x}\right)^2 \dot{x} + R_v\left(\frac{\partial \Delta l_v}{\partial x}\right)\left(\frac{\partial \Delta l_v}{\partial y}\right)\dot{y} + K_h x + K_v\Delta l_v\frac{\partial \Delta l_v}{\partial x} = 0 \\
m\ddot{y} + R_v\left(\frac{\partial \Delta l_v}{\partial x}\right)\left(\frac{\partial \Delta l_v}{\partial y}\right)\dot{x} + R_v\left(\frac{\partial \Delta l_v}{\partial y}\right)^2 \dot{y} + K_v\Delta l_v\frac{\partial \Delta l_v}{\partial y} = -P
\end{cases} \tag{1.152}$$

namely, by keeping account of (1.149):

$$\begin{cases}
m\ddot{x} + R_h\dot{x} + R_v\left(\frac{x}{l}\right)^2 \dot{x} + R_v\left(\frac{x(l_s + y)}{l^2}\right)\dot{y} + K_h x + K_v\frac{l - l_s}{l}x = 0 \\
m\ddot{y} + R_v\left(\frac{x(l_s + y)}{l^2}\right)\dot{x} + R_v\left(\frac{(l_s + y)}{l}\right)^2 \dot{y} + K_v\frac{l - l_s}{l}(l_s + y) = -P
\end{cases} \tag{1.153}$$

By keeping account of expressions (1.148), these equations enable us to describe the motion in the large of the system, they are differential to the total nonlinear derivatives and, for this reason, must be integrated using step-by-step numerical methods [1, 19].

In order to obtain the static equilibrium equations of the system in question, it is necessary to solve the expressions:

$$\begin{cases}
K_h x + K_v\frac{l - l_s}{l}x = 0 \\
K_v\frac{l - l_s}{l}(l_s + y) = -P
\end{cases}
=
\begin{cases}
\left(K_h + K_v\frac{l - l_s}{l}\right)x = 0 \\
K_v\frac{y}{l}(l - l_s) + K_v\frac{l_s}{l}(l - l_s) = -P
\end{cases} \tag{1.154}$$

As is generally the case, these algebraic equations are nonlinear, meaning that they can thus be solved by using iterative numerical methods of the Newton-Raphson type [1, 19]. In the case in question, they are simple and linear with solution:

$$\begin{cases} x_o = 0 \\ K_v \dfrac{y_o}{l_s + y_o}(l_s + y_o - l_s) + K_v \dfrac{l_s}{l_s + y_o}(l_s + y_o - l_s) = -P \end{cases} \Rightarrow \begin{cases} x_o = 0 \\ y_o^2 + l_s y_o + \dfrac{P}{K_v}(l_s + y_o) = 0 \end{cases}$$
$$\begin{cases} x_o = 0 \\ y_o^2 + y_o\left(l_s + \dfrac{P}{K_v}\right) + \dfrac{P}{K_v} l_s = 0 \end{cases} \Rightarrow \begin{cases} x_o = 0 \\ y_o = -\dfrac{P}{K_v} \end{cases} \tag{1.155}$$

The length of the vertical springs in the static equilibrium position are therefore:

$$l_o = l_s + y_o = l_s - \frac{P}{K_v} \tag{1.156}$$

In order to obtain linearized equations of motion in the neighbourhood of the static equilibrium position defined by expressions (1.155), it is necessary to expand the various forms of energy (1.150) in the neighbourhood of this position up to the second derivatives:

$$\begin{aligned} E_c &= \frac{1}{2} m\dot{x}^2 + \frac{1}{2} m\dot{y}^2 \\ V &= V_o + \left(\frac{\partial V}{\partial x}\right)_o (x - x_o) + \left(\frac{\partial V}{\partial y}\right)_o (y - y_o) \\ &\quad + \frac{1}{2}\left(\frac{\partial^2 V}{\partial x^2}\right)_o (x - x_o)^2 + \frac{1}{2}\left(\frac{\partial^2 V}{\partial y^2}\right)_o (y - y_o)^2 + \left(\frac{\partial^2 V}{\partial y \partial x}\right)_o (y - y_o)(x - x_o) \\ D &= \frac{1}{2} R_h \left(\left(\frac{\partial \Delta l_h}{\partial x}\right)_o \dot{x} + \left(\frac{\partial \Delta l_h}{\partial y}\right)_o \dot{y}\right)^2 + \frac{1}{2} R_v \left(\left(\frac{\partial \Delta l_v}{\partial x}\right)_o \dot{x} + \left(\frac{\partial \Delta l_v}{\partial y}\right)_o \dot{y}\right)^2 \\ \delta * L &= 0 \end{aligned} \tag{1.157}$$

By introducing the independent variables $\bar{x}$ and $\bar{y}$ which define the perturbed motion in the neighbourhood of the static equilibrium position:

$$\begin{aligned} x &= x_o + \bar{x} \\ y &= y_o + \bar{y} \end{aligned} \tag{1.158}$$

these expressions (1.157) become:

$$
\begin{aligned}
E_c &= \frac{1}{2}m\dot{\bar{x}}^2 + \frac{1}{2}m\dot{\bar{y}}^2 \\
V &= V_o + \left(\frac{\partial V}{\partial x}\right)_o \bar{x} + \left(\frac{\partial V}{\partial y}\right)_o \bar{y} + \frac{1}{2}\left(\frac{\partial^2 V}{\partial x^2}\right)_o \bar{x}^2 + \frac{1}{2}\left(\frac{\partial^2 V}{\partial y^2}\right)_o \bar{y}^2 + \left(\frac{\partial^2 V}{\partial y \partial x}\right)_o \overline{yx} \\
D &= \frac{1}{2}\left(R_h\left(\frac{\partial \Delta l_h}{\partial x}\right)_o^2 + R_v\left(\left(\frac{\partial \Delta l_v}{\partial x}\right)^2\right)_o\right)\dot{\bar{x}}^2 + \frac{1}{2}\left(R_h\left(\frac{\partial \Delta l_h}{\partial y}\right)_o^2 + R_v\left(\left(\frac{\partial \Delta l_v}{\partial y}\right)^2\right)_o\right)\dot{\bar{y}}^2 \\
&\quad + \left(R_h\left(\frac{\partial \Delta l_h}{\partial x}\right)_o\left(\frac{\partial \Delta l_h}{\partial y}\right)_o + R_v\left(\frac{\partial \Delta l_v}{\partial x}\right)_o\left(\frac{\partial \Delta l_v}{\partial y}\right)_o\right)\overline{\dot{x}\dot{y}} \\
\delta * L &= 0
\end{aligned}
\tag{1.159}
$$

To evaluate these quadratic expressions, it is necessary to calculate the Jacobians and the Hessians of the physical variables as a function of the independent variables: by keeping account of expressions (1.148) and (1.149) we obtain:

$$
\begin{aligned}
&\frac{\partial \Delta l_h}{\partial x} = 1; \quad \frac{\partial \Delta l_h}{\partial y} = 0 \\
&\frac{\partial \Delta l_v}{\partial x} = \frac{x}{\sqrt{x^2 + (l_s + y)^2}} = \frac{x}{l}; \quad \frac{\partial \Delta l_v}{\partial y} = \frac{(l_s + y)}{\sqrt{x^2 + (l_s + y)^2}} = \frac{(l_s + y)}{l} \\
&\frac{\partial h}{\partial x} = 0; \quad \frac{\partial h}{\partial y} = 1
\end{aligned}
\tag{1.160}
$$

$$
\begin{aligned}
&\frac{\partial^2 \Delta l_h}{\partial x^2} = \frac{\partial^2 \Delta l_h}{\partial y \partial x} = \frac{\partial^2 \Delta l_h}{\partial y^2} = 0 \\
&\frac{\partial^2 \Delta l_v}{\partial x^2} = \frac{1}{\sqrt{x^2 + (l_s + y)^2}} - \frac{1}{2}\frac{1}{\left(x^2 + (l_s + y)^2\right)^{\frac{3}{2}}} 2x^2 \\
&\frac{\partial^2 \Delta l_v}{\partial y \partial x} = \frac{\partial^2 \Delta l_v}{\partial x \partial y} = -\frac{1}{2}\frac{1}{\left(x^2 + (l_s + y)^2\right)^{\frac{3}{2}}} 2x(l_s + y) \\
&\frac{\partial^2 \Delta l_v}{\partial y^2} = \frac{1}{\sqrt{x^2 + (l_s + y)^2}} - \frac{1}{2}\frac{1}{\left(x^2 + (l_s + y)^2\right)^{\frac{3}{2}}} 2\,(l_s + y)^2 \\
&\frac{\partial^2 h}{\partial x^2} = \frac{\partial^2 h}{\partial y \partial x} = \frac{\partial^2 h}{\partial y^2} 0
\end{aligned}
$$

These functions, calculated in the neighbourhood of the static equilibrium position (keeping account of expressions (1.155)), become:

$$\begin{aligned}
&\left(\frac{\partial \Delta l_h}{\partial x}\right)_o = 1; \quad \left(\frac{\partial \Delta l_h}{\partial y}\right)_o = 0\\
&\left(\frac{\partial \Delta l_v}{\partial x}\right)_o = 0; \quad \left(\frac{\partial \Delta l_v}{\partial y}\right)_o = \frac{(l_s + y_o)}{l_o} = \frac{l_o}{l_o} = 1\\
&\left(\frac{\partial h}{\partial x}\right)_o = 0; \quad \left(\frac{\partial h}{\partial y}\right)_o = 1
\end{aligned} \tag{1.161}$$

$$\begin{aligned}
&\left(\frac{\partial^2 \Delta l_h}{\partial x^2}\right)_o = \left(\frac{\partial^2 \Delta l_h}{\partial y \partial x}\right)_o = \left(\frac{\partial^2 \Delta l_h}{\partial y^2}\right)_o = 0\\
&\left(\frac{\partial^2 \Delta l_v}{\partial x^2}\right)_o = \frac{1}{\sqrt{(l_s + y_o)^2}} = \frac{1}{l_o}; \left(\frac{\partial^2 \Delta l_v}{\partial y \partial x}\right)_o = \left(\frac{\partial^2 \Delta l_v}{\partial x \partial y}\right)_o = 0\\
&\left(\frac{\partial^2 \Delta l_v}{\partial y^2}\right)_o = \frac{1}{\sqrt{(l_s + y_o)^2}} - \frac{1}{2}\frac{1}{\left((l_s + y_o)^2\right)^{\frac{3}{2}}} 2\,(l_s + y_o)^2 = \frac{1}{l_o} - \frac{1}{l_o^3} l_o^2 = 0\\
&\left(\frac{\partial^2 h}{\partial x^2}\right)_o = \left(\frac{\partial^2 h}{\partial y \partial x}\right)_o = \left(\frac{\partial^2 h}{\partial y^2}\right)_o = 0
\end{aligned} \tag{1.162}$$

The quadratic expression of potential energy (1.157) can be expressed as:

$$\begin{aligned}
&\left(\frac{\partial V}{\partial x}\right)_o = \left(\frac{\partial V}{\partial \Delta l_h}\right)_o \left(\frac{\partial \Delta l_h}{\partial x}\right)_o + \left(\frac{\partial V}{\partial \Delta l_v}\right)_o \left(\frac{\partial \Delta l_v}{\partial x}\right)_o + \left(\frac{\partial V}{\partial h}\right)_o \left(\frac{\partial h}{\partial x}\right)_o = (K_h \Delta l_h)_o \left(\frac{\partial \Delta l_h}{\partial x}\right)_o + (K_v \Delta l_v)_o \left(\frac{\partial \Delta l_v}{\partial x}\right)_o\\
&\left(\frac{\partial V}{\partial y}\right)_o = \left(\frac{\partial V}{\partial \Delta l_h}\right)_o \left(\frac{\partial \Delta l_h}{\partial y}\right)_o + \left(\frac{\partial V}{\partial \Delta l_v}\right)_o \left(\frac{\partial \Delta l_v}{\partial y}\right)_o + \left(\frac{\partial V}{\partial h}\right)_o \left(\frac{\partial h}{\partial y}\right)_o = (K_h \Delta l_h)_o \left(\frac{\partial \Delta l_h}{\partial y}\right)_o + (K_v \Delta l_v)_o \left(\frac{\partial \Delta l_v}{\partial y}\right)_o + P\\
&\left(\frac{\partial^2 V}{\partial x^2}\right)_o = K_h \left(\frac{\partial \Delta l_h}{\partial x}\right)_o^2 + (K_h \Delta l_h)_o \left(\frac{\partial^2 \Delta l_h}{\partial x^2}\right)_o + K_v \left(\frac{\partial \Delta l_v}{\partial x}\right)_o^2 + (K_v \Delta l_v)_o \left(\frac{\partial^2 \Delta l_v}{\partial x^2}\right)_o\\
&\left(\frac{\partial^2 V}{\partial y^2}\right)_o = K_h \left(\frac{\partial \Delta l_h}{\partial y}\right)_o^2 + (K_h \Delta l_h)_o \left(\frac{\partial^2 \Delta l_h}{\partial y^2}\right)_o + K_v \left(\frac{\partial \Delta l_v}{\partial y}\right)_o^2 + (K_v \Delta l_v)_o \left(\frac{\partial^2 \Delta l_v}{\partial y^2}\right)_o\\
&\left(\frac{\partial^2 V}{\partial x \partial y}\right)_o = \left(\frac{\partial^2 V}{\partial y \partial x}\right)_o = K_h \left(\frac{\partial \Delta l_h}{\partial y}\right)_o \left(\frac{\partial \Delta l_h}{\partial x}\right)_o + (K_h \Delta l_h)_o \left(\frac{\partial^2 \Delta l_h}{\partial x \partial y}\right)_o + K_v \left(\frac{\partial \Delta l_v}{\partial y}\right)_o \left(\frac{\partial \Delta l_v}{\partial x}\right)_o + (K_v \Delta l_v)_o \left(\frac{\partial^2 \Delta l_v}{\partial x \partial y}\right)_o
\end{aligned} \tag{1.163}$$

By keeping account of expressions (1.161) and (1.162), the various terms of (1.163) are reduced to the expressions:

$$\begin{aligned}
&\left(\frac{\partial V}{\partial x}\right)_o = K_h \Delta l_h; \quad \left(\frac{\partial V}{\partial y}\right)_o = (K_v \Delta l_v)_o + P\\
&\left(\frac{\partial^2 V}{\partial x^2}\right)_o = K_h + (K_v \Delta l_v)_o \frac{1}{l_o} = K_h - \frac{P}{l_o}; \quad \left(\frac{\partial^2 V}{\partial y^2}\right)_o = K_v; \quad \left(\frac{\partial^2 V}{\partial x \partial y}\right)_o = \left(\frac{\partial^2 V}{\partial y \partial x}\right)_o = 0
\end{aligned} \tag{1.164}$$

where P:

$$P = -(K_v \Delta l_v)_o \tag{1.165}$$

is the static preload of the vertical springs

By keeping account of expressions (1.163), (1.164) and (1.147), expressions (1.159) can be rewritten as:

$$\begin{aligned} E_c &= \frac{1}{2} m \dot{\bar{x}}^2 + \frac{1}{2} m \dot{\bar{y}}^2 \\ V &= (K_h \Delta l_h)_o \bar{x} + (K_v \Delta l_v + P)_o \bar{y} + \frac{1}{2}\left(K_h - \frac{P}{l_o}\right)_o \bar{x}^2 + \frac{1}{2} K_v \bar{y}^2 \\ D &= \frac{1}{2} R_h \dot{\bar{x}}^2 + \frac{1}{2} R_h \dot{\bar{y}}^2 \\ \delta * L &= 0 \end{aligned} \tag{1.166}$$

By applying Lagrange's equations to the energy forms that have been reduced in quadratic forms, the following equations of motion are obtained:

$$\begin{aligned} m\ddot{\bar{x}} + R_h \dot{\bar{x}} + \left(K_h - \frac{P}{l_o}\right)\bar{x} &= 0 \\ m\ddot{\bar{y}} + R_v \dot{\bar{y}} + K_v \bar{y} &= 0 \end{aligned} \tag{1.167}$$

where $\left(-\frac{P}{l_o}\right)$ represents the negative elastic restoring force due to the preload of the vertical spring (generally defined as $(K_j \Delta l_j)_o \left(\frac{\partial^2 \Delta l_{ji}}{\partial q_i^2}\right)_o$ (1.127) which, as already mentioned, is different from zero in the case in which the link between Δl_j and generic coordinate q_i is nonlinear. In the case analysed, spring K_v preloaded with P, for a displacement x, generates a horizontal force $\frac{P}{l_o}x$, where $\frac{x}{l_o}$ is the angle formed with the vertical.[12]

By analysing the equivalent stiffness $\left(K_h - \frac{P}{l_o}\right)$ in the horizontal direction of the system shown in Fig. 1.8, linearized in the neighbourhood of the static equilibrium position, it is possible to observe how the negative term tends to reduce the natural frequency of the system or even justify it, in the case in which:

$$\left(K_h - \frac{P}{l_o}\right) < 0 \Rightarrow K_h < \frac{P}{l_o} \tag{1.168}$$

[12]If they were taken as independent variables, the elongations of the springs, "the gravitational term" would be derived from the potential energy associated with gravitational force, being in this case a non-linear relationship between the displacement of the center of gravity and the independent variables.

is a static instability, typically of a divergence type, as will be explained in detail in Chap. 2.

The analysis of systems with several degrees of freedom will be dealt with in the next chapter.

1.6 Multi-Body Systems

A more general approach to analyse these systems is traditionally referred to as *Multibody System Dynamics*, [1, 2, 20]. These methods were created for the nonlinear analysis of complex mechanical systems in space (land vehicles, aeroplanes, helicopters, space structures and robots) and were developed from a mathematical point of view for the use on modern digital computers in order to provide accurate analyses of the structures undergoing dynamic loads. These methods are based on the concept of substituting the real system with an equivalent model consisting of a series of rigid or flexible bodies, which are generally subjected to motion in the large, connected by linear and nonlinear elastic and dissipation elements. The equations (of the order of hundreds or thousands) governing the motion of these systems are obviously differential and highly nonlinear. For this reason, in almost all cases, they cannot be solved in a closed analytical form but need to be solved numerically, using step-by-step integration methods. In this text it is not possible to give an in-depth explanation of this methodology (reference to the above can be found in the widespread specialist bibliography available on this subject [1, 2, 5, 9, 7, 13]: for this reason, only a few references, based on several consolidated techniques, will be made, in order to allow readers to have a better understanding of the logic underlying this approach.

As an example, Fig. 1.10 shows several mechanical systems that can be modelled as multi-body systems. In general, a multi-body system can be defined as a set of sub-systems termed bodies, components or substructures. The motion of these sub-systems is generally constrained and, as previously mentioned, each sub-system may be subjected to translations and rotations in the large. The configuration of a multi-body system can be described in terms of displacements, velocity and accelerations: these kinematic quantities can be defined with respect to a system of coordinates, usually right handed, with orthogonal axes $X_k - Y_k - Z_k$ and an origin O_k. These systems can be:

- inertial, meaning that it is possible, for example, to define the absolute trajectory of the various bodies composing the system;
- embedded in the generic rigid body;
- connected to a specific point of the generic rigid body, with axes oriented in such a way as to permit the simple, immediate introduction of the applied forces.

Fig. 1.10 Examples of multi-body systems

Among the many possibilities, the following are often used as independent variables:

- the displacement of the centre of gravity of each rigid body G_k (often made to coincide with the $O_k = G_k$ of a reference frame connected to the body itself, with axes parallel to the principal axes of inertia), referred to an absolute reference frame, unique for the entire mechanical system;
- displacement of the origin of a moveable reference frame with reference to another generic reference frame, in turn, in motion;
- absolute or relative rotations of a reference fame that is connected to the generic body referred to an inertial/moveable reference frame etc.

The equations of motion [1, 2] can be obtained either by using the *cardinal equations of Dynamics* or energy approaches such as, for example, *Lagrange's equations*. When writing the equations of motion, various calculation methodologies, based on matrix algorithms, have been developed.

In the explanation that follows, to simplify matters and bearing in mind the objectives of the text itself, reference will be made to the following work hypotheses (to learn more about the various methodologies and calculation strategies developed and differentiated by various authors, please refer to the bibliography at the end of this chapter):

- analysis of multi-body systems with rigid bodies;
- rh cartesian reference frames;
- use of an absolute reference frame and one that is connected to each body;
- writing of equations using the Lagrangian approach.

In order to define the equations of motion of a generic multi-body system, it is necessary to make some references based on matrix algebra algorithms, which are extremely convenient and efficient for setting up the equations of motion of the system.

1.6.1 Vector Analysis

First and foremost, let us analyse how it is possible to use a vector representation of the quantities in play (displacements, velocity, accelerations, forces etc.) using a matrix approach.

Let us consider (see Fig. 1.11) the generic reference system of the axes $X_k - Y_k - Z_k$ and origin O_k: with regard to this system, as is known, it is possible to define a generic geometric vector (P–O) as:

$$(P - O) = \overrightarrow{PO} = \vec{i}_k\, x_{PO,k} + \vec{j}_k\, y_{PO,k} + \vec{k}_k\, z_{PO,k} \tag{1.169}$$

Further on in this text, in order to ensure a better understanding, reference will be made to a displacement vector. However, it goes without saying that this explanation can and will be extended to any other vector quantity. By introducing a column matrix $\underline{\vec{h}}_k$ which contains the three unit vectors of the reference frame to which reference is made and a column matrix $\underline{X}_{PO,k}$ which contains the components of the same geometric vector:

$$\underline{\vec{h}}_k = \begin{Bmatrix} \vec{i}_k \\ \vec{j}_k \\ \vec{k}_k \end{Bmatrix}; \quad \underline{X}_{PO,k} = \begin{Bmatrix} x_{PO,k} \\ y_{PO,k} \\ z_{PO,k} \end{Bmatrix} \tag{1.170}$$

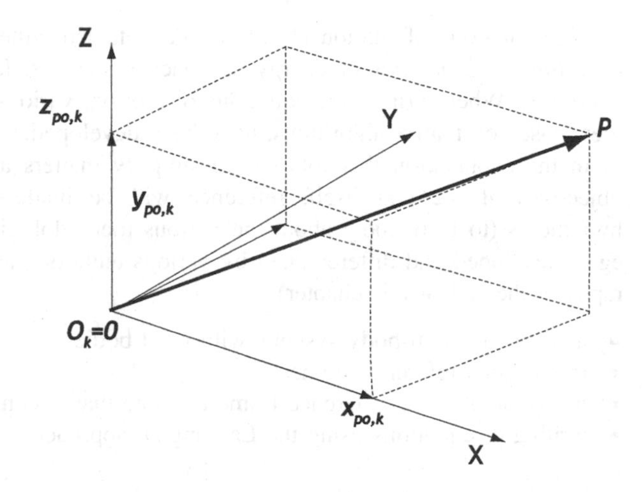

Fig. 1.11 Vectorial representation in space

the same vector can be defined in matrix form as:

$$(P-O)=\underline{\vec{h}}_k^T\,\underline{X}_{PO,k} \tag{1.171}$$

The meaning of the subscripts is of fundamental importance to understand the next part of the text and to clarify, unequivocally, the meaning of single quantities. To be more specific we will use:

- $\underline{Z}$ to indicate a vector, i.e. a column matrix;
- $[A]$ to indicate a matrix;
- $\vec{i}_k$ to indicate the generic versor, unit vector that defines the direction of the generic axis of a generic reference system $(O_k-X_k-Y_k-Z_k)$;
- $x_{PO,k}$ to indicate the generic component of the generic vector $(P-O)$, defined in the generic reference system $(O_k-X_k-Y_k-Z_k)$.

By keeping account of the above, the generic vector $(P-O)$ can be defined in two different reference systems $(O_i-X_i-Y_i-Z_i)$ and $(O_j-X_j-Y_j-Z_j)$ as:

$$\begin{aligned}(P-O)&=\vec{i}_i\,x_{PO,i}+\vec{j}_i\,y_{PO,i}+\vec{k}_i\,z_{PO,i}\\(P-O)&=\vec{i}_j\,x_{PO,j}+\vec{j}_j\,y_{PO,j}+\vec{k}_j\,z_{PO,j}\end{aligned} \tag{1.172}$$

or in matrix form as:

$$(P-O)=\underline{\vec{h}}_i^T\,\underline{X}_{PO,i}=\underline{\vec{h}}_j^T\,\underline{X}_{PO,j} \tag{1.173}$$

By pre-multiplying[13] the two terms of Eq. (1.173) in scalar form for the column matrix $\underline{\vec{h}}_i$ which contains the unit vectors of the nth triad:

$$\underline{\vec{h}}_i\times\underline{\vec{h}}_i^T\,X_{PO,i}=\underline{\vec{h}}_i\times\underline{\vec{h}}_j^T\,\underline{X}_{PO,j}\;\Rightarrow\;\underline{X}_{PO,i}=[\Lambda_{ji}]\,\underline{X}_{PO,j} \tag{1.174}$$

it is possible to calculate the relation that correlates the components of the same vector with respect to the two systems of different coordinates.

Relation (1.174) represents a transformation of coordinates and matrix $[\Lambda_{ji}]$ is the transformation matrix of same or the matrix of the direction cosines matrix:

$$[\Lambda_{ji}]=\underline{\vec{h}}_i\times\underline{\vec{h}}_j^T=\begin{bmatrix}\vec{i}_i\times\vec{i}_j & \vec{i}_i\times\vec{j}_j & \vec{i}_i\times\vec{k}_j\\ \vec{j}_i\times\vec{i}_j & \vec{j}_i\times\vec{j}_j & \vec{j}_i\times\vec{k}_j\\ \vec{k}_i\times\vec{i}_j & \vec{k}_i\times\vec{j}_j & \vec{k}_i\times\vec{k}_j\end{bmatrix} \tag{1.175}$$

[13] In the text we will use $\times$ to indicate the scalar product between the two vectors and Λ to indicate the vectorial product.

The columns of this matrix contain the projections of the unit vectors of the jth reference frame on the ith reference frame, while the rows contain projections of the third i on the j reference frame. If the two reference systems are equal, this matrix obviously become the identity matrix:

$$\underline{\vec{h}}_i \times \underline{\vec{h}}_i^T = [I] \tag{1.176}$$

By collecting the terms of this matrix by columns:

$$[\Lambda_{ji}] = \begin{bmatrix} \underline{n}_{ji} & \underline{b}_{ji} & \underline{t}_{ji} \end{bmatrix} \tag{1.177}$$

and by bearing in mind that this contains the unit vectors of the ith and jth cartesian reference frames, it is possible to keep account of the following properties:

$$\begin{aligned} \underline{n}_{ji}^T \times \underline{t}_{ji} = \underline{t}_{ji}^T \times \underline{b}_{ji} = \underline{b}_{ji}^T \times \underline{n}_{ji} = 0 \\ \underline{n}_{ji}^T \times \underline{n}_{ji} = \underline{t}_{ji}^T \times \underline{t}_{ji} = \underline{b}_{ji}^T \times \underline{b}_{ji} = 1 \end{aligned} \tag{1.178}$$

By keeping account of the 6 relations (1.178), the transformation matrix of the coordinates, which contains 9 terms (1.175), can therefore be expressed by only 3 independent coordinates: in terms of independent coordinates, it is possible to assume the 3 Cardan angles, the Euler 3 angles, etc. [1, 2]. This direction cosines matrix is orthogonal, i.e. its inverse coincides with the transposed matrix, meaning that

$$\begin{aligned} &\underline{X}_{PO,i} = [\Lambda_{ji}]\,\underline{X}_{PO,j} \Rightarrow [\Lambda_{ji}]^{-1}\,\underline{X}_{PO,i} = [\Lambda_{ji}]^{-1}[\Lambda_{ji}]\,\underline{X}_{PO,j} \\ &\underline{X}_{PO,j} = [\Lambda_{ji}]^{-1}\,\underline{X}_{PO,i} = [\Lambda_{ij}]\;\underline{X}_{PO,i} \\ &[\Lambda_{ij}] = [\Lambda_{ji}]^{-1} = [\Lambda_{ji}]^T \\ &\underline{X}_{PO,j} = [\Lambda_{ji}]^T\;\underline{X}_{PO,i} \end{aligned} \tag{1.179}$$

We will now go on to explain how it is possible to describe the kinematics of a generic rigid body, using the matrix approach that has just been introduced.

1.6.2 Kinematic Analysis of the Rigid Body

For reasons of simplicity, let us consider a generic rigid body which composes the overall multi-body system (Fig. 1.12): let us consider an inertial (absolute) reference frame $(O - X_o - Y_o - Z_o)$ and a moveable reference frame which is connected to the body considered, with origin $(O - X_o - Y_o - Z_o)$ coinciding with centre of gravity G_1 of the body itself and axes parallel to the principal axes of same. Let us consider a generic point P of the body in question. The absolute

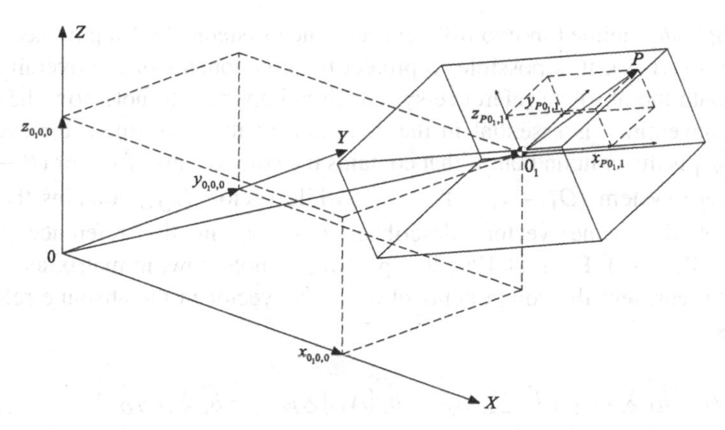

Fig. 1.12 Kinematic analysis of the generic rigid body

position of generic point P can be expressed, as previously described, by the following relation:

$$(P-O)=(P-O_1)+(O_1-O)=\underline{\vec{h}}_1^T\,\underline{X}_{PO_1,1}+\underline{\vec{h}}_o^T\,\underline{X}_{O_1O,O} \tag{1.180}$$

Where the column matrices that respectively contain the unit vectors of the inertial reference frame and the moveable reference frame have been defined by $\underline{\vec{h}}_o$ and $\underline{\vec{h}}_1$:

$$\underline{\vec{h}}_o=\begin{Bmatrix}\vec{i}_o\\ \vec{j}_o\\ \vec{k}_o\end{Bmatrix}\quad \underline{\vec{h}}_1=\begin{Bmatrix}\vec{i}_1\\ \vec{j}_1\\ \vec{k}_1\end{Bmatrix} \tag{1.181}$$

and by $\underline{X}_{PO_1,O}$ and $\underline{X}_{O_1O,1}$:

$$\underline{X}_{PO_1,O}=\begin{Bmatrix}x_{PO_1,O}\\ y_{PO_1,O}\\ z_{PO_1,O}\end{Bmatrix}\underline{X}_{O_1O,1}=\begin{Bmatrix}x_{O_1O,1}\\ y_{O_1O,1}\\ z_{O_1O,1}\end{Bmatrix} \tag{1.182}$$

the components that respectively describe the position of the centre of gravity of the body (i.e. of the origin of the moveable reference frame) with respect to the absolute reference system and the position of point P with respect to the centre of gravity of the body. The two vectors are defined with respect to the absolute reference system and with respect to the moveable reference system: by using this approach, components $\underline{X}_{PO_1,1}$ (i.e. the components of vector $(P-O_1)$ defined in the moveable reference system) are easily definable and coincide with the coordinates of point P (constants) with respect to the centre of gravity of the same body. The two components of vector $(P-O)$ of (1.182) cannot obviously be added up directly on account of being defined in two different reference systems: by keeping account of expressions (1.179), it is possible to project the components of the overall vector $(P-O)$ onto the absolute reference system (it is important to note how the use of suitable conventions is essential in the definition of the subscripts): for example, vector $\underline{X}_{PO,1}$ is the column matrix that contains the components of vector $(P-O)$ in the reference system $(O_1-X_1-Y_1-Z_1)$, while vector $\underline{X}_{PO,O}$ contains the components of the same vector, described, however in the reference system $(O-X_0-Y_0-Z_0)$. From (1.179) it is possible to note how, in matrix terms, it is possible to represent the components of a generic vector in the absolute reference system as:

$$\begin{aligned}(P-O) &= \underline{\vec{h}}_1^T \underline{X}_{PO_1,1} + \underline{\vec{h}}_o^T \underline{X}_{O_1O,O} = \underline{\vec{h}}_o^T [\Lambda_{1o}] \underline{X}_{PO_1,1} + \vec{h}_o^T \underline{X}_{O_1O,O} \\ &= \underline{\vec{h}}_o^T \underline{X}_{PO_1,O} + \underline{\vec{h}}_o^T \underline{X}_{O_1O,O} = \underline{\vec{h}}_o^T \{\underline{X}_{PO_1,o} + \underline{X}_{O_1O,O}\} = \underline{\vec{h}}_o^T \underline{X}_{PO,O}\end{aligned} \tag{1.183}$$

where:

$$\underline{X}_{PO,O} = [\Lambda_{1O}] \underline{X}_{PO_1,1} + \underline{X}_{O_1O,O} \tag{1.184}$$

By keeping account of expressions (1.180) and (1.179) in the moveable reference system, the same vector can be defined by means of the following relations:

$$\begin{aligned}(P-O) &= \underline{\vec{h}}_1^T \underline{X}_{PO_1,1} + \underline{\vec{h}}_o^T \underline{X}_{O_1O,O} = \underline{\vec{h}}_1^T \underline{X}_{PO_1,1} + \underline{\vec{h}}_1^T [\Lambda_{1O}]^T \underline{X}_{O_1O,O} \\ &= \underline{\vec{h}}_1^T \underline{X}_{PO_1,1} + \underline{\vec{h}}_1^T \underline{X}_{O_1O,1} = \underline{\vec{h}}_1^T \{\underline{X}_{PO_1,1} + \underline{X}_{O_1O,1}\} = \underline{\vec{h}}_1^T \underline{X}_{PO,1}\end{aligned} \tag{1.185}$$

where

$$\underline{X}_{PO,1} = \underline{X}_{PO_1,1} + [\Lambda_{1O}]^T \underline{X}_{O_1O,O} \tag{1.186}$$

Starting from the definition of the vector expressed in matrix form by means of (1.183), i.e. with its components projected on the absolute reference system, it is possible, by deriving the velocity and acceleration vector with respect to time, to symbolically obtain:

$$\begin{aligned}(P-O) &= \underline{\vec{h}}_o^T \underline{X}_{PO,O} = \underline{\vec{h}}_o^T \{[\Lambda_{1O}] \underline{X}_{PO_1,1} + \underline{X}_{O_1O,O}\} \\ \vec{V}_P &= \frac{d(P-O)}{dt} = \underline{\vec{h}}_o^T \left\{\left[\dot{\Lambda}_{1O}\right] \underline{X}_{PO_1,1} + \dot{\underline{X}}_{O_1O,O}\right\} \\ \vec{A}_P &= \frac{d\vec{V}_P}{dt} = \underline{\vec{h}}_o^T \left\{\left[\ddot{\Lambda}_{1O}\right] \underline{X}_{PO_1,1} + \ddot{\underline{X}}_{O_1O,O}\right\}\end{aligned} \tag{1.187}$$

By considering the classic approach of Applied Mechanics, these terms represent the absolute velocity and acceleration of point P, which is assumed to be rigidly fixed to the relative coordinate system. This approach is easy to understand and imitates, in space, the method generally adopted in the basic courses of Applied Mechanics [8, 16, 17].

With the idea of using kinematic with a view to writing the equations of motion of a complex system in space, we will now introduce an alternative approach to obtain the velocity (and, consequently, the accelerations) of a generic point. We would like to underline that, compared to the classic approach described above, this approach only has the advantage of permitting a more immediate and simple writing of the equations of motion of a generic body in space.

When using this alternative method, velocity is evaluated by deriving the displacement with respect to time, defined by relation

$$(P-O)=\underline{\vec{h}}_1^T \underline{X}_{PO_1,1}+\underline{\vec{h}}_o^T \underline{X}_{O_1O,O} \tag{1.188}$$

before projecting all vectors onto one single reference system. Thus:

$$\vec{V}_P=\frac{d(P-O)}{dt}=\frac{d\underline{\vec{h}}_1^T}{dt}\underline{X}_{PO_1,1}+\underline{\vec{h}}_o^T \dot{\underline{X}}_{O_1O,O} \tag{1.189}$$

In this case, the first term of velocity is connected to the fact that the column matrix that contains the unit vectors of the moveable reference frame must be derived with respect to time, in that the unit vectors themselves, while obviously maintaining a constant (unitary) module, change direction. The matrix term $\frac{d\underline{\vec{h}}_1^T}{dt}$ of (1.188) can be divided up into three components:

$$\frac{d\underline{\vec{h}}_1^T}{dt}\underline{X}_{PO_1,1}=\frac{d\vec{i}_1}{dt}x_{PO_1,1}+\frac{d\vec{j}_1}{dt}y_{PO_1,1}+\frac{d\vec{k}_1}{dt}z_{PO_1,1} \tag{1.190}$$

By recalling the Poisson formulae [3, 10], the generic term of expressions (1.189) can be rewritten as:

$$\begin{aligned}\frac{d\vec{i}_1}{dt}x_{PO_1,1}&=\vec{\omega}_1\,\Lambda\,\vec{i}_1\,x_{PO_1,1}=\left(\vec{i}_1\,\omega_{x1,1}+\vec{j}_1\,\omega_{y1,1}+\vec{k}_1\,\omega_{z1,1}\right)\Lambda\,\vec{i}_1\,x_{PO_1,1}\\&=-\vec{k}_1\,\omega_{y1,1}x_{PO_1,1}+\vec{j}_1\,\omega_{z1,1}x_{PO_1,1}\end{aligned} \tag{1.191}$$

having used $\vec{\omega}_1$ to indicate the vector of angular velocity to which the body is subjected, i.e. the angular velocity of the moveable reference frame. Based on the matrix approach used, this geometric vector can be represented as:

$$\vec{\omega}_1=\underline{\vec{h}}_1^T\,\underline{\omega}_{1,1}=\underline{\vec{h}}_1^T\begin{Bmatrix}\omega_{x1,1}\\\omega_{y1,1}\\\omega_{z1,1}\end{Bmatrix} \tag{1.192}$$

As will be shown in Sect. 1.7.5, having used $\omega_{x1,1}$, $\omega_{y1,1}$ and $\omega_{z1,1}$ to indicate the three components of the angular velocity of the body projected onto the moveable reference frame (as previously mentioned, with an origin coinciding with the centre of gravity of the body and axes parallel to the principal axes of inertia) this choice simplifies the writing of the kinetic energy associated with the generic body, thus allowing for the introduction of the constant and diagonal inertia tensor, regardless of the position that the body gradually assumes in its motion in the large in space [3, 10]. This will be demonstrated in Sect. 1.7.5.

Steps similar to (1.190) can also be performed for the other two components, from which we obtain[14]

$$\begin{aligned}\frac{d\underline{\vec{h}}_1^T}{dt}\underline{X}_{PO_1,1} &= \vec{i}_1\left(-\omega_{z1,1}y_{PO_1,1}+\omega_{y1,1}z_{PO_1,1}\right)+\vec{j}_1\left(\omega_{z1,1}x_{PO_1,1}-\omega_{x1,1}z_{PO_1,1}\right)\\ &+\vec{k}_1\left(-\omega_{y1,1}x_{PO_1,1}+\omega_{x1,1}y_{PO_1,1}\right)=\underline{\vec{h}}_1^T\left\{\begin{array}{c}-\omega_{z1,1}y_{PO_1,1}+\omega_{y1,1}z_{PO_1,1}\\ \omega_{z1,1}x_{PO_1,1}-\omega_{x1,1}z_{PO_1,1}\\ -\omega_{y1,1}x_{PO_1,1}+\omega_{x1,1}y_{PO_1,1}\end{array}\right\}\\ &=\underline{\vec{h}}_1^T\left[\omega_{1,1}\right]\underline{X}_{PO_1,1}=\underline{\vec{h}}_1^T\left[X_{PO_1,1}\right]\underline{\omega}_{1,1}\end{aligned} \tag{1.193}$$

Having used $\left[\omega_{1,1}\right]$ to indicate the matrix

$$\left[\omega_{1,1}\right]=\begin{bmatrix}0 & -\omega_{z1,1} & \omega_{y1,1}\\ \omega_{z1,1} & 0 & -\omega_{x1,1}\\ -\omega_{y1,1} & \omega_{x1,1} & 0\end{bmatrix} \tag{1.194}$$

And $\left[X_{PO_1,1}\right]$ to indicate the matrix:

$$\left[X_{PO_1,1}\right]=\begin{bmatrix}0 & z_{PO_1,1} & -y_{PO_1,1}\\ -z_{PO_1,1} & 0 & x_{PO_1,1}\\ y_{PO_1,1} & -x_{PO_1,1} & 0\end{bmatrix} \tag{1.195}$$

The two different formulations of Eq. (1.192) are obviously totally equivalent: in the event of our wishing to use the kinematics calculation to facilitate automatic writing of the equations of motion, as will be explained further on, the second expression is

[14]The same relation can be obtained directly in matrix form by means of:

$$\begin{aligned}\frac{d\underline{\vec{h}}_1^T}{dt}\underline{x}_{PO_1,1} &= \vec{\omega}l_1\Lambda\underline{\vec{h}}_1^T\underline{x}_{PO_1,1}=\underline{\vec{h}}_1\underline{\omega}_{1,1}\Lambda\underline{\vec{h}}_1^T\underline{x}_{PO_1,1}=\underline{\omega}_{1,1}^T\underline{\vec{h}}_1\Lambda\underline{\vec{h}}_1^T\underline{x}_{PO_1,1}\\ &=\underline{\omega}_{1,1}^T\begin{bmatrix}0 & \vec{k}_1 & -\vec{j}_1\\ -\vec{k}_1 & 0 & \vec{i}_1\\ \vec{j}_1 & -\vec{i}_1 & 0\end{bmatrix}\underline{x}_{PO_1,1}=\underline{\vec{h}}_1^T\left[\omega_{1,1}\right]\underline{X}_{PO_1,1}=\underline{\vec{h}}_1^T\left[X_{PO_1,1}\right]\underline{\omega}_{1,1}\end{aligned} \tag{1.14.1}$$

preferable. In fact, this approach will make it easier to isolate the velocity terms of the independent variables in the kinetic energy expression and, in particular, in the velocity expression. By now keeping account of Eq. (1.192) in the form:

$$\frac{d\underline{\vec{h}}_1^T}{dt}\underline{X}_{PO_1,1} = \underline{\vec{h}}_1^T\left[X_{PO_1,1}\right]\underline{\omega}_{1,1} \tag{1.196}$$

the velocity of generic point P becomes:

$$\begin{aligned}\vec{V}_P &= \underline{\vec{h}}_1^T\left[X_{PO_1,1}\right]\underline{\omega}_{1,1} + \underline{\vec{h}}_o^T\dot{\underline{X}}_{O_1O,O} = \underline{\vec{h}}_o^T[\Lambda_{1O}]\left[X_{PO_1,1}\right]\underline{\omega}_{1,1} + \underline{\vec{h}}_o^T\dot{\underline{X}}_{O_1O,O}\\ &= \underline{\vec{h}}_o^T\left\{\dot{\underline{X}}_{O_1O,O} + [\Lambda_{1O}]\left[X_{PO_1,1}\right]\underline{\omega}_{1,1}\right\}\end{aligned} \tag{1.197}$$

By comparing Eq. (1.197) with Eq. (1.187) and by keeping account of Eq. (1.193), we obtain:

$$\left[\dot{\Lambda}_{1O}\right] = [\Lambda_{1O}]\left[\omega_{1,1}\right]. \tag{1.198}$$

1.6.3 Rotations and Angular Velocity of the Rigid Body

As is known, the rotations cannot be represented as vectors [1, 3, 12]. Hence, it is necessary not only to choose specific angles as independent variables but also to establish, right from the very beginning, the sequence of same. Among the many options available in the bibliography to describe the rotation of the generic rigid body of the mechanical system schematized by means of multi-body techniques, in this particular text, we will adopt the Cardan angles:

$$\underline{q}_{\vartheta ij} = \begin{Bmatrix}\rho_{ij}\\ \beta_{ij}\\ \sigma_{ij}\end{Bmatrix} = \begin{Bmatrix}\rho\\ \beta\\ \sigma\end{Bmatrix} \tag{1.199}$$

These angles represent the rotations, in the order around axis Z, axis X and axis Y belonging to different intermediate reference frames as shown in Fig. 1.13.

To introduce these angles, for reasons of convenience, we will refer to the two reference frames $(O - X_i - Y_i - Z_i)$ and $(O - X_j - Y_j - Z_j)$ in relative motion one with respect to the other on account of having the same origin. The order defined is as follows:

- snaking (or yaw) rotation around axis Z_i: a first intermediate reference frame $(O - X_I - Y_I - Z_I)$ subjected to this rotation has axis Z_I in common with the previous reference frame, while unit vectors X_I and Y_I are rotated by σ_{ij} with respect to the initial reference frame;

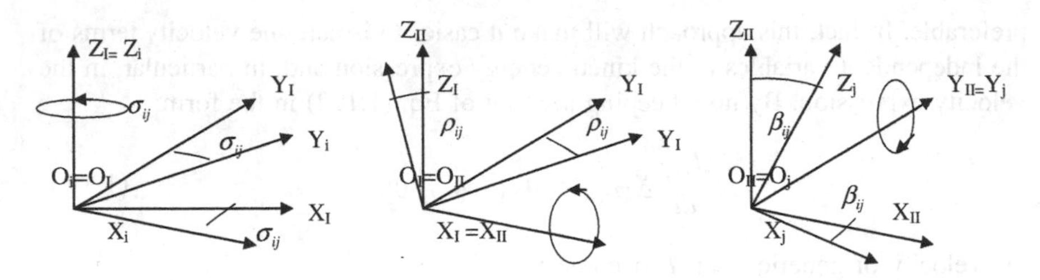

Fig. 1.13 Definition of the cardan angles

- roll rotation around axis X_I of the first intermediate reference frame: a second intermediate reference frame $(O - X_{II} - Y_{II} - Z_{II})$, subjected to this rotation has axis X_{II} in common with the previous reference frame, while unit vectors $\vec{k}_I$ and $\vec{j}_I$ are rotated by ρ_{ij} r with respect to the previous reference frame;
- pitch rotation around axis Y_{II} of the second intermediate reference frame: the final reference frame $(O - X_j - Y_j - Z_j)$, subjected to this rotation, has axis Y_j in common with the previous reference frame, while unit vectors $\vec{k}_I$ and $\vec{i}_I$ are rotated by β_{ij} with respect to the previous reference frame.

By using the matrix algorithm, the corresponding angular velocities:

$$\underline{\dot{q}}_{\vartheta ij} = \begin{Bmatrix} \dot{\rho}_{ij} \\ \dot{\beta}_{ij} \\ \dot{\sigma}_{ij} \end{Bmatrix} = \begin{Bmatrix} \dot{\rho} \\ \dot{\beta} \\ \dot{\sigma} \end{Bmatrix} \tag{1.200}$$

given the definition of the Cardan angles assumed and given the order established for these rotations, can be expressed as:

$$\begin{aligned}
\vec{\dot{\sigma}} &= \vec{k}_i\,\dot{\sigma} = \vec{k}_I\,\dot{\sigma} = \begin{Bmatrix} \vec{i}_I \\ \vec{j}_I \\ \vec{k}_I \end{Bmatrix}^T \begin{Bmatrix} 0 \\ 0 \\ \dot{\sigma} \end{Bmatrix} = \underline{\vec{h}}_I^T\,\underline{\dot{\sigma}}_I \\
\vec{\dot{\rho}} &= \vec{i}_I\,\dot{\rho} = \vec{i}_{II}\,\dot{\rho} = \begin{Bmatrix} \vec{i}_{II} \\ \vec{j}_{II} \\ \vec{k}_{II} \end{Bmatrix}^T \begin{Bmatrix} \dot{\rho} \\ 0 \\ 0 \end{Bmatrix} = \underline{\vec{h}}_{II}^T\,\underline{\dot{\rho}}_{II} \\
\vec{\dot{\beta}} &= \vec{j}_{II}\,\dot{\beta} = \vec{j}_j\,\dot{\beta} = \begin{Bmatrix} \vec{i}_j \\ \vec{j}_j \\ \vec{k}_j \end{Bmatrix}^T \begin{Bmatrix} 0 \\ \dot{\beta} \\ 0 \end{Bmatrix} = \underline{\vec{h}}_j^T\,\underline{\dot{\beta}}_j
\end{aligned} \tag{1.201}$$

The relative angular velocity vector of reference frame "j" with respect to reference frame "i" can thus be defined as:

$$\begin{aligned}\vec{\omega}_{ij} &= \vec{\dot{\sigma}} + \vec{\dot{\rho}} + \vec{\dot{\beta}} = \underline{\vec{h}}_I^T \underline{\dot{\sigma}}_I + \underline{\vec{h}}_{II}^T \underline{\dot{\rho}}_{II} + \underline{\vec{h}}_j^T \underline{\dot{\beta}}_j = \underline{\vec{h}}_j^T \underline{\omega}_{ij} \\ \vec{\omega}_{ij} &= \underline{\vec{h}}_j^T \left\{ [\Lambda_{j,II}]^T [\Lambda_{II,I}]^T \underline{\dot{\sigma}}_I + [\Lambda_{j,II}]^T \underline{\dot{\rho}}_{II} + \underline{\dot{\beta}}_j \right\}\end{aligned} \tag{1.202}$$

1.6.4 The Transformation Matrix of the Coordinates in Terms of Cardan Angles

Let us now define the transformation matrix of the coordinates as a function of the type of independent variables assumed to define the rotation of the bodies in space (Cardan angles) and of the sequence adopted for same (yaw, roll and pitch). Let us consider a generic vector (in this specific case, for reasons of convenience, we will continue to make reference to a displacement vector) and let us consider two generic reference frames $(O_i - X_i - Y_i - Z_i)$ and $(O_j - X_j - Y_j - Z_j)$ with a common origin $O_i = O_j$ (Fig. 1.13). The generic vector can be defined with respect to reference frame $(O_i - X_i - Y_i - Z_i)$ or reference frame $(O_I - X_I - Y_I - Z_I)$ which has been subjected to a yawing rotation with respect to the initial one:

$$(P - O) = \underline{\vec{h}}_i^T \underline{X}_{PO,i} = \underline{\vec{h}}_I^T \underline{X}_{PO,I} \tag{1.203}$$

By keeping account of the explanation given in the previous paragraph we obtain:

$$\underline{\vec{h}}_i \times \underline{\vec{h}}_i^T \underline{X}_{PO,i} = \underline{\vec{h}}_i \times \underline{\vec{h}}_I^T \underline{X}_{PO,I} \Rightarrow \quad \underline{X}_{PO,i} = [\Lambda_{I,i}] \underline{X}_{PO,I} \tag{1.204}$$

This relation enables us to correlate the components of the same vector in two different reference systems: in this case, the transformation matrix of coordinates $[\Lambda_{I,i}]$ is:

$$[\Lambda_{I,i}] = \underline{\vec{h}}_i \times \underline{\vec{h}}_I^T = \begin{bmatrix} \vec{i}_i \times \vec{i}_I & \vec{i}_i \times \vec{j}_I & \vec{i}_i \times \vec{k}_I \\ \vec{j}_i \times \vec{i}_I & \vec{j}_i \times \vec{j}_I & \vec{j}_i \times \vec{k}_I \\ \vec{k}_i \times \vec{i}_I & \vec{k}_i \times \vec{j}_I & \vec{k}_i \times \vec{k}_I \end{bmatrix} = \begin{bmatrix} \cos\sigma & -\sin\sigma & 0 \\ \sin\sigma & \cos\sigma & 0 \\ 0 & 0 & 1 \end{bmatrix} \tag{1.205}$$

The same vector $(P - O)$ can be expressed once again also in relation to the intermediate reference system $(O_{II} - X_{II} - Y_{II} - Z_{II})$ which has been subjected to a roll rotation with respect to system $(O_I - X_I - Y_I - Z_I)$:

$$\begin{aligned}(P - O) &= \underline{\vec{h}}_I^T \underline{X}_{PO,I} = \underline{\vec{h}}_{II}^T \underline{X}_{PO,II} \\ \underline{\vec{h}}_I \times \underline{\vec{h}}_I^T \underline{X}_{PO,I} &= \underline{\vec{h}}_I \times \underline{\vec{h}}_{II}^T \underline{X}_{PO,II} \Rightarrow \quad \underline{X}_{PO,I} = [\Lambda_{II,I}] \underline{X}_{PO,II}\end{aligned} \tag{1.206}$$

where $\left[\Lambda_{II,I}\right]$ is the transformation matrix of the coordinates between the two reference systems:

$$\left[\Lambda_{II,I}\right] = \underline{\vec{h}}_I \times \underline{\vec{h}}_{II}^T = \begin{bmatrix} \vec{i}_I \times \vec{i}_{II} & \vec{i}_I \times \vec{j}_{II} & \vec{i}_I \times \vec{k}_{II} \\ \vec{j}_I \times \vec{i}_{II} & \vec{j}_I \times \vec{j}_{II} & \vec{j}_I \times \vec{k}_{II} \\ \vec{k}_I \times \vec{i}_{II} & \vec{k}_I \times \vec{j}_{II} & \vec{k}_I \times \vec{k}_{II} \end{bmatrix} \tag{1.207}$$

The same considerations and steps can be developed by considering the second intermediate reference frame and the final reference frame $\left(O_j - X_j - Y_j - Z_j\right)$, from which:

$$\begin{gathered} (P - O) = \underline{\vec{h}}_{II}^T \underline{X}_{PO,II} = \underline{\vec{h}}_J^T \underline{X}_{PO,j} \\ \underline{\vec{h}}_{II} \times \underline{\vec{h}}_{II}^T \underline{X}_{PO,II} = \underline{\vec{h}}_{II} \times \underline{\vec{h}}_j^T \underline{X}_{PO,j} \Rightarrow \quad \underline{X}_{PO,II} = \left[\Lambda_{j,II}\right] \underline{X}_{PO,j} \\ \left[\Lambda_{j,II}\right] = \underline{\vec{h}}_{II}^T \times \underline{\vec{h}}_j^T = \begin{bmatrix} \vec{i}_{II}\, x \vec{i}_j & \vec{i}_{II} \times \vec{j}_j & \vec{i}_{II} \times \vec{k}_j \\ \vec{j}_{II}\, x \vec{i}_j & \vec{j}_{II} \times \vec{j}_j & \vec{j}_{II} \times \vec{k}_j \\ \vec{k}_{II}\, x \vec{i}_j & \vec{k}_{II} \times \vec{j}_j & \vec{k}_{II} \times \vec{k}_j \end{bmatrix} = \begin{bmatrix} \cos\beta & 0 & \sin\beta \\ 0 & 1 & 0 \\ -\sin\beta & 0 & \cos\beta \end{bmatrix} \end{gathered} \tag{1.208}$$

By keeping account of Eqs. (1.204), (1.206) and (1.207), at this point, Eq. (1.203) can be rewritten as:

$$\begin{aligned} \underline{X}_{PO,i} &= \left[\Lambda_{I,i}\right] \underline{X}_{PO,I} = \left[\Lambda_{I,i}\right]\left[\Lambda_{II,I}\right] \underline{X}_{PO,II} \\ &= \left[\Lambda_{I,i}\right]\left[\Lambda_{II,I}\right]\left[\Lambda_{j,II}\right] \underline{X}_{PO,j} = \left[\Lambda_{j,i}\right] \underline{X}_{PO,j} \end{aligned} \tag{1.209}$$

where $\left[\Lambda_{j,i}\right]$ is the transformation matrix between the two reference systems $(O_i - X_i - Y_i - Z_i)$ and $\left(O_j - X_j - Y_j - Z_j\right)$, defined as:

$$\begin{aligned} \left[\Lambda_{j,i}\right] &= \left[\Lambda_{I,i}\right]\left[\Lambda_{II,I}\right]\left[\Lambda_{j,II}\right] \\ &= \begin{bmatrix} \cos\sigma\cos\beta - \sin\beta\sin\sigma\sin\rho & -\sin\sigma\cos\rho & \cos\sigma\sin\beta + \sin\sigma\sin\rho\cos\beta \\ \sin\sigma\cos\beta + \cos\sigma\sin\rho\sin\beta & \cos\rho\cos\sigma & \sin\beta\sin\sigma - \cos\sigma\sin\rho\cos\beta \\ -\sin\beta\cos\rho & \sin\rho & \cos\rho\cos\beta \end{bmatrix} \end{aligned} \tag{1.210}$$

This matrix thus proves to be a nonlinear function of the same independent variables used to describe the angular position of one reference frame with respect to another, in other words, in the case proposed, as a function of the Cardan angles, bearing in mind the sequence order foreseen.

1.6.5 Relationship Between the Angular Velocities and the Velocities in Terms of Cardan Angles

At this point it is possible to explain the link between the components of the angular velocity geometric vector of the generic rigid body (Eq. 1.195) and the independent variables assumed to define the angular position of same, i.e. the Cardan angles in space. This link was previously introduced as:

$$\vec{\omega}_{ij} = \underline{\vec{h}}_j^T \left\{ \left[\Lambda_{j,II}\right]^T \left[\Lambda_{II,I}\right]^T \underline{\dot{\sigma}} + \left[\Lambda_{j,II}\right]^T \underline{\dot{\rho}} + \underline{\dot{\beta}} \right\} \tag{1.211}$$

which, by keeping account of Eqs. (1.205), (1.207) and (1.208), becomes:

$$\underline{\omega}_{ji,j} = \begin{Bmatrix} \cos\beta\,\dot{\rho} - \sin\beta\cos\rho\,\dot{\sigma} \\ \sin\rho\,\dot{\sigma} + \dot{\beta} \\ \sin\beta\,\dot{\rho} + \cos\beta\cos\rho\,\dot{\sigma} \end{Bmatrix} \tag{1.212}$$

This relation can be rewritten in a more compact matrix form as:

$$\underline{\omega}_{ji,j} = \left[A_{\vartheta ij}\right] \underline{\dot{q}}_{\vartheta ij} \tag{1.213}$$

Where the matrix is also a nonlinear function of the same Cardan angles:

$$\left[A_{\vartheta ij}\right] = \begin{bmatrix} \cos\beta & 0 & -\sin\beta\cos\rho \\ 0 & 1 & \sin\rho \\ \sin\beta & 0 & \cos\beta\cos\rho \end{bmatrix}. \tag{1.214}$$

1.7 The Dynamics of a Rigid Body

Let us consider a generic multi-body system with rigid bodies in space. The independent variables necessary to describe motion are $6xnc$ which can be formally grouped in a vector $\underline{q}$ defined as:

$$\underline{q}^T = \left\{ \underline{q}_1^T \quad \underline{q}_2^T \quad \cdots \quad \underline{q}_{nc-1}^T \quad \underline{q}_{nc}^T \right\} \tag{1.215}$$

on account of there being 6 independent coordinates $\underline{q}_i$ which allow for the description of motion of each body:

$$\underline{q}_i^T = \left\{ x_{0i0,0} \quad y_{0i0,0} \quad z_{0i0,0} \quad \rho_i \quad \beta_i \quad \sigma_i \right\} = \left\{ \underline{q}_{xi}^T \quad \underline{q}_{\vartheta i}^T \right\} \tag{1.216}$$

The equations of motion of the system can be obtained by means of Lagrange's equations:

$$\frac{d}{dt}\left\{\frac{\partial E_c}{\partial \underline{\dot{q}}}\right\}^T - \left\{\frac{\partial E_c}{\partial \underline{q}}\right\}^T + \left\{\frac{\partial D}{\partial \underline{\dot{q}}}\right\}^T + \left\{\frac{\partial V}{\partial \underline{q}}\right\}^T = \underline{Q} \tag{1.217}$$

where E_c, D and V are respectively the overall kinetic energy of the system, the dissipation function and the potential energy. In Eq. (1.217), $\underline{Q}$ is the vector of the generalized forces acting on the system which can be defined by the virtual work of same by means of the known relation:

$$\delta^* L = \underline{Q}^T \, \delta^* \underline{q} \tag{1.218}$$

E_c, D, V and $\delta * L$ must be expressed as a function of the independent variables $\underline{q}$: these energies are generally defined as a function of the physical variables (i.e. of those variables which are the more convenient to define the same energy functions) and it is, thus, necessary to find and impose the link between the physical variables and those of $\underline{q}$ themselves.

In this explanation, generally only of an introductory nature as with the multi-body techniques, our objective is not to give a detailed definition of all the computational and operative aspects of this approach: in fact, to simplify matters, without losing sight of the main objectives, we will make reference to the single body, to the single force and to the single connecting element. For a broader overview of the subject, readers should refer to the widespread bibliography available on the subject [1, 2, 5, 9, 13, 21, 22].

Let us now analyse the motion of a generic rigid body in space (Fig. 1.14), in general motion (translation and rotation) with respect to an absolute reference frame $(O - Z_o - Y_o - X_o)$ of unit vectors $\vec{i}_o, \vec{j}_o$ and $\vec{k}_o$ (grouped in the column matrix $\underline{\vec{h}}_o$): we will associate a moveable reference frame,$(O_i - Z_i - Y_i - X_i)$ with origin O_i

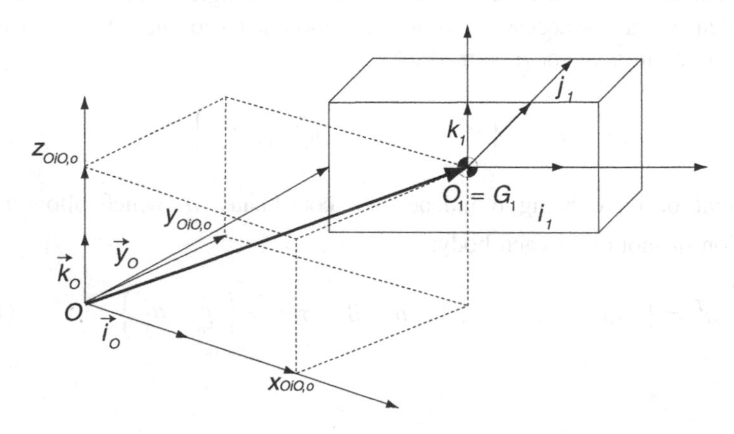

Fig. 1.14 Analysis of the rigid body in space

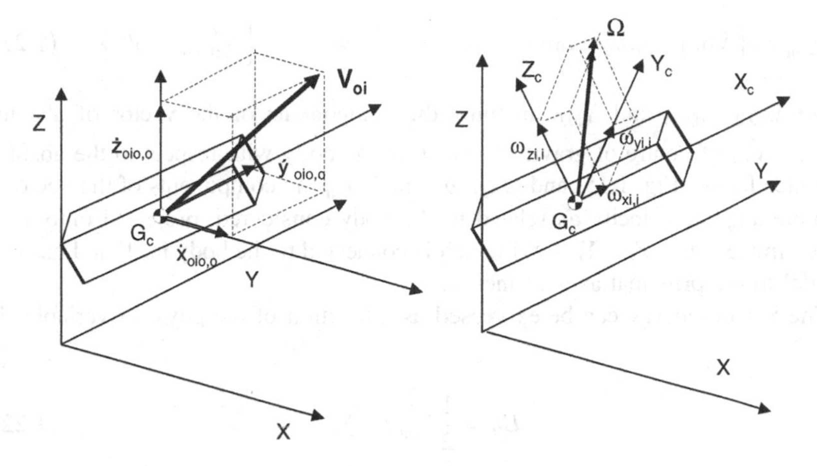

Fig. 1.15 Analysis of a rigid body in space

coinciding with the centre of gravity of body G_i and unit vectors $\vec{i}_i$, $\vec{j}_i$ and $\vec{k}_i$ (base $\underline{\vec{h}}_i$) which define three axes that are parallel to the main axes of inertia.

It goes without saying that the body in space has 6 d.o.f.: as previously mentioned, let us assume the displacement components $(O_i - O)$ of the centre of gravity of the body as independent variables defined with respect to the absolute reference system and the Cardan angles, previously defined, to distinguish the spatial angular position (Fig. 1.13, Eq. 1.200).

Let us now analyse separately the single inertial, elastic and dissipative contributions and the work of the external forces.

1.7.1 Inertial Terms

To calculate the inertial terms according to Lagrange, it is necessary to calculate the overall kinetic energy E_c of the system:

$$E_c = \sum_{i=1}^{nc} E_{ci} \tag{1.219}$$

where E_{ci} is the kinetic energy associated with the generic i-nth body, which can be expressed as a function of the physical variables $\underline{\dot{Y}}_{mi}$ which respectively define:

- the components, with respect to the fixed reference frame, $\underline{\dot{X}}_{0i0,0}$ of the centre of gravity velocity $\vec{V}_{Oi}$;
- the components $\underline{\omega}_{i,i}$ of the vector of the absolute velocity of body $\vec{\omega}_i$ projected onto the reference frame $(O_i - Z_i - Y_i - X_i)$ which is connected to the body and has axes parallel to the principal inertia axes, or rather:

$$\underline{\dot{Y}}_{mi}^T = \left\{ \dot{x}_{0i0,0} \quad \dot{y}_{0i0,0} \quad \dot{z}_{0i0,0} \quad \omega_{xi,i} \quad \omega_{yi,i} \quad \omega_{zi,i} \right\} = \left\{ \underline{\dot{X}}_{0i0,0}^T \quad \underline{\omega}_{i,i}^T \right\} \tag{1.220}$$

where $\dot{x}_{0i0,0}$, $\dot{y}_{0i0,0}$ and $\dot{z}_{0i0,0}$ are thus the components of the vector of absolute velocity $\vec{V}_{Oi}$ of centre of gravity $G_i = O_i$ of the body with respect to the absolute reference frame (Fig. 1.15) and $\omega_{xi,i}$, $\omega_{yi,i}$ and $\omega_{zi,i}$ the components of the vector of absolute angular velocity $\vec{\omega}_i$ related to the body considered, projected onto a reference frame $(O_i - Z_i - Y_i - X_i)$ which is connected to the body itself and has axes parallel to the principal axes of inertia.

The kinetic energy can be expressed as a function of the physical variables [3, 10, 23]:

$$E_{ci} = \frac{1}{2} \underline{\dot{Y}}_{mi}^T [m_i] \, \underline{\dot{Y}}_{mi} \tag{1.221}$$

where $[m_i]$ is thus a diagonal matrix containing, in an orderly way, mass m of the body and the principal inertia moments J_{xi}, J_{yi} and J_{zi}:

$$[m_i] = \begin{bmatrix} m_i & 0 & 0 & 0 & 0 & 0 \\ 0 & m_i & 0 & 0 & 0 & 0 \\ 0 & 0 & m_i & 0 & 0 & 0 \\ 0 & 0 & 0 & I_{xi} & 0 & 0 \\ 0 & 0 & 0 & 0 & I_{yi} & 0 \\ 0 & 0 & 0 & 0 & 0 & I_{zi} \end{bmatrix} = \begin{bmatrix} [m_{mi}] & [0] \\ [0] & [J_{mi}] \end{bmatrix} \tag{1.222}$$

By keeping account of the link between the physical variables $\underline{Y}_{mi}$ and the independent variables $\underline{q}_i$:

$$\underline{Y}_{mi} = \underline{Y}_{mi}\left(\underline{q}_i\right) \tag{1.223}$$

Equation (1.221) becomes:

$$E_{ci} = \frac{1}{2} \underline{\dot{q}}_i^T \, [\Lambda_{mi}]^T [m_i][\Lambda_{mi}] \, \underline{\dot{q}}_i = \frac{1}{2} \underline{\dot{q}}_i^T [M_i] \, \underline{\dot{q}}_i \tag{1.224}$$

where:

$$[\Lambda_{mi}] = \left[\frac{\partial \underline{Y}_{mi}}{\partial \underline{q}_i} \right] \tag{1.225}$$

is the matrix, i.e. the Jacobian, obtained as a derivative of a column matrix $\underline{Y}_{mi}$ which contains the generic physical variables, as opposed to another column matrix $\underline{q}$ which contains the independent variables: this matrix has 6 rows, i.e. as many as the physical variables considered, and 6 columns, i.e. as many as the independent variables q_i assumed. It is now necessary to define the kinematics of the body in order to clarify the aforementioned Jacobian matrix. The position of the centre of gravity of body $(O_i - O)$ (Fig. 1.15) is defined by the following vectorial expression, in the usual matrix formulation:

$$(O_i - O) = \underline{\vec{h}}_0^T \underline{X}_{0i0,0} = \underline{\vec{h}}_0^T \begin{Bmatrix} x_{0i0,0} \\ y_{0i0,0} \\ z_{0i0,0} \end{Bmatrix} \tag{1.226}$$

The velocity of the centre of gravity is thus given by:

$$\vec{V}_{0i} = \frac{d(O_i - O)}{dt} = \underline{\vec{h}}_0^T \dot{\underline{X}}_{0i0,0} = \underline{\vec{h}}_0^T \begin{Bmatrix} \dot{x}_{0i0,0} \\ \dot{y}_{0i0,0} \\ \dot{z}_{0i0,0} \end{Bmatrix} \tag{1.227}$$

from which it is possible to note how, in this case, the physical variables $\dot{\underline{X}}_{0i0,0}$ coincide with the independent variables $\dot{\underline{q}}_{xi}$:

$$\begin{aligned} \dot{\underline{X}}_{0i0,0} &= \left[\frac{\partial \underline{X}_{0i0,0}}{\partial \underline{q}_{xi}} \right] \dot{\underline{q}}_{xi} = [I] \dot{\underline{q}}_{xi} = \dot{\underline{q}}_{xi} \\ \vec{V}_{0i} &= \underline{\vec{h}}_0^T \dot{\underline{q}}_{xi} \end{aligned} \tag{1.228}$$

Having grouped Eq. (1.216) in vector $\underline{q}_{\vartheta i}$ the independent variables Cardan angles which define the absolute angular position in the space assumed by the reference frame $(O_i - Z_i - Y_i - X_i)$ which is connected to the body, the link between the physical variables components $\underline{\omega}_{1,1}$ of the absolute angular velocity $\vec{\omega}_1$ projected onto the reference frame that is connected to the body (see Sect. 1.6.5, Eq. 1.213) and the independent variables $\underline{q}_{\vartheta 1}$, is given by:

$$\underline{\omega}_{i,i} = [A_{\vartheta i}] \dot{\underline{q}}_{\vartheta i} = \begin{bmatrix} \cos\beta_i & 0 & -\sin\beta_i \cos\rho_i \\ 0 & 1 & \sin\rho_i \\ \sin\beta_i & 0 & \cos\beta_i \cos\rho_i \end{bmatrix} \begin{Bmatrix} \dot{\rho}_i \\ \dot{\beta}_i \\ \dot{\sigma}_i \end{Bmatrix} \tag{1.229}$$

By using this notation, matrix $[A_{\vartheta i}] = \left[A_{\vartheta i}\left(\underline{q}_{\vartheta i} \right) \right]$ is the Jacobian matrix that permits us to express the physical variables $\underline{\omega}_{1,1}$ as a function of the independent variables $\dot{\underline{q}}_{\vartheta i}$.

By keeping account of Eqs. (1.228) and (1.229), Eq. (1.223) can be rewritten as:

$$\dot{\underline{Y}}_{mi} = \begin{Bmatrix} \dot{\underline{X}}_{0i0,0} \\ \underline{\omega}_{i,i} \end{Bmatrix} = [\Lambda_{mi}] \dot{\underline{q}}_i = \begin{bmatrix} [I] & \\ & [A_{\vartheta i}] \end{bmatrix} \dot{\underline{q}}_i \tag{1.230}$$

By imposing Eq. (1.230) in Eq. (1.224), it is possible to calculate the mass matrix of the generic i-nth rigid body as a function of the independent variables assumed:

$$[M_i] = [\Lambda_{mi}]^T [m_i][\Lambda_{mi}] = \begin{bmatrix} [m_{mi}] & [0] \\ [0] & [A_{\vartheta i}]^T [J_i][A_{\vartheta i}] \end{bmatrix} \quad (1.231)$$

By suitably assembling the matrices of each single body constituting the multi-body system, it is possible to evaluate the expression of kinetic energy in the independent coordinates:

$$E_c = \sum_{i=1}^{nc} E_{ci} = \frac{1}{2} \underline{\dot{q}}^T [M] \underline{\dot{q}} \quad (1.232)$$

and to apply Lagrange's Equation (1.217) to this expression.

1.7.2 External Excitation Forces

To introduce the external forces applied to the multi-body system, it is necessary to calculate the work of same:

$$\delta^* L = \sum_{i=1}^{nf} \delta^* L_i \quad (1.233)$$

where $\delta^* L_i$ is the work performed by the generic i-nth force.

Let us now analyse, always as an example, how it is possible to define the Lagrangian components of a generic excitation force $\vec{F}_i$ applied to a generic point P_i which is connected to the body considered (Fig. 1.16).

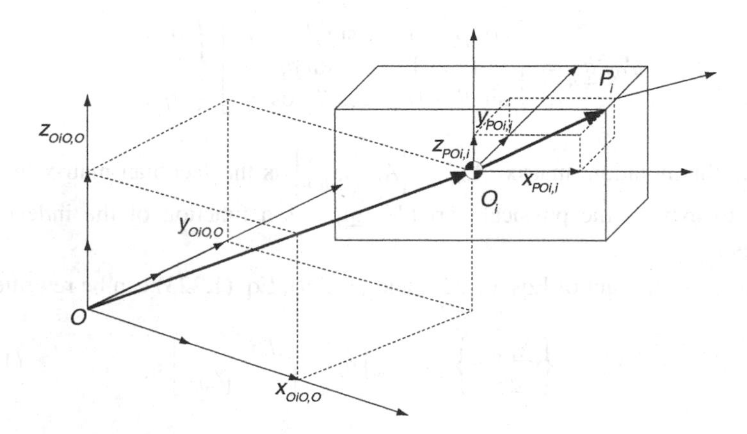

Fig. 1.16 Generic excitation force

The virtual work performed by the generic force $\vec{F}_i$ is defined by:

$$\delta^* L_i = \vec{F}_i \times \delta^* \vec{X}_{fi} \tag{1.234}$$

By rewriting the two vectors in matrix form:

$$\begin{aligned} \vec{F}_i &= \underline{h}_0^T \underline{F}_{i,0} \\ \vec{X}_{fi} &= \underline{h}_0^T \underline{X}_{fi,0} \end{aligned} \tag{1.235}$$

where $\underline{F}_{i,0}$ is the column matrix that contains the components of the vector $\vec{F}_i$ and $\underline{X}_{fi,0}$ the column matrix that contains the components of vector $\vec{X}_{fi}$, both referring to the absolute reference system, we obtain:

$$\delta^* L_i = \underline{F}_{i,0}^T \, \delta^* \, \underline{X}_{fi,0} \tag{1.236}$$

By remembering the dependence between the physical variables $\underline{X}_{fi,0}$ and the independent variables $\underline{q}_i$:

$$\underline{X}_{fi,0} = \underline{X}_{fi,0}\left(\underline{q}_i\right) \tag{1.237}$$

Equation (1.236) can be rewritten as:

$$\delta * L_i = \underline{F}_{i,0}^T \left[\frac{\partial \underline{X}_{fi,0}}{\partial \underline{q}_i} \right] \delta * \underline{q}_i = \underline{F}_{i,0}^T [\Lambda_{fi}] \delta * \underline{q}_i = \underline{Q}_i^T \, \delta * \underline{q}_i \tag{1.238}$$

where:

$$\underline{Q}_i^T = [\Lambda_{fi}]^T \, \underline{F}_{i,0} \tag{1.239}$$

In order to evaluate the generalized force $\underline{Q}_i$ it is, thus, necessary to evaluate the Jacobian $[\Lambda_{fi}]$: for this purpose, it is convenient to start once again from vector $(P_i - O)$ which defines the position of point P_i, the application point of the excitation force considered:

$$\begin{aligned} (P_i - O) &= (P_i - O_i) + (O_i - O) = \underline{h}_i^T \underline{X}_{fi,i} + \underline{h}_0^T \underline{X}_{0i0,0} \\ &= \underline{h}_0^T \{ [\Lambda_{i0}] \underline{X}_{fi,i} + \underline{X}_{0i0,0} \} = \underline{h}_0^T \underline{X}_{fi,0} \end{aligned} \tag{1.240}$$

The absolute velocity of point P is the first total derivative, with respect to time, of the displacement vector, which, by keeping account of Eqs. (1.193) and (1.213), can be expressed as:

$$\vec{V}_{fi} = \frac{d(P_i - O)}{dt} = \underline{h}_i^T [X_{fi,i}] \underline{\omega}_{i,i} + \underline{h}_0^T \dot{\underline{X}}_{0i0,0} = \underline{h}_i^T [X_{fi,i}] [A_{\vartheta i}] \dot{\underline{q}}_{\vartheta i} + \underline{h}_0^T \dot{\underline{X}}_{0i0,0}$$
$$= \underline{h}_0^T \left\{ [\Lambda_{i0}] [X_{fi,i}] [A_{\vartheta i}] \dot{\underline{q}}_{\vartheta i} + \dot{\underline{q}}_{xi} \right\} = \underline{h}_0^T \dot{\underline{X}}_{fi,0} \quad (1.241)$$

where:

$$\dot{\underline{X}}_{fi,0} = \left[\frac{\partial \underline{X}_{fi,0}}{\partial \underline{q}_i} \right] \dot{\underline{q}}_i = [\Lambda_{fi}] \dot{\underline{q}}_i \quad (1.242)$$

By recalling Eq. (1.193) it is possible to obtain the Jacobian $[\Lambda_{fi}]$ (1.241) necessary to evaluate the Lagrangian components of the force applied to the body:

$$\left[\frac{\partial \underline{X}_{fi,0}}{\partial \underline{q}_i} \right] = [\Lambda_{fi}] = [[I] | [\Lambda_{i0}] [X_{fi,i}] [A_{\vartheta i}]]_{3x6} \quad (1.243)$$

where

$$[X_{fi,i}] = \begin{bmatrix} 0 & z_{fi,i} & -y_{fi,i} \\ -z_{fi,i} & 0 & x_{fi,i} \\ y_{fi,i} & -x_{fi,i} & 0 \end{bmatrix} \quad (1.244)$$

The vector of the Lagrangian components of the generic force applied to a point P of a body of a generic multi-body system can, for this reason, be evaluated by means of the following relation

$$\underline{Q}_i^T = \underline{F}_{i,0}^T [\Lambda_{fi}] \quad \Rightarrow \quad \underline{Q}_i = \begin{bmatrix} [I] \\ [A_{\vartheta i}]^T [X_{fi,i}]^T [\Lambda_{i0}]^T \end{bmatrix}_{6x3} \underline{F}_{i,0} \quad (1.245)$$

By suitably assembling the generic vector of the Lagrangian components $\underline{Q}_i$ on each body constituting the multi-body system, it is possible to evaluate the expression of the overall work of the external forces in the independent coordinates, formally speaking:

$$\delta * L = \underline{Q}^T \delta * \underline{q} \quad (1.246)$$

and to apply the Lagrange Equation (1.217) to this expression.

1.7.3 Elastic and Gravitational Forces

The introduction of the field of elastic and gravitational forces is performed according to Lagrange's equations by means of potential energy *V*:

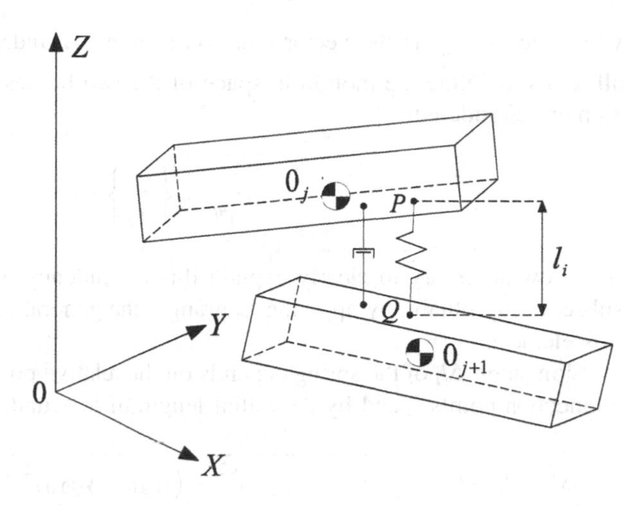

Fig. 1.17 Generic elastic and dissipative element

$$V = \sum_{i=1}^{nv} V_i = \sum_{i=1}^{nk} V_{ki} + \sum_{i=1}^{ng} V_{gi} \tag{1.247}$$

associated with the elastic elements present in the system, with the same gravitational field[15]

$$\begin{aligned} V_{ki} &= \frac{1}{2} \Delta l_i^T \, k_i \, \Delta l_i \\ V_{gi} &= m_i \, g \, h_i \end{aligned} \tag{1.248}$$

where k_i is the stiffness of the generic k-nth elastic element, Δl_i the elongation of the ends of the spring itself, $m_i g$ the weight of the generic body and h_i the elevation of the corresponding centre of gravity.

In order to calculate the terms associated with the elastic field according to the Lagrange equation (1.217), it is convenient to calculate the potential energy V_{ki} (1.248) associated with the generic elastic element as a function of the most convenient physical variable, in this case, elongation Δl_i of same. Since the elongation depends on the motion of the two attachment points P and Q of the spring (Fig. 1.17) and these on the independent variables $\underline{q}_{ipq}$ which define the motion of the connected bodies:

$$\Delta l_i = \Delta l_i \left(\underline{q}_{ipq} \right) \tag{1.249}$$

[15]Consider here also the gravitational field.

where vector $\underline{q}_{ipq}$ is the vector that contains the 12 independent coordinates that allow us to define the motion in space of the two bodies connected by the elastic element considered:

$$\underline{q}_{ipq} = \begin{Bmatrix} \underline{q}_p \\ \underline{q}_q \end{Bmatrix} \quad (1.250)$$

It is now necessary to clearly explain this dependency in Eq. (1.249) in order to subsequently obtain, by applying Lagrange, the generalized forces associated with this element.

Elongation Δl_i of the spring depends on the relative position assumed by the two connection points l_i and by the initial length of the rundown spring l_{si}:

$$\Delta l_i = l_i - l_{si} = \sqrt{\left(x_{P0,0} - x_{Q0,0}\right)^2 + \left(y_{P0,0} - y_{Q0,0}\right)^2 + \left(z_{P0,0} - z_{Q0,0}\right)^2} - l_{si} \quad (1.251)$$

By defining the column matrices that respectively contain the components of the displacement vector of points P and Q in the absolute reference system with $\underline{X}_{P0,0}$ and $\underline{X}_{Q0,0}$:

$$\underline{X}_{P0,0} = \begin{Bmatrix} x_{P0,0} \\ y_{P0,0} \\ z_{P0,0} \end{Bmatrix}; \quad \underline{X}_{Q0,0} = \begin{Bmatrix} x_{Q0,0} \\ y_{Q0,0} \\ z_{Q0,0} \end{Bmatrix} \quad (1.252)$$

vector $\underline{X}_{PQ,0}$, which contains the relative displacement components between points P and Q in the absolute reference system, can be defined as:

$$\underline{X}_{PQ,0} = \underline{X}_{P0,0} - \underline{X}_{Q0,0} \quad (1.253)$$

The elongation of the spring can thus be evaluated as:

$$\Delta l_i = \sqrt{\underline{X}_{PQ,0}^T \underline{X}_{PQ,0}} - l_{si} \quad (1.254)$$

The displacements of the ends of spring $\underline{X}_{PQ,0}$ depend on the independent variables which, in turn, define the motion of the two bodies. In the hypothesis of a reference frame that is connected to each body and only one absolute reference frame, these independent variables are:

$$\underline{q}_i = \begin{Bmatrix} \underline{q}_{xj} \\ \underline{q}_{\vartheta j} \\ \underline{q}_{xj+1} \\ \underline{q}_{\vartheta j+1} \end{Bmatrix} \quad (1.255)$$

Vector $\underline{q}_i$ contains:

- components $\underline{q}_{xj}$ of the displacement vector of the centre of gravity O_j of body j with respect to the absolute reference frame (vector $(O_j - O)$) expressed in the absolute reference system;
- components $\underline{q}_{xj+1}$ of the displacement vector of centre of gravity O_{j+1} of body $j+1$ with respect to the absolute reference frame (vector $(O_{j+1} - O)$) expressed in the absolute reference system;
- the Cardan angles $\underline{q}_{\vartheta j}$ which define the angular position of body j with respect to the absolute reference frame;
- the Cardan angles $\underline{q}_{\vartheta j+1}$ which define the angular position of body $j+1$ with respect to the absolute reference frame.

In this case, vector $\underline{X}_{PQ,0}$, which defines the components of the relative displacement of the ends of the elastic element (1.253) can easily be defined by keeping account of equations:

$$\begin{aligned}(P-O) &= \underline{h}_p^T \underline{X}_{POp,p} + \underline{h}_0^T \underline{X}_{Op0,0} = \underline{h}_0^T \left\{ [\Lambda_{p0}] \underline{X}_{POp,p} + \underline{X}_{Op0,0} \right\} = \underline{h}_0^T \underline{X}_{P0,0} \\ (P-O) &= \underline{h}_q^T \underline{X}_{QOq,q} + \underline{h}_0^T \underline{X}_{Oq0,0} = \underline{h}_0^T \left\{ [\Lambda_{q0}] \underline{X}_{QOq,q} + \underline{X}_{Oq0,0} \right\} = \underline{h}_0^T \underline{X}_{Q0,0}\end{aligned} \tag{1.256}$$

from which:

$$\underline{X}_{PQ,0} = \underline{X}_{P0,0} - \underline{X}_{Q0,0} = \left\{ \left[\Lambda_{p0}\left(\underline{q}_{\vartheta j}\right)\right] \underline{X}_{POp,p} + \underline{q}_{xj} - \left[\Lambda_{q0}\left(\underline{q}_{\vartheta j+1}\right)\right] \underline{X}_{QOq,q} - \underline{q}_{xj+1} \right\} \tag{1.257}$$

By substituting this relation in Eqs. (1.254) and (1.248), it is, thus, possible to define the term of potential energy associated with the generic elastic element as a function of the independent variables.

To calculate the terms associated with the gravitational field according to Lagrange, it is necessary to express the potential energy V_{gi} associated with the generic rigid body:

$$V_{gi} = m_i g h_i \tag{1.258}$$

as a function of the most convenient physical variable, in this case, elevation h_i of the centre of gravity of same (Fig. 1.20): this elevation depends on the independent variables $\underline{q}_i$ which define the motion of same:

$$h_i = h_i\left(\underline{q}_i\right) \tag{1.259}$$

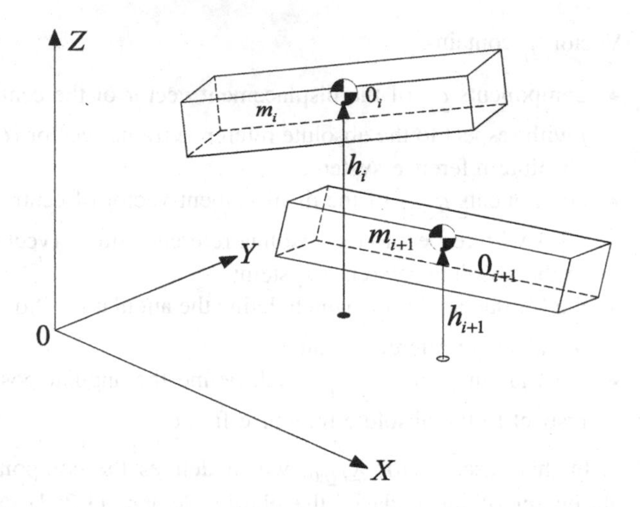

Fig. 1.18 Introduction of the gravitational field

In this case too, it is necessary to clearly express this dependency in Eq. (1.259) in order to subsequently apply Lagrange's equations. Using the conventions shown in Fig. 1.18, the elevation of the centre of gravity can be defined as:

$$h_i = \underline{u}^T \underline{q}_i \tag{1.260}$$

where $\underline{q}_i$ is the vector that contains the 6 independent coordinates that describe the motion of the body considered and:

$$\underline{u}^T = \{0 \quad 0 \quad 1 \quad 0 \quad 0 \quad 0\} \tag{1.261}$$

from which:

$$V_{gi} = m_i \, g \, \underline{u}^T \underline{q}_i \tag{1.262}$$

1.7.4 Dissipation Forces

To calculate the terms associated with the dissipation field according to Lagrange, it is necessary to introduce the dissipation function of the system analysed:

$$D = \sum_{i=1}^{ns} D_i \tag{1.263}$$

where D_i is the dissipation function associated with the generic damping element of constant r_i (Fig. 1.18):

$$D_i = \frac{1}{2}\dot{\Delta l}_i^T r_i \dot{\Delta l}_i \tag{1.264}$$

as a function of the elongation velocity $\dot{\Delta l}_i$ of same. Owing to the fact that the elongation (Fig. 1.17) depends on the motion of the two attachment points P and Q and the latter on the independent variables $\underline{q}_i$:

$$\underline{q}_i = \begin{Bmatrix} \underline{q}_{xj} \\ \underline{q}_{\vartheta j} \\ \underline{q}_{xj+1} \\ \underline{q}_{\vartheta j+1} \end{Bmatrix} \tag{1.265}$$

which define the motion of the connected bodies, it is necessary to clearly explain this dependency in Eq. (1.264):

$$\dot{\Delta l}_i = \left[\frac{\partial \Delta l_i}{\partial \underline{q}_{iPQ}}\right]\dot{\underline{q}}_i = \left[\frac{\partial \Delta l_i}{\partial \underline{X}_{PQ,0}}\right]\left[\frac{\partial \underline{X}_{PQ,0}}{\partial \underline{q}_i}\right]\dot{\underline{q}}_i = \left[\frac{\partial \Delta l_i}{\partial \underline{X}_{PQ,0}}\right][\Lambda_{ri}]\,\dot{\underline{q}}_i \tag{1.266}$$

in order to subsequently apply Lagrange. The Jacobian

$$[\Lambda_{ri}] = \left[\frac{\partial \underline{X}_{PQ,0}}{\partial \underline{q}_i}\right] \tag{1.267}$$

present in Eq. (1.267) can be expressed, by keeping account of Eq. (1.257), starting from the expression of the velocities of the two connection points:

$$\begin{aligned}
\vec{V}_P &= \underline{h}_p^T[X_{P0p,p}]\underline{\omega}_{p,p} + \underline{h}_0^T\dot{\underline{X}}_{Op0,0} = \underline{h}_0^T\left\{\left[\Lambda_{10}\left(\underline{q}_{\vartheta j}\right)\right][X_{P0p,p}]\left[A_{\vartheta j}\left(\underline{q}_{\vartheta j}\right)\right]\dot{\underline{q}}_{\vartheta j} + \dot{\underline{q}}_{xj}\right\} \\
&= \underline{h}_0^T\dot{\underline{X}}_{P0,0} \\
\vec{V}_Q &= \underline{h}_q^T[X_{Q0q,q}]\underline{\omega}_{q,q} + \underline{h}_0^T\dot{\underline{X}}_{Oq0,0} = \underline{h}_0^T\left\{\left[\Lambda_{20}\left(\underline{q}_{\vartheta j+1}\right)\right][X_{Q0q,q}]\left[A_{\vartheta j+1}\left(\underline{q}_{\vartheta j+1}\right)\right]\dot{\underline{q}}_{\vartheta j+1} + \dot{\underline{q}}_{xj+1}\right\} \\
&= \underline{h}_0^T\dot{\underline{X}}_{Q0,0}
\end{aligned} \tag{1.268}$$

The link between the relative velocities of the two points and the independent variables that describe the motion of same becomes:

$$\begin{aligned}
\dot{\underline{X}}_{PQ,0} = \dot{\underline{X}}_{P0,0} - \dot{\underline{X}}_{Q0,0} &= \left\{\left[\Lambda_{j0}\left(\underline{q}_{\vartheta j}\right)\right][X_{P0p,p}]\left[A_{\vartheta j}\left(\underline{q}_{\vartheta j}\right)\right]\dot{\underline{q}}_{\vartheta j} + \dot{\underline{q}}_{xj}\right. \\
&\left. - \left[\Lambda_{j+10}\left(\underline{q}_{\vartheta j+1}\right)\right][X_{Q0q,q}]\left[A_{\vartheta j+1}\left(\underline{q}_{\vartheta j+1}\right)\right]\dot{\underline{q}}_{\vartheta j+1} - \dot{\underline{q}}_{xj+1}\right\} = \left[\frac{\partial \underline{X}_{PQ,0}}{\partial \underline{q}_i}\right]\dot{\underline{q}}_i
\end{aligned} \tag{1.269}$$

from which it is possible to calculate the Jacobian of Eq. (1.267):

$$[\Lambda_{ri}] = \left[\frac{\partial \underline{X}_{PQ,0}}{\partial \underline{q}_i}\right] = \left[[I] \quad \left[\left[\Lambda_{j0}\left(\underline{q}_{\vartheta j}\right)\right][X_{P0p,p}]\left[A_{\vartheta j}\left(\underline{q}_{\vartheta j}\right)\right]\right] \quad -[I] \right. \\ \left. -\left[\left[\Lambda_{j+10}\left(\underline{q}_{\vartheta j+1}\right)\right][X_{Q0q,q}]\left[A_{\vartheta j+1}\left(\underline{q}_{\vartheta j+1}\right)\right]\right]\right] \tag{1.270}$$

On the contrary, the partial derivative of elongation $\left[\frac{\partial \Delta l_i}{\partial \underline{X}_{PQ,0}}\right]$ with respect to displacements (1.266) is simple:

$$\left[\frac{\partial \Delta l_i}{\partial \underline{X}_{PQ,0}}\right] = \left[\frac{\partial l_i}{\partial \underline{X}_{PQ,0}}\right] = \left[\frac{\partial l_i}{\partial x_{PQ,0}} \quad \frac{\partial l_i}{\partial y_{PQ,0}} \quad \frac{\partial l_i}{\partial z_{PQ,0}}\right] = \left[\frac{1}{l_i}x_{PQ,0} \quad \frac{1}{l_i}y_{PQ,0} \quad \frac{1}{l_i}z_{PQ,0}\right] \\ = \frac{1}{l_i}\underline{X}^T_{PQ,0} \tag{1.271}$$

Thus, by keeping account of Eqs. (1.270) and (1.271), the dissipation function of the generic interconnection element of the multi-body system (1.264) can thus be expressed as a function of the independent variables assumed as:

$$D_i = \frac{1}{2}\underline{\dot{q}}_i^T\left[[\Lambda_{ri}]^T \underline{X}_{PQ,0}\frac{1}{l_i}r_i\frac{1}{l_i}\underline{X}^T_{PQ,0}[\Lambda_{ri}]\right]\underline{\dot{q}}_i = \frac{1}{2}\underline{\dot{q}}_i^T[R_i]\underline{\dot{q}}_i \tag{1.272}$$

having used:

$$[R_i] = \left[[\Lambda_{ri}]^T \underline{X}_{PQ,0}\frac{1}{l_i}\, r_i\, \frac{1}{l_i}\underline{X}^T_{PQ,0}[\Lambda_{ri}]\right] \tag{1.273}$$

to indicate the generalized damping matrix in the independent coordinates $\underline{\dot{q}}_i$ of the system. In this way, it is possible to apply Lagrange and to obtain the generalized forces caused by the dissipation terms.

1.7.5 Definition of Kinetic Energy

In Sect. 1.7.1 we introduced the kinetic energy expression of a body in space by introducing the tensor of diagonal inertia. By using the multi-body systems approach it is possible to demonstrate its expression by using the matrix methodology proposed.

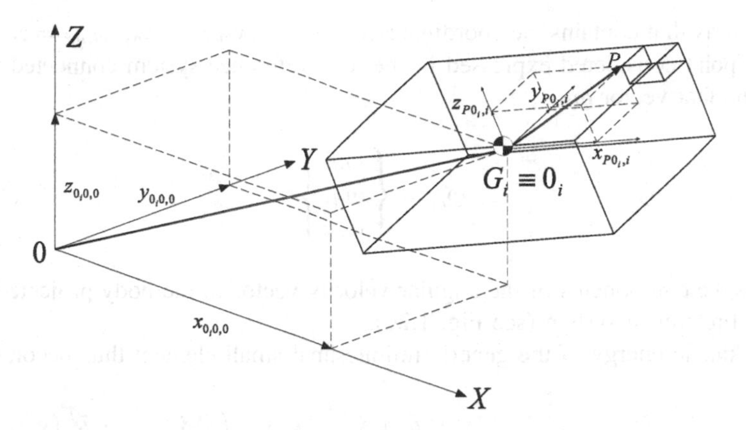

Fig. 1.19 Kinematic analysis of the generic i-nth rigid body

Let us consider the generic i-nth rigid body (Fig. 1.19) in space whose kinetic energy we wish to evaluate. The kinetic energy of the infinitesimal element (point P) belonging to the rigid body [1, 3, 10, 23][16]:

$$dE_{ci} = \frac{1}{2} V_P^2 dm = \frac{1}{2} \vec{V}_P^T \text{x } \vec{V}_P dm \tag{1.274}$$

where $dm = \rho dV_{ol}$ is the infinitesimal mass, ρ the mass per unit of volume and finally, $\vec{V}_P$ the velocity of the same point. The kinetic energy of the rigid body will be the integral extended to the volume of the body itself of Eq. (1.274).

To define the velocity of the point, it is necessary to derive vector position *(P–O)* with respect to time, in order to obtain, through use of the methodology proposed, the following expressions:

$$\begin{aligned} (P - O) &= (G_i - O) + (P - G_i) = \underline{\vec{h}}_o^T X_{GiO,O} + \underline{\vec{h}}_i^T \underline{X}_{PiGi,i} = \underline{\vec{h}}_o^T \underline{X}_{GiO,O} + \underline{\vec{h}}_i^T \underline{X}_{PiGi,i} \\ \vec{V}_P &= \frac{d(P - O)}{dt} = \underline{\vec{h}}_o^T \underline{\dot{X}}_{GiO,O} + \frac{d\underline{\vec{h}}_1^T}{dt} \underline{X}_{PiGi,i} = \underline{\vec{h}}_o^T \underline{\dot{X}}_{GiO,O} + \underline{\vec{h}}_i^T [X_p] \underline{\omega}_{i,i} \end{aligned} \tag{1.275}$$

in which we recall that:

$$[X_p] = [X_{PiGi,i}] = \begin{bmatrix} 0 & z_{PiGi,i} & -y_{PiGi,i} \\ -z_{PiGi,i} & 0 & x_{PiGi,i} \\ y_{PiGi,i} & -x_{PiGi,i} & 0 \end{bmatrix} = \begin{bmatrix} 0 & z_p & -y_p \\ -z_p & 0 & x_p \\ y_p & -x_p & 0 \end{bmatrix} \tag{1.276}$$

[16]The two relations in Eq. (1.274) are totally similar: in this section we prefer the second expression which allows for an easier matrix formulation of the kinetic energy.

is the matrix that contains the coordinates $x_{PiGi,i} = x_p, y_{PiGi,i} = y_p, z_{PiGi,i} = z_p$ of the generic point considered expressed in the local reference system connected to the body and that vector $\underline{\omega}_{i,i}$:

$$\underline{\omega}_{i,i} = \begin{Bmatrix} \omega_{xi,i} \\ \omega_{yi,i} \\ \omega_{zi,i} \end{Bmatrix} \tag{1.277}$$

contains the components of the angular velocity vector of the body projected onto the local reference system (see Fig. 1.20).

The kinetic energy of the generic infinitesimal small element thus becomes:

$$\begin{aligned} dE_{ci} &= \frac{1}{2}\vec{V}_P{}^T \times \vec{V}_P dm = \frac{1}{2}\left(\underline{\omega}_{i,i}^T [X_p]^T \vec{\underline{h}}_i + \underline{\dot{X}}_{GiO,O}^T \vec{\underline{h}}_o\right) \times \left(\vec{\underline{h}}_o^T \underline{\dot{X}}_{GiO,O} + \vec{\underline{h}}_i^T [X_p] \underline{\omega}_{i,i}\right) dm \\ &= \frac{1}{2}\underline{\omega}_{i,i}^T [X_p]^T \vec{\underline{h}}_i \times \vec{\underline{h}}_o^T \underline{\dot{X}}_{GiO,O}\, dm + \frac{1}{2}\underline{\omega}_{i,i}^T [X_p]^T \vec{\underline{h}}_i \times \vec{\underline{h}}_i^T [X_p] \underline{\omega}_{i,i} dm \\ &+ \frac{1}{2}\underline{\dot{X}}_{Gi,o}^T \vec{\underline{h}}_o \times \vec{\underline{h}}_o^T \underline{\dot{X}}_{GiO,O}\, dm + \frac{1}{2}\underline{\dot{X}}_{GiO,o}^T \vec{\underline{h}}_o \times \vec{\underline{h}}_i^T [X_p] \underline{\omega}_{i,i} dm \end{aligned} \tag{1.278}$$

keeping account of equations:

$$\begin{aligned} \vec{\underline{h}}_o^T \underline{\dot{X}}_{GiO,O} &= \vec{\underline{h}}_i^T [\Lambda_{Oi}] \underline{\dot{X}}_{GiO,O} \\ \vec{\underline{h}}_i \times \vec{\underline{h}}_i^T &= [I] \end{aligned} \tag{1.279}$$

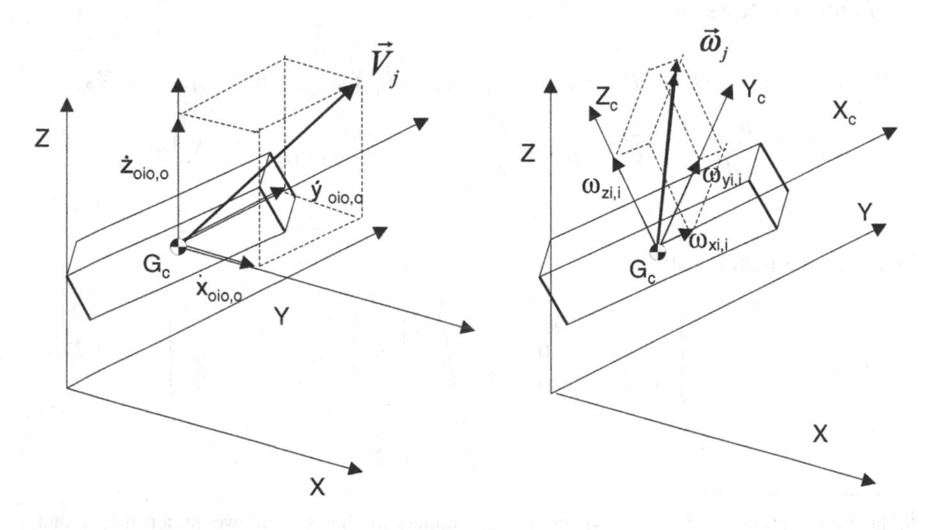

Fig. 1.20 Definition of the velocity components $\vec{V}_i$ and $\vec{\omega}_i$ for a generic rigid body

and keeping account that the kinetic energy is a scalar term, (1.278) can be reduced to the following expression:

$$dE_{ci} = \underline{\omega}_{i,i}^T [X_p]^T [\Lambda_{oi}] \dot{\underline{X}}_{GiO,O} dm + \frac{1}{2} \underline{\omega}_{i,i}^T [X_p]^T [X_p] \underline{\omega}_{i,i} dm + \frac{1}{2} \dot{\underline{X}}_{GiO,O}^T [I] \dot{\underline{X}}_{GiO,O} dm \quad (1.280)$$

The kinetic energy associated with the entire rigid body considered can, at this point, be calculated as:

$$\begin{aligned} E_{ci} = \int_{Vol} dE_{ci} = \underline{\omega}_{i,i}^T \left[\int_{Vol} [X_p]^T \rho \, dV_{ol} \right] [\Lambda_{oi}] \dot{\underline{X}}_{GiO,O} \\ + \frac{1}{2} \underline{\omega}_{i,i}^T \left[\int_{Vol} [X_p]^T [X_p] \rho \, dV_{ol} \right] \underline{\omega}_{i,i} + \frac{1}{2} \dot{\underline{X}}_{GiO,O}^T \left[\int_{Vol} [I] \rho dV_{ol} \right] \dot{\underline{X}}_{GiO,O} \end{aligned} \quad (1.281)$$

the term:

$$\left[\int_{Vol} [I] \rho \, dV_{ol} \right] = \begin{bmatrix} m_i & & \\ & m_i & \\ & & m_i \end{bmatrix} = [m_{mi}] \quad (1.282)$$

is the portion of the mass matrix associated with the translation terms. Having considered the reduction pole corresponding to the centre of gravity of the body itself, the term:

$$\left[\int_{Vol} [X_p]^T \rho d_{Vol} \right] = \begin{bmatrix} 0 & \int_{Vol} z_p \rho d_{Vol} & -\int_{Vol} y_p \rho d_{Vol} \\ -\int_{Vol} z_p \rho \, d_{Vol} & 0 & \int_{Vol} x_p \rho d_{Vol} \\ \int_{Vol} y_p \rho \, d_{Vol} & -\int_{Vol} x_p \rho d_{Vol} & 0 \end{bmatrix} = [0] \quad (1.283)$$

is null and, finally, it is possible to evaluate the inertia tensor defined in the reference system connected to body $[J_i]$ and, thus, for this reason, constant in time:

$$\begin{aligned} [J_i] &= \left[\int_{Vol} [X_p]^T [X_p] \rho dV_{ol} \right] = \left[\int_{Vol} \begin{bmatrix} 0 & -z_p & y_p \\ z_p & 0 & -x_p \\ -y_p & x_p & 0 \end{bmatrix} \begin{bmatrix} 0 & z_p & -y_p \\ -z_p & 0 & x_p \\ y_p & -x_p & 0 \end{bmatrix} \rho dV_{ol} \right] \\ &= \begin{bmatrix} \int_{Vol} \left(z_p^2 + y_p^2 \right) \rho \, dV_{ol} & -\int_{Vol} y_p x_p \rho \, dV_{ol} & -\int_{Vol} z_p x_p \rho \, dV_{ol} \\ -\int_{Vol} x_p y_p \rho \, dV_{ol} & \int_{Vol} \left(z_p^2 + x_p^2 \right) \rho \, dV_{ol} & -\int_{Vol} z_p y_p \rho \, dV_{ol} \\ -\int_{Vol} z_p x_p \rho \, dV_{ol} & -\int_{Vol} y_p z_p \rho \, dV_{ol} & \int_{Vol} \left(y_p^2 + x_p^2 \right) \rho \, dV_{ol} \end{bmatrix} = \begin{bmatrix} J_{xi} & -J_{xyi} & -J_{xzi} \\ -J_{xyi} & J_{yi} & -J_{yzi} \\ -J_{xzi} & -J_{yzi} & J_{zi} \end{bmatrix} \end{aligned} \quad (1.284)$$

If the reference system, connected to the body, proves to have axes that are parallel to the main axes of inertia, the same tensor becomes diagonal [3, 4, 10]:

$$[J_i] = \begin{bmatrix} J_{xi} & 0 & 0 \\ 0 & J_{yi} & 0 \\ 0 & 0 & J_{zi} \end{bmatrix} \tag{1.285}$$

By keeping account of Eqs. (1.282), (1.283) and (1.284), the kinetic energy of the generic rigid body (1.281) thus becomes:

$$E_{ci} = \frac{1}{2}\underline{\omega}_{i,i}^T[J_i]\underline{\omega}_{i,i} + \frac{1}{2}\underline{\dot{X}}_{GiO,O}^T[m_{mi}]\underline{\dot{X}}_{GiO,O} \tag{1.286}$$

Having reunited the two vectors:

$$\underline{\dot{Y}}_{mi} = \begin{Bmatrix} \underline{\dot{X}}_{GiO,O} \\ \underline{\omega}_{i,i} \end{Bmatrix} \tag{1.287}$$

the kinetic energy can, thus, be defined in compact matrix form as:

$$E_{ci} = \frac{1}{2}\underline{\dot{Y}}_{mi}^T[m_i]\underline{\dot{Y}}_{mi} = \frac{1}{2}\underline{\dot{Y}}_{mi}^T\begin{bmatrix} [m_{mi}] & [0] \\ [0] & [J_i] \end{bmatrix}\underline{\dot{Y}}_{mi} \tag{1.288}$$

Finally, by keeping account of the link between the physical variables and the independent variables (centre of gravity displacements and Cardan angles), Eqs. (1.228) and (1.229):

$$\begin{aligned} \underline{\dot{X}}_{GiO,O} &= \underline{\dot{q}}_{xi} \\ \underline{\omega}_{i,i} &= \left[A\left(\underline{q}_{\theta i}\right)\right]\underline{\dot{q}}_{\theta i} \end{aligned} \tag{1.289}$$

we obtain:

$$\underline{\dot{Y}}_{mi} = [\Lambda_{mi}]\begin{Bmatrix} \underline{\dot{q}}_{xi} \\ \underline{\dot{q}}_{\theta i} \end{Bmatrix} = \begin{bmatrix} [I] & [0] \\ [0] & \left[A\left(\underline{q}_{\theta i}\right)\right] \end{bmatrix}\begin{Bmatrix} \underline{\dot{q}}_{xi} \\ \underline{\dot{q}}_{\theta i} \end{Bmatrix} = [\Lambda_{mi}]\underline{\dot{q}}_i \tag{1.290}$$

from which:

$$E_{ci} = \frac{1}{2}\underline{\dot{q}}_i^T[\Lambda_{mi}]^T[m_i][\Lambda_{mi}]\underline{\dot{q}}_i = \frac{1}{2}\underline{\dot{q}}_i^T[M_i]\underline{\dot{q}}_i \tag{1.291}$$

where:

$$[M_i] = [\Lambda_{mi}]^T[m_i][\Lambda_{mi}] = \begin{bmatrix} [m_i] & [0] \\ [0] & \left[A\left(\underline{q}_{\theta i}\right)\right]^T[J_i]\left[A\left(\underline{q}_{\theta i}\right)\right] \end{bmatrix} \tag{1.292}$$

The mass matrix in independent coordinates of the generic rigid body.

1.7.6 Writing the Equations of Motion

By keeping account of the expressions of the various forms of kinetic energy E_c (1.232), potential V (1.247) and (1.248), of the dissipation function D (1.272) and, finally, of the work of the active external forces $\delta * L$ (1.246), explained in the independent variables, it is possible to apply Lagrange's Equation (1.217).

The equations obtained are formally of the type:

$$[M(\underline{q})]\underline{\ddot{q}} = -[\dot{M}(\underline{q}, \underline{\dot{q}})]\underline{\dot{q}} + \underline{Q}_c(\underline{q}, \underline{\dot{q}}) - \underline{Q}_p(\underline{q}) - [R(\underline{q})]\underline{\dot{q}} \tag{1.293}$$

where $[M(\underline{q})]$ has been used to indicate the mass matrix of the system which is dependent on the position assumed by same, terms $-[\dot{M}(\underline{q}, \underline{\dot{q}})]\underline{\dot{q}} + \underline{Q}_c(\underline{q}, \underline{\dot{q}})$ contain the gyroscopic effects and the Coriolis components, term $\underline{Q}_c(\underline{q}, \underline{\dot{q}})$ derives, in particular, from term $-\left\{\frac{\partial E_c}{\partial \underline{q}}\right\}^T$, $[R(\underline{q})]\underline{\dot{q}}$ represents the contribution of the dissipation elements and $\underline{Q}_p(\underline{q})$ contains the elastic terms.

In order to obtain a more detailed description of the methodology used for the automatic definition of the equations of motion, starting from Lagrange's equations, the reader should refer to specialized texts [1, 2, 24].

In general, Eq. (1.293) should be numerically integrated, using suitable numeric step-by-step integration methods, more widely described in [1, 9, 14, 18, 19].

More often than not, in engineering problems, it is sufficient to evaluate the response of the mechanical system being examined, linearized in the neighbourhood of a certain static equilibrium position (idle or steady state): in these cases, it is possible to linearize the Eqs. (1.293) in the neighbourhood of this position (as previously seen for systems with *1* or *2* d.o.f. and as widely described [2, 4, 23] for systems that generically have n d.o.f.., thus obtaining the following equation:

$$[M]_o\underline{\ddot{\overline{q}}} + [R]_o\underline{\dot{\overline{q}}} + [K]_o\underline{\overline{q}} = \underline{\overline{Q}}(t) \tag{1.294}$$

in which $\underline{\overline{q}}$ represents the vector of the degrees of freedom of the linearized system in the neighbourhood of the equilibrium position considered, $[M]_o$, $[R]_o$ and $[K]_o$ are the relative matrices of mass, damping and stiffness and, finally, $\underline{\overline{Q}}(t)$ is the vector of the Lagrangian components.

The next Chap. 2 will be entirely dedicated to the analysis of these linear or linearized system and to the description of the methodologies necessary to evaluate the dynamic response either in the presence or absence of external excitation forces.

1.7.7 The Cardinal Equations of Dynamics

As previously mentioned, writing of the equations of motion can also be done by using the cardinal equations of dynamics [1, 2, 5]. Let us consider a generic ith rigid body in space (see Figs. 1.19 and 1.20) subjected to "*nf*" forces $\vec{F}_k$ and "*nm*" applied moments: the Newton-Euler equations [1, 2] are:

$$\sum_{k=1}^{nf} \vec{F}_k = \frac{d(\vec{Q}_i)}{dt}$$
$$\sum_{k=1}^{nm} \vec{M}_{gk} = \frac{d\left(\vec{\Gamma}_i\right)}{dt} \tag{1.295}$$

Having used $\vec{Q}_i$ and $\vec{\Gamma}_i$ to respectively indicate the momentum and the angular momentum [3, 10]. These amplitudes associated with a generic infinitesimal small element of the body considered can be written as (see Fig. 1.21)[17]:

[17] By generally considering the vectorial product between the two generic vectors $\vec{a}$ and $\vec{b}$ expressed in just as a generic reference system; this product can be expressed in matrix form as:

$$\begin{aligned}\vec{c} = \vec{a}\Lambda\vec{b} &= \left(\vec{i}a_x + \vec{j}a_y + \vec{k}a_z\right)\Lambda\left(\vec{i}b_x + \vec{j}b_y + \vec{k}b_z\right)\\ &= \vec{k}a_xb_y - \vec{j}a_xb_z - \vec{k}a_yb_x + \vec{i}a_yb_z + \vec{j}a_zb_x - \vec{i}a_zb_y\\ &= +\vec{i}\left(a_yb_z - a_zb_y\right) + \vec{j}\left(a_zb_x - a_xb_z\right) + \vec{k}\left(a_xb_y - a_yb_x\right)\\ &= \underline{\vec{h}}^T \begin{Bmatrix} a_yb_z - a_zb_y \\ a_zb_x - a_xb_z \\ a_xb_y - a_yb_x \end{Bmatrix} = \underline{\vec{h}}^T[a]\underline{b} = \underline{\vec{h}}^T[b]\underline{a}\end{aligned} \tag{1.17.1}$$

where

$$[a] = \begin{bmatrix} 0 & -a_z & a_y \\ a_z & 0 & -a_x \\ -a_y & a_x & 0 \end{bmatrix}; \quad [b] = \begin{bmatrix} 0 & b_z & -b_y \\ -b_z & 0 & b_x \\ b_y & -b_x & 0 \end{bmatrix}; \quad \underline{a} = \begin{Bmatrix} a_x \\ a_y \\ a_z \end{Bmatrix}; \quad \underline{b} = \begin{Bmatrix} b_x \\ b_y \\ b_z \end{Bmatrix} \tag{1.17.2}$$

By directly introducing the matrix approach proposed to define the vectors:

$$\vec{a} = \underline{\vec{h}}^T \underline{a}; \quad \vec{b} = \underline{\vec{h}}^T \underline{b}; \quad \vec{c} = \underline{\vec{h}}^T \underline{c} \tag{1.17.3}$$

The same vectorial product (1.16.1) can also be defined as:

$$\vec{c} = \vec{a}\Lambda\vec{b} = \vec{a}^T\Lambda\vec{b} = \underline{a}^T\underline{\vec{h}}\Lambda\underline{\vec{h}}^T\underline{b} = \underline{a}^T \begin{bmatrix} 0 & \vec{k} & -\vec{j} \\ -\vec{k} & 0 & \vec{i} \\ \vec{j} & -\vec{i} & 0 \end{bmatrix} \underline{b} = \underline{\vec{h}}^T[a]\underline{b} = \underline{\vec{h}}^T[b]\underline{a} \tag{1.17.4}$$

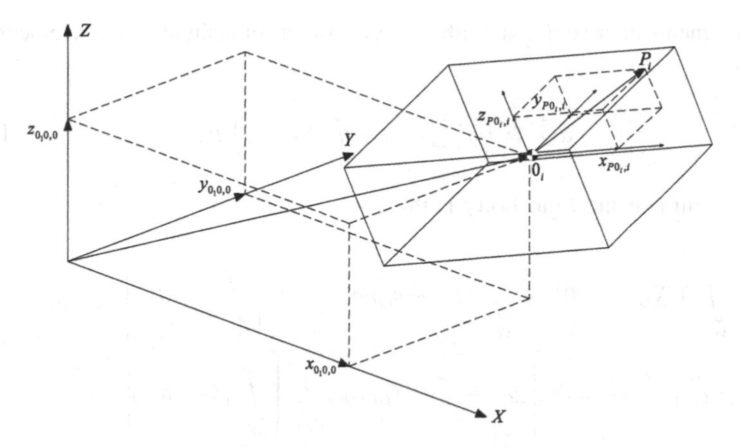

Fig. 1.21 Kinematic analysis of the generic ith rigid body

$$
\begin{aligned}
d\vec{Q}_i &= \vec{V}_P \, dm \\
d\vec{\Gamma}_i &= (P - G)\Lambda \vec{V}_P \, dm = (P - G)^T \Lambda \vec{V}_P \, dm
\end{aligned}
\tag{1.296}
$$

As usual, the velocity of point P can be calculated by means of the following expression:

$$
\begin{aligned}
(P - O) &= (P - G) + (G - P) = \underline{\vec{h}}_o^T \underline{X}_{GiO,O} + \underline{\vec{h}}_i^T \underline{X}_{PiGi,i} \\
\vec{V}_P &= \frac{d(P - O)}{dt} = \underline{\vec{h}}_o^T \underline{\dot{X}}_{GiO,O} + \underline{\vec{h}}_i^T [X_P] \underline{\omega}_{i,i}
\end{aligned}
\tag{1.297}
$$

where (1.276):

$$
[X_p] = \begin{bmatrix} 0 & z_p & -y_p \\ -z_p & 0 & x_p \\ y_p & -x_p & 0 \end{bmatrix}
\tag{1.298}
$$

is the matrix that contains the coordinates of the generic point considered, expressed in the local reference system connected to the body and vector $\underline{\omega}_{i,i}$ (1.277):

$$
\underline{\omega}_{i,i} = \begin{Bmatrix} \omega_{xi,i} \\ \omega_{yi,i} \\ \omega_{zi,i} \end{Bmatrix}
\tag{1.299}
$$

contains the components of the angular velocity vector of the body projected onto the local reference system (see Fig. 1.20).

The momentum associated with the generic infinitesimal small element thus becomes:

$$d\vec{Q}_i = \left(\underline{\vec{h}}_o^T \underline{\dot{X}}_{GiO,O} + \underline{\vec{h}}_i^T [X_P] \underline{\omega}_{i,i}\right) dm \tag{1.300}$$

The momentum of the rigid body is thus:

$$\begin{aligned}\vec{Q}_i &= \int_{Vol} \underline{\vec{h}}_o^T \underline{\dot{X}}_{GiO,O}\, \rho\, dV_{ol} + \int_{Vol} \underline{\vec{h}}_i^T [X_P] \underline{\omega}_{i,i} \rho dV_{ol} = \underline{\vec{h}}_o^T \left[\int_{Vol} [I]\, \rho\, dV_{ol}\right] \underline{\dot{X}}_{GiO,O} \\ &+ \underline{\vec{h}}_i^T \left[\int_{Vol} [X_P] \rho dV_{ol}\right] \underline{\omega}_{i,i} = \underline{\vec{h}}_o^T [m] \underline{\dot{X}}_{GiO,O} + \underline{\vec{h}}_i^T \left[\int_{Vol} [X_P]\, \rho\, dV_{ol}\right] \underline{\omega}_{i,i}\end{aligned} \tag{1.301}$$

Having considered the reduction pole corresponding to the centre of gravity of the body itself, the term

$$\left[\int_{Vol} [X_P] \rho dV_{ol}\right] = \begin{bmatrix} 0 & \int_{Vol} z_p \rho dV_{ol} & -\int_{Vol} y_p \rho dV_{ol} \\ -\int_{Vol} z_p \rho dV_{ol} & 0 & \int_{Vol} x_p \rho dV_{ol} \\ \int_{Vol} y_p \rho dV_{ol} & -\int_{Vol} x_p \rho dV_{ol} & 0 \end{bmatrix} = [0] \tag{1.302}$$

is null, and, by keeping account of Eqs. (1.302), (1.301) becomes:

$$\vec{Q}_i = \underline{\vec{h}}_o^T [m_{mi}] \underline{\dot{X}}_{GiO,O} \tag{1.303}$$

from which Eq. (1.295) becomes:

$$\sum_{k=1}^{nf} \vec{F}_k = \frac{d(\vec{Q}_i)}{dt} \Rightarrow \underline{\vec{h}}_o^T \underline{F}_o = \frac{d(\vec{Q}_i)}{dt} = \underline{\vec{h}}_o^T [m_{mi}] \underline{\ddot{X}}_{GiO,O} \Rightarrow \underline{F}_o = [m_{mi}] \underline{\ddot{X}}_{GiO,O} \tag{1.304}$$

Similarly, the angular momentum of the infinitesimal small element can be defined as:

$$\begin{aligned} d\vec{\Gamma} &= (P-G)^T \Lambda \vec{V}\, dm = \underline{X}_p^T \underline{\vec{h}}_i \Lambda \left(\underline{\vec{h}}_o^T \underline{\dot{X}}_{GiO,O} + \underline{\vec{h}}_i^T [X_P] \underline{\omega}_{i,i}\right) dm \\ &= \underline{X}_p^T \underline{\vec{h}}_i \Lambda \underline{\vec{h}}_o^T \underline{\dot{X}}_{GiO,O}\, dm + \underline{X}_p^T \underline{\vec{h}}_i \Lambda \underline{\vec{h}}_i^T [X_P] \underline{\omega}_{i,i}\, dm \\ &= \underline{X}_p^T \underline{\vec{h}}_i \Lambda \underline{\vec{h}}_i^T [\Lambda_{Oi}] \underline{\dot{X}}_{GiO,O} dm + \underline{X}_p^T \underline{\vec{h}}_i \Lambda \underline{\vec{h}}_i^T [X_P] \underline{\omega}_{i,i} dm \end{aligned} \tag{1.305}$$

By keeping account of the following relations:

$$\underline{X}_p^T \underline{\vec{h}}_i \Lambda \underline{\vec{h}}_i^T = \underline{X}_p^T \begin{Bmatrix} \vec{i}_i \\ \vec{j}_i \\ \vec{k}_i \end{Bmatrix} \{ \vec{i}_i \quad \vec{j}_i \quad \vec{k}_i \}$$

$$= \{ x_p \quad y_p \quad z_p \} \begin{bmatrix} 0 & \vec{k}_i & -\vec{j}_i \\ -\vec{k}_i & 0 & \vec{i}_i \\ \vec{j}_i & -\vec{i}_i & 0 \end{bmatrix} \tag{1.306}$$

$$= \{ -\vec{k}_i y_p + \vec{j}_i z_p \quad \vec{k}_i x_p - \vec{i}_i z_p \quad -\vec{j}_i x_p + \vec{i}_i y_p \} = \underline{\vec{h}}_i^T [X_P]^T$$

the angular momentum (1.10) becomes:

$$d\vec{\Gamma} = \underline{\vec{h}}_i^T [X_P]^T [\Lambda_{Oi}] \dot{\underline{X}}_{GiO,O}\, dm + \underline{\vec{h}}_i^T [X_P][X_P]^T \underline{\omega}_{i,i} dm \tag{1.307}$$

and for the entire rigid body:

$$\vec{\Gamma} = \int_{Vol} d\vec{\Gamma} = \underline{\vec{h}}_i^T \left[\int_{Vol} [X_P]^T \rho dV_{ol} \right] [\Lambda_{O1}] \dot{\underline{X}}_{GiO,O} + \underline{\vec{h}}_i^T \left[\int_{Vol} [X_P][X_P]^T \rho\, dV_{ol} \right] \underline{\omega}_{i,i} \tag{1.308}$$

Since relation (1.302) is valid, and by indicating the inertia tensor (1.284) by $[J_i]$, the relation which defines the angular momentum (1.308) can be rewritten as:

$$\vec{\Gamma} = \underline{\vec{h}}_i^T [J_i] \underline{\omega}_{i,i} \tag{1.309}$$

from which:

$$\sum_{k=1}^{nm} \vec{M}_{gk} = \underline{\vec{h}}_i^T \underline{M}_i = \frac{d(\vec{\Gamma})}{dt} = \frac{d\left(\underline{\vec{h}}_i^T [J_i] \underline{\omega}_{i,i}\right)}{dt} = \underline{\vec{h}}_i^T [\omega_{i,i}] [J_i] \underline{\omega}_{i,i} + \underline{\vec{h}}_i^T [J_i] \dot{\underline{\omega}}_{i,i} \tag{1.310}$$

having used matrix $[\omega_{i,i}]$ to indicate:

$$[\omega_{i,i}] = \begin{bmatrix} 0 & -\omega_{zi,i} & \omega_{yi,i} \\ \omega_{zi,i} & 0 & -\omega_{xi,i} \\ -\omega_{yi,i} & \omega_{xi,i} & 0 \end{bmatrix} \tag{1.311}$$

In short, we have[18]:

$$\begin{cases} \underline{F}_o = [m] \underset{GiO,O}{\ddot{\underline{X}}} \\ \underline{M}_i = [\omega_{i,i}] [J_i] \underline{\omega}_{i,i} + [J_i] \underline{\dot{\omega}}_{i,i} \end{cases} \tag{1.312}$$

A relation in which it is then possible to substitute the relationship between the physical variables and the independent variables:

$$\begin{cases} \ddot{\underline{X}}_{GiO,O} = \ddot{\underline{q}}_x \\ \underline{\omega}_{i,i} = \left[A(\underline{q}_\theta)\right] \dot{\underline{q}}_\theta \end{cases} \tag{1.313}$$

Summary This chapter of the book describes the large motion non-linear dynamics of multi-body discrete 1, 2 and "n" degree-of-freedom systems using scalar and matrix methodologies (Lagrange's equations) to write the related equations of motion. Concepts regarding the degrees of freedom of mechanical systems (associated with constraints and their schematization), physical variables and independent coordinates are introduced. The equations of "motion in large" and their linearization in the neighborhood of the equilibrium position (static or steady state) are introduced and described. For 3D motions, the basic concepts of multi-body methods are shown. Several examples are shown in the text.

References

1. Cheli F, Pennestrì E (2006) Cinematica e dinamica dei sistemi multibody. Casa Editrice Ambrosiana, Milano
2. Shabana AA (2005) Dynamics of multibody systems, 3rd edn. Cambridge University Press, Cambridge
3. Finzi B (1991) Meccanica razionale. Zanichelli, Bologna

[18]In the case in which the axes of the reference system are the principal inertia axes, the inertial tensor $[J_i]$ becomes

$$[J_i] = \begin{bmatrix} J_{xi} & 0 & 0 \\ 0 & J_{yi} & 0 \\ 0 & 0 & J_{zi} \end{bmatrix} \tag{1.18.1}$$

and the components of the $\bar{M}_i$ vector become:

$$\begin{aligned} M_{xi} &= J_{xi}\,\dot{\omega}_{xi,i} - \omega_{y,i}\omega_{zi,i}J_{yi} + \omega_{yi,i}\omega_{zi,i}J_{zi} \\ M_{yi} &= J_{yi}\,\dot{\omega}_{yi,i} - \omega_{x,i}\omega_{zi,i}J_{zi} + \omega_{xi,i}\omega_{zi,i}J_{xi} \\ M_{zi} &= J_{zi}\,\dot{\omega}_{zi,i} - \omega_{x,i}\omega_{yi,i}J_{xi} + \omega_{xi,i}\omega_{yi,i}J_{yi} \end{aligned} \tag{1.18.2}$$

4. Rao A (2006) Dynamics of particles and rigid bodies. Cambridge University Press, Cambridge
5. Coutinho MG (2001) Dynamics simulations of multibody systems. Springer, New York
6. Ambròsio JAC, Kleiber M (2001) Computational aspects of non linear systems with large rigid body motion. IOS Press, Amsterdam
7. Nikravesh P (1988) Computer-Aided analysis of mechanical systems. Prentice Hall, Upper Saddle River
8. Bachschmid N, Bruni S, Collina A, Pizzigoni B, Resta F (2003) Fondamenti di meccanica teorica ed applicata. McGraw Hill, New York
9. Shabana AA (2001) Computational dynamics. Wiley Interscience, New York
10. Cecignani C (1987) Spazio, tempo e movimento. Bologna, Zanichelli
11. Diana G, Resta F (2006) Controllo dei sistemi meccanici. Polipress, Milano
12. Udwadia ET, Kalaba RE (1996) Analytical mechanics: a new approach. Cambridge University Press, Cambridge
13. Schielen W, Journal of multibody system dynamics. Springer, New York
14. Meirovitch L (1980) Computational methods in structural dynamics. The Sijthoff & Noordhoff International Publishers, The Hague
15. Meirovitch L (1970) Methods of analytical dynamics. McGraw-Hill, New York
16. Sesini O (1964) Meccanica applicata alle macchine. Casa Editrice Ambrosiana, Milano
17. Diana G et al (1985) Appunti di meccanica applicata alle macchine. Edizione Spiegel, Milano
18. Argyris J, Mleinek HP (1991) Texts on computational mechanics, vol v, dynamic of structures. Elsevier, New York
19. Bathe KJ, Wilson EL (1976) Numerical methods in finite element method. Prentice-Hall, Upper Saddle River
20. Schielen W (1997) Multibody system dynamics: roots and perspectives, multibody system dynamics. Kluwer Academic Publishers, Dordrecht
21. Blundell M, Harty D (2004) The multibody systems approach to vehicle dynamics. Butterworth-Heinemann, Oxford
22. Geradin M, Cardona A (2001) Flexible multibody dynamics: a finite element approach. Wiley, New York
23. Wittenburg J (1997) Dynamic of systems of rigid bodies. Teubner, Stuttgart, B.G
24. Diana G, Cheli F (1997) Cinematica e dinamica dei sistemi multi-corpo. Edizioni Spiegel, Milano

Chapter 2
The Dynamic Behaviour of Discrete Linear Systems

2.1 Introduction

After describing the methodology to define the equations of motion of a generic nonlinear discrete system in Chap. 1, generally n-degree-of-freedom (d.o.f.) systems, in this chapter the case of linear or linearized systems about the static equilibrium position will be analysed. From a strictly engineering point of view, this procedure not only permits the simplified modelling of machinery and structures but also, more specifically, the resolution of dynamic problems related to mechanical vibrations, or rather to the small oscillations to which the latter might be subjected during standard operating procedures. In this chapter, a rigorous, rapid methodology, based on matrix notation, will be used in order to enable us to directly write the linearized equations of motion. Subsequently, an analytical-numerical procedure will also be provided in order to allow us to calculate the solution of these equations not only in the case of free motion (response of the system to an initial disturbance starting from the static equilibrium position) but also in the case of forced motion.

2.2 Writing Equations of Motion

The methodology used to write the equations of motion for discrete systems consisting of rigid bodies, with several linear or linearized d.o.f. about the static equilibrium position (static position) may be generalized, as shown in the Chap. 1, by adopting a matrix method that is particularly suitable for implementation on computers.

This approach allows us to obtain the equations of motion of the system using both Euler's (cardinal equations of dynamics) and Lagrange's equations (a methodology that will also be used when expounding on this particular procedure).

F. Cheli and G. Diana, *Advanced Dynamics of Mechanical Systems*,
DOI 10.1007/978-3-319-18200-1_2

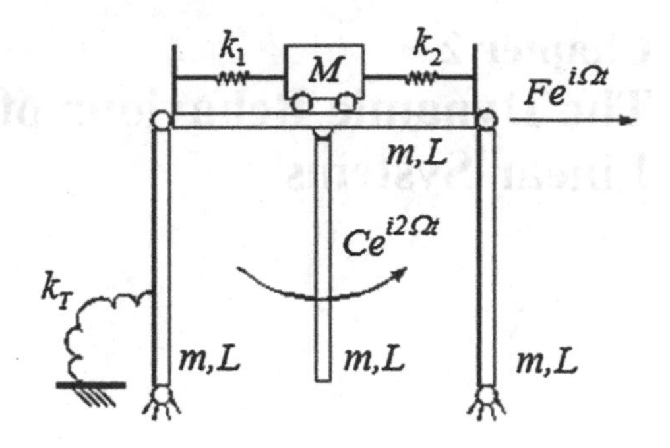

Fig. 2.1 Typical multi-body system

Let us consider a generic n d.o.f. system (Fig. 2.1), consisting of n_c rigid bodies, mutually interconnected either by means of ideal kinematic constraints or elastic and damping elements, subjected to external forces.

Let us use:

- $\underline{x}$ to define the vector of n independent coordinates that define the generic motion of the system where:

$$n = 6n_c - n_v \tag{2.1}$$

  and n_v is the number of the constraint equations;
- $\overline{\underline{x}}$ to define the vector of n independent coordinates that define the perturbed motion

$$\overline{\underline{x}} = \underline{x} - \underline{x}_0 \tag{2.2}$$

  about the static equilibrium position:

$$\underline{x} = \underline{x}_0 \tag{2.3}$$

- $\underline{Y}$ to define the vector of the physical coordinates that are useful to define the various forms of energy that appear in the Lagrange's equations. As seen in Chap. 1, in terms of physical coordinates $\underline{Y}$, it is possible to assume, for example, the absolute displacements of the centres of gravity of the single bodies and the absolute rotations of the bodies themselves (variables that are convenient in terms of defining the kinetic energy E_c of the system), the relative displacements of the extremities of the elastic and damping elements (amplitudes that are suitable to define potential energy V and dissipation function D) and the displacements of the application points of the external forces in order to evaluate their virtual work $\delta^* L$.

As already widely described in the Chap. 1, the equations of motion of the system can be obtained from Lagrange's equations which, in matrix form, assume the following expression:

$$\left\{\frac{d}{dt}\left(\frac{\partial E_C}{\partial \underline{\dot{x}}}\right)\right\}^T - \left\{\frac{\partial E_C}{\partial \underline{x}}\right\}^T + \left\{\frac{\partial D}{\partial \underline{\dot{x}}}\right\}^T + \left\{\frac{\partial V}{\partial \underline{x}}\right\}^T = \underline{Q} \tag{2.4}$$

in which E_c, D and V respectively indicate the kinetic energy, the dissipation function and the potential energy of the system, while the term $\underline{Q}$ represents a column vector whose generic ith element Q_i coincides with the ratio between the virtual work $\delta^* L_i$ performed by the active forces due to the sole generic virtual displacement $\delta^* \overline{x}_i$ and the displacement itself:

$$Q_i = \frac{\delta^* L_i}{\delta^* \overline{x}_i} \quad (i = 1, 2, \ldots, n) \tag{2.5}$$

To obtain the linearized equations of motion about the static equilibrium position, it is necessary to make the various forms of energy quadratic.

For this purpose, it is necessary:

- to define as a function of physical coordinates $\underline{Y}$ and, in matrix form, kinetic energy E_c, potential V, dissipative D and virtual work $\delta^* L$ of the active external forces;
- to define the links between physical coordinates $\underline{Y}$ and independent coordinates $\underline{x}$ through functions of the type $\underline{Y} = \underline{Y}(\underline{x})$;
- to replace these links and to make the various forms of energy quadratic, by evaluating the Jacobian and Hessian matrices that link the physical variables to the independent variables in the static equilibrium position considered.

2.2.1 Kinetic Energy

To express kinetic energy E_c, in view of the fact that we are here dealing with rigid bodies, the simplest approach, as previously mentioned, is to refer to their centre of gravity, by writing (see Fig. 2.2):

- their contribution to translation as a function of components $\dot{x}_j, \dot{y}_j, \dot{z}_j$ of the absolute velocity of the centre of gravity $\vec{v}_j$ expressed in the absolute reference system;
- their contribution to rotation by expressing the absolute velocity vector of rotation $\vec{\omega}_j$ of the body as a function of the 3 components $\dot{\theta}_{j1}, \dot{\theta}_{j2}, \dot{\theta}_{j3}$ expressed in a Cartesian reference system $\xi_{j1} - \xi_{j2} - \xi_{j3}$ connected to the body itself of origin O_j in the centre of gravity of body G_j and axes parallel to the main axes of inertia; in this way, the inertia tensor is diagonal [13, 18].

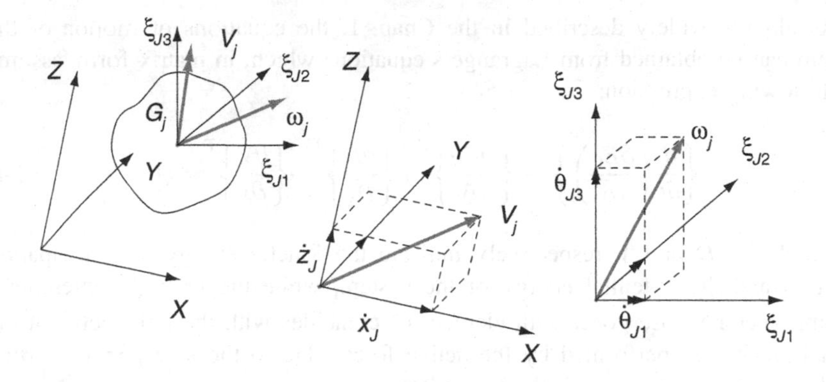

Fig. 2.2 Definition of velocity components for the generic rigid body

Using this approach, the total kinetic energy E_c of the system is given by the sum of contributions E_{cj} of n_c rigid bodies constituting the system itself:

$$E_c = \sum_{j=1}^{n_c} E_{cj} = \sum_{j=1}^{n_c} \frac{1}{2} m_j v_j^2 + \frac{1}{2}\left(J_{j1}\dot{\theta}_{j1}^2 + J_{j2}\dot{\theta}_{j2}^2 + J_{j3}\dot{\theta}_{j3}^2\right) \tag{2.6}$$

which, having used M_{jk} to generically indicate both the real mass of each body m_j, and the generic moment of inertia J_{jk} with respect to a main axis, can be expressed, in a more general formulation, as:

$$E_c = \sum_{j=1}^{n_c} \sum_{k=1}^{6} \frac{1}{2} M_{jk} \dot{y}_{jk}^2 \tag{2.7}$$

And then, in a more compact matrix form, as:

$$E_c = \frac{1}{2} \underline{\dot{Y}}_m^T [M_y] \underline{\dot{Y}}_m \tag{2.8}$$

having used:

$$\underline{\dot{Y}}_m = \begin{Bmatrix} \dot{y}_{11} \\ \dot{y}_{12} \\ \cdot \\ \cdot \\ \dot{y}_{n_c 6} \end{Bmatrix} \tag{2.9}$$

to generically define the vector of physical variables (generally a vector composed of $6 \times n_c$ elements) and $[M_y]$ the mass matrix as a function of the same physical coordinates which, by taking into account (2.7), is diagonal:

$$[M_y] = \begin{bmatrix} M_{11} & 0 & \dots & 0 & \dots & 0 \\ 0 & M_{12} & \dots & 0 & \dots & 0 \\ \vdots & \vdots & \dots & \vdots & \dots & \vdots \\ 0 & 0 & \dots & 0 & \dots & M_{n_c6} \end{bmatrix} = diag\{M_{11}\, M_{12} \dots M_{n_c6}\} \quad (2.10)$$

Generally speaking, the physical coordinates y_{jk} do not coincide with the free coordinates x_i of the entire system: the link between the physical coordinates and the free coordinates, which represent holonomic constraint conditions, of the type:

$$y_{jk} = y_{jk}(x_1, x_2, \dots, x_n) = y_{jk}(\underline{x}) \quad (j = 1, \dots, m;\ k = 1, \dots, 6) \quad (2.11)$$

are generally nonlinear, having collected the independent variables x_i in the vector $\underline{x}$. Equation (2.11) can briefly be rewritten in matrix form as:

$$\underline{Y}_m = \underline{Y}_m(\underline{x}) \quad (2.12)$$

Each component of the velocity vector can be calculated by the time derivative of (2.11) in scalar form, as:

$$\dot{y}_{jk} = \frac{\partial y_{jk}}{\partial x_1}\dot{x}_1 + \frac{\partial y_{jk}}{\partial x_2}\dot{x}_2 + \dots + \frac{\partial y_{jk}}{\partial x_n}\dot{x}_n \quad (j = 1, \dots, \mathrm{n_c};\ k = 1, \dots, 6) \quad (2.13)$$

i.e. by vector:

$$\dot{y}_{jk} = \left\{\frac{\partial y_{jk}}{\partial \underline{x}}\right\}\dot{\underline{x}} \quad (2.14)$$

It is then possible to define vector $\underline{Y}_m$ of the velocity components in matrix form, by taking into account that:

$$\dot{\underline{Y}}_m = \left[\frac{\partial \underline{Y}_m}{\partial \underline{x}}\right]\dot{\underline{x}} = \left[\frac{\partial \underline{Y}_m}{\partial \underline{x}}\right]\dot{\underline{x}} \quad (2.15)$$

and remembering that the derivative of a vector with respect to another vector is a matrix, we have:

$$\dot{\underline{Y}}_m = \left[\frac{\partial \underline{Y}_m}{\partial \underline{x}}\right]\dot{\underline{x}} = \begin{bmatrix} \frac{\partial y_{11}}{\partial x_1} & \frac{\partial y_{11}}{\partial x_2} & \dots & \frac{\partial y_{11}}{\partial x_n} \\ \dots & \dots & \dots & \dots \\ \frac{\partial y_{jk}}{\partial x_1} & \frac{\partial y_{jk}}{\partial x_2} & \dots & \frac{\partial y_{jk}}{\partial x_n} \\ \dots & \dots & \dots & \dots \\ \frac{\partial y_{n_c6}}{\partial x_1} & \frac{\partial y_{n_c6}}{\partial x_2} & \dots & \frac{\partial y_{n_c6}}{\partial x_{n1}} \end{bmatrix}\dot{\underline{x}} = [\Lambda(\underline{x})]_m \dot{\underline{x}} \quad (2.16)$$

Matrix $[\Lambda(\underline{x})]_m$ is termed *Jacobian* and, for nonlinear systems, depends on the same independent variables $\underline{x}$. For perturbed systems about the static equilibrium position, in order to obtain linear equations of motion it is necessary, as previously mentioned, to make the various forms of energy quadratic: as regards kinetic energy, this can be reduced to a quadratic form by evaluating the Jacobian $[\Lambda(\underline{x})]_m$ in correspondence to the equilibrium position, or rather for $\bar{x}_i = 0$ (i = 1, n):

$$[\Lambda(\underline{x})]_m = \left[\frac{\partial \underline{Y}_m}{\partial \underline{x}}\right]_0 = [\Lambda_m] = const. \tag{2.17}$$

By now introducing the link between the physical variables and the independent variables (2.15) and (2.17) into (2.8), for the generic linear or linearized system, we obtain:

$$\begin{aligned} E_c &= \frac{1}{2}\dot{\underline{\bar{x}}}^T \left[\frac{\partial \underline{Y}_m}{\partial \underline{x}}\right]_0^T [M_y] \left[\frac{\partial \underline{Y}_m}{\partial \underline{x}}\right]_0 \dot{\underline{\bar{x}}} \\ &= \frac{1}{2}\dot{\underline{\bar{x}}}^T [\Lambda_m]^T [M_y] [\Lambda_m] \dot{\underline{\bar{x}}} = \frac{1}{2}\dot{\underline{\bar{x}}}^T [M_0] \dot{\underline{\bar{x}}} \end{aligned} \tag{2.18}$$

The product:

$$[M_0] = [\Lambda_m]^T [M_y] [\Lambda_m] \tag{2.19}$$

is a square matrix of $n \times n$ order and is the mass matrix of the system, generalized with the d.o.f. of vector $\underline{x}$.

2.2.2 Dissipation Function

In the event of only considering concentrated viscous dampers with a constant r_j (all types of damping are attributable, either directly or approximately, to this type of damping), dissipative function D of the entire system is expressed in a formal manner similar to that of potential energy:

$$D = \frac{1}{2}\sum_{j=1}^{n_r} r_j \Delta\dot{l}_j^2 = \frac{1}{2}\Delta\dot{\underline{l}}_r^T [r] \Delta\dot{\underline{l}}_r \tag{2.20}$$

where $[r]$ is a diagonal matrix which contains the damping constants r_j of the individual dampers:

$$[r] = \begin{bmatrix} r_1 & 0 & \dots & 0 & \dots & 0 \\ 0 & r_2 & \dots & 0 & \dots & 0 \\ \dots & \dots & \dots & \dots & \dots & \dots \\ 0 & 0 & \dots & 0 & \dots & r_{n_r} \end{bmatrix} \tag{2.21}$$

while vector $\dot{\underline{\Delta l}}_r$ contains the corresponding elongation speeds $\dot{\Delta l}_j$:

$$\dot{\underline{\Delta l}}_r^T = \{ \dot{\Delta l}_1 \quad \dot{\Delta l}_2 \quad \dots \quad \dots \quad \dot{\Delta l}_{n_r} \} \tag{2.22}$$

Constants r_j (2.20) can be obtained from the hysteresis curve of the elastic dissipative element or from the characteristic curves that simulate the link between the force transmitted by the damping element and the relative speed $\dot{\Delta l}_j$ of its extremities. The values obtained from r_j can thus be constant or functions of variables $\underline{x}$ and $\dot{\underline{x}}$ themselves: in the pages that follow we will analyse the case of constant values of r_j. Elongations Δl_j are generally expressed as a nonlinear function of independent coordinates x_i:

$$\underline{\Delta l}_r = \underline{\Delta l}_r(\underline{x}) = \begin{Bmatrix} \Delta l_1(x_1, x_2, \dots, x_n) \\ \Delta l_2(x_1, x_2, \dots, x_n) \\ \dots \\ \Delta l_{nr}(x_1, x_2, \dots, x_n) \end{Bmatrix} \tag{2.23}$$

The time derivatives of the elongations, or rather the relative speeds between the damper extremities, are:

$$\dot{\underline{\Delta l}}_r = \left[\frac{\partial \underline{\Delta l}_r}{\partial \underline{x}}\right]\dot{\underline{x}} = [\Lambda(\underline{x})]_d \dot{\underline{x}} = [\Lambda(\underline{x})]_d \dot{\bar{\underline{x}}} \tag{2.24}$$

If the links between the physical variables $\underline{\Delta l}_r$ and the independent variables $\underline{x}$ are linear or if the same equations are linearized about thc static equilibrium position, then

$$[\Lambda(x)]_d = \left[\frac{\partial \underline{\Delta l}_r}{\partial \underline{x}}\right] = \begin{bmatrix} \frac{\partial \Delta l_1}{\partial x_1} & \frac{\partial \Delta l_1}{\partial x_2} & \dots & \frac{\partial \Delta l_1}{\partial x_n} \\ \dots & \dots & \dots & \dots \\ \frac{\partial \Delta l_j}{\partial x_1} & \frac{\partial \Delta l_j}{\partial x_2} & \dots & \frac{\partial \Delta l_j}{\partial x_n} \\ \dots & \dots & \dots & \dots \\ \frac{\partial \Delta l_{nr}}{\partial x_1} & \frac{\partial \Delta l_{nr}}{\partial x_2} & \dots & \frac{\partial \Delta l_{nr}}{\partial x_n} \end{bmatrix} \tag{2.25}$$

becomes a constant matrix:

$$[\Lambda(\underline{x})]_d = \left[\frac{\partial \underline{\Delta}_l}{\partial \underline{x}}\right]_0 = [\Lambda_d] = const. \tag{2.26}$$

For linear or linearized systems about the static equilibrium position, dissipation function D can thus be expressed as:

$$D = \frac{1}{2}\dot{\underline{\Delta l}}_r^T [r] \dot{\underline{\Delta l}}_r = \frac{1}{2}\dot{\bar{\underline{x}}}^T [\Lambda_d]^T [r][\Lambda_d] \dot{\bar{\underline{x}}} = \frac{1}{2}\dot{\bar{\underline{x}}}^T [R_0] \dot{\bar{\underline{x}}} \tag{2.27}$$

where $[R]$ is the generalized damping matrix of the overall system.

2.2.3 Potential Energy

Let us now consider, on a general basis, the contribution made not only by the elastic elements k_j that connect the various bodies, either to each other or to the outside world, but also the gravitational field to the potential energy. By using Δl_j to define the elongations of the generic jth elastic element, starting from the position of the unloaded spring, where p_j and h_j are respectively the weight and elevation of the centre of gravity of the general body that constitutes the mechanical system under consideration, potential energy V of the system can be so expressed as:

$$V = \frac{1}{2}\sum_{j=1}^{n_k} k_j \Delta l_j^2 + \sum_{j=1}^{n_P} p_j h_j = \frac{1}{2}\underline{\Delta l}_k^T [k] \underline{\Delta l}_k + \underline{p}^T \underline{h} \tag{2.28}$$

having collected the elongations Δl_j of the n_k system elements in vector $\underline{\Delta l}_k$:

$$\underline{\Delta l}_k^T = \{ \Delta l_1 \quad \Delta l_2 \quad \ldots \quad \ldots \quad \Delta l_{n_k} \} \tag{2.29}$$

with $[k]$ thus being a diagonal matrix containing stiffnesses k_j of the individual springs:

$$[k] = \begin{bmatrix} k_1 & 0 & \ldots & 0 & \ldots & 0 \\ 0 & k_2 & \ldots & 0 & \ldots & 0 \\ \ldots & \ldots & \ldots & \ldots & \ldots & \ldots \\ 0 & 0 & \ldots & 0 & \ldots & k_{n_k} \end{bmatrix} \tag{2.30}$$

In (2.28) we also collected the weight and elevations of the centres of gravity in the following vectors

$$\begin{aligned} \underline{p}^T &= \{ p_1 \quad p_2 \quad \ldots \quad \ldots \quad p_{n_p} \} \\ \underline{h}^T &= \{ h_1 \quad h_2 \quad \ldots \quad \ldots \quad h_{n_p} \} \end{aligned} \tag{2.31}$$

Spring constants k_j are obtained from the linearization of the curve—generally nonlinear—which correlates the displacement impressed on the extremities of the generic elastic jth element and the force that is transmitted as a consequence. Elongations Δl_j and the elevations of the centres of gravity h_j (2.28) are generally expressed as a function of the independent coordinates x_i in scalar form as:

$$\begin{cases} \Delta l_j = \Delta l_j(x_1, x_2, \ldots, x_n) = \Delta l_j(\underline{x}) \\ h_j = h_j(x_1, x_2, \ldots, x_n) = h_j(\underline{x}) \end{cases} \tag{2.32}$$

or rather, by collecting spring elongations Δl_j in the sole vector $\underline{\Delta l}_k$ and the elevation of the centres of gravity h_j in the single vector $\underline{h}$,

$$
\underline{\Delta l}_k = \underline{\Delta l}_k(\underline{x}) = \begin{Bmatrix} \Delta l_1(x_1, x_2, \ldots, x_n) \\ \Delta l_2(x_1, x_2, \ldots, x_n) \\ \ldots \\ \Delta l_{n_k}(x_1, x_2, \ldots, x_n) \end{Bmatrix}
$$
$$
\underline{h} = \underline{h}(\underline{x}) = \begin{Bmatrix} h_1(x_1, x_2, \ldots, x_n) \\ h_2(x_1, x_2, \ldots, x_n) \\ \ldots \\ h_{n_p}(x_1, x_2, \ldots, x_n) \end{Bmatrix} \tag{2.33}
$$

the potential energy V associated with the elastic elements is:

$$
V = \frac{1}{2} \underline{\Delta l}_k(\underline{x})^T [k] \, \underline{\Delta l}_k(\underline{x}) + \underline{p}^T \underline{h}(\underline{x}) \tag{2.34}
$$

It is necessary to underline the fact that in this chapter we hypothesized a linear behaviour of the elastic elements ($[k]$ = constant). Only the link between the elongations of the extremities and the independent variables (geometrical nonlinearity) might possibly prove to be nonlinear.

In the event of our wishing to linearize the equations of motion, it is necessary to first make the potential energy expression quadratic:

$$
V = V(\underline{x}_0) + \left\{ \frac{\partial V}{\partial \underline{x}} \right\}_{\underline{x}=\underline{x}_0} \overline{\underline{x}} + \frac{1}{2} \overline{\underline{x}}^T \left[\frac{\partial}{\partial \underline{x}} \left\{ \frac{\partial V}{\partial \underline{x}} \right\} \right]_{\underline{x}=\underline{x}_0}
$$
$$
\overline{\underline{x}} = V(\underline{x}_0) + \left\{ \frac{\partial V}{\partial \underline{x}} \right\}_0 \overline{\underline{x}} + \frac{1}{2} \overline{\underline{x}}^T [K_0] \overline{\underline{x}} \tag{2.35}
$$

where $[K_0]$ is the stiffness matrix of the overall system, expressed as a function of the independent variables $\underline{x}$.

The first derivative of potential energy evaluated in the static equilibrium position can be rewritten in matrix form as:

$$
\left\{ \frac{\partial V}{\partial \underline{x}} \right\}_0 = \{\underline{\Delta l}_k\}_0^T [k] \left[\frac{\partial \underline{\Delta l}_k}{\partial \underline{x}} \right]_0 + \underline{p}^T \left[\frac{\partial \underline{h}}{\partial \underline{x}} \right]_0 \tag{2.36}
$$

and, as a consequence, the second derivative, always evaluated about this position, is:

$$
\left[\frac{\partial}{\partial \underline{x}} \left\{ \frac{\partial V}{\partial \underline{x}} \right\} \right]_0 = [K_0] = \left[\frac{\partial \underline{\Delta l}_k}{\partial \underline{x}} \right]_0^T [k] \left[\frac{\partial \underline{\Delta l}_k}{\partial \underline{x}} \right]_0 + \left[\left[\frac{\partial}{\partial \underline{x}} \left[\frac{\partial \underline{\Delta l}_k}{\partial \underline{x}} \right]^T \right] \right]_0 [k] \{\underline{\Delta l}_k\}_0 + \left[\left[\frac{\partial}{\partial \underline{x}} \left[\frac{\partial \underline{h}}{\partial \underline{x}} \right]^T \right] \right]_0 \underline{p} \tag{2.37}
$$

where $\left[\left[\frac{\partial}{\partial \underline{x}}\left[\frac{\partial \Delta l_k}{\partial \underline{x}}\right]\right]\right]_0$ *and* $\left[\left[\frac{\partial}{\partial \underline{x}}\left[\frac{\partial h}{\partial \underline{x}}\right]\right]\right]_0$ are three-dimensional matrices.[1]

In the case in which the physical variables Δl_j and h_j are linearly dependent on independent coordinates $\underline{x}$, Eq. (2.37) reduces to:

$$\left\{\frac{\partial}{\partial \underline{x}}\left\{\frac{\partial V}{\partial \underline{x}}\right\}\right\}_0 = [K_0] = \left[\frac{\partial \Delta l_k}{\partial \underline{x}}\right]_0^T [k] \left[\frac{\partial \Delta l_k}{\partial \underline{x}}\right]_0 \tag{2.38}$$

In (2.35) $V(\underline{x}_0)$ is a constant, $\underline{Q}_0 = \left\{\frac{\partial V}{\partial \underline{x}}\right\}_0$ is null by definition of the static equilibrium position, while the second derivative of potential energy V with respect to vector $\underline{x}$ represents the stiffness matrix $[K_0]$: this matrix is symmetric because, by definition of conservative system, the rotor of the forces that admit a potential [13, 18] is null and the second mixed derivatives of the potential energy are equal.

In the case in which the perturbations about the static equilibrium position are analysed, and the static equilibrium position is stable, $[K_0]$ is also positive definite (V calculated in $\underline{x} = \underline{x}_0$ has a relative minimum).

The aforementioned approach which refers to the use of three-dimensional matrices can be replaced with a method that uses planar matrices: for this purpose (2.36) can be rewritten as:

$$\left\{\frac{\partial V}{\partial \underline{x}}\right\}_0 = \left\{\sum_{j=1}^{n_k} k_j \Delta l_j \left\{\frac{\partial \Delta l_j}{\partial \underline{x}}\right\} + \sum_{j=1}^{n_P} p_j \left\{\frac{\partial h_j}{\partial \underline{x}}\right\}\right\}_0 \tag{2.39}$$

and (2.37) as:

$$\begin{aligned}\left[\frac{\partial}{\partial \underline{x}}\left\{\frac{\partial V}{\partial \underline{x}}\right\}\right]_0 = \Bigg[&\sum_{j=1}^{n_k}\left[\left\{\frac{\partial \Delta l_j}{\partial \underline{x}}\right\}^T k_j \left\{\frac{\partial \Delta l_j}{\partial \underline{x}}\right\}\right]_0 \\ &+ \sum_{j=1}^{n_k}\left[k_j \Delta l_j \frac{\partial}{\partial \underline{x}}\left\{\frac{\partial \Delta l_j}{\partial \underline{x}}\right\}\right]_0 + \sum_{j=1}^{n_P}\left[p_j \frac{\partial}{\partial \underline{x}}\left\{\frac{\partial h_j}{\partial \underline{x}}\right\}\right]_0\Bigg]\end{aligned} \tag{2.40}$$

[1]The two aforementioned matrices are three-dimensional ($m \times n \times n$), where m is the number of physical variables (number of elastic elements of the system or number of bodies that compose it) and n is the number of d.o.f. in $\underline{x}$. In this way, the generic term of this matrix is $\hat{k}(i,j,l) = \frac{\partial^2 \Delta l_i}{\partial x_j \partial x_l}$ or $\hat{k}(i,j,l) = \frac{\partial^2 h_i}{\partial x_j \partial x_l}$.

where:

$$\left[k_j \Delta l_j \frac{\partial}{\partial \underline{x}}\left\{\frac{\partial \Delta l_j}{\partial \underline{x}}\right\}\right]_0 = k_j \Delta l_j \begin{bmatrix} \frac{\partial^2 \Delta l_j}{\partial x_1 \partial x_1} & \cdots & \frac{\partial^2 \Delta l_j}{\partial x_1 \partial x_i} & \cdots & \frac{\partial^2 \Delta l_j}{\partial x_1 \partial x_n} \\ \cdots & \cdots & \cdots & \cdots & \cdots \\ \frac{\partial^2 \Delta l_j}{\partial x_i \partial x_1} & \cdots & \frac{\partial^2 \Delta l_j}{\partial x_i \partial x_i} & \cdots & \frac{\partial^2 \Delta l_j}{\partial x_i \partial x_n} \\ \cdots & \cdots & \cdots & \cdots & \cdots \\ \frac{\partial^2 \Delta l_j}{\partial x_n \partial x_1} & \cdots & \frac{\partial^2 \Delta l_j}{\partial x_n \partial x_i} & \cdots & \frac{\partial^2 \Delta l_j}{\partial x_n \partial x_n} \end{bmatrix}_0 = [K_{jII}]$$

$$\left[p_j \frac{\partial}{\partial \underline{x}}\left\{\frac{\partial h_j}{\partial \underline{x}}\right\}\right]_0 = p_j \begin{bmatrix} \frac{\partial^2 h_j}{\partial x_1 \partial x_1} & \cdots & \frac{\partial^2 h_j}{\partial x_1 \partial x_i} & \cdots & \frac{\partial^2 h_j}{\partial x_1 \partial x_n} \\ \cdots & \cdots & \cdots & \cdots & \cdots \\ \frac{\partial^2 h_j}{\partial x_i \partial x_1} & \cdots & \frac{\partial^2 h_j}{\partial x_i \partial x_i} & \cdots & \frac{\partial^2 h_j}{\partial x_i \partial x_n} \\ \cdots & \cdots & \cdots & \cdots & \cdots \\ \frac{\partial^2 h_j}{\partial x_n \partial x_1} & \cdots & \frac{\partial^2 h_j}{\partial x_n \partial x_i} & \cdots & \frac{\partial^2 h_j}{\partial x_n \partial x_n} \end{bmatrix}_0 = [K_{jIII}] \tag{2.41}$$

Taking into account (2.40) and (2.41), (2.37) can be rewritten as:

$$\begin{aligned} [K_0] &= \left[\left[\frac{\partial \Delta l_k}{\partial \underline{x}}\right]_0^T [k] \left[\frac{\partial \Delta l_k}{\partial \underline{x}}\right]_0 + \sum_{j=1}^{n_k} [K_{jII}]_0 + \sum_{j=1}^{n_P} [K_{jIII}]_0\right] \\ &= \left[\left[\frac{\partial \Delta l_k}{\partial \underline{x}}\right]_0^T [k] \left[\frac{\partial \Delta l_k}{\partial \underline{x}}\right]_0\right] + \left[\sum_{j=1}^{n_k} [K_{jII}]_0 + \sum_{j=1}^{n_P} [K_{jIII}]_0\right] \\ &= \left[\left[\frac{\partial \Delta l_k}{\partial \underline{x}}\right]_0^T [k] \left[\frac{\partial \Delta l_k}{\partial \underline{x}}\right]_0\right] + [K_g]_0 \end{aligned} \tag{2.42}$$

where $[K_g]_0 = \left[\sum_{j=1}^{n_k} [K_{jII}]_0 + \sum_{j=1}^{n_P} [K_{jIII}]_0\right]$ is the stiffness-gravitational reference matrix of the system.

2.2.4 Virtual Work of Active Forces

For the calculation of the Lagrangian component of the active forces, not contemplated in terms of kinetic, potential energy or the dissipation function previously defined, we use f_j to denote the component of the external generic force (or torque) and y_{fj} the respective displacement (or rotation) component of the relative point of application of the force. We group the n_f components of the f_j forces applied to the system in vector $\underline{f}$:

$$\underline{f}^T = \{f_1 \quad f_2 \quad \dots \quad f_i \quad \dots \quad f_{n_f}\} \tag{2.43}$$

and with $\delta^* \underline{Y}_f$ the vector that contains the n_f virtual displacement components $\delta^* y_{fj}$ of the force application points in the direction of the force themselves:

$$\delta^* \underline{Y}_f^T = \{\delta^* y_{f1} \quad \delta^* y_{f2} \quad \dots \quad \delta^* y_{fj} \quad \dots \quad \delta^* y_{fn_f}\} \tag{2.44}$$

The displacement (or rotation) of the generic application point of the forces (or torques) is generally a nonlinear function of the independent coordinates $\underline{x}$:

$$\underline{Y}_f = \underline{Y}_f(\underline{x}) \tag{2.45}$$

The virtual work can thus be expressed in scalar and matrix form as:

$$\delta^* L = \sum_{j=1}^{n_f} f_j \delta^* y_{fj} = \underline{f}^T \delta^* \underline{Y}_f \tag{2.46}$$

The virtual variation $\delta^* \underline{Y}_f$ as a consequence of a virtual variation of the free coordinates $\delta^* \underline{x}$ can be evaluated as:

$$\delta^* \underline{Y}_f = \left[\frac{\partial \underline{Y}_f}{\partial \underline{x}}\right] \delta^* \underline{x} = [\Lambda(x)]_f \delta^* \underline{x} \tag{2.47}$$

where $[\Lambda(x)]_f$ is the Jacobian matrix:

$$[\Lambda(\underline{x})]_f = \begin{bmatrix} \frac{\partial y_{f1}}{\partial x_1} & \frac{\partial y_{f1}}{\partial x_2} & \dots & \frac{\partial y_{f1}}{\partial x_n} \\ \dots & \dots & \dots & \dots \\ \frac{\partial y_{fj}}{\partial x_1} & \frac{\partial y_{fj}}{\partial x_2} & \dots & \frac{\partial y_{fj}}{\partial x_n} \\ \dots & \dots & \dots & \dots \\ \frac{\partial y_{fn_f}}{\partial x_1} & \frac{\partial y_{fn_f}}{\partial x_2} & \dots & \frac{\partial y_{fn_f}}{\partial x_n} \end{bmatrix} \tag{2.48}$$

If the dependence of $\delta^* \underline{Y}_f$ from $\underline{x}$ is linear, this matrix is constant, while if nonlinear (similarly to what was seen in the previous sections) this matrix still proves to be a function of the same independent variables $\underline{x}$.

To define the response of the vibrating system about the static equilibrium position it is possible, as usual, to linearize this relationship about the same position:

$$[\Lambda(\underline{x})]_f = \left[\frac{\partial \underline{Y}_f}{\partial \underline{x}}\right]_0 = [\Lambda_f] = const. \tag{2.49}$$

Based on this hypothesis the Jacobian $[\Lambda_f]$ calculated for $\underline{x} = \underline{x}_0$ becomes a constant matrix. In the case of linear or linearized systems about the static equilibrium position, the virtual work of forces thus becomes:

$$\delta^* L = \underline{f}^T \delta^* \underline{Y}_f = \underline{f}^T [\Lambda_f] \delta^* \overline{\underline{x}} = \underline{Q}^T \delta^* \overline{\underline{x}} \tag{2.50}$$

where $\underline{Q}$ is the vector of the generalized force defined as:

$$\underline{Q} = [\Lambda_f]^T \underline{f} \tag{2.51}$$

2.2.5 Equations of Motion

Having now defined all the forms of energy:

- kinetic E_c (2.18);
- dissipation function D (2.27);
- potential V (2.35);

in quadratic form and having reduced the virtual work of the external active forces $\delta^* L$ as shown in (2.50), it is possible, for a linear or linearized system about the static equilibrium position, to define, by applying Lagrange's equations (2.4), the individual terms appearing in the equation of motion of the system itself. In particular, for kinetic energy, the term:

$$\left\{ \frac{d}{dt} \left(\frac{\partial E_C}{\partial \dot{\overline{x}}_i} \right) \right\} - \left\{ \frac{\partial E_C}{\partial \overline{x}_i} \right\} \tag{2.52}$$

is reduced, in linear systems, to the first term only, where term $\left\{ \frac{\partial E_C}{\partial \overline{x}_i} \right\}$ is different from zero only for nonlinear systems on account of the fact that, in this case, the Jacobian $[\Lambda(\underline{x})]_m$ (2.16) is a function of the same independent variables $\underline{x}$. In linear or linearized systems, the mass matrix $[M]$ of the system is symmetric and constant. Therefore, based on (2.19) the derivation according to Lagrange [22, 34] leads to the inertia term:

$$\begin{aligned} \left\{ \frac{d}{dt} \left(\frac{\partial E_C}{\partial \dot{\overline{\underline{x}}}} \right) \right\}^T &= \left\{ \frac{d}{dt} \left\{ \frac{1}{2} \dot{\overline{\underline{x}}}^T [M] + \frac{1}{2} \dot{\overline{\underline{x}}}^T [M]^T \right\} \right\}^T \\ &= [M] \ddot{\overline{\underline{x}}} = [\Lambda_m]^T [m] [\Lambda_m] \ddot{\overline{\underline{x}}} \end{aligned} \tag{2.53}$$

The terms due to forces that allow potential (2.35) can be obtained by means of:

$$\left\{\frac{\partial V}{\partial \underline{x}}\right\}^T = \left\{\frac{1}{2}\underline{x}^T[K] + \frac{1}{2}\underline{x}^T[K]^T\right\}^T = [K]\underline{x} = [\Lambda_k]^T[k][\Lambda_k]\underline{x} \tag{2.54}$$

which are valid in the linear case, and with the same applying to the dissipation function (2.27):

$$\left\{\frac{\partial D}{\partial \dot{\underline{x}}}\right\}^T = \left\{\frac{1}{2}\dot{\underline{x}}^T[R] + \frac{1}{2}\dot{\underline{x}}^T[R]^T\right\}^T = [R]\dot{\underline{x}} = [\Lambda_r]^T[r][\Lambda_r]\dot{\underline{x}} \tag{2.55}$$

Finally, the Lagrangian components $\underline{Q}$ of the active forces along the d.o.f. $\underline{x}$ can be obtained by means of the definition (2.50) shown below:

$$\underline{Q} = [\Lambda_f]^T\underline{f} \tag{2.56}$$

The final form of the equation of motion is thus:

$$[M]\ddot{\underline{x}} + [R]\dot{\underline{x}} + [K]\underline{x} = \underline{Q} \tag{2.57}$$

As already demonstrated, the correct methodology to linearize the nonlinear equations of motion of a system about an equilibrium position is the one outlined in Sects. 2.2.1–2.2.4: more specifically, as regards the potential energy associated with elastic and gravitational terms (Sect. 2.2.3), the static preloads of the springs and the presence of the weight force are taken in account in the linearization through the *stiffness-gravitational* matrix

$$[K_g]_0 = \left[\sum_{j=1}^{n_k}[K_{jII}]_0 + \sum_{j=1}^{n_P}[K_{jIII}]_0\right] \tag{2.57a}$$

If the system is linear, if the Hessians present in (2.42) are null or if these contributions (due to the preload of the springs and gravitational effects) are negligible and therefore need not to be considered, the potential energy (Eq. 2.34) can be defined as:

$$V = \frac{1}{2}\underline{\Delta l}_k(\underline{x})^T[k]\,\underline{\Delta l}_k(\underline{x}) \tag{2.58}$$

In this instance it is possible, as in the case of the kinetic energy [Eqs. (2.15)–(2.18)] and the dissipative function [Eqs. (2.24)–(2.27)], to linearize the link between the physical variables and the independent variables:

$$\Delta \underline{l}_k = \left[\frac{\partial \Delta l_k}{\partial \underline{x}} \right] \underline{x} = [\Lambda]_0 \underline{x} \tag{2.59}$$

which, when substituted in (2.34) gives:

$$V = \frac{1}{2} \underline{x}^T [\Lambda]_0^T [k] [\Lambda]_0 \underline{x} \tag{2.60}$$

In this case, the equations of motion of the linearized system thus always prove to be (2.57):

$$[M]\ddot{\underline{x}} + [R]\dot{\underline{x}} + [K]\underline{x} = \underline{Q} \tag{2.61}$$

on account of simply being [Eqs. (2.47)–(2.50), (2.23)–(2.27), (2.15)–(2.19) and (2.34)]:

$$\begin{aligned} [K] &= [\Lambda]_0^T [k] [\Lambda]_0 \\ [R] &= [\Lambda_d]^T [r] [\Lambda_d] \\ [M] &= [\Lambda_m]^T [M_y] [\Lambda_m] \\ \underline{Q} &= [\Lambda_f]^T \underline{f} \end{aligned} \tag{2.62}$$

2.3 Some Application Examples

2.3.1 One-Degree-of-Freedom Systems

Let us now consider the system of Fig. 2.3. constituted by a roller which rolls without sliding on a smooth surface, constrained by a spring of constant k and a viscous damper of constant r. We will use m to indicate disc mass and J_G to indicate the mass moment of inertia of the same and finally R to indicate its radius.

By only considering planar motion and the pure rolling condition, the system has only 1 d.o.f. $n = 3xn_c - n_v = 3x1 - 2 = 1$. As an independent variable, let us consider the displacement x of the centre of gravity of the disc and, as a

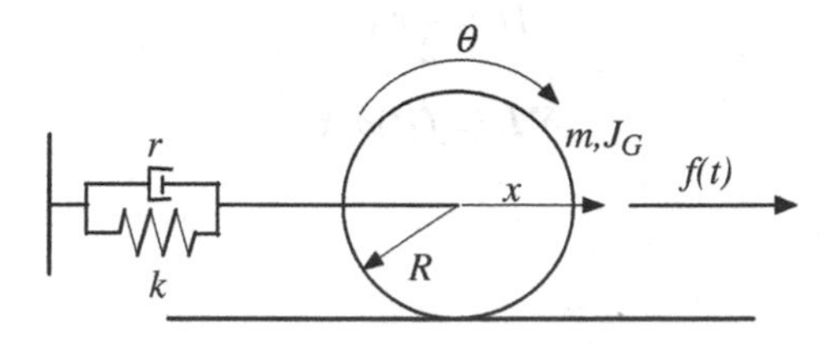

Fig. 2.3 Vibrating 1 d.o.f. system

consequence, rotation θ, elongation Δl of the spring (conventionally considered positive in extension) and the relative speed $\dot{\Delta l}$ between the extremities of the damper as physical variables. The expression of kinetic energy, as a function of the physical variables, is the following:

$$E_c = \frac{1}{2} m \dot{x}^2 + \frac{1}{2} J_G \dot{\theta}^2 \tag{2.63}$$

Potential Energy is due to the only spring of stiffness k:

$$V = \frac{1}{2} k \, \Delta l^2 \tag{2.64}$$

due to the fact that the centre of gravity h of the disc does not perform any vertical movement. Dissipative energy is due to the presence of the damper and is equal to half of the power developed by the viscous force, changed in sign or rather:

$$D = \frac{1}{2} r \, \dot{\Delta l}^2 \tag{2.65}$$

The virtual work performed by external force $f(t)$ is:

$$\delta^* L = f(t) \delta^* * x \tag{2.66}$$

By now introducing the links between the physical variables and the independent variable x into the various forms of energy:

$$\begin{aligned} \theta &= \frac{x}{R} \\ \Delta l &= x \end{aligned} \tag{2.67}$$

it is possible to rewrite the same expressions as a function of x alone:

$$E_c = \frac{1}{2} m \ddot{x}^2 + \frac{1}{2} \frac{J_G}{R^2} \dot{x}^2 \tag{2.68}$$

$$V = \frac{1}{2} k x^2 \tag{2.69}$$

$$D = \frac{1}{2} r \dot{x}^2 \tag{2.70}$$

$$\delta^* L = f(t) \delta^* x \tag{2.71}$$

By means of Lagrange's equation (2.4) and by keeping account of (2.68)–(2.71), we obtain:

$$\left(m+\frac{J_G}{R^2}\right)\ddot{x}+r\dot{x}+kx=f(t) \tag{2.72}$$

The term $(m+\frac{J_G}{R^2})$ is defined as a generalized mass according to the independent Lagrangian coordinate x, r the generalized damper and finally k the generalized stiffness: these generalized amplitudes may coincide (as in the case analysed for k and r) with physical terms or represent equivalent amplitudes from a point of view of energy.

2.3.2 *Two-Degree-of-Freedom Systems*

From a didactic point of view, before proceeding with a generalized approach to dissipative n d.o.f. systems, perturbed about the static equilibrium position, it would be advisable to first discuss 2 d.o.f. systems. The diagram of a 2 d.o.f. system can assume a number of different forms depending on the specific problem that one wishes to simulate. By following the same outline already shown for 1 d.o.f. systems, we will analyse an extremely simple system like the one illustrated in Fig. 2.4, writing the equations of motion both with dynamic equilibriums (Sect. 2.2.1), and Lagrange's equations (Sect. 2.2.2). More specifically, we will consider linear systems in which the physical coordinates (elevations of centres of gravity, relative elongations of the extremities of the elastic and damping elements etc.) are linked to the free coordinates, regardless of their linear relationships.

Later on (Sect. 2.4), we will demonstrate a systematic approach, based on Lagrange's equations, to express the equations of motion in matrix form for systems consisting of mutually coupled rigid bodies: more specifically, we will address the problem regarding the definition of the various forms of energy as a function of the physical coordinates of the system and, subsequently, the definition of the link between the physical and independent coordinates chosen to describe the system itself. Finally we will resort (Sect. 2.4.2) to the methodology required to linearize the equations of motion of the system in the case of nonlinear links between physical and independent variables.

2.3.2.1 Dynamic Equilibrium Equation

As a first example, let us consider a vibrating two d.o.f. system (see Fig. 2.4) which is linear and subjected to two generic external forces $f_1=f_1(t)$ and $f_2=f_2(t)$.

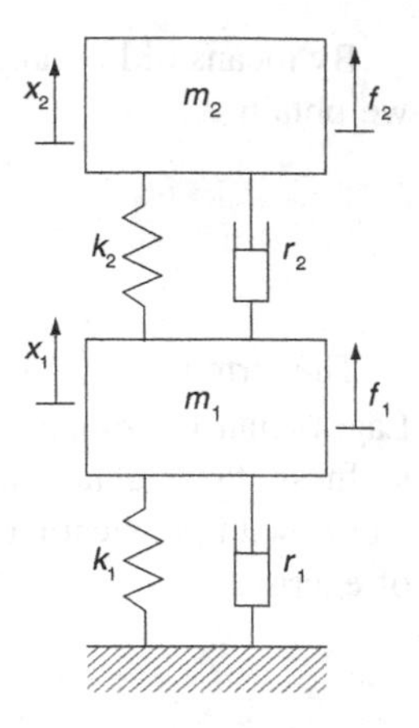

Fig. 2.4 Linear two d.o.f. system

As independent variables, we will choose the absolute displacements x_1 and x_2 evaluated starting from the static equilibrium position (defined by $x_1 = x_2 = 0$): x_1 and x_2 thus represent the perturbed motion about this position and, for this reason, the weight forces[2] will not appear in the equations of motion. Furthermore, for the sake of convention, let us assume that the forces are positive in the same directions of the displacements. The two equations of motion of the system can be obtained by expressing dynamic equilibrium equations [1, 13, 18, 33] at translation of the two masses, considered as isolated bodies and highlighting the forces that the outside world exerts on same (free body diagram of Fig. 2.5). The forces acting on the first mass are:

- external force f_1;
- inertia force f_{in1};
- elastic forces f_{k1} and f_{k2} due to the presence of the elastic elements of constants k_1 and k_2;
- viscous forces f_{r1} and f_{r2} due to the dampers of constants r_1 and r_2:

$$\begin{cases} f_{in1} = -m_1\ddot{x}_1 \\ f_{r1} = -r_1\dot{x}_1 \\ f_{r2} = -r_2(\dot{x}_1 - \dot{x}_2) \\ f_{k1} = -k_1x_1 \\ f_{k2} = -k_2(x_1 - x_2) \end{cases} \tag{2.73}$$

[2] As seen in Chap. 1, Sect. 1.5, the effect of the weight force and of the static preload of the springs does not appear in the equations of motion because, in this particular case, the link between the elevation of the centre of gravity of the single masses, the spring elongations and the independent variables is linear.

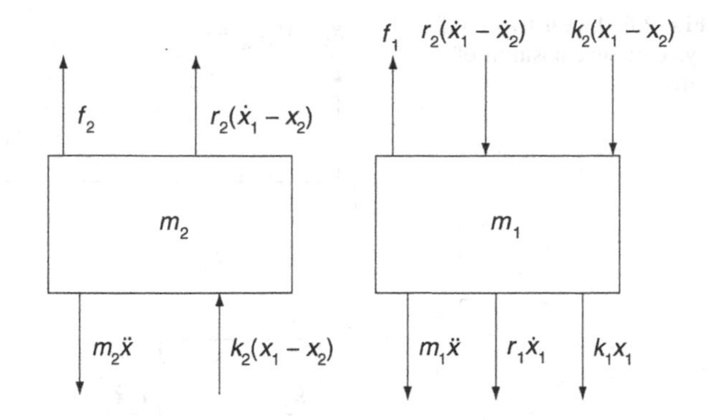

Fig. 2.5 Linear two d.o.f. system (Fig. 2.4): expression of the equations of motion using dynamic equilibriums: forces acting on the single masses

Similarly, the following forces:

$$\begin{cases} f_{in2} = -m_2\ddot{x}_2 \\ f_{r2} = -r_2(\dot{x}_2 - \dot{x}_1) \\ f_{k2} = -k_2(x_2 - x_1) \end{cases} \tag{2.74}$$

will act on the second mass, considered isolated.

By expressing the balance of the forces on the individual masses and by bringing the applied forces to the right hand side, a 2 equation system is obtained (Fig. 2.6):

$$\begin{cases} m_1\ddot{x}_1 + (r_1 + r_2)\dot{x}_1 - r_2\dot{x}_2 + (k_1 + k_2)x_1 - k_2x_2 = f_1 \\ m_2\ddot{x}_2 - r_2\dot{x}_1 + r_2\dot{x}_2 - k_2x_1 + k_2x_2 = f_2 \end{cases} \tag{2.75}$$

It is possible to rewrite the equations of motion (2.75) in matrix form as:

$$[M]\underline{\ddot{x}} + [R]\underline{\dot{x}} + [K]\underline{x} = \underline{f} \tag{2.76}$$

corresponding, for a generic system with 2 d.o.f., to the following generic expression in scalar form:

$$\begin{cases} m_{11}\ddot{x}_1 + m_{12}\ddot{x}_2 + r_{11}\dot{x}_1 + r_{12}\dot{x}_2 + k_{11}x_1 + k_{12}x_2 = f_1 \\ m_{21}\ddot{x}_1 + m_{22}\ddot{x}_2 + r_{21}\dot{x}_1 + r_{22}\dot{x}_2 + k_{21}x_1 + k_{22}x_2 = f_2 \end{cases} \tag{2.77}$$

In this case the subscripts associated with the different amplitudes define the location occupied by same in the corresponding matrices (respectively rows and columns). As previously seen, expression in matrix form (2.76) is convenient for a synthetic approach to the n d.o.f. systems because we are reduced to the expression of only one matrix equation, corresponding to n scalar equations. As will be seen later on, matrices $[M]$, $[R]$ and $[K]$ of Eq. (2.76) can also be directly obtained thanks to the application of Lagrange's equations defined in matrix form. Given the hypothesis of a dissipative system perturbed about its static and stable equilibrium position, matrices $[M]$, $[R]$ and $[K]$ are symmetrical and positive definite. In fact,

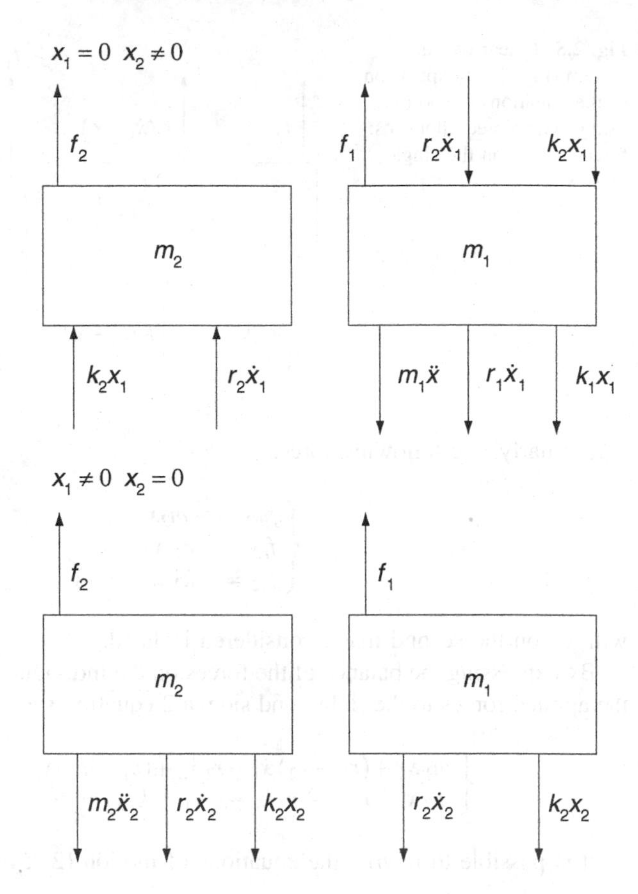

Fig. 2.6 Linear two d.o.f. system: superposition of effects

they are associated with energy forms (kinetic energy E_c, dissipation function D and potential energy V) represented by quadratic forms which are positive definite. A positive definite symmetrical matrix is associated with a quadratic form which is also positive definite: this is always true if the matrices are obtained using an energy approach, while the same might not apply when using dynamic equilibriums approach, due to specific choices of independent coordinates. Finally we note that:

- $[M]$ is the matrix which, when multiplied by acceleration vector $\underline{\ddot{x}}$, defines the inertia forces acting on the system according to the various d.o.f. Due to the specific choice of coordinates (absolute) adopted in the example, the system is inertially uncoupled and the mass matrix therefore proves to be diagonal. The generic term m_{ij} represents the inertia force, changed in sign, acting on the generic ith d.o.f. for a unitary acceleration assigned to the generic jth d.o.f;
- $[R]$, multiplied by the velocity vector $\underline{\dot{x}}$, represents the viscous forces acting on the system;
- $[K]$, multiplied by the displacement vector, provides the elastic forces which always refer to the system's $\underline{x}$ d.o.f.

As previously mentioned, a precise physical meaning corresponds to each element of the generic matrix: for example, generic element k_{ij} of the stiffness matrix represents the elastic force (changed in sign) acting on the ith d.o.f. for an unitary displacement of the jth d.o.f., on account of the fact that the other displacements are null. Terms k_{ij} are denoted as stiffness constants of the vibrating system.

2.3.2.2 Lagrange's Equations

The same equations of motion (2.77) can obviously be obtained by adopting a Lagrangian approach in scalar form [8, 13, 18], i.e. by using the following expression:

$$\frac{d}{dt}\left(\frac{\partial E_c}{\partial \dot{x}_i}\right) - \frac{\partial E_c}{\partial x_i} + \frac{\partial D}{\partial \dot{x}_i} + \frac{\partial V}{\partial x_i} = \frac{\delta^* L}{\delta^* x_i} = Q_i \quad (i = 1, n) \tag{2.78}$$

after expressing kinetic energy E_c, potential energy V, dissipation function D and virtual work $\delta^* L$ of the non-conservative forces from which it is possible to evaluate the generic Lagrangian component Q_i. In the case of the example in Fig. 2.5, the various energy forms can be directly expressed as a function of the independent coordinates (having conventionally assumed the elongations of the elastic and damping elements as positive):

$$E_c = \frac{1}{2} m_1 \dot{x}_1^2 + \frac{1}{2} m_2 \dot{x}_2^2 \tag{2.79}$$

$$V = \frac{1}{2} k_2 (x_2 - x_1)^2 + \frac{1}{2} k_1 x_1^2 \tag{2.80}$$

$$D = \frac{1}{2} r_2 (\dot{x}_2 - \dot{x}_1)^2 + \frac{1}{2} r_1 \dot{x}_1^2 \tag{2.81}$$

$$\delta^* L = \delta^* L_1 + \delta^* L_2 - f_1 \delta^* x_1 + f_2 \delta^* x_2 \tag{2.82}$$

To express the 2 equations of motion, the various energy forms (2.79)–(2.82) are derived according to Lagrange (2.4) with respect to coordinate x_1:

$$\frac{d}{dt}\left(\frac{\partial E_c}{\partial \dot{x}_1}\right) = m_1 \ddot{x}_1 \tag{2.83}$$

$$\frac{\partial E_c}{\partial x_1} = 0$$

$$\frac{\partial V}{\partial x_1} = -k_2 (x_2 - x_1) + k_1 x_1 \tag{2.84}$$

$$\frac{\partial D}{\partial \dot{x}_1} = -r_2(\dot{x}_2 - \dot{x}_1) + r_1 \dot{x}_1$$

$$Q_1 = \frac{\delta^* L_1}{\delta^* x_1} = f_1$$

and to coordinate x_2:

$$\begin{aligned}
&\frac{d}{dt}\left(\frac{\partial E_c}{\partial \dot{x}_2}\right) = m_2 \ddot{x}_2 \\
&\frac{\partial E_c}{\partial x_2} = 0 \\
&\frac{\partial V}{\partial x_2} = k_2(x_2 - x_1) \\
&\frac{\partial D}{\partial \dot{x}_2} = r_2(\dot{x}_2 - \dot{x}_1) \\
&Q_2 = \frac{\delta^* L_2}{\delta^* x_2} = f_2
\end{aligned} \tag{2.85}$$

assumed as Lagrangian coordinates. By bearing in mind Eqs. (2.83)–(2.85), it is once again possible to obtain the same system (2.77) of two equations in coordinates x_1 and x_2 which, as previously seen, can be rewritten in matrix form as:

$$[M]\underline{\ddot{x}} + [R]\underline{\dot{x}} + [K]\underline{x} = \underline{f} \tag{2.86}$$

or rather:

$$\begin{aligned}
&\left[\begin{pmatrix} m_1 & 0 \\ 0 & m_2 \end{pmatrix}\right]\begin{Bmatrix} \ddot{x}_1 \\ \ddot{x}_2 \end{Bmatrix} + \left[\begin{pmatrix} r_1 + r_2 & -r_2 \\ -r_2 & r_2 \end{pmatrix}\right]\begin{Bmatrix} \dot{x}_1 \\ \dot{x}_2 \end{Bmatrix} \\
&+ \left[\begin{pmatrix} k_1 + k_2 & -k_2 \\ -k_2 & k_2 \end{pmatrix}\right]\begin{Bmatrix} x_1 \\ x_2 \end{Bmatrix} = \begin{Bmatrix} f_1 \\ f_2 \end{Bmatrix}
\end{aligned} \tag{2.87}$$

Given that the system is dissipative and perturbed about the stable equilibrium position, matrices $[M]$, $[R]$ and $[K]$ are symmetrical and defined as positive. Additionally, the associated energy forms (quadratic forms defined as positive) can be expressed in matrix form as:

$$E_c = \frac{1}{2}\underline{\dot{x}}^T [M]\underline{\dot{x}}$$

$$D = \frac{1}{2}\underline{\dot{x}}^T [R]\underline{\dot{x}} \tag{2.88}$$
$$V = \frac{1}{2}\underline{x}^T [K]\underline{x}$$
$$\delta^* L = \underline{f}^T \delta^* \underline{x}$$

However, it is necessary to remember that the equations of motion expressed in terms of dynamic equilibriums are not always the same as those obtained when using a Lagrangian approach.

2.3.3 *An Additional Example of Two-Degree-of-Freedom Systems*

In order to express the equations of motion, we will now apply the methods outlined in the previous sections, using not only dynamic equilibriums but also Lagrange's equations for the vibrating system shown in the Fig. 2.7 which represents a simplified model designed to study the vertical motion of a vehicle excited, for example, by the unbalanced inertia forces of the engine. We will subsequently resort to the same example to illustrate the general method of expressing equations of motions for systems with n d.o.f

The system - considering only motion in the vertical plane - is schematized by means of a two d.o.f. model: we will use m to define body mass and J_G to define the inertia moment with respect to centre of gravity G. The elastic elements, of constants k_1 and k_2, and the dampers, of constants r_1 and r_2, connect the system to the ground and are respectively placed at distances l_1 and l_2 from the centre of gravity G of same.

The system is finally subjected to an applied force $f = f(t)$ applied in point B and positioned at distance b from the centre of gravity G (as conventions for displacements, rotations and forces, we assume those shown in Fig. 2.7).

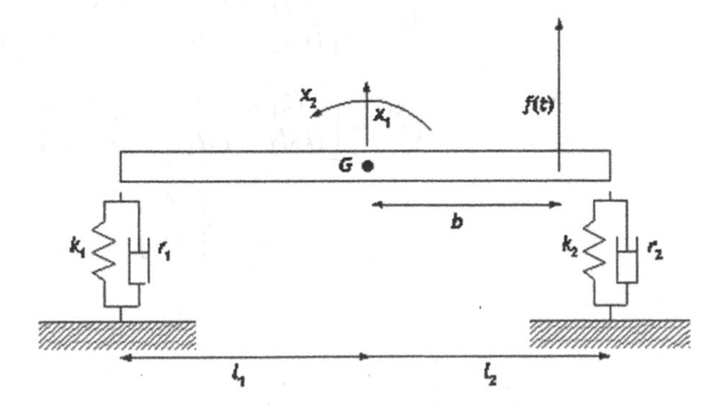

Fig. 2.7 Linear two d.o.f. system: simplified vehicle model

The independent coordinates chosen to describe the system are the absolute vertical displacement of the centre of gravity x_1 and the absolute rotation x_2 of the rigid body defined with respect to the horizontal reference line: both variables are defined starting from the static equilibrium position, i.e. they only define the variations due to perturbed motion (this means that the effect of the weight and static deformations of the springs will not appear in the equations of the dynamic equilibrium or in the Lagrange's equations—see the Sect. 1.5 of Chap. 1).

It is therefore necessary, given the initial conditions, to determine the differential equations which, once integrated, allow us to define system motion.

2.3.3.1 Expression Through Dynamic Equilibrium Equations

By considering the body as an isolated system - highlighting the forces that the outside world exerts on same - it is possible to write the equations of motion:

$$M\ddot{x}_1 + (r_1 + r_2)\,\dot{x}_1 + (r_2 l_2 - r_1 l_1)\,\dot{x}_2 + (k_1 + k_2)x_1 + (k_2 l_2 - k_1 l_1)x_2 = f \quad (2.89)$$

$$J_G\ddot{x}_2 + (r_2 l_2 - r_1 l_1)\,\dot{x}_1 + (r_1 l_1^2 + r_2 l_2^2)\,\dot{x}_2 + (k_2 l_2 - k_1 l_1)x_1 + +(k_1 l_1^2 + k_2 l_2^2)x_2 = fb$$

Having organized the vector of independent variables x as:

$$\underline{x} = \begin{Bmatrix} x_1 \\ x_2 \end{Bmatrix} \quad (2.90)$$

Equations (2.89) can be rewritten in the usual matrix form:

$$[M]\underline{\ddot{x}} + [R]\underline{\dot{x}} + [K]\underline{x} = \underline{f} \quad (2.91)$$

which, in this case, is:

$$[M] = \begin{bmatrix} m & 0 \\ 0 & J_G \end{bmatrix}$$

$$[R] = \begin{bmatrix} (r_1 + r_2) & (r_2 l_2 - r_1 l_1) \\ (r_2 l_2 - r_1 l_1) & (r_2 l_2^2 + r_1 l_1^2) \end{bmatrix} \quad (2.92)$$

$$[K] = \begin{bmatrix} (k_1 + k_2) & (k_2 l_2 - k_1 l_1) \\ (k_2 l_2 - k_1 l_1) & (k_2 l_2^2 + k_1 l_1^2) \end{bmatrix}$$

$$\underline{f} = \begin{Bmatrix} f \\ f\,b \end{Bmatrix}$$

As can be noted:

- Stiffness matrix $[K]$ and mass matrix $[M]$ are symmetrical (analytical evidence corresponding to the conservative field of the elastic and inertia forces);
- The damping matrix $[R]$, which is symmetrical and defined as positive, corresponds to a force field depending on the dissipative speed;
- Matrix $[K]$ is symmetrical and defined as positive, i.e. its equilibrium position is stable.

2.3.3.2 Using Lagrange's Equations in Scalar Form

The same equations of motion (2.89) or (2.91) can obviously be obtained adopting a Lagrangian approach in scalar form. In the case of the example shown in Fig. 2.7, the various energy forms can be expressed directly, as a function of the physical coordinates, as:

$$
\begin{aligned}
E_c &= \frac{1}{2}m\dot{x}_1^2 + \frac{1}{2}J_G\dot{x}_2^2 \\
V &= \frac{1}{2}k_2\Delta l_2^2 + \frac{1}{2}k_1\Delta l_1^2 \\
D &= \frac{1}{2}r_2\Delta \dot{l}_2^2 + \frac{1}{2}r_1\Delta \dot{l}_1^2 \\
\delta^* L &= f\delta^* x_f
\end{aligned}
\tag{2.93}
$$

where Δl_1 and Δl_2 are the relative elongations to which the two springs (positive according to the conventions of Fig. 2.7) are subjected and x_f is the displacement of the point of application of the applied force in the direction of the force itself: these amplitudes are the physical variables adopted in this example. The link between the physical variables and the independent variables can easily be expressed by:

$$
\begin{aligned}
\Delta l_1 &= x_1 - x_2 l_1 \\
\Delta l_2 &= x_1 + x_2 l_2 \\
x_f &= x_1 + x_2 b \rightarrow \delta^* x_f = \frac{\partial x_f}{\partial x_1}\delta^* x_1 + \frac{\partial x_f}{\partial x_2}\delta^* x_2 = \delta^* x_1 + \delta^* x_2 b
\end{aligned}
\tag{2.94}
$$

By substituting these links (2.94) in (2.93) the various energy forms can be expressed as a function of the only independent coordinates:

$$
E_c = \frac{1}{2}m\dot{x}_1^2 + \frac{1}{2}J_G\dot{x}_2^2
$$

$$V = \frac{1}{2}k_2(x_1 + x_2\,l_2)^2 + \frac{1}{2}k_1(x_1 - x_2\,l_1)^2$$
$$D = \frac{1}{2}r_2(\dot{x}_1 + \dot{x}_2\,l_2)^2 + \frac{1}{2}r_1(\dot{x}_1 - \dot{x}_2\,l_1)^2 \tag{2.95}$$
$$\delta^*L = f\delta^*x_1 + f\,b\,\delta^*x_2$$

By applying Lagrange's equations (2.78) and subsequently considering $i = 1$ and $i = 2$, we once again obtain the equations of motion of system (2.93) which can also be rewritten in matrix form (2.91).

In Sect. 2.3.5, an alternative methodology will be described in order to define—directly in matrix form—Eq. (2.91) of a linear vibrating system with n d.o.f. without using the scalar notation shown here.

As a form of exercise, we will now describe the same vibrating system shown in Fig. 2.7. However, in this case, we will use another set of independent coordinates, for example, vertical displacement q_1 of a point O lying at a distance of a from the centre of gravity of the body and rotation q_2 around the same pole O, see Fig. 2.8. The various energy forms, as a function of the physical variables, vertical displacement of the centre of gravity x_1 and rotation x_2, are the same as those shown in (2.79)–(2.82): the link between the physical and independent variables now becomes:

$$x_1 = q_1 + q_2\,a \rightarrow \dot{x}_1 = \dot{q}_1 + \dot{q}_2\,a$$
$$x_2 = q_2 \rightarrow \dot{x}_2 = \dot{q}_2$$
$$\Delta l_1 = q_1 - q_2(l_1 - a) = q_1 - q_2\,\bar{l}_1$$
$$\Delta l_2 = q_1 + q_2(l_2 + a) = q_1 + q_2\,\bar{l}_2$$
$$x_f = q_1 + q_2\,(b + a) \rightarrow \delta^*x_f = \frac{\partial x_f}{\partial q_1}\,\delta^*q_1 + \frac{\partial x_f}{\partial q_2}\delta^*q_2 = \delta^*q_1 + \delta^*q_2\,(b + a) \tag{2.96}$$

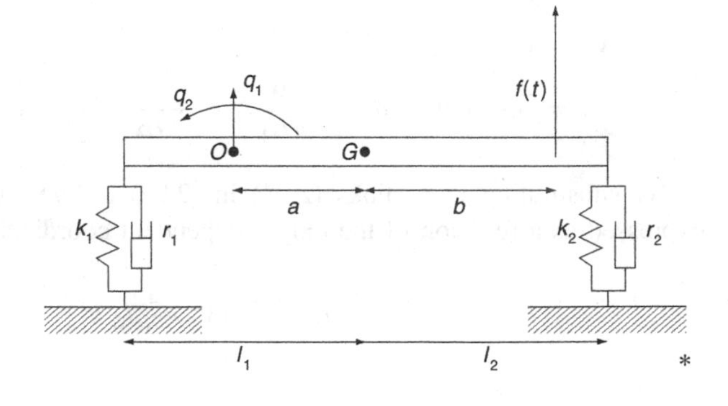

Fig. 2.8 Linear two d.o.f. system (of Fig. 2.7) and a different set of free coordinates

By substituting these links (2.96) in (2.79)–(2.82) it is possible to express the various energy forms as a function of the only new independent coordinates q_1 and q_2:

$$\begin{aligned} E_c &= \frac{1}{2} m\,(\dot{q}_1 + a\,\dot{q}_2)^2 + \frac{1}{2} J_G\,\dot{q}_2^2 \\ V &= \frac{1}{2} k_2 (q_1 + q_2\,\bar{l}_2)^2 + \frac{1}{2} k_1 (q_1 - q_2\,\bar{l}_1)^2 \\ D &= \frac{1}{2} r_2 (\dot{q}_1 + \dot{q}_2\,\bar{l}_2)^2 + \frac{1}{2} r_1 (\dot{q}_1 - \dot{q}_2\,\bar{l}_1)^2 \\ \delta^* L &= f\delta^* q_1 + f(b+a)\delta^* q_2 \end{aligned} \tag{2.97}$$

By applying Lagrange's equations (2.78) and subsequently considering $i = 1$ and $i = 2$, we obtain the equations of motion in the new independent variables:

$$\begin{aligned} & m\ddot{q}_1 + ma\ddot{q}_2 + (r_1 + r_2)\dot{q}_1 - (r_1\bar{l}_1 - r_2\bar{l}_2)\dot{q}_2 \\ & \quad + (k_1 + k_2) q_1 - (k_1\bar{l}_1 - k_2\bar{l}_2) q_2 = f \\ & ma\ddot{q}_1 + (J_G + ma^2)\ddot{q}_2 - (r_1\bar{l}_1 - r_2\bar{l}_2)\dot{q}_1 + (r_1\bar{l}_1^2 + r_2\bar{l}_2^2)\dot{q}_2 \\ & \quad - (k_1\bar{l}_1 - k_2\bar{l}_2) q_1 + (k_1\bar{l}_1^2 + k_2\bar{l}_2^2) q_2 = f(b+a) \end{aligned} \tag{2.98}$$

Having used $\underline{q}$ to define the vector of the new independent variables:

$$\underline{q} = \begin{Bmatrix} q_1 \\ q_2 \end{Bmatrix} \tag{2.99}$$

Equation (2.98) can be rewritten in matrix form as:

$$[M_q]\underline{\ddot{q}} + [R_q]\underline{\dot{q}} + [K_q]\underline{q} = \underline{f}_q \tag{2.100}$$

where:

$$\begin{aligned} [M_q] &= \begin{bmatrix} m & (ma) \\ (ma) & (ma^2 + J_G) \end{bmatrix} \\ [R_q] &= \begin{bmatrix} (r_1 + r_2) & (r_2\,\bar{l}_2 - r_1\,\bar{l}_1) \\ (r_2\,\bar{l}_2 - r_1\,\bar{l}_1) & (r_2\,\bar{l}_2^2 + r_1\,\bar{l}_1^2) \end{bmatrix} \\ [K_q] &= \begin{bmatrix} (k_1 + k_2) & (k_2\,\bar{l}_2 - k_1\,\bar{l}_1) \\ (k_2\,\bar{l}_2 - k_1\,\bar{l}_1) & (k_2\,\bar{l}_2^2 + k_1\,\bar{l}_1^2) \end{bmatrix} \end{aligned} \tag{2.101}$$

$$\underline{f}_q = \begin{Bmatrix} f \\ f(b+a) \end{Bmatrix}$$

Unlike the previous case, mass matrix $[M_q]$ is no longer a diagonal matrix, meaning that, in this case, the equations are also inertially coupled.

2.3.3.3 Expressing Lagrange's Equations in Matrix Form

We will once again analyse the example of a linear vibrating two d.o.f. system shown in Fig. 2.7: as was the case in Sect. 2.3.3.1, here too we will consider the displacement of the centre of gravity and rotation (2.76) as variables x_1 and x_2. In this case, total kinetic energy E_c (2.97) can be expressed as:

$$E_c = \frac{1}{2}\dot{\underline{Y}}_m^T [M_y] \dot{\underline{Y}}_m \tag{2.102}$$

where $\underline{Y}_m$ is the vector containing the displacement of the centre of gravity and body rotation:

$$\underline{Y}_m = \begin{Bmatrix} x_1 \\ x_2 \end{Bmatrix} \tag{2.103}$$

meaning, therefore, that, in this case, $[M_y]$ is:

$$[M_y] = \begin{bmatrix} m & 0 \\ 0 & J_G \end{bmatrix} \tag{2.104}$$

The vector of physical variables $\underline{Y}_m$ (2.103), used to define the kinetic energy of the body, coincides with the vector of independent variables $\underline{x}$ so that, from (2.91), the Jacobian matrix $[\Lambda_m]$ proves not only to be constant but also unitary:

$$\underline{Y}_m = [I]\,\underline{x} \tag{2.105}$$

Kinetic energy (2.102) thus becomes:

$$E_c = \frac{1}{2}\dot{\underline{Y}}_m^T [M_y]\dot{\underline{Y}}_m = \frac{1}{2}\dot{\underline{x}}^T [I]^T [M_y]\,[I]\dot{\underline{x}} = \frac{1}{2}\dot{\underline{x}}^T [M]\dot{\underline{x}} \tag{2.106}$$

and the relative mass matrix, defined in the independent coordinates, is:

$$[M] = [M_y] = \begin{bmatrix} m & 0 \\ 0 & J_G \end{bmatrix} \tag{2.107}$$

The potential energy associated with the system in question [see general expression in (2.34)] can be expressed as:

$$V = \frac{1}{2} \underline{\Delta l}_k^T [k] \underline{\Delta l}_k \tag{2.108}$$

Having used $[k]$ (see Fig. 2.5), to indicate matrix:

$$[k] = \begin{bmatrix} k_1 & 0 \\ 0 & k_2 \end{bmatrix} \tag{2.109}$$

and $\underline{\Delta l}_k$ to indicate the relative displacement vector of the extremities of its springs, functions in turn of independent coordinates:

$$\begin{aligned} \underline{\Delta l}_k &= \begin{Bmatrix} \Delta l_1 \\ \Delta l_2 \end{Bmatrix} = \underline{\Delta l}_k(\underline{x}) = \begin{Bmatrix} \Delta l_1(\underline{x}) \\ \Delta l_2(\underline{x}) \end{Bmatrix} = \begin{Bmatrix} x_1 - x_2 l_1 \\ x_1 + x_2 l_2 \end{Bmatrix} \\ &= [\Lambda_k]\underline{x} = \begin{bmatrix} 1 & -l_1 \\ 1 & l_2 \end{bmatrix} \underline{x} \end{aligned} \tag{2.110}$$

Potential energy is thus:

$$V = \frac{1}{2} \underline{\Delta l}_k^T [k] \underline{\Delta l}_k = \frac{1}{2} \underline{x}^T [\Lambda_k]^T [k] [\Lambda_k] \underline{x} = \frac{1}{2} \underline{x}^T [K] \underline{x} \tag{2.111}$$

where $[K]$ is the total matrix defined by:

$$[K] = \begin{bmatrix} k_1 + k_2 & -l_1 k_1 + l_2 k_2 \\ -l_1 k_1 + l_2 k_2 & k_1 l_1^2 + k_2 l_2^2 \end{bmatrix} \tag{2.112}$$

By operating in a similar manner, the dissipation function can be expressed as:

$$D = \frac{1}{2} \underline{\Delta l}_r^T [r] \underline{\Delta l}_r = \frac{1}{2} \begin{Bmatrix} \dot{\Delta l}_1 \\ \dot{\Delta l}_2 \end{Bmatrix}^T \begin{bmatrix} r_1 & 0 \\ 0 & r_2 \end{bmatrix} \begin{Bmatrix} \dot{\Delta l}_1 \\ \dot{\Delta l}_2 \end{Bmatrix} \tag{2.113}$$

By bearing in mind Eq. (2.94) we have:

$$\dot{\underline{\Delta l}}_r = [\Lambda_d]\dot{\underline{x}} = \begin{bmatrix} 1 & -l_1 \\ 1 & l_2 \end{bmatrix} \dot{\underline{x}} \tag{2.114}$$

and, by substituting Eq. (2.114) in Eq. (2.113), the dissipation function is thus given by:

$$D = \frac{1}{2}\underline{\dot{x}}^T [\Lambda_d]^T [r] \, [\Lambda_d]\underline{\dot{x}} = \frac{1}{2}\underline{\dot{x}}^T [R] \, \underline{\dot{x}}$$
$$= \frac{1}{2}\begin{Bmatrix} \dot{x}_1 \\ \dot{x}_2 \end{Bmatrix}^T \begin{bmatrix} r_1 + r_2 & -l_1 r_1 + l_2 r_2 \\ -l_1 r_1 + l_2 r_2 & r_1 l_1^2 + r_2 l_2^2 \end{bmatrix} \begin{Bmatrix} \dot{x}_1 \\ \dot{x}_2 \end{Bmatrix} \quad (2.115)$$

A force f is applied at a point B of the system's body (Fig. 2.8): y_f is denoted as the displacement of the application point of the force itself (physical variable), meaning that the virtual work can thus be expressed as:

$$\delta^* L = f^T \delta^* y_f \quad (2.116)$$

y_f proves to be correlated to independent variables $\underline{x}$ by means of a function defined as:

$$y_f = y_f(\underline{x}) = x_1 + bx_2 \Rightarrow \delta^* y_f = \left\{\frac{\partial y_f}{\partial \underline{x}}\right\}\delta^* \underline{x}$$
$$\delta^* y_f = \{\Lambda_f\}\delta^* \underline{x} = \{1 \quad b\}\delta^* \underline{x} \quad (2.117)$$

that is:

$$\delta^* L = f^T \{\Lambda_f\}\delta^* \underline{x} = \begin{Bmatrix} f \\ bf \end{Bmatrix}^T \delta^* \underline{x} = \underline{f}^T \delta^* \underline{x} \quad (2.118)$$

The equations of motion of the system thus become:

$$[M]\underline{\ddot{x}} + [R]\underline{\dot{x}} + [K]\underline{x} = \underline{f} \quad (2.119)$$

and are more or less similar to those obtained by other means (scalar method) in Sect. 2.3.3.1, Eq. (2.91).

Always considering the two d.o.f. example, subjected generically, on the contrary, to a set of forces $f_1, f_2, \ldots, f_{nf}$ (Fig. 2.9), the relationships between physical variables y_{fj} and independent variables x_i, are:

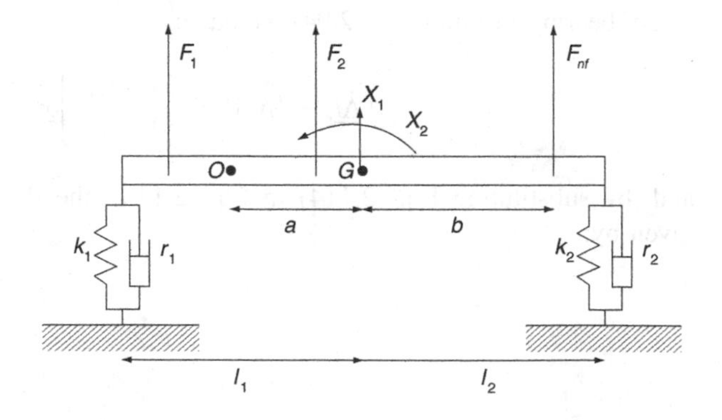

Fig. 2.9 Vibrating system with 2 d.o.f. of Fig. 2.7, with different forces applied

$$\begin{cases} y_{f1} = x_1 - a_1 x_2 \\ y_{f2} = x_1 - a_2 x_2 \\ \dots \\ y_{fnf} = x_1 + a_{nf} x_2 \end{cases} \tag{2.120}$$

In this case, matrix $[\Lambda_f]$ proves to be a rectangular matrix having n_f rows, i.e. the same number as the application points of the forces and as many columns as those of the d.o.f.:

$$[\Lambda_f] = \begin{bmatrix} 1 & -a_1 \\ 1 & -a_2 \\ \dots & \dots \\ 1 & +a_{nf} \end{bmatrix} \tag{2.121}$$

As once again shown in Fig. 2.5, we will now analyze the same vibrating system, by considering vertical displacement z_1 and z_2 of the spring connection points as independent variables to describe motion (Fig. 2.10).

$$\underline{z} = \begin{Bmatrix} z_1 \\ z_2 \end{Bmatrix} \tag{2.122}$$

In this case, unlike the case described in the previous section, the vector of physical variables $\underline{Y}_m$, used to define kinetic energy (2.106), does not coincide with independent variable vector $\underline{z}$ where:

$$\underline{Y}_m = [\Lambda_m]\,\underline{z} = \begin{bmatrix} \frac{l_2}{l_1+l_2} & \frac{-l_1}{l_1+l_2} \\ \frac{1}{l_1+l_2} & \frac{1}{l_1+l_2} \end{bmatrix} \underline{z} \tag{2.123}$$

Thus, with the new set of independent variables $\underline{z}$, kinetic energy becomes:

$$E_c = \frac{1}{2}\dot{\underline{z}}^T [\Lambda_m]^T [M_y] [\Lambda_m] \dot{\underline{z}} = \frac{1}{2}\dot{\underline{z}}^T [M_z] \dot{\underline{z}} \tag{2.124}$$

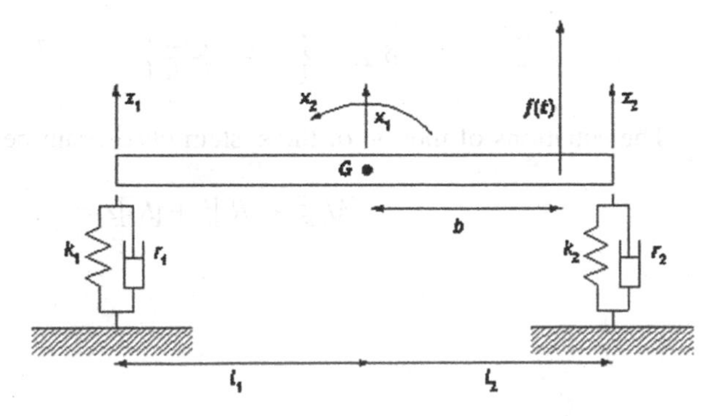

Fig. 2.10 Linear system with 2 d.o.f. (also see Fig. 2.7), with new set of independent coordinates

and the relative mass matrix becomes:

$$[M_z] = \frac{1}{(l_1 + l_2)^2} \begin{bmatrix} ml_2^2 + J_G & -ml_1 l_2 + J_G \\ -ml_1 l_2 + J_G & ml_1^2 + J_G \end{bmatrix} \underline{z} \tag{2.125}$$

The potential energy associated with the system in question can be directly expressed as a function of the independent variables, where $\underline{\Delta l}_k$ can be directly expressed by:

$$\underline{\Delta l}_k = [I]\underline{z} = \underline{z} \tag{2.126}$$

from which:

$$V = \frac{1}{2}\underline{z}^T [I]^T [k][I]\underline{z} = \frac{1}{2}\underline{z}^T [K_z]\underline{z} = \frac{1}{2}\begin{Bmatrix} z_1 \\ z_2 \end{Bmatrix}^T \begin{bmatrix} k_1 & 0 \\ 0 & k_2 \end{bmatrix} \begin{Bmatrix} z_1 \\ z_2 \end{Bmatrix} \tag{2.127}$$

Working in a similar manner, the damping matrix, expressed as a function of the new independent variables $\underline{z}$, becomes:

$$[R_z] = \begin{bmatrix} r_1 & 0 \\ 0 & r_2 \end{bmatrix} \tag{2.128}$$

Finally, by analysing the contribution of applied force f and using y_f to indicate the displacement of the application point of the force itself, in this case we obtain:

$$y_f = y_f(\underline{z}) = -\left\{ \frac{l_1 z_1 + l_2 z_2}{l_1 + l_2} - b\frac{z_1 - z_2}{l_1 + l_2} \right\} \Rightarrow \delta^* y_f = \left\{ \frac{\partial y_f}{\partial \underline{z}} \right\} \delta^* \underline{z} \tag{2.129}$$

$$\delta^* y_f = \{\Lambda_f\}^T \delta^* \underline{z} = \left\{ \tfrac{l_1 - b}{l_1 + l_2} \quad \tfrac{l_2 + b}{l_1 + l_2} \right\} \delta^* \underline{z} \tag{2.130}$$

Virtual work $\delta^* L$ can thus be expressed as:

$$\delta^* L = f^T \delta^* y_f = f^T \left\{ \frac{\partial y_f}{\partial \underline{z}} \right\} \delta^* \underline{z} = f^T \{\Lambda_f\}^T \delta^* \underline{z} \tag{2.131}$$

$$\delta^* L = \left\{ \tfrac{l_1 - b}{l_1 + l_2} \quad \tfrac{l_2 + b}{l_1 + l_2} \right\} \delta^* \underline{z} = \underline{f}_z^T \delta^* \underline{z} \tag{2.132}$$

The equations of motion of the system once again become:

$$[M_z]\ddot{\underline{z}} + [R_z]\dot{\underline{z}} + [K_z]\underline{z} = \underline{f}_z. \tag{2.133}$$

2.3.4 A Further Example of a Two-Degree-of-Freedom Systems

Let us now consider, as another example, the vibrating system consisting of two 2 rigid rods, of mass M and mass moment of inertia J with respect to centre of gravity J, linked one to another and to ground as shown in Fig. 2.11. The system has two d. o.f.: as independent variables let us consider the absolute rotation of the first rod θ_1 and the absolute rotation θ_2 of the second rod:

$$\underline{x} = \begin{Bmatrix} \theta_1 \\ \theta_2 \end{Bmatrix} \tag{2.134}$$

To introduce the terms associated with total kinetic energy E_c, 4 physical variables, Y_1, Y_2, θ_1, and θ_2, which respectively represent the vertical displacements of the centre of gravity of the first and second rod and the corresponding absolute rotations are introduced:

$$\underline{Y}_m = \begin{Bmatrix} \underline{Y}_1 \\ - \\ \underline{Y}_2 \end{Bmatrix} = \begin{Bmatrix} Y_1 \\ \theta_1 \\ - \\ Y_2 \\ \theta_2 \end{Bmatrix} \tag{2.135}$$

In this way, kinetic Energy can be defined as:

$$E_c = \frac{1}{2} \dot{\underline{Y}}_m^T [M_y] \dot{\underline{Y}}_m \tag{2.136}$$

where, in this case:

$$[M_y] = \begin{bmatrix} m & 0 & 0 & 0 \\ 0 & J & 0 & 0 \\ 0 & 0 & m & 0 \\ 0 & 0 & 0 & J \end{bmatrix} \tag{2.137}$$

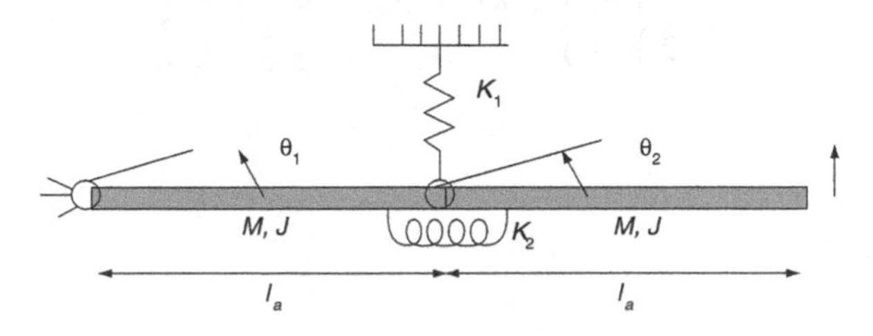

Fig. 2.11 Two d.o.f. system: system for technical multi-body applications

The link between physical variables and independent variables can be expressed as:

$$\underline{\dot{Y}}_m = [\Lambda_m]\,\underline{\dot{x}} = \begin{bmatrix} \frac{l_a}{2} & 0 \\ 1 & 0 \\ l_a & \frac{l_a}{2} \\ 0 & 1 \end{bmatrix} \underline{\dot{x}} \tag{2.138}$$

meaning that total kinetic energy becomes:

$$E_c = \frac{1}{2}\underline{\dot{x}}^T [\Lambda_m]^T [M_y] [\Lambda_m] \underline{\dot{x}} = \frac{1}{2}\underline{\dot{x}}^T [M] \underline{\dot{x}} \tag{2.139}$$

and the relative mass matrix is:

$$[M] = \begin{bmatrix} \frac{ml_a^2}{4} + J + ml_a^2 & \frac{ml_a^2}{2} \\ \frac{ml_a^2}{2} & \frac{ml_a^2}{4} + J \end{bmatrix} \tag{2.140}$$

The potential energy associated with the system in question can be expressed as:

$$V = \frac{1}{2}\underline{\Delta l}_k^T [k] \underline{\Delta l}_k = \frac{1}{2} \begin{Bmatrix} \Delta l_1 \\ \Delta \Phi_2 \end{Bmatrix}^T \begin{bmatrix} k_1 & 0 \\ 0 & k_2 \end{bmatrix} \begin{Bmatrix} \Delta l_1 \\ \Delta \Phi_2 \end{Bmatrix} \tag{2.141}$$

Elongations $\underline{\Delta l}_k$, considered positive in extension, are functions of the independent coordinates by means of functions $\Delta l_1(\underline{x})$ and $\Delta l_2(\underline{x})$:

$$\underline{\Delta l}_k = \underline{\Delta l}_k(\underline{x}) = \begin{Bmatrix} \Delta l_1(\underline{x}) \\ \Delta \Phi_2(\underline{x}) \end{Bmatrix} = [\Lambda_k] = \begin{bmatrix} l_a & 0 \\ -1 & 1 \end{bmatrix} \underline{x} \tag{2.142}$$

Potential energy V is thus:

$$\begin{aligned} V &= \frac{1}{2}\underline{x}^T [\Lambda_k]^T [k] [\Lambda_k] \underline{x} = \frac{1}{2}\underline{x}^T [K] \underline{x} \\ &= \frac{1}{2}\underline{x}^T \begin{bmatrix} l_a & -1 \\ 0 & 1 \end{bmatrix}^T [k] \begin{bmatrix} l_a & 0 \\ -1 & 1 \end{bmatrix} \underline{x} \\ &= \frac{1}{2} \begin{Bmatrix} \theta_1 \\ \theta_2 \end{Bmatrix}^T \begin{bmatrix} k_2 + k_1 l_a^2 & -k_2 \\ -k_2 & k_2 \end{bmatrix} \begin{Bmatrix} \theta_1 \\ \theta_2 \end{Bmatrix} \end{aligned} \tag{2.143}$$

A force f is applied at point B of the second rod of the system (see Fig. 2.11): to define the virtual work performed by the same force, let us consider, as physical variable, the displacement of application point y_f correlated to independent variables $\underline{x}$ by:

$$y_f = y_f(\underline{x}) = \theta_1 l_a + \theta_2 l_a = \{\, l_a \quad l_a \,\} \begin{Bmatrix} \theta_1 \\ \theta_2 \end{Bmatrix} = \{\Lambda_f\}^T \underline{x} \tag{2.144}$$

The virtual work can thus be expressed as:

$$\delta^* L = f^T \delta^* y_f = f^T \left\{ \frac{\partial y_f}{\partial \underline{x}} \right\} \delta^* \underline{x} = f^T \{\Lambda_f\}^T \delta^* \underline{x} = \begin{Bmatrix} l_a f \\ l_a f \end{Bmatrix}^T \delta^* \underline{x} \tag{2.145}$$

2.3.5 n-Degree-of-Freedom Systems

Let us now extend what we have learnt regarding two d.o.f. systems to systems generically having n d.o.f.: from a didactic point of view it is advisable to use a defined model in order to make the explanation clearer.

By thus considering a linear system formed by a shaft with n keyed discs (Fig. 2.12): neglecting the inertia moments of the shaft sections between 2 discs

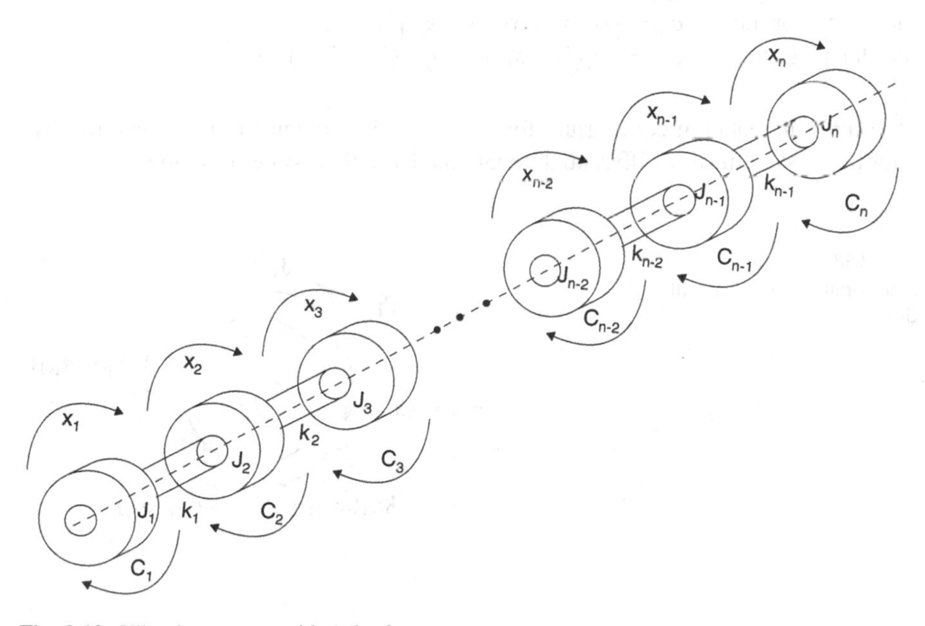

Fig. 2.12 Vibrating system with n d.o.f

with respect to the inertia moments J_i (i = 1, ..., n) of the discs themselves, we obtain a concentrated parameter model of the system. Let:

$$k_j = \frac{GJ_{pj}}{l_j} \quad (j = 1,\ n - 1) \tag{2.146}$$

be the torsional stiffness of the individual sections of the shaft (of length l_j) to which we associate a proportional viscous damping r_j due to the elastic structural hysteresis or other dissipation phenomena [35]. We hypothesize that generic torques c_i act on the discs and we assume, as independent variables, the n absolute rotations x_i of the single discs. First and foremost, we express the equations of motion in scalar form using two different approaches: dynamic equilibriums and Lagrange's equations. We subsequently analyse a general systematic approach for the automatic expression of the equations of motion in vibrating systems with n d.o.f., always basing our reasoning on the example shown in Fig. 2.12.

2.3.5.1 Dynamic Equilibrium Equations

To express the dynamic equilibrium equations, we isolate the generic ith disc (Fig. 2.13) by applying all the forces that the outside world exerts on same (the conventions for rotations and torques are shown in the same figure): more specifically, the following will act on the disc:
the external torque c_i
the inertia torque $c_{ini} = -J_i\ddot{x}_i$
the elastic torque $c_{ki} = -k_j(x_i - x_{i+1}) - k_{j-1}(x_i - x_{i-1})$
the damping torque $c_{ri} = -r_j(\dot{x}_i - \dot{x}_{i+1}) - r_{j-1}(\dot{x}_i - \dot{x}_{i-1})$

Expressing rotation equilibrium for each discs it is obtained, in scalar form, by adopting a system of n differential equations that define system motion:

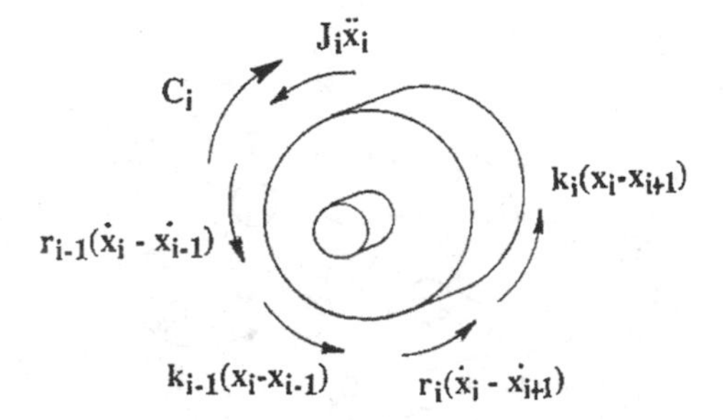

Fig. 2.13 Dynamic equilibrium of a generic ith disc

$$\begin{cases} J_1\ddot{x}_1 + r_1(\dot{x}_1 - \dot{x}_2) + k_1(x_1 - x_2) = c_1 \\ J_2x_2 + r_1(\dot{x}_2 - \dot{x}_1) + r_2(\dot{x}_2 - \dot{x}_3) + k_1(x_2 - x_1) + k_2(x_2 - x_3) = c_2 \\ \dots \\ J_i\ddot{x}_i + r_{i-1}(\dot{x}_i - \dot{x}_{i-1}) + r_i(\dot{x}_i - \dot{x}_{i+1}) + k_{i-1}(x_i - x_{i-1}) + k_i(x_i - x_{i+1}) = c_i \\ \dots \\ J_n\ddot{x}_n + r_{n-1}(\dot{x}_n - \dot{x}_{n-1}) + k_{n-1}(x_n - x_{n-1}) = c_n \end{cases} \tag{2.147}$$

2.3.5.2 Lagrange's Equations—Scalar Approach

As seen in the previous sections, expressing the equations of motion of the same system by means of Lagrange's equations requires the definition of the various energy forms as a function of independent variables x_i:

- kinetic energy E_c:

$$E_c = \sum_{i=1}^{n} E_{ci} = \frac{1}{2}\sum_{i=1}^{n} J_i\dot{x}_i^2 \tag{2.148}$$

- elastic potential energy V:

$$V = \sum_{j=1}^{n-1} V_j = \frac{1}{2}\sum_{j=1}^{n-1} k_j(x_j - x_{j+1})^2 \tag{2.149}$$

- dissipation function D:

$$D = \sum_{j=1}^{n-1} D_j = \frac{1}{2}\sum_{j=1}^{n-1} r_j(\dot{x}_j - \dot{x}_{j+1})^2 \tag{2.150}$$

- the virtual work performed by non-conservative forces:

$$\delta^*L = \sum_{i=1}^{n} \delta^*L_i = \sum_{i=1}^{n} c_i\delta^*x_i = \sum_{i=1}^{n} Q_i\delta^*x_i \tag{2.151}$$

On account of the example analysed being particularly simple, these expressions are already expressed as a function of independent variables X_i and it is therefore possible to apply the Lagrange's equations directly which, in scalar form, we recall as being:

$$\frac{d}{dt}\left(\frac{\partial E_C}{\partial \dot{x}_i}\right) - \frac{\partial E_C}{\partial x_i} + \frac{\partial D}{\partial \dot{x}_i} + \frac{\partial V}{\partial x_i} = \frac{\delta^* L}{\delta^* x_i} = Q_i \quad (i = 1, 2, \ldots, n) \tag{2.152}$$

By applying (2.152) and taking into account (2.148)–(2.151), we once again obtain the equations of motion of the system previously shown in (2.147): the same equations of motion, after grouping the independent variables in vector $\underline{x}$:

$$\underline{x}^T = \{x_1 \quad x_2 \quad \ldots \quad x_i \ldots \quad x_n\} \tag{2.153}$$

can subsequently be rewritten in matrix form as:

$$[M]\ddot{\underline{x}} + [R]\dot{\underline{x}} + [K]\underline{x} = \underline{f} \tag{2.154}$$

where $[M]$ is the mass matrix which, in this case, having assumed the absolute rotations as independent variables, is diagonal (this corresponds to the fact that the equations are inertially decoupled):

$$[M] = \begin{bmatrix} J_1 & 0 & \ldots & 0 & \ldots & 0 \\ 0 & J_2 & \ldots & 0 & \ldots & 0 \\ 0 & 0 & \ldots & J_i & \ldots & 0 \\ 0 & 0 & \ldots & 0 & \ldots & 0 \\ 0 & 0 & \ldots & 0 & \ldots & J_n \end{bmatrix} = diag\{J_1 \quad \ldots \quad J_i \quad \ldots \quad J_n\} \tag{2.155}$$

In (2.154), $[K]$ is the stiffness matrix: in this particular case, this matrix is banded, symmetric and non-negative definite:

$$[K] = \begin{bmatrix} k_1 & -k_1 & 0 & \ldots & 0 & 0 & 0 & \ldots & 0 \\ -k_1 & k_1 + k_2 & -k_2 & \ldots & 0 & 0 & 0 & \ldots & 0 \\ \ldots & \ldots & \ldots & \ldots & \ldots & \ldots & \ldots & \ldots & \ldots \\ 0 & 0 & 0 & \ldots & k_{i-2} + k_{i-1} & -k_{i-1} & 0 & \ldots & 0 \\ 0 & 0 & 0 & \ldots & -k_{i-1} & k_{i-1} + k_i & -k_i & \ldots & 0 \\ 0 & 0 & 0 & \ldots & 0 & -k_i & k_{i+1} + k_i & \ldots & 0 \\ \ldots & \ldots & \ldots & \ldots & \ldots & \ldots & \ldots & \ldots & \ldots \\ 0 & 0 & 0 & \ldots & 0 & 0 & 0 & \ldots & k_n \end{bmatrix} \tag{2.156}$$

Similarly, matrix $[R]$ is:

$$[R] = \begin{bmatrix} r_1 & -r_1 & 0 & \dots & 0 & 0 & 0 & \dots & 0 \\ -r_1 & r_1 + r_2 & -r_2 & \dots & 0 & 0 & 0 & \dots & 0 \\ \dots & \dots & \dots & \dots & \dots & \dots & \dots & \dots & \dots \\ 0 & 0 & 0 & \dots & r_{i-2} + r_{i-1} & -r_{i-1} & 0 & \dots & 0 \\ 0 & 0 & 0 & \dots & -r_{i-1} & r_{i-1} + r_i & -r_i & \dots & 0 \\ 0 & 0 & 0 & \dots & 0 & -r_i & r_{i+1} + r_i & \dots & 0 \\ \dots & \dots & \dots & \dots & \dots & \dots & \dots & \dots & \dots \\ 0 & 0 & 0 & \dots & 0 & 0 & 0 & \dots & r_n \end{bmatrix} \tag{2.157}$$

and the vector of the applied forces:

$$\underline{f}^T = \{f_1 \quad f_2 \quad \dots \quad f_i \quad \dots \quad f_n\} \tag{2.158}$$

2.3.5.3 Lagrange's Equations—Matrix Approach

Instead of directly using the matrix approach total kinetic energy E_c can be defined as:

$$E_c = \frac{1}{2}\underline{\dot{Y}}_m^T [M_y] \underline{\dot{Y}}_m \tag{2.159}$$

having used

$$\underline{Y}_m^T = \{x_1 \quad x_2 \quad \dots \quad x_i \quad \dots \quad x_n\} \tag{2.160}$$

to indicate the vector of physical variables that are more convenient to describe this energy form (the absolute rotations of the single discs) and being:

$$[M_y] = \begin{bmatrix} J_1 & 0 & \dots & 0 & \dots & 0 \\ 0 & J_2 & \dots & 0 & \dots & 0 \\ 0 & 0 & \dots & J_i & \dots & 0 \\ 0 & 0 & \dots & 0 & \dots & 0 \\ 0 & 0 & \dots & 0 & \dots & J_n \end{bmatrix} \tag{2.161}$$

the mass matrix in physical coordinates.

In this case the transformation between physical and independent variables results simply to be

$$\underline{\dot{Y}}_m = [\Lambda_m]\underline{\dot{x}} = [I]\underline{\dot{x}} \tag{2.162}$$

For which the total kinetic energy becomes:

$$E_c = \frac{1}{2}\underline{\dot{x}}^T [\Lambda_m]^T [M_y][\Lambda_m]\underline{\dot{x}} = \frac{1}{2}\underline{\dot{x}}^T [M]\underline{\dot{x}} \tag{2.163}$$

The potential energy associated with the system in question can be expressed as:

$$V = \frac{1}{2}\underline{\Delta l}_k^T [k]\underline{\Delta l}_k = \frac{1}{2}\begin{Bmatrix} \Delta\theta_1 \\ \dots \\ \Delta\theta_j \\ \dots \\ \Delta\theta_{n-1} \end{Bmatrix}^T \begin{bmatrix} k_1 & & & & \\ & \dots & & & \\ & & k_j & & \\ & & & \dots & \\ & & & & k_{n-1} \end{bmatrix} \begin{Bmatrix} \Delta\theta_1 \\ \dots \\ \Delta\theta_j \\ \dots \\ \Delta\theta_{n-1} \end{Bmatrix} \tag{2.164}$$

Having used $\underline{\Delta l}_k$ to denote the relative rotation vector $\Delta\theta_j$ of the extremities of the single rotor sections and $[k]$ to denote the stiffness matrix (diagonal) in the physical coordinates.

The link between physical variables and independent coordinates can be expressed by means of a compact matrix formula of the type:

$$\begin{aligned}\underline{\Delta l}_k = \underline{\Delta l}_k(\underline{x}) &= \begin{Bmatrix} \Delta\theta_1(\underline{x}) \\ \dots \\ \Delta\theta_j(\underline{x}) \\ \dots \\ \theta_{n-1}(\underline{x}) \end{Bmatrix} = \begin{Bmatrix} \theta_1 - \theta_2 \\ \dots \\ \theta_{j-1} - \theta_j \\ \dots \\ \theta_{n-1} - \theta_n \end{Bmatrix} \\ &= [\Lambda_k]\underline{x} = \begin{bmatrix} 1 & -1 & & & & \\ & 1 & -1 & & & \\ & & \dots & & & \\ & & & 1 & -1 & \\ & & & & 1 & -1 \end{bmatrix} \underline{x}\end{aligned} \tag{2.165}$$

Potential energy V thus proves to be:

$$V = \frac{1}{2}\underline{x}^T[\Lambda_k]^T[k][\Lambda_k]\underline{x} = \frac{1}{2}\underline{x}^T[K]\underline{x}$$

$$= \frac{1}{2}\underline{x}^T \begin{bmatrix} 1 & -1 & & & \\ & 1 & -1 & & \\ & & \cdots & & \\ & & & 1 & -1 \\ & & & & 1 & -1 \end{bmatrix}^T \begin{bmatrix} k_1 & & & & \\ & \cdots & & & \\ & & k_j & & \\ & & & \cdots & \\ & & & & k_{n-1} \end{bmatrix} \begin{bmatrix} 1 & -1 & & & \\ & 1 & -1 & & \\ & & \cdots & & \\ & & & 1 & -1 \\ & & & & 1 & -1 \end{bmatrix} \underline{x} =$$

$$= \frac{1}{2}\{x_1 \;\; \ldots \;\; x_i \;\; \ldots \;\; x_n\} \begin{bmatrix} k_1 & -k_1 & 0 & \ldots & 0 & 0 & 0 & \ldots & 0 \\ -k_1 & k_1+k_2 & -k_2 & \ldots & 0 & 0 & 0 & \ldots & 0 \\ \ldots & \ldots & \ldots & \ldots & \ldots & \ldots & \ldots & \ldots & \ldots \\ 0 & 0 & 0 & \ldots & k_{i-2}+k_{i-1} & -k_{i-1} & 0 & \ldots & 0 \\ 0 & 0 & 0 & \ldots & -k_{i-1} & k_{i-1}+k_i & -k_i & \ldots & 0 \\ 0 & 0 & 0 & \ldots & 0 & -k_i & k_{i+1}+k_i & \ldots & 0 \\ \ldots & \ldots & \ldots & \ldots & \ldots & \ldots & \ldots & \ldots & \ldots \\ 0 & 0 & 0 & \ldots & 0 & 0 & 0 & \ldots & k_n \end{bmatrix} \begin{Bmatrix} x_1 \\ \ldots \\ x_i \\ \ldots \\ x_n \end{Bmatrix}$$

A similar approach can obviously also be used for the dissipation function, which can be expressed as:

$$D = \frac{1}{2}\underline{\Delta\dot{l}}_r^T[r]\underline{\Delta\dot{l}}_r = \frac{1}{2}\begin{Bmatrix} \Delta\dot{\theta}_1 \\ \ldots \\ \Delta\dot{\theta}_j \\ \ldots \\ \Delta\dot{\theta}_{n-1} \end{Bmatrix}^T \begin{bmatrix} r_1 & & & \\ & \ldots & & \\ & & r_j & \\ & & & \ldots & \\ & & & & r_{n-1} \end{bmatrix} \begin{Bmatrix} \Delta\dot{\theta}_1 \\ \ldots \\ \Delta\dot{\theta}_j \\ \ldots \\ \Delta\dot{\theta}_{n-1} \end{Bmatrix} \tag{2.166}$$

Having used $\underline{\Delta\dot{l}}_k$ to denote the vector of the relative rotation speeds $\Delta\dot{\theta}_j$ of the extremities of the single rotor sections and $[r]$ the stiffness matrix (diagonal) in the physical coordinates.

The link between physical variables and independent coordinates can be expressed in a compact matrix form of the type:

$$\begin{aligned} \underline{\Delta\dot{l}}_r = \underline{\Delta\dot{l}}_r(\underline{\dot{x}}) &= \begin{Bmatrix} \Delta\dot{\theta}_1(\underline{\dot{x}}) \\ \ldots \\ \Delta\dot{\theta}_j(\underline{\dot{x}}) \\ \ldots \\ \Delta\dot{\theta}_{n-1}(\underline{\dot{x}}) \end{Bmatrix} = \begin{Bmatrix} \dot{\theta}_1 - \dot{\theta}_2 \\ \ldots \\ \dot{\theta}_{j-1} - \dot{\theta}_j \\ \ldots \\ \dot{\theta}_{n-1} - \dot{\theta}_n \end{Bmatrix} \\ &= [\Lambda_r]\underline{\dot{x}} = \begin{bmatrix} 1 & -1 & & & \\ & 1 & -1 & & \\ & & \ldots & & \\ & & & 1 & -1 \\ & & & & 1 & -1 \end{bmatrix} \underline{\dot{x}} \end{aligned} \tag{2.167}$$

Dissipation functions D thus proves to be:

$$D = \frac{1}{2}\underline{\dot{x}}^T[\Lambda_r]^T[r][\Lambda_r]\underline{\dot{x}} = \frac{1}{2}\underline{\dot{x}}^T[R]\underline{\dot{x}}$$

$$= \frac{1}{2}\underline{\dot{x}}^T\begin{bmatrix} 1 & -1 & & & \\ & 1 & -1 & & \\ & & \cdots & & \\ & & & 1 & -1 \\ & & & 1 & -1 \end{bmatrix}^T \begin{bmatrix} r_1 & & & & \\ & \cdots & & & \\ & & r_j & & \\ & & & \cdots & \\ & & & & r_{n-1} \end{bmatrix} \begin{bmatrix} 1 & -1 & & & \\ & 1 & -1 & & \\ & & \cdots & & \\ & & & 1 & -1 \\ & & & 1 & -1 \end{bmatrix}\underline{\dot{x}}$$

$$= \frac{1}{2}\{\dot{x}_1 \quad \ldots \quad \dot{x}_i \quad \ldots \quad \dot{x}_n\}\begin{bmatrix} r_1 & -r_1 & 0 & \ldots & 0 & 0 & 0 & \ldots & 0 \\ -r_1 & r_1+r_2 & -r_2 & \ldots & 0 & 0 & 0 & \ldots & 0 \\ \ldots & \ldots & \ldots & \ldots & \ldots & \ldots & \ldots & \ldots & \ldots \\ 0 & 0 & 0 & \ldots & r_{i-2}+r_{i-1} & -r_{i-1} & 0 & \ldots & 0 \\ 0 & 0 & 0 & \ldots & -r_{i-1} & r_{i-1}+r_i & -r_i & \ldots & 0 \\ 0 & 0 & 0 & \ldots & 0 & -r_i & r_{i+1}+r_i & \ldots & 0 \\ \ldots & \ldots & \ldots & \ldots & \ldots & \ldots & \ldots & \ldots & \ldots \\ 0 & 0 & 0 & \ldots & 0 & 0 & 0 & \ldots & r_n \end{bmatrix}\begin{Bmatrix} \dot{x}_1 \\ \ldots \\ \dot{x}_i \\ \ldots \\ \dot{x}_n \end{Bmatrix} \tag{2.168}$$

The virtual work performed by the torques can be expressed in matrix form through relation:

$$\delta^* L = \underline{c}^T \delta^* \underline{Y}_f = \{c_1 \quad \ldots \quad c_2 \quad \ldots \quad c_3\}\begin{Bmatrix} \delta^*\theta_1 \\ \ldots \\ \delta^*\theta_i \\ \ldots \\ \delta^*\theta_n \end{Bmatrix} \tag{2.169}$$

having used $\underline{Y}_f$ to denote the vector of the most convenient physical variables (the absolute rotations of the discs) to introduce this term.

Even in this case the transformation between physical and independent variables is simply:

$$\delta^* \underline{Y}_f = [\Lambda_f]\delta^* \underline{x} = [I]\delta^* \underline{x} \tag{2.170}$$

so that the virtual work becomes:

$$\delta^* L = \underline{c}^T \delta^* \underline{Y}_f = \underline{c}^T [\Lambda_f]\delta^* \underline{x} = \underline{f}^T \delta^* \underline{x} \tag{2.171}$$

where:

$$\underline{f} = [\Lambda_f]\underline{c} = \underline{c} \tag{2.172}$$

is the vector of the generalized forces.

The equations of motion can obviously once again be rewritten in matrix form as:

$$[M]\underline{\ddot{x}} + [R]\underline{\dot{x}} + [K]\underline{x} = \underline{f} \tag{2.173}$$

2.4 Solving the Equations of Motion

2.4.1 One-Degree-of-Freedom System

As seen in Sect. 2.3.1, the equations of motion of a generic 1 d.o.f. system, linear or linearized about the static equilibrium position is:

$$m\ddot{x} + r\dot{x} + kx = f(t) \tag{2.175}$$

having used x to indicate the generic independent variable with which the motion of the system under examination is described, m, r and k to respectively indicate generalized mass, damping and stiffness and, finally, $f(t)$ to indicate the generalized generic force.

We will now analyse the methodologies used to evaluate the dynamic response of the system (2.175) in various situations (free motion and forced motion).

2.4.1.1 Undamped Motion

In the absence of damping and of external applied forces, expression (2.175) becomes a linear homogeneous differential equation of the second order with constant coefficients:

$$m\ddot{x} + kx = 0 \tag{2.176}$$

A particular integral of Eq. (2.176) is of the type [2, 28, 37]:

$$x = \overline{X}e^{\lambda t} \tag{2.177}$$

where λ is a parameter, generally complex, that needs to be determined and $\overline{X}$ represents the amplitude, also generically complex. By substituting expression (2.177) in (2.176) we obtain a parametric, homogenous, algebraic equation in λ:

$$(\lambda^2 m + k)\overline{X} = 0 \tag{2.178}$$

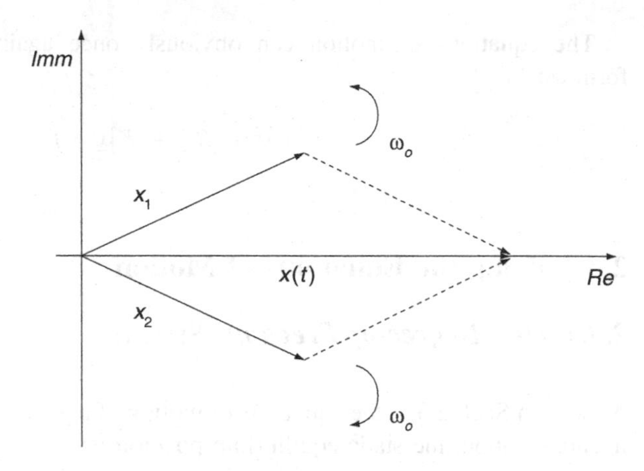

Fig. 2.14 Counter-rotating vectors

which, as a non-trivial solution, admits that which annuls the characteristic polynomial:

$$\lambda^2 m + k = 0 \tag{2.179}$$

The solutions of (2.179) are:

$$\lambda_{1,2} = \sqrt{-\frac{k}{m}} = \pm i\sqrt{\frac{k}{m}} = \pm i\omega_0 \tag{2.180}$$

where i is the imaginary unit ($i = \sqrt{-1}$), while $\omega_0 = \sqrt{k/m}$ is the frequency of the undamped system.

In view of the fact that (2.179) has two separate solutions, Eq. (2.176) will have two linearly independent particular integrals $\overline{X}_1 e^{\lambda_1 t}$ and $\overline{X}_2 e^{\lambda_2 t}$, to which a general integral corresponds, given by their linear combination:

$$x(t) = x_G(t) = \alpha \overline{X}_1 e^{\lambda_1 t} + \beta \overline{X}_2^{\lambda_2 t} = X_1 e^{\lambda_1 t} + X_2^{\lambda_2 t} \tag{2.181}$$

where X_1 and X_2 are defined up to a constant and having included in same the coefficients of the linear combination. In the case under examination, by keeping account of (2.180), the general integral becomes:

$$x(t) = x_G = X_1 e^{i\omega_0 t} + X_2^{-i\omega_0 t} \tag{2.182}$$

In (2.182) $x(t)$ is real and, as a consequence, X_1 and X_2 prove to be complex conjugate constants whose value depends on initial conditions.[3]

The general integral, or rather the response of the free system to an initial disturbance about the equilibrium position, can thus be calculated (in Gauss's complex plane) as the vector sum of two complex counter-rotating conjugate vectors of constant modulus (Fig. 2.14).

By developing (2.182) by means of Euler's formula [2], the same equation can also be rewritten in the following form:

$$x(t) = X_1(\cos\omega_0 t + i\sin\omega_0 t) + X_2(\cos\omega_0 t - i\sin\omega_0 t) \tag{2.183}$$

or rather, by gathering the terms in sine and cosine:

$$x(t) = (X_1 + X_2)\cos\omega_0 t + i(X_1 - X_2)\sin\omega_0 t \tag{2.184}$$

By bearing in mind the fact that X_1 and X_2 are complex conjugates, their sum is a real number, while their difference is a purely imaginary number; expression (2.184) thus becomes:

$$x(t) = A\cos\omega_0 t + B\sin\omega_0 t \tag{2.185}$$

Based on the aforementioned statement, A and B are real constants, that need to be determined by imposing the initial conditions of displacement and velocity:

---

[3]In (2.182) X_1 and X_2 can be defined by means of the initial conditions. By imposing, at time $t = 0$:

$$\begin{aligned} x(0) &= x_0 \\ \dot{x}(0) &= \dot{x}_0 \end{aligned} \tag{2.3.1}$$

It is possible to obtain a system of equations of the type:

$$\begin{aligned} x_0 &= X_1 + X_2 \\ \dot{x}_0 &= i\omega_0 X_1 - i\omega_0 X_2 \end{aligned} \tag{2.3.2}$$

From which:

$$\begin{aligned} X_1 &= \frac{x_0}{2} - i\frac{\dot{x}_0}{2\omega_0} \\ X_2 &= \frac{x_0}{2} + i\frac{\dot{x}_0}{2\omega_0} \end{aligned} \tag{2.3.3}$$

As can be noted, X_1 and X_2 are complex conjugates.

$$\begin{aligned} x(0) &= x_0 \\ \dot{x}(0) &= \dot{x}_0 \end{aligned} \tag{2.186}$$

Expression (2.185), derived with respect to time, provides the velocity trend:

$$\dot{x}(t) = -\omega_0 A \sin \omega_0 t + \omega_0 B \cos \omega_0 t \tag{2.187}$$

By imposing the initial conditions in (2.185) and (2.187) we obtain:

$$x(0) = x_0 = A \tag{2.188}$$

$$\dot{x}(0) = \dot{x}_0 = B\omega_0 \tag{2.189}$$

The value of constants A and B as a function of the initial displacement and velocity of the system thus become:

$$A = x_0 \tag{2.190}$$

$$B = \frac{\dot{x}_0}{\omega_0} \tag{2.191}$$

The generic expression of vibration of the undamped free system perturbed about the equilibrium position is thus:

$$x(t) = x_0 \cos \omega_0 t + \frac{\dot{x}_0}{\omega_0} \sin \omega_0 t \tag{2.192}$$

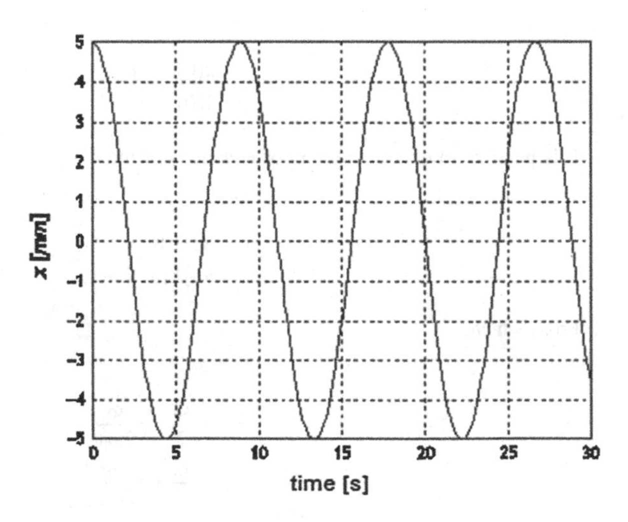

Fig. 2.15 System with 1 d.o. f.: free undamped vibration

An additional alternative expression to (2.185), can be obtained by performing a change of variables:

$$A = |X|\cos(\varphi) \tag{2.193}$$

$$B = -|X|\sin(\varphi) \tag{2.194}$$

which, when substituted in (2.185) give:

$$x(t) = |X|\cos(\varphi)\cos\omega_0 t - |X|\sin(\varphi)\sin\omega_0 t \tag{2.195}$$

It is possible to note that expression (2.195) is a development of the following formula:

$$x(t) = |X|\cos(\omega_0 t + \varphi) \tag{2.196}$$

This Eq. (2.196) suggests how vibration $x(t)$ can also be considered as the projection on the real axis of complex vector X, modulus $|X|$ and phase φ, rotating in the complex plane at speed ω_0:

$$x(t) = Re\left(Xe^{i\omega_0 t}\right) = Re\left(|X|e^{i\varphi}e^{i\omega_0 t}\right) = |X|\cos(\omega_0 t + \varphi) \tag{2.197}$$

as shown in Fig. 2.15. The links between formulations (2.185) and (2.197) are expressed by:

$$|X| = \sqrt{A^2 + B^2} \tag{2.198}$$

$$\varphi = arctg\left(-\frac{B}{A}\right) \tag{2.199}$$

To summarize, vibration $x(t)$ which defines the undamped free motion in the absence of external forces (transient conditions) can be expressed in one of the following three forms (2.182), (2.185) and (2.197). The general integral of an undamped one d.o.f. system is thus represented by a harmonic oscillation with frequency ω_0 (measured in rad/s), characterized (see Fig. 2.15) by a period T_0 (expressed in seconds) equal to:

$$T_0 = \frac{2\pi}{\omega_0} \tag{2.200}$$

Relationship:

$$f_0 = \frac{\omega_0}{2\pi} = \frac{1}{T_0} \tag{2.201}$$

which defines the number of cycles (oscillations) that the system completes in a unit of time is defined as natural frequency f_0 (expressed in Hz).

2.4.1.2 Damped Free Motion

In real systems, the condition of free motion with an amplitude that is constant in time is not verified because, in all vibrating systems, there is always a certain amount of energy dissipation. This dissipation can be attributed to the elastic hysteresis of the material, to Coulomb friction or to the presence of an actual damper. In analytic models we keep account of this circumstance by introducing equivalent damping elements of a viscous type, which exercise a force that is proportional to the speed and in the opposite direction to same. Even though this type of schematization does not correspond to the actual dissipation mechanism, it is possible to determine an equivalent viscous damping coefficient whereby, in each oscillation cycle, we have the same amount of dissipated energy as in the real system.[4]

By keeping account of the damping, the equation of motion (2.175) becomes:

$$m\ddot{x} + r\dot{x} + kx = 0 \tag{2.202}$$

or rather, once again a differential equation of the second order, with total derivatives, having constant coefficients. Given:

$$\omega_0^2 = \frac{k}{m} \tag{2.203}$$

expression (2.202) can be rewritten as:

$$\ddot{x} + \frac{r}{m}\dot{x} + \omega_0^2 x = 0 \tag{2.204}$$

By imposing (2.177) in Eq. (2.204), it is possible to obtain non-trivial solutions by setting the characteristic polynomial to zero:

$$\lambda^2 + \frac{r}{m}\lambda + \omega_0^2 = 0 \tag{2.205}$$

whose roots are:

$$\lambda_{1,2} = -\frac{r}{2m} \pm \sqrt{\left(\frac{r}{2m}\right)^2 - \omega_0^2} \tag{2.206}$$

In this case, we have two separate roots of the characteristic equation: the general integral, which defines the perturbed motion of the system, is given by the combination of the two particular integrals of expression (2.202):

[4]Concerning this we suggest referring both to Sect. 2.4.1.3 and Chap. 8, Sect. 8.2 relative to *Techniques of modal identification*, where, by means of the analysis of experimental tests, we will outline a procedure to identify the various terms of the equation.

$$x(t) = x_G(t) = X_1 e^{\lambda_1 t} + X_2 e^{\lambda_2 t} \tag{2.207}$$

where X_1 and X_2 are constants that need to be determined on the basis of the initial conditions. The discussion of the solutions given by (2.206) can be traced back to the discussion of the sign of the discriminant:

$$\Delta = \left(\frac{r}{2m}\right)^2 - \omega_0^2 \tag{2.208}$$

We define "critical damping" r_c of system as that value of damping r which annuls discriminant (2.208):

$$r_c = 2m\omega_0 \tag{2.209}$$

By using h to name the relationship between the actual damping r of the system and critical damping r_c:

$$h = \frac{r}{r_c} \tag{2.210}$$

and by keeping account of (2.203) and (2.209), the solutions can be expressed as follows:

$$\lambda_{1,2} = -\frac{r\omega_0}{2m\omega_0} \pm \sqrt{\left(\frac{r\omega_0}{2m\omega_0}\right)^2 - \omega_0^2} \tag{2.211}$$

By keeping account of the definition of critical damping (2.209) and of (2.210), (2.211) can be redefined, in non-dimensional form, as:

$$\lambda_{1,2} = \omega_0\left(-h \mp \sqrt{h^2 - 1}\right) \tag{2.212}$$

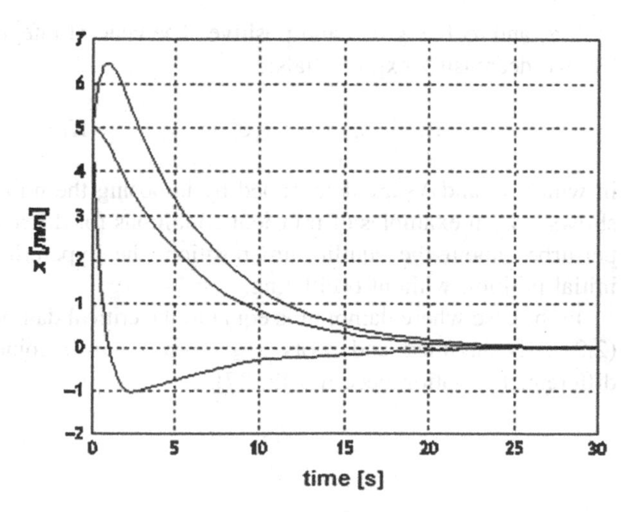

Fig. 2.16 System with 1 d.o.f.: free motion of a damped system with $h > 1$ for different initial conditions

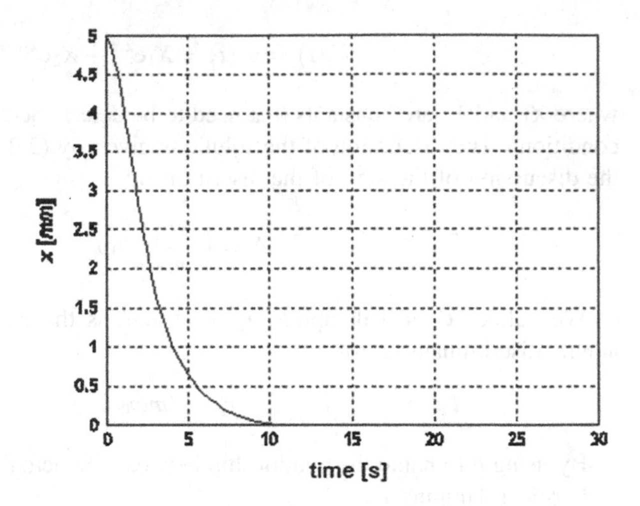

Fig. 2.17 System with 1 d.o.f.: free motion of a damped system with h = 1

The discussion is thus reduced to the possible relations:

$$\begin{aligned} h<1 &\Rightarrow r<r_c \\ h=1 &\Rightarrow r=r_c \\ h>1 &\Rightarrow r>r_c \end{aligned} \tag{2.213}$$

In the event of the damping being greater than the critical damping (hypercritical system) $h > 1$, roots $\lambda_{1,2}$ (2.211) are both real and negative seeing that the radicand is positive and its minor root, in absolute value, is h:

$$\lambda_1 = -\alpha_1 < 0 \tag{2.214}$$

$$\lambda_2 = -\alpha_2 < 0 \tag{2.215}$$

with α_1 and α_2 being real and positive. The general integral (2.207) will be defined by two decreasing exponentials:

$$x(t) = x_G(t) = X_1 e^{\lambda_1 t} + X_2 e^{\lambda_2 t} = X_1 e^{-\alpha_1 t} + X_2 e^{-\alpha_2 t} \tag{2.216}$$

in which X_1 and X_2 are determined by imposing the initial conditions. Figure 2.16 shows several examples of transient conditions for different initial conditions: once perturbed about the equilibrium position, the hypercritical system returns to its initial position without oscillating.

In the case where damping is equal to the critical damping $h = 1$, the radicand of (2.212) is annulled and roots $\lambda_{1,2} = -\alpha = -\omega_0 h$ coincide: the solution of the differential equation becomes [2, 37]:

$$x(t) = x_G(t) = X_1 e^{-\alpha t} + X_2 t e^{-\alpha t} \tag{2.217}$$

This solution (Fig. 2.17) is the limit between non-oscillating solutions ($h > 1$) and harmonic oscillating solutions ($h < 1$). A system with damping equal to critical value ($h = 1$), disturbed about the rest position, returns to the rest position in less time both with respect to a hypercritical system ($h > 1$) and a system with a lower damping than the critical one ($h < 1$) [3, 31, 35].

In the case of a system with $h < 1$ (with a lower damping than the critical value), the radicand of (2.211) is negative and its root is thus a pure imaginary number. For this reason, the two solutions $\lambda_{1,2}$ are complex conjugates with a real negative part:

$$\lambda_{1,2} = \omega_0\left(-h \pm i\sqrt{1-h^2}\right) = -\alpha \pm i\omega \tag{2.218}$$

in which:

$$\begin{aligned} \omega &= \omega_0\sqrt{1-h^2} \\ \alpha &= h\omega_0 \end{aligned} \tag{2.219}$$

where ω is the natural frequency of the damped system. The solution of the equation of motion (2.207) becomes:

$$x(t) = x_G(t) = X_1 e^{-\alpha t + i\omega t} + X_2 e^{-\alpha t - i\omega t} \tag{2.220}$$

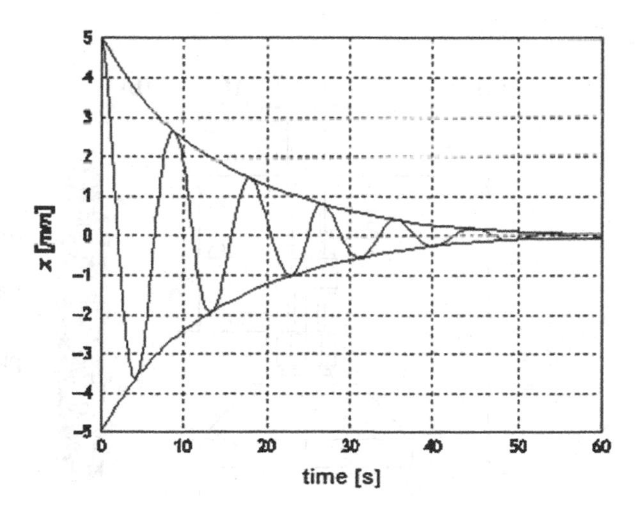

Fig. 2.18 System with 1 d.o.f.: free motion of a damped system with $h < 1$

And by gathering the exponential term:

$$x(t) = x_G(t) = \mathrm{e}^{-\alpha t}\left(X_1 \mathrm{e}^{+i\omega t} + X_2 \mathrm{e}^{-i\omega t}\right) \tag{2.221}$$

The term in parentheses of (2.221) has the same form as the solution of free motion in the absence of damping (2.182), meaning that the same transformations seen in Sect. 2.2 can be applied to it, thus obtaining alternative expressions of the type:

$$x(t) = x_G(t) = \mathrm{e}^{-\alpha t}(A\cos\omega t + B\sin\omega t) \tag{2.222}$$

$$x(t) = x_G(t) = \mathrm{e}^{-\alpha t}(|X|\cos(\omega t + \varphi)) \tag{2.223}$$

Expressions (2.222) and (2.223) evidence how, in the presence of a lower damping than that of the critical damping, motion is represented by an oscillation with frequency ω and a decreasing amplitude (Fig. 2.18) due to the presence of a negative exponential. Constants A and B ($|X|$ and φ) present in expressions (2.222) and (2.223) are determined, using the same procedure seen for the free undamped system, by imposing the initial conditions. In the case of a system with a low damping value ($h \ll 1$), frequency ω of the damped motion and that of the undamped motion ω_0 practically coincide:

$$\omega = \omega_0\sqrt{1 - h^2} \approx \omega_0 \tag{2.224}$$

Similarly to what we have seen for the undamped free system, we define the relationship

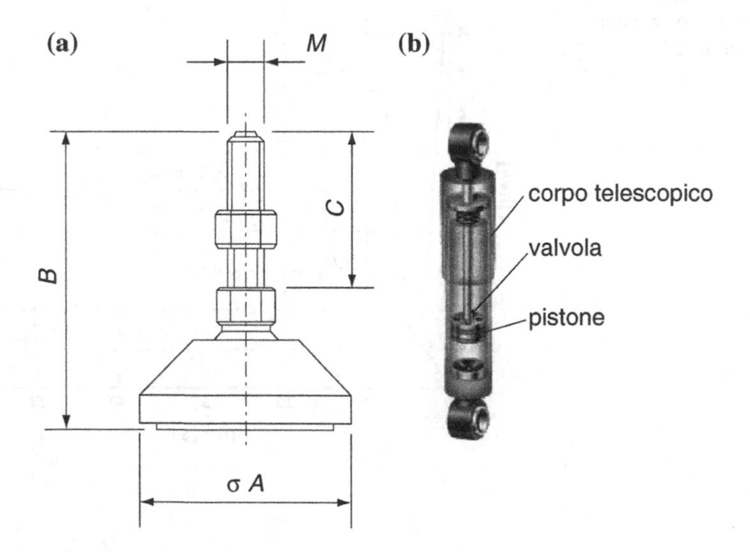

Fig. 2.19 Connection elements: **a** rubber element, **b** viscous damper

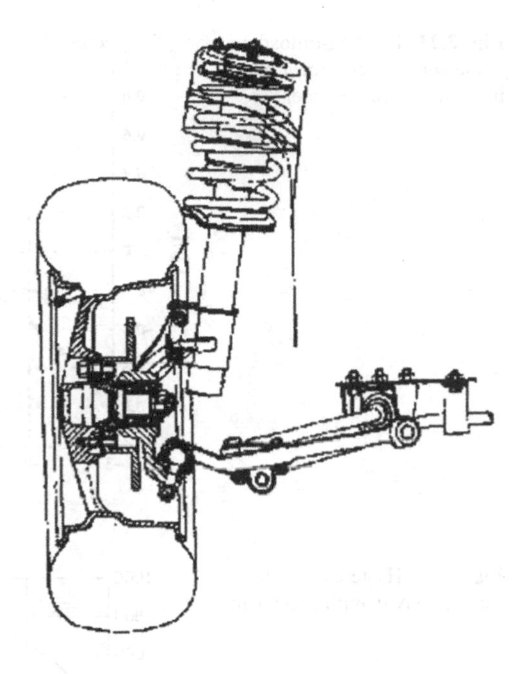

Fig. 2.20 Motorcar suspension: we note the presence of a viscous damper

$$T = \frac{2\pi}{\omega} \tag{2.225}$$

as the natural period of damped motion T, and the relationship:

$$f = \frac{\omega}{2\pi} = \frac{1}{T} \tag{2.226}$$

as the natural frequency f (Hz).

2.4.1.3 Identification of Damping

The problem of identification of the parameters of a system is fundamental for a correct simulation of the behaviour of the system itself. This subject will be the object of Chap. 8, dedicated to *Identification Techniques*. In this instance we wish to introduce the subject for preparatory purposes through the simple application of one d.o.f. system, limiting the discussion to the only damping parameter r: as previously mentioned, term $r\dot{x}$ represents the dissipative effects of the system. Furthermore, let us consider the case in which the dissipative effects are due to only one element that can be physically isolated from the system itself, such as, for example, the hysteresis of a spring or a rubber element (Fig. 2.19a) or an actual viscous damper (Figs. 2.19b and 2.20). In this case, it is possible, thanks to the use of experimental tests, to trace the dissipation of energy E_d due to the element itself

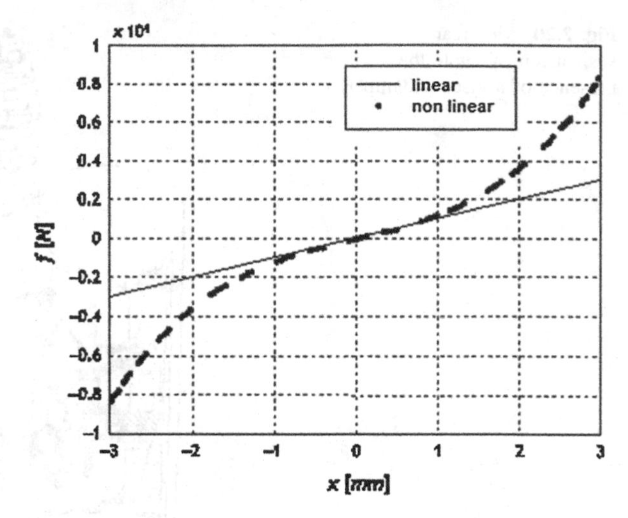

Fig. 2.21 Loading-offloading cycle for a purely linear and nonlinear elastic element

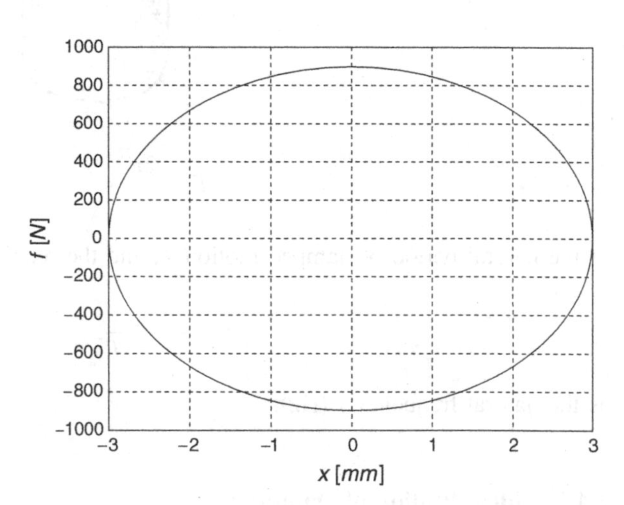

Fig. 2.22 Hysteresis cycle for a purely damping element

and, based on this, the value of the equivalent viscous damping r [5, 8, 15, 21, 22, 31, 35]. By subjecting the element to a harmonically variable elongation x:

$$x = x_0 \cos \Omega t \tag{2.227}$$

by measuring the force $f = f(t)$ necessary to impose this displacement and finally by plotting this force as a function of displacement x we obtain a diagram of the type shown in Fig. 2.21 (relative to the case of linear and non linear elastic element), in Fig. 2.22 (relative to the case of a purely damping element) or in Fig. 2.23 (relative to the case of a linear elastic element which dissipates energy by hysteresis): area A enclosed within the cycle is proportionate to dissipated energy E_d. The form of the cycle depends on the type of dissipation actually present in the system:

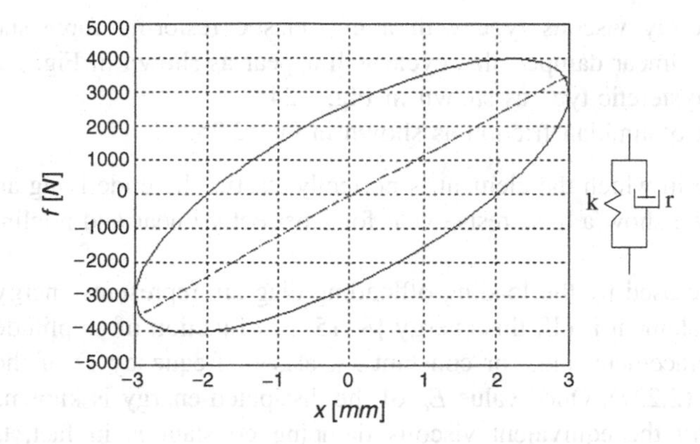

Fig. 2.23 Hysteresis cycle for a linear elastic element equipped with hysteretic dissipation: the slope of the *dotted line* represents the stiffness of the element considered (k), the underlying area is proportional to the damping introduced by same (r)

Fig. 2.24 Hysteresis cycle for a nonlinear elastic element equipped with hysteretic dissipation

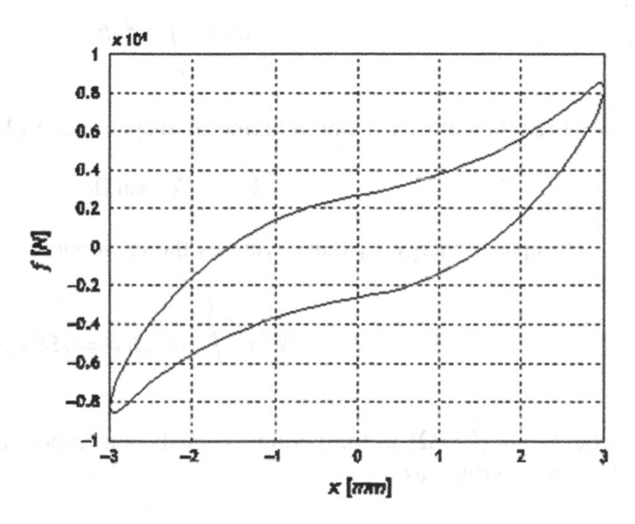

Fig. 2.25 Hysteresis cycle for a spring element with Coloumbian friction

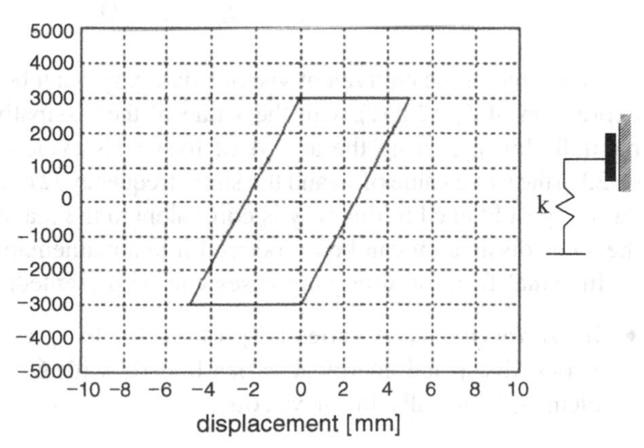

- If of a purely viscous type with a null elastic restoring force such as, for example, a linear damper, this cycle will appear as shown in Fig. 2.22;
- if of the hysteretic type as shown in Fig. 2.24;
- if due to Coloumbian friction as shown in Fig. 2.25.

In the case in which the element is perfectly elastic, the underlying area is null: in Fig. 2.21 we show a hysteresis cycle for a perfectly linear and nonlinear elastic element.

Area A enclosed by the loading-offloading diagram represents energy E_d dissipated by the element itself: this energy [8, 15] is a function of amplitude x_0 of the imposed displacement and, for constant x_0, also of frequency Ω of the imposed displacement (2.227). Once value E_d of the dissipated energy is known, it is possible to obtain the equivalent viscous damping constant r: in fact, the energy dissipated in a cycle by the equivalent viscous element subjected to a generic elongation x, with velocity $\dot{x}$, is worth:

$$E_d = \int_0^T r\dot{x}^2\,dt \tag{2.228}$$

in which the velocity, being a harmonic displacement (2.227), is:

$$\dot{x} = -\Omega x_0 \sin \Omega t \tag{2.229}$$

Dissipated energy E_d thus, assumes the expression:

$$E_d = r\Omega^2 x_0^2 \int_0^T \sin^2 \Omega t dt = r\Omega^2 x_0^2 \frac{T}{2} \tag{2.230}$$

where $T = (2\pi/\Omega)$ is the period of oscillation imposed. Expression (2.230) can thus be rewritten as:

$$E_d = r\pi x_0^2 \Omega \tag{2.231}$$

The value of the equivalent viscous damping r can be obtained by equalizing the expression of E_d (2.231) with the value of the dissipative energy obtained experimentally by measuring the area A of hysteresis cycle and by placing in the same (2.231) the same value of x_0 and the same frequency Ω of the test: the value of viscous damping r obtained in this way is equivalent to the real damping, i.e., it is such that the same dissipation can be introduced into a mathematical model, in one cycle.

In actual fact, the following cases can be presented:

- if, by varying x_0, A varies proportionately to x_0^2 as in (2.231) and if, when Ω varies, dissipated energy A varies linearly with Ω (2.231), then the dissipative element is actually linear viscous;

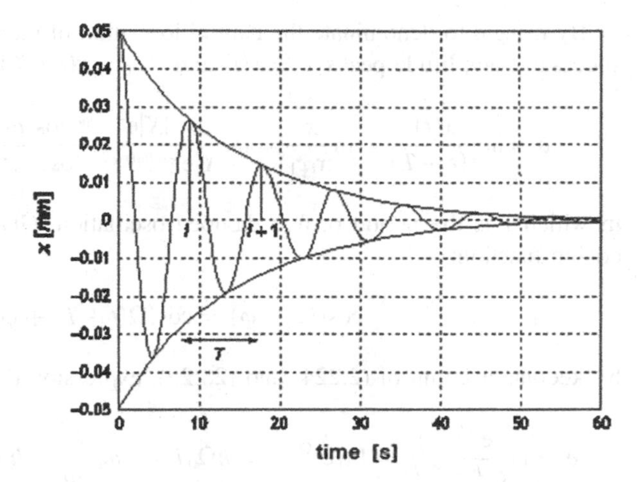

Fig. 2.26 Measurement of critical damping using the logarithmic decrement method

- conversely, if A depends on x_0 with a quadratic law, but is independent from Ω, then the dissipative element is termed hysteretic and value r obtained from (2.231) proves once again to depend on frequency Ω: in this case, it is thus necessary to introduce a constant value of equivalent damping r which is different according to the frequency considered;
- if finally, keeping the same frequency and by only varying x_0, dissipated energy A does not depend quadratically on x_0, the damping element is not linear and, as a consequence, coefficient r is a function of the amplitude: in this case, it would be necessary to have values that differ from the equivalent viscous damping depending on the amplitude of oscillation.

Everything stated until now is based on the assumption of using a linear model to simulate the complex mechanism responsible for the dissipation of energy. This dissipation, as previously mentioned, is associated, for the most part, with the elastic hysteresis of the material which does not work in a purely elastic field. In the event of our wishing to simulate this complex mechanism adequately, it is necessary to resort to nonlinear elements in terms of both velocity and displacement. Concerning this, models known as "rheological", capable of simulating various types of hysteresis cycles [17, 22, 23], exist.

In the event of it not being possible to perform tests of the above mentioned type, as an alternative to the method previously described, it is possible to adopt a method known as the *logarithmic decrement* method [35], which can be applied to vibrating systems with a damping that is much lower than the critical damping $h \ll 1$: this approach is based on the measurement of the damped free motion (referred to as *decay*). Based on the assumptions made, the free motion of a damped system is described by (2.222):

$$x(t) = |X|e^{-h\omega_0 t}\cos(\omega t + \varphi) \tag{2.232}$$

By using δ to denominate the natural logarithm of the relationship between two successive amplitude peaks $x_i = x(t)$ and $x_{i+1} = x(t+T)$ (Fig. 2.26), we obtain:

$$\delta = \ln\frac{x(t)}{x(t+T)} = \ln\frac{x_i}{x_{i+1}} = \ln\frac{|X|\mathrm{e}^{-h\omega_0 t}\cos(\omega t+\varphi)}{|X|\mathrm{e}^{-h\omega_0 (t+T)}\cos(\omega(t+T)+\varphi)} \tag{2.233}$$

in which T is the period of free damped oscillation. Given the periodicity of the cosine function:

$$\cos(\Omega t+\varphi) = \cos(\Omega(t+T)+\phi) \tag{2.234}$$

by keeping account of (2.224) and (2.225), expression (2.233) becomes:

$$\delta = \ln\frac{\mathrm{e}^{-h\omega_0 t})}{\mathrm{e}^{-h\omega_0 (t+T)}} = \ln\left(\mathrm{e}^{h\Omega_0 T}\right) = h\Omega_0 T = h\omega_0\frac{2\pi}{\omega} = h\omega_0\frac{2\pi}{\omega_0\sqrt{1-h^2}} \tag{2.235}$$

and by assuming that $h \ll 1$:

$$\delta = \ln\frac{x_i}{x_{i+1}} = 2\pi h \tag{2.236}$$

By obtaining the temporal trend $x(t)$ of the vibration and by measuring two successive peaks x_i and x_{i+1}, it is possible, by means of (2.236), to determine the logarithmic decrement δ and, from (2.235), the value of the non-dimensional damping h of the system compared with critical value. By measuring natural period T (Fig. 2.26) and, therefore, frequency ω of same, once mass m of the system is known, it is possible to determine the value of damping constant r by means of:

$$r = h2m\omega_0 \tag{2.237}$$

This procedure can be repeated between any two pair of peaks: if the system behaves as a linear system, the value of δ does not change, neither during the course of the single time history nor by imposing different initial conditions. Conversely, a change of δ with the amplitude of oscillations is indicative of the nonlinear behaviour of the system. Damping coefficient r obtained using this methodology represents an equivalent damping from a point of view of energy, owing to the fact that, in one cycle, it produces the same dissipation of energy (in the form of a kinetic energy loss) as the real dissipation present in the system.

2.4.1.4 Forced Motion

In this chapter we will be analysing the dynamic behaviour of linear systems for which, as is known, the principle of the superposition of effects [13, 18] can be applied. In this section, in particular, we will deal with excitation forces depending solely on time, leaving to Chap. 5 a more general explanation, in which the forces can also depend on independent variables and where we will deal with force fields.

Forces which are solely explicit functions of time are generally:

- constant forces (such as, for example, those due to gravitational fields);
- sinusoidal forces (e.g. such as those due, in rotors, to unbalances);
- periodic forces [32];
- aperiodic excitation forces (deterministic, but not periodic);
- random excitation forces [12]: the response of dynamic systems to aperiodic or random excitation forces will be dealt with in Chap. 7.

In the explanation dealing with linear systems, we will only analyse the response to excitation forces depending explicitly on time, in that the presence of constant forces only modifies the particular integral of a constant term: later on in the book, always with regard to linear systems, unless the contrary is expressly requested, we will analyse the response of the system starting from the static equilibrium position defined by the presence of these constant forces.[5] By analysing linear systems, it is possible, by using a Fourier analysis [2, 12, 27, 37], to trace the periodic excitation forces to a linear combination of harmonic functions (trigonometric series or Fourier series) whose frequencies are integer multiples of a fundamental frequency Ω_0:

---

[5]The equation of motion of a generic 1 d.o.f. linear system subjected to a force as a function of time $f = f(t)$ and to a constant force, e.g. weight, is generically of the type:

$$m\ddot{x} + r\dot{x} + kx = p + f(t) \tag{2.5.1}$$

I.e. a differential linear equation with constant coefficients. The steady state solution (2.3.1) (i.e. once the transient has been exhausted [see Sect 2.4.1.2)], since the superposition of effects is valid for the linear system, is given by the sum of the two terms:

- a constant particular integral (static solution of the system subjected, for example, to the gravitational field) due to the sole weight force $x_0 = {}^{p}/_{k}$;
- a particular integral $x_p(t)$ that describes the oscillation about static value x_0.

By considering, as seen in Chap. 1, the displacement of system $\bar{x}$, starting from the static equilibrium position $x = x_0$, as a new independent variable:

$$x = x_0 + \bar{x}; \quad \dot{x} = \dot{\bar{x}}; \quad x = \ddot{\bar{x}} \tag{2.5.2}$$

The equation of motion (2.3.1) becomes:

$$m\ddot{\bar{x}} + r\dot{\bar{x}} + k\bar{x} + kx_0 = p + f(t) \Rightarrow m\ddot{\bar{x}} + r\dot{\bar{x}} + k\bar{x} = f(t) \tag{2.5.3}$$

This equation of motion defines the dynamic equilibrium conditions of the vibrating system analysed starting from the static configuration.

$$f(t) = \sum_{n=0}^{\infty} |F_n| \cos(n\Omega_0 t + \theta_n) \tag{2.238}$$

where Ω_0 is the natural frequency of the fundamental harmonic ($\Omega_0 = 2\pi/T_0$, where T_0 is the period of the periodic function), F_n is the generic nth harmonic (coefficient of the series or Fourier coefficient), i.e. the complex generic term of the development (2.238) of modulus F_{no} and phase θ_n:

$$F_n = F_n(n\Omega_0) = \frac{2}{T_0} \int_{-\frac{T_0}{2}}^{\frac{T_0}{2}} f(t) e^{-in\Omega_0 t} dt = |F_n| e^{i\theta_n} \tag{2.239}$$

$|F_0|$ in (2.239) represents the average value of $f(t)$. By using the superposition of effects principle, the dynamic response $x(t)$ of a linear system subjected to a periodic excitation force can be obtained as the sum of the responses of the same system to each harmonic $F_n(t)$ in which the excitation force $f(t)$ has been decomposed. For these reasons, the study of the response of a linear system subjected to a single harmonic excitation forces proves to be of fundamental importance: the response to a harmonic excitation force is usually defined as a "harmonic transfer function" (see Sect. 2.4.1.4.1).

In the case of a aperiodic excitation force $f(t)$, the period tends towards the infinity $T_0 \to \infty$, the discrete variable $n\Omega_0$ becomes a continuous variable Ω so that (2.239) becomes:

$$F(\Omega) = \frac{1}{2\pi} \int_{-\infty}^{+\infty} f(t) e^{-i\Omega t} dt \tag{2.240}$$

Expression (2.240), which represents a continuous spectrum, generally goes by the name of a Fourier transform of $f(t)$. Thanks to this transform, excitation force $f(t)$ is seen as formally constituted by infinite excitation forces $F(\Omega)d\Omega$ distributed with continuity in the domain $O < \Omega < \infty$ so that (2.238), in the case of aperiodic excitation forces [2, 31], is transformed into:

$$f(t) = \int_{-\infty}^{+\infty} \frac{1}{2\pi} F(\Omega) e^{i\Omega t} d\Omega \tag{2.241}$$

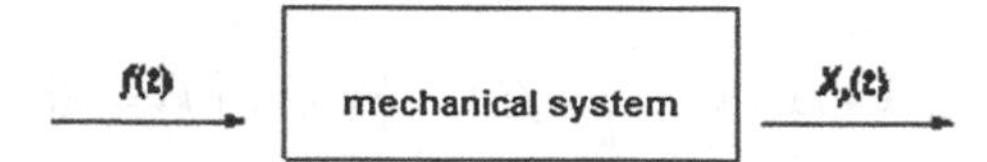

Fig. 2.27 Typical approach of control systems

defined as an inverse Fourier transform or a Fourier integral. An alternative, always in the case of aperiodic excitation forces, is to use the so-called convolution integral [31, 35] which evaluates the response of the system $x(t)$ as the sum of the responses to subsequent impulses with which the method envisages the same function $f(t)$ as being decomposed. It is possible to demonstrate that the response to the impulse is nothing else but the inverse transform (or Fourier integral) of the harmonic transfer function [24].

2.4.1.4.1 Harmonic Transfer Function

The response of a 1 d.o.f. system to a known excitation force can be more generally contextualized in the response of a system to a generic excitation, through algorithms typical of the control discipline [10, 11, 16, 24, 26].

From this point of view, the excitation force represents the input magnitude and the response of the system the output magnitude. The problem is thus ascribable to that represented in Fig. 2.27, where the vibrating system is considered like a box, on which, as an input magnitude, we can impose a generic applied force $f(t)$ and from which, as an "output magnitude" we can obtain displacement $x(t)$ of the system.

This representation can also easily be adopted for systems, even extremely complex ones, differing from a 1 d.o.f. system, as long as it is possible to identify an input and an output. The behaviour of the system in the box can be represented by the harmonic transfer function (defined in this section) or by the transfer function in complex variable s, using the Laplace transform (Sect. 2.4.1.4.6). The harmonic transfer function $H(\Omega)$ is used when one wishes to pass from a time domain to a frequency domain, using the Fourier transform.

Let us now consider a linear vibrating 1 d.o.f. system, excited by an harmonic applied force of modulus F_0, phase θ null and natural frequency Ω:

$$m\ddot{x} + r\dot{x} + kx = f(t) = F_0 \cos \Omega t \tag{2.242}$$

Solution $x(t)$ of (2.242) is the sum of the general integral $x_G(t)$ of the associated homogenous, already analysed in the previous sections, and of the particular integral $x_P(t)$ due to the known term, i.e. excitation force $f(t)$:

$$x(t) = x_G(t) + x_P(t) \tag{2.243}$$

The general integral $x_G(t)$ of the associated homogenous which describes the initial transient, in the case of real damping ($r > 0$) defines a damped motion tending to zero; conversely the steady state solution is described by the particular integral. The most convenient approach to calculate the steady state response $x_P(t)$ of a system with a harmonic excitation is to use a complex vector representation of the excitation and of the response. Thanks to this method, the actual force $F_0 \cos \Omega t$ can be considered as a projection on the real axis of a vector (complex number) $F = F_0 e^{i\Omega t}$ rotating in a Gauss plane at angular speed Ω (see Fig. 2.28):

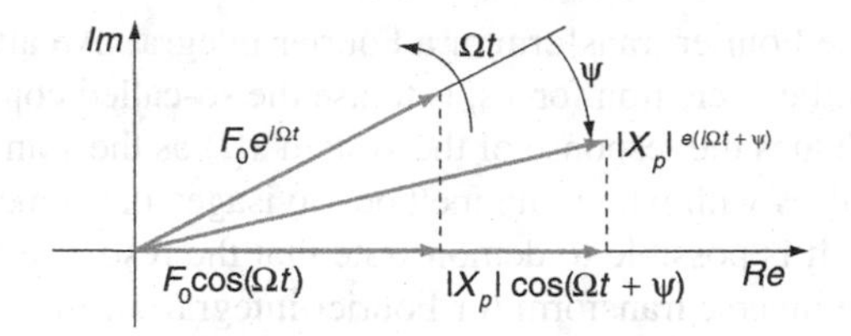

Fig. 2.28 Vector representation of a harmonic excitation force and a steady state response

$$F_0 \cos \Omega t = Re\left(F_0 e^{i\Omega t}\right) \tag{2.244}$$

The equation of motion (2.242) can be rewritten, in a complex notation, as:

$$m\ddot{x} + r\dot{x} + kx = F_0 e^{i\Omega t} \tag{2.245}$$

In a complex field, the solution of (2.245) under steady state conditions, is given by:

$$x_P = X_P e^{i\Omega t} = |X_P| e^{i(\Omega t + \psi)} \tag{2.246}$$

where X_P is a complex number of modulus $|X_P|$ and phase ψ. The actual solution $x_P(t)$[6] is represented by the projection on the real axis of the complex rotating vector $X_P e^{i\Omega t}$ (2.246):

$$x_P(t) = Re\left(X_P e^{i\Omega t}\right) = Re\left(|X_P| e^{i(\Omega t + \psi)}\right) = |X_P| \cos(\Omega t + \psi) \tag{2.247}$$

[6]In order to justify using a complex notation, it is possible to rewrite an auxiliary equation in parallel to the real Eq. (2.242):

$$\begin{aligned} m\ddot{x} + r\dot{x} + kx &= F_0 \cos(\Omega t) \\ m\ddot{y} + r\dot{y} + ky &= F_0 \sin(\Omega t) \end{aligned} \tag{2.6.1}$$

By multiplying the second of the Eq. (2.6.1), by imaginary unit $i = \sqrt{-1}$, adding it up member by member and assuming a new independent variable $z = x + iy$, the system of Eq. (2.6.1) can be rewritten (always keeping account of Euler's relation [2] as:

$$m\ddot{z} + r\dot{z} + kz = F_0 e^{i\Omega t} \tag{2.6.2}$$

The real solution $x(t)$ represents the real part of the complex solution $z(t)$: traditionally, during the transit from the real Eq. (2.6.1) to the equivalent complex one (2.6.2), formally speaking we retain the same name of the variable, thus obtaining expression (2.245).

where $|X_P|$ represents the vibration amplitude and ψ the phase between the vibration and the applied force. By substituting (2.246) and its derivatives with respect to time in (2.245) we obtain[7]:

$$\left(-m\Omega^2 + i\Omega r + k\right)X_P e^{i\Omega t} = F_0 e^{i\Omega t} \tag{2.248}$$

By using this approach it is possible to eliminate with ease the dependence on time and to obtain the expression of the complex solution:

$$X_P = \frac{F_0}{(k - m\Omega^2) + i\Omega r} \tag{2.249}$$

By dividing both terms of (2.249) by F_0 and dividing the numerator and the denominator of the term on the right of the equal sign by k, the same equation can be rewritten as:

$$\frac{X_P}{F_0} = \frac{1/k}{\left(1 - \frac{m\Omega^2}{k}\right) + i\frac{\Omega r}{k}} \tag{2.250}$$

and, by using:

- $\omega_0^2 = k/m$ to indicate the natural frequency of the undamped system;
- $a = \frac{\Omega}{\omega_0}$ the non-dimensional ratio of the frequencies;
- $h = \frac{r}{r_c}$ the non-dimensional damping ratio (2.210);

expression (2.249) can be rewritten as follows:

$$H(\Omega) = \frac{X_P}{F_0} = \frac{1/k}{(1 - a^2) + i2ah} \tag{2.251}$$

[7]In this expression (2.248), the term:

$$k_c = (k + i\Omega r) \tag{2.7.1}$$

can be considered as a complex stiffness, also inclusive of the dissipative term. Using this formality, in the frequency domain, i.e. by analysing only the steady state response of the system to a harmonic applied force, the same Eq. (2.248) can be rewritten as:

$$\left(-\Omega^2 m + k_c\right)x_p = F_0 \tag{2.7.2}$$

This representation will be convenient to introduce the hysteretic damping, in Sect. 3.7.1.

where $H(\Omega)$ is the so-called harmonic transfer function, i.e. a function that represents the response of a vibrating system to a harmonic applied force of a unitary modulus.

To obtain the modulus of (2.251), it is possible to proceed with the rationalization of the expression itself:

$$\begin{aligned} H(\Omega) &= \frac{1/k}{(1-a^2)+i2ah}\frac{(1-a^2)-i2ah}{(1-a^2)-i2ah} \\ &= \frac{1}{k}\frac{(1-a^2)-i2ah}{(1-a^2)^2+(2ah)^2} = |H(\Omega)|e^{i\psi} \end{aligned} \tag{2.252}$$

By separating the real part $\mathrm{Re}(H(\Omega))$ and the imaginary part $\mathrm{Im}(H(\Omega))$ of this relationship:

$$\mathrm{Re}(H(\Omega)) = \frac{1}{k}\frac{(1-a^2)}{(1-a^2)^2+(2ah)^2} \tag{2.253}$$

$$\mathrm{Im}(H(\Omega)) = -\frac{1}{k}\frac{i2ah}{(1-a^2)^2+(2ah)^2} \tag{2.254}$$

we obtain the modulus of the steady state response, as the composition of both the real and imaginary parts of the response:

$$|H(\Omega)| = \sqrt{\mathrm{Re}(H(\Omega))^2+\mathrm{Im}(H(\Omega))^2} = \frac{1/k}{\sqrt{(1-a^2)^2+(2ha)^2}} \tag{2.255}$$

while the phase is:

$$\tan\psi = \frac{\mathrm{Im}(H(\Omega))}{\mathrm{Re}(H(\Omega))} = -\frac{2ha}{1-a^2} \Rightarrow \psi = \psi(\Omega) = a\tan\left(\frac{-2ha}{1-a^2}\right) \tag{2.256}$$

The frequency response of a generic vibrating system is used to define the diagram showing gain $|H(\Omega)|$ of the harmonic transfer function (2.255) and phase diagram $\psi(\Omega)$ (2.256) both as a function of the frequency of excitation force Ω. This function, which will be analysed in more detail in Chap. 8, Sect. 8.2, relative to *Modal identification techniques*, represents the response of the system to a unitary harmonic excitation force. Conversely, we define the set of curves that, as a function of frequency Ω (in logarithmic scale) show the gain in decibels $\mathrm{Im}(|H(\Omega)|) = 20\log|H(\Omega)|$ and the phase angle ψ as a Bode diagram. The harmonic transfer function is used to calculate the response of a system with one harmonic excitation force (and, as will be seen further on, also a periodic excitation force that can be developed in a Fourier series).

Diagrams $|H(\Omega)|$ and $\psi(\Omega)$ are indicative of the dynamic behaviour of each 1 d.o.f. system. The modulus and phase diagrams can be divided into three areas depending on the value of relationship $a = (\Omega/\omega_0)$:

- Quasi-static area with $a = {}^{\Omega}/_{\omega_0} < 1$;
- resonance area with $a = {}^{\Omega}/_{\omega_0} = 1$;
- seismographic area with $a = {}^{\Omega}/_{\omega_0} > 1$.

The first area (defined as a quasi-static area), in which $a < 1$ (an area in which frequency Ω of the excitation force is much smaller than the natural frequency ω_0 of the system) is characterized by a harmonic transfer function $|H(\Omega)|$ near to the static value $1/k$. In this area, the vibration amplitudes assume values near to those that would be obtained by applying the excitation force statically. In a quasi-static area, phase ψ remains almost null: the force and displacement therefore prove to be in-phase. For n d.o.f. systems (Sect. 2.4.3), the area in which Ω is much smaller than the first natural frequency of ω_1 is considered almost static. To evaluate the stress state induced by a sinusoidal excitation force with $a \ll 1$, static calculations can be used, even if it is necessary to bear in mind that the stress is alternated meaning that a material fatigue problem exists.

The resonance area is characterized by a frequency of excitation force Ω near to the natural frequency of system ω_0: the harmonic transfer function $|H(\Omega)|$ for $a = (\Omega/\omega_0) = 1$, see (2.252), is given by:

$$|H(\Omega)|_{a=1} = \frac{1}{k}\frac{1}{2h} \tag{2.257}$$

The vibration amplitudes in resonance prove to be inversely proportionate to damping $h = (r/r_c)$ and the width of the resonance peak increases with damping. The harmonic transfer function $|H(\Omega)|$ is lower than the value assumed in the quasi-static zone only in the case of $h > 0.5$: this damping value is not easily achieved for structures or machine elements, where usual values of h are of the order of a few percent and, as a consequence, the harmonic transfer function in resonance conditions $H(\Omega)_{a=1}$ can show an amplification coefficient of the order of ten. Under these conditions, phase ψ is always equal to $-90°$ regardless of the damping value. For a null damping value h, expression (2.252) shows that the vibration amplitudes $|X_P|$ tend to the infinity. In reality, for various reasons, this type of situation never occurs:

- while for small amplitudes it is possible to neglect the damping effect, this is no longer allowable if the amplitudes become high, because, under these conditions, generally speaking, nonlinear effects introduce an increased dissipation of energy;

- to reach infinite vibration amplitudes, an infinite time is required: it can, in fact, be shown [3, 27, 36] that when there is a state of resonance and a lack of damping, the vibration follows the law:

$$x_P(t) = \frac{F_0}{2m\omega} t \sin \Omega t \tag{2.258}$$

Therefore, under resonance conditions, the dynamic effect on the deformation state can no longer be neglected: the stress state proves to be considerably amplified with respect to the value calculated statically.

In the third area, termed seismographic, characterized by $a = \Omega/\omega_0 > 1$, the harmonic transfer function $|H(\Omega)|$ decreases with the increase of ratio (Ω/ω_0) itself: the dynamic effect therefore reduces the oscillation amplitude. Phase ψ between vibration X_P and force F tends to $-180°$.

The harmonic transfer function $H(\Omega)$, defined as:

$$H(\Omega) = \frac{X_P(\Omega)}{F_0(\Omega)} \tag{2.259}$$

is thus, as previously mentioned, a complex function that allows us to calculate the response of the system in the frequency domain, once the excitation force in complex form $F = F(\Omega)$ is known. The response is obtained as a product of excitation force $F(\Omega)$ by $H(\Omega)$, a function of frequency Ω of the excitation force itself:

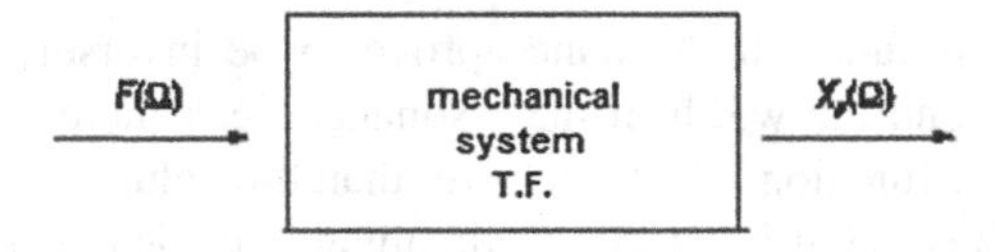

Fig. 2.29 Input and output variables: transfer function

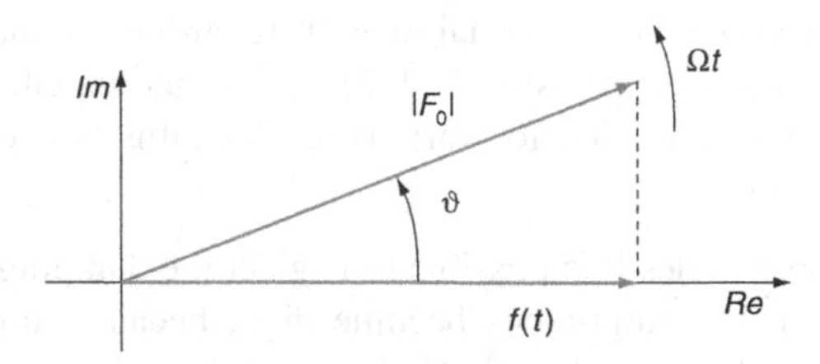

Fig. 2.30 Vector representation of a harmonic excitation force

$$X_P(\Omega) = H(\Omega)\,F(\Omega) \tag{2.260}$$

From this viewpoint (see Fig. 2.29), the system is seen as a block having one input, excitation force $F(\Omega)$, and one output, excitation response $X_P(\Omega)$: based on the response in complex form, it is then possible to evaluate the real part X_{PR} and the imaginary part X_{PI} and thus calculate modulus $|X_P|$ and phase ψ. In conclusion, the particular integral $x_P(t)$, i.e. the steady state solution for a system subjected to a harmonic excitation force is the projection on the real axis of vector $X_P e^{i(\Omega t+\psi)}$ i.e.:

$$x_P(t) = |X_P|\cos(\Omega t + \psi) \tag{2.261}$$

In (2.261) $|X_P|$ represents the amplitude of sinusoidal harmonic motion $x_P(t)$ displaced by ψ with respect to the excitation force: thus, at steady state the system vibrates at the same frequency as the excitation force.

Let us now generically consider (Fig. 2.30) a harmonic excitation force of modulus F_0 and phase θ, defined as a projection on the real axis of a complex vector $F = F_0 e^{i(\Omega t+\theta)}$:

$$f(t) = |F_0|\cos(\Omega t + \theta) = \mathrm{Re}(F) = \mathrm{Re}\left(|F_0| e^{i(\Omega t+\theta)}\right) \tag{2.262}$$

The steady state response of the system can be obtained as a product of the harmonic transfer function $H(\Omega)$ for the complex vector $F = F_0 e^{i\Omega t}$:

$$x(t) = \mathrm{Re}(H(\Omega)F(\Omega)) = \mathrm{Re}\left(|H(\Omega)| e^{i\psi} |F_0| e^{i(\theta+\Omega t)}\right) \tag{2.263}$$

i.e., by recalling De Moivre's theorem [2] on the product of complex numbers:

$$x(t) = \mathrm{Re}\left(|H(\Omega)||F_0| e^{i(\psi+\theta)} e^{i\Omega t}\right) \tag{2.264}$$

$$= |H(\Omega)||F_0|\cos(\Omega t + \psi + \theta) = |X_P|\cos(\Omega t + \psi + \theta) \tag{2.265}$$

where θ represents the force phase, dependent on the value of force phase $f(t)$ at initial instant $t = 0$, while ψ represents the phase lag of the response with respect to the excitation force.

2.4.1.4.2 Representation in a Complex Plane

To better explain the meaning of the three zones shown in the diagram of transfer function $H(\Omega)$, it is possible to resort to the complex representation of the equation of motion in the real variable $x(t)$:

$$-m\ddot{x} - r\dot{x} - kx + F_0 \cos \Omega t = 0 \tag{2.266}$$

Expression (2.266) can be interpreted as the dynamic equilibrium equation of the forces acting on the system:

$$\begin{array}{ll} \text{inertia forces} & f_i = -m\ddot{x} \\ \text{damping forces} & f_d = -r\dot{x} \\ \text{elastic forces} & f\_\{k\} = -kx \\ \text{external forces} & f(t) = F_0 \cos \Omega t \end{array} \tag{2.267}$$

The same relation can be rewritten by adopting a complex notation: by considering all the components of the forces as projections on the real axis of force vectors rotating in the Gauss plane (it is necessary to underline that this step is possible in that all the vectors rotate at the same angular speed Ω) and, remembering that the same expression (2.266) represents the sum of all the forces acting on mass m, it is, in fact, possible to rewrite the equation of motion of the one d.o.f. system subjected to a harmonic excitation as:

$$\text{Re}(F_i + F_d + F_k + F) = 0 \Rightarrow F_i + F_d + F_k + F = 0 \tag{2.268}$$

Excitation force $f(t) = F_0 \cos \Omega t$ can be seen as the projection on the real axis of a rotating vector F in the complex plane:

$$F = F_0 \cos(\Omega t) \Rightarrow F = F_0 \text{e}^{i\Omega t} \tag{2.270}$$

By bearing in mind that the particular integral of (2.266) can be expressed, in a complex field [see (2.247)] as:

$$x_P = X_P \text{e}^{i\Omega t} = |X_P| \text{e}^{i\psi} \text{e}^{i\Omega t} = |X_P| \text{e}^{i(\Omega t + \psi)} \tag{2.271}$$

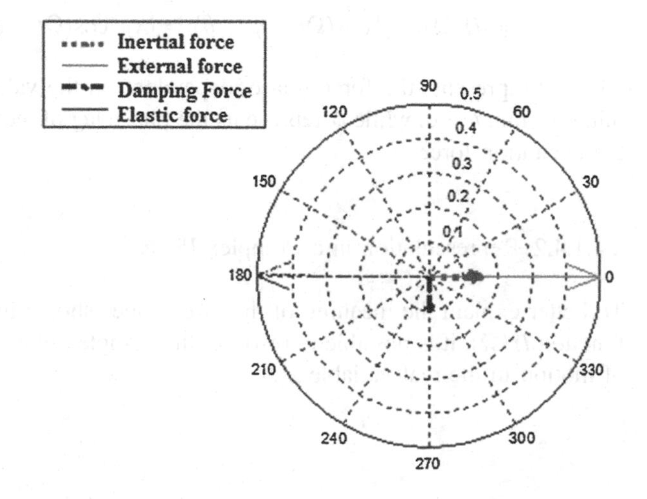

Fig. 2.31 One d.o.f. system: vector diagram of the forces for a ≪ 1

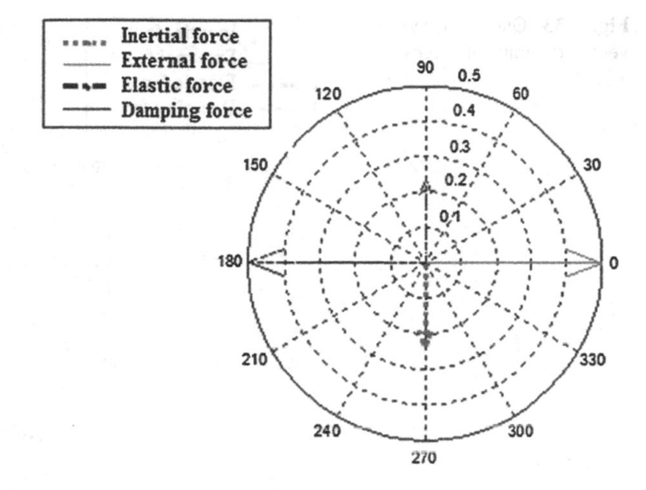

Fig. 2.32 One d.o.f. system: vector diagram of forces for a = 1

and, as a consequence, that the velocity and acceleration can be expressed as:

$$\dot{x}_P = i\Omega X_P e^{i\Omega t} = i\Omega |X_P| e^{i(\Omega t+\psi)} \tag{2.272}$$

$$\ddot{x}_P = -\Omega^2 X_P e^{i\Omega t} = -\Omega^2 |X_P| e^{i(\Omega t+\psi)} \tag{2.273}$$

the expression of the inertia force F_i (2.267) thus becomes:

$$f_i = -m\ddot{x}_p \Rightarrow F_i = m\Omega^2 X_P e^{i\Omega t} = m\Omega^2 x_P \tag{2.274}$$

i.e. it is represented by a rotating vector having the same direction as the complex vector x_P; on the contrary, the elastic force has the opposite direction of x_P:

$$f_k = -kx_P \Rightarrow F_k = -kX_P e^{i\Omega t} = -kx_P \tag{2.275}$$

and, lastly, the viscous force has a direction that is perpendicular to x_P and has a 90° phase lag on the displacement itself:

$$f_d = -r\dot{x}_p \Rightarrow F_d = -ir\Omega X_P e^{i\Omega t} = -ir\Omega x_P \tag{2.276}$$

The vector equation (2.268) is graphically represented in Fig. 2.31. This representation allows us to better evidence the peculiar characteristics of the three zone in the diagram of the transfer function already seen in Sect. 2.4.1.4.1.

In particular, for the first zone, ($a = (\Omega/\omega_0) < 1$), inertia force F_i and viscous force F_d are small, on account of frequency Ω being small and, furthermore, phase lag ψ is practically null. Thus, ultimately, it results that external excitation force F is balanced by the only elastic force F_k (Fig. 2.31). Under this condition, the

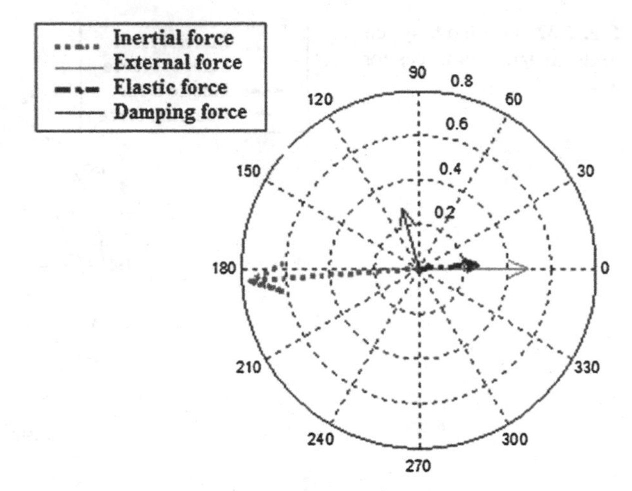

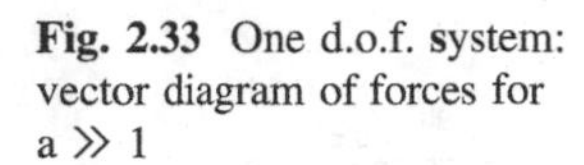
Fig. 2.33 One d.o.f. system: vector diagram of forces for a ≫ 1

excitation force is offloaded onto the external constraints by means of the interposed elastic element.

In the second zone in which $a = 1$ (i.e. $\Omega = \omega_0$) elastic force F_k balances inertia force F_i, on account of having the same modulus and an opposite direction. Therefore, external force F is balanced by the only viscous force F_d since phase lag ψ is equal to 90° (see Fig. 2.32). By equalling the modulus of external force $|F|$ with the modulus of viscous force F_d (2.267), we once again obtain the steady state vibration amplitude in resonance:

$$|X_P| = \frac{F_0}{r\omega_0} = \frac{F_0}{r\Omega} = \frac{F_0/k}{2h} \tag{2.277}$$

As previously seen, as the damping increases the amplitude of oscillation decreases.

In the third zone, termed the seismographic zone, defined by parameter $a > 1$, inertia force F_i, which depends on the square of frequency Ω of the excitation force, is much greater than elastic force F_k and viscous force F_d. In this zone, phase ψ between response x_P and external excitation force F is worth approximately 180°, regardless of the value assumed by damping r. External force F is therefore balanced almost exclusively by the only inertia force F_i (Fig. 2.33) and the amplitudes of oscillation tend to annul themselves by Ω tending towards infinity. Already for the values of ratio $a = 2$ the elastic forces F_k and the viscous ones F_d are negligible with respect to those of inertia F_i and the mass can therefore be seen as free in space, i.e. devoid of external constraints. The fact of being able to consider the system as if it were devoid of ground constraints is exploited in problems related to isolation of foundations: by creating a system having a very low natural frequency ω_0 with respect to the frequency of excitation force Ω, so that functioning takes place above the natural frequency of the system: in this case, the force transmitted by the spring and by the damper to the external constraints decreases. This property is, in fact, exploited in isolation of vibrations by means of an elastic foundation (see Sect. 2.4.1.4.1).

2.4.1.4.3 Coefficient of Dynamic Amplification

In some cases, the frequency response can also be represented in terms of a dynamic amplification coefficient: by remembering that the natural frequency of the undamped system is $\omega_0^2 = k/m$, by introducing the non-dimensional relationship $a = \frac{\Omega}{\omega_0}$ and after defining the non-dimensional damping ratio $h = \frac{r}{r_c}$, it is possible to rewrite (2.249) in the following form:

$$X_P = \frac{(F_0/k)}{(1 - a^2) + i2ah} \tag{2.278}$$

By now using X_{st} to define the displacement to which the system is subjected due to the effect of static application (i.e. with null frequency Ω) of the same force F_0:

$$X_{st} = \frac{F_0}{k} \tag{2.279}$$

expression (2.278) can thus be rewritten as[8]:

$$\frac{X_P}{X_{st}} = \frac{1}{(1 - a^2) + i2ah} \tag{2.280}$$

The term *dynamic amplification coefficient* $A(a)$ is used to define the modulus of relationship (2.280):

$$A(a) = A\left(\frac{\Omega}{\omega_0}\right) = \left|\frac{X_P}{X_{st}}\right| = \frac{1}{(1 - a^2)^2 + (2ah)^2} \tag{2.281}$$

[8]Function ${}^{X_P}/_{X_{st}}$ can be obtained from $H(i\Omega)$ simply divided by ${}^{1}/_{k}$.

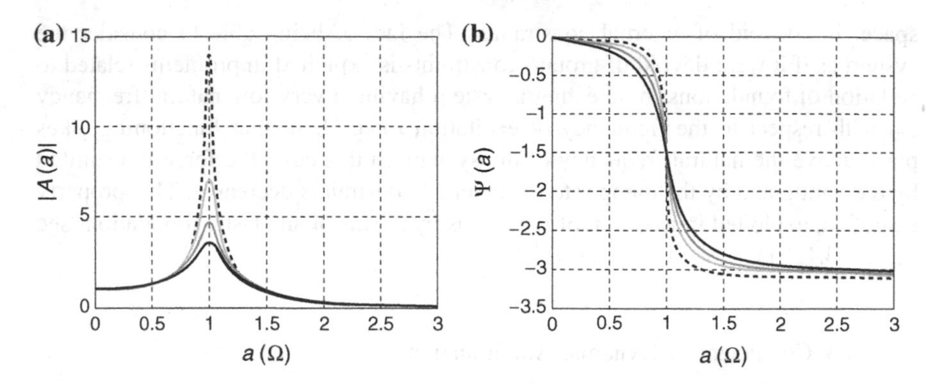

Fig. 2.34 Forced 1 d.o.f. system: **a** dynamic amplification coefficient $A(a)$; **b** phase $\psi(a)$ of the response with respect to a harmonic excitation force

and the corresponding phase is given by:

$$tan\psi = -\frac{2ha}{1-a^2} \tag{2.282}$$

This relationship provides amplification of the response amplitude of the system with respect to the static case: calculation of the steady state amplitude can be performed by simply multiplying the static deformation X_{st} by amplification coefficient $A(a)$ given by (2.281).

To highlight the dynamic effects of the application of the harmonic excitation force, it is customary, as a function of non-dimensional parameter $a = \Omega/\omega_0$, to plot the dynamic amplification coefficient $A(a)$ (2.281) and the relative phase ψ (2.282): function $A(a)$ thus represents the vibration amplitude, compared with the static displacement, as a function of the relationship between frequency Ω of the excitation force and natural frequency ω_0. Function $\psi = \psi(a)$ (Fig. 2.34) as previously mentioned expresses the angular phase lag between excitation force $F = F_0 e^{i\Omega t}$ and vibration $X_P = |X_P| e^{i(\Omega t + \psi)}$ (Fig. 2.34); on the contrary, the temporal phase lag is equal to:

$$\Delta t = \frac{\psi}{\Omega} \tag{2.283}$$

Diagrams $A(a)$ and $\psi(a)$ are representative of the dynamic behaviour of each 1 d.o.f. system. The diagrams of modulus and phase can be divided into three zones depending on the value of the ratio $a = (\Omega/\omega_0)$ similar to the situation involving the harmonic transfer function, so that we obtain:

- a quasi-static zone with $a = {}^{\Omega}/_{\omega_0} < 1$;
- a resonance zone with $a = {}^{\Omega}/_{\omega_0} = 1$;
- a seismographic zone with $a = {}^{\Omega}/_{\omega_0} > 1$.

The first zone (defined as a quasi-static zone), in which $a<1$ (a zone in which frequency Ω of the excitation force is much smaller than the natural frequency ω_0 of the system) is characterized by a dynamic amplification coefficient $A(a)$ near to unity. In this zone, the vibration amplitudes assume values near to those that would be obtained by applying the excitation force statically. In the almost static zone, phase ψ remains almost null: the force and displacement thus prove to be in phase.

The resonance zone is characterized by an excitation force frequency Ω near to the natural frequency of system ω_0: the dynamic amplification $A(a)$ for $a = (\Omega/\omega_0) = 1$ [see (2.281)] is worth:

$$A(a) = \frac{|X_P|}{X_{st}} = \frac{1}{2h} \tag{2.284}$$

The vibration amplitudes in resonance are inversely proportionate to damping $h = (r/r_c)$ and the width of the resonance peak base increases with damping. Amplification coefficient $A(a)$ is smaller than unity only in the case of $h > 0.5$: this damping value is difficult to achieve for structures or machine elements, where usual values of h are of the order of a few percent and, as a consequence, the dynamic amplification coefficient in resonance $A(a)_{a=1}$ can become of the order of some tens. Under these conditions, phase ψ, is always equal to $-90°$, regardless of the damping value (2.282). For a null damping value h, expression (2.284) shows that vibration amplitudes $|X_P|$ tend towards the infinity. In the third zone, termed seismographic, characterized by $a > 1$, the dynamic amplification coefficient $A(a)$ is smaller than unity and decreases as relationship (Ω/ω_0) itself increases: the dynamic effect thus reduces the amplitude of oscillation. Phase ψ between vibration X_P and force F tends towards $-180°$.

2.4.1.4.4 An Example of Application: Isolation of Vibrations by Means of a Foundation

In theory, there is often the problem of reducing the level of vibrations both of the machine as well as the surrounding environment: a classic example is the case of a machine mounted on a base. While operating, the machine transmits a force to the base (an inertia force in an alternative machine, an unbalance in a rotating machine, etc.) which, generally speaking, is periodic and therefore can be developed, by means of a Fourier series, into harmonics with frequency $n\Omega_0$ proportional to fundamental frequency Ω_0 (let us remember that n is used to define the order of the harmonic). In this case, the aim of the foundation is to limit the value of the vibrations of the base and the value of the force transmitted to the constraints: depending on the applications, the first or second aspect will be predominant.

A first approach to the problem is to use a simple diagram with one d.o.f., in which m represents the mass of the base plus that of the machine, k the stiffness of the foundation and r its damping. Let us consider the vibrating system as being subjected to a harmonic force $f(t) = F_0 \cos \Omega t$: the equations of motion, in complex

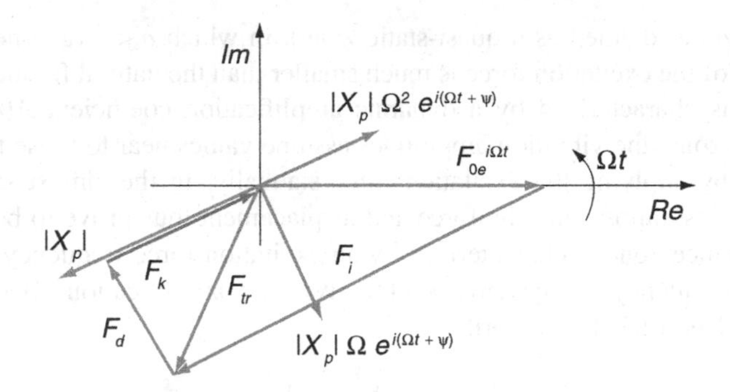

Fig. 2.35 Vector representation of the steady state situation: displacements and forces acting on a vibrating one d.o.f. system

form, that allow us to define the steady state response of the system, have been defined in Sect. 2.4.1.4, Eq. (2.242) and the solution has been obtained by using the vector notation in (2.247). The force transmitted by vibrating system f_{tr} is offloaded onto the constraints by means of spring k and damper r: under steady state conditions, by resorting to the complex vector notation, by keeping account of (2.271) (2.276), f_{tr} therefore simply becomes:

$$f_{tr} = \mathrm{Re}(F_{tr}) = kx + r\dot{x} \Rightarrow F_{tr} = kx_P + r\,\dot{x}_P \tag{2.285}$$

i.e.:

$$F_{tr} = (k + i\Omega r)X_P \tag{2.286}$$

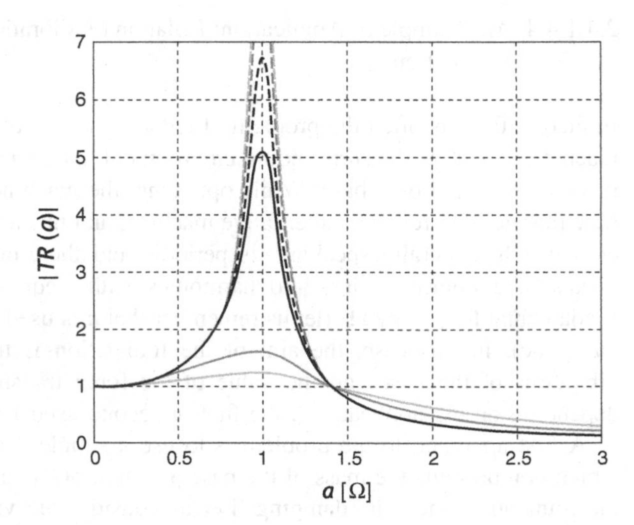

Fig. 2.36 Transmissivity for various structural damping values

The modulus of the force transmitted is:

$$|F_{tr}| = \sqrt{(k|X_P|)^2 + (\Omega r|X_P|)^2} = |X_P|\sqrt{k^2 + (\Omega r)^2} \quad (2.287)$$

Figure 2.35 shows a steady-state vector representation of:

- displacement X_P, velocity $\dot{X}_P$, acceleration $\ddot{X}_P$, (2.272) and (2.273), of mass m;
- excitation force $F = F_0 e^{i\Omega t}$ acting on the system;
- the corresponding elastic force F_k, viscous force F_d and inertia force F_i (2.267);
- transmitted force F_{tr} (2.286).

We note how the vector notation enables us to easily evidence the fact that viscous force F_d foreruns the elastic force F_k.

By recalling the expression of the steady state vibration amplitude $|X_P|$ (2.278), the modulus of transmitted force $|F_{tr}|$ becomes:

$$|F_{tr}| = \frac{1}{k}\frac{F_0}{(1-a^2)^2 + (2ah)^2}\sqrt{k^2 + (\Omega r)^2} = \frac{F_0\sqrt{1 + (2ha)^2}}{(1-a^2)^2 + (2ah)^2} \quad (2.288)$$

The relationship between the modulus of the force transmitted to constraint $|F_{tr}|$ and the modulus of the excitation force applied to vibrating system $|F_0|$ is defined as a *TR* transmissivity relationship (Fig. 2.36):

$$TR = \frac{|F_{tr}|}{|F_0|} = \frac{\sqrt{1 + (2ha)^2}}{\sqrt{(1-a^2)^2 + (2ah)^2}} \quad (2.289)$$

In this way, modulus $|F_{tr}|$ of the force transmitted can be rewritten as a product of the impressed force modulus $|F_0|$ multiplied by transmissibility coefficient *TR*:

$$|F_{tr}| = |F_0|\ TR \quad (2.290)$$

In the diagram showing the trend of the *TR* transmissivity ratio between the modulus of the force transmitted to the constraint and excitation force (Fig. 2.36), it is possible to observe a first zone (corresponding to $\Omega/\omega_0 < 1$) in correspondence to which the entire excitation force F_0 is transmitted to the constraints (in this case, we refer to a rigid foundation) and a third zone (seismographic zone) in which the *TR* ratio is lower than unity and tends to annul itself as ratio $a = (\Omega/\omega_0)$ increases. As can be noted, for values of $a = \sqrt{2}$ the whole of force F_0 is transmitted to the ground, while for $a > \sqrt{2}$ the force tends to decrease, as frequency Ω of the excitation force increases. It is interesting to note how, in a seismographic zone, as damping h increases so does the force transmitted. To isolate a machine, it is therefore necessary to design the foundation so that its frequency ω_0 is much lower than the frequency of excitation force Ω, with a damping that, though possibly limited is, nevertheless, compatible with passing through resonance. Furthermore, it is necessary to bear in mind that, in addition to the periodic (or harmonic) force, there is also a noise factor, associated with the random excitation forces (Chap. 7; [14]) resulting from uncontrollable phenomena that act on the machine. Generally speaking, the noise covers a wide frequency band and is also capable of exciting the natural frequency of the base: for this reason, it is necessary that same is sufficiently damped. From a technical point of view, the largest problems are encountered when creating low values of ω_0, which can be obtained either with low stiffness values k, or high mass values m: however, from a practical point of view, this solution is not always an option. In fact, by rewriting the relation that describes the natural frequency of a 1 d.o.f. system as:

$$\omega_0 = \sqrt{\frac{k}{m}} = \sqrt{\frac{kg}{mg}} = \sqrt{\frac{g}{\delta_{st}}} \tag{2.291}$$

where g is the acceleration of gravity and δ_{st} is the static drawdown sustained by the mass of the system due to the effect of the natural weight, it is possible to observe that, in order to obtain low frequencies, it is necessary to create a highly deformable suspension. Expression (2.291) shows how as natural frequency ω_0 of the foundation decreases, the value of static drawdown δ_{st}, required by the elastic element, increases and that this value is not easily achievable in non-theoretical situations: for example, to obtain a natural frequency equal to 1 rad/s (0.16 Hz), it is necessary to have a static deformation of as much as 9.81 m! Furthermore, it is important to note that a high value of δ_{st}, i.e. a low value of k, implies that the system, subjected to generic external actions, sustains considerable displacements that are not always compatible with the correct functioning of the machine which is bound to the outside world either to receive or supply energy.

2.4.1.4.5 Response to a Periodic Excitation Force

In Sect. 2.4.1.4.1 we showed how a vibrating system, subjected to a harmonic excitation force of frequency Ω, oscillates at steady-state conditions having the same frequency as the excitation force with amplitudes that depend on the relationship [see Eqs. (2.247) and (2.278)] between the frequency of excitation force Ω and the natural frequency ω of the system. In the previous Sect. 2.4.1.4.1 we saw how a representation that is similar to dynamic amplification coefficient $A(a)$ (Sect. 2.4.1.4.3, (2.281), Fig. 2.34), can be provided by the modulus of the harmonic transfer function $|H(i\Omega)|$ defined, in the frequency domain, as the relationship between the complex vector $X_P e^{i\Omega t}$ (2.259), whose projection on the real axis provides the actual vibration $x_P(t)$, and the complex vector $F = F_0 e^{i\Omega t}$, whose projection on the real axis represents the actual excitation force $f(t)$ (see Footnote 6). If the system is linear, then the principle of the superposition of effects applies to it: the dynamic response to several excitation forces is thus equal to the sum of responses to each single excitation force. As previously mentioned in Sect. 2.4.1.4.1, this property is useful when one wishes to analyse the dynamic behaviour of a linear system, subjected to a periodic excitation force. We consider a generic periodic excitation force $f(t)$ whereby relationship $f(t) = f(t + T_0)$ applies, where T_0 is the period of the same function. This function can be represented as:

$$f(t) = \sum_{n=0}^{\infty} \mathrm{Re}\left(F_n e^{in\Omega_0 t}\right) = \sum_{n=0}^{\infty} |F_n| \cos(n\Omega_0 t + \theta_n) \tag{2.292}$$

where $F_n = |F_n|e^{i\theta_n}$ is the generic complex harmonic component and $\Omega_n = n\Omega_0$ is the generic frequency, a multiple of the fundamental harmonic Ω_0 [rad/s] associated with fundamental period T_0:

$$\Omega_0 = \frac{2\pi}{T_0} = 2\pi f_0 \tag{2.293}$$

Expression (2.292) can also be rewritten as:

$$f(t) = \sum_{n=0}^{\infty} F_{An} \cos(n\Omega_0 t) + F_{Bn} \sin(n\Omega_0 t) \tag{2.294}$$

The generic harmonic F_n of (2.292) can be evaluated by means of (see also [2, 12, 27]):

$$F_n = |F_n|e^{i\theta_n} = \frac{2}{T_0} \int_{-To/2}^{To/2} f(t) e^{-in\Omega_0 t} dt \tag{2.295}$$

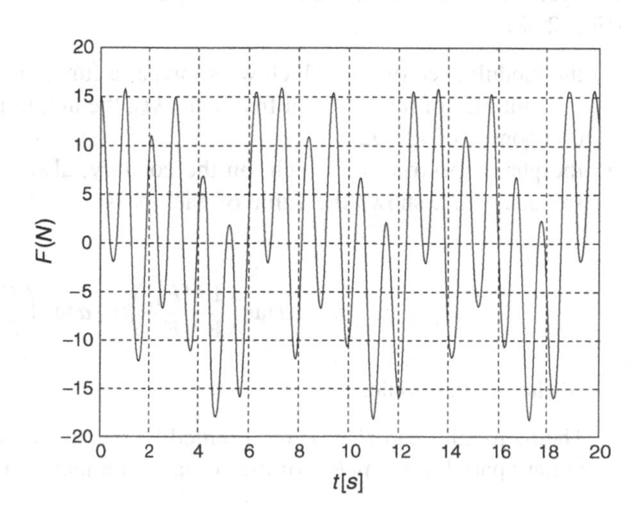

Fig. 2.37 Periodic excitation force: time history

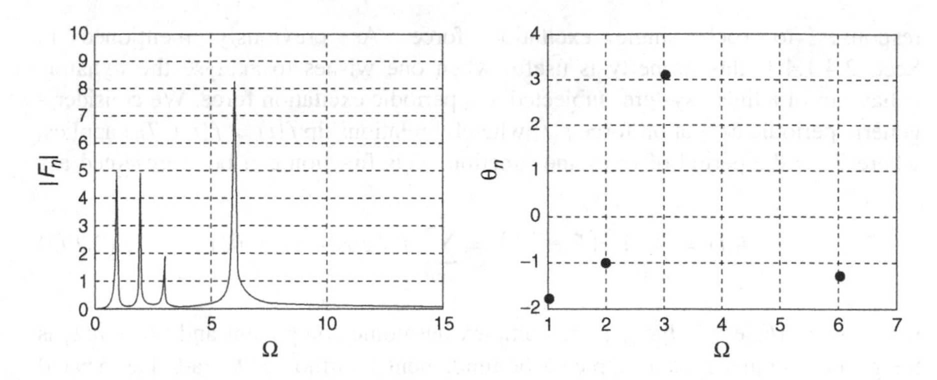

Fig. 2.38 Periodic excitation force: spectrum($|F_n|$ amplitude, θ_n phase)

while, on the contrary, for formulation (2.294) components F_{An} and F_{Bn} prove to be:

$$F_{An} = \frac{2}{T_0} \int_{-T_0/2}^{T_0/2} f(t) \cos(n\Omega_0 t) dt \qquad (2.296)$$

$$F_{Bn} = \frac{2}{T_0} \int_{-T_0/2}^{T_0/2} f(t) \sin(n\Omega_0 t) dt \qquad (2.297)$$

Relationships (2.296) and (2.297) are nothing other than a variable transformation that allow us to pass from a time domain to a frequency domain: the representation of generic amplitude $f(t)$ in the time domain is obviously given by the same time history (Fig. 2.37), while its representation in the frequency domain is given by a discrete spectrum $F_n = F(\Omega_n)$ (with $n \to \infty$), i.e. by two diagrams (Fig. 2.38):

- the moduli spectrum in which we show, as a function of frequencies $\Omega_n = n\Omega_0$, all multiples of fundamental frequency Ω_0, the amplitude $|F_n|$ of each harmonic component of (2.292);
- the phase spectrum θ_n which, on the contrary, always as a function of generic frequency Ω_n, shows the value of each phase:

$$\theta_n = a\tan\left(\frac{\mathrm{Im}(F_n)}{\mathrm{Re}(F_n)}\right) = a\tan\left(\frac{F_{nI}}{F_{nR}}\right) \qquad (2.298)$$

of the nth harmonic.

The harmonics can also be represented by means of real part $F_{nR} = \mathrm{Re}(F_n)$ and imaginary part $F_{nI} = \mathrm{Im}(F_n)$ of the complex number F_n in (2.295). Representation of a periodic excitation force by means of the corresponding frequency spectrum proves to be extremely advantageous and immediate in numerous representations, as will be seen in more depth later on.

Once the excitation force $f(t)$ has been broken up into single harmonic components F_n (2.292), should one wish to analyse the behaviour of a linear system, it is possible to calculate the steady-state response X_{Pn} to each component F_n of the excitation force using the harmonic transfer function $H(\Omega_n)$, Sect. 2.4.1.4.1, (2.259):

$$X_{Pn} = |H(i\Omega)|F_n \tag{2.299}$$

In theory, all the infinite harmonic of $f(t)$ defined by expression (2.292) should be considered: from an engineering viewpoint, only the most significant harmonics are considered. This approximation is justified by three considerations:

- first and foremost, whatever the physical phenomenon that gives rise to a periodic excitation force $f(t)$, the harmonic content (i.e. the amplitude of the various harmonics) of the latter tends to annul itself at high frequencies, due to the fact that the energy associated with the physical phenomenon under examination is proportional to the sum of the squares of the harmonics; since this energy cannot be infinite, only a frequency band limited in the upper part must be significant, as is the case in the spectrum shown in Fig. 2.38;
- furthermore, in the case of n d.o.f. systems, the influence of the harmonics of an order higher than the highest natural frequency becomes negligible. This phenomenon will be explained more clearly in Sect. 2.5, with reference to the modal approach;
- damping increases as the frequency increases.

Thus, for these reasons, the Fourier series (2.292) is truncated to the first N terms, since N is a function of the specific problem considered:

$$f(t) = \sum_{n=0}^{N} \mathrm{Re}\left(F_n e^{in\Omega_0 t} = \sum_{n=0}^{N} |F_n| \cos n\Omega_0 t + \theta_n \right) \tag{2.300}$$

The complex response can be obtained by recomposing the single X_{Pn} responses using the Fourier series:

$$\begin{aligned} x(t) &= \sum_{n=0}^{N} \mathrm{Re}\left(X_{Pn} e^{i(n\Omega_0 t)} \right) = \sum_{n=0}^{N} \mathrm{Re}\left(H(n\Omega_0) F_n e^{i(n\Omega_0 t)} \right) \\ &= \sum_{n=0}^{N} \mathrm{Re}\left(|X_{Pn}| e^{i(n\Omega_0 t + \theta_n + \psi_n)} \right) = \sum_{n=0}^{N} \left(|X_{Pn}| \cos(n\Omega_0 t + \theta_n + \psi_n) \right) \\ &= X_{st} + \sum_{n=1}^{N} |X_{Pn}| \cos(n\Omega_0 t + \theta_n + \psi_n) \end{aligned} \tag{2.301}$$

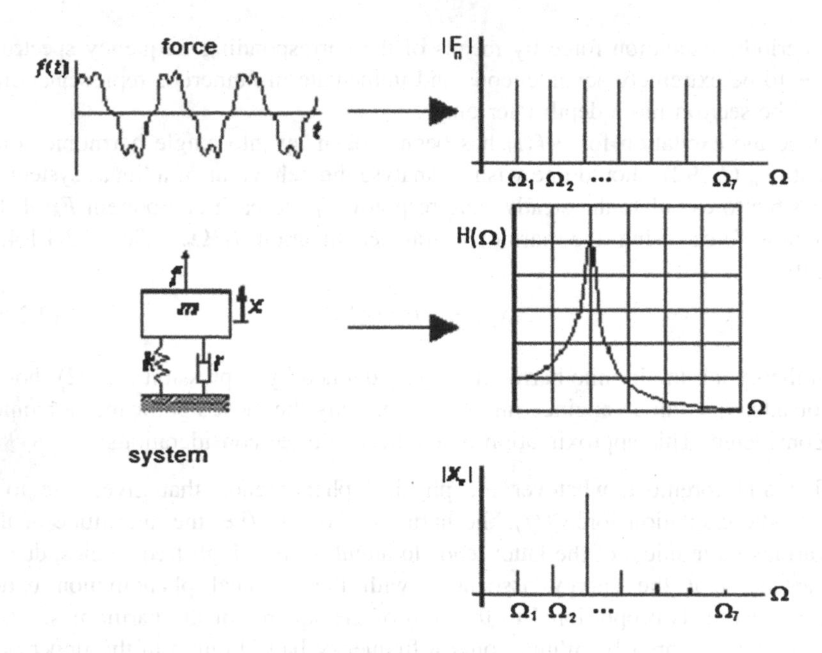

Fig. 2.39 System response to a periodic excitation force: response spectrum

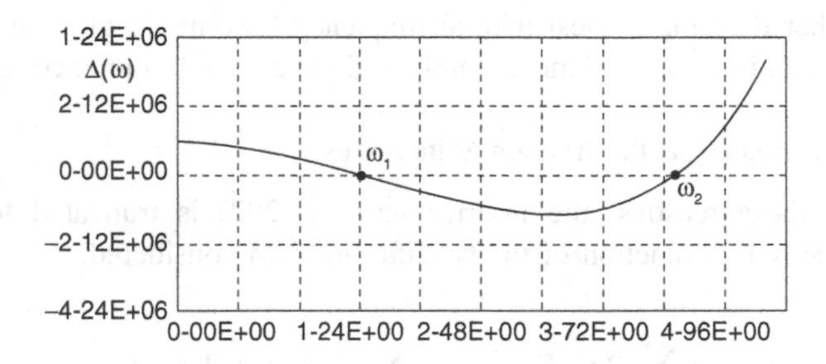

Fig. 2.40 Trend of the determinant of the coefficient matrix of problem (2.85) for the system in Fig. 2.4 (numerical data shown in Table 2.2)

having used X_{st} to indicate the static response of the system, i.e. the response to the constant component of the force (corresponding to $n = 0$).

As seen in Sect. 2.3.3, depending on the relationship between frequency Ω_n of the generic harmonic F_n of excitation force $f(t)$ and natural frequency ω of the system, the dynamic amplification coefficient $A(a)$ ((2.281), Fig. 2.34) can assume widely differing values:

- equal to approximately 1 in the quasi-static zone ($a_n = (\Omega_n/\omega) \ll 1$);
- much larger than one in the resonance zone (corresponding to $a_n = (\Omega_n/\omega) = 1$);
- lower than one (tending to zero) in the seismographic zone ($a_n \gg 1$).

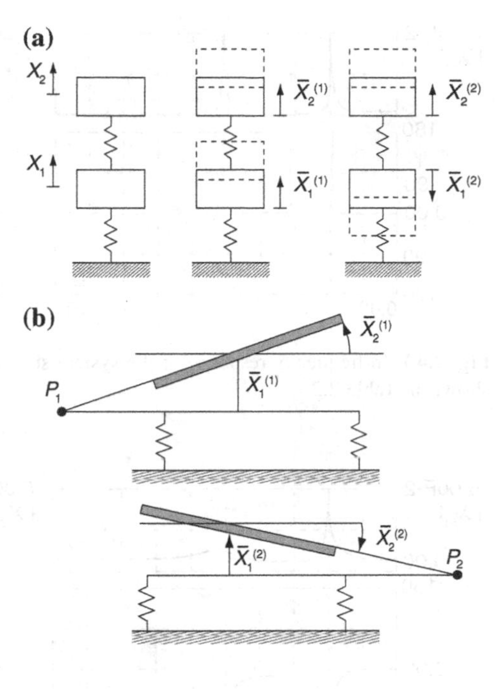

Fig. 2.41 **a** Vibration modes of the system in Fig. 2.4 (data shown in Table 2.2). **b** Vibration modes of the system in Fig. 2.7 (data shown in Table 2.3)

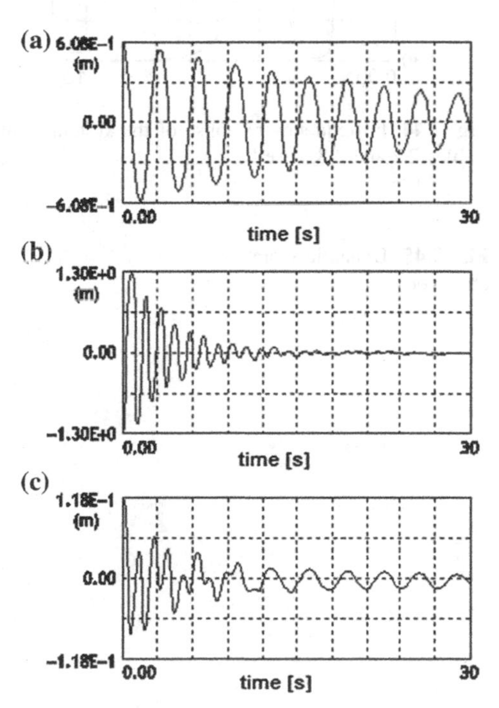

Fig. 2.42 Two d.o.f. system: examples of free motion for different initial conditions

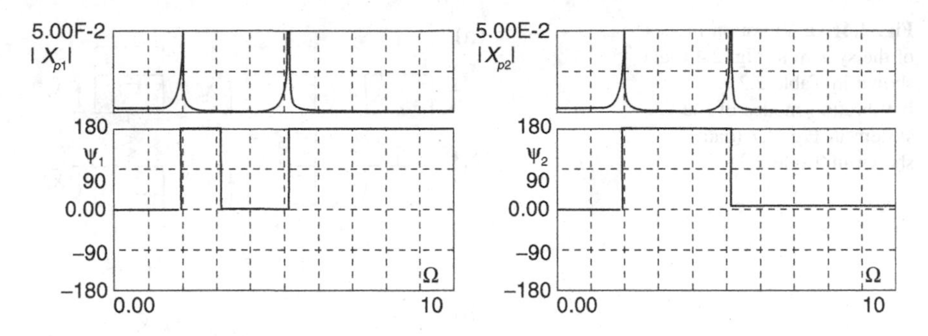

Fig. 2.43 In-frequency response of the system shown in Fig. 2.4 without damping: relative data shown in Table 2.2

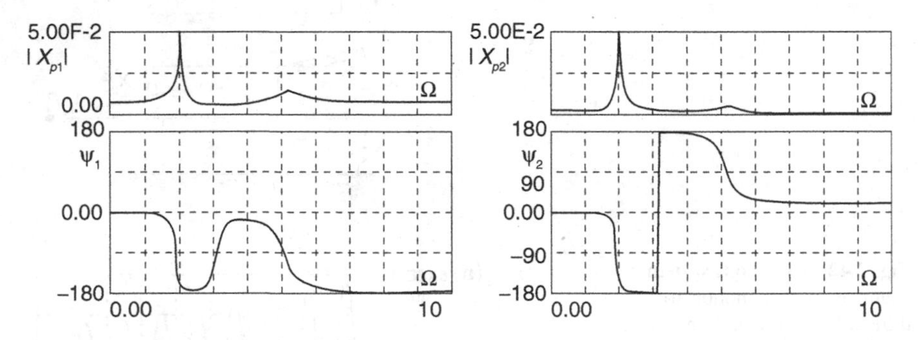

Fig. 2.44 In-frequency response of the system shown in Fig. 2.4 with damping: data shown in Tables 2.2 and 2.8, case b

Fig. 2.45 Dynamic vibration absorber

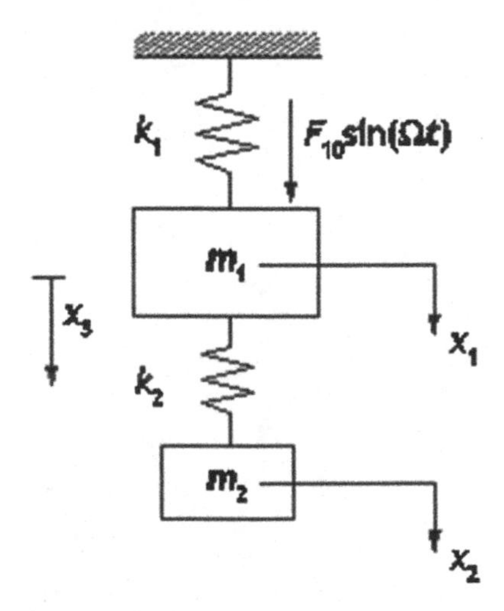

Even phase lag ψ_n between the response and excitation force is a function of relationship a_n. As an example, in the frequency domain, Fig. 2.39 shows the response of a 1 d.o.f. system subjected to an excitation in which the first seven harmonics, of equal amplitude $|F_n|$ (with n = 1, 2, ... ,7), were considered as representative: as can be seen, transfer function $|H(\Omega)|$ of the 1 d.o.f. system modifies the response spectrum and, as a consequence, the time history which can be defined by means of (2.301). From this point of view, the 1 d.o.f. system can thus be considered as a mechanical filter. If the non-dimensional damping ratio $h = (r/r_c)$ is very small, only the harmonics near to the resonance will be amplified and the system will thus operate as a low pass filter, filtering all the harmonics with frequency Ω_n which is much greater than natural frequency ω of the system.

2.4.1.4.6 Laplace Transform

To complete the subject, in this section we will analyse an alternative approach to solve the linear differential equations that define the forced motion of a generic discrete 1 d.o.f. system. This approach is used, above all, within the context of a control theory [31, 35] and is based on Laplace transforms: for a more in-depth analysis of the problem see [10, 11, 16]. In particular, we will show how the harmonic transfer function can be obtained from the transfer function in complex variable s. Let us consider a generic function $f(t)$, defined for $t = 0$, if a number b is real and positive so that product $e^{-bt}f(t)$ tends to a finite value for $t \to \infty$, then the Laplace transform of function $f(t)$ exists:

$$F(s) = L[f(t)] = \int_0^\infty f(t)e^{-st}dt \tag{2.302}$$

where s represents a generically complex variable. This transformation, which, to simplify matters, we will refer to hereinafter with the symbol *TdL*, allows us to pass from a function $f(t)$ of the real variable t to a function $F(s)$ of the complex variable s (which is named after the Laplace operator) and, as will be seen, allows us to transform a linear differential equation into an algebraic equation, thus facilitating the search for a general solution. Let us briefly recall the main properties of this transformation [24, 26, 28, 29, 31, 37]:

- the Laplace transform is a linear transformation, i.e.:

$$L[A_1f_1(t) + A_2f_2(t)] = A_1F_1(s) + A_2F_2(s) \tag{2.303}$$

  where A_1 and A_2 are two constants and $F_1(s)$ and $F_2(s)$ are respectively the Laplace transforms of functions $f_1(t)$ and $f_2(t)$;

Table 2.1 Laplace transform of several characteristic functions

f(*t*)	*F*(*s*)
$\delta(t)$ *impulse*	1
$u(t)$ *step*	$\frac{1}{s}$
$t^n \quad n = 1, 2, \ldots$	$\frac{n!}{s^{n+1}}$
$e^{-\omega t}$	$\frac{1}{s+\omega}$
$te^{-\omega t}$	$\frac{1}{(s+\omega)^2}$
$\cos \omega t$	$\frac{s}{s^2+\omega^2}$
$\sin \omega t$	$\frac{\omega}{s^2+\omega^2}$
$\cosh \omega t$	$\frac{s}{s^2-\omega^2}$
$\sinh \omega t$	$\frac{\omega}{s^2-\omega^2}$
$1-e^{-\alpha t}$	$\frac{\omega}{s(s+\omega)}$
$1-\cos \omega t$	$\frac{\omega^2}{s(s^2+\omega^2)}$
$\omega t - \sin \omega t$	$\frac{\omega^3}{s^2(s^2+\omega^2)}$
$\omega t \cos \omega t$	$\frac{\omega(s^2-\omega^2)}{(s^2+\omega^2)^2}$
$\omega t \sin \omega t$	$\frac{2\omega^2 s}{(s^2+\omega^2)^2}$
$\frac{1}{(1-\xi^2)^{1/2}\omega} e^{-\xi\omega t} \sin(1-\xi^2)^{1/2}\omega t$	$\frac{1}{s^2+2\xi\omega s+\omega^2}$
$e^{-\xi\omega t}\left[\cos(1-\xi^2)^{1/2}\omega t + \frac{\xi}{(1-\xi^2)^{1/2}} \sin(1-\xi^2)^{1/2}\omega t\right]$	$\frac{s+2\xi\omega}{s^2+2\xi\omega s+\omega^2}$

- the Laplace transform of the derivative of order r with respect to the time of a function $f(t)$ is given by:

$$L\left[\frac{d^r f(t)}{dt^r}\right] = s^r F(s) - s^{r-1} f(0) - s^{r-2} \frac{d^r f(0)}{dt^r} - \ldots - \frac{d^{r-1} f(0)}{dt^{r-1}} \tag{2.304}$$

where $f(0)$ and $\frac{d^r f(0)}{dt^r}$ respectively represent function $f(t)$ and the rth derivative of $f(t)$ calculated at time $t = 0$: if, for example, we assume that these values are null for all the initial conditions, then expression (2.304) simply becomes:

$$L\left[\frac{d^r f(t)}{dt^r}\right] = s^r F(s) \tag{2.305}$$

i.e., the derivative operation corresponds, in the Laplace domain, to a multiplication for variable s.

- the inverse transform, that allows us to perform an inverse step, is given by [15, 26, 30]:

$$L^{-1}[F(s)] = f(t) = \frac{1}{2\pi i} \int_{c-i\infty}^{c+i\infty} F(s)e^{st} ds \tag{2.306}$$

where $i = \sqrt{-1}$ represents the imaginary unit and c is a real constant, termed abscissa of convergence, greater than the real part of all the singular points of $F(s)$ [26, part IV]. Determination of $f(t)$, starting from $F(s)$ by means of (2.306), usually proves to be extremely laborious: in the event of it not being possible to perform an inverse Laplace transform, by taking advantage of Table 2.1 (which, as an example, shows the Laplace transforms and the inverse transforms of several characteristic functions) we commonly resort to other methods such as, for example, the partial fractions method.

Let us now consider the equation of motion of a 1 d.o.f. linear vibrating system:

$$m\ddot{x} + r\dot{x} + kx = f(t) \tag{2.307}$$

where, according to this approach $x = x(t)$ and can be seen as an output variable and $f = f(t)$ is considered as an input variable. By keeping account of properties (2.302) and (2.303) and assuming null initial conditions, the transform of this equation becomes:

$$L[m\ddot{x} + r\dot{x} + kx] = L[f(t)] \tag{2.308}$$

$$L[m\ddot{x}] + L[r\dot{x}] + L[kx] = L[f(t)] \tag{2.309}$$

$$ms^2X(s) + rsX(s) + kX(s) = F(s) \tag{2.310}$$

where $X(s)$ and $F(s)$ are the Laplace transform of the system's response (output) and of the excitation force (input): as can be noted, the transformation has allowed for the transition from a linear differential equation (2.307) to an algebraic equation in variable s (2.308) which can be rewritten as:

$$X(s) = H(s)F(s) \tag{2.311}$$

in which $H(s)$ is the transfer function:

$$H(s) = \frac{1}{ms^2 + rs + k} \tag{2.312}$$

which enables us to obtain transform $X(s)$ of the response (in this case steady-state) of the system to excitation force $f(t)$, simply by multiplying transform $F(s)$ of the excitation force by $H(s)$. In order to then have response $x(t)$ in the time domain, it is necessary to perform the inverse transform of $X(s)$ (see Table 2.1).

As can be noted from (2.312), the transfer function depends univocally on the parameters of the system: it remains unvaried regardless of the variations over time of the input amplitude, i.e. the force. More generally speaking, for a linear n d.o.f. system, expression (2.312) can be expressed [24, 26] as the ratio between two polynomials in s, with numerator of degree m and denominator of degree n (where, for physical systems $n > m$):

$$H(s) = \frac{X(s)}{F(s)} = K\frac{(s - z_1)(s - z_2) \cdot (s - z_m)}{(s - p_1)(s - p_2) \cdot (s - p_n)} \tag{2.313}$$

where $z_1, z_2, \ldots, z_m$ are the zeros of transfer function $H(s)$ and $p_1, p_2, \ldots, p_n$ define the poles.

In the case analysed, i.e. a 1.d.o.f. linear vibrating system, formulation (2.313) of $H(s)$ is given by:

$$H(s) = \frac{1}{ms^2 + rs + k} = \frac{1}{(s - p_1)(s - p_2)} \tag{2.314}$$

Poles p_1 and p_2 therefore correspond to solutions λ_i $(i = 1, 2)$ of the characteristic equation (see Sect. 2.4.1.2, (2.205):

$$p_1 = -\alpha + i\omega \tag{2.315}$$

$$p_2 = -\alpha - i\omega \tag{2.316}$$

As an example, we will now analyse the case of an impulsive excitation force, i.e. [31, 35] of a force f_{imp} applied in a time $\varepsilon \rightarrow 0$ so that:

$$\int_0^e f_{imp}(t)dt = 1 \tag{2.317}$$

By referring to Table 2.1, the Laplace transform of this quantity proves to be:

$$L[f_{imp}(t)] = F_{imp}(s) = 1 \tag{2.318}$$

and by keeping account of (2.311) and (2.312) the response in the Laplace domain is thus simply equal to the transfer function:

$$X_{imp}(s) = H(s)F_{imp}(s) = H(s) = \frac{1}{ms^2 + rs + k} \tag{2.319}$$

By performing the inverse transform (2.319) [see Table 2.1 penultimate function and Eq. (2.317)] and by recalling that $\alpha = r/(2m)$, we obtain the response of the system defined in the time domain:

$$x(t) = L^{-1}[X_{imp}(s)] = \frac{1}{\omega m} e^{-\alpha t} \sin(\omega t) \tag{2.320}$$

It is possible to demonstrate [10, 16, 24, 26, 31] that the harmonic transfer function $H(\Omega)$ of a linear system (Sect. 2.4.1.4) is nothing else but the transfer function, the Laplace transform of the response to impulse $X_{imp}(s)$, by assuming $s = i\Omega$ as a Laplace variable, i.e.:

$$H(\Omega) = H(s)_{s=i\Omega} = X_{imp}(s) = \frac{1}{-m\Omega^2 + i\Omega r + k} \tag{2.321}$$

$$= \frac{1}{m(-\Omega^2 + i2h\omega_0\Omega + \omega_0^2)} = \frac{1/k}{\left(1 - \frac{\Omega^2}{\omega_0^2}\right) - i\frac{\Omega r}{k}} = \frac{1/k}{(1 - a^2) - i2ha} \tag{2.322}$$

2.4.1.5 Solution of Generic Motion

The complete solution of the equation of motion (2.175) of a vibrating system is given by the sum:

- of the general integral x_G of the associated homogenous (Sect. 2.4.1.2), which describes the free motion;
- of the particular integral x_P (Sect. 2.4.1.4) due to excitation force $f(t)$ acting on the system, which describes the steady-state motion.

If the system [always see expression (2.175)] has an equivalent viscous damping of constant $r > 0$, the general integral of the associated homogenous tends to zero as time t elapses, meaning that, in steady-state conditions, the solution is practically given by the particular integral x_P only. In any case, if the damping is greater than the critical damping, the general integral of the associated homogenous is given by two decreasing exponential functions (2.217):

$$x_G(t) = X_1 e^{-\alpha_1 t} + X_2 e^{-\alpha_2 t} \tag{2.323}$$

while, if the damping is lower than the critical damping, integral x_G describes an oscillation with a decreasing amplitude (2.222):

$$x_G(t) = e^{-\alpha t}(A \cos \omega t + B \sin \omega t) \tag{2.324}$$

Let us consider Eq. (2.175) in the case of $r > 0$ but lower than the critical damping and forced by a harmonic excitation force with frequency Ω:

$$m\ddot{x} + r\dot{x} + kx = f(t) = F_0 \cos(\Omega t) \tag{2.325}$$

The complete solution is the sum of the general integral which, in this case, describes an oscillation with a frequency equal to natural frequency ω of the damped system ($h = r/r_c < 1$), (2.324), and of the particular integral which, in the case of the harmonic excitation force, has a frequency Ω equal to the frequency of the excitation force [Sect. 2.4.1.4, (2.247)]:

$$x(t) = x_G(t) + x_P(t) = e^{-\alpha t}(A \cos \omega t + B \cos \omega t) + |X_P| \cos(\Omega t + \psi) \tag{2.326}$$

Let us now analyse the case in which the damping constant is negative ($r < 0$): this situation can arise in mechanical systems subjected to force fields (as we will see in Chap. 5), while it does not pertain in the dissipative systems analysed in this chapter. Under these conditions, the general integral defines an expansive solution of the type (2.222):

$$x_G(t) = e^{\alpha t}(A \cos \omega t + B \sin \omega t) \tag{2.327}$$

which will thus be predominant with respect to the particular integral. Constants A and B are determined by imposing the initial position and velocity conditions $x(0) = x_0$ and $\dot{x}(0) = \dot{x}_0$ on the complete solution $x(t)$ (2.243).

2.4.2 Two-Degree-of-Freedom Systems

We propose to study the system of differential equations which describes the perturbed motion of a two d.o.f. system, linear or linearized about the static equilibrium position, written in the more general matrix form (using e.g. one of the methods outlined in Chap. 1, Sect. 1.5 and in Sect. 2.3.2) as:

$$[M]\underline{\ddot{x}} + [R]\underline{\dot{x}} + [K]\underline{x} = \underline{f}(t) \tag{2.328}$$

By following the same outline already seen for one d.o.f. systems, we start analysing, first and foremost, the free undamped and damped free motion, (Sects. 2.4.2.1 and 2.4.2.2), and, finally, the forced motion (Sect. 2.3.2.3).

2.4.2.1 Undamped Free Motion

2.4.2.1.1 Calculating Natural Frequencies and Vibration Modes

In the absence of damping and applied external excitation forces, the vibrating system becomes conservative and (2.328) is reduced to:

$$[M]\underline{\ddot{x}} + [K]\underline{x} = \underline{0} \tag{2.329}$$

If, as in the cases analysed until now, matrices $[M]$ and $[K]$ are symmetrical and positive definite (a condition always verified for a disturbed motion of a conservative system about the stable static equilibrium position), solution [2, 20] is of the type[9]:

$$\overline{x} = \overline{\underline{X}}e^{i\omega t} = \begin{Bmatrix} X_1 \\ X_2 \end{Bmatrix} e^{i\omega t} \rightarrow \underline{\dot{x}} = i\omega\overline{\underline{X}}e^{i\omega t} \rightarrow \underline{\ddot{x}} = -\omega^2\overline{\underline{X}}e^{i\omega t} \tag{2.330}$$

where $\overline{\underline{X}}$ is generally complex or rather:

$$\begin{aligned} \overline{X}_1 &= |\overline{X}_1|e^{i\psi_1} \\ \overline{X}_2 &= |\overline{X}_2|e^{i\psi_2} \end{aligned} \tag{2.331}$$

By substituting (2.239) in the differential equation (2.330) and simplifying term $e^{i\omega t}$, always different from zero, we obtain:

$$\left[-\omega^2[M] + [K]\right]\overline{\underline{X}} = \underline{0} \tag{2.332}$$

which represents a homogenous linear algebraic system in the unknown $\overline{\underline{X}}$ (i.e. in the two scalar unknowns $\overline{X}_1$ and $\overline{X}_2$). As is known, this system will accept a different solution from the trivial one if and only if the determinant of the coefficient matrix is equal to zero [2, 20], i.e.:

$$det\left|-\omega^2[M] + [K]\right| = 0 \tag{2.333}$$

By keeping account of the general expression of the mass $[M]$ and stiffness $[K]$ matrices for a generic two d.o.f. system, (2.333) can be rewritten as:

$$\Delta(\omega^2) = det\left|\begin{pmatrix} (-\omega^2 m_{11} + k_{11}) & (-\omega^2 m_{12} + k_{12}) \\ (-\omega^2 m_{21} + k_{21}) & (-\omega^2 m_{22} + k_{22}) \end{pmatrix}\right| = 0 \tag{2.334}$$

[9]The general solution of (2.325) has the form $\underline{x} = \overline{\underline{X}}e^{\lambda t}$, but, since $[K]$ and $[M]$ are symmetric and positive definite, roots λ_i are always imaginary [2, 21] i.e. $\lambda_i = \pm i\omega_i$.

i.e. the determinant:

$$\Delta(\omega^2) = \omega^4(m_{11}m_{22} - m_{12}m_{21}) - \omega^2(m_{11}k_{22} + k_{11}m_{22} - m_{12}k_{21} - k_{12}m_{21}) \\ + (k_{11}k_{22} - k_{12}k_{21}) = 0 \quad (2.334a)$$

thus proves to be a polynomial in ω of order 4 (generally speaking, for a system with n d.o.f. of the 2n order) in ω^2 of the second order of the type:

$$\Delta(\omega^2) = \omega^4 a + \omega^2 b + c = 0 \quad (2.335)$$

where coefficients a, b and c [by comparing (2.334) with (2.335)] are functions of physical parameters of the system, represented by the elements of the mass and stiffness matrices. The (2.335) represents a biquadratic equation and its solution is therefore of the type:

$$\omega_{I,II}^2 = \frac{-b \pm \sqrt{b^2 - 4ac}}{2a} \quad (2.336)$$

with ω_I^2 and ω_{II}^2 always real and positive. There are four values of ω for which the determinant $\Delta(\omega^2)$ is zero, (2.335) in correspondence to which the algebraic system has a non-trivial solutions:

$$\omega_{1,3} = \pm\sqrt{\omega_I^2} \quad (2.337)$$

$$\omega_{2,4} = \pm\sqrt{\omega_{II}^2} \quad (2.338)$$

By subsequently substituting in (2.335) the four solutions we obtain:

$$\left[-\omega_I^2[M] + [K]\right]\underline{X}^{(1)} = 0 \rightarrow \left[-\omega_I^2[M] + [K]\right]\underline{X}^{(3)} = \underline{0} \quad (2.339)$$

$$\left[-\omega_{II}^2[M] + [K]\right]\underline{X}^{(2)} = 0 \rightarrow \left[-\omega_{II}^2[M] + [K]\right]\underline{X}^{(4)} = \underline{0} \quad (2.340)$$

Based on (2.333), the matrices of coefficients $-\omega_I^2[M] + [K]$ and $-\omega_{II}^2[M] + [K]$ prove to be singular: this condition corresponds to the fact that one of the two equations, both in (2.339), and (2.340), is a linear combination of the other.[10] For this reason, the generic eigenvector $\underline{X}^{(1)}$, $\underline{X}^{(3)}$, $\underline{X}^{(2)}$, $\underline{X}^{(4)}$ (calculated respectively

[10]Actually, for an n d.o.f system the number of rows presenting linear combination with others is equal to the number of coincident solutions of the generic ω_i; in case of a single solution, the row linear combination of the others is the one that, once eliminated, a minor with denominators different from zero are determined [2, 21]. As far as multiple roots ω_i are concerned, see [29, 31].

for $\omega = \omega_1, \omega = \omega_3, \omega = \omega_2$ *and* $\omega = \omega_4$) can be defined, up to a constant, by using only one of the equations of (2.339) and (2.340), for example, the first:

$$\begin{aligned} (-\omega_I^2 m_{11} + k_{11})\overline{X}_1^{(1)} + (-\omega_I^2 m_{12} + k_{12})\overline{X}_2^{(1)} &= 0 \\ (-\omega_{II}^2 m_{11} + k_{11})\overline{X}_1^{(2)} + (-\omega_{II}^2 m_{12} + k_{12})\overline{X}_2^{(2)} &= 0 \end{aligned} \tag{2.341}$$

The solution to the problem therefore univocally defines the relationship between the two values of the solution (2.339), a relationship which can be obtained immediately from (2.340) and (2.341):

$$\begin{aligned} \mu_1 &= \frac{\overline{X}_1^{(1)}}{\overline{X}_2^{(1)}} = \frac{|\overline{X}_1^{(1)}|e^{i\psi_1^{(1)}}}{|\overline{X}_2^{(1)}|e^{i\psi_2^{(1)}}} = -\frac{(-\omega_I^2 m_{12} + k_{12})}{(-\omega_I^2 m_{11} + k_{11})} \\ \mu_2 &= \frac{\overline{X}_1^{(2)}}{\overline{X}_2^{(2)}} = \frac{|\overline{X}_1^{(2)}|e^{i\psi_1^{(2)}}}{|\overline{X}_2^{(2)}|e^{i\psi_2^{(2)}}} = -\frac{(-\omega_{II}^2 m_{12} + k_{12})}{(-\omega_{II}^2 m_{11} + k_{11})} \end{aligned} \tag{2.342}$$

where:

$$\begin{aligned} \frac{\overline{X}_1^{(3)}}{\overline{X}_2^{(3)}} &= \frac{|\overline{X}_1^{(3)}|e^{i\psi_1^{(3)}}}{|\overline{X}_2^{(3)}|e^{i\psi_2^{(3)}}} = \mu_1 \\ \frac{\overline{X}_1^{(4)}}{\overline{X}_2^{(4)}} &= \frac{|\overline{X}_1^{(4)}|e^{i\psi_1^{(4)}}}{|\overline{X}_2^{(4)}|e^{i\psi_2^{(4)}}} = \mu_2 \end{aligned} \tag{2.343}$$

By analysing (2.343) it is possible to note how these characteristic relationships are real, thus enabling us to obtain:

$$\begin{aligned} \psi_1^{(1)} &= \psi_2^{(1)} = \psi^{(1)} \\ \psi_1^{(2)} &= \psi_2^{(2)} = \psi^{(2)} \\ \psi_1^{(3)} &= \psi_2^{(3)} = \psi^{(3)} \\ \psi_1^{(4)} &= \psi_2^{(4)} = \psi^{(4)} \end{aligned} \tag{2.344}$$

In the following Sect. 2.4.2.1.2, we will show, in greater detail, how a pair of counter-rotating vectors in the complex plane corresponds to a pair of complex conjugate solutions $\pm\omega_i$, which as noted for a 1 d.o.f. system, Sect. 2.4.1.1, can be reduced to a projection on the real axis of a single rotating vector having an angular speed of ω_i. In this way, in a two d.o.f. system, we find ourselves faced with only two solutions: ω_1 and ω_2. The two relationships μ_1 and μ_2 obtained in this way, or the two pairs of solutions $(\overline{X}_1^{(1)}; \overline{X}_2^{(1)})$ or $(\overline{X}_1^{(2)}; \overline{X}_2^{(2)})$ defined up to a constant, define the deformed shape of the two d.o.f. system, in the event of this vibrating respectively with frequency ω_1 or ω_2, i.e. with one of its natural frequencies. The two deformed shapes of the system are termed vibration modes of the undamped system or rather principal vibration modes. In a vibrating system with n d.o.f. we will obviously have n vibration modes, i.e. a vibration mode for each natural frequency, each defined up to a constant, since only the relationships between an independent coordinate and the remaining $(n-1)$ are univocally fixed. Generally speaking, the eigenvectors $\underline{\overline{X}}^{(i)}$ [Eqs. (2.330), (2.342)] are numerically evaluated by assigning a real arbitrary value to one of the coordinates $\overline{X}_j^{(i)}$ and obtaining the other from the characteristic relationship μ_i (2.342): this operation is termed normalization of the eigenvector (by means of this normalization, as a consequence, the generic eigenvector $\underline{\overline{X}}^{(i)}$ proves to be real). In the case of the system with two d.o.f. analysed in this section, in the hypothesis, for example, of normalizing the generic eigenvector $\underline{\overline{X}}^{(i)}$ by assuming $\overline{X}_2^{(i)} = 1$, these, by considering (2.342), become:

$$\underline{\overline{X}}^{(1)} = \begin{Bmatrix} \mu_1 \\ 1 \end{Bmatrix} \qquad \underline{\overline{X}}^{(2)} = \begin{Bmatrix} \mu_2 \\ 1 \end{Bmatrix} \tag{2.345}$$

From a purely didactic point of view, it could be useful to solve the problem of the free motion of an undamped system in scalar form: though easy when operating on a system having only 2 d.o.f., it becomes more laborious in the case of a n d.o.f. system. In the absence of external excitation forces and damping, the system of differential equations that describe the motion of the generic two d.o.f. system (2.175) becomes:

$$\begin{aligned} m_{11}\ddot{x}_1 + m_{12}\ddot{x}_2 + k_{11}x_1 + k_{12}x_2 = 0 \\ m_{21}\ddot{x}_1 + m_{22}\ddot{x}_2 + k_{21}x_1 + k_{22}x_2 = 0 \end{aligned} \tag{2.346}$$

Based on what has gone before, by imposing the harmonic solution:

$$\begin{aligned} x_1 = \overline{X}_1 e^{i\omega t} \\ x_2 = \overline{X}_2 e^{i\omega t} \end{aligned} \tag{2.347}$$

we obtain a homogenous linear algebraic system, parametric in ω^2, with unknowns $\overline{X}_1$ and $\overline{X}_2$, corresponding to the matrix form (2.332):

$$\begin{aligned}(-\omega^2 m_{11} + k_{11})\overline{X}_1 + (-\omega^2 m_{12} + k_{12})\overline{X}_2 = 0 \\ (-\omega^2 m_{21} + k_{21})\overline{X}_1 + (-\omega^2 m_{22} + k_{22})\overline{X}_2 = 0\end{aligned} \tag{2.348}$$

The annulment condition of the determinant of the matrix of coefficients obviously leads to (2.334a).

An alternative approach to the annulment method of the determinant is to consider the problem of the evaluation of the natural frequencies ω_i and relative vibration modes $\underline{\overline{X}}^{(i)}$ as an eigenvalues-eigenvectors problem. It is important to remember that we define eigenvalue λ of a generic matrix $[A]$ as that value which, when subtracted from the principal diagonal of same, annuls the determinant, or rather [2, 20]:

$$[-\lambda[I] + [A]]\underline{V} = \underline{0} \rightarrow det|[A] - \lambda[I]| = 0 \tag{2.349}$$

and that there exist as many eigenvalues as the rank of the square matrix $[A]$. After imposing a solution of the type (2.329), we have seen that the equations of motion of the free system (2.330), have been attributed (2.332) to a system of parametric homogenous algebraic linear equations in ω of the type:

$$\left[-\omega^2[M] + [K]\right]\underline{\overline{X}} = \underline{0} \tag{2.350}$$

By pre-multiplying the matrix of coefficients by $[M]^{-1}$, which undoubtedly exists because hypothetically $[M]$ is symmetric and positive definite (and therefore not singular), we obtain:

$$\left[-\omega^2[M]^{-1}[M] + [M]^{-1}[K]\right]\underline{\overline{X}} = \underline{0} \rightarrow \left[-\omega^2[I] + [M]^{-1}[K]\right]\underline{\overline{X}} = \underline{0} \tag{2.351}$$

where $[M]^{-1}[M] = [I]$ is the identity matrix. It is thus possible to attribute the calculation of the natural frequencies ω_i of the free undamped vibrating system (2.350) to the calculation of the eigenvalues λ_i of matrix $[M]^{-1}[K]$, the two amplitudes ω_i and λ_i are linked by the relationship:

$$\lambda_i = \omega_i^2 \tag{2.352}$$

and the vibration modes that coincide with the eigenvectors of the same matrix $[M]^{-1}[K]$:

$$\underline{\overline{X}}^{(i)} = \underline{V}^{(i)} \tag{2.353}$$

This factor is extremely convenient because subprograms are available for the numerical calculation of the eigenvalues of a generic matrix. Therefore, usually, once matrix $[M]^{-1}[K]$ has been calculated, calculation of the ω_i can generally be performed without having to develop the determinant or calculate the solution of the characteristic algebraic equation obtained: one of the methods used is to diagonalize $[M]^{-1}[K]$ because, by definition, the eigenvalues of a diagonal matrix are the diagonal elements themselves.

2.4.2.1.2 Calculation of the Response to Initial Conditions

Having used the methods described above to define natural frequencies ω_i and the main vibration modes $\underline{\overline{X}}^{(i)}$, it is now possible to define the free motion of the system, in the absence of damping, or rather its response to an initial assigned disturbance. This motion is defined analytically by the general integral [2, 37] of the equation of motion (2.346) of the vibrating system: in the hypothetical case of distinct eigenvalues, this general integral is given by the linear combination of the particular integrals (2.348) of the homogenous itself:

$$\underline{x}(t) = \sum_{i=1}^{n=4} a_i \underline{\overline{X}}^{(i)} e^{\lambda_i t} \tag{2.354}$$

i.e. in scalar form:

$$\begin{aligned} x_1(t) &= a_1 X_1^{(1)} e^{i\omega_1 t} + a_2 X_1^{(2)} e^{i\omega_2 t} + a_3 X_1^{(3)} e^{-i\omega_1 t} + a_4 X_1^{(4)} e^{-i\omega_2 t} \\ x_2(t) &= a_1 X_2^{(1)} e^{i\omega_1 t} + a_2 X_2^{(2)} e^{i\omega_2 t} + a_3 X_2^{(3)} e^{-i\omega_1 t} + a_4 X_2^{(4)} e^{-i\omega_2 t} \end{aligned} \tag{2.355}$$

Using similar steps to those implemented for one d.o.f. systems [Sect. 2.4.1, Eq. (2.311)] the (2.355) can be rewritten as:

$$\begin{aligned} x_1(t) &= \alpha \overline{X}_1^{(1)} \cos(\omega_1 t + \psi^{(1)}) + \beta \overline{X}_1^{(2)} \cos(\omega_2 t + \psi^{(2)}) \\ x_2(t) &= \alpha \overline{X}_2^{(1)} \cos(\omega_1 t + \psi^{(1)}) + \beta \overline{X}_2^{(2)} \cos(\omega_2 t + \psi^{(2)}) \end{aligned} \tag{2.356}$$

In order to define these expressions numerically, it is necessary to evaluate the 2 unknowns α and β and the two phases $\psi^{(1)}$ and $\psi^{(2)}$, since the two natural frequencies ω_1 and ω_2 are calculated by means of Eq. (2.350) and because the modes are normalized as shown in (2.345). To evaluate the four unknowns, it is necessary to impose the four initial conditions:

$$\begin{aligned} x_1(t)_{t=0} &= x_{10} \quad \dot{x}_1(t)_{t=0} = \dot{x}_{10} \\ x_2(t)_{t=0} &= x_{20} \quad \dot{x}_2(t)_{t=0} = \dot{x}_{20} \end{aligned} \tag{2.357}$$

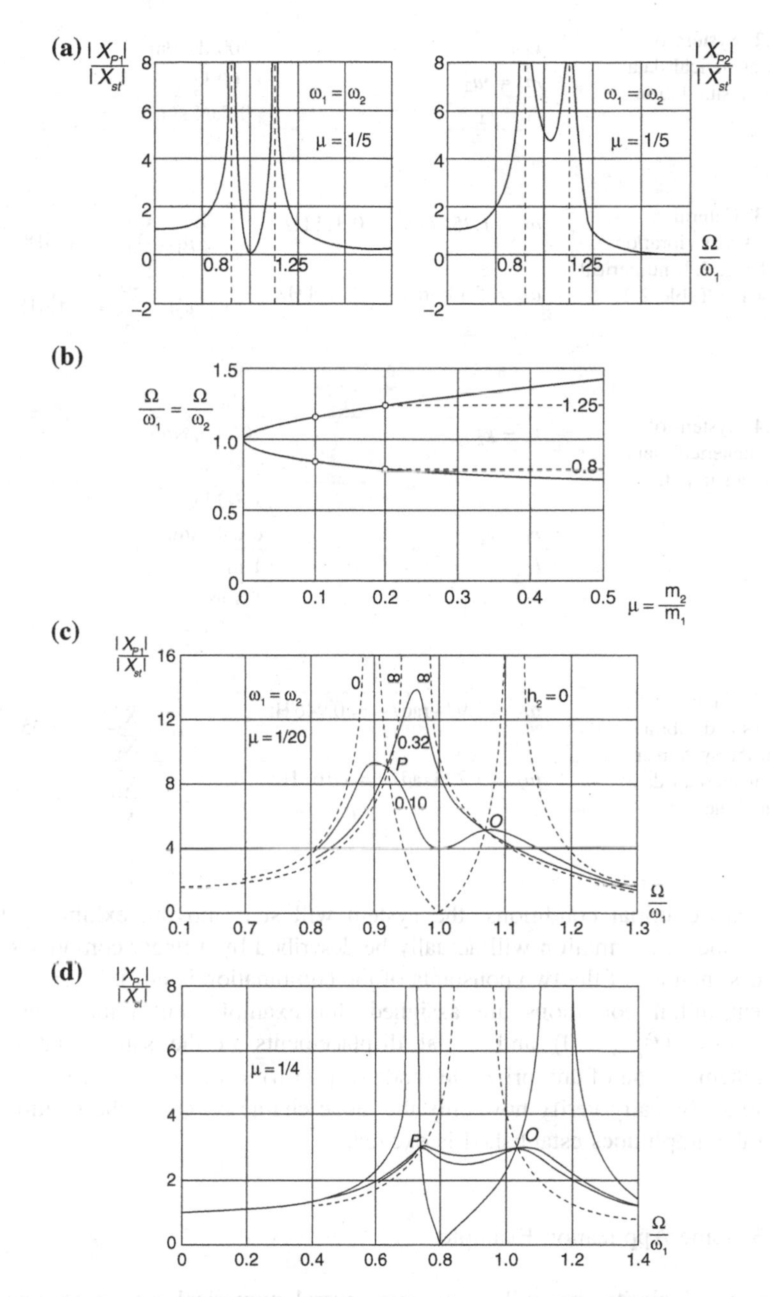

Fig. 2.46 **a** Dynamic absorber: in-frequency response (μ = 1/5). **b** Dynamic vibration absorber: complete frequency system of Fig. 2.41 (with $\omega_1 = \omega_2$) as a function of the relationship of masses μ. **c** Dynamic vibration absorber: in-frequency response of mass m_1 (with $\omega_1 = \omega_2 = \Omega$) versus non-dimensional damping value h_2. **d** Dynamic vibration absorber: optimal in-frequency response of mass m_1 (with $\omega_1 = \omega_2 = \Omega$ and optimal non-dimensional damping h_2)

Table 2.2 System of Fig. 2.4, numerical data relative to a simulation

$k_1 = k_2$	100 daN/m
$m_1 = m_2$	100 kg
$r_1 = r_2$	0 daN s/m

Table 2.3 Natural frequencies and vibration modes of Fig. 2.4, numerical data shown in Table 2.2

$\omega_1 = 1.954\,\text{rad/s} = 0.311\,\text{Hz}$	$\mu_1 = \frac{\overline{X}_1^{(1)}}{\overline{X}_2^{(1)}} = +0.618$
$\omega_2 = 5.116\,\text{rad/s} = 0.814\,\text{Hz}$	$\mu_2 = \frac{\overline{X}_1^{(2)}}{\overline{X}_2^{(2)}} = -1.618$

Table 2.4 System of Fig. 2.7, numerical data relative to a simulation

$k_1 = k_2$	2000 daN/m
m	950 kg
J_G	1400 kg m^2
$r_1 = r_2$	0 daN s/m
l_1	1 m
l_2	1.5 m

Table 2.5 Natural frequencies and vibration modes of the system in Fig. 2.7, numerical data shown in Table 2.4

$\omega_1 = 5.944\,\text{rad/s} = 0.946\,\text{Hz}$	$\mu_1 = \frac{\overline{X}_1^{(1)}}{\overline{X}_2^{(1)}} = +1.55$
$\omega_2 = 7.294\,\text{rad/s} = 1.161\,\text{Hz}$	$\mu_2 = \frac{\overline{X}_1^{(2)}}{\overline{X}_2^{(2)}} = -0.948$

For generic initial conditions, the system will start moving exhibiting both vibration modes, i.e. motion will actually be described by a linear combination of two modes (neither of the two constants of the combination is null). If, vice versa, convenient initial conditions are assigned (for example null initial velocities $(\dot{x}_1(t)_{t=0} = 0, \dot{x}_2(t)_{t=0} = 0)$ and initial displacements in the same relationship between them as one of any principle mode $((x_{10}/x_{20}) = \mu_1$ or $(x_{10}/x_{20}) = \mu_2))$, it is easy to analytically verify how only the mode characterized by the relationship between the amplitudes established is excited.

2.4.2.1.3 Some Application Examples

For reasons of clarity, we will now give several numerical results obtained in particular cases. Let us consider the vibrating linear system in Fig. 2.46, whose equations of motion are shown in (2.85), with the numerical values outlined in Table 2.2. Figure 2.51 shows the trend of determinant $D(\omega)$ (2.334) as a function of ω: as can be noted, $D(\omega)$ becomes zero in correspondence to the two values ω_1 and

Table 2.6 System of Fig. 2.7, numerical data relative to a simulation

$k_1 = k_2$	2000 daN/m
M	950 kg
J_G	1400 kg m^2
$r_1 = r_2$	0 daN s/m
$l_1 = l_2$	1.5 m

Table 2.7 Natural frequencies and vibration modes of the system shown in Fig. 2.7, case of symmetry: numerical data shown in Table 2.6

$\omega_1 = 5.944\,\mathrm{rad/s}$	$\mu_1 = \frac{\overline{X}_1^{(1)}}{\overline{X}_2^{(1)}} = 1.55$
$\omega_2 = 7.294\,\mathrm{rad/s}$	$\mu_2 = \frac{\overline{X}_1^{(2)}}{\overline{X}_2^{(2)}} = -0.948$

ω_2 corresponding to the two natural frequencies of the system, shown in Table 2.3 with their relative vibration modes represented, for the sake of convenience, by assigning an arbitrary value to one of the two variables, $\overline{X}_2^{(1)} = 1$ and $\overline{X}_2^{(2)} = 1$ and thus defining the other which, as a consequence, is $\overline{X}_1^{(1)} = \mu_1$ and $\overline{X}_1^{(2)} = \mu_2$. Figure 2.40 graphically shows these deformed shapes: the first vibration mode corresponds to an in-phase motion of the two masses (relationship μ_1 is positive), while in the second vibration mode the two masses oscillate in counter-phase (μ_2 negative). As can be seen further on, recognition of the type of modal deformed shape is important in the event of one wishing to intervene and modify the natural frequencies of the system. Always as an example, in Table 2.5, we show the natural frequencies and relative vibration modes of the system shown in Fig. 2.4 (2.87), for the particular numeric values shown in Table 2.4.

In this example, the generic vibration mode defined by relation:

$$\bar{X}_1^{(i)} = \mu_i \bar{X}_2^{(i)} \quad (i = 1, 2) \tag{2.358}$$

represents the rotation of a rigid body around a point P_i lying at a distance of μ_i from centre of gravity G of the body itself, see Fig. 2.41 where, if we consider small displacements, rotation $\bar{X}_2^{(i)}$ around point P_i causes a displacement $\bar{X}_1^{(i)}$ of centre of gravity G equal to $\bar{X}_1^{(i)} = \mu_i \bar{X}_2^{(i)}$. Thus, in the generic mode it is as if there were an ideal hinge in point P_i. Generally speaking, the two points P_i are positioned on opposite ends with respect to centre of gravity G. Obviously, in the event of us assigning null velocities and a configuration that satisfies the generic characteristic relationship μ_i to the system as initial conditions, the system will rotate around generic point P_i, or rather the ideal generic hinge: under these conditions, as has been seen, only one vibration mode is excited and the system thus behaves as if it only had one d.o.f. Furthermore, by analysing the modes of the system, it is possible to observe how both vertical motion and rotating motion are coupled in them.

Among other things, always by analysing the vibrating system in Fig. 2.7, it is possible to observe how, when the system is symmetric with respect to the gravity centre (Table 2.6), the stiffness $[K]$ and mass $[M]$ matrices prove to be diagonal (2.90): in this case, the two d.o.f. are uncoupled or rather the extra diagonal terms, which represent the forces that one d.o.f. exercises on the other, are null: the result is that the two vibration modes (Table 2.7) prove to be uncoupled, i.e. pure vertical motion (position of P_1 to the infinite) and pure torsional motion ($P_2 = G$) exist. By generalizing the question regarding a n d.o.f. system, as will be described in-depth further on (Sect. 2.4.3), there is always an n-set of independent coordinates, termed principal coordinates, meaning that the n differential equations are always uncoupled (diagonal matrices).

2.4.2.2 Damped Free Motion

2.4.2.2.1 Calculating Natural Frequencies and Vibration Modes

In reality, all vibrating systems have their own damping, in other words, during vibration, energy is dissipated in different forms. It is thus necessary to keep account of the presence of this damping by referring, as previously mentioned, to a viscous damping (i.e. proportionate to the vibration velocity). In the absence of excitation forces, the equations of motion (2.328), can be traced back to:

$$[M]\underline{\ddot{x}} + [R]\underline{\dot{x}} + [K]\underline{x} = \underline{0} \tag{2.359}$$

The solution of (2.359) is of the type:

$$\underline{x} = \underline{\overline{X}}e^{\lambda t} = \begin{Bmatrix} \overline{X}_1 \\ \overline{X}_2 \end{Bmatrix} e^{\lambda t} \tag{2.360}$$

where, generally speaking, both the generic eigenvalues λ_i, and the generic vibration mode $\underline{X}^{(i)}$ are complex:

$$\begin{aligned} \overline{X}_1^{(i)} &= |\overline{X}_1^{(i)}|e^{i\psi_1} \\ \overline{X}_2^{(i)} &= |\overline{X}_2^{(i)}|e^{i\psi_2} \end{aligned} \tag{2.361}$$

By substituting (2.360) in (2.359) we obtain:

$$[\lambda^2[M] + \lambda[R] + [K]]\underline{\overline{X}} = 0 \tag{2.362}$$

i.e. a system of homogenous linear algebraic equations in unknowns $\underline{X}$. The system (2.362) admits non-trivial solutions if the determinant of the coefficient matrix is null, i.e. for:

$$\Delta(\lambda) = det\left|\lambda^2[M] + \lambda[R] + [K]\right| = 0 \tag{2.363}$$

which, in the case of a two d.o.f. system, Eq. (2.346), is worth:

$$\Delta(\lambda) = \begin{vmatrix} \lambda^2 m_{11} + \lambda r_{11} + k_{11} & \lambda^2 m_{12} + \lambda r_{12} + k_{12} \\ \lambda^2 m_{21} + \lambda r_{21} + k_{21} & \lambda^2 m_{22} + \lambda r_{22} + k_{22} \end{vmatrix} = 0 \tag{2.364}$$

i.e. it can be associated with a polynomial of the fourth order in λ of the type:

$$\Delta(\lambda) = a\lambda^4 + b\lambda^3 + c\lambda^2 + d\lambda + e = 0 \tag{2.365}$$

The solutions in λ of this polynomial will be two by two complex conjugates, of the type (assuming that matrices $[M]$, $[R]$ and $[K]$ are symmetrical and positive definite):

$$\begin{aligned} \lambda_{1,3} &= -\alpha_1 \pm i\omega_1 \\ \lambda_{2,4} &= -\alpha_2 \pm i\omega_2 \end{aligned} \tag{2.366}$$

With α_i and ω_i real and positive: ω_i $(i = 1, 2)$ represents the natural frequency of the damped system and α_i the corresponding exponential coefficient. To solve this polynomial, in the case of two d.o.f. systems, it is possible to resort to analytical methods. Conversely, for systems with several d.o.f., it is necessary to use numerical methods. By substituting the generic value of λ_i $(i = 1, 2, 3, 4)$ in one of the two equations of system (2.362) (since the other is a linear combination of the first) we obtain:

$$(\lambda_i^2 m_{11} + \lambda_i r_{11} + k_{11})\overline{X}_1^{(i)} + (\lambda_i^2 m_{12} + \lambda r_{12} + k_{12})\overline{X}_2^{(i)} = 0 \tag{2.367}$$

thus making it possible to define, in a way that is similar to that shown in the previous section, the characteristic relationship:

$$\mu_i = \frac{\overline{X}_1^{(i)}}{\overline{X}_2^{(i)}} = -\frac{\lambda_i^2 m_{12} + \lambda_i r_{12} + k_{12}}{\lambda_i^2 m_{11} + \lambda_i r_{11} + k_{11}} \tag{2.368}$$

which is complex because λ_i is complex. In this case, due to the presence of damping, in addition to relationship μ_i even the generic vibration mode $\underline{X}^{(i)}$ will prove to be complex and the same vibration modes will prove to be two by two complex conjugates:

$$\begin{aligned} \underline{\bar{X}}^{(3)} &= coniug\left\{\underline{\bar{X}}^{(1)}\right\} \\ \underline{\bar{X}}^{(4)} &= coniug\left\{\underline{\bar{X}}^{(2)}\right\} \end{aligned} \tag{2.369}$$

The resulting motion due to a disturbance of the initial static conditions (as will be seen in Sect. 2.4.1.3), is defined by means of a linear combination of the vibration modes considered. An alternative approach to the annulment method of determinant (2.364) is to consider the problem of the evaluation of the λ_i s and relative vibration modes $\underline{X}^{(i)}$ as an eigenvalue (eigenvectors) problem. For this reason, a dummy identity is added to the system with the equations of motion of the system (2.363) to obtain:

$$\begin{aligned} &[M]\underline{\ddot{x}} + [R]\underline{\dot{x}} + [K]\underline{x} = \underline{0} \\ &[M]\underline{\dot{x}} = [M]\underline{\dot{x}} \end{aligned} \tag{2.370}$$

Given the dummy added identity, this step does not alter the equations of the initial system and, in particular, it does not change the natural frequencies of same. The reason justifying the choice of type of identity will be given further one. By gathering the variables in a new vector $\underline{z}$ defined as:

$$\underline{z} = \begin{Bmatrix} \underline{\dot{x}} \\ \underline{x} \end{Bmatrix} \tag{2.371}$$

it is possible to rewrite expressions (2.370) as:

$$[B]\underline{\dot{z}} + [C]\underline{z} = \underline{0} \tag{2.372}$$

where:

$$\begin{aligned} [B] &= \begin{bmatrix} [M] & [0] \\ [0] & [M] \end{bmatrix} \\ [C] &= \begin{bmatrix} [R] & [K] \\ -[M] & [0] \end{bmatrix} \end{aligned} \tag{2.373}$$

Expression (2.372), in a more compact form, can be rewritten as:

$$\underline{\dot{z}} - [A]\underline{z} = \underline{0} \tag{2.374}$$

Having used $[A]$ to indicate matrix:

$$[A] = -[B]^{-1}[C] \tag{2.375}$$

This form of writing will lead us to a system of differential equations of the first order: matrix $[A]$ is generally defined as a state matrix. The solutions of expression (2.374) are of the type:

$$\underline{z} = \underline{\overline{Z}}e^{\lambda t} \tag{2.376}$$

By substituting expressions (2.376) in (2.374) we obtain:

$$[[A] - \lambda[I]]\underline{\overline{Z}} = \underline{0} \tag{2.376a}$$

Let us now recall the definition of an eigenvector of a matrix [2, 20], from (2.375) we can ascertain that there are as many λ as the rank of matrix $[A]$, i.e. $2n$, and that they coincide with the eigenvalues λ_i of matrix $[A]$ itself. The vibration modes coincide with the eigenvectors, in this case complex, of the same matrix $[A]$. In the systems with n d.o.f. analysed until now, i.e. whose matrices $[K]$, $[R]$ and $[M]$ are symmetric and defined as positive, solutions λ_i are always two by two complex conjugates with a real negative part (stable motion).

2.4.2.2.2 Calculation of the Response to Given Initial Conditions

The response of the free damped motion of a two d.o.f. system is given by the general integral, obtainable as a linear combination of the particular integrals of the homogenous equation:

$$\begin{aligned} x_1(t) &= a_1\overline{X}_1^{(1)}e^{\lambda_1 t} + a_2\overline{X}_1^{(2)}e^{\lambda_2 t} + a_3\overline{X}_1^{(3)}e^{\lambda_3 t} + a_4\overline{X}_1^{(4)}e^{\lambda_4 t} \\ x_2(t) &= a_1\overline{X}_2^{(1)}e^{\lambda_1 t} + a_2\overline{X}_2^{(2)}e^{\lambda_2 t} + a_3\overline{X}_2^{(3)}e^{\lambda_3 t} + a_4\overline{X}_2^{(4)}e^{\lambda_4 t} \end{aligned} \tag{2.377}$$

where:

$$\begin{aligned} \lambda_{1,2} &= -\alpha_1 \pm i\omega_1 \\ \lambda_{3,4} &= -\alpha_2 \pm i\omega_2 \end{aligned} \tag{2.377a}$$

by keeping account of expression (2.377a), expressions (2.377) can be rewritten as:

$$\begin{aligned} x_1(t) &= e^{-\alpha_1 t}\left(a_1\overline{X}_1^{(1)}e^{i\omega_1 t} + a_2\overline{X}_1^{(2)}e^{-i\omega_1 t}\right) + e^{-\alpha_2 t}\left(a_3\overline{X}_1^{(3)}e^{i\omega_2 t} + a_4\overline{X}_1^{(4)}e^{-i\omega_2 t}\right) \\ x_2(t) &= e^{-\alpha_1 t}\left(a_1\overline{X}_2^{(1)}e^{i\omega_1 t} + a_2\overline{X}_2^{(2)}e^{-i\omega_1 t}\right) + e^{-\alpha_2 t}\left(a_3\overline{X}_2^{(3)}e^{i\omega_2 t} + a_4\overline{X}_2^{(4)}e^{-i\omega_2 t}\right) \end{aligned} \tag{2.378}$$

This expression is similar to the previous expression (2.356) evaluated, always for a generic two d.o.f. system, in the absence of damping. As previously mentioned, complex conjugate eigenvectors correspond to couples of complex conjugate eigenvalues (2.377a): each couple of displacements corresponds to two vectors in the complex plane, counter-rotating with an angular speed ω_i and, therefore, as already noted for one d.o.f. systems, equivalent to the projection of only one rotating vector on the real axis. Using steps that are similar in all respects to those used in the previous section, it is possible to rewrite expressions (2.378) as:

$$\begin{aligned} x_1(t) &= \alpha e^{-\alpha_1 t}|\overline{X}_1^{(1)}|\cos\left(\omega_1 t + \psi^{(1)} + \Psi_{12}^{(1)}\right) + \beta e^{-\alpha_2 t}|\overline{X}_1^{(2)}|\cos\left(\omega_2 t + \psi^{(2)} + \Psi_{12}^{(2)}\right) \\ x_2(t) &= \alpha e^{-\alpha_1 t}|\overline{X}_2^{(1)}|\cos\left(\omega_1 t + \psi^{(1)}\right) + \beta e^{-\alpha_2 t}|\overline{X}_1^{(2)}|\cos\left(\omega_2 t + \psi^{(2)}\right) \end{aligned} \tag{2.379}$$

This expression is similar in all respects to (2.356) defined for undamped systems except for:

- the exponential term, which, based on the dissipative systems analysed until now, introduces a reduction of the oscillation amplitudes $e^{-\alpha_i t}$ (i = 1,2) which are superimposed on harmonic motion $\alpha_i > 0$;
- relative phase $\Psi_{12}^{(i)}$ (i = 1, 2) since, in the presence of damping, the characteristic relationship of the generic mode proves to be complex.

To numerically define the expression of the general integral (2.379), or rather to define the free damped motion of a vibrating two d.o.f. system, it is thus necessary to determine the four unknowns $\alpha, \beta, \psi^{(1)}, \psi^{(2)}$ after normalizing, for example, the vibration modes as follows:

$$\begin{aligned} \overline{X}_2^{(1)} &= 1 \Rightarrow \overline{X}_1^{(1)} = \mu_{10} e^{i\Psi_{12}^{(1)}} \Rightarrow |\overline{X}_1^{(1)}| = \mu_{10} \\ \overline{X}_2^{(2)} &= 1 \Rightarrow \overline{X}_1^{(2)} = \mu_{20} e^{i\Psi_{12}^{(2)}} \Rightarrow |\overline{X}_1^{(2)}| = \mu_{20} \end{aligned} \tag{2.380}$$

Using the two relations defined by the characteristic relationships of the modes. The 4 unknowns $\alpha, \beta, \psi^{(1)}, \psi^{(2)}$ are thus defined by the four initial conditions:

$$x_1(t)_{t=0} = x_{10} \quad \dot{x}_1(t)_{t=0} = \dot{x}_{10}$$

Table 2.8 System of Fig. 2.4, numerical data relative to a simulation

	$k_1 = k_2$	100 daN/m
	$m_1 = m_2$	100 kg
case (a)	$r_1 = r_2$	2 daN s/m
case (b)	$r_1 = r_2$	20 daN s/m
case (c)	$r_1 = r_2$	200 daN s/m

Table 2.9 Natural frequencies and vibration modes of the system shown in Fig. 2.4, numerical data shown in Table 2.8

Case a		
$\omega_1 = 1.95\,\text{rad/s} = 0.31\,\text{Hz}$	$\alpha_1 = -0.038$	$\mu_1 = \frac{\overline{X}_1^{(1)}}{\overline{X}_2^{(1)}} = 0.618e^{i0.00}$
$\omega_2 = 5.11\,\text{rad/s} = 0.81\,\text{Hz}$	$\alpha_2 = -0.261$	$\mu_2 = \frac{\overline{X}_1^{(2)}}{\overline{X}_2^{(2)}} = 1.47e^{i3.14}$
Case b		
$\omega_1 = 1.92\,\text{rad/s} = 0.305\,\text{Hz}$	$\alpha_1 = -0.382$	$\mu_1 = \frac{\overline{X}_1^{(1)}}{\overline{X}_2^{(1)}} = 0.618e^{i0.00}$
$\omega_2 = 4.39\,\text{rad/s} = 0.699\,\text{Hz}$	$\alpha_2 = -2.618$	$\mu_2 = \frac{\overline{X}_1^{(2)}}{\overline{X}_2^{(2)}} = 1.47e^{i3.14}$
Case c		
$\omega_1 = 0.00\,\text{rad/s} = 0.00\,\text{Hz}$	$\alpha_1 = -0.504$	$\mu_1 = \frac{\overline{X}_1^{(1)}}{\overline{X}_2^{(1)}} = 1.62e^{i3.14}$
$\omega_2 = 0.00\,\text{rad/s} = 0.00\,\text{Hz}$	$\alpha_2 = -0.538$	$\mu_2 = \frac{\overline{X}_1^{(2)}}{\overline{X}_2^{(2)}} = 1.47e^{i0.00}$
$\omega_1 = 0.00\,\text{rad/s} = 0.00\,\text{Hz}$	$\alpha_1 = -7.101$	$\mu_2 = \frac{\overline{X}_1^{(2)}}{\overline{X}_2^{(2)}} = 0.618e^{i0.00}$
$\omega_1 = 0.00\,\text{rad/s} = 0.00\,\text{Hz}$	$\alpha_1 = -51.856$	$\mu_2 = \frac{\overline{X}_1^{(2)}}{\overline{X}_2^{(2)}} = 1.62e^{i3.14}$

$$x_2(t)_{t=0} = x_{20} \quad \dot{x}_2(t)_{t=0} = \dot{x}_{20} \tag{2.381}$$

2.4.2.2.3 Some Application Examples

As an example, Table 2.9 shows the natural frequencies and the relative vibration modes of the system in Fig. 2.4 regarding a simulation with the data shown in Table 2.8. As can be noted, phase $\Psi_{12}^{(i)}$ of the characteristic relationship $\mu_{i0}e^{i\Psi_{12}^{(i)}}$ is always or 0 or π. More generally speaking, it is possible to show [29] that this occurs when damping matrix $[R]$ is proportionate to stiffness matrix $[K]$. When damping r_1 and r_2 increase, the real part α_i of eigenvalues $\lambda_i = -\alpha_i \pm i\omega_i$ always remains negative though its modulus increases. For high damping values (case *c*), solutions are no longer complex conjugate but purely real and negative, an indication of a stable hypercritical non-vibrating system (the eigenvalues represent modes with $(r/r_c) > 1$). For the same cases, Fig. 2.42 shows the time histories of free motion of the system analysed, i.e. the solution of the equations of damped motion (see Table 2.9b), respectively imposing:

- initial conditions such as to excite the first mode;
- initial conditions such as to excite the second mode;
- any initial condition.

2.4.2.3 Forced Motion

The equation of motion (2.328) complete with excitation term is:

$$[M]\ddot{\underline{x}} + [R]\dot{\underline{x}} + [K]\underline{x} = \underline{f}(t) \tag{2.382}$$

The complete solution of (2.382) is given by the sum of the general integral, which defines the transient to which the system is subjected (defined by the linear combination of the integrals of the homogenous equation (see previous section) and by the particular integral of the complete equation, which defines steady-state motion. The aim of this section is to evaluate the particular integral of (2.382). More specifically, we will analyse the case of the harmonic excitation force because, as previously mentioned in Sect. 2.4.1.4.1, once the response to a harmonic excitation force is known, it is possible to obtain the response to any arbitrary excitation forces, bearing in mind the Fourier series development for periodic excitation forces and the Fourier transform for any excitation forces. A more general explanation, also including random excitation forces, will be resumed in Chap. 7.

2.4.2.3.1 Forced Motion in the Absence of Damping

Let us begin by analysing the response of an undamped vibrating 2 d.o.f. system in Fig. 2.4 with two excitation forces applied respectively to mass m_1 and mass m_2: the general equation of motion in scalar form is:

$$\begin{aligned} m_{11}\ddot{x}_1 + m_{12}\ddot{x}_2 + k_{11}x_1 + k_{12}x_2 &= F_1 \cos \Omega t \\ m_{21}\ddot{x}_1 + m_{22}\ddot{x}_2 + k_{21}x_1 + k_{22}x_2 &= F_2 \cos \Omega t \end{aligned} \tag{2.383}$$

Let us assume:

$$\underline{x}_P = \begin{Bmatrix} X_{P1} \\ X_{P2} \end{Bmatrix} \cos \Omega t \tag{2.384}$$

as a particular integral. By substituting this solution in (2.383) and collecting the terms proportionate to $\cos \Omega t$, since the solution being searched for is independent from time, by equalizing the terms on the left and right of the equations proportionate to $\cos \Omega t$, we will obtain:

$$\begin{aligned} (k_{11} - m_{11}\Omega^2)X_{P1} + (k_{12} - m_{12}\Omega^2)X_{P2} &= F_1 \\ (k_{21} - m_{21}\Omega^2)X_{P1} + (k_{22} - m_{22}\Omega^2)X_{P2} &= F_2 \end{aligned} \tag{2.385}$$

By imposing solution (2.384) in the differential equations of motion we obtained a linear algebraic system with real coefficients and known non-null terms. The basic difference between this system of equations and the corresponding one for the vibrating system devoid of excitation forces (2.362) lies in the presence of a known term and the fact that in (2.385) the forcing frequency x_i is known, (for free motion, frequency ω was an unknown parameter, on account of being one of the natural frequencies of the system). The solutions of (2.385) can be determined by applying, for example Cramer's rule [2, 20] or rather:

$$
\begin{aligned}
X_{P1} &= \frac{\Delta_1}{\Delta} = \frac{\begin{vmatrix} F_1 & (k_{12} - m_{12}\Omega^2) \\ F_2 & (k_{22} - m_{22}\Omega^2) \end{vmatrix}}{\begin{vmatrix} (k_{11} - m_{11}\Omega^2) & (k_{12} - m_{12}\Omega^2) \\ (k_{21} - m_{21}\Omega^2) & (k_{22} - m_{22}\Omega^2) \end{vmatrix}} \\
X_{P2} &= \frac{\Delta_2}{\Delta} = \frac{\begin{vmatrix} (k_{11} - m_{11}\Omega^2) & F_1 \\ (k_{21} - m_{21}\Omega^2) & F_2 \end{vmatrix}}{\begin{vmatrix} (k_{11} - m_{11}\Omega^2) & (k_{12} - m_{12}\Omega^2) \\ (k_{21} - m_{21}\Omega^2) & (k_{22} - m_{22}\Omega^2) \end{vmatrix}}
\end{aligned}
\quad (2.386)
$$

where Δ is the determinant (real) of the matrix of coefficients of the associated homogeneous and Δ_i is the determinant of the matrix in which, instead of the generic i-nth column, we substituted the vector of known terms (i = 1, 2). In the presence of two generic real forces [in-phase with each other, as in the case of (2.386)] determinant Δ_i is real and therefore the same applies to solutions X_{P1} and X_{P2}. From a physical point of view, as already noted in 1 d.o.f. systems this means that the response can have:

- phases ψ_1 and ψ_2 null with respect to the excitation force (if X_{P1} or X_{P2} are positive), i.e. the vibration is in-phase with respect to the excitation force itself;
- phases ψ_1 and ψ_2 with respect to the excitation force equal to 180° (X_{P1} or X_{P2} are negative) i.e. the vibration is in counter-phase with respect to the excitation force.

In relation to the example given in Fig. 2.4 and the data shown in Table 2.2, Fig. 2.43 shows the frequency response of two d.o.f. system in terms of both modulus $|X_{P1}|$ and $|X_{P2}|$ and phase ψ_1 and ψ_2 for a single unitary modulus force F_1 applied to d.o.f. x_1: these diagrams (also termed Bode diagrams) thus represent the transfer functions $H_{11}(\Omega) = |X_{P1}|e^{i\psi_1}$ and $H_{21}(\Omega) = |X_{P2}|e^{i\psi_2}$ of the undamped system under examination where, as previously mentioned, F_1 is unitary. When frequency Ω of the excitation forces approaches one of the two natural frequencies ω_1 and ω_2 of the system, determinant Δ tends towards zero (2.386) and therefore solutions X_{P1} and X_{P2} tend towards infinite, i.e. we have two resonance conditions. More generally speaking, we can say that a 1 d.o.f. system has a resonance while a vibrating 2 d.o.f. system has two resonances and a generic n d.o.f. system has

n resonances. All the assertions and considerations made until now for an undamped 2 d.o.f. system, forced with a harmonic excitation force, can be generalized for undamped n d.o.f. systems, whose equations, as seen previously, are in matrix form[11]:

$$[M]\underline{\ddot{x}} + [K]\underline{x} = \underline{F}_0 e^{i\Omega t} \tag{2.387}$$

We impose a solution of the type:

$$\underline{x}_P = \underline{X}_P e^{i\Omega t} \tag{2.388}$$

in the complex domain and, by resorting to a matrix notation, we obtain:

$$\left[-\Omega^2[M] + [K]\right]\underline{X}_P = \underline{F}_0 \Rightarrow [D(\Omega)]\underline{X}_P = \underline{F}_0 \tag{2.389}$$

where matrix $[D(\Omega)]$ has real coefficients depending on the characteristics of the system and on frequency Ω of the excitation force. Having named the matrix obtained from $[D]$ with $[D_i]$ by substituting the i-nth column with the column of known terms $\underline{F}_0$, it is possible to once again use Cramer to solve the system (2.389) by means of:

$$X_{Pi} = \frac{|D_i|}{|D|} \quad (i = 1, 2, \ldots, n) \tag{2.390}$$

For Ω tending to generic ω_i, the denominator of (2.390) tends to zero and therefore the corresponding solutions X_{Pi} tend to infinite. In resonance conditions, expression (2.390), just as the similar scalar expression (2.386), are not valid any longer, because determinant $[D]$ goes to zero. A more efficient method of solving (2.389) is to pre-multiply the terms of same by $[D(\Omega)]^{-1}$ to obtain:

$$[D(\Omega)]^{-1}[D(\Omega)]\underline{X}_P = [D(\Omega)]^{-1}\underline{F}_0 \Rightarrow \underline{X}_P = [D(\Omega)]^{-1}\underline{F}_0 \tag{2.391}$$

i.e. the particular integral solution in a direct form (obviously the inversion of matrix $[D(\Omega)]$ is only possible if $\Omega \neq \omega_i$, i.e. if resonance conditions do not apply).

---

[11] As already seen for 1 d.o.f. systems, whenever an excitation force has the form $\underline{F} = \underline{F}_0 cos\Omega t$ it is useful to substitute it with the complex $\underline{F} = \underline{F}_0 e^{i\Omega t}$. This allows us an easier solution of the problem (particularly avoiding doubling the equations of motion) since it facilitates the elimination of the time dependence from the equations of motion. The solution found has to be projected on the real axis. These passages are implied in what follows (and will not be shown) and excitation force will be expressed as $\underline{F} = \underline{F}_0 e^{i\Omega t}$ and the same will be done for the response $\underline{x} = \underline{X}_P e^{i\Omega t}$.

2.4.2.3.2 Forced Motion in the Presence of Damping

Let us now consider the steady-state response of a vibrating 2 d.o.f. system (see Fig. 2.4) whose damping is subjected to sinusoidal excitation forces: the general equation of motion in scalar form is:

$$\begin{aligned} m_{11}\ddot{x}_1 + m_{12}\ddot{x}_2 + r_{11}\dot{x}_1 + r_{12}\dot{x}_2 + k_{11}x_1 + k_{12}x_2 &= F_1 \cos \Omega t \\ m_{21}\ddot{x}_1 + m_{22}\ddot{x}_2 + r_{21}\dot{x}_1 + r_{22}\dot{x}_2 + k_{21}x_1 + k_{22}x_2 &= F_2 \cos \Omega t \end{aligned} \tag{2.392}$$

When damping is present, the solution can be obtained more easily (Sect. 2.4.1.4.1) by passing to the complex domain, i.e. by considering the following substitution:

$$F_1 \cos \Omega t \Rightarrow F_1 e^{i\Omega t}; F_2 \cos \Omega t \Rightarrow F_2 e^{i\Omega t} \tag{2.393}$$

so that the equation of motion (2.392) becomes:

$$\begin{aligned} m_{11}\ddot{x}_1 + m_{12}\ddot{x}_2 + r_{11}\dot{x}_1 + r_{12}\dot{x}_2 + k_{11}x_1 + k_{12}x_2 &= F_1 e^{i\Omega t} \\ m_{21}\ddot{x}_1 + m_{22}\ddot{x}_2 + r_{21}\dot{x}_1 + r_{22}\dot{x}_2 + k_{21}x_1 + k_{22}x_2 &= F_2 c^{i\Omega t} \end{aligned} \tag{2.394}$$

A particular integral of (2.394) is given by:

$$\underline{X}_P = \begin{Bmatrix} X_{P1} \\ X_{P2} \end{Bmatrix} e^{i\Omega t} \tag{2.395}$$

By substituting this solution in (2.394) and simplifying the term $e^{i\Omega t}$ (always different from zero) we will obtain:

$$\begin{aligned} \left(k_{11} + ir_{11}\Omega - m_{11}\Omega^2\right)X_{P1} + \left(k_{12} + ir_{12}\Omega - m_{12}\Omega^2\right)X_{P2} &= F_1 \\ \left(k_{21} + ir_{21}\Omega - m_{21}\Omega^2\right)X_{P1} + \left(k_{22} + ir_{22}\Omega - m_{22}\Omega^2\right)X_{P2} &= F_2 \end{aligned} \tag{2.396}$$

By imposing solution (2.395) in differential equations (2.392) we obtained a linear algebraic system with non-homogenous complex coefficients. The solutions of (2.396) can be determined by once again applying Cramer's rule, i.e.:

$$X_{P1} = \frac{\Delta_1}{\Delta} = \frac{\begin{vmatrix} F_1 & (k_{12} + ir_{12}\Omega - m_{12}\Omega^2) \\ F_2 & (k_{22} + ir_{22}\Omega - m_{22}\Omega^2) \end{vmatrix}}{\begin{vmatrix} (k_{11} + ir_{11}\Omega - m_{11}\Omega^2) & (k_{12} + ir_{12}\Omega - m_{12}\Omega^2) \\ (k_{21} + ir_{21}\Omega - m_{21}\Omega^2) & (k_{22} + ir_{22}\Omega - m_{22}\Omega^2) \end{vmatrix}}$$
$$X_{P2} = \frac{\Delta_2}{\Delta} = \frac{\begin{vmatrix} (k_{11} + ir_{11}\Omega - m_{11}\Omega^2) & F_1 \\ (k_{21} + ir_{21}\Omega - m_{21}\Omega^2) & F_2 \end{vmatrix}}{\begin{vmatrix} (k_{11} + ir_{11}\Omega - m_{11}\Omega^2) & (k_{12} + ir_{12}\Omega - m_{12}\Omega^2) \\ (k_{21} + ir_{21}\Omega - m_{21}\Omega^2) & (k_{22} + ir_{22}\Omega - m_{22}\Omega^2) \end{vmatrix}} \tag{2.397}$$

where Δ is the determinant (complex) of the matrix of coefficients and Δ_i is the determinant of the matrix in which, instead of the generic ith column, we substituted the vector of known terms (i = 1, 2).

When frequency Ω of the excitation force tends to one of the two natural frequencies ω_1 and ω_2 of the system, due to the presence of damping terms, determinant Δ no longer tends to zero (2.397), but will have a minimum.

Therefore, under resonance conditions, although solutions X_{P1} and X_{P2} have a maximum, they no longer tend towards the infinite. Since both determinant Δ and determinant Δ_i are complex, solutions, X_{P1} and X_{P2} are also complex:

$$\underline{X}_P = \begin{Bmatrix} |X_{P1}|e^{i\psi 1} \\ |X_{P2}|e^{i\psi 2} \end{Bmatrix} e^{i\Omega t} \tag{2.398}$$

where

$$\begin{aligned} X_{P1} &= \mathrm{Re}(X_{P1}) + i\mathrm{Im}(X_{P1}) = X_{P1R} + iX_{P1I} \\ X_{P2} &= \mathrm{Re}(X_{P2}) + i\mathrm{Im}(X_{P2}) = X_{P2R} + iX_{P2I} \end{aligned} \tag{2.399}$$

and

$$\begin{aligned} |X_{P1}| &= \sqrt{X_{P1R}^2 + X_{P1I}^2}; \quad \psi_1 = arctg\left(\frac{X_{P1I}}{X_{P1R}}\right) \\ |X_{P2}| &= \sqrt{X_{P2R}^2 + X_{P2I}^2}; \quad \psi_2 = arctg\left(\frac{X_{P2I}}{X_{P2R}}\right) \end{aligned} \tag{2.400}$$

The actual solution, i.e. the steady-state response of the 2 d.o.f. system to the harmonic excitation force of frequency Ω, bearing in mind (2.393), becomes:

$$\begin{aligned} x_1(t) &= |X_{P1}| \cos(\Omega t + \psi_1) \\ x_2(t) &= |X_{P2}| \cos(\Omega t + \psi_2) \end{aligned} \tag{2.401}$$

i.e. the response is generally out-of-phase with respect to the excitation force: this will prove to be more or less in phase with the force for $\Omega < \omega_1$ and at 180° for $\Omega > \omega_2$. Always as an example of Fig. 2.4 and the data shown in Table 2.2 and in Table 2.8 case b, Fig. 2.44 shows the in-frequency response of two d.o.f. systems in terms of modulus $|X_{P1}|$ and $|X_{P2}|$ and of phase ψ_1 and ψ_2 (2.400) for only one force F_1 of the unitary modulus applied to the d.o.f. x_1: these diagrams (Bode diagrams) thus represent the transfer functions $H_{11}(\Omega) = |X_{P1}|e^{i\psi_1}$ and $H_{21}(\Omega) = |X_{P2}|e^{i\psi_2}$ of the system being examined in the presence of damping. These transfer functions connect the input (the force on the first d.o.f.) to the outputs (displacements x_1 and x_2). In the event of our wishing to define the overall response of the system, it would be necessary to obtain matrix $[D]^{-1}$ (2 × 2) (2.389) containing transfer functions $H_{ij}(\Omega)$ (i = 1, 2, j = 1, 2): this matrix thus contains the response of the system, or rather outputs x_1 and x_2 (first and second line) due to inputs F_1 and F_2 (first and second column). All the statements and considerations made until now for a damped 2 d.o.f. system, forced with a harmonic excitation force, can be generalized for n d.o.f. systems forced in the presence of damping, whose equations, in matrix forms, are [see (2.394)]:

$$[M]\ddot{\underline{x}} + [R]\dot{\underline{x}} + [K]\underline{x} = F_0 e^{i\Omega t} \tag{2.402}$$

By imposing a solution of the type:

$$\underline{x}_P = \underline{X}_P e^{i\Omega t} \tag{2.403}$$

and making recourse to a matrix notation, we obtain:

$$\left[-\Omega^2[M] + i\Omega[R] + [K]\right]\underline{X}_P = \underline{F}_0 \Rightarrow [D(\Omega)]\,\underline{X}_P = \underline{F}_0 \tag{2.404}$$

where, this time, matrix $[D(\Omega)]$ has complex coefficients, always, obviously, dependent on the characteristics of the system and the frequency of the excitation force. This equation can be solved once again by using Cramer, or, by pre-multiplying the terms of same for $[D(\Omega)]^{-1}$ thus directly obtaining the particular integral solution:

$$[D(\Omega)]^{-1}[D(\Omega)]\underline{X}_P = [D(\Omega)]^{-1}\underline{F}_0 \Rightarrow \underline{X}_P = [D(\Omega)]^{-1}\underline{F}_0 = [H(i\Omega)]\underline{F}_0 \tag{2.405}$$

Due to the presence of damping terms, the inversion of matrix $[D(\Omega)]$ is now possible; in fact, even for $\Omega = \omega_i$. $[H(i\Omega)]$ is the harmonic transfer matrix in which the various terms $\left[H_{jk}(i\Omega)\right]$ represent the transfer function between the jth point in which the system's response is evaluated and the kth point where the excitation force is applied.

2.4.2.4 One Application: The Dynamic Vibration Absorber

Let us consider a mechanical system, modelled as having 1 d.o.f., on which a harmonic excitation force $f_1 = F_{10}\text{sen}\Omega t$ of frequency Ω, assigned and constant, acts. Let us assume that this frequency is near to the natural frequency $\omega_1 = \sqrt{k_1/m_1}$ of the system itself and, furthermore, that the damping is small: under these conditions, the oscillations prove to be high and unacceptable for the seamless functioning of the system. In order to reduce the vibration amplitudes, it is possible to:

- limit the excitation force: for example, in the case where this is due to an eccentric rotating mass, by balancing the rotor;
- to change either the mass or stiffness of the system;
- to increase damping.

At times, these remedies can be onerous or even impossible to apply. Another option is to mount a dynamic vibration absorber (first introduced by Frahm in 1909, see [15] and [37]). This absorber consists of a vibrating 1 d.o.f. system, relatively small with respect to the initial system, to be added to the principal mass m_1 (Fig. 2.53): the natural frequency ω_2 of this auxiliary system is chosen to coincide with the frequency of excitation force:

$$\omega_2 = \sqrt{\frac{k_2}{m_2}} = \Omega \tag{2.406}$$

We will demonstrate that, thanks to this absorber, machine mass m_1 no longer oscillates, while the same absorber oscillates so that the force transmitted by spring k_2 is equal instant by instant and opposed to external force $f_1 = F_{10}\text{sen}\Omega t$ acting on the machine. In actual fact, this proposition is theoretical because it presupposes the absence of damping, which actually is always present, and the equality of the excitation frequency Ω with natural frequency ω_2. To better interpret the problem, we will write the equations of motion of the system in Fig. 2.45, obtainable from the general expression found in (2.396), neglecting damping for the time being:

$$\begin{aligned} m_1\ddot{x}_1 + (k_1 + k_2)x_1 - k_2x_2 &= F_{10}\sin\Omega t \\ m_2\ddot{x}_2 - k_2x_1 + k_2x_2 &= 0 \end{aligned} \tag{2.407}$$

As always, by imposing a steady-state solution of the type:

$$\begin{aligned} x_{P1} &= X_{P1}\sin\Omega t \\ x_{P2} &= X_{P2}\sin\Omega t \end{aligned} \tag{2.408}$$

we obtain:

$$\begin{aligned} -\Omega^2 m_1 + (k_1 + k_2)X_{P1} - k_2 X_{P2} &= F_{10} \\ -k_2 X_{P1} + \left(-\Omega^2 m_2 + k_2\right)X_{P2} &= 0 \end{aligned} \tag{2.409}$$

For an easier and more direct interpretation of the results, we divide the first equation by k_1 and the second equation by k_2: furthermore, we use:

- $X_{ST} = \frac{F_{10}}{k_1}$ to denominate the static deflection of the main system;
- $\mu = \frac{m_1}{m_2}$ the relationship between the two masses;

thus obtaining:

$$\begin{aligned} \left(-\Omega^2 \frac{m_1}{k_1} + \left(1 + \frac{k_2}{k_1}\right)\right)X_{P1} - \frac{k_2}{k_1}X_{P2} = \frac{F_{10}}{k_1} \Rightarrow \left(1 + \frac{k_2}{k_1} - \frac{\Omega^2}{\omega_1^2}\right)X_{P1} - \frac{k_2}{k_1}X_{P2} = X_{ST} \\ -X_{P1} + \left(-\Omega^2 \frac{m_2}{k_2} + 1\right)X_{P2} = 0 \Rightarrow X_{P1} = \left(1 - \frac{\Omega^2}{\omega_2^2}\right)X_{P2} \end{aligned} \tag{2.410a}$$

from which, by solving in X_{P1} and X_{P2} we obtain:

$$\begin{aligned} \left(\left(1 + \frac{k_2}{k_1} - \frac{\Omega^2}{\omega_1^2}\right)\left(1 - \frac{\Omega^2}{\omega_2^2}\right) - \frac{k_2}{k_1}\right)X_{P2} = X_{ST} \\ \Rightarrow \frac{X_{P2}}{X_{ST}} = \frac{1}{\left(\left(1 + \frac{k_2}{k_1} - \frac{\Omega^2}{\omega_1^2}\right)\left(1 - \frac{\Omega^2}{\omega_2^2}\right) - \frac{k_2}{k_1}\right)} \end{aligned} \tag{2.410b}$$

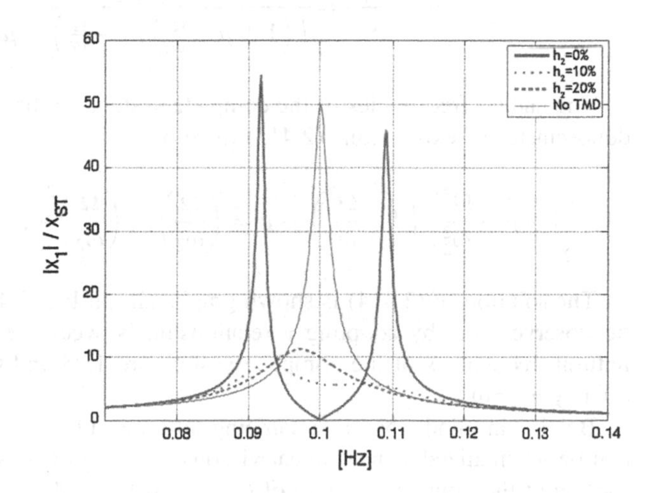

Fig. 2.47 Dynamic absorber $\mu = m_1/m_2 = 3\,\%$ frequency response of the main system for different values of the non-dimensional damping h_2

$$\frac{X_{P1}}{X_{ST}} = \frac{\left(1 - \frac{\Omega^2}{\omega_2^2}\right)}{\left(\left(1 + \frac{k_2}{k_1} - \frac{\Omega^2}{\omega_1^2}\right)\left(1 - \frac{\Omega^2}{\omega_2^2}\right) - \frac{k_2}{k_1}\right)} \tag{2.410c}$$

By analysing (2.410c) we note that amplitude X_{p1} of steady-state mass m_1 is null, or rather that the machine no longer oscillates. Always bearing in mind the same relationship (2.410c), from (2.410b) we obtain:

$$\frac{X_{P2}}{X_{ST}} = \frac{1}{-\frac{k_2}{k_1}} = \frac{-k_1}{k_2} \Rightarrow X_{P2} = -\frac{X_{ST}k_1}{k_2} \Rightarrow k_2 X_{P2} = -F_{10} \tag{2.411}$$

or rather, instant by instant, the force transmitted by spring k_2 balances the external excitation force. To give an example, in Fig. 2.46 we show the frequency response of the main mass m_1 and of the dynamic absorber m_2 as a function of the frequency, in the case of $\mu = {}^1/_5$ and $\omega_1 = \omega_2$: as can be noted, there are obviously two resonance peaks (the two d.o.f. system) and, in correspondence to the frequency of the excitation force, as previously mentioned, amplitudes X_{p1} are null. Furthermore, by imposing relationship:

$$\omega_1 = \omega_2 \Rightarrow \frac{k_1}{m_1} = \frac{k_2}{m_2} \Rightarrow \frac{k_2}{k_1} = \frac{m_2}{m_1} = \mu \tag{2.412}$$

expressions (2.410c) and (2.411) become:

$$\frac{X_{P2}}{X_{ST}} = \frac{1}{\left(\left(1 + \mu - \frac{\Omega^2}{\omega_2^2}\right)\left(1 - \frac{\Omega^2}{\omega_2^2}\right) - \mu\right)} \tag{2.413}$$

$$\frac{X_{P1}}{X_{ST}} = \frac{\left(1 - \frac{\Omega^2}{\omega_2^2}\right)}{\left(\left(1 + \mu - \frac{\Omega^2}{\omega_2^2}\right)\left(1 - \frac{\Omega^2}{\omega_2^2}\right) - \mu\right)}$$

The natural frequencies of the complete system can be obtained by equating the denominator of expressions (2.413) to zero:

$$\left(1 + \mu - \frac{\Omega^2}{\omega_2^2}\right)\left(1 - \frac{\Omega^2}{\omega_2^2}\right) - \mu = \left(\frac{\Omega}{\omega_2}\right)^4 - \left(\frac{\Omega}{\omega_2}\right)^2 (2 + \mu) + 1 = 0 \tag{2.414}$$

The solution of (2.414) is shown graphically in Fig. 2.47: as an example, it can be observed that by adopting a relationship between the masses $\mu = 2$, the two natural frequencies of the complete system are 1.25 and 0.8 times the initial frequency $\omega_1 = \omega_2$.

Bearing in mind a possible damping introduced by the dynamic absorber, which can be schematized using a linear viscous damping of constant r_2, the equations of motion of the complete system of (2.409) of Fig. 2.45 becomes:

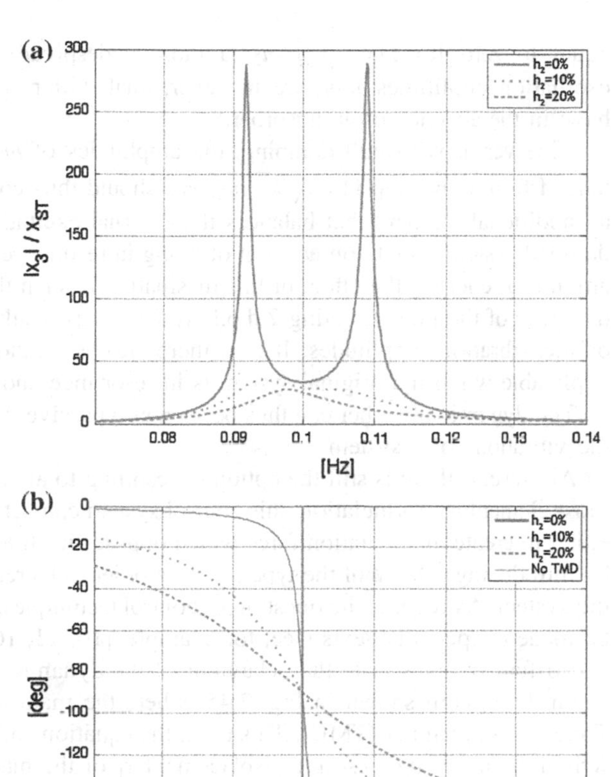

Fig. 2.48 **a** Dynamic absorber $\mu = m_1/m_2 = 3\,\%$ frequency response of the secondary system—magnitude—for different values of the non-dimensional damping h_2. **b** Dynamic absorber $\mu = m_1/m_2 = 3\,\%$ frequency response of the secondary system—phase φ_{31}—for different values of the non-dimensional damping h_2

$$
\begin{aligned}
m_1\ddot{x}_1 + r_2\dot{x}_1 - r_2\dot{x}_2 + (k_1 + k_2)x_1 - k_2x_2 &= F_{10}e^{i\Omega t}\\
m_2\ddot{x}_2 - r_2\dot{x}_1 + r_2\dot{x}_2 - k_2x_1 + k_2x_2 &= 0
\end{aligned}
\tag{2.415}
$$

Figure 2.47 shows the frequency response of mass m_1 (with $\omega_1 = \omega_2 = \Omega$ and $\mu = 1/20$) with the variation of non-dimensional damping $h_2 = r_2/(2m_2\omega_2)$. The continuous line indicates the null damping curve ($h_2 = 0$), while the dotted line is used to show the same amplitude in the presence of damping. When the damping increases, the two peaks decrease to once again reunite in a peak tending towards infinite as $h_2 \to \infty$. The optimal damping value to be assigned to the dynamic absorber can be obtained [15, 36] by exploiting the property whereby all the curves pass through points P and Q of Fig. 2.47 regardless of the value of h_2 and by making sure that the frequency response passes through these two points with a null tangent, in order to obtain a diagram like the one shown in Fig. 2.48 (in the case of $\mu = 1/4$). The presence of damping means that the mass amplitude m_1 is no longer equal to zero for $\Omega = \sqrt{k_2/m_2}$: a more widespread field exists in which the oscillation amplitudes of m_1 are, however, small with respect to those that we would have in the absence of an absorber.

Vice versa, with null damping, the amplitudes of m_1 are amplified when distanced from condition $\Omega = \sqrt{k_2/m_2}$. We should thus consider the absorber not as an additional element that balances the external excitation force but as a suitably damped system which, on account of being in resonance, dissipates a considerable amount of energy: the effect of this dissipation is such that it increases the overall damping of the new vibrating 2 d.o.f. system, thus resulting in an overall decrease of the vibration amplitudes. It can, therefore, be deduced that this is effectively applicable when the original system is in resonance and has a small damping.

The dynamic absorber can thus be seen as a passive system to control (decrease) the vibrations of a system.

At present, there is still the option of resorting to an active control system: in its most elementary formulation, this control system can act on the mechanical system either to create an excitation force which opposes itself to the external excitation or by introducing a force of the type $(-r\dot{x})$ in order to increase the overall damping of the system. As regards the question of control techniques, we suggest that reference be made to specialist texts (see, for example, [10, 11, 16, 30]).

To further investigate the behaviour of the dynamic absorber we will consider again the system shown in Fig. 2.45, where the main system is equipped with a Tuned Mass Damper (TMD). This time, the equations of motion (2.415) can be rewritten by using the absolute displacement x_1 of the main system and the relative displacement x_3 of the TDM as new d.o.f.'s. According to the sign conventions shown in the same figure, we have:

$$\begin{cases} (m_1 + m_2)\ddot{x}_1 + m_2\ddot{x}_3 + r_1\dot{x}_1 + k_1x_1 = f_1 \\ m_2\ddot{x}_1 + m_2\ddot{x}_3 + r_2\dot{x}_3 + k_2x_3 = 0 \end{cases} \tag{2.416}$$

Instead of elastic and damping connections between the TMD and the main system, in this case, system (2.416) is coupled by means of the TMD mass.

By analysing the first equation of system (2.416), we can see how the absorber effect increases the main system mass, which becomes $m_1 + m_2$, where, generally speaking, mass m_2 is negligible with respect to m_1.

The damping introduced by TMD can thus be calculated, in non-dimensional form, by means of relationship (see Sect. 2.4.1.3):

$$h_{eq} = \frac{E_d}{4\pi E_{c\,max}} \tag{2.417a}$$

where $E_{c\,max}$ is the maximum value of the total kinetic energy of the system and E_d is the energy dissipated by the TMD.

The energy dissipated in one period T by the TMD, and in particular by the linear viscous damping of constant r_2, can be expressed as:

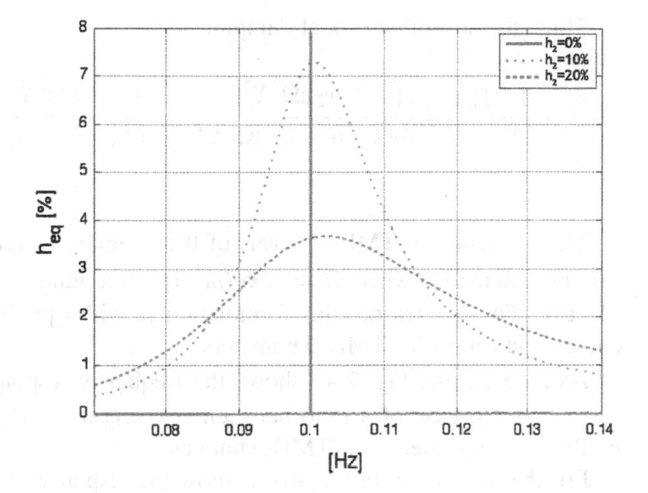

Fig. 2.49 Dynamic absorber $\mu = m_1/m_2 = 3\,\%$ damping h_{eq} for different values of the non-dimensional damping h_2

$$E_d = \int_0^T r_2 \dot{x}_3^2 dt = r_2 \pi \Omega^2 X_{p3}^2 \tag{2.417b}$$

Having used X_{p3} to define the magnitude of the relative motion of the TMD. The kinetic energy of the system, considering the two variables x_1 and x_3, is:

$$E_c = \frac{1}{2} m_1 \dot{x}_1^2 + \frac{1}{2} m_2 (\dot{x}_1 + \dot{x}_3)^2 \tag{2.417c}$$

By analysing the second equation of (2.416), when $\Omega = \sqrt{k_2/m_2}$ the ratio between the magnitude of the relative motion of the TMD X_{p3} and the magnitude of the dynamic response of the main system X_{p1} is:

$$\frac{X_{p3}}{X_{p1}} = \frac{1}{2h} \tag{2.417d}$$

where $h = r_2/(2m_2\omega_2)$ and $\omega_2 = \sqrt{k_2/m_2}$, and, moreover, X_{p3} and X_{p1} are in quadrature. Considering that m_1 is typically much greater than m_2 and considering that X_{p3} and X_{p1} are in quadrature, the maximum value of the kinetic energy in one period T for the system is:

$$E_{cmax} = \frac{1}{2}(m_1 + m_2) X_{p1}^2 \Omega^2 \tag{2.417e}$$

Thus, the non-dimensional damping factor is:

$$h_{eq} = \frac{1}{4\pi}\frac{E_d}{E_{cmax}} = \frac{1}{4\pi}\frac{r_2\pi\Omega^2 X_{p3}^2}{\frac{1}{2}(m_1+m_2)\Omega^2 X_{p1}^2} = \frac{1}{4\pi}\frac{r_2\pi\Omega^2 X_{p1}^2\left(\frac{1}{2h}\right)^2}{\frac{1}{2}(m_1+m_2)\Omega^2 X_{p1}^2} = \frac{1}{4h}\frac{m_2}{(m_1+m_2)} \tag{2.417f}$$

The efficiency of TMD in terms of the damping increment of the main system can be evaluated, when $\Omega = \sqrt{k_2/m_2}$ (perfect tuning condition), by means of (2.417f). For this reason, this formulation provides preliminary indications in the choice and setting of TMD parameters.

As an example, Fig. 2.47 shows the frequency response $|X_{p1}|/X_{st}$ of the main system, in the case of a ratio between $\mu = m_1/m_2 = 3\,\%$, as damping ratio $h_2 = r_2/2m_2\omega_2$ associated with TMD, changes.

For the same case, Fig. 2.48a, b show the response of the secondary system, in magnitude $|X_{p2}|/X_{st}$ and the relative phase φ_{31}.

The damping trend introduced into the main system as a function of damping h_2, in a parametric form, is shown in Fig. 2.49.

Once again, these results show how TMD efficiency depends, first and foremost, on the ratio between added mass and main system mass and how the damping introduced also depends on the transfer function between the relative motion of the added mass and the motion of the system to be damped.

The dynamic absorber can thus be seen as a passive system to control (decrease) the vibrations of a system.

At present, there is still the option of resorting to an active control system: in its most elementary formulation, this control system can act on the mechanical system either to create an excitation force which opposes itself to the external excitation or by introducing a force of the type $(-r\dot{x})$ in order to increase the overall damping of the system. As regards the question of control techniques, we suggest that reference be made to specialist texts (see, for example, [10, 11, 16, 30]).

2.4.3 n-Degree-of-Freedom System

2.4.3.1 Undamped Free Motion

2.4.3.1.1 Calculation of Natural Frequencies and Vibration Modes

As noted, the equation in matrix form for a system with n d.o.f. is:

$$[M]\underline{\ddot{x}} + [R]\underline{\dot{x}} + [K]\underline{x} = \underline{F}(t) \tag{2.418}$$

where $\underline{F}$ is the vector of the generalized forces depending on the d.o.f. (independent coordinates $\underline{x}$) pre-chosen to describe the system. To calculate the natural frequencies, as in the case of 2 d.o.f. systems (Sect. 2.4.2.1.1), in the absence of damping, we consider equation:

$$[M]\underline{\ddot{x}} + [K]\underline{x} = \underline{0} \tag{2.419}$$

As seen in Sect. 2.4.2.1, for the 2 d.o.f. system, when $[M]$ and $[K]$ are symmetric and defined as positive, the solution of expression (2.419) is the following:

$$\underline{x} = \underline{X}e^{i\omega t} \tag{2.420}$$

which when placed in the same equation leads to:

$$[-\omega^2[I] + [M]^{-1}[K]\underline{X} = \underline{0} \tag{2.421}$$

The natural frequencies of system ω_i can be evaluated by calculating the eigenvalues λ_i of matrix $[A] = [M]^{-1}[K]$, since relationship:

$$\omega_i^2 = \lambda_i \tag{2.422}$$

is valid. Conversely, the generic principal vibration mode $\underline{X}^{(i)}$ corresponds to the relative eigenvector:

$$\underline{X} = \begin{Bmatrix} X_1^{(i)} \\ X_2^{(i)} \\ \ldots \\ X_n^{(i)} \end{Bmatrix} \tag{2.423}$$

These eigenvectors are defined up to a constant, or rather $n \times (n - 1)$ can be defined as characteristic relationships of the type:

$$\mu_{jk}^{(i)} = \frac{X_j^{(i)}}{X_k^{(i)}} \quad (j = 2, n\ i = 1, n) \tag{2.424}$$

which define the relationship (in relation to the generic ith vibrating mode) between the generic jth coordinate $X_j^{(i)}$ and the kth coordinate $X_k^{(i)}$ chosen as a reference. The eigenvectors (2.424) can be normalized, i.e. numerically defined, by assigning e.g. an arbitrary value to a variable chosen as a reference $X_k^{(i)}$: for example, by imposing (for k = 1) $X_1^{(i)} = 1$ and keeping account of (2.424), the eigenvector relative to the generic natural frequency ω_i will be defined numerically as:

$$\underline{X}^{(i)} = \begin{Bmatrix} 1 \\ \mu_{21}^{(i)} \\ \cdots \\ \mu_{n1}^{(i)} \end{Bmatrix} \tag{2.425}$$

There are obviously countless ways of normalizing the eigenvectors, several of which have proven to be particularly convenient for subsequent manipulations of the same equations of motion: this subject will be resumed further on in Sect. 2.5, when discussing the modal approach.

As previously mentioned in Sect. 2.4.2.1, relationships $\mu_{jk}^{(i)}$ (2.424) which define the generic eigenvector, in the case where no damping is present, are real: this means that the motions of the various d.o.f. of each vibrating mode are only in phase or in counter-phase one with the other (depending on the sign of the various components of eigenvector $X_j^{(i)}$). One particular and useful property of the eigenvectors of the free, undamped system, where matrices $[K]$ and $[M]$ are symmetrical and positive definite, is the orthogonality (in the broad sense of the term [20, 22, 34]) with respect to mass $[M]$ and stiffness $[K]$ matrices:

$$\begin{aligned} \underline{X}^{(i)T}[M]\underline{X}^{(j)} &= \delta_{ij}m_{ij} \\ \underline{X}^{(i)T}[K]\underline{X}^{(j)} &= \delta_{ij}k_{ij} \end{aligned} \tag{2.426}$$

where δ_{ij} is the Kroneker delta ($\delta_{ij} = 1$ by $i = j$ and $\delta_{ij} = 0$ by $i \neq j$) and m_{ij} and k_{ij} two real values. This property will be demonstrated later on and exploited in the modal approach (Sect. 2.5). To calculate these eigenvectors it is necessary, most of all for a n d.o.f. system, to use numeric algorithms, using the calculation subroutines already existing in scientific calculation software, which use different numerical algorithms, including, for example, the QR method [29, 31, 37].

2.4.3.1.2 Calculation of the Response to Initial Conditions

The free motion of an undamped vibrating n d.o.f. system, subjected to an initial disturbance is described by the general integral (as seen for 2 d.o.f. systems, Sect. 2.4.2), i.e. by the linear combination of the particular integrals of the associated homogenous (2.419):

$$\underline{x}(t) = C_1\underline{X}^{(1)}\cos(\omega_1 t + \phi_1) + \cdots + C_n\underline{X}^{(n)}\cos(\omega_n t + \phi_n) \tag{2.427}$$

i.e. in a more convenient form:

$$\begin{aligned}\underline{x}(t) &= \underline{X}^{(1)}(A_1\cos\omega_1 t + B_1\sin\omega_1 t) + \cdots + \underline{X}^{(n)}(A_n\cos\omega_n t + B_n\sin\omega_n t)\\ &= [\Phi]\begin{Bmatrix} A_1\cos\omega_1 t + B_1\sin\omega_1 t\\ \dots\\ A_n\cos\omega_n t + B_n\sin\omega_n t\end{Bmatrix}\end{aligned} \tag{2.428}$$

where $[\Phi]$ is the matrix, termed modal (this matrix will be referred to again in Sect. 2.5, in relation to the modal approach), having n vibration modes as columns, i.e. the n eigenvectors $\underline{X}^{(i)}$, systematically placed side by side as shown below:

$$[\Phi] = \begin{bmatrix}\underline{X}^{(1)} & \underline{X}^{(2)} & \dots & \underline{X}^{(n)}\end{bmatrix} = \begin{bmatrix} X_1^{(1)} & X_1^{(2)} & \dots & X_1^{(n)}\\ X_2^{(1)} & X_2^{(2)} & \dots & X_2^{(n)}\\ \dots & \dots & \dots & \dots\\ X_n^{(1)} & X_n^{(2)} & \dots & X_n^{(n)}\end{bmatrix} \tag{2.429}$$

By deriving (2.428) with respect to time, we obtain:

$$\begin{aligned}\dot{\underline{x}}(t) &= \underline{X}^{(1)}(-\omega_1 A_1\sin\omega_1 t + \omega_1 B_1\cos\omega_1 t) + \cdots + \underline{X}^{(n)}(-\omega_n A_n\sin\omega_n t + \omega_n B_n\cos\omega_n t)\\ &= [\Phi][\omega]\begin{Bmatrix} -A_1\sin\omega_1 t + B_1\cos\omega_1 t\\ \dots\\ -A_n\sin\omega_n t + B_n\cos\omega_n t\end{Bmatrix}\end{aligned} \tag{2.430}$$

where $[\omega]$ is the diagonal matrix:

$$[\omega] = \begin{bmatrix} \omega_1 & 0 & \dots & 0\\ 0 & \omega_2 & \dots & 0\\ \dots & \dots & \dots & \dots\\ 0 & 0 & \dots & \omega_n\end{bmatrix} \tag{2.431}$$

Thus, by imposing the initial conditions and keeping account of expressions (2.428) and (2.430), we obtain:

$$\underline{x}_0 = \underline{x}(0) = [\Phi]\begin{Bmatrix} A_1 \\ A_2 \\ \ldots \\ A_n \end{Bmatrix} \Rightarrow [\Phi]\underline{A} = \underline{x}_0$$
$$\dot{\underline{x}}_0 = \dot{\underline{x}}(0) = [\Phi][\omega]\begin{Bmatrix} B_1 \\ B_2 \\ \ldots \\ B_n \end{Bmatrix} \Rightarrow [\Phi][\omega]\underline{B} = \dot{\underline{x}}_0 \tag{2.432}$$

equations that enable us to obtain constants $\underline{A}$ and $\underline{B}$ which are necessary to define the free motion of the system (2.419).

2.4.3.2 Damped Free Motion

2.4.3.2.1 Calculation of the Natural Frequencies and Vibration Modes

Let us now consider the presence of damping: the matrix equation that describes the free damped motion of a generic n d.o.f. system is given by (on a pair with 2 d.o.f. systems, Sect. 2.4.2):

$$[M]\ddot{\underline{x}} + [R]\ddot{\underline{x}} + [K]\underline{x} = \underline{0} \tag{2.433}$$

We will look for a solution of the type:

$$\underline{x} = \underline{X}e^{\lambda t} \tag{2.434}$$

which, when substituted in (2.433), results in a homogenous linear system in the unknown $\underline{X}$:

$$\left(\lambda^2[M] + \lambda[R] + [K]\right)\underline{X} = \underline{0} \tag{2.435}$$

This system accepts a solution that is different from the trivial solution only if the determinant of the coefficient matrix is null:

$$det\left|\lambda^2[M] + \lambda[R] + [K]\right| = 0 \tag{2.436}$$

Development of the determinant leads to a complete polynomial expression of order 2 in λ which, generally speaking, will have complex solutions (purely real or complex conjugate, because the coefficient matrix is real [2, 8, 9, 20, 21]. For n d.o.f. systems the analytical development of the characteristic polynomial and the zero setting of (2.436) is extremely laborious: therefore, as noted for 2 d.o.f. systems (Sect. 2.4.2.1.1), we prefer to calculate the natural frequencies of a damped system by solving an eigenvalues/eigenvectors problem. As previously seen, for this purpose, alongside the equation of motion (2.370), we consider an auxiliary identity:

$$\begin{aligned} [M]\ddot{\underline{x}} + [R]\dot{\underline{x}} + [K]\underline{x} &= \underline{0} \\ [M]\dot{\underline{x}} &= [M]\dot{\underline{x}} \end{aligned} \tag{2.437}$$

and we perform the following change of variables:

$$\underline{z} = \begin{Bmatrix} \dot{\underline{x}} \\ \underline{x} \end{Bmatrix} = \begin{Bmatrix} \underline{y} \\ \underline{x} \end{Bmatrix} \Rightarrow \dot{\underline{z}} = \begin{Bmatrix} \ddot{\underline{x}} \\ \dot{\underline{x}} \end{Bmatrix} = \begin{Bmatrix} \dot{\underline{y}} \\ \dot{\underline{x}} \end{Bmatrix} \tag{2.438}$$

In this way, system (2.437) can be rewritten as:

$$\begin{aligned} [M]\dot{\underline{y}} + [R]\underline{y} + [K]\underline{x} &= \underline{0} \\ [M]\dot{\underline{x}} - [M]\underline{y} &= \underline{0} \end{aligned} \tag{2.439}$$

By defining matrices $[B]$ and $[C]$:

$$\begin{aligned} [B] &= \begin{bmatrix} [M] & [0] \\ [0] & [M] \end{bmatrix} \\ [C] &= \begin{bmatrix} [R] & [K] \\ -[M] & [0] \end{bmatrix} \end{aligned} \tag{2.440}$$

expression (2.439) is synthetically expressed as:

$$[B]\dot{\underline{z}} + [C]\underline{z} = \underline{0} \tag{2.441}$$

where matrix $[B]$ proves to be symmetrical and positive definite on account of having been obtained by suitably assembling mass matrix $[M]$. Starting from (2.441), it is possible to obtain an expression that is frequently adopted in the systems control field [10, 11, 16, 19, 24, 28]:

$$\dot{\underline{z}} = [A]\underline{z} \tag{2.442}$$

having used $[A]$ to define matrix:

$$[A] = -[B]^{-1}[C] \tag{2.443}$$

In (2.443) matrix $[B]$ is definitely invertible. The solution of expression (2.442) is of the type:

$$\underline{z} = \underline{Z}e^{\lambda t} \tag{2.444}$$

which, when substituted in (2.442) leads to a homogenous linear algebraic system in Z with real coefficients, parametric in λ[12]:

$$[[A] - \lambda[I]]\underline{Z} = \underline{0} \Rightarrow \underline{Z} = \left\{ \begin{array}{c} \underline{Y} \\ \underline{X} \end{array} \right\} = \left\{ \begin{array}{c} \lambda\underline{X} \\ \underline{X} \end{array} \right\} \tag{2.445}$$

which accepts a solution that is different from the banal solution if the determinant of the matrix of coefficients is null. The $2n$ values of which annul the determinant are, by definition [see (2.445)], the eigenvalues of matrix $[A]$. In dissipative systems, i.e. those having symmetrical matrices and defined as positive, and small structural damping, the eigenvalues are two by two complex conjugates of the type:

$$\lambda_{i,i+1}C = -\alpha_i \pm i\omega_i \tag{2.446}$$

with α_i and ω_i being real positives. In (2.446), the imaginary part ω_i (rad/s) represents the natural frequency of the damped system, while the real part α_i provides information about the damping of the system: more specifically, since matrices $[K]$, $[R]$ and $[M]$ are symmetric and positive definite, for dissipative systems perturbed about the stable rest position the real parts α_i of the eigenvalues are positive, i.e. free motion is damped. As seen in Sect. 2.4.2.1.1 for the free motion of damped two d.o. f. systems, unlike that which occurs for undamped systems, the eigenvectors $\underline{Z}^{(i)}$ are no longer real but complex, involving non-orthogonality in a generalized sense (i.e. with respect matrix $[M]$ or $[K]$) of the vibration modes themselves. Eigenvectors $\underline{Z}^{(i)}$ of expression (2.445) are also two by two complex conjugates.

2.4.3.2.2 Calculation of the Response to Initial Conditions

Having given an initial generic condition to the damped system, defined in terms of state variables $\underline{z}$:

$$\underline{z}(t)_{t=0} = \underline{z}_0 \tag{2.447}$$

[12]Keeping account of (2.438) we have:

$$\dot{\underline{x}} = \underline{y} \tag{2.12.1}$$

and, from (2.444) we obtain:

$$\underline{x} = \underline{X}e^{\lambda t} \Rightarrow \underline{y} = \underline{Y}e^{\lambda t} = \lambda\underline{X}e^{\lambda t}. \tag{2.12.2}$$

the motion of the system is defined by the combination of particular integrals (2.444) of the associated homogenous (2.441):

$$\underline{z}(t) = \sum_{i=1}^{2n} c_i \underline{Z}^{(i)} e^{\lambda_i t} \tag{2.448}$$

where c_i (i = 1, 2n) are complex constants that can be defined by imposing the initial conditions (2.447) which, by keeping account of (2.448), become:

$$\underline{z}_0 = \sum_{i=1}^{2n} c_i \underline{Z}^{(i)} \tag{2.449}$$

By pre-multiplying expression (2.449) by matrix $[B]$, see (2.440), and by the complex generic vibration mode $\underline{Z}^{(k)T}$ we obtain:

$$\underline{Z}^{(k)T}[B]\underline{z}_0 = \underline{Z}^{(k)T}[B] \sum_{i=1}^{2n} c_i \underline{Z}^{(i)} \tag{2.450}$$

By keeping account of the properties of orthogonality of the eigenvectors (Sect. 2.5.2.1, [2, 20, 29] and, in particular, Argiris, 1991, Sect. 9.1) all the terms on the right of equal sign of (2.450) are null with the exception of the one with index $j = k$: from this it is easy to obtain the value of the generic complex constant c_k of the general integral (2.448):

$$c_k = \left[\underline{Z}^{(k)T}[B]\underline{Z}^{(k)}\right]^{-1}\left\{\underline{Z}^{(k)T}[B]\underline{z}_0\right\} \tag{2.451}$$

2.4.3.3 Forced Motion

In the case of a sinusoidal excitation force and by using the complex exponential notation (as seen for 1 d.o.f. systems, Sect. 2.4.1 and in 2 d.o.f. systems, Sect. 2.4.2), the equation of motion of a generic n d.o.f. system (2.328) can be rewritten:

$$[M]\ddot{\underline{x}} + [R]\dot{\underline{x}} + [K]\underline{x} = \underline{f}(t) = \underline{F}_0 e^{i\Omega t} \tag{2.452}$$

As a particular integral, expression (2.452) will have:

$$\underline{x}_P = \underline{X}_P e^{i\Omega t} \tag{2.453}$$

where $\underline{X}_P$ is generally complex. By introducing (2.453) into the equation of motion (2.452) we obtain a complete linear algebraic system with complex coefficients (i.e. have an unknown non-null term):

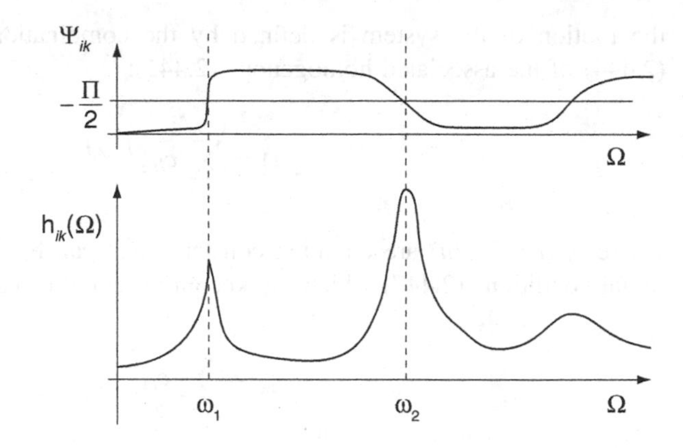

Fig. 2.50 Vibrating n d.o.f. system: example of a harmonic transfer function

$$\left[-\Omega^2[M]+i\Omega[R]+[K]\right]\underline{X}_P=\underline{F}_0 \tag{2.454}$$

i.e.:

$$[D(\Omega)]\,\underline{X}_P=\underline{F}_0 \Rightarrow \underline{X}_P=[D(\Omega)]^{-1}\underline{F}_0 \tag{2.455}$$

The response:

$$x_j(t)=X_{Pj}e^{i\Omega t}=\left|X_{Pj}\right|\cos\left(\Omega t+\psi_j\right) \quad (j=1,n) \tag{2.456}$$

is generally out-of-phase with respect to the excitation force. The known term $\underline{F}_0$ in (2.455) is real if all the excitation forces acting on different d.o.f. of the system are all in-phase or in counter phase; conversely, if the excitation forces on various d.o.f. are generally out-of-phase in time, the known term $\underline{F}_0$ of (2.455) is complex. As with the 1 d.o.f. system (Sect. 2.4.1.4.1), it is possible to define as harmonic transfer function $H_{jk}(\Omega)$ the ratio:

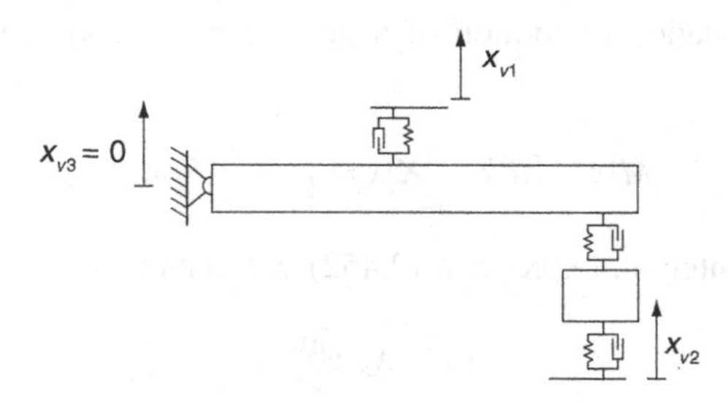

Fig. 2.51 Vibrating n d.o.f. system: motion imposed to constraints

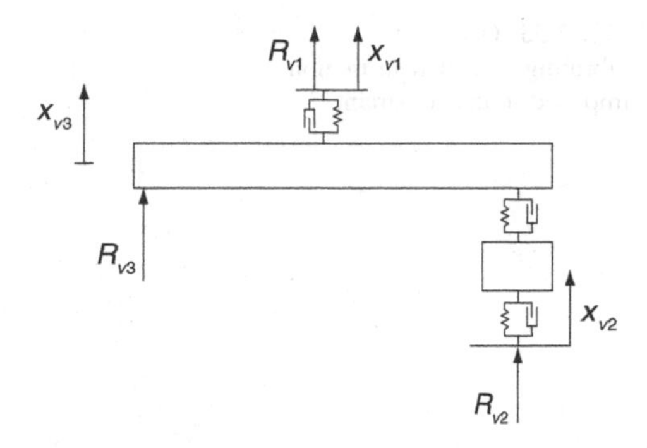

Fig. 2.52 Vibrating n d.o.f. system: applied forces on the system free in space

$$H_{jk}(\Omega) = \frac{X_{Pjk}(\Omega)}{F_{0k}(\Omega)} \tag{2.457}$$

between the response of the generic jth d.o.f.:

$$x_{Pjk} = X_{Pjk}e^{i\Omega t} = |X_{Pjk}|e^{i\psi_j}e^{i\Omega t} \tag{2.458}$$

which can be obtained by solving (2.456) where we impose:

$$\underline{F}_0 = \begin{Bmatrix} 0 \\ \cdots \\ F_{0k} \\ \cdots \\ 0 \end{Bmatrix} \tag{2.459}$$

and the generic harmonic excitation force applied according to the kth d.o.f. of the system:

$$f_k = F_{0k}e^{i\Omega t} \tag{2.460}$$

The harmonic transfer function $H_{jk}(\Omega)$ is, obviously, complex, since the response of the system (2.457) is generically complex: this function is usually represented, as a function of excitation frequency Ω, in modulus and phase as shown, as an example, in Fig. 2.50. In damped and forced n d.o.f. systems as seen in 1 and 2 d.o.f. systems (Sects. 2.4.1.4.1 and 2.4.2.3.2), when frequency Ω of the excitation force coincides with one of the natural frequencies of system ω_i, there will be an increase of steady-state amplitudes which, unlike undamped systems, will not reach infinite values. More specifically, as the damping increases, the resonance peaks are lowered and the frequencies of the undamped system are more strongly diversified than those of the same damped system.

Finally, let us remember that, generally speaking the solution to expression (2.455) is obtained numerically, using suitable algorithms in the complex field (e.g. Gauss-Jordan, Cholesky and others, see, for example [4, 7, 8, 26]) implemented as

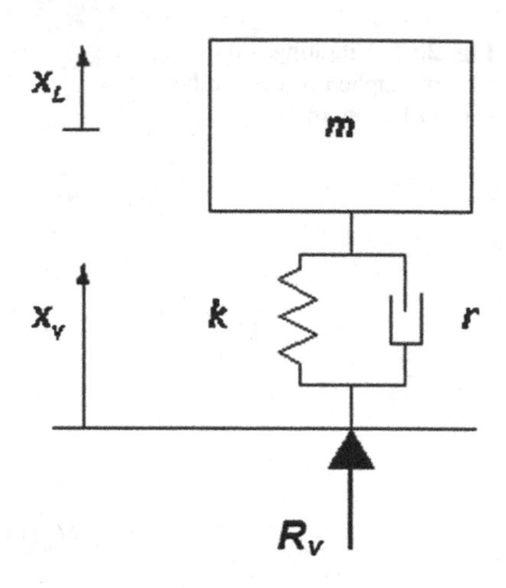

Fig. 2.53 One d.o.f. vibrating system with motion imposed to the constraint

subprograms of automatic calculation in various scientific software packages (IMSL etc.).

2.4.3.4 Motion Imposed on Constraints

Let us now consider a generic vibrating n d.o.f. system constrained to the outside world with a certain number of constraints that can either be fixed or in motion with an assigned law, see Fig. 2.51; $\underline{x} = \underline{x}_L$ is the vector of the independent variables that describe the behaviour of the system and $\underline{x}_V$ the vector of the constraint displacements:

$$\underline{x}_L = \begin{Bmatrix} x_1 \\ \cdots \\ x_i \\ \cdots \\ x_n \end{Bmatrix}; \quad \underline{x}_V = \begin{Bmatrix} x_{V1} \\ x_{V2} \\ \cdots \\ x_{Vnv} \end{Bmatrix} \tag{2.461}$$

A general methodology used to define the equations of motion of this system consists in writing the same equations by considering as independent coordinates $\underline{x}$ not only free coordinates $\underline{x}_L$ but also coordinates $\underline{x}_V$ imposed by the law of motion:

$$\underline{x} = \begin{Bmatrix} \underline{x}_L \\ \underline{x}_V \end{Bmatrix} \tag{2.462}$$

When using this approach, the system proves to be free in space (since the imposed constraints were eliminated during this first phase) and, therefore, in addition to forces $\underline{f}_L$ acting on effectively free d.o.f., as external forces it also has constraint reactions $\underline{R}_v$ (Fig. 2.52). The equations of motion are obtained later on by using, for example Lagrange's equations: these equations will contain as unknowns, by suitably transferring to the right hand side those terms depending on coordinates $\underline{x}_V$ (imposed laws of motion), the displacements associated with the actual d.o.f.'s $\underline{x}_L$ as well as the constraint reactions $\underline{R}_v$: these reactions are those necessary to maintain the imposed law of motion. It is tacitly understood that this approach can also be used to calculate constraint reactions $\underline{R}_v$ when the constraints $\underline{x}_V = \underline{0}$ are fixed, thus representing the particular situation whereby the imposed law of motion is constant and null. Let us consider, as a first example, a vibrating system with 1 d.o.f. with an imposed motion $\underline{x}_V = \underline{x}_V(t)$ to the constraint (see Fig. 2.53).

By adopting a Lagrangian approach, we will evaluate the various forms of energy, also considering, in addition to the free variable $\underline{x}_L$, $\underline{x}_V$ as an independent variable and assigning the term $\underline{R}_V$ to the constraint reaction:

$$\begin{gathered} E_C = \frac{1}{2} m \dot{x}_L^2; \quad V = \frac{1}{2} k (x_L - x_V)^2 \\ D = \frac{1}{2} r (\dot{x}_L - \dot{x}_V)^2; \quad \delta^* L = R_V \, \delta^* x_V \end{gathered} \tag{2.463}$$

To write the two equations of motion, as is customary, the forms of energy are derived according to Lagrange, with respect to coordinate $\underline{x}_L$.

$$\begin{aligned} &\frac{d}{dt}\left(\frac{\partial E_C}{\partial \dot{x}_L}\right) = m\ddot{x}_L; \quad \frac{\partial E_C}{\partial x_L} = 0 \\ &\frac{\partial V}{\partial x_L} = k(x_L - x_V); \quad \frac{\partial D}{\partial \dot{x}_L} = r(\dot{x}_L - \dot{x}_V); \quad \frac{\delta^* L}{\delta^* x_L} = 0 \end{aligned} \tag{2.464}$$

and coordinate $\underline{x}_V$:

$$\begin{aligned} &\frac{d}{dt}\left(\frac{\partial E_C}{\partial \dot{x}_V}\right) = 0; \quad \frac{\partial E_C}{\partial x_V} = 0 \\ &\frac{\partial V}{\partial x_V} = -k(x_L - x_V); \quad \frac{\partial D}{\partial \dot{x}_V} = -r(\dot{x}_L - \dot{x}_V); \quad \frac{\delta^* L}{\delta^* x_V} = R_V \end{aligned} \tag{2.465}$$

which leads to the two equation system in coordinates x_V and x_L:

$$\begin{gathered} m\ddot{x}_L + r\dot{x}_L - r\dot{x}_V + kx_L - kx_V = 0 \\ -r\dot{x}_L + r\dot{x}_V - kx_L + kx_V = R_V \end{gathered} \tag{2.466}$$

By bearing in mind that x_V is, in actual fact, an assigned function of time $x_V = x_V(t)$, the two equations (2.466) can be rewritten as:

$$\begin{aligned} m\ddot{x}_L + r\dot{x}_L + kx_L &= r\dot{x}_V + kx_V = \widetilde{f}_L \\ -r\dot{x}_L + r\dot{x}_V - kx_L + kx_V &= R_V \end{aligned} \tag{2.467}$$

The second member of the first equation $\widetilde{f}_L = r\dot{x}_V + kx_V$ constitutes a known equivalent excitation force, due to imposed displacement x_V: after solving the first equation, the second equation provides the reaction $R_V = R_V(t)$ to be applied to the constraint to obtain the imposed law $x_V = x_V(t)$. The procedure outlined can be generalized for a vibrating n d.o.f. system in which, generally speaking there are n_v constrains having a known, imposed motion.

By using $\underline{x}$ (2.469) to define the vector containing both the sub-vector $\underline{x}_L$ of the n independent variables and sub-vector x_V containing the constrained d.o.f., by applying the Lagrange's equations, we obtain the equation of motion of the system (see Sect. 2.4) which, in matrix form, can be written as:

$$[M]\ddot{\underline{x}} + [R]\dot{\underline{x}} + [K]\underline{x} = \underline{f} \tag{2.468}$$

By ordering vector $\underline{x}$ so that it proves to be:

$$\underline{x} = \begin{Bmatrix} \underline{x}_L \\ \underline{x}_V \end{Bmatrix} \tag{2.469}$$

and, as a consequence, by reordering matrices $[M]$, $[R]$ and $[K]$ and partitioning same into 4 sub-matrices:

$$[M] = \begin{bmatrix} [M_{LL}] & [M_{LV}] \\ [M_{VL}] & [M_{VV}] \end{bmatrix}; \; [R] = \begin{bmatrix} [R_{LL}] & [R_{LV}] \\ [R_{VL}] & [R_{VV}] \end{bmatrix}; \; [K] = \begin{bmatrix} [K_{LL}] & [K_{LV}] \\ [K_{VL}] & [K_{VV}] \end{bmatrix} \tag{2.470}$$

expressions (2.468) can be rewritten as two matrix equations:

$$\begin{aligned} &[M_{LL}]\,\ddot{\underline{x}}_L + [M_{LV}]\,\ddot{\underline{x}}_V + [R_{LL}]\,\dot{\underline{x}}_L + [R_{LV}]\,\dot{\underline{x}}_V + [K_{LL}]\,\underline{x}_L + [K_{LV}]\,\underline{x}_V = \underline{f}_L \\ &[M_{VL}]\,\ddot{\underline{x}}_L + [M_{VV}]\,\ddot{\underline{x}}_V + [R_{VL}]\,\dot{\underline{x}}_L + [R_{VV}]\,\dot{\underline{x}}_V + [K_{VL}]\,\underline{x}_L + [K_{VV}]\,\underline{x}_V = \underline{R}_V \end{aligned} \tag{2.471}$$

In expressions (2.471) $\underline{f}_L$ represents the vector containing the active external excitation forces acting on actual d.o.f., while $\underline{R}_V$ represents the vector of the generalized reactions applied to the constrained d.o.f. By keeping account of the fact that vector $\underline{x}_V$, and its derivatives with respect to time, are vectors of known functions, the first of the two equations (2.471) can be rewritten as:

$$\begin{aligned} [M_{LL}]\ddot{\underline{x}}_L + [R_{LL}]\dot{\underline{x}}_L + [K_{LL}]\underline{x}_L &= -[M_{LV}]\ddot{\underline{x}}_V \\ &- [R_{LV}]\dot{\underline{x}}_V - [K_{LV}]\underline{x}_V + \underline{f}_L = \tilde{\underline{f}}_L + \underline{f}_L = \tilde{\tilde{\underline{f}}}_L \end{aligned} \tag{2.472}$$

in which the second member $\underline{\tilde{f}}_L$ is composed solely of known terms, i.e. the effective external excitation forces $\underline{f}_L$ and the equivalent excitation forces:

$$\underline{\tilde{f}}_L = -[M_{LV}]\underline{\ddot{x}}_V - [R_{LV}]\underline{\dot{x}}_V - [K_{LV}]\underline{x}_V \tag{2.473}$$

due to the motion imposed on constraints $\underline{x}_V$. Expression (2.472) thus represents a system of n equations, the same number as the $\underline{x}_L$ unknowns, which when solved can be used to calculate the motion of the vibrating system. Conversely, the equation relative to constrained d.o.f. [based on the matrix equation of expressions (2.471)] allows us to obtain the constraint reactions $\underline{R}_V$, once expressions (2.472) have been integrated and the values of $\underline{x}_L$ and their derivatives obtained.

2.5 Modal Approach for Linear *n*-Degree-of-Freedom Systems

In the previous sections, the methods most commonly used to write the equations of motion of discrete 2-n d.o.f. systems were introduced. Generally speaking, the equations obtained are coupled i.e. the corresponding matrices of mass $[M]$, damping $[R]$ and stiffness $[K]$ are full: for systems with a high number of d.o.f. this involves a complex calculation process to obtain the solution of same (see Sects. 2.3 and 2.4). In this chapter, we will demonstrate a particular type of approach, termed the "modal approach" which, under certain conditions and thanks to a suitable, preordained choice of a specific set of independent coordinates, enables us to uncouple the equations of motion of a generic vibrating 2-n d.o.f. system: instead of n coupled equations, it allows us to solve n independent equations with 1 d-o-f., thus saving a considerable amount of time and occupation of space on the computer and, even more important, to use a model that allows us to have a better understanding of the behaviour of the system. For teaching purposes, as in the past, we will analyse this technique applied, first and foremost, to 2 d.o.f. systems (Sect. 2.5.1), to subsequently extend the procedure to n d.o.f. systems (Sect. 2.5.2).

2.5.1 Modal Approach for Two-Degree-of-Freedom Systems

Let us resume the example of a vibrating 2 d.o.f. system shown in Fig. 2.4: for the same system as the one in Sect. 2.3.2.1, using two different approaches, we defined the equations of motion of a free undamped system (2.85) which, for the sake of convenience, is shown in a compact matrix form:

$$[M]\underline{\ddot{x}} + [K]\underline{x} = \underline{0} \tag{2.474}$$

having used $\underline{x}$ to indicate the vector of independent coordinates:

$$\underline{x} = \begin{Bmatrix} x_1 \\ x_2 \end{Bmatrix} \tag{2.475}$$

and $[K]$ and $[M]$ to indicate the matrices of stiffness and damping of the same system. For the system in question, in Sect. 2.3.1 we subsequently obtained the natural frequencies, i.e. the two eigenvalues ω_i $(i = 1, 2)$ and the corresponding vibration modes, i.e. eigenvectors $\overline{\underline{X}}^{(i)}$:

$$\begin{aligned} \omega = \omega_1 \Rightarrow \overline{\underline{X}}^{(1)} &= \begin{Bmatrix} \overline{X}_1^{(1)} \\ \overline{X}_2^{(1)} \end{Bmatrix} \\ \omega = \omega_2 \Rightarrow \overline{\underline{X}}^{(2)} &= \begin{Bmatrix} \overline{X}_1^{(2)} \\ \overline{X}_2^{(2)} \end{Bmatrix} \end{aligned} \tag{2.476}$$

We can now define a generic change of variables, by moving from a couple of independent variables x_1 and x_2 to a new couple of variables q_1 and q_2 and imposing:

$$\begin{aligned} x_1 &= \overline{X}_1^{(1)} q_1 + \overline{X}_1^{(2)} q_2 \\ x_2 &= \overline{X}_2^{(1)} q_1 + \overline{X}_2^{(2)} q_2 \end{aligned} \tag{2.477}$$

This change of variables is determined univocally once the eigenvectors $\overline{X}_j^{(i)}$ have been defined numerically thanks to the use of a normalization process. The new variables q_1 and q_2 can be interpreted as multiplicative coefficients of the vibration modes of the system. This change of variables, introduced progressively by means of expressions (2.477) can be rewritten in matrix form as:

$$\underline{x} = [\Phi]\underline{q} \tag{2.478}$$

Having ordered the new independent variables in $\underline{q}$:

$$\underline{q} = \begin{Bmatrix} q_1 \\ q_2 \end{Bmatrix} \tag{2.479}$$

and the vibration modes organized by columns in matrix $[\Phi]$:

$$[\Phi] = \left[\underline{X}^{(1)} \quad \underline{X}^{(2)}\right] = \begin{bmatrix} \overline{X}_1^{(1)} & \overline{X}_2^{(1)} \\ \overline{X}_1^{(2)} & \overline{X}_2^{(2)} \end{bmatrix} \tag{2.480}$$

This matrix is referred to as a modal matrix: given the particular change of variables proposed, $[\Phi]$ is defined by placing the eigenvectors of problem (2.480) one next to the other, i.e., in this particular case, the two eigenvectors (2.476). For a generic n d.o.f. system, if the number of vibration modes taken into consideration is n (i.e. the same as the number of d.o.f. of the system), $[\Phi]$ is a $n \times n$ square matrix, and its columns contain the n vibration modes of the structure. By using a Lagrange's equation (Sect. 2.2.5), all the matrix relations, expressed previously as a function of the independent coordinates $\underline{x}$, can be rewritten in the new variables $\underline{q}$. Potential energy V (2.80) becomes:

$$V = \frac{1}{2}\underline{x}^T[K]\underline{x} = \frac{1}{2}\underline{q}^T[\Phi]^T[K][\Phi]\underline{q} = \frac{1}{2}\underline{q}^T\left[\overline{K}\right]\underline{q} \tag{2.481}$$

having used:

$$\left[\overline{K}\right] = [\Phi]^T[K][\Phi] \tag{2.482}$$

to indicate the stiffness matrix in the new principal independent coordinates $\underline{q}$. It is important to note how, by applying the new set of independent variables, i.e. by writing the potential energy V as a function of $\underline{q}$, the formal writing of the same form of energy does not change its appearance. By applying a change of variables (2.478) in the expression of kinetic energy E_c [Sect. 2.4.2, (2.79)] we obtain:

$$E_c = \frac{1}{2}\underline{\dot{x}}^T[M]\underline{\dot{x}} = \frac{1}{2}\underline{\dot{q}}^T[\Phi]^T[M][\Phi]\underline{\dot{q}} = \frac{1}{2}\underline{\dot{q}}^T\left[\overline{M}\right]\underline{\dot{q}} \tag{2.483}$$

having used:

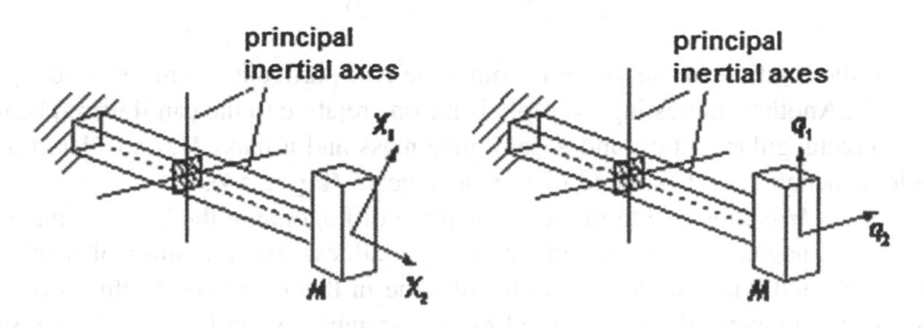

Fig. 2.54 Physical meaning of the principal coordinates

$$\left[\overline{M}\right] = [\Phi]^T [M][\Phi] \tag{2.484}$$

to indicate the mass matrix in the new principal independent coordinates $\underline{q}$. The equations of motion (2.474), in the new independent coordinates $\underline{q}$, thus become:

$$\left[\overline{M}\right]\ddot{\underline{q}} + \left[\overline{K}\right]\underline{q} = \underline{0} \tag{2.485}$$

It is possible to demonstrate that the new independent coordinates $\underline{q}$, obtainable explicitly, in the case of a squared modal matrix $[\Phi]$, from relation:

$$\underline{q} = [\Phi]^{-1}\underline{x} \tag{2.486}$$

have, on account of the way in which they are defined, the ability to uncouple the n equations of motion of the system (2.474), i.e. they are such as to have in each equation only one unknown. As a consequence of this property, the $\underline{q}$ variables thus allow us to diagonalize matrices $[M]$ and $[K]$ of a discrete n d.o.f. system: for this reason, the new coordinates are termed "principal" or "orthogonal" (an example of this property will be given in Sect. 2.5.2.1). If $[\Phi]$ were an ordinary matrix, i.e. not a modal matrix, matrices $\left[\overline{K}\right]$ and $\left[\overline{M}\right]$ shown in expressions (2.482) and (2.484) would be full: if, on the contrary, among all the possible variable changes, we decided to actually assume the modal matrix of (2.480) as a transformation matrix, then $\left[\overline{K}\right]$ and $\left[\overline{M}\right]$ are diagonal and the equations of motion (2.485) expressed as a function of the new independent variables $\underline{q}$ prove to be uncoupled. Generally speaking, the principal independent coordinates $\underline{q}$ for a vibrating system do not have a specific physical meaning. However, in some particular cases, this physical meaning can emerge like, for example, for the system shown in Fig. 2.7 (Sect. 2.3.3), defined by means of the two independent coordinatesx_1, the absolute vertical displacement of the centre of gravity and x_2 the absolute rotation of the rigid body: the two principal coordinates q_1 and q_2 which define the coordinate transformation (see Sect. 2.4.2.1.3 and, in particular, (2.478) and Fig. 2.41b):

$$\begin{aligned} x_1 &= \overline{X}_1^{(1)} q_1 + \overline{X}_1^{(2)} q_2 \\ x_2 &= \overline{X}_2^{(1)} q_1 + \overline{X}_2^{(2)} q_2 \end{aligned} \tag{2.487}$$

represent the rotation of the system around the two particular points P_1 and P_2 of Fig. 2.41b. Another interesting example is the one relative to the cantilevered beam, having a rectangular section and a negligible mass and a mass M, considered as a particle, concentrated at one end of the same beam (Fig. 2.54).

This mass M is thus free to move in the plane orthogonal to the longitudinal axis of the beam, neglecting, on account of their smallness, the longitudinal displacements associated with the deformability of same in this direction. In this way, the mass proves to have 2 d.o.f. identified by two variables x_1 and x_2 which represent

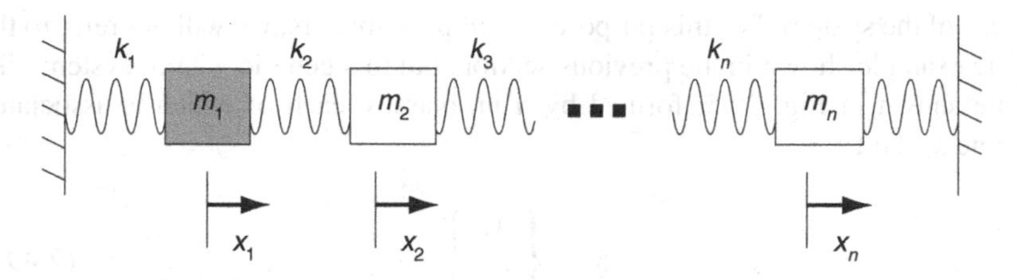

Fig. 2.55 Generic vibrating n d.o.f. system

the displacements in two generic mutually orthogonal directions (Fig. 2.54). In general, the equations of dynamic equilibrium which define the free motion of mass M, in the absence of damping, according to the two generic directions x_1 and x_2 are of the following type:

$$\begin{aligned} M\ddot{x}_1 + k_{11}x_1 + k_{12}x_2 &= 0 \\ M\ddot{x}_2 + k_{21}x_1 + k_{22}x_2 &= 0 \end{aligned} \tag{2.488}$$

i.e. they are two equations which are not uncoupled, due to the presence of mixed coefficients k_{ij} of stiffness, i.e. the elastic restoring forces which arise in the generic ith ($i = 1, 2$) direction due to the effect of a displacement in the other direction j (we find ourselves in unsymmetrical bending conditions). If, among the infinite set of variables, as free coordinates of the system we choose displacements q_1 and q_2 (Fig. 2.54), which define the motion of mass in the directions parallel to the principal axes of the central inertia ellipse of the beam section (Fig. 2.54), the equations are reduced, as explained by the Science of Constructions (straight bending conditions, in which one displacement in one direction does not generate a force in the other and vice versa), to:

$$\begin{aligned} M\ddot{q}_1 + k_{q1}q_1 &= 0 \\ M\ddot{q}_2 + k_{q2}q_2 &= 0 \end{aligned} \tag{2.489}$$

In other words, for the vibrating system shown in Fig. 2.54, the principal coordinates $\underline{q}$ (i.e. the coordinates that uncouple the equations of motion) represent the displacements according to the main directions of the axes of the central inertia ellipse.

2.5.2 Modal Approach for n-Degree-of-Freedom Systems

We now wish to demonstrate how a change of coordinates, expressed in scalar form in (2.477) or in matrix form in (2.478), uncouples the equations of motion of a vibrating n d.o.f. system, i.e. results in the diagonalization of the stiffness and mass matrices of the system. For this purpose, to simplify matters, we will not refer to the specific example shown in the previous section, but to a generic n d.o.f. system, like the one shown in Fig. 2.55, formed by n m_i masses, each of which is associated with one x_i d.o.f.:

$$\underline{x} = \begin{Bmatrix} x_1 \\ \cdots \\ x_n \end{Bmatrix} \tag{2.490}$$

For example, the equations of motion can be written by using the equations of dynamic equilibrium relative to the generic mass: since variables $\underline{x}$ define the disturbed motion about the static equilibrium configuration and since the link between the elongations Δl_j of the springs is linear with the independent variables x_i themselves, the equations of equilibrium will be linear and of the following type:

$$\begin{cases} m_1 \ddot{x}_1 + (k_1 + k_2)x_1 - k_2 x_2 = 0 \\ \cdots \\ m_i \ddot{x}_i - k_i x_{i-1} - k_{i+1} x_{i+1} + (k_i + k_{i+1})x_i = 0 \\ \cdots \\ m_n \ddot{x}_n - k_n x_{n-1} + (k_n + k_{n+1})x_n = 0 \end{cases} \tag{2.491}$$

The same system of equations can obviously be obtained using an energy-related approach, exploiting, for example, Lagrange's equations by defining potential energy V and kinetic energy E_c of the system as:

$$V = \sum_j^{n+1} \frac{1}{2} k_j \Delta_{lj}^2 \tag{2.492}$$

$$E_c = \sum_i^{n} \frac{1}{2} m_i \dot{x}_i^2 \tag{2.493}$$

having used:

$$\begin{cases} \Delta l_1 = x_1 \\ \cdots \\ \Delta l_i = x_i - x_{i-1} \\ \cdots \\ \Delta l_{n+1} = x_n \end{cases} \tag{2.494}$$

to indicate the elongations of the elastic elements (positive by convention if producing traction). By applying Lagrange's form to expressions (2.492) and (2.493) we thus obtain a system of equations, similar to that obtained with dynamic equilibriums, which in matrix form can be expressed as usual by the following relation:

$$[M]\underline{\ddot{x}} + [K]\underline{x} = \underline{0} \tag{2.495}$$

having used $\underline{x}$ to indicate the vector of the independent variables (2.490) and due to the fact that $[K]$ and $[M]$ are respectively the matrices of stiffness and mass of the system under examination:

$$[K] = \begin{bmatrix} k_1 + k_2 & -k_2 & \dots & 0 \\ -k_2 & k_2 + k_3 & \dots & \dots \\ \dots & \dots & \dots & \dots \\ 0 & \dots & -k_n & k_n + k_{n+1} \end{bmatrix}$$
$$[M] = \begin{bmatrix} m_1 & 0 & \dots & 0 \\ 0 & m_2 & \dots & \dots \\ \dots & \dots & \dots & \dots \\ 0 & \dots & 0 & m_n \end{bmatrix} \tag{2.496}$$

We now note that, since the conservative system analysed is perturbed about a stable equilibrium configuration, $[K]$ and $[M]$ are positive definite symmetrical matrices (in this particular example, given the specific type of independent variables assumed, the mass matrix $[M]$ is also diagonal, while the stiffness matrix $[K]$ is a three diagonal band matrix) and the solution is of the following type (Sect. 2.3.1):

$$\underline{x} = \underline{X}e^{i\omega t} \tag{2.497}$$

which, when substituted in expressions (2.495), results in a homogenous linear algebraic system in $\underline{X}$ and parametric in ω:

$$(-\omega^2[M] + [K])\underline{X} = \underline{0} \tag{2.498}$$

The non-trivial solutions can be ascribed (Sect. 2.4.2.1.1) to a problem of eigenvalues eigenvectors on account of having defined matrix

$$[A] = [M]^{-1}[K] \tag{2.499}$$

with $[A]$ i.e.:

$$([A] - \omega^2[I])\underline{X} = \underline{0} \tag{2.500}$$

The eigenvalues of matrix $[A]$ correspond to n values of ω^2 which satisfy expression (2.500) and which represent the natural frequencies of the system. By imposing the generic value ω_i in system (2.498) on natural frequency ω, it is possible to calculate the generic *i*th principal vibration mode $\underline{X}^{(i)}$ of the system, whose shape will be defined, up to a constant:

$$\omega = \omega_i \Rightarrow \underline{X}^{(i)} = \begin{Bmatrix} \overline{X}_1^{(1)} \\ \cdots \\ \overline{X}_n^{(i)} \end{Bmatrix} \quad (i = 1, 2, \ldots, n) \tag{2.501}$$

Let us now perform a change of variables. The new coordinates $\underline{q}$:

$$\underline{q} = \begin{Bmatrix} q_1 \\ \cdots \\ q_n \end{Bmatrix} \tag{2.502}$$

will now be defined as multiplicative coefficients of the deformation shapes of n principal vibration modes: the generic deformation shape $\underline{x}$ is thus represented as a linear combination of the eigenvectors (2.501) i.e. in scalar form:

$$\begin{cases} x_1 = \overline{X}_1^{(1)} q_1 + \overline{X}_1^{(2)} q_2 + \cdots + \overline{X}_1^{(i)} q_i + \cdots + \overline{X}_1^{(n)} q_n \\ x_2 = \overline{X}_2^{(1)} q_1 + \overline{X}_2^{(2)} q_2 + \cdots + \overline{X}_2^{(i)} q_i + \cdots + \overline{X}_2^{(n)} q_n \\ \cdots \\ x_n = \overline{X}_n^{(1)} q_1 + \overline{X}_n^{(2)} q_2 + \cdots + \overline{X}_n^{(i)} q_i + \cdots + \overline{X}_n^{(n)} q_n \end{cases} \tag{2.503}$$

where the various mode components should be considered numerically fixed, i.e. the generic principal vibration mode $\underline{X}^{(i)}$ should be considered normalized, i.e. for each mode the relative constant multiplicative has been fixed arbitrarily a priori. The aforementioned change of variables expressed in scalar form in (2.503) can be rewritten in a more compact matrix form as:

$$\underline{x} = [\Phi]\underline{q} \tag{2.504}$$

having defined, as in the previous section in relation to 2 d.o.f. systems (2.478), the modal matrix $[\Phi]$ having as columns n vibration modes, i.e. n eigenvectors $\underline{X}^{(i)}$, ordered by columns:

$$[\Phi] = \begin{bmatrix} \underline{X}^{(1)} & \underline{X}^{(2)} & \ldots & \underline{X}^{(n)} \end{bmatrix} = \begin{bmatrix} \overline{X}_1^{(1)} & \overline{X}_2^{(1)} & \ldots & \overline{X}_n^{(1)} \\ \overline{X}_1^{(2)} & \overline{X}_2^{(2)} & \ldots & \overline{X}_n^{(2)} \\ \ldots & \ldots & \ldots & \ldots \\ \overline{X}_1^{(n)} & \overline{X}_2^{(n)} & \ldots & \overline{X}_n^{(n)} \end{bmatrix} \tag{2.505}$$

Bearing in mind that:

$$\underline{x}^T = \left\{[\Phi]\underline{q}\right\}^T = \underline{q}^T [\Phi]^T \tag{2.506}$$

it is thus possible, in matrix form, to express kinetic energy E_c and potential energy V of the system analysed as a function of the new variables $\underline{q}$ as:

$$V = \frac{1}{2}\underline{q}^T [\Phi]^T [K][\Phi]\underline{q} = \frac{1}{2}\underline{q}^T \left[\overline{K}\right]\underline{q} \tag{2.507}$$

$$E_c = \frac{1}{2}\underline{\dot{q}}^T [\Phi]^T [M][\Phi]\underline{\dot{q}} = \frac{1}{2}\underline{\dot{q}}^T \left[\overline{M}\right]\underline{\dot{q}} \tag{2.508}$$

Let us now show how matrices $\left[\overline{K}\right]$ and $\left[\overline{M}\right]$ are diagonal, i.e. how the undamped vibration modes of the system have the property of uncoupling the equations of motion.

2.5.2.1 The Orthogonality of Vibration Modes

As seen, potential and kinetic energy do not formally change their appearance when the independent coordinates, chosen to describe the motion of the vibrating system, vary: in order to ensure that the change of variables uncouples the equations of motion, matrices $\left[\overline{K}\right]$ and $\left[\overline{M}\right]$ must prove to be diagonal. We will now show how the modal matrix $[\Phi]$ defined in (2.505) makes these matrices diagonal. For this purpose, we will consider two ordinary generic modes $\underline{X}^{(r)}$ and $\underline{X}^{(s)}$ corresponding to two generic natural frequencies ω_r and ω_s, i.e. any two ordinary columns of the modal matrix $[\Phi]$. Obviously the generic eigenvalue ω_i and the corresponding eigenvector $\underline{X}^{(i)}$ satisfy (2.498):

$$\begin{aligned} -\omega_s^2[M]\,\underline{X}^{(s)} + [K]\,\underline{X}^{(s)} &= \underline{0} \Rightarrow [K]\,\underline{X}^{(s)} = \omega_s^2[M]\,\underline{X}^{(s)} \\ -\omega_r^2[M]\,\underline{X}^{(r)} + [K]\,\underline{X}^{(r)} &= \underline{0} \Rightarrow [K]\,\underline{X}^{(r)} = \omega_r^2[M]\,\underline{X}^{(r)} \end{aligned} \tag{2.509}$$

By pre-multiplying the first of the Eq. (2.509) by $\underline{X}^{(s)\,T}$ and the second by $\underline{X}^{(r)\,T}$ we obtain:

$$\begin{aligned} \underline{X}^{(r)T}[K]\underline{X}^{(s)} &= \omega_s^2\underline{X}^{(r)T}[M]\underline{X}^{(s)} \\ \underline{X}^{(s)T}[K]\underline{X}^{(r)} &= \omega_r^2\underline{X}^{(s)T}[M]\underline{X}^{(r)} \end{aligned} \tag{2.510}$$

By transposing the first equation of expressions (2.510), the same equations can be rewritten as:

$$\begin{aligned} \underline{X}^{(s)T}[K]\underline{X}^{(r)} &= \omega_s^2\underline{X}^{(s)T}[M]\underline{X}^{(r)} \\ \underline{X}^{(s)T}[K]\underline{X}^{(r)} &= \omega_r^2\underline{X}^{(s)T}[M]\underline{X}^{(r)} \end{aligned} \tag{2.511}$$

where, as we recall, the transpose of a matrix product can be obtained by transposing the single matrices and inverting their order in the product itself [2, 20] and where matrices $[K]$ and $[M]$ are symmetrical, so that the following relations apply:

$$\begin{aligned} \left\{\underline{X}^{(s)T}\right\}^T &= \underline{X}^{(s)} \\ [K]^T &= [K] \\ [M]^T &= [M] \end{aligned} \tag{2.512}$$

Let us note that the first members of expressions (2.511) are equal: by equalling the second members we obtain:

$$(\omega_s^2 - \omega_r^2)\underline{X}^{(s)T}[M]\underline{X}^{(r)} = 0 \tag{2.513}$$

If we now consider two distinct modes (i.e. corresponding to two non-coincident eigenvalues), the following relations must most certainly apply:

$$\begin{aligned} r \neq s \Rightarrow \omega_s^2 \neq \omega_r^2 \Rightarrow \underline{X}^{(s)T}[M]\underline{X}^{(r)} = 0 \\ r = s \Rightarrow \omega_s^2 = \omega_r^2 \Rightarrow \underline{X}^{(s)T}[M]\underline{X}^{(r)} = m_{ss} \neq 0 \end{aligned} \tag{2.514}$$

The two conditions (2.514) obtained for the eigenvectors enable us to define the same as orthogonal in the broadest sense of the term[13] [2, 4, 20] because they satisfy the generic condition:

$$\begin{aligned} \underline{X}^{(s)T}[M]\underline{X}^{(r)} = m_{sr} \Rightarrow m_{sr} = 0 \quad (r \neq s) \\ \underline{X}^{(s)T}[M]\underline{X}^{(r)} = m_{sr} \Rightarrow m_{sr} \neq 0 \quad (r = s) \end{aligned} \tag{2.515}$$

where the generic term m_{ss} represents the "generalized mass" relative to the sth vibration mode.

From expressions (2.509) and (2.510) it also results that:

$$\begin{aligned} \underline{X}^{(s)T}[K]\underline{X}^{(r)} = 0 \quad (r \neq s) \\ \underline{X}^{(s)T}[K]\underline{X}^{(r)} = k_{sr} \Rightarrow k_{sr} \neq 0 \quad (r = s) \end{aligned} \tag{2.516}$$

[13]Let us recall that (2.516) represents orthogonality condition in broad sense, while condition:

$$\underline{X}^{(r)T}\underline{X}^{(s)} = x_{rs}\delta_{rs} \tag{2.13.1}$$

being δ_{rs} the Kroneker's δ ($\delta_{rs} = 0$ for r $\neq$ s and $\delta_{rs} = 1$ for r = s), is intended in strict sense.

Having obtained these results, let us return to the problem of showing that a change of variables (2.504) diagonalizes the mass and stiffness matrices. For this purpose, let us consider the modal matrix $[\Phi]$ defined in partitioned form (2.505) where each vector column $\overline{\underline{X}}^{(i)}$ (corresponding to the generic ith eigenvector) is constituted by $n\,\overline{X}_j^{(i)}$ components. As can be seen, in this way $[\Phi]$ formally assumes the appearance of a row vector having column vectors as elements:

$$[\Phi] = \left[\overline{\underline{X}}^{(1)} \quad \overline{\underline{X}}^{(2)} \quad \ldots \quad \overline{\underline{X}}^{(n)}\right] \tag{2.517}$$

If we now consider $[\Phi]^T$, this can be seen as a column vector having, as elements, row vectors neatly defined as the transposed of the eigenvectors themselves:

$$[\Phi]^T = \begin{bmatrix} \overline{\underline{X}}^{(1)T} \\ \overline{\underline{X}}^{(2)T} \\ \ldots \\ \overline{\underline{X}}^{(n)T} \end{bmatrix} \tag{2.518}$$

By using $[\Phi]$ in this compact form it is possible to calculate the two products found in expressions (2.507) and (2.508):

$$\begin{aligned} [\Phi]^T[K][\Phi] &= \left[\overline{K}\right] \\ [\Phi]^T[M][\Phi] &= \left[\overline{M}\right] \end{aligned} \tag{2.519}$$

Let us consider the first product: first of all we will calculate product $[M][\Phi]$ which, by keeping account of (2.505), becomes a row vector whose terms are, in turn, represented by column vectors $[[M]\,\underline{X}^{(i)}]$:

$$[M][\Phi] = \left[[M]\,\overline{\underline{X}}^{(1)} \quad [M]\,\overline{\underline{X}}^{(2)} \quad \ldots \quad [M]\,\overline{\underline{X}}^{(n)}\right] \tag{2.520}$$

Let us now pre-multiply product $[[M][\Phi]]$ (2.520) by matrix $[\Phi]^T$ in the form of a column vector (2.505), formally executing a product of a "row vector" $[\Phi]^T$ by a "column vector" $[M][\Phi]$. The final result of this operation is a square matrix of $n \times n$ order, whose elements are constituted by matrix products and, more precisely:

$$[\Phi]^T[M][\Phi] = \begin{bmatrix} \overline{\underline{X}}^{(1)T} \\ \overline{\underline{X}}^{(2)T} \\ \ldots \\ \overline{\underline{X}}^{(n)T} \end{bmatrix} \left[[M]\,\overline{\underline{X}}^{(1)} \quad [M]\,\overline{\underline{X}}^{(2)} \quad \ldots \quad [M]\,\overline{\underline{X}}^{(n)}\right] \tag{2.521}$$

by performing the products of (2.521) we obtain:

$$[\Phi]^T[M][\Phi]\dot{\underline{q}} = [\overline{M}] = \begin{bmatrix} \underline{X}^{(1)T}[M]\underline{X}^{(1)} & \underline{X}^{(1)T}[M]\underline{X}^{(2)} & \dots & \underline{X}^{(1)T}[M]\underline{X}^{(n)} \\ \underline{X}^{(2)T}[M]\underline{X}^{(1)} & \underline{X}^{(2)T}[M]\underline{X}^{(2)} & \dots & \underline{X}^{(2)T}[M]\underline{X}^{(n)} \\ \dots & \dots & \dots & \dots \\ \underline{X}^{(n)T}[M]\underline{X}^{(1)} & \underline{X}^{(n)T}[M]\underline{X}^{(2)} & \dots & \underline{X}^{(n)T}[M]\underline{X}^{(n)} \end{bmatrix} \tag{2.522}$$

By keeping account of the relationships of orthogonality of the principal vibration modes (2.515), we find that in the previous matrix $[\overline{M}]$ all the extra-diagonal terms are null (since the indices relative to the order of the modes are different, $r \neq s$), while only the diagonal terms (where the two eigenvectors have the same index) are different from zero:

$$[\overline{M}] = \begin{bmatrix} m_{11} & 0 & \dots & 0 \\ 0 & m_{22} & \dots & 0 \\ \dots & \dots & \dots & \dots \\ 0 & 0 & \dots & m_{nn} \end{bmatrix} \qquad (2.523))$$

As previously mentioned, the generic terms of the principal diagonal m_{ii} is defined as a generalized mass relative to the generic ith principal vibrating mode. The numeric value associated with the generalized mass depends on the type of normalization adopted to define, always numerically, the corresponding eigenvector: to be more specific, if, for example, we choose the arbitrary multiplicative constant of the mode so that:

$$[\overline{M}] = [I] \tag{2.524}$$

the mass matrix in principal coordinates is an identity matrix. By using exactly the same method, from (2.346) we also show that the product:

$$[\Phi]^T[K][\Phi] = [\overline{K}] = \begin{bmatrix} k_{11} & 0 & \cdots & 0 \\ 0 & k_{22} & \cdots & 0 \\ \dots & \dots & \dots & \dots \\ 0 & 0 & \cdots & k_{nn} \end{bmatrix} \tag{2.525}$$

becomes a diagonal matrix (always due to the orthogonality of the eigenvectors): the generic term k_{ii}, similar to the definition of the generalized mass m_{ii}, is defined as generalized stiffness. If the elements of the eigenvectors are normalized in the form shown in (2.354), the diagonal elements of $[\overline{K}]$ are tidily given by $\omega_1^2, \omega_2^2, \dots, \omega_n^2$ i.e. by the square of the eigenvalues of problem (2.498):

$$\left[\overline{K}\right] = \begin{bmatrix} \omega_1^2 & 0 & \dots & 0 \\ 0 & \omega_2^2 & \dots & 0 \\ \dots & \dots & \dots & \dots \\ 0 & 0 & \dots & \omega_n^2 \end{bmatrix} \tag{2.526}$$

where $\omega_i^2 = k_{ii} m_{ii}$. The equations of motion of the system in principal coordinates with normalization (2.524) can be rewritten as:

$$\left[\overline{M}\right]\ddot{\underline{q}} + \left[\overline{K}\right]\underline{q} = \underline{0} \tag{2.527}$$

i.e.:

$$\begin{bmatrix} 1 & 0 & \dots & 0 \\ 0 & 1 & \dots & 0 \\ \dots & \dots & \dots & \dots \\ 0 & 0 & \dots & 1 \end{bmatrix} \begin{Bmatrix} \ddot{q}_1 \\ \ddot{q}_2 \\ \dots \\ \ddot{q}_n \end{Bmatrix} + \begin{bmatrix} \omega_1^2 & 0 & \dots & 0 \\ 0 & \omega_2^2 & \dots & 0 \\ \dots & \dots & \dots & \dots \\ 0 & 0 & \dots & \omega_n^2 \end{bmatrix} \begin{Bmatrix} q_1 \\ q_2 \\ \dots \\ q_n \end{Bmatrix} = \begin{Bmatrix} 0 \\ 0 \\ 0 \\ 0 \end{Bmatrix} \tag{2.528}$$

or:

$$[I]\ddot{\underline{q}} + \left[diag(\omega_i^2)\right]\underline{q} = \underline{0} \tag{2.529}$$

Given the particular structure of matrices $\left[\overline{M}\right]$ and $\left[\overline{K}\right]$ (2.524) and (2.526), these equations will obviously each contain one single unknown: in extended scalar form the system of differential equations, generally coupled, which describes the motion of a generic vibrating n d.o.f. system, see (2.529) is reduced to principal coordinates with n uncoupled equations of the type shown below:

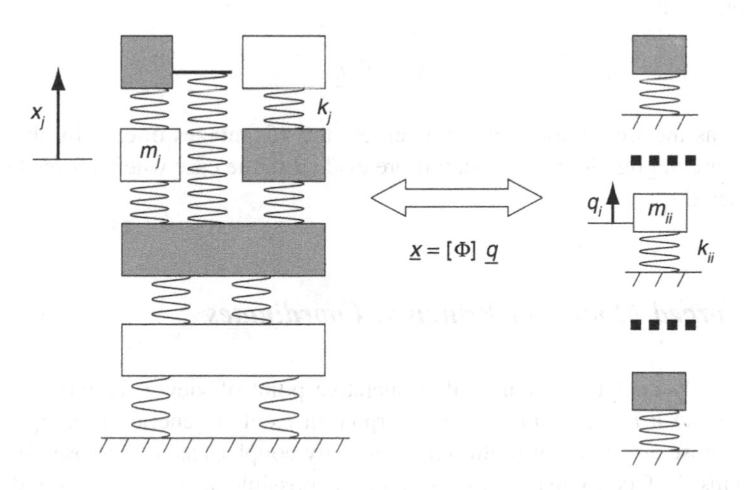

Fig. 2.56 Logic diagram of the modal approach

$$\begin{cases} \ddot{q}_1 + \omega_1^2 q_1 = 0(911) \\ \ddot{q}_2 + \omega_2^2 q_2 = 0(912) \\ \ldots\ldots\ldots\ldots(913) \\ \ddot{q}_n + \omega_n^2 q_n = 0(914) \end{cases} \tag{2.530}$$

Conversely, by adopting a generic normalization of the eigenvectors we obtain:

$$\begin{cases} m_{11}\ddot{q}_1 + k_{11}q_1 = 0 \\ m_{22}\ddot{q}_2 + k_{22}q_2 = 0 \\ \ldots \\ m_{nn}\ddot{q}_n + k_{nn}q_n = 0 \end{cases} \tag{2.531}$$

where the relationship $\omega_i^2 = k_{ii}m_{ii}$ is valid and where, as we recall, m_{ii} and k_{ii} are respectively the generalized mass of the generic ith vibration mode and the corresponding generalized stiffness. At this point, Eqs. (2.530) or (2.531) can be integrated: although the analytical advantage of operating in principal coordinates is a plus, it is also extremely important to highlight the physical aspects of the use of these coordinates. Basically speaking, the fact of operating on n uncoupled equations of the type (2.531) means interpreting the vibrating n d.o.f. system as being composed by n separate systems, each with 1 d.o.f., whose mass is the generalized mass of the mode considered m_{ii} and whose stiffness is the generalized stiffness k_{ii}: in other words, each of the n Eq. (2.531) in principal coordinates expresses the movement of the system as if it only had one d.o.f., i.e. the ith one, characterized by the natural frequency ω_i (Fig. 2.56).

The generic motion of the system is thus given by a combination of the n vibration modes. Once the motion of the defined system, expressed in principal coordinates $\underline{q}$, is known, it is possible (see diagram in Fig. 2.56) to obtain same as a function of the free coordinates $\underline{x}$ by means of the transformation relationship of coordinates:

$$\underline{x} = [\Phi]\underline{q} \tag{2.532}$$

As far as the free motions are concerned, the advantages offered by the modal approach are negligible, being much more evident in the case when a forced system is considered.

2.5.3 Forced Motion in Principal Coordinates

From a purely analytical though also operative point of view it is extremely convenient to use a modal approach to interpret an excited generic vibrating n d.o.f. system (whose equations of motion are generally coupled) as if composed by n 1 d. o.f. systems. In fact, by using this method, it is possible, as will be seen in detail in this section, to see the same excitation force distributed individually on each vibration mode and to study the generic vibration mode subjected to this generalized excitation force separately. This analysis can be conducted exactly as if the generalized excitation force were applied to a system having only 1 d.o.f. so that the well-known diagram of dynamic response as a function of $\left|{}^{X_p}/_{X_{st}}\right|$ related to a 1 d.o.f. system can be used (Chap. 1, Sect. 1.3.3). It is clear that in order to obtain the response of the vibrating system in terms of independent coordinates $\underline{x}$ (the only ones with which, generally speaking, a specific physical meaning is associated) it is necessary to superimpose the effect of the single modes. More specifically, if the excitation force has a frequency Ω coinciding with one of the natural frequencies of system ω_i, the approach in principal coordinates, i.e. the modal distribution of the excitation force, is very interesting: in fact, under these conditions, the ith vibration mode finds itself in resonance conditions while all the others, despite being excited, are far removed from this condition. If the damping of the system is null, the generic ith resonance mode has infinite amplitudes q_i, meaning that, under this condition, the contribution of the other modes is negligible: under this condition the system behaves exactly like a 1 d.o.f. system and the deformation shape of the system coincides with that of resonance mode $\underline{X}^{(i)}$. In the presence of small damping (r/r_c < 2–3 %, as is usually the case in structures and machines), the generic ith mode reaches finite though elevated amplitudes (10–30 times that of the static deformation shape), while all the other principal coordinates q_j will remain far removed from resonance condition. In other words, when a vibrating n d.o.f. system is excited with $\Omega = \omega_i$ the contribution relative to the resonance mode is strongly amplified and predominates over all the other modes: it is clear that, under this condition, the motion of the system is prevalently determined, even in the presence of damping, by the resonance mode, since the contributions of all the other modes are negligible and expression (2.503) thus becomes:

$$\Omega = \omega_i \Rightarrow \begin{cases} x_1 = \overline{X}_1^{(i)} q_i \\ x_2 = \overline{X}_2^{(i)} q_i \\ \dots \\ x_n = \overline{X}_n^{(i)} q_i \end{cases} \tag{2.533}$$

The statement ascertaining that when a vibrating n d.o.f. system is excited with a coincident frequency with one of its natural frequencies it actually behaves like a 1 d.o.f. system and assumes as a deformation shape the specific one of the resonant vibration mode [in other words, its deformation shape is no longer a linear combination of all the modes, in that only the resonant one is predominant (2.533)] is justified. This behaviour is operatively exploited experimentally, in that set of methodologies, referred to as Modal Identification Techniques (described in detail in Chap. 8) that allow us to experimentally obtain the characteristics of complex vibrating systems.

What has been described until now, in qualitative terms, regarding excitation in principal coordinates, can be analytically described by writing the equations of motion of the excited system in principal coordinates. Let us now reconsider the generic excited vibrating system in Fig. 2.55, whose equations of motion, as a function of principal independent coordinates $\underline{q}$:

$$\underline{q} = \begin{Bmatrix} q_1 \\ \dots \\ q_n \end{Bmatrix} \tag{2.534}$$

can be obtained, as mentioned several times, by applying, for example Lagrange's equations in matrix form (Sect. 2.4.1.2):

$$\left\{\frac{d}{dt}\left\{\frac{\partial E_c}{\partial \underline{\dot{q}}}\right\}\right\}^T - \left\{\frac{\partial E_c}{\partial \underline{q}}\right\}^T + \left\{\frac{\partial D}{\partial \underline{\dot{q}}}\right\}^T + \left\{\frac{\partial V}{\partial \underline{q}}\right\}^T = \underline{Q} \tag{2.535}$$

where $\underline{Q}$ is the vector of the Lagrangian components of the active (non-conservative) forces defined by relation:

$$\delta * L = \underline{Q}^T \delta * \underline{q} \tag{2.536}$$

Using the usual methodology, it is necessary to express the various forms of energy found in (2.535) as a function of physical coordinates Y, to find the $\underline{Y} = \underline{Y}(\underline{x})$ link between the physical coordinates and the independent coordinates $\underline{x}$ and, finally, to impose the transformation of coordinates (2.504).

As regards this first example (Fig. 2.55), kinetic energy and potential energy have already been developed in (2.492) and (2.493): on the contrary, it is necessary to develop these steps with the dissipative function D and the virtual work $\delta * L$ of active non-conservative forces. In the example given, it is easy to immediately obtain the expression of these quantities directly in independent coordinates $\underline{x}$ by writing:

$$D = \sum_{j=1}^{n+1} \frac{1}{2} r_j \Delta \dot{l}_j^2 \tag{2.537}$$

having used $\Delta \dot{l}_j = (\dot{x}_i - \dot{x}_{i-1})$ to indicate the relative speeds of the ends of the dissipative elements. Using similar steps to those used for potential energy, we obtain:

$$D = \frac{1}{2} \underline{\dot{x}}^T [R] \underline{\dot{x}} \tag{2.538}$$

where $[R]$ is the damping matrix:

$$[R] = \begin{bmatrix} r_1 + r_2 & -r_2 & \dots & 0 \\ -r_2 & r_2 + r_3 & \dots & 0 \\ \dots & \dots & \dots & \dots \\ 0 & 0 & \dots & r_n + r_{n+1} \end{bmatrix} \tag{2.539}$$

and similarly the virtual work performed by the external non-conservative forces f_j is:

$$\delta * L = \sum_{j=1}^{n} f_j \, \delta * x_j = \underline{f}^T \, \delta * \underline{x} \tag{2.540}$$

having ordered vector $\underline{f}$ as:

$$\underline{f} = \begin{Bmatrix} f_1 \\ \dots \\ f_j \\ \dots \\ f_n \end{Bmatrix} \tag{2.541}$$

Having now imposed the modal transformation (2.532) where:

$$\delta * \underline{x} = [\Phi] \delta * \underline{q} \tag{2.542}$$

expressions (2.538) and (2.540) can be rewritten as:

$$D = \frac{1}{2} \underline{\dot{x}}^T [R] \underline{\dot{x}} = \frac{1}{2} \underline{\dot{q}}^T [\Phi]^T [R] [\Phi] \underline{\dot{q}} = \frac{1}{2} \underline{\dot{q}}^T \left[\overline{R}\right] \underline{\dot{q}} \quad \left[\overline{R}\right] = [\Phi]^T [R] [\Phi] \tag{2.543}$$

$$\delta * L = \underline{f}^T [\Phi] \delta * \underline{q} = \underline{Q}^T \, \delta * \underline{q} \tag{2.544}$$

where $\left[\overline{R}\right]$ is the damping matrix in principal coordinates and $\underline{Q}$ is the vector of the generalized forces:

$$\underline{Q}^T = \underline{f}^T [\Phi] \Rightarrow \underline{Q} = [\Phi]^T \underline{f} \tag{2.545}$$

The generic ith component Q_i of vector $\underline{Q}$ is thus worth:

$$Q_i = \underline{\overline{X}}^{(i)T} \underline{f} = \sum_{j=1}^{n} f_j \overline{X}_j^{(i)} \quad (i = 1, n) \tag{2.546}$$

Basically speaking, this component represents the work performed by external forces f_j for one displacement corresponding to the generic ith vibrating mode: from this viewpoint, it is clear how component Q_{ij} due to only one force f_j:

$$Q_{ij} = f_j \overline{X}_j^{(i)} \tag{2.547}$$

is annulled in the event of the force being applied in a nodal point of the mode considered, since the displacement of the application point of the force itself is null. In the case in which the generic component (2.547) is null, the system of applied excitation forces is incapable of introducing energy into the system, i.e. for no reason whatsoever will the excitation force considered excite that natural vibration mode, even under resonance conditions. Generally speaking, the matrix of damping in principal coordinates $[\overline{R}]$ is not diagonal: conversely, it is diagonal if, and only if, matrix $[R]$ can be expressed as a linear combination of $[M]$ and $[K]$:

$$[R] = \alpha[M] + \beta[K] \tag{2.548}$$

In fact, by pre-multiplying expression (2.548) by $[\Phi]^T$, post-multiplying the same by $[\Phi]$ and recalling expressions (2.519), by using a number of simple steps we obtain:

$$\begin{aligned} [\overline{R}] &= [\Phi]^T[R][\Phi] = \alpha\,[\Phi]^T[M][\Phi] + \beta\,[\Phi]^T[K][\Phi] \\ &= \alpha[\overline{M}] + \beta[\overline{K}] \end{aligned} \tag{2.549}$$

Therefore, in this case, the matrix is diagonal and, on the principal diagonal contains the generalized damping coefficients r_{ii} defined as:

$$r_{ii} = \alpha m_{ii} + \beta k_{ii} \tag{2.550}$$

By operating the normalization of the eigenvectors with respect to the matrix of mass based on expression (2.524) we obtain:

$$[\overline{R}] = \alpha[I] + \beta\left[diag(\omega_i^2)\right] \Rightarrow r_{ii} = \alpha + \beta\omega_i^2 \tag{2.551}$$

By keeping account of expressions (2.519) and (2.549), applying Lagrange's equations and considering principal coordinates $\underline{q}$ as independent coordinates, we obtain the equations of motion of the damped and excited system:

$$[\overline{M}]\underline{\dot{q}} + [\overline{R}]\underline{\dot{q}} + [\overline{K}]\underline{q} = \underline{Q} \tag{2.552}$$

where the matrices of mass $[\overline{M}]$ and stiffness $[\overline{K}]$ are always diagonal and contain the generalized masses m_{rr} (2.515) and the generalized stiffnesses k_{rr} (2516) relative to the various vibration modes considered. In (2.552), generally speaking, the matrix of damping $[\overline{R}]$ is full: this is diagonal if relation (2.552) applies and, in this case, contains the generalized dampings r_{rr} defined in (2.551). Finally, in Eq. (2.552), vector $\underline{Q}$ of the active force components in principal coordinates is obtained from relation (2.545). In the hypothesis that the damping matrix is diagonal, the system of Eq. (2.552) can be rewritten as n uncoupled equations, each of which is formally similar to the equation of motion of a vibrating 1 d.o.f. system:

$$m_{ii}\ddot{q}_i + r_{ii}\dot{q}_i + k_{ii}q_i = Q_i \quad (i = 1, n) \tag{2.553}$$

where m_{ii}, r_{ii} and k_{ii} are respectively the generalized mass, damping and stiffness of the generic vibration mode. By now analysing expression (2.552) it is possible to demonstrate how, in steady state conditions, if the excitation forces applied to the system are such as to annul the generic Lagrangian component Q_i, as previously ascertained, the excitation force considered will not excite the mode considered, even under resonance conditions. In this case, the ith vibration mode will not bring any contribution to the global motion of the system, since the latter is always described as a linear combination of the principal modes. Let us now analyse the damping in greater detail: if the modal matrix $[\Phi]$ also makes the damping matrix $\boxed{R}$ diagonal, we are still able to see the n d.o.f. system, by reasoning in principal coordinates, as n 1 d.o.f. systems which can be studied separately, expressing each one of the n vibration modes (evaluated in the absence of damping) of the system itself. Conversely, in the case in which matrix $\boxed{R}$ does not become diagonal, Eq. (2.552) remain coupled in the velocity terms, in spite of obviously being uncoupled in $\underline{\dot{q}}$ and $\underline{q}$, i.e., in scalar for we will have:

$$m_{ii}\ddot{q}_i + \sum_{j=1}^{n}\left(r_{ij}\dot{q}_j\right) + k_{ii}q_i = Q_i \quad (i = 1, n) \tag{2.554}$$

It is clear that, in this case, one of the advantages of the modal approach ceases to exist, because there is no longer the option of uncoupling the equations of motion and separately analysing the various vibration modes. In actual fact, it often happens that the damping matrix $[R]$ is a linear combination of $[M]$ and $[K]$. If the damping is hysteretic (Sect. 2.4.1.3), this proves to be associated, like elastic phenomena, with deformations; the energy dissipated by same is most certainly a fraction of the similar elastic potential energy, hence the proportionality between $[R]$ and $[K]$:

$$[R] = \beta[K] \Rightarrow r_{rr} = \beta k_{rr} \tag{2.555}$$

In the case in which $\boxed{R}$ is not diagonalized by modal matrix $[\Phi]$, if the extra-diagonal terms of matrix $\boxed{R}$ are small with respect to those on the principal diagonal, it is possible to neglect these and apply the modal analysis methodology, by once again considering the uncoupled Eq. (2.552): in this case, it is clear that the solution of motion defined in this way is approximated. Finally, it is important to remember that the modal approach allows for the introduction of structural damping measured experimentally, as will be seen later on, mode by mode, otherwise not definable analytically, by directly introducing diagonal matrix $\left[\overline{R}\right]$. The equations of motion of the system in principal coordinates (2.552), on account of not being coupled one to the other, can be integrated separately one at a time, by following the usual analytical method inherent in 1 d.o.f. systems (Sect. 2.4.1).

2.5.3.1 Harmonic Excitation Forces in Principal Coordinates

Let us consider the case of sinusoidal excitation forces of frequency Ω:

$$\underline{f}(t) = \underline{F}_0\, e^{i\Omega t} = \begin{Bmatrix} F_{o1} \\ F_{oj} \\ F_{on} \end{Bmatrix} e^{i\Omega t} \tag{2.556}$$

By keeping account of (2.545) the generic component of active non-conservative forces becomes:

$$\underline{Q} = [\Phi]^T \underline{f} = [\Phi]^T \underline{F}_0\, e^{i\Omega t} = \underline{Q}_0\, e^{i\Omega t} \tag{2.557}$$

i.e.:

$$m_i\, \ddot{q}_i + r_i\, \dot{q}_i + k_i q_i = Q_{oi} e^{i\Omega t} \tag{2.558}$$

having respectively used m_i, r_i and k_i to indicate the generalized mass, the generalized damping and the generalized stiffness relative to the generic ith vibration mode. In the event of our wanting to evaluate the steady-state response, i.e. the response of the system once the initial transient motion has finished, the solution, in complex form, of expressions (2.558) is, as known, of the following type:

$$q_i = q_{oi} e^{i\Omega t} \quad (i = 1, n) \tag{2.559}$$

which, when placed in the same equation of motion (2.558), gives:

$$q_{oi} = \frac{Q_{oi}}{(-\Omega^2 m_i + i\Omega r_i + k_i)} \quad (i = 1, n) \tag{2.560}$$

i.e. in modulus and in phase:

$$q_{oi} = |q_{oi}| e^{i\Psi_i} = \frac{Q_{oi}}{\sqrt{(-\Omega^2 m_i + k_i)^2 + (\Omega r_i)^2}} e^{i\Psi_i} \tag{2.561}$$

where

$$\Psi_i = a\tan\left(\frac{-i\Omega r_i}{-\Omega^2 m_i + k_i}\right) \tag{2.562}$$

For an improved interpretation of results, it is convenient to put these amplitudes (see see Sect. 1.3.3) in non-dimensional form, by defining the generic non-dimensional frequency a_i and the non-dimensional damping (compared with critical) h_i, both relative to the generic ith vibration mode:

$$\begin{aligned} a_i &= \frac{\Omega}{\omega_i} \\ h_i &= \frac{r_i}{2m_i\omega_i} \end{aligned} \tag{2.563}$$

so that expressions (2.561) become:

$$\begin{aligned} q_{oi} = |q_{oi}|e^{i\Psi_i} &= \frac{Q_{oi}/k_i}{\sqrt{(-a_i+1)^2+(2h_i a_i)^2}} e^{i\Psi_i} \\ &= \frac{q_{ist}}{\sqrt{(-a_i+1)^2+(2h_i a_i)^2}} e^{i\Psi_i} \end{aligned} \tag{2.564}$$

where:

$$\Psi_i = a\tan\left(\frac{-2h_i a_i}{-a_i+1}\right) \tag{2.565}$$

In (2.564):

$$q_{ist} = Q_{oi}/k_i \tag{2.566}$$

was used to indicate the value of the response of the generic ith vibration mode, based on the assumption that the vibrating system is excited by an excitation force of modulus Q_{oi} but of a null frequency $\Omega = 0$. In the more specific case of normalization of the eigenvectors with respect to the matrix of mass (2.524), expressions (2.564) and (2.565) become:

$$q_{oi} = |q_{oi}|e^{i\Psi_i} = \frac{Q_{oi}/\omega_i^2}{\sqrt{(-a_i+1)^2+(2h_i a_i)^2}} e^{i\Psi_i} \tag{2.567}$$

$$\Psi_i = a\tan\left(\frac{-2h_i a_i}{-a_i+1}\right) \quad (i = 1, n) \tag{2.568}$$

Having obtained the amplitude of the vibration of the ith mode through (2.567), it is possible to determine the dynamic amplification coefficient of the same mode defined by ratio:

$$\frac{q_{oi}}{Q_{oi}/k_i} = \frac{1}{\sqrt{(-a_i+1)^2+(2h_i a_i)^2}} e^{i\Psi_i} \quad (i = 1, n) \tag{2.569}$$

which defines the extent to which the dynamic response of the system (i.e. when the force is applied dynamically with a certain frequency Ω) is modified with respect to the static case. Obviously the study of this function will lead to conclusions, already widely analysed in the case of vibrating 1 d.o.f. systems (Sect. 2.4.1): the diagram showing the dynamic amplification (in modulus and in phase, because the same amplitude is complex) is shown for the sake of convenience in Fig. 2.57. As already mentioned in Sect. 2.4.1.4 this diagram can be divided into 3 separate zones:

- $a_i \ll 1 \Rightarrow \Omega \ll \omega_i$ a quasi-static zone: in this case the excitation force is prevalently balanced by the elastic force and the response of the mode is in-phase with this force;
- $a_i = 1 \Rightarrow \Omega = \omega_i$ a resonance zone: in this zone the vibration amplitude increases, tending towards the infinite if the damping is null;
- $a_i \gg 1 \Rightarrow \Omega \gg \omega_i$ a seismographic zone: the vibration amplitude decreases and the external force is prevalently balanced by the inertia force, the phase tends towards 180°.

In the event of a vibrating system being excited by a harmonic excitation force with frequency $\Omega = \omega_i$, the ith mode finds itself in resonance conditions and the corresponding amplitude q_{oi} is just as high as damping h_i is small: on account of not being near to resonance condition $\Omega \neq \omega_j$ the amplitudes relative to the other vibration modes q_{oj} $(j \neq i)$ remain small. Now, it is always true that the motion of

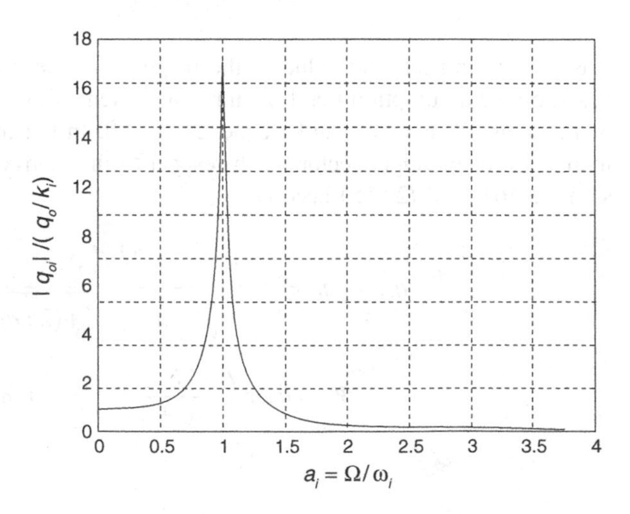

Fig. 2.57 Diagram of the dynamic amplification coefficient of the generic ith vibration mode

the vibrating system is described, using the modal approach, as a linear combination of vibration modes:

$$\begin{cases} x_1 = \left(q_{o1}\overline{X}_1^{(1)} + \cdots + q_{oi}\overline{X}_1^{(i)} + \cdots + q_{on}\overline{X}_1^{(n)}\right)e^{i\Omega t} \\ x_2 = \left(q_{o1}\overline{X}_2^{(1)} + \cdots + q_{oi}\overline{X}_2^{(i)} + \cdots + q_{on}\overline{X}_2^{(n)}\right)e^{i\Omega t} \\ \quad\cdots \\ x_n = \left(q_{o1}\overline{X}_n^{(1)} + \cdots + q_{oi}\overline{X}_n^{(i)} + \cdots + q_{on}\overline{X}_n^{(n)}\right)e^{i\Omega t} \end{cases} \quad (2.570)$$

In the event of a mode being excited in resonance, contribution q_{oi} relative to the resonance mode which becomes predominant over the other modes q_{oj} $(j \neq i)$ is strongly amplified: this assertion is even truer the smaller the damping of the resonance mode. It is thus evident that, under this condition, the movement of the system is prevalently determined by the resonance mode, since the contributions of all the other mode are negligible, i.e.:

$$\Omega = \omega_i \Rightarrow \begin{cases} x_1 \approx q_{oi}\overline{X}_1^{(i)}e^{i\Omega t} \\ x_2 \approx q_{oi}\overline{X}_2^{(i)}e^{i\Omega t} \\ \quad\cdots \\ x_n \approx q_{oi}\overline{X}_n^{(i)}e^{i\Omega t} \end{cases} \quad (2.571)$$

As a result, by operating in principal coordinates, i.e. by describing a generic vibrating n d.o.f. system through its principal vibration modes:

- it is possible to consider the same system as n vibrating systems with only 1 d.o. f. each (2.564);
- if one of the frequencies of the excitation forces is near to one of the natural frequencies ω_i of the system itself during actual motion it is possible to neglect the contribution of all the non-resonant modes (2.571) if the dampings are low and the natural frequencies are distant.

In other words, the vibrating system, excited in resonance, behaves as if it were a 1 d.o.f. system. In the event of the system being excited in resonance, despite the corresponding Lagrangian component of the active stresses Q_{oi} being null, the excitation force cannot supply energy to excite the resonant mode and therefore its response is null: conversely, in this situation, the contribution of the non-resonant modes though small becomes important.

We now show everything that has been said until now in graphic form and more specifically diagram modulus $\left|X_{jo}\right|$ of the response of the generic jth d.o.f. x_j in terms of modal reconstruction (2.570):

$$\begin{aligned} x_j = X_{jo}e^{i\Omega t} &= \left(q_{o1}\,\overline{X}_j^{(1)} + \cdots + q_{oi}\,\overline{X}_j^{(i)} + \cdots + q_{on}\,\overline{X}_j^{(n)}\right)e^{i\Omega t} \\ &= \left(X_{jo}^{(1)} + \cdots + X_{jo}^{(i)} + \cdots + X_{jo}^{(n)}\right)e^{i\Omega t} \end{aligned} \quad (2.572)$$

having used:

$$\left|X_{jo}^{(i)}\right| = \left|q_{oi}\overline{X}_j^{(i)}\right| \tag{2.573}$$

to indicate the generic term, the contribution to the overall response $|X_{jo}|$ of the generic *i*th vibrating mode. Figure 2.58 shows the qualitative trend of the single terms $\left|X_{jo}^{(i)}\right|$ as a function of the frequency of the excitation force and the overall response $|X_{jo}|$.

This representation allows us to highlight another advantage, possibly the most important, of the modal approach, i.e. that of being able to reduce the number of d.o.f. that allow us to describe the dynamic behaviour of a generic vibrating mechanical *n* d.o.f. system, all the while maintaining high precision of the solution. The figure highlights several peculiarities of the system:

- the generic contribution of the generic i-nth vibration mode $\left|X_{jo}^{(i)}\right|$ (2.573) varies with the variation of the frequency of excitation force Ω, passing from the static zone in which:

$$\Omega < \omega_i \Rightarrow \frac{|q_{oi}|}{q_{st}} = 1 \tag{2.574}$$

to the resonance zone in which:

$$\Omega = \omega_i \Rightarrow \frac{|q_{oi}|}{q_{ist}} \rightarrow \infty \ (without{-}dampiong) \tag{2.575}$$

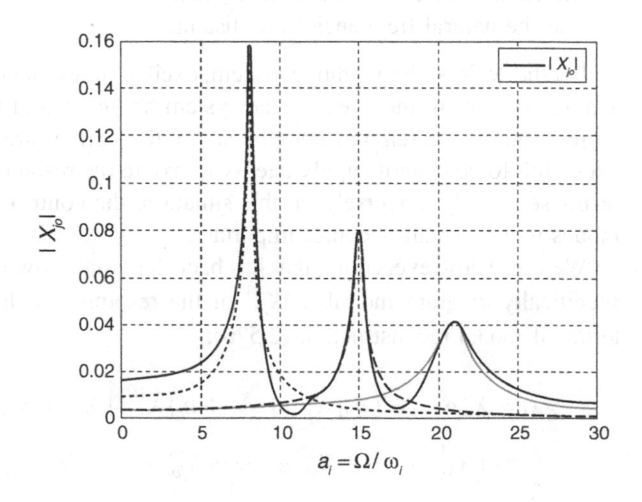

Fig. 2.58 Response of a *n* d.o.f. system: modal approach

to the seismographic zone in which:

$$\Omega > \omega_i \Rightarrow \frac{|q_{oi}|}{q_{st}} \to 0 \tag{2.576}$$

where q_{ist} represents the deformation shape assumed statically by q_i i.e. for $\Omega = 0$, i.e.:

$$q_{ist} = \frac{Q_{oi}}{\omega_i^2} \tag{2.577}$$

- as the number of order of the vibration mode considered increases, i.e. with the increase of natural frequency ω_i, static component q_{ist} of (2.577) decreases, with the consequent decrease of the corresponding value of $|q_{oi}|$;
- this means, with the same Lagrangian component Q_{oi} of active forces, a decrease in the importance of the contribution to the overall response $\left|X_{jo}^{(i)}\right|$ of the modes with a high number of order, which, for this reason, can thus be neglected

In other words, if we consider a generic frequency $\overline{\Omega}$ of excitation force, the contribution of the modes with frequencies near to $\overline{\Omega}$ is definitely important: the modes in the static zone (i.e. the modes having a frequency ω_i that is almost 2 times higher than the frequency of excitation force $\overline{\Omega}$) are negligible, while those corresponding to natural frequency ω_i lower than the frequency of excitation force $\overline{\Omega}$, despite being in a seismographic zone nevertheless present a not negligible contribution. Furthermore, it is important to point out that:

- external excitation forces f_j remaining unaltered, the Lagrangian component Q_{oi}, generally speaking, decreases when the order of the considered vibration mode increases;
- it is possible to verify, experimentally, that, generally speaking, structural damping h_i increases when frequency ω_i increases.

These considerations make more realistic the hypothesis of considering all the modes having a frequency lower than 2 times the frequency $\overline{\Omega}$ of the excitation force as vibration modes participating in the response of the system. Having thus defined a certain frequency range of interest $0 < \Omega < \Omega_{\max}$, the modal transformation can be performed by only considering all the p modes ($p < n$) having a natural frequency $\omega_i < 2\Omega_{\max}$ $(i = 1, 2, p)$, i.e. by considering a reduced modal matrix $[\Phi_r]$ (with n rows equalling the d.o.f. of the system and p columns equalling the vibration modes considered) from (2.504) and (2.517):

$$[\Phi_r] = \begin{bmatrix} \underline{X}^{(1)} & \underline{X}^{(2)} & \ldots & \underline{X}^{(p)} \end{bmatrix} \tag{2.578}$$

from which the transformation in principal coordinates becomes:

$$\underline{x} = [\Phi_r]\underline{q} \tag{2.579}$$

and the equations of motion of the system in principal coordinates (2.558) can be retraced to:

$$m_i\ddot{q}_i + r_i\dot{q}_i + k_iq_i = Q_{oi}e^{i\Omega t} \quad (i = 1, 2, \ldots, p) \tag{2.580}$$

This results in a reduction, often considerable, of the d.o.f. of the system and the need to define a limited number of natural frequencies and principal vibration modes with respect to the total d.o.f. number n.[14]

2.5.3.2 An Example

Let us now consider, as an applicative example of everything described in this section, the 2 d.o.f. system in Fig. 2.7, whose equations of motion (2.91) have already been defined in Sect. 2.3.3.1: the independent coordinates chosen to describe the motion of the system are the absolute vertical displacement of centre of gravity x_1 and absolute rotation x_2 of the rigid body, collected in vector $\underline{x}$:

$$x = \begin{Bmatrix} x_1 \\ x_2 \end{Bmatrix} \tag{2.581}$$

By using Lagrange's equations, the various forms of energy in scalar form (2.95) are:

$$\begin{cases} E_c = \frac{1}{2}m\dot{x}_1^2 + \frac{1}{2}J_G\dot{x}_2^2 \\ V = \frac{1}{2}k_2(x_1 + x_2l_2)^2 + \frac{1}{2}k_1(x_1 - x_2l_1)^2 \\ D = \frac{1}{2}r_2(\dot{x}_1 + \dot{x}_2l_2)^2 + \frac{1}{2}r_1(\dot{x}_1 - \dot{x}_2l_1)^2 \\ \delta * L = f(\delta * x_1 + \delta * x_2b) = f\delta * x_1 + fb\delta * x_2 \end{cases} \tag{2.582}$$

[14]Computer program packages which implement mathematical algorithms allowing us to calculate the first p natural frequencies of vibrating systems in a certain pre-established frequency range are already available at scientific libraries.

and, in matrix form, as:

$$\begin{aligned} E_c &= \frac{1}{2}\dot{\underline{x}}^T[M]\dot{\underline{x}} = \frac{1}{2}\dot{\underline{x}}^T\begin{bmatrix} m & 0 \\ 0 & J_G \end{bmatrix}\dot{\underline{x}} \\ V &= \frac{1}{2}\underline{x}^T[K]\underline{x} = \frac{1}{2}\underline{x}^T\begin{bmatrix} k_1+k_2 & -k_1l_1+k_2l_2 \\ -k_1l_1+k_2l_2 & k_1l_1^2+k_2l_2^2 \end{bmatrix}\underline{x} \\ D &= \frac{1}{2}\dot{\underline{x}}^T[R]\dot{\underline{x}} = \frac{1}{2}\dot{\underline{x}}^T\begin{bmatrix} r_1+r_2 & -r_1l_1+r_2l_2 \\ -r_1l_1+r_2l_2 & r_1l_1^2+r_2l_2^2 \end{bmatrix}\dot{\underline{x}} \\ \delta * L &= \underline{f}^T\delta * \underline{x} = \begin{Bmatrix} f \\ fb \end{Bmatrix}^T \delta * \underline{x} \end{aligned} \tag{2.583}$$

and the relative equations of motion become:

$$[M]\ddot{\underline{x}} + [R]\dot{\underline{x}} + [K]\underline{x} = \underline{f} \tag{2.584}$$

In order to adopt a modal approach, it is necessary to define the natural frequencies and the relative vibration modes by analysing the equations of free motion of the system in absence of damping:

$$[M]\ddot{\underline{x}} + [K]\underline{x} = \underline{0} \tag{2.585}$$

evaluating, for example, as described in Sect. 2.4.2.1, the eigenvalues and eigenvectors of matrix $[A] = [M]^{-1}[K]$. Having defined natural frequencies ω_i and vibration modes $\underline{X}^{(i)}$ in this way:

$$\begin{aligned} \omega = \omega_1 &\Rightarrow \underline{\overline{X}}^{(1)} = \begin{Bmatrix} \overline{X}_1^{(1)} \\ \overline{X}_2^{(1)} \end{Bmatrix} \\ \omega = \omega_2 &\Rightarrow \underline{\overline{X}}^{(2)} = \begin{Bmatrix} \overline{X}_2^{(1)} \\ \overline{X}_2^{(2)} \end{Bmatrix} \end{aligned} \tag{2.586}$$

In this case, modal matrix $[\Phi]$ is:

$$[\Phi] = \begin{bmatrix} \underline{\overline{X}}^{(1)} & \underline{\overline{X}}^{(2)} \end{bmatrix} = \begin{bmatrix} \overline{X}_1^{(1)} & \overline{X}_2^{(1)} \\ \overline{X}_2^{(1)} & \overline{X}_2^{(2)} \end{bmatrix} \tag{2.587}$$

These eigenvectors must be normalized, for example, with respect to the matrix of mass $[M]$ imposing that[15]:

$$\begin{aligned} \underline{\overline{X}}^{(1)T}[M]\underline{\overline{X}}^{(1)} &= 1 \\ \underline{\overline{X}}^{(2)T}[M]\underline{\overline{X}}^{(2)} &= 1 \end{aligned} \tag{2.588}$$

At this point it is possible to impose a change of variables:

$$\underline{x} = [\Phi]\underline{q} \tag{2.589}$$

thus obtaining:

$$\begin{cases} E_c = \frac{1}{2}\underline{\dot{q}}^T[\Phi]^T[M][\Phi]\underline{\dot{q}} = \frac{1}{2}\underline{\dot{q}}^T\left[\overline{M}\right]\underline{\dot{q}} \\ V = \frac{1}{2}\underline{q}^T[\Phi]^T[K][\Phi]\underline{q} = \frac{1}{2}\underline{q}^T\left[\overline{K}\right]\underline{q} \\ D = \frac{1}{2}\underline{\dot{q}}^T[\Phi]^T[R][\Phi]\underline{\dot{q}} = \frac{1}{2}\underline{\dot{q}}^T\left[\overline{R}\right]\underline{\dot{q}} \\ \delta * L = \underline{f}^T[\Phi]^T\delta * \underline{q} = \underline{Q}^T\delta * \underline{q} \end{cases} \tag{2.590}$$

[15]Actually, convenient normalizing procedure with respect to mass matrix are performed by:

- normalizing the eigenvectors $\underline{\overline{X}}^{(1)}$ and $\underline{\overline{X}}^{(2)}$ in any chosen way (for instance equating to one the maximum value);
- calculating the generalized masses obtained from:

$$\begin{aligned} \underline{\overline{X}}^{(1)T}[M]\underline{\overline{X}}^{(1)} &= m_{11} \\ \underline{\overline{X}}^{(2)T}[M]\underline{\overline{X}}^{(2)} &= m_{22} \end{aligned} \tag{2.15.1}$$

- obtaining:

$$\begin{aligned} \underline{X}^{(1)} &= \frac{1}{\sqrt{m_{11}}}\underline{\overline{X}}^{(1)} \\ \underline{X}^{(2)} &= \frac{1}{\sqrt{m_{22}}}\underline{\overline{X}}^{(2)} \end{aligned} \tag{2.15.2}$$

This procedure can be naturally extended to a n d.o.f. systems, so that:

$$\underline{X}^{(i)} = \frac{1}{\sqrt{m_{ii}}}\underline{\overline{X}}^{(i)} \tag{2.15.3}$$

being $\underline{\overline{X}}^{(i)}$ a generic vibration mode, normalized initially in any chosen way.

where vector $\underline{Q}$ is defined as:

$$\underline{Q} = \begin{Bmatrix} Q_1 \\ Q_2 \end{Bmatrix} = \begin{Bmatrix} f\,\overline{X}_1^{(1)} - bf\,\overline{X}_2^{(1)} \\ f\,\overline{X}_1^{(2)} - bf\,\overline{X}_2^{(2)} \end{Bmatrix} \tag{2.591}$$

We immediately note that if $f = f(t)$ is applied at point P_1 so that $b = \mu_1 = \overline{X}_1^{(1)}\overline{X}_2^{(1)}$, the centre of rotation for the first vibration mode (see (2.358), Sect. 2.4.2.1.3), we obtain:

$$\begin{aligned} Q_1 &= f\overline{X}_1^{(1)} - bf\overline{X}_2^{(1)} = f\left(\overline{X}_1^{(1)} - b\overline{X}_2^{(1)}\right) \\ &= f\overline{X}_1^{(1)}\left(1 - b\frac{\overline{X}_2^{(1)}}{\overline{X}_1^{(1)}}\right) = f\overline{X}_1^{(1)}\left(1 - \frac{b}{\mu_1}\right) \end{aligned} \tag{2.592}$$

Similarly if this were applied at P_2 (see Fig. 2.41b), the Lagrangian component Q_2 of the second mode would be null.

Summary This chapter outlines the dynamic behaviour of discrete systems described by linear or linearized equations of motion. From an engineering point of view, this approach simplifies the modelling of machines and mechanical systems subject to vibrations. One, two up to n-degree-of-freedom (d.o.f.) systems are considered, by writing the related equations of motion both in scalar and matrix form. Solution methods for the resulting equations are described for both free and forced motion. Several numerical examples are shown. At the end of the chapter. the modal approach in principal coordinates for discrete n d.o.f. systems is illustrated.

References

1. AA VV (1985) Appunti di meccanica applicata alle macchine. Edizioni Spiegel, Milano
2. Amerio L (1977) Analisi matematica. UTET, Torino
3. Andronov AA, Viu AA, Khaikin SE (1966) Theory of oscillators. Dover Publications, Inc., New York
4. Argyris J, Mleinek HP (1991) Texts on computational mechanics. Dynamic of structures, vol V. Elsevier Science Publishers, New York
5. ASME Meeting (1959) Structural damping. Pergamon Press, Atlanta
6. Bathe KJ (1982) Finite element procedures in engineering analysis. Prentice-Hall, Englewood Cliffs
7. Bathe KJ, Wilson EL (1976) Numerical methods in finite element method. Prentice-Hall, Englewood Cliffs
8. Bishop RED, Johnson DC (1960) The mechanics of vibrations. Cambridge University Press, London

9. Bishop RED, Gladwell GML, Michaelson S (1965) The matrix analysis of vibration. Cambridge University Press, London
10. Bittanti S, Schiavoni N (1982) Automazione e regolazione. Clup, Milano
11. Bittanti S, Schiavoni N (1982) Modellistica e controllo. Clup, Milano
12. Brigham EO (1974) The fast fourier transform. Prentice-Hall Inc., Englewood Cliffs, New Jersey
13. Cercignani C (1987) Spazio, tempo e movimento. Zanichelli, Bologna
14. Crandall SH, Mark WD (1963) Random vibration in mechanical systems. Academic Press, New York
15. Den Hartog JP (1956) Mechanical vibrations, 4th edn. McGraw-Hill Co., New York
16. Diana G, Resta F (2006) Controllo di sistemi meccanici. Polipress, Milano
17. Diana G, Mimmi G, Sirtori S (1982) Un modello di materiale quasi elasto- plastico sensibile alla velocità di deformazione. VI Congresso Nazionale AIMET A, Genova
18. Finzi B (1978) Meccanica razionale. Zanichelli, Bologna
19. Friedmann P, Hammond CE (1977) Efficient numerical treatment of periodic systems with application to stability problems. Int J Numer Methods Eng 11:1117–1136
20. Gasapina U (1979) Algebra delle matrici. Edizioni La Viscontea, Milano
21. Ghigliazza R, Galletti A (1989) Meccanica applicata alle macchine. UTET, Torino
22. Harris CM, Crede CE (1980) Shock and vibration handbook. McGraw-Hill Book Co., New York
23. Hayashi C (1964) Non-linear oscillations in physical systems. McGraw Hill Company, New York
24. Hinton E et al (1982) Non-linear computational mechanics. Pineridge Press Limited, Swansea
25. Holzer SH (1985) Computer analysis of structures. Elsevier, New York
26. Jacazio G, Piombo B (1977) Meccanica applicata alle macchine. Levrotto e Bella, Torino
27. Marple SL (1987) Digital spectral analysis with applications. Prentice-Hall Inc., Englewood Cliffs
28. Meirovitch L (1967) Analytical methods in vibrations. The Macmillan Co., New York
29. Meirovitch L (1970) Methods of analytical dynamics. McGraw-Hill Book Co., New York
30. Meirovitch L (1980) Computational methods in structural dynamics. The Sijthoff & Noordhoff International Publishers, The Netherlands
31. Meirovitch L (1986) Elements of vibration analysis. Mc Graw-Hill Book Co., New York
32. Pandit SM, Wu SM (1983) Time series and system analysis with applications. Wiley, New York
33. Sesini O (1964) Meccanica applicata alle macchine. Casa Editrice Ambrosiana, Milano
34. Shabana AA (1989) Dynamics of multibody systems. Wiley, New York
35. Thomson WT (1974) Vibrazioni meccaniche -teoria ed applicazioni. Tamburini Editore, Milano
36. Timoshenko SP, Young DH, Weaver W (1974) Vibration problems in engineering, 4th edn. Wiley, New York
37. Vernon JB (1967) Linear vibration theory—general properties and numerical methods. Wiley, New York

Chapter 3
Vibrations in Continuous Systems

3.1 Introduction

The analysis of a continuous system can be seen as an extrapolation, for n tending to infinity, of the analysis of discrete systems with n d.o.f. (see Chap. 2): the problem is that of analytically implementing this formal step. Equations related to continuous systems, unlike those of discrete systems, will be partial derivatives, since the amplitudes, that define the motion of the system in this case, depend both on time t and on space. All real systems should, in actual fact, be studied as continuous systems: the exact solution is only obtained in particularly simple cases. In complex structures, the analytical solution cannot be obtained using the natural equations of the continuous system. Therefore, in these situations, it becomes essential to return to discrete schemes, using suitable methodologies: concentrated parameters [3, 16], transfer matrices [3, 7, 20], finite elements (see Chap. 4) and, last but not least, finite differences [14, 16]. The study that will be conducted on a continuous system will therefore assume a prevalently didactic/educational aspect, with the same applying to the description of discretization methods. Furthermore, we will limit ourselves to particularly simple cases, for which an analytical solution in closed form is possible.

3.2 Transverse Vibrations of Cables

Some simple hypotheses will be assumed for these elements:

(a) Cables have a constant mass per unit length $m(\xi) = \rho(\xi)A(\xi) = m$, where ξ is the abscissa along the cable.
(b) Tension $S = S_o$, obtained by axially preloading the cable, is constant along the length of the cable (i.e. independent from space ξ). This hypothesis assumes that the catenary, representing the shape of the cable in static equilibrium

F. Cheli and G. Diana, *Advanced Dynamics of Mechanical Systems*,
DOI 10.1007/978-3-319-18200-1_3

condition, has a very small deflection, thereby allowing us to disregard the variations in the tension. Therefore, by assuming that the catenary has a small deflection it is possible, for a heavy cable, to approximate its static equilibrium configuration with a straight line.

(c) The motion of the cable will be described by means of a function $w(\xi, t)$ of time t and the current coordinate ξ, which indicates the generic section of the continuous structure; in the example that we will develop, as constraint conditions at both ends we will assume the same support conditions as those shown in Fig. 3.1. We will use function $w(\xi, t)$ to describe the transverse displacement of the generic section of the cable in a plane p defined by axes x–z of a right hand cartesian reference frame, positioned according to Fig. 3.1. This means that both the longitudinal motions (i.e. parallel to axis x) and transverse motions in the plane perpendicular to plane p will be assumed as null.

(d) Another simplified hypothesis regarding motion is that this occurs in any plane containing the axis of the cable: for this purpose, we must assume that the cable has a polar symmetry (i.e. a symmetry with respect to its axis); the motion will thus prove to be independent from plane p considered (all planes are equal in polar symmetry); any spatial motion may thus be described by means of 2 of its components on 2 orthogonal planes: in the hypotheses given, the two equations that govern the two component plane motions are equal and uncoupled, so that even the study of motion in space can always be traced back to the study of two plane motions that are independent from each other.

Having established these hypotheses and simplifications, the writing of the equations of motion can be obtained by defining the dynamic equilibrium equations for a generic infinitesimal element dx of the cable, placed in a generic position x. An inertia force f_{in}, parallel to the direction of motion will act on this element. By using the conventions assumed and shown in Fig. 3.1, this can be expressed as

$$f_{in} = -m(x)dx\frac{\partial^2 w}{\partial t^2} \tag{3.1}$$

where

- $m(\xi) = m$ is the mass per unit of length of the cable, assumed to be constant;
- $m\, d\xi$ is the mass of the infinitesimal element of length $d\xi$;

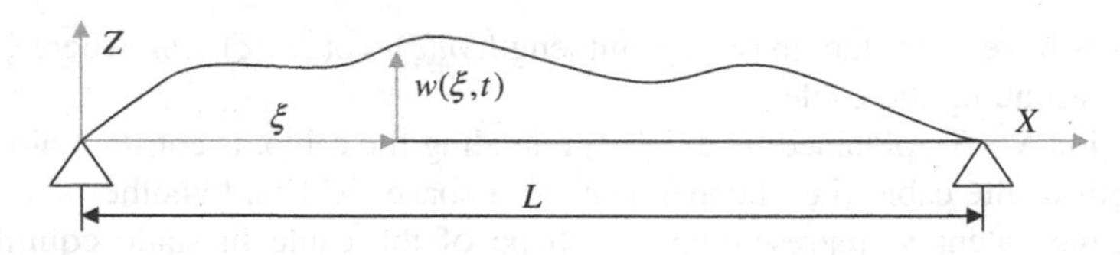

Fig. 3.1 Taut cable: simply supported conditions

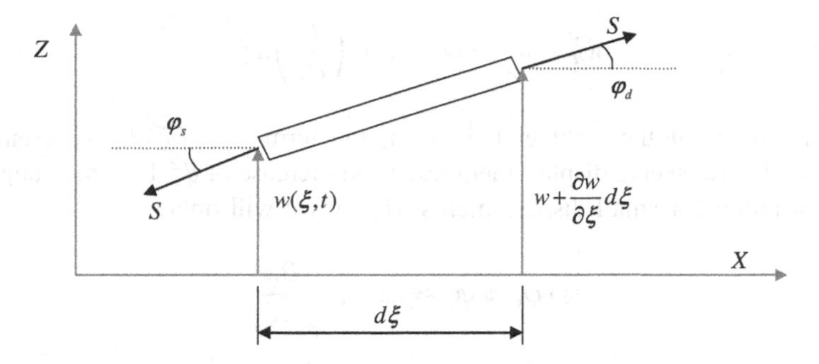

Fig. 3.2 Taut cable: undefined equations of equilibrium

- $\partial^2 w/\partial t^2$ is the transverse component of absolute acceleration to which the infinitesimal element is subjected.

Tension $S = S_o$ acting on the element is assumed constant along the cable and tangent at every point of the cable itself. By projecting tension force S (Fig. 3.2), in correspondence to the left and right end of the element, in direction z (i.e. in the direction along which the dynamic equilibrium is being written) and by keeping account of the conventions adopted (forces considered as positive if in agreement with the transverse displacements w) we obtain:

$$S_s = -S \sin \varphi_s \tag{3.2}$$

$$S_d = S \sin \varphi_d \tag{3.3}$$

where φ_s and φ_d are the angles formed by the tangent to the cable respectively in x and in $x + dx$ with respect to the undeformed configuration (i.e. horizontally), and with generally (3.2):

$$\varphi_s \neq \varphi_d \tag{3.4}$$

Given the initial hypotheses (disturbed motion in the about the static equilibrium configuration, assumed to be rectilinear, in order to neglect the weight and the coupling of the axial motion with the transverse one), no other forces should appear in the dynamic equilibrium equations. The equation of motion thus becomes:

$$-m(\xi)d\xi \frac{\partial^2 w}{\partial t^2} - S \sin \varphi_s + S \sin \varphi_d = 0 \tag{3.5}$$

If we wish to analyse the small vibrations about the static equilibrium configuration it is possible to linearize the problem. Having used $w_s = w(\xi, t)$ to define the displacement of the left end of the generic infinitesimal element considered and

$$w_d = w + dw = w + \left(\frac{\partial w}{\partial \xi}\right) d\xi \tag{3.6}$$

the displacement of the right end, in (3.6) the term $(\partial w/\partial \xi)d\xi$ represents the increase of the transverse displacement due to an increase of $d\xi$. For small angles φ_s and φ_d, or rather for small displacements $w(\xi, t)$, we will obtain:

$$\sin \varphi_s \approx \varphi_s \approx \tan \varphi_s = \frac{\partial w}{\partial \xi} \tag{3.7}$$

and furthermore:

$$\sin \varphi_d \approx \varphi_d \approx \tan \varphi_d = \frac{\partial}{\partial \xi}\left(w + \frac{\partial w}{\partial \xi} d\xi\right) = \frac{\partial w}{\partial \xi} + \left(\frac{\partial^2 w}{\partial \xi^2}\right) d\xi \tag{3.8}$$

By substituting Eqs. (3.2) and (3.3) in the equation that defines the motion of the generic infinitesimal small element of cable (3.5) and using suitable simplifications, we obtain

$$-m(\xi)d\xi\left(\frac{\partial^2 w}{\partial t^2}\right) + S\left(\frac{\partial^2 w}{\partial \xi^2}\right) d\xi = 0 \tag{3.9}$$

Where, as far as tension S is concerned, the only contribution to the dynamic equilibrium associated with the variation along $d\xi$ of angle φ appears. By furthering simplifying Eq. (3.9) we obtain:

$$m\left(\frac{\partial^2 w}{\partial t^2}\right) = S\left(\frac{\partial^2 w}{\partial \xi^2}\right) \tag{3.10}$$

This is the partial derivatives equation of second order with constant coefficients which defines the transverse motion of the cable, i.e. the trend of function $w(\xi, t)$. Similar equations are often used in the study of vibrations in continuous systems, e.g. torsional vibrations of bars (which will be analysed in Sect. 3.5) or longitudinal vibrations of beams (Sect. 3.4). More generally in physics, this equation, known as the "wave equation", is used to express the propagation of a wave-like motion. It goes without saying that this equation defines the perturbed motion of the cable, i.e. the free motion of same as determined by initial conditions. Finally, it is necessary to point out that this Eq. (3.10) is valid in any part of the cable in which the hypotheses indicated in the writing of same (mass m and tension S constant, absence of distributed or concentrated forces) hold true.

3.2.1 Propagative Solution

A general integral of Eq. (3.10) is the function[1]:

$$w(\xi,t) = f_1(\xi - Ct) + f_2(\xi + Ct) \tag{3.11}$$

[1]It is easy to verify how (3.11) is actually the solution of (3.9) regardless of the shape of functions f_1 and f_2 themselves: therefore, for reasons of convenience, let us assume that:

$$\begin{aligned} a &= \xi - Ct \\ b &= \xi + Ct \end{aligned} \tag{3.1.1}$$

By substituting (3.11) in (3.9) and expressing the partial derivatives as:

$$\begin{aligned} \frac{\partial f_1}{\partial t} &= \frac{\partial f_1}{\partial a}\frac{\partial a}{\partial t} \\ \frac{\partial^2 f_1}{\partial t^2} &= \frac{\partial^2 f_1}{\partial a^2}\left(\frac{\partial a}{\partial t}\right)^2 + \frac{\partial f_1}{\partial a}\frac{\partial^2 a}{\partial t^2} \end{aligned} \tag{3.1.2}$$

we obtain:

$$\begin{aligned} & m\left(\frac{\partial^2 f_1}{\partial a^2}\left(\frac{\partial a_1}{\partial t}\right)^2 + \frac{\partial f_1}{\partial a}\frac{\partial^2 a}{\partial t^2} + \frac{\partial^2 f_2}{\partial b^2}\left(\frac{\partial b}{\partial t}\right)^2 + \frac{\partial f_2}{\partial b}\frac{\partial^2 b}{\partial t^2}\right) \\ & = S\left(\frac{\partial^2 f_1}{\partial a^2}\left(\frac{\partial a_1}{\partial \xi}\right)^2 + \frac{\partial f_1}{\partial a}\frac{\partial^2 a}{\partial \xi^2} + \frac{\partial^2 f_2}{\partial b^2}\left(\frac{\partial b}{\partial \xi}\right)^2 + \frac{\partial f_2}{\partial b}\frac{\partial^2 b}{\partial \xi^2}\right) \end{aligned} \tag{3.1.3}$$

where, therefore, from (3.1.1):

$$\begin{aligned} \frac{\partial a}{\partial t} &= -C \Rightarrow \frac{\partial^2 a}{\partial t^2} = 0 \\ \frac{\partial b}{\partial t} &= C \Rightarrow \frac{\partial^2 b}{\partial t^2} = 0 \\ \frac{\partial a}{\partial \xi} &= 1 \Rightarrow \frac{\partial^2 a}{\partial \xi^2} = 0 \\ \frac{\partial b}{\partial \xi} &= 1 \Rightarrow \frac{\partial^2 b}{\partial \xi^2} = 0 \end{aligned} \tag{3.1.4}$$

Equation (3.1.4) becomes:

$$m\left(\frac{\partial^2 f_1}{\partial a^2}C^2 + \frac{\partial^2 f_2}{\partial b^2}C^2\right) = S\left(\frac{\partial^2 f_1}{\partial a^2} + \frac{\partial^2 f_2}{\partial b^2}\right) \tag{3.1.5}$$

or rather Eq. (3.11) is the solution to (3.9) if constant C applies:

$$C = \sqrt{\frac{S}{m}}. \tag{3.1.6}$$

where functions f_1 and f_2 are functions to be suitably defined, while constant C is defined as $C = \sqrt{S/m}$ and represents the longitudinal propagation velocity of the deformation wave. These functions both represent the motion of a wave travelling at velocity C along the cable and, more specifically:

- f_1 represents a displacement wave travelling in a direction corresponding to ξ positive, without altering either the shape or the profile of the wave: this profile is explicitly defined by the shape of function f_1 itself;
- f_2 is similar to f_1, but represents a wave travelling in the opposite direction.

As an example, let us consider function $f_1(\xi - Ct)$ (similar considerations can be made for function $f_2(\xi + Ct)$: its value is determined by argument $(\xi - Ct)$ and therefore by a series of correlated values of ξ and t. The value of function $f_1(\xi - Ct)$ obviously remains unaltered if the argument $(\xi - Ct)$ is the same and, therefore, a generic increase $d\xi$ leads to the same value of the argument if the time is increased by $\Delta t = \Delta x/C$. In other words, the transverse displacement $w(\xi, t)$, in direction z, reaches the same value w in different points $\xi_1, \xi_2, \ldots, \xi_n$ in different instants of time $t_1, t_2, \ldots, t_n$ correlated by $\Delta t = \Delta x/C$:

$$\overline{w} = f_1(\xi_1 - Ct_1) = f_1(\xi_2 - Ct_2) = f_1((\xi_1 + \Delta\xi) - C(t_1 + \Delta\xi)) \tag{3.12}$$

Physically, this means that function $f_1(\xi - Ct)$ represents a phenomenon that is generally propagative, i.e. it describes the profile of the wave moving in a positive direction along axis ξ with velocity C (Fig. 3.3).

In the same way we show how function $f_2(\xi - Ct)$ represents a wave moving in a negative direction along axis ξ always with, however, a propagation velocity C. A generic disturbance which starts from the left end of the cable, runs along the cable itself with velocity C:

$$C = \sqrt{\frac{S}{m}} \tag{3.13}$$

For ascending ξ, when this disturbance reaches the right end, it is reflected and then turns back, always with velocity C, towards the left end ([12], Sect. 5.10). This phenomenon is attenuated as a function of the dissipations found in the cable or at its ends. This consideration can be used to measure tension S experimentally, once the mass m of the cable is known: in fact, it is sufficient to measure at which velocity C an impulsive disturbance is propagated along the cable. If L is the length of the cable and $\bar{t}$ the time taken by the disturbance to travel from one extremity to the other, we will obtain:

$$C = \left(\frac{L}{\bar{t}}\right) \tag{3.14}$$

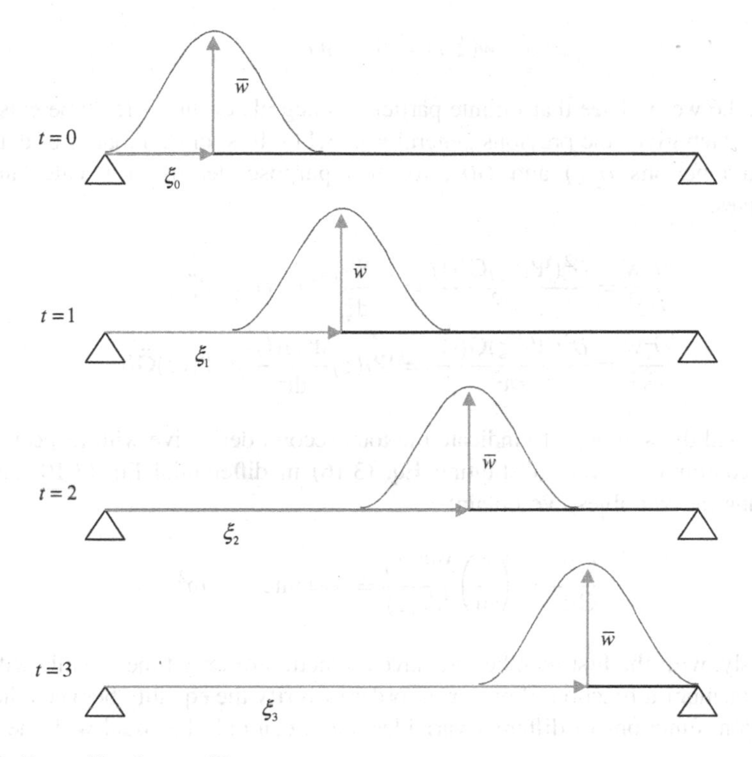

Fig. 3.3 Taut cable: the travelling wave

and subsequently tension S from Eq. (3.13). The previous particular integral (3.11) is useful, above all, to describe the transient motion of the cable, i.e. the motion following a generic initial perturbation.

3.2.2 Stationary Solution

Conversely, should we wish to study the steady-state vibrations of the cable subjected to a generic excitation force (which, as usual, we will consider sinusoidal), we will find ourselves dealing with a phenomenon that is no longer propagative but stationary: the generic section of the cable oscillates in a transverse direction $w(\xi, t)$ with a given amplitude and a certain frequency. Therefore, under stationary conditions, the shape of the profile $\psi_f(\xi)$ of the deformation does not change, but the amplitudes in various points of the continuous system change, moment by moment, governed by a single time function $G(t)$: this type of motion, referred to as "stationary wave", can be described by means of a particular integral of the equation of motion in the form:

$$w(\xi,t) = \Psi_f(\xi)G(t) \tag{3.15}$$

In Sect. 3.6 we will see that infinite particular integrals of the (3.15) type exist, the sum of which gives the previous general Eq. (3.11). It is, thus, a matter of defining the two functions $\psi_f(\xi)$ and $G(t)$; for this purpose, let us now calculate the derivatives:

$$\begin{aligned} \frac{\partial^2 w}{\partial \xi^2} &= \frac{\partial^2(\Psi_f(\xi)G(t))}{\partial \xi^2} = \frac{d^2\Psi_f(\xi)}{d\xi^2}G(t) = \Psi_f''(\xi)G(t) \\ \frac{\partial^2 w}{\partial t^2} &= \frac{\partial^2(\Psi_f(\xi)G(t))}{\partial t^2} = \Psi_f(\xi)\frac{d^2G(t)}{dt^2} = \Psi_f(\xi)\ddot{G}(t) \end{aligned} \tag{3.16}$$

having used the symbol $''$ to indicate the total second derivative with respect to the spatial coordinate ξ. By substituting Eq. (3.16) in differential Eq. (3.10) and by separating the variables, we obtain:

$$\frac{\ddot{G}(t)}{G(t)} = \left(\frac{S}{m}\right)\frac{\Psi_f''(\xi)}{\Psi_f(\xi)} = \text{costante} = -\omega^2 \tag{3.17}$$

Obviously, with the first member we have a function of only time t, while with the second member a function of only x: in order to verify the equality between the two expression, functions of different variables, these can only be equal with the same constant:

$$\frac{\ddot{G}(t)}{G(t)} = \left(\frac{S}{m}\right)\frac{\Psi_f''(\xi)}{\Psi_f(\xi)} = \text{costante} = -\omega^2 \tag{3.18}$$

As will be explained further on, constant $(-\omega^2)$ is defined in such a way as to be negative. We can now consider the two differential equations with total derivatives (3.18) separately, in that functions $\psi_f(\xi)$ and $G(t)$ are respectively only functions of space ξ and of time t. By considering the first equation of (3.18) we obtain:

$$\ddot{G}(t) + \omega^2 G(t) = 0 \tag{3.19}$$

i.e. a differential equation with constant coefficients similar to those previously analysed for systems with 1 d.o.f. (Sect. 1.3.1). The general integral of Eq. (3.19) is given by:

$$G(t) = A\sin\omega t + B\cos\omega t = D\cos(\omega t + \varphi) \tag{3.20}$$

where A and B (or D and φ) are the integration constants to be defined by means of the initial conditions (assigned) and ω is the frequency (still to be defined) with which the generic section of the cable vibrates.

By analysing solution (3.20) of (3.19) we thus justify the choice made with regard to constant $(-\omega^2)$ of relation (3.18): owing to the fact that in this chapter we are analysing a conservative system perturbed about the stable static equilibrium position, the steady-state motion, over time, obviously has to be sinusoidal, not expansive (see Chap. 1, Sect. 1.3.1). To obtain this result, it is necessary that the coefficient multiplying $G(t)$ in (3.19) is definitely positive, i.e. it is absolutely essential that the constant in Eq. (3.18) is negative.

The second differential equation in (3.18) allows for the definition of the spatial function $\psi_f(\xi)$ which describes how the motion varies along the cable itself; this equation can be developed as:

$$\frac{S}{m}\Psi_f''(\xi) + \omega^2\Psi_f(\xi) = 0 \tag{3.21}$$

formally identical, except for the constants, to the previous relation (3.19). We will thus have:

$$\Psi_f''(\xi) + \left(\frac{m}{S}\omega^2\right)\Psi_f(\xi) = 0 \tag{3.22}$$

which by assuming:

$$\gamma^2 = \frac{m}{S}\omega^2 \Rightarrow \gamma = \sqrt{\frac{m}{S}}\omega \tag{3.23}$$

gives:

$$\Psi_f''(\xi) + \gamma^2\Psi_f(\xi) = 0 \tag{3.24}$$

Relation (3.24) admits:

$$\Psi_f(\xi) = F_1 \sin\gamma\xi + F_2\cos\gamma\xi = F_1\sin\frac{\omega}{C}\xi + F_2\cos\frac{\omega}{C}\xi \tag{3.25}$$

as a particular integral where γ will be determined by fixing ω (or vice versa) and F_1 and F_2 must be defined as a function of the spatial boundary conditions (functions that are independent from time). The following can be imposed as boundary conditions:

- congruence conditions;
- equilibrium conditions.

In the case in question (a cable supported at both end see Fig. 3.1) the boundary conditions are congruence conditions, in that the generic space-time function $w(\xi, t)$ must satisfy the fact that the transverse displacement is null at the extremities of the cable. This condition must be verified in every instant of time t, i.e. regardless of time, or rather regardless of the value of $G(t)$:

$$\begin{aligned} w(0,t) = \Psi_f(0) \quad G(t) = 0 \quad \Rightarrow \quad \Psi_f(0) = 0 \\ w(L,t) = \Psi_f(L) \quad G(t) = 0 \quad \Rightarrow \quad \Psi_f(L) = 0 \end{aligned} \tag{3.26}$$

where L is the length of the cable. From (3.26) and (3.24) we thus obtain:

$$\Psi_f(0) = 0 \quad \Rightarrow \quad \mathbf{F}_2 = 0 \tag{3.27}$$

so that (3.15) becomes:

$$w(\xi, t) = F_1 \sin\gamma\xi(A \sin\omega t + B \cos\omega t) \tag{3.28}$$

Since F_1, A and B are all constants that need to be determined, it is possible to assume:

$$\overline{A} = F_1 A \quad \overline{B} = F_1 B \tag{3.29}$$

and thus obtain:

$$w(\xi, t) = \sin\gamma\xi\left(\overline{A} \sin\omega t + \overline{B} \cos\omega t\right) \tag{3.30}$$

where

$$\Psi_f(\xi) = \sin\gamma\xi = \sin\frac{\omega}{C}\xi \tag{3.31}$$

Let us now impose the second condition:

$$\Psi_f(L) = \sin\gamma L = 0 \quad \Rightarrow \quad \gamma_n L = n\pi \quad (n = 1, 2, \ldots\infty) \tag{3.32}$$

At this point, by recalling (3.23), relation (3.32) imposes a condition on frequency ω of the motion of the various sections:

$$\gamma_n = \frac{n\pi}{L} \quad \Rightarrow \quad \omega_n = \frac{n\pi}{L}\sqrt{\frac{S}{m}} \quad (n = 1, 2, \ldots\infty) \tag{3.33}$$

and, from this relation, we obtain infinite values ω_n, each associated with a specific value assigned to the entire parameter n. It is interesting to note how, in this case, the ω_n all prove to be multiples of a fundamental frequency:

$$\omega_1 = \frac{\pi}{L}\sqrt{\frac{S}{m}}\infty \tag{3.34}$$

In short, for the normal system, schematized as a continuous system, they are defined as infinite natural frequencies ω_n, as an extrapolation of the "n" natural frequencies of the discrete systems with "n" d.o.f., since the normal system has infinite d.o.f. A normal mode $(\psi_f(\xi))_n$ defined by the spatial deformation $(\psi_f(\xi))_{\omega=\omega n}$ assumed by the system in correspondence to the generic ω_n associated with it corresponds to each natural frequency ω_n:

$$\phi^{(n)}(\xi) = \sin\frac{\omega_n}{C}\xi = \sin\frac{n\pi}{L}\xi = \sin\frac{2\pi}{\lambda_n}\xi \tag{3.35}$$

Having used λ_n to define the wave length corresponding to the generic nth normal mode, intended as the distance between homologous points in successive spatial periods of deformation:

$$\lambda_n = \frac{2L}{n} \tag{3.36}$$

The deformation of the generic normal mode $\Phi^{(n)}(\xi)$ is now described, in the case analysed, by a function of the sinusoidal type, and not by a finite number of terms contained in eigenvalue $\underline{X}^{(i)}$ as previously seen in the discrete systems (Chap. 2, Sect. 2.4.3.1), on account of the fact that we are dealing with a continuous system, i.e. with infinite d.o.f. As an example, let us consider the deformation of the first mode:

$$\phi^{(n)}(\xi) = \sin\frac{n\pi}{L}\xi \Rightarrow n = 1 \Rightarrow \phi^{(1)}(\xi) = \sin\frac{\pi}{L}\xi \tag{3.37}$$

We note that this only has one maximum (or antinode) in the midspan, where the wavelength λ_1 of the first mode (see (3.36)) is:

$$\lambda_1 = 2L \tag{3.38}$$

Therefore, for the generic nth normal mode we will have:

$$\omega_n = \frac{n\pi}{L}\sqrt{\frac{S}{m}} \Rightarrow \phi^{(n)}(\xi) = \sin\frac{n\pi}{L} = \sin\frac{2\pi}{\lambda_n}\xi \tag{3.39}$$

So that the deformation assumed by the cable for the nth mode will be described by a harmonic wavelength function $\boldsymbol{\lambda_n}$ equal to:

$$\lambda_n = \frac{2L}{n} \tag{3.40}$$

Relation (3.39) shows how the deformation of the nth mode has n antinodes and $n + 1$ nodes, i.e. points with a null displacement (see [3]). The previous definition of wavelength λ_n can be included in the writing of the natural frequency of system ω_n (3.33):

$$\omega_n = \frac{n\pi}{L}\sqrt{\frac{S}{m}} = \frac{2n\pi}{2L}\sqrt{\frac{S}{m}} = \frac{2\pi}{\lambda_n}\sqrt{\frac{S}{m}} \Rightarrow \gamma_n = \frac{2\pi}{\lambda_n} \tag{3.41}$$

We note that the value of the generic frequency ω_n, in addition to increasing with tension S and decreasing with mass m per unit length of the cable, also proves to be inversely proportional to the wavelength of the associated mode $\Phi^n(\xi)$. By examining a continuous system of any type whatsoever, we will thus always have infinite natural frequencies ω_n and, at the same time, infinite normal modes. Furthermore, in the continuous system, it is possible to describe the generic deformation of steady-state motion as a linear combination of the normal modes $\Phi^i(\xi)$, in this case, defined by (3.35). This combination can be seen as an in-series development of the space-time function $w(\xi, t)$ in variable ξ:

$$w(\xi, t) = \sum_{i=1}^{\infty} \phi^{(i)}(\xi) q_i \tag{3.42}$$

where it is necessary to sum infinite deformations $\Phi^i(\xi)$, each of which is a function of ξ and of a multiplicative coefficient q_i. Relation (3.42) can be considered as a transformation of coordinates: multiplicative coefficients q_i are the infinite new coordinates (termed principal or orthogonal) which can be used to describe the motion of the system itself.

The q_i s thus define the extent of the contribution of the generic ith normal mode associated with the motion of the system: we will see (Sect. 3.7) how, for these, orthogonal coordinates, similar peculiarities considered in the discrete system, are valid. Needless to say, in case we want to schematize the continuous system by means of "n" d.o.f., i.e. in discrete form, in order to define the generic deformation $w(\xi, t)$ it is only necessary to consider the first "n" terms of the previous in-series development (3.42):

$$w(\xi, t) = \sum_{i=1}^{n} \phi^{(i)}(\xi) q_i \tag{3.43}$$

As an example, Fig. 3.4 shows the first five natural frequencies ω_i and relative deformations $\Phi^i(\xi)$ of a taut cable using the data given in Table 3.1.

3.3 Transverse Vibrations in Beams

In this paragraph we will analyse the small transverse oscillations of a beam in the neighbourhood of its static equilibrium condition (Fig. 3.5). We will assume that the beam is homogenous, i.e. having transverse section A, bending stiffness EJ and density ρ all constant along the beam; furthermore we will also assume the absence of axial loads S. We will then analyse motion in a transverse direction z, coinciding

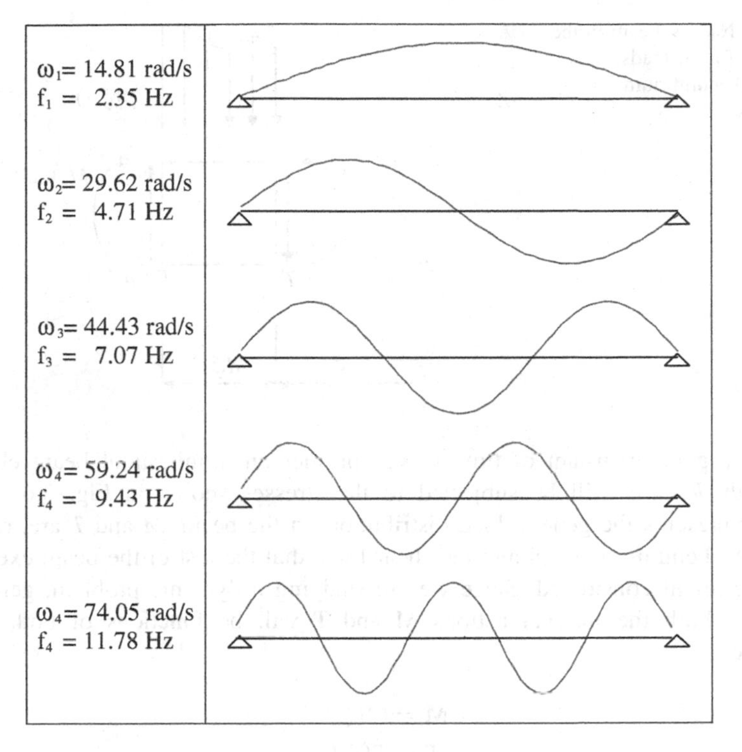

Fig. 3.4 Taut cable: natural frequencies and normal modes (data shown in Table 3.1)

Table 3.1 Taut cable: inertial geometric data

Length	L = 30 m
Tension	S = 30,000 N
Mass by unit of length	m = 1.5 kg/m

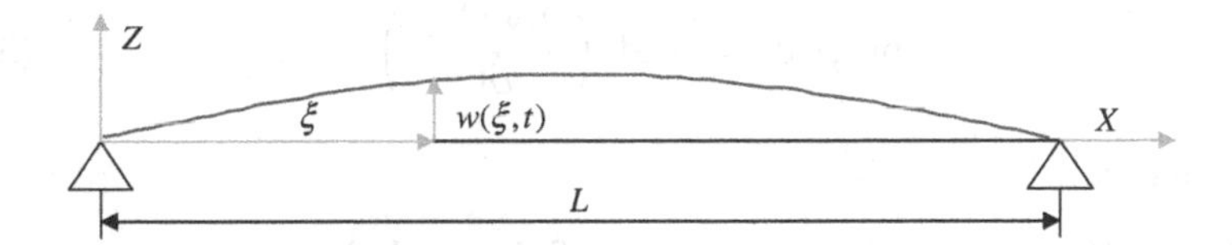

Fig. 3.5 Transverse vibration in a narrow beam

with the main inertia direction of the beam. Let $w(\xi, t)$ be the displacement along z of the generic section of abscissa ξ along the beam.

Furthermore, let us assume that:

- even under dynamic conditions, the beam undergoes always to bending in a plane of simmetry;
- the sections perpendicular to the axis remain plane.

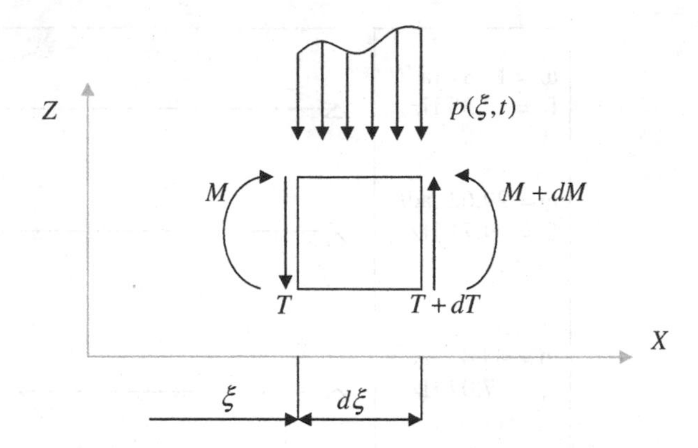

Fig. 3.6 Narrow beam in the absence of axial loads: undefined equilibrium equations

If in a generic instant of time t, we consider an infinitesimal beam element of length $d\xi$, this will be subjected to the stresses shown in Fig. 3.6, where $p(\xi, t)$ represents the generic load distribution on the beam, M and T are, respectively, the bending moment and the shear force that the rest of the beam exercises on the element considered. Since we are studying a dynamic problem, generally speaking, both the internal actions M and T will be functions of both space and time.

$$\begin{aligned} \mathrm{M} &= \mathrm{M}(\xi, \mathrm{t}) \\ \mathrm{T} &= \mathrm{T}(\xi, \mathrm{t}) \end{aligned} \tag{3.44}$$

Furthermore, by referring to the static equilibrium configuration, assumed as stable, or rather by analysing the perturbed motion about the static equilibrium configuration, the natural weight does not formally appear in the dynamic equilibrium equations. For this reason, function $p(\xi, t)$ is only defined by the inertia force which, according to the conventions assumed, becomes:

$$\mathrm{p}(\xi, \mathrm{t})\mathrm{d}\xi = \rho \mathrm{Ad}\xi\left(\frac{\partial^2 \mathrm{w}(\xi, \mathrm{t})}{\partial \mathrm{t}^2}\right) \tag{3.45}$$

in which the term:

- $\rho A d\xi = m d\xi$ represents the mass of the infinitesimal element;
- $\partial^2 \mathrm{w}(\xi, \mathrm{t})/\partial \mathrm{t}^2$ is the acceleration to which the element itself is subjected.

Two dynamic equilibrium equations can be written for the same infinitesimal element. The first equation is the one relative to a translation in a vertical direction z which, by neglecting the higher order infinitesimals, becomes:

$$\frac{\partial \mathrm{T}}{\partial \xi}\mathrm{d}\xi = \mathrm{p}(\xi, \mathrm{t})\mathrm{d}\xi = \rho \mathrm{Ad}\xi \frac{\partial^2 \mathrm{w}}{\partial \mathrm{t}^2} \tag{3.46}$$

a well known relation: the variation of the shear force equals the distributed load. The second equation that we will write is the dynamic equilibrium equation at rotation, from which, by neglecting the higher order infinitesimals, we obtain:

$$\frac{\partial M(\xi, t)}{\partial \xi} = -T(\xi, t) \tag{3.47}$$

another well-known link between the derivative of the bending moment M and shear force T. Thus, these differential relationships, i.e. (3.46) and (3.47), do not lose validity even in the dynamic case, but are expressed as partial derivatives forms, being functions of independent variables represented by space ξ and time t. By keeping account of the relation (3.46), the equilibrium equation at rotation (3.47) can be expressed as:

$$\frac{\partial^2 M}{\partial \xi^2} = -\frac{\partial T}{\partial \xi} = -\rho A \frac{\partial^2 w}{\partial t^2} \tag{3.48}$$

We can now use the elastic line equation, the Euler-Bernoulli relation [2, 3, 13, 16], also valid in the dynamic case. As is known, this relation correlates the bending moment M with the curvature of the deformation line $w(\xi, t)$, i.e. with the second derivative of same with respect to the current abscissa ξ; thus, by disregarding the deformation due to the shear force (narrow beam hypothesis) we obtain:

$$EJ\left(\frac{\partial^2 w}{\partial \xi^2}\right) = M \tag{3.49}$$

in which E represents the modulus of normal elasticity of the material (Young's modulus), while J is the inertia moment of the section with respect to the neutral axis: in the hypothesis of a uniform beam with a constant section, EJ is constant along the beam By substituting this latter link in the previous relation, we obtain:

$$\frac{\partial^2}{\partial \xi^2}\left(EJ\frac{\partial^2 w}{\partial \xi^2}\right) = -\rho A \frac{\partial^2 w}{\partial t^2} \Rightarrow EJ\frac{\partial^4 w}{\partial \xi^4} = -\rho A \frac{\partial^2 w}{\partial t^2} \tag{3.50}$$

This differential equation of the fourth order with partial derivatives defines the transverse motion $w(\xi, t)$ of the beam analysed: relation (3.50) is defined as an *undefined equilibrium equation* and is valid for each section of the beam in which the hypotheses made (homogeneous beam, absence of distributed loads other than inertia loads) have been verified. In this case too, as with the cable (Sect. 3.2), in the event of our wishing to study free steady-state motion, i.e. to look for a possible stationary solution of (3.50), we assume a particular integral of this equation written in the form of:

$$w(\xi, t) = \Psi_b(\xi) G(t) \tag{3.51}$$

By substituting (3.51) in (3.50) and by re-ordering the terms, we obtain:

$$\frac{EJ}{\Psi_b(\xi)} \frac{d^4\Psi_b(\xi)}{d\xi^4} = -\frac{\rho A}{G(t)} \frac{d^2G(t)}{dt^2} \tag{3.52}$$

In (3.52) we note the substitution of the partial derivatives of the space-time $w(\xi, t)$ function with the total derivatives of functions $\psi_b(\xi)$ and $G(t)$. Thus, the terms on the left and on the right of Eq. (3.52) are one a function of the only spatial coordinate ξ and the other only of time t: for this reason, both terms will be equal if, and only if, they are equal to the same constant K. This enables us to define two separate differential ordinary equations; we can write the first one, relative to $G(t)$, as:

$$-\frac{1}{G(t)} \frac{d^2G(t)}{dt^2} = K \tag{3.53}$$

Given that the system is conservative and perturbed about a static position, of stable equilibrium, it is necessary to impose condition:

$$K = \omega^2 \tag{3.54}$$

so that the solution represents a non-expansive harmonic motion:

$$G(t) = A \sin \omega t + B \cos \omega t = D \cos(\omega t + \varphi) \tag{3.55}$$

Where constants A and B (or amplitude D and phase φ) depend on the initial conditions assigned to the system at time t = 0. The second ordinary differential equation, relative to $\psi_b(\xi)$, will be:

$$\frac{EJ}{\rho A} \frac{1}{\Psi_b(\xi)} \left(\frac{d^4\Psi_b(\xi)}{d\xi^4} \right) = \omega^2 \Rightarrow \left(\frac{d^4\Psi_b(\xi)}{d\xi^4} \right) - \gamma^4 \Psi_b(\xi) = 0 \tag{3.56}$$

where

$$\gamma^4 = \omega^2 \frac{\rho A}{EJ} \tag{3.57}$$

In terms of a solution, (3.56) admits:

$$\Psi_b(\xi) = \bar{F} e^{\lambda \xi} \tag{3.58}$$

which, by being substituted in (3.56) itself, gives an algebraic, homogeneous, parametric equation in λ:

$$(\lambda^4 - \gamma^4)\overline{\mathrm{F}} = 0 \tag{3.59}$$

which admits nontrivial solutions for:

$$(\lambda^4 - \gamma^4) = 0 \tag{3.60}$$

i.e. for:

$$\begin{aligned} \lambda_{1,2} &= \pm\sqrt{-\gamma^2} = \pm i\gamma \\ \lambda_{3,4} &= \pm\sqrt{\gamma^2} = \pm i\gamma \end{aligned} \tag{3.61}$$

For this reason, the general integral of (3.56) can be expressed as:

$$\Psi_b(\xi) = \overline{\mathrm{F}}_1 e^{\lambda_1 \xi} + \overline{\mathrm{F}}_2 e^{\lambda_2 \xi} + \overline{\mathrm{F}}_3 e^{\lambda_3 \xi} + \overline{\mathrm{F}}_4 e^{\lambda_4 \xi} = \overline{\mathrm{F}}_1 e^{i\gamma\xi} + \overline{\mathrm{F}}_2 e^{i\gamma\xi} + \overline{\mathrm{F}}_3 e^{i\gamma\xi} + \overline{\mathrm{F}}_4 e^{i\gamma\xi} \tag{3.62}$$

which can also be expressed as:

$$\Psi_b(\xi) = \mathrm{F}_1 \sinh\gamma\xi + \mathrm{F}_2 \sinh\gamma\xi + \mathrm{F}_3 \sinh\gamma\xi + \mathrm{F}_4 \sinh\gamma\xi \tag{3.63}$$

The stationary solution which represents the free steady-state motion of the beam is thus given by the relation:

$$\mathrm{w}(\xi, \mathrm{t}) = (\mathrm{F}_1 \sin\gamma\xi + \mathrm{F}_2 \cos\gamma\xi + \mathrm{F}_3 \sinh\gamma\xi + \mathrm{F}_4 \cosh\gamma\xi)(\mathrm{A}\sin\omega\mathrm{t} + \mathrm{B}\cos\omega\mathrm{t}) \tag{3.64}$$

This equation defines the transverse motion of a beam (homogenous and devoid of distributed loads), whatever the boundary or initial conditions. In (3.64) constants F_1, F_2, F_3 and F_4, depend on the boundary conditions, i.e. on the constraints assigned to the extremities of the beam: by varying the latter, the solution is modified.

A particularly simple, though significant example, is given by the beam which is simply supported at both extremities (Fig. 3.7): in this case, regardless of the instant considered, the vertical displacement at the ends must be null, or rather, by keeping account of (3.64), the boundary conditions become simple:

$$\begin{aligned} [\mathrm{w}(\xi, \mathrm{t})]_{\xi=0} &= 0(\text{at any t}) \Rightarrow \Psi_b(0) = 0 \\ [\mathrm{w}(\xi, \mathrm{t})]_{\xi=\mathrm{L}} &= 0(\text{at any t}) \Rightarrow \Psi_b(\mathrm{L}) = 0 \end{aligned} \tag{3.65}$$

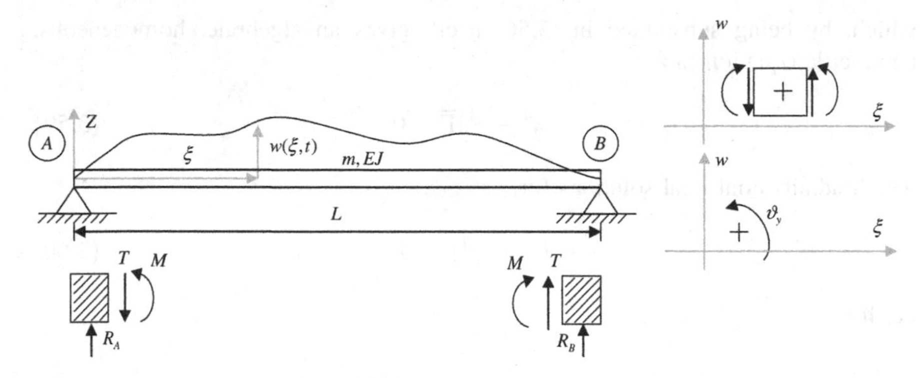

Fig. 3.7 Simply supported beam: transverse vibrations

Furthermore, regardless of time t, the bending moment M must be null in the same points: by recalling the link given by the elastic line (bending moment M proportional to the curvature) this gives us:

$$
\begin{aligned}
&[\mathrm{M}]_{\xi=0} = \mathrm{EJ}\left[\frac{\partial^2 \mathrm{w}}{\partial \xi^2}\right]_{\xi=0} = 0 \Rightarrow \left[\frac{\mathrm{d}^2 \Psi_\mathrm{b}(\xi)}{\mathrm{d}\xi^2}\right]_{\xi=0} = 0 \\
&[\mathrm{M}]_{\xi=\mathrm{L}} = \mathrm{EJ}\left[\frac{\partial^2 \mathrm{w}}{\partial \xi^2}\right]_{\xi=\mathrm{L}} = 0 \Rightarrow \left[\frac{\mathrm{d}^2 \Psi_\mathrm{b}(\xi)}{\mathrm{d}\xi^2}\right]_{\xi=\mathrm{L}} = 0
\end{aligned}
\tag{3.66}
$$

By imposing the boundary conditions which, as we remember, must generally be equilibrium (e.g. relations (3.66)) or congruence (3.65) conditions, it is possible to define the equations which must satisfy the spatial function $\psi_b(\xi)$ and thus define the integration constants of (3.64). in the specific case analysed, we obtain:

$$
\begin{aligned}
&\frac{\mathrm{d}\Psi_\mathrm{b}(\xi)}{\mathrm{d}\xi} = \gamma(\mathrm{F}_1 \cos\gamma\xi - \mathrm{F}_2 \sin\gamma\xi + \mathrm{F}_3 \cosh\gamma\xi + \mathrm{F}_4 \sinh\gamma\xi) \\
&\frac{\mathrm{d}^2\Psi_\mathrm{b}(\xi)}{\mathrm{d}\xi^2} = \gamma^2(-\mathrm{F}_1 \sin\gamma\xi - \mathrm{F}_2 \cos\gamma\xi + \mathrm{F}_3 \sinh\gamma\xi + \mathrm{F}_4 \cosh\gamma\xi)
\end{aligned}
\tag{3.67}
$$

By imposing the boundary conditions (3.65) and (3.66):

$$
\begin{aligned}
&\Psi_\mathrm{b}(0) = 0 \Rightarrow \mathrm{F}_2 + \mathrm{F}_4 = 0 \\
&\left[\frac{\mathrm{d}^2\Psi_\mathrm{b}(\xi)}{\mathrm{d}\xi^2}\right]_{\xi=0} = 0 \Rightarrow -\mathrm{F}_2 + \mathrm{F}_4 = 0
\end{aligned}
\tag{3.68}
$$

we immediately obtain:

$$F_2 = F_4 = 0 \tag{3.69}$$

so that, in this case, $\psi_b(\xi)$ is reduced to the following expression

$$\Psi_b(\xi) = F_1 \sin\gamma\xi + F_3 \sinh\gamma\xi \tag{3.70}$$

If we now impose the last two conditions on $\psi_b(\xi)$ (3.70), we obtain an algebraic system, homogenous in the two unknowns F_1 and F_3 and parametric in γ:

$$\begin{gathered}\Psi_b(L) = 0 \Rightarrow F_1 \sin\gamma L + F_3 \sinh\gamma L = 0\\ \left[\frac{d^2\Psi_b(\xi)}{d\xi^2}\right]_{\xi=L} = 0 \Rightarrow -\gamma^2 F_1 \sin\gamma L + \gamma^2 F_3 \sinh\gamma L = 0\end{gathered} \tag{3.71}$$

i.e. in matrix form, having organized the unknowns in a vector $\underline{Z}$ as:

$$\underline{Z} = \begin{Bmatrix} F_1 \\ F_3 \end{Bmatrix} \Rightarrow [H(\gamma)]\underline{Z} = 0 \quad (\text{con } \gamma = \gamma(\omega)) \tag{3.72}$$

where

$$[H(\gamma)] = \begin{bmatrix} \sin\gamma L & \sinh\gamma L \\ -\gamma^2 \sin\gamma L & \gamma^2 \sinh\gamma L \end{bmatrix} \tag{3.73}$$

This system will only admit nontrivial solutions, i.e. with at least one of the unknowns being non-null (the trivial case $\underline{Z} = \underline{0}$ corresponds to absence of motion, a case almost devoid of interest) only if the determinant of the coefficient matrix $[H(\gamma)]$ is annulled, i.e. for:

$$\begin{gathered}\det|H(\gamma)| = 0\\ \gamma^2 \sin\gamma L \sinh\gamma L + \gamma^2 \sin\gamma L \sinh\gamma L = 2\gamma^2 \sin\gamma L \sinh\gamma L = 0\end{gathered} \tag{3.74}$$

now obviously being:

$$L \neq 0 \quad \text{and} \quad \gamma \neq 0 \Rightarrow \sinh\gamma L \neq 0 \tag{3.75}$$

Equation (3.74) becomes:

$$\sin\gamma L = 0 \tag{3.76}$$

from which:

$$\gamma L = n\pi \quad (n = 1, 2, \ldots \infty) \Rightarrow \gamma_n = \frac{n\pi}{L} \tag{3.77}$$

Therefore, infinite values of the parameter $\gamma = \gamma_n$, which satisfy the relation (3.76), exist; a generic natural frequency ω_n corresponds to each γ_n parameter defined, see (3.56), as:

$$\gamma_n^4 = \omega_n^2 \frac{\rho A}{EJ} \Rightarrow \omega_n = \gamma_n^2 \sqrt{\frac{EJ}{\rho A}} \Rightarrow \omega_n = \frac{n^2 \pi^2}{L^2} \sqrt{\frac{EJ}{\rho A}} \quad (n = 1, 2, \ldots \infty) \tag{3.78}$$

Having annulled the determinant of the coefficient matrix of the system (3.72) in order to calculate the nontrivial solutions, the same matrix $[H(\gamma)]$ evaluated by $\gamma = \gamma_n$ becomes singular, or rather one of the lines of the algebraic system (3.72) is linearly dependent on the other: therefore, as previously observed in discrete systems, vector $\underline{Z}$ of the unknowns can be multiplied by any arbitrary constant. In the case in question, this is translated into the fact that the constants F_1 and F_3 of (3.70) and (3.72) can be obtained from one of the two Eq. (3.71), e.g. the second:

$$-\gamma_n^2 F_1^{(n)} \sin \gamma_n L + \gamma_n^2 F_3^{(n)} \sinh \gamma_n L = 0 \quad (n = 1, 2, \ldots \infty) \tag{3.79}$$

By keeping account of (3.76), (3.79) is reduced to:

$$\gamma_n^2 F_3^{(n)} \sinh \gamma_n L = 0 \quad (n = 1, 2, \ldots \infty) \tag{3.80}$$

which, with:

$$\sinh \gamma_n L \neq 0 \quad (n = 1, 2, \ldots \infty) \tag{3.81}$$

imposes the relation:

$$F_3^{(n)} = 0 \quad (n = 1, 2, \ldots \infty) \tag{3.82}$$

By keeping account of (3.82), (3.70) is reduced to:

$$(\Psi_b(\xi))_{\omega=\omega_n} = F_1^{(n)} \sin \gamma_n \xi \quad (n = 1, 2, \ldots \infty) \tag{3.83}$$

Since $F_1^{(n)}$ can be scaled by means of an arbitrary constant, it is possible to assume, for example:

$$F_1^{(n)} = 1 \quad (n = 1, 2, \ldots \infty) \tag{3.84}$$

By keeping account of (3.83), (3.84), (3.77) and (3.78), the spatial function $(\psi_b(\xi))_{\omega=\omega n} = \Phi^{(n)}(\xi)$, which defines the deformation assumed by the beam, considered in correspondence to the generic natural frequency ω_n, thus becomes:

$$(\Psi_b(\xi))_{\omega=\omega_n} = \phi^{(n)}(\xi) = \sin\left[\frac{n\pi}{L}\xi\right] = \sin\gamma_n\xi \quad (n = 1, 2, \ldots\infty) \tag{3.85}$$

This expression thus represents the deformation of the system (the simply supported beam in Fig. 3.7) in correspondence to the natural frequency mode ω_n: let us use $\Phi^{(n)}(\xi)$ to define the generic normal mode of the system. The generic particular integral of (3.52) is thus of the following type:

$$\begin{aligned} w_n(\xi, t) &= (\Psi_b(\xi))_{\omega=\omega_n} G_n(t) \\ w_n(\xi, t) &= \sin\left(\frac{n\pi}{L}\right)\xi D_n \cos(\omega_n t + \varphi_n) = \sin\gamma_n\xi(A_n \sin\omega_n t + B_n \cos\omega_n t) \end{aligned} \tag{3.86}$$

Among the infinite normal modes that are allowed (in keeping with the fact that the continuous system has infinite degrees of freedom.), $w_n(\xi, t)$ represents the generic nth mode. Naturally, the actual free motion of the beam, due to any initial disturbance, is described by the general integral of (3.52), or rather by a linear combination of the infinite normal modes. Given the particular contraint conditions analysed (see Fig. 3.7; Eqs. (3.65) and (3.66)) in all the normal modes $w_n(\xi, t)$ of the system analysed (the simply supported beam) there is a null deformation in correspondence to the extremities of the beam, where the restraint prevents displacements in a transverse direction. Always in the case in question, as an example of the taut cable in Sect. 3.3, on length L, the generic nth normal mode, shows "n" semi-wave lengths, as shown in Fig. 3.8 for a numeric example (the relevant data are shown in Table 3.2). Finally, as regards the simply supported beam, we notice that the relationship between the two subsequently ordered natural frequencies ω_n and $\omega_n + 1$ is equal to (3.78):

$$\frac{\omega_n}{\omega_{n+1}} = \frac{n^2}{(n+1)^2} \tag{3.87}$$

When the boundary conditions vary, the conditions to be imposed on the spatial function that defines the transverse deformation of the beam $\psi_b(\xi)$ will also obviously vary and, as a consequence, so will the natural frequencies of the system ω_n and the relative normal modes $\Phi^{(n)}(\xi)$. For example, in the case of a restrained beam in correspondence to the extremities (Fig. 3.9) the restraint will prevent both the transverse displacements w and rotations Θ_y: the corresponding boundary conditions, in this case, congruence conditions, become:

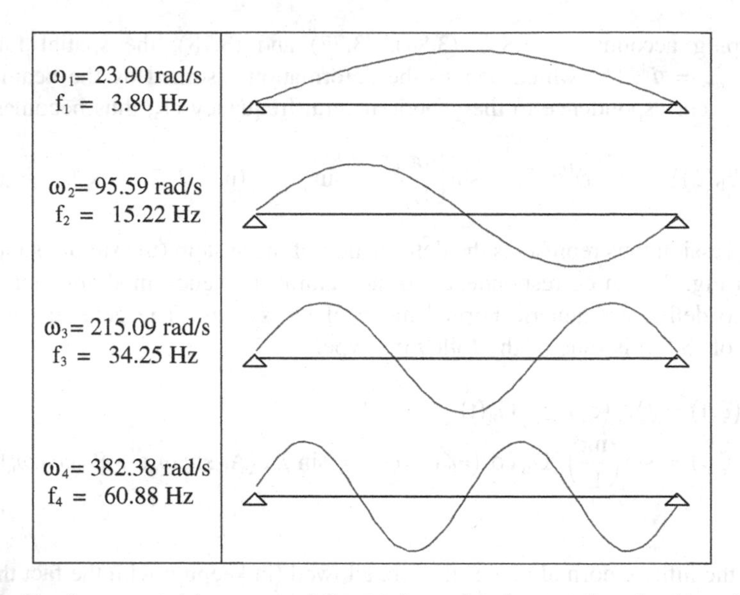

Fig. 3.8 Simply supported beam natural frequencies and main normal modes (data shown in Table 3.2)

Table 3.2 Beam: inertia geometry data

Length	$L = 30$ m
Bending stiffness	$EJ = 1.9 \times 10^8$ Nm^2
Mass per unit of length	$m = 40$ kg/m

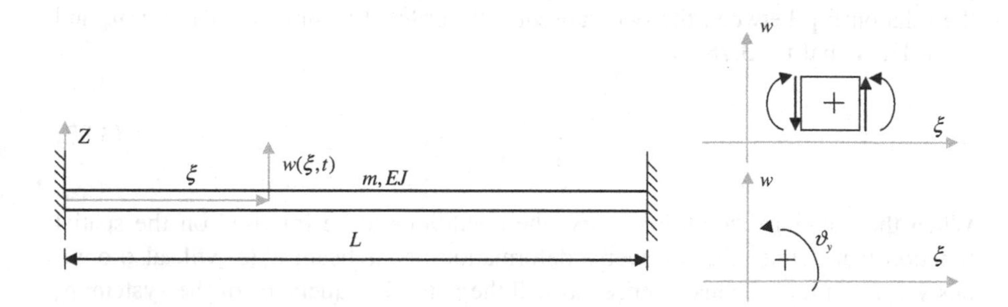

Fig. 3.9 Simply restrained beam

$$
\begin{aligned}
&[\mathrm{w}(\xi,\mathrm{t})]_{\xi=0} = 0\,(\text{at any t}) \Rightarrow \Psi_\mathrm{b}(0) = 0 \\
&[w(\xi,t)]_{\xi=L} = 0\,(\text{at any t}) \Rightarrow \Psi_\mathrm{b}(\mathrm{L}) = 0 \\
&[\Theta_\mathrm{y}]_{\xi=0} = \left[\frac{\partial \mathrm{w}}{\partial \xi}\right]_{\xi=0} = 0 \Rightarrow \left[\frac{\mathrm{d}\Psi_\mathrm{b}(\xi)}{\mathrm{d}\xi}\right]_{\xi=0} = 0 \\
&[\Theta_\mathrm{y}]_{\xi=\mathrm{L}} = \left[\frac{\partial \mathrm{w}}{\partial \xi}\right]_{\xi=\mathrm{L}} = 0 \Rightarrow \left[\frac{\mathrm{d}\Psi_\mathrm{b}(\xi)}{\mathrm{d}\xi}\right]_{\xi=\mathrm{L}} = 0
\end{aligned}
\tag{3.88}
$$

By substituting these boundary conditions with the generic relation (3.63) and defining the vector of the unknowns, or rather of the constants F_1, F_2, F_3, F_4, with $\underline{Z}$, we obtain the generic homogenous linear algebraic equation, parametric in γ and generally of the type:

$$[\mathrm{H}(\gamma)]\underline{\mathrm{Z}} = \underline{0} \tag{3.89}$$

The matrix terms of coefficients $[H(\gamma)]$ obviously change, when the boundary conditions vary. This system will admit nontrivial solutions only if the determinant of the coefficient matrix $[H(\gamma)]$ is annulled, that is for:

$$\det|\mathrm{H}(\gamma)| = 0 \tag{3.90}$$

In simpler cases (such as for example in the case analysed, i.e. a simply supported beam) the infinite roots γ_n of this expression can be obtained analytically or, in more complex cases, numerically. By means of the relation:

$$\gamma_\mathrm{n}^4 = \omega_\mathrm{n}^2 \frac{\rho \mathrm{A}}{\mathrm{EJ}} \tag{3.91}$$

it is thus possible to define the generic natural frequency ω_n of the system. Always in the case of the beam, the relative normal mode $\Phi^{(n)}(\xi)$ is generally of the following type:

$$\phi^{(\mathrm{n})}(\xi) = \mathrm{F}_1^{(\mathrm{n})} \sin\gamma_\mathrm{n}\xi + \mathrm{F}_2^{(\mathrm{n})} \cos\gamma_\mathrm{n}\xi + \mathrm{F}_3^{(\mathrm{n})} \sinh\gamma_\mathrm{n}\xi + \mathrm{F}_4^{(\mathrm{n})} \cosh\gamma_\mathrm{n}\xi \tag{3.92}$$

and can be scaled by means of any constant, by imposing an arbitrary value on one of the amplitudes $F_1^{(n)}$, $F_2^{(n)}$, $F_3^{(n)}$ or $F_4^{(n)}$, for example $F_1^{(n)} = 1$, and obtaining the others by means of:

$$[\mathrm{h}(\gamma_\mathrm{n})]\underline{\mathrm{z_n}} = \underline{\mathrm{N_n}} \tag{3.93}$$

where

- $\underline{z}_n$ is the vector that contains the non-defined constants:

$$\underline{\mathrm{z_n}} = \begin{Bmatrix} \mathrm{F}_2^{(\mathrm{n})} \\ \mathrm{F}_3^{(\mathrm{n})} \\ \mathrm{F}_4^{(\mathrm{n})} \end{Bmatrix} \tag{3.94}$$

- $[h(\gamma_n)]$ is the matrix obtained from $[H(\gamma_n)]$ reduced by eliminating the linear combination line of the others and the column that multiplies the constant whose value has been assigned (in the case in question, the first column);
- is vector $\underline{N}_n$ obtained by multiplying the column corresponding to the constant assigned for the same constant whose sign has changed.

Depending on the different boundary conditions, we will thus have different natural frequencies and, consequently, different normal modes. As regards the transverse motions of a narrow, homogenous beam with a bending stiffness EJ and density ρ, the value of the natural frequencies can be calculated using a simple formula [10, 16] which is always valid, regardless of the boundary conditions:

$$\omega_\mathrm{n} = \mathrm{K_n}\left(\frac{\mathrm{n}^2\pi^2}{\mathrm{L}^2}\right)\sqrt{\frac{\mathrm{EJ}}{\rho \mathrm{A}}} \tag{3.95}$$

where for coefficient K_n it depends on the specific boundary conditions (see Table 3.3).

The bibliography contains the tabulated values of both the natural frequencies and the normal modes in correspondence to the most common boundary conditions [10, 16].

Table 3.3 Coefficient K_n Eq. (3.95) for a homogenous beam as a function of the boundary conditions

Extremity A	Extremity B	K_1	K_2	K_3	K_4
Simply supported	Simply supported	1	1	1	1
Restrained	Restrained	2.27	1.56	1.36	1.26
Restrained	Simply supported	1.56	1.26	1.17	1.13
Restrained	Free	0.36	0.56	0.69	0.77
Free	Free	2.27	1.56	1.36	1.26

3.3.1 Transverse Vibrations in Beams Subjected to an Axial Load (Tensioned Beam)

The transverse vibrations of a beam are influenced by the presence of an axial load: to be more explicit, both the natural frequencies and the relative normal modes change when axial load S varies. A tensile load increases the frequencies while, on the contrary, a compression load reduces them. Furthermore, a certain compression load value (referred to as a critical load, see [13]) exists whereby the beam becomes statically unstable. In order to justify this phenomenon analytically, in this paragraph we will analyse the small transverse oscillations around the static equilibrium condition of a homogenous beam (Fig. 3.10) which, though identical to the one analysed in the previous paragraph, is subjected to an axial load S. We will always only analyse motion in the *z direction*, in the plane transversal to its own longitudianl axis ξ (abscissa running along the beam).

If in a generic instant of time t, we consider an infinitesimal element of the beam along $d\xi$, this will be subjected to the stresses outlined in Fig. 3.11, where

$$p(\xi,t)d\xi = \rho A d\xi\left(\frac{\partial^2 w(\xi,t)}{\partial t^2}\right) \tag{3.96}$$

represents the only inertia force due to the motion of the element itself while M, T and S are respectively the bending action, the shear and the axial action that the remainder of the beam exercises on the element considered.

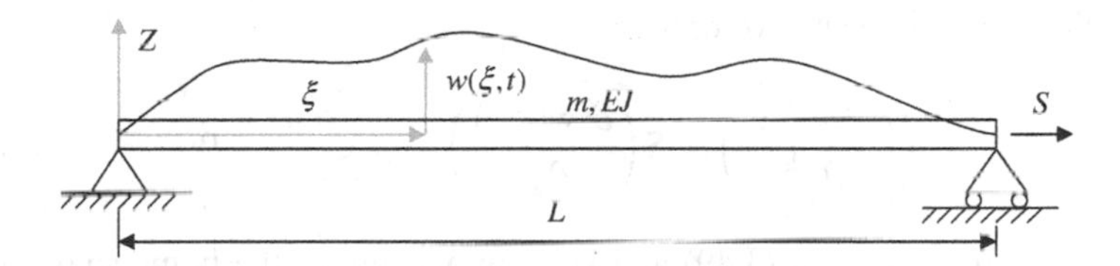

Fig. 3.10 Tensioned beam

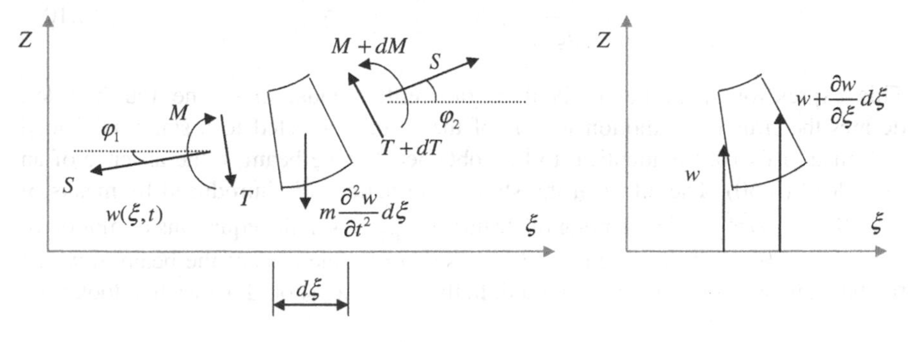

Fig. 3.11 Tensioned beam: undefined equilibrium equations

Since we are studying a dynamic problem, the internal actions *M*, *T* and *S* will generally be functions of both space and time:

$$\mathrm{M}=\mathrm{M}(\xi,\mathrm{t});\quad \mathrm{T}=\mathrm{T}(\xi,\mathrm{t});\quad \mathrm{S}=\mathrm{S}(\xi,\mathrm{t}) \tag{3.97}$$

The equilibrium equation with translation in a vertical direction *z*, disregarding the higher order infinitesimals, is given by:

$$\begin{aligned}&-\mathrm{T}-\rho\mathrm{A}d\xi\frac{\partial^2\mathrm{w}}{\partial\mathrm{t}^2}+\mathrm{T}+d\mathrm{T}-\mathrm{S}\sin\varphi_1+\mathrm{S}\sin\varphi_2=0\\&\Rightarrow\frac{\partial\mathrm{T}}{\partial\xi}d\xi-\rho\mathrm{A}\frac{\partial^2\mathrm{w}}{\partial\mathrm{t}^2}d\xi+\mathrm{S}\frac{\partial^2\mathrm{w}}{\partial\xi^2}d\xi=0\end{aligned} \tag{3.98}$$

The dynamic equilibrium equation at rotation around pole O, or rather around the left side of the infinitesimal generic element becomes, by disregarding the higher order infinitesimals

$$\mathrm{T}(\xi,\mathrm{t})=-\left(\frac{\partial\mathrm{M}(\xi,\mathrm{t})}{\partial\xi}\right) \tag{3.99}$$

By deriving (3.99):

$$\frac{\partial\mathrm{T}(\xi,\mathrm{t})}{\partial\xi}=-\left(\frac{\partial^2\mathrm{M}(\xi,\mathrm{t})}{\partial\xi^2}\right) \tag{3.100}$$

and substituting it in (3.98) we obtain:

$$-\left(\frac{\partial^2\mathrm{M}(\xi,\mathrm{t})}{\partial\xi^2}\right)+\mathrm{S}\left(\frac{\partial^2\mathrm{w}(\xi,\mathrm{t})}{\partial\xi^2}\right)-\rho\mathrm{A}\frac{\partial^2\mathrm{w}}{\partial\mathrm{t}^2}=0 \tag{3.101}$$

which, by recalling relation (3.49) and keeping account of the homogenous beam hypothesis (*EJ* = *cost*), becomes:

$$-\mathrm{EJ}\left(\frac{\partial^4\mathrm{w}}{\partial\xi^4}\right)+\mathrm{S}\left(\frac{\partial^2\mathrm{w}}{\partial\xi^2}\right)=\rho\mathrm{A}\frac{\partial^2\mathrm{w}}{\partial\mathrm{t}^2} \tag{3.102}$$

This expression, a partial derivatives differential equation of the fourth order, defines the transverse motion $w(\xi, t)$ of the beam subjected to a static axial load *S*. Equation (3.102) is identical to that obtained for the beam in the absence of an axial load (3.50). The effect of the static axial action *S* is introduced by means of term $\mathrm{S}(\partial^2\mathrm{w}/\partial\xi^2)$ which, among other things, appears in the equations of motion of the cable (3.9) which can be considered as a borderline case of the beam in which the bending stiffness *EJ* is disregarded. In this case too, should we wish to look for a possible stationary solution to (3.102), we will assume a particular integral of the same equation in the following form:

$$w(\xi, t) = \Psi_c(\xi)G(t) \tag{3.103}$$

which, when placed in (3.102), due to the intrinsic structure of functions $\psi_c(\xi)$ and $G(t)$, gives us a differential equation with total derivatives:

$$EJ\Psi_c''''(\xi)G(t) - S\Psi_c''(\xi)G(t) + \rho A\Psi_c(\xi)\ddot{G}(t) = 0 \tag{3.104}$$

where, for the sake of brevity, we have used the symbols Ψ'' and Ψ'''' to indicate respectively the second derivative and the fourth derivative totals of the spatial function $\Psi_c(\xi)$:

$$\Psi_c''(\xi) = \frac{d^2\Psi_c(\xi)}{d\xi^2}; \quad \Psi_c''''(\xi) = \frac{d^4\Psi_c(\xi)}{d\xi^4} \tag{3.105}$$

As with the previous cases, (3.104) by separating the variables, i.e.:

$$\left(\frac{EJ}{\rho A}\right)\frac{\Psi_c''''(\xi)}{\Psi_c(\xi)} - \left(\frac{S}{\rho A}\right)\frac{\Psi_c''(\xi)}{\Psi_c(\xi)} = -\frac{\ddot{G}(t)}{G(t)} = \omega^2 \tag{3.106}$$

we are able to define two distinct differential ordinary equations; the first of (3.106), relative to $G(t)$ can be written as:

$$-\frac{1}{G(t)}\frac{d^2G(t)}{dt^2} = \omega^2 \Rightarrow G(t) = A\sin\omega t + B\cos\omega t = D\cos(\omega t + \varphi) \tag{3.107}$$

while, in this case, the second differential ordinary equation, relative to $\Psi_c(\xi)$ will be:

$$EJ\Psi_c''''(\xi) - S\Psi_c''(\xi) - \rho A\omega^2\Psi_c(\xi) = 0 \tag{3.108}$$

By imposing the generic solution:

$$\Psi_c(\xi) = \overline{F}e^{\lambda\xi} \tag{3.109}$$

in differential equation (3.108), we obtain an algebraic, linear, homogenous and parametric system in λ of the type:

$$(\lambda^4 EJ - \lambda^2 S - \rho A\omega^2)\overline{F} = 0 \tag{3.110}$$

which admits non-trivial solutions for:

$$\lambda^4 \mathrm{EJ} - \lambda^2 \mathrm{S} - \rho \mathrm{A}\omega^2 = 0 \tag{3.111}$$

from which:

$$\begin{aligned} \lambda_{\mathrm{I}}^2 &= \frac{\mathrm{S} - \sqrt{\mathrm{S}^2 + 4\mathrm{EJ}\rho\mathrm{A}\omega^2}}{2\mathrm{EJ}} \Rightarrow \lambda_{1,2} = \pm\sqrt{\lambda_{\mathrm{I}}^2} \\ \lambda_{\mathrm{II}}^2 &= \frac{\mathrm{S} + \sqrt{\mathrm{S}^2 + 4\mathrm{EJ}\rho\mathrm{A}\omega^2}}{2\mathrm{EJ}} \Rightarrow \lambda_{3,4} = \pm\sqrt{\lambda_{\mathrm{II}}^2} \end{aligned} \tag{3.112}$$

On account of the fact that λ_{I}^2 is always negative and λ_{II}^2 always positive since the root in (3.112) is always higher than S, the solution thus becomes of the following type:

$$\Psi_{\mathrm{c}}(\xi) = \mathrm{F}_1 \sin\gamma_1\xi + \mathrm{F}_2 \cos\gamma_1\xi + \mathrm{F}_3 \sinh\gamma_2\xi + \mathrm{F}_4 \cosh\gamma_2\xi \tag{3.113}$$

where

$$\gamma_1 = \sqrt{-\lambda_{\mathrm{I}}^2}; \quad \gamma_2 = \sqrt{\lambda_{\mathrm{II}}^2} \tag{3.114}$$

That stationary solution, which represent free motion in a steady-state transverse direction of the beam subjected to an axial load S, is thus given by the relation:

$$\mathrm{w}(\xi, \mathrm{t}) = (\mathrm{F}_1 \sin\gamma_1\xi + \mathrm{F}_2 \cos\gamma_1\xi + \mathrm{F}_3 \sinh\gamma_2\xi + \mathrm{F}_4 \cosh\gamma_2\xi)(\mathrm{A}\sin\omega\mathrm{t} + \mathrm{B}\cos\omega\mathrm{t}) \tag{3.115}$$

In (3.115) constants F_1, F_2, F_3 and F_4, depend on the boundary conditions, i.e. on the constraints assigned to the extremities of the beam. A particularly simple, though significant, case is given by the beam simply supported on both extremities (Fig. 3.10); in this case, the boundary conditions are:

$$\begin{aligned} &[\mathrm{w}(\xi, \mathrm{t})]_{\xi=0} = 0\,(\text{at any t}) \Rightarrow \Psi_{\mathrm{c}}(0) = 0 \\ &[\mathrm{w}(\xi, \mathrm{t})]_{\xi=\mathrm{L}} = 0\,(\text{at any t}) \Rightarrow \Psi_{\mathrm{c}}(\mathrm{L}) = 0 \\ [\mathrm{M}]_{\xi=0} = \mathrm{EJ}&\left[\frac{\partial^2 \mathrm{w}}{\partial\xi^2}\right]_{\xi=0} = 0\,(\text{at any t}) \Rightarrow \left[\frac{\mathrm{d}^2\Psi_{\mathrm{c}}(\xi)}{\mathrm{d}\xi^2}\right]_{\xi=0} = 0 \\ [\mathrm{M}]_{\xi=\mathrm{L}} = \mathrm{EJ}&\left[\frac{\partial^2 \mathrm{w}}{\partial\xi^2}\right]_{\xi=\mathrm{L}} = 0\,(\text{at any t}) \Rightarrow \left[\frac{\mathrm{d}^2\Psi_{\mathrm{c}}(\xi)}{\mathrm{d}\xi^2}\right]_{\xi=\mathrm{L}} = 0 \end{aligned} \tag{3.116}$$

From which, as in the case outlined in the previous paragraph, we obtain:

$$\sin\gamma_1\mathrm{L} = 0 \tag{3.117}$$

from which:

$$\gamma_1 L = n\pi \quad (n = 1, 2, \ldots \infty) \Rightarrow \gamma_{1n} = \frac{n\pi}{L} \tag{3.118}$$

$$(\Psi_c(\xi))_{\omega=\omega_n} = \phi^{(n)}(\xi) = \sin\left[\left(\frac{n\pi}{L}\right)\xi\right] = \sin\gamma_{1n}\xi \quad (n = 1, 2, \ldots \infty) \tag{3.119}$$

This expression thus represents the deformation (or rather the generic normal mode $\Phi_{(n)}(\xi)$) of the system analysed, in correspondence to the mode with a natural frequency ω_n which can be evaluated using simple steps such as:

$$\omega_n^2 = \left(\frac{n^2\pi^2}{\rho A L^2}\right)\left(S + EJ\left(\frac{n^2\pi^2}{L^2}\right)\right) \Rightarrow \omega_n = \left(\frac{n\pi}{L}\right)\sqrt{\frac{S + EJ\left(\frac{n^2\pi^2}{L^2}\right)}{m}} \tag{3.120}$$

If the axial load is of tensile type (i.e. S is positive according to the conventions shown in Fig. 3.11) the natural frequency of system ω_n increases; conversely, if the static axial load is of the compression type ($S < 0$), the natural frequency, as shown in the graph in Fig. 3.12, decreases.

In the case the load is:

$$S_{crit} = -\frac{EJ\pi^2}{L^2} \tag{3.121}$$

the natural is annulled: S has reached the limit value S_{crit}, or rather Eulero's critical load [13]. If compression load S increases above this value, the beam becomes statically unstable. From a dynamic point of view, the critical load S_{crit} can thus be defined as that limit load value due to which the first frequency of the beam tends towards zero.

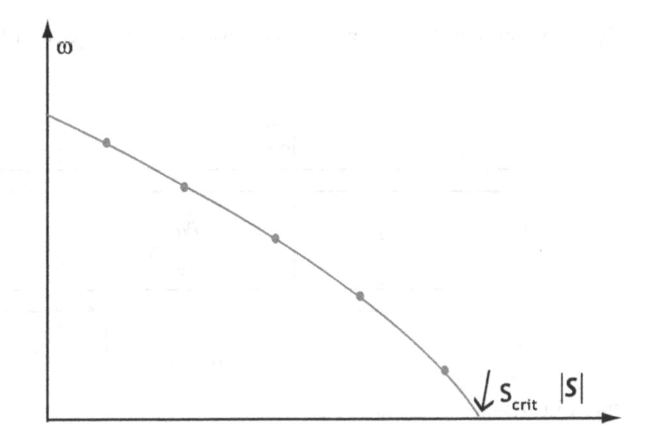

Fig. 3.12 Tensioned beam: effect of axial compression S on the first natural frequency ω1 of the system

The trend of the frequency resulting from an increase in the compressive load shown in Fig. 3.12 can be used to evaluate the critical load without reaching it. In fact, if by axially loading a beam we measure the first natural frequency as the load itself increases, it is possible to extrapolate the value of the critical load without reaching it, as briefly outlined in Fig. 3.13.

Let us now analyse the longitudinal vibrations about the static equilibrium condition in bars, i.e. beams whose longitudinal dimension is predominant with respect to the others (Fig. 3.14), by assuming, as in the previous paragraphs, a homogenous beam, i.e. with transverse section A, axial stiffness EA and density ρ all constant. Let us consider a generic infinitesimal beam element of length $d\xi$ placed at a distance ξ from the left extremity of the beam itself: let us use $u = u(\xi)$ to define the displacement of the generic section. The longitudinal deformation to which the element is subjected, at any time t, can be expressed as:

$$\varepsilon = \frac{u + \frac{\partial u}{\partial \xi} d\xi - u}{d\xi} = \frac{\partial u}{\partial \xi} \tag{3.122}$$

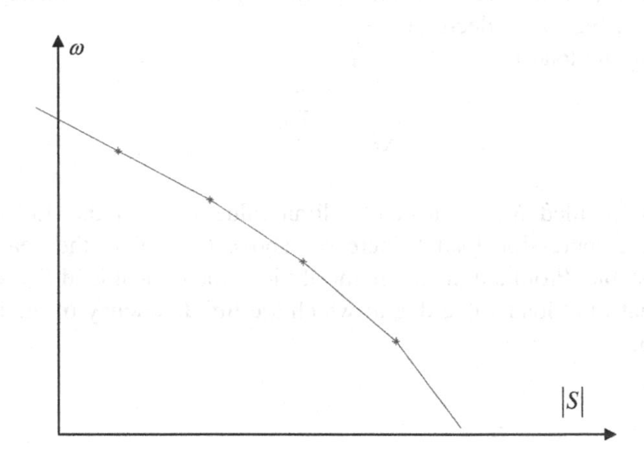

Fig. 3.13 Tensioned beam: experimental evaluation of the critical load

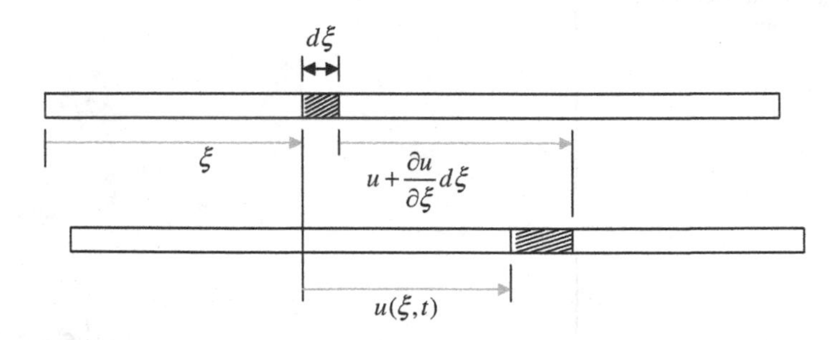

Fig. 3.14 Beam: axial vibrations

According to Hooke's law [13] stress σ found in the section is thus:

$$\sigma = \frac{S}{A} = E\varepsilon = E\frac{\partial u}{\partial \xi} \tag{3.123}$$

where S is the axial action present in the beam. By deriving (3.123) we obtain:

$$\frac{\partial S}{\partial \xi} = AE\frac{\partial^2 u}{\partial \xi^2} \tag{3.124}$$

Since we are studying a dynamic problem, generally speaking, internal axial action S will be a function of both space and of time:

$$S = S(\xi, t) \tag{3.125}$$

With regard to the same infinitesimal element, it is possible to write the dynamic equilibrium equation at translation in a longitudinal direction ξ (Fig. 3.15) as:

$$-S + S + \frac{\partial S}{\partial \xi}d\xi - \rho A\frac{\partial^2 u}{\partial t^2}d\xi = 0 \Rightarrow \frac{\partial S}{\partial \xi} = \rho A\frac{\partial^2 u}{\partial t^2} \tag{3.126}$$

and by keeping account of (3.124):

$$AE\frac{\partial^2 u}{\partial \xi^2} = \rho A\frac{\partial^2 u}{\partial t^2} \tag{3.127}$$

Having used

$$C = \sqrt{\frac{AE}{\rho A}} = \sqrt{\frac{E}{\rho}} \tag{3.128}$$

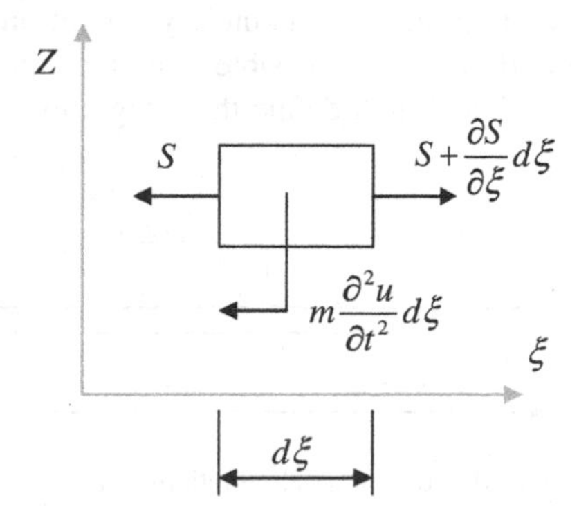

Fig. 3.15 Beam: axial vibrations—undefined equilibrium equations

to indicate the propagation velocity in the longitudinal direction (velocity of sound), we thus obtain the undefined equilibrium equation for longitudinal motion expressed in the form of

$$\frac{\partial^2 u}{\partial \xi^2} = \frac{1}{C^2} \frac{\partial^2 u}{\partial t^2} \tag{3.129}$$

an expression similar to the one already seen for the transverse vibrations of the cable. This equation, a second order partial derivatives differential equation, defines the longitudinal motion $u(\xi, t)$ of the beam analysed. In this case too, as with the cable (Sect. 3.2), if we wish to study a possible stationary solution of (3.128) we will assume a particular integral of the equation itself in the form of:

$$u(\xi, t) = \Psi_u(\xi) G(t) \tag{3.130}$$

By using an identical procedure to the one described in the previous paragraphs, we will obtain the stationary solution:

$$\begin{aligned} u(\xi, t) &= \left(F_1 \sin\left(\frac{\omega}{C}\xi\right) + F_2 \cos\left(\frac{\omega}{C}\xi\right) \right) (A \sin \omega t + B \cos \omega t) \\ u(\xi, t) &= \left(F_1 \sin\left(\frac{\omega}{C}\xi\right) + F_2 \cos\left(\frac{\omega}{C}\xi\right) \right) D \cos(\omega t + j) \end{aligned} \tag{3.131}$$

This equation defines the longitudinal motion of a homogenous beam, regardless of the boundary conditions or the initial conditions. In (3.131) constants F_1 and F_2 depend on the boundary conditions, i.e. on the constraints assigned to the beam extremities: by varying the latter, the solution is obviously modified.

A particularly simple example is given by the beam constrained at its extremities (see Fig. 3.16); in this case, the boundary conditions are the following:

$$\begin{aligned} [u(\xi, t)]_{\xi=0} = 0 \Rightarrow \Psi_u(0) = 0 \\ S_{\xi=L} = AE \left[\frac{\partial u(\xi, t)}{\partial \xi} \right]_{\xi=L} = 0 \Rightarrow \left[\frac{\partial \Psi_u}{\partial \xi} \right]_{\xi=L} = 0 \end{aligned} \tag{3.132}$$

By imposing the boundary conditions, respectively congruence and equilibrium conditions, it is possible to define equations which must satisfy the spatial function $\Psi_u(\xi)$ and thus define the integration constants of (3.131):

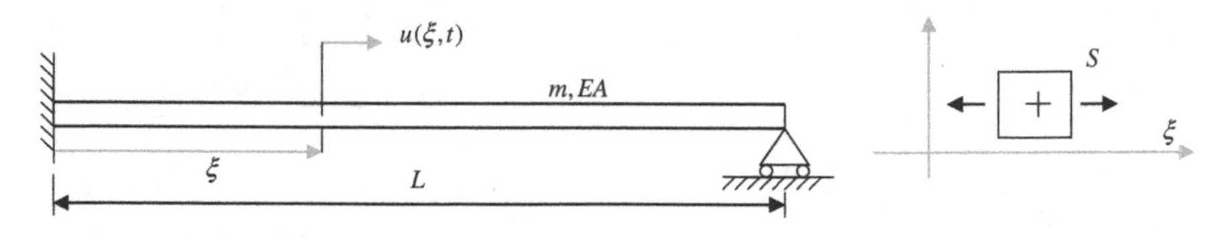

Fig. 3.16 Beam: axial vibrations—hinge-cart constraint configuration

$$\begin{aligned} \Psi_u(0) = 0 \Rightarrow F_2 = 0 \\ \left[\frac{\partial \Psi_u}{\partial \xi}\right]_{\xi=L} = 0 \Rightarrow \frac{\omega}{C} F_1 \cos\left(\frac{\omega}{C} L\right) = 0 \end{aligned} \tag{3.133}$$

As usual, the second equation is of the linear, algebraic homogenous and parametric type in ω. The same equation will only admit nontrivial solutions if:

$$\cos\left(\frac{\omega}{C} L\right) = 0 \Rightarrow \frac{\omega}{C} L = \frac{\pi}{2} + (n-1)\pi \Rightarrow \omega_n = \left(\frac{\pi}{2} + (n-1)\pi\right)\frac{C}{L} \tag{3.134}$$

By keeping account of (3.131), (3.133) and (3.134), the spatial function $\Psi_u(\xi)$, which defines the deformation assumed by the beam, considered in correspondence to the generic natural frequency ω_n, thus becomes:

$$\Psi_u(\xi)_{\omega=\omega_n} = \phi^{(n)}(\xi) = \sin\left[\left(\frac{\frac{\pi}{2} + (n-1)\pi}{L}\right)\xi\right] \quad (n = 1, 2, \ldots \infty) \tag{3.135}$$

3.4 Torsional Vibrations in Beams

In this paragraph we intend to analyse the equations of motion to describe the torsional vibrations in beams, whose dimension, namely the longitudinal one, is predominant with respect to the others (Fig. 3.17), by assuming, as in the previous paragraphs, a homogenous beam, i.e. with a transverse section with constant torsional stiffness GJ_p and density ρ. We assume that the centre of torsion coincides with the centroid of the beam section so that the torsion can be considered uncoupled from bending. We will introduce variable $\Theta_x(\xi, t)$ as an independent variable to describe this motion. This variable describes the torsional time-space deformation assumed by the beam itself. Let us once again consider the generic infinitesimal element of the beam of length $d\xi$ placed at distance ξ from the left extremity of the same beam subjected to a torsional moment M_t; the relative rotation

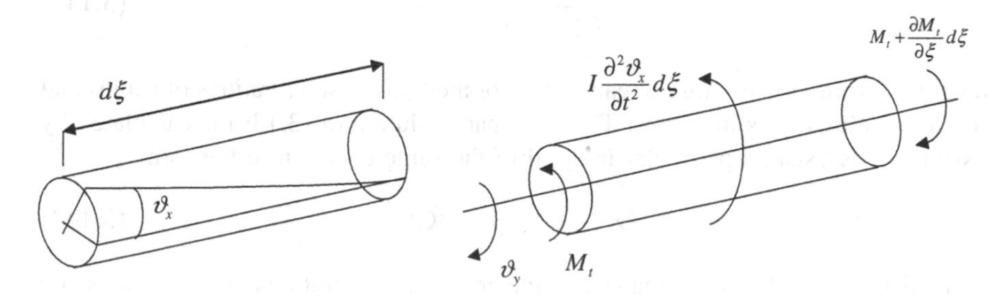

Fig. 3.17 Narrow beam: torsional vibrations

$d\Theta_x$ between the two sections of the extremity of the small element will be linked to the same torsional moment by the well-known relation [13, 20]:

$$d\Theta_x = \chi \frac{M_t d\xi}{GJ_p} \tag{3.136}$$

where J_p represents the polar area moment of the section, G is the shear modulus and finally χ is the torsion factor: this factor [2, 11] depends, as is known, on the type of section ($\chi = 1$ for a circular section).

By deriving relation (3.136) with respect to space ξ, we obtain:

$$\frac{\partial M_t}{\partial \xi} = \frac{1}{\chi} GJ_p \frac{\partial^2 \Theta_x}{\partial \xi^2} \tag{3.137}$$

The dynamic equilibrium equation at rotation for the generic infinitesimal small element becomes:

$$-M_t + M_t + \frac{\partial M_t}{\partial \xi} d\xi - I \frac{\partial^2 \Theta_x}{\partial t^2} d\xi = 0 \Rightarrow \frac{1}{\chi} GJ_p \frac{\partial^2 \Theta_x}{\partial \xi^2} = I \frac{\partial^2 \Theta_x}{\partial t^2} \tag{3.138}$$

where I is the mass moment of inertia by unit of length (for homogenous beams $I = \rho J_p$ [10]), from which:

$$\frac{\partial^2 \Theta_x}{\partial t^2} = \frac{1}{\chi} \frac{G}{\rho} \frac{\partial^2 \Theta_x}{\partial \xi^2} \tag{3.139}$$

Having used

$$C = \sqrt{\frac{1}{\chi} \frac{G}{\rho}} \tag{3.140}$$

To indicate the propagation velocity of the torsion wave, we thus obtain the undefined equilibrium equation for torsional motions expressed in the usual form:

$$\frac{\partial^2 \Theta_x}{\partial \xi^2} = \frac{1}{C^2} \frac{\partial^2 \Theta_x}{\partial t^2} \tag{3.141}$$

a similar expression to those already seen for the transverse vibrations of cables and the longitudinal ones in beams. The stationary solution of (3.139) is calculated by assuming, as usual, a particular integral of the same equation in the form:

$$\Theta_x(\xi, t) = \Psi_\theta(\xi) G(t) \tag{3.142}$$

and all the considerations made for the transverse vibrations of the cables are still valid.

3.5 Analysis of the General Integral of the Equation of Motion in Continuous Systems

In Sects. 3.2–3.4, for the different types of vibrations (transverse, longitudinal and axial) to which the continuous systems analysed could be subjected (beams and cables), we calculated the natural frequencies ω_n and the normal modes $\phi^{(n)}(\xi)$.[2] Each of these normal modes represents a particular integral of the equations of motion. The most generic normal mode of the system is given by the linear combination of the same normal modes, namely:

$$y(\xi,t) = \sum_{n=1}^{\infty} \phi^{(n)}(\xi) D_n \cos(\omega_n t + \varphi_n) = \sum_{n=1}^{\infty} \phi^{(n)}(\xi)(A_n \cos\omega_n t + B_n \sin\omega_n t) \tag{3.143}$$

Since we are dealing with a continuous system, the initial conditions will be defined by two spatial functions which define the initial deformation imposed on the system with time t = 0 and a possible spatial distribution of velocity:

$$\begin{aligned} y_0(\xi) &= (y(\xi,t))_{t=0} \\ \dot{y}_0(\xi) &= \left(\frac{\partial y(\xi,t)}{\partial t}\right)_{t=0} \end{aligned} \tag{3.144}$$

By keeping account of (3.143) and (3.144), regardless:

- of the type of vibrations analysed (transverse, axial or torsional);
- of the type of continuous system analysed (cable or beam);
- of the constraint conditions which, as mentioned several times, only modify the expression of the natural frequencies and the relative normal modes, we obtain:

$$y_0(\xi) = \sum_{n=1}^{\infty} \phi^{(n)}(\xi) A_n \tag{3.145}$$

[2] As previously mentioned, in this chapter we will only consider continuous systems from which we can easily obtain the analytical solution in closed form and, for which, the spatial deformation is a function of the sole coordinate ξ. More complex continuous systems, such as circular or rectangular plates, are also dealt with on a widespread basis in literature and for these too, analytical solutions can be found, in which, however, the deformation is a function of two spatial variables ξ and η. For more in-depth information regarding this problem, please refer to [12, 16, 17, 18].

$$\dot{y}_0(\xi) = \left[\sum_{n=1}^{\infty} \phi^{(n)}(\xi)(-\omega_n A_n \sin \omega_n t + \omega_n B_n \cos \omega_n t)\right]_{t=0} = \sum_{n=1}^{\infty} \phi^{(n)}(\xi)\omega_n B_n \tag{3.146}$$

These expressions are nothing else but in-series developments, in variable ξ, of the two functions $\mathbf{y}_0(\boldsymbol{\xi})$ and $\dot{\mathbf{y}}_0(\boldsymbol{\xi})$ in the normal modes considered. Constants A_i and B_i (i = 1, n) can be obtained [1, 15] by means of:

$$A_i = \int_0^L y_0(\xi)\phi^{(i)}(\xi)d\xi; \quad B_i = \frac{1}{\omega_i}\int_0^L \dot{y}_0(\xi)\phi^{(i)}(\xi)d\xi \tag{3.147}$$

In the specific case of the simply supported cable (Sect. 3.2), the natural frequencies ω_n and the relative vibratine modes $\phi^{(n)}(\xi)$ are (see (3.33) and (3.35)):

$$\omega_n = \left(\frac{n\pi}{L}\right)\sqrt{\frac{S}{m}} \Rightarrow \phi^{(n)}(\xi) = \sin\left(\frac{n\pi}{L}\right)\xi \quad (n = 1, 2, \ldots\infty) \tag{3.148}$$

The more generic free motion of the system $w(\xi, t)$, i.e. the response of same to an initial perturbation, is given by the general integral defined by the linear combination of all the infinite normal modes (3.143):

$$w(\xi, t) = \sum_{n=1}^{\infty} \sin\left(\frac{n\pi}{L}\xi\right)(A_n \cos(\omega_n t) + B_n \sin(\omega_n t)) \tag{3.149}$$

In other words, (3.149) expresses the more generic free motion of the continuous system as the sum of the contribution of all the modes, weighted differently. In order to obtain a complete definition of this motion, it is necessary to define all the constants A_n and B_n, functions of the initial conditions assigned to the system. As seen in the previous paragraph, the initial conditions are defined by two spatial functions which, with time t = 0, provide the initial deformation imposed on the system and its initial motion:

$$\begin{aligned} w_0(\xi) &= (w(\xi, t))_{t=0} \\ \dot{w}_0(\xi) &= \left(\frac{\partial w(\xi, t)}{\partial t}\right)_{t=0} \end{aligned} \tag{3.150}$$

By keeping account of (3.148) and (3.147), constants A_i and B_i ($i = 1, n$) (3.149) become:

$$A_i = \frac{2}{L}\int_0^L w_o(\xi)\sin\left(\frac{i\pi\xi}{L}\right)d\xi$$
$$B_i = \frac{2}{L\omega_i}\int_0^L \dot{w}_o(\xi)\sin\left(\frac{i\pi\xi}{L}\right)d\xi \tag{3.151}$$

In this specific case, therefore, the constants are multiplicative coefficients of the in-series developments of Fourier [1, 4] of functions $w_0(\xi)$ and $\dot{w}_0(\xi)$. Equation (3.151) allows us to calculate explicitly the constants present in the general integral of the system's free motion (3.149). If, for example, we assign the following initial conditions:

$$w_0(\xi) = \sin\left(\frac{\pi}{L}\xi\right);\quad \dot{w}_0(\xi) = 0 \tag{3.152}$$

i.e. if initially the cable is positioned exactly as its first mode and kept still, the constants of the free motion of the system subjected to the initial condition defined by (3.149) are:

$$B_1 = B_2 = \cdots = B_n = 0$$
$$A_n = \frac{2}{L}\int_0^L \sin\left(\frac{\pi}{L}\xi\right)\sin\left(\frac{n\pi}{L}\xi\right)d\xi \Rightarrow A_1 = 1;\ A_n(n \neq 1) = 0 \tag{3.153}$$

Therefore, by adopting the initial conditions (3.152), the general integral is reduced to:

$$w(\xi, t) = \sin\left(\frac{\pi}{L}\xi\right)\cos\omega_1 t \tag{3.154}$$

By assigning to the system, as initial conditions, those corresponding to the first normal mode (this obviously applies for the generic normal mode) and null initial velocities, the system vibrates in steady-state free motion (stationary motion) exciting only the first (or generic) mode, i.e. maintaining the initial deformation. When subjected to these initial conditions, the system will vibrate with its first natural frequency ω_1 (or with the generic ω_n). Conversely, if the initial conditions are generic, all the normal modes will be excited: these modes will appear with non-null coefficients in the writing the general integral (3.149): thus motion is generally seen as a combination of normal modes of the system itself. The more general solution will thus be aperiodic with beats[3] [12, 16].

3.6 Analysis of the Particular Integral of Forced Motion

In the case of harmonic concentrated excitation forces of frequency Ω, the solution can be found by imposing these excitation forces as the boundary conditions of the domains with which it is necessary to divide the continuous system itself.

In the event, for example, of analysing a cable of length L (Fig. 3.18) excited by an excitation force applied at a distance "a" from the left extremity, it is necessary to divide the cable into two domains for which Eq. (3.10) with valid partial derivatives for free motion are valid. However, should we wish to obtain the only particular integral associated with the harmonic forced motion, the solution will be harmonic and therefore generally of the following type:

$$w(\xi, t) = \Psi(\xi)G(t) \tag{3.155}$$

[3]When a mechanical system is excited by several excitation forces, harmonic with a similar amplitude and slightly different frequencies [14, 19], the answer is of the type shown in the following figure:

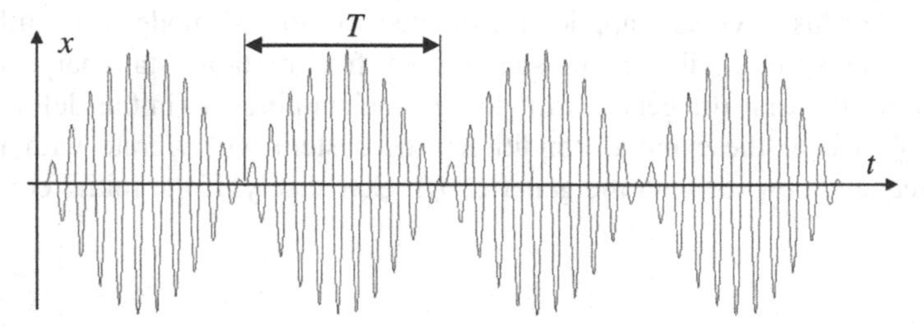

Example of beat

This particular type of motion is defined as "beat". Each time that the amplitude has a maximum we say that a beat has occurred, while the beat frequency is determined by two maximum consecutive amplitudes.

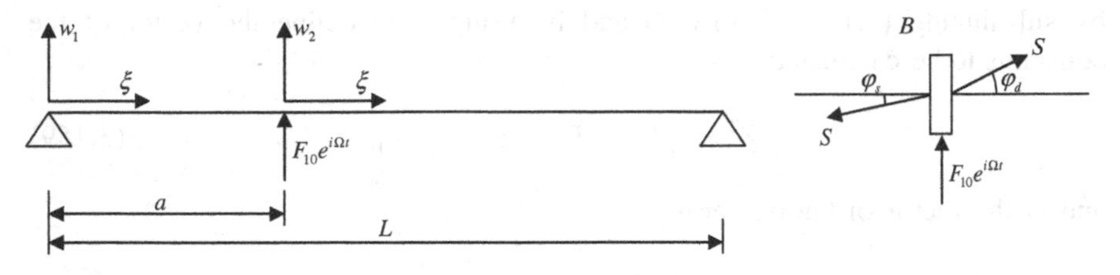

Fig. 3.18 Cable undergoing to a harmonic excitation force

with[4]:

$$G(t) = G_0 e^{i\Omega t} \tag{3.156}$$

In the example cited (Fig. 3.18) in the two domains, the deformation can be expressed as:

$$\begin{aligned} w_1(\xi,t) &= \Psi_1(\xi)G_{01}(t) = \left(F_{11}\sin\frac{\Omega}{C}\xi + F_{21}\cos\frac{\Omega}{C}\xi\right)e^{i\Omega t} \\ w_2(\xi,t) &= \Psi_2(\xi)G_{02}(t) = \left(F_{12}\sin\frac{\Omega}{C}\xi + F_{22}\cos\frac{\Omega}{C}\xi\right)e^{i\Omega t} \end{aligned} \tag{3.157}$$

By imposing the boundary conditions (which, as previously mentioned, can be equilibrium and congruence conditions):

- condition I (congruence) $[w_1(\xi,t)]_{\xi=0} = 0$
- condition II (congruence) $[w_2(\xi,t)]_{\xi=(L-a)} = 0$
- condition III (congruence) $[w_1(\xi,t)]_{\xi=a} = [w_2(\xi,t)]_{\xi=0}$ (3.158)
- condition IV (equilibrium at B)

$$F_0e^{i\Omega t} - S\sin\varphi_s + S\sin\varphi_d = 0 \Rightarrow F_0e^{i\Omega t} - S\left[\frac{\partial w_1}{\partial\xi}\right]_{\xi=a} + S\left[\frac{\partial w_2}{\partial\xi}\right]_{\xi=0} = 0$$

[4]The real force acting on the system iso f the type $f_j = F_j\cos\Omega t$. As already seen in systems with 1 d.o.f. and 2 d.o.f. (Chap. 2), an effective resolving formalism to evaluate the steady-state response of the system being examined is the one that uses complex numbers so that:

$$\mathbf{f_j = F_j\cos\Omega t \Rightarrow f_j = F_j e^{i\Omega t}} \tag{4.4.1}$$

and, by using this approach, solution $y(\xi,t)$ will obviously prove to be complex:

$$Y_o(\xi)e^{i\Omega t} \tag{4.4.2}$$

The real response of the system will be defined by the projection on the real axis of (4.4.2):

$$y(\xi,t) = \mathrm{Re}\left(Y_o(\xi)e^{i\Omega t}\right). \tag{4.4.3}$$

by substituting (3.157) in (3.158) and by using $\underline{Z}$ to define the vector of the constants to be calculated:

$$\underline{Z}^T = \{ F_{11} \quad F_{12} \quad F_{21} \quad F_{22} \} \tag{3.159}$$

and $\underline{\mathbf{N}}$ the vector of known terms:

$$\underline{N}^T = \{ 0 \quad 0 \quad 0 \quad -F_0 \} \tag{3.160}$$

we thus obtain a non-homogenous linear algebraic system of as many equations as there are constants to define $\underline{\mathbf{Z}}$:

$$[H(\Omega)]\underline{Z}^T = \underline{N} \tag{3.161}$$

where, in this case, the matrix of the real coefficients:

$$[H(\Omega)] = \begin{bmatrix} 0 & 1 & 0 & 0 \\ 0 & 0 & \sin\left(\frac{\Omega(L-a)}{C}\right) & \cos\left(\frac{\Omega(L-a)}{C}\right) \\ \sin\left(\frac{a\Omega}{C}\right) & \cos\left(\frac{a\Omega}{C}\right) & 0 & -1 \\ -\frac{S\Omega}{C}\cos\left(\frac{a\Omega}{C}\right) & \frac{S\Omega}{C}\cos\left(\frac{a\Omega}{C}\right) & \frac{S\Omega}{C} & 0 \end{bmatrix} \tag{3.162}$$

In the case of several external excitation forces, the domains with which the continuous systems will be divided increase and, as consequence so do the number of constants to be defined and the boundary conditions to be imposed. As a consequence, the order of vectors $\underline{Z}$ and $\underline{N}$ and of matrix $[H(\Omega)]$ increases. However, it is necessary to point out that in order to evaluate the steady-state response of a continuous system to harmonic excitation forces this approach cannot be used if:

- the excitation forces are distributed (we would need an infinite number of domains and, consequently, of spatial-temporal functions $w_k(\xi, t)$ which describe the deformation of the system);
- the excitation forces move their application point along the continuous system like, for example, in the case of mobile loads (in this case, it is not possible to impose boundary conditions in correspondence to the generic travelling load, in that these vary in time).

Conversely, by using this approach, it is possible to keep account of springs or dampers applied to the continuous system itself. For example, in case we want to analyse the taut cable in Fig. 3.19, it is necessary to divide the continuous system into 3 domains inside which the deformation is described by:

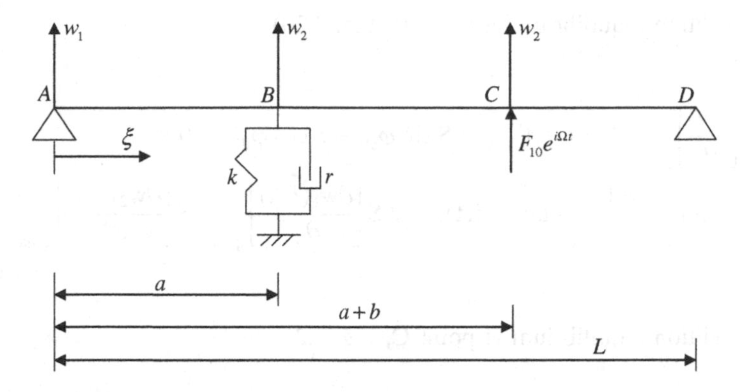

Fig. 3.19 Forced continuous system (taut cable example)

- domain I $(0<\xi<a)$ $\Rightarrow$ $w_1(\xi,t)=\left(F_{11}\sin\frac{\Omega}{C}\xi+F_{21}\cos\frac{\Omega}{C}\xi\right)e^{i\Omega t}$
- domain II $(a<\xi<(a+b))$ $\Rightarrow$ $w_2(\xi,t)=\left(F_{12}\sin\frac{\Omega}{C}\xi+F_{22}\cos\frac{\Omega}{C}\xi\right)e^{i\Omega t}$
- domain III $((a+b)<\xi<L)$ $\Rightarrow$ $w_3(\xi,t)=\left(F_{13}\sin\frac{\Omega}{C}\xi+F_{23}\cos\frac{\Omega}{C}\xi\right)e^{i\Omega t}$

(3.163)

In this case, the boundary conditions (Fig. 3.20) will be:

- condition I (congruence) $[w_1(\xi,t)]_{\xi=0}=0$
- condition II (congruence) $[w_3(\xi,t)]_{\xi=L}=0$
- condition III (congruence) $[w_1(\xi,t)]_{\xi=a}=[w_2(\xi,t)]_{\xi=a}$
- condition IV (congruence) $[w_2(\xi,t)]_{\xi=a+b}=[w_3(\xi,t)]_{\xi=a+b}$

(3.164)

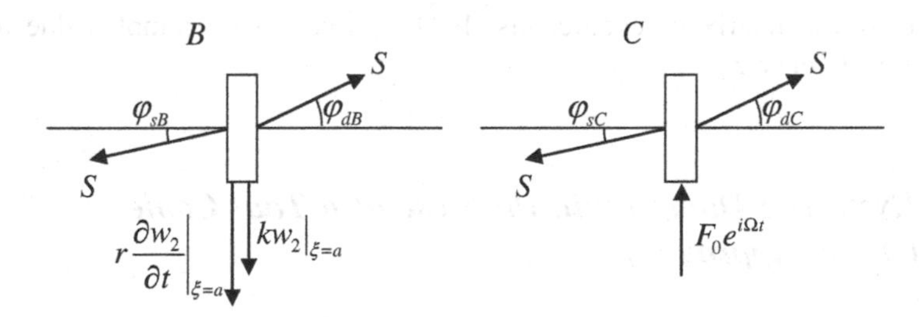

Fig. 3.20 Definition of the boundary conditions

- IV condition (equilibrium at point B, Fig. 3.20)

$$\begin{aligned}
& r\left[\frac{\partial w_2}{\partial t}\right]_{\xi=a} + k[w_2(\xi,t)]_{\xi=a} + S\sin\varphi_{sB} - S\cos\varphi_{dB} = 0 \\
& \Rightarrow r\left[\frac{\partial w_2}{\partial t}\right]_{\xi=a} + k[w_2(\xi,t)]_{\xi=a} + S\left[\frac{\partial w_1(\xi,t)}{\partial \xi}\right]_{\xi=a} - S\left[\frac{\partial w_2(\xi,t)}{\partial \xi}\right]_{\xi=a} = 0
\end{aligned} \tag{3.165}$$

- VI condition (equilibrium at point C, Fig. 3.20)

$$\begin{aligned}
& -F_0 e^{i\Omega t} + S\sin\varphi_{sC} - S\cos\varphi_{dC} = 0 \\
& \Rightarrow -F_0 e^{i\Omega t} + S\left[\frac{\partial w_2(\xi,t)}{\partial \xi}\right]_{\xi=a+b} - S\left[\frac{\partial w_3(\xi,t)}{\partial \xi}\right]_{\xi=a+b} = 0
\end{aligned} \tag{3.166}$$

By substituting the solution imposed (3.163) in the 6 boundary conditions (3.164), (3.165) and (3.166) and by eliminating term $e^{i\Omega t}$ on account of its being common in all the equations, we obtain a system of algebraic equations, complete:

$$[H(\Omega)]\underline{Z}^T = \underline{N} \tag{3.167}$$

in the 6 unknowns $\underline{Z}$:

$$\underline{Z}^T = \{F_{11} \quad F_{21} \quad F_{12} \quad F_{22} \quad F_{13} \quad F_{23}\} \tag{3.168}$$

where $\underline{N}$ is the vector of the known terms:

$$\underline{N}^T = \{0 \quad 0 \quad 0 \quad 0 \quad 0 \quad F_0\} \tag{3.169}$$

In this case, the matrix of coefficients $[H(\Omega)]$ proves to be complex due to the presence of damper **r**.

3.6.1 *Hysteretic Damping in the Case of a Taut Cable (Direct Approach)*

As seen in discrete systems, and in particular in systems with 1 d.o.f. (see Chap. 1, Sect. 1.2.3.5) hysteretic damping can be introduced into the equations of motion of a system only in the event of our wishing to evaluate the steady-state solution with an external sinusoidal excitation force $F = F_0 e^{i\Omega t}$ considering a complex stiffness [8, 9]. As regards continuous systems and, in particular, with reference to the example of a taut cable, hysteretic damping can be introduced, in the same way as it was in Sect. 1.2.3.5, by considering an overall tension S_c:

$$S_c = (S + iS_i) \tag{3.170}$$

where **S** represents the actual tension acting on the cable and $\mathbf{S_i}$ repersents the hysteretic term, proportional to the elastic forces. The partial derivatives differential equation which defines the transverse motions of a taut cable, using this approach (or rather for the study of the steady-state response of the system) is:

$$m\left(\frac{\partial^2 w}{\partial t^2}\right) = S_c\left(\frac{\partial^2 w}{\partial \xi^2}\right) \tag{3.171}$$

The solution of (3.171) thus becomes of the type (see Sect. 3.2.2):

$$w(\xi, t) = \Psi_f(\xi)G(t) = \Psi_f(\xi)G_0^{i\Omega t} \tag{3.172}$$

where Ω is the frequency of the excitation force and therefore, at steady-state conditions, of the induced vibration. By substituting (3.172) in (3.171) and by using similar steps to those shown in Sect. 3.2.2, we obtain:

$$\Psi_f''(\xi) + \gamma^2\Psi_f(\xi) = 0 \tag{3.173}$$

having used γ to indicate the complex term [3]:

$$\gamma = \Omega\sqrt{\frac{m}{S_c}} \tag{3.174}$$

This explanation reflects the one already seen in Sect. 3.2 with the use, in this case, of complex algebra. As an example, we will analyse the cable in Fig. 3.18 considering, this time, the hysteretic damping introduced by means of (3.170). The generic deformation $w(\xi, t)$, expressed in complex terms, is described by two different spatial-temporal functions:

$$\begin{aligned} w_1(\xi, t) &= \left(F_{11}e^{i\gamma\xi} + F_{12}e^{-i\gamma\xi}\right)e^{i\Omega t} \\ w_2(\xi, t) &= \left(F_{21}e^{i\gamma\xi} + F_{22}e^{-i\gamma\xi}\right)e^{i\Omega t} \end{aligned} \tag{3.175}$$

Similar to the operation performed previously, by imposing these four conditions on the boundary, it is possible to obtain a system of algebraic equations, complete in the four complex unknowns F_{11}, F_{12}, F_{21} and F_{22}:

$$[H(S_c, \gamma)]\underline{Z} = \underline{N}. \tag{3.176}$$

having used $\underline{Z}$, as usual, to indicate the vector of unknowns (which, this time, unlike the example without damping, are complex) and $\underline{N}$ the vector of the known terms. The response of the system will be give by the projection on the real axis of (3.175):

$$\begin{aligned} w_1(\xi, t) &= Re\left(\left(F_{11}e^{i\gamma\xi} + F_{12}e^{-i\gamma\xi}\right)e^{i\Omega t}\right) \\ w_2(\xi, t) &= Re\left(\left(F_{21}e^{i\gamma\xi} + F_{22}e^{-i\gamma\xi}\right)e^{i\Omega t}\right) \end{aligned} \tag{3.177}$$

With regard to, by using the same procedure as that used for the taut cable, we will introduce a complex stiffness EJ_c of the type:

$$EJ_c = (EJ + iEJ_i) \tag{3.178}$$

where EJ is the real bending stiffness of the beam, whereas EJ_i represents the hysteretic term, proportional to the elastic forces.

3.7 Approach in Principal Coordinates

A more "hands on" approach to handling the equations of motion of continuous systems in the presence:

- of concentrated or distributed excitation forces;
- of structural type damping

consists in using the Lagrange method to write the same equations in principal coordinates. Although, as will be seen further on, this approach does not have any advantages over the direct method as far as the calculation of the response of the free system is concerned, it is fairly advantageous when it comes to handling both forced problems as well as problems connected to damped systems. As with discrete systems (Chap. 2, Sect. 2.5), even in continuous systems it is possible to perform a transformation of the coordinates by passing from the physical coordinate $y(\xi, t)$ which defines the motion (transverse, longitudinal and torsional) of the generic continuous system using the principal coordinates $q_i(t)$ through transformation ((3.43), Sect. 3.2.2):

$$y(\xi, t) = \sum_{i=1}^{n} \phi^{(i)}(\xi) q_i(t) \tag{3.179}$$

where $q_i(t)$ represent the multiplicative coefficients of the normal modes $\phi^{(i)}(\xi)$ with which deformation $\mathbf{y}(\boldsymbol{\xi}, \mathbf{t})$ is approximated. By developing this summation we obtain:

$$y(\xi,t) = \phi^{(1)}(\xi)q_1(t) + \cdots + \phi^{(i)}(\xi)q_i(t) + \cdots + \phi^{(n)}(\xi)q_n(t) \qquad (3.180)$$

In continuous systems, in which the generic normal mode $\phi^{(i)}(\xi)$ is expressed by a function of space, this expression is similar to that used for discrete systems with "n" d.o.f. (Sect. 2.5) where the same change of variables was expressed by:

$$x_k(t) = X_k^{(1)}q_1(t) + \cdots + X_k^{(i)}q_i(t) + \cdots + X_k^{(n)}q_n(t) \quad (k = 1, n) \qquad (3.181)$$

where $X_k^{(i)}$ is the generalized displacement of the generic *k-n*th d.o.f. in correspondence to the generic *i-n*th normal mode: in matrix form, relation (3.181), became of the type:

$$\underline{x} = [\Phi]\underline{q} \qquad (3.182)$$

where $[\Phi]$ is the modal matrix, whose columns contain the normalized normal modes of the system (therefore, this matrix contains constant terms). Even in continuous systems it is convenient to express (3.180) in matrix form as:

$$y(\xi,t) = \underline{\phi}^T(\xi)\underline{q}(t) \qquad (3.183)$$

where by $\underline{\phi}(\xi)$ we intend a vector of "n" terms (as many as the normal modes considered), each containing a function of space corresponding to the generic normal mode:

$$y(\xi,t) = \underline{\phi}^T(\xi)\underline{q}(t) \qquad (3.184)$$

where $\underline{q}(t)$ is the vector that contains the "n" principal coordinates:

$$\underline{q}(t) = \begin{Bmatrix} q_1 \\ \cdots \\ q_i \\ \cdots \\ q_n \end{Bmatrix} \qquad (3.185)$$

In this way, the independent variables are only function of time t, in that the variable space ξ is contained in the deformations of the natural normal modes $\phi^{(i)}(\xi)$. Therefore, we move from a system with infinite degrees of freedom to a system with "n" degrees of freedom. The discretization introduced by the modal approach thus concerns the number of modes considered, while the physical description of the spatial deformations is performed by means of the continuous functions represented by the same normal modes $\phi^{(i)}(\xi)$. Here, in relation to the choice of the number of modes, the same considerations made for systems with "n" degrees of freedom (Chap. 2, Sect. 2.5) are valid: obviously, in this case, the modal approach provides the exact solution if infinite modes are considered. In terms of principal coordinates the equations prove to be uncoupled and this is due to the known property of orthogonality of the normal modes (see for example [15]): in actual fact, the equations will undoubtedly be uncoupled as far as elastic and inertia terms are concerned while this does not necessarily apply to dissipation terms. In order to write the equations of motion of the system in terms of principal coordinates we will use Lagrange equations with which it is easy, by means of a proper dissipation function D and the $Q_i (i = 1, n)$ components of nonconservative forces, to introduce both the damping (concentrated or hysteretic) and excitation force terms. As usual, first and foremost, it is necessary to define the various forms of kinetic E_c, potential V and dissipative energy D and the virtual work performed by the nonconservative forces $\delta^* L$ as a function of variable $y(\xi, t)$, to be considered in this case a physical variable, or rather of a generic spatial-temporal function which describes the deformation assumed by the continuous system. We will then perform the coordinate transformation defined in matrix form in (3.183) in order to express the same forms of energy as a function of the independent coordinates q. As an example, we will now develop this approach in the simple case of a simply supported taut cable (already analysed in Sect. 3.2.2).

3.7.1 Taut Cable Example

Let us consider the taut cable in Fig. 3.1, assuming, as a first step, the absence of damping and external applied excitation forces. In order to write the equations of free motion of the system in principal coordinates, in this case, we must define the kinetic energy function E_c and the potential energy function V first and foremost as a function of the physical variable $w(\xi, t)$ which defines the transverse displacement to which the infinitesimal generic element of the cable itself has been subjected. Having used m to define the mass per unit length and $\dot{w}(\xi, t) = (\partial w / \partial t)$ the absolute velocity in the transverse direction of the general infinitesimal element of length $d\xi$, the kinetic energy dE_c associated with the same small element is:

$$dE_c = \frac{1}{2} m d\xi \dot{w}^2 \tag{3.186}$$

where $md\xi$ is thus the mass associated with the same small element. The energy associated with the entire cable will thus be:

$$E_c = \int_0^L dE_c = \frac{1}{2} \int_0^L m \dot{w}^2 d\xi = \frac{1}{2} \int_0^L m \dot{w}^T \dot{w} d\xi \tag{3.187}$$

By keeping account of link (3.183) between the physical variable $w(\xi,t)$ and the independent variables principle coordinates $\underline{q}$, we will obtain:

$$\dot{w} = \frac{\partial w}{\partial t} = \frac{\partial\left(\underline{\phi}^T\underline{q}\right)}{\partial t} = \underline{\phi}^T\underline{\dot{q}} \tag{3.188}$$

where vector $\underline{\phi}$ of the normal modes (3.35) is a function of space only. In the case of the taut cable $\underline{\phi}$ becomes:

$$\underline{\phi}(\xi) = \begin{Bmatrix} \sin\left(\frac{\pi\xi}{L}\right) \\ \dots \\ \sin\left(\frac{i\pi\xi}{L}\right) \\ \dots \\ \sin\left(\frac{n\pi\xi}{L}\right) \end{Bmatrix} \tag{3.189}$$

expression (3.187), by keeping account of (3.188), becomes:

$$E_c = \frac{1}{2}\int_0^L m\underline{\dot{q}}^T\underline{\phi}\underline{\phi}^T\underline{\dot{q}}d\xi \tag{3.190}$$

and bearing in mind that vector $\underline{\dot{q}}$ does not depend on the space variable (integration variable):

$$E_c = \frac{1}{2}\underline{\dot{q}}^T\left[\int_0^L m\underline{\phi}\underline{\phi}^T d\xi\right]\underline{\dot{q}} = \frac{1}{2}\underline{\dot{q}}^T[M]\underline{\dot{q}} \tag{3.191}$$

where [M] is the mass matrix, expressed in principle coordinates, and defined as:

$$[M] = \begin{bmatrix} \int_0^L m\phi^{(1)}\phi^{(1)}d\xi & \dots & \int_0^L m\phi^{(1)}\phi^{(i)}d\xi & \dots & \int_0^L m\phi^{(1)}\phi^{(n)}d\xi \\ \dots & \dots & \dots & \dots & \dots \\ \int_0^L m\phi^{(i)}\phi^{(1)}d\xi & \dots & \int_0^L m\phi^{(i)}\phi^{(i)}d\xi & \dots & \int_0^L m\phi^{(i)}\phi^{(n)}d\xi \\ \dots & \dots & \dots & \dots & \dots \\ \int_0^L m\phi^{(n)}\phi^{(1)}d\xi & \dots & \int_0^L m\phi^{(n)}\phi^{(i)}d\xi & \dots & \int_0^L m\phi^{(n)}\phi^{(n)}d\xi \end{bmatrix} \tag{3.192}$$

This matrix is diagonal on account of the eigensolution property which, in the specific case of the simply supported cable and by keeping account of (3.189), is expressed by the orthogonality of the sine functions integrated on a period or a multiple of the period so that:

$$\int_0^L m\phi^{(r)}\phi^{(s)}d\xi = m_{ss} = m\frac{1}{2}\ (\text{per } s = r); \quad \int_0^L m\phi^{(r)}\phi^{(s)}d\xi = 0\ (\text{per } s \neq r) \tag{3.193}$$

The mass matrix is thus:

$$[M] = m\frac{L}{2}[I] \tag{3.194}$$

where [I] is the unit matrix. The generic term of the principle diagonal of the mass matrix matrix:

$$m_{ii} = m\frac{L}{2} \tag{3.195}$$

is defined as a generalized mass relative to the generic *i*th normal mode and retains, obviously even in the case of continuous systems, the meaning assumed for discrete systems. As regards potential energy, this will be defined as the work performed by tension S (which represents the only elastic restoring force in the cable) due to the elongation effect of the cable itself associated with the transverse displacement $w(\xi, t)$. Let us consider an initial generic small element of length $d\xi$: when the cable is deformed in generic position $w(\xi, t)$, this becomes of length dl (see Fig. 3.21):

$$dl = \sqrt{d\xi^2 + \left(\frac{\partial w}{\partial \xi}d\xi\right)^2} \tag{3.196}$$

work dl performed by tension S thus becomes:

$$dL = S(dl - d\xi) = S\left(\sqrt{d\xi^2 + \left(\frac{\partial w}{\partial \xi}d\xi\right)^2} - d\xi\right) \tag{3.197}$$

where $(dl - d\xi)$ is the elongation of section $d\xi$. Let us observe how in this work expression, written for a constant tension S, coefficient $(1/2)$ which normally appears in the work of the elastic forces does not appear here: in fact, in the generic case the elastic forces are not constant, as in this case, but vary linearly with elongations.[5] Total potential energy V can be obtained by integrating the expression

[5]Assuming the pull to be constant means neglecting the variations to which it is subjected due to the effect of the transverse deformation $w(\xi, t)$ and this is allowable if pull S is sufficiently high and if the transverse displacements themselves are small.

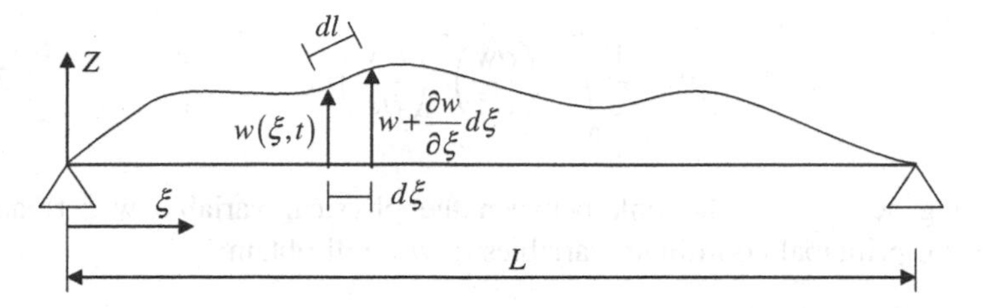

Fig. 3.21 Calculation of potential energy: evaluation of the elongation of the cable due to the effect of a transverse displacement

relative to the generic small element (3.197) along the entire longitudinal length of the cable:

$$V = \int_0^L dL = \int_0^L S\left(\sqrt{d\xi^2 + \left(\frac{\partial w}{\partial \xi}d\xi\right)^2} - d\xi\right) = \int_0^L S\left(\sqrt{1 + \left(\frac{\partial w}{\partial \xi}\right)^2} - 1\right)d\xi \tag{3.198}$$

As can be seen, this expression of potential energy V is not a quadratic form and, therefore, the corresponding Lagrange term $\left(\partial V/\partial \underline{q}\right)$ will prove to be non-linear. In the event of us wanting to obtain linear equations of motion, in the hypothesis of small oscillations in the neighbourhood of the static equilibrium position, it is possible to develop the term $\sqrt{1 + (\partial w/\partial \xi)^2}$ which appears in the in-series expression of V (3.198) up to the quadratic term, i.e.:

$$\begin{aligned} V &= \int_0^L S\left(\sqrt{1 + \left(\frac{\partial w}{\partial \xi}\right)^2} - 1\right)d\xi = \int_0^L S\left(1 + \frac{1}{2}\left(\frac{\partial w}{\partial \xi}\right)^2 - 1\right)d\xi \\ &= \frac{1}{2}\int_0^L S\left(\frac{\partial w}{\partial \xi}\right)^2 d\xi \end{aligned} \tag{3.199}$$

Bearing in mind the fact that term $(\partial w/\partial \xi)$ is a scalar term and that therefore relation:

$$\left(\frac{\partial w}{\partial \xi}\right)^T = \left(\frac{\partial w}{\partial \xi}\right) \tag{3.200}$$

applies to it, this expression can be rewritten, according to the same procedure used for kinetic energy, as:

$$V = \frac{1}{2}\int_0^L S\left(\frac{\partial w}{\partial \xi}\right)^T\left(\frac{\partial w}{\partial \xi}\right)d\xi \tag{3.201}$$

By keeping account of the link between the physical variable $w(\xi, t)$ and the independent principal coordinate variables $\underline{q}$, we will obtain:

$$\frac{\partial w}{\partial \xi} = \frac{\partial\left(\underline{\phi}^T\underline{q}\right)}{\partial \xi} = \underline{\phi}'^T\underline{q} \tag{3.202}$$

where $\underline{\phi}'$ is the vector containing the derivatives of the spatial functions that define the normal modes derived with respect to the spatial variable, i.e. in the taut cable case analysed:

$$\underline{\phi}'(\xi) = \begin{Bmatrix} \frac{\pi}{L}\cos\left(\frac{\pi\xi}{L}\right) \\ \dots \\ \frac{i\pi}{L}\cos\left(\frac{i\pi\xi}{L}\right) \\ \dots \\ \frac{n\pi}{L}\cos\left(\frac{n\pi\xi}{L}\right) \end{Bmatrix} \tag{3.203}$$

Thus, the expression of potential energy V, in principle coordinates, becomes:

$$V = \frac{1}{2}\int_0^L S\underline{q}^T\underline{\phi}'(\xi)\underline{\phi}'(\xi)^T\underline{q}d\xi \tag{3.204}$$

or rather, bearing in mind that vector $\underline{q}$ does not depend on the space variable ξ (integration variable):

$$V = \frac{1}{2}\underline{q}^T\left[\int_0^L S\underline{\phi}'(\xi)\underline{\phi}'(\xi)^T d\xi\right]\underline{q} = \frac{1}{2}\underline{q}^T[K]\underline{q} \tag{3.205}$$

where [K] is the stiffness matrix expressed in principle coordinates defined as:

$$[K] = \begin{bmatrix} \int_0^L S\phi^{(1)'}\phi^{(1)'}d\xi & \dots & \int_0^L S\phi^{(1)'}\phi^{(i)'}d\xi & \dots & \int_0^L S\phi^{(1)'}\phi^{(n)'}d\xi \\ \dots & \dots & \dots & \dots & \dots \\ \int_0^L S\phi^{(i)'}\phi^{(1)'}d\xi & \dots & \int_0^L S\phi^{(i)'}\phi^{(i)'}d\xi & \dots & \int_0^L S\phi^{(i)'}\phi^{(n)'}d\xi \\ \dots & \dots & \dots & \dots & \dots \\ S\phi^{(n)'}\phi^{(1)'}d\xi & \dots & S\phi^{(n)'}\phi^{(i)'}d\xi & \dots & S\phi^{(n)'}\phi^{(n)'}d\xi \end{bmatrix} \tag{3.206}$$

This matrix proves to be diagonal on account of the property of eigensolutions which, in the specific case of the simply supported cable, is expressed by the orthogonal nature of the cosine functions integrated on a period or a multiple of the period so that, as is known:

$$\begin{aligned} &\int_0^L S\left(\frac{i\pi}{L}\right)\cos\left(\frac{i\pi\xi}{L}\right)\left(\frac{j\pi}{L}\right)\cos\left(\frac{j\pi\xi}{L}\right)d\xi = S\left(\frac{i\pi}{L}\right)^2\frac{L}{2} \quad (\text{con } i = j)\\ &\int_0^L S\left(\frac{i\pi}{L}\right)\cos\left(\frac{i\pi\xi}{L}\right)\left(\frac{j\pi}{L}\right)\cos\left(\frac{j\pi\xi}{L}\right)d\xi = 0 \quad (\text{con } i \neq j) \end{aligned} \tag{3.207}$$

By keeping account of (3.198) the stiffness matrix is thus:

$$[K] = S\frac{L}{2}\begin{bmatrix} \left(\frac{\pi}{L}\right)^2 & \dots & 0 & \dots & 0\\ \dots & \dots & \dots & \dots & \dots\\ 0 & \dots & \left(\frac{i\pi}{L}\right)^2 & \dots & 0\\ \dots & \dots & \dots & \dots & \dots\\ 0 & \dots & 0 & \dots & \left(\frac{n\pi}{L}\right)^2 \end{bmatrix} = S\left(\frac{\pi^2}{2L}\right)[i^2] \tag{3.208}$$

where $[i^2]$ is a diagonal matrix whose generic element is given by the square of the order of the mode considered. The generic term of the principle diagonal of the stiffness matrix:

$$k_{ii} = \int_0^L S\left(\frac{i\pi}{L}\right)^2\cos^2\left(\frac{i\pi\xi}{L}\right)d\xi = S\left(\frac{i\pi}{L}\right)^2\frac{L}{2} = m\frac{L}{2}\omega_i^2 \tag{3.209}$$

is defined as a generalized stiffness relative to the generic *n*th normal mode and obviously retains, even in the case of continuous systems, the meaning assumed for discrete systems. Thus, for a continuous system, by considering a certain "n" number of independent coordinates, in this instance all the properties and characteristics of the modal approach already developed for discrete systems with 2-n d.o.f., are valid. By applying the Langrange equations, always in the case of an undamped free system:

$$\left(\frac{d}{dt}\left(\frac{\partial E_c}{\partial \dot{\underline{q}}}\right)\right)^T + \left(\frac{\partial V}{\partial \underline{q}}\right)^T = 0 \tag{3.210}$$

it is possible to obtain, directly in matrix form, the equations of motion of the continuous system in principle coordinates which will thus be:

$$[M]\ddot{\underline{q}} + [K]\underline{q} = 0 \qquad (3.211)$$

which are translated into "n" uncoupled equations, each of the ordinary differential type, in "n" unknown principal coordinates. In the case of the simply supported taut cable, by recalling (3.195) and (3.209) these equations can be rewritten as:

$$\left(\frac{mL}{2}\right)\ddot{q}_i + \left(i^2\frac{S\pi^2}{2L}\right)q_i = 0 \quad (i = 1, 2, \ldots, n) \qquad (3.212)$$

By integrating relations (3.212) we thus obtain the equation that defines the free motion of the system in the absence of damping in the principal coordinates:

$$q_i(t) = A_i \sin\omega_i t + B_i \cos\omega_i t \quad (i = 1, 2, \ldots, n) \qquad (3.213)$$

where $\boldsymbol{\omega}_i$ is the frequency of the generic normal mode:

$$\omega_i = \sqrt{\frac{\left(i^2\frac{S\pi^2}{2L}\right)}{\left(\frac{mL}{2}\right)}} = \left(\frac{i\pi}{L}\right)\sqrt{\frac{S}{m}} \qquad (3.214)$$

or rather it is still the natural frequency already defined by the integration of partial derivative equations (3.10) by means of the particular integral:

$$w_i(\xi, t) = (\Psi(\xi))_{\omega=\omega_i} G_i(t) = \phi^{(i)}(\xi) G_i(t) \qquad (3.215)$$

Thus, in the case of undamped free motion, there is a coincidence between the generic *i*th principal coordinate $q_i(t)$ and the corresponding temporal function $\mathbf{G_i(t)}$ which defines the temporal motion of the continuous system using the direct approach: for this reason, the study of free motion of the system in principal coordinates is of little use. Conversely, this approach is more convenient when dealing with the equations of motion of forced continuous systems in which damping is also present.

3.7.1.1 Forced Vibrations in Principal Coordinates

In the case of forced vibrations, the equations of motion in principal coordinates become:

$$[M]\ddot{\underline{q}} + [K]\underline{q} = \underline{Q} \qquad (3.216)$$

It is necessary to calculate the Lagrangian components Q_i $(i = 1, n)$ of the non-conservative excitation forces applied to the system, relative to the normal modes considered in the transformation of the coordinates (3.183). In the specific case of the taut cable, the generic equation of motion (always uncoupled from the others) becomes:

$$\left(\frac{mL}{2}\right)\ddot{q}_i + \left(i^2\frac{S\pi^2}{2L}\right)q_i = Q_i \quad (i = 1, 2, \ldots, n) \tag{3.217}$$

As usual we will have to evaluate the virtual work of the external forces applied to the system and expressed first and foremost as a function of the physical coordinate $w(\xi, t)$:

$$\delta^* L = \delta^* L(w(\xi, t)). \tag{3.218}$$

in order to then impose the transformation of coordinates (3.183) to achieve an expression of virtual work as a function of the independent coordinates, formally defined by a matrix relation of the type:

$$\delta^* L = \underline{Q}^T \delta^* \underline{q} \tag{3.219}$$

where $\delta^* \underline{q}$ is the vector that contains the virtual displacements associated with the principal coordinates chosen in order to define, in discrete form, the motion of the system, $\underline{Q}$ is the vector of the Langrangian components whose generic term Q_i is given by the relation:

$$Q_i = \left(\frac{\delta^* L}{\delta^* q_i}\right) \quad (i = 1, 2, \ldots, n) \tag{3.220}$$

where $\delta^* L_i$ is the work performed by the excitation forces for a virtual displacement $\delta^* q_i$.

3.7.1.1.1 Forced Vibrations in Principal Coordinates: Concentrated Excitation Forces

First and foremost, let us analyse the case of the taut cable excited by a single harmonic force of the type:

$$f_1(t) = F_{10} e^{i\Omega t} \tag{3.221}$$

applied at a point of coordinate ξ_1 (see Fig. 3.22). To calculate its Langragian component it is necessary, as previously mentioned, to express the virtual work performed by this force, first and foremost as a function of the physical coordinate $w(\xi, t)$, a space-time function which defines the generic deformation assumed by the system:

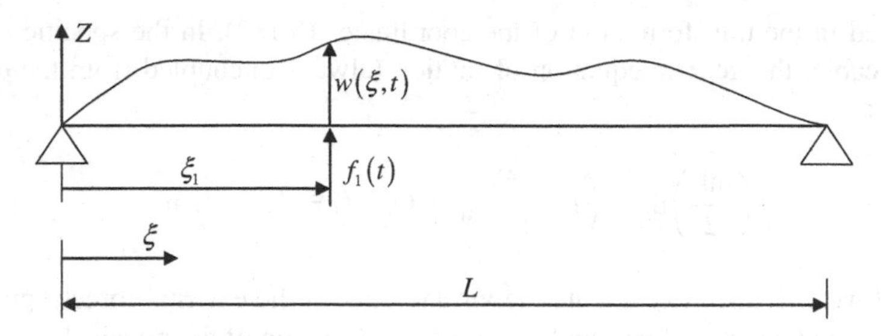

Fig. 3.22 Continuous system forced by a single excitation force: modal approach

$$\delta^* L = f_1^T(t)\delta^* w(\xi_1, t) \tag{3.222}$$

By recalling the transformation of coordinates (3.183) the virtual displacement expressed as a function of the new independent principle coordinates becomes:

$$\delta^* w(\xi_1, t) = \left[\frac{\partial w(\xi, t)}{\partial \underline{q}}\right]_{\xi=\xi_1} \delta^* \underline{q} = \underline{\phi}(\xi_1)^T \delta^* \underline{q} \tag{3.223}$$

so that (3.222) can be expressed as:

$$\delta^* L = f_1^T(t)\underline{\phi}(\xi_1)^T \delta^* \underline{q} = \underline{Q}^T \delta^* \underline{q} \tag{3.224}$$

where, in this case the vector of the Lagrangian components is defined as:

$$\underline{Q} = \underline{\phi}(\xi_1) f_1(t) = \begin{Bmatrix} \sin\left(\frac{\pi\xi_1}{L}\right) \\ \cdots \\ \sin\left(\frac{i\pi\xi_1}{L}\right) \\ \cdots \\ \sin\left(\frac{n\pi\xi_1}{L}\right) \end{Bmatrix} f_1(t) = \begin{Bmatrix} Q_1(t) \\ \cdots \\ Q_i(t) \\ \cdots \\ Q_n(t) \end{Bmatrix} \tag{3.225}$$

In physical terms, the generic Lagrangian component $Q_i(t) = f_1(t)\sin(i\pi\xi_1/L)$ represents the work that the excitation force $f_1(t)$ is capable of introducing into the generic *i*th normal mode for a displacement of the only coordinate q_i. As Q_i increases, the work introduced in correspondence to the generic normal mode also increases: the amplitude of Q_i thus also represents the excitation force's ability to excite the *i*th mode. On the contrary, small Q_i (or, at most, null, if force f_1 is applied in a vibration node, meaning that we have $\sin(i\pi\xi_1/L) = 0$) means that the same applied force will only excite marginally (or not at all) the mode considered. In the example analysed (Fig. 3.23), if the excitation force is applied in a vibration node, the generic mode considered will not be excited, not even in resonance for $\Omega = \omega_i$ in that the displacement of the application point of the excitation force is null, i.e. $Q_i = 0$.

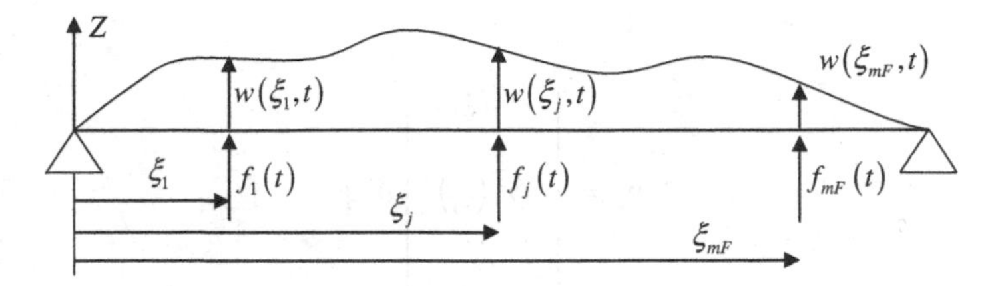

Fig. 3.23 Continuous system forced by discrete excitation forces: modal approach

By generalizing the previous writing, if the cable is forced by "m_F" excitation forces $f_j(t)(j = 1, F)$ applied at points ξ_j (Fig. 3.23) we will obtain:

$$\delta^* L = \sum_{j=1}^{mF} f_j(t)\delta^* w(\xi, t) \tag{3.226}$$

An expression which in matrix form:

- by organizing in vector $\underline{f}(t)$ the applied excitation forces

$$\underline{f}(t) = \begin{Bmatrix} f_1 \\ \cdots \\ f_j \\ \cdots \\ f_{mF} \end{Bmatrix} \tag{3.227}$$

- by defining a vector $\underline{w}$ containing the displacements of the application points of the excitation forces:

$$\underline{w} = \begin{bmatrix} w(\xi_1, t) \\ \cdots \\ w(\xi_j, t) \\ \cdots \\ w(\xi_{mF}, t) \end{bmatrix} = \begin{bmatrix} \underline{\phi}^T(\xi_1) \\ \cdots \\ \underline{\phi}^T(\xi_j) \\ \cdots \\ \underline{\phi}^T(\xi_{mF}) \end{bmatrix} \underline{q} \tag{3.228}$$

can be expressed as:

$$\delta^* L = \underline{f}^T(t)\delta^* \underline{w} \tag{3.229}$$

where

$$\delta^* \underline{w} = \begin{bmatrix} \underline{\phi}^T(\xi_1) \\ \cdots \\ \underline{\phi}^T(\xi_j) \\ \cdots \\ \underline{\phi}^T(\xi_{mF}) \end{bmatrix} \delta^* \underline{q} \tag{3.230}$$

Equation (3.229) thus becomes:

$$\delta^* L = \begin{Bmatrix} f_1 \\ \cdots \\ f_j \\ \cdots \\ f_{mF} \end{Bmatrix}^T \begin{bmatrix} \underline{\phi}^T(\xi_1) \\ \cdots \\ \underline{\phi}^T(\xi_j) \\ \cdots \\ \underline{\phi}^T(\xi_{mF}) \end{bmatrix} \delta^* \underline{q} = \underline{f}^T \left[w_q \right] \delta^* \underline{q} \tag{3.231}$$

matrix $\left[w_q\right]$, a rectangular matrix of "m_F" lines, the same number as the excitation forces applied to the system and "n" columns, the same number as the principle coordinates assumed to define the motion of the system itself, correlates the physical variables $\underline{w}$ with the independent variables. Virtual work (3.231) can be expressed in compact form as:

$$\delta^* L = \underline{f}^T(t) \left[w_q \right] \delta^* \underline{q} = \underline{Q}^T(t) \delta^* \underline{q} \tag{3.232}$$

and the Lagrangian component $\underline{Q}$ of the external forces is thus:

$$\underline{Q} = \left[w_q \right]^T \underline{f}(t) = \left[\underline{\phi}(\xi_1) \quad \cdots \quad \underline{\phi}(\xi_j) \quad \cdots \quad \underline{\phi}(\xi_{mF}) \right] \begin{Bmatrix} f_1 \\ \cdots \\ f_j \\ \cdots \\ f_{mF} \end{Bmatrix} \tag{3.233}$$

3.7.1.1.2 Forced Vibrations in Principal Coordinates: Distributed Excitation Force

In the case in which the system is forced (see Fig. 3.24) by a distributed excitation force $f(\xi, t)$ (force per unit of length in correspondence to generic abscissa ξ), the stress acting on the generic infinitesimal small element of length $d\xi$ will be given by $f(\xi, t)d\xi$. The work $\delta^* L_{\Delta\xi}$ performed by the excitation force distributed on the infinitesimal generic small element will therefore be:

$$\delta^* L_{\Delta\xi} = f(\xi, t) \delta^* w(\xi, t) d\xi \tag{3.234}$$

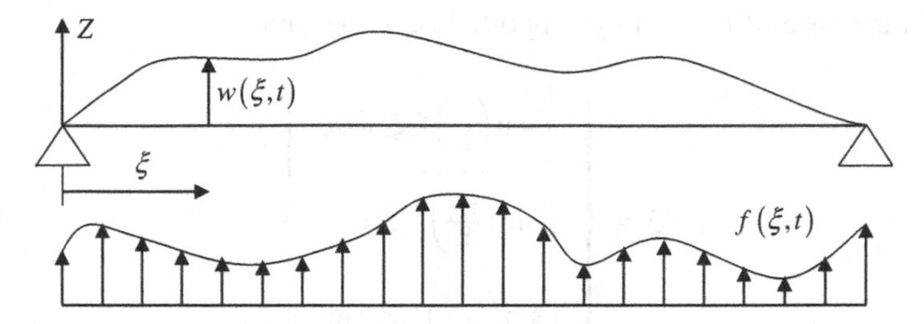

Fig. 3.24 Continuous system forced by a distributed excitation force: modal approach

The overall work as a function of the physical variable $w(\xi, t)$ will therefore be:

$$\delta^* L = \int_0^L f(\xi, t)\delta^* w(\xi, t) d\xi \tag{3.235}$$

By recalling relation (3.223), the same virtual work can be expressed as a function of the principle coordinates in matrix form as:

$$\begin{aligned} \delta^* L &= \int_0^L f(\xi, t)\left(\frac{\partial w(\xi, t)}{\partial \underline{q}}\right)\delta^* \underline{q} d\xi = \int_0^L f(\xi, t)\underline{\phi}^T(\xi)\delta^* \underline{q} d\xi \\ &= \left\{\int_0^L f(\xi, t)\underline{\phi}^T(\xi) d\xi\right\}\delta^* \underline{q} \end{aligned} \tag{3.236}$$

that is:

$$\delta^* L = \underline{Q}^T \delta^* \underline{q} \tag{3.237}$$

where, in this case, the vector of the Langrangian components $\underline{Q}$ is equal to:

$$\underline{Q} = \begin{Bmatrix} Q_1(t) \\ \cdots \\ Q_i(t) \\ \cdots \\ Q_n(t) \end{Bmatrix} = \left\{\int_0^L \underline{\phi}(\xi) f(\xi, t) d\xi\right\} = \begin{Bmatrix} \int_0^L \phi^{(1)}(\xi) f(\xi, t) d\xi \\ \cdots \\ \int_0^L \phi^{(i)}(\xi) f(\xi, t) d\xi \\ \cdots \\ \int_0^L \phi^{(n)}(\xi) f(\xi, t) d\xi \end{Bmatrix} \tag{3.238}$$

which, in the case of the simply supported cable becomes:

$$\underline{Q} = \left\{ \begin{array}{c} \int_0^L \sin\left(\frac{\pi\xi}{L}\right) f(\xi,t) d\xi \\ \cdots \\ \int_0^L \sin\left(\frac{i\pi\xi}{L}\right) f(\xi,t) d\xi \\ \cdots \\ \int_0^L \sin\left(\frac{n\pi\xi}{L}\right) f(\xi,t) d\xi \end{array} \right\} \tag{3.239}$$

Thus, in the hypothesis of a simply supported cable, if the distributed excitation force is of the type:

$$f(\xi,t) = p_0 e^{i\Omega t} \tag{3.240}$$

(or rather the force per unit length po is constant with respect to space ξ), the integrals of (3.238), given the particular trend of the eigenfunctions (number of antinodes equal to "i"), become:

$$Q_i = \int_0^L \sin\left(\frac{i\pi\xi}{L}\right) p_0 e^{i\Omega t} d\xi \tag{3.241}$$

i.e. the generic Lagrangian component Q_i thus proves to be different from zero only for modes with an odd number of antinodes: in fact, the work performed in correspondence to the modes with "i" even number of antinodes becomes null overall. On the contrary, in correspondence to an odd mode (that is if the number of antinodes is odd) the work proves to be different from zero and equal, with the exception of term $p_0 e^{i\Omega t}$, to the area underlying the last antinode (positive or negative), see Fig. 3.25.

A constant distributed force in space, always introduces less energy into the system as the frequency increases, in that (always only for odd modes) the work introduced for the generic mode (proportional to the integral on only one semi-period) decreases. Thus with the same amplitude this work will always be smaller

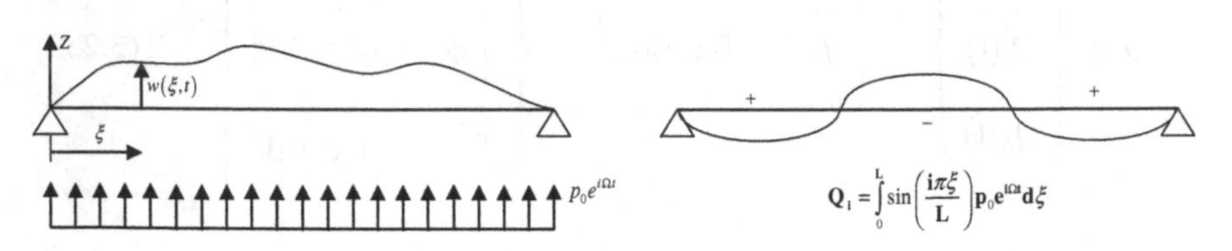

Fig. 3.25 Continuous system excited by a distributed excitation force $f(\xi,t) = p_0 e^{i\Omega t}$ modal approach—physical meaning of lagrangian component Q_i

as the mode considered increases, since the wavelength decreases and, as a consequence, the corresponding integral on the last semi-period. A constant distributed force in space thus only excites the odd modes with an intensity that gradually decreases as order "i" of the normal mode considered increases. Conversely, in the event of the distributed force assuming a sinusoidal trend in ξ with a wavelength equal to $(2L/k)$ (with k (entire) $= 1, n$):

$$f(\xi, t) = p_0 \sin\left(\frac{k\pi}{L}\xi\right) e^{i\Omega t} \tag{3.242}$$

i.e. in the event of $f(\xi, t)$ proving to be sinusoidal in space and in time, it is evident that the Lagrangian components Q_i present in (3.238):

$$Q_i = \int_0^L \sin\left(\frac{i\pi\xi}{L}\right) p_0 \sin\left(\frac{k\pi}{L}\xi\right) e^{i\Omega t} d\xi \tag{3.243}$$

are all null (due to the known property of orthogonality of the sine functions integrated on the natural period) with the exception of the one corresponding to the generic *k*th normal mode meaning that we would obtain:

$$Q_k = \int_0^L \sin\left(\frac{k\pi\xi}{L}\right) p_0 \sin\left(\frac{k\pi}{L}\xi\right) e^{i\Omega t} d\xi = p_0 \frac{L}{2} e^{i\Omega t} \tag{3.244}$$

In this case the only *k*th mode will prove to be excited by the external excitation force. Whatever the structure of the generic Lagrangian component Q_i, the equations of motion of the taut cable forced in the absence of damping will thus be of the type:

$$\left(\frac{mL}{2}\right) \ddot{q}_i + \left(i^2 \frac{S\pi^2}{2L}\right) q_i = Q_i \quad (i = 1, 2, \ldots, n) \tag{3.245}$$

where the modal mass m_{ii}:

$$m_{ii} = \left(\frac{mL}{2}\right) \tag{3.246}$$

is independent from the normal mode considered, while the term:

$$k_{ii} = \left(i^2 \frac{S\pi^2}{2L}\right) \tag{3.247}$$

represents generalized stiffness. If the generic Lagrangian component Q_i is harmonic:

$$Q_i = Q_{i0}e^{i\Omega t} \tag{3.248}$$

with frequency $\Omega = \omega_i$, the generic main coordinate $q_i(t)$ proves to be harmonic, with amplitudes that tend towards the infinite, for $\mathbf{t} \to \infty$. Generally speaking, the explanation given with particular reference to the simple case of a taut cable can be extended to a generic continuous system, with any boundary conditions: in principal coordinates will always obtain a system of "n" equations (the same number as the d.o.f. considered in the transformation of uncoupled coordinates (3.183)) of the type:

$$m_{ii}\ddot{q}_i + k_{ii}q_i = Q_i \quad (i = 1, 2, \ldots, n) \tag{3.249}$$

Having solved this equation and having calculated the q_i by means of relation (3.249), it is possible to evaluate the response of the system as a function of the physical variable $w(\xi, t)$ which defines the space-time deformation assumed by the system subjected to the excitation forces considered using:

$$w(\xi, t) = \underline{\phi}^T(\xi)\underline{q} \tag{3.250}$$

3.7.1.1.3 Forced Vibrations in Principal Coordinates: Moving Load

Let us now assume the presence of a constant moving load P along the cable translating with a constant velocity V (Fig. 3.26). This case, like the one relative to distributed loads along the continuous system, cannot be solved by using the direct approach, in that the extremities of the domains into which the cable can be divided with the presence of the load itself vary as a function of time. For this reason, the response to the moving load can be dealt with using the modal approach (or, as will be seen further on, the finite element method).

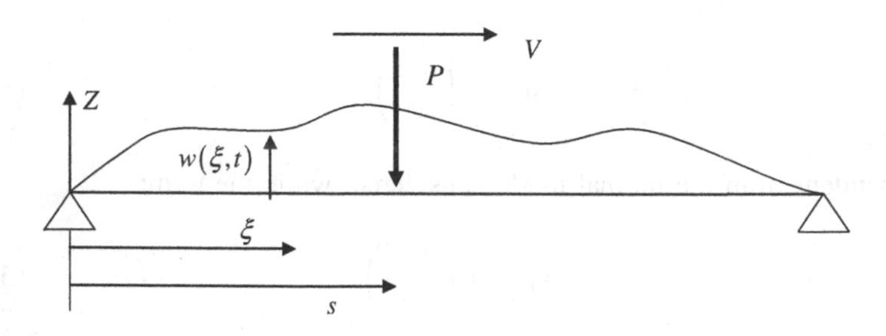

Fig. 3.26 Continuous system forced by a moving load

Let us thus consider the generic cable in Fig. 3.26, subjected to the moving load P which moves at velocity V (constant) and occupies position s = Vt: the equations of motion in principal coordinates, in the absence of structural damping, thus once again prove to be of the type:

$$m_{ii}\ddot{q}_i + k_{ii}q_i = Q_i \quad (i = 1, 2, \ldots, n) \tag{3.251}$$

To evaluate the Lagrangian component associated with the moving load, it is necessary to express work $\boldsymbol{\delta * L}$ performed by same as a function of the transverse displacement physical coordinate **w** and subsequently, by exploiting the transformation in independent coordinates (3.110) $\underline{\mathbf{q}}$ as a function of the latter:

$$\delta * L = -P\delta * w(s,t)|_{s=Vt} = -P\underline{\phi}^T(s)\delta * \underline{q} = \underline{Q}^T(Vt)\delta * \underline{q} \tag{3.252}$$

having indicated the vector of Lagrangian components with $\underline{\mathbf{Q}}$:

$$\underline{Q}(Vt) = -P\underline{\phi}(Vt)\delta * L = \underline{Q} = \underline{Q}(t) == -P\begin{Bmatrix} \sin\left(\frac{\pi Vt}{L}\right) \\ \cdots \\ \sin\left(\frac{i\pi Vt}{L}\right) \\ \cdots \\ \sin\left(\frac{n\pi Vt}{L}\right) \end{Bmatrix} \tag{3.253}$$

Thus, in this case, the generic Lagrangian component becomes:

$$Q_i = -P\sin\left(\frac{i\pi Vt}{L}\right) \Rightarrow \Omega_i = \frac{i\pi V}{L} \Rightarrow Q_i = -P\sin(\Omega_i t) \tag{3.254}$$

In principal coordinates this component thus proves to be a generic harmonic excitation force of frequency $\Omega_i = \frac{i\pi V}{L}$ different from zero only in time interval $0 \le t \le T;\ T = L/V$, in which excitation force P is present on the cable: the temporal histories of the generalized forces on the nodes have the trends shown in Fig. 3.27.

In this case, the generic generalized force is incapable of exciting the cable in resonance since, as previously mentioned, it is different from zero only in the interval $0 - T$.

The response to the moving load cannot therefore be calculated in terms of a steady state response, but must be calculated in the time domain by numerically integrating the equations of motion (3.251) and reconstructing the response of the system in any point of the continuous system to generic time using the relation:

$$w(\xi, t) = \underline{\phi}(\xi)^T\underline{q}(t) = \sum_{i=1}^{nm} \sin\left(\frac{i\pi\xi}{L}\right) q_i(t) \tag{3.255}$$

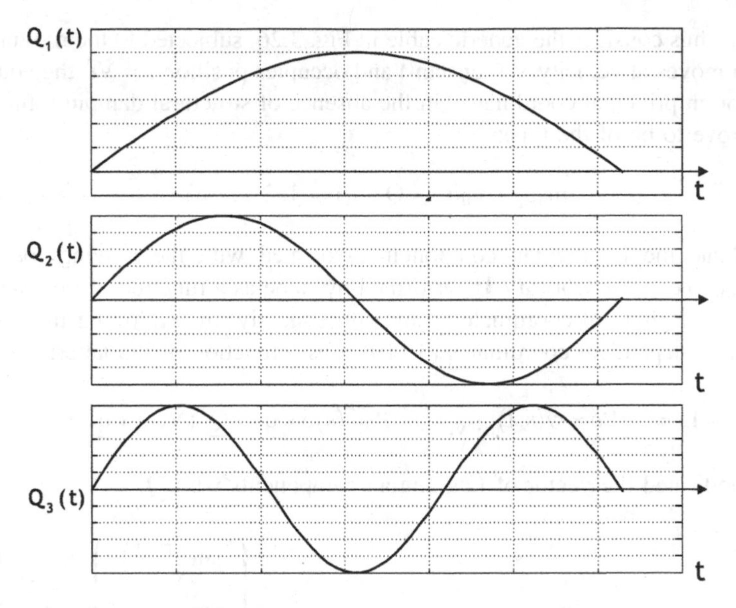

Fig. 3.27 Trend of the generalized forces in modal coordinates due to the effect of the transit of a moving load

By keeping account of (3.254), it is possible to define a critical velocity V_{cr} in which the frequency of the excitation force coincides with the natural frequency of the system:

$$\Omega_i = \omega_i \Rightarrow \frac{i\pi V}{L} = \sqrt{\frac{k_{ii}}{m_{ii}}} = \left(\frac{i\pi}{L}\right)\sqrt{\frac{S}{m}} \tag{3.256}$$

from which:

$$V_{cr} = \sqrt{\frac{S}{m}} \tag{3.257}$$

In the case in which the continuous system is excited by a sequence of moving loads at distance d (Fig. 3.28) the temporal phase displacement between the loads is:

$$\Delta t = \frac{d}{V} \tag{3.258}$$

where

$$T = \frac{1}{V} \tag{3.259}$$

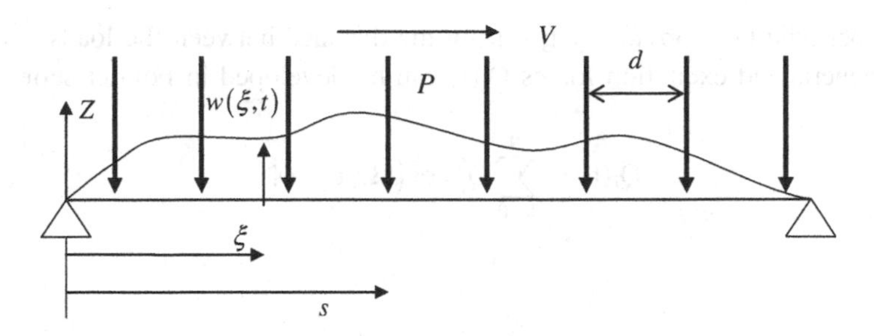

Fig. 3.28 Continuous system forced by a sequence of moving loads

is the stay time of the single excitation force. By assuming an infinite series of equally spaced moving loads and by keeping account of (3.254), in principal coordinates we will obtain generalized periodic excitation forces $Q_i(t)$ whose trend is shown in Fig. 3.29.

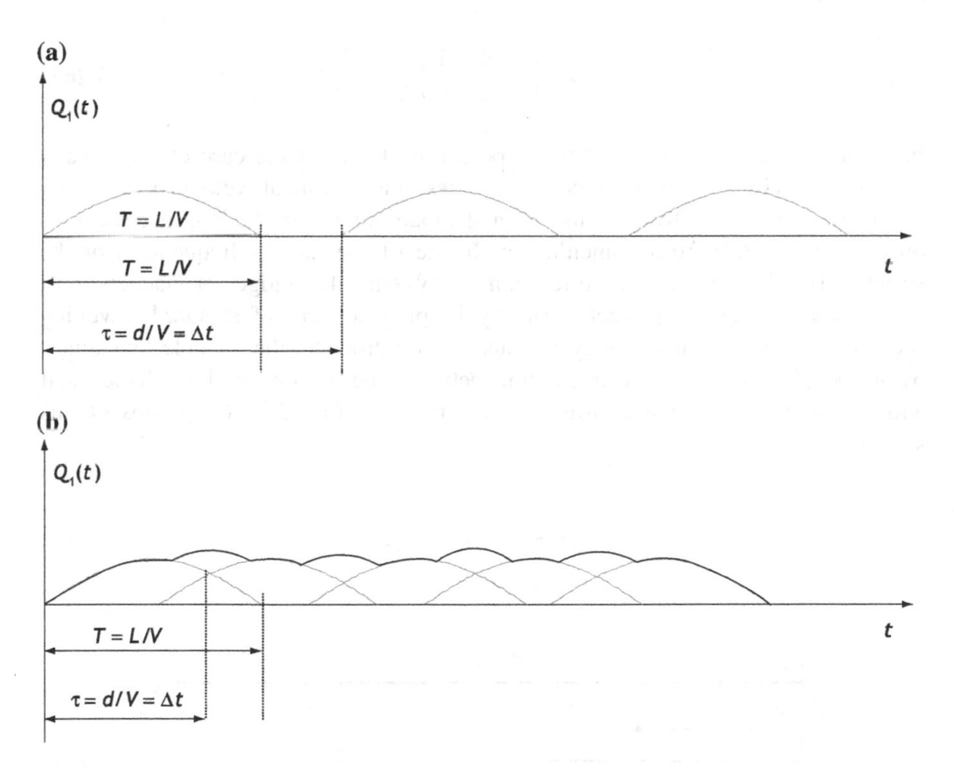

Fig. 3.29 Lagrangian components for a sequence of moving loads: with d > L (**a**) and with d < L (**b**)

The periodicity is obviously given by the distance between the loads $\tau = {}^{d}\!/_{V}$. These generalized excitation forces $Q_i(t)$ can be developed in Fourier series:

$$Q_i(t) = \sum_{k=0}^{N} Q_i^k \cos\left(k\Omega_o t + \varphi_i^k\right) \tag{3.260}$$

where

$$\Omega_o = \frac{2\pi}{\tau} = \frac{2\pi V}{d} \tag{3.261}$$

In this case we will have a resonance condition each time that:

$$\Omega_i = k\Omega_o = \omega_i \Rightarrow k\frac{2\pi V}{d} = \sqrt{\frac{k_{ii}}{m_{ii}}} = \left(\frac{i\pi}{L}\right)\sqrt{\frac{S}{m}} \tag{3.262}$$

from which:

$$(V_{cr})_i^k = \frac{d}{2k}\left(\frac{i}{L}\right)\sqrt{\frac{S}{m}} \tag{3.263}$$

In reality, critical conditions of this type can be found in the case of long heavy convoys travelling over viaducts or bridges; these critical velocities are only potentially dangerous for the first normal modes of the deck. It is necessary to prevent Ω_o (3.262) from coinciding with one of the natural frequencies of the structure ($i = 1$). To avoid these resonance problems, the bridges or viaducts must be checked during the project phase by keeping account of standard travelling speeds. A more realistic analysis must also introduce the vehicle dynamics, by introducing the dynamic interaction between the motion of the vehicle itself and the motions of the infrastructures, as shown in Fig. 3.30 (a' propos of this see [5, 6]).

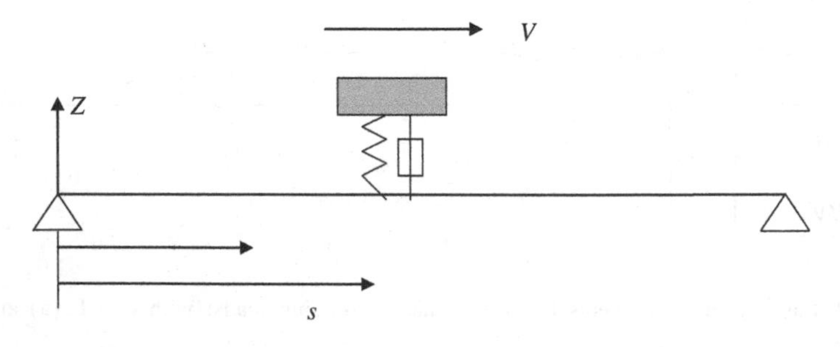

Fig. 3.30 Model of vehicle-infrastructure system

3.7.1.2 Damping in Principle Coordinates

In the hypothesis of structural damping such as a hysteretic or similar one, the equations of motion remain uncoupled even in principal coordinates: it is, in fact, possible to define a modal $\mathbf{r_{ii}}$ damping term associated with each normal mode $\phi^{(i)}(\xi)$. By using this approach, the equations of motion of the generic mode thus become:

$$m_{ii}\ddot{q}_i + r_{ii}\dot{q}_i + k_{ii}q_i = Q_i \quad (i = 1, 2, \ldots, n) \tag{3.264}$$

or rather in matrix form once again of the type:

$$[M]\ddot{\underline{q}} + [R]\dot{\underline{q}} + [K]\underline{q} = \underline{Q} \tag{3.265}$$

where [M], [R] and [K] are the diagonal matrices of mass, damping and stiffness and respectively contain, on the principle diagonal, generic modal mass m_{ii}, modal damping r_{ii} and modal stiffness k_{ii}, similar to what encountered with discrete systems (Chap. 2, Sect. 2.5). The value of modal damping r_{ii}, always referring to what has been seen in discrete systems, is obtained using suitable modal identification techniques (Chap. 8). In the case of continuous systems, it is also possible to obtain this value by analysing the free motion of the system itself: in the example given for the cable if, as is usually the case, the structural damping is small, the following procedure can be adopted:

- the cable is excited in resonance in correspondence to the generic *i*th normal mode;
- the excitation force is eliminated;
- the free motion of the system is analysed (Fig. 3.31).

Since the damping is small, the free motion prevalently contains the excited mode and will be of the type shown in Fig. 3.31:

$$w(\xi, t) = \underline{\phi}^T(\xi)\underline{q} = \sum_{j=1}^{n} \phi^{(j)}(\xi)q_j(t) \approx \phi^{(i)}(\xi)q_i(t) \tag{3.266}$$

In this case, the value of logarithmic decrement $\boldsymbol{\delta_i}$ (Chap. 2, Sect. 2.4.1.3, (2.235)) is defined by the relation:

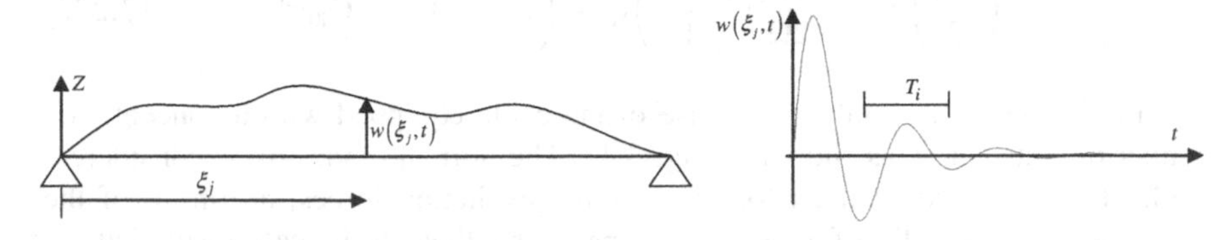

Fig. 3.31 Definition of the modal damping from the free motion of the system

$$\delta_i = \ln\left(\frac{w(\xi_j, t)}{w(\xi_j, t + T_i)}\right) = \ln\left(\frac{q_i(t)}{q_i(t + T_i)}\right) \tag{3.267}$$

where $w(\xi_i, t)$ and $w(\xi_i, t + T_i)$ are the amplitudes corresponding to two successive peaks evaluated in a generic point of the cable of abscissa ξ_i (see Fig. 3.31) and T_i is the oscillation period corresponding to the excited normal mode:

$$T_i = \frac{2\pi}{\omega_i} \tag{3.268}$$

From (3.267) it is possible to obtain the nondimensional damping h_i relative to the generic *i*th normal mode by means of:

$$h_i = \left(\frac{r}{r_c}\right)_i = \frac{\delta_i}{2\pi} \tag{3.269}$$

from which:

$$r_{ii} = 2m_{ii}\omega_i h_i = 2m_{ii}\omega_i \frac{\delta_i}{2\pi} \tag{3.270}$$

If, on the contrary, concentrated dissipative elements, mounted on the structure, exist, it is possible to introduce their effect, in terms of principal coordinates, by means of the corresponding dissipative function D or, in exactly the same way as the external excitation forces f_j were introduced, by means of the virtual work δ^*L performed by the dissipative forces. In the case of hysteretic damping, similar to the procedure used both in discrete systems (Chap. 1, Sect. 1.2.3.5) and in continuous systems using the direct approach (Sect. 3.7), if we wish to evaluate the steady state response of the system subjected to sinusoidal excitation $Q_i = Q_{i0}e^{i\Omega t}$, the equation of motion of the generic mode (3.264) can be rewritten by introducing a complex stiffness such as ([3], Sect. 5.10):

$$m_{ii}\ddot{q}_i + (k_{ii} + i\gamma k_{ii})q_i = Q_{i0}e^{i\Omega t} \quad (i = 1, 2, \ldots n) \tag{3.271}$$

By analysing the taut cable, and in the only case of harmonic excitation, we will have:

$$\left(\frac{mL}{2}\right)\ddot{q}_i + \left(\frac{\gamma S}{2\Omega}\right)\left(\frac{i^2\pi^2}{L}\right)\dot{q}_i + \left(i^2\frac{S\pi^2}{2L}\right)q_i = Q_{i0}e^{i\Omega t} \tag{3.272}$$

As an example, let us analyse the case of the cable equipped with a concentrated damper of constant $\mathbf{r}_j$, as shown in Fig. 3.32. The normal modes to be considered, similar to the procedure used for the external excitation forces, are those of the system without excitation forces and dampers, or rather, in the case examined:

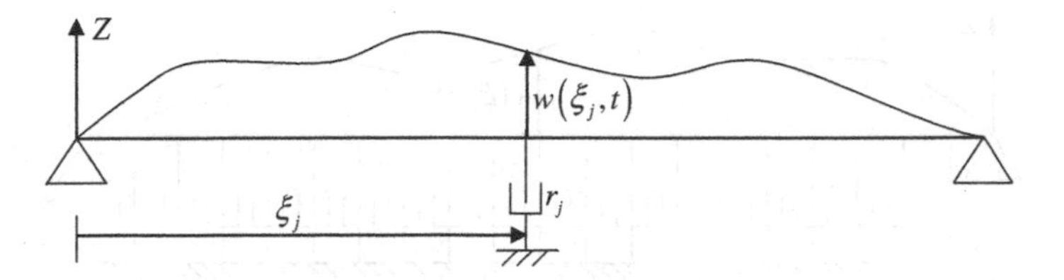

Fig. 3.32 Continuous system with concentrated damper

$$\phi^{(i)} = \sin\frac{i\pi\xi}{L} \tag{3.273}$$

The dissipative function of the damper can be directly defined as a function of the physical variable $\mathbf{w}(\xi_j, t)$ which defines the displacement of the point of the cable in which the same damper is applied:

$$D_j = \frac{1}{2} r_j \dot{w}^2(\xi_j, t) = \frac{1}{2} r_j \dot{w}^T(\xi_j, t) \dot{w}(\xi_j, t) = \frac{1}{2} r_j \left(\frac{\partial w_j}{\partial t}\right)^T \left(\frac{\partial w_j}{\partial t}\right) \tag{3.274}$$

By keeping account of the transformation of modal coordinates:

$$w(\xi_j, t) = \underline{\phi}^T(\xi_j) \underline{q}(t) \tag{3.275}$$

$\mathbf{D_j}$ can be expressed as a function of the independent variables represented by modal displacements by means of:

$$D_j = \frac{1}{2} r_j \underline{\dot{q}}^T(t) \underline{\phi}(\xi_j) \underline{\phi}^T(\xi_j) \underline{\dot{q}}(t) = \frac{1}{2} \underline{\dot{q}}^T(t) \left[r_j \underline{\phi}(\xi_j) \underline{\phi}^T(\xi_j)\right] \underline{\dot{q}}(t) = \frac{1}{2} \underline{\dot{q}}^T(t) [R_j] \underline{\dot{q}}(t) \tag{3.276}$$

where $[R_j]$ is the damping matrix in principle coordinates. This matrix is square of order "n" as the number of the normal modes considered in (3.276) and is generally full:

$$[R_j] = \left[r_j \underline{\varphi}(\xi_j) \underline{\varphi}^T(\xi_j)\right] = \begin{bmatrix} \varphi^{(1)}(\xi_j)\varphi^{(1)}(\xi_j) & \dots & \varphi^{(1)}(\xi_j)\varphi^{(i)}(\xi_j) & \dots & \varphi^{(1)}(\xi_j)\varphi^{(n)}(\xi_j) \\ \dots & \dots & \dots & \dots & \dots \\ \varphi^{(i)}(\xi_j)\varphi^{(1)}(\xi_j) & \dots & \varphi^{(i)}(\xi_j)\varphi^{(i)}(\xi_j) & \dots & \varphi^{(i)}(\xi_j)\varphi^{(n)}(\xi_j) \\ \dots & \dots & \dots & \dots & \dots \\ \varphi^{(n)}(\xi_j)\varphi^{(1)}(\xi_j) & \dots & \varphi^{(n)}(\xi_j)\varphi^{(i)}(\xi_j) & \dots & \varphi^{(n)}(\xi_j)\varphi^{(n)}(\xi_j) \end{bmatrix} \tag{3.277}$$

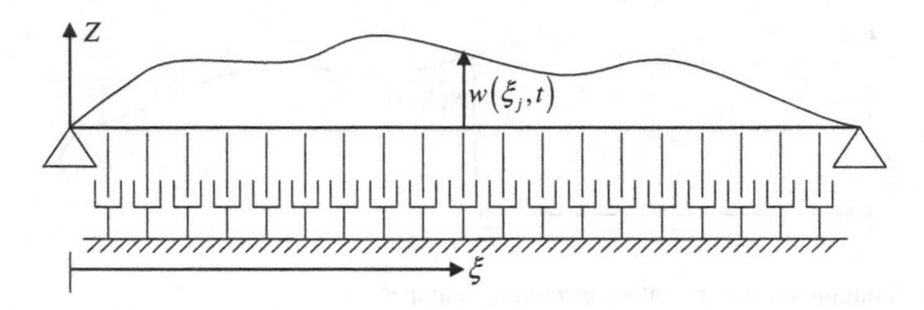

Fig. 3.33 Continuous system with distributed damping

Finally, let us analyse the case of the cable with distributed damping of constant $r(\xi) = r$ constant, as shown in Fig. 3.33.

The normal modes to be considered, similar to the procedure used for the external excitation forces, are those of the system devoid of excitation forces and dampers (outlined in (3.273)). The dissipative function of the damper can be directly defined as a function of the physical variable $\mathbf{w}(\boldsymbol{\xi}, \mathbf{t})$ which defines the displacement of a generic point of the cable:

$$D = \frac{1}{2}\int_0^L r\dot{w}^2(\xi, t)d\xi = \frac{1}{2}\int_0^L r\dot{w}^T\dot{w}d\xi \tag{3.278}$$

or rather, by keeping account of the transformation of modal coordinates (3.275):

$$D = \frac{1}{2}\underline{\dot{q}}^T(t)\left[\int_0^L r\underline{\phi}(\xi)\underline{\phi}^T(\xi)d\xi\right]\underline{\dot{q}}(t) = \frac{1}{2}\underline{\dot{q}}^T(t)[R]\underline{\dot{q}}(t) \tag{3.279}$$

where $[\mathbf{R}]$ is the damping matrix in principle coordinates: this matrix is diagonal and, in the case of uniformly distributed damping, the generic term is:

$$R_{ii} = r\frac{L}{2} \quad (i = 1, 2, \ldots, n) \tag{3.280}$$

Summary In this chapter, continuous body models (i.e. systems with infinite degrees of freedom, governed by partial differential equations, functions of both time and space) are introduced. As is known, closed form solutions can only be obtained in some simple applications. For this reason, the study was conducted mainly for educational purposes, in preparation for the discretization methods described in the following chapter. In particular, vibrations in continuous systems, using both the direct (the "propagative" solution) and modal approach in principal coordinates (the "stationary" solution), have been studied. At the end of the chapter, the bending vibrations of cables, beams and "taut beams" are shown as examples.

References

1. Amerio L (1977) Analisi matematica. UTET
2. Belloni G, Bemasconi G (1975) Sforzi e deformazioni e loro legami. Tamburini Editore, Milano
3. Bishop RED, Iohnson DC (1960) The mechanics of vibrations. Cambridge University Press, Cambridge
4. Brigham EO (1974) The fast fourier transform. Prentice-Hall, Englewood Cliffs
5. Bruni S, Cheli F, Collina A, Diana G (1985) Train-track interaction: a comparison between a numerical model and full scale experiments. In: Heavy vehicle system. Int J Veh Des 6(1–4):115–146
6. Bruni S, Collina A, Corradi R (2005) Train-track-bridge interaction: influence of track topology on structural dynamic performances. In: Proceedings of EURODYN 2005, Paris, France
7. Den Hartog IP (1956) Mechanical vibrations, 4th edn. McGraw-Hill, New York
8. Diana G, Claren R (1966) Vibrazioni nei conduttori ricerche teoriche e sperimentalifi - vibrazioni trasversali di un conduttore tesato con ammortizzatore. L'Energia Elettrica XLIII
9. Diana G, Claren R (1966) Vibrazioni nei conduttori ricerche teoriche e sperimentalifi - vibrazioni trasversali di un conduttore tesato. L'Energia Elettrica XLIII
10. Harris CM, Crede CE (1980) Shock and vibration handbook. McGraw-Hill, New York
11. Massa E (1981) Costruzioni di macchine. Masson, Milano
12. Meirovitch L (1986) Elements of vibration analysis. Mc-Graw-Hill, New York
13. Politecnico di Milano (1986) Lezioni di scienza delle Costruzioni. Clup, Milano
14. Roark RJ, Young WC (1975) Formulas for stress and strain. McGraw-Hill, New York
15. Shabana AA (1997) Theory of vibration. Springer, Berlin
16. Thomson WT, Dahleh MD (1997) Theory of vibration with applications, 5th edn. Prentice Hall, Englewood Cliffs
17. Timoshenko SP, Gere JM (1961) Theory of elastic stability. McGraw-Hill, New York
18. Timoshenko SP, Krieger SW (1965) Theory of plates and shells. McGraw-Hill, New York
19. Timoshenko SP, Young DH (1965) Theory of structures. McGraw-Hill, New York
20. Timoshenko SP, Young DH, Weaver W (1974) Vibration problems in engineering, 4th edn. Wiley, New York

Chapter 4
Introduction to the Finite Element Method

4.1 Introduction

The increase in complexity of structures and the need for more rigorous, in-depth structural analyses have made the development of systematic methodologies, allowing for the discretization of continuous systems, necessary. In fact, all real systems can be represented as continuous systems, with infinite degree-of-freedom (d.o.f.): however, in the best case scenario, working on a continuous system means dealing with partial derivatives differential equations (see Chap. 3) to be integrated into different domains. If the elastic and inertia characteristics are differentiated and/or the geometry of the system to be analysed is complex, the number of domains necessary and the necessary relative boundary conditions are such as to immediately cause considerable difficulties in terms of calculation or, in any event, the approach does not lend itself to a generalization of the procedure for all analysable systems (for each system it is necessary to redefine the equations of motion of same). It is thus possible to discretize the continuous system, or rather to make the transition from a continuous system with infinite d.o.f. (the real system) to an approximate discrete model with n d.o.f. (the corresponding mathematical model), which adequately approximates its behaviour. Various techniques and approaches can be adopted to discretize the continuous system.

- Schematization with *concentrated parameters* ([5, 6, 14, 17, 19,23], Chaps. 1 and 2);
- A *modal approach* (Chap. 3, Sect. 3.7);
- The *transfer matrix* method [11, 23, 24];
- The *finite differences* method [1];
- the *finite element* method;
- the *boundary element* method (see, for example, [7]).

In the past, the methodology most widely used was the *concentrated parameters* one. This method concentrates masses, dampings and stiffnesses, in actual fact

F. Cheli and G. Diana, *Advanced Dynamics of Mechanical Systems*,
DOI 10.1007/978-3-319-18200-1_4

distributed in a continuous form on the real system, in geometric points of the system. By adopting this scheme, it is possible to obtain discrete models with n d.o.f. whose equations of motion can be obtained by means of dynamic equilibrium equations, influence coefficients or Lagrange's equations (see Chaps. 1 and 2).

As an example, in Fig. 4.1, we show the scheme that can be used to simulate the dynamic behaviour of a rotating shaft equipped with flywheel masses: for this real system, it is possible to obtain a discrete n d.o.f. model by concentrating the distributed masses of the connecting sections on the flywheels and associating them with the elasticity of the shaft while considering the stiffness of the flywheels themselves as infinite.

Similarly, a continuous beam (real system) can be divided (Fig. 4.2) into sectors, by attributing, to the generic sector, the sole bending stiffness and by concentrating the mass of each sector in the centre of gravity of same (discretized model).

The *concentrated parameters* technique is not particularly valid because there are not strict criteria to concentrate the parameters themselves and this methodology does not allow for a systematic approach (i.e. one that can be retraced to logic diagrams and standardized procedures).

At present, there is no doubt that the most effective and powerful method is the finite element one that lends itself particularly well to the study of various problems (static and dynamic structural calculation, fluid-dynamic, heat transmission, etc.) whose phenomenology can be described analytically by partial derivatives differential equations. The success of this method is due to the rapid development of digital calculators and to the considerable evolution of numeric methods for the

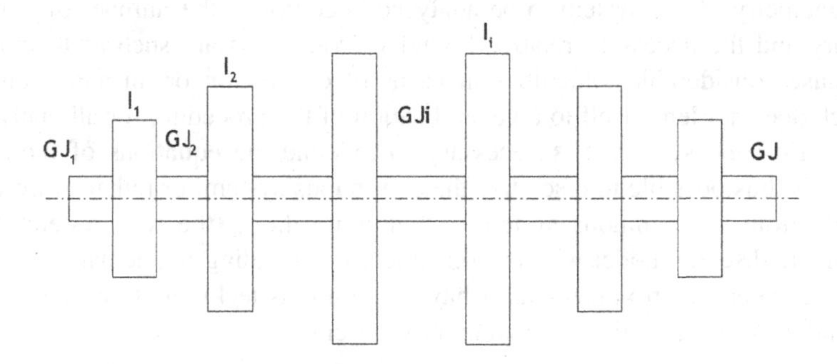

Fig. 4.1 Concentrated parameters schematization of a shaft with mounted flywheel masses

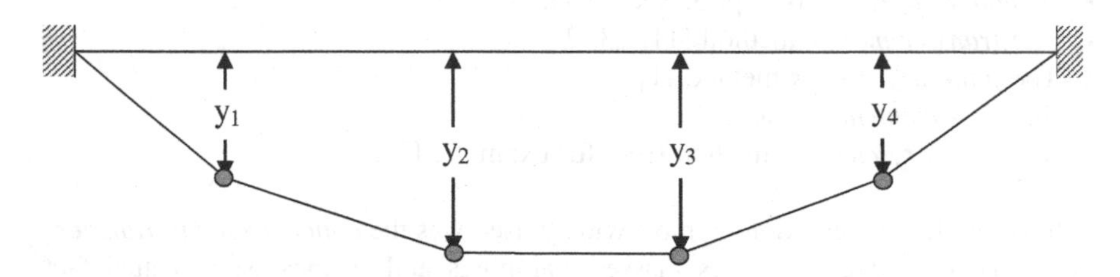

Fig. 4.2 Schematization showing the concentrated parameters of a beam for the study of bending vibrations

resolution of higher order systems of algebraic equations. Furthermore, its versatility and power is increased by the fact that any structure, complex in any case, can be seen as a set of simple elements (*finite elements*) whose elastic-inertia characteristics are defined in parametric form. From a point of view of the analysis of mechanical systems, by using a discrete n d.o.f. model of the continuous real system, this method permits:

- determination of the static equilibrium position of the structure;
- calculation of the natural frequencies ω_i and of the related normal modes $\underline{X}^{(i)}$;
- evaluation of the response (static and/or dynamic) with assigned external excitation forces.

As general independent coordinates, the Finite Element Method (indicated hereinafter as F.E.M.) considers the displacement of geometric points of the continuous system referred to as nodes (Fig. 4.3). In the more general case, (i.e. in the case of a spatial system) if n are the nodes of the discrete mathematical model, which define the displacements of the system, the d.o.f. will be $3 \times n$, if three d.o.f. are associated with the displacement of the node in space. This method considers the deformation between two nodes inside the finite element as a variable dependent on the nodal displacements: this deformation is, in fact, described by means of analytical functions dependent on the node displacements, termed Shape Functions (referred to hereinafter as *s.f.*). Once the nodal displacements have been calculated, these functions allow for the definition of the deformation assumed by the continuous system between two nodes. The structure is, thus, divided into *finite elements* (see Fig. 4.3), which definite the grid constituted by the nodes of the mathematical model. In this way, the F.E.M. reduces the continuous system to a finite number of d.o.f., where the number of independent variables assumed is finite, while maintaining the distribution properties of mass and elasticity of the continuous system unaltered by means of the *s.f.* themselves.

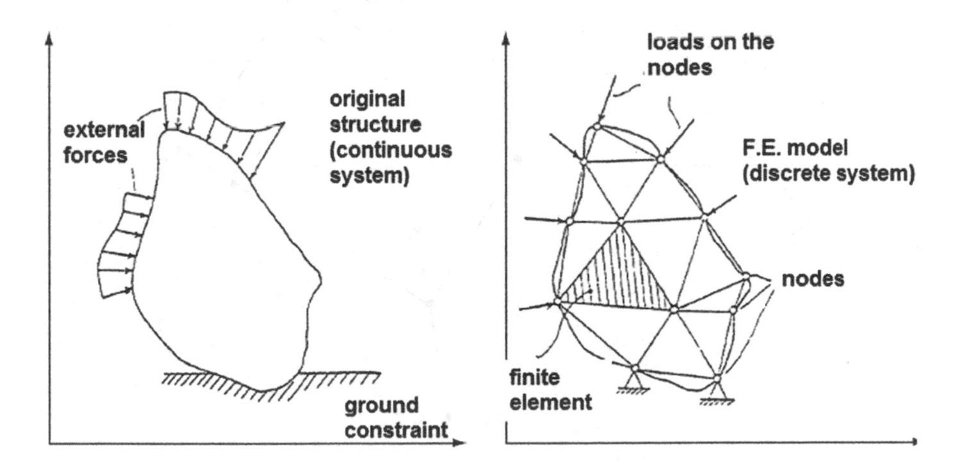

Fig. 4.3 Finite element schematization of a structure

4.2 The Shape Function

Let us consider a continuous system that occupies an assigned position in space and let us define (using *O–X–Y–Z*) a left-to-right Cartesian absolute reference systems: where *x*, *y* and *z* (Fig. 4.4) are the coordinates of a generic point P belonging to the continuous system. Let us indicated with $u = \vec{u}(x, y, z, t)$ the vector that defines the displacement of the point starting from the reference position defined previously. By using this approach, vector $\vec{u}(x, y, z, t)$ represents the independent variable used to describe the motion of the continuous system.

In the model discretized using the finite element method, deformation $\vec{u}(x, y, z, t)$ inside the generic finite element is, on the contrary, a function dependent on the node displacements. The problem in this case is to choose a suitable morphology of the *s.f.* which allows us to define the deformations of the single finite elements and to approximate, in the best way possible, the real deformation of the continuous system. To facilitate calculation, the *s.f.* is defined with respect to a local reference system (O_j–X_j–Y_j–Z_j) (see Fig. 4.4) that is integral with the *j*th undeformed finite element and displacement *u* is defined with respect to the local reference system. The behaviour of the points inside each finite element can be formulated according to different form functions:

- polynominal *s.f.*;
- trigonometric *s.f.*;
- *s.fs.* that are linear combinations of several polynomials etc.

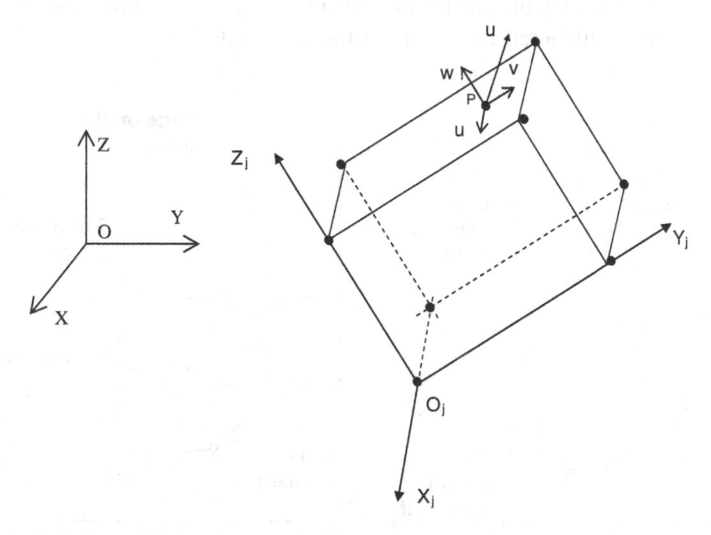

Fig. 4.4 Definition of global and local reference system, definition of the position (*x*, *y*, *z*) of the generic node and its displacement components $\vec{u}(x, y, z, t)$

which are calculated by leaving, first and foremost, a certain number of undefined parameters, subsequently determined by imposing the nodal displacement relative to the finite element considered on the *s.fs.* themselves, as boundary conditions.

Consequently, the number of parameters to be determined must necessarily be equal to the number of d.o.f. associated with each finite element, equal to the product between the number of interconnecting nodes with the rest of the model and the number of d.o.f. of each node.

Since the F.E.M uses shape functions that are clearly an approximation of the actual distribution of internal point displacements, it offers an approximate solution to the problem.

To satisfy all the above mentioned requirements, in terms of *s.fs.*, if analytically known, we generally use the static deformation of the single finite element, seen as a subject with forces concentrated on the nodes since, otherwise, we would obtain a static deformation, also a function of the load distribution inside the element and not only of the type of element.

Assuming that we adopt a good *s.f.* (for some continuous systems these are known and analytically defined while, for others (see Sect. 4.6) such as plates and membranes this does not occur) it is obvious that the good quality of the F.E.M. schematization increases as the number of nodes considered increase, that is with the thickening of the mesh. To facilitate calculation, during a first phase, the nodal displacements for the generic finite element are defined in the local reference system (O_j–Z_j–Y_j–X_j), which is different from element to element. Only subsequently are all the displacement expressed as a function of a single global reference system (O–Z–Y–X).

In the paragraphs that follow, mainly for didactic-explanatory purposes, we will show the definition of the *s.f.s* assumed for the taut cable finite element (Sect. 4.2.1) and for the beam finite element (Sect. 4.2.2). Finally, in Sect. 4.6 we will describe the procedure for a generic two-dimensional or three-dimensional finite element.

4.2.1 Shape Function for the Taut Cable Element

In a first approximation, a cable is devoid of bending and torsional stiffness, i.e. it does not react to applied external torques.[1]

Based on these assumptions, it is not necessary to introduce, into the mathematical model, the section rotations as independent variables.

The *taut cable* finite element is thus limited by only two nodes: in the spatial case (Fig. 4.5) each node has 3 d.o.f. (the three displacement components) and each element, considered rectilinear, thus has 6 d.o.f. and its deformation between two

[1]In actual fact, a real cable has a bending stiffness EJ and a torsional stiffness GJx. If, however, the static tension S_o is high and the area of the transverse section is limited, these effects are, in actual fact, negligible. If this were not the case, as a finite element it would be necessary to use the taut beam which will be described in Sect. 4.5.

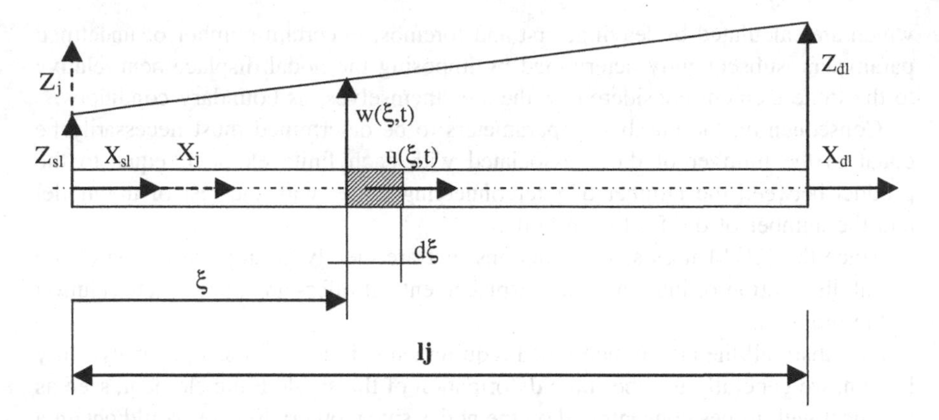

Fig. 4.5 Taut cable finite element

nodes will be defined through the 6 nodal displacement independent coordinates. By initially limiting, for didactic purposes, the study of motion to only one plane, each node of the generic *j*th element only has 2 d.o.f. and its deformation is associated with 4 independent coordinates (Fig. 4.5). The local reference system (O_j–Z_j–Y_j–X_j) is defined by a reference frame of Cartesian axes where axis X_j coincides with the longitudinal axis that connects the two end nodes and axis Z_j is perpendicular to X_j, lying in the plane considered and, finally, Y_j perpendicular to the two previous ones oriented in such a way that the local reference frame proves to be right handed. The physical coordinates that define the displacements of the generic section *P* of the cable, placed at a distance ξ measured starting from the left end of the finite element, are functions $w_j(\xi, t) = w(\xi, t)$ which define the transverse displacement components of the section in direction Z_j (a function of space and time, as in the continuous system) and function $u_j(\xi, t) = u(\xi, t)$ which defines the axial displacement of the same section in direction X_j.[2]

As independent variables, we choose the local displacement of the two nodes in longitudinal directions (X_{sl} and X_{dl}) and in a direction that is normal to the point of connection of the nodes themselves (Z_{sl} and Z_{dl})[3]: these displacements (independent variables) are grouped in a vector $\underline{X}_{jl}$:

[2]As already seen in Sect. 3.2, relative to the analysis of the cables schematized as continuous elements, the transverse behaviour of the cable is independent from the plane considered: for this reason, further one, we will consider the plane problem, or rather we will only consider plane *Zj*–*Xj*.

[3]Both here and further on, in order not to make the formalism and the writing of the various equations more complex, often for amplitudes relative to the generic *j*th finite element (different from element to element and thus identifiable with the subscript *j*) this index will be omitted. For example, in relation (4.1), we will omit $X_{sl} = X_{sjl}$, $Y_{sl} = Y_{sjl}$, $\underline{X}_{jl} = \underline{X}_l$, etc.

$$\underline{X}_{jl} = \underline{X}_l = \begin{Bmatrix} X_{sl} \\ Z_{sl} \\ X_{dl} \\ Z_{dl} \end{Bmatrix} \tag{4.1}$$

It is now necessary to define the link between the deformation assumed by the continuous system inside the generic finite element defined by the two components $u(\xi,t) = u_j(\xi,t)$ and $w(\xi,t) = w_j(\xi,t)$ and the vector of the nodal displacements in the local reference system $\underline{X}_l = \underline{X}_{jl}$ (4.1). The static deformation of a cable subjected to loads concentrated on nodes is a broken line: to describe this configuration, it is possible to assume a linear *s.f.* for the single taut cable finite element both for the transverse displacements and the longitudinal displacements:

$$\begin{aligned} w(\xi,t) &= a_1\xi + b_1 \\ u(\xi,t) &= a_2\xi + b_2 \end{aligned} \tag{4.2}$$

These *s.f.* contain the four undefined parameters a_1, a_2, b_1 and b_2 which are determined by imposing the boundary conditions on the single finite element, i.e. by imposing that the functions $u(\xi, t)$ and $w(\xi, t)$ coincide, in correspondence to the end nodes, with the displacements sustained by the $\underline{X}_l$ themselves (4.1), variables assumed regardless of the method:

$$\begin{aligned} w(\xi,t)|_{\xi=0} &= Z_{sl}(t); \quad w(\xi,t)|_{\xi=l_j} = Z_{dl}(t) \\ u(\xi,t)|_{\xi=0} &= X_{sl}(t); \quad u(\xi,t)|_{\xi=l_j} = X_{dl}(t) \end{aligned} \tag{4.3}$$

where l_j is the length of the generic finite element.

By solving the system we obtain:

$$a_1 = \frac{Z_{dl} - Z_{sl}}{l_j}; \quad a_2 = \frac{X_{dl} - X_{sl}}{l_j}; \quad b_1 = Z_{sl}; \quad b_2 = X_{sl} \tag{4.4}$$

The complete writing of the physical variables $w(\xi, t)$ and $u(\xi, t)$ (4.2) which describe the motion of the continuous system as a function of the independent variables $\underline{X}_l$ (4.1) becomes, in matrix form and keeping account of (4.4):

$$\begin{aligned} w(\xi,t) &= \underline{f}_w^T(\xi)\underline{X}_l(t) \Rightarrow w_j(\xi,t) = \underline{f}_{wj}^T(\xi)\underline{X}_{jl}(t) \\ u(\xi,t) &= \underline{f}_u^T(\xi)\underline{X}_l(t) \Rightarrow u_j(\xi,t) = \underline{f}_{uj}^T(\xi)\underline{X}_{jl}(t) \end{aligned} \tag{4.5}$$

having used $\underline{f}_w(\xi)$ and $\underline{f}_u(\xi)$ to indicate the shape functions:

$$\underline{f}_w(\xi) = \begin{Bmatrix} 0 \\ 1 - \frac{\xi}{l_j} \\ 0 \\ \frac{\xi}{l_j} \end{Bmatrix}; \quad \underline{f}_u(\xi) = \begin{Bmatrix} 1 - \frac{\xi}{l_j} \\ 0 \\ \frac{\xi}{l_j} \\ 0 \end{Bmatrix} \tag{4.6}$$

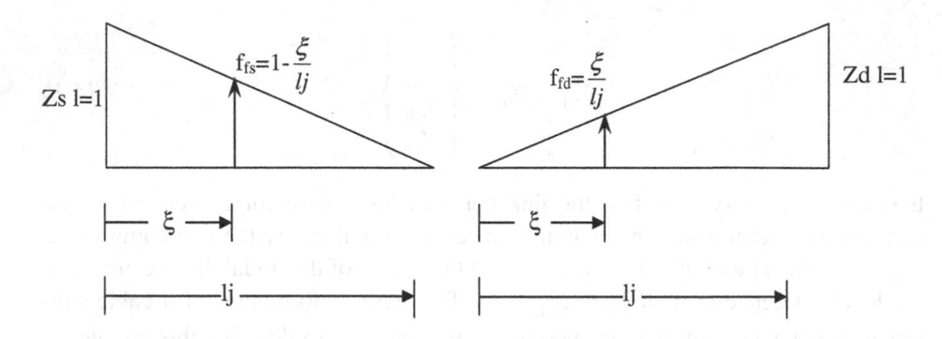

Fig. 4.6 Meaning of the shape function $f_w(\xi)$ for the taut cable element

The shape function depends parametrically on the length of the finite element, see (4.5), and on the spatial generic coordinate ξ. Each element of the vectors $f_w(\xi)$ and $f_u(\xi)$ represents the internal deformation assumed for a unit displacement assigned to the generic component of the nodal displacement (Fig. 4.6). By exploiting (4.5), once the displacement of the end nodes $\underline{X}_l$ are known, it is possible to trace the origin of the internal deformation.

4.2.2 The Shape Function for the Beam Element

Let us now consider a finite beam element, i.e. a finite element that schematizes the inertial elastic behaviour of a narrow, rectilinear beam with a constant transverse section. To simplify our explanation, in this first phase we will assume:

- uncoupled bending, torsional and axial motions (shear centre coinciding with the centroid of the section);
- stress state due to pure bending (bending moment vector parallel to a main axis of inertia of the section);
- effects of order higher than the first are negligible (like the effect of an axial force on the bending behaviour).

Furthermore, let us consider the specific case of bending motion in the plane (Fig. 4.7). As in the case of the taut beam, we assume a local reference system defined by a reference frame of Cartesian axes where axis X_j coincides with the longitudinal axis that connects the two end nodes and axis Z_j is perpendicular to X_j lying in the plane considered and, finally, Y_j is perpendicular to the two previous one oriented in such a way that the reference frame (O_j–X_j–Y_j–Z_j) proves to be right handed. The motion of the generic section, placed at distance ξ from the left end, can be described by the only physical variable $w_j(\xi, t) = w(\xi, t)$ (a function, as in the continuous system, of space ξ and of time t) which defines the transverse displacements (Fig. 4.7) of the beam itself.

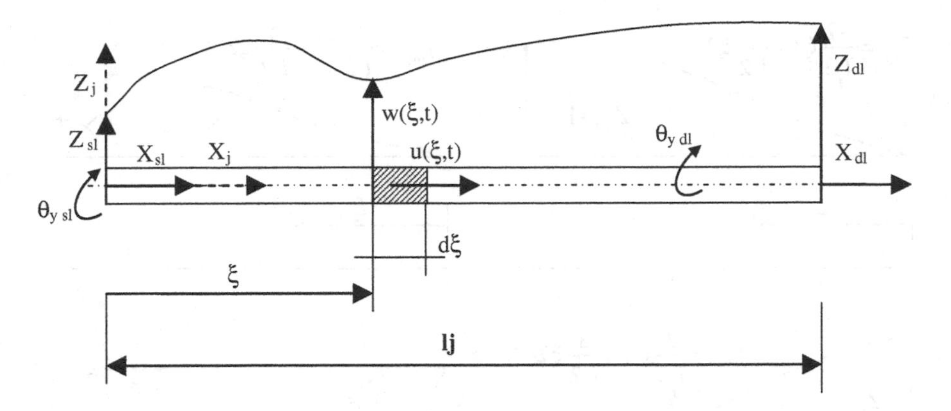

Fig. 4.7 Finite beam element

The local transverse displacements of the two end nodes Z_{sl} and Z_{dl} and the rotations θ_{ysl} and θ_{ydl} of the nodal sections of the beam are chosen as independent variables: these displacements and rotations are grouped in a vector $\underline{X}_{jl} = \underline{X}_j$:

$$\underline{X}_{jl} = \underline{X}_l = \begin{Bmatrix} Z_{sl} \\ \theta_{ysl} \\ Z_{dl} \\ \theta_{ydl} \end{Bmatrix} \tag{4.7}$$

It is now necessary to define the link between the transverse deformation $w(\xi, t)$ assumed by the continuous system inside the generic finite element and the vector of the nodal displacement of the finite beam element in the local reference system X_l. The shape function chosen is a cubic one, since this is the shape of the static deformation of a beam that is not subjected to distributed or concentrated loads between node and node [10, 26]:

$$w(\xi, t) = a_1\xi^3 + a_2\xi^2 + a_3\xi + a_4 \tag{4.8}$$

Constants a_1, a_2, a_3 and a_4 can be determined by imposing the boundary conditions (bearing in mind the sign conventions adopted, Fig. 4.7) on deformation $w(\xi, t)$:

$$\begin{aligned} w(\xi, t)|_{\xi=0} &= Z_{sl}(t); \quad \left.\frac{\partial w(\xi, t)}{\partial \xi}\right|_{\xi=0} = -\theta_{ysl}(t) \\ w(\xi, t)|_{\xi=lj} &= Z_{dl}(t); \quad \left.\frac{\partial w(\xi, t)}{\partial \xi}\right|_{\xi=lj} = -\theta_{ydl}(t) \end{aligned} \tag{4.9}$$

where l_j is the length of the generic finite element.

The expression of the physical variables that describe the motion of the continuous system as a function of the independent variables that describe the nodal displacements thus becomes (in matrix form and keeping account of (4.7), (4.8), (4.9)):

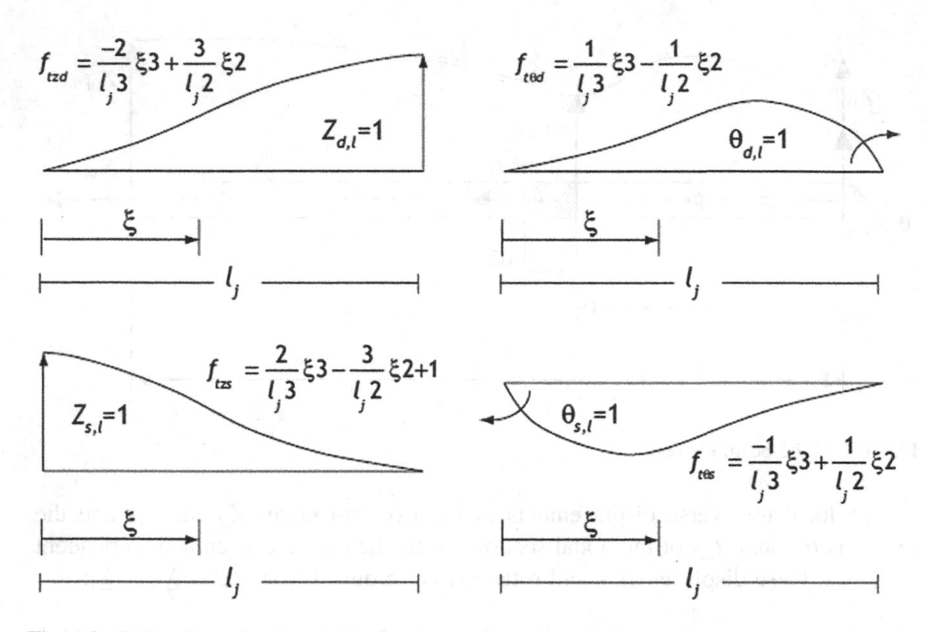

Fig. 4.8 Cubic shape function in the bending deformation of a beam element

$$\mathrm{w}(\xi,\mathrm{t}) = \underline{\mathrm{f}}_{\mathrm{w}}^{\mathrm{T}}(\xi)\underline{\mathrm{X}}_{\mathrm{l}}(\mathrm{t}) \Rightarrow \mathrm{w}_{\mathrm{j}}(\xi,\mathrm{t}) = \underline{\mathrm{f}}_{\mathrm{wj}}^{\mathrm{T}}(\xi)\underline{\mathrm{X}}_{\mathrm{jl}}(\mathrm{t}) \tag{4.10}$$

having indicated the shape function for the beam element and for the bending motions in plane X_j–Z_j with $\underline{\mathrm{f}}_{\mathrm{w}}(\xi)$:

$$\underline{\mathrm{f}}_{\mathrm{w}}(\xi) = \underline{\mathrm{f}}_{\mathrm{wj}}(\xi) = \begin{Bmatrix} \frac{2}{\mathrm{l}_{\mathrm{j}}^{3}}\xi^{3} - \frac{3}{\mathrm{l}_{\mathrm{j}}^{2}}\xi^{2} + 1 \\ -\frac{1}{\mathrm{l}_{\mathrm{j}}^{2}}\xi^{3} + \frac{2}{\mathrm{l}_{\mathrm{j}}}\xi^{2} - \xi \\ -\frac{2}{\mathrm{l}_{\mathrm{j}}^{3}}\xi^{3} + \frac{3}{\mathrm{l}_{\mathrm{j}}^{2}}\xi^{2} \\ \frac{1}{\mathrm{l}_{\mathrm{j}}^{2}}\xi^{3} - \frac{1}{\mathrm{l}_{\mathrm{j}}}\xi^{2} \end{Bmatrix} \tag{4.11}$$

Each term of this vector $\underline{\mathrm{f}}_{\mathrm{w}}$ represents the internal deformation of the finite element for a unit displacement associated with the generic d.o.f. (Fig. 4.8).

4.2.3 Shape Function for Generic Finite Elements

Given a generic finite element with n_o nodes (Fig. 4.9), by assuming an arbitrary local reference system of axes X_j–Y_j–Z_j and by considering vector $\underline{\mathrm{X}}_{\mathrm{jl}}$ of the nodal displacements defined in the local reference system:

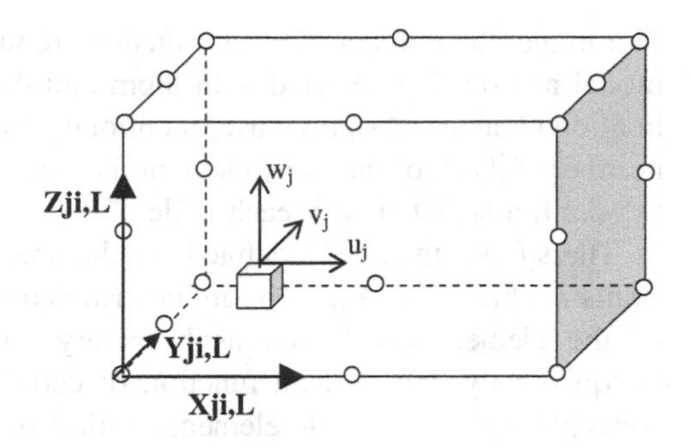

Fig. 4.9 Generic finite element

$$\underline{X}_{jl} = \begin{Bmatrix} X_{j1,1} \\ Y_{j1,1} \\ Z_{j1,1} \\ X_{j2,1} \\ Y_{j2,1} \\ Z_{j2,1} \\ \cdots \\ \cdots \\ \cdots \\ X_{jn_o,1} \\ Y_{jn_o,1} \\ Z_{jn_o,1} \end{Bmatrix} \tag{4.12}$$

it is possible to write the three functions that provide the displacement in three directions in the local reference system of generic point *P* inside the finite element. These are:

$$\begin{aligned} u_j(\xi,\eta,\zeta,t) &= \underline{f}_{uj}^T(\xi,\eta,\zeta)\underline{X}_{jl}(t) \\ w_j(\xi,\eta,\zeta,t) &= \underline{f}_{wj}^T(\xi,\eta,\zeta)\underline{X}_{jl}(t) \\ v_j(\xi,\eta,\zeta,t) &= \underline{f}_{vj}^T(\xi,\eta,\zeta)\underline{X}_{jl}(t) \end{aligned} \tag{4.13}$$

where $f_{uj}(\xi,\eta,\zeta)$, $f_{wj}(\xi,\eta,\zeta)$ and $f_{vj}(\xi,\eta,\zeta)$ are the shape functions that describe the deformations between two nodes.

As previously mentioned various *s.f.* can be used: among these, the most common is the polynomial one so that expressions (4.13) can be expressed as:

$$\begin{aligned} u_j(\xi,\eta,\zeta,t) &= a_0 + a_1\xi + b_1\eta + c_1\zeta + a_2\xi^2 + b_2\eta^2 + c_2\zeta^2 + \cdots + a_i\eta\zeta + \cdots + c_n\zeta^n \\ w_j(\xi,\eta,\zeta,t) &= d_0 + d_1\xi + e_1\eta + f_1\zeta + d_2\xi^2 + e_2\eta^2 + f_2\zeta^2 + \cdots + d_i\eta\zeta + \cdots + f_n\zeta^n \\ v_j(\xi,\eta,\zeta,t) &= g_0 + g_1\xi + h_1\eta + p_1\zeta + g_2\xi^2 + h_2\eta^2 + p_2\zeta^2 + \cdots + g_i\eta\zeta + \cdots + p_n\zeta^n \end{aligned} \tag{4.14}$$

The higher the number of terms considered, the greater the number of nodes in the model and d.o.f. associated with them and thus, obviously, the better the approximation obtained. As previously mentioned, the number of constants is equal to the number of d.o.f. of the finite element, i.e. to the product of the number of nodes for the d.o.f. associated with each node.

The s.*f.* are then fully defined. In the specific case of polynomial *s.f.* the constants a_1, b_1, c_1, ..., c_n, f_n, p_n are determined by assuming the nodal displacements of the element considered as boundary conditions: in this way, the *s.f.* are unequivocally defined as a function of coordinates ξ, η and ζ. In Sect. 4.6 these concepts, applied to finite elements with 2 or 3 dimensions, will be resumed.

4.3 The Equations of Motion of the System

As previously mentioned, the finite element approach is of the Lagrangian type: displacements $\underline{X}_t$ of the nodes of schematization (Fig. 4.3) are chosen as independent coordinates (Lagrangian coordinates). As a function of $\underline{X}_t$ it is possible to define the displacements of the continuous system $u_j(\xi, \eta, \zeta, t)$, $w_j(\xi, \eta, \zeta, t)$ and $v_j(\xi, \eta, \zeta, t)$ inside the *j*th finite element by means of known links constituted by the shape functions $\underline{f}_{wj}(\xi,\eta,\zeta)$, $\underline{f}_{uj}(\xi,\eta,\zeta)$, and $\underline{f}_{vj}(\xi,\eta,\zeta)$ (4.13). The various forms of energy that appear in the Lagrange's equations, expressed as a function of the independent coordinates nodal displacements $\underline{X}_t$, can be obtained from the deformation of the continuous system to obtain the equation of motion:

$$\left\{\frac{d}{dt}\left(\frac{\partial E_{ctot}}{\partial \underline{\dot{X}}_t}\right)\right\}^T - \left\{\frac{\partial E_{ctot}}{\partial \underline{X}_t}\right\}^T + \left\{\frac{\partial D_{tot}}{\partial \underline{\dot{X}}_t}\right\}^T + \left\{\frac{\partial V_{tot}}{\partial \underline{X}_t}\right\}^T = \underline{F}_t \tag{4.15}$$

with:

$$\delta^* L_{tot} = \underline{F}_t^T \delta^* \underline{X}_t \tag{4.16}$$

where $\delta^* L_{tot}$ is the virtual work performed by the external forces acting on the system which are neither inertial nor elastic nor viscous, for a virtual displacement $\delta^* \underline{X}_t$. In (4.15) E_{ctot} and V_{tot} are respectively the kinetic energy and the potential energy of the complete system, defined as the sum of contributions associated with the *j*th finite element E_{cj} and $V_j (j = 1, 2, \ldots, m)$:

$$E_{ctot} = \sum_{j=1}^{m} E_{cj}; \quad V_{tot} = \sum_{j=1}^{m} V_j \tag{4.17}$$

and D_{tot} is the corresponding dissipation function. By applying Lagrange's equations (4.15), based on the hypothesis, often valid in reality, of linearizing the problem relative to the inertial and elastic terms, we obtain the equations of motion of the mathematical model of the structure and finite elements in the form:

$$[M_{tot}]\ddot{\underline{X}}_t + [R_{tot}]\dot{\underline{X}}_t[K_{tot}]\underline{X}_t = \underline{F}_t \tag{4.18}$$

where, let us remember, $[M_{tot}]$ is the mass matrix of the complete finite element model, $[K_{tot}]$ the relative stiffness matrix, $[R_{tot}]$ the damping matrix and, finally, $\underline{F}_t$ is the vector of the generalized forces. Examples of a nonlinear approach will be described further on (Sect. 4.7). We will now analyse, making reference for the sake of convenience to specific application examples, the procedure used to obtain the equations of motion of the complete system formally defined in (4.18).

4.4 Taut Cable Finite Element (an Application Example)

As a first example, let us consider the F.E.M. of a span, fitted out with a conductor for aerial power lines (Fig. 4.10), whose main data are shown in Table 4.1.

This structure can be schematized with finite elements of the taut cable type whose characteristics and relative shape function have already been described in Sect. 4.2.1. In the analysis we will only consider motion in the plane, i.e. transverse and axial vibrations of the conductor about the static equilibrium configuration, assumed to have been assigned.[4] We will outline the procedure to be followed to schematize any type of structure using the finite element method. However, the considerations made will be integrated, at each step, with references to the guide example in Fig. 4.10.

4.4.1 Discretization of the Structure

The first step is the choice of the type of finite elements to be used and the number of nodes of the schematization (i.e. the definition of the *mesh* of the model). In the case proposed, by assuming that the bending stiffness is negligible, one can adopt the taut cable finite element, previously described in Sect. 4.2.1.

Having chosen a suitable type (or types of finite elements) to represent the real behaviour of the system to be analysed, it is then necessary to fix a global reference *O–X–Y–Z* and divide the structure in finite elements, identifying the position of the nodes, i.e. the connecting points between adjacent elements. The choice of the number of elements and, therefore, of the number of nodes of the mesh must guarantee that the schematization used does not have any flaws, i.e. it must ensure

[4]In Sect. 4.7 the methodologies required to solve the static problem (normally not linear) to define this configuration will be analyzed.

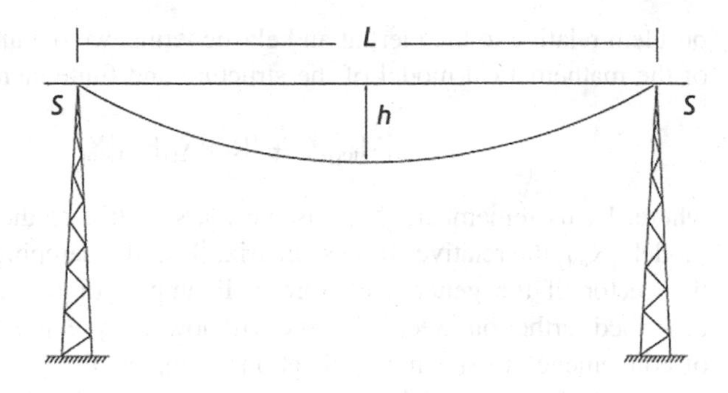

Fig. 4.10 Span of an electric line constituted by a single conductor

Table 4.1 Data relative to an electric line constituted by a single conductor

Span length	L = 200 m
Linear mass	m = 2.0 kg/m
Horizontal stretch component	S = 30,000 N

that the hypotheses adopted in the definition of the single finite element are satisfied:

- use of a specific shape form (for the taut cable element, this function, see Sect. 4.2.1, is linear and, for this reason, valid in the static field);
- use of constant elastic and inertial characteristics inside the single element (for the taut cable this involves assuming as constant, inside the *j*th finite element, tension S_j, mass by unit of length m_j and the area of the transverse section A_j).

In order to ensure that these hypotheses are able to satisfy the above mentioned requisites, it will be necessary to use sufficiently small finite elements in order to reproduce the variations of these parameters along the cable. In this example, the variable parameter is the tension. Furthermore, the length of the single element must be such to ensure that, during the vibrations of the cable, it maintains an almost static behaviour.[5] It is, thus, necessary that the first natural frequency of the generic element ω_{elj} is distinctly higher than the field of frequencies involved in the analysis performed. As a first approximation formula to calculate ω_{elj} it is possible to use the one that defines the natural frequency of a taut cable restrained at the ends by two supports (already described in Chap. 3, Sect. 3.2.2):

$$\omega_{elj} = \frac{\pi}{l_j}\sqrt{\frac{S_j}{m_j}} \tag{4.19}$$

[5]The shape function adopted is the static one: for this reason, the behavior of the finite element model will be similar to that of the real system, all the more so because the dynamic behavior of the single finite element will be similar to the static one.

By assuming, for example, a field of analysis included between 0 and Ω_{max} = 10 rad/s we will obtain:

$$\omega_{elj} \geq 1.5\Omega_{max} = 15\,\text{rad/s} \Rightarrow l_j \leq \frac{\pi}{15}\sqrt{\frac{S_j}{m_j}} = 25.65\,\text{m} \tag{4.20}$$

By making reference to the structure in Fig. 4.10, it is thus possible to consider 9 nodes and to schematize the structure as shown in Fig. 4.11. In reality, during the vibrations, tension S_j varies with respect to static value S_{oj}: in fact, axial deformations are associated to the longitudinal and transverse displacements of the cable and, therefore, variations of tension ΔS_j present in the *j*th finite element ($S_j = S_{oj} + \Delta S_j$) and, additionally longitudinal and transverse displacements are coupled, making the problem nonlinear, with serious analytical complication (in Sect. 4.7 you will see how it is possible to consider these effects nonlinear). If however [2, 10, 12, 25, 26], the initial static tension $S_{oj} = S_o$ is sufficiently high, the variation of tension ΔS_j only relative to the transverse behaviour of the cable, can be disregarded and it is thus possible, we repeat only for the transverse motions, to consider the tension constant.

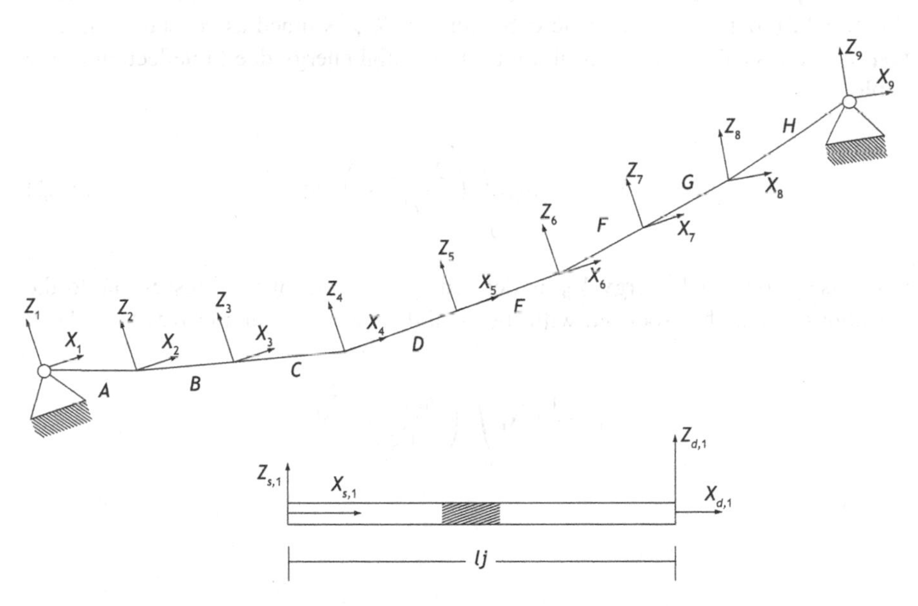

Fig. 4.11 Schematization with taut cable finite elements: absolute reference system, local reference systems

4.4.2 Definition of the Stiffness $[K_j]$ and Mass $[M_j]$ Matrix of the Taut Cable Finite Element in the Local Reference System

Let us express the potential energy V_j and kinetic energy E_{cj} associated with the generic jth cable finite element (recalling that these energies will be expressed as a function of the independent coordinates that define the disturbed motion about the static equilibrium configuration. Having assumed, for the only transverse component of the displacements, tension S_j constant and equal to the static value S_{oj} inside the jth element and not considering the elastic couplings between the longitudinal motions and the transverse motions, the calculation diagram becomes linear and it is possible to apply the superposition of the effects. The potential energy relative to the jth finite element V_j can be evaluated as the sum of two contributions:

$$V_j = V_{jtr} + V_{ja} \tag{4.21}$$

where V_{jtr} is associated with the transverse displacements $w_j(\xi, t)$ (normal to the longitudinal axis of the cable) and V_{ja} is associated with the longitudinal displacements $u_j(\xi, t)$ (Fig. 4.5). V_{jtr} is the potential energy, already defined for the cable subject to transverse motion alone, analysed with the continuous systems (Sect. 3.2.2), depending on static cable tension S_{oj}, assumed as constant, and corresponds physically to the contribution of potential energy due to deflection of the cable[6]:

$$V_{jtr} = \frac{1}{2} S_{0j} \int_0^{l_j} \left(\frac{\partial w_j(\xi, t)}{\partial \xi} \right)^2 d\xi \tag{4.22}$$

Conversely, potential energy $\mathbf{V_{ja}}$ is defined by the work of the stresses due to the variations in length associated with the axial displacements and proves to be [10]:

$$V_{ja} = \frac{1}{2} EA_j \int_0^{l_j} \left(\frac{\partial u_j(\xi, t)}{\partial \xi} \right)^2 d\xi \tag{4.23}$$

[6]We suppose a linear stress-strain relationship, otherwise potential energy would not have a quadratic form and, consequently, equations of motion would be nonlinear.

The potential energy associated with the generic *j*th finite element is thus[7]:

$$\begin{aligned}
V_j = V_{jtr} + V_{ja} &= \frac{1}{2} S_{oj} \int_0^{l_j} \left(\frac{\partial w_j(\xi, t)}{\partial \xi} \right)^2 d\xi + \frac{1}{2} EA_j \int_0^{l_j} \left(\frac{\partial u_j(\xi, t)}{\partial \xi} \right)^2 d\xi \\
&= \frac{1}{2} S_{oj} \int_0^{l_j} \left(\frac{\partial w_j(\xi, t)}{\partial \xi} \right)^T \left(\frac{\partial w_j(\xi, t)}{\partial \xi} \right) d\xi + \frac{1}{2} EA_j \int_0^{l_j} \left(\frac{\partial u_j(\xi, t)}{\partial \xi} \right)^T \left(\frac{\partial u_j(\xi, t)}{\partial \xi} \right) d\xi
\end{aligned} \tag{4.24}$$

Similar to the case of potential energy V_j, kinetic energy E_{cj} associated with the generic finite element proves to be defined by the sum of the two contributions, corresponding respectively to transverse motion $w_j(\xi, t)$ and axial motion $u_j(\xi, t)$:

$$\begin{aligned}
E_{cj} = E_{cjtr} + E_{cja} &= \frac{1}{2} m_j \int_0^{l_j} \left(\frac{\partial w_j(\xi, t)}{\partial \xi} \right)^2 d\xi + \frac{1}{2} m_j \int_0^{l_j} \left(\frac{\partial u_j(\xi, t)}{\partial \xi} \right)^2 d\xi \\
&= \frac{1}{2} m_j \int_0^{l_j} \left(\frac{\partial w_j(\xi, t)}{\partial \xi} \right)^T \left(\frac{\partial w_j(\xi, t)}{\partial \xi} \right) d\xi + \frac{1}{2} m_j \int_0^{l_j} \left(\frac{\partial u_j(\xi, t)}{\partial \xi} \right)^T \left(\frac{\partial u_j(\xi, t)}{\partial \xi} \right) d\xi
\end{aligned} \tag{4.25}$$

in which $w_j(\xi, t)$ and $u_j(\xi, t)$ represent the two components of absolute velocity of the generic infinitesimal element in plane X_j–Z_j and $m_j(\xi) = m_j$ constant is the mass by unit of the length of the cable (m_j dξ represents the mass associated with the generic infinitesimal section). Having assumed the shape functions $\underline{f}_{wj}$ and $\underline{f}_{uj}$ which correlate the transverse $w_j(\xi, t)$ and axial $u_j(\xi, t)$ displacements (to which the generic point inside the generic finite element has been subjected) to the displacements of the end nodes of the same:

$$\begin{aligned}
w_j(\xi, t) &= \underline{f}_{wj}^T(\xi) \underline{X}_{jl} \\
u_j(\xi, t) &= \underline{f}_{uj}^T(\xi) \underline{X}_{jl}
\end{aligned} \tag{4.26}$$

[7]Given a generic scalar quantity **p** it is always possible to express its square as:

$$p^2 = p^T p \tag{4.7.1}$$

This will be useful in the matrix form expressions that will follow, as already seen for the continuous systems (Chap. 3).

potential energy V_j (4.24) and kinetic energy E_{cj} can be expressed as a function of the independent variables $\underline{X}_{jl}$, which represent the nodal displacements to which the end nodes of the generic cable element in the local reference system have been subjected as:

$$\begin{aligned} E_{cj} &= \frac{1}{2}\dot{\underline{X}}_{jl}^T \left[m_j \int_0^{l_j} \underline{f}_{wj}\underline{f}_{wj}^T d\zeta + m_j \int_0^{l_j} \underline{f}_{uj}\underline{f}_{uj}^T d\zeta \right] \dot{\underline{X}}_{jl} = \frac{1}{2}\dot{\underline{X}}_{jl}^T [M_{jl}] \dot{\underline{X}}_{jl} \\ V_j &= \frac{1}{2}\underline{X}_{jl}^T \left[S_{oj} \int_0^{l_j} \underline{f}'_{wj}\underline{f}'^T_{wj} d\zeta + EA_j \int_0^{l_j} \underline{f}'_{uj}\underline{f}'^T_{uj} d\zeta \right] \underline{X}_{jl} = \frac{1}{2}\underline{X}_{jl}^T [K_{jl}] \underline{X}_{jl} \end{aligned} \tag{4.27}$$

having indicated with:

$$\underline{f}'_{wj} = \frac{\partial \underline{f}_{wj}}{\partial \xi} = \begin{Bmatrix} 0 \\ -\frac{1}{l_j} \\ 0 \\ \frac{1}{l_j} \end{Bmatrix}; \quad \underline{f}'_{uj} = \frac{\partial \underline{f}_{uj}}{\partial \xi} = \begin{Bmatrix} -\frac{1}{l_j} \\ 0 \\ \frac{1}{l_j} \\ 0 \end{Bmatrix} \tag{4.28}$$

In (4.27) $[M_{jl}]$ is the mass matrix in the local reference system of the generic taut cable finite element:

$$\begin{aligned} [M_{jl}] &= \left[\int_0^{l_j} m_j \underline{f}_{wj}\underline{f}_{wj}^T d\xi + \int_0^{l_j} m_j \underline{f}_{uj}\underline{f}_{uj}^T d\xi \right] \\ &= m_j \begin{bmatrix} \int_0^{l_j}\left(1-\frac{\xi}{l_j}\right)^2 d\xi & 0 & \int_0^{l_j}\left(1-\frac{\xi}{l_j}\right)\left(\frac{\xi}{l_j}\right) d\xi & 0 \\ 0 & \int_0^{l_j}\left(1-\frac{\xi}{l_j}\right)^2 d\xi & 0 & \int_0^{l_j}\left(1-\frac{\xi}{l_j}\right)\left(\frac{\xi}{l_j}\right) d\xi \\ \int_0^{l_j}\left(\frac{\xi}{l_j}\right)\left(1-\frac{\xi}{l_j}\right) d\xi & 0 & \int_0^{l_j}\left(1-\frac{\xi}{l_j}\right)^2 d\xi & 0 \\ 0 & \int_0^{l_j}\left(\frac{\xi}{l_j}\right)\left(1-\frac{\xi}{l_j}\right) d\xi & 0 & \int_0^{l_j}\left(1-\frac{\xi}{l_j}\right)^2 d\xi \end{bmatrix} \\ &= l_j m_j \begin{bmatrix} \frac{1}{3} & 0 & \frac{1}{6} & 0 \\ 0 & \frac{1}{3} & 0 & \frac{1}{6} \\ \frac{1}{6} & 0 & \frac{1}{3} & 0 \\ 0 & \frac{1}{6} & 0 & \frac{1}{3} \end{bmatrix} \end{aligned} \tag{4.29}$$

$[K_{jl}]$ is the relative stiffness matrix, always defined in the local reference system:

$$
\begin{aligned}
[\mathrm{K_{jl}}] &= \left[\mathrm{S_{0j}} \int_0^{l_j} \underline{\mathrm{f}}'_{wj} \underline{\mathrm{f}}'^{T}_{wj} d\xi + \mathrm{EA_j} \int_0^{l_j} \underline{\mathrm{f}}'_{uj} \underline{\mathrm{f}}'^{T}_{uj} d\xi \right] \\
&= \begin{bmatrix}
\mathrm{EA_j} \int_0^{l_j} \left(-\frac{1}{l_j}\right)^2 d\xi & 0 & \mathrm{EA_j} \int_0^{l_j} \left(-\frac{1}{l_j}\right)\left(\frac{1}{l_j}\right) d\xi & 0 \\
0 & \mathrm{S_{0j}} \int_0^{l_j} \left(-\frac{1}{l_j}\right)^2 d\xi & 0 & \mathrm{S_{0j}} \int_0^{l_j} \left(-\frac{1}{l_j}\right)\left(\frac{1}{l_j}\right) d\xi \\
\mathrm{EA_j} \int_0^{l_j} \left(\frac{1}{l_j}\right)\left(-\frac{1}{l_j}\right) d\xi & 0 & \mathrm{EA_j} \int_0^{l_j} \left(\frac{1}{l_j}\right)^2 d\xi & 0 \\
0 & \mathrm{S_{0j}} \int_0^{l_j} \left(\frac{1}{l_j}\right)\left(-\frac{1}{l_j}\right) d\xi & 0 & \mathrm{S_{0j}} \int_0^{l_j} \left(\frac{1}{l_j}\right)^2 d\xi
\end{bmatrix} \\
&= \begin{bmatrix}
\frac{\mathrm{EA_j}}{l_j} & 0 & -\frac{\mathrm{EA_j}}{l_j} & 0 \\
0 & \frac{\mathrm{S_{0j}}}{l_j} & 0 & -\frac{\mathrm{S_{0j}}}{l_j} \\
-\frac{\mathrm{EA_j}}{l_j} & 0 & \frac{\mathrm{EA_j}}{l_j} & 0 \\
0 & -\frac{\mathrm{S_{0j}}}{l_j} & 0 & \frac{\mathrm{S_{0j}}}{l_j}
\end{bmatrix}
\end{aligned}
\tag{4.30}
$$

As it was possible to foresee, by studying a conservative system, $[K_{jl}]$ is symmetric and singular (corresponding to the real situation of the free *taut cable* element in space).

Let us observe that the product between the ith line of matrix $[K_{jl}]$ and vector $\underline{X}_{jl}$ expresses the Lagrangian component of the elastic forces of the system according to the ith local free coordinate, while the generic term $K_{jl}(r,c)$ (rth line and cth column) represents the generalized elastic force that is created according to the rth d.o.f. due to a unitary displacement relative to the cth d.o.f. The fact that $[M_{jl}]$ and $[K_{jl}]$ are 4×4 matrices depends on the d.o.f. attributed to each node. If, for example, we consider a cable element in space, we have to attribute 3 d.o.f. to each of the two end nodes: in this case, we would have 6×6 matrices. In the case of the beam, where rotations $[M_{jl}]$ and $[K_{jl}]$ are also assumed as independent variables, this would have 12×12 size.

4.4.3 Transformation of Coordinates: Local Reference System, Absolute Reference System

Having defined the schematization of the structure (or *mesh*) according to the criteria outlined in Sect. 4.4.1 and having evaluated (as shown in Sect. 4.4.2) the stiffness $[K_{jl}]$ and mass $[M_{jl}]$ matrices of the taut cable finite element in the local reference system, it is now necessary to define the stiffness $[K_{jl}]$ and mass $[M_{jl}]$ matrices of the single finite elements in the sole global reference system and to assemble suitably the matrices of all the elements in order to define the equations of motion of the complete system. The position of the nodes (Fig. 4.11) is defined by the static equilibrium configuration assumed by the cable which, as is known, is a catenary (this static configuration must, in any case, be determined previously). In this way, each finite element assumes a different spatial orientation, as can clearly be seen in Fig. 4.11, and each finite element will be subjected to a different static tension S_{oj} (this means that the stiffness matrices $[K_{jl}]$ of the various finite elements are different). To write the equation of motion, a Lagrangian approach is adopted and it will, thus, be necessary to define the total kinetic energy E_{ctot} and the total potential elastic energy V_{tot} of the structure, given by the sum of the single contributions E_{cj} and V_j of all the m finite elements that make up the structure itself. These functions, keeping account of (4.27), become:

$$
\begin{aligned}
E_{ctot} &= \sum_{j}^{m} E_{cj} = \frac{1}{2}\sum_{j}^{m} \underline{\dot{X}}_{jl}^{T}[M_{jl}]\underline{\dot{X}}_{jl} \\
V_{tot} &= \sum_{j}^{m} V_j = \frac{1}{2}\sum_{j}^{m} \underline{X}_{jl}^{T}[K_{jl}]\underline{X}_{jl}
\end{aligned}
\tag{4.31}
$$

In these calculations, the local coordinates relative to a generic node are different, depending on whether they belong to the right or left finite element of the node itself (see Fig. 4.12). As previously mentioned, it is thus convenient to perform a coordinate transformation by passing from a generic system of local coordinates (O_j–X_j–Y_j–Z_j) to a system of absolute coordinates (O–X–Y–Z), so that only one set of coordinates that define the displacement exist for each node. Let

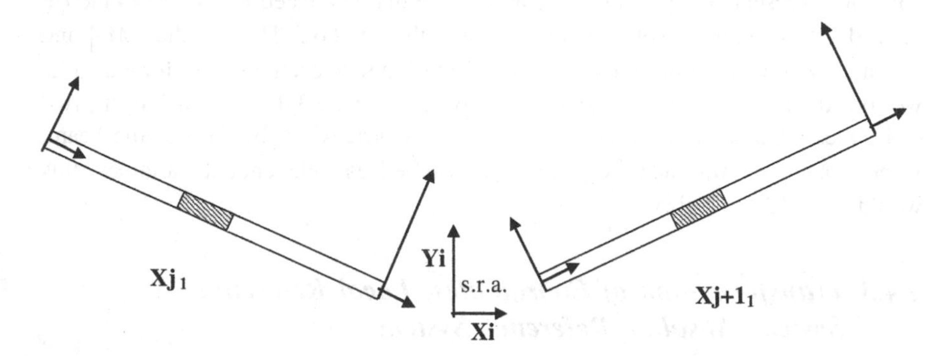

Fig. 4.12 Different systems of local coordinates for the generic node

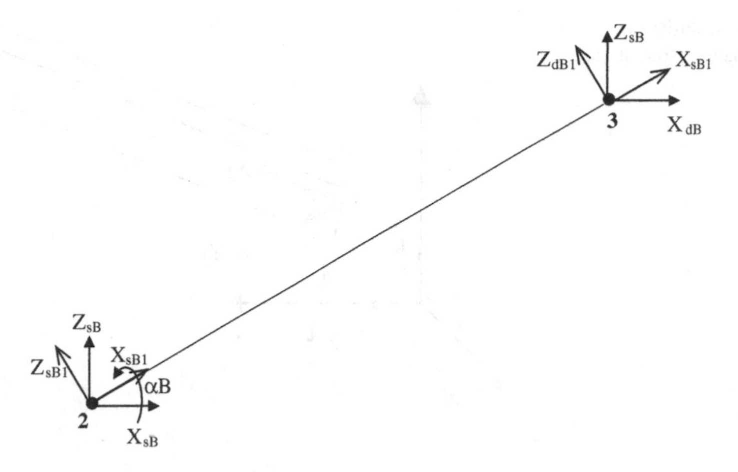

Fig. 4.13 Coordinate transformation: plane case

- $\underline{X}_{jl}$ be the displacements of the end nodes of the generic finite element in the local reference system (Fig. 4.13):

$$\underline{X}_{jl}^{T} = \{X_{sl}\ Z_{sl}\ X_{dl}\ Z_{dl}\} \tag{4.32}$$

- $\underline{X}_{j}$ the displacements expressed in the absolute reference system:

$$\underline{X}_{j}^{T} = \{X_{s}\ Z_{s}\ X_{d}\ Z_{d}\} \tag{4.33}$$

In general, the link between the displacements $\underline{X}_{jl}$ of the end nodes of a *j*th element in the local reference system and the corresponding displacements $\underline{X}_{j}$ expressed in the global absolute reference system:

$$\underline{X}_{jl} = [\Lambda_{j}]\underline{X}_{j} \tag{4.34}$$

is given by the coordinate transformation matrix $[\Lambda_j]$ formed by directory cosines in the local reference frame with respect to the global one.

In the case of the cable element in space equipped with six d.o.f. if $\vec{i}_L, \vec{j}_L$ and $\vec{k}_L$ are unit vectors of the local reference frame of the generic element (Fig. 4.14) and $\vec{i}, \vec{j}$, and $\vec{k}$ are the unit vectors relative to the global absolute reference frame, the transformation matrix $[\Lambda_j]$ [2, 10, 26] is formed by two sub-matrices $[\lambda_j]$ which, in their lines, have the direction cosines of the unit vectors of the local reference frame $\vec{i}_L$, $\vec{j}_L$, and $\vec{k}_L$, with respect to the global reference frame:

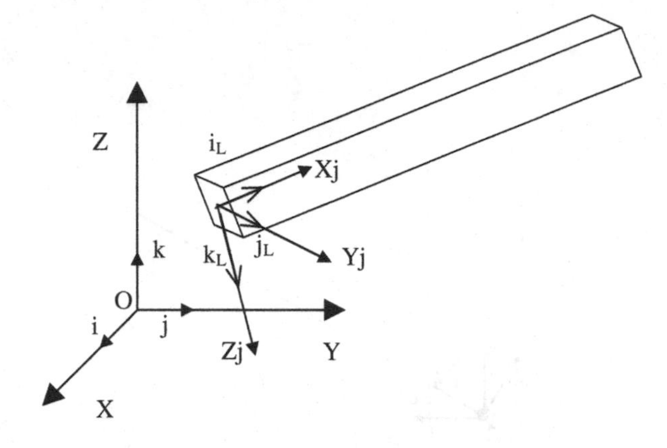

Fig. 4.14 Coordinate transformation: spatial case

$$[\Lambda_j] = \begin{bmatrix} [\lambda_j] & [0] \\ [0] & [\lambda_j] \end{bmatrix} \rightarrow [\lambda_j] = \begin{bmatrix} l_i & m_i & n_i \\ l_j & m_j & n_j \\ l_k & m_k & n_k \end{bmatrix} \tag{4.35}$$

where

- l_i, m_i and n_i are the direction cosines of unit vector $\vec{i}_L$ with respect to the axes of the global reference frame;
- l_j, m_j and n_j are the direction cosines of unit vector $\vec{j}_L$;
- l_k, m_k and n_k are the direction cosines of unit vector $\vec{k}_L$.

In the plane case analysed, this matrix is easily evaluated: let us consider, for example, finite element B of Fig. 4.13 (the same considerations could be made for all the other finite elements in the diagram). Let us look for the geometric link that enables us to correlate the independent generalized coordinates $\underline{\mathbf{X}}_{\mathbf{Bl}}$, that express the end node displacements in the local reference system (O_B–X_B–Y_B–Z_B), with the generalized coordinates $\underline{\mathbf{X}}_{\mathbf{B}}$ expressed in the global reference system O–X–Y–Z:

$$\underline{X}_{Bl} = \begin{Bmatrix} X_{sB,l} \\ Z_{sB,l} \\ X_{dB,l} \\ Z_{dB,l} \end{Bmatrix}; \quad \underline{X}_B = \begin{Bmatrix} X_{sB} \\ Z_{sB} \\ X_{dB} \\ Z_{dB} \end{Bmatrix} = \begin{Bmatrix} X_2 \\ Z_2 \\ X_3 \\ Z_3 \end{Bmatrix} \tag{4.36}$$

From an analysis of Fig. 4.13, it is possible to easily express this link in scalar form as:

$$\begin{aligned} X_{sB,1} &= X_2 \cos\alpha_B + Z_2 \sin\alpha_B \\ X_{sB,1} &= -X_2 \sin\alpha_B + Z_2 \cos\alpha_B \Rightarrow \underline{X}_{Bl} = [\Lambda_B]\underline{X}_B \\ X_{dB,1} &= X_3 \cos\alpha_B + Z_3 \sin\alpha_B \\ X_{dB,1} &= -X_3 \sin\alpha_B + Z_3 \cos\alpha_B \end{aligned} \tag{4.37}$$

where $[\Lambda_B]$ is the coordinate transformation matrix of element B defined as:

$$[\Lambda_B] = \begin{bmatrix} [\lambda_B] & [0] \\ [0] & [\lambda_B] \end{bmatrix} = \begin{bmatrix} \cos\alpha_B & \sin\alpha_B & 0 & 0 \\ -\sin\alpha_B & \cos\alpha_B & 0 & 0 \\ 0 & 0 & \cos\alpha_B & \sin\alpha_B \\ 0 & 0 & -\sin\alpha_B & \cos\alpha_B \end{bmatrix} \tag{4.38}$$

Similarly it is possible to obtain the coordination transformation matrix $[\Lambda_j]$ relative to the jth finite element constituting the overall model of the structure. The terms of $[\lambda_B]$ are the direction cosines of the local reference frame with respect to the absolute reference frame.

4.4.4 *Definition of the Stiffness [K_j] and Mass [M_j] Matrix of the Taut Cable Element in the Global Reference System*

By introducing the coordinate transformation (4.34) into the expression of kinetic energy E_{ctot} and potential energy V_{tot} (4.31) we obtain:

$$E_{ctot} = \sum_{j}^{m} \frac{1}{2} \underline{\dot{X}}_j^T [\Lambda_j]^T [M_{jl}] [\Lambda_j] \underline{\dot{X}}_j = \sum_{j}^{m} \frac{1}{2} \underline{\dot{X}}_j^T [M_j] \underline{\dot{X}}_j \tag{4.39}$$

$$V_{tot} = \sum_{j}^{m} \frac{1}{2} \underline{X}_j^T [\Lambda_j]^T [K_{jl}] [\Lambda_j] \underline{X}_j = \sum_{j}^{m} \frac{1}{2} \underline{X}_j^T [K_j] \underline{X}_j \tag{4.40}$$

having respectively used $[M_j]$ and $[K_j]$ to indicate the matrices of mass and stiffness of the finite element referring to the global coordinates:

$$\begin{aligned} [M_j] &= [\Lambda_j]^T [M_{jl}] [\Lambda_j] \\ [K_j] &= [\Lambda_j]^T [K_{jl}] [\Lambda_j] \end{aligned} \tag{4.41}$$

In (4.39) and (4.40) vectors $\underline{X}_j$ contain the 4 d.o.f. of the nodes of the jth element defined in the global reference system.

4.4.5 *Assembly of the Complete Structure*

Let us now indicate with $\underline{X}_t$ the vector that contains all the d.o.f. of the structure analysed, assumed, for the time being, as free in space. In the example analysed, (see Fig. 4.11) we will have:

$$\underline{X}_t = \begin{Bmatrix} X_1 \\ Z_1 \\ \vdots \\ X_i \\ Z_i \\ \vdots \\ X_9 \\ Z_9 \end{Bmatrix} \tag{4.42}$$

The generic term of kinetic energy E_{cj} or potential energy V_j can thus be redefined as:

$$\begin{aligned} E_{cj} &= \frac{1}{2}\dot{\underline{X}}_j^T [M_j] \dot{\underline{X}}_j = \frac{1}{2}\dot{\underline{X}}_t^T [\overline{M}_j] \dot{\underline{X}}_t \\ V_j &= \frac{1}{2}\underline{X}_j^T [K_j] \underline{X}_j = \frac{1}{2}\underline{X}_t^T [\overline{K}_j] \underline{X}_t \end{aligned} \tag{4.43}$$

this procedure is defined as "expansion to the dimensions of the complete model". In expressions (4.43) the matrices $[\overline{M}_j]$ or $[\overline{K}_j]$ are singular matrices except in correspondence to the terms present in the corresponding reduced matrices $[M_j]$ and $[K_j]$ as shown in Fig. 4.15.

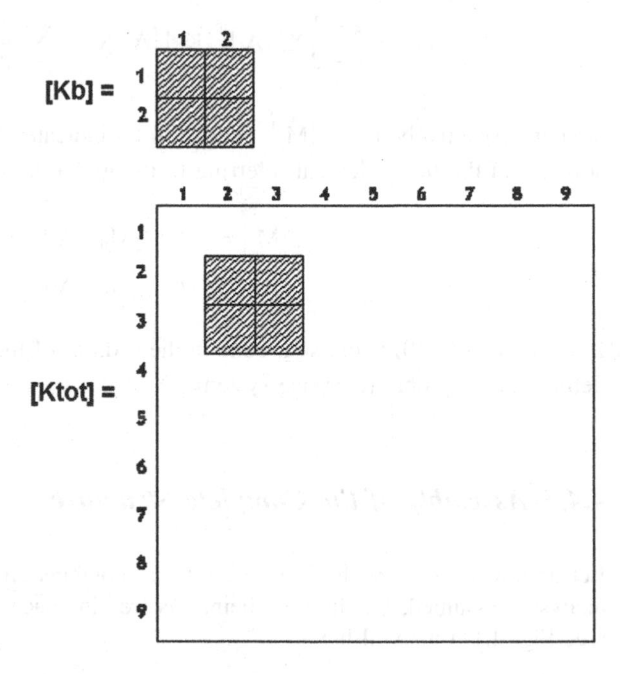

Fig. 4.15 Expansion procedure of the matrices to total dimensions

In other words, with the expression shown in (4.43), the generic form of energy of the generic finite element is expressed as a function of displacements $\underline{X}_t$ relative to all the nodes of the overall model. The kinetic and overall potential energy (4.39) and (4.40) can, in this way, keeping account of the formalism shown in (4.43), be expressed as:

$$\begin{aligned} E_{ctot} &= \frac{1}{2}\sum_{j}^{m} \dot{\underline{X}}_t^T [\overline{M}_j] \dot{\underline{X}}_t = \frac{1}{2}\dot{\underline{X}}_t^T \left[\sum_{j}^{m} [\overline{M}_j] \right] \dot{\underline{X}}_t = \frac{1}{2}\dot{\underline{X}}_t^T [M_{tot}] \dot{\underline{X}}_t \\ V_{tot} &= \frac{1}{2}\sum_{j}^{m} \underline{X}_t^T [\overline{K}_j] \underline{X}_t = \frac{1}{2}\underline{X}_t^T \left[\sum_{j}^{m} [\overline{K}_j] \right] \underline{X}_t = \frac{1}{2}\underline{X}_t^T [K_{tot}] \underline{X}_t \end{aligned} \tag{4.44}$$

$$[M_{tot}] = \left[\sum_{j}^{m} [\overline{M}_j] \right]; \quad [K_{tot}] = \left[\sum_{j}^{m} [\overline{K}_j] \right] \tag{4.45}$$

respectively the matrices of mass and stiffness of the complete structure (still devoid of constraints): these matrices are thus defined through the assembly operation, i.e. the sum of the expanded matrices relative to the single elements.

To obtain the complex matrices $[K_{tot}]$ and $[M_{tot}]$ it is not, however, necessary, nor even convenient, to use the complete expression of kinetic energy E_{ctot} and potential energy V_{tot} according to formulation (4.44). By analysing the assembly procedure shown in Fig. 4.16, it is easy to observe how the generic term $M_{tot}(r,c)$, $K_{tot}(r,c)$, of the global mass $[M_{tot}]$ and stiffness $[K_{tot}]$ matrices of (4.45)

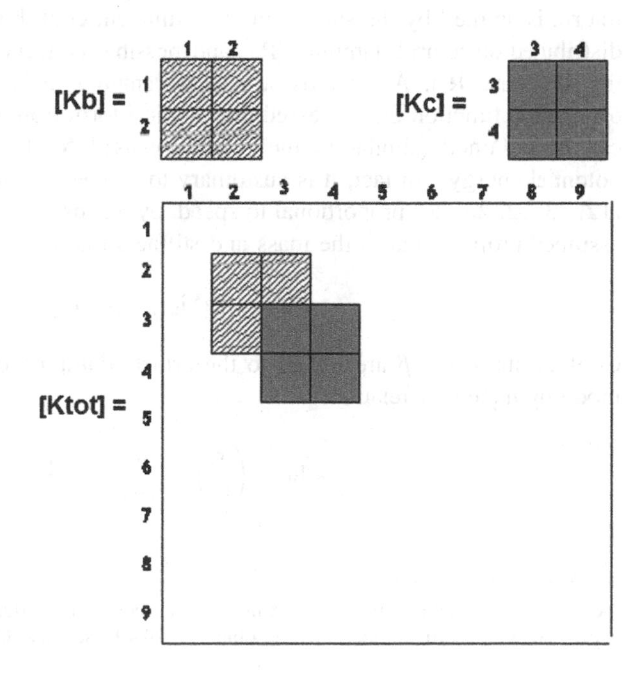

Fig. 4.16 Procedure of operative assembly

is obtained by adding up the contribution $M_j(r,c)$ and $K_j(r,c)$ due to the single matrices $[M_j]$ and $[K_j]$ relative to the same d.o.f. r and c:

$$M_{tot}(r,c) = \sum_{j=1}^{m} M_j(r,c); \quad K_{tot}(r,c) = \sum_{j=1}^{m} K_j(r,c) \tag{4.46}$$

It is thus clear how the matrices $[\mathrm{K_{tot}}]$ and $[\mathrm{M_{tot}}]$ of the complete model can be easily obtained by means of a suitable assembly of the single matrices $[\mathrm{K_j}]$ and $[\mathrm{M_j}]$ relative to the single finite elements (defined in the global reference system) without resorting to the expansion of these matrices in $[\mathrm{K_j}]$ and $[\mathrm{M_j}]$. The assembly operation is thus performed by algebraically adding up the single contributions $\mathrm{K_j(i,l)}$ and $\mathrm{M_j(i,l)}$ directly in the final matrices of the complete structure $[\mathrm{K_{tot}}]$ and $[\mathrm{M_{tot}}]$, as shown in Fig. 4.16.

4.4.5.1 Damping Matrix $[R_{tot}]$

Damping is introduced into a finite element model as seen in n d.o.f. systems (Chap. 2) through the dissipation function:

$$D_{tot} = D_d + D_c = \frac{1}{2}\underline{\dot{X}}_t^T[R_d]\underline{\dot{X}}_t + \frac{1}{2}\underline{\dot{X}}_t^T[R_c]\underline{\dot{X}}_t = \frac{1}{2}\underline{\dot{X}}_t^T[R_{tot}]\underline{\dot{X}}_t \tag{4.47}$$

where $[\mathrm{R_{tot}}]$ is the damping matrix of the complete system. Generally speaking this matrix is defined by the sum of the two different contributions associated with the distributed structural damping $[\mathrm{R_d}]$ and possible concentrated dampers present in the structure $[\mathrm{R_c}]$. As regards structural damping, it is not possible to define a dissipation function $\mathrm{D_d}$ expressed in analytical form from which the damping terms can be obtained (similar to the procedure used for the kinetic energy and the potential energy). In fact, it is customary to define an equivalent viscous damping [12, 13, 22, 25, 26] proportional to speed, by means of matrix $[\mathrm{R_{tot}}]$. This matrix is assumed proportional to the mass and stiffness matrix[8]:

$$[\mathrm{R_d}] = \alpha[\mathrm{M_{tot}}] + \beta[\mathrm{K_{tot}}] \tag{4.48}$$

Coefficients α and β are linked to the critical damping of the generic ith normal mode by means of relation:

$$\mathrm{h_i} = \left(\frac{\mathrm{r}}{\mathrm{r_c}}\right)_\mathrm{i} = \frac{\alpha}{2\omega_\mathrm{i}} + \frac{\beta\omega_\mathrm{i}}{2} \tag{4.49}$$

[8]According to this, the damping matrix in principal coordinates is diagonal. Its generic term can be expressed as $\mathrm{r_{ii}} = \alpha \mathrm{m_{ii}} + \beta \mathrm{k_{ii}}$, thereby obtaining (4.49), see also Chap. 2.

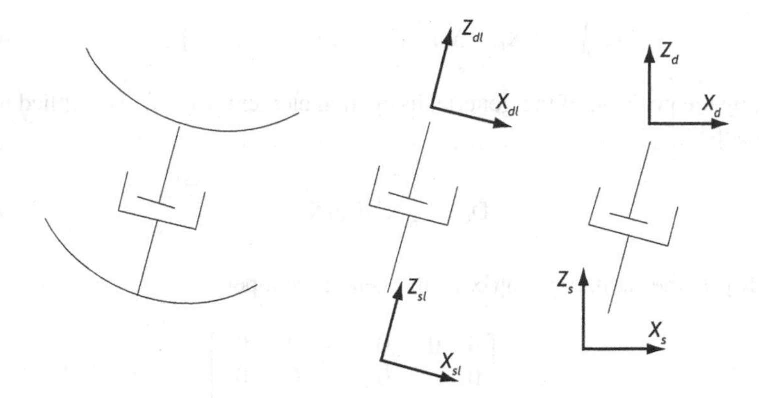

Fig. 4.17 Introduction of a dissipation element in a finite element diagram

It is, thus, possible to determine α and β:

- in a deterministic manner, i.e. by imposing the critical damping value (obtained experimentally) in correspondence to two specific natural frequencies;
- by imposing, in the frequency field of interest $0 < \omega_i < \Omega_{max}$, different values of h_i as a function of the natural frequencies ω_i, thus obtaining coefficients α and β using a least squares approach.[9]

Conversely, as regards the dissipative function D_c associated with concentrated elements, this is expressed by using a procedure similar to the one used for the elastic and kinetic energy relative to the finite elements. Dissipative function D_{cj} relative to the generic damper (Fig. 4.17):

$$D_{cj} = \frac{1}{2} r_j \dot{\Delta l}_{rj}^2 = \frac{1}{2} r_j \left(\dot{Z}_{sl}^2 - \dot{Z}_{dl}^2 \right) \tag{4.50}$$

is expressed, first and foremost, in a system of local coordinates $\underline{X}_{jl}$ in which Z is oriented following the longitudinal axis of the damper (Fig. 4.17):

[9]In this case, coefficients α and β are defined minimizing the root mean square error function:

$$f(\alpha, \beta) = \sum_{1}^{nf} \left(h_i - \frac{\alpha}{2\omega_i} + \frac{\beta \omega_i}{2} \right)^2 = \min \tag{4.9.1}$$

i.e. by imposing:

$$\frac{\partial f}{\partial \alpha} = 0; \quad \frac{\partial f}{\partial \beta} = 0. \tag{4.9.2}$$

$$\underline{X}_{jl}^{T} = \{\, X_{sl} \quad Y_{sl} \quad Z_{sl} \quad X_{dl} \quad Y_{dl} \quad Z_{dl} \,\} \tag{4.51}$$

Defined by the position of the generic dissipation element (Fig. 4.17), applied to the nodal points:

$$D_{cj} = \frac{1}{2}\dot{\underline{X}}_{jl}^{T}[R_{sl}]\dot{\underline{X}}_{jl} \tag{4.52}$$

where $[R_{sl}]$ is the damping matrix of the generic damper:

$$[R_{sl}] = \begin{bmatrix} 0 & 0 & 0 & 0 & 0 & 0 \\ 0 & 0 & 0 & 0 & 0 & 0 \\ 0 & 0 & r_j & 0 & 0 & -r_j \\ 0 & 0 & 0 & 0 & 0 & 0 \\ 0 & 0 & 0 & 0 & 0 & 0 \\ 0 & 0 & -r_j & 0 & 0 & r_j \end{bmatrix} \tag{4.53}$$

Subsequently, the coordination transformation is performed from the local reference system (i.e. from variables X_{sl} of (4.50) to the global one coordinates X_s):

$$\underline{X}_{j}^{T} = \{X_s\, Y_s\, Z_s\, X_d\, Y_d\, Z_d\} \tag{4.54}$$

Through the coordinates transformation matrix $[\Lambda_{sj}]$:

$$\underline{X}_{jl} = [\Lambda_{sj}]\underline{X}_{j} = \begin{bmatrix} [\lambda_j] & [0] \\ [0] & [\lambda_j] \end{bmatrix}\underline{X}_{j} \tag{4.55}$$

Formed by directory cosines in the local reference frame with respect to the global one. By introducing the coordinates transformation (4.55) into the dissipation function expression $\mathbf{D}_{cj}$ (4.51) we obtain:

$$D_{cj} = \frac{1}{2}\dot{\underline{X}}_{j}^{T}[\Lambda_{sj}]^{T}[R_{sl}][\Lambda_{sj}]\dot{\underline{X}}_{j} = \frac{1}{2}\dot{\underline{X}}_{j}^{T}[R_s]\dot{\underline{X}}_{j} \tag{4.56}$$

where $[R_s]$ is the damping matrix in global coordinates:

$$[R_s] = [\Lambda_{sj}]^{T}[R_{sl}][\Lambda_{sj}] \tag{4.57}$$

The overall dissipation function D_c associated with m_r concentrated elements is thus given by the sum of the single contributions:

$$\begin{aligned} D_c &= \frac{1}{2}\sum_{s=1}^{mr} \underline{\dot{X}}_j^T [R_s] \underline{\dot{X}}_j = \frac{1}{2}\sum_{s=1}^{mr} \underline{\dot{X}}_t^T \left[\overline{R_s}\right] \underline{\dot{X}}_t \\ &= \frac{1}{2}\underline{\dot{X}}_t^T \left[\sum_{s=1}^{mr} \left[\overline{R_s}\right]\right] \underline{\dot{X}}_t = \frac{1}{2}\underline{\dot{X}}_t^T [R_c] \underline{\dot{X}}_t \end{aligned} \tag{4.58}$$

4.4.6 Calculation of the Generalized Forces

To define the vector of the generalized forces $\underline{F}_t$ due to external forces (4.18), it is necessary to evaluate the virtual work $\delta^* L_{tot}$ performed by same for a virtual displacement of the independent variables $\delta^* \underline{X}_t$ which, in this case, are the node displacements of the finite element schematization. By following the systematic approach, already used both for the discrete systems (Chap. 2, Sect. 2.2), first and foremost, we will define the virtual work as a function of the virtual displacement of the application point of the excitation force, a displacement to be considered as a physical variable.

We will subsequently correlate this physical variable with the local independent variables, i.e. nodal displacements $\underline{X}_{jl}$, by means of the natural shape functions of the generic finite element used, and, therefore, with the nodal displacements $\underline{X}_j$ through the transformation matrices $[\Lambda_j]$ (4.55). The treatment that follows is independent of whether it is a question of dealing with a static (in which the forces do not depend on time) or dynamic problem (in which, conversely, the forces depend on time). The type of excitation force (static or dynamic) will only influence the techniques necessary to solve the resulting equations of motion. Finally, it is necessary to remember that since the structure is still free in space, it will also be necessary to express the virtual work of the constraint reactions $\underline{F}_V$.

4.4.6.1 Forces Concentrated at the Nodes

First and foremost, let us examine the simpler case of m_F concentrated forces applied in the nodal points of the structure analysed (Fig. 4.18). In this case, the virtual work performed by these forces is:

$$\delta^* L_{tot} = \sum_{k=1}^{mF} \delta^* L_k \tag{4.59}$$

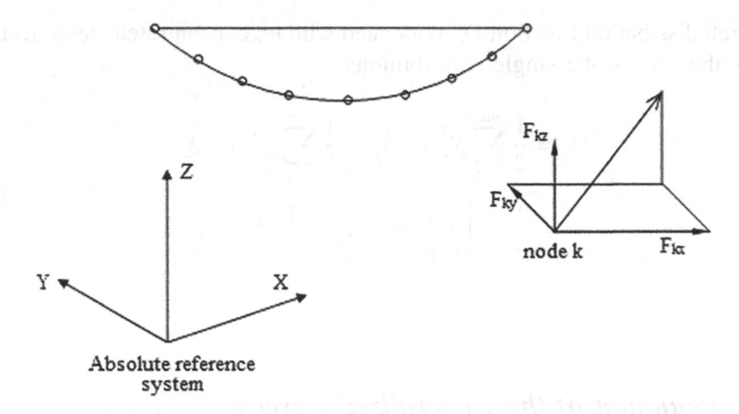

Fig. 4.18 Concentrated force in the generic *k*th node

where $\delta^{*}L_{k}$ is the virtual work of the generic concentrated force $\vec{F}_{k}(k = 1, m_{F})$ applied in generic node *k* of the schematization of the span for the virtual displacement $\delta^{*}w_{k}$ of the application point of the same force:

$$\delta^{*}L_{k} = F_{k}\,\delta^{*}w_{k} \tag{4.60}$$

Let us consider vector $\underline{F}_{k}$, containing the components of the generic *k*th force $\vec{F}_{k}$ in the absolute reference (Fig. 4.16):

$$\underline{F}_{k} = \begin{Bmatrix} F_{kx} \\ F_{ky} \\ F_{kz} \end{Bmatrix} \Rightarrow \underline{F}_{k} = \underline{i}F_{kx} + \underline{j}F_{ky} + \underline{k}F_{kz} \tag{4.61}$$

Having used $\underline{X}_{k}$:

$$\underline{X}_{k} = \begin{Bmatrix} X_{kx} \\ X_{ky} \\ X_{kz} \end{Bmatrix} \Rightarrow \underline{w}_{k} = \underline{i}X_{x} + \underline{j}X_{y} + \underline{k}X_{z} \tag{4.62}$$

to indicate the vector containing the 3 displacement components of the generic *k*th node, defined by the absolute reference system, work $\delta^{*}L_{k}$ of the external force acting on the generic *k*th node can be expressed as:

$$\delta^{*}L_{k} = \underline{F}_{k}^{T}\delta^{*}\underline{X}_{k} \tag{4.63}$$

For the sake of convenience, similar to the procedure adopted for the assembly of the total stiffness $[K_{tot}]$ and $[M_{tot}]$ matrices, let us consider a vector $\overline{\underline{F}}_{k}$ having the same dimensions as the number of d.o.f. of the system, consisting of the elements of vector $\underline{F}_k$ placed in correspondence to the d.o.f. relative to the kth node, where the other terms of the same vector are null:

$$\underline{\overline{F}}_k = \begin{Bmatrix} \underline{0} \\ \underline{0} \\ \cdots \\ \underline{F}_k \\ \cdots \\ \underline{0} \\ \underline{0} \end{Bmatrix} \tag{4.64}$$

The virtual work performed by the generic kth concentrated force acting on the generic kth node can be rewritten as:

$$\delta^* L_k = \underline{F}_k^T \delta^* \underline{X}_k = \underline{\overline{F}}_k^T \delta^* \underline{X}_t \tag{4.65}$$

in which $\underline{X}_t$ is the vector that contains all the d.o.f. of the structure (4.42). Let us consider all the mF concentrated forces applied to the structure, the total work is the sum of contributions of each element affected by the concentrated forces:

$$\delta^* L_{tot} = \sum_{k=1}^{mF} \delta^* L_k = \sum_{k=1}^{mF} \underline{\overline{F}}_k^T \delta^* \underline{X}_t = \left\{ \sum_{k=1}^{mF} \underline{\overline{F}}_k^T \right\} \delta^* \underline{X}_t = \underline{F}_t^T \delta^* \underline{X}_t \tag{4.66}$$

Having used $\underline{F}_t$ to define the vector as a sum of the single contributions:

$$\underline{F}_t = \sum_{k=1}^{mF} \underline{\overline{F}}_k \tag{4.67}$$

As seen in Sect. 4.4.5 in relation to the mass and stiffness matrices (Eq. (4.45); Fig. 4.16), this procedure is anything but convenient. The complete vector $\underline{F}_t$ of the forces is defined by directly adding up the single contributions $\underline{\overline{F}}_k$ in correspondence to the d.o.f. relative to the corresponding node (Fig. 4.19), using the methodology already described in Sect. 4.4.5 and in Fig. 4.18, without passing through the expansion of $\underline{F}_k$ in vector $\underline{\overline{F}}_k$.

4.4.6.2 Distributed Forces

In the case in which the finite element is subjected to the distributed forces $\vec{p}(\xi)$ of components $p_v(\xi)$, $p_u(\xi)$, and $p_w(\xi)$ (projected in the local reference system (O_j–X_j–Y_j–Z_j), see Fig. 4.20, the virtual work performed by these forces can be defined as:

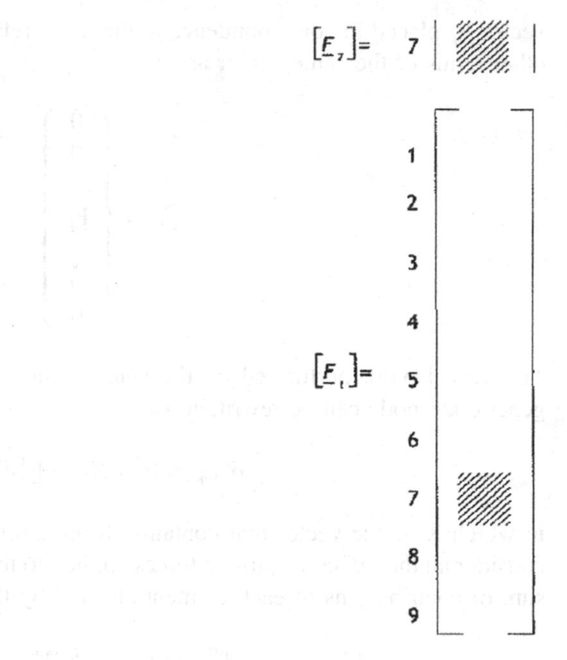

Fig. 4.19 Operative assembly procedure of the vector of the generalized forces $\underline{F}_t$

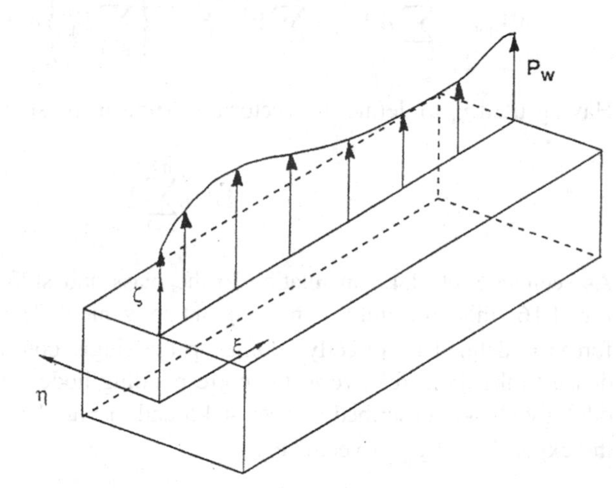

Fig. 4.20 Distributed forces p_w in local reference

$$\delta * L = \int_0^{l_j} p_u(\xi)\delta^* u \, d\xi + \int_0^{l_j} p_v(\xi)\delta^* v \, d\xi + \int_0^{l_j} p_w(\xi)\delta^* w \, d\xi \tag{4.68}$$

By means of the shape form vectors (4.6), (4.5), it is possible to express the virtual displacements in the local reference, as a function of the d.o.f. of the element, always in the local reference:

$$\begin{aligned} \delta^* u &= \underline{f}_u^T(\xi)\delta^*\underline{X}_{jl} \\ \delta^* v &= \underline{f}_v^T(\xi)\delta^*\underline{X}_{jl} \\ \delta^* w &= \underline{f}_w^T(\xi)\delta^*\underline{X}_{jl} \end{aligned} \tag{4.69}$$

Let us now calculate the work performed by the distributed forces as a function of the independent variables nodal displacements $\underline{X}_{jl}$ of the generic *j*th finite element in the local reference system:

$$\delta^* L_j = \left[\int_0^{l_j} p_u(\xi)\underline{f}_u^T(\xi)d\xi + \int_0^{l_j} p_v(\xi)\underline{f}_v^T(\xi)d\xi + \int_0^{l_j} p_w(\xi)\underline{f}_w^T(\xi)d\xi\right]\delta^*\underline{X}_{jl} = \underline{F}_{jl}^T\delta^*\underline{X}_{jl} \tag{4.70}$$

Let us recall that the vectors of the shape functions have as many elements as the d.o.f. of the finite element so that, as a consequence, vector $\underline{F}_{jl}$ defined in (4.70) has the same number of elements. The elements of vector $\underline{F}_{jl}$ assume the meaning of generalized nodal forces (in the local reference system), energetically equivalent to the distributed forces applied to the element itself. By means of the coordinate transformation matrix $[\Lambda_j]$ (4.35) which correlates the independent coordinates relative to the *j*th generic finite element in the local reference system and those in the absolute reference system $\underline{X}_j$ (4.34), the virtual work expressed as a function of the independent coordinates in the global reference is:

$$\delta^* L_j = \underline{F}_{jl}^T\delta^*\underline{X}_{jl} = \underline{F}_{jl}^T[\Lambda_j]\delta^*\underline{X}_{jl} = \underline{F}_j^T\delta^*\underline{X}_j \tag{4.71}$$

where $\underline{F}_j$ is a vector of the dimensions equal to the number of d.o.f. of the generic finite element that contains the generalized forces concentrated at the nodes of the discretized finite element model defined in the global reference system:

$$\underline{F}_j = [\Lambda_j]^T\underline{F}_{jl} \tag{4.72}$$

As already seen in Sect. 4.4.6.1, (4.66), by considering a vector $\overline{\underline{F}}_j$ having the same dimensions as the number of d.o.f. of the system, consisting only of elements of vector $\underline{F}_j$ placed in correspondence to the d.o.f. relative to the *j*th element, in this way, the total work is the sum of the contributions of each element affected by the distributed forces:

$$\delta^* L_t = \sum_{j=1}^{m}\overline{\underline{F}}_j^T\delta^*\underline{X}_t = \left\{\sum_{j=1}^{m}\overline{\underline{F}}_j^T\right\}\delta^*\underline{X}_t = \underline{F}_t^T\delta^*\underline{X}_t \tag{4.73}$$

As already observed in the two previous paragraphs, the assembly of these forces $\underline{\overline{F}}_j$ in the overall vector of the generalized external excitation forces $\underline{F}_t$ is performed by algebraically adding up the single terms of vector $\underline{\overline{F}}_j$ in correspondence to the relative d.o.f. The extension to the calculation of the generalized forces due to concentrated forces applied inside the generic finite element is a particular application of the more general case of distributed forces.

4.4.6.3 Moving Loads

Let us now consider the case, already tackled using the modal method, applied to continuous systems (Chap. 3, Sect. 3.7.1.1.3) of a moving load acting on the structure. To simplify matters, we will refer to a simply supported beam traversed by a moving load of constant module P, travelling with a constant velocity V (Fig. 4.21). The assembly of the mass, stiffness and damping matrix of the system considered is obviously taken for granted, while we will now analyse the question regarding calculation of the Lagrangian component of the moving force. At the generic instant of time t, the load is found inside the jth finite element (see Fig. 4.21): for this reason, the vector of the generalized forces on the free nodes of the structure will obviously be completely null, except in correspondence to the d.o.f. of the end nodes of the generic jth finite element.

For this reason, the work performed by the moving load can be evaluated as a function of the physical variable $w_j(\xi = Vt - \Delta l_{j-1})$ which represents the displacement of the application point of the force itself:

$$\delta^* L = -P\,\delta^* w_j(\overline{\xi} = Vt - \Delta l_{j-1}) \quad (0 \leq \overline{\xi} \leq l_j) \tag{4.74}$$

evaluated at $\overline{\xi}$ (which changes in time),

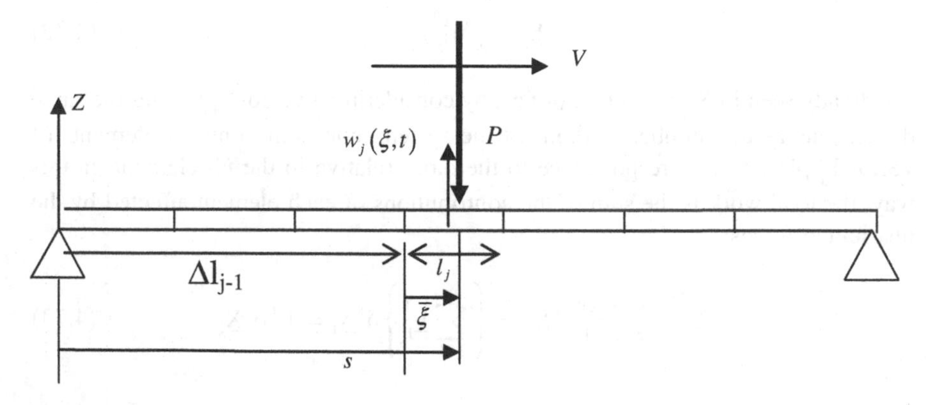

Fig. 4.21 Moving loads on a beam schematized with finite elements

$$w_j(\overline{\xi}) = \underline{f}_w^T(\overline{\xi})\underline{X}_{jl}(t) \tag{4.75}$$

By keeping account of the link between the physical variable w_j and displacement $\underline{X}_{jl}$ of the end nodes of the finite element considered in the local reference system (defined by means of the shape function $\underline{f}_w^T$, see (4.75), Fig. 4.22) (4.74) can be rewritten as:

$$\delta^*L = -P\underline{f}_w^T(\overline{\zeta} = Vt - \Delta l_{j-1})\delta^*\underline{X}_{jl} = \underline{F}_{jl}^T(t)\delta^*\underline{X}_{jl} \quad (0 \leq \overline{\zeta} \leq l_j) \tag{4.76}$$

having used:

$$\underline{F}_{jl}(t) = -P\underline{f}_w(\overline{\zeta}) \tag{4.77}$$

to indicate the vector of the generalized forces in the local reference system.

By keeping account of the transformation from a local reference to a global system (4.78) defined by the coordinate transformation matrix $[\Lambda_j]$(in this particular case an identity matrix):

$$\underline{X}_{jl} = [\Lambda_j]\underline{X}_{jg} \tag{4.78}$$

(4.76) becomes:

$$\delta^*L = \underline{F}_{jl}^T(t)\delta^*\underline{X}_{jl} = \underline{F}_{jl}^T(t)[\Lambda_j]\delta^*\underline{X}_{jg} = \underline{F}_{jg}^T(t)\,\delta^*\underline{X}_{jg} \quad (0 \leq \zeta \leq l_j) \tag{4.79}$$

having used

$$\underline{F}_{jg} = [\Lambda_j]^T\underline{F}_{jl} \tag{4.80}$$

to indicate the vector of the generalized forces with the generic instant of time.

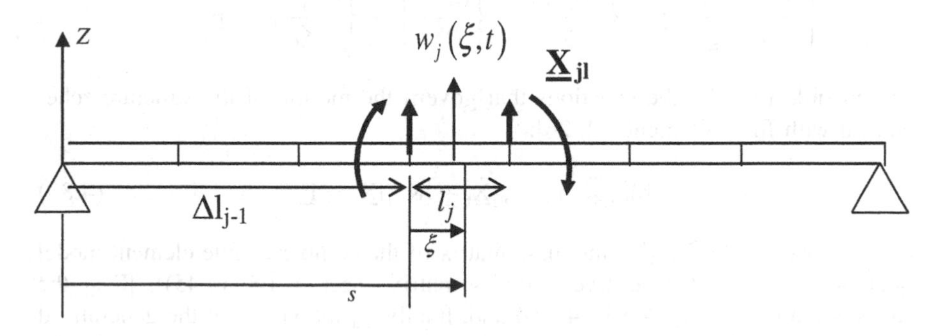

Fig. 4.22 d.o.f. $\underline{X}_{jl}$ of the generic finite element

This vector must subsequently be assembled with vector $\underline{F}_{tot}$ of the overall system. Obviously, because the moving load is travelling along the beam, instant by instant, this will affect the different finite elements, and the vector of the generalized forces, relative to the *j*th generic finite element in which the load finds itself, will also vary over time. The numerical integration of the equations of motion of the system will provide the dynamic response of same to the moving load. Particular attention [8, 9] should be paid in terms of correctly locate the finite elements of the structure, keeping account of the introduction of a fictitious periodicity of the generalized force introduced by the very discretization of the system in finite elements. As far as the approach that uses finite elements is concerned, all the considerations made in Chap. 3, Sect. 3.7.1.1.3 obviously apply, due to the need to often simulate the moving load (in the event of this representing a road or rail vehicle) using a dynamic discrete model.

4.4.7 Imposition of Constraints (Boundary Conditions)

Thus, as shown in Sect. 4.4.5, having obtained the stiffness $[K_{tot}]$, mass $[M_{tot}]$ and damping $[R_{tot}]$ matrices of the total structure and having defined the vector of the Lagrangian components of non-conservative forces $\underline{F}_t$, as shown in Sect. 4.4.6, it is possible to express the various forms of energy as a function of the sole independent coordinates represented by nodal displacements:

$$\begin{aligned} E_{ctot} &= \frac{1}{2}\dot{\underline{X}}_t^T [M_{tot}]\dot{\underline{X}}_t \\ V_{tot} &= \frac{1}{2}\underline{X}_t^T [K_{tot}]\underline{X}_t \\ D_{tot} &= \frac{1}{2}\dot{\underline{X}}_t^T [R_{tot}]\dot{\underline{X}}_t \end{aligned} \quad (4.81)$$

In this way, by applying the Lagrange equations (Sect. 4.3, (4.15)):

$$\left\{\frac{d}{dt}\left(\frac{\partial E_{ctot}}{\partial \dot{\underline{X}}_t}\right)\right\}^T - \left\{\frac{\partial E_{ctot}}{\partial \underline{X}_t}\right\}^T + \left\{\frac{\partial D_{tot}}{\partial \dot{\underline{X}}_t}\right\}^T + \left\{\frac{\partial V_{tot}}{\partial \underline{X}_t}\right\}^T = \underline{F}_t \quad (4.82)$$

it is possible to write the equations that govern the motion of the structure schematized with finite elements [2, 26]:

$$[M_{tot}]\ddot{\underline{X}}_t + [R_{tot}]\dot{\underline{X}}_t + [K_{tot}]\underline{X}_t = \underline{F}_t \quad (4.83)$$

where, as we recall, $[M_{tot}]$ is the mass matrix of the complete finite element model (Sect. 4.4.5), $[K_{tot}]$ the relative stiffness matrix (Sect. 4.4.5, (4.45)), $[R_{tot}]$ the damping matrix (Sect. 4.4.5.1, (4.47)) and, finally, $\underline{F}_t$ the vector of the generalized external forces of the discretized finite element model (Sects. 4.4.6.1 and 4.4.6.2).

Equations (4.83) represent the equations of motion of the system analysed, discretized with finite elements, still free in space, since, as yet, no constraint condition has been imposed: for this reason, for example, the overall stiffness matrix $[K_{tot}]$ is singular, in that, at this point, rigid motions of the structure are still possible. Having chosen the constraints of the structure towards the outside world, it is necessary to re-number the equations of the system (4.83) by reorganizing the vector of the generalized coordinates $\underline{X}_t$ in a form divided as follows:

$$\underline{X}_t = \begin{Bmatrix} \underline{X}_L \\ \underline{X}_V \end{Bmatrix} \tag{4.84}$$

having used:

- $\underline{X}_L$ to indicate the vector relative to the actual d.o.f. of the structure;
- $\underline{X}_V$ to indicate the vector relative to the "constrained" d.o.f., i.e. of those d.o.f. whose displacement value is known on account of having been imposed.

The displacements of the constraints, defined by the terms present in vector $\underline{X}_V$, can be specifically:

- null, in the case of structures with fixed constraints;
- different from zero in the case of motion imposed on the constraints as, for example, occurs in simulations of the behaviour of a structure with seismic excitation.

Given the new order imposed on the independent variables in vector $\underline{X}_t$ (4.84), the global matrices $[M_{tot}]$, $[R_{tot}]$, and $[K_{tot}]$ must be re-ordered in such a way that the system of initial equations (4.83) coincides with that obtained after the renumbering of the variable inside vector $\underline{X}_t$: this can be obtained by re-ordering the lines and the columns of the initial matrices with the same new order associated with vector $\underline{X}_t$ (4.84). In this way, the re-ordered matrices can be assumed, in turn, as divided into sub-matrices and the initial equations (4.83) can be rewritten as[10] [2, 10, 13, 22]:

$$\begin{bmatrix} [M_{LL}] & [M_{LV}] \\ [M_{VL}] & [M_{VV}] \end{bmatrix} \begin{Bmatrix} \ddot{\underline{X}}_L \\ \ddot{\underline{X}}_V \end{Bmatrix} + \begin{bmatrix} [R_{LL}] & [R_{LV}] \\ [R_{VL}] & [R_{VV}] \end{bmatrix} \begin{Bmatrix} \dot{\underline{X}}_L \\ \dot{\underline{X}}_V \end{Bmatrix} + \begin{bmatrix} [K_{LL}] & [K_{LV}] \\ [K_{VL}] & [K_{VV}] \end{bmatrix} \begin{Bmatrix} \underline{X}_L \\ \underline{X}_V \end{Bmatrix} = \begin{Bmatrix} \underline{F}_L \\ \underline{F}_V + \underline{R}_V \end{Bmatrix} = \begin{Bmatrix} \underline{F}_L \\ \underline{F}_V \end{Bmatrix} \tag{4.85}$$

[10] From a practical operative point of view, the assembly of the different structural matrices and of the vector of the generalized external forces is, in actual fact, always performed directly in the split form of Eq. (4.96) without passing through vector $\underline{X}_t$.

In (4.85):

- matrix $[M_{LL}]$ thus represents the inertia forces on free d.o.f. $\underline{X}_L$ due to a unitary acceleration $\ddot{\underline{X}}_L$ of the d.o.f. themselves;
- matrix $[M_{LV}]$ represents the inertia forces on free d.o.f. $\underline{X}_L$ due to a unit acceleration $\ddot{\underline{X}}_V$ imposed on the constrained d.o.f.;
- matrix $[M_{VL}]$ represents the inertia forces on the constrained d.o.f. $\underline{X}_V$ due to a unit acceleration $\ddot{\underline{X}}_L$;
- matrix $[M_{VV}]$ represents the inertia forces on the constrained d.o.f. $\underline{X}_V$ due to a unit acceleration $\ddot{\underline{X}}_V$ and so forth.

In the same equation:

- $\underline{F}_L$ is used to indicate the vector (known) of the generalized forces acting on the free d.o.f. due to active external stresses;
- $\underline{F}_V$ is used to indicate the vector (known) of the generalized forces acting on the constrained d.o.f. due to the active external stresses (these forces will be off-loaded directly on the constraints);
- $\underline{R}_V$ is used to indicate the vector (unknown) of the restrained reactions.

Using this formalism, it is possible to highlight the unknowns of the problem, i.e. the displacement of the independent d.o.f. $\underline{X}_L$ (and their derivatives) and the restrained forces $\underline{R}_V$ contained in vector $\underline{F}_V$ together with the part of generalized forces relative to the constrained d.o.f. $\underline{F}_V$. For a better manipulation of these equations it is convenient to consider this system as decomposed into two matrix subsystems.

$$[M_{LL}]\ddot{\underline{X}}_L + [M_{LV}]\ddot{\underline{X}}_V + [R_{LL}]\dot{\underline{X}}_L + [R_{LV}]\dot{\underline{X}}_V + [K_{LL}]\underline{X}_L + [K_{LV}]\underline{X}_V = \underline{F}_L \quad (4.86)$$

$$[M_{VL}]\ddot{\underline{X}}_L + [M_{VV}]\ddot{\underline{X}}_V + [R_{VL}]\dot{\underline{X}}_L + [R_{VV}]\dot{\underline{X}}_V + [K_{VL}]\underline{X}_L + [K_{VV}]\underline{X}_V = \underline{F}_V + \underline{R}_V \quad (4.87)$$

The product of vector $\underline{X}_V$ and its derivatives for the respective matrices $[K_{LV}]$, $[R_{LV}]$, and $[M_{LV}]$, which appear in (4.86), represent the forces acting on the unrestrained d.o.f. of the structure due to the motion imposed on the constraints: these forces are known in that the displacements of constraints $\underline{X}_V$ themselves are known. For this reason, expressions (4.86) and (4.87) can be rewritten as:

$$[M_{LL}]\ddot{\underline{X}}_L + [R_{LL}]\dot{\underline{X}}_L + [K_{LL}]\underline{X}_L = \underline{F}_L - [M_{LV}]\ddot{\underline{X}}_V - [R_{LV}]\dot{\underline{X}}_V - [K_{LV}]\underline{X}_V = \overline{\underline{F}}_L \quad (4.88)$$

$$\underline{R}_V = [M_{VL}]\ddot{\underline{X}}_L + [M_{VV}]\ddot{\underline{X}}_V + [R_{VL}]\dot{\underline{X}}_L + [R_{VV}]\dot{\underline{X}}_V + [K_{VL}]\underline{X}_L + [K_{VV}]\underline{X}_V - \underline{F}_V \quad (4.89)$$

where $\overline{\underline{F}}_L$ is the vector containing the overall forces acting on the structure, inclusive of both the applied external forces $\underline{F}_L$ and the effects of the displacements of the constraints (this result was already highlighted in the systems with 2 d.o.f., Chap. 2, Sect. 2.4.3.4, subjected to constraint displacements). In Eq. (4.88) $[M_{LL}]$, $[R_{LL}]$, and $[K_{LL}]$ represent the overall matrices of the unrestrained d.o.f.: these matrices, if referred to perturbed motion about the static equilibrium position, assumed stable, of a dissipation system, prove to be symmetric and are positive definite. Having assigned the imposed motion to constraints $\underline{X}_V$ and having evaluated the generalized forces $\underline{F}_L$ acting on the structure, by solving Eq. (4.88), it is possible to calculate the response of the system in terms of displacements relative to unrestrained d.o.f. $\underline{X}_L$. Having introduced this solution into (4.89) and since the motion imposed on the constraints is known, it is possible to evaluate the constraint reactions $\underline{R}_V$. The two matrix Eqs. (4.88) and (4.89) obtained in this way are nothing other but the generalized equations of dynamic equilibrium relative to the free nodes (4.88) and to the constrained nodes (4.89). Let us now return to the example considered (Fig. 4.11): in this case, the d.o.f. relative to the constraints are:

$$\underline{X}_V^T = \{X_1 \quad Z_1 \quad X_9 \quad Z_9\} \tag{4.90}$$

while those of the free nodes are the remaining ones:

$$\underline{X}_L^T = \{X_2 \quad Z_2 \quad \dots \quad X_7 \quad Z_7 \quad X_8 \quad Z_8\} \tag{4.91}$$

It is, thus, possible to renumber vector $\underline{X}_t$ (4.42) as:

$$\underline{X}_t = \begin{Bmatrix} \underline{X}_L \\ - \\ \underline{X}_V \end{Bmatrix} \underline{X}_L^T = \begin{Bmatrix} X_2 \\ Z_2 \\ \dots \\ X_8 \\ Z_8 \\ - \\ X_1 \\ Z_1 \\ X_9 \\ Z_9 \end{Bmatrix} \tag{4.92}$$

The overall stiffness matrix of the free structure in space, already assembled in Sect. 4.4.5, ((4.45); Fig. 4.15), must be re-ordered $[K_{tot}]$, keeping account of the new order of variables in the divided vector $\underline{X}_t$ (4.92) as shown in Fig. 4.23. The same division must be made to obtain the damping $[R_{tot}]$ and mass $[M_{tot}]$ matrix and the vector of the generalized forces $\underline{F}_t$.

It goes without saying that the assembly of the single matrices of stiffness and mass, as well as the generalized forces on the nodes, can be performed by directly assembling the matrices of the single finite elements $[M_j]$, $[K_j]$, and $\underline{F}_j$ (defined in

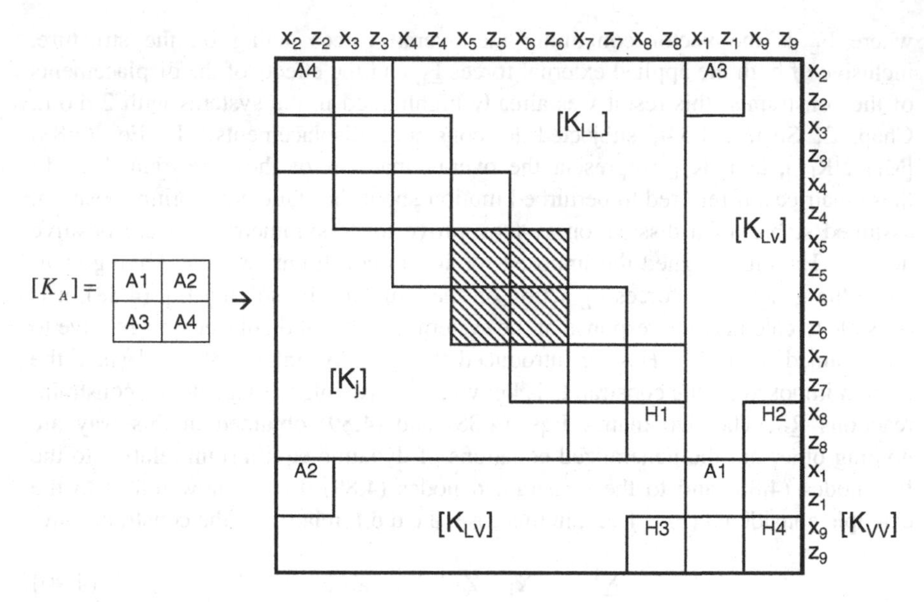

Fig. 4.23 Division of the stiffness matrix

the absolute reference system, Sect. 4.4.4, (4.39) and (4.40) in matrices $[M_{tot}]$, $[K_{tot}]$, and in vector $\underline{F}_t$ by making direct reference to the new numbering of vector $\underline{X}_t$ of (4.92).

4.4.8 Solving the Equations of Motion

Having defined the methods for the calculation of the generalized forces, in the most usual cases (concentrated and distributed forces), we will now consider all the different types of linear analysis that can be tackled using the finite element method and more specifically:

- Static linear analyses (Sect. 4.4.8.1);
- Dynamic analyses
  - (a) calculation of natural frequencies and normal modes (Sects. 4.4.8.2.1 and 4.4.8.2.2);
  - (b) steady-state response of a structure with an applied external excitation force (Sect. 4.4.8.2.3);
  - (c) steady-state response of a structure with impressed constraint displacements.

The analysis of the nonlinear, static and dynamic type, will be dealt with in Sect. 4.7. In particular, we will show how to tackle static nonlinear problems while for nonlinear dynamic analyses we suggest that you refer to Chap. 5, relative to nonconservative force fields (see also [2, 4, 18, 21]).

4.4.8.1 Static Linear Analyses

In order to analyse the static behaviour of the structure schematized with finite elements subjected to constant external forces $\underline{F}_L = \underline{F}_0$, it is necessary to impose the constraints with the outside world on same:

$$\ddot{\underline{X}}_V = \dot{\underline{X}}_V = \underline{X}_V = 0 \tag{4.93}$$

and, since we are carrying out a static analysis, we will also have[11]:

$$\begin{aligned} \ddot{\underline{X}}_L &= \ddot{\underline{X}} = 0 \\ \dot{\underline{X}}_L &= \dot{\underline{X}} = 0 \end{aligned} \tag{4.94}$$

By keeping account of (4.94) and (4.95), the equations that allow us to evaluate the response of the structure and the relative constraint reactions (4.88), (4.89) are reduced, in this case, to:

$$\begin{aligned} [K_{LL}]\underline{X}_0 &= \underline{F}_0 \Rightarrow [K]\underline{X}_0 = \underline{F}_0 \Rightarrow \underline{X}_0 = [K]^{-1}\underline{F}_0 \\ \underline{R}_{0V} &= -\underline{F}_{0V} + [K_{VL}]\underline{X}_0 \end{aligned} \tag{4.95}$$

Once the generalized forces F_0 applied to the free nodes are known, the first matrix equation enables us to obtain the static displacements $\underline{X}_0$ to which this is subjected, while the second, having obtained displacements $\underline{X}_0$ from (4.95), enables us to determine the constraint reactions $\underline{R}_{0V}$. Let us remember that if, as a shape function, we have assumed the static deformation of the element subject to forces applied only at the nodes, the static solution shown here is rigorous when the forces are applied only at the nodal points.

4.4.8.2 Dynamic Analyses

4.4.8.2.1 Calculation of the Natural Frequencies and Normal Modes

To study the undamped free motion of the structure it is necessary to trace back the complete equations (4.86), as seen for discrete *n* d.o.f. systems, to the characteristic homogenous equation:

[11] Further on and unless otherwise indicated, in order not to complicate the symbology, we will omit the subscripts meaning that we will thus use $\underline{X}_L = \underline{X}$, $\underline{F}_L = \underline{F}$, $[K_{LL}] = [K]$, $[M_{LL}] = [M]$, etc.

$$[\mathrm{M}]\ddot{\underline{\mathrm{X}}} + [\mathrm{K}]\underline{\mathrm{X}} = \underline{0} \tag{4.96}$$

The calculation of the natural frequencies of system ω_i and of the relative normal modes $\underline{\mathrm{X}}^{(\mathrm{i})}$ is traced back, as already seen for 2-n d.o.f. systems, Chap. 2, Sect. 2.3.1, to the calculation of the eigenvalues and the eigenvectors of matrix $[\mathrm{M}]^{-1}[\mathrm{K}]$. In the event of our wishing to study the damped free motion of the structure, it is necessary to evaluate the solutions of the characteristic homogenous equation:

$$[\mathrm{M}]\ddot{\underline{\mathrm{X}}} + [\mathrm{R}]\dot{\underline{\mathrm{X}}} + [\mathrm{K}]\underline{\mathrm{X}} = \underline{0} \tag{4.97}$$

in other words, it is necessary to obtain the natural frequencies of system $\lambda_i = \alpha_i + i\omega_i$ (complex conjugates) and the relative normal modes $\underline{\mathrm{X}}^{(\mathrm{i})}$. It is possible to trace this problem back (always see Chap. 2, Sect. 2.4.2.2.1) to the calculation of the eigenvalues and eigenvectors of matrix $[\mathrm{A}] = -[\mathrm{B}]^{-1}[\mathrm{C}]$, having used $[\mathrm{B}]$ and $[\mathrm{C}]$ to define:

$$[\mathrm{B}] = \begin{bmatrix} [\mathrm{M}] & [\mathrm{R}] \\ [0] & [\mathrm{I}] \end{bmatrix}; \quad [\mathrm{C}] = \begin{bmatrix} [0] & [\mathrm{K}] \\ -[\mathrm{I}] & [0] \end{bmatrix} \tag{4.98}$$

4.4.8.2.2 Response of the Structure to External Applied Excitation Forces

Let us now consider the case of a structure, with fixed constraints (i.e. with $X_v = 0$), subjected to external concentrated or distributed excitation forces generally of time: in this case it is necessary to once again return to (4.86) in its complete form [3]:

$$[\mathrm{M}]\ddot{\underline{\mathrm{X}}} + [\mathrm{R}]\dot{\underline{\mathrm{X}}} + [\mathrm{K}]\underline{\mathrm{X}} = \underline{\mathrm{F}}(\mathrm{t}) \tag{4.99}$$

$$[\mathrm{M}_{\mathrm{VL}}]\ddot{\underline{\mathrm{X}}} + [\mathrm{R}_{\mathrm{VL}}]\dot{\underline{\mathrm{X}}} + [\mathrm{K}_{\mathrm{VL}}]\underline{\mathrm{X}} - \underline{F}_{\mathrm{V}}(t) = \underline{\mathrm{R}}_{\mathrm{V}} \tag{4.100}$$

The solution of (4.99) can be obtained analytically or, more frequently, numerically: having thus defined the values of $\underline{\mathrm{X}}$, $\dot{\underline{\mathrm{X}}}$, and $\ddot{\underline{\mathrm{X}}}$, the constraint reactions $\underline{\mathrm{R}}_{\mathrm{V}}$ can be obtained from (4.100). In the specific case of applied harmonic excitation forces:

$$\underline{F} = \underline{F}(t) \Rightarrow \underline{F} = \underline{F}_o e^{i\Omega t}; \quad \underline{F}_V = \underline{F}_V(t) \Rightarrow \underline{F}_V = \underline{F}_{Vo} e^{i\Omega t} \tag{4.101}$$

the steady-state solution will be of the following type:

$$\underline{\mathrm{X}} = \underline{\mathrm{X}}_0 e^{\mathrm{i}\Omega \mathrm{t}} \tag{4.102}$$

which when placed in (4.98) gives:

$$\underline{X}_0 = \left[-\Omega^2[M] + i\Omega[R] + [K]\right]^{-1}\underline{F}_0 \tag{4.103}$$

By substituting (4.102) in (4.99) it is possible to evaluate the constraint reactions which, in complex terms, are given by the following relation:

$$\left\{\left[-\Omega^2[M_{VL}] + i\Omega[R_{VL}] + [K_{VL}]\right]\underline{X}_0 - \underline{F}_{V0}\right\}e^{i\Omega t} = \underline{R}_{V0}e^{i\Omega t} \tag{4.104}$$

Solution $\underline{X}_0$ of (4.103) is a vector of complex numbers: the module of each element represents the vibration amplitude of the generic d.o.f., while the phase represents the phase lag with respect to the excitation force.

4.4.8.2.3 Response of the Structure to Imposed Constraint Displacements

If we assume that the constraints (or some of them) are subjected to imposed displacements, i.e. $\underline{X}_V \neq \underline{0}, \dot{\underline{X}}_V \neq \underline{0}, \ddot{\underline{X}}_V \neq \underline{0}$, from (4.86) and (4.87) we will obtain:

$$[M]\ddot{\underline{X}} + [R]\dot{\underline{X}} + [K]\underline{X} = -[M_{LV}]\ddot{\underline{X}}_V - [R_{LV}]\dot{\underline{X}}_V - [K_{LV}]\underline{X}_V = \overline{\underline{F}}_L \tag{4.105}$$

$$\underline{R}_V = [M_{VL}]\ddot{\underline{X}} + [M_{VV}]\ddot{\underline{X}}_V + [R_{VL}]\dot{\underline{X}} + [R_{VV}]\dot{\underline{X}}_V + [K_{VL}]\underline{X} + [K_{VV}]\underline{X}_V - \underline{F}_V \tag{4.106}$$

where the terms $\overline{\underline{F}}_L$ on the right hand side of the equal sign in the first matrix equation are all known since the displacements $\underline{X}_V$ imposed on the constraints are known. The solution of these equations can thus be traced back to the solutions of the equations of the previous case.

4.4.9 A Numerical Example

For the mathematical model, discretized with finite elements, of a span of a high tension line already described in Sect. 4.4, Figs. 4.10 and 4.11, using the procedures previously described, it is thus possible to calculate the natural frequencies ω_i and the response of the system to an external excitation.

The first natural frequencies are shown in Fig. 4.24 and a marked continuous line is used to show the associated modes (in the vertical plane) evaluated in the absence of damping. As can be seen, the first normal mode does not show an antinode in the middle: this is due to the stiffening caused by the particular geometric configuration assumed by the cable in the static equilibrium position (or rather by the presence of the coordinate transformation matrices $[\Lambda_j]$ of the single finite elements that introduce axial stiffness effect of the cable into the vertical plane). The same model was

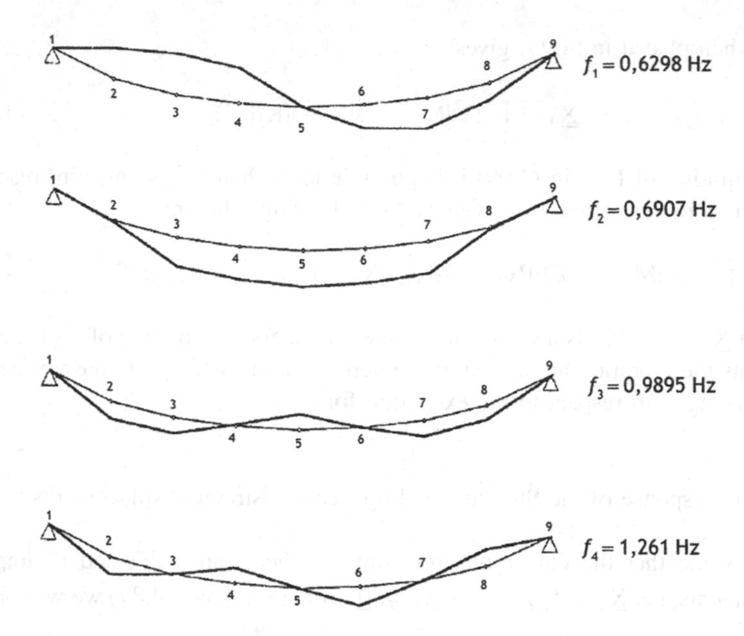

Fig. 4.24 Deformations of the first four vertical normal modes calculated

subjected to a sinusoidal excitation force $F_7 = |F_7|e^{i\Omega t}$ with a constant amplitude ($|F_7| = 100$ da N) applied to node 7 of the model (see Fig. 4.11) in a field of frequencies $0\,\text{Hz} \le {}^{\Omega}/_{2\pi} \le 1.5\,\text{Hz}$. The damping matrix, defined as reported in (4.48) was assumed in order to obtain a critical damping ${}^{r}/_{r_c}$ equal to approximately 1–2 % in the field of frequencies analysed. As an example, Fig. 4.25 shows the trend, resulting from the variation of the excitation frequency Ω, of module $|Z_3|$ and of phase ϕ_3 of the vertical displacement of node 3 where $Z_3(t) = |Z_3|\cos(\Omega t + \phi_3)$: the frequency response shows three resonance peaks, corresponding to the first three natural frequencies excited by the force applied in the field of frequencies considered (note the simultaneous shift to 90° of the phase angle).

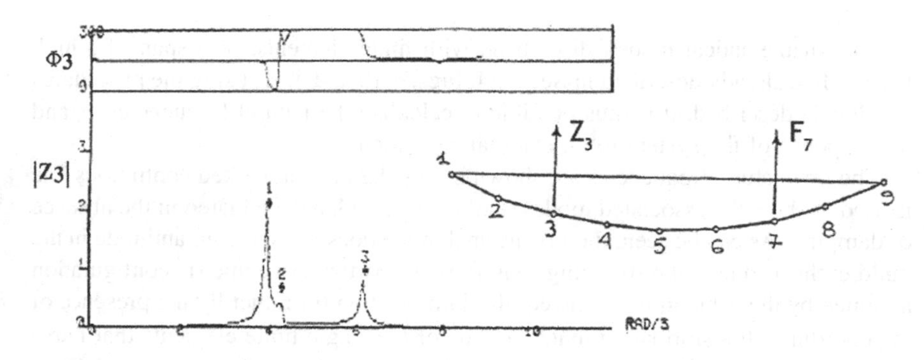

Fig. 4.25 Frequency response of the vertical d.o.f.: node 3 of the structure

Figure 4.26 shows the trend of module $|Z_5|$ and of the relative displacement phase ϕ_5 of the central point (node 5, see Fig. 4.11) when the frequency varies. As it was possible to foresee from an analysis of the natural frequencies and, above all, of the relative normal modes of the structure shown in Fig. 4.24, only two peaks are visible for point 5, that correspond to the second and third natural frequency.

Conversely, a peak in correspondence to the first frequency is not shown in the graph because as far as this is concerned point 5 is a nodal point (see Fig. 4.24). Always for the forced case considered, Fig. 4.27 shows the trend of module $|Z_2|$ and $|Z_8|$ and of phases ϕ_2 and ϕ_8 (relative to the excitation force) of the vertical

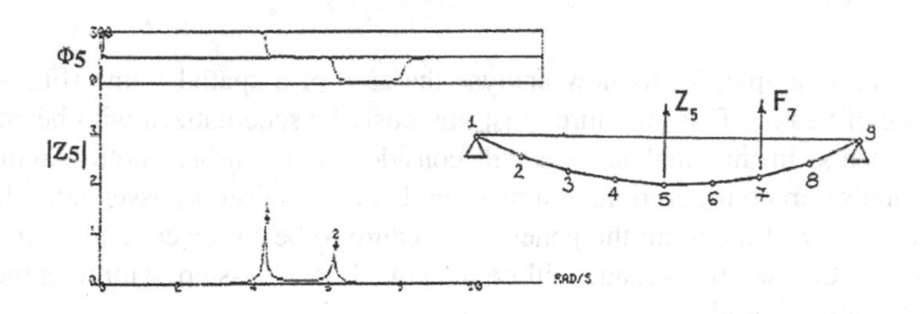

Fig. 4.26 Frequency response of the vertical d.o.f.: node 5 of the structure

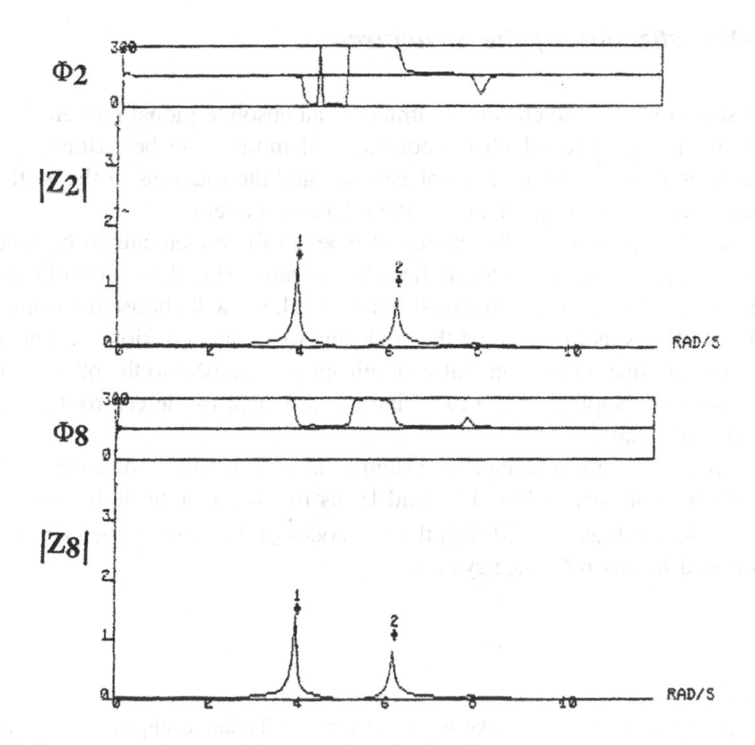

Fig. 4.27 Frequency response of the vertical d.o.f. of nodes 2 and 8 of the structure

displacements of node 2 and 8. The transition through the first natural frequency occurs with a relative phase shift $\phi_r = \phi_2 - \phi_8$ of approximately 180°, while at the second resonance (corresponding to the third normal mode) the relative phase ϕ_r is null: as can be verified, this is congruent with the deformations of the normal modes shown in Fig. 4.24: in fact, in the first normal mode Z_2 and Z_8 are in counter-phase ($\phi_r = 180°$), while in the third mode Z_2 and Z_8 are in phase with each other ($\phi_r = 0$).

4.5 An Application Example: Finite Beam Element

As a second example, let us now analyse the case of a spatial frame (Fig. 4.28), consisting of beams. This structure must obviously be schematized with beam-type finite elements. In this analysis, we will consider the perturbed motion about the static equilibrium configuration, assumed, in this application, as assigned.[12] In this paragraph, we will illustrate the general procedure to be followed, and to simplify matters, the considerations made will be integrated, step-by-step, with references to the pilot example used.

4.5.1 Discretization of the Structure

The first step concerns the choice (arbitrary) of an absolute global reference system (X–Y–Z) with respect to which the equations of motion can be written: as independent variables we assume the displacements and the rotations of the sections in correspondence to the single nodes in the reference system.

The second step concerns the choice of types of finite elements to be used and the choice of the number of nodes in the schematization (i.e. the choice of the mesh of the model). In this case, as previously mentioned, we will choose the beam finite element that allows us to represent the static and dynamic behaviour of a prismatic rod (a body with one dimension that is dominant with respect to the other two) that reacts at moments applied in the two planes orthogonal to the centroid axis, both torsionally and axially.

Let us consider a local reference system of axes x, h and z connected with the generic beam as shown in Fig. 4.29 and let us use $\underline{X}_{jl}$ to indicate the vector that contains the 12 d.o.f. associated with the end nodes of the beam (dependent on time alone) defined in this reference system:

[12]Suitable procedures to define an unknown initial static equilibrium configuration of a structure will be illustrated in Sect. 4.7, valid for structures of linear or nonlinear behaviour.

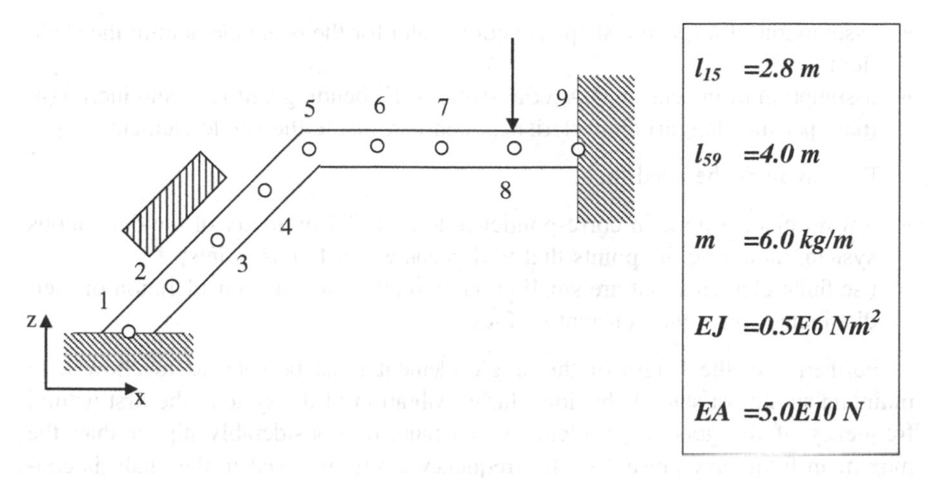

Fig. 4.28 Spatial frame: finite element schematization

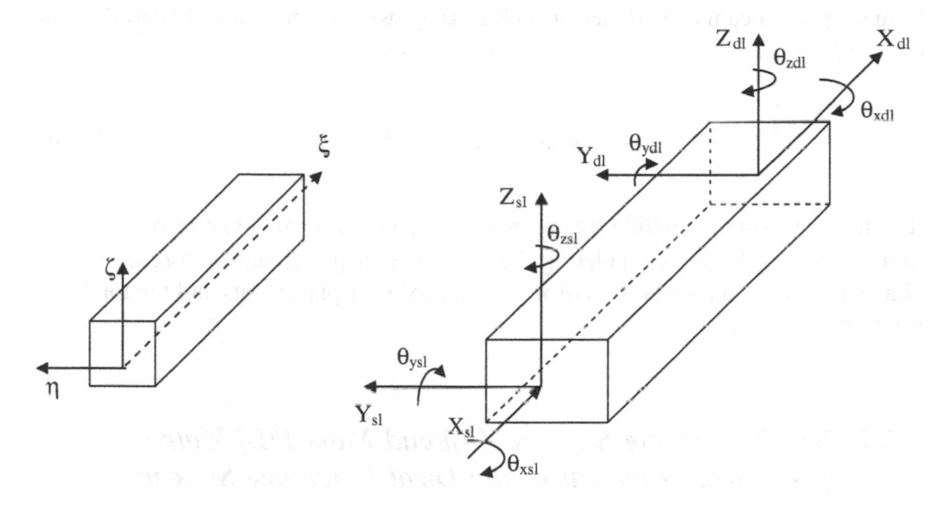

Fig. 4.29 Nodal d.o.f. of the beam element

$$\underline{X}_{jl}^{T} = \left\{ X_{sl} \quad Y_{sl} \quad Z_{sl} \quad \theta_{xsl} \quad \theta_{ysl} \quad \theta_{zsl} \quad X_{dl} \quad Y_{dl} \quad Z_{dl} \quad \theta_{xdl} \quad \theta_{ydl} \quad \theta_{zdl} \right\} \tag{4.107}$$

It will then, subsequently, be necessary to divide the structure into finite elements, identifying the position of the nodes, i.e. of the connecting points between adjacent elements. The choice of the number of elements and thus of the number of nodes of the mesh must be such to ensure that the schematization used does not have any shortcomings, i.e. the hypotheses, adopted in the definition of the single finite elements, must be satisfied. These can be summarized as:

- assumption of a specific shape function, valid for the beam element in the static field;
- assumption of the elastic (EA axial stiffness, EJ bending stiffness) and inertia (m mass per unit length) characteristics, constant inside the single element.

This involves the need to:

- always place a node in correspondence to each discontinuity of the continuous system, including the points that will become constraints points;
- use finite elements that are small enough to allow for the consideration of their distributed mass and constant stiffnesses.

Furthermore, the length of the single element must be such as to allow it to maintain an almost static behaviour during vibration of the system: the first natural frequency of the generic *j*th element ω_{elj} must be considerably higher than the maximum frequency present in the frequency range involved in the analysis conducted $0 \leq \Omega \leq \Omega_{max}$. In the case of the beam (and for bending motion), as a first approximation formula, it is possible to use the one that defines the first natural frequency of a beam constrained at both ends by two supports (already described in Sect. 3.3):

$$\omega_{\mathrm{elj}} = \left(\frac{\pi}{\mathrm{l_j}}\right)^2 \sqrt{\frac{\mathrm{EJ_j}}{\mathrm{m_j}}} \tag{4.108}$$

It is now necessary to define the stiffness $[K_{jl}]$ and mass $[M_{jl}]$ matrix of the beam finite element in space, considering the transverse displacements according to two planes parallel to the main axes of inertia, the axial displacements and the torsional rotations.[13]

4.5.2 Definition of the Stiffness $[K_l]$ and Mass $[M_l]$ Matrix of the Beam Element in the Local Reference System

In order to define the stiffness $[K_{jl}] = [K_l]$ and mass $[M_{jl}] = [M_l]$ matrix of the generic beam finite element it is first necessary to define a local reference system: as previously mentioned, for reasons of convenience we will assume a specific reference system, with axis x parallel to the longitudinal axis of the beam itself and axes η and ζ perpendicular to the first and parallel to the main axes of inertia of the section (Fig. 4.29). This choice is conditioned by being able to consider the bending

[13] In the part that follows, as with the beam, in order not to complicate the matrix notation, when referring to amplitudes relative to the generic *j-n*th beam, we will use the subscript "j" also necessary to highlight the fact that these same amplitudes change when the element considered varies (for example $\mathrm{l_j = l; EA_j = EA; \ldots; \underline{X}_{jl} = \underline{X}_l; \ldots; [K_{jl}] = [K]}$ etc.).

separately in the two planes ξ–η and ξ–ζ. Additionally, we will also consider the case of the centroid of the section coincident with the bending centre in order to consider the bending motions uncoupled from the torsional motions. Finally, by assuming small displacements and rotations, we will also assume the axial motions uncoupled. By choosing this reference system and based on these assumptions it is possible to consider separately:

- the two bending motions in two planes (Fig. 4.30) defined respectively by two space-time functions $w(\xi, t) = w_j(\xi, t)$ and $v(\xi, t) = v_j(\xi, t)$;
- the torsional motion described by function $\theta_x(x, t) = \theta_{xj}(x, t)$;
- the axial motion, defined by function $u(\xi, t) = u_j(\xi, t)$.

To simplify matters we will subdivide vector $\underline{X}_{jl} = \underline{X}_j$ (4.1) which contains the twelve d.o.f. of the two end nodes of the beam, into four sub-vectors: let us thus use $\underline{X}_w$ to define the vector relative to only those d.o.f. affected by transverse displacements in plane (ξ–ζ) (Fig. 4.30):

$$\underline{X}_w = \begin{Bmatrix} Z_{sl} \\ \theta_{ysl} \\ Z_{dl} \\ \theta_{ydl} \end{Bmatrix} \tag{4.109}$$

$\underline{X}_v$ the vector relative to only those d.o.f. affected by the transverse displacements in plane (ξ–η) (Fig. 4.30):

$$\underline{X}_v = \begin{Bmatrix} Y_{sl} \\ \theta_{zsl} \\ Y_{dl} \\ \theta_{zdl} \end{Bmatrix} \tag{4.110}$$

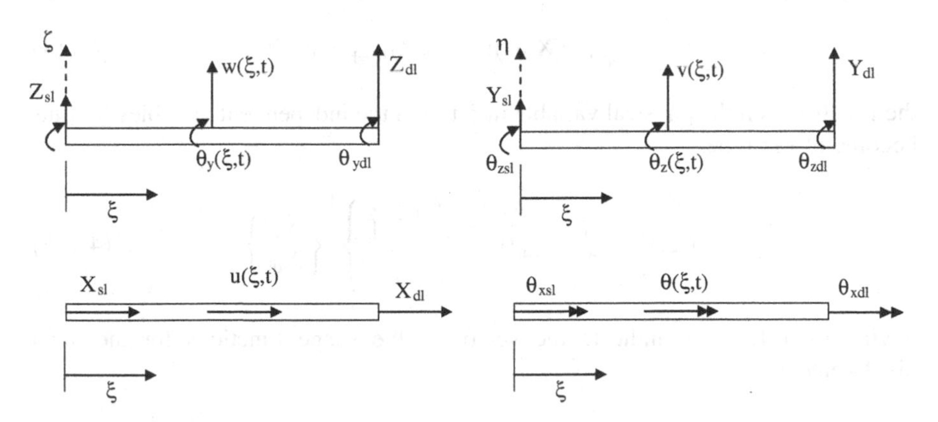

Fig. 4.30 Definition of the d.o.f. of the beam in different directions

$\underline{X}_u$ the vector of the axial displacements (Fig. 4.30):

$$\underline{X}_u = \begin{Bmatrix} X_{sl} \\ X_{dl} \end{Bmatrix} \tag{4.111}$$

and $\underline{X}_{\theta x}$ the vector of the torsional rotations around axis x (Fig. 4.30):

$$\underline{X}_{\theta x} = \begin{Bmatrix} \theta_{xsl} \\ \theta_{xdl} \end{Bmatrix} \tag{4.112}$$

First and foremost, it is necessary to define the link between the physical variables $w(\xi,t)$, $v(\xi,t)$, $\theta_x(\xi,t)$ and $u(\xi,t)$, dependent on both space and time, that define the generic deformation (axial, torsional and bending) inside the finite element as a function of the displacements of the end nodes $\underline{X}_{jl}$ assumes as independent variables.

4.5.2.1 Shape Function

4.5.2.1.1 Shape Functions for Axial Motion $\underline{f}_u(\xi)$

Similar to the procedure used for the cable, Sect. 4.2.1, in the case of the beam, in order to describe axial motion we assumed the physical coordinate $u(\xi,t)$, while X_{sl} and X_{dl} represent the longitudinal displacements of respectively the left and right node of the generic element grouped together in a vector $\underline{X}_u$ (4.111) (see Fig. 4.31). The deformation assumed by the continuous system inside the generic finite element is still of the type (4.2):

$$u(\xi,t) = a_2\xi + b_2 \tag{4.113}$$

By imposing the boundary conditions, of the type (4.3):

$$u(\xi,t)|_{\xi=0} = X_{sl}(t); \quad u(\xi,t)|_{\xi=l_j} = X_{dl}(t) \tag{4.114}$$

the link between the physical variable $u(\xi,t)$ and the independent variables $\underline{X}_u$ thus become of the type:

$$u(\xi,t) = \underline{f}_u^T(\xi)\underline{X}_u(t) = \begin{Bmatrix} 1 - \frac{\xi}{l_j} \\ \frac{\xi}{l_j} \end{Bmatrix}^T \begin{Bmatrix} X_{sl} \\ X_{dl} \end{Bmatrix} \tag{4.115}$$

having used $\underline{f}_u(\xi)$ to indicate the vector of the shape functions for the axial displacements.

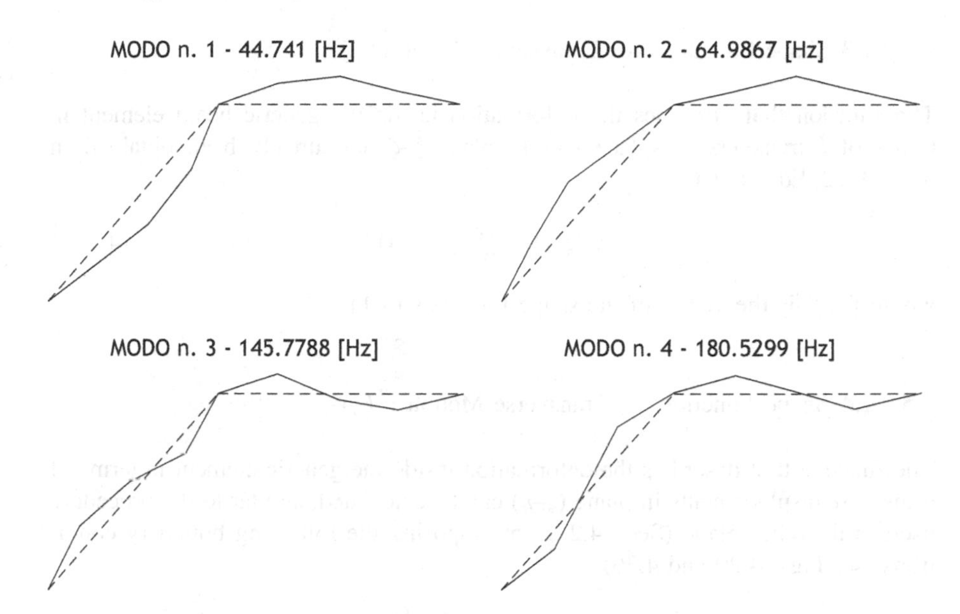

Fig. 4.31 Deformations of the first four vertical normal modes calculated

4.5.2.1.2 Shape Functions for Torsional Motion $\underline{f}_{\theta x}^{T}(\xi)$

The torsional rotations to which the generic section of the beam, positioned at ξ, is subjected are defined by function $\theta_{xj}(\xi, t) = \theta_x(\xi, t)$: this function is assumed to be linear, or rather of the type:

$$\theta_x(\xi, t) = b_1\xi + b_2 \tag{4.116}$$

The boundary conditions must be such that $\theta_x(\xi, t)$ coincides, in correspondence to the end nodes of the generic finite element, with rotations θ_{xsl} and θ_{xdl} assumed as independent variables (Fig. 4.34):

$$\theta_x(\xi, t)|_{\xi=0} = \theta_{xsl}(t); \quad \theta_x(\xi, t)|_{\xi=lj} = \theta_{xdl}(t) \tag{4.117}$$

The link of the physical variables $\theta_x(\xi, t)$ and the independent variables $\underline{X}_{\theta x}$ (torsional rotations of the end nodes) thus become in matrix form:

$$\theta_x(\xi, t) = \underline{f}_{\theta x}^{T}(\xi)\underline{X}_{\theta x}(t) \tag{4.118}$$

having used $\underline{f}_{\theta x}(\xi)$ to indicate:

$$\underline{f}_{\theta x}(\xi) = \left\{ \begin{array}{c} 1 - \frac{\xi}{l_j} \\ \frac{\xi}{l_j} \end{array} \right\} \tag{4.119}$$

4.5.2.1.3 Shape Functions for Transverse Motion $\underline{f}_w^T(\xi)$

The function that describes the deformation inside the generic beam element in terms of a transverse displacement in plane ξ–ζ has already been obtained in Sect. 4.2.2, Eq. (4.10):

$$w(\xi, t) = \underline{f}_w^T(\xi)\underline{X}_w(t) \tag{4.120}$$

where $\underline{f}_w(\xi)$ is the vector of the shape functions (4.11).

4.5.2.1.4 Shape Functions of Transverse Motion $\underline{f}_v^T(\xi)$

The function that describes the deformation inside the generic element in terms of transverse displacements in plane (ξ–η) can be calculated, similar to the procedure used in the other plane (Sect. 4.2.2), by imposing the following boundary conditions (see Figs. 4.29 and 4.30)

$$v(\xi, t)|_{\xi=0} = Y_{sl}(t); \quad \left.\frac{\partial v(\xi, t)}{\partial \xi}\right|_{\xi=0} = \theta_{zsl}(t) \tag{4.121a}$$

$$v(\xi, t)|_{\xi=lj} = Y_{dl}(t); \quad \left.\frac{\partial v(\xi, t)}{\partial \xi}\right|_{\xi=lj} = \theta_{zdl}(t) \tag{4.121b}$$

on function $v(\xi, t)$, which describes the transverse displacement in the plane considered. The link between the physical variable $v(\xi, t)$, which describes the motion of the continuous system inside the generic element and the independent variables (displacements and rotations of the end nodes of the same element) thus becomes, in matrix form:

$$v(\xi, t) = \underline{f}_v^T(\xi)\underline{X}_v(t) \tag{4.122}$$

where $\underline{f}_v(\xi)$ is the vector shape function:

$$\underline{f}_v(\xi) = \begin{Bmatrix} \frac{2}{l_j^3}\xi^3 - \frac{3}{l_j^2}\xi^2 + 1 \\ -\left(-\frac{1}{l_j^2}\xi^3 + \frac{2}{l_j}\xi^2 - \xi\right) \\ -\frac{2}{l_j^3}\xi^3 + \frac{3}{l_j^2}\xi^2 \\ -\frac{1}{l_j^2}\xi^3 + \frac{1}{l_j}\xi^2 \end{Bmatrix} \tag{4.123}$$

With respect to the shape function defined in plane ξ–ζ, only the terms depending on the rotation proved to have changed their sign, because unlike what happens in plane ξ–ζ, the positive convention for θ_{zs} and θ_{zd} is in agreement with the derivative of displacement $v(\xi, t)$, with respect to the current coordinate ξ.

4.5.2.2 Definition of the Mass and Stiffness Matrices in Local Coordinates

To define the stiffness $[K_l] = [K_{jl}]$ and mass $[M_l] = [M_{jl}]$ matrix of the generic *j*th complete final beam element, it is necessary to define the potential energy V_j and the kinetic energy E_{cj} associated with the same element. In particular, as regards the elastic potential energy, the sum of the single contributions [10, 21, 26]:

$$
\begin{aligned}
V &= V_u + V_w + V_v + V_{\theta x} \\
&= \frac{1}{2}\int_0^{l_j} EA\left(\frac{\partial u}{\partial \xi}\right)^T\left(\frac{\partial u}{\partial \xi}\right)d\xi + \frac{1}{2}\int_0^{l_j} EJ_y\left(\frac{\partial^2 w}{\partial \xi^2}\right)^T\left(\frac{\partial^2 w}{\partial \xi^2}\right)d\xi \\
&+ \frac{1}{2}\int_0^{l_j} EJ_z\left(\frac{\partial^2 v}{\partial \xi^2}\right)^T\left(\frac{\partial^2 v}{\partial \xi^2}\right)d\xi + \frac{1}{2}\int_0^{l_j} GJ_x\left(\frac{\partial \theta_x}{\partial \xi}\right)^T\left(\frac{\partial \theta_x}{\partial \xi}\right)d\xi
\end{aligned}
\tag{4.124}
$$

where J_x, J_y and J_z are the inertia moments of the area of the section with respect to axes ξ–η–ζ, A the area of the section and E and G respectively the module of normal and tangent elasticity. Where kinetic energy is in question, both the translational kinetic energy and that due to torsional motion are considered [10]:

$$
\begin{aligned}
E_c &= E_{cu} + E_{cw} + E_{cv} + E_{c\theta x} \\
&= \frac{1}{2}\int_0^{l_j} m\left(\frac{\partial u}{\partial \xi}\right)^T\left(\frac{\partial u}{\partial \xi}\right)d\xi + \frac{1}{2}\int_0^{l_j} m\left(\frac{\partial w}{\partial \xi}\right)^T\left(\frac{\partial w}{\partial \xi}\right)d\xi \\
&+ \frac{1}{2}\int_0^{l_j} m\left(\frac{\partial v}{\partial \xi}\right)^T\left(\frac{\partial v}{\partial \xi}\right)d\xi + \frac{1}{2}\int_0^{l_j} I_x\left(\frac{\partial \theta_x}{\partial \xi}\right)^T\left(\frac{\partial \theta_x}{\partial \xi}\right)d\xi
\end{aligned}
\tag{4.125}
$$

where I_x is the mass moment of inertia per unit length with respect to longitudinal axis x_j. Let us now separately examine the single contributions associated with the axial, bending and torsional displacements of the potential energy and the kinetic energy.

4.5.2.2.1 Axial Motion

Once the links given by the shape functions (4.115) have been introduced, the expression of elastic potential energy associated with the axial deformation of the beam becomes:

$$V_u = \frac{1}{2}\int_0^{l_j} EA\left(\frac{\partial u}{\partial \xi}\right)^T\left(\frac{\partial u}{\partial \xi}\right)d\xi = \frac{1}{2}\int_0^{l_j} EA\underline{X}_u^T\underline{f}_u'(\xi)\underline{f}_u'^T(\xi)\underline{X}_u d\xi \tag{4.126}$$

by removing the vectors of the axial d.o.f. that only depend on time from the integral sign, we highlight the stiffness matrix according to the components of the same vector:

$$V_u = \frac{1}{2}\underline{X}_u^T\left[\int_0^{l_j} EA\underline{f}_u'(\xi)\underline{f}_u'^T(\xi)d\xi\right]\underline{X}_u = \frac{1}{2}\underline{X}_u^T[K_u]\underline{X}_u \tag{4.127}$$

In (4.127) vector $f_u'(\xi)$ contains the first derivatives of the shape functions with respect to the current coordinate ξ:

$$f_u'(\xi) = \left\{\frac{\partial \underline{f}_u(\xi)}{\partial \xi}\right\} = \left\{\begin{matrix} -\frac{1}{l_j} \\ \frac{1}{l_j}\end{matrix}\right\} \tag{4.128}$$

The stiffness matrix $[K_u]$ is of order 2 and similar to that obtained for the cable (4.30), Sect. 4.4.2:

$$[K_u] = EA\begin{bmatrix} \frac{1}{l_j} & -\frac{1}{l_j} \\ -\frac{1}{l_j} & \frac{1}{l_j}\end{bmatrix} \tag{4.129}$$

By introducing the links given by the shape functions, the kinetic energy proves to be:

$$\begin{aligned} E_{cu} &= \frac{1}{2}\int_0^{l_j} m\dot{\underline{X}}_u^T\underline{f}_u(\xi)\underline{f}_u^T(\xi)\dot{\underline{X}}_u d\xi \\ &= \frac{1}{2}\dot{\underline{X}}_u^T\left[\int_0^{l_j} m\underline{f}_u(\xi)\underline{f}_u^T(\xi)d\xi\right]\dot{\underline{X}}_u = \frac{1}{2}\dot{\underline{X}}_u^T[M_u]\dot{\underline{X}}_u \end{aligned} \tag{4.130}$$

where

$$[M_u] = m\begin{bmatrix} \frac{l_j}{3} & \frac{l_j}{6} \\ \frac{l_j}{6} & \frac{l_j}{3} \end{bmatrix} \tag{4.131}$$

4.5.2.2.2 Torsional Motion

Once the links given by the shape functions have been introduced, the expression of elastic potential energy $V_{\theta x}$ associated with the beam torsion, becomes:

$$\begin{aligned} V_{\theta x} &= \frac{1}{2}\int_0^{l_j} GJ_x\left(\frac{\partial\theta_x}{\partial\xi}\right)^T\left(\frac{\partial\theta_x}{\partial\xi}\right)d\xi = \frac{1}{2}\int_0^{l_j} GJ_x\underline{X}_{\theta x}^T\underline{f}'_{\theta x}(\xi)\underline{f}'^T_{\theta x}(\xi)\underline{X}_{\theta x}d\xi \\ &= \frac{1}{2}\underline{X}_{\theta x}^T\left[GJ_x\int_0^{l_j}\underline{f}'_{\theta x}(\xi)\underline{f}'^T_{\theta x}(\xi)d\xi\right]\underline{X}_{\theta x} = \frac{1}{2}\underline{X}_{\theta x}^T[K_{\theta x}]\underline{X}_{\theta x} \end{aligned} \tag{4.132}$$

In (4.132) $\underline{f}'_{\theta x}$ represents the vector that contains the first derivatives with respect to the current coordinate ξ of vector $\underline{f}_{\theta x}$ and $[K_{\theta x}]$ is the stiffness matrix for the torsion:

$$[K_{\theta x}] = GJ_x\begin{bmatrix} \frac{1}{l_j} & -\frac{1}{l_j} \\ -\frac{1}{l_j} & \frac{1}{l_j} \end{bmatrix} \tag{4.133}$$

By introducing the shape functions (4.118), kinetic energy $E_{c\theta x}$, proves to be:

$$E_{c\theta x} = \frac{1}{2}\dot{\underline{X}}_{\theta x}^T\left[\int_0^{l_j} I_x\underline{f}_{\theta x}(\xi)\underline{f}_{\theta x}^T(\xi)d\xi\right]\dot{\underline{X}}_{\theta x} = \frac{1}{2}\dot{\underline{X}}_{\theta x}^T[M_{\theta x}]\dot{\underline{X}}_{\theta x} \tag{4.134}$$

$[M_{\theta x}]$ in (4.134) represents the mass matrix relative to the torsional d.o.f.:

$$[M_{\theta x}] = I_x\begin{bmatrix} \frac{l_j}{3} & \frac{l_j}{6} \\ \frac{l_j}{6} & \frac{l_j}{3} \end{bmatrix} \tag{4.135}$$

4.5.2.2.3 Bending Motion

Let us now consider the transverse vibrations $w(\xi, t)$ in plane ξ–ζ of the beam, defined about the static equilibrium configuration (Fig. 4.30). To determine the stiffness matrix of the finite element for transverse motion only, it is necessary to define the potential energy V_w associated with it which, as already seen, proves to be defined as:

$$V_w = \frac{1}{2} EJ_y \int_0^{l_j} \left(\frac{\partial^2 w}{\partial \xi^2}\right)^2 d\xi = \frac{1}{2} EJ_y \int_0^{l_j} \left(\frac{\partial^2 w(\xi, t)}{\partial \xi^2}\right)^T \left(\frac{\partial^2 w(\xi, t)}{\partial \xi^2}\right) d\xi \tag{4.136}$$

Similarly, as regards the kinetic energy $E_{cw} = E_{cwj}$ of the generic beam element associated with the bending motion in the plane considered, by neglecting the inertias associated with the rotations of the sections (these terms will be introduced subsequently in Sect. 4.5.2.3.1), this will be:

$$E_{cw} = \frac{1}{2} m_j \int_0^{l_j} \left(\frac{\partial w(\xi, t)}{\partial t}\right)^2 d\xi = \frac{1}{2} EJ_y \int_0^{l_j} \left(\frac{\partial w(\xi, t)}{\partial t}\right)^T \left(\frac{\partial w(\xi, t)}{\partial t}\right) d\xi \tag{4.137}$$

where $m = m_j$ is the mass per unit length of the beam, meaning that the product m_j $d\xi$ represents the mass associated with the generic infinitesimal section. By keeping account of the transformation of coordinates (4.10) and by thus introducing the shape function vector $\underline{f}_w(\xi)$ (4.11), we obtain:

$$\begin{aligned} E_{cw} &= \frac{1}{2} \dot{\underline{X}}_w^T \left[\int_0^{l_j} m \underline{f}_w(\xi) \underline{f}_w^T(\xi) d\xi \right] \dot{\underline{X}}_w = \frac{1}{2} \dot{\underline{X}}_w^T [M_w] \dot{\underline{X}}_w \\ V_w &= \frac{1}{2} \underline{X}_w^T \left[EJ_y \int_0^{l_j} \underline{f}_w''(\xi) \underline{f}_w''^T(\xi) d\xi \right] \underline{X}_w = \frac{1}{2} \underline{X}_w^T [K_w] \underline{X}_w \end{aligned} \tag{4.138}$$

having used $\underline{f}_w''$ to indicate the vector of the second derivatives of $\underline{f}_w$ with respect to the current coordinate ξ along the beam:

$$\underline{f}_w'' = \left\{ \frac{d^2 \underline{f}_w(\xi)}{d\xi^2} \right\} = \left\{ \begin{array}{c} \frac{12}{l_j^3} \xi - \frac{6}{l_j^2} \\ -\frac{6}{l_j^2} \xi + \frac{4}{l_j} \\ -\frac{12}{l_j^3} \xi + \frac{6}{l_j^2} \\ \frac{6}{l_j^2} \xi - \frac{2}{l_j} \end{array} \right\} \tag{4.139}$$

As regards the transverse motion in plane ζ–ξ, in (4.138) $[M_w]$ is the mass matrix, defined in the local reference system of the generic finite element beam

$$[M_w] = m\left[\int_0^{l_j} \underline{f}_w(\xi)\underline{f}_w^T(\xi)d\xi\right] = ml_j\begin{bmatrix} \frac{13}{35} & -\frac{11}{210}l_j & \frac{9}{70} & \frac{13}{420}l_j \\ -\frac{11}{210}l_j & \frac{l_j^2}{105} & -\frac{13}{420}l_j & -\frac{l_j^2}{140} \\ \frac{9}{70} & -\frac{13}{420}l_j & \frac{13}{35} & \frac{11}{210}l_j \\ \frac{13}{420}l_j & -\frac{l_j^2}{140} & \frac{11}{210}l_j & \frac{l_j^2}{105} \end{bmatrix} \tag{4.140}$$

Conversely, in (4.138) $[K_w]$ is the relative stiffness matrix always defined in the local reference system:

$$[K_w] = \left[EJ_y\int_0^{l_j} \underline{f}_w''(\xi)\underline{f}_w''^T(\xi)d\xi\right] = EJ_y\begin{bmatrix} \frac{12}{l_j^3} & -\frac{6}{l_j^2} & -\frac{12}{l_j^3} & -\frac{6}{l_j^2} \\ -\frac{6}{l_j^2} & \frac{4}{l_j} & \frac{6}{l_j^2} & \frac{2}{l_j} \\ -\frac{12}{l_j^3} & \frac{6}{l_j^2} & \frac{12}{l_j^3} & \frac{6}{l_j^2} \\ -\frac{6}{l_j^2} & \frac{2}{l_j} & \frac{6}{l_j^2} & \frac{4}{l_j} \end{bmatrix} \tag{4.141}$$

As foreseeable, by studying a conservative system, $[K_w]$ proves to be symmetrical and singular (corresponding to the real situation of the beam element which is still free in space: this condition will only disappear when the constraints are imposed on the fully assembled structure). Matrices $[K_v]$ and $[M_v]$ relative to plane ξ–η, are the same as $[K_w]$ and $[M_w]$ defined in plane ξ–ζ: given the sign difference present in some terms of the shape functions $\underline{f}_v(\xi)$ (4.122) and $\underline{f}_w(\xi)$ (4.26) matrices $[K_v]$ and $[M_v]$ will differ from the same ones only in terms of the signs of the mixed terms which depend on the rotations θ_{zl}.

4.5.2.3 Definition of the Overall Mass and Stiffness Matrices

By resuming (4.124) and (4.125) we can write the elastic potential and kinetic energy in matrix form as the sum of bending, axial and torsional contributions:

$$\begin{aligned} V = V_w + V_v + V_u + V_{\theta x} &= \frac{1}{2}X_w^T[K_w]X_w + \frac{1}{2}X_v^T[K_v]X_v \\ &+ \frac{1}{2}X_u^T[K_u]X_u + \frac{1}{2}X_{\theta x}^T[K_{\theta x}]X_{\theta x} \end{aligned} \tag{4.142}$$

$$\begin{aligned} E_c = E_{cw} + E_{cv} + E_{cu} + E_{c\theta x} &= \frac{1}{2}\dot{X}_w^T[M_w]\dot{X}_w + \frac{1}{2}\dot{X}_v^T[M_v]\dot{X}_v \\ &+ \frac{1}{2}\dot{X}_u^T[M_u]\dot{X}_u + \frac{1}{2}\dot{X}_{\theta x}^T[M_{\theta x}]\dot{X}_{\theta x} \end{aligned} \tag{4.143}$$

By gathering the sub-vectors $\underline{X}_w, \underline{X}_v, \underline{X}_u$ and $\underline{X}_{\theta x}$ in one single vector $\underline{Y}_l$:

$$\underline{Y}_l = \begin{Bmatrix} \underline{X}_w \\ \underline{X}_v \\ \underline{X}_u \\ \underline{X}_{\theta x} \end{Bmatrix} \tag{4.144}$$

and, a consequence, by ordering the stiffness matrices relative to each sub-vector, we obtain the overall 12 × 12 stiffness $[K_Y]$ and mass $[M_Y]$ matrix of the beam element, relative to all the d.o.f. assumed:

$$[K_Y] = \begin{bmatrix} [K_w] & & & \\ & [K_v] & & \\ & & [K_u] & \\ & & & [K_{\theta x}] \end{bmatrix} \quad [M_Y] = \begin{bmatrix} [M_w] & & & \\ & [M_v] & & \\ & & [M_u] & \\ & & & [M_{\theta x}] \end{bmatrix} \tag{4.145}$$

By using this change of variables, kinetic energy $E_{cj} = E_c$ and potential energy $V_j = V$ of the generic jth finite element can be expressed as:

$$\begin{aligned} E_c &= \frac{1}{2}\dot{\underline{Y}}_l^T [M_Y] \dot{\underline{Y}}_l \\ V &= \frac{1}{2}\underline{Y}_l^T [K_Y] \underline{Y}_l \end{aligned} \tag{4.146}$$

In general, however, it is preferable to order the d.o.f. of the beam element according to vector $\underline{X}_l$ (4.107), or rather by gathering, in an orderly way, the displacements and rotations of the first node in the first six locations and the amplitudes relative to the second node of the beam in the subsequent locations:

$$\underline{Y}_l = [B]\underline{X}_l \tag{4.147}$$

where [B] is the following matrix:

$$[B] = \begin{bmatrix} 0 & 0 & 1 & 0 & 0 & 0 & 0 & 0 & 0 & 0 & 0 & 0 \\ 0 & 0 & 0 & 0 & 1 & 0 & 0 & 0 & 0 & 0 & 0 & 0 \\ 0 & 0 & 0 & 0 & 0 & 0 & 0 & 0 & 1 & 0 & 0 & 0 \\ 0 & 0 & 0 & 0 & 0 & 0 & 0 & 0 & 0 & 0 & 1 & 0 \\ 0 & 1 & 0 & 0 & 0 & 0 & 0 & 0 & 0 & 0 & 0 & 0 \\ 0 & 0 & 0 & 0 & 0 & 1 & 0 & 0 & 0 & 0 & 0 & 0 \\ 0 & 0 & 0 & 0 & 0 & 0 & 0 & 1 & 0 & 0 & 0 & 0 \\ 0 & 0 & 0 & 0 & 0 & 0 & 0 & 0 & 0 & 0 & 0 & 1 \\ 1 & 0 & 0 & 0 & 0 & 0 & 0 & 0 & 0 & 0 & 0 & 0 \\ 0 & 0 & 0 & 0 & 0 & 0 & 1 & 0 & 0 & 0 & 0 & 0 \\ 0 & 0 & 0 & 1 & 0 & 0 & 0 & 0 & 0 & 0 & 0 & 0 \\ 0 & 0 & 0 & 0 & 0 & 0 & 0 & 0 & 0 & 1 & 0 & 0 \end{bmatrix} \tag{4.148}$$

The kinetic and potential energy can thus be defined as:

$$\begin{aligned} E_{cj} &= E_c = \frac{1}{2}\dot{X}_l^T[B]^T[M_Y][B]\dot{X}_l = \frac{1}{2}\dot{X}_{jl}^T[M_{jl}]\dot{X}_{jl} \\ V_j &= V = \frac{1}{2}X_l^T[B]^T[K_Y][B]X_l = \frac{1}{2}X_{jl}^T[K_{jl}]X_{jl} \end{aligned} \tag{4.149}$$

4.5.2.3.1 Inertia Effects Associated with Rotation Inertia

By keeping account of the rotational inertia (neglected in the explanation given until now), the kinetic energy of the beam in its complete form becomes:

$$\begin{aligned} E_c = &\frac{1}{2}\int_0^{l_j} m\left(\frac{\partial u}{\partial t}\right)^T\left(\frac{\partial u}{\partial t}\right)d\xi + \frac{1}{2}\int_0^{l_j} m\left(\frac{\partial w}{\partial t}\right)^T\left(\frac{\partial w}{\partial t}\right)d\xi + \frac{1}{2}\int_0^{l_j} m\left(\frac{\partial v}{\partial t}\right)^T\left(\frac{\partial v}{\partial t}\right)d\xi \\ &+ \frac{1}{2}\int_0^{l_j} I_x\left(\frac{\partial \theta_x}{\partial t}\right)^T\left(\frac{\partial \theta_x}{\partial t}\right)d\xi + \frac{1}{2}\int_0^{l_j} I_y\left(\frac{\partial \theta_y}{\partial t}\right)^T\left(\frac{\partial \theta_y}{\partial t}\right)d\xi + \frac{1}{2}\int_0^{l_j} I_z\left(\frac{\partial \theta_z}{\partial t}\right)^T\left(\frac{\partial \theta_z}{\partial t}\right)d\xi \end{aligned} \tag{4.150}$$

where

$$\theta_y = \left(\frac{\partial}{\partial t}\left(\frac{\partial w}{\partial \xi}\right)\right) = \underline{f}_w'^T(\xi)\underline{\dot{X}}_w(t) \tag{4.151}$$

$$\theta_z = \left(\frac{\partial}{\partial t}\left(\frac{\partial v}{\partial \xi}\right)\right) = \underline{f}_v'^T(\xi)\underline{\dot{X}}_v(t) \tag{4.152}$$

By keeping account of these relations, the two additional terms in the kinetic energy expression become:

$$\begin{aligned} E_{c\theta y} &= \frac{1}{2}\underline{\dot{X}}_w^T\left[\int_0^{l_j} I_y\underline{f}_w'(\xi)\underline{f}_w'^T(\xi)d\xi\right]\underline{\dot{X}}_w = \frac{1}{2}\underline{\dot{X}}_w^T[M_{\theta y}]\underline{\dot{X}}_w \\ E_{c\theta z} &= \frac{1}{2}\underline{\dot{X}}_v^T\left[\int_0^{l_j} I_z\underline{f}_v'(\xi)\underline{f}_v'^T(\xi)d\xi\right]\underline{\dot{X}}_v = \frac{1}{2}\underline{\dot{X}}_v^T[M_{\theta z}]\underline{\dot{X}}_v \end{aligned} \tag{4.153}$$

These matrices add up, according to the order of vector $\underline{Y}_l$ (4.144), in the global matrix $[M_Y]$ (4.145) which thus becomes:

$$[M_Y] = \begin{bmatrix} [M_{\theta y}] + [M_w] & & & \\ & [M_{\theta z}] + [M_v] & & \\ & & [M_u] & \\ & & & [M_{\theta x}] \end{bmatrix} \tag{4.154}$$

Table 4.2 shows the mass matrix $\left[M_{jl}\right]$ of the beam element in its most complete formulation.

4.5.2.3.2 Effect of the Static Axial Pre-load in the Beam: Taut Beam Finite Element

In some structures, the single beam, in the static equilibrium position, proves to be axially loaded by a force S_0 (of either traction or compression). As with the example of the cable (Sect. 4.4.2, (4.24)), this axial load changes the terms of potential energy for bending motions [26]: in this case, the potential elastic energy V_j becomes:

Table 4.2 Elements of the mass matrix $[M_{jl}]$ in complete form

$$\left[M_{jl}\right] = A\rho l \begin{bmatrix}
\frac{1}{3} & & & & & & & & & & & \\
0 & \frac{13}{35}+\frac{6I_z}{5A\rho l^2} & & & & & & & & & & \\
0 & 0 & \frac{13}{35}+\frac{6I_y}{5A\rho l^2} & & & & & & & & & \\
0 & 0 & 0 & \frac{I_x}{3A\rho} & & & & & \text{sym} & & & \\
0 & 0 & -\frac{11l}{210}-\frac{I_y}{10A\rho l} & 0 & \frac{l^2}{105}+\frac{2I_y}{15A\rho} & & & & & & & \\
0 & \frac{11l}{210}+\frac{I_z}{10A\rho l} & 0 & 0 & 0 & \frac{l^2}{105}+\frac{2I_z}{15A\rho} & & & & & & \\
\frac{1}{6} & 0 & 0 & 0 & 0 & 0 & \frac{1}{3} & & & & & \\
0 & \frac{9}{70}-\frac{6I_z}{5A\rho l^2} & 0 & 0 & 0 & \frac{13l}{420}-\frac{I_z}{10A\rho l} & 0 & \frac{13}{35}+\frac{6I_z}{5A\rho l^2} & & & & \\
0 & 0 & \frac{9}{70}-\frac{6I_y}{5A\rho l^2} & 0 & -\frac{13l}{420}+\frac{I_y}{10A\rho l} & 0 & 0 & 0 & \frac{13}{35}+\frac{6I_y}{5A\rho l^2} & & & \\
0 & 0 & 0 & \frac{I_x}{3A\rho} & 0 & 0 & 0 & 0 & 0 & \frac{I_x}{3A\rho} & & \\
0 & 0 & \frac{13l}{420}-\frac{I_y}{10A\rho l} & 0 & -\frac{l^2}{140}-\frac{I_y}{30A\rho} & 0 & 0 & 0 & \frac{11l}{210}+\frac{I_y}{10A\rho l} & 0 & \frac{l^2}{105}+\frac{2I_y}{15A\rho} & \\
0 & -\frac{13l}{420}+\frac{I_z}{10A\rho l} & 0 & 0 & 0 & -\frac{l^2}{140}-\frac{I_z}{30A\rho} & 0 & -\frac{11l}{210}-\frac{I_z}{10A\rho l} & 0 & 0 & 0 & \frac{l^2}{105}+\frac{2I_z}{15A\rho}
\end{bmatrix}$$

$$
\begin{aligned}
V_j &= \frac{1}{2}\int_0^{l_j} EA\left(\frac{\partial u}{\partial \xi}\right)^T\left(\frac{\partial u}{\partial \xi}\right)d\xi + \frac{1}{2}\int_0^{l_j} S_o\left(\frac{\partial w}{\partial \xi}\right)^T\left(\frac{\partial w}{\partial \xi}\right)d\xi \\
&+ \frac{1}{2}\int_0^{l_j} S_o\left(\frac{\partial v}{\partial \xi}\right)^T\left(\frac{\partial v}{\partial \xi}\right)d\xi + \frac{1}{2}\int_0^{l_j} EJ_y\left(\frac{\partial^2 w}{\partial \xi^2}\right)^T\left(\frac{\partial^2 w}{\partial \xi^2}\right)d\xi \\
&+ \frac{1}{2}\int_0^{l_j} EJ_z\left(\frac{\partial^2 v}{\partial \xi^2}\right)^T\left(\frac{\partial^2 v}{\partial \xi^2}\right)d\xi + \frac{1}{2}\int_0^{l_j} GJ_x\left(\frac{\partial^2 \theta_x}{\partial \xi^2}\right)^T\left(\frac{\partial^2 \theta_x}{\partial \xi^2}\right)d\xi
\end{aligned}
\tag{4.155}
$$

The additional terms due to the presence of axial action S_o can, using similar steps to those performed in Sect. 4.4.2, be defined as:

$$
\begin{aligned}
V_{Sw} &= \frac{1}{2}\int_0^{l_j} S_o\left(\frac{\partial w}{\partial \xi}\right)^T\left(\frac{\partial w}{\partial \xi}\right)d\xi = \frac{1}{2}\underline{X}_w^T[K_{Sw}]\underline{X}_w \\
V_{Sv} &= \frac{1}{2}\int_0^{l_j} S_o\left(\frac{\partial v}{\partial \xi}\right)^T\left(\frac{\partial v}{\partial \xi}\right)d\xi = \frac{1}{2}\underline{X}_v^T[K_{Sv}]\underline{X}_v
\end{aligned}
\tag{4.156}
$$

where $[K_{Sw}]$ and $[K_{Sv}]$ are the stiffness matrices, always defined in the local reference system, which define the linearized elastic forces due to the presence of the axial action in the transverse motion. Potential energy V_j of the generic taut beam is thus:

$$
\begin{aligned}
V &= V_u + V_w + V_{Sw} + V_v + V_{Sv} + V_{\theta x} \\
&= \frac{1}{2}\underline{X}_u^T[K_u]\underline{X}_u + \frac{1}{2}\underline{X}_w^T[K_w]\underline{X}_w + \frac{1}{2}\underline{X}_w^T[K_{Sw}]\underline{X}_w \\
&+ \frac{1}{2}\underline{X}_v^T[K_v]\underline{X}_v + \frac{1}{2}\underline{X}_v^T[K_{Sv}]\underline{X}_v + \frac{1}{2}\underline{X}_{\theta x}^T[K_{\theta x}]\underline{X}_{\theta x}
\end{aligned}
\tag{4.157}
$$

According to vector $\underline{Y}_1$ (4.144), the stiffness matrix $[K_Y]$ (4.145), in its complete form, thus becomes:

$$
[K_Y] = \begin{bmatrix} [K_w]+[K_{Sw}] & & & \\ & [K_v]+[K_{Sv}] & & \\ & & [K_u] & \\ & & & [K_{\theta x}] \end{bmatrix}
\tag{4.158}
$$

In Table 4.3 we show the complete stiffness matrix $[K_{jl}]$ of the taut beam element (i.e. the beam element in its most general configuration). As can be noted, the effect of an axial static pre-load of traction (S_o positive) increases the bending stiffness of the beam, while an axial action of compression (S_o negative) reduces the stiffness

Table 4.3 Elements of the taut beam stiffness stiffness $[K_{jl}]$ matrix

$$
[K_{jl}]=\begin{bmatrix}
\frac{EA}{l} & & & & & & & & & & & \\
0 & \frac{12EJ_z}{l^3}+\frac{6S_o}{5l} & & & & & & & & & & \\
0 & 0 & \frac{12EJ_y}{l^3}+\frac{6S_o}{5l} & & & & & & & & & \\
0 & 0 & 0 & \frac{GJ_x}{l} & & & & & \text{sym} & & & \\
0 & 0 & -\frac{6EJ_y}{l^3}-\frac{S_o}{10} & 0 & \frac{4EJ_y}{l}+\frac{2lS_o}{15} & & & & & & & \\
0 & \frac{6EJ_z}{l^2}+\frac{S_o}{10} & 0 & 0 & 0 & \frac{4EJ_z}{l}+\frac{2lS_o}{15} & & & & & & \\
-\frac{EA}{l} & 0 & 0 & 0 & 0 & 0 & \frac{EA}{l} & & & & & \\
0 & -\frac{12EJ_z}{l^3}-\frac{6S_o}{5l} & 0 & 0 & 0 & -\frac{6EJ_z}{l^2}-\frac{S_o}{10} & 0 & \frac{12EJ_z}{l^3}+\frac{6S_o}{5l} & & & & \\
0 & 0 & -\frac{12EJ_y}{l^3}-\frac{6S_o}{5l} & 0 & \frac{6EJ_y}{l^2}-\frac{S_o}{10} & 0 & 0 & 0 & \frac{12EJ_y}{l^3}+\frac{6S_o}{5l} & & & \\
0 & 0 & 0 & -\frac{GJ_x}{l} & 0 & 0 & 0 & 0 & 0 & \frac{GJ_x}{l} & & \\
0 & 0 & -\frac{6EJ_y}{l^3}-\frac{S_o}{10} & 0 & \frac{2EJ_y}{l}-\frac{lS_o}{30} & 0 & 0 & 0 & \frac{6EJ_y}{l^2}+\frac{S_o}{10} & 0 & \frac{4EJ_y}{l}+\frac{2lS_o}{15} & \\
0 & \frac{6EJ_z}{l^2}+\frac{S_o}{10} & 0 & 0 & 0 & \frac{2EJ_z}{l}-\frac{lS_o}{30} & 0 & -\frac{6EJ_z}{l^2}-\frac{S_o}{10} & 0 & 0 & 0 & \frac{4EJ_z}{l}+\frac{2lS_o}{15}
\end{bmatrix}
$$

(see (4.158); Table 4.3). In the chapter that follows, we will always refer, unless otherwise stated, to the beam element with two complete mass (Table 4.2) and stiffness (Table 4.3) matrices. Having thus defined the mass and stiffness matrices of the generic beam expressed in the local reference system, we will now return to the initial example (Fig. 4.28) and proceed with the writing of the equations of motion.

4.5.3 Definition of the Stiffness $[K_j]$ and Mass $[M_j]$ Matrix of the Beam Element in the Global Reference System

Having defined the schematization of the structure according to the criteria described in Sect. 4.5.1 and having calculated the stiffness $\left[K_{jl}\right]$ and mass $\left[M_{jl}\right]$ matrices of the single finite elements of the local reference system (Sect. 4.5.2), it is possible to define the total kinetic energy E_{ctot} and the total elastic potential energy V_{tot} of the structure given by the sum of the single contributions E_{cj} and V_j of all the *m* finite elements that make up the structure itself [2, 10]:

$$
\begin{aligned}
E_{ctot} &= \sum_j^m E_{cj} = \frac{1}{2}\sum_j^m \underline{\dot{X}}_{jl}^T \left[M_{jl}\right] \underline{\dot{X}}_{jl} \\
V_{tot} &= \sum_j^m V_j = \frac{1}{2}\sum_j^m \underline{X}_{jl}^T \left[K_{jl}\right] \underline{X}_{jl}
\end{aligned}
\tag{4.159}
$$

In order to sum the effect of each finite element on the generic node of the model, it is necessary to perform a transformation of the coordinates that enable us to express the link between the displacements $\underline{X}_{jl}$ of the end nodes of the generic finite beam element in the local reference system (Fig. 4.28) and the displacements $\underline{X}_j$ expressed in the absolute reference system:

$$\underline{X}_j^T = \{ X_s \quad Y_s \quad Z_s \quad \theta_{xs} \quad \theta_{ys} \quad \theta_{zs} \quad X_d \quad Y_d \quad Z_d \quad \theta_{xd} \quad \theta_{yd} \quad \theta_{zd} \} \tag{4.160}$$

The link between displacements $\underline{X}_{jl}$ of the end nodes of a generic *j*th beam element in the local reference system and the corresponding displacements $\underline{X}_j$ expressed in the absolute global reference system is given by the transformation matrix of coordinates $[\Lambda_j]$:

$$\underline{X}_{jl} = [\Lambda_j]\underline{X}_j \tag{4.161}$$

where

$$[\Lambda_j] = \begin{bmatrix} [\lambda_j] & & & \\ & [\lambda_j] & & \\ & & [\lambda_j] & \\ & & & [\lambda_j] \end{bmatrix} \tag{4.162}$$

being $[\lambda_j]$ (see (4.35)) the matrix (3 × 3) formed by the directory cosines of the local reference system with respect to the global one [10, 21, 26]. By introducing the transformation of coordinates (4.161) into the expression of kinetic E_{ctot} and potential V_{tot} (4.159) energy:

$$\begin{aligned} E_{ctot} &= \frac{1}{2}\sum_j^m \dot{\underline{X}}_j^T [\Lambda_j]^T [M_{jl}] [\Lambda_j] \dot{\underline{X}}_j = \frac{1}{2}\sum_j^m \dot{\underline{X}}_j^T [M_j] \dot{\underline{X}}_j \\ V_{tot} &= \frac{1}{2}\sum_j^m \underline{X}_{jl}^T [\Lambda_j]^T [K_{jl}] [\Lambda_j] \underline{X}_{jl} = \frac{1}{2}\sum_j^m \underline{X}_j^T [K_j] \underline{X}_{jl} \end{aligned} \tag{4.163}$$

having respectively used $[M_j]$ and $[K_j]$ to indicate the matrices of mass and stiffness of the generic *j*th finite element referring to the global coordinates:

$$\begin{aligned} [M_j] &= [\Lambda_j]^T [M_{jl}] [\Lambda_j] \\ [K_j] &= [\Lambda_j]^T [K_{jl}] [\Lambda_j] \end{aligned} \tag{4.164}$$

In (4.163) vectors $\underline{X}_j$ contain the 12 d.o.f. of the nodes of the generic *j*th element, defined in the global reference system.

4.5.4 Writing of the Equations of Motion and Their Solution

The writing of the equations of motion of the structure schematized with finite beam element obviously follows the same procedure as that performed for the cable element: matrices $[M_j]$ and $[K_j]$ of (4.164) are assembled in the overall matrices $[K_{tot}]$ and $[M_{tot}]$ as shown in Sect. 4.4.5. In this case too, the equations of motion are of the type:

$$[M_{tot}]\ddot{\underline{X}}_t + [R_{tot}]\dot{\underline{X}}_t + [K_{tot}]\underline{X}_t = \underline{F}_t \tag{4.165}$$

Having used $\underline{X}_t$ to indicate the vector that contains all the d.o.f. of the structure analysed, assumed, for the time being, as free in space, organized in a distributed form as follows:

$$\underline{X}_t = \begin{Bmatrix} \underline{X}_L \\ \underline{X}_V \end{Bmatrix} \tag{4.166}$$

where $\underline{X}_L$ is once again the vector relative to the actual d.o.f. of the structure and $\underline{X}_V$ the vector relative to the constrained d.o.f. The method used to solve these same Eq. (4.165) is widely described in Sect. 4.4.8.

4.5.5 A Numerical Example

As regards the discretized mathematical model with finite elements analysed, Fig. 4.28, using the procedures described in the previous paragraphs the first natural frequencies ω_i (Fig. 4.31) were calculated. The same model was subjected to a sinusoidal excitation force $F_8 = |F_8|e^{i\Omega t}$ with a constant amplitude ($|F_8| = 5 \cdot 10^4$ daN) applied to node 8 of the model (see Fig. 4.28) in a frequency range $1\,\text{Hz} \le \Omega \le 200\,\text{Hz}$. As an example, in Fig. 4.32 you can see the trend of the module $|Z_7|$ and of the phase Φ_7 relative to the excitation force of the vertical displacement of node 7, while Fig. 4.33 shows the trend of module $|Z_8|$ and of phase Φ_8 (always relative to the excitation force) of the displacement of node 8 when the frequency varies.

4.6 Two-Dimensional and Three-Dimensional Finite Elements (Brief Outline)

The cable and beam finite elements, analysed in Sects. 4.4 and 4.5, can only simulate the behaviour of the structural elements in which one dimension predominates with respect to the others. Generally speaking, the real structures also

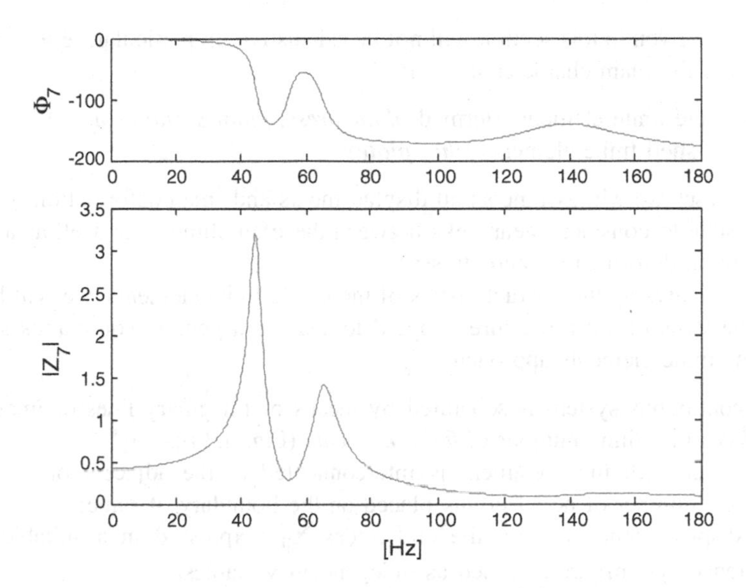

Fig. 4.32 Frequency response of the vertical d.o.f. of node 7 of the structure

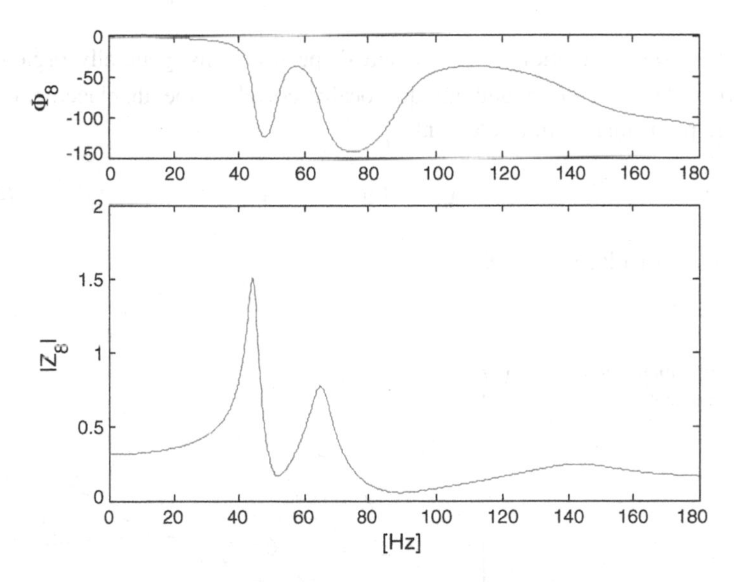

Fig. 4.33 Frequency response of the vertical d.o.f. of node 8 of the structure

have elements with physical dimensions that are comparable in all directions. The aim, prevalently for study purposes, of this paragraph is to give a general overview of the subject to facilitate the reading of numerous specialized texts on the subject (Sect. 4.8, for example [2, 10, 21, 26]) by introducing some basic notions about two-dimensional and three-dimensional finite elements. For obvious reasons, the information given in this section will not be exhaustive. In particular, reference will be made to the main characteristics of:

- membrane finite elements (termed *plain stress*, *plain strain element*);
- plate or shell finite elements (*thin plate*);

In all cases we will assume small displacements and small deformations (so that it is possible to consider linear links between these amplitudes) as well as a linear link between deformations and stresses.

Before analysing the characteristics of the single finite elements, we will briefly recall the steps of the procedure adopted to analyse a generic continuous system using the finite elements approach:

- the continuous system is separated by means of imaginary lines or imaginary surfaces in a finite number of *finite elements* (Fig. 4.34);
- the generic *j*th finite element is interconnected to the adjacent ones with a discrete number of nodal points placed on the boundary of same;
- the displacements of only the end nodes $\underline{X}_{jl}$ (expressed in a suitable local reference system) are assumed as independent variables;

$$\underline{X}_{jl}^{T} = \{ X_{11} \quad Y_{11} \quad Z_{11} \quad \ldots \quad \ldots \} \tag{4.167}$$

- a suitable set of functions $[f_j]$ (termed shape functions, generally organized in matrix $[f_j]$) are chosen and unequivocally correlate the displacement of the generic point inside finite element $\underline{u}_j$:

$$\underline{u}_j^{T} = \{ u \quad v \quad w \} \tag{4.168}$$

with nodal displacements $\underline{X}_{jl}$;

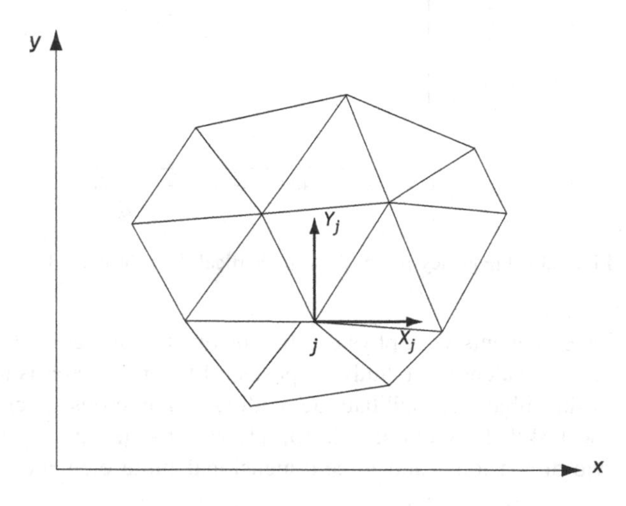

Fig. 4.34 Schematization with finite elements

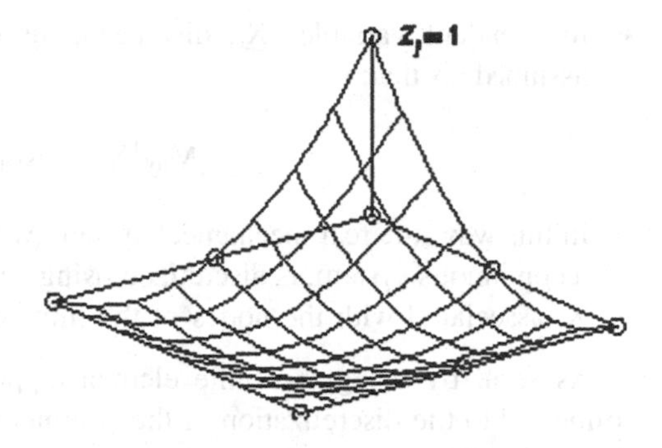

Fig. 4.35 Shape function meaning

- the shape functions $[f_j]$ thus unequivocally define the state of deformation inside the generic finite element (Fig. 4.35):

$$\underline{u}_j = [f_j]\underline{X}_{jl} \tag{4.169}$$

Subsequently, a Lagrangian approach is used: first and foremost, the various forms of energy (kinetic E_c, potential V) are defined and the virtual work δ^*L performed by the external forces applied to the system as a function of the physical variables u_j;

- a link is imposed between the physical variables u_j (4.169) and the independent variables $\underline{X}_{jl}$ (4.167) nodal displacements thus allowing us to define the various forms of energy, relative to the same element, as:

$$\begin{aligned} V_j &= \frac{1}{2}\underline{X}_{jl}^T[K_{jl}]\underline{X}_{jl} \\ E_{cj} &= \frac{1}{2}\dot{\underline{X}}_{jl}^T[M_{jl}]\dot{\underline{X}}_{jl} \\ \delta^*L_j &= \underline{F}_{jl}^T\delta^*\underline{X}_{jl} \end{aligned} \tag{4.170}$$

- finally a coordinate transformation is performed to express this energy in one single absolute reference system:

$$\underline{X}_{jl} = [\Lambda_j]\underline{X}_j \tag{4.171}$$

- we assemble the various stiffness $[K_j]$ and mass $[M_j]$ matrices and the vector of the Lagrangian components of the applied external forces $\underline{Q}_j$ expressed in the absolute reference system:

$$\begin{aligned} [K_j] &= [\Lambda_j]^T[K_{jl}][\Lambda_j] \\ [M_j] &= [\Lambda_j]^T[M_{jl}][\Lambda_j] \\ \underline{F}_j &= [\Lambda_j]^T\underline{F}_{jl} \end{aligned} \tag{4.172}$$

- independent variables $\underline{X}_t$, displacements of all the nodes of the model, are assumed so that:

$$[M_{tot}]\ddot{\underline{X}}_t + [K_{tot}]\underline{X}_t = \underline{F}_t \tag{4.173}$$

in this way, the real mechanical system, with infinite d.o.f., on account of being a continuous system, is discretized using a n d.o.f., i.e. the same number of d.o.f. $\underline{X}_t$ associated with the nodes of the finite element model adopted.

As seen, by using the finite element approach, we introduce approximations connected to the discretization of the continuous system adopted and to the type of shape function assumed: having concentrated the generalized forces (elastic, external and inertial) on the nodes, the equilibrium conditions are only satisfied in a general sense, i.e. they might not be respected inside each element or on the boundary of same:

- in the choice, among other things arbitrary, of the shape functions, it is not always possible to respect the equations of continuity between adjacent elements along the boundary.

In order to reduce, as far as is possible, the errors introduced using this method, it is necessary to increase the number of finite elements or to use more refined shape functions. The choice of the shape functions must, in any case, satisfy the convergence conditions that may be summarized as follows [3, 26]:

- $[f_j]$ must be such in order to define, within the generic finite element, a continuous field of displacements;
- the shape functions $[f_j]$ must be such to ensure that no deformations are introduced inside the generic finite element in the event of this being subjected to a rigid motion;
- the shape functions $[f_j]$ must be such as to allow us to reproduce a constant deformation state;
- the elements must be compatible, i.e. there should not be any slacks or overlapping of nodes (no sudden variations in slope must be allowed either in the beams or in the plates) in the beams and in the plates.

4.6.1 Definition of the Generic Shape Function

To evaluate matrix $[f_j]$ (4.169) it is necessary:

- to define a certain shape function, usually a polynomial containing an n number of multiplicative constants to be determined as equal to the d.o.f. associated with the end nodes of the generic finite element analysed (i.e. equal to the number of elements of vector $\underline{\mathbf{X}}_{jl}$):

$$\begin{cases} \mathrm{u_j = u_j(x, y, z) = a_0 + a_1 x + a_2 y + \cdots} \\ \mathrm{v_j = v_j(x, y, z) = b_0 + b_1 x + b_2 y + \cdots} \\ \mathrm{w_j = w_j(x, y, z) = c_0 + c_1 x + c_2 y + \cdots} \end{cases} \quad (4.174)$$

or rather in matrix form:

$$\underline{\mathrm{u}}_{\mathrm{j}} = [\mathrm{P_j}]\underline{\mathrm{a}}_{\mathrm{j}} \quad (4.175)$$

where vector $\underline{\mathrm{a}}_{\mathrm{j}}$ contains the n constants to be determined:

$$\underline{\mathrm{a}}_{\mathrm{j}}^{\mathrm{T}} = \{\mathrm{a_0} \quad \mathrm{b_0} \quad \mathrm{c_0} \quad \ldots \quad \ldots \quad \ldots\} \quad (4.176)$$

- to impose the congruence of these functions, which define the displacements inside the generic finite element, in correspondence to the end nodes:

$$\underline{\mathrm{X}}_{\mathrm{jl}} = [\mathrm{C_j}]\underline{\mathrm{a}}_{\mathrm{j}} \quad (4.177)$$

  where $[\mathrm{C_j}]$ is a known matrix that contains the coordinates relative to the single nodes of the finite element considered.

Having used (4.177) to evaluate the constants of the functions (4.174), it is possible to express (4.175) in matrix form as:

$$\underline{\mathrm{u}}_{\mathrm{j}} = [\mathrm{P_j}][\mathrm{C_j}]^{-1}\underline{\mathrm{X}}_{\mathrm{jl}} = [\mathrm{f_j}]\underline{\mathrm{X}}_{\mathrm{jl}} \quad (4.178)$$

where $[\mathrm{f_j}] = [\mathrm{P_j}][\mathrm{C_j}]^{-1}$ is the matrix ($3 \times n$) of the shape functions being looked for [26].

4.6.2 *General Definition of the Stiffness and Mass Matrices of the Generic Three-Dimensional Finite Element*

At this point, it is convenient to introduce a generalization in terms of the methodology necessary to define the stiffness $[K_{jl}]$ and mass $[M_{jl}]$ matrices of a generic three-dimensional finite element and of the relative generalized forces $\underline{F}_{jl}$ (expressed in the local reference system). For this purpose, we will briefly recall the basic concepts relative to the stress-deformation and deformation-displacement links. Let us consider the generic jth three-dimensional finite element in space and, more specifically, an infinitesimal generic element of volume dV inside this same finite element: the relation that will correlate the position of the deformed $\underline{R}_j$ and undeformed $\underline{R}_{j0}$ element is given by:

$$\underline{R}_j = \begin{Bmatrix} x \\ y \\ z \end{Bmatrix} = \underline{R}_{j0} + \underline{u}_j = \begin{Bmatrix} x_0 \\ y_0 \\ z_0 \end{Bmatrix} + \begin{Bmatrix} u \\ v \\ w \end{Bmatrix} \tag{4.179}$$

where u_j is the vector containing the displacement components along the three axes of the reference system assumed. As is known [4, 15, 20, 21, 26] the link between strains and displacements is given, assuming small deformations, by the relations:

$$\begin{aligned} &\varepsilon_{xx} = \frac{\partial u}{\partial x}; \quad \varepsilon_{yy} = \frac{\partial v}{\partial y}; \quad \varepsilon_{zz} = \frac{\partial w}{\partial z}; \quad \gamma_{xy} = \frac{\partial u}{\partial y} + \frac{\partial v}{\partial x}; \\ &\gamma_{yz} = \frac{\partial v}{\partial z} + \frac{\partial u}{\partial y}; \quad \gamma_{zx} = \frac{\partial u}{\partial z} + \frac{\partial w}{\partial x} \end{aligned} \tag{4.180}$$

Having used $\underline{\varepsilon}$ to define the vector that contains the deformation components:

$$\underline{\varepsilon}^T = \left\{ \varepsilon_x \quad \varepsilon_y \quad \varepsilon_z \quad \gamma_{xy} \quad \gamma_{yz} \quad \gamma_{zx} \right\} \tag{4.181}$$

and [d] a 6 × 3 matrix of differential operators defined as:

$$[d] = \begin{bmatrix} \frac{\partial}{\partial x} & 0 & 0 \\ 0 & \frac{\partial}{\partial y} & 0 \\ 0 & 0 & \frac{\partial}{\partial z} \\ \frac{\partial}{\partial y} & \frac{\partial}{\partial x} & 0 \\ 0 & \frac{\partial}{\partial z} & \frac{\partial}{\partial y} \\ \frac{\partial}{\partial z} & 0 & \frac{\partial}{\partial x} \end{bmatrix} \tag{4.182}$$

Equation (4.180) can be rewritten in matrix form as:

$$\underline{\varepsilon} = [d]\underline{u} \tag{4.183}$$

The relation, for isotropic material, between stresses and strains, becomes, in matrix form [3, 26]:

$$\underline{\sigma} = [E]\underline{\varepsilon} \tag{4.184}$$

where $\underline{\sigma}$ is the vector of the stresses:

$$\underline{\sigma}^T = \left\{ \sigma_x \quad \sigma_y \quad \sigma_z \quad \gamma_{xy} \quad \gamma_{yz} \quad \gamma_{zx} \right\} \tag{4.185}$$

and [E] is the matrix that correlates the stresses to the strains:

$$[\mathrm{E}] = \frac{\mathrm{E}}{(1+v)(1-2v)}\begin{bmatrix} (1-v) & 0 & 0 & 0 & 0 & 0 \\ 0 & (1-v) & 0 & 0 & 0 & 0 \\ 0 & 0 & (1-v) & 0 & 0 & 0 \\ 0 & 0 & 0 & \left(\frac{1-2v}{2}\right) & 0 & 0 \\ 0 & 0 & 0 & 0 & \left(\frac{1-2v}{2}\right) & 0 \\ 0 & 0 & 0 & 0 & 0 & \left(\frac{1-2v}{2}\right) \end{bmatrix} \tag{4.186}$$

being E the Young's modulus of the material and v the Poisson's ratio. Keeping account of Eqs. (4.178), (4.183) and (4.184), stresses and strains can be directly expressed as functions of nodal displacements, i.e.:

$$\underline{\varepsilon} = [\mathrm{d}][\mathrm{f_j}]\underline{\mathrm{X}}_{\mathrm{jl}} = [\mathrm{B_j}]\underline{\mathrm{X}}_{\mathrm{jl}} \tag{4.187}$$

$$\underline{\sigma} = [\mathrm{E}]\underline{\varepsilon} = [\mathrm{E}][\mathrm{B_j}]\underline{\mathrm{X}}_{\mathrm{jl}} \tag{4.188}$$

where $[\mathrm{B_j}]$ is a 6 × n matrix containing the space derivatives of the shape functions $[\mathrm{f_j}]$ (*n* is the number of d.o.f of the generic finite element in use). It is then possible to define, for the generic three-dimensional finite element (of volume Vol) the potential energy $\mathrm{V_j}$ related to elastic forces, the kinetic energy $\mathrm{E_{cj}}$ and the virtual work $\delta * \mathrm{L_j}$ performed by external forces $p(x, y, z, t)$, generally distributed on the finite element, as:

$$\begin{aligned} V_j &= \frac{1}{2}\int_{Vol} \underline{\sigma}^T \underline{\varepsilon} dVol \\ E_{cj} &= \frac{1}{2}\int_{Vol} \rho \underline{\dot{u}}_j^T \underline{\dot{u}}_j dVol \\ \delta * L_j &= \int_{Vol} \underline{p}_j(x,y,z,t)^T \delta * \underline{u}_j dVol \end{aligned} \tag{4.189}$$

being ρ the material density, so that $\rho\, dVol$ is the mass of the infinitesimal element. It has to be observed that $\underline{\dot{u}}_j$ is a column matrix containing the three components of the geometric velocity vector of the infinitesimal element of mass $\rho\, dVol$. In (4.189) $\delta^* \underline{\mathrm{u}}_\mathrm{j}$ is the vector containing the virtual displacement components of the generic point of application of a force:

$$\delta^* \underline{u}_\mathrm{j}^\mathrm{T} = \{\delta^*\mathrm{u} \quad \delta^*\mathrm{w} \quad \delta^*\mathrm{v}\} \tag{4.190}$$

and $\underline{p}_j$ is the vector of the corresponding force components:

$$\underline{p}_j^T = \{ p_x \quad p_y \quad p_z \} \tag{4.191}$$

Equation (4.189) can be expressed, keeping account of (4.188) and (4.187), as functions of nodal displacement coordinates:

$$\begin{aligned} V_j &= \frac{1}{2}\underline{X}_{jl}^T \left[\int_{Vol} [B_j]^T [E][B_j] dVol \right] \underline{X}_{jl} = \frac{1}{2}\underline{X}_{jl}^T [K_{jl}] \underline{X}_{jl} \\ E_{cj} &= \frac{1}{2}\dot{\underline{X}}_{jl}^T \left[\int_{Vol} \rho [f_j]^T [f_j] dVol \right] \dot{\underline{X}}_{jl} = \frac{1}{2}\dot{\underline{X}}_{jl}^T [M_{jl}] \dot{\underline{X}}_{jl} \\ \delta^* L_j &= \left\{ \int_{Vol} \underline{p}_j(x,y,z,t)^T [f_j(x,y,z)] dVol \right\} \delta^* \underline{X}_{jl} = \underline{F}_{jl}^T \delta^* \underline{X}_{jl} \end{aligned} \tag{4.192}$$

where $[K_{jl}]$:

$$[K_{jl}] = \left[\int_{Vol} [B_j]^T [E][B_j] dVol \right] \tag{4.193}$$

is the generic stiffness matrix; $[M_{jl}]$ is the mass matrix:

$$[M_{jl}] = \left[\int_{Vol} \rho [f_j]^T [f_j] dVol \right] \tag{4.194}$$

and, finally, $\underline{F}_{jl}$ is the vector of the generalized forces:

$$\underline{F}_{jl} = \left\{ \int_{Vol} \underline{p}_j(x,y,z,t)^T [f_j(x,y,z)] dVol \right\} \tag{4.195}$$

The dimensions of these matrices obviously change when the number of d.o.f. associated with the generic element varies, despite their morphology remaining unchanged. When the type of finite element varies, the matrices of the shape functions $[f_j]$ change. Let us now analyse some of the main types of two-dimensional and three-dimensional finite elements.

4.6.3 Two-Dimensional Elements (Membrane)

4.6.3.1 The Triangular Flat Element

Figure 4.36 shows a classical triangular element (defined as a linear triangular element with 3 nodes), in which, as independent variables, we assume the 6 displacements of the 3 end nodes:

$$\underline{X}_{j1}^{T} = \{X_1 \quad Y_1 \quad Z_1 \quad X_2 \quad Y_2 \quad Z_2\} \tag{4.196}$$

By using this element, it is possible to simulate the behaviour of a system subjected to plane stresses and strains. The displacement components u and v of the generic infinitesimal element inside the generic finite element are defined unequivocally by these 6 displacements: a simpler representation is given by linear shape functions of the type:

$$\begin{aligned} u &= a_1 + a_2 x + a_{3y} \\ v &= b_1 + b_2 x + b_{3y} \end{aligned} \tag{4.197}$$

or rather in matrix form:

$$\underline{u} = \begin{Bmatrix} u \\ v \end{Bmatrix} = [P_{t3}]\underline{a}_{t3} = \begin{bmatrix} 1 & x & y & 0 & 0 & 0 \\ 0 & 0 & 0 & 1 & x & y \end{bmatrix} \begin{Bmatrix} a_1 \\ a_2 \\ a_3 \\ b_1 \\ b_2 \\ b_3 \end{Bmatrix} \tag{4.198}$$

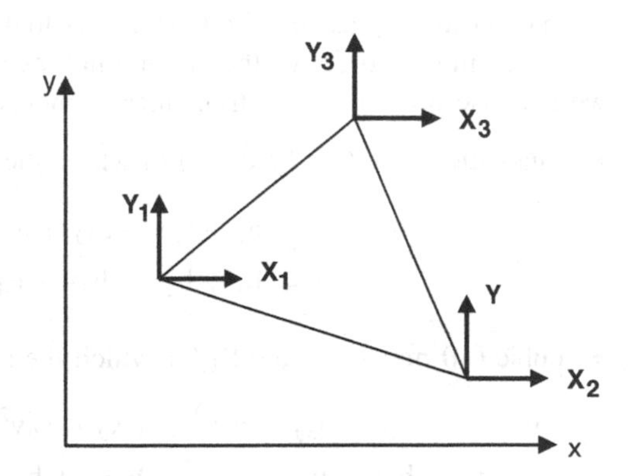

Fig. 4.36 Linear triangular element with a constant deformation

The 6 constants in (4.198) can be defined by 6 boundary conditions:

$$\begin{aligned} X_1 &= u(x_1, y_1) = a_1 + a_2 x_1 + a_3 y_1 \\ Y_1 &= v(x_1, y_1) = b_1 + b_2 x_1 + b_3 y_1 \\ X_2 &= u(x_2, y_2) = a_1 + a_2 x_2 + a_3 y_2 \\ &\ldots \end{aligned} \tag{4.199}$$

which, always in matrix form, become:

$$\underline{X}_{jl} = \begin{Bmatrix} X_1 \\ Y_1 \\ X_2 \\ Y_2 \\ X_3 \\ Y_3 \end{Bmatrix} = [C_{t3}]\underline{a}_{t3} = \begin{bmatrix} 1 & x_1 & y_1 & 0 & 0 & 0 \\ 0 & 0 & 0 & 1 & x_1 & y_1 \\ 1 & x_2 & y_2 & 0 & 0 & 0 \\ 0 & 0 & 0 & 1 & x_2 & y_2 \\ 1 & x_3 & y_3 & 0 & 0 & 0 \\ 0 & 0 & 0 & 1 & x_3 & y_3 \end{bmatrix} \begin{Bmatrix} a_1 \\ a_2 \\ a_3 \\ b_1 \\ b_2 \\ b_3 \end{Bmatrix} \tag{4.200}$$

By imposing relations (4.198), using the same procedure as that used for the finite beam and cable element, it is possible to define the generic displacement vector of the generic infinitesimal element inside the finite element as:

$$\underline{u} = \begin{Bmatrix} u \\ v \end{Bmatrix} = [P_{t3}][C_{t3}]^{-1}\underline{X}_{jl} = [f_{t3}(x, y)]\underline{X}_{jl} \tag{4.201}$$

where $[f_{t3}(x, y)]$ is a matrix consisting of 2 lines and 6 columns, a function of only the spatial coordinates x and y. Using the general formulation shown in (4.193), this expression enables us to evaluate the stiffness $[K_{t3}]$ and mass $[M_{t3}]$ matrix of the same triangular finite element with a constant strain. From an operative point of view, in order to define $[f_{t3}]$ the approach presently defined is not convenient. On the contrary, it is advisable to use another set of coordinates, termed *area coordinates* (see [26]).

In order to decrease the d.o.f. necessary to define the behaviour of a real structure by increasing the order of the polynomial relative to the shape function, in practice we also use other types of triangular elements (Fig. 4.37):

- quadratic (6 nodes, 12 d.o.f.) in which the shape function is of the type:

$$\begin{aligned} u &= a_1 + a_2 x + a_3 y + a_4 x^2 + a_5 xy + a_6 y^2 \\ v &= b_1 + b_2 x + b_3 y + b_4 x^2 + b_5 xy + b_6 y^2 \end{aligned} \tag{4.202}$$

- cubic (10 nodes, 20 d.o.f.) for which the shape function becomes:

$$\begin{aligned} u &= a_1 + a_2 x + a_3 y + a_4 x^2 + a_5 xy + a_6 y^2 + a_7 x^3 + a_8 x^2 y + a_9 xy^2 + a_{10} y^3 \\ v &= b_1 + b_2 x + b_3 y + b_4 x^2 + b_5 xy + b_6 y^2 + b_7 x^3 + b_8 x^2 y + b_9 xy^2 + b_{10} y^3 \end{aligned} \tag{4.203}$$

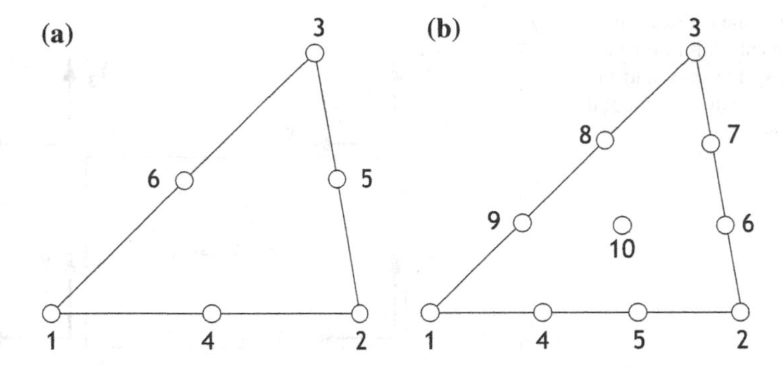

Fig. 4.37 Other types of triangular plane elements: **a** triangular quadratic element; **b** triangular cubic element

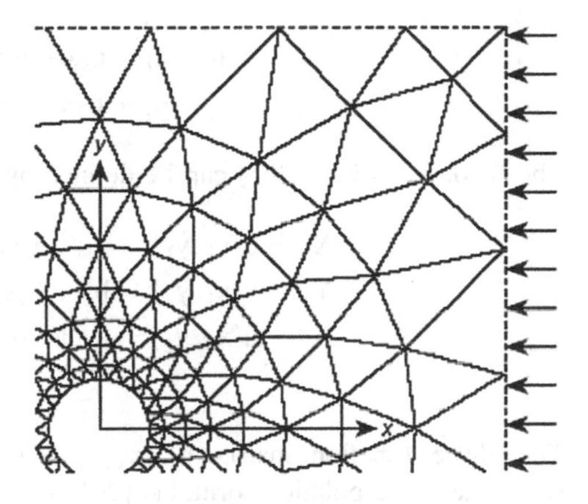

Fig. 4.38 Triangular finite element: an application example

As an example, Fig. 4.38 shows a finite element model to simulate the strain and stress state (static and dynamic) in the neighbourhood of a hole in the event of plane stresses.

4.6.3.2 Rectangular Plane Element (Membrane)

Another two-dimensional finite element is the linear rectangular one with 4 nodes (Fig. 4.39) for which, as independent variables, the displacement of the 4 end nodes:

$$\underline{X}_{jl}^{T} = \{X_1 \quad Y_1 \quad X_2 \quad Y_2 \quad X_3 \quad Y_3 \quad X_4 \quad Y_4\} \tag{4.204}$$

The components of the generic displacement u and v of the generic infinitesimal element inside the generic element are defined unequivocally by these 8 displacements with bilinear functions of the type:

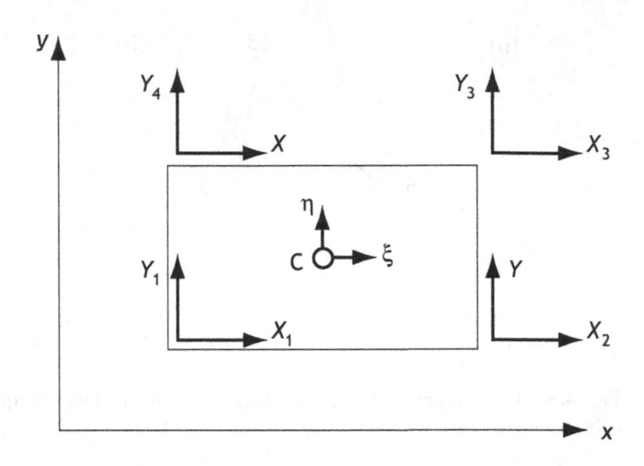

Fig. 4.39 Linear rectangular finite element: definition in reference system x–y and in reference system ξ–η (natural coordinates)

$$\begin{aligned} u &= a_1 + a_2 x + a_3 y + a_4 xy \\ v &= b_1 + b_2 x + b_3 y + b_4 xy \end{aligned} \tag{4.205}$$

The 8 constants in (4.205) can be defined by the 8 boundary conditions:

$$\begin{aligned} X_1 &= u(x_1, y_1) = a_1 + a_2 x_1 + a_3 y_1 + a_4 x_1 y_1 \\ Y_1 &= v(x_1, y_1) = b_1 + b_2 x_1 + b_3 y_1 + b_4 x_1 y_1 \\ X_2 &= u(x_2, y_2) = a_1 + a_2 x_2 + a_3 y_2 + a_4 x_2 y_2 \\ &\ldots \end{aligned} \tag{4.206}$$

The shape function matrix $[f_{r4}(x, y)]$ can be obtained directly by using the Lagrange's interpolation formulae [26] and a local reference system, of axes ξ, η and origin in the centre of rectangle x_c and y_c (Fig. 4.39):

$$\begin{aligned} \xi &= \frac{x - x_c}{b} \\ \eta &= \frac{y - y_c}{c} \end{aligned} \tag{4.207}$$

These new coordinates ξ and η, termed *natural* or *intrinsic*, allow us (in the same way as the coordinates of area defined for the triangular element and the current coordinate ξ for the finite beam or cable element) to define the characteristics of the finite element, regardless of its spatial orientation: furthermore, the use of natural coordinates allows for an easy extension to the treatment of curved isoparametric elements [3, 26]. By imposing relations (4.207) in (4.205) it is possible to express the displacements *u* and *v* as a function of the new natural coordinates ξ and η: in this way, the functions $u(\xi, \eta, t)$ and $v(\xi, \eta, t)$ prove to be independent from the dimension and spatial position assumed by the generic finite element.

By imposing the boundary conditions (4.206) in the new variables, the shape function matrix $[\mathbf{f}_{r4}]$ can immediately be evaluated as ([26], Sect. 5.4):

$$[\mathrm{f}_{r4}] = \begin{bmatrix} \mathrm{f}_1 & 0 & \mathrm{f}_2 & 0 & \mathrm{f}_3 & 0 & \mathrm{f}_4 & 0 \\ 0 & \mathrm{f}_1 & 0 & \mathrm{f}_2 & 0 & \mathrm{f}_3 & 0 & \mathrm{f}_4 \end{bmatrix} \tag{4.208}$$

where

$$\begin{aligned} f_1 &= \frac{1}{4}(1-\xi)(1-\eta) \\ f_2 &= \frac{1}{4}(1+\xi)(1-\eta) \\ f_3 &= \frac{1}{4}(1+\xi)(1+\eta) \\ f_4 &= \frac{1}{4}(1-\xi)(1+\eta) \end{aligned} \tag{4.209}$$

4.6.4 Three-Dimensional Elements (Brick Elements)

Analysis of the three-dimensional stress obviously encompasses all cases of practical interest although a two-dimensional analysis often provides a suitable and *economic* approximation. Similar to the triangular element in the two-dimensional analysis, the simpler three-dimensional finite element is a tetrahedron (Fig. 4.40): in order to define accurately the actual stress and strain state in a generic continuous system it is necessary to use a large number of these elements, thereby involving problems related to the occupation of storage space and calculation times. For this reason, we introduced other types of three-dimensional finite elements, with more complex shape functions, which do, however, allow us to reduce the overall number of d.o.f.

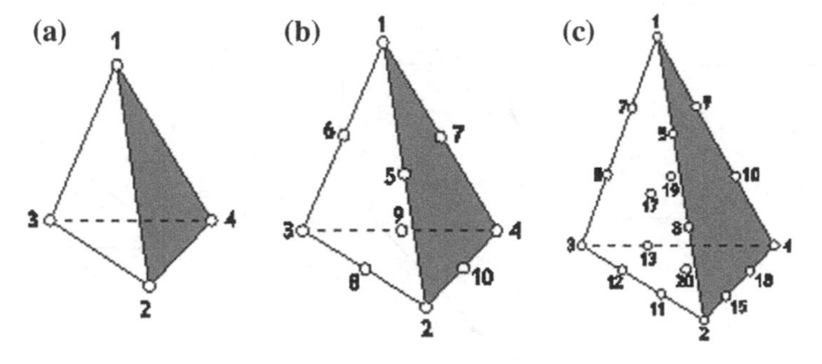

Fig. 4.40 Three-dimensional tetrahedral finite elements: **a** linear (4 nodes); **b** quadratic tetrahedral finite elements (8 nodes); **c** cubic tetrahedral element (20 nodes)

4.6.4.1 Tetrahedral Finite Element

In the linear tetrahedron with 4 nodes (Fig. 4.40), the displacements of the 4 end nodes are assumed as independent variables:

$$X_{jl}^{T} = \{ x_1 \quad y_1 \quad z_1 \quad x_2 \quad y_2 \quad z_2 \quad x_3 \quad y_3 \quad z_3 \quad x_4 \quad y_4 \quad z_4 \} \tag{4.210}$$

Components *u*, *v* and *w* of the displacement of the infinitesimal generic element inside the tetrahedron generic element are defined completely by these 12 displacements: in this case, as a shape function, we assume a linear shape function of the type:

$$\begin{aligned} u &= a_1 + a_2 x + a_3 y + a_4 z \\ v &= b_1 + b_2 x + b_3 y + b_4 z \\ w &= c_1 + c_2 x + c_3 y + c_4 z \end{aligned} \tag{4.211}$$

Other tetrahedral-type finite elements can have (see Figs. 4.40 and 4.41) 9 nodes (quadratic elements) and 20 nodes (cubic elements), with shape functions, obviously, of a higher order.

4.6.4.2 Brick Elements Consisting of Tetrahedrons with 8 Nodes

A more convenient representation (from a point of view of visualization of the mesh and numeration of the nodes) could be that in which the real continuous system is

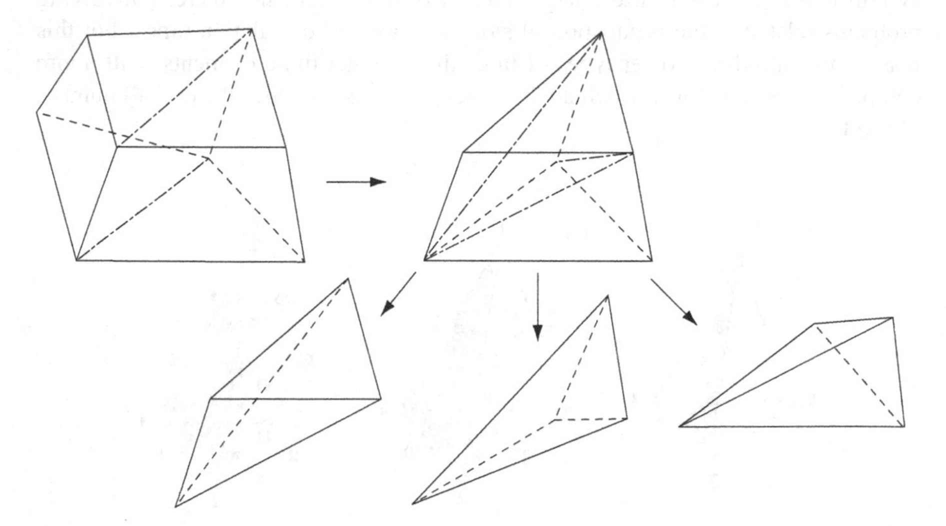

Fig. 4.41 A systematic approach of dividing the continuous system into brick elements with 8 nodes composed by several tetrahedrons

divided into brick elements with 8 nodes, each of which can subsequently be divided into 2 or more tetrahedral-type elements as shown in Fig. 4.41: this approach is useful to implement automatic schematization procedures.

4.6.4.3 Rectangular Prisms with 8 Nodes

Another three-dimensional finite element that can be used to analyse a generic state of spatial stress is the rectangular prismatic element (Fig. 4.42a) with 8 nodes. Other rectangular prismatic-type elements can have (see Fig. 4.42b) 20 nodes (quadratic elements) and 32 nodes (cubic elements), with shape functions of a higher order. As an example, in Fig. 4.43 we show other possible types of triangular prismatic-type finite elements, respectively linear (with 6 nodes), quadratic (with 8 nodes) and cubic (with 26 nodes).

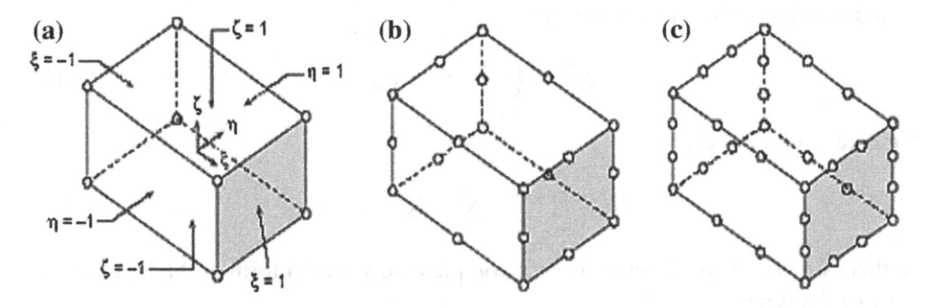

Fig. 4.42 **a** Linear prism finite element with 8 nodes; **b** quadratic element; **c** cubic element

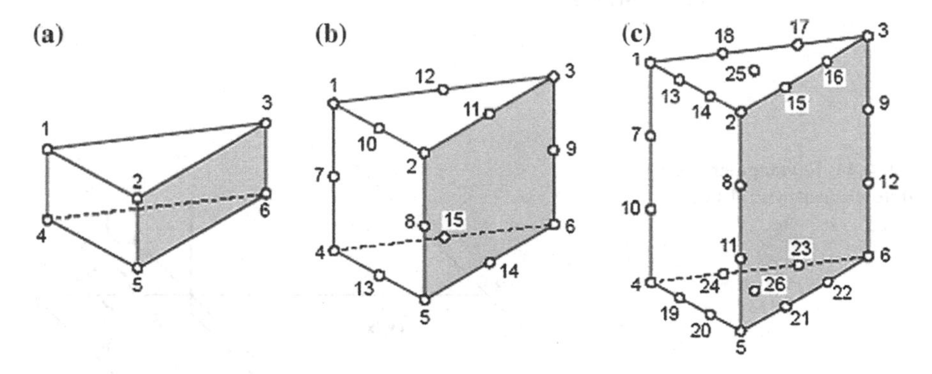

Fig. 4.43 Triangular prism finite elements: **a** linear, **b** quadratic and **c** cubic

4.6.5 Plate Elements and Shell Elements

In the previous paragraphs, the relations between stresses and strains were given in their exact form: conversely, in the classic plate approach [3, 26] we introduce approximations to simplify the problem (in reality three-dimensional) to a problem with two dimensions: the hypothesis normally introduced is that of considering the stress and strain variation linear in the normal direction to the plane of the plate and this hypothesis is valid for narrow plates and small strains. The strain state can thus be defined by the only transverse displacement $\mathbf{w}(\mathbf{x}, \mathbf{y}, \mathbf{z})$ of the medium plane of the plate, also imposing continuity on the derivative of this amplitude.

4.6.5.1 Rectangular Plate Element

Figure 4.44 shows the plate finite element with 4 nodes: associated with each node are 3 d.o.f. corresponding to the vertical displacement w, to rotation θ_{xi} around axis x and to rotation θ_{yi} around axis y:

$$\underline{X}_i^T = \{ w_i \quad \theta_{xi} \quad \theta_{yi} \} \tag{4.212}$$

for a total of 12 d.o.f.:

$$\underline{X}_{ji}^T = \{ \underline{X}_1^T \quad \underline{X}_2^T \quad \underline{X}_3^T \quad \underline{X}_4^T \} \tag{4.213}$$

in this case, the shape function is an incomplete polynomial form of the fourth order [26] of the type:

$$\begin{aligned} w(x, y) = {} & a_1 + a_2 x + a_3 y + a_4 x^2 + a_5 xy + a_6 y^2 + a_7 x^3 + a_8 x^2 y + a_9 xy^2 \\ & + a_{10} y^3 + a_{11} x^3 y + a_{12} xy^3 \end{aligned} \tag{4.214}$$

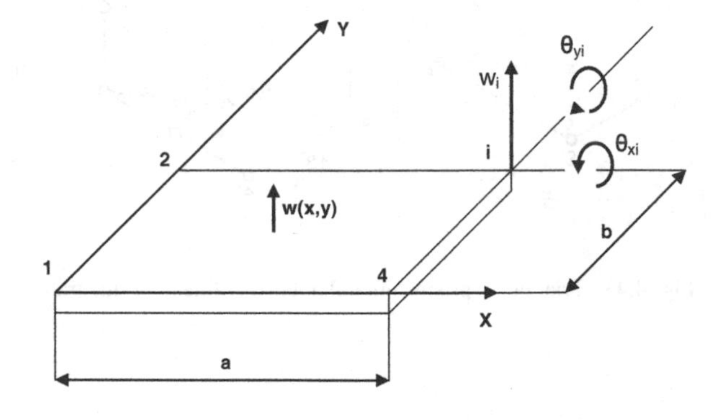

Fig. 4.44 Rectangular plate finite element with 4 nodes

As can be seen, the displacement with x = constant or y = constant is cubic and, in particular, on the boundaries: since a cubic function, as with the beam, is defined by 4 constants that are completely defined by imposing displacements and rotations on the ends, the 12 constants can be defined by imposing the 12 boundary conditions on the nodes:

$$\begin{aligned} w_i &= a_1 + a_2 x_i + a_3 y_i + \cdots \\ -\left(\frac{\partial w}{\partial y}\right)_i &= \theta_{xi} = -a_3 - a_5 x_i + \cdots \\ -\left(\frac{\partial w}{\partial x}\right)_i &= \theta_{yi} = -a_2 - 2a_4 x_i + \cdots \end{aligned} \tag{4.215}$$

The relative stiffness $\left[K_p\right]$ and mass $\left[M_p\right]$ matrix of the plate element can, at this point, be obtained from (4.193) and (4.194).

4.6.5.2 Shell Element (Shell)

A shell structure has both a *membrane* (Sect. 4.5.3.2) and *plate-type structure* in that it reacts both to in-plane forces and to bending stresses. The shell element is thus used to simulate the behaviour of these structures which, from a geometric point of view, have curved surfaces. To this purpose, it is possible to use:

- a plane shell element (4 nodes with 5 d.o.f. per node, Figs. 4.45 and 4.46)

$$\underline{X}_k^T = \left\{ x_k \quad y_k \quad z_k \quad \theta_{xk} \quad \theta_{yk} \quad \theta_{zk} \right\} \quad (k = 1, 2, 3, 4) \tag{4.216}$$

  a combination of a plate element and a membrane element: although this shell has a simple formulation it does not offer a high level of accuracy;
- a curved shell element, deriving from the theory of shells: when using this approach, we introduce many d.o.f. often with derivatives of a higher order than the first;
- a solid (or degenerated solid) element similar to the plate elements or isoparametric plate elements [3, 26].

4.6.6 Isoparametric Elements

The increase in the number of the nodes of the generic finite element causes a refinement of the shape function of same and therefore necessitates a lower number of d.o.f. in order to schematize suitably the generic continuous system. In order to simulate a relatively complex shape, using a limited number of finite elements, the

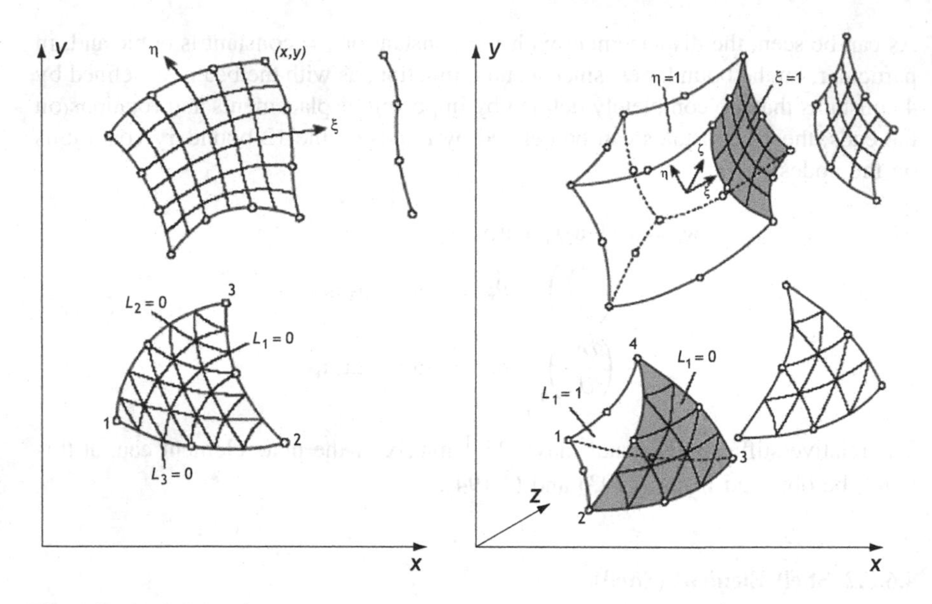

Fig. 4.45 Shell finite elements

Fig. 4.46 Shell finite elements: an application

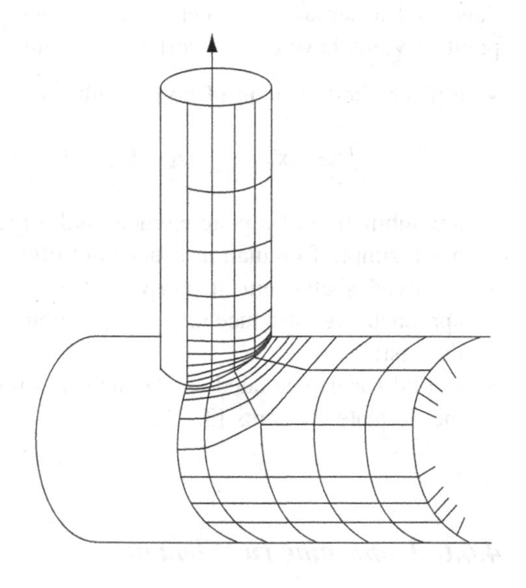

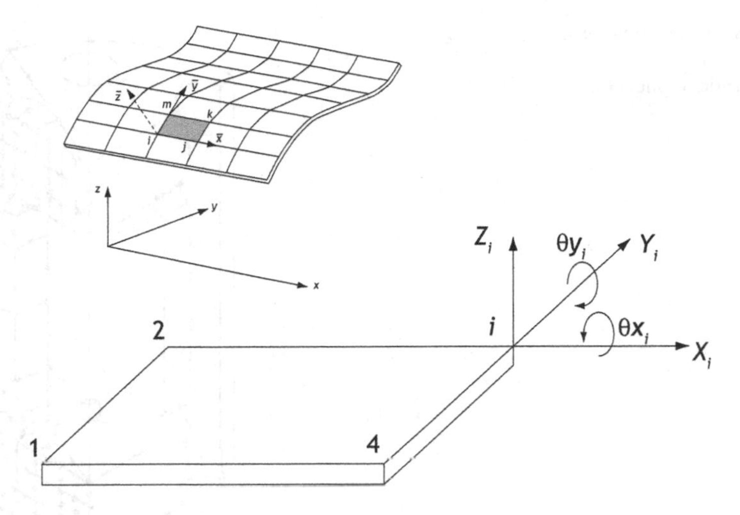

Fig. 4.47 Isoparametric elements

use of isoparametric elements (Fig. 4.47)[14] might prove to be necessary. Their main characteristic is the fact that they have curvilinear boundaries that are able to clearly define solids with curved boundaries and reduce or thicken the mesh, e.g. in those areas that have high stress gradient values (Fig. 4.48). For more details reference should be made to the bibliography.

[14]The coordinates of a generic point inside a generic finite element can be defined as a function of the coordinates of the end nodes of same as:

$$\underline{x}_p = \begin{Bmatrix} x_p \\ y_p \\ z_p \end{Bmatrix} = [N_p]\underline{C} = [N_p]\begin{Bmatrix} x_1 \\ y_1 \\ z_1 \\ \dots \\ x_m \\ y_m \\ z_m \end{Bmatrix} \tag{4.14.1}$$

This expression is formally the same as that used to define the link between the displacements of the same point P as a function of displacements $\underline{X}_{jl}$ of the end nodes (4.95). In relations (4.14.1) both $[N_p]$, and the shape function matrix $[f_j]$ prove to be dependent on the spatial coordinates: an element is termed "isoparametric" if $[N_p]$ and $[f_j]$ are identical [10].

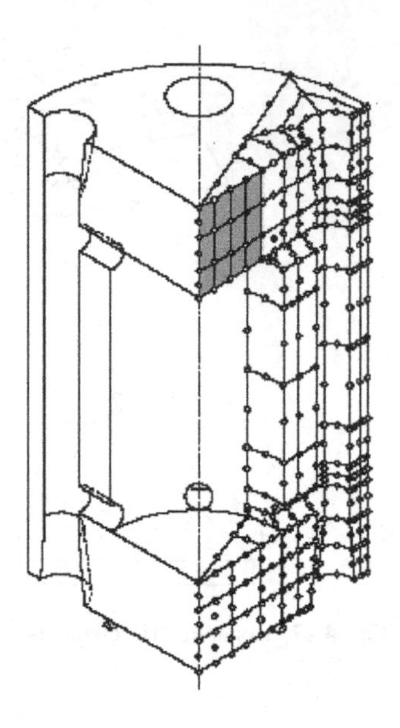

Fig. 4.48 An example of the application of an isoparametric application

4.7 Nonlinear Analysis in Structures Using the Finite Element Method (Brief Outline)

Until now, we have analysed the dynamic behaviour of structures in the linear field, in other words we assumed a linear relationship between stresses and strains and we analysed small oscillations about a pre-established equilibrium position: when these assumptions are not verified, the equations of motion obviously become nonlinear. Several of the behaviours of the structure can be classified as nonlinear:

- nonlinearity of material, caused by a specific link between stresses and strains (both dependent and not dependent on time);
- nonlinearity correlated to the effect of large displacements or rotations;
- nonlinearity due to variations in constraint conditions; in particular parts of the structure can slide relatively and clearances in connections may open or close.

In the latter two cases we generally talk about geometric nonlinearity. Often, the most common structural problems can be solved by using linear algorithms in that, generally speaking, materials work in an elastic field and the strains that can be supported by mechanical systems are small: furthermore, small nonlinearities do not basically change a basic project defined in a linear field. Conversely, when nonlinearities become significant, a nonlinear analysis, which is also more complicated from a computational point of view, is necessary. This is also due to the fact that, in this case, the numerical procedures have not yet been completely standardized. Often, one of the disadvantages of this approach is the difficulty of interpreting results and the fact that it does not allow for analyses of either sensitivity or optimization in *engineering* times. The aim of this paragraph is simply to provide a brief outline of the main problems related to nonlinear (static and dynamic) analysis, with an application targeted at nonlinear techniques in finite elements: in this case it is necessary to point out that, in Sects. 1.5 and 2.3.2, the methodologies for the analysis of nonlinear 1 or *n* d.o.f. systems, consisting of mutually coupled rigid bodies, have already been described. The explanation given in this paragraph is not exhaustive and we therefore suggest that reference is made to the detailed biography on the subject [2, 3, 10, 18, 21, 26]. In chapter,[15] we will analyse, more specifically, geometric nonlinearities. In order to simplify matters, we will base our explanation on an application example.

4.7.1 Introduction to the Non-linear Problem

In the previous paragraphs, we wrote the equations of motion of a generic system discretized with finite elements, based on the assumption of the linear behaviour of same and obtaining a matrix equation of the type:

$$[M]\ddot{\underline{X}} + [R]\dot{\underline{X}} + [K]\underline{X} = \underline{F}(t) \tag{4.217}$$

Relation (4.217) was obtained by assuming that:

- the material always works in a linear field, in other words a linear link between stresses σ and strains ε is assumed;
- displacements $\underline{X}$ of the complete model of the structure which, as previously mentioned, represent a disturbed motion about the static equilibrium position, are small and thus incapable of introducing geometric nonlinearities;
- the constraint conditions remain unaltered during the application of loads.

---

[15] In relation to the finite element method, the Newton-Raphson method was developed in a particular form based on two different approaches [2, 10, 26]:

- A Lagrangian approach (termed "stationary Lagrangian" or "total Lagrangian";
- A Eulerian approach (defined as "updated Lagrangian").

In the first approach, the reference system remains the absolute one regardless of how big the rotations or displacements become: as the displacements gradually become bigger, it is necessary to add stresses and strains to the link, nonlinear terms that give rise to stiffness matrices, additional linear, quadratic, etc. functions of the same independent variables. Conversely, in the Eulerian approach, a local reference system, connected to each finite element (and, therefore, different from element to element) is used. The position of each local reference system is re-updated as a function of the new configuration $\underline{\mathbf{X}}^{(i)}$ reached by the system.

The advantages of one method with respect to the other mainly depend on the type of application and there are no set rules for the choice of one or other algorithm. In the explanation that follows, we will only refer to the Eulerian formula ("updated Lagrangian") which, generally speaking, lends itself better to the analysis of geometric nonlinearities.

When these conditions are not verified, the behaviour of the system is nonlinear. In this case, the equations of motion formally become[16]:

$$\underline{F}_i(\underline{X},\dot{\underline{X}},\ddot{\underline{X}}) + \underline{F}_s(\underline{X},\dot{\underline{X}}) + \underline{F}_e(\underline{X}) = \underline{F}(t) \tag{4.218}$$

where

- $\underline{F}(t)$ is the vector of the external forces;
- $\underline{F}_i$ is the vector of the generalized inertia forces, nonlinear functions of displacements $\underline{X}$, velocity $\dot{\underline{X}}$ and accelerations $\ddot{\underline{X}}$ of the system;
- $\underline{F}_s$ is the vector of the generalized dissipation forces, nonlinear functions of displacements $\underline{X}$ and of velocity $\dot{\underline{X}}$ of the system;
- $\underline{F}_e$ finally, is the vector of the elastic forces, also generally nonlinear functions of displacements $\underline{X}$.

The direct integration of these equations (4.218) is performed by using step-by-step numerical integration methods: this approach to the problem is necessary when one needs to evaluate the motion in large of the nonlinear system. Often the motion of the system occurs as the result of small displacements about the static equilibrium position. In this case, it is, however, necessary to first evaluate the static equilibrium position $\underline{X}_0$ by solving the nonlinear static problem:

$$\underline{F}_e(\underline{X}_0) = \underline{F}_c \tag{4.219}$$

where $\underline{F}_c$ was used to indicate the constant part of the constant external forces applied to the system. About this position it is subsequently possible to linearize the equations of motion (4.218). After adding and subtracting $\underline{F}_c$ we obtain:

$$[M_0]\ddot{\underline{X}} + [R_0]\dot{\underline{X}} + \underline{F}_e(\underline{X}_0) + [K_0]\{\underline{X} - \underline{X}_0\} = \{\underline{F}(t) - \underline{F}_c\} + \underline{F}_c \tag{4.220}$$

where, respectively:

$$[M_0] = \left[\frac{\partial \underline{F}_i}{\partial \ddot{\underline{X}}}\right]_{X=X_0}; \quad [R_0] = \left[\frac{\partial \underline{F}_s}{\partial \dot{\underline{X}}}\right]_{X=X_0}; \quad [K_0] = \left[\frac{\partial \underline{F}_e}{\partial \underline{X}}\right]_{X=X_0} \tag{4.221}$$

$[M_0]$ is the mass matrix of the system, evaluated about the static equilibrium position, $[R_0]$ is the relative damping matrix and finally $[K_0]$ is the tangent stiffness matrix. By keeping account of the static equilibrium equation (4.219) and performing a change of variables:

$$\underline{X} = \underline{X}_0 + \overline{\underline{X}} \tag{4.222}$$

[16]These equations can be obtained by means of Lagrange's equations: in particular, it is important to remember that the terms will be nonlinear if the various forms of energy are not quadratic functions (a propos of this, see Chap. 2, Sect. 2.4.2).

having used $\underline{\overline{X}}$ to indicate the vector that defines the disturbed motion about the static equilibrium position $\underline{X}_0$, relation (4.220) can be rewritten as:

$$[M_0]\ddot{\underline{\overline{X}}} + [R_0]\dot{\underline{\overline{X}}} + [K_0]\underline{\overline{X}} = \{\underline{F}(t) - \underline{F}_c\} = \underline{\overline{F}}(t) \tag{4.223}$$

This equation proves to be linear.

4.7.2 *Linearization of the Equations of Motion About the Equilibrium Position*

The methodology necessary to linearize the equations of motion about the static equilibrium position has already been described in depth in Chaps. 1 and 2, with particular emphasis on discrete systems generally with n d.o.f. We will now describe the methodology necessary to define the static equilibrium position of a nonlinear system, with particular reference to the finite element technique.

4.7.2.1 Nonlinear Static Analysis

Equation (4.219) which defines the configuration assumed by the system under static conditions can be obtained from the Lagrange's equations, by annulling the terms of velocity and acceleration. Generally speaking, this equation is nonlinear and its solution must be looked for using specific numerical procedures: among the methods used, the most common is the Newton-Raphson method, with its numerous variants. From a didactic point of view, in order to simplify the explanation, we will now refer to the highly frequent and particularly simple example of constant external forces $\underline{F}(t) = \underline{P}$ from which we obtain:

$$\underline{F}_e(\underline{X}) = \underline{P} \tag{4.224}$$

We will analyse this method applied, first and foremost, to a system with only one d.o.f. (for which it is also possible to give a significant geometric interpretation of the algorithm), in order to subsequently (Sect. 4.7.2.1.2) generalize the approach for generic n d.o.f. systems.

4.7.2.1.1 The Newton-Raphson Method Applied to One-Degree-of-Freedom Systems

The equation that defines the static equilibrium position in a nonlinear one d.o.f. system, can thus be rewritten as:

$$F_e(x) = P \tag{4.225}$$

in which we have also kept account of the fact that the external forces P are constant and where the only free coordinate of the system has been indicated by x. An iterative approach can be used when looking for the solution, for example, by using the Newton-Raphson method, expanding in a Taylor series, about an initial trial solution (defined by prime$^{(0)}$) Eq. (4.225):

$$F_e\left(x^{(0)}\right) + \left(\frac{\partial F_e}{\partial x}\right)_{x=x^{(0)}} \left(x - x^{(0)}\right) = P \tag{4.226}$$

Equation (4.226) can be rewritten as:

$$\left(\frac{\partial F_e}{\partial x}\right)_{x=x^{(0)}} \left(x - x^{(0)}\right) = P - F_e\left(x^{(0)}\right) \Rightarrow K^{(0)} \Delta x^{(0)} = \Delta R^{(0)} \tag{4.227}$$

where the term:

$$\Delta R^{(0)} = P - F_e\left(x^{(0)}\right) \tag{4.228}$$

can be interpreted as a static unbalance associated with the fact that the position of the first attempt does not represent the real static solution (otherwise, this term would, obviously, be null). In (4.227) the term:

$$K^{(0)} = \left(\frac{\partial F_e}{\partial x}\right)_{x=x^{(0)}} = \left(\frac{\partial^2 V}{\partial x^2}\right)_{x=x^{(0)}} \tag{4.229}$$

represents the tangent stiffness evaluated about the chosen first trial position: this term generally proves to be variable with the chosen $x^{(0)}$ position (its geometric meaning is highlighted in Fig. 4.49). Finally, the term:

$$\Delta x^{(0)} = \left(x - x^{(0)}\right) \tag{4.230}$$

represents the difference between the first trial position $x^{(0)}$ and solution **x** which derives from the linearized expression of $F_e(x)$: this amplitude is an index of error in the estimation of the equilibrium position. This method foresees the calculation of $\Delta x^{(0)}$ by solving (4.230) and by assuming, as the next new position (i = 1):

$$x^{(1)} = x^{(0)} + \Delta x^{(0)} \tag{4.231}$$

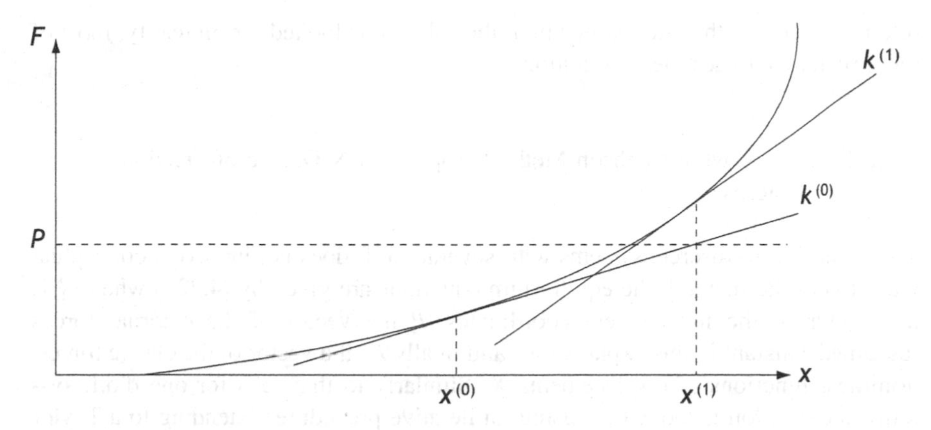

Fig. 4.49 Nonlinear one d.o.f. system: Newton-Raphson method

by once again solving:

$$\left(\frac{\partial F_e}{\partial x}\right)_{x=x^{(1)}}\left(x-x^{(1)}\right)=P-F_e\left(x^{(1)}\right)\;\Rightarrow\;K^{(1)}\Delta x^{(1)}=\Delta R^{(1)} \tag{4.232}$$

where $K^{(1)}$ represents the tangent stiffness, unlike the previous case on account of being evaluated in correspondence to a different $x^{(1)}$ strain. The procedures continues in an iterative way by using the generic *i*th step to define the trial solution as:

$$x^{(i)}=x^{(i-1)}+\Delta x^{(i-1)}=x^{(0)}+\sum_{k=0}^{i-1}\Delta x^{(k)} \tag{4.233}$$

and solving:

$$\left(\frac{\partial F_e}{\partial x}\right)_{x=x^{(i)}}\left(x-x^{(i)}\right)=P-F_e\left(x^{(i)}\right)\Rightarrow K^{(i)}\Delta x^{(i)}=\Delta R^{(i)} \tag{4.234}$$

From the first Eq. (4.233) we obtain a value of $x^{(i)}$ which, after being inserted into the second Eq. (4.234), provides a better approximation of the solution been looked for. The solution is found when, for a certain value of $x^{(i)}$, $\Delta x^{(i)}$ proves to be null: this condition is not normally reached and, in actual fact, the iterative procedure (whose geometric meaning, for a 1 d.o.f. system, is described in Fig. 4.49) is halted when this condition is verified:

$$\Delta x^{(i)}<\varepsilon \tag{4.235}$$

where ε is the error, as small as is willed, considered acceptable. This procedure proves to be convergent for a monotonous function without maximum or minimum relative values, in the interval in which the solution is looked for: in reality, most of the structure satisfies these conditions.

4.7.2.1.2 The Newton-Raphson Method Applied to N-Degree-of-Freedom Systems

The extension to discrete systems with several d.o.f. does not involve a conceptual variation of the method: the equilibrium equations are given by (4.224) where $\underline{X}$ is the vector of the independent coordinates, $\underline{P}$ the vector of the external forces (assumed constant in this explanation) and finally $\underline{F}_e$ the vector of the elastic forces, nonlinear functions of displacements $\underline{X}$. Similarly to that seen for one d.o.f. systems, the solution is looked for using an iterative procedure, extending to a Taylor series, about a trial solution (defined by means of prime$^{(i)}$) Eq. (4.224) and triggering an iterative procedure that can be summarized by the following equations:

$$\underline{X}^{(i)} = \underline{X}^{(i-1)} + \Delta\underline{X}^{(i-1)} = \underline{X}^{(0)} + \sum_{k=0}^{i-1} \Delta\underline{X}^{(k)} \tag{4.236}$$

$$\left[\frac{\partial F_e}{\partial \underline{X}}\right]_{x=x^{(i)}} \left\{\underline{X} - \underline{X}^{(i)}\right\} = \underline{P} - \underline{F}_e\left(\underline{X}^{(i)}\right) \Rightarrow \left[K^{(i)}\right]\Delta\underline{X}^{(i)} = \Delta\underline{R}^{(i)} \tag{4.237}$$

The iterative procedure stops when condition:

$$\left|\Delta\underline{X}^{(i)}\right| < \varepsilon \tag{4.238}$$

is verified or rather when norm $\left|\Delta\underline{X}^{(i)}\right|$ of the difference between the displacements between the generic ith iteration and the previous one is lower than a certain maximum, prefixed error. In Eq. (4.237) vector:

$$\Delta\underline{R}^{(i)} = \underline{P} - \underline{F}_e\left(\underline{X}^{(i)}\right) \tag{4.239}$$

can be interpreted as the static unbalance due to the fact that the generic $\underline{X}^{(i)}$ position does not represent the real static solution (otherwise this terms would obviously be a rigorously null vector).

Matrix $\left[K^{(i)}\right]$:

$$\left[K^{(i)}\right] = \left[\frac{\partial F_e}{\partial \underline{X}}\right]_{x=x^{(i)}} = \left[\frac{\partial^2 V}{\partial \underline{X}^2}\right]_{x=x^{(i)}} \tag{4.240}$$

represents the tangent stiffness matrix evaluated about the generic ith trial position. This matrix coincides with the stiffness matrix calculated using the linear approach developed in the previous paragraphs and is a function of the $\underline{X}^{(i)}$ position about which the same is evaluated. The vector of the elastic forces $\underline{F}_e\left(\underline{X}^{(i)}\right)$, a generally nonlinear function of displacements $\underline{X}^{(i)}$, is formally evaluated using the relation:

$$\underline{F}_e\left(\underline{X}^{(i)}\right) = \left\{\frac{\partial V}{\partial \underline{X}}\right\}^{(i)} \tag{4.241}$$

In Eq. (4.237) vector:

$$\Delta\underline{X}^{(i)} = \left\{\underline{X} - \underline{X}^{(i)}\right\} \tag{4.242}$$

represents the difference between the attempt position $\underline{X}^{(i)}$ and the real solution $\underline{X}$: this amplitude is an index of error in the estimation of the equilibrium position. Finally, let us remember that, often, to improve the convergence of the method, the external load $\underline{P}$ is not applied in a single solution, but is divided into fractions which are subsequently applied, calculating, each time, the intermediate equilibrium position reached by following the procedure outlined above.

In this way, the gradual application of a load is simulated, the aim being to avoid the first attempt solution being too far off from the final one, which could lead to convergence problems of the method.

4.7.2.1.3 An Application Example

As an example, let us now analyse the span of a high tension electric line previously analysed in Sect. 4.4, Fig. 4.10. Let us consider the schematization already adopted in Sect. 4.4.1 (Fig. 4.11), or rather a model of the span divided into 8 finite elements and nine nodes, and, furthermore, let us only study its behaviour in the vertical plane. We will now evaluate the static equilibrium position reached by the cable subjected to the initial tension S_o and its own weight (Fig. 4.50). The vector of the independent variables $\underline{X}_t$, already divided into free d.o.f. $\underline{X}_L$ and constraint d.o.f. $\underline{X}_V$ has already been defined in (4.84):

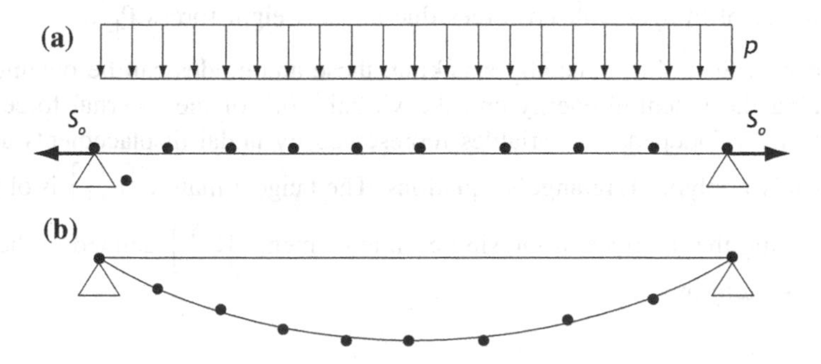

Fig. 4.50 Taut cable: **a** initial configuration $\underline{X}^{(o)}$, **b** generic configuration $\underline{X}^{(i)}$

$$\underline{X}_t = \begin{Bmatrix} \underline{X}_L \\ \underline{X}_V \end{Bmatrix} \tag{4.243}$$

where, as we recall,

$$\begin{aligned} \underline{X}_V^T &= \{ X_1 \quad Y_1 \quad X_9 \quad Y_9 \} \\ \underline{X}_L^T &= \{ X_2 \quad Y_2 \quad X_3 \quad Y_3 \quad X_4 \quad Y_4 \quad X_5 \quad Y_5 \quad X_6 \quad Y_6 \quad X_7 \quad Y_7 \quad X_8 \quad Y_8 \} \end{aligned} \tag{4.244}$$

The expressions that define the linearized equations about a generic static configurations are (4.85)

$$\begin{bmatrix} [K_{LL}] & [K_{LV}] \\ [K_{VL}] & [K_{VV}] \end{bmatrix} \begin{Bmatrix} \underline{X}_L \\ \underline{X}_V \end{Bmatrix} = \begin{Bmatrix} \underline{F}_L \\ \underline{F}_V + \underline{R}_V \end{Bmatrix} = \begin{Bmatrix} \underline{F}_L \\ \underline{F}_V \end{Bmatrix} \tag{4.245}$$

The iterative procedure used to determine this configuration, already defined in the previous paragraph (4.236), (4.237), is shown below for reasons of convenience:

$$\underline{X}_L^{(i)} = \underline{X}_L^{(i-1)} + \Delta\underline{X}_L^{(i-1)} = \underline{X}_L^{(0)} + \sum_{k=0}^{i-1} \Delta\underline{X}_L^{(k)} \tag{4.246}$$

$$\left[K_{LL}^{(i)}\right] \Delta\underline{X}_L^{(i)} = \Delta\underline{R}_L^{(i)} \tag{4.247}$$

$$\Delta\underline{R}_L^{(i)} = \underline{P}_L^{(i)} - \underline{F}_{eL}\left(\underline{X}_L^{(i)}\right) \tag{4.248}$$

In order to solve these equations using an iterative method and having assigned the generic trial configuration $\underline{X}_L^{(i)}$, it is necessary to define:

- the tangent stiffness matrix $\left[K_{LL}^{(i)}\right]$;
- the vector of the elastic forces $\underline{F}_{eL}\left(\underline{X}_L^{(i)}\right)$;
- the vector of the generalized forces due to the weight forces $\underline{P}_L^{(i)}$.

As seen in Sect. 4.6, generally speaking, these amplitudes can be obtained, by determining the potential energy and the virtual work of the external forces as a function of the independent variables represented by nodal displacements and by subsequently applying Lagrange's equations. The tangent matrix $\left[K_{LL}^{(i)}\right]$ is obtained by assembling the matrices of the single finite elements $\left[K_j^{(i)}\right]$ defined in the local reference system by:

$$\left[K_j^{(i)}\right] = \left[\Lambda_j^{(i)}\right]^T \left[K_{jl}^{(i)}\right] \left[\Lambda_j^{(i)}\right] \tag{4.249}$$

where $\left[K_{jl}^{(i)}\right]$ is the stiffness matrix of the finite element taut cable in the plane (Sect. 4.4.2, (4.30)):

$$\left[K_{jl}^{(i)}\right] = \begin{bmatrix} \frac{EA_j}{l_j} & 0 & -\frac{EA_j}{l_j} & 0 \\ 0 & \frac{S_j^{(i)}}{l_j} & 0 & -\frac{S_j^{(i)}}{l_j} \\ -\frac{EA_j}{l_j} & 0 & \frac{EA_j}{l_j} & 0 \\ 0 & -\frac{S_j^{(i)}}{l_j} & 0 & \frac{S_j^{(i)}}{l_j} \end{bmatrix} \tag{4.250}$$

In Eq. (4.250) tension $S_j^{(i)}$ can be defined by adding the initial tension $S_j^{(0)} = S_o$ to the increase of the axial action due to strain:

$$S_j^{(i)} = S_j^{(0)} + \frac{EA_j}{l_j} \Delta X_{jl}^{(i)} \tag{4.251}$$

in which $\Delta X_{jl}^{(i)}$ is the elongation to which the generic finite element is subjected with respect to the reference configuration (Fig. 4.51):

$$\Delta X_{jl}^{(i)} = \Delta X_{jls}^{(i)} - \Delta X_{jld}^{(i)} \tag{4.252}$$

where $\Delta X_{jls}^{(i)}$ and $\Delta X_{jld}^{(i)}$ are the displacements of the end nodes in the axial direction of the local reference system of the element that can be obtained from vector $\underline{X}_{jl}^{(i)}$:

$$\underline{X}_{jl}^{(i)} = \left[\Lambda_j^{(i)}\right] \underline{X}_j^{(i)} \tag{4.253}$$

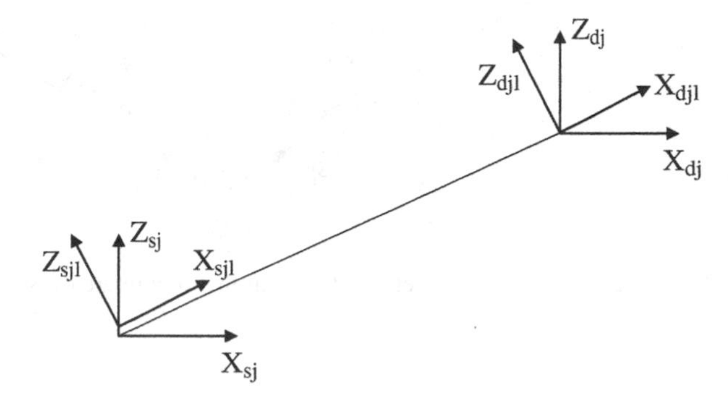

Fig. 4.51 Absolute and local reference system of the generic finite element

where $\left[\Lambda_j^{(i)}\right]$ is the coordinate transformation matrix (already introduced in Sect. 4.4.3, (4.35)):

$$\left[\Lambda_j^{(i)}\right] = \begin{bmatrix} \left[\lambda_j^{(i)}\right] & 0 \\ 0 & \left[\lambda_j^{(i)}\right] \end{bmatrix} \Rightarrow \left[\Lambda_j^{(i)}\right] = \begin{bmatrix} \cos \alpha_j^{(i)} & \sin \alpha_j^{(i)} \\ -\sin \alpha_j^{(i)} & \cos \alpha_j^{(i)} \end{bmatrix} \tag{4.254}$$

$\alpha_j^{(i)}$ the angle formed with respect to the global reference system by the generic finite element in the configuration corresponding to the attempt solution $\underline{X}_j^{(i)}$ (Fig. 4.51):

$$\alpha_j^{(i)} = \arctan\left(\frac{Z_{dj}^{(i)} - Z_{sj}^{(i)}}{X_{dj}^{(i)} - X_{sj}^{(i)}}\right) \tag{4.255}$$

Vector $\underline{F}_{eL}\left(\underline{X}_L^{(i)}\right)$ of the elastic forces, that can be obtained from Eq. (4.241), represents the generalized forces applied, in this case, to the nodes of the structure that here coincide with the components according to the global directions of the elastic forces applied to the nodes. In this case it is, therefore, simpler to define these forces by evaluating them in the local reference system of the *j*th finite element in the generic position (Fig. 4.52):

$$\underline{F}_{ejl}\left(\underline{X}^{(i)}\right) = -\begin{Bmatrix} S_j^{(i)} \\ 0 \\ -S_j^{(i)} \\ 0 \end{Bmatrix} \tag{4.256}$$

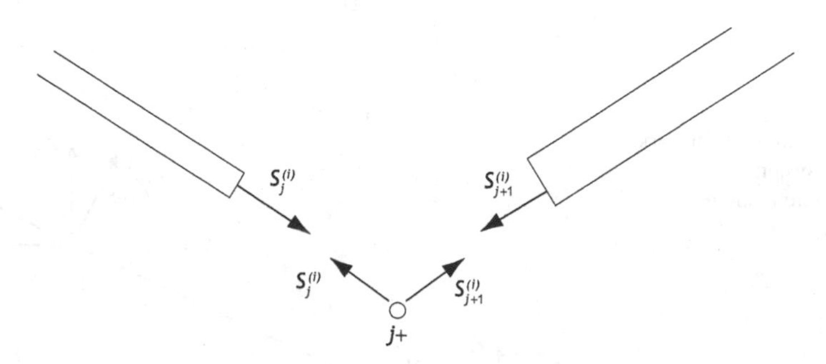

Fig. 4.52 Definition of the elastic forces in the local reference system

which, when projected into the global reference system, become:

$$\underline{F}_{ej}^{(i)} = \left[\Lambda_j\left(\underline{X}^{(i)}\right)\right]^T \underline{F}_{ejl}\left(\underline{X}^{(i)}\right) \tag{4.257}$$

These forces are assembled in the global vector $\underline{F}_{eL}\left(\underline{X}_L^{(i)}\right)$, (4.248) (Fig. 4.53).

As regards the system in question, the vector of the external forces $\underline{P}_L^{(i)}$ (4.248) depends solely on the weight force: the components of the weight force per unit length in the local reference system of the *j*th finite element are (see Fig. 4.51):

$$\begin{aligned} p_{uj}^{(i)} &= -mg\sin\alpha_j \\ p_{wj}^{(i)} &= -mg\cos\alpha_j \end{aligned} \tag{4.258}$$

The virtual work performed by these forces can be defined as (Sect. 4.4.6.2):

$$\delta^* L_j = \int_0^{l_j} p_{uj}^{(i)}(\xi)\delta^* u_j d\xi + \int_0^{l_j} p_{wj}^{(i)}(\xi)\delta^* w_j d\xi \tag{4.259}$$

By recalling that (Sect. 4.2.1, (4.5)):

$$u_j(\xi, t) = \underline{f}_u^T(\xi)\underline{X}_{jl} \tag{4.260}$$

$$w_j(\xi, t) = \underline{f}_w^T(\xi)\underline{X}_{jl} \tag{4.261}$$

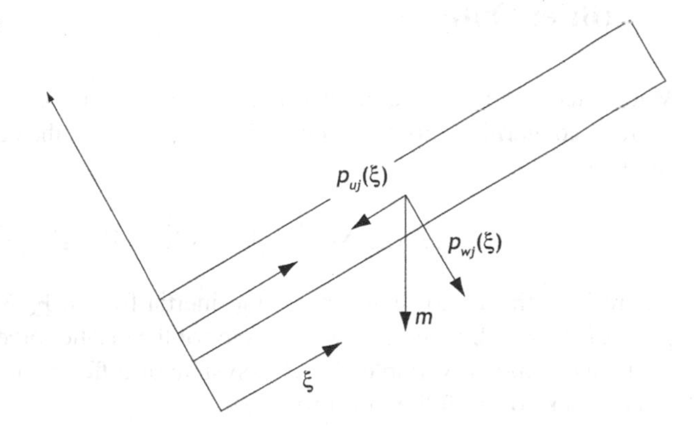

Fig. 4.53 Decomposition of the weight force in the local reference system

$\delta^* L_j$ can be expressed as:

$$\delta^* L_j = \left\{ \int_0^{l_j} p_{uj}^{(i)}(\xi) \underline{f}_u^T(\xi) d\xi + \int_0^{l_j} p_{wj}^{(i)}(\xi) \underline{f}_w^T(\xi) d\xi \right\} \delta^* \underline{X}_{jl} = \underline{P}_{jl}^{(i)T} \delta^* \underline{X}_{jl} \quad (4.262)$$

where $\underline{P}_{jl}^{(i)}$ is a vector, whose dimensions are equal to the number of d.o.f. of the generic finite element, which contains the generalized forces concentrated at the nodes of the finite element model. The virtual work expressed as a function of the independent coordinates in the global reference system is:

$$\delta^* L_j = \underline{P}_{jl}^{(i)T} \left[\Lambda_j^{(i)} \right] \delta^* \underline{X}_j = \underline{P}_j^{(i)T} \delta^* \underline{X}_j \quad (4.263)$$

where $\underline{P}_j^{(i)}$ is a vector whose dimensions are equal to the number of d.o.f. of the generic finite element, which contains the generalized forces concentrated at the nodes of the model, defined in the global reference system:

$$\underline{P}_j^{(i)} = \left[\Lambda_j^{(i)} \right]^T \underline{P}_{jl}^{(i)} \quad (4.264)$$

By assembling vectors $\underline{P}_j^{(i)}$ of the excitation forces for each element, as already seen in Sect. 4.4.6.1, we obtain the definition of the generalized forces $\underline{P}_L^{(i)}$ for all the d.o. f. of the system. In this way it is possible to solve Eqs. (4.246), (4.247) and (4.248) using an iterative approach in order to find the static equilibrium position.

4.8 Numerical Integration of the Equations of Motion (Brief Outline)

With reference to the equations obtained in the case of nonlinear 1, 2 or *n* d.o.f. system (concerning this see Sects. 1.5, 2.4.2 and 4.7), the equations of motion are of the type:

$$\underline{F}_i(\underline{X}, \dot{\underline{X}}, \ddot{\underline{X}}) + \underline{F}_s(\dot{\underline{X}}, \ddot{\underline{X}}) + \underline{F}_e(\underline{X}) = \underline{F}(t) \quad (4.265)$$

where $\underline{F}_i$ is the vector that defines the inertia forces, $\underline{F}_s$ is the vector of the dissipation forces and, finally, $\underline{F}_e$ is the vector of the elastic forces all nonlinear functions of the independent variables $\underline{X}$. This system of differential equations can always be traced back to the following form [10]:

$$[M(\underline{X})]\ddot{\underline{X}} = \underline{F}_r(\underline{X}, \dot{\underline{X}}, t) \tag{4.266}$$

having collected all the force contributions not directly proportional to acceleration $\ddot{\underline{X}}$ in vector $\underline{F}_r$. In the case of linear systems (always see Sects. 2.4 and 4.3) the equations of motion can be traced back to the known form:

$$[M]\ddot{\underline{X}} = -[R]\dot{\underline{X}} - [K]\underline{X} - \underline{F}(t) \tag{4.267}$$

These equations can be integrated directly, as already mentioned several times, in numerical form: in this paragraph we will make a brief reference to the methodologies of numerical integration of differential equations with total derivatives of the type shown in (4.266) and (4.267) most frequently used, with particular reference to courses on Numerical Calculation and specific texts on the subject [2, 16, 18].

All the integration methods, both in a linear and nonlinear field, are based on two different ideas:

- satisfying Eqs. (4.266) and (4.267) not in a continuous manner, but only in particular instants of time t_o, $(t_o + \Delta t)$, $(t_o + 2\Delta t)$ etc.;
- imposing, in approximated form, the value of acceleration $\ddot{\underline{X}}(\tau)$ inside the integration step Δt $(0 \leq \tau \leq \Delta t)$ and deducing both speed and displacement from this during the next step:

$$\begin{aligned}
\dot{\underline{X}}(t + \Delta t) &= \dot{\underline{X}}(t) + \Delta\dot{\underline{X}} = \dot{\underline{X}}(t) + \left\{ \int_0^{\Delta t} \ddot{\underline{X}}(\tau) d\tau \right\} \\
\underline{X}(t + \Delta t) &= \underline{X}(t) + \Delta\underline{X} = \underline{X}(t) + \left\{ \Delta t \dot{\underline{X}}(t) + \int_0^{\Delta t} \left\{ \int_0^{\Delta t} \ddot{\underline{X}}(\tau) d\tau \right\} d\tau \right\}
\end{aligned} \tag{4.268}$$

The choice of the acceleration trend X(t) inside the step determines the accuracy, the numeric stability and the burden of the solution method. In this way, the generic integration procedure proves to be iterative:

- at time $t = 0$ the initial conditions $\underline{X}(0)$ and $\dot{\underline{X}}(0)$ are known, because they are imposed;
- it is possible to evaluate all the terms on the right of the equal sign of Eq. (4.266) or (4.267) and to thus obtain acceleration $\ddot{\underline{X}}(0)$;
- $\underline{X}(\Delta t)$ and $\dot{\underline{X}}(\Delta t)$ are calculated by (4.268);
- the procedure is developed, starting from time $(t + \Delta t)$ and subsequently the amplitudes at time $(t + 2\Delta t)$ are obtained and so on.

The methods are divided into explicit and implicit methods:

- explicit (or predictor) methods are those for which neither displacements $\underline{X}(t + \Delta t)$, nor velocity $\dot{\underline{X}}(t + \Delta t)$ are made to depend on acceleration $\ddot{\underline{X}}(t + \Delta t)$ at step $t + \Delta t$, or rather, it is possible to perform the integrations only by knowing the accelerations $\ddot{\underline{X}}(t)$ at step t; these methods can be applied directly and indifferently both in linear or nonlinear problems;
- implicit (or corrector) methods are those in which, conversely, displacements and speeds depend on accelerations in two instances t and $t + \Delta t$; since when using these approaches it is necessary to evaluate all the forces acting on the system even at time $t + \Delta t$ [2], these methods can only be directly applied to linear problems (the extension of these algorithms in a nonlinear field involves the introduced, at a frozen time, of the iterative procedures of the predictor corrector type [1]).

The explicit methods are conditionally stable, that is they require small integration steps Δt (smaller than the smallest oscillation period present in the system to be analysed): if integration steps that are too large are used, the integration method diverges. The implicit methods are unconditionally stable, i.e. any type of integration step Δt can be used for them: in this case, the choice of integration step only depends on the accuracy desired in the results (these methods filter the frequencies for which the integration step adopted is too high). Among the most widely used integration methods, worth particular note are:

- the Runge-Kutta fourth order (explicit method) method which, despite having an error of the order of Δt^5 nevertheless requires 4 solutions of the equations for each step and the integration step must be sufficiently small (otherwise there would be numerical instability);
- the Newmark method, which will be described in the next paragraph; or the Houbolt [2] method which, however, damps the modes with a T period lower than $20\Delta t$ (this method is often preferred to the Newmark method in nonlinear problems);
- the Park-method [1] which appears to be better than the Houbolt method for nonlinear systems;
- the central differences method [2];
- the Hughes method, the Taylor method [1];
- the Hilber method which represents and extension of the Newmark method.

Out of all the above mentioned integration methodologies that can be used, in this part, reference will be made to the algorithms based on the Newmark method. As regards other types of approaches, reference can be made to the specialized texts on the subject [1, 2].

4.8.1 The Newmark Method in a Linear Field

This method can be considered as an extension of the linear acceleration method: the following assumptions, based on the hypothesis of constant acceleration in step Δt, are introduced:

$$\begin{aligned}\dot{\underline{X}}(t+\Delta t) &= \dot{\underline{X}}(t) + \left[(1-\delta)\ddot{\underline{X}}(t) + \delta\ddot{\underline{X}}(t+\Delta t)\right]\Delta t \\ \underline{X}(t+\Delta t) &= \underline{X}(t) + \Delta t\dot{\underline{X}}(t) + \left[\left(\frac{1}{2}-\alpha\right)\ddot{\underline{X}}(t) + \alpha\ddot{\underline{X}}(t+\Delta t)\right]\Delta t^2\end{aligned} \tag{4.269}$$

The equations of motion for the generic linear system are written at instant $t+\Delta t$:

$$[M]\ddot{\underline{X}}(t+\Delta t) + [R]\dot{\underline{X}}(t+\Delta t) + [K]\underline{X}(t+\Delta t) = \underline{F}(t+\Delta t) \tag{4.270}$$

By obtaining acceleration $\ddot{\underline{X}}(t+\Delta t)$ from the second equation of (4.269) and substituting it in the first relation, we obtain:

$$\begin{aligned}\ddot{\underline{X}}(t+\Delta t) &= \frac{1}{\alpha\Delta t^2}\left\{\underline{X}(t+\Delta t) - \underline{X}(t) - \Delta t\dot{\underline{X}}(t) - \Delta t^2\left[\left(\frac{1}{2}-\alpha\right)\ddot{\underline{X}}(t)\right]\right\} \\ \dot{\underline{X}}(t+\Delta t) &= \dot{\underline{X}}(t) + \Delta t(1-\delta)\ddot{\underline{X}}(t) + \delta\Delta t\frac{1}{\alpha\Delta t^2}\left\{\underline{X}(t+\Delta t) - \underline{X}(t) - \Delta t\dot{\underline{X}}(t) - \Delta t^2\left[\left(\frac{1}{2}-\alpha\right)\ddot{\underline{X}}(t)\right]\right\}\end{aligned} \tag{4.271}$$

By substituting relations (4.271) in Eq. (4.270) we obtain an algebraic equation in $\underline{X}(t+\Delta t)$:

$$\begin{aligned}&\left[[M]\frac{1}{\alpha\Delta t^2} + [R]\frac{\delta}{\alpha\Delta t} + [K]\right]\underline{X}(t+\Delta t) = \underline{F}(t+\Delta t) \\ &\quad + \left[[M]\frac{1}{\alpha\Delta t^2} + [R]\frac{\delta}{\alpha\Delta t}\right]\underline{X}(t) + \left[[M]\frac{1}{\alpha\Delta t} - [R] + [R]\frac{\delta}{\alpha}\right]\dot{\underline{X}}(t) \\ &\quad + \left[[M]\frac{1}{\alpha}\left(\frac{1}{2}-\alpha\right) - [R]\Delta t(1-\delta) + [R]\left(\frac{\delta\Delta t}{\alpha}\left(\frac{1}{2}-\alpha\right)\right)\right]\ddot{\underline{X}}(t)\end{aligned} \tag{4.272}$$

or rather:

$$[\bar{K}]\underline{X}(t+\Delta t) = \underline{R}(t+\Delta t) \tag{4.273}$$

where $[\bar{K}]$ is the matrix of reduced stiffness:

$$[\bar{K}] = \left[[M]\frac{1}{\alpha\Delta t^2} + [R]\frac{\delta}{\alpha\Delta t} + [K]\right] \tag{4.274}$$

and $\underline{R}(t + \Delta t)$ is the vector of the reduced forces:

$$\begin{aligned}\underline{R}(t+\Delta t) = \underline{F}(t+\Delta t) &+ \left[[M]\frac{1}{\alpha\Delta t^2} + [R]\frac{\delta}{\alpha\Delta t}\right]\underline{X}(t) \\ &+ \left[[M]\frac{1}{\alpha\Delta t} + [R]\left(\frac{\delta}{\alpha} - 1\right)\right]\underline{\dot{X}}(t) \\ &+ \left[[M]\frac{1}{\alpha}\left(\frac{1}{2} - \alpha\right) - [R]\Delta t(1-\delta) + [R]\left(\frac{\delta\Delta t}{\alpha}\left(\frac{1}{2} - \alpha\right)\right)\right]\underline{\ddot{X}}(t)\end{aligned} \tag{4.275}$$

Having assigned a certain integration step Δt and having assigned the values of constants α and δ it is possible to calculate the reduced stiffness matrix $[\bar{K}]$ from Eq. (4.274). Having assigned the initial conditions (for $t = 0$) it is possible to evaluate the vector of the fictitious forces $\underline{R}(\Delta t)$ at time Δt. By solving Eq. (4.273) it is possible, at the same time, to obtain the displacements and, by substituting these values in Eq. (4.271), it is possible to calculate velocity and accelerations. At this point, time is increased and the procedure is repeated. As can be seen, this method is implicit and it can be shown that for:

$$\delta \geq 0.5; \quad \alpha \geq \frac{(0.5+\delta)^2}{4} \tag{4.276}$$

this is unconditionally stable [1, 3]: as the integration step increases, the effect introduced by the numerical integration is that of filtering the high frequencies (associated with periods that are lower than the integration step Δt used). This characteristic is useful in the case of structures modelled with finite elements, where high fictitious frequencies are introduced by the same schematization used.

4.8.2 The Newmark Method in a Nonlinear Field

As noted, the Newmark method requires an understanding of the real forces at step $t + \Delta t$ meaning that, therefore, this cannot be used for nonlinear systems in which the forces depend on displacements speeds and accelerations not known in the next step. In this case, it is necessary to modify this methodology (see [3]) by introducing, independently of time, an iteration that will converge to the equilibrium at each step by introducing a dynamic equilibrium equation of the forces which must be verified at the end of each integration step.

4.9 Summary

This chapter deals with the study of the dynamics of continuous mechanical systems using the approximated finite element discretization approach. The general methodology used to define the matrix equations of motion in the case of vibration problems is described. Cable, beam and "taut beam" finite elements are analysed as examples of application; notions regarding two-dimensional and three-dimensional finite elements are shown. The basics concepts regarding the finite element method approach for nonlinear systems and methods of numerical integration of the nonlinear equations of motion are introduced.

References

1. Argyris J, Mleinek HP (1991) Texts on computational mechanics, dynamic of structures. Elsevier, Amsterdam
2. Bathe KJ (1982) Finite element procedures in engineering analysis. Prentice-Hall, Englewood Cliffs
3. Bathe KJ, Wilson EL (1976) Numerical methods in finite element method. Prentice-Hall, Englewood Cliffs
4. Belloni G, Bemasconi G (1975) Sforzi e deformazioni e loro legami. Tamburini Editore, Milano
5. Bishop RED, Johnson DC (1960) The mechanics of vibrations. Cambridge University Press, Cambridge
6. Bishop RED, Gladwell GML, Michaelson S (1965) The matrix analysis of vibration. Cambridge University Press, Cambridge
7. Brebbia CA (1991) Boundary element techniques—theory and application in engineering. McGraw-Hill, New York
8. Bruni S, Cheli F, Collina A, Diana G (1985) Train-track interaction: a comparison between a numerical model and full scale experiments. In: Heavy vehicle system. Int J Veh Des 6(1–4):115–146
9. Bruni S, Collina A, Corradi R (2005) Train-track-bridge interaction: influence of track topology on structural dynamic performances. In: Proceedings of EURODYN 2005, Paris, France
10. Cook RD (1981) Concepts and applications of finite element analysis. Wiley, New York
11. Den Hartog JP (1956) Mechanical vibrations, 4th edn. McGraw-Hill, New York
12. Desay CS, Abel JF (1972) Introduction to the finite element method: a numerical method for engineering analysis. Van Nostrand Reinhold Company
13. Gallager H (1973) Finite elementanalysis fundamentals. Prentice-Hall, Englewood Cliffs
14. Holzer SH (1985) Computer analysis of structures. Elsevier, Amsterdam
15. Massa E (1981) Costruzioni di machine. Masson Italia, Milano
16. Meirovitch L (1967) Analytical methods in vibrations. The Macmillan, New York
17. Meirovitch L (1970) Methods of analytical dynamics. McGraw-Hill, New York
18. Meirovitch L (1980) Computational methods in structural dynamics. The Sijthoff & Noordhoff Intemational Publishers
19. Meyers VJ (1983) Matrix analysis of structures. Harper and Row, New York
20. Politecnico di Milano (1986) Lezioni di scienza delle Costruzioni. Clup, Milano
21. Przemieniecki JS (1968) Theory of matrix structural analysis. McGraw-Hill, New York

22. Ross CTF (1985) Finite element methods in structural mechanics. Ellis Horwood Limited, Wiley, New York
23. Thomson WT, Dahleh MD (1997) Theory of vibration with applications, 5th edn. Prentice Hall, Englewood Cliffs
24. Timoshenko SP, Young DH, Weaver W (1974) Vibration problems in engineering, 4th edn. Wiley, New York
25. Weaver WW, Johnston PR (1987) Structural dynamics by finite elements. Prentice Hall Inc, Englewood Cliffs
26. Zienkiewicz OC (1987) The finite element method, 4th edn. McGraw-Hill, New York

Chapter 5
Dynamical Systems Subjected to Force Fields

5.1 Introduction

In order to analyse the dynamics of a mechanical system we must address an issue which, in general, is very complex. In fact, there are systems that allow a position of static equilibrium and systems that, in general, do not allow such positions and are, on the contrary, designed to create a certain motion, such as machines in general. As regards systems that allow a position of static equilibrium, we have already examined the class corresponding to *dissipative systems* where the forms of energy addressed are elastic, gravitational and the dissipative function associated with non-conservative forms arising from hysteresis, friction or, in any case, from dissipation in the elastic elements themselves. In reality, these systems are subject to other force fields in addition to the elastic and gravitational ones: consider, for example, the case where a part, or the whole system, comes into contact with a fluid that exerts actions that depend on the relative motion between fluid and object, actions that are expressed, therefore, as force fields. In the case of a system with two elements that come into contact with each other, these are subject to actions that can again be expressed as a function of relative motion and, therefore, can be defined once more through a force field. An electromagnetic force acting on a part or an entire system can again be defined as a force field. The presence of force fields can alter the static and dynamic behaviour of the actual system in a more or less substantial manner. These cases are referred to as fluid-elastic or aeroelastic or magnetoelastic systems, depending on whether the force field results from a fluid in general, from the action of air or from an electromagnetic field. These force fields can be:

- conservative;
- non-conservative.

Conservative force fields, similar to the elastic force field (conservative by definition), overlap the latter, altering, as we will see later, natural frequencies and

F. Cheli and G. Diana, *Advanced Dynamics of Mechanical Systems*,
DOI 10.1007/978-3-319-18200-1_5

normal modes of vibration: these fields may give rise to static type instability (static divergence), but cannot generate instability problems of a dynamic nature. In reality, purely conservative force fields do not actually exist. The elastic force field is also seen as conservative since the non-conservative part associated with it, due to structural damping, is introduced with dissipative terms, proportional to velocity: these systems have been defined as dissipative. Only those systems where their non-conservative nature cannot be attributed to the dissipative phenomena mentioned above (typically hysteresis or viscous damping), but is an intrinsic characteristic of the force field, should be defined as non-conservative systems. Therefore, if the force field is non-conservative and non-dissipative, any energy introduced by the field into the system may give rise to:

- unstable motions;
- motions that increase over time (dynamic instability), motions which may also lead to the collapse of the system.

As mentioned, force fields may act:

- on systems that allow a position of static equilibrium;
- on systems which, in general, do not allow such positions, for example, machines.

In this chapter we will discuss those mechanical systems subject to force fields:

- perturbed about rest, where motion only results from distributed and/or concentrated elasticity (Sects. 5.2 and 5.3);
- perturbed about a state of motion (Sect. 5.4).

A steady-state solution will be obtained for these systems, if one exists, and, in the event to analyse its stability we will consider the perturbed motion about the steady-state or about another solution (if one exists). In the discussion we will show that, in the case of systems perturbed about a situation of non-rest, conditions of instability may also be observed in cases where these systems are subject to conservative force fields. For example, in Sect. 6.6 we will analyse the phenomena of instability that may arise in a rotating shaft with different degrees of bending stiffnesses: in this case the energy is supplied by external forces, needed to maintain the steady-state or the motion, in general, of the system. These forces become, in such a system, functions of the perturbed motion and so they themselves are a non-conservative force field: for this reason the whole system is considered to be non-conservative. In this chapter, we will illustrate some typical examples of such forms of instability, for example, flutter instability in aerofoils (Sects. 5.2 and 5.3.1) or oil film instability in journal bearings with hydrodynamic lubrication (Sect. 5.3.2.2) and other vibratory phenomena induced, in general, by a fluid influencing a body. Subsequently, further examples of systems where the force field is generated by the interaction of two bodies coming into contact will also be analysed: the actions of contact that are exchanged between two bodies can, in fact, give rise to various phenomena such as flutter instability which may occur above a certain speed in railway cars or the so-called wheel *shimmy* in cars [94, 102]. In the dynamics of interconnected bodies we will consider the various bodies that make up the system, either rigid or deformable, and apply to them the actions caused by external force fields, i.e.:

- known external actions $F(t)$;
- actions arising from systems that interact with them.

Due to the constraints that interconnect the bodies, the term "constraint reactions" is generally used to describe these actions. If an energetic approach is used to write the equations of motion, for example the Lagrange equations, these reactions do not perform work and so cannot be calculated. The constraints that exist between the various bodies are translated into n_c constraint equations that reduce the degree-of-freedom of the entire system. In the event that the constraints are holonomic, if n are the coordinates required to define the motion of the generic system[1] then the number of independent variables drops to $n - n_c$. In actual fact, this approach is only theoretical, or applies in a first approximation, since the coupling of various bodies:

- due to the interposition of any lubricant between the contact surfaces;
- due to the effect of the contact mechanisms (micro-slip, deformation);

is not equivalent, in general, to the imposition of kinematic constraints.

For this reason, the n_c constraint equations cannot be used but, in relation to the type of contact, must be defined as a function of the independent variables that define the forces exchanged between the bodies themselves. To further clarify this point, we can use the following example: we will consider two bodies in planar motion (in this case the total degree-of-freedom are $n = 6$), coupled in the journal with a hinge made up of a lubricated journal bearing coupling. If we neglect the motion between journal and bearing we can write $n_c = 2$ constraint equations that require that the centre of the journal considered to belong to body A undergoes the same displacements as the centre of the bearing, belonging to body B (Fig. 5.1). In this case, the effective degree-of-freedom of the system are reduced from $n = 6$ to $(n - n_c) = 4$.

If, however, we consider a case where there is oil film, it is not possible to write the $n_c = 2$ constraint equations, it is necessary to define the forces exchanged in the coupling. These forces are functions of the displacements and of the relative velocities between journal centre and bearing and the angular velocities of the two bodies A and B: the same forces can, therefore, be estimated as a function of the independent variables that define the motion of the two bodies A and B (the same procedure can be applied in the case where mechanisms of local deformability are under consideration). In this case, it is not possible to write holonomic constraint equations, which reduce the number of degree-of-freedom, but, once the link

[1]In a discrete system with rigid bodies, as we know, $n = 6 \cdot m_c$ (in the general case of motion in space) or $n = 3 \cdot m_c$ (in the case of planar motion), m_c being the number of bodies that make up the actual system.

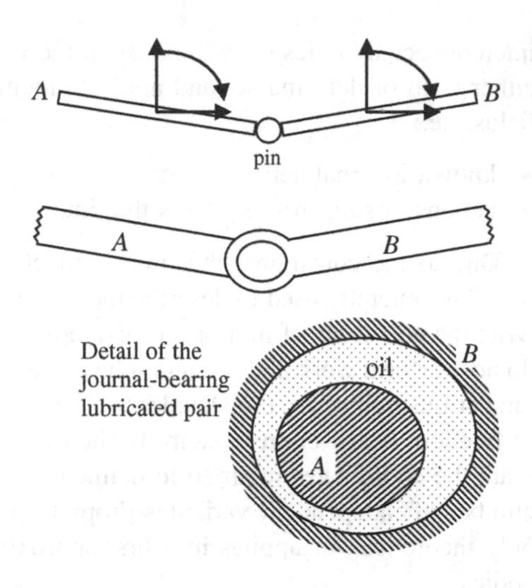

Fig. 5.1 Constraints between bodies: lubricated journal-bearing coupling

between the forces exchanged and the variables that define the system's motion have been defined, the same forces must appear in the equations of motion of the overall system (described by all 6 coordinates for each body). The expressions of these forces, which also include the derivatives of the variables that define the system's motion, are, ultimately, relations of non-holonomic constraint. With this approach to the problem, the rigid or deformable bodies that make up the system will come to have all the degree-of-freedom of the various bodies, as if they were free: the introduction of constraints occurs through the force fields that represent the interaction between the bodies themselves. This method, which replaces the traditional one [1, 6]:

- is closer to reality;
- requires a number of coordinates greater than the actual number of degree-of-freedom to describe the motion of a system, since it does not introduce the constraint equations; furthermore, it increases the complexity of the problem.

This approach should be considered as a refinement and a generalisation of the approach where the interaction between the bodies making up the system can be described by directly imposing the constraint conditions. We should point out that the general discussion here can also be applied to controlled mechanical systems, since the action of control is exerted via forces that are generally functions of the state variables i.e. variables associated with the degree-of-freedom of the system, so they can still be considered to be systems subject to field actions. Traditionally, on the other hand, specific techniques were developed for these systems. For example, in the event that we wish to judge the stability of the system, they are based on the Laplace transform: these techniques, already referred to in Chap. 1, Sect. 1.3.5.2, will not be discussed here; the reader should refer to specialised texts on the subject [3, 4, 86]. For mainly didactic reasons, and similarly to what was done for dissipative systems, in the first part of this chapter we will analyse the behaviour of vibrating systems, subject to force fields, perturbed about rest (small motions). Firstly we will consider those with a one degree-of-freedom (Sect. 5.2), subsequently those with two degree-of-freedom (Sect. 5.3) and, lastly, those with n or ∞ degree-of-freedom (Sect. 5.4). In the second part of the chapter we will consider systems subject to force fields, whose motion is defined as being large, focusing on systems with one, two and n degree-of-freedom. The general discussion will be accompanied by several explanatory examples, relating to the different issues being analysed: during the discussion, it will be necessary to refer to specific force fields and we will briefly illustrate their characteristics. In particular, in Sect. 5.2.1.1 we will illustrate the characteristics of the force field caused by a fluid interacting with a fixed or vibrating body while in Sect. 5.5.2.1.1 we will illustrate the characteristics of the force field that are generated in contacts between wheel and road or wheel and track.

5.2 Vibrating Systems with 1 DOF Perturbed Around the Position of Equilibrium

We will consider the vibrating system with one degree-of-freedom in Fig. 5.2, made up of a mass m_s suspended elastically on a spring with elastic constant k_s and also constrained to the ground by a damper of constant r_s: We can describe the horizontal displacement of the mass using independent variable x.

We will assume that the system under consideration is subject to a known force as a time function $F(t)$ and the same is subject to a force field, generically functions of displacement x, of velocity $\dot{x}$ and acceleration $\ddot{x}$: we will define with $F_x(x, \dot{x}, \ddot{x})$ the component of these forces depending on the degree-of-freedom x. The motion equation of the vibrating system analysed is:

$$m_s\ddot{x} + r_s\dot{x} + k_s x = F_x(x, \dot{x}, \ddot{x}) + F(t) \tag{5.1}$$

This differential equation is non-linear and so cannot generally be integrated analytically: the solution of (5.1) can be estimated with approximate methods of step-by-step numerical integration [2, 5]. To analyse the response of the system for small oscillations about the position of static equilibrium or to estimate the stability it is possible to linearise (5.1), once the position of static equilibrium has been defined, if it exists. In correspondence to this solution, $F(t)$ is zero or it is equal to a

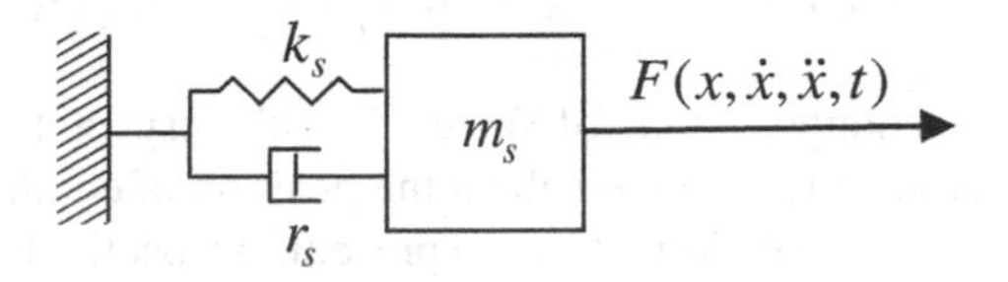

Fig. 5.2 System with one degree-of-freedom subject to a force field

constant. The position of static equilibrium x_o is such that, with $F(t) = 0$, the following relation is satisfied:

$$k_s x_0 = F_x(x_0, 0, 0) \tag{5.2a}$$

or, more generally, the function $f(x)$ is zero:

$$f(x_0) = k_s x_0 - F_x(x_0, 0, 0) = 0 \tag{5.2b}$$

Equation (5.2a) is a non-linear algebraic equation that can be solved, to obtain x_0, analytically in simple cases and, more generally, using numerical procedures [2, 5]. To linearise (5.1) about the position of static equilibrium x_0, the force function F_x must be expressed in the Taylor series:

$$m_s\ddot{x} + r_s\dot{x} + k_s x = F_x(x_0, 0, 0) + \left(\frac{\partial F_x}{\partial x}\right)_0 (x - x_0) + \left(\frac{\partial F_x}{\partial \dot{x}}\right)_0 \dot{x} + \left(\frac{\partial F_x}{\partial \ddot{x}}\right)_0 \ddot{x} \tag{5.3}$$

For convenience, it is possible to introduce a coordinate transformation that makes it possible to describe the motion of the system not as a function of independent variable x but as a function of a new variable $\bar{x}$, which defines the perturbed motion of the system about the position of static equilibrium x_0 defined by (5.2a) and (5.2b), as:

$$\bar{x} = x - x_0 \tag{5.4}$$

By replacing (5.4) in the linearised equation of motion (5.3) we obtain:

$$m_s\ddot{\bar{x}} + r_s\dot{\bar{x}} + k_s\bar{x} + k_s x_0 = F_x(x_0, 0, 0) + \left(\frac{\partial F_x}{\partial x}\right)_0 \bar{x} + \left(\frac{\partial F_x}{\partial \dot{x}}\right)_0 \dot{\bar{x}} + \left(\frac{\partial F_x}{\partial \ddot{x}}\right)_0 \ddot{\bar{x}} \tag{5.5a}$$

Taking into account (5.2a), (5.5a) is reduced to:

$$m_s\ddot{\bar{x}} + r_s\dot{\bar{x}} + k_s\bar{x} = \left(\frac{\partial F_x}{\partial x}\right)_0 \bar{x} + \left(\frac{\partial F_x}{\partial \dot{x}}\right)_0 \dot{\bar{x}} + \left(\frac{\partial F_x}{\partial \ddot{x}}\right)_0 \ddot{\bar{x}} \tag{5.5b}$$

Now we can define with

$$m_F = -\left(\frac{\partial F_x}{\partial \ddot{x}}\right)_o \quad r_F = -\left(\frac{\partial F_x}{\partial \dot{x}}\right)_o \quad k_F = -\left(\frac{\partial F_x}{\partial x}\right)_o \tag{5.6}$$

the derivatives, with changed sign, of force F_x with respect to displacement x, velocity $\dot{x}$ and acceleration $\ddot{x}$, measured about the position of equilibrium defined by $x = x_o$ and $\dot{x} = \ddot{x} = 0$: these derivatives represent, respectively, the equivalent mass, damping and stiffness, caused by the specific force field being analysed.

Taking into account the definition introduced in (5.6), the motion Eqs. (5.5a) and (5.5b) can be rewritten as:

$$(m_s + m_F)\ddot{\bar{x}} + (r_s + r_F)\dot{\bar{x}} + (k_s + k_F)\bar{x} = 0 \tag{5.7}$$

which after having defined the constants:

$$m_T = m_s + m_F \quad r_T = r_s + r_F \quad k_T = k_s + k_F \tag{5.8}$$

is reduced to:

$$m_T\ddot{\bar{x}} + r_T\dot{\bar{x}} + k_T\bar{x} = 0 \tag{5.9}$$

As we can see, the force field modifies the dynamic behaviour of the vibrating system, since it modifies the inertial, elastic and damping characteristics. In fact, the natural frequency of the system, with $r_T = 0$, subject to the force field is:

$$\omega_0 = \sqrt{\frac{k_T}{m_T}} \tag{5.10}$$

and is therefore modified, as the following are modified:

- the overall stiffness k_T which depends (5.8) on the structural stiffness k_s, which by definition is always positive, and on equivalent stiffness k_F of the force field (5.6), which can, however, assume both positive and negative values.
- damping r_T which is the sum, in turn, of structural damping r_s (always positive) and of the term r_F; the latter may be positive or negative (5.6).

The term of equivalent mass m_F is, however, normally always positive and does not modify the stability of the system. Analysis of the linearised system also makes it possible to judge the stability of small motions (or incipient instability) of the non-linear system: in fact, by making use of Lyapunov's theorem, it is possible to confirm that, if the linearised system is stable (or unstable), then so is the non-linear system, provided that we take into consideration small perturbations about the position of equilibrium. We can talk about asymptotic stability of the system, in cases where the system perturbed about the position of static equilibrium x_0, returns to the initial conditions for $t \rightarrow \infty$. In the event that we wish to consider the stability of large motions, i.e. for not small movements, then linearisation is not valid and we must consider non-linear complete equations. In the region of the small oscillations, by definition, the behaviour of the linear system and that of the non-linear system do not differ, while it may be different for large amplitudes. It is typical of systems perturbed about a configuration of unstable equilibrium, when we consider extensive motion, i.e. we take into account the non-linearity, the occurrence of an increase in the amplitudes of vibration that tend to a finite limit cycle (Fig. 5.3a). It may also occur that a system that is stable with small motions is

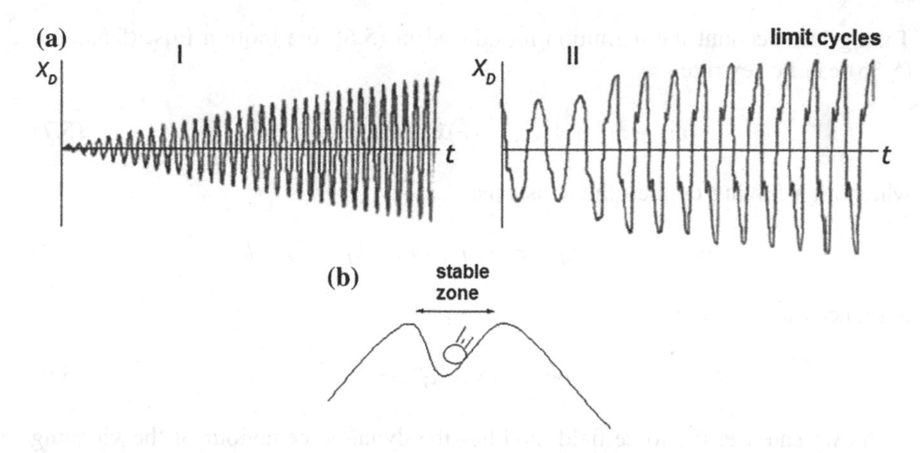

Fig. 5.3 **a** System with one degree-of-freedom placed in a force field: case relating to a system with incipient instability: (I) response of the linearised system (unstable); (II) response of the non-linear system (limit cycle). **b** System with one degree-of-freedom placed in a force field: a case of stable system with small motion and unstable with large motion. Non-trivial solutions to (5.12) are obtained in correspondence to the values of l which cancel out the characteristic equation

no longer stable when motion is large: a typical example is shown in Fig. 5.3b. Now we will examine the possible behaviours of a linearised system. A particular integral of (5.9) is, as we know, given by:

$$\bar{x} = \overline{X}e^{\lambda t} \tag{5.11}$$

which, placed in the actual equation (5.9), brings us to:

$$\left(\lambda^2 m_T + \lambda r_T + k_T\right)\overline{X} = 0 \tag{5.12}$$

$$\lambda^2 m_T + \lambda r_T + k_T = 0 \tag{5.13}$$

Therefore, the solutions are:

$$\lambda_{1,2} = -\frac{r_T}{2m_T} \pm \sqrt{\left(\frac{r_T}{2m_T}\right)^2 - \frac{k_T}{m_T}} \tag{5.14}$$

Now we will discuss the possible solutions (5.14). First of all we will consider the case where the total damping r_T is positive:

- if $k_T > 0$ and $\frac{k_T}{m_T} > \frac{r_T}{4m_T}$ the roots $\lambda_{1,2}$ of (5.14) are complex conjugates:

$$\lambda_{1,2} = \alpha \pm i\omega \tag{5.15a}$$

with $\alpha = -\frac{r_T}{2m_T}$ real and negative: the values (5.15a) correspond to a stable oscillating solution of the form:

$$\bar{x}(t) = \bar{X}e^{\alpha t}\cos(\omega t + \phi) \tag{5.15b}$$

- if $k_T > 0$ and $\frac{k_T}{m_T} < \frac{r_T}{4m_T}$, the roots $\lambda_{1,2}$ are both real and negative:

$$\lambda_1 = \alpha_1; \quad \lambda_2 = \alpha_2 \tag{5.15c}$$

with α_1 and α_2 real and negative: i.e. the corresponding solution is stable, but not oscillating:

$$\bar{x}(t) = \overline{X}_1 e^{\alpha_1 t} + \overline{X}_2 e^{\alpha_2 t} \tag{5.15d}$$

- if $k_T < 0$, then at least one solution is real positive ($\lambda = \alpha > 0$) and we have static divergence.

If, however, damping r_T is negative:

- if $k_T > 0$, with $k_T > \frac{r_T^2}{4m_T}$, we have an oscillating expanding solution ($\alpha > 0$);
- if $k_T > 0$, with $k_T < \frac{r_T^2}{4m_T}$, (5.14) will give rise to a supercritical solution ($\lambda_1 = \alpha_1 < 0$) and a non-oscillating expanding solution ($\lambda_2 = \alpha_2 > 0$);
- a negative value of k_T, even in this case of negative damping, still gives rise to instability of a static nature (divergence).

Therefore, to sum up, we can confirm that:

- if the eigenvalues of the linearised system, both have negative real parts, then the system is asymptotically stable and the resulting motion $x(t)$ can be decreasingly exponential or decreasingly oscillating: the system, in this condition, once perturbed, returns to its initial state for $t \to \infty$;
- if the eigenvalues have a positive real part then the system is asymptotically unstable (with solution that is merely exponential or oscillating exponential);
- if one of the eigenvalues of the linearised system has a zero real part, then stability is non-asymptotic.

These characteristics of stability also apply, obviously, with little motion, to the non-linear system. Lastly, as far as the problem of linearisation of the forces is concerned, the development of the force derivatives, starting from the analytical expressions of the forces, is not always easy. When function F_x, which defines the force, is the result of a product of several functions $F_x = f \cdot g \cdot h$ we can, without loss of generality, develop the single functions f, g e h that make up the product in series around the position of equilibrium, up to the linear term. Subsequently, the product of the linearised functions is performed, bearing in mind just the linear terms in the product: it is possible to demonstrate that this approach leads to the same results obtained when calculating the derivative of the product function of the single functions. This approach can also be extended to the case of a composite function $F_x = F_x(y(x))$: by develojournalg function F_x with respect to the intermediate variable y, and the latter with respect to the independent variable x and substituting the linearised expression of $y(x)$ in the linearised expression of $F_x(y(x))$, we obtain the same result that we would obtain by directly develojournalg $F_x(y(x))$ with respect to variable x. This way of dealing with the problem of linearisation makes it possible to carry out calculations that are, in general, more simple than development carried out directly on the function to be linearised. Below we will examine some significant examples of systems with one degree-of-freedom subject to force fields, with particular reference to the aerodynamic forces exchanged between a fluid and a body.

5.2.1 System with 1 DOF Placed in an Aerodynamic Force Field

Now we will analyse, as a first example, the vibrating system shown in Fig. 5.4, i.e. a symmetrical aerofoil, constrained in such a way as to create a vibrating system with just one torsional degree-of-freedom, k_s being the constants of the elastic return springs, r_s the constants associated with the concentrated dampings (reproducing structural damping) and lastly J the moment of inertia of the body around its centre of gravity G. The aerofoil is hit by a fluid stream, with speed U constant in time and space. In order to write the equations of motion we must estimate the generalised forces that act on the wing due to the force field induced by the fluid stream affecting it.

5.2.1.1 Defining the Aerodynamic Forces on a Rigid Body

In this section, for a more complete discussion, we will provide some references relating to the aerodynamic actions that are exerted on a body hit by a fluid stream. As we know (see the Bibliography, section "Aerodynamic forces acting on

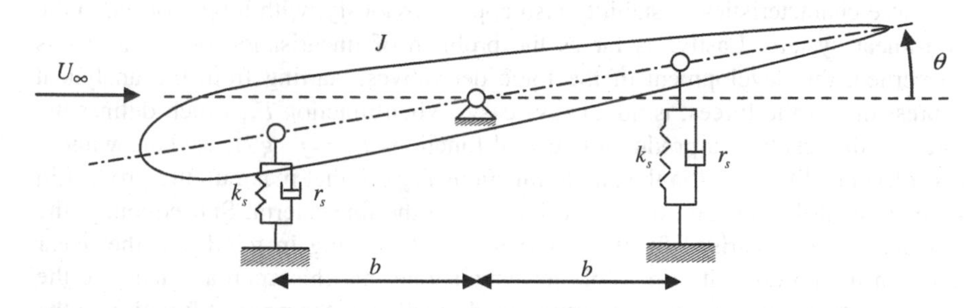

Fig. 5.4 System with 1 torsional DOF: aerofoil hit by a fluid stream

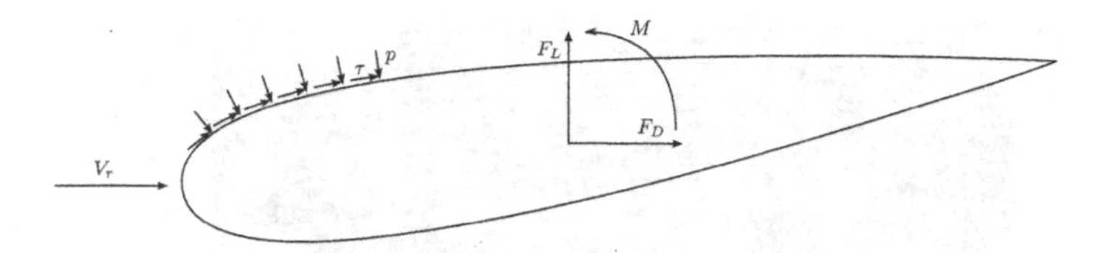

Fig. 5.5 Forces acting on a body hit by a fluid stream

structures" [9–39], a rigid body hit by a fluid in a two-dimensional stream[2] is subject, on the separation surface between the fluid and the actual body, to tangential τ and normal p actions which, as an overall effect, can be traced back to a resultant applied to a given point, known as the centre of pressure, or, considering the centre of gravity as the reduction pole, to an equivalent system of forces made up of (Fig. 5.5):

- a drag force F_D, acting in the same direction as the relative velocity V_r of the stream with respect to the actual body (in the following, it can also be referred to as F_R);
- a lift force F_L, acting in an orthogonal direction to the former; or by an aerodynamic torque M, the value of which depends on the distance between the centre of pressure (i.e. the point of application of the lift and drag forces) and the reduction point of the forces themselves (in the following, it can also be referred to as F_P);

These forces depend on the type of flow that is established around the object, a function, in turn, of the shape of the aerofoil. Aerofoils that are elongated in the direction of the incoming stream present a low drag force and, depending on the direction of the incoming stream, may have a high lift. For these aerofoils (Fig. 5.6a) the boundary layer adheres to the object on the entire surface, there is no separation of the stream and, as a consequence, there is no formation of vortices. In this case, the tangential actions contribute significantly to the drag force, while the contribution of the normal actions p to the drag force (form drag) is small.

Vice versa, bluff-bodies, such as, for example, a cylinder with a circular cross-section, are subject to high drag force: in these bodies, for Reynolds numbers $Re > 800$–1000 there is formation and shedding of vortices (Fig. 5.6b). The flow separation that is established on these bluff bodies creates a wake of alternating vortices that produces drag and lift forces that are variable in time even if the velocity of the blowing stream has constant magnitude and direction. The forces that are variable in time due to the vortex shedding in the case of a stationary body have random characteristics (see Chap. 7) with power spectral density that shows a peak (Fig. 5.6c) in correspondence to a particular frequency f_s, defined by the Strouhal relation:

[2]In this case the body must be cylindrical with its axis being perpendicular to the influencing flow.

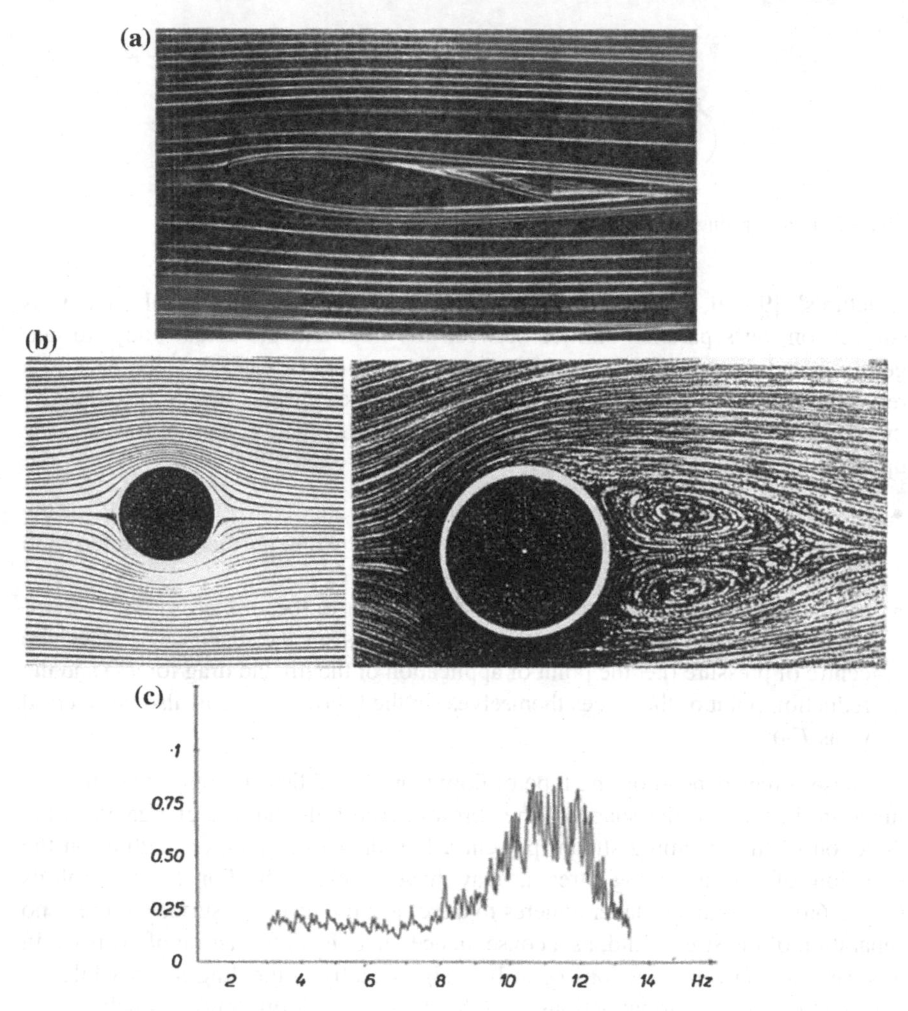

Fig. 5.6 **a** Flow conditions in an elongated aerofoil (wing). **b** Flow conditions in a bluff bodies: (a) condition for *Re* < 800–1000; (b) condition for *Re* > 800–1000. **c** Power spectral density (stationary profile) due to vortex shedding. **d** Values of Cs for different profiles. **e** Fixed cylinder: alternating vortices

$$f_s = \frac{c_s V}{D} \tag{5.16a}$$

where V is the fluid velocity, c_s a constant, characteristic of the body and lastly D a dimension characteristic of the same: in the case of a cylinder with a circular cross-section, for example, we have $c_s = 0.2$, D being the diameter. Figure 5.6d shows the values of c_s for different profiles.

It is customary to reproduce vortex shedding with an alternating lift force with a frequency equal to the frequency of Strouhal f_s (5.16a) and with drag force (also

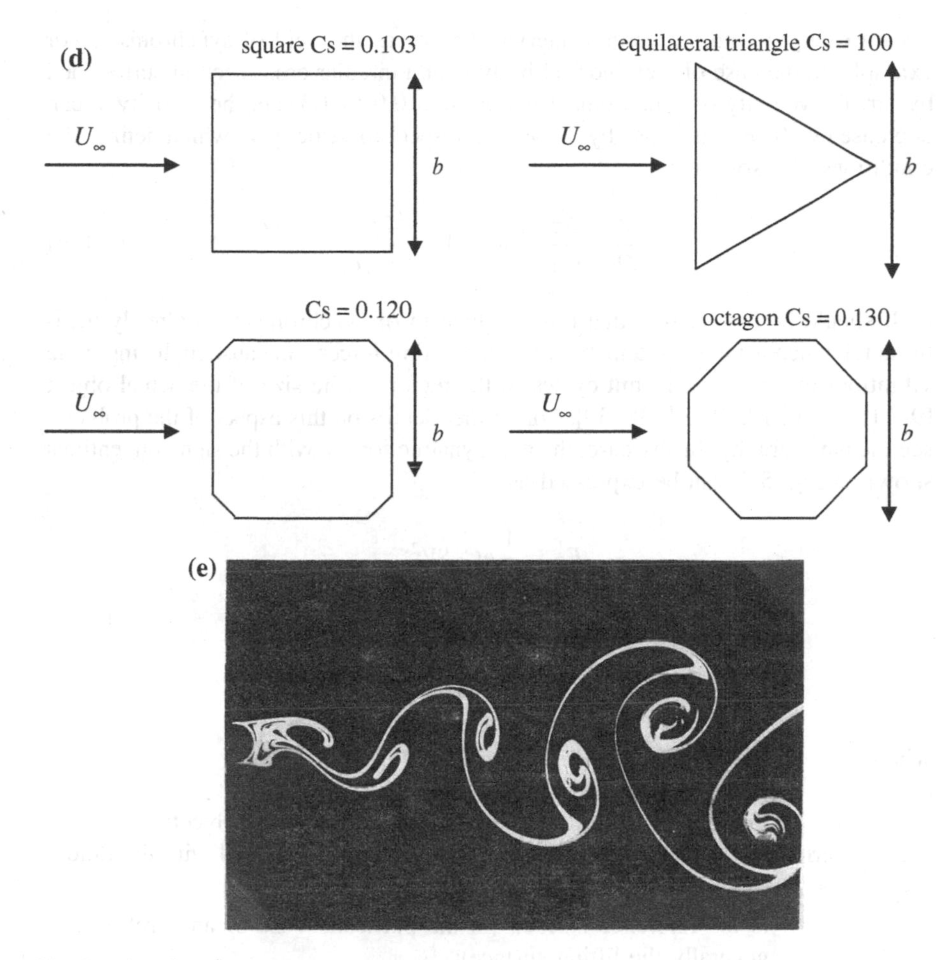

Fig. 5.6 (continued)

alternating) with double frequency $2f_s$. Figure 5.6e shows the trend of a series of alternating vortices in the case of a fixed cylinder with a circular cross-section: the vortex shedding frequency is equal to frequency f_s.

When the object is free to vibrate in an orthogonal direction to the stream, if the natural vibrating frequency of the object $f_o = \frac{\omega_o}{2\pi}$ coincides with the Strouhal frequency f_s, the object is excited in resonance and begins to oscillate. In these conditions the vortex shedding, which occurs for the fixed object in a disorderly manner in the various sections along the object giving rise to random forces, as a result of the vibration it synchronises with the actual vibration to generate a force that is no longer random but almost harmonic with a frequency that is attuned with the actual vibration. Therefore, in these conditions, a phenomenon of synchronism occurs between the vortex shedding and the vibration that increases the intensity of the alternating lift force. Synchronism is maintained even if the velocity V of the stream changes within a certain interval, known as the field of synchronism. For example, in the case of a cylindrical body with a circular cross-section surrounded by air, the velocity of synchronism varies from 0.9 to 1.3, i.e. the velocity V can decrease by 10 % or increase by 30 % with respect to velocity V_s which defines the conditions of resonance:

$$f_s = c_s \frac{V}{D} = \frac{2\pi}{\omega_0} = f_0 \Rightarrow V_s = \frac{2\pi D}{\omega_0 c_s} = f_0 \frac{D}{c_s} \tag{5.16b}$$

It should be emphasised that this mechanism of synchronism, as already mentioned, by increasing the intensity of the alternating forces can cause high amplitude vibrations that can reach limit cycles in the region of the size of the actual object [9–11, 13, 16, 17, 26–34, 38, 39]. For further details on this aspect of the problem, see the bibliography. In any case, the aerodynamic forces, with the sign conventions shown in Fig. 5.7, can be expressed as:

$$\begin{aligned} F_D &= \frac{1}{2}\rho C_D S V_r^2 \\ F_L &= \frac{1}{2}\rho C_L S V_r^2 \\ M &= \frac{1}{2}\rho C_m S C V_r^2 \end{aligned} \tag{5.17}$$

where:

ρ	is the density of the fluid;
V_r	the relative velocity of the stream relative to the object;
the term $1/2\rho V_r^2$	is the kinetic energy per unit of volume associated with the fluid;
S	is the reference surface (in the particular case of an aerofoil S is generally the lifting surface);
C, lastly	is a characteristic linear dimension, the chord line in the case of the aerofoil (Fig. 5.7)

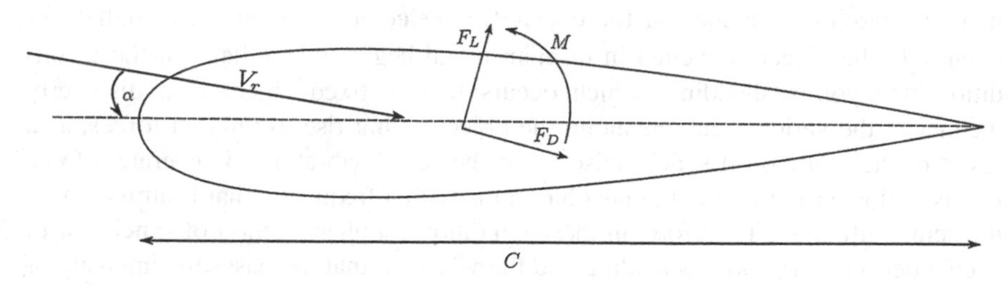

Fig. 5.7 Sign conventions: aerodynamic forces, angle of attack α

In the expressions (5.17) C_D, C_L and C_m represent non-dimensional aerodynamic coefficients: usually these coefficients only refer to the mean value of the aerodynamic forces. In large fields of the Reynolds number [9, 22, 33, 34, 38], these coefficients can be considered constant, merely functions of the type of profile analysed and of the angle of attack of the influencing flow (Fig. 5.7), i.e. the angle formed by the direction of the relative velocity V_r with a reference direction that is integral to the body. In fact, the aerodynamic coefficients are a function of the Reynolds number: Fig. 5.8 shows, again in the case of a cylinder with a circular cross-section hit by a flow that is perpendicular to the axis of the actual cylinder, the trend of the drag coefficient C_r as a function of the Reynolds number (*Re*). As we can see:

- at low Re ($Re < 10^3$) there is laminar flow and so drag is mainly caused by tangential actions;
- for $10^3 < Re < 10^5$ the flow regime becomes turbulent with a laminar boundary layer: in this situation the drag resistance is equal to $C_D = 1.2$;
- for values of $Re > 10^5$ (critical Reynolds number), as a function of the roughness of the cylinder, the boundary limit becomes turbulent and there is a decrease in drag: at the same time the vortex shedding diminishes greatly compared to that which occurs in the previous zone;
- for $Re > 10^6$, the drag coefficient C_D begins to rise again: in this region the phenomenon of vortex shedding occurs again.

These coefficients can be obtained analytically by integrating the Stokes-Navier equations or equations deriving from them. In general, for all types of profiles, we

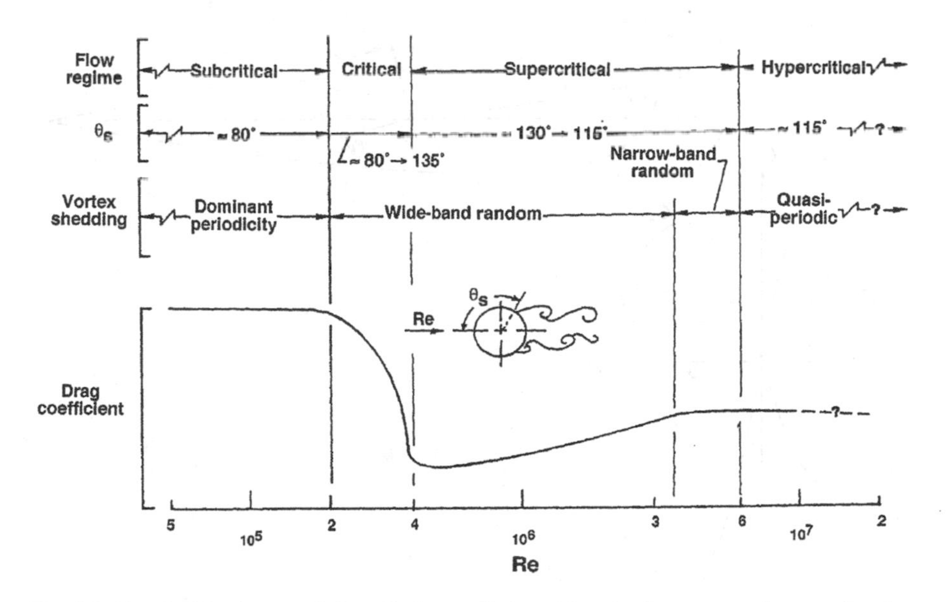

Fig. 5.8 Trend of the drag coefficient C_D for a cylinder with a circular cross-section as a function of *Re*

can also use experimental methods to statically measure the forces acting on physical scale models placed in wind tunnels or hydraulic channels, upon variation of the angle of attack of the flow α. In the case of a symmetrical profile, for example, the trend as a function of the angle of attack (a) of the coefficients, with the conventions of Fig. 5.7, is shown in Fig. 5.9. The curves C_L and C_m grow almost linearly with α about the origin and have maximum value in correspondence to an angle of incidence (as) of approximately 10°–15°. Upon increasing the angle of incidence above this value, lift and torque decrease, while drag increases. Angle α in correspondence to the maximum is the stall angle of the profile where there is detachment of the flow from the profile and formation of vortices with consequent increase in drag and decrease in lift. Profiles that do not detach the stream, as already mentioned, have low drag coefficients; they are generally profiles that are elongated in the direction of the flow. Non-elongated profiles, with a strong front surface, generally detach the stream and are accompanied by a high drag coefficient.

The (5.17) relations apply in conditions of stationary flow, i.e. when the velocity of the stream relative to the object forms an angle that is constant in time. If, due to the motion of the object or to changes in the direction or also in the intensity of the fluid velocity, the relative velocity V_r changes between stream and object, the (5.17) relations can still be considered valid if the motion of the object or the change in direction and velocity of the fluid occur slowly, i.e. almost statically. In quantitative terms this is true if the value of the reduced frequency of the fluid-elastic system in question, defined as:

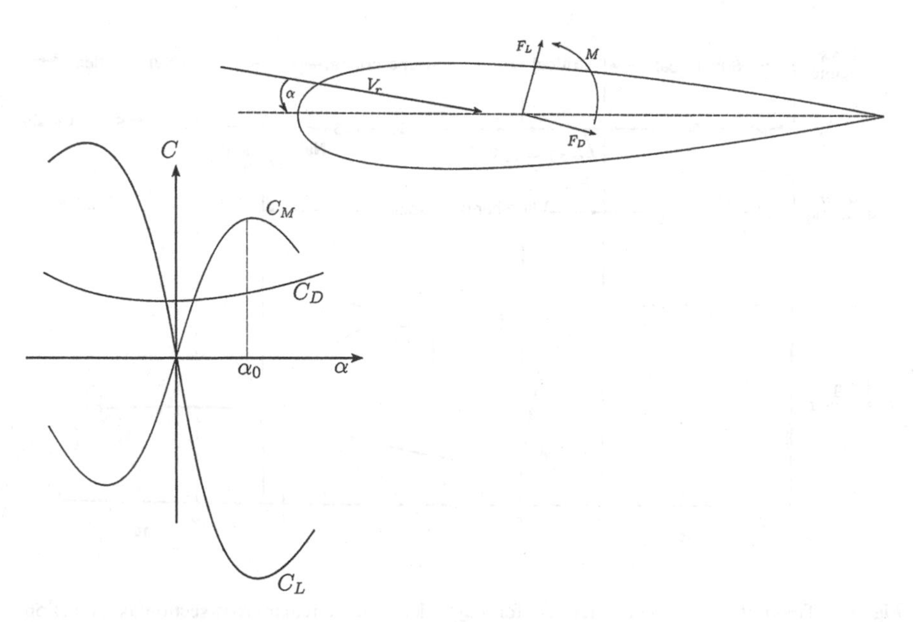

Fig. 5.9 Aerodynamic coefficients for a symmetrical profile

$$f_r = \frac{f_0}{U/C} \tag{5.17a}$$

is small, $f_0 = \frac{\omega_o}{2\pi}$ being the frequency of oscillation of the vibrating system (or of oscillation of the velocity or of the direction of the fluid), U being the average velocity of the fluid and C the chord line. The ratio U/C represents a frequency correlated to the time required for a fluid particle to cross the region occupied by the body. In fact, if the body in question does not vibrate ($f_0 = 0$), the relative reduced frequency f_r is zero: the coefficients measured statically in the wind tunnel on a physical model, placed in the (5.17) relations, make it possible to precisely calculate the aerodynamic forces acting on the actual body. As the value of f_r increases, the aerodynamic force field comes to depend increasingly on the oscillations of the body: coefficients C_r, C_p and C_m obtained experimentally, must be *corrected* as a function of the reduced frequency (correct quasi-static theory [21, 22, 35–37]. Often, to define aerodynamic coefficients in relation to the quasi-static theory, *corrected* if necessary, we can also refer to another non-dimensional parameter, called reduced velocity V_{rid}, defined as:

$$V_{rid} = \frac{U}{f_0 \cdot C} \tag{5.17b}$$

which corresponds to the inverse of reduced frequency f_r (5.17a). In the following discussion we will assume, for simplicity's sake, that the quasi-static theory applies and we will assume that the values of the aerodynamic coefficients are known, while, for further details, we refer the reader to the Bibliography shown in the section "Aerodynamic forces acting on structures" [12, 14–27, 35–37].

5.2.1.2 Torsional Vibrating System Subjected to a Fluid Stream

Now we will go back to considering the vibrating system shown in Fig. 5.3, i.e. a symmetric profile, constrained in such a way as to create a vibrating system with just one torsional degree-of-freedom, having defined the constants of the elastic restoring spring with k_s, the constants associated with the concentrated dampers (reproducing structural damping) with r_s and lastly the moment of inertia of the body around its centre of gravity G with J. We will assume that the profile is hit by an incoming air flow by velocity U that is constant in time and space. To describe the motion of the one d.o.f. system in question, we assume, as an independent variable, rotation θ. We write the equations of motion using, for example, the Lagrange equations: the kinetic energy of the system E_c is:

$$E_c = \frac{1}{2} J \dot{\theta}^2 \tag{5.18a}$$

Assuming that the potential energy V of the actual system in the static equilibrium position, in the absence of the aerodynamic force field, is zero, the same can be expressed as:

$$V = \frac{1}{2} 2k_s b^2 \theta^2 \quad (5.18b)$$

and similarly the dissipative function D is:

$$D = \frac{1}{2} 2r_s b^2 \dot{\theta}^2 \quad (5.18c)$$

The work $\delta^* L$ carried out by the external forces applied (the aerodynamic forces) for the only virtual displacement $\delta^* \theta$ that the body is permitted by the constraints equals:

$$\delta^* L = M \delta^* \theta = \frac{1}{2} \rho C_m(\alpha) SCV_r^2 \delta^* \theta \quad (5.18d)$$

where M represents the aerodynamic torque acting on the profile already defined in the previous section, and is dependent on the force field generated by the incoming flow. Upon applying the Lagrange equations to the various forms of energy defined previously (5.18a)–(5.18d), the equation of motion of the system becomes:

$$J\ddot{\theta} + 2r_s b^2 \dot{\theta} + 2k_s b^2 \theta = \frac{1}{2} \rho C_m(\alpha) SCV_r^2 \quad (5.19)$$

The generalised aerodynamic torque M depends (5.17) on the relative velocity V_r, and on the angle of incidence α, i.e. on the motion of the body. Let us consider the generic position of the body in question: the relative velocity V_r must be evaluated as the vector difference of the absolute velocity U of the stream and the velocity V_t of a point belonging to the body:

$$V_r = U - V_t \quad (5.20a)$$

Obviously, this term changes upon variation of the generic point P_i considered on the body (Fig. 5.10), i.e. from the distance of the rotation axis: this considerably complicates the definition of the aerodynamic forces. In fact, the relative velocity is different at every point on the body thus making it impossible to apply the quasi-static theory, which assumes that the relative velocity is the same for all the points of the body. Nevertheless, it is possible to demonstrate that it is still legitimate to resort to the quasi-static theory if we refer to a specific point P_1 of the body located at a certain distance b_1 from the rotation axis: in the particular case of the aerofoil, this point P_1 is, in general, close to the leading edge (Fig. 5.10), distant C/2 from the geometric centre of the section. The value of b_1 is different depending on the geometry of the profile being considered and can be defined experimentally by

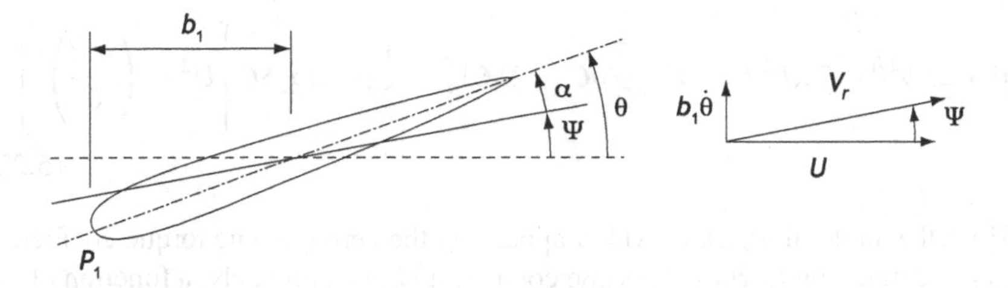

Fig. 5.10 Calculation of the relative velocity V_r for calculation of the aerodynamic forces (5.16a and 5.16b)

making the profile oscillate in a wind tunnel and measuring the response of the actual profile [18, 20–24]. In other words, once point P_1 has been defined, the relative velocity is calculated as if all the points of the profile had a relative velocity equal to that of P_1.

The angle ψ formed by the relative velocity with the horizontal axis equals (Fig. 5.10):

$$\psi = \operatorname{atan}\left(\frac{-V_t}{U}\right) = \operatorname{atan}\left(\frac{-b_1\dot{\theta}}{U}\right) \tag{5.20b}$$

The angle of attack α, between the straight line that is integral to the body (which is horizontal when the body is stationary) and the relative velocity V_r, with the conventions of Fig. 5.7,[3] is given by:

$$\alpha = \theta - \psi = \theta - \frac{b_1\dot{\theta}}{U} \tag{5.20c}$$

θ being the rotation of the body. The relative velocity, in magnitude, is equal to:

$$V_r^2 = U^2 + \left(\frac{b_1\dot{\theta}}{U}\right)^2 \tag{5.21}$$

Taking into account the relations (5.20a)–(5.20c) and (5.21), the equation of motion (5.19) of the vibrating system with 1 DOF, hit by a fluid stream (Fig. 5.3), is non-linear in the independent variable θ and in its derivatives:

[3]Rotating the direction of the velocity in an anti-clockwise direction, taking into account the sign conventions for the positive angles shown in Fig. 5.7, is equal to assuming that the angles of the body's clockwise rotation are positive.

$$J\ddot{\theta} + 2r_s b^2\dot{\theta} + 2k_s b^2\theta = M = \frac{1}{2}\rho C_m(\alpha)SCV_r^2 = \frac{1}{2}\rho C_m(\alpha)SC\left[U^2 + \left(\frac{b_1\dot{\theta}}{U}\right)^2\right] \tag{5.22}$$

Since the angle of attack α, which appears in the aerodynamic torque coefficient $C_m(\alpha)$, is defined by (5.20c), the same coefficient C_m is, ultimately, a function of the rotation θ and the velocity $\dot{\theta}$:

$$C_m(\alpha) = C_m\left(\theta - a\tan\left(\frac{b_1\dot{\theta}}{U}\right)\right) = C_m\left(\theta, \dot{\theta}\right) \tag{5.23}$$

Integration of Eq. (5.22) can only be carried out numerically, while updating at every step of integration, the generalised aerodynamic torque value M as a function of the current values of θ and $\dot{\theta}$. To analyse the conditions of incipient instability of the system (or the small oscillations around the position of static equilibrium) it is possible, using the Lyapunov theorem, to linearise (5.22): to do this, we must express the generalised aerodynamic force in a Taylor series, around the position of static equilibrium. The equation that makes it possible to define the position of static equilibrium is the following[4]:

$$2k_s b^2\theta_0 = \frac{1}{2}\rho C_m(\theta_0, 0)SCU^2 \tag{5.24a}$$

Which is obtained from (5.22) setting $\dot{\theta}$ as zero. Equation (5.24a) accepts as a solution, assuming a flow in the direction of the axis of symmetry of the profile and for the particular trend of the assumed aerodynamic torque coefficient of the symmetric aerofoil (Fig. 5.9):

$$\theta_0 = 0 \Rightarrow \alpha_0 = 0 \tag{5.24b}$$

Now to linearise the equation of motion (5.22) around the position of static equilibrium, it is convenient, as already mentioned, to linearise the single terms of the product $\frac{1}{2}\rho C_m(\theta_0, 0)SCU^2$ i.e.:

[4]More generally, the aerodynamic field force affects not only the dynamic behaviour of the system, but also the actual position of static equilibrium.

$$C_m(\alpha) \approx C_m(\alpha_0) + \left(\frac{\partial C_m}{\partial \alpha}\right)_0 (\alpha - \alpha_0)$$
$$\alpha \approx \theta - \left(\frac{b_1\dot{\theta}}{U}\right) \tag{5.25a}$$
$$U^2 + \left(\frac{b_1\dot{\theta}}{U}\right)^2 \approx U^2$$

The aerodynamic torque $C_m(\alpha)$ is thus linearised around the position of static equilibrium defined by the pair of values $\theta_o = 0$ and $\dot{\theta}_o = 0$. We can use K_{mo} to define the derivative of the torque coefficient estimated around the position of static equilibrium:

$$K_{m0} = \left(\frac{\partial C_m}{\partial \alpha}\right)_0 \tag{5.25b}$$

The right-hand member of (5.22) (i.e. the aerodynamic moment M acting on the vibrating system), taking into account (5.25a) and (5.25b), becomes:

$$M = \frac{1}{2}\rho\left(C_m(\alpha_0) + \left(\frac{\partial C_m}{\partial \alpha}\right)_0 (\alpha - \alpha_0)\right)SCU^2 \tag{5.25c}$$

Since the position of static equilibrium, in the case being analysed, is defined by a value of the angle of attack (5.24b) $\alpha_o = 0$ and that $C_{mo}(\alpha_o) = 0$ (see Fig. 5.9), Eq. (5.25c) becomes:

$$M = \left(\frac{1}{2}\rho SCK_{m0}U^2\right)\theta - \left(\frac{1}{2}\rho SCUK_{m0}b_1\right)\dot{\theta} \tag{5.25d}$$

Lastly, assuming:

$$a = \frac{1}{2}\rho S \tag{5.25e}$$

the equation of motion (5.22) can be rewritten, bringing to the left-hand member the equivalent linearised terms of damping and stiffness of (5.25d):

$$J\ddot{\theta} + \left(2r_s b^2 + aK_{m0}Ub_1C\right)\dot{\theta} + \left(2k_s b^2 - aU^2K_{m0}C\right)\theta = 0 \tag{5.26}$$

In this case the variable θ that defines the perturbed motion around the position of static equilibrium, taking into account that $\theta_o = 0$, is coincident with the initial independent variable:

$$\overline{\theta} = \theta - \theta_0 = \theta \tag{5.26a}$$

In the particular case of the aerofoil, as mentioned, b_1 is practically equal to $C/2$. Since K_{mo} (5.25b) and Fig. 5.9 is greater than zero, the equivalent elastic term K_F due to the force field (5.26):

$$K_F = -aU^2 K_{m0} C \tag{5.27a}$$

is negative; in other words, there is a stream velocity U_{lim} above which the equivalent elastic restoring term of the aerodynamic forces K_F is equal to or greater in magnitude than the structural elastic restoring term $2k_s b^2$:

$$aU_{\text{lim}}^2 K_{m0} C \geq 2k_s b^2 \tag{5.27b}$$

For velocities of the incoming stream greater than this value U_{lim}, the overall stiffness K_T of the aero-elastic system:

$$K_T = \left(2k_s b^2\right) + K_F \tag{5.27c}$$

is negative, i.e. the system is statically unstable (that is, there is a static divergence). The equivalent damping term r_F however:

$$r_F = aK_{m0} U b_1 C \tag{5.27d}$$

is positive for the aerofoil: the aeroelastic system cannot be dynamically unstable, on the contrary, as the velocity of the incoming stream increases, the overall damping r_T (structural $2r_s b^2$ and aerodynamic r_F) increases. If another type of profile had been analysed, for example, the box profile in Fig. 5.11 or the same aerofoil but considering an angle of incidence α_o greater than the stall angle Fig. 5.9, the resulting equations (5.22) would have been the same. In this case, however, as we can see from the same Fig. 5.11, the derivative of torque coefficient K_{mo} is negative: therefore, for the specific profile or for small oscillations around the

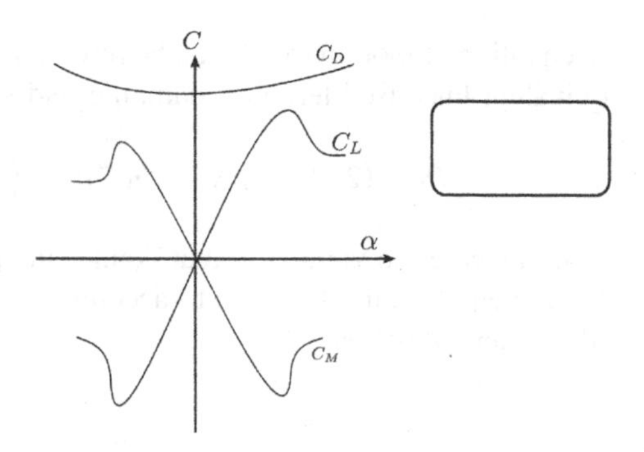

Fig. 5.11 Aerodynamic coefficients of a box profile

position of static equilibrium, instability of a static nature cannot occur. However, it is possible that a dynamic type of instability may be established (with oscillations expanding over time) when the term r_F (5.27d) (always, in this case, less than zero) becomes, in magnitude, greater than structural damping and this occurs for a velocity $U_{\lim}$ equal to:

$$U_{\lim} = \left| \frac{2r_s b^2}{aK_{m0}b_1 C} \right| \tag{5.28}$$

The value of this velocity, also known as instability velocity, increases as structural damping *rs* increases and as the absolute value of the slope of the curve of the aerodynamic torque, i.e. of K_{mo} decreases (K_{mo} being negative). Equation (5.26), obtained for the simple case of a torsional vibrating system with one degree-of-freedom subject to an aerodynamic force field, already shows the possible alternatives for eliminating any possible forms of instability (static and/or dynamic) present in a generic vibrating system:

- modifying the shape of the profile (i.e. changing its aerodynamic coefficients);
- introducing greater structural damping into the system or increasing the structural stiffness.

5.2.1.3 Translating Vibrating System Subjected to a Fluid Stream

Now we will consider a case where the aerofoil is constrained in such a way as to create a translating vibrating system with one degree-of-freedom, as shown in Fig. 5.12. We will consider vertical displacement x as a generalised independent variable, to describe the motion of the system: the different terms of energy, in this case, can be evaluated as:

$$E_c = \frac{1}{2}m\dot{x}^2 \quad V = \frac{1}{2}k_s x^2 \quad D = \frac{1}{2}r_s\dot{x}^2 \quad \delta^* L = F_x \delta^* x \tag{5.29a}$$

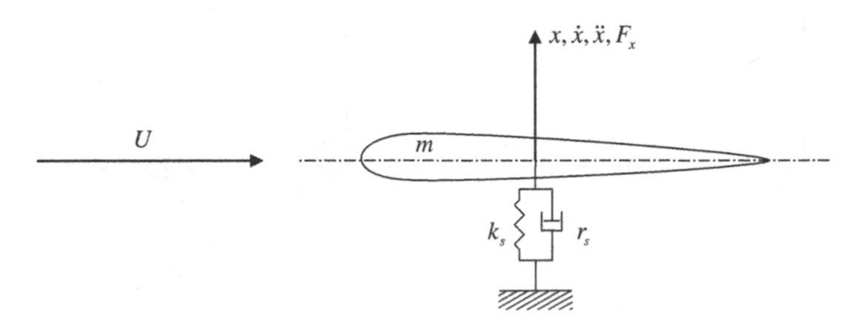

Fig. 5.12 Translating system hit by a fluid stream

where F_x represents the component in the vertical plane of the aerodynamic forces acting on the body (with this arrangement of constraints, both the horizontal component of the forces, and the aerodynamic torque are balanced by the constraint reactions).

The equation of motion of the aero-elastic system becomes:

$$m\ddot{x} + r_s\dot{x} + k_s x = F_x \tag{5.29b}$$

Now we need to estimate generalised force F_x. We assign a generic displacement x to the body in question, with an assigned velocity $\dot{x}$: the forces that act on the system (Fig. 5.13) are only the drag force F_r and lift force F_p. The relative velocity V_r is given by the vector difference of the absolute velocity of the stream U and the velocity (by dragging) of the object, see Fig. 5.13, while the square of the magnitude of V_r equals:

$$V_r^2 = U^2 + \dot{x}^2 \tag{5.30a}$$

The angle ψ formed by vector V_r with the horizontal (see Fig. 5.13), velocity x being small compared to that of the incoming stream U, equals:

$$\psi = a\tan\left(\frac{\dot{x}}{U}\right) \approx \frac{\dot{x}}{U} \tag{5.30b}$$

The angle of attack α, formed by the relative velocity V_r with respect to a straight line that is integral to the body (assumed to have the same direction as the undisturbed flow U), is thus given by:

$$\alpha = \psi = \frac{\dot{x}}{U} \tag{5.30c}$$

Since the drag and lift forces act, respectively, in the same direction and orthogonally to the direction of the relative velocity, the vertical resultant F_x (Fig. 5.12) can be expressed as:

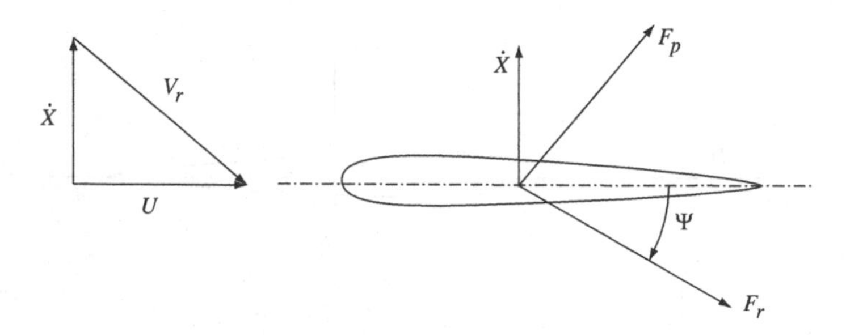

Fig. 5.13 Definition of the aerodynamic actions acting on a translating profile

$$F_x = -F_r \sin\psi + F_p \cos\psi \tag{5.30d}$$

Taking into account (5.30a)–(5.30d) the equation of motion (5.29b) of the translating system surrounded by a fluid stream becomes non-linear in the independent variable $\dot{x}$ and x:

$$m\ddot{x} + r_s\dot{x} + k_s x = \frac{1}{2}\rho S\left(U^2 + \dot{x}^2\right)\left(-C_r(\alpha)\sin\left(\frac{\dot{x}}{U}\right) + C_p(\alpha)\sin\left(\frac{\dot{x}}{U}\right)\right) \tag{5.31}$$

In particular, (5.31) is non-linear in the velocity $\dot{x}$: for this reason, it can be said that the aerodynamic force field, in this case of a purely translating profile, once linearised will only give rise to equivalent damping terms:

$$F_x(\dot{x}) = F_x(0) + \left(\frac{\partial F_x}{\partial \dot{x}}\right)_0 \dot{x} \tag{5.32a}$$

If we consider, as an example, the profile whose characteristic curves (polar) are shown in Fig. 5.9, the position of static equilibrium is defined by $x_o = 0$, the value of the lift coefficient being zero $C_p(\alpha)$ by $\alpha_o = 0$. In this case too, we will linearise function F_x by develojournalg the single terms of the product that makes up F_x in a Taylor series:

$$F_x = \frac{1}{2}\rho S\left(U^2 + \dot{x}^2\right)\left(-C_r(\alpha)\sin\left(\frac{\dot{x}}{U}\right) + C_p(\alpha)\sin\left(\frac{\dot{x}}{U}\right)\right) \tag{5.32b}$$

i.e.:

$$\begin{aligned}
&\alpha 0 = 0; \quad \alpha \approx \frac{\dot{x}}{U} \\
&\sin\left(\frac{\dot{x}}{U}\right) \approx \left(\frac{\dot{x}}{U}\right); \quad \cos\left(\frac{\dot{x}}{U}\right) \approx 1; \quad V_r^2 = U_2 \\
&C_r(\alpha) \approx C_{r0} + \left(\frac{\partial C_r}{\partial \alpha}\right)_0 (\alpha - \alpha_0) = C_{r0} + K_{r0}(\alpha - \alpha_0) \\
&C_p(\alpha) \approx C_{p0} + \left(\frac{\partial C_p}{\partial \alpha}\right)_0 (\alpha - \alpha_0) = C_{p0} + K_{p0}(\alpha - \alpha_0)
\end{aligned} \tag{5.32c}$$

Taking into account (5.32c), assuming:

$$a = \frac{1}{2}\rho S \tag{5.32d}$$

and neglecting the non-linear terms, force F_x at the right-hand member of (5.31), (5.32b) can be linearised as:

$$F_x = -(a \cdot U \cdot C_{r0})\dot{x} + (a \cdot U^2 \cdot C_{p0}) + (a \cdot U \cdot K_{p0})\dot{x} \tag{5.33a}$$

Since in the example analysed, see Fig. 5.9, $C_{po}(\alpha) = C_p(\alpha_o) = C_p(0) = 0$ and $K_{ro} = \left(\partial C_r/\partial\alpha\right)_o$, the linearised expression of the generalised aerodynamic force is reduced to the form:

$$F_x = a \cdot U(K_{p0} - C_{r0})\dot{x} \tag{5.33b}$$

where K_{po} and C_{ro} indicate (5.33a), respectively, the derivate of the lift coefficient C_p estimated at the origin and the drag coefficient C_r estimated in correspondence to the actual angle of attack $\alpha_o = 0$. The linearised equation of the system's motion thus becomes:

$$m\ddot{x} + (r_s + r_F)\dot{x} + k_s x = 0 \tag{5.34}$$

having assumed:

$$r_F = -a \cdot U(K_{p0} - C_{r0}) = a \cdot U(C_{r0} - K_{p0}) \tag{5.34a}$$

In this case too, the variable x that defines the perturbed motion around the position of static equilibrium, taking into account the specific type of profile being analysed (Fig. 5.9), defined by $x_0 = 0$, proves to be coincident with the initial independent variable:

$$\overline{x} = x - x_0 = x \tag{5.34b}$$

As already mentioned, the aerodynamic force field in this case only affects the terms of velocity: C_{ro} being always positive and K_{po} negative, in the example of the aerofoil, their contribution in the equivalent damping term r_F is purely dissipative, being always positive. Therefore, the force field introduces actual damping which adds to the structural damping: this characteristic is exploited by using aerofoils with the only degree-of-freedom of translation like in the case of the dynamic stabilisers adopted on vessels (anti-roll stabilisers). If, however, we consider profiles that have an important front surface (Fig. 5.11), or the same aerofoil for angles α_o greater than the stall angle, then K_{po} may be positive. In this case, if K_{po} is, in absolute value, greater than C_{ro}, then r_F proves to be negative: if, in this case, the magnitude of damping caused by the force field $|r_F|$ is greater than the structural damping r_s, this determines an instability in the 1 o.d.f. model, corresponding to a flexural expanding oscillating motion. Since r_F (5.34a) depends on U, there is a critical velocity U_{crit}, in correspondence to which r_F assumes a negative value and, in absolute value, greater than r_s. In correspondence to this velocity instability of the system occurs, which persists for values of velocity $U > U_{crit}$, characterised by a coefficient of instability $\delta = \left(\alpha/\omega\right)^5$ which becomes increasingly higher as the velocity of the stream increases above the critical value.

5.3 Vibrating Systems with 2 d.o.f. Perturbed Around the Position of Equilibrium

Now we will extend our analysis of the dynamic behaviour of mechanical systems subject to force fields in relation to two-degree-of-freedom systems. Naturally, the method for analysing these systems is the same applied in the previous section to systems with one degree-of-freedom:

- finding the position of static equilibrium; or linearisation of the motion equations;
- study of the limited stability.

In two-degree-of-freedom systems, in addition to those seen for systems with one degree-of-freedom, other forms of instability are also possible, such as, for example, flutter instability, where the unstable motion is defined by combining two vibration modes, as will be illustrated in the following sections. Another difference lies in the fact that while in systems with one degree-of-freedom it is always possible to develop the discussion of stability by simply analysing the coefficients of the differential equation of motion, in the case of two-degree-of-freedom systems this is only possible in the case of a force field as a function of just the position. In the case of a force field that also depends on velocity, analysis of the stability is not so immediate, even if it is possible, however, to outline sufficient criteria for stability that concern the structure of the matrices.

Now we will illustrate the case of the vibrating two-degree-of freedom system in Fig. 5.14,[6] placed in a force field that is a function of the position, velocity and acceleration of the system, whose equations can thus be written in scalar form:

---

[5]Coefficient of instability defines the relation between the real part α (positive if there is instability) and the imaginary part ω of the generic root λ = α + iω of the characteristic equation corresponding to the solution. This relation is defined in this case by:

$$\delta = \left(\frac{\alpha}{\omega}\right) = -\frac{\frac{(r_F+r_s)}{2m}}{\omega_0\sqrt{(1-h^2)}} = -\frac{\frac{(r_F+r_s)\omega_0}{2m\omega_0}}{\omega_0\sqrt{(1-h^2)}} = -\frac{h}{\sqrt{(1-h^2)}} \approx -h \tag{5.4.1}$$

In stable systems (con $r_T = (r_F + r_S) > 0$) δ is negative, while for unstable systems ($r_T = (r_F + r_S) < 0$) it is positive.

[6]This system represents the simplest system with 2 DOF that can be considered: the discussion can be extended to any system with 2 DOF by analysing it with a modal approach, i.e. in principle

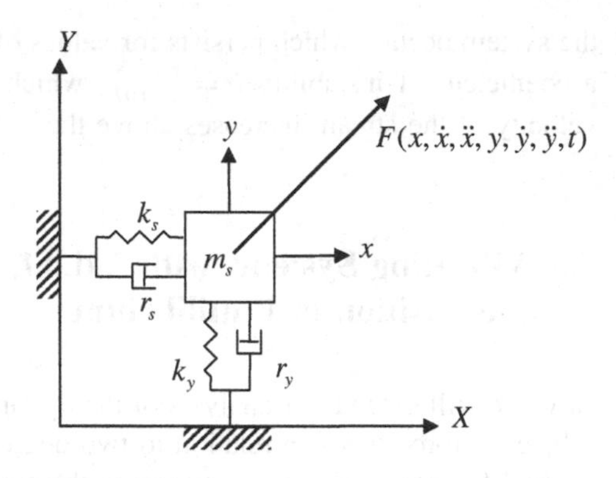

Fig. 5.14 Vibrating two-degree-of-freedom system placed in a force field

$$
\begin{aligned}
m\ddot{x} + r_x\dot{x} + k_x x &= F_x(x, y, \dot{x}, \dot{y}, \ddot{x}, \ddot{y}) \\
m\ddot{y} + r_y\dot{y} + k_y y &= F_y(x, y, \dot{x}, \dot{y}, \ddot{x}, \ddot{y})
\end{aligned}
\tag{5.35}
$$

The terms F_x and F_y due to the force field are, in general, non-linear functions of the displacements x and y and of their derivatives: therefore, integration of the Eq. (5.35) cannot generally be done analytically and in these cases it is necessary to resort to numerical integration methods for the differential equations [2, 5, 7]. Equation (5.35) can be rewritten in matrix form, as:

$$
\begin{bmatrix} m & 0 \\ 0 & m \end{bmatrix} \begin{Bmatrix} \ddot{x} \\ \ddot{y} \end{Bmatrix} + \begin{bmatrix} r_x & 0 \\ 0 & r_y \end{bmatrix} \begin{Bmatrix} \dot{x} \\ \dot{y} \end{Bmatrix} + \begin{bmatrix} k_x & 0 \\ 0 & k_y \end{bmatrix} \begin{Bmatrix} x \\ y \end{Bmatrix} = \begin{Bmatrix} F_x\left(x, y, \dot{x}, \dot{y}, \ddot{x}, \ddot{y}\right) \\ F_y\left(x, y, \dot{x}, \dot{y}, \ddot{x}, \ddot{y}\right) \end{Bmatrix}
\tag{5.36}
$$

Having defined the following vectors:

$$
\underline{z} = \begin{Bmatrix} x \\ y \end{Bmatrix} \quad \underline{F} = \begin{Bmatrix} F_x \\ F_y \end{Bmatrix}
\tag{5.36a}
$$

(5.36) becomes, with a more compact formulation:

$$
[M]\underline{\ddot{z}} + [R]\underline{\dot{z}} + [K]\underline{z} = \underline{F}(\underline{z}, \underline{\dot{z}}, \underline{\ddot{z}})
\tag{5.36b}
$$

Following the procedure already used for systems with one degree-of-freedom, first we search for the position of static equilibrium defined by a pair of values x_0, y_0.[7] Then we must solve the system of non-linear equations:

(Footnote 6 continued)
coordinates (Sect. 2.5) and, therefore, with structurally decoupled equations, of the same type as those considered in the example analysed here.

[7] We should remember that, as the equations are non-linear, there may be more than one position of equilibrium.

$$K_x x_0 = F_x(x_0, y_0)$$
$$K_y y_0 = F_y(x_0, y_0) \tag{5.37a}$$

i.e.:

$$[K]\underline{z}_0 = \underline{F}(\underline{z}_0, \underline{0}, \underline{0}) \tag{5.37b}$$

The solution can be found, for example, with the Newton-Raphson method, for systems of equations ([2, 5] and Sect. 4.7). Once the position of equilibrium $\underline{z}_0$, if it exists, has been obtained, it is possible to linearise the system around the position of equilibrium found, develojournalg the non-linear forces in series up to the first order:

$$\underline{F}(\underline{z}, \dot{\underline{z}}, \ddot{\underline{z}}) = \underline{F}(\underline{z}_0, \underline{0}, \underline{0}) + \left[\frac{\partial \underline{F}}{\partial \underline{z}}\right]_0 (\underline{z} - \underline{z}_0) + \left[\frac{\partial \underline{F}}{\partial \dot{\underline{z}}}\right]_0 \dot{\underline{z}} + \left[\frac{\partial \underline{F}}{\partial \ddot{\underline{z}}}\right]_0 \ddot{\underline{z}} + \ldots \tag{5.38a}$$

Remembering that the derivative of a vector (in this case $\underline{F}$) with respect to another vector (in this case $\underline{z}$), represents a matrix defined as:

$$\left[\frac{\partial \underline{F}}{\partial \underline{z}}\right]_0 = \begin{bmatrix} \frac{\partial F_1}{\partial x_1} & \frac{\partial F_1}{\partial x_2} \\ \frac{\partial F_2}{\partial x_1} & \frac{\partial F_2}{\partial x_1} \end{bmatrix}_0 \quad \left[\frac{\partial \underline{F}}{\partial \dot{\underline{z}}}\right]_0 = \begin{bmatrix} \frac{\partial F_1}{\partial \dot{x}_1} & \frac{\partial F_1}{\partial \dot{x}_2} \\ \frac{\partial F_2}{\partial \dot{x}_1} & \frac{\partial F_2}{\partial \dot{x}_2} \end{bmatrix}_0 \quad \left[\frac{\partial \underline{F}}{\partial \ddot{\underline{z}}}\right]_0 = \begin{bmatrix} \frac{\partial F_1}{\partial \ddot{x}_1} & \frac{\partial F_1}{\partial \ddot{x}_2} \\ \frac{\partial F_2}{\partial \ddot{x}_1} & \frac{\partial F_2}{\partial \ddot{x}_2} \end{bmatrix}_0 \tag{5.38b}$$

considering the coordinate transformation:

$$\underline{z} = \underline{z}_0 + \bar{\underline{z}} \tag{5.38c}$$

having used $\bar{\underline{z}}$ to indicate the vector of the independent variables that define the perturbed motion around the position of static equilibrium $\underline{z}_0$:

$$\bar{\underline{z}} = \begin{Bmatrix} \bar{x} \\ \bar{y} \end{Bmatrix} = \begin{Bmatrix} x - x_0 \\ y - y_0 \end{Bmatrix} \tag{5.38d}$$

then Eq. (5.36b) can be rewritten, while also taking into account the solution regarding static equilibrium (5.37b) as:

$$[[M] + [M_F]]\ddot{\bar{\underline{z}}} + [[R] + [R_F]]\dot{\bar{\underline{z}}} + [[K] + [K_F]]\bar{\underline{z}} = \underline{0} \tag{5.39a}$$

thus corresponding to the equation of free motion, in the presence of the linearised force field. The system of Eq. (5.39a) can be rewritten as:

$$[M_t]\ddot{\bar{\underline{z}}} + [R_t]\dot{\bar{\underline{z}}} + [K_t]\bar{\underline{z}} = \underline{0} \tag{5.39b}$$

Where $[M_t] = [M] + [M_F]$ is the matrix of the system's mass taking into account the force field, $[R_t] = [R] + [R_F]$ is the matrix of overall damping and lastly $[K_t] = [K] + [K_F]$ the matrix of total stiffness. The matrices of mass $[M_F]$, damping $[R_F]$ and stiffness $[K_F]$ equivalent to the force field are defined as:

$$-[M_F] = -\begin{bmatrix} \frac{\partial F_x}{\partial \ddot{x}} & \frac{\partial F_x}{\partial \ddot{y}} \\ \frac{\partial F_y}{\partial \ddot{x}} & \frac{\partial F_y}{\partial \ddot{y}} \end{bmatrix}_o \quad [R_F] = -\begin{bmatrix} \frac{\partial F_x}{\partial \dot{x}} & \frac{\partial F_x}{\partial \dot{y}} \\ \frac{\partial F_y}{\partial \dot{x}} & \frac{\partial F_y}{\partial \dot{y}} \end{bmatrix}_o \quad [K_F] = -\begin{bmatrix} \frac{\partial F_x}{\partial x} & \frac{\partial F_x}{\partial y} \\ \frac{\partial F_y}{\partial x} & \frac{\partial F_y}{\partial y} \end{bmatrix}_o \tag{5.39c}$$

As we saw previously for systems with one degree-of-freedom, (5.39a) highlights how the force field can, in general, change the characteristics of mass, damping and stiffness of the mechanical system, by changing the natural frequencies, damping and, therefore, affecting its stability. The stability of the system is, in actual fact, conditioned (5.39a), (5.39c) by the form of the matrices $[M_F]$, $[R_F]$ and $[K_F]$. In fact, we can make the following observations:

- if the matrices $[M_F]$, $[R_F]$e $[K_F]$ are symmetrical and positive definite (and therefore the matrices $[M_T]$, $[R_T]$ and $[K_T]$ prove to be the same) the system is always stable and admits eigenvalues $\lambda_i = \alpha_i \pm i\omega_i$ with a negative real part α_i. In these conditions, the system surrounded by the force field still behaves like a dissipative system. The matrix $[R_F]$ is symmetrical and positive definite if the velocity field is dissipative, while $[K_F]$ is symmetrical and defined positive if the field of positional forces is conservative: this matrix adds to the system's elastic matrix $[K]$, which is definitely symmetrical and positive definite.
- if $[K_F]$ is not symmetrical flutter instability may occur: these conditions will be discussed in detail in the following section;
- if $[K_F]$ is non-positive definite and assumes such values that $[K_T]$ also is also non-positive definite then divergence type instability occurs (see Sect. 5.3.1.1.2);
- if $[R_F]$ is not symmetrical the force field of velocity is not dissipative and could give rise to instability;
- if $[R_F]$ is non-positive definite in such a way that $[R_T]$ is also non-positive definite, then dynamic instability occurs.

The same also applies for two-degree-of-freedom but also more generally for n-degree-of-freedom systems. To analyse stability in quantitative terms we must examine the solution of free motion, obtained by introducing a particular integral of the homogeneous equation:

$$\underline{z}(t) = \underline{Z}e^{\lambda t} \tag{5.40}$$

where λ is generally complex. By replacing (5.40) in (5.39a), we obtain a homogeneous linear equation in $\underline{Z}$:

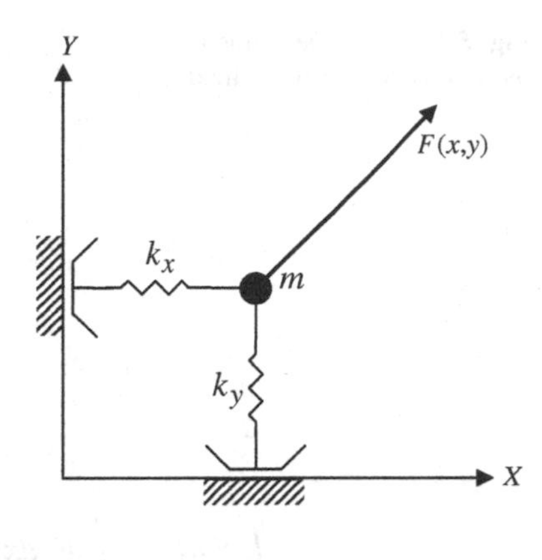

Fig. 5.15 Vibrating two-degree-of freedom system surrounded by a field of positional forces

$$\left[\lambda^2[M_t] + \lambda[R_t] + [K_t]\right]\underline{Z} = 0 \tag{5.40a}$$

By setting the determinant of the coefficient matrix (i.e. the characteristic polynomial which, as we know, is of degree $2n$ in λ) equal to zero, we obtain $2n$ values of λ, n being the number of degree-of-freedom of the system. As n increases, to find the values of λ as solutions of an eigenvalue problem is preferable, as explained previously in Sect. 2.3.1. This procedure will be referred to again in the section on systems generically with n-degree-of-freedom. Now we will study stability around the position of rest, of two-degree-of-freedom systems subject to different force fields, by firstly examining a generic field of forces that are purely positional and, subsequently, also taking into account the terms of velocity and acceleration.

5.3.1 Two-Degree-of-Freedom System with Placed in a Field of Purely Positional Forces

Now we will analyse the behaviour of two-degree-of-freedom systems with by firstly introducing some observations on purely positional force fields. We will refer to the simple example in Fig. 5.15, where m is a mass that is free to move in the plane, elastically constrained by two springs with constants k_x and k_y, according to the two directions of the reference axes, where we deliberately neglect the dissipative terms associated with the elastic hysteresis of the springs. $F(x,y)$ is a force acting on the mass, a function of the position x, y of the mass itself: the force defines a force field that can be conservative or non conservative. We should remember that a force field is conservative if the work of the force for a generic trajectory does not depend on the path, but only on the point of departure and arrival, i.e. (Fig. 5.16)

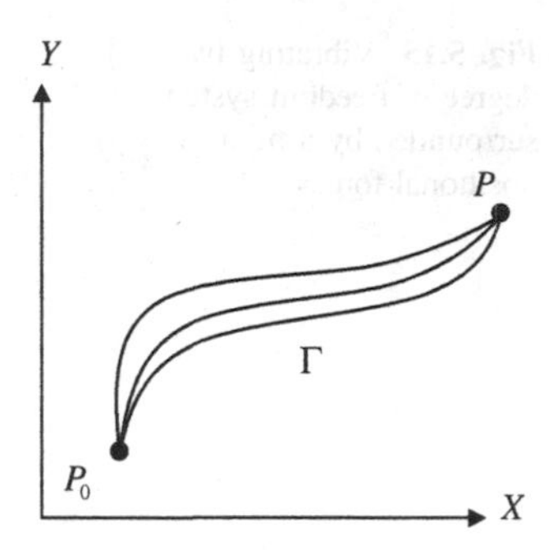

Fig. 5.16 Work performed by a conservative force field

$$L = \int_{\Gamma} F\,dP = \int_{\Gamma} F_x\,dx + \int_{\Gamma} F_y\,dy = f(x,y) - f(x_0,y_0) \tag{5.41}$$

The work of force F, of components F_x and F_y, if the force field is conservative does not depend on the path Γ, but only on the extremes P_o and P. If the line is closed then the work performed by the conservative force field will be zero.

In the case of conservative systems we should remember that the force F is defined as the gradient of a scalar $f(x,y)$ which represents the potential of the force field: the expression that appears in (5.41) is then an exact differential and, as a consequence, the components of the forces according to x and y are:

$$F_x = \frac{\partial f}{\partial x}; \quad F_y = \frac{\partial f}{\partial y} \tag{5.42}$$

Again, for a conservative force field, the mixed derivatives of the components of the forces with respect to the independent variables are equal, i.e.:

$$\frac{\partial F_x}{\partial y} = \frac{\partial F_y}{\partial x} \tag{5.42a}$$

In fact, f is a continuous function:

$$\frac{\partial^2 f}{\partial x \partial y} = \frac{\partial^2 f}{\partial y \partial x} \tag{5.42b}$$

that is, the rotor of F is zero [1, 5, 6]:

$$rot\,F = 0 \tag{5.42c}$$

and for (5.42) we have (5.42a). In the case of a non-conservative system, there is no potential f and the mixed derivatives of the components of the force F are different:

$$\frac{\partial F_x}{\partial y} \neq \frac{\partial F_y}{\partial x} \tag{5.43a}$$

that is:

$$rot\, F \neq 0 \tag{5.43b}$$

The work L performed by the force field along a closed path Γ is, in this case, different to zero: so a non-conservative force field for a closed trajectory gives or subtracts energy from the system. We will once more consider the two-degree-of-freedom system in Fig. 5.15, subject, as already mentioned, to a purely positional force field and we will write the equations of dynamic equilibrium in the two orthogonal directions x and y:

$$\begin{aligned} m\ddot{x} + k_x x &= F_x(x, y) \\ m\ddot{y} + k_y y &= F_y(x, y) \end{aligned} \tag{5.44}$$

The terms $F_x(x, y)$ and $F_y(x, y)$ are non-linear functions of the displacements x and y: however, the integration of these equations is not generally possible analytically and in these cases we must resort, as already mentioned, to numerical methods of integration of the above mentioned differential equations [2, 5, 7] or to approximate analytical methods. As usual (5.44) can be rewritten in matrix form as:

$$\begin{bmatrix} m & 0 \\ 0 & m \end{bmatrix} \begin{Bmatrix} \ddot{x} \\ \ddot{y} \end{Bmatrix} + \begin{bmatrix} k_x & 0 \\ 0 & k_y \end{bmatrix} \begin{Bmatrix} x \\ y \end{Bmatrix} = \begin{Bmatrix} F_x(x, y) \\ F_y(x, y) \end{Bmatrix} \tag{5.45}$$

Note that the system, which was initially uncoupled, is now coupled by the presence of the force field, a function of both the variables x and y. If the purpose of analysis is to estimate the small oscillations around the position of static equilibrium or to study the stability of small motions of the actual system, it is possible, as already mentioned, to linearise the motion equations around that position. The position of static equilibrium is defined by a pair of values x_o and y_o that satisfy the equations:

$$\begin{aligned} K_x x_0 &= F_x(x_0, y_0) \\ K_y y_0 &= F_y(x_0, y_0) \end{aligned} \tag{5.46}$$

which are obtained from (5.45) for zero values of accelerations x, y and velocity $\dot{x}$, $\dot{y}$. The unknowns x_o and y_o appear again in non-linear form in the expressions of the positional forces of the field: the solution of (5.46) must be calculated with specific numerical procedures for non-linear problems (for example, the Newton Raphson method referred to in Chap. 4, Sect. 4.7). Once the pair of values x_o and y_o has been defined, it is possible to carry out a coordinate transformation, introducing two new variables x and y, which define the perturbed motion around the position of static equilibrium, defined by:

$$\bar{x} = x - x_0; \quad \bar{y} = y - y_0 \tag{5.47}$$

Now, to analyse the motion of the vibrating system in a small neighbourhood around the position of equilibrium, we can use the Lyapunov theorem to linearise the actual equations, that is, the non-linear terms F_x and F_y in a Taylor series in the neighbourhood of x_o and y_o, up to the linear terms (using subscript $_o$ to indicate the value of the generic quantity estimated in the position of static equilibrium). Taking into account (5.46) and (5.47), (5.45) becomes:

$$\begin{aligned} m\ddot{\bar{x}} + k_x\bar{x} + k_x x_0 &= F_x(x_0, y_0) + \left(\frac{\partial F_x}{\partial x}\right)_0 (x - x_0) + \left(\frac{\partial F_x}{\partial y}\right)_0 (y - y_0) + \cdots \\ m\ddot{\bar{y}} + k_y\bar{y} + k_y y_0 &= F_y(x_0, y_0) + \left(\frac{\partial F_y}{\partial x}\right)_0 (x - x_0) + \left(\frac{\partial F_y}{\partial y}\right)_0 (y - y_0) + \cdots \end{aligned} \tag{5.48}$$

Taking into account the equation of static equilibrium (5.46) and bringing linear elements of the development of the force field to the left-hand member, we obtain:

$$\begin{aligned} m\ddot{\bar{x}} + k_x\bar{x} - \left(\frac{\partial F_x}{\partial x}\right)_0 \bar{x} - \left(\frac{\partial F_x}{\partial y}\right)_0 \bar{y} = 0 \\ m\ddot{\bar{y}} + k_y\bar{y} - \left(\frac{\partial F_y}{\partial x}\right)_0 \bar{x} + \left(\frac{\partial F_y}{\partial y}\right)_0 \bar{y} = 0 \end{aligned} \tag{5.48a}$$

In this way we have reduced to a differential equations system, linear and homogeneous, in the two variables x and y, variables which, as mentioned, describe the perturbed motion of the system around the position of static equilibrium: in matrix form the Eq. (5.48a) can be rewritten as:

$$[M]\ddot{\underline{z}} + [[K] + [K_F]]\underline{z} = 0 \tag{5.49}$$

having used $\underline{z}$ to indicate the vector containing the displacements x and y:

$$\underline{z} = \left\{ \begin{matrix} \bar{x} \\ \bar{y} \end{matrix} \right\} \tag{5.49a}$$

and with $[M]$ the mass matrix of the system. In (5.49) $[K]$ represents the system's elastic matrix, while $[K_F]$ is an equivalent elastic matrix, due to the presence of positional force field F where the vibrating system is located:

$$[K_F] = -\begin{bmatrix} \left(\frac{\partial F_x}{\partial x}\right)_0 & \left(\frac{\partial F_x}{\partial y}\right)_0 \\ \left(\frac{\partial F_y}{\partial x}\right)_0 & \left(\frac{\partial F_y}{\partial y}\right)_0 \end{bmatrix} \quad (5.49b)$$

The Eq. (5.49) is similar in form to (5.39a) with matrices $[M_F]$ and $[R_F]$ being null, since the force field is considered to be merely positional. When analysing (5.49) we can immediately see how the field of positional forces changes the dynamic behaviour of the vibrating system, namely, its natural frequencies and relative vibration modes: in fact, in the absence of the force field the same equations can be reduced to:

$$\begin{aligned} m\ddot{\bar{x}} + k_x\bar{x} &= 0 \\ m\ddot{\bar{y}} + k_y\bar{y} &= 0 \end{aligned} \quad (5.50a)$$

which are uncoupled. The natural pulsations and relative vibration modes, for the system not subject to a force field, are:

$$\begin{aligned} \omega_{1_0} &= \sqrt{\frac{k_x}{m}} \Rightarrow \underline{Z}^{(1_0)} = \begin{Bmatrix} 1 \\ 0 \end{Bmatrix} \\ \omega_{2_0} &= \sqrt{\frac{k_y}{m}} \Rightarrow \underline{Z}^{(2_0)} = \begin{Bmatrix} 0 \\ 1 \end{Bmatrix} \end{aligned} \quad (5.50b)$$

As we can see, the relations between the elements of the eigenvectors (vibration modes) are jointly real and orthogonal (Fig. 5.17) as x and y are already principal coordinates.

If, however, we consider the force field, once the matrix $[K_T]$ has been defined as the sum of the contribution of the elastic forces and the linearised force field:

$$[K_T] = [K] + [K_F] = \begin{bmatrix} k_x & 0 \\ 0 & k_y \end{bmatrix} - \begin{bmatrix} \left(\frac{\partial F_x}{\partial x}\right)_0 & \left(\frac{\partial F_x}{\partial y}\right)_0 \\ \left(\frac{\partial F_y}{\partial x}\right)_0 & \left(\frac{\partial F_y}{\partial y}\right)_0 \end{bmatrix} = \begin{bmatrix} K_{xx} & K_{xy} \\ K_{xy} & K_{yy} \end{bmatrix} \quad (5.51)$$

(5.49) can be rewritten as:

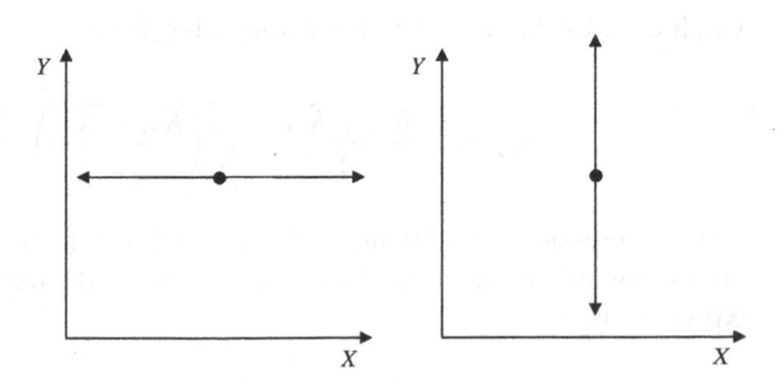

Fig. 5.17 Vibrating system with 2 d.o.f.: vibration modes in the absence of a force field

$$[M]\ddot{\underline{z}} + [K_T]\underline{z} = \underline{0} \tag{5.52}$$

The solution of (5.52), which represents the equation of free motion in the presence of the linearised force field, is, as usual, of the form:

$$\underline{z} = \underline{Z}e^{\lambda t} \tag{5.52a}$$

which leads to:

$$[\lambda^2[M]\underline{Z} + [K_T]]\underline{Z} = \underline{0} \tag{5.52b}$$

The system thus obtained, admits non-trivial solutions when the determinant Δ of the matrix of the coefficients is zero:

$$\Delta = \Delta(\lambda) = \left|\lambda^2[M] + [K_T]\right| = 0 \tag{5.52c}$$

i.e.:

$$\Delta(\lambda) = \begin{vmatrix} \lambda^2 m + K_{xx} & K_{xy} \\ K_{yx} & \lambda^2 m + K_{xx} \end{vmatrix} = 0 \tag{5.52d}$$

By develojournalg the determinant (5.52d) we obtain the characteristic polynomial of degree $2n$ in λ (in this case, $n = 2$ being of the fourth order)

$$\Delta(\lambda) = \lambda^4 m^2 + \lambda^2 \left(mK_{xx} + mK_{yy}\right) + \left(K_{xx}K_{yy} - K_{xy}K_{yx}\right) = 0 \tag{5.52e}$$

By dividing (5.52e) by m^2 and assuming:

$$\overline{K}_{xx} = \frac{K_{xx}}{m}; \quad \overline{K}_{yy} = \frac{K_{yy}}{m}; \quad \overline{K}_{xy} = \frac{K_{xy}}{m}; \quad \overline{K}_{yx} = \frac{K_{yx}}{m} \tag{5.52f}$$

we obtain as the solution of the polynomial (5.52e):

$$\lambda^2_{I,II} = -\frac{\overline{K}_{xx} + \overline{K}_{yy}}{2} \pm \sqrt{\left(\frac{\overline{K}_{xx} + \overline{K}_{yy}}{2}\right)^2 - \left(\overline{K}_{xx}\overline{K}_{yy} - \overline{K}_{xy}\overline{K}_{yx}\right)} \tag{5.53a}$$

which can also be rewritten in the following form:

$$\lambda^2_{I,II} = -\frac{\overline{K}_{xx} + \overline{K}_{yy}}{2} \pm \sqrt{\left(\frac{\overline{K}_{xx} - \overline{K}_{yy}}{2}\right)^2 + \overline{K}_{xy}\overline{K}_{yx}} \tag{5.53b}$$

The expressions (5.53a) and (5.53b), which will be used below in discussing the various possible cases, depending on the nature of the force field, can also be briefly expressed as:

$$\lambda_{I,II}^2 = -\gamma \pm \sqrt{\beta} = -\gamma \pm \Delta \tag{5.53c}$$

Now we will discuss the possible solutions (5.53a)–(5.53c), bearing in mind the conservative or non-conservative nature of the force field that the system in Fig. 5.15 is subject to.

5.3.1.1 Conservative Force Field

If the force field is conservative, then the matrix of stiffness $[K_T]$ is symmetrical (the extra-diagonal terms of the stiffness matrix are equal): in this case, the term under the square root of (5.53b) is positive definite, as it is the sum of two squares.

5.3.1.1.1 Stiffness Matrix $[K_T]$ Positive Definite

If the overall stiffness matrix in addition to being symmetrical is also positive definite, then the terms K_{xx}, K_{yy}, $K_{xx}K_{yy} - K_{xy}$, K_{yx}, which represent the principal minors, are positive, and from (5.53a), it can easily be demonstrated[8] how the values of $\lambda_{I,II}^2$ are negative, (5.53c) being $\Delta < |\gamma|$. The values of $\lambda_{1,2,3,4}$ (equal to the square root of $\lambda_{I,II}^2$) are, therefore, purely imaginary:

$$\begin{aligned} \lambda_{1,2} &= \pm\sqrt{\lambda_I^2} = \pm\sqrt{-\gamma + \Delta} = \pm i\omega_1 \quad (essendo\,(\Delta - \gamma) < 0) \\ \lambda_{3,4} &= \pm\sqrt{\lambda_{II}^2} = \pm\sqrt{-\gamma - \Delta} = \pm i\omega_2 \quad (essendo\,(-\Delta - \gamma) < 0) \end{aligned} \tag{5.54}$$

In this case, as the real part of the eigenvalues is zero, the linearised system is stable, but not asymptotically. The pulsations ω_1 and ω_2 of the linearised system, represent the pulsations of the small oscillations of the non-linear system around the position of equilibrium. This shows that if the matrices $[M]$ and $[K_T]$ are symmetrical and positive definite the solution to the free problem (5.52) is a pure

[8] As the matrix is $[K_T]$ symmetrical, by analysing (5.53b) it is easy to see how the radicand β (5.53c) is certainly positive (being the sum of two squares) hence:

$$\Delta = \sqrt{\beta} = reale \tag{5.8.1}$$

and $\left(\overline{K}_{xx}\overline{K}_{yy} - \overline{K}_{xy}\overline{K}_{yx}\right) > 0$ being, Δ in magnitude is lower than $\gamma = \frac{\overline{K}_{xx}+\overline{K}_{yy}}{2}$ hence:

$$\lambda_{I,II}^2 = -\gamma \pm \sqrt{\beta} = -\gamma \pm \Delta < 0. \tag{5.8.2}$$

harmonic motion: the eigenvalues λ are purely imaginary (5.54) and the pulsations ω_1 and ω_2 are different compared to the values that we saw in the absence of the force field (5.50b).

5.3.1.1.2 Stiffness Matrix $[K_T]$ not Positive Definite

If the matrix $[K_T]$ is symmetrical (so always representative of a conservative force field) but is not positive definite, at least one of the principal minors is negative. In particular, if the determinant is negative, then it is always, based on (5.53a):

$$\Delta > |\gamma| \tag{5.55a}$$

and therefore (5.53c), regardless of the sign of γ, just one of the $\lambda_{I,II}^2$ is positive, so:

$$\begin{aligned} \lambda_{1,2} &= \pm\sqrt{\lambda_I^2} = \pm\sqrt{-\gamma+\Delta} = \pm\alpha \quad (\textit{being}\,(\Delta-\gamma)<0) \\ \lambda_{3,4} &= \pm\sqrt{\lambda_{II}^2} = \pm\sqrt{-\gamma-\Delta} = \pm i\omega_1 \quad (\textit{being}\,(-\Delta-\gamma)<0) \end{aligned} \tag{5.55b}$$

One of the roots, $\lambda_1 = \alpha$, therefore, is real and positive and this corresponds to the onset of static instability (divergence). The solution $\lambda_2 = -\alpha$ is, however, negative (corresponding to a motion that vanishes over time). On the other hand, the second solution $\lambda_{3,4} = \pm i\omega_1$ corresponds to a harmonic motion.

In the case of a stiffness matrix $[K_T]$ that is not positive definite, but with a determinant greater than zero, then the only possibility[9] is to have negative terms of the principal diagonal to result in $\gamma<0$. As the determinant is positive, this results in (5.53a):

$$\Delta<|\gamma| \tag{5.56a}$$

so both the values $\lambda_{I,II}^2$ are positive, as the sign γ prevails:

$$\begin{aligned} \lambda_{1,2} &= \pm\sqrt{\lambda_I^2} = \pm\sqrt{-\gamma+\Delta} = \pm\alpha_1 \quad (\textit{essendo}\,(\Delta-\gamma)<0) \\ \lambda_{3,4} &= \pm\sqrt{\lambda_{II}^2} = \pm\sqrt{-\gamma-\Delta} = \pm\alpha_2 \quad (\textit{essendo}\,(-\Delta-\gamma)<0) \end{aligned} \tag{5.56b}$$

[9]If the determinant of the matrix is positive, we have:

$$\overline{K}_{xx}\overline{K}_{yy} > \overline{K}_{xy}\overline{K}_{yx} \tag{5.9.1}$$

If the force field is conservative, the product $\overline{K}_{xy}\overline{K}_{yx}$ is positive and, therefore, the terms $\overline{K}_{xx}$ and $\overline{K}_{yy}$ must have the same sign.

The four roots $\lambda_{1,2,3,4}$ are all real, two negatives and two positives, corresponding to two instabilities of a static nature, that is, to two different modes of static divergence. To sum up, we can say that:

- in cases where the matrix $[K_T]$ is symmetrical and positive definite there is stability;
- if $[K_T]$, although symmetrical, is not positive definite, then at least one solution has positive exponent, with divergence type static instability.

To calculate the eigenvectors corresponding to the solutions $\lambda_{1,2,3,4}$ found, we will reconsider the linearised equations of motion (5.52a):

$$\begin{aligned} \ddot{\overline{x}} + \overline{K}_{xx}\overline{x} + \overline{K}_{xy}\overline{y} = 0 \\ \ddot{\overline{y}} + \overline{K}_{yy}\overline{y} + \overline{K}_{yx}\overline{x} = 0 \end{aligned} \tag{5.57a}$$

Replacing in (5.57a) the solution of the form:

$$\underline{\overline{z}} = \underline{\overline{Z}}e^{\lambda t} = \{array*20c\overline{X}\overline{Y}\}e^{\lambda t} \tag{5.57b}$$

the same system of differential equations is reduced to a system of linear algebraic equations of the form:

$$\begin{aligned} \lambda^2\overline{X} + \overline{K}_{xx}\overline{X} + K_{xy}\overline{Y} = 0 \\ \lambda^2\overline{Y} + \overline{K}_{yy}\overline{Y} + \overline{K}_{yx}\overline{X} = 0 \end{aligned} \tag{5.57c}$$

If l corresponds to one of the eigenvalues calculated in the two cases analysed in this section, (5.55b) and (5.56b), then the two Eq. (5.57c) are linear combinations of each other: from the first of (5.57c) it is possible, for example, to define the relation between the amplitudes in correspondence to the solutions of the characteristic equation:

$$\mu_{I,II} = \left(\frac{\overline{Y}}{\overline{X}}\right)_{I,II} = -\frac{\lambda_{I,II}^2 + \overline{K}_{xx}}{\overline{K}_{xy}} \tag{5.57d}$$

Since, in any case, for conservative systems the values of $\lambda_{I,II}^2$ are always real, it follows that the characteristic ratios $\mu_{I,II}$ will be real. Furthermore, given that the overall elastic matrix $[K_T]$ is symmetrical, having assumed a conservative force field, the eigenvectors $\underline{\overline{Z}}^{(I,II)}$:

$$\underline{\overline{Z}}^{(I,II)} = \left\{ \begin{array}{c} \overline{X}^{(I,II)} \\ \overline{Y}^{(I,II)} \end{array} \right\} \tag{5.57e}$$

will also be orthogonal in respect to the overall stiffness matrix $[K_T]$ and to the mass matrix $[M]$ (see Sect. 2.5.2.1), despite being different, obviously, to those $\underline{\overline{Z}}^{(1o)}$, $\underline{\overline{Z}}^{(2o)}$ that are obtained in the absence of the force field (5.50b).

5.3.1.2 Non-conservative Force Field

Now we will analyse the most interesting case, namely, the case where the force field is non- conservative: as the matrix $[K_T]$, in this case, is no longer symmetrical, the extra-diagonal terms are different and their product can be both positive and negative. We will analyse the two cases separately.

5.3.1.2.1 Negative Radicand β (Flutter Instability)

The radicand β of (5.53b) is negative if:

- $\overline{K}_{xy}$ is of the opposite sign to $\overline{K}_{yx}$:

$$\overline{K}_{xy}\overline{K}_{yx} < 0 \tag{5.58a}$$

- And if the following relation exists:

$$\left|\overline{K}_{xy}\overline{K}_{yx}\right| > \left(\frac{\overline{K}_{xx} - \overline{K}_{yy}}{2}\right)^2 \tag{5.58b}$$

In this case the values of $\lambda_{I,II}^2$ are complex conjugates:

$$\lambda_{I,II}^2 = -\gamma \pm i\sqrt{|\beta|} = a \pm ib = \sqrt{a^2 + b^2} \cdot e^{\pm i\phi} = M \cdot e^{\pm i\phi} \tag{5.58c}$$

and can be represented by a pair of numbers with equal magnitude M and phase ϕ equal and opposite in sign. The values of $\lambda_{1,2,3,4}$, which are obtained as the square root of $\lambda_{I,II}^2$, are also complex conjugates, in pairs, with the same imaginary part, as also shown in Fig. 5.18:

$$\begin{aligned}
\lambda_{1,2} &= \pm\sqrt{\lambda_I^2} = \pm\sqrt{a + ib} = \pm\sqrt{M} \cdot e^{i\frac{\phi}{2}} \quad \Rightarrow \lambda_1 = \alpha + i\omega; \quad \lambda_2 = -\alpha - i\omega \\
\lambda_{3,4} &= \pm\sqrt{\lambda_{II}^2} = \pm\sqrt{a - ib} = \pm\sqrt{M} \cdot e^{-i\frac{\phi}{2}} \quad \Rightarrow \lambda_3 = \alpha - i\omega; \quad \lambda_4 = -\alpha + i\omega
\end{aligned} \tag{5.58d}$$

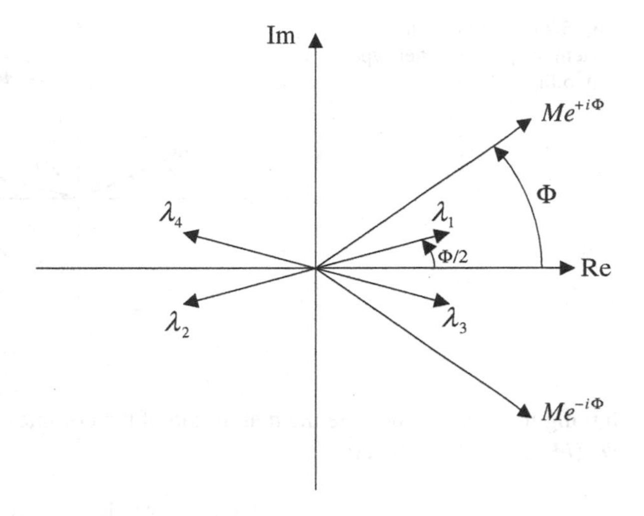

Fig. 5.18 Solutions in the case of non-conservative field, with b < 0

Each pair of conjugate roots ($\lambda_{1,3}$, $\lambda_{2,4}$) as we have often seen, provides the same solution in the real field, so we will have as a general integral in the real field:

$$\begin{aligned} \underline{z} &= Re\left(\underline{\overline{Z}}^{(1)} e^{\lambda_1 t}\right) + Re\left(\underline{\overline{Z}}^{(2)} e^{\lambda_2 t}\right) = Re\left(\underline{\overline{Z}}^{(1)} e^{\alpha t} \cdot e^{i\omega t}\right) + Re\left(\underline{\overline{Z}}^{(2)} e^{-\alpha t} \cdot e^{i\omega t}\right) \\ &= e^{\alpha t} \cdot Re\left(\underline{\overline{Z}}^{(1)} \cdot e^{i\omega t}\right) + e^{-\alpha t} \cdot Re\left(\underline{\overline{Z}}^{(2)} \cdot e^{i\omega t}\right) \end{aligned} \tag{5.59}$$

i.e. two isofrequential vibration modes, one of which is expanding and the other damped. The eigenvectors $\underline{\overline{Z}}^{(1)}$ and $\underline{\overline{Z}}^{(2)}$ are, in general, complex and the corresponding motion will be of an elliptical nature. In fact, the characteristic relation between the amplitudes (5.57d):

$$\mu_{I,II} = \left(\frac{\overline{Y}}{\overline{X}}\right)_{I,II} = -\frac{\lambda_{I,II}^2 + \overline{K}_{xx}}{\overline{K}_{xy}} = -\frac{Me^{\pm i\phi} + \overline{K}_{xx}}{\overline{K}_{xy}} \tag{5.60a}$$

in this case is complex as the value of $\lambda_{I,II}^2$ is a complex number: $\overline{Y}^{(1)}$ and $\overline{X}^{(1)}$ and $\overline{Y}^{(2)}$ and $\overline{X}^{(2)}$, which form the vectors $\underline{\overline{Z}}^{(1)}$ and $\underline{\overline{Z}}^{(2)}$ of (5.59), are therefore correlated by the relations μ_I and μ_{II}:

$$\begin{aligned} \frac{\overline{Y}^{(1)}}{\overline{X}^{(1)}} &= \mu_I = -\frac{\lambda_I^2 + \overline{K}_{xx}}{\overline{K}_{xy}} \quad \Rightarrow \overline{Y}^{(1)} = \overline{X}^{(1)} |\mu_I| e^{i\psi_I} \\ \frac{\overline{Y}^{(2)}}{\overline{X}^{(2)}} &= \mu_{II} = -\frac{\lambda_{II}^2 + \overline{K}_{xx}}{\overline{K}_{xy}} \quad \Rightarrow \overline{Y}^{(2)} = \overline{X}^{(2)} |\mu_{II}| e^{i\psi_{II}} \end{aligned} \tag{5.60b}$$

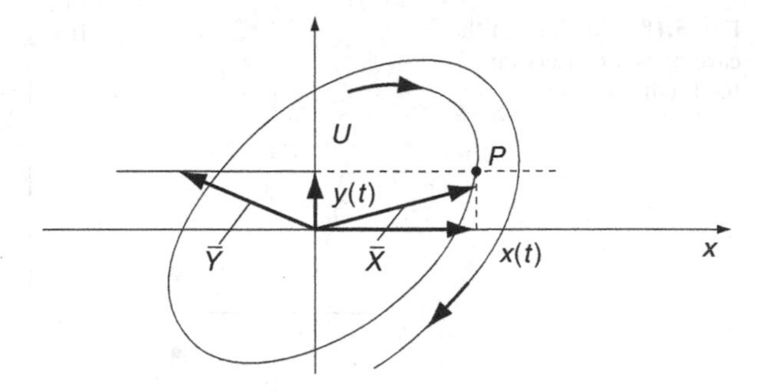

Fig. 5.19 Motion of the system subject to flutter type instability

having used $|\mu_i|$ to indicate the magnitude of the complex number μ_i $(i = I, II)$ and ψ_i $(i = I, II)$ the subject:

$$\begin{aligned}\psi_I &= \operatorname{arct} g\left(\frac{M \sin\phi}{-M\cos\phi - \overline{K}_{xx}}\right)\\ \psi_I &= -\psi_{II}\end{aligned} \tag{5.60c}$$

(5.59) developed in the components x and y become:

$$\begin{aligned}\overline{x}(t) &= Re\left(C_1\overline{X}^{(1)}e^{\lambda_1 t} + C_2\overline{X}^{(2)}e^{\lambda_2 t}\right) = e^{\alpha t}\cdot Re\left(C_1\overline{X}^{(1)}e^{i\omega t}\right) + e^{-\alpha t}\cdot Re\left(C_2\overline{X}^{(2)}e^{i\omega t}\right)\\ \overline{y}(t) &= Re\left(C_1\overline{Y}^{(1)}e^{\lambda_1 t} + C_2\overline{Y}^{(2)}e^{\lambda_2 t}\right) = e^{\alpha t}\cdot Re\left(C_1\overline{Y}^{(1)}e^{i\omega t}\right) + e^{-\alpha t}\cdot Re\left(C_2\overline{Y}^{(2)}e^{i\omega t}\right)\\ &= e^{\alpha t}\cdot Re\left(C_1\overline{X}^{(1)}|\mu_I|e^{i\Psi_I}e^{i\omega t}\right) + e^{-\alpha t}\cdot Re\left(C_2\overline{X}^{(2)}|\mu_{II}|e^{i\Psi_{II}}e^{i\omega t}\right)\end{aligned} \tag{5.60d}$$

where C_1 and C_2 (complex) can be determined from the initial conditions. The trajectory of the mass (Fig. 5.19) is defined by two counter rotating elliptical motions, as the vibrations along X and Y are harmonic and out of phase with each other: one motion is exponentially expanding $\lambda_1 = \alpha + i\omega$ and the other exponentially decreasing $\lambda_4 = -\alpha + i\omega$.

The generic trajectory, cleansed of the term $e^{\pm\alpha t}$, is elliptical since, as already mentioned, it is defined by two sinusoidal components along the X and Y axes (projections of the vectors $\overline{X}^{(1,2)}$ and $\overline{Y}^{(1,2)}$ on the respective real axes (Figs. 5.19 and 5.60d), out of phase with each other as the relation $\mu_{I,II} = \left(\overline{Y}/\overline{X}\right)^{(1,2)}$ (5.60b), (5.60c) is complex.[10]

---

[10]For ease of discussion, we will consider the contribution to vibration caused by λ_1, for example, by cleansing it of the exponential term. The trajectory described by the mass m is defined by a generic relation of the form (5.60d):

$$\begin{aligned} x &= X\cos(\omega t + \theta) \\ y &= Y\cos(\omega t + \gamma) \end{aligned} \quad (5.10.1)$$

In order to define the type of trajectory, it is possible to express (5.9.1) with a shift of time $t = \tau - (\theta/\omega)$ so that the same expressions can be defined as:

$$\begin{aligned} x &= X\cos(\omega t + \theta) = X\cos\left(\omega\tau - \omega\frac{\theta}{\omega} + \theta\right) = X\cos(\omega\tau) \\ y &= Y\cos(\omega t + \gamma) = Y\cos\left(\omega\tau - \omega\frac{\theta}{\omega} + \gamma\right) = Y\cos(\omega\tau + (\gamma - \theta)) = Y\cos(\omega\tau + \Psi) \end{aligned} \quad (5.10.2)$$

where ψ represents the relative angle between x and y, see (5.60c). By making the cosine term in the second equation explicit and substituting the first, we have

$$\begin{aligned} \frac{y}{Y} &= \cos(\omega\tau + \Psi) = \cos(\omega\tau)\cos(\Psi) - \sin(\omega\tau)\sin(\Psi) \\ &= \cos(\omega\tau)\cos(\Psi) - \left(\sqrt{1-\cos^2(\omega\tau)}\right)\sin(\Psi) = \frac{x}{X}\cos(\Psi) - \left(\sqrt{1-\left(\frac{x}{X}\right)^2}\right)\sin(\Psi) \end{aligned} \quad (5.10.3)$$

By squaring this expression we have:

$$\left(\frac{y}{Y}\right)^2 + \left(\frac{x}{X}\right)^2 - 2\left(\frac{y}{Y}\right)\left(\frac{x}{X}\right)\cos\Psi = \sin^2\Psi \quad (5.10.4)$$

This equation represents, in general, an elliptical. Assuming that the relative angle ψ is zero (i.e. $\theta = \gamma$), we have:

$$\left(\frac{y}{Y}\right)^2 + \left(\frac{x}{X}\right)^2 - 2\left(\frac{y}{Y}\right)\left(\frac{x}{X}\right) = 0 \Rightarrow \left(\frac{y}{Y} - \frac{x}{X}\right)^2 = 0 \Rightarrow y = x\left(\frac{Y}{X}\right)^2 \quad (5.10.5)$$

which represents a rectilinear motion. In the particular case where $X = Y = A$ and $\Psi = 90°$, (5.10.4) becomes:

$$\left(\frac{y}{A}\right)^2 + \left(\frac{x}{A}\right)^2 - 2\left(\frac{y}{Y}\right) = 1 \Rightarrow y^2 + x^2 = A^2 \quad (5.10.6)$$

which represents a circular motion. To define the direction the elliptical trajectory travels, it can be described in polar coordinates:

$$z = \rho \cdot e^{i\beta} \quad (5.10.7)$$

The direction of travel on the trajectory causing the body to absorb energy from the force field is, therefore, defined by solution λ_1 (that with the positive real part), while the other solution corresponds to a motion (the direction of travel of the relative trajectory is opposite to the previous) that transfers energy to the force field (the motion gradually dampens). In vibrating systems subject to conservative fields of positional forces, the relations between the elements of the generic eigenvector (generally complex) are always real (Sect. 2.3.1) and so the vibrations associated with the different elements are always either in phase or in counter phase: for this reason, in these systems the motion, represented in the plane x–y and for a two-degree-of freedom system is always a segment of a straight line. The instability that can be seen in non-conservative systems surrounded by fields of purely positional forces is, on the other hand, characterised by the fact that two natural frequencies become equal and the resulting motion is of an elliptical nature. This is a characteristic that is peculiar to unstable non-conservative systems, with expanding vibrations generated by the action of the force field. The force field makes the two frequencies of the system equal, which in the absence of the force field would be different, and makes it possible for the mass to describe an elliptical trajectory and to absorb the energy from the field. Returning, in fact, to the example considered in Fig. 5.15), in the absence of the force field, the frequency would be:

$$\begin{aligned} \omega_x &= \sqrt{\frac{k_x}{m}} = \omega_{10} \\ \omega_y &= \sqrt{\frac{k_y}{m}} = \omega_{20} \end{aligned} \tag{5.61}$$

(Footnote 10 continued)
z being a complex vector (of magnitude ρ and anomaly β) which describes the trajectory of the mass m:

$$\begin{aligned} \rho &= \sqrt{X^2\cos^2(\omega t+\theta) + Y^2\sin^2(\omega t+\gamma)} \\ \beta &= a\tan\left(\frac{Y\sin(\omega t+\gamma)}{X\cos(\omega t+\theta)}\right) \end{aligned} \tag{5.10.8}$$

The direction of travel can be defined by the sign of the:

$$\begin{aligned} \frac{d\beta}{dt} &= \frac{1}{1-\beta^2}\cdot\frac{\omega YX\cos(\omega t+\gamma)\cos(\omega t+\theta) + \omega YX\sin(\omega t+\gamma)\sin(\omega t+\theta)}{X^2\cos^2(\omega t+\theta)} = \\ &= \frac{1}{1-\beta^2}\cdot\frac{\omega YX\sin(\gamma-\theta)}{X^2\cos^2(\omega t+\theta)} \end{aligned} \tag{5.10.9}$$

The sign of the term depends, therefore, on the sign of $\sin(\theta-\gamma) = -\sin(\Psi)$, i.e. on the value assumed by the two phases Ψ_I and Ψ_{II} (5.60c).

with, in general, ω_x differing from ω_y. The non-conservative field of positional forces modifies such frequencies through $[K_F]$ and these, at the onset of instability, converge to the same value of ω given by (5.58b). This physical aspect of the problem suggests a way to control these forms of instability by acting appropriately on the value of ω_x and ω_y: in fact, if, for example, ω_x and ω_y were very different, the force field through the terms of $[K_F]$ would be less likely to make them equal, thus distancing the system from phenomena of instability. Generalising the discussion to n-degree-of freedom systems, the purely inertial and elastic system is characterised by natural frequencies that are generally different: the presence of $[K_F]$, equivalent elastic matrix of the non-conservative force field, changes the natural frequencies of the system and when instability occurs with expanding vibrations (conditions given by (5.58a), two of these frequencies become equal (complex conjugate roots with positive real part). This type of instability is defined as flutter instability. Examples of this form of instability will be shown later, in particular relating to an aerofoil with two degree-of-freedom (Sect. 5.3.2.1), to a railway wheelset and to the motion of a journal in a bearing with hydrodynamic lubrication (Sect. 5.3.2.2).

5.3.1.2.2 β Positive Radicand

We will now analyse the second case, where the radicand $\beta > 0$: this condition always occurs if $\overline{K}_{xy}$ and $\overline{K}_{yx}$, although having different magnitudes, have the same sign or if the magnitude of their product satisfies:

$$\left|\overline{K}_{xy}\overline{K}_{yx}\right| < \left(\frac{\overline{K}_{xx} + \overline{K}_{yy}}{2}\right)^2 \tag{5.62}$$

as highlighted in (5.53b). In this condition, see (5.53c), the two values of λ_I^2 and λ_{II}^2 are real: at this point, we need to distinguish the possible cases on the basis of the sign of the matrix's $[K_T]$ determinant. In the case of a determinant that is less than zero (matrix not positive definite), as seen previously in the case of conservative systems, the result is always:

$$\Delta > |\gamma|. \tag{5.63a}$$

And then, once again, similarly to what happens to systems surrounded by fields of conservative forces with $[K_T]$ not positive definite (sect. 5.3.1.1.2, (5.55b)), one solution λ_I^2 is positive while the other λ_{II}^2 proves to be negative, so we will have:

$$\begin{aligned} \lambda_{1,2} &= \pm\sqrt{\lambda_I^2} = \pm\sqrt{(-\gamma + \Delta)} = \pm\alpha \\ \lambda_{3,4} &= \pm\sqrt{\lambda_{II}^2} = \pm\sqrt{(-\gamma - \Delta)} = \pm i\omega_1 \end{aligned} \tag{5.63b}$$

giving rise to static divergence. In the case of a determinant that is greater than zero, however, the result is always:

$$\Delta < |\gamma| \tag{5.64a}$$

and so the sign of $\lambda^2_{I,II}$ is determined by the sign of γ. If the terms $\overline{K}_{xx}$ and $\overline{K}_{yy}$ are positive, then the matrix $[K_T]$ is positive definite, as the determinant is also positive. A positive value of γ, provides two negative values of $\lambda^2_{I,II}$, so the roots $\lambda_{1,2,3,4}$ will all be imaginary, i.e. the linearised system is stable:

$$\begin{aligned} \lambda_{1,2} &= \pm\sqrt{\lambda_I^2} = \pm\sqrt{(-\gamma+\Delta)} = \pm i\omega_1 \\ \lambda_{3,4} &= \pm\sqrt{\lambda_{II}^2} = \pm\sqrt{(-\gamma-\Delta)} = \pm i\omega_2 \end{aligned} \tag{5.64b}$$

The ratios of the elements of the eigenvectors are real but, unlike the similar case obtained for the conservative force field, they will no longer be orthogonal, since the matrix $[K_T]$ is no longer symmetrical (Sect. 2.5.2.1). So to summarise, we can say that:

- in the case of a non-conservative field of positional forces, $[K_T]$ non-symmetrical, under suitable conditions, given by (5.58a) and (5.58b) the onset of flutter type instability is possible, being typical of non-conservative systems.
- furthermore, if the matrix of overall stiffness $[K_T]$ (including the elastic terms and the force field equivalent terms $[K_F]$) is not positive definite, we will, once again, find the divergence type static instabilities already seen in conservative systems.

5.3.1.3 An Example: Instability in a Cutting Tool

We will analyse, as an example of a mechanical system subject to a field of positional forces perturbed around the position of rest, that relating to a cutting tool: more specifically, in this section, we will analyse the possible forms of instability introduced by the force field due to the contact with the piece to be cut, instabilities which, as we know, cause the tool to jump around, with the consequent unwanted irregularities in the process. The problem is complex as the description of the actual force field that acts on the *tool* system is complex. Given the purely didactic purpose of the discussion, we will adopt suitable simplifications that will, however, in no way alter the specific characteristics of the phenomenon to be analysed.

The longitudinal axis of the tool (axis X in Fig. 5.20) is inclined, in relation to the normal to the plane of the object to be processed, with an angle θ: k_x and k_y are the two stiffnesses in direction X and Y, Y being normal to X. The force P transmitted from the piece to the tool can be, in a first approximation, considered proportional to the thickness of the swarf thickness S removed, i.e.:

$$P = k_s S \tag{5.65}$$

while the inclination of this force is defined by a certain angle ψ (Fig. 5.20), function of the different processing parameters. Assuming, for simplicity's sake, that the tool is rigid and neglecting the rotations, it is possible to assume, as independent variables, the displacement x and y of the actual tool, according to the directions X and Y (Fig. 5.20) defined in relation to the position of equilibrium: these variables describe the perturbed motion around this position. The angle of inclination of the force transmitted to the tool is considered independent from x and y and from the velocity $\dot{x}$ and $\dot{y}$, but depends only on the cutting speed . Its equations of motion, supposing that the dissipative terms can be neglected and using m to indicate the mass of the tool, are:

$$\begin{aligned} -m\ddot{x} - k_x x + P_x &= 0 \\ -m\ddot{y} - k_y y + P_y &= 0 \end{aligned} \tag{5.66}$$

P_x and P_y being the two components of the force transmitted by contact P in direction x and y, defined by the relations:

$$\begin{aligned} P_x &= P\cos(\psi - \theta) \\ P_y &= -P\sin(\psi - \theta) \end{aligned} \tag{5.66a}$$

The displacement of the tool, in its two components x and y, also defines (Fig. 5.20) the thickness of the swarf removed P using the simple relations:

$$S = -y \cdot \sin\theta - x \cdot \cos\theta \tag{5.67}$$

Taking into account (5.66a) and (5.67) the motion equations of the system can be rewritten in explicit form as:

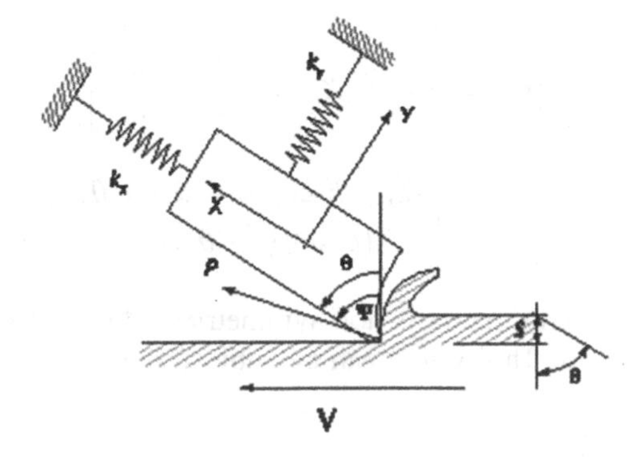

Fig. 5.20 Definition of the tool model

$$\begin{aligned} m\ddot{x} + k_x x - P\cos(\psi - \theta) = 0 \Rightarrow m\ddot{x} + k_x x - k_s S\cos(\psi - \theta) = 0 \\ m\ddot{y} + k_y y + P\sin(\psi - \theta) = 0 \Rightarrow m\ddot{y} + k_y y - k_s S\sin(\psi - \theta) = 0 \end{aligned} \tag{5.68}$$

i.e.:

$$\begin{aligned} -m\ddot{x} - k_x x - k_s(-y \cdot \sin\theta - x \cdot \cos\theta)\cos(\psi - \theta) = 0 \\ -m\ddot{y} - k_y y + k_s(-y \cdot \sin\theta - x \cdot \cos\theta)\sin(\psi - \theta) = 0 \end{aligned} \tag{5.69}$$

Given the particular choice of the independent variables, these directly define the perturbed motion, i.e. with the symbology adopted:

$$\begin{aligned} x = \overline{x} \\ y = \overline{y} \end{aligned} \tag{5.70a}$$

and (5.68) become:

$$\begin{aligned} m\ddot{\overline{x}} + k_x\overline{x} - k_s(-\overline{y} \cdot \sin\theta - \overline{x} \cdot \cos\theta)\cos(\psi - \theta) = 0 \\ m\ddot{\overline{y}} + k_y\overline{y} + k_s(-\overline{y} \cdot \sin\theta - \overline{x} \cdot \cos\theta)\sin(\psi - \theta) = 0 \end{aligned} \tag{5.70b}$$

The vector of the independent variables is defined with $\underline{z}$:

$$\underline{z} = \begin{Bmatrix} \overline{x} \\ \overline{y} \end{Bmatrix} \tag{5.71}$$

(5.70b) can be rewritten in matrix form as:

$$[M]\ddot{\underline{z}} + [[K_S] + [K_F]]\underline{z} = [M]\ddot{\underline{z}} + [K_T]\underline{z} = \underline{0} \tag{5.72}$$

$[M]$ and $[K_T]$ being the matrices of mass and overall stiffness (sum of the structural term $[K_S]$ and of the term due to the force field $[K_F]$) defined as:

$$[M] = \begin{bmatrix} m & 0 \\ 0 & m \end{bmatrix} \tag{5.72a}$$

$$\begin{aligned} [K_T] = [K_S] + [K_F] = \begin{bmatrix} K_{xx} & K_{xy} \\ K_{yx} & K_{yy} \end{bmatrix} \\ = \begin{bmatrix} (k_x + k_s\cos(\psi - \theta)\cos\theta) & (k_s\cos(\psi - \theta)\sin\theta) \\ -(k_s\sin(\psi - \theta)\cos\theta) & (k_y - k_s\sin(\psi - \theta)\sin\theta) \end{bmatrix} \end{aligned} \tag{5.72b}$$

This matrix is non-symmetrical, due to the non-conservative nature of the force field. The system can be subject, therefore, to dynamic type instability when:

$$K_{xy}K_{yx} < 0 \Rightarrow -k_s^2 \cos(\psi - \theta) \sin\theta \sin(\psi - \theta) \cos\theta < 0 \tag{5.73a}$$

and the magnitude of their product:

$$\begin{gathered}
|K_{xy}K_{yx}| > \frac{K_{xx}^2 - K_{yy}^2}{4} \Rightarrow \\
|k_s^2 \cos(\psi - \theta) \sin\theta \sin(\psi - \theta) \cos\theta| > \\
\frac{(k_x + k_s \cos(\psi - \theta) \cos\theta)^2 - (k_y - k_s \sin(\psi - \theta) \sin\theta)^2}{4}
\end{gathered} \tag{5.73b}$$

So a necessary condition to have dynamic instability becomes:

$$\sin(\psi - \theta) > 0 \Rightarrow \psi > 0 \tag{5.74a}$$

The same system can be subject to static instability (divergence) when one of the principal minors of the stiffness matrix is negative, i.e. when:

$$k_y < k_s \sin(\psi - \theta) \sin\theta \tag{5.74b}$$

5.3.2 *Two-Degrees-of-Freedom System Placed in a Field of Position and Velocity Dependent Forces*

We consider it useful, for mainly didactic purposes, not to directly introduce the general discussion on the stability of a generic two-degree-of-freedom system placed in field of forces that are depending on position and velocity, but firstly we will refer to two specific examples (aerofoil hit by a confined flow [Sect. 5.3.2.1) and journal bearing lubricated pair (Sect. 5.3.2.2)], to then return to the general discussion in the context of the examples and the discussion relating to systems with n DOF (Sect. 5.4).

5.3.2.1 An Example: An Aerofoil with Two Degree-of-Freedom Hit by a Confined Flow

As an example of a two-degree-of-freedom system placed in a field of forces that are depending on both position and velocity, we will consider the aerofoil described in Fig. 5.21: this body (with mass m and moment of inertia J around the centre of gravity) can translate on a vertical plane and rotate, constrained by two vertical springs, each with elastic constant $k_x/2$. To describe its motion we can assume, as independent variables, the vertical displacement x and the rotation θ.

As we saw in Sect. 5.2.1.1, the aerodynamic actions of drag and lift that act on the body are, respectively, parallel and normal to the direction of the relative velocity V_r. This relative velocity is calculated in relation to an observer that is integral to the aerofoil at a characteristic point at a distance b_1 from the centre of gravity of the aerofoil (Sect. 5.2.2) and it is inclined at angle ψ with the horizontal (Fig. 5.22). The equations of motion, therefore, are the following:

$$\begin{aligned} m\ddot{x} + r_x\dot{x} + k_x x &= F_p \cos\psi - F_r \sin\psi \\ J\ddot{\theta} + r_\theta\dot{\theta} + k_\theta\theta &= M \end{aligned} \tag{5.75}$$

Having used k_θ to define the torsional stiffness $k_\theta = k_x l^2$ and r_θ for the torsional damping $r_\theta = r_x l^2$. By defining the vector $\underline{z}$ formed by the independent coordinates x and θ as:

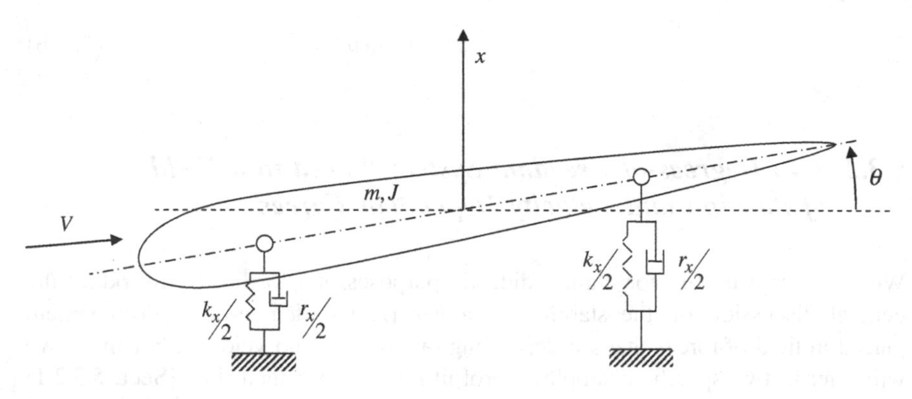

Fig. 5.21 Two-degree-of-freedom aerofoil

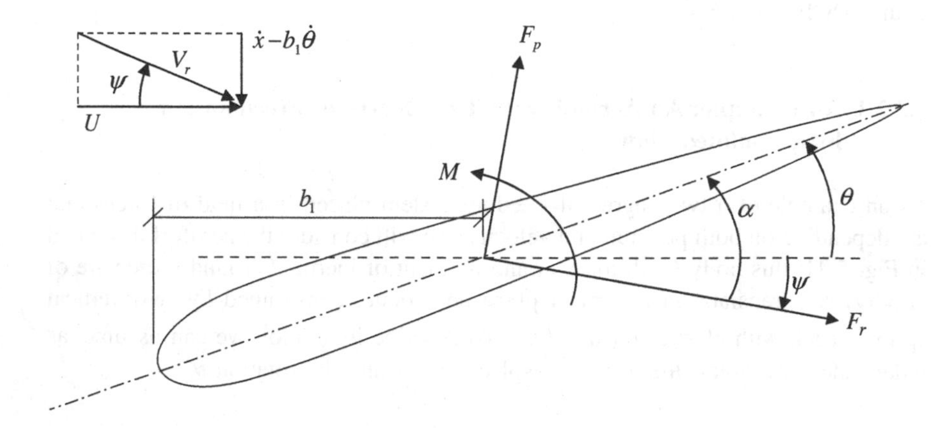

Fig. 5.22 Calculating relative velocity

$$\underline{z} = \begin{Bmatrix} x \\ \theta \end{Bmatrix} \tag{5.75a}$$

(5.75) can be rewritten in matrix form as:

$$\begin{bmatrix} m & 0 \\ 0 & J \end{bmatrix} \begin{Bmatrix} \ddot{x} \\ \ddot{\theta} \end{Bmatrix} + \begin{bmatrix} r_x & 0 \\ 0 & r_\theta \end{bmatrix} \begin{Bmatrix} \dot{x} \\ \dot{\theta} \end{Bmatrix} + \begin{bmatrix} k_x & 0 \\ 0 & k_\theta \end{bmatrix} \begin{Bmatrix} x \\ \theta \end{Bmatrix} = \begin{Bmatrix} F_p \cos\psi - F_r \sin\psi \\ M \end{Bmatrix} \tag{5.75b}$$

or:

$$[M_s]\ddot{\underline{z}} + [R_s]\dot{\underline{z}} + [K_s]\underline{z} = \underline{F}(\underline{z}, \dot{\underline{z}}) \tag{5.75c}$$

As we saw in Sect. 5.2.1.1, the aerodynamic actions of drag, lift and torque can be expressed as:

$$\begin{aligned} F_r &= \frac{1}{2}\rho S C_r(\alpha) V_r^2 \\ F_p &= \frac{1}{2}\rho S C_p(\alpha) V_r^2 \\ M &= \frac{1}{2}\rho S C_m(\alpha) V_r^2 \end{aligned} \tag{5.76}$$

where the aerodynamic coefficients are functions of the angle α formed between the relative velocity V_r and a reference line that is integral to the body (Fig. 5.22). Using the quasi-static theory, which is valid, as mentioned, in the case of reduced speed $V_{rid} > 20–30$, (see Sect. 5.2.1.1, (5.17) about this), and the angle ψ between the relative velocity V_r and the absolute reference being small, we can assume:

$$\tan\psi = \frac{\dot{x} - b_1\dot{\theta}}{U} \approx \psi \tag{5.77a}$$

The angle of attack between the relative velocity and the body is, in this case, equal to:

$$\alpha = \theta + \psi \tag{5.77b}$$

(Fig. 5.22) being:

$$V_r^2 = U^2 + \left(\dot{x} - b_1\dot{\theta}\right)^2 \tag{5.77c}$$

The expressions of the generalised forces according to the degree-of-freedom of the system are, therefore, non-linear functions of the actual variables and of their derivatives in relation to time: to linearise the motion equations and examine the stability of the system, we proceed as seen in Sect. 5.3.1, i.e. by linearising each term that appears in the expressions of the generalised forces and bearing in mind just the linear terms (so neglecting those of a higher order that appear in the developments in series). We will consider, in particular, the case of a symmetrical aerofoil (Fig. 5.9) whose lift and torque coefficients are zero at the origin:

$$C_{p0} = 0; \quad C_{m0} = 0 \tag{5.78a}$$

The position of static equilibrium, considering the direction of the velocity of the free stream coinciding with the axis of symmetry of the profile, and bearing in mind (5.78a), thus becomes:

$$\begin{aligned} x_0 &= 0 \\ \theta_0 &= \alpha_0 = 0 \end{aligned} \tag{5.78b}$$

The relative velocity V_r and its values of $\cos\psi$ and $\sin\psi$, developed around the position of static equilibrium defined by (5.78b), can be expressed as:

$$\begin{aligned} &V_r^2 = U^2 + \ldots \approx U^2; \quad \psi \approx \frac{\dot{x} - b_1\dot{\theta}}{U} \\ &\cos\psi = 1 + \frac{\psi^2}{2} + \ldots \approx 1; \quad \sin\psi = \psi + \ldots \approx \psi \end{aligned} \tag{5.78c}$$

The aerodynamic coefficients $C_r(\alpha)$, $C_p(\alpha)$ and $C_m(\alpha)$, as we saw in Sect. 5.2.1, can be developed in Taylor series around the position of static equilibrium defined by $\alpha_o = 0$ (5.78b). The forcing terms $\underline{F}(\underline{z}, \dot{\underline{z}})$ linearised in (5.75b) by expanding the products and considering just the linear terms of the product obtained, are reduced to the following expressions:

$$\begin{aligned} F_p(\alpha)\cos\psi &= aV_r^2 C_p \cos\psi \\ &= aU^2\left(\frac{\partial C_p}{\partial \alpha}\right)_0 (\alpha - \alpha_0) = aU^2 K_{p0}\left(\theta + \frac{\dot{x} - b_1\dot{\theta}}{U}\right) \\ F_r(\alpha)\sin\psi &= aV_r^2 C_r \sin\psi = aU^2 C_r \psi = aU^2 C_{r0}\left(\frac{\dot{x} - b_1\dot{\theta}}{U}\right) \\ M(\alpha) &= aCV_r^2 C_m = aCU^2\left(\frac{\partial C_m}{\partial \alpha}\right)_0 (\alpha - \alpha_0) = aCU^2 K_{m0}\left(\theta + \frac{\dot{x} - b_1\dot{\theta}}{U}\right) \end{aligned} \tag{5.79}$$

having indicated, as always, with:

$$a = \frac{1}{2}\rho S; \quad C_{r0} = (C_r)_{\alpha=\alpha_0}$$
$$K_{p0} = \left(\frac{\partial C_p}{\partial \alpha}\right)_0 = \left(\frac{\partial C_p}{\partial \alpha}\right)_{\alpha=\alpha_0}; \quad K_{m0} = \left(\frac{\partial C_m}{\partial \alpha}\right)_0 = \left(\frac{\partial C_m}{\partial \alpha}\right)_{\alpha=\alpha_0} \tag{5.79a}$$

The motion Eq. (5.75) of the vibrating two-degree-of-freedom system analysed while placed in a force field (Fig. 5.21) can be rewritten, having linearised the forces and by ordering the terms according to the degree-of-freedom of the system, as:

$$\begin{bmatrix} m & 0 \\ 0 & J \end{bmatrix}\begin{Bmatrix} \ddot{x} \\ \ddot{\theta} \end{Bmatrix} + \begin{bmatrix} r_x & 0 \\ 0 & r_\theta \end{bmatrix}\begin{Bmatrix} \dot{x} \\ \dot{\theta} \end{Bmatrix} + \begin{bmatrix} k_x & 0 \\ 0 & k_\theta \end{bmatrix}\begin{Bmatrix} x \\ \theta \end{Bmatrix}$$
$$= \begin{Bmatrix} aU(K_{p0} - C_{r0})\dot{x} + aU(C_{r0} - K_{p0})b_1\dot{\theta} + aU^2K_{p0}\theta \\ aUCK_{m0}\dot{x} - aUCb_1K_{m0}\dot{\theta} + aU^2CK_{m0}\theta \end{Bmatrix} \tag{5.80a}$$

Also in this case, bearing in mind the particular type of profile analysed in Fig. 5.9, the variables x and θ that define the perturbed motion around the position of static equilibrium, defined by $x_o = 0$ and $\theta_o = 0$, coincide with the actual independent variables initially chosen:

$$\begin{aligned} \overline{x} &= x - x_0 = x \\ \overline{\theta} &= \theta - \theta_0 = \theta \end{aligned} \tag{5.80b}$$

Having used $\underline{z}$ to indicate the vector of the independent variables:

$$\underline{z} = \begin{Bmatrix} x \\ \theta \end{Bmatrix} \tag{5.80c}$$

By bringing to the left of the equals sign the second-member terms, which represent the linearised aerodynamic forces, it is possible to define the Eq. (5.80a) in matrix terms as:

$$[M]\ddot{\underline{z}} + [[R_s] + [R_F]]\dot{\underline{z}} + [[K_s] + [K_F]]\underline{z} = \underline{0} \tag{5.80d}$$

having used $[K_F]$ and $[R_F]$ to indicate the equivalent aerodynamic matrices of stiffness and of damping:

$$[K_F] = aU^2\begin{bmatrix} 0 & -K_{p0} \\ 0 & -CK_{m0} \end{bmatrix}$$
$$[R_F] = aU\begin{bmatrix} C_{r0} - K_{p0} & (K_{p0} - C_{r0})b_1 \\ CK_{m0} & Cb_1K_{m0} \end{bmatrix} \tag{5.80e}$$

By summing the matrices due to linearisation of the force field $[K_F]$ and $[R_F]$ to the structural matrices $[K_S]$ and $[R_S]$ we obtain the overall matrices of stiffness $[K_T]$ and damping $[R_T]$:

$$
\begin{aligned}
[K_T] &= [K_s] + [K_F] = \begin{bmatrix} k_x & -aU^2K_{p0} \\ 0 & k_\theta - aU^2CK_{m0} \end{bmatrix} \\
[R_T] &= [R_s] + [R_F] = \begin{bmatrix} r_x + aU(C_{r0} - K_{p0}) & aU(K_{p0} - C_{r0})b_1 \\ -aUCK_{m0} & r_\theta - aUCb_1K_{m0} \end{bmatrix}
\end{aligned} \tag{5.80f}
$$

whereas the matrix of mass remains unaltered as, in this case, the aerodynamic actions do not depend on accelerations. It is important to note that:

- the aerodynamic matrices $[K_F]$ and $[R_F]$ couple the equations that, initially, with the absence of fluid dynamic forces, prove to be uncoupled;
- the two matrices $[K_F]$ and $[R_F]$ are not symmetrical: so the force field is non-conservative for the positional terms and non-dissipative for those of velocity: in fact, since the equivalent matrix of damping is not symmetrical, it can both dissipate energy and introduce energy into the system;
- the terms of the equivalent matrices of the force field increase as the velocity U of the flow increases.

We will now review the possible types of instability linked to the aerodynamic force field that may occur. These are, basically:

- static divergence of the torsional degree-of-freedom;
- instability with one torsional degree-of-freedom;
- instability with one flexural degree-of-freedom;
- flutter instability.

5.3.2.1.1 Torsional Static Divergence

Static divergence occurs when the direct term of equivalent torsional stiffness $-aU^2CK_{m0}$ (5.80d), in addition to being negative, is in absolute value greater than the corresponding term of structural stiffness k_θ (global matrix $[K_T]$ not positive definite):

$$
\begin{aligned}
& -\mathrm{aU^2CK_{m0}} < 0 \\
& \left|\mathrm{aU^2CK_{m0}}\right| > \mathrm{k_\theta}
\end{aligned} \tag{5.81}
$$

This instability occurs when the values of the derivative $\mathrm{K_{m0}}$ of the torque coefficient are positive, as, for example, occurs in aerofoils (Fig. 5.11). To overcome this type of instability, it is necessary, at equal velocity U, to either increase the torsional stiffness k_θ of the system or decrease the value of the derivative of the torque coefficient $\mathrm{K_{m0}}$.

5.3.1.2.2 Dynamic Instability with One-Degree-of-Freedom

Dynamic instability with one-degree-of-freedom, either torsional or flexural, occurs when the corresponding direct term (either vertical or torsional) of the equivalent damping matrix $[R_F]$ of the force field is (5.80c) negative (matrix $[R_F]$ not positive definite) and, in absolute value, predominant compared to the corresponding term of structural damping:

$$\begin{array}{l} aU\left(C_{r0} - K_{p0}\right) < 0 \\ r_x < \left|aU\left(C_{r0} - K_{p0}\right)\right|. \end{array} \tag{5.82a}$$

$$\begin{array}{l} aUCb_1K_{m0} < 0 \\ r_\theta < \left|aUCb_1K_{m0}\right| \end{array} \; \mathrm{n\_(t)} = 6 \tag{5.82b}$$

In this situation the system oscillates while expanding, by receiving energy via the degree-of-freedom affected by the instability. More specifically, aerofoils (with one-degree-of-freedom), for which the derivative of the lift coefficient K_{p0} is negative, while the derivative of the torque coefficient K_{m0} is positive, are always stable. This instability with one-degree-of-freedom occurs, on the other hand, in non-aerofoil profiles, with a large front surface (for example, beams with simple and double-T cross-sections or similar sections).

5.3.1.2.3 Flutter Instability

Aerofoil type profiles that are stable if constrained in such a way as to create a vibrating system with one degree-of-freedom can be subject, if constrained so that a two-degree-of-freedom system is created, to flutter instability, due to the simultaneous action of positional and velocity parts of the force field. The positional part of the force field is characterised (5.80a)–(5.80f) by an extra-diagonal zero term K_{yx}, so the positional terms alone cannot be responsible for the instability, since the necessary condition can never exist (5.58a and 5.58b) (with the absence of damping) to justify this phenomenon (Sect. 5.3.1.2) being:

$$K_{xy}K_{yx} = 0 \tag{5.82c}$$

On the other hand however, the positional part of the force field modifies, via the term $k_\theta - aU^2CK_{m0}$, the torsional frequency of the system and, in particular, for positive values of the torque coefficient derivative K_{m0}, the torsional frequency ω_θ decreases. If, as is usually the case, the frequency ω_x relating to vertical motion is less than the torsional frequency ω_θ, as the velocity U of the incident flow increases, the positional part tends to draw the two frequencies together, synchronising them. In this situation ($[R_F]$ is not symmetrical), above a certain value of velocity, known as the flutter velocity, the terms of velocity are able to make the system unstable with amplitudes that increase over time affecting both the vertical and torsional motion. Flutter instability is typical of structures such as the wings of aircraft and the decks of suspension bridges, where the natural torsional frequency ω_θ is generally greater than the flexural frequency ω_x and the almost wing-like form of the profile is such as to present a positive derivative of the torque coefficient K_{m0} and negative derivative of the lift coefficient K_{p0}. To avoid the danger of flutter instability, in the decks of suspension bridges, for example, we can:

- modify the form of the profile, seeking to decrease the value of the derivatives of the coefficients of lift K_{p0} and torque K_{m0}, in order to reduce the influence of the aerodynamic force field;
- seek to space out the natural vertical ω_x and torsional ω_θ frequencies, to raise the flutter velocity (though this is not always possible on some structures).

We will examine the phenomenon of flutter in more detail while bearing in mind, as mentioned, that:

- the terms directly responsible for introducing energy into the system are those of velocity;
- the tendency towards vertical and torsional synchronisation is essential to allow energy to be introduced into the system.

Under conditions of flutter, as already mentioned in Sect. 5.3.1.2.1, the solutions become of the form $\lambda_{1,2} = \pm\alpha \pm i\omega$, i.e. with equal pulsations: so, under this condition, the profile translates and rotates with the same frequency ω, synchronously: the two motions are also out of phase, with phases that vary depending on the velocity above the threshold of instability. When the index of instability reaches the maximum values, the relative phase between the translational and rotational motion becomes $\pi/2$. This implies that:

- as the profile passes through $x = 0$ (with $\dot{x}$ facing downwards), it has the maximum rotation with its upstream part facing downwards;
- however, as the profile rises, again in $x = 0$ (but with $\dot{x}$ facing upwards), it has its upstream part facing upwards (Fig. 5.23).

For this reason, at any given time the velocities of the profile, under these conditions, have the same direction as the lift force: so, the actual force performs positive work at every cycle, i.e. it introduces energy into the system: the synchronisation of the two vertical and torsional moments mean that this condition is maintained over time, so the introduction of energy continues over time generating instability.

The above makes it possible to describe, in qualitative terms, the phenomenon of flutter instability that can affect a particular profile. To define the actual phenomenon in quantitative terms, we must, as mentioned, integrate (5.80d) by imposing a generic solution:

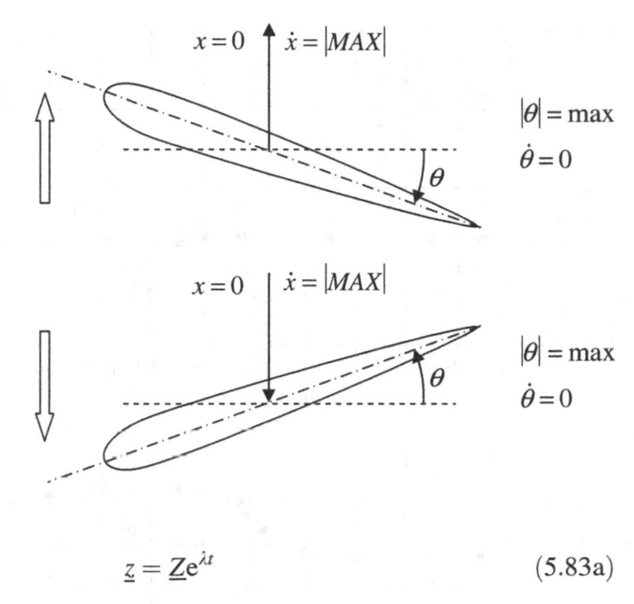

Fig. 5.23 Subsequent positions assumed by an aerofoil subject to flutter type instability

$$\underline{z} = \underline{Z}e^{\lambda t} \tag{5.83a}$$

obtaining, as usual, a homogeneous algebraic equation of the form:

$$\left[\lambda^2[M_T] + \lambda[R_T] + [K_T]\right]\underline{Z} = 0 \tag{5.83b}$$

By setting the determinant of the coefficient matrix to zero we can calculate the roots λ_i. Figure 5.24 shows, as a function of the velocity of the incident flow U, an example of the trend of the initial natural pulsations (ω_θ torsional and flexural ω_x) of a section of suspension bridge deck and of the instability coefficient α/ω: the same figure also shows the values adopted for calculation. With the presence of the equivalent damping terms of the force field $[R_F]$ the onset of instability is possible even when the two frequencies are not strictly coincident.

5.3.2.2 Instability of a Journal in a Bearing with Hydrodynamic Lubrication

Now we will address the problem concerning the interaction between a journal and a bearing with natural hydrodynamic lubrication: this issue will be resumed in Chap. 6, dedicated to the dynamics of rotors, so please refer to that chapter for further discussion on the problems of lubrication. Here we will briefly summarise the key aspects of the problem while referring the reader to specialised texts and lecture notes for more detailed discussion (see the bibliography and, in particular, [41, 50, 52, 53]. We will be considering a lubricated cylindrical bearing (Fig. 5.25) subject to a vertical load Q: the centre of the journal O_p is placed in an eccentric position e_c with respect to the geometric centre O_c of the bearing, thus forming a meatus with variable thickness that determines a distribution of pressure that

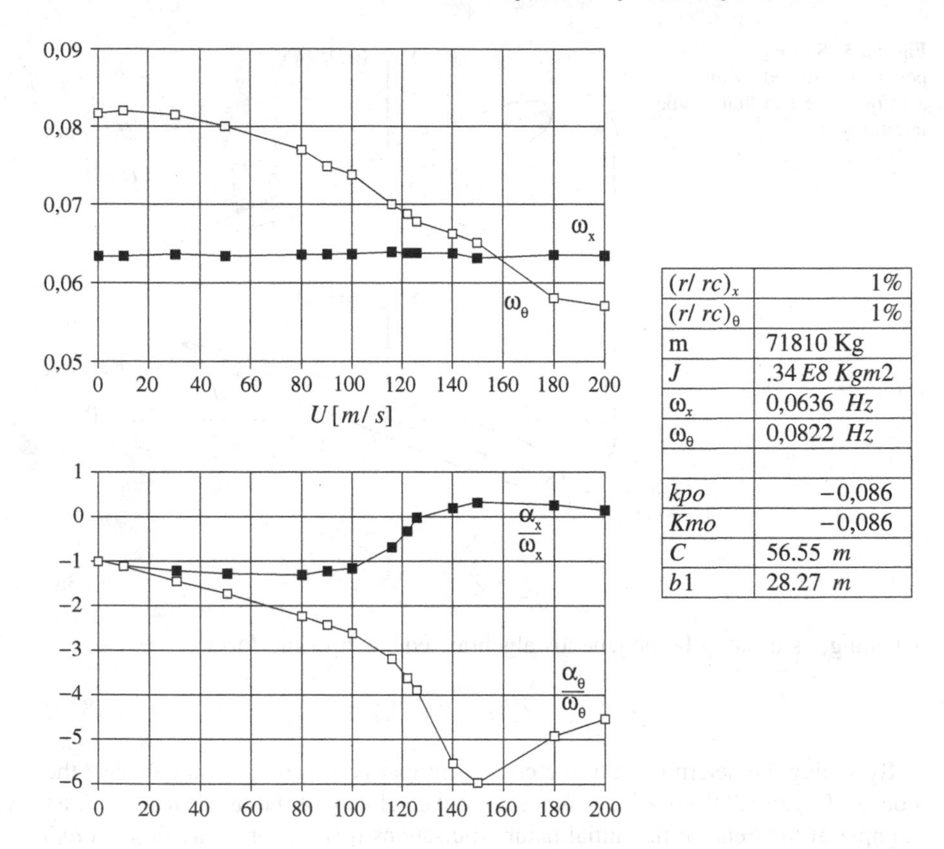

$(r/rc)_x$	1%
$(r/rc)_\theta$	1%
m	71810 Kg
J	.34 *E8 Kgm2*
ω_x	0,0636 *Hz*
ω_θ	0,0822 *Hz*
kpo	−0,086
Kmo	−0,086
C	56.55 *m*
*b*1	28.27 *m*

Fig. 5.24 Plot of the torsional ω_θ and flexural ω_x pulsations and of the instability coefficient $^{\alpha}/_{\omega}$, as a function of the velocity of the incident flow U, of a typical section of suspension bridge deck

balances the external load acting on the actual journal. The value of the magnitude $|e_c|$ of the vector $O_c - O_p$, that defines the position of the centre of the journal inside the bearing and its anomaly β depend on the load Q acting on the journal, on its angular speed Ω, as well as the geometry of the bearing and the characteristics of the lubricant. More in general [41, 53, 54], we can say that, inside the meatus the presence of an oil film creates a force F_p, sustaining the journal, as a function:

- of the relative position of the centre of the journal O_p with respect to the geometric centre of the bearing O_c, defined by the coordinates x_c and y_c;
- of the components $\dot{x}_c$ and $\dot{y}_c$ of the approach speed of the journal with respect to the bearing.

This force F_p can be divided, according to the two axes X and Y, into two components F_x and F_y the values of which are functions of the coordinates x_c, y_c and their first derivatives, $\dot{x}_c$ and $\dot{y}_c$:

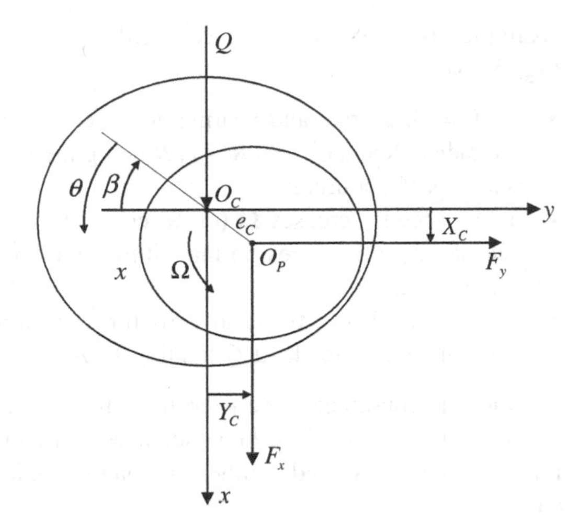

Fig. 5.25 Position of the journal inside the bearing

$$F_x = F_x(x_c, y_c, \dot{x}_c, \dot{y}_c)$$
$$F_y = F_y(x_c, y_c, \dot{x}_c, \dot{y}_c) \quad (5.84)$$

These Eq. (5.84) indicate that the journal is subject to a field of forces that depend on position and velocity, depending on the characteristics of the bearing and the speed of rotation Ω. To define this force field we use the lubrication theory which is based on the Reynolds equation (see Bibliography, section "Lubrication and bearings with hydrodynamic lubrication" [40–54]. In particular, once the position x_c, y_c and the velocity $\dot{x}_c$ and $\dot{y}_c$ of the journal inside the bearing have been assigned, it is possible to solve the Reynolds equation using numerical or semi-analytical methods to define the trend of the pressures in the oil film and, as a consequence, the forces F_x and F_y that act on the journal.

In this discussion we will use the term *steady-state conditions* to define those conditions where the journal rotates in the bearing with a constant angular speed Ω, without performing oscillations within the bearing, i.e. with $\dot{x}_c = 0$ and $\dot{y}_c = 0$: the position of the centre of the journal O_p with respect to the centre of the bearing O_c is defined by the equilibrium between the constant external load Q and the resultant of the pressures generated by the oil film (Fig. 5.25). The eccentricity e_c that is created between the axes of the journal and the bearing depends, therefore, on both the rotational speed Ω and the intensity of the load Q on the journal:

$$e_c = e_c(\Omega, Q) \quad (5.85)$$

The term *load locus* is used to define the locus of the points described by the centre of the journal O_p with variable Ω and constant load Q, or vice versa. The trend of the load locus can be defined analytically by integrating the Reynolds equations (Chap. 6, Sect. 6.2.2) or by taking experimental measurements. Obviously, the load locus is a function of the type of bearing being analysed.

Examples of this locus, for two different types of bearing, are illustrated in Fig. 5.26a:

- for $\Omega = 0$, journal and bearing are in contact and the eccentricity e_c is equal to the radial clearance $\delta = R - r$, R being the inner radius of the bearing and r the radius of the journal;
- as the speed increases Ω (or, at constant speed, as the load decreases Q), the eccentricity e_c changes, in magnitude and direction, to produce the film needed for support;
- the eccentricity e_c tends to zero for Ω tending to infinity, or, in the case of constant speed, for load Q tending to zero.

Another important parameter in defining the dynamic characteristics of the bearing is the so-called load coefficient $A(\chi)$ which, once the different parameters in play have been assigned, defines the load per unit of width N that the fluid film can sustain:

$$N = \frac{Q}{b} = A(\chi)\Omega r \frac{r^2}{\delta^2}\mu \tag{5.86}$$

$\chi = {}^{e_c}/_{\delta}$ being the non-dimensional eccentricity, b the width of the bearing, r the radius of the journal and μ the viscosity of the lubricant. The trend of $A(\chi)$ is of the type illustrated in Fig. 5.26b: the function $A(\chi)$ can be determined by numerically integrating the Reynolds equations ([41, 53, 54] and Chap. 6, Sect. 6.2.2). Knowledge of the function $A(\chi)$ is necessary as, once the load acting on bearing Q is known, and therefore the load per unit of width N, it is possible to obtain the value of $A(\chi)$ from (5.86) and, entering the curve $A = A(\chi)$ of Fig. 5.26b, establish χ: once χ is known it is possible to estimate the eccentricity e_c required by the journal to support the actual load. Once the eccentricity e_c and the load locus are known, it is also possible to establish the direction of the eccentricity vector, by

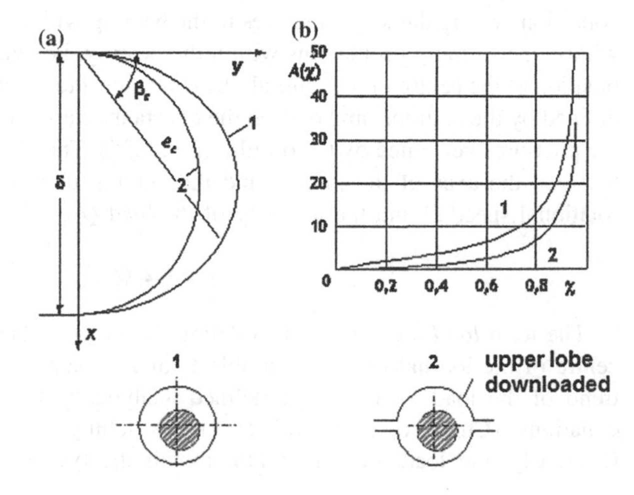

Fig. 5.26 **a** Load locus, **b** load coefficient $A(\chi)$; (*1*) cylindrical bearing; (*2*) single lobe bearing

means of anomaly β_c, thus being able to define, for the particular type of bearing being analysed, the geometry of the meatus for the assumed values of Q and Ω. The position of the journal, therefore, can be established, in static conditions, with the procedure described above: when the rotor vibrates, however, the relative position between journal and bearing changes dynamically and the forces F_x and F_y, non-linear expressions in x_c, y_c, $\dot{x}_c$ and $\dot{y}_c$, are given by (5.84). Considering with x_{co} and y_{co} the position of the centre of the journal with respect to the centre of the bearing under steady-state conditions, i.e. with $\dot{x}_c = 0$ and $\dot{y}_c = 0$:

$$\begin{aligned} x_{c0} &= |e_{c0}| \sin \beta_{c0} \\ y_{c0} &= |e_{c0}| \cos \beta_{c0} \end{aligned} \tag{5.87}$$

in order to analyse the perturbed motion of the journal around this steady-state position,[11] it is possible to linearise the expression of the forces F_x and F_y (5.84) by developing them in series, i.e.:

$$\begin{aligned} F_x = F_x(x_{c0}, y_{c0}, 0, 0) + \left(\frac{\partial F_x}{\partial x_c}\right)_{x_{c0}, y_{c0}} (x_c - x_{c0}) + \left(\frac{\partial F_x}{\partial y_c}\right)_{x_{c0}, y_{c0}} (y_c - y_{c0}) + \\ = \left(\frac{\partial F_x}{\partial \dot{x}_c}\right)_{x_{c0}, y_{c0}} \dot{x}_c + \left(\frac{\partial F_x}{\partial \dot{y}_c}\right)_{x_{c0}, y_{c0}} \dot{y}_c + \cdots \end{aligned} \tag{5.88a}$$

$$\begin{aligned} F_y = F_y(x_{c0}, y_{c0}, 0, 0) + \left(\frac{\partial F_y}{\partial x_c}\right)_{x_{c0}, y_{c0}} (x_c - x_{c0}) + \left(\frac{\partial F_y}{\partial y_c}\right)_{x_{c0}, y_{c0}} (y_c - y_{c0}) + \\ = \left(\frac{\partial F_y}{\partial \dot{x}_c}\right)_{x_{c0}, y_{c0}} \dot{x}_c + \left(\frac{\partial F_y}{\partial \dot{y}_c}\right)_{x_{c0}, y_{c0}} \dot{y}_c + \ldots \end{aligned} \tag{5.88b}$$

where $F_x(x_{c0}, y_{c0}, 0, 0)$ and $F_y(x_{c0}, y_{c0}, 0, 0)$ represent the forces exerted by the oil-film under steady-state conditions (or static equilibrium) and thus satisfy the equation:

$$\begin{aligned} F_x(x_{c0}, y_{c0}, 0, 0) + Q_x = 0 \\ F_y(x_{c0}, y_{c0}, 0, 0) + Q_y = 0 \end{aligned} \tag{5.89}$$

[11]Normally, and perhaps improperly, this steady-state situation, defined by a constant position of the journal centre x_{c0}, y_{c0}, by the journal constant rotational speed Ω and by a constant load Q, is defined as the *position of static equilibrium of the journal*: the issue can be covered in the section on systems perturbed in the neighbourhood of rest as perturbed motion is considered to be that defined by the variables $\overline{x} = x - x_{c0}$ and $\overline{y} = y - y_{c0}$ not in the angular speed Ω which is assumed to be constant.

The linearised dynamic forces exchanged between journal and bearing, due to perturbed motion around the position of static equilibrium, taking into account (5.88a), (5.88b) and (5.89) and introducing a change of coordinates that makes it possible to highlight the perturbed motion:

$$\begin{aligned}\bar{x} &= x_c - x_{c0}\\ \bar{y} &= y_c - y_{c0}\end{aligned} \tag{5.90}$$

become:

$$\begin{aligned}F_{xdin} &= F_x - F_x(x_{c0}, y_{c0}, 0, 0) = -K_{xx}\bar{x} - K_{xy}\bar{y} - R_{xx}\dot{\bar{x}} - R_{xy}\dot{\bar{y}}\\ F_{ydin} &= F_y - F_y(x_{c0}, y_{c0}, 0, 0) = -K_{yx}\bar{x} - K_{yy}\bar{y} - R_{yx}\dot{\bar{x}} - R_{yy}\dot{\bar{y}}\end{aligned} \tag{5.91}$$

i.e. in matrix form:

$$\begin{Bmatrix} F_{xdin} \\ F_{ydin} \end{Bmatrix} = -\begin{bmatrix} K_{xx} & K_{xy} \\ K_{yx} & K_{yy} \end{bmatrix}\begin{Bmatrix} \bar{x} \\ \bar{y} \end{Bmatrix} - \begin{bmatrix} R_{xx} & R_{xy} \\ R_{yx} & R_{yy} \end{bmatrix}\begin{Bmatrix} \dot{\bar{x}} \\ \dot{\bar{y}} \end{Bmatrix} \tag{5.92}$$

having used the constants K_{xx}, K_{xy}, K_{yx} and K_{yy} to indicate the partial derivatives, with a change in the sign, of the functions F_x and F_y with respect to the variables x_c and y_c, estimated around the position of static equilibrium x_{c0}, y_{c0}; these derivatives can also be expressed in non-dimensional form as:

$$\begin{aligned}K_{xx} &= -\left(\frac{\partial F_x}{\partial x_c}\right)_{x_{c0}, y_{co}} = \frac{Q}{\delta} C_{xx}\\ K_{xy} &= -\left(\frac{\partial F_x}{\partial y_c}\right)_{x_{c0}, y_{co}} = \frac{Q}{\delta} C_{xy}\\ K_{yx} &= -\left(\frac{\partial F_y}{\partial x_c}\right)_{x_{c0}, y_{co}} = \frac{Q}{\delta} C_{yx}\\ K_{yy} &= -\left(\frac{\partial F_y}{\partial y_c}\right)_{x_{c0}, y_{co}} = \frac{Q}{\delta} C_{yy}\end{aligned} \tag{5.93}$$

as $\delta = R - r$, Q being the applied load and C_{xx}, C_{xy}, C_{yx} and C_{yy} non-dimensional stiffnesses (functions of χ). In the case of cylindrical bearings, for example, these stiffnesses assume the trend shown in Fig. 5.27.

The constants R_{xx}, R_{xy}, R_{yx} and R_{yy} in (5.92) represent the partial derivatives of F_x and of F_y with respect to the components of velocity: the same quantities can be expressed in non-dimensional form as:

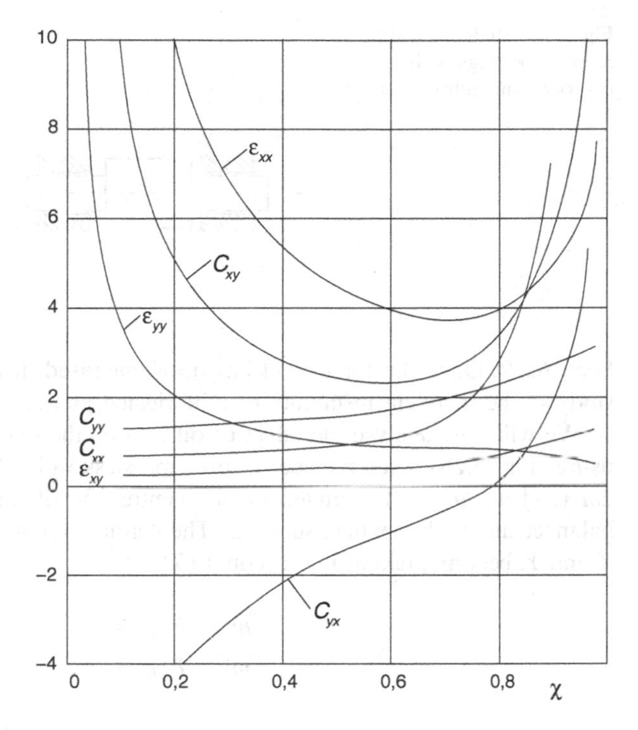

Fig. 5.27 Non-dimensional stiffnesses C_{ij} and dampings ε_{ij} for a cylindrical bearing

$$
\begin{aligned}
R_{xx} &= -\left(\frac{\partial F_x}{\partial \dot{x}_c}\right)_{x_{c0},y_{co}} = \frac{Q}{\Omega\delta}\varepsilon_{xx} \\
R_{xy} &= -\left(\frac{\partial F_x}{\partial \dot{y}_c}\right)_{x_{c0},y_{co}} = \frac{Q}{\Omega\delta}\varepsilon_{xy} \\
R_{yx} &= -\left(\frac{\partial F_y}{\partial \dot{x}_c}\right)_{x_{c0},y_{co}} = \frac{Q}{\Omega\delta}\varepsilon_{yx} \\
R_{yy} &= -\left(\frac{\partial F_y}{\partial \dot{y}_c}\right)_{x_{c0},y_{co}} = \frac{Q}{\Omega\delta}\varepsilon_{yy}
\end{aligned}
\tag{5.94}
$$

where ε_{xx}, ε_{xy}, ε_{yx} and ε_{yy} are called coefficients of non-dimensional damping: these coefficients, always in the case of cylindrical bearings, show the trend as a function of χ that is shown in the diagram in Fig. 5.27. The values of ε_{xy} are ε_{yx} are coincident and, for this reason, the equivalent damping matrix, due to the oil-film $[R_F]$, is symmetrical: it is also always positive definite. With this approach, the effect of the force field can be attributed to two equivalent matrices, respectively elastic and viscous: the physical non-conservative nature of the field of positional forces, will, once again, be reflected by the non-symmetry of the equivalent elastic matrix. The numerical values of the single elements of the equivalent matrices can be defined either experimentally or analytically, as will be described in Chap. 6,

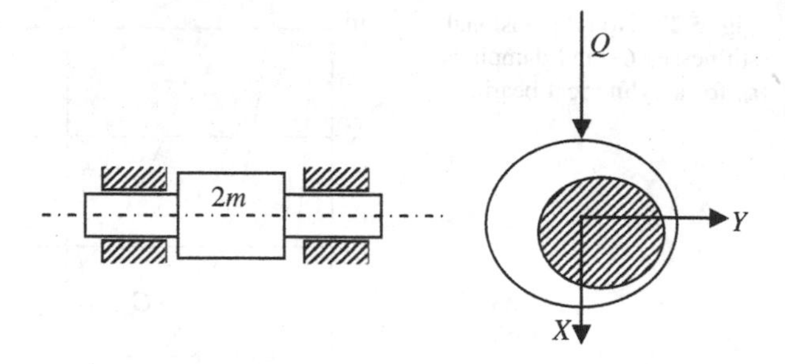

Fig. 5.28 Model of a stiff rotor on bearings with hydrodynamic lubrication

Sect. 6.2.2. Once the force field has been linearised, it is possible to address the study of the incipient instability of a lubricated journal.

We will now analyse the effect of oil-film on the dynamic behaviour of rotors, using (Fig. 5.28) a rotor made up of a massless rigid shaft, with a disk of mass $2m$ (and weight $2Q$) mounted on the centre line of the two supports, perfectly balanced and with identical supports. The perturbed motion equations, in directions X and Y, become, taking into account (5.92):

$$\begin{aligned} m\ddot{\overline{x}} - F_{xdin} &= 0 \\ m\ddot{\overline{y}} - F_{ydin} &= 0 \end{aligned} \tag{5.95}$$

i.e.:

$$\begin{aligned} m\ddot{\overline{x}} + \frac{Q}{\Omega\delta}\varepsilon_{xx}\dot{\overline{x}} + \frac{Q}{\Omega\delta}\varepsilon_{xy}\dot{\overline{y}} + \frac{Q}{\delta}C_{xx}\overline{x} + \frac{Q}{\delta}C_{xy}\overline{y} = 0 \\ m\ddot{\overline{y}} + \frac{Q}{\Omega\delta}\varepsilon_{yx}\dot{\overline{x}} + \frac{Q}{\Omega\delta}\varepsilon_{yy}\dot{\overline{y}} + \frac{Q}{\delta}C_{yx}\overline{x} + \frac{Q}{\delta}C_{yy}\overline{y} = 0 \end{aligned} \tag{5.95a}$$

We will first analyse the effect of just the positional terms. While neglecting the terms of velocity and assuming for convenience:

$$\omega_0 = \sqrt{\frac{g}{\delta}} \tag{5.96a}$$

and since $Q = mg$, Eq. (5.95a) are reduced to:

$$\begin{bmatrix} 1 & 0 \\ 0 & 1 \end{bmatrix} \begin{Bmatrix} \ddot{\overline{x}} \\ \ddot{\overline{y}} \end{Bmatrix} + \omega_0^2 \begin{bmatrix} C_{xx} & C_{yx} \\ C_{xy} & C_{yy} \end{bmatrix} \begin{Bmatrix} \overline{x} \\ \overline{y} \end{Bmatrix} = \begin{Bmatrix} 0 \\ 0 \end{Bmatrix} \tag{5.96b}$$

Using the discussion already seen in Sect. 5.3.1.2, on purely positional force fields, we can say (5.58a) and (5.58b) that flutter instability will arise in the event that the following occur:

$$C_{xy}C_{yx} < 0; \quad |C_{xy}C_{yx}| > \left(\frac{C_{xx} - C_{yy}}{2}\right)^2 \tag{5.97}$$

By observing the trend of the non-dimensional stiffness coefficients (Fig. 5.27), we can see that the term C_{xy} is always positive, while the term C_{yx} changes sign for a value of χ equals, in cylindrical bearings, $\chi_{\lim} = 0.78$. For a value of non-dimensional eccentricity χ less than the limit value $\chi_{\lim}$, that is, if the journal decreases its eccentricity e_c (which happens with low loads Q or high rotational speeds Ω) the extra-diagonal terms of the stiffness matrix C_{xy} and C_{yx} are of opposite signs and the equations in (5.97) are verified. In these conditions the journal is unstable, that is, once it is perturbed around the configuration of steady-state equilibrium, it is subject to expanding elliptical motion (see Sect. 5.3.1.2 on this) with pulsation ω and coefficient of expansion α (Fig. 5.29). The value of $\chi = 0.78$, so the real part of the solution is zero ($\alpha = 0$), corresponding to the value that separates the zone of stability from that of instability: so the non-dimensional coefficient (α/ω) if positive, provides an index of the system's instability (Fig. 5.30) but if negative, an index of stability.

On the other hand, with an equivalent elastic matrix of the oil-film not positive definite we would see divergence type unstable conditions: in the case in question this condition does not occur in any situation as the terms of stiffness are always such that the actual matrix of stiffness of the force field is always positive definite, that is, the position of the journal is always statically stable. We will now also analyse the effect of the terms of velocity, while always referring to the simplified model with two DOF in Fig. 5.28: the linearised equations, in this case, having again assumed (5.96a), equal:

$$\begin{bmatrix} 1 & 0 \\ 0 & 1 \end{bmatrix} \begin{Bmatrix} \ddot{\bar{x}} \\ \ddot{\bar{y}} \end{Bmatrix} + \frac{\omega_0^2}{\Omega} \begin{bmatrix} \varepsilon_{xx} & \varepsilon_{yx} \\ \varepsilon_{xy} & \varepsilon_{yy} \end{bmatrix} \begin{Bmatrix} \dot{\bar{x}} \\ \dot{\bar{y}} \end{Bmatrix} + \omega_0^2 \begin{bmatrix} C_{xx} & C_{yx} \\ C_{xy} & C_{yy} \end{bmatrix} \begin{Bmatrix} \bar{x} \\ \bar{y} \end{Bmatrix} = \begin{Bmatrix} 0 \\ 0 \end{Bmatrix} \tag{5.98a}$$

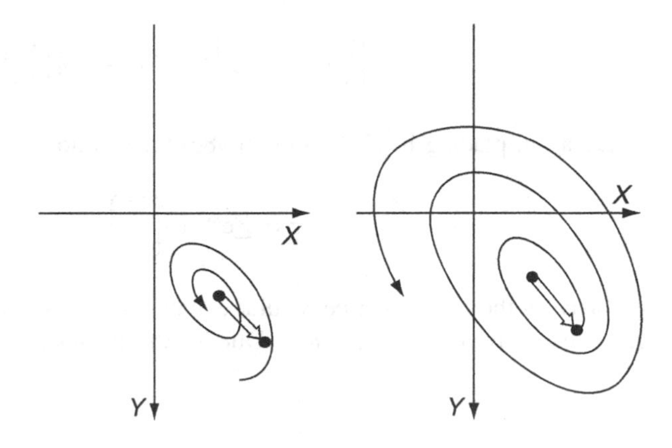

Fig. 5.29 Stable motion and unstable motion

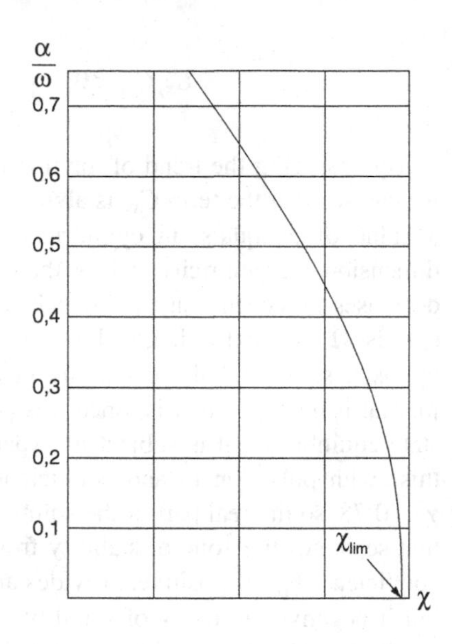

Fig. 5.30 Stability of a cylindrical bearing with just the positional terms: trend of the ratio $^{\alpha}/_{\omega}$ as a function of non-dimensional eccentricity χ

Having used the vector $\underline{z}$ to indicate:

$$\underline{z} = \begin{Bmatrix} \overline{x} \\ \overline{y} \end{Bmatrix} \tag{5.98b}$$

the vector that contains the independent variables that define the perturbed motion of the journal around the position of static equilibrium, (5.98a) can be rewritten as:

$$[M]\ddot{\underline{z}} + [R_F]\dot{\underline{z}} + [K_F]\underline{z} = \underline{0} \tag{5.98c}$$

having used $[R_F]$ and $[K_F]$ to indicate, respectively the equivalent matrices of the oil-film:

$$[R_F] = \frac{\omega_0^2}{\Omega}\begin{bmatrix} \varepsilon_{xx} & \varepsilon_{yx} \\ \varepsilon_{xy} & \varepsilon_{yy} \end{bmatrix}; \quad [K_F] = \omega_0^2\begin{bmatrix} C_{xx} & C_{yx} \\ C_{xy} & C_{yy} \end{bmatrix} \tag{5.98d}$$

By again placing in (5.98a) or (5.98c) the solution:

$$\underline{z} = \underline{Z}e^{\lambda t} = \begin{Bmatrix} \overline{X} \\ \overline{Y} \end{Bmatrix} e^{\lambda t} \tag{5.98e}$$

we arrive at the fourth degree secular equation in l (Sect. 2.3.2), which when solved gives the 4 roots $\lambda_i = \alpha_i \pm i\omega_i$. Due to the presence of terms of velocity, the stability of the system will not depend solely on the equivalent elastic matrix $[K_F]$ of the force field which, as seen previously, is a function of χ, but also on the equivalent damping matrix $[R_F]$ due to the actual force field. This matrix is not only a function of χ, via the coefficients ε_{ij}, but also of the angular speed Ω of the rotor. The terms of velocity are stabilising as $[R_F]$ is symmetrical and positive definite. More specifically [43, 44], two of these solutions λ_i are real and negative or complex conjugates with a strongly negative real part (corresponding to a stable solution). The other two values of λ_i are complex conjugates with a real part that can be positive or negative: we can use α to define the real part and ω for the imaginary part of these solutions. Figures 5.31a and 5.31b show, in relation to the pair of unstable solutions only, the trend of ratio $^{\alpha}/_{\omega}$ and of ratio $^{\Omega}/_{\omega}$ as a function of the non-dimensional eccentricity χ, again for cylindrical bearing and stiff shaft.

$$\begin{aligned} \chi &< \chi_{\lim} \\ \Omega &> 2\overline{\omega} \end{aligned} \tag{5.98f}$$

Since instability is dynamic or caused by flutter, we have expanding oscillatory motion in the plane x–y with expanding elliptical trajectory that has a frequency equal to the natural pulsation ω of the system. Lastly, we should remember that the linearised solution applies for small oscillations around the position of equilibrium. As the amplitude of the oscillations increases, then the non-linear effects become significant, causing the establishment of limit cycles with oscillation amplitudes that reach maximum values close to the boundary conditions of contact between journal and bearing.

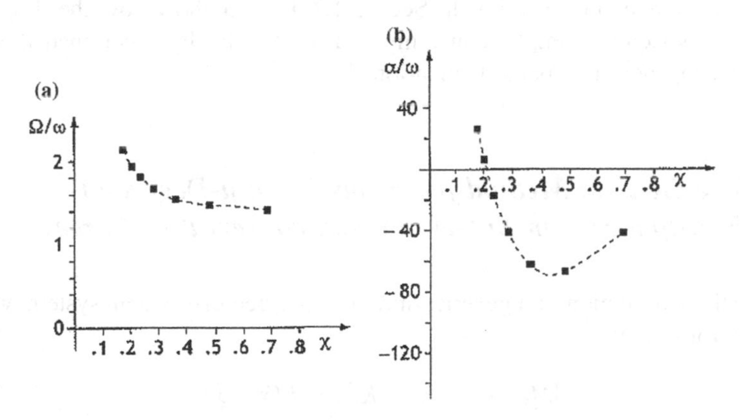

Fig. 5.31 Stability of a cylindrical bearing with velocity and positional terms: **a** trend of the ratio $^{\alpha}/_{\omega}$ as a function of non-dimensional eccentricity χ; **b** trend of the ratio $^{\Omega}/_{\omega}$ as a function of non-dimensional eccentricity ec

5.4 Multi-Degree-of-Freedom Vibrating Systems Perturbed Around the Position of Equilibrium

The reality that surrounds us is always made up of continuous systems, that is, systems with ∞ degree-of-freedom: often, however, we can reduce them to discrete models with 1, 2 or n-degree-of-freedom depending on the schematization adopted and the level of refinement required in the results. Usually, the schematization used is the simplest possible, but is such that it can correctly reproduce the characteristics, at least qualitative, of the phenomenon being analysed. In problems relating to systems subject to non-conservative force fields, for a qualitative analysis of the phenomena we usually reduce to simple models with one or two-degree-of-freedom, depending on the problem being analysed (see previous sections). If we wish to reproduce all the characteristics of the phenomenon in question as faithfully as possible, including in quantitative terms, we need to resort to more sophisticated models with n or ∞ degree-of-freedom. Basically, there are two approaches that can be used for this purpose:

- the modal approach, Chap. 3, Sect. 3.8;
- the finite elements method, Chap. 4.

The difference in using the first or second approach lies purely in the way the equations are obtained, but leads, in any case, to a discrete system with n degree-of-freedom. In this section we will illustrate the methods used to analyse a generic mechanical system with n degree-of-freedom surrounded by a force field that is a function of the position, velocity and acceleration of the actual system, with a discussion which is clearly a logical extension of that already seen for two-degree-of-freedom systems in Sect. 5.3. In Sect. 5.4.2 we will show how the discussion applies to a specific example, concerning an aerofoil hit by a confined flow and analysed using both the methods mentioned.

5.4.1 The General Method for Analysing a n-Degree-of-Freedom System Subject to Non-conservative Forces

The equations of motion of a generic vibrating n-degree-of-freedom system, written in matrix form, are:

$$[M_s]\ddot{\underline{x}} + [R_s]\dot{\underline{x}} + [K_s]\underline{x} = \underline{F}(\ddot{\underline{x}}, \dot{\underline{x}}, \underline{x}) \tag{5.99}$$

Following what has already been done for systems with one and two degree-of-freedom, first we must find the position of static equilibrium (we should remember that as the equations are non-linear, the positions of equilibrium may be more than one). For this purpose, we must solve a system of non-linear equations of the form:

$$[K_s]\underline{x}_0 = \underline{F}(\underline{0}, \underline{0}, \underline{x}_0) \tag{5.100}$$

Which can be solved, for example, using the Newton-Raphson method, for systems of equations [2, 5, 8]. Having obtained a position of equilibrium $\underline{x}_o$, if one exists, it is possible to linearise the system's equations of motion around the position of static equilibrium found, developing the non-linear forces $\underline{F}(\underline{\ddot{x}}, \underline{\dot{x}}, \underline{x})$ in a Taylor series up to the first order:

$$\underline{F}(\underline{\ddot{x}}, \underline{\dot{x}}, \underline{x}) = \underline{F}(\underline{0}, \underline{0}, \underline{x}_0) + \left[\frac{\partial \underline{F}}{\partial \underline{x}}\right]_0 (\underline{x} - \underline{x}_0) + \left[\frac{\partial \underline{F}}{\partial \underline{\dot{x}}}\right]_0 \underline{\dot{x}} + \left[\frac{\partial \underline{F}}{\partial \underline{\ddot{x}}}\right]_0 \underline{\ddot{x}} + \ldots \tag{5.101}$$

The derivatives of the vector $\underline{F}$ with respect to vector $\underline{x}$ (with the sign changed) formally represent the matrices of equivalent stiffness $[K_F]$, damping $[R_F]$ and mass $[M_F]$, respectively, due to the force field:

$$[K_F] = -\left[\frac{\partial \underline{F}}{\partial \underline{x}}\right]_0 = -\begin{bmatrix} \frac{\partial F_1}{\partial x_1} & \frac{\partial F_1}{\partial x_2} & \cdots & \frac{\partial F_1}{\partial x_n} \\ \cdots & \cdots & \cdots & \cdots \\ \frac{\partial F_n}{\partial x_1} & \frac{\partial F_n}{\partial x_2} & \cdots & \frac{\partial F_n}{\partial x_n} \end{bmatrix}_o$$

$$[R_F] = -\begin{bmatrix} \frac{\partial F_1}{\partial \dot{x}_1} & \frac{\partial F_1}{\partial \dot{x}_2} & \cdots & \frac{\partial F_1}{\partial \dot{x}_n} \\ \cdots & \cdots & \cdots & \cdots \\ \frac{\partial F_n}{\partial \dot{x}_1} & \frac{\partial F_n}{\partial \dot{x}_2} & \cdots & \frac{\partial F_n}{\partial \dot{x}_n} \end{bmatrix}_o \qquad [M_F] = -\begin{bmatrix} \frac{\partial F_1}{\partial \ddot{x}_1} & \frac{\partial F_1}{\partial \ddot{x}_2} & \cdots & \frac{\partial F_1}{\partial \ddot{x}_n} \\ \cdots & \cdots & \cdots & \cdots \\ \frac{\partial F_n}{\partial \ddot{x}_1} & \frac{\partial F_n}{\partial \ddot{x}_2} & \cdots & \frac{\partial F_n}{\partial \ddot{x}_n} \end{bmatrix}_o \tag{5.101b}$$

Equation (5.99) can thus be rewritten, also taking into account (5.101), as:

$$\begin{aligned} [[M_s] + [M_F]]\underline{\ddot{\overline{x}}} + [[R_s] + [R_F]]\underline{\dot{\overline{x}}} + [[K_s] + [K_F]]\underline{\overline{x}} = \underline{0} \\ [M_T]\underline{\ddot{\overline{x}}} + [R_T]\underline{\dot{\overline{x}}} + [K_T]\underline{\overline{x}} = \underline{0} \end{aligned} \tag{5.102}$$

having indicated, as usual, with:

$$\underline{\overline{x}} = \underline{x} - \underline{x}_0 \tag{5.102a}$$

the variable that describes the perturbed motion around the position of static equilibrium. So Eq. (5.102) is the free motion equation of the system, with the presence of the linearised force field: this equation, relating to any system with n degree-of-freedom, can obviously also refer to:

- a continuous system discretized using the finite elements method;
- a continuous system discretized using the modal approach.

As we have already seen for systems with one or two degree-of-freedom, Eq. (5.102) highlights how the force field can, in general, change the system's characteristics of mass, damping and stiffness, thus changing its natural frequencies and influencing its stability.

5.4.1.1 Stability Analysis

As we saw in both the previous examples, it is also possible to achieve linearisation of the non-conservative forces that act on a system with n degree-of-freedom, in order to analyse the stability of the linearised system. One method used to estimate the stability of a system consists in analysing the free motion solution, obtained by finding a particular integral of the homogeneous (5.102):

$$\overline{\underline{x}}(t) = \overline{\underline{X}} e^{\lambda t} \tag{5.103}$$

where λ can generally be complex. By replacing the solution (5.103) in (5.102), we obtain a homogeneous linear equation in $\overline{\underline{X}}$:

$$\left[\lambda^2 [M_T] + \lambda [R_T] + [K_T]\right] \overline{\underline{X}} = 0 \tag{5.104}$$

By annulling the determinant of the coefficient matrix, we obtain $2n$ values of λ, n being the number of degree-of-freedom of the system. As already stated, as the degree-of-freedom increase, this procedure becomes less suitable, so we reduce the search for values of λ to the solution of an eigenvalue problem, a problem which can be solved using ordinary calculation codes. For this purpose, as already seen (Sect. 2.4.3.1), we consider an auxiliary identity alongside of the motion Eq. (5.102):

$$\begin{aligned} [M_T]\ddot{\overline{\underline{x}}} + [R_T]\dot{\overline{\underline{x}}} + [K_T]\overline{\underline{x}} &= \underline{0} \\ [M_T]\dot{\overline{\underline{x}}} &= [M_T]\dot{\overline{\underline{x}}} \end{aligned} \tag{5.105}$$

And we perform the following change in variables:

$$\overline{\underline{z}} = \begin{Bmatrix} \dot{\overline{\underline{x}}} \\ \overline{\underline{x}} \end{Bmatrix} = \begin{Bmatrix} \overline{\underline{y}} \\ \overline{\underline{x}} \end{Bmatrix} \Rightarrow \dot{\overline{\underline{z}}} = \begin{Bmatrix} \ddot{\overline{\underline{x}}} \\ \dot{\overline{\underline{x}}} \end{Bmatrix} = \begin{Bmatrix} \dot{\overline{\underline{y}}} \\ \dot{\overline{\underline{x}}} \end{Bmatrix} \tag{5.105a}$$

The system (5.105) can thus be rewritten as:

$$\begin{aligned} [M_T]\dot{\overline{\underline{y}}} + [R_T]\overline{\underline{y}} + [K_T]\overline{\underline{x}} &= \underline{0} \\ [M_T]\overline{\underline{y}} &= [M_T]\dot{\overline{\underline{x}}} \end{aligned} \tag{5.105b}$$

Defining matrices $[B]$ and $[C]$:

$$[B] = \begin{bmatrix} [M_T] & 0 \\ 0 & [M_T] \end{bmatrix} [C] = \begin{bmatrix} [R_T] & [K_T] \\ -[M_T] & 0 \end{bmatrix} \tag{5.106a}$$

(5.106b) is briefly expressed as:

$$[B]\dot{\underline{z}} + [C]\underline{z} = \underline{0} \tag{5.106b}$$

For convenience (5.106b) can be written as:

$$\dot{\underline{z}} = [A]\underline{z} \tag{5.106c}$$

being:

$$[A] = -[B]^{-1}[C] \tag{5.106d}$$

By imposing for (5.106c) a solution of the form:

$$\underline{z}(t) = \underline{Z}e^{\lambda t} \tag{5.107a}$$

we reduce to the eigenvalue problem:

$$[[A] - \lambda[I]]\underline{Z} = \underline{0} \tag{5.107b}$$

The values of $\lambda_i = \alpha_i \pm i\omega_i$ are the eigenvalues of the matrix $[A]$, also known as the state transition matrix transition matrix. Examination of the real part of these eigenvalues λ_i makes it possible to establish whether the system is asymptotically stable or not: a positive value of the real part indicates instability, a negative value indicates stability. The equivalent matrices of the force field $[M_F]$, $[R_F]$ and $[K_F]$ are, in general, functions of a parameter p that is characteristic of the phenomenon being analysed: for example, in the event that we wish to analyse the behaviour of a system placed in a confined flow, this parameter is the velocity U of the actual flow. For this reason, bearing in mind (5.102), the overall matrices $[M_T]$, $[R_T]$ and $[K_T]$ of the system surrounded by the force field are functions of this parameter:

$$[M_T] = [M_T(p)]; \quad [R_T] = [R_T(p)]; \quad [K_T] = [K_T(p)] \tag{5.108}$$

The roots λ of the system analysed (5.104) are, in turn, functions of the parameter p:

$$[\lambda^2[M_T(p)] + \lambda[R_T(p)] + [K_T(p)]]\underline{X} = 0 \Rightarrow \lambda = \lambda(p) \tag{5.108a}$$

We define the value above which a generic eigenvalue $\lambda = \lambda(p)$ has a positive real part as the critical value of parameter p_{crit} (in the case of the aerofoil critical velocity U_{crit}), i.e. where $p = \overline{p} > p_{crit}$ there is at least one eigenvalue λ_k:

$$\lambda_k = \lambda_k(\overline{p}) = \alpha_k(\overline{p}) + i\omega_k(\overline{p}) \quad con\, \alpha_k(\overline{p}) > 0 \tag{5.108b}$$

Alongside calculation of the eigenvalues of the linearised system, which can sometimes present problems of a numerical nature, we can use the *forced method* which offers the advantage of greater numerical stability [43].

5.4.2 An Example: An Aerofoil Hit by a Confined Flow (n-Degree-of-Freedom System)

We will now analyse the fluid-elastic vibrations of the wing on an aircraft which moves in a fluid in constant pressure (with the absence of turbulence): for simplicity's sake, we will assume, as shown in Fig. 5.32, that the wing has one end wedged into the ground, although in reality it would actually be connected to the fuselage. We should actually analyse the global wing + fuselage system, but the discussion would become unnecessarily complicated for the educational requirements that we seek to satisfy. In reality, the wing moves through the air at a certain velocity U: in general, the issue is dealt with while assuming that the wing is fixed and the fluid has equal velocity (in magnitude and direction) but in the opposite direction. This problem is similar, for example, to that of fluid-elastic vibrations in the decks of suspension bridges or cable-stayed bridges, to the vibrations of blades in turbo-alternators hit by fluid, to the vibrations caused in helicopter rotor blades etc.

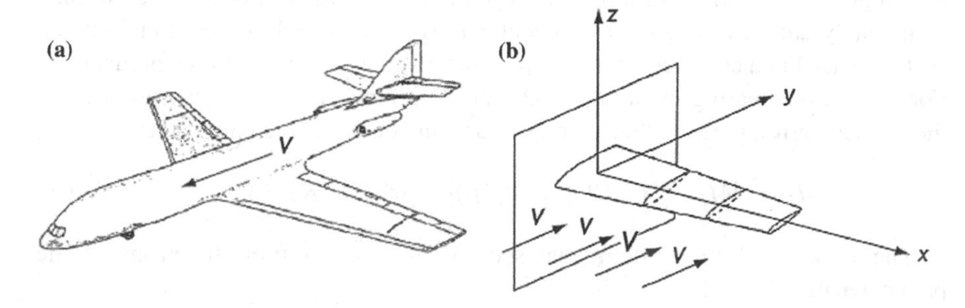

Fig. 5.32 The wing hit by a confined flow

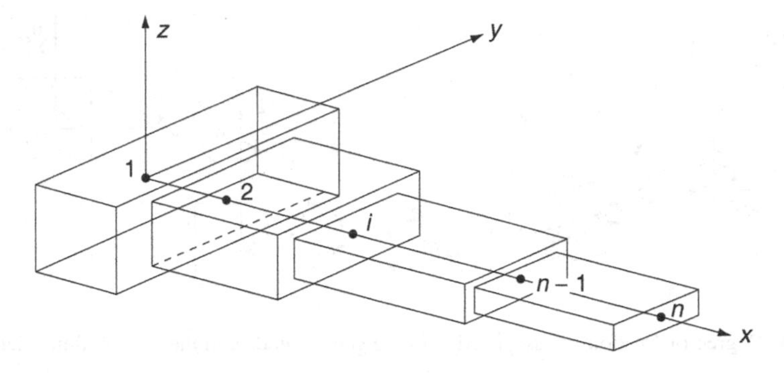

Fig. 5.33 The wing schematised with beam-type finite elements

5.4.2.1 Model with Finite Elements

First of all we will examine a case where the structure is schematised with finite elements (Fig. 5.33) by adopting beam-type finite elements (Sect. 4.5). This schematisation applies as long as the wing is sufficiently extended in the longitudinal direction and so the various cross sections of the wing can be considered rigid: in this case it is possible to schematise the wing with finite beam-type elements, for which just the flexural and torsional characteristics are defined.[12] As always, the number of finite elements adopted to reproduce the dynamic behaviour of the wing, in addition to being such as to reproduce the variations of section that can be encountered along the longitudinal axis of the actual wing, must be adjusted to the range of frequencies that we intend to analyse.

We use $\underline{x}$ to define the vector that contains the degree-of-freedom relative to the nodes of the model with finite elements (Fig. 5.33) and, more specifically, the degree-of-freedom that are actually free (Chap. 4, Sect. 4.5) $\underline{x} = \underline{x}_L$:

$$\underline{x} = \begin{Bmatrix} \underline{x}_2 \\ \underline{x}_3 \\ \cdots \\ \underline{x}_i \\ \cdots \\ \underline{x}_n \end{Bmatrix} \tag{5.109}$$

$\underline{x}_i$ being the vector (Fig. 5.34) that contains the degree-of-freedom of the generic ith node:

[12]If the wing were an assembled structure, and so difficult to define from a structural point of view, a more sophisticated schematisation of the actual structure would be necessary, introducing, for example, other finite elements such as plates (Sect. 4.6). Often, experimental testing is necessary on a prototype to better define the elasto-inertial of the structure (for more information see Chap. 8 on Techniques of identification.

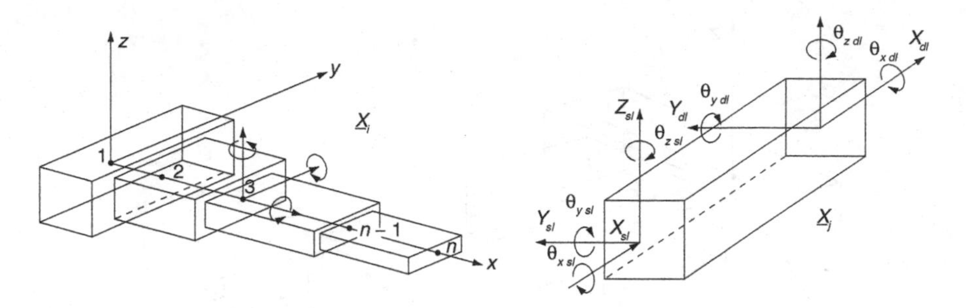

Fig. 5.34 Degree-of-freedom x_i associated with the generic node and the generic finite element x_j

$$\underline{x}_i = \begin{Bmatrix} x_i \\ y_i \\ z_i \\ \theta_{xi} \\ \theta_{yi} \\ \theta_{zi} \end{Bmatrix} \tag{5.110}$$

The equations that govern the motion of the structure, with the presence of aerodynamic actions are, therefore:

$$[M_s]\underline{\ddot{x}} + [R_s]\underline{\dot{x}} + [K_s]\underline{x} = \underline{Q}_x(\underline{x}, \underline{\dot{x}}) \tag{5.111}$$

where the structural matrices $[M_s]$and $[K_s]$ are the matrices relating solely to the degree-of-freedom that are actually free:

$$[M_s] = [M_{LL}]; \quad [K_s] = [K_{LL}] \tag{5.111a}$$

Defined with the known assembly and partition procedures (for more information see Chap. 4, Sect. 4.5) starting from the matrices of mass $[M_j]$ and stiffness $[K_j]$ of the single finite elements (Fig. 5.35). In (5.111) the damping matrix $[R_s]$ is obtained as a linear combination of the global matrices $[M_s]$ and $[K_s]$:

$$[R_s] = \alpha[M_s] + \beta[K_s] \tag{5.111b}$$

α and β being two coefficients that are appropriately defined to reproduce the structural damping of the system in the range of frequencies concerned. Now we can write the expression of the generalised aerodynamic forces acting on the structure: for this purpose, we consider a very small generic infinitesimal elemental area of length $d\xi$ belonging to the jth finite element (Fig. 5.36).

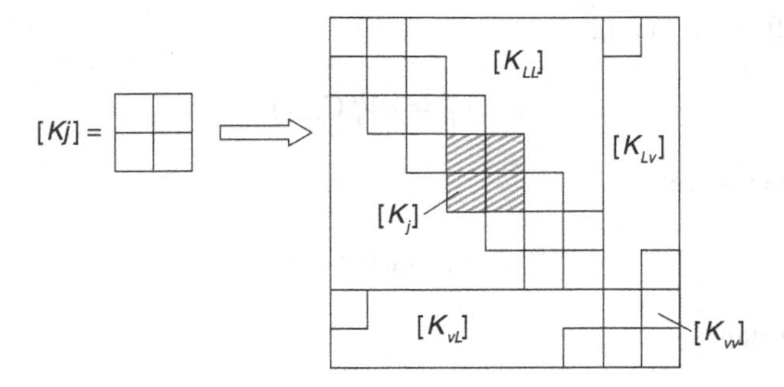

Fig. 5.35 Stiffness matrix assembly $[K_s]$

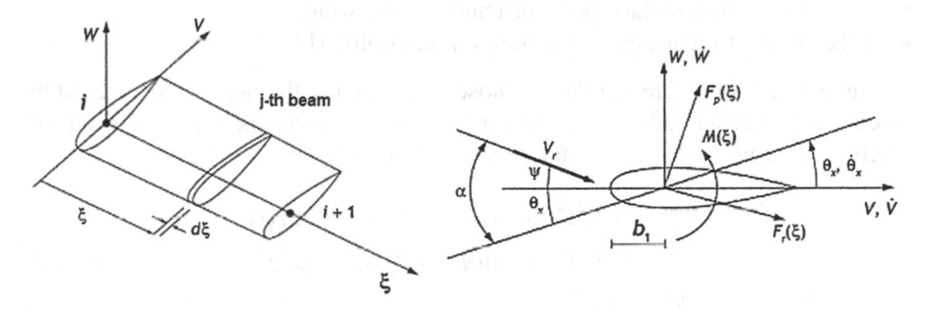

Fig. 5.36 Calculating the generalised aerodynamic forces

The presence of the incident fluid, with velocity U assumed constant in space and in time, generates aerodynamic forces[13]: these forces per unit of length can be expressed, if projected in a vertical and transversal direction and taking into account the nomenclature adopted in Sect. 5.2.1 (where there is the full discussion of the quasi-static theory used) as:

$$\begin{aligned} F_w &= \left(F_p \cos\psi - F_r \sin\psi\right) \\ F_v &= \left(F_p \sin\psi + F_r \cos\psi\right) \\ M &= aCV_r^2 C_m(\alpha) \end{aligned} \tag{5.112a}$$

being:

[13]In this discussion we will neglect the effect of turbulent wind which would add to the velocity of the aircraft, giving rise to velocity U that is no longer constant, but variable over time in magnitude and direction.

- F_p the lift force (5.66):

$$F_p = aV_r^2 C_p(\alpha) \tag{5.112b}$$

- F_r the drag force:

$$F_r = aV_r^2 C_r(\alpha) \tag{5.112c}$$

- a the constant:

$$a = \frac{1}{2}\rho S \tag{5.112d}$$

- S is the reference surface per unit length of the wing;
- α the angle of incidence of the flow on the profile (Fig. 5.36).

These expressions are similar to those obtained for the aerofoil with 2 DOF (Sect. 5.3.2), having added the resultant F_v of the aerodynamic forces, per unit of length, according to the horizontal direction (Fig. 5.32)[14]:

$$\begin{aligned} F_w &= \left(aV_r^2 C_p(\alpha)\cos\psi - aV_r^2 C_r(\alpha)\sin\psi\right) \\ F_v &= \left(aV_r^2 C_p(\alpha)\sin\psi + aV_r^2 C_r(\alpha)\cos\psi\right) \\ M &= aCV_r^2 C_m(\alpha) \end{aligned} \tag{5.113}$$

In this case, the angles α and ψ (5.77a)–(5.77c) and the relative velocity V_r are functions of the current abscissa ξ (varying the motion of the different sections with ξ):

$$\alpha = \alpha(\xi) = \theta_x(\xi) + \psi(\xi) \tag{5.113a}$$

$$\psi = \psi(\xi) = arctg\left(\frac{\dot{w}(\xi)}{U - \dot{v}} - \frac{b_1\dot{\theta}_x(\xi)}{U - \dot{v}}\right) \tag{5.113b}$$

$$V_r^2 = V_r^2(\xi) = \left(b_1\dot{\theta}_x(\xi) - \dot{w}(\xi)\right)^2 + (U - \dot{v})^2 \tag{5.113c}$$

where w, v are θ_x represent, respectively, the vertical and lateral displacement and torsional rotation of the generic infinitesimal element placed at distance ξ from the far-left of the generic jth finite element (Fig. 5.36).

---

[14]In this section, we use, for ease of discussion, the correct quasi-static theory which only applies for very high reduced velocities V_r: we refer the reader to [14, 20, 21, 24, 25] for more sophisticated analysis.

Once the position of static equilibrium $\underline{x}_o$ has been defined using (5.100), as we saw in Sect. 5.3.2, it is possible to linearise the expressions of the aerodynamic forces, per unit of length, around this position. Assuming:

$$\begin{gathered} V_r^2 = U^2; \quad \cos\psi = 1; \quad \sin\psi = \psi \\ \alpha(\xi) = \theta_x(\xi) + \psi(\xi); \quad \psi(\xi) = \frac{\dot{w}(\xi)}{U} - \frac{b_1\dot{\theta}_x(\xi)}{U} \end{gathered} \tag{5.114a}$$

and defining with:

$$\begin{gathered} C_{r0} = (C_r(\alpha))_{\alpha=\alpha_0}; \quad C_{p0} = (C_p(\alpha))_{\alpha=\alpha_0}; \quad C_{m0} = (C_m(\alpha))_{\alpha=\alpha_0} \\ K_{r0} = \left(\frac{\partial C_r(\alpha)}{\partial\alpha}\right)_{\alpha=\alpha_0}; \quad K_{p0} = \left(\frac{\partial C_p(\alpha)}{\partial\alpha}\right)_{\alpha=\alpha_0}; \quad K_{m0} = \left(\frac{\partial C_m(\alpha)}{\partial\alpha}\right)_{\alpha=\alpha_0} \end{gathered} \tag{5.114b}$$

the aerodynamic forces linearised on the generic section Eq. (5.113) can be rewritten as:

$$\begin{aligned} F_w &= aU^2K_{p0}\overline{\alpha} - aU^2C_{r0}\overline{\psi} - aU^2K_{r0}\overline{\alpha}\,\overline{\psi} \\ F_v &= aU^2C_{p0}\overline{\psi} + aU^2K_{p0}\overline{\alpha}\,\overline{\psi} + aU^2K_{r0}\overline{\alpha} \\ M &= aCU^2K_{m0}\overline{\alpha} \end{aligned} \tag{5.114c}$$

having used $\overline{\alpha}$ and $\overline{\psi}$ to indicate the perturbed motion around the position of static equilibrium. By subsequently eliminating the higher order terms from (5.114c), we obtain:

$$\begin{aligned} F_w &= aU^2K_{p0}\overline{\alpha} - aU^2C_{r0}\overline{\psi} \\ F_v &= aU^2C_{p0}\overline{\psi} + aU^2K_{r0}\overline{\alpha} \\ M &= aCU^2K_{m0}\overline{\alpha} \end{aligned} \tag{5.114d}$$

Taking into account (5.114a) in (5.114d), it is possible to express the aerodynamic forces linearised in correspondence to the generic section of the wing as a function of the displacements and the velocities that it is subject to. Having used $\underline{x}_\xi$ to indicate the vector containing the lateral $v(\xi)$ and vertical $w(\xi)$ displacement coordinates and the rotation $\theta_x(\xi)$ of the generic section of the wing (Fig. 5.36), which represent the perturbed motion around the position of static equilibrium (in the case in question, Fig. 5.9), this position is defined by $v_o(\xi) = w_o(\xi) = \theta_{xo}(\xi) = 0$:

$$\underline{x}_{\xi}(\xi) = \begin{Bmatrix} w(\xi) - w_0(\xi) \\ v(\xi) - v_0(\xi) \\ \theta_x(\xi) - \theta_{x0}(\xi) \end{Bmatrix} = \begin{Bmatrix} \overline{w}(\xi) \\ \overline{v}(\xi) \\ \overline{\theta}_x(\xi) \end{Bmatrix} = \begin{Bmatrix} w(\xi) \\ v(\xi) \\ \theta_x(\xi) \end{Bmatrix} \tag{5.115a}$$

The expression of the aerodynamic forces acting on the generic section of the wing $\underline{F}_{\xi}$ per unit of length, cleansed of the constant terms and linearised (5.114d), can be written as:

$$\begin{aligned} F_w &= aU^2 K_{p0}\left(\theta_x + \frac{\dot{w}}{U} - \frac{b_1\dot{\theta}_x}{U}\right) - aU^2 C_{r0}\left(\frac{\dot{w}}{U} - \frac{b_1\dot{\theta}_x}{U}\right) \\ F_v &= aU^2 C_{p0}\left(\frac{\dot{w}}{U} - \frac{b_1\dot{\theta}_x}{U}\right) + aU^2 K_{r0}\left(\theta_x + \frac{\dot{w}}{U} - \frac{b_1\dot{\theta}_x}{U}\right) \\ M &= aCU^2 K_{m0}\left(\theta_x + \frac{\dot{w}}{U} - \frac{b_1\dot{\theta}_x}{U}\right) \end{aligned} \tag{5.115b}$$

and, in matrix form:

$$\underline{F}_{\xi} = \begin{Bmatrix} F_w \\ F_v \\ M \end{Bmatrix} = -[K_{A\xi}]\underline{x}_{\xi} - [R_{A\xi}]\dot{\underline{x}}_{\xi} \tag{5.115c}$$

where the matrices $[K_{A\xi}]$ and $[R_{A\xi}]$ assume the following expressions:

$$\begin{aligned} [R_{A\xi}] &= aU\begin{bmatrix} (-K_{p0} + C_{r0}) & 0 & (K_{p0} - C_{r0})b_1 \\ (-C_{p0} - K_{r0}) & 0 & C_{p0}b_1 + K_{r0}b_1 \\ -CK_{m0} & 0 & -CK_{m0}b_1 \end{bmatrix} \\ [K_{A\xi}] &= aU^2\begin{bmatrix} 0 & 0 & -K_{p0} \\ 0 & 0 & K_{r0} \\ 0 & 0 & -CK_{m0} \end{bmatrix} \end{aligned} \tag{5.115d}$$

In (5.115d) we should remember:

- C_{r0}, C_{p0} and C_{m0} (5.114b) are, respectively, the coefficients of drag, lift and torque of the profile calculated in the position of static equilibrium, i.e. for $\alpha = \alpha_o$;

- K_{r0}, K_{p0} and K_{m0} (5.114b) are the respective derivatives with respect to the angle of attack calculated, again, around the position of static equilibrium;
- b_1 is the reference distance for calculating the relative velocity of the flow with respect to the profile;
- C is the quantity of reference of the profile (the chord in the case of the aerofoil).

As we know, the generalised forces must be estimated by expressing the virtual work performed by the real forces $\underline{F}_\xi$ as a function of the independent variables, in this case the displacements $\underline{x}_i$ of the nodes of the finite element model. We can then calculate the virtual work $\delta^* L_\xi$ of the aerodynamic forces acting on a generic infinitesimal element of a beam

$$\delta^* L_\xi = \{\underline{F}_\xi d\xi\}^T \delta^* \underline{x}_\xi = -\left\{\underline{x}_\xi^T [K_{A\xi}]^T d\xi + \underline{\dot{x}}_\xi^T [R_{A\xi}]^T d\xi\right\} \delta^* \underline{x}_\xi \tag{5.116a}$$

The work relative to the generic section of the wing must be then integrated along the length l_j of the jth finite element:

$$\delta^* L_j = \int_0^{l_j} \{\underline{F}_\xi d\xi\}^T \delta^* \underline{x}_\xi = \left\{\int_0^{l_j} \left\{-\underline{x}_\xi^T [K_{A\xi}]^T d\xi - \underline{\dot{x}}_\xi^T [R_{A\xi}]^T d\xi\right\}\right\} \delta^* \underline{x}_\xi \tag{5.116b}$$

This virtual work $\delta^* L_j$ must be estimated, subsequently, as a function of the system's free coordinates, i.e. of the nodal displacements of the model $\underline{x}$. The quantities contained in the vector $\underline{x}_\xi$ are linked to the nodal coordinates relating to the generic jth finite element $\underline{x}_j$, by shape functions, previously defined in the chapter regarding the finite element method (Chap. 4, Sect. 4.3):

$$\underline{x}_\xi = \begin{Bmatrix} w(\xi) \\ v(\xi) \\ \theta_x(\xi) \end{Bmatrix} = \begin{Bmatrix} \underline{f}_{w_j}^T(\xi) \\ \underline{f}_{v_j}^T(\xi) \\ \underline{f}_{\theta_{xj}}^T(\xi) \end{Bmatrix} \underline{x}_j = \left[\underline{f}_j(\xi)\right]^T \underline{x} \tag{5.116c}$$

$\left[\underline{f}_j(\xi)\right]$ being the matrix of the functions of form which, in this case, has three lines (as many as the variables $\underline{x}_\xi$ of the section affected by the aerodynamic forces) and twelve columns (as many as the degree-of-freedom $\underline{x}_j$ associated with the end nodes of the generic beam element):

$$\left[\underline{f}_j(\xi)\right]^T = \begin{Bmatrix} \underline{f}_{w_j}^T(\xi) \\ \underline{f}_{v_j}^T(\xi) \\ \underline{f}_{\theta_{xj}}^T(\xi) \end{Bmatrix} \tag{5.117a}$$

being:

$$\underline{f}_{\theta_{xj}}(\xi) = \begin{Bmatrix} 0 \\ 0 \\ 0 \\ (1-\xi/l_j) \\ 0 \\ 0 \\ 0 \\ 0 \\ 0 \\ (\xi/l_j) \\ 0 \\ 0 \end{Bmatrix}; \quad \underline{f}_{w_j}(\xi) = \begin{Bmatrix} 0 \\ 0 \\ \left(\frac{2}{l_j^3}\xi^3 - \frac{3}{l_j^2}\xi^2 + 1\right) \\ 0 \\ \left(-\frac{1}{l_j^2}\xi^3 + \frac{2}{l_j}\xi^2 - \xi\right) \\ 0 \\ 0 \\ 0 \\ \left(-\frac{2}{l_j^3}\xi^3 + \frac{3}{l_j^2}\xi^2\right) \\ 0 \\ \left(\frac{1}{l_j^2}\xi^3 - \frac{1}{l_j}\xi^2\right) \\ 0 \end{Bmatrix}; \tag{5.117b}$$

$$\underline{f}_{v_j}(\xi) = \begin{Bmatrix} 0 \\ \left(\frac{2}{l_j^3}\xi^3 - \frac{3}{l_j^2}\xi^2 + 1\right) \\ 0 \\ 0 \\ 0 \\ \left(\frac{1}{l_j^2}\xi^3 - \frac{2}{l_j}\xi^2 + \xi\right) \\ 0 \\ \left(-\frac{2}{l_j^3}\xi^3 + \frac{3}{l_j^2}\xi^2\right) \\ 0 \\ 0 \\ 0 \\ \left(-\frac{1}{l_j^2}\xi^3 + \frac{1}{l_j}\xi^2\right) \end{Bmatrix}$$

We will now calculate the virtual work $\delta^* L_j$ of the aerodynamic forces as a function of the independent variables, by introducing the link (5.116c):

$$\delta^* L_j = \left\{ \int_0^{l_j} \left\{ -\underline{x}_j^T [f_j(\xi)] [K_{A\xi}]^T - \underline{\dot{x}}_j^T [f_j(\xi)] [R_{A\xi}]^T d\xi \right\} \right\} [f_j(\xi)]^T \delta^* \underline{x}_j = \underline{Q}_{xj}^T \delta^* \underline{x}_j \tag{5.118a}$$

$\underline{Q}_{xj}$ being the vector of the linearised generalised aerodynamic forces acting on the jth finite element:

$$
\begin{aligned}
\underline{Q}_{xj} &= \int_0^{l_j} \left\{ -[f_j(\xi)][K_{A\xi}][f_j(\xi)]^T \underline{x}_j - [f_j(\xi)][R_{A\xi}][f_j(\xi)]^T \dot{\underline{x}}_j \right\} d\xi \\
&= \left[\int_0^{l_j} [f_j(\xi)][K_{A\xi}][f_j(\xi)]^T d\xi \right] \underline{x}_j - \left[\int_0^{l_j} [f_j(\xi)][R_{A\xi}][f_j(\xi)]^T d\xi \right] \dot{\underline{x}}_j \\
&= -[K_{Aj}]\underline{x}_j - [R_{Aj}]\dot{\underline{x}}_j
\end{aligned}
\tag{5.118b}
$$

In (5.118b) $[K_{Aj}]$ and $[R_{Aj}]$ are two equivalent matrices (of stiffness and of damping) of the aerodynamic force field relating to the single jth finite element:

$$
\begin{aligned}
[K_{Aj}] &= \left[\int_0^{l_j} [f_j(\xi)][K_{A\xi}][f_j(\xi)]^T d\xi \right] \\
[R_{Aj}] &= \left[\int_0^{l_j} [f_j(\xi)][R_{A\xi}][f_j(\xi)]^T d\xi \right]
\end{aligned}
\tag{5.118c}
$$

By assembling the single terms $\underline{Q}_{xj}$, we obtain the vector $\underline{Q}_x$ of the Lagrangian components of the aerodynamic forces, according to all the degree-of-freedom of the structure:

$$
\underline{Q}_x = -[K_F]\underline{x} - [R_F]\dot{\underline{x}}
\tag{5.118d}
$$

The matrices $[K_F]$ and $[R_F]$ represent the overall equivalent matrices of stiffness and damping, obtained by assembling the single equivalent matrices of stiffness $[K_{Aj}]$ and damping $[R_{Aj}]$ of the single finite elements, with a process that is very similar to that used for structural matrices (Fig. 5.35). By bringing $\underline{Q}_x$ to the first member of the Eq. (5.111), we obtain the equation of motion with the linearised forces:

$$
[M_s]\ddot{\underline{x}} + [[R_s] + [R_F(U)]]\dot{\underline{x}} + [[K_s] + [K_F(U)]]\underline{x} = \underline{0}
\tag{5.119}
$$

Thus we have reduced the problem of a continuous system, subject to a non-conservative force field, to a discrete system with n-degree-of-freedom, linearised, where we can apply the methods already illustrated in the previous section to study its stability. The matrices of the linearised system are a function of the force field and, in particular, in the case of the wing, they are a function of the velocity U of the incident flow (5.119). By calculating the eigenvalues:

$$\lambda_i = \lambda_i(U) = \alpha_i + i\omega_i = \alpha_i(U) + i\omega_i(U) \tag{5.119a}$$

as the velocity U of the flow varies, it is possible to calculate the threshold value (critical velocity U_{crit}) above which instability will start to occur:

$$\alpha_i(U_{crit}) > 0 \tag{5.119b}$$

In practical cases this kind of approach may present problems of a numerical nature, if the number of degree-of-freedom is high: in this case, the modal approach, reducing analysis to just the modes affected by the phenomenon of instability, may be advantageous compared to that of finite elements; it is, however, accompanied by the drawback of having to choose, in advance, the vibration modes affected by instability.

5.4.2.2 Modal Approach

Now we will deal with the same problem, concerning the wing hit by a confined flow (Fig. 5.32), using a modal approach where the main advantages lie in the possibility of reducing the number of degree-of-freedom and in the possibility of inserting damping in a rigorous way (Sect. 2.8). Assuming that we will again schematise the structure with a finite element model (Fig. 5.33), the equations of motion of the wing, with the presence of aerodynamic actions, are, once again, (5.111):

$$[M_s]\underline{\ddot{x}} + [R_s]\underline{\dot{x}} + [K_s]\underline{x} = \underline{Q}_x(\underline{x}, \underline{\dot{x}}) \tag{5.120}$$

If we wish to adopt a modal approach, we must define the vibration modes $\underline{X}^{(i)}$ of the structure, with the absence of damping and applied external forces, estimated by solving the following equation with the usual techniques (Sect. 2.3 and Chap. 4, Sect. 4.5):

$$[M_s]\underline{\ddot{x}} + [K_s]\underline{x} = \underline{0} \tag{5.121a}$$

By organising the single eigenvectors defined in this way into the modal matrix $[\Phi]$, taking into account only the first p[15] modes relating to the range of frequencies considered:

$$[\Phi] = \left[\underline{X}^{(1)}\underline{X}^{(2)}\ldots\underline{X}^{(i)}\ldots\underline{X}^{(p)}\right] \tag{5.121a}$$

and performing the transformation of coordinates:

[15]Usually, in these problems of instability, the lowest modes are those that are involved.

$$\underline{x} = [\Phi]\underline{q} \tag{5.121b}$$

it is possible to rewrite (5.120) in terms of principal coordinates:

$$[m]\underline{\ddot{q}} + [r]\underline{\dot{q}} + [k]\underline{q} = [\Phi]^T \underline{Q}_x(\underline{x}, \underline{\dot{x}}) \tag{5.122}$$

In (5.122), $[m]$, $[r]$ and $[k]$ re used to indicate the matrices, respectively, of mass, damping and stiffness in principal coordinates (we should remember that they are diagonal):

$$[m] = [\Phi]^T[M_s][\Phi]; \quad [r] = [\Phi]^T[R_s][\Phi]; \quad [k] = [\Phi]^T[K_s][\Phi] \tag{5.122a}$$

The term $[\Phi]^T \underline{Q}_x(\underline{x}, \underline{\dot{x}})$ of (5.122), taking into account the transformation of coordinates (5.121b) and considering the linearised terms of the aerodynamic forces (5.118d), becomes:

$$\begin{aligned} [\Phi]^T \underline{Q}_x(\underline{x}, \underline{\dot{x}}) &= -[\Phi]^T[K_F]\underline{x} - [\Phi]^T[R_F]\underline{\dot{x}} \\ &= -[\Phi]^T[K_F][\Phi]\underline{q} - [\Phi]^T[R_F][\Phi]\underline{\dot{q}} = -[k_F]\underline{q} - [r_F]\underline{\dot{q}} \end{aligned} \tag{5.122b}$$

$[k_F]$ and $[r_F]$ being full matrices (pxp) that define, in principal coordinates, the linearised generalised aerodynamic forces:

$$\begin{aligned} [k_F] &= [k_F(U)] = [\Phi]^T[K_F(U)][\Phi] \\ [r_F] &= [r_F(U)] = [\Phi]^T[R_F(U)][\Phi] \end{aligned} \tag{5.122c}$$

and having used $\underline{q}$ to indicate the vector containing the principal coordinates that represent the perturbed motion around the position of static equilibrium. The linearised equations of motion (5.122) thus become:

$$[m]\underline{\ddot{q}} + [[r] + [r_F]]\underline{\dot{q}} + [[k] + [k_F]]\underline{q} = \underline{0} \tag{5.122d}$$

In this equation it is again possible to see how the aerodynamic forces couple the vibration modes using the full matrices equivalent to the field $[k_F]$ and $[r_F]$. The modal method can be applied once the vibration modes are known. These modes can be obtained:

- with a finite elements schematisation (the approach described above);
- using the equations of continuos systems (Chap. 3);
- from experimental testing on the structure, using techniques of modal identification (Chap. 8).

By using the last two approaches, we have a continuous or discrete function that defines the generic ith vibration mode $\underline{\Phi}^{(i)}(\zeta)$ of the wing in question:

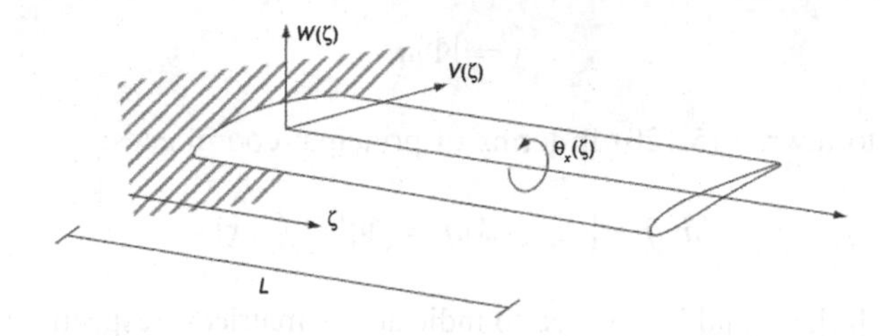

Fig. 5.37 Displacements $\underline{x}_\zeta$ of the generic section of the wing

$$\underline{\Phi}^{(i)}(\zeta) = \begin{Bmatrix} \Phi_w^{(i)}(\zeta) \\ \Phi_v^{(i)}(\zeta) \\ \Phi_{\theta_x}^{(i)}(\zeta) \end{Bmatrix} \tag{5.123a}$$

where $\Phi_w^{(i)}(\zeta)$, $\Phi_v^{(i)}(\zeta)$ and $\Phi_{\theta_x}^{(i)}(\zeta)$ are the components of the mode analysed according to the vertical, lateral and rotational directions. It is possible then, in this case too, to use a modal approach, without using a finite elements schematisation and thus arrive at an equation of motion similar to (5.122):

$$[m]\underline{\ddot{q}} + [r]\underline{\dot{q}} + [k]\underline{q} = \underline{Q}_q \tag{5.123b}$$

where $\underline{Q}_q$ is the vector of the generalised aerodynamic forces.

The approach to define this vector $\underline{Q}_q$ is conceptually similar to that seen previously: however, in this case the displacements $\underline{x}_\zeta$ sustained by the generic section of the wing (Fig. 5.37) are linked to the generalised coordinates $\underline{q}$ via a matrix $[\Phi(\zeta)]$ of order *3xp* (*p* being the vibration modes considered):

$$\begin{aligned} \underline{x}_\zeta = \begin{Bmatrix} w(\zeta) \\ v(\zeta) \\ \theta_x(\zeta) \end{Bmatrix} &= \left[\underline{\Phi}^{(1)}(\zeta)\underline{\Phi}^{(2)}(\zeta)\ldots\underline{\Phi}^{(i)}(\zeta)\ldots\underline{\Phi}^{(p)}(\zeta)\right]\underline{q} \\ &= \begin{bmatrix} \Phi_w^{(1)}(\zeta)\Phi_w^{(2)}(\zeta)\ldots\Phi_w^{(i)}(\zeta)\ldots\Phi_w^{(p)}(\zeta) \\ \Phi_v^{(1)}(\zeta)\Phi_v^{(2)}(\zeta)\ldots\Phi_v^{(i)}(\zeta)\ldots\Phi_v^{(p)}(\zeta) \\ \Phi_{\theta_x}^{(1)}(\zeta)\Phi_{\theta_x}^{(2)}(\zeta)\ldots{}_{\theta_x}^{(i)}(\zeta)\ldots\Phi_{\theta_x}^{(p)}(\zeta) \end{bmatrix}\underline{q} = [\Phi(\zeta)]\underline{q} \end{aligned} \tag{5.123c}$$

which contain the lateral, vertical and rotational components of the *p* modes considered. The virtual work of the aerodynamic forces acting on the generic section of coordinate ζ and on an infinitesimal portion $d\zeta$, taking into account (5.123a) and (5.123), thus equals:

$$\delta^* L_\zeta = \underline{F}_\zeta^T \cdot d\zeta \cdot \delta^* \underline{x}_\zeta \tag{5.124}$$

as, similar to what we have seen previously, the aerodynamic force $\underline{F}_\zeta$ per unit of length is given by:

$$\underline{F}_\zeta = \begin{Bmatrix} F_w \\ F_v \\ M \end{Bmatrix} = -[K_{A\zeta}]\underline{x}_\zeta - [R_{A\zeta}]\underline{\dot{x}}_\zeta \tag{5.124a}$$

The work relating to the generic section must then be integrated along the length L of the wing:

$$\delta^* L = \int_0^L \{\underline{F}_\zeta d\zeta\}^T \delta^* \underline{x}_\zeta = \int_0^L \left\{-\underline{x}_\zeta^T [K_{A\zeta}]^T - \underline{\dot{x}}_\zeta^T [R_{A\zeta}]^T\right\} d\zeta \delta^* \underline{x}_\zeta \tag{5.124b}$$

Taking into account the transformation of coordinates (5.123a), (5.124b) can be rewritten as:

$$\begin{aligned}
\delta^* L &= \left\{ \int_0^L \left\{-\underline{q}^T [\phi(\zeta)]^T [K_{A\zeta}]^T - \underline{\dot{q}}^T [\phi(\zeta)]^T [R_{A\zeta}]^T\right\} d\zeta [\phi(\zeta)] \right\} \delta^* \underline{q} \\
&= -\underline{q}^T \left[\int_0^L [\phi(\zeta)]^T [K_{A\zeta}]^T [\phi(\zeta)] d\zeta \right] \delta^* \underline{q} - \underline{\dot{q}}^T \left[\int_0^L [\phi(\zeta)]^T [R_{A\zeta}]^T [\phi(\zeta)] d\zeta \right] \delta^* \underline{q} \\
&= \underline{Q}_q^T \delta^* \underline{q}
\end{aligned} \tag{5.124c}$$

$\underline{Q}_q$ being the vector of the linearised generalised aero elastic forces, expressed in principal coordinates:

$$\begin{aligned}
\underline{Q}_q &= -\left[\int_0^L [\phi(\zeta)]^T [K_{A\zeta}]^T [\phi(\zeta)] d\zeta \right] \underline{q} - \left[\int_0^L [\phi(\zeta)]^T [R_{A\zeta}]^T [\phi(\zeta)] d\zeta \right] \underline{\dot{q}} \\
&= -[k_F]\underline{q} - [r_F]\underline{\dot{q}}
\end{aligned} \tag{5.124d}$$

In (5.124d) the matrices $[k_F]$ and $[r_F]$ are the equivalent aerodynamic matrices of stiffness and damping of the force field, linearised and expressed in modal coordinates:

$$[k_F] = \left[\int_0^L [\phi(\zeta)]^T [K_{A\zeta}]^T [\phi(\zeta)] d\zeta \right]$$
$$[r_F] = \left[\int_0^L [\phi(\zeta)]^T [R_{A\zeta}]^T [\phi(\zeta)] d\zeta \right] \tag{5.124e}$$

By bringing the linearised forcing terms to the first member of the Eq. (5.124), we obtain:

$$[m]\underline{\ddot{q}} + [[r] + [r_F]]\underline{\dot{q}} + [[k] + [k_F]]\underline{q} = \underline{0} \tag{5.125}$$

similar to that obtained in (5.119).

5.5 Systems Perturbed Around the Steady-State Position

We will now address the study of stability in systems, again, placed in force fields, perturbed around a steady-state motion: in some ways this study is addressed in a similar way to what we have already seen for systems perturbed in the position of rest.

5.5.1 Systems with 1 d.o.f

For simplicity's sake we will be considering a dynamic one-degree-of-freedom system subject to forces that are generically functions of displacement and velocity, as well as time:

$$m\ddot{x} = F(x, \dot{x}, t) \tag{5.126}$$

To study the stability of the system, first we must find, if it exists, the steady-state solution $x_r(t)$ of the equation of motion (5.126)[16]:

$$m\ddot{x}_r = F(x_r, \dot{x}_r, t) \tag{5.127}$$

In particular $x_r(t)$ may be of the form $x_r = x_{ro}$, $\dot{x}_r = 0$, i.e. a position of rest, or $\dot{x}_r = V_o$ constant (absolute steady-state), i.e. a condition of uniform motion with constant speed, or even a periodic solution (periodic steady-state). Similarly to what

[16]The case of a system perturbed around the position of rest has already been considered in the previous Sect. 5.4.

was done for studying stability, we can now define the perturbed motion x of the system around the steady-state solution x_r as:

$$\bar{x} = x - x_r; \quad \bar{\dot{x}} = \dot{x} - \dot{x}_r; \quad \bar{\ddot{x}} = \ddot{x} - \ddot{x}_r \tag{5.128}$$

At this stage, if we wish to analyse the small oscillations around the steady-state solution, we must develop the force $F = F(x_r, \dot{x}_r, t)$ in a Taylor series, around the steady-state solution x_r, by calculating the derivatives of F with respect to the independent variables:

$$F \approx F(x_r, \dot{x}_r, t) + \left(\frac{\partial F(x, \dot{x}, t)}{\partial x}\right)_{x_r, \dot{x}_r}(x - x_r) + \left(\frac{\partial F(x, \dot{x}, t)}{\partial \dot{x}}\right)_{x_r, \dot{x}_r}(\dot{x} - \dot{x}_r) \tag{5.129}$$

where the derivatives must be calculated in correspondence to the steady-state solution x_r. Remembering the definition given of perturbed motion (5.128) and the development of the force (5.129), the equation of motion (5.126) is rewritten as:

$$m(\bar{\ddot{x}} + \ddot{x}_r) = F(x_r, \dot{x}_r, t) + \left(\frac{\partial F(x, \dot{x}, t)}{\partial x}\right)_{x_r, \dot{x}_r}(x - x_r) + \left(\frac{\partial F(x, \dot{x}, t)}{\partial \dot{x}}\right)_{x_r, \dot{x}_r}(\dot{x} - \dot{x}_r) \tag{5.130}$$

Assuming:

$$K_x = -\left(\frac{\partial F(x, \dot{x}, t)}{\partial x}\right)_{x_r, \dot{x}_r} \qquad R_V = -\left(\frac{\partial F(x, \dot{x}, t)}{\partial \dot{x}}\right)_{x_r, \dot{x}_r} \tag{5.131a}$$

and taking into account the fact that $x_r(t)$ is the solution of the equation of motion (5.127), we obtain, after appropriate simplification, an equation in the variable x:

$$m\bar{\ddot{x}} + R_v\bar{\dot{x}} + K_x\bar{x} = 0 \tag{5.131b}$$

If the derivatives K_x and R_v are constants, we find ourselves with a differential equation with constant coefficients that we can study with the methods already seen for systems perturbed around the state of rest. If the derivatives are explicit functions of time and the solution is periodic, then K_x and R_v can be periodic: in this case the resulting equation is linear, but with periodically variable coefficients. In the specific case where these coefficients vary harmonically, (5.131b) can be brought back, see Sect. 1.5, to a Mathieu equation: analysis of the stability, in this case, can be done with specific methods for the type of equation being considered, for example the perturbation method.

5.5.1.1 An Example: The Stability of the Torsional Motion of a Rotor

We will consider, as the first example (Fig. 5.38), a motor and a user coupled on the same shaft with moment of inertia J: the drive torque M_m and resistant torque M_r are known and we assume they are functions of the angular speed $\dot{\theta}$ only (Fig. 5.39). The system's equation of motion, with rotation θ assumed to be the free coordinate, is the following:

$$J\ddot{\theta} = M_m\left(\dot{\theta}\right) - M_r\left(\dot{\theta}\right) \tag{5.132}$$

Firstly, we must find the steady-state solutions, which can be sought as the intersection of the characteristic curve of the motor $M_m = M_m\left(\dot{\theta}\right)$ with that of the user $M_r = M_r\left(\dot{\theta}\right)$ (Fig. 5.39). In the particular case being analysed there are two intersections, that is, two possible steady-state solutions, respectively, for angular speeds ω_1 and ω_2. For both the solutions, the steady-state velocity $\dot{\theta}_r = \omega_r$ satisfies the equation:

$$M_m(\omega_r) - M_r(\omega_r) = 0; \quad \ddot{\theta} = 0 \tag{5.133a}$$

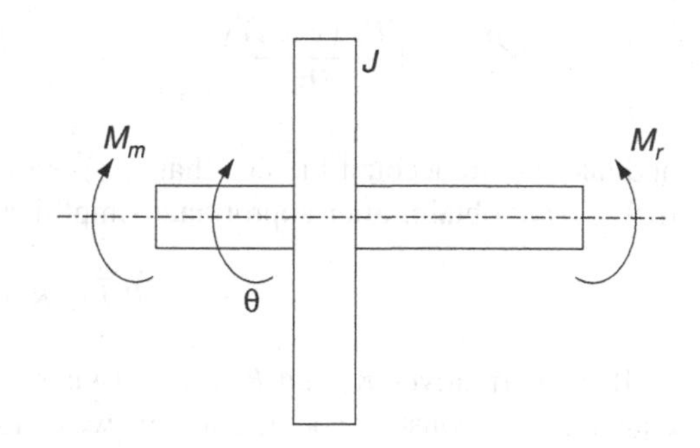

Fig. 5.38 Schematisation of a.c. motor and user

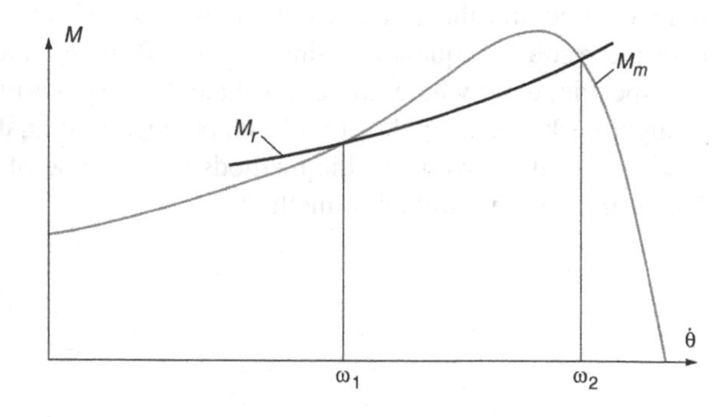

Fig. 5.39 Characteristic curves of a.c. motor and user

So we have a case of absolute steady-state motion at a constant speed. For the sake of convenience we can define the values of the drive torque and the resistance torque under steady-state conditions as:

$$\overline{M}_m = M_m(\omega_r); \quad \overline{M}_r = M_r(\omega_r) \tag{5.133b}$$

We will now analyse the perturbed motion, around the steady-state solution, defined by the variable $\overline{\theta}$, together with its derivatives:

$$\overline{\theta} = \theta - \theta_r; \quad \overline{\dot{\theta}} = \dot{\theta} - \dot{\theta}_r; \quad \overline{\ddot{\theta}} = \ddot{\theta} - \ddot{\theta}_r \tag{5.134}$$

By developing the non-linear terms of the equation of motion (5.132), around the generic steady-state situation ($\dot{\theta}_r = \omega$, $\ddot{\theta}_r = 0$, with r = 1, 2), corresponding to motion with constant angular speed up to the linear terms only of the development, we have:

$$J\overline{\ddot{\theta}} = \overline{M}_m + \left(\frac{\partial M_m}{\partial \dot{\theta}}\right)_{\dot{\theta}=\omega_r} \left(\dot{\theta} - \dot{\theta}_r\right) - \overline{M}_r - \left(\frac{\partial M_r}{\partial \dot{\theta}}\right)_{\dot{\theta}=\omega_r} \left(\dot{\theta} - \dot{\theta}_r\right) \tag{5.135}$$

Taking into account the steady-state solution (5.133a) and the definition of perturbed motion (5.134), Eq. (5.135) of the perturbed motion around the steady-state solution can be rewritten, with suitable simplifications:

$$J\overline{\ddot{\theta}} + \left[-\left(\frac{\partial M_m}{\partial \dot{\theta}}\right)_{\dot{\theta}=\omega_r} + \left(\frac{\partial M_r}{\partial \dot{\theta}}\right)_{\dot{\theta}=\omega_r}\right]\overline{\dot{\theta}} = 0 \tag{5.135a}$$

Using K_m and K_r to define the derivatives of the engine and resistance torque with respect to the angular speed, estimated in the steady-state position:

$$K_m = \left(\frac{\partial M_m}{\partial \dot{\theta}}\right)_{\dot{\theta}=\omega_r}; \quad K_r = \left(\frac{\partial M_r}{\partial \dot{\theta}}\right)_{\dot{\theta}=\omega_r} \tag{5.135b}$$

and assuming, as a variable, the angular speed $\overline{\omega}$:

$$\overline{\omega} = \overline{\dot{\theta}} \tag{5.135c}$$

we have:

$$J\overline{\ddot{\theta}} + [K_r - K_m]\overline{\dot{\theta}} = 0 \tag{5.136}$$

Equation (5.136) is a linear equation with constant coefficients, of the first order in the variable $\overline{\omega}$. Replacing a solution of the form:

$$\overline{\omega} = \overline{\omega}_o e^{\lambda t} \tag{5.137a}$$

in the Eq. (5.136), we obtain:

$$(J\lambda + [K_r - K_m])\overline{\omega_o} e^{\lambda t} = 0 \tag{5.137b}$$

and, by solving the characteristic equation:

$$(J\lambda + [K_r - K_m]) = 0 \tag{5.137c}$$

we obtain the real value of λ:

$$\lambda = \frac{[K_m - K_r]}{J} \tag{5.137d}$$

which corresponds to:

- an exponentially decreasing solution (corresponding to stable motion) for $\lambda < 0$;
- an exponentially increasing solution (corresponding to unstable motion) for $\lambda > 0$.

In this case the stability of the system is dictated by the sign of the difference $K_m - K_r$, J always being positive. In particular, upon examining the characteristic curves we can see that in correspondence to the solution $\dot{\theta}_r = \omega_2$, K_m being negative and K_r positive, the system is definitely stable. Physically, this corresponds to the fact that, for example, a perturbation of the positive angular speed $+\Delta\omega$, leads to a reduction in the drive torque and an increase in the resistance torque: under this condition the system has a tendency to self-adjust. Vice versa, in correspondence to the solution $\dot{\theta}_r = \omega_1$, the signs of the derivatives K_r and K_m are both positive, but with $K_m > K_r$. Bearing in mind (5.122d), the coefficient $K_m - K_r$ is positive: therefore, the steady-state situation is unstable. In fact, an increase in angular speed $+\Delta\omega$ corresponds to an increase in the drive torque C_m and a lesser increase in that of resistance C_r: the angular speed, therefore, tends to increase.

5.5.2 Systems with 2 d.o.f

We will now examine systems with two degree-of-freedom, again perturbed around the steady-state position: more specifically, we will examine:

- the lateral vibrations and the yaw rotation of a railway wheelset, perturbed around the configuration of straight running at constant velocity (Sect. 5.5.2.2);
- the stability of a road vehicle driving in a straight line at constant velocity (Sect. 5.5.2.3).

5.5.2.1 Introduction to Studying the Dynamics of a Land Vehicle

Both railway and road vehicles are important examples of mechanical systems surrounded by a force field: in these cases the force field is mainly due to actions that are exerted in the contact zone between wheel and runway. Analysing the stability of a railway vehicle, either running in a straight line or through a bend, along with the similar problem for a road vehicle, offers an important analysis of the stability of a system perturbed around a steady-state solution. In this case, the steady-state (or particular integral) is defined by the rectilinear motion of the vehicle or by the motion on a curvilinear trajectory with the known law of motion, these being the possible particular solutions of the problem. The discussion, for simplicity's sake, will be limited to the case study of perturbed motion around the configuration assumed by the system running at a steady-state and in a straight line. We will be considering models that have been greatly simplified yet still reproduce, at least qualitatively, the behaviour of the real system and the problems linked to it. Here we will only touch briefly on more complex models while referring the reader to specialist publications for further details (see bibliography). The problems present in a railway vehicle and in a road vehicle are basically similar, with some obvious differences:

- in a road vehicle, the definition of its trajectory is controlled by the driver, whereas in a railway vehicle it is defined by the track layout;
- the problems for defining the contact forces generated between tyre and road and wheel and track are also different.

Although it is certainly not necessary to provide a description of a road vehicle (Fig. 5.40), as it is widely recognised by everyone, it may be useful to briefly recall the main morphological characteristics of a railway vehicle. Rolling stock, in its standard configuration (Fig. 5.41), is made up of an carbody, two or more bogies and their wheelsets (meaning the axis and pair of wheels):

- the wheelsets can be rigid or with independent wheels (for example, for ease of control on bends with low radii of curvature);

Fig. 5.40 A road vehicle

Fig. 5.41 A railway vehicle

- the wheelset is connected to the bogie via the so-called vertical, transverse and longitudinal primary suspensions (created physically with elastic elements like helical springs and deformable connections like, for example, in the case of the bushings between wheelset and bogie);
- the bogies may be self-steering.
- vertical, transversal and longitudinal secondary suspensions, created physically with helical springs and torsional return bars, are positioned between carbody and bogie;
- in the secondary stage of suspension viscous or friction dampers are also considered.

A correct and complete analysis of the behaviour of a rail vehicle must take into account the mutual interaction between the vehicle and the infrastructure, that is, the connection that constrains the rail to the outside world: in fact, where rail vehicles are concerned, we cannot ignore, if not in a first approximation, the dynamic effect of the rail which can be schematised as a beam constrained with elastic elements to the crossbeams which, in turn, usually rest on a ballast, to filter the forces transmitted to the ground and limit noise emissions.

For road vehicles, however, the road can be considered rigid, given the relationship between the stiffness of the tyre and that of the actual runway. Unlike the rail vehicle, the road vehicle has just one stage of suspensions between chassis and wheel (Fig. 5.40), but the deformability of the tyre is far higher than that of the wheel on a wheelset. In the road vehicle, assuming that the road is horizontal, the contact normal always runs in a vertical direction (Fig. 5.42), while the contact normal in the case of wheel-rail, given the particular morphology of the rail and the wheel rim (Fig. 5.43a), changes as the relative wheelset-track position varies. When the wheelset is centred, the contact is on a truncated conical area of the wheel-rim (known as the rolling surface), with low conicity γ ($\gamma = 0.05 - 0.1$ rad); as the relative displacement Y_{rel} between wheelset and track increases (Fig. 5.43b) the contact normal of the wheel whose edge approaches the track rises up to a value of

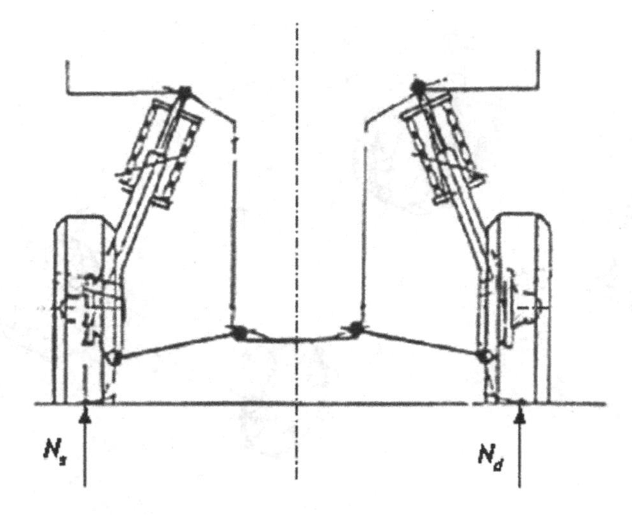

Fig. 5.42 Road vehicle: direction of the contact normals

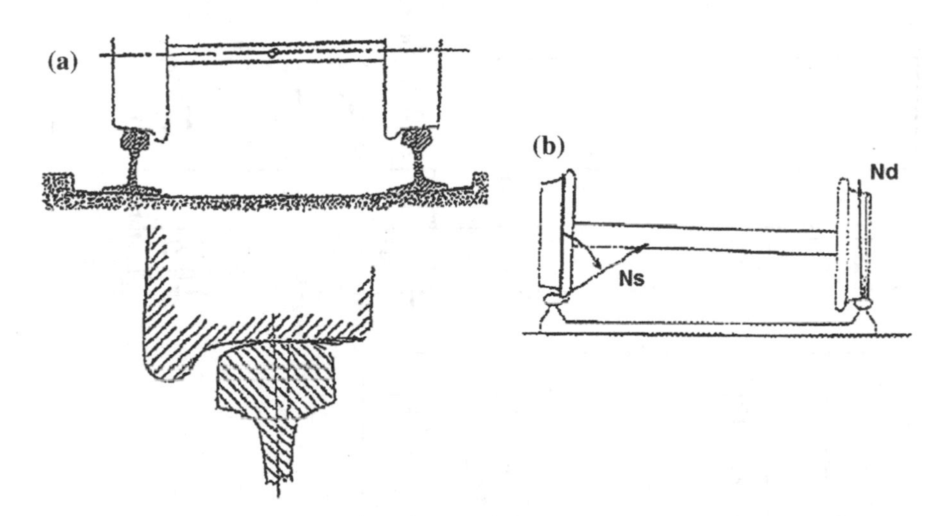

Fig. 5.43 Railway vehicle. **a** Geometry of the wheel-rail contact; **b** direction of the contact normal in the case of wheel-rail

$\gamma_{max} = 60^{\circ} - 70^{\circ}$, while the contact normal on the other wheel remains more or less constant. Therefore, the particular morphology of the wheel rim and the rail (Fig. 5.43) introduces two conditions of constraint as the contact normal component of the relative velocity between each wheel and the corresponding rail must be zero: these constraints reduce the degree-of-freedom of the actual wheelset.

In order to reproduce the behaviour of the real system (Figs. 5.40 and 5.41), we must correctly schematise the vehicle and the actions of contact so as to define suitable mathematical models. These model (Figs. 5.44 and 5.45) are made up of a

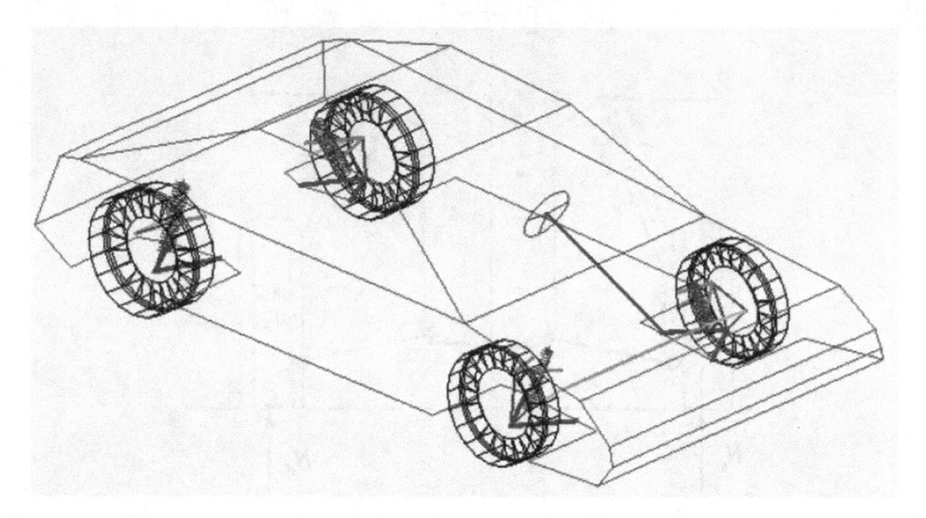

Fig. 5.44 ADAMS model of a road vehicle

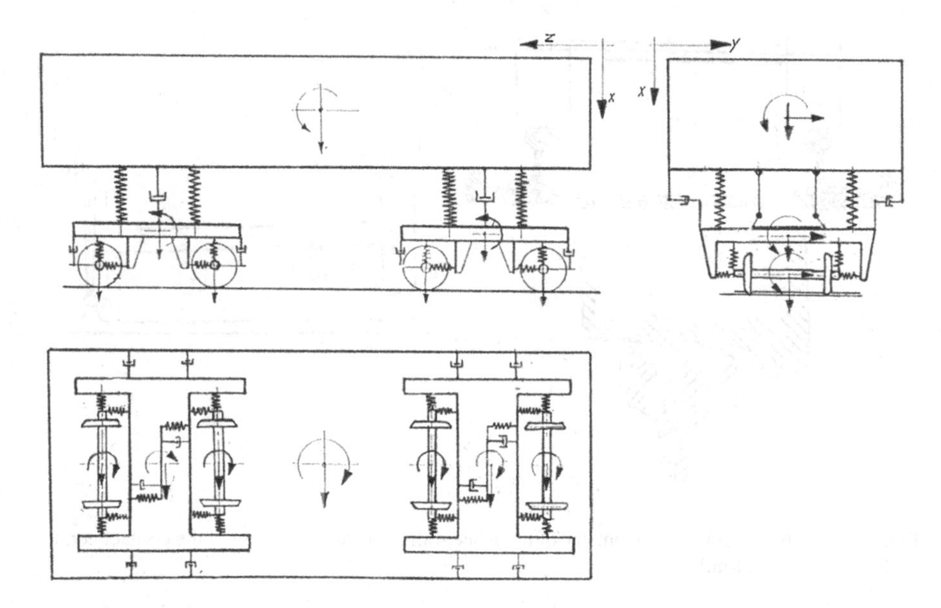

Fig. 5.45 Model of a railway vehicle

set of rigid (or deformable) bodies, interconnected with elastic and damping elements, both linear and non-linear: external forces such as aerodynamics and forces due to wheel-pathway contact act on the vehicle. Figure 5.44 shows a possible 14 degree-of-freedom model for the road vehicle (6 for the chassis, 4 associated with the suspension wishbones, 4 associated with the rolling of the wheels). A classic model adopted for the rail vehicle is of the type shown in Fig. 5.45: in particular, it is a model with 23 degree-of-freedom for studying the lateral and vertical dynamics of the rolling stock. When making use of models like this, which often involve a high number of degree-of-freedom, studying the dynamic behaviour of the vehicle becomes rather complicated and requires the use of specific algorithms for writing the equations of motion (*Multi-body methods*, [5, 7, 8, 56–84, 86–96, 98–109] and appropriate numerical methods for integrating the actual equations of motion [2, 5, 56, 64, 69–71].

To define the dynamic behaviour of rail and road vehicles we need to analyse what happens during wheel-runway contact, that is, analyse the quantities that influence the actions exchanged in the contact zone:

- in Sect. 5.5.2.1.1 below, we will go over the basic concepts that make it possible to define the forces exchanged between wheel and pathway and, more specifically, we will analyse the problems linked to the forces exchanged between wheel and rail (section "The Contact Forces Between Wheel and Rail") and between tyre and road (section "The Contact Forces Between Tyre and Road");
- in the sections immediately below we will use these notions to analyse the dynamics of the wheelset (Sect. 5.5.2.2) of the road vehicle (Sect. 5.5.2.3), while studying, in particular and mainly for didactic purposes, the stability of the rectilinear steady-state motion.

5.5.2.1.1 The Contact Forces Between Wheel and Pathway

To introduce the main problems involved when defining the contact forces exchanged between wheel and pathway, we must consider, for example, the motion of a drive wheel which moves forward by rolling along a flat surface: in the traditional approach of Applied Mechanics [1, 6] the product of the normal action N by the coefficient of static friction f_s is greater than the tangential action T, this motion is schematised as pure rolling of the wheel (Fig. 5.46a).

With this view, the wheel is considered to be rigid and the contact between wheel and plane is punctiform or linear. The contact point C has velocity $\mathbf{V_c}$ equal

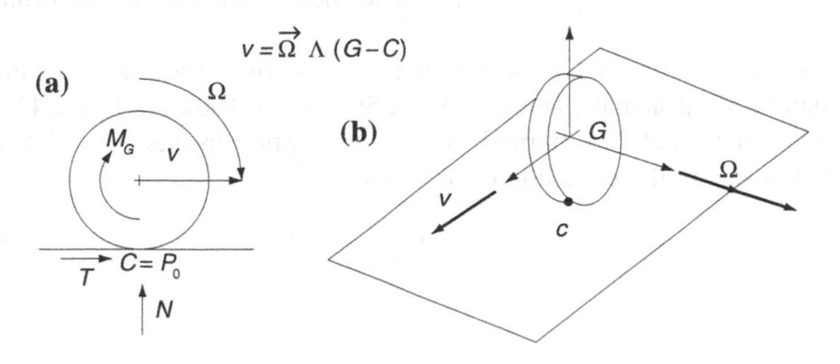

Fig. 5.46 Pure rolling hypothesis

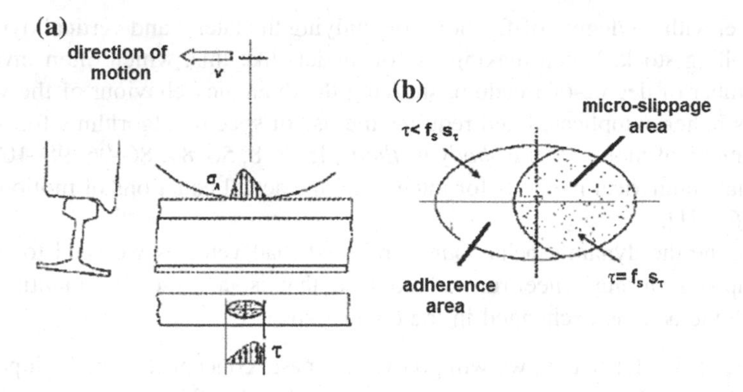

Fig. 5.47 Railway wheel: **a** contact patch, **b** distribution of normal σ_x and shear τ stresses

to zero, so it becomes the centre of instantaneous rotation P_o. In this case, the tie between the forward velocity $\mathbf{V}$ of the wheel centre G and the angular velocity $\mathbf{\Omega}$ of rotation (Fig. 5.46b) is[17]:

$$\mathbf{V} = \mathbf{\Omega} \times (G - C) \tag{5.138}$$

If this occurs, vector $\mathbf{\Omega}$ is directed like the axis of the wheel while vector $\mathbf{V}$ is orthogonal to the axis of the wheel and to the vector $(C - G)$. In reality, however, the bodies are deformable and contact is no longer a point, but extends to an area with a roughly elliptical shape: Fig. 5.47 shows this situation in the case of a railway wheel. In this situation there is no longer pure rolling and (5.138) no longer holds true, as:

$$\mathbf{V} \neq \mathbf{\Omega} \times (G - C) \tag{5.139}$$

The generic infinitesimal element on the external circumference of the wheel is subject to variations in peripheral velocity, due to the deformations that the actual wheel undergoes. Away from the contact patch, the peripheral velocity (i.e. the velocity relating to a system of reference, translating with the wheel centre) is constant, and the element appears to be undeformed (if we exclude the centrifugal field).

In the contact patch normal σ_x and shear τ stresses arise, associated with the state of deformation of the actual element [56, 58, 59, 84, 98–100], see Fig. 5.47. In the front area of the patch the normal and shear elastic stresses (associated with deformations) satisfy the condition of adhesion:

$$\tau \leq f_s \sigma_x \tag{5.140}$$

[17]We should remember that in this discussion geometric vectors are highlighted in bold.

where f_s is the coefficient of static friction. In the rear area of the patch, small local slipping motions take place, without macroscopic slipping of the whole wheel occurring. Basically, we have a similar mechanism to that of a belt on a pulley [1], where one area of the belt's winding arc is the site of micro-slip, while in the other area there is adhesion. Upon increase of the torque (braking or drive) acting on the wheel, or rather, the tangential reaction T required by wheel-terrain contact, the area affected by micro-slip extends, until it covers the entire area of contact: above this limit real macroscopic slipping actually begins. The trend of the overall reaction transmitted from the ground to the wheel (i.e. the integral of tangential actions τ extended to the entire contact patch) is normally defined as the product of the normal action N for a suitable friction coefficient. This reaction tends to a maximum value $T_{max} = \mu_{max} N$, to then decrease more or less rapidly tending asymptotically to the value corresponding to total macroscopic slipping: in these conditions the entire area is the site of slipping.

As mentioned, due to the deformability of the wheel and the rail and due to the presence of micro-slip, the velocity $\mathbf{V}$ of the wheel centre is different to the vector product (5.139), as it would be in the case of pure rolling. In addition to the braking or drive torque, which give rise to longitudinal actions, in general the wheel is also affected by transversal forces (consider, for example, the centrifugal force in a vehicle going around a bend). In this way, the micro-slip zone is generally affected both by transversal actions and longitudinal actions. The centre of the wheel, in the presence of transversal forces, has a component of velocity in a direction that is perpendicular to the plane of the actual wheel (Fig. 5.48). To define the contact forces we need to define the distribution of the normal σ_x and shear τ forces in the contact zone, correlating it to the kinematic quantities in play, i.e. to the velocity of the wheel centre and to its angular velocity Ω: solving this problem is very complex. On the other hand, it is possible to demonstrate both experimentally and analytically [99, 100, 102] that the contact forces can be defined as a function of a single kinematic parameter $\boldsymbol{\varepsilon}_{\mathbf{tot}}$ defined as *creepage* in the railway sector and *slip* in

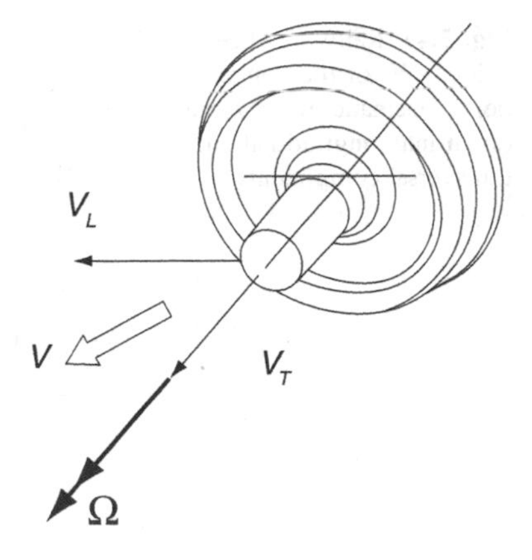

Fig. 5.48 Components of the velocity $\mathbf{V}$ of the wheel centre in a generic case

the automotive sector. So we introduce the vector quantity of creepage $\boldsymbol{\varepsilon}_{\mathbf{tot}}$, defined as the ratio of the absolute velocity $\mathbf{V}_c$ of the contact point kinematic entity (corresponding to the centre of the contact patch), calculated as if that point where rigidly constrained to the wheel, i.e.:

$$\mathbf{V}_c = \mathbf{V} + \boldsymbol{\Omega} \times (C - G) = \mathbf{V} - \boldsymbol{\Omega} \times (G - C) \tag{5.141a}$$

and the module of the wheel centre velocity $|\mathbf{V}|$:

$$\boldsymbol{\varepsilon}_{tot} = \frac{\mathbf{V} - \boldsymbol{\Omega} \times (G - C)}{|\mathbf{V}|} \tag{5.141b}$$

The difference between the effective velocity of the wheel centre $\mathbf{V}$ and the quantity $\boldsymbol{\Omega} \times (G - C)$ is a measure of how the wheel micro-slips: this difference also equals the absolute velocity of the contact patch's centre, considered to be rigidly constrained to the wheel. In other words, if we consider a translating reference system with origin in the wheel centre, $\mathbf{V}$ represents a "drag" velocity (i.e. velocity of non-rotating wheel) and $\boldsymbol{\Omega} \times (G - C)$ is the relative velocity of point C. The velocity $\mathbf{V}_c$ of the contact patch's centre represents a kinematic entity and not the actual velocity of the point of the wheel that occupies the position at the centre of the contact patch. In fact, the actual velocity of this point should be calculated while also taking into account the terms of relative velocity with respect to a reference system that is integral with the hub, due to the deformability of the wheel in the contact patch. To define the forces that are established in the contact patch, without introducing the deformability of bodies in contact, we use total creepage $\boldsymbol{\varepsilon}_{\mathbf{tot}}$, defined by (5.141b): with this approach, the contact forces, as we will illustrate below, come to depend on the velocity $\mathbf{V}_c$ of the contact point kinematic entity. The geometric creepage vector $\boldsymbol{\varepsilon}_{\mathbf{tot}}$ is usually broken down into two components (Fig. 5.49) ε_L and ε_T, respectively, in the longitudinal (defined by unit

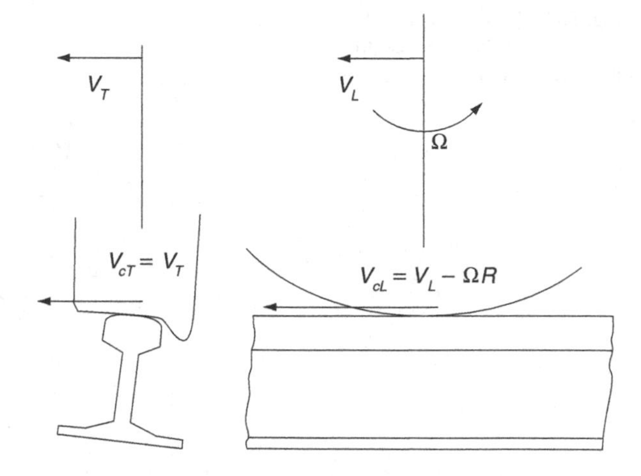

Fig. 5.49 Railway wheel: velocity $\mathbf{V}_c$ of the contact point kinematic entity broken down into longitudinal and transversal components

vector **l**) and the transversal direction (defined by unit vector **t**), with respect to the axis of the wheel:

$$\boldsymbol{\varepsilon}_{tot} = \frac{\mathbf{V}_c}{|\mathbf{V}|} = \mathbf{l}\varepsilon_L + \mathbf{t}\varepsilon_T \tag{5.142a}$$

being, respectively:

$$\varepsilon_L = \frac{V_L - \Omega R}{|V|}; \quad \varepsilon_T = \frac{V_T}{|V|} \tag{5.142b}$$

where V_L and V_T represent the components of the velocity **V** of the wheel centre, in longitudinal **l** and transversal **t** direction (Fig. 5.49), respectively.

The contact force is made to depend on the total creepage (5.142a) and, more precisely, the longitudinal F_L and transversal F_T components of the contact force are non-linear functions of the longitudinal ε_L and transversal ε_T components of creepage (5.142b) and of the normal component N of the contact force in the form [94, 97, 100, 102, 103]:

$$\begin{aligned} F_L &= -\mu_L(\varepsilon_L, \varepsilon_T, N)N \\ F_T &= -\mu_T(\varepsilon_T, \varepsilon_L, N)N \end{aligned} \tag{5.142c}$$

where μ_L and μ_T are said to be coefficients of friction. As we can see from (5.142b), creepages ε_L and ε_T are functions of the components of the wheel centre velocity V_L and V_T and of the wheel's rotational speed Ω: these quantities are functions of the independent variables that make it possible to define the motion of the actual wheel: so for these reasons, (5.142c) can be considered as force field that are functions, in this case, of the derivatives with respect to time of the independent variables that define the motion of bodies in contact. This approach to defining the contact forces is used both in the case of wheel-rail and in the case of tyre-road with different formulations and nomenclatures: for example, as already mentioned in the introduction to the section, in the case of wheel-rail, we must also bear in mind the particular morphology assumed by the wheel flange and the rail (Fig. 5.43a). For these reasons, we will now address a discussion that differs for the two cases:

- wheel-rail (section "The Contact Forces Between Wheel and Rail");
- tyre-road (section "The Contact Forces Between Tyre and Road").

The Contact Forces Between Wheel and Rail

As seen in the previous section, the contact force is made to depend on the creepage, more precisely the longitudinal and transversal components, F_L and F_T, of the contact force are non-linear functions of the longitudinal ε_L and transversal ε_T components of the total creepage $\vec{\varepsilon}_{tot}$ and of the normal component N of the contact force in the form of (5.142c): the same equation highlights how the friction

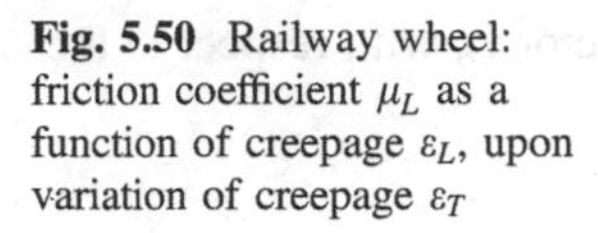

Fig. 5.50 Railway wheel: friction coefficient μ_L as a function of creepage ε_L, upon variation of creepage ε_T

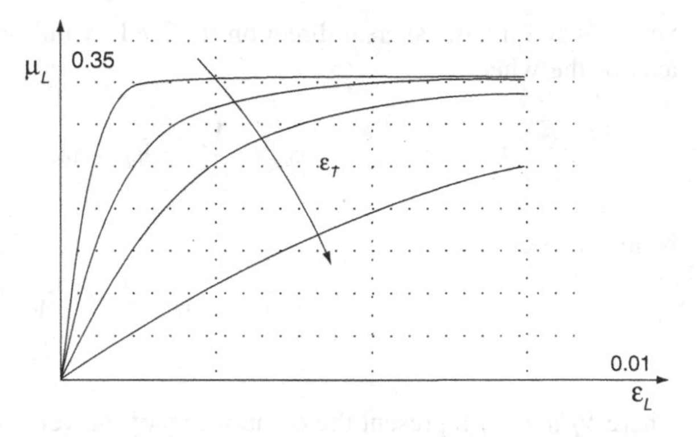

coefficient μ, and thus the corresponding contact force component in one direction, also depends on the creepage in the other direction. Figure 5.50 shows the trend of the longitudinal friction coefficient μ_L as a function of the longitudinal creepage ε_L upon variation of the lateral creepage coefficient ε_T. Upon variation of ε_L and when $\varepsilon_T = 0$, μ_L increases in a non-linear form until, for values of ε_L higher than 0.005, it reaches a maximum value $\mu_{L\max}$. In this situation the constraint is saturated, that is, it cannot provide greater longitudinal forces. Taking into account the simultaneous presence of transversal creepage ε_T, the maximum value that can be reached by the friction coefficient μ_L decreases. This figure, and those following it (relative to the railway wheel) do not show the subsequent reduction in the value of μ which, as already mentioned, tends asymptotically to the value of the macro-*slipping* friction coefficient. Function $\mu_L = \mu_L(\varepsilon_L, \varepsilon_T)$ (and likewise $\mu_T = \mu_T(\varepsilon_T, \varepsilon_L)$ is odd with respect to ε_L and even with respect to ε_T: the subsequent figures will only show the quarter corresponding to positive ε_L and ε_T. Figure 5.51 uses a three-dimensional representation to show how, as transversal creepage ε_L increases (we have the same graphs for coefficient μ_T, as parameter ε_L varies) friction coefficient decreases and so force F_L, which develops from contact in a longitudinal direction, is reduced. At worst, if the creepage in the transversal direction is so high that it involves the entire contact area with transversal creepage ε_T, force F_L is cancelled out: in this case all the adhesion has already been fully exploited in the perpendicular direction and the creepage zone affects the entire contact patch (the constraint has reached saturation point). Figure 5.52 shows the trend of the longitudinal friction coefficient μ_L as a function of longitudinal creepage ε_L upon variation of the normal load N. The normal load, when the dimensions of the contact patch change, also affects the mechanism that defines the links between the various quantities in play. Defining the various quantities shown in the curves (Figs. 5.50, 5.51 and 5.52) can be done:

- experimentally [102, 107];
- using suitable mathematical models, developed mainly by two researchers: Kalker [99, 100] and Johnson [98]

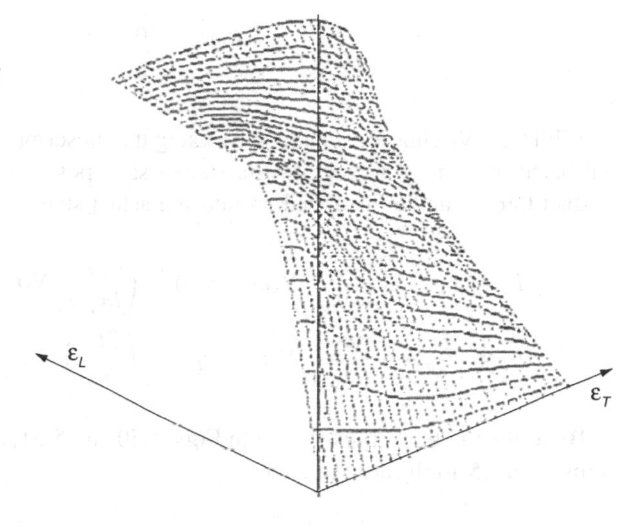

Fig. 5.51 Railway wheel: friction coefficient μ_L as a function of creepage ε_L and of creepage ε_T

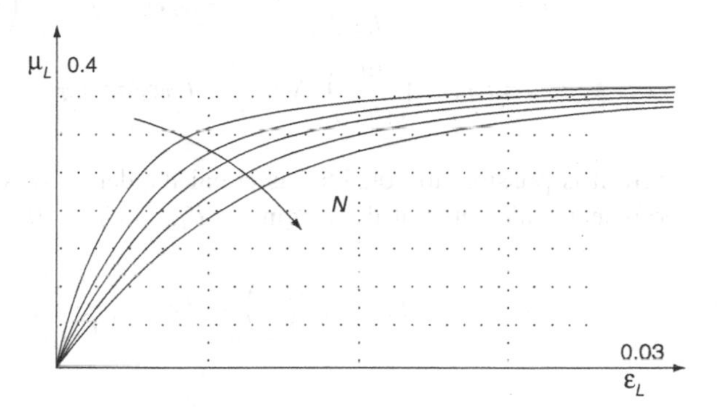

Fig. 5.52 Railway wheel: friction coefficient μ_L as a function of creepage ε_L, upon variation of the normal load N

If we wish to simulate the generic motion of the vehicle, we must use the complete expressions of the contact forces (5.142c) and introduce them into the mathematical model of the vehicle: the resulting equations are, as a consequence, non-linear and so must be numerically integrated using split-step methods. At the generic instant of time, the velocities of the individual bodies and in particular of the wheelset are known, either from the initial conditions or from the previous integration step: it is possible, for this reason, to use (5.142b) to estimate the creepages ε_L and ε_T and so, using (5.142c), the contact forces F_L and F_T needed to proceed in integrating them [56–60, 62–64, 80–84, 87–91, 93, 95, 96, 108].

To analyse the stability of the vehicle affected by the actions of contact described above, it is sufficient to linearise (5.142b) around the steady-state position. When analysing the perturbed motion around the configuration defined by running in a straight line at constant speed, if we assume that the wheel is being pulled and that the longitudinal forces associated with the mechanism of rolling friction can be neglected, and we take into account (5.142c) and the definition of creepage (5.142b) we will have:

$$\begin{aligned} \varepsilon_{To} &= 0 \\ \varepsilon_{Lo} &= 0 \end{aligned} \tag{5.143a}$$

With load N being constant, and by using the subscript "o" to define the quantities estimated in correspondence to the steady-state position defined by (5.143a), the contact forces can be linearised around the actual steady-state position using:

$$\begin{aligned} F_L &= F_{Lo} - \left(\frac{\partial \mu_L}{\partial \varepsilon_L}\right)_o N(\varepsilon_L - \varepsilon_{LO}) - \left(\frac{\partial \mu_L}{\partial \varepsilon_T}\right)_o N(\varepsilon_T - \varepsilon_{TO}) \\ F_T &= F_{To} - \left(\frac{\partial \mu_T}{\partial \varepsilon_L}\right)_o N(\varepsilon_L - \varepsilon_{LO}) - \left(\frac{\partial \mu_T}{\partial \varepsilon_T}\right)_o N(\varepsilon_T - \varepsilon_{TO}) \end{aligned} \tag{5.143b}$$

By analysing the curves shown in Figs. 5.50 and 5.51, and the mixed derivatives being zero, (5.143b) are reduced to:

$$\begin{aligned} F_L &= F_{Lo} - \left(\frac{\partial \mu_L}{\partial \varepsilon_L}\right)_o N\varepsilon_L = -f_{oL}\varepsilon_L \Rightarrow f_{oL} = \left(\frac{\partial \mu_L}{\partial \varepsilon_L}\right)_o \\ F_T &= F_{To} - \left(\frac{\partial \mu_T}{\partial \varepsilon_T}\right)_o N\varepsilon_T = -f_{oT}\varepsilon_T \Rightarrow f_{oT} = \left(\frac{\partial \mu_T}{\partial \varepsilon_T}\right)_o \end{aligned} \tag{5.143c}$$

where it is possible to consider as equal the derivatives f_{oL} and f_{oT} of the friction coefficients calculated at the origin, as $F_{Lo} = F_{To} = 0$:

$$f_{oL} = \left(\frac{\partial \mu_L}{\partial \varepsilon_L}\right)_o = f_{oT} = \left(\frac{\partial \mu_T}{\partial \varepsilon_T}\right)_o = f_o \tag{5.143d}$$

i.e.:

$$\begin{aligned} F_L &= -f_o\varepsilon_L \\ F_T &= -f_o\varepsilon_T \end{aligned} \tag{5.144}$$

These definitions will be applied below in studying the stability of a railway wheelset running in a straight line and perturbed around the steady-state situation at constant V. We should also remember that, as mentioned previously, in the lateral dynamics of the railway vehicle, in addition to tangential actions, normal actions also appear, modifying the direction upon variation of angle γ formed by the tangent to the contact plane, defined by the coupled profiles of wheel and rail (Fig. 5.43) [56–60, 62–64, 66, 80–84, 93, 95, 96, 108].

The Contact Forces Between Tyre and Road

In order to define the behaviour of a generic road vehicle we must analyse what happens during contact between tyre and terrain, that is, analyse the quantities that

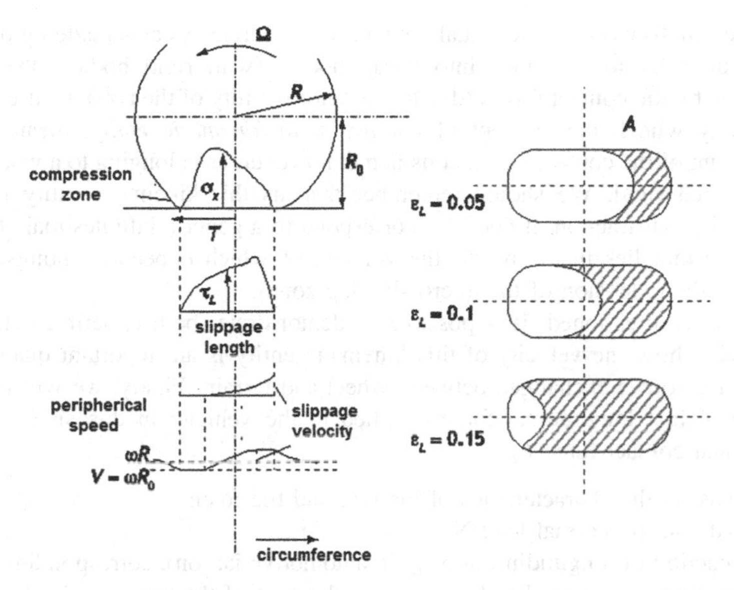

Fig. 5.53 The tyre: contact patch and induced stresses (drive wheel)

affect the actions exchanged in the contact patch. The tyre is a highly deformable element: for this reason contact develops in an area that is much larger than that of the railway vehicle: in the case of the railway vehicle, the size of the contact patch is about ten millimetres, whereas in the case of the tyre it is about ten centimetres.

We will analyse Fig. 5.53 which shows the case of a drive wheel, with no transversal forces (similar considerations can be conducted with a braking wheel): the figure shows the trend, in the contact patch, of the normal stresses σ_x and shear stresses τ_L. The peripheral velocity, meaning the velocity of a small element of tread seen by an observer integral to the centre of the wheel, changes upon variation of its position in the contact patch:

- there is a first zone of complete adhesion where the peripheral velocity is equal and opposite to the velocity of the wheel centre;
- there is a second zone where there is local micro-slipping (similar to what we saw for the railway wheel), where the peripheral velocity (Fig. 5.53) is greater (in the case of the drive wheel) or lower (case of the trailer wheel) than the forward speed.

Similarly to what we saw for the railway wheel, to correctly reproduce the tangential forces exchanged between wheel and terrain we must take into account the deformability of the tyre and this involves the need to write partial derivative equations as is the case for continuous systems: as we know, these equations are complicated, difficult to integrate analytically and so would involve particularly laborious calculations. On the other hand, one model usually adopted to describe the dynamic behaviour of the vehicle is based on *multibody* methods [5, 7, 69–72], i.e. on schematisations of the actual vehicle with discrete systems made up of rigid bodies. In order to introduce into these models (with rigid bodies) the effect introduced by the contact forces (due to the deformability of the tyre) we use, as for the railway wheel, the concept of *contact point kinematic entity*, meaning the central point of the contact patch, considered, however, as belonging to a wheel that is considered rigid. We should remember that, as this kinematic entity is pure mathematical abstraction, it does not correspond to a generic infinitesimal element of the tyre that lies in the patch (the velocity of which depends, amongst other things, on the extension of the micro-slipping zone).

As already mentioned, it is possible to demonstrate, both experimentally and analytically, how the velocity of this kinematic entity is an important quantity in defining the forces exchanged between wheel and terrain. Firstly, we will analyse the case of just longitudinal forces applied to the vehicle: in this situation, the longitudinal contact force F_L:

- depends on the characteristics of the tyre and the road;
- depends on the normal load N;
- is a function of longitudinal slip ε_L (in automotive jargon), corresponding to the longitudinal creepage already defined in the case of the railway wheel.

So, as we saw in the previous section, ε_L is defined as the relationship between the longitudinal component of the absolute velocity of the kinematic entity contact point V_C and the module of the forward speed $|V|$ of the wheel centre:

$$\varepsilon_L = \frac{V_L - \Omega R}{|V|} \tag{5.145}$$

So, if a longitudinal force F_L, resulting from driving or braking torque, is applied to the tyre then longitudinal slip ε_L occurs. The link between the longitudinal force F_L, in non-dimensional terms, and ε_L can be expressed by:

$$\frac{F_L}{N} = -\mu_L(\varepsilon_L, N, c_p, c_s) \tag{5.146}$$

where N, as already mentioned, is the normal load, c_p is a parameter that defines the characteristics of the tyre and c_s is a parameter that defines the characteristics of the road. Figure 5.54a shows an example of the trend of μ_L as a function of longitudinal slip ε_L (without transversal forces): for limited ε_L, μ_L depends linearly on the actual slip (creepage), while, as this quantity increases, the link becomes strongly non-linear. In the area where the curve reaches an asymptote, virtually the entire area of contact is the site of slip; for even higher values of creepage, slipping is macroscopic and this situation is not stable (dashed curve). The maximum value $\mu_{L\max}$ and the minimum value $\mu_{L\min}$ also depend, in turn, on the forward speed V of the wheel centre. By comparing this curve with the similar curve in Fig. 5.47 (relating to the railway wheel), we can see that the two phenomena are quite similar: the only significant difference concerns the value of creepage $\varepsilon_L = 0.005$ in correspondence

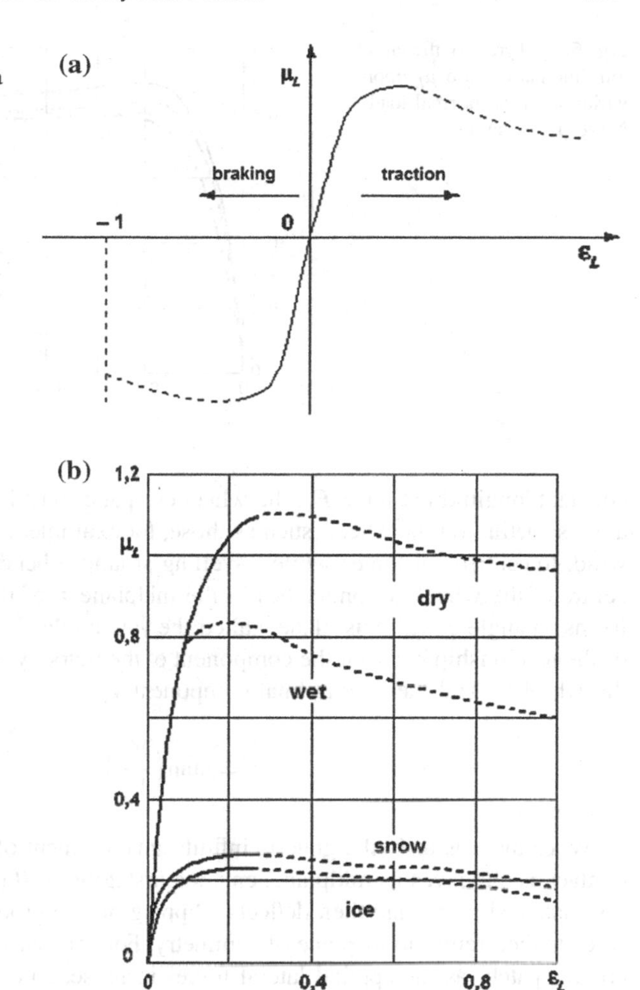

Fig. 5.54 **a** Tyre: coefficient of longitudinal friction μ_L as a function of slip ε_L (zero yaw angle). **b** Tyre: coefficient of longitudinal friction μ_L upon variation of the road surface conditions

to which the constraint reaches saturation in the case of the railway wheel (Fig. 5.47), while in the case of the tyre the slip value ε_L in correspondence to which this phenomenon occurs is equal to approximately $\varepsilon_L = 0.3$ (Fig. 5.54a). Figure 5.54b shows how μ_L varies as environmental conditions vary, i.e. dry or wet road, presence of snow or ice. As we can see, the slipping friction coefficient μ_L can vary enormously (from 1 to 0.2) as these conditions vary. The presence of water also reduces μ_L: in this regard, we should also remember the phenomenon of aquaplaning, i.e. floating of the vehicle, (function of speed V) due to a film of water [94] build under its wheels. In Fig. 5.55 we can see the dependency of the friction coefficient μ_L on the normal load N: the effect of the non-linear dependency of μ_L on the load N is lower than the effect of longitudinal force F_L due to the direct linear dependence (5.146): the overall result is such that a heavily loaded wheel, with the other conditions being equal, has lower friction coefficients μ_L and so, to provide a

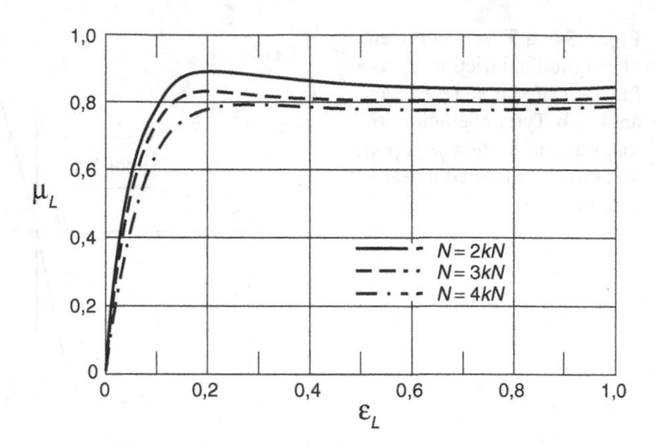

Fig. 5.55 Tyre: coefficient of longitudinal friction μ_L upon variation of the normal load N (zero yaw angle)

constant longitudinal force F_L, the wheel creepage must be higher. If lateral forces are also acting on the wheel (such as those, for example, that act on vehicles due to wind, to centrifugal forces while travelling around a bend etc.) the velocity of the centre of the wheel no longer lies in the midplane π of the wheel (Fig. 5.56), but forms an angle α with this plane, called the yaw angle,[18] defined by the arctangent of the relationship between the component of the velocity transversal to the plane of the wheel V_T and the longitudinal component V_L:

$$\alpha = \mathrm{atan}\left(\frac{V_T}{V_L}\right) \tag{5.147}$$

When there is drift, the generic infinitesimal element of the tread that enters the contact patch leaves its midplane, enters a first zone *A–B* (Fig. 5.56) where there is no relative slipping and then deflects, slipping on the ground until the exit, section *B–C*, to then return to its plane of symmetry. For this reason, side-slip occurs in the contact patch: as the applied lateral forces increase, so does the drift and the area where local slipping occurs between tread and terrain.[19] In the presence of external

[18] In this discussion we refer to the typical nomenclature of those who are involved with road vehicles, even if this may appear at first glance to be a different approach in defining the contact forces in the case of road vehicles compared to the case of railway vehicles. This difference, in actual fact, does not exist, even if the two approaches foresee, historically, slightly different formalisms: the yaw angle used in road vehicles actually corresponds to the lateral creepage adopted when analysing railway vehicles.

[19] In the event that there are no lateral forces or torques (driving or braking) applied to the wheel, there is no micro-slip ε_L and the yaw angle α is zero: in these conditions the forward velocity of the wheel centre, the direction of which in this case is parallel to the midplane of the actual wheel, satisfies (5.138): pure rolling is thus established only under these conditions, conditions, on the other hand, that have never occurred in reality (in fact there is always an aerodynamic force, the hysteresis of the material causes the non-symmetry of the diagrams of pressure σ_x that gives rise to rolling resistance, etc.).

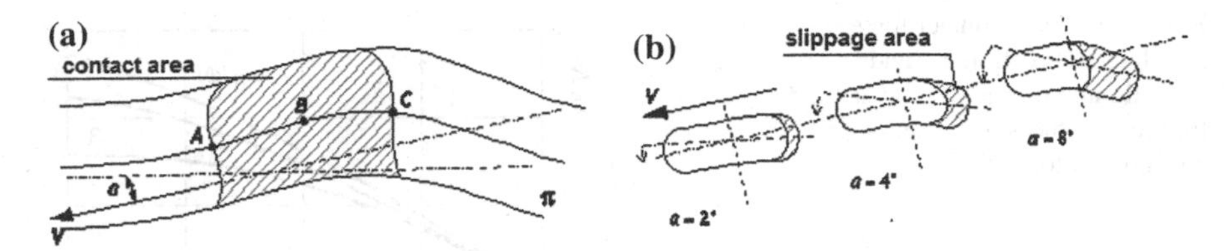

Fig. 5.56 Tyre: contact patch with lateral forces

transversal forces only, the components of transversal contact force F_T then prove to be non-linear functions:

- of the normal load acting on the wheel N;
- of the characteristics of the tyre c_p and the road surface c_s;
- of the yaw angle α.

Similarly F_T (odd function of α) can be expressed in non-dimensional form using:

$$\frac{F_T}{N} = -\mu_T(\alpha, N, c_p, c_s) \tag{5.148}$$

Figure 5.57 shows, upon variation of the module of the yaw angle α and of the normal load N, respectively, the module of the F_T (5.148) and the module of self-aligning torque M_{aut} (normal vector at the contact patch): this torque is established in the contact patch due to the effect of non-symmetry, in a transversal direction, of the distribution of the shear stresses τ_T.[20] F_T and M_{aut} again show a linear trend for small yaw angles, while they become strongly non-linear for large α. There is also a lateral mechanism of constraint saturation (the curves tend to a horizontal asymptote after which they decrease).

Compared with the approach used traditionally in the rail sector to define the transversal contact forces F_T, estimated as a function of transversal creepage ε_T, in the automotive sector we resort to the definition of the yaw angle: however, this difference is purely formal as both these quantities represent the relationship between the transversal and longitudinal components of the wheel's velocity. Although the two approaches are therefore conceptually quite similar, in any case it is more convenient to use the different symbols and different procedures that are now well established in the different road and rail sectors. Lastly, we will analyse the more general case of external forces, both longitudinal and transversal, that effect the vehicle: just like in the railway wheel, the contact forces F_T and F_L are coupled, in both directions, by longitudinal slip ε_L and by the yaw angle α: the dependence of the generic contact force on both microslip actions can be explained

[20]This self-aligning torque M_{aut} tends to straighten the wheel and is discharged on the steering wheel. In the case of the railway wheel, there is a similar term (neglected in this discussion due to its limited size for rectilinear motions), defined in railway jargon as *spin momentum* [99, 100].

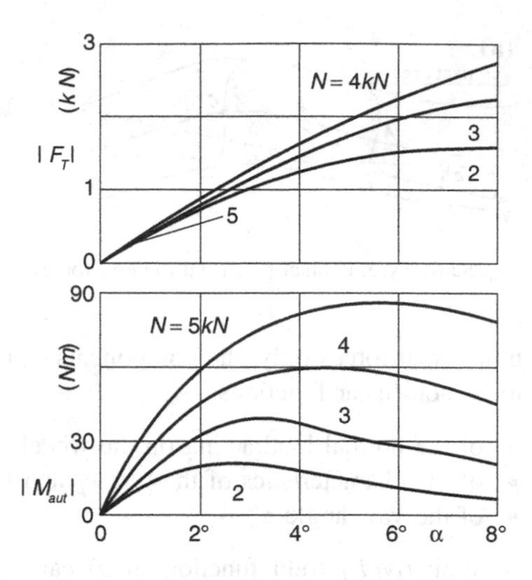

Fig. 5.57 Tyre: contact force in a lateral direction F_T and self-aligning torque M_{aut} as a function of the slip-angle α and normal load N

qualitatively by bearing in mind the fact that the combination of traction and lateral forces generate a state of stress τ_T and τ_L which add together, thus increasing the slipping zone and obviously affecting both the components of the contact force. In other words, the adherence used in one direction, reduces the availability of adherence in the other. Thus the components of the longitudinal F_L and transversal F_T contact force are non-linear functions:

- of the normal load acting on the wheel N;
- of the characteristics of the tyre and the road surface;
- of the longitudinal slip ε_L and the yaw angle α;

$$\begin{aligned} F_L &= -\mu_L(\varepsilon_L, \alpha, N, c_p, c_s)N \\ F_T &= -\mu_T(\varepsilon_L, \alpha, N, c_p, c_s)N \end{aligned} \quad (5.149a)$$

Figure 5.58 shows the total force F exchanged between tyre and road:

$$\mathbf{F} = \mathbf{l}F_L + \mathbf{t}F_T \quad (5.149b)$$

for a pre-established yaw angle α: as we can see, similarly to what we saw in the railway wheel (Figs. 5.42, 5.43 and 5.44), the maximum lateral force can be obtained when longitudinal forces are not required and vice versa. Figure 5.59 shows the same polar diagram for different yaw angles.

If a traction force (or a lateral force) is such as to create extensive slip ε_L in the forward direction (or transversal direction) the same constraint cannot provide force in a perpendicular direction, i.e. very similarly to what we saw in the previous section, the constraint becomes saturated:

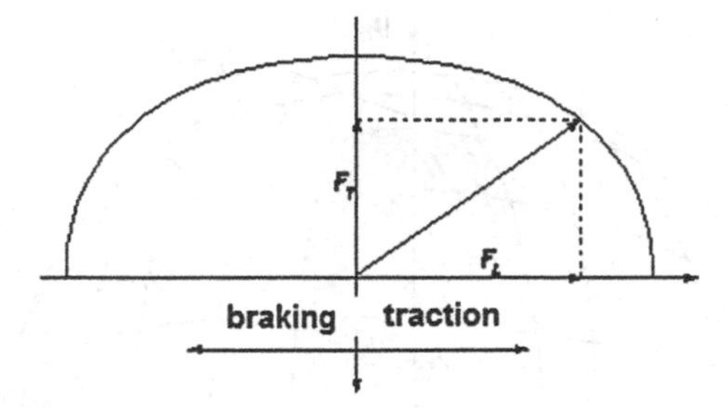

Fig. 5.58 Tyre: polar diagram of the force exchanged at a constant yaw angle

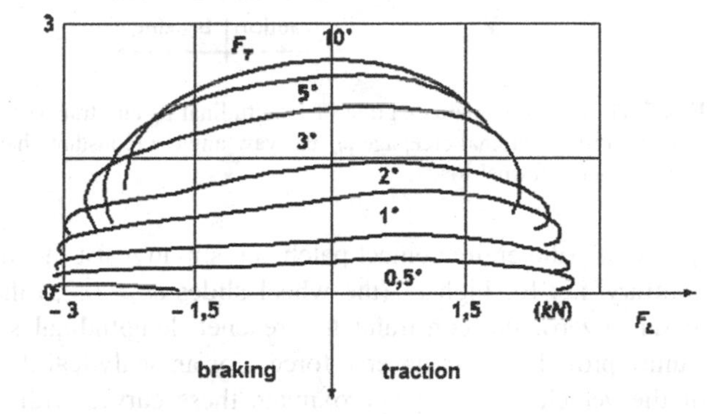

Fig. 5.59 Tyre: polar diagram of the force exchanged for different yaw angles

- we can, for example, think of a car that is trying to set off on snow; the driving torque applied to the wheels makes them slide (correspondingly the longitudinal slip ε_L is extensive, i.e. the entire patch is slipping): in these conditions a small lateral force is enough to move the vehicle sideways;
- another well known case may be where a vehicle enters a bend while braking sharply: the longitudinal slipping becomes extensive (the wheels could even lock) the patch cannot provide extensive lateral forces (with the wheels locked it cannot provide any lateral force). Instead of going into the curvilinear trajectory, the vehicle will proceed in a straight line, regardless of the steering angle δ given to the wheels.

Lastly, Fig. 5.60 shows a curve that defines, in full, the characteristics of wheel-terrain contact for a given load N, a certain type of tyre and road situation. The inputs to this curve are the longitudinal slip ε_L and yaw angle α (i.e. side-slip): there are curves with constant ε_L and curves with constant α. The intersection of a curve for a certain value of ε_L with the curve corresponding to a certain value of α, assumed to be known, defines a point where the abscissa and ordinate define, respectively, the friction coefficient μ_L in the longitudinal and μ_T in transversal directions, from which it is possible to obtain the contact forces transmitted in the patch area using the formulae (5.149a). This graph clearly highlights the fact that for example, if the longitudinal slip ε_L is zero, the side friction coefficient μ_T is

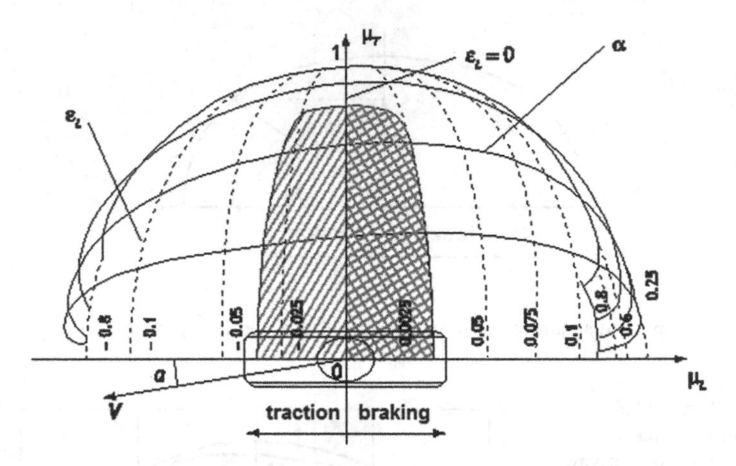

Fig. 5.60 Tyre: complete diagram of longitudinal μ_L and transversal μ_T friction coefficients as a function of longitudinal creepage ε_L and yaw angle α (constant characteristics of the tyre and the road, constant normal load)

maximum (the entire contact patch can slip to resist the lateral forces only). On the contrary, i.e. for high ε_L (the wheel slides $\varepsilon_L > 0.25$) the side-slip coefficient μ_T tends to zero: the constraint has reached longitudinal saturation and therefore it cannot provide any transverse force. For an analytical discussion on the behaviour of the vehicle we must approximate these curves with simple functions: a good approximation could be that shown in Fig. 5.61 where the various functions are approximated with ellipses.[21]

Summarising the results highlighted in the last two sections, regarding wheel-rail contact and tyre-road contact, the main characteristics of the phenomenon are:

- the condition of pure rolling does not actually exist and, in any case, is a primary simplistic model of the real phenomenon;
- in the contact zone longitudinal F_L and transversal F_T forces are established (5.149a) as functions of longitudinal ε_L or transversal ε_T creepage (or yaw angle α);
- the actual velocity distribution of a generic infinitesimal element integral with the wheel is difficult to define, as is the real state of deformation and stress in the contact zone;
- we therefore refer back to definition of the forces F_L and F_T via estimation of the velocity of the contact area centre kinematic entity, considered integral with the wheel which is assumed to be rigid: this entity is representative of the actual phenomenon.

[21]With a constant yaw angle α, the various curves can be interpolated, with good approximation from curves such as:

$$\left(\frac{\mu_L}{\mu_{L\,\max}}\right)^2+\left(\frac{\mu_T}{\mu_{T\,\max}}\right)^2=1. \tag{5.21.1}$$

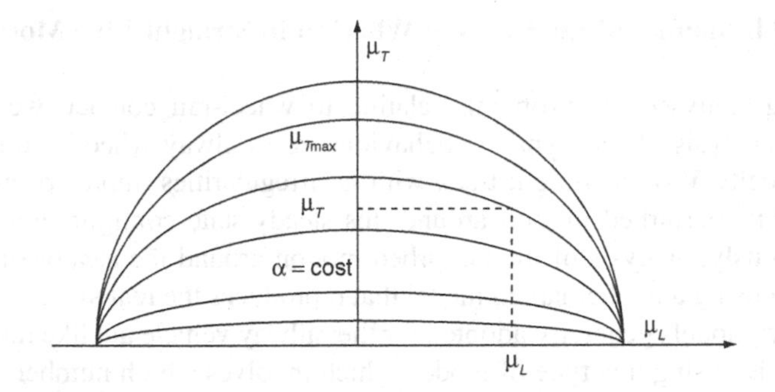

Fig. 5.61 Tyre: simplified diagram of transversal friction coefficients μ_T as a function of the longitudinal creepage ε_L and yaw angle α (constant characteristics of the tyre and the road, constant normal load)

With this approach, the contact forces are modelled like a force field as the same friction forces F_L and F_T are made to depend not only on the normal load N and the type and conditions of the contact surface, but also:

- on the creepages ε_L and ε_T, in the case of the railway wheel;
- on the longitudinal slip ε_L and the yaw angle α (as we have seen, this quantity is similar to transversal creepage ε_T), in the case of the tyre.
- the condition of pure rolling does not actually exist and, in any case, is a primary simplistic model of the real phenomenon;
- in the contact zone longitudinal F_L and transversal F_T (5.149a) forces are established as functions of longitudinal ε_L or transversal ε_T creepage (or yaw angles α).

With this approach, the contact forces are modelled like a force field as the same friction forces F_L and F_T are made to depend not only on the normal load N and the type and conditions of the contact surface:

- but also on the creepages ε_L and ε_T, in the case of the railway wheel;
- on the longitudinal slip ε_L and the yaw angle α (as we have seen, this quantity is similar to transversal creepage ε_T), in the case of the tyre.

These slips or creepages are a function, in turn, of the independent variables that define the motion of the vehicle, or rather, the motion of the generic wheel. Thus a field of non-linear forces is defined that could be linearised around a steady-state configuration, giving rise to a matrix of equivalent damping and stiffness $[R_F]$ and $[K_F]$ which could result, for example, in problems linked to the stability of the actual vehicle. Now we will apply these concepts to analysing the dynamics of the railway wheelset and the road vehicle.

5.5.2.2 The Dynamics of the Railway Wheelset in Straight Line Motion

After having analysed the problems relating to wheel-rail contact, we will now address the analysis of the dynamic behaviour of a railway wheelset travelling at constant velocity **V** on a straight track with no irregularities: more specifically we will analyse its perturbed motion around this steady-state configuration. As mentioned previously, analysis of the perturbed motion around the steady-state should be addressed using a mathematical model that reproduces the real system in the best possible way: models generally adopted for the railway vehicle are like those shown in Fig. 5.45. By using this type of model, which involves a high number of degree-of-freedom (23–31 dof), studying stability becomes complex and requires the use of numerical methods to calculate the eigenvalues of the matrices of the linearised system. Alternatively, and mainly for educational purposes, we can make a simplified qualitative study of stability by analysing just the wheelset (Fig. 5.62). This simplified model still allows us to shed light on the nature of the problem and to highlight how the non-conservative force field, due to the actions of contact, can introduce energy into the system and thus induce unstable motions. More specifically, the simplified model is made up of a wheelset connected to a bogie with springs and dampers (corresponding to primary suspensions). The bogie has purely rectilinear translatory motion at constant velocity **V** and so its perturbed motion is not taken into consideration. The wheelset, with mass M and moment of inertia J around a vertical axis, shows steady-state forward motion in a straight line with constant velocity **V**: the equations of perturbed motion are linearised around this steady-state situation. The degree-of-freedom of the system are the lateral displacement y and the yaw rotation σ, with the conventions shown in Fig. 5.62. We assume the wheelset in the generic deformed position (Fig. 5.63) according to a lateral displacement y and a yaw rotation σ[22]: the equations of dynamic equilibrium in lateral direction and on yaw are the following (Figs. 5.62 and 5.63):

$$
\begin{aligned}
m\ddot{y} + 2r_y\dot{y} + 2k_y y &= F_y = (F_{Ts} + F_{Td})\cos\sigma - (F_{Ls} + F_{Ld})\sin\sigma \\
J\ddot{\sigma} + 2b^2 r_x\dot{\sigma} + 2b^2 k_x\sigma &= M_\sigma = (F_{Ld} - F_{Ls})S
\end{aligned}
\tag{5.150a}
$$

where k_y and k_x represent the lateral and longitudinal stiffnesses of the wheelset-bogie connection, r_y and r_x are the corresponding dissipative terms and S represents the wheelset semi-gauge. As we can see, the equations, as far as the structural part is concerned, are uncoupled and coupling between the degrees of freedom lateral y and yaw σ is introduced by the terms:

$$
\begin{aligned}
F_y &= (F_{Ts} + F_{Td})\cos\sigma - (F_{Ls} + F_{Ld})\sin\sigma \\
M_\sigma &= (F_{Ld} - F_{Ls})S
\end{aligned}
\tag{5.150b}
$$

[22] In this discussion we assume that the motion of the wheelset occurs in the plane containing the two tracks.

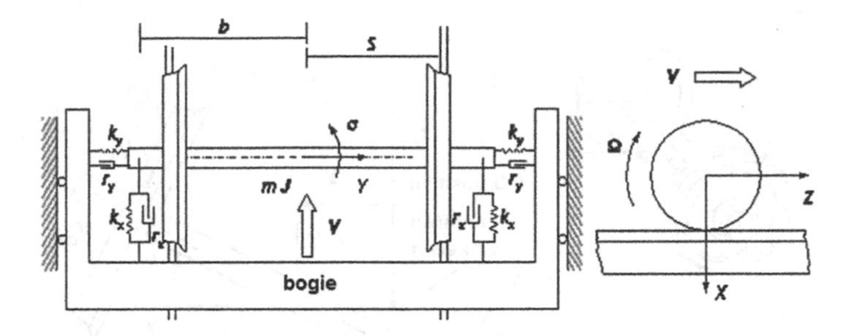

Fig. 5.62 Two-degree-of-freedom model for studying the stability of the railway wheelset

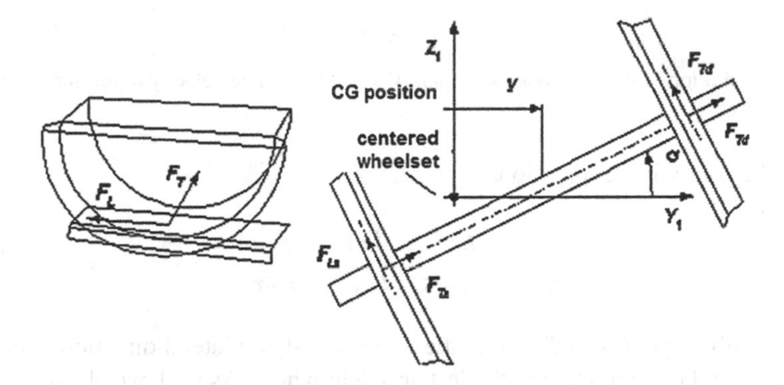

Fig. 5.63 Contact forces acting on the wheelset

which represent the components of the contact forces which, as we will see, are functions of the degree-of-freedom of the system.

Before proceeding to calculate the contact forces we need to define:

- the velocity $\mathbf{V}_c$ of the kinematic entity contact point;
- the creepages ε_L and ε_T in the longitudinal and transversal direction (unit vectors $\mathbf{l}$ and $\mathbf{t}$) with respect to the plane of the wheel (i.e. with respect to a system of reference integral with the wheelset).

The contact forces, if considered applied to the wheels, will be directed in the opposite direction to the lateral and longitudinal creepages in the reference of the wheelset (Fig. 5.64) and, if linearised around the steady-state position defined by:

$$\begin{aligned} y_o &= 0 \\ \sigma_o &= 0 \end{aligned} \tag{5.151a}$$

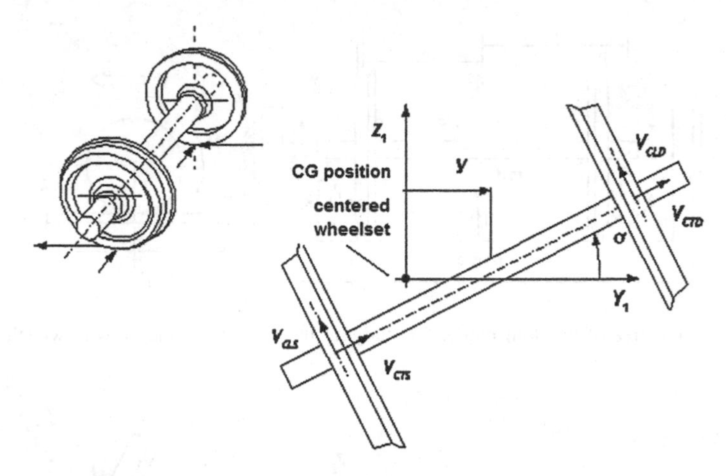

Fig. 5.64 Definition of the components V_{cL} and V_{cT} of the absolute velocity of the generic contact point

assume the following expressions:

$$\begin{aligned} F_{Ls} &= -f_o\varepsilon_{Ls}; \quad F_{Ld} = -f_o\varepsilon_{Ld} \\ F_{Ts} &= -f_o\varepsilon_T; \quad F_{Td} = -f_o\varepsilon_{Td} \end{aligned} \tag{5.151b}$$

where the subscripts L and T denote the longitudinal and lateral direction, while the subscripts s and d are relative to the left and right wheel. As we have already seen in (5.143b) and (5.143d), f_o represents the tangent to the origin of the curve of the forces $F_L = -\mu_L N$ and $F_T = -\mu_T N$, i.e. the derivative of them with respect to creepages ε_L and ε_T, N being the normal load on the single wheel with centred wheelset and μ_L and μ_T the friction coefficients as functions of the actual creepages [Figs. 5.50 and 5.51; Eqs. (5.143a)–(5.143d) and (5.144)].

Now we will proceed to calculate the longitudinal ε_L and transversal ε_T creepages that are functions of the projections V_{cL} and V_{cT}, on a system of reference integral with the generic wheel, of the velocity $\mathbf{V}_c$ of the contact point (Fig. 5.64), of the left-hand wheel and the right-hand wheel, divided by the forward speed V, (5.142b):

$$\begin{aligned} \varepsilon_{Ls} &= \frac{V_{cLs}}{|V|}; \quad \varepsilon_{Ts} = \frac{V_{cTs}}{|V|} \\ \varepsilon_{Ld} &= \frac{V_{cLd}}{|V|}; \quad \varepsilon_{Ts} = \frac{V_{cTd}}{|V|} \end{aligned} \tag{5.152}$$

To estimate the components of velocity of the kinematic entity contact point in the reference system integral with the generic wheel, it is best to use several reference systems of reference:

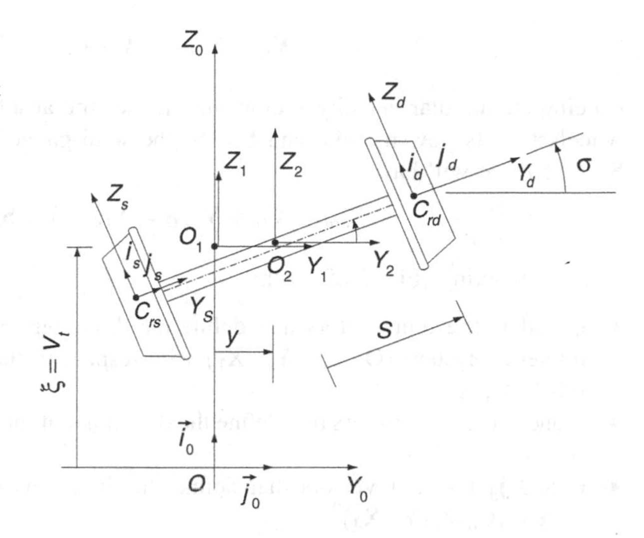

Fig. 5.65 Reference reference systems assumed for analysing the contact point kinematics

- a translating reference system (O_1-Z_1-Y_1-X_1) (Fig. 5.65) with velocity **V** and origin O_1 placed, on the centreline of the track on a level with the rail plane, in correspondence to the abscissa $s = Vt$, in correspondence to the centre of gravity of the centred wheelset;
- a reference system (C_d-Z_d-Y_d-X_d) integral with the right-hand wheel (except for the rotation Ω) and centre C_d coinciding with the geometric centre of the actual wheel;
- a reference system (C_s-Z_s-Y_s-X_s) integral with the left-hand wheel (except for the rotation Ω) and centre C_s coinciding with the geometric centre of the actual wheel.

The velocity $\mathbf{V}_{cs}$ of the contact point on the left wheel can thus be estimated (Fig. 5.65) as the sum of the velocity $\mathbf{V}_{crs}$ of the wheel centre (to be considered as "dragging" velocity) plus the relative velocity $\mathbf{V}_{cs,crs}$ of the contact point with respect to the reference system (C_s-Z_s-Y_s-X_s) integral with the wheel:

$$\mathbf{V}_{cs} = \mathbf{V}_{crs} + \mathbf{V}_{cs,crs} = \mathbf{V}_{crs} + \Omega \times \mathbf{R}_s \tag{5.153a}$$

Ω being the vector of the angular velocity of rolling and $\mathbf{R}_s$ the vector representing the radius of the left wheel. Similarly, for the right wheel, we will have:

$$\mathbf{V}_{cd} = \mathbf{V}_{crd} + \mathbf{V}_{cd,crd} = \mathbf{V}_{crd} + \Omega \times \mathbf{R}_d \tag{5.153b}$$

The velocity of the centre of the left wheel $\mathbf{V}_{crs}$ can be, in turn, estimated as the sum of the velocity $\mathbf{V}_G$ of the centre of gravity G of the wheelset (meaning the "dragging" velocity), plus the relative velocity $\mathbf{V}_{cs,G}$ of the wheel centre with respect to the centre of gravity (relative velocity):

$$\mathbf{V}_{crs} = \mathbf{V}_G + \mathbf{V}_{cs,G} = \mathbf{V}_G + \dot{\sigma} \times \mathbf{S}_s \tag{5.154a}$$

$\dot{\sigma}$ being the angular velocity vector, that defines the absolute angular speed of the wheelset in its yaw motion, and $\mathbf{S} = \mathbf{S}_s$ the semi-gauge. For the right wheel (as $\mathbf{S} = -\mathbf{S}_d$) we will have:

$$\mathbf{V}_{crd} = \mathbf{V}_G + \mathbf{V}_{cd,G} = \mathbf{V}_G + \dot{\sigma} \times \mathbf{S}_d \tag{5.154b}$$

Denominating (Fig. 5.65) with:

- $\mathbf{i}_o$ and $\mathbf{j}_o$ the unit vectors that define the directions of the axes of translating reference system (O_1 -Z_1 -Y_1 -X_1) with respect to the fixed reference system (O-Z_o-Y_o-X_o);
- $\mathbf{i}_s$ and $\mathbf{j}_s$ the unit vectors that define the directions of the axes of reference system (C_s -Z_s -Y_s -X_s);
- $\mathbf{i}_d$ and $\mathbf{j}_d$ the unit vectors that define the directions of the axes of reference system (C_d-Z_d-Y_d-X_d)[23];

the absolute velocity of the centre of the left and the right wheel, taking into account (5.154a) and (5.154b), becomes:

$$\begin{aligned}\mathbf{V}_{crs} &= \mathbf{i}_o V + \mathbf{j}_o \dot{y} - \mathbf{i}_s \dot{\sigma} S = \mathbf{i}_s V \cos\sigma - \mathbf{i}_s \dot{y} \sin\sigma + \mathbf{j}_s \dot{y} \cos\sigma + \mathbf{j}_s V \sin\sigma - \mathbf{i}_s \dot{\sigma} S \\ &= \mathbf{i}_s (V \cos\sigma - \dot{y}\sin\sigma - \dot{\sigma} S) + \mathbf{j}_s (V \sin\sigma + \dot{y}\cos\sigma)\end{aligned} \tag{5.155a}$$

$$\begin{aligned}\mathbf{V}_{crd} &= \mathbf{i}_o V + \mathbf{j}_o \dot{y} + \mathbf{i}_d \dot{\sigma} S = \mathbf{i}_d V \cos\sigma - \mathbf{i}_d \dot{y} \sin\sigma + \mathbf{j}_d \dot{y} \cos\sigma + \mathbf{j}_d V \sin\sigma y + \mathbf{i}_d \dot{\sigma} S \\ &= \mathbf{i}_d (V \cos\sigma - \dot{y}\sin\sigma + \dot{\sigma} S) + \mathbf{j}_d (V \sin\sigma + \dot{y}\cos\sigma)\end{aligned} \tag{5.155b}$$

The absolute velocities of the contact points (in the reference system integral with the wheel) estimated on the right and left wheel can be expressed as, taking into account (5.153a and 5.153b) and (5.155a and 5.155b), see Fig. 5.65:

$$\begin{aligned}\mathbf{V}_{cs} &= \mathbf{i}_s (V\cos\sigma - \dot{y}\sin\sigma - \dot{\sigma} S) + \mathbf{j}_s (V \sin\sigma + \dot{y}\cos\sigma) - \mathbf{i}_s \Omega R_s \\ &= \mathbf{i}_s (V\cos\sigma - \dot{y}\sin\sigma - \dot{\sigma} S - \Omega R_s) + \mathbf{j}_s (V\sin\sigma + \dot{y}\cos\sigma) = \mathbf{i}_s V_{Ls} + \mathbf{j}_s V_{Ts} \\ \mathbf{V}_{cd} &= \mathbf{i}_d (V\cos\sigma - \dot{y}\sin\sigma + \dot{\sigma} S) + \mathbf{j}_d (V\sin\sigma + \dot{y}\cos\sigma) - \mathbf{i}_d \Omega R_d \\ &= \mathbf{i}_d (V\cos\sigma - \dot{y}\sin\sigma + \dot{\sigma} S - \Omega R_d) + \mathbf{j}_d (V\sin\sigma + \dot{y}\cos\sigma) = \mathbf{i}_d V_{Ld} + \mathbf{j}_d V_{Td}\end{aligned} \tag{5.156}$$

Taking into account (5.156), for the left wheel and the right wheel, the longitudinal creepages (Fig. 5.66) are, therefore:

[23] If we only consider the unit vectors lying in the plane containing the two tracks.

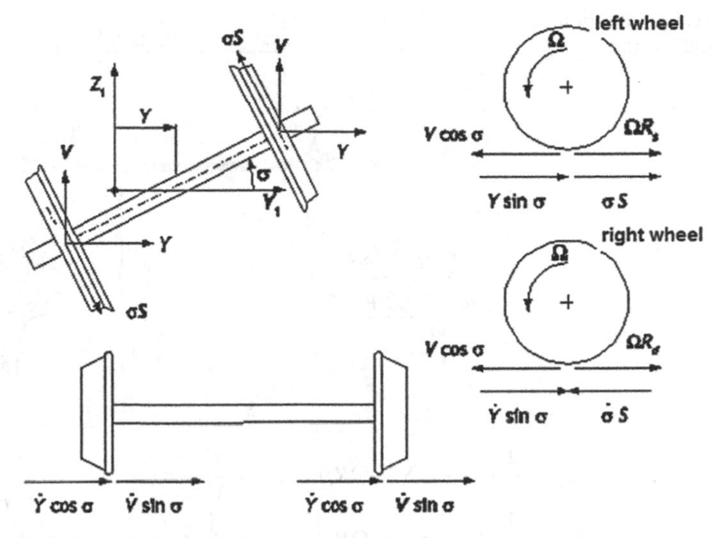

Fig. 5.66 Definition of longitudinal and transversal creepages

$$\begin{aligned} \varepsilon_{Ls} &= \frac{1}{V}(V\cos\sigma - \dot{y}\sin\sigma - \dot{\sigma}S - \Omega R_s) \\ \varepsilon_{Ld} &= \frac{1}{V}(V\cos\sigma - \dot{y}\sin\sigma + \dot{\sigma}S - \Omega R_d) \end{aligned} \tag{5.157a}$$

whereas the corresponding transversal creepages ((5.152b), Fig. 5.66) become:

$$\begin{aligned} \varepsilon_{Ts} &= \frac{1}{V}(V\sin\sigma + \dot{y}\cos\sigma) \\ \varepsilon_{Td} &= \frac{1}{V}(V\sin\sigma + \dot{y}\cos\sigma) \end{aligned} \tag{5.157b}$$

These expressions are non-linear in the independent variables y and σ. If we wish to linearise the motion equations around the steady-state configuration, corresponding to the position of centred wheelset (5.151a and 5.151b) we must develop the terms $\cos\sigma$, $\sin\sigma$ and the other non-linear terms around this position ($y_o = 0,\ \ \sigma_o = 0$):

$$\begin{aligned} \cos\sigma &\approx \cos\sigma_o + (-\sin\sigma_o)(\sigma - \sigma_o) = 1 \\ \sin\sigma &\approx \sin\sigma_o + (\cos\sigma_o)(\sigma - \sigma_o) = \sigma \\ \dot{y}\cos\sigma &\approx (\cos\sigma_o)\dot{y} - (\dot{y}\sin\sigma_o)(\sigma - \sigma_o) = \dot{y} \\ V\sin\sigma &\approx V(\cos\sigma_o)(\sigma - \sigma_o) = V\sigma \end{aligned} \tag{5.158}$$

and taking into account, in the products, of the linear terms only, we arrive at the linearised expression of the longitudinal and transversal creepages as a function of the system's coordinates:

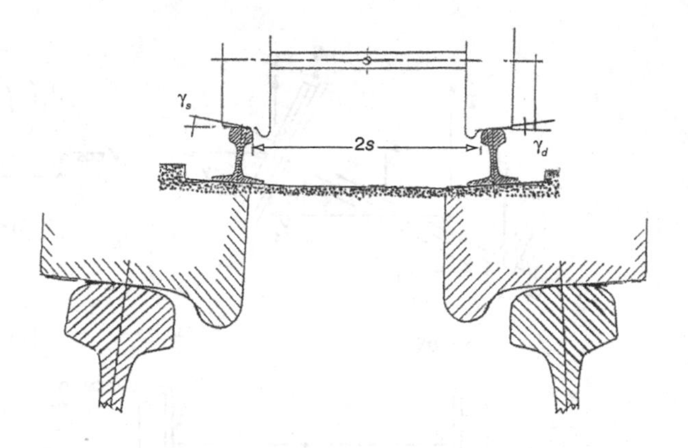

Fig. 5.67 Real profiles of the wheel rim and the rail

$$\begin{aligned} \varepsilon_{Ls} &= \frac{1}{V}(V - \dot{\sigma} S - \Omega R_s) \quad \varepsilon_{Ts} = \frac{\dot{y}}{V} + \sigma \\ \varepsilon_{Ld} &= \frac{1}{V}(V + \dot{\sigma} S - \Omega R_d) \quad \varepsilon_{Td} = \frac{\dot{y}}{V} + \sigma \end{aligned} \tag{5.159}$$

In (5.159), we should remember that R_s and R_d indicate the rolling radii of left wheel and the right wheel: in this regard, as mentioned in the introduction to this section, we must bear in mind the actual profile of the wheel flange and the rail (Fig. 5.67). For a positive lateral displacement y of the wheelset the right-hand wheel moves close to the flange, while the left-hand wheel moves away from it: the rolling radii of the two wheels are, therefore, a non-linear function of lateral displacement. In the event that we analyse small oscillations around the position of centred wheelset, contact remains in the truncated conical area of the wheel rim, called the plane: in this area the tangent in contact remains constant and equal to the conicity γ (equal for both wheels). It follows that a lateral displacement y of the wheelset (Fig. 5.67) results in a radius value equal to:

$$\begin{aligned} R_s &= R - y\gamma \\ R_d &= R + y\gamma \end{aligned} \tag{5.160}$$

R being the rolling radius in the centre wheelset position. When introducing these expressions into the definition of longitudinal creepage and remembering that $V = \Omega R$, we have:

$$\begin{aligned} \varepsilon_{Ls} &= \frac{1}{V}(V - \dot{\sigma} S - \Omega(R - y\gamma)) = \frac{y\gamma}{R} - \frac{\dot{\sigma} S}{V} \\ \varepsilon_{Ld} &= \frac{1}{V}(V + \dot{\sigma} S - \Omega(R + y\gamma)) = -\frac{y\gamma}{R} + \frac{\dot{\sigma} S}{V} \end{aligned} \tag{5.161}$$

The linearised contact forces (5.150a), taking into account (5.159) and (5.161) can now be expressed as a function of the free coordinates of the system:

$$
\begin{aligned}
F_{Ls} &= -f_o \varepsilon_{Ls} = -f_o \left(\frac{y\gamma}{R} - \frac{\dot{\sigma} S}{V} \right) \\
F_{Ld} &= -f_o \varepsilon_{Ld} = -f_o \left(-\frac{y\gamma}{R} + \frac{\dot{\sigma} S}{V} \right) \\
F_{Ts} &= -f_o \varepsilon_T = -f_o \left(\frac{\dot{y}}{V} + \sigma \right) \\
F_{Td} &= -f_o \varepsilon_{Td} = -f_o \left(\frac{\dot{y}}{V} + \sigma \right)
\end{aligned}
\tag{5.162}
$$

Note the expressions of the longitudinal and transversal components of the contact forces, it is now possible to calculate components F_y and M_σ (5.150a) of these forces according to the free coordinates of the wheelset:

$$
\begin{aligned}
F_y &= (F_{Ts} + F_{Td}) \cos\sigma - (F_{Ls} + F_{Ld}) \sin\sigma = -2f_o \left(\frac{\dot{y}}{V} + \sigma \right) \cos\sigma \\
M_\sigma &= (F_{Ld} - F_{Ls}) S = -2f_o \left(-\frac{y\gamma}{R} + \frac{\dot{\sigma} S}{V} \right) S
\end{aligned}
\tag{5.163}
$$

These components are still non-linear: it is possible to linearise these expressions around the steady-state position by assuming:

$$
\begin{aligned}
\cos\sigma &= 1 \\
\sin\sigma &= \sigma
\end{aligned}
\tag{5.164a}
$$

and, neglecting the higher order terms, we obtain:

$$
\begin{aligned}
F_y &= -2f_o \frac{\dot{y}}{V} - 2f_o \sigma \\
M_\sigma &= 2f_o \frac{yS}{R} \gamma - 2f_o \frac{\dot{\sigma} S^2}{V}
\end{aligned}
\tag{5.164b}
$$

At this point, it is more convenient, in the wheelset motion equation (5.150a) and (5.150b), to bring the terms to first member (5.164a and 5.164b) due to the linearised contact forces:

$$
\begin{aligned}
m\ddot{y} + 2r_y \dot{y} + 2k_y y + 2f_o \frac{\dot{y}}{V} + 2f_o \sigma &= 0 \\
J\ddot{\sigma} + 2b^2 r_x \dot{\sigma} + 2b^2 k_x \sigma - 2f_o \frac{yS}{R} \gamma + 2f_o \frac{\dot{\sigma} S^2}{V} &= 0
\end{aligned}
\tag{5.165}
$$

Now using $\underline{z}$ to define the vector of the free coordinates, ordered as:

$$\underline{z} = \begin{Bmatrix} y \\ \sigma \end{Bmatrix} \tag{5.166}$$

it is possible to rewrite the equation of motion (5.165) in matrix form, highlighting the contributions due to the linearisation of the force field:

$$[M]\ddot{\underline{z}} + [[R_s] + [R_F]]\dot{\underline{z}} + [[K_s] + [K_F]]\underline{z} = \underline{0} \tag{5.167}$$

where $[M]$, $[R_s]$ and $[K_s]$ are the matrices of structural mass, damping and stiffness, respectively:

$$[M] = \begin{bmatrix} M & 0 \\ 0 & J \end{bmatrix}; \quad [R_s] = \begin{bmatrix} 2r_y & 0 \\ 0 & 2r_x b^2 \end{bmatrix}; \quad [K_s] = \begin{bmatrix} 2k_y & 0 \\ 0 & 2k_x b^2 \end{bmatrix}; \tag{5.167a}$$

In (5.167), however, the equivalent matrices $[R_F]$ and $[K_F]$, due to linearisation of the force field of contact between wheel and rail, have the following expressions:

$$[R_F] = \begin{bmatrix} 2\frac{f_o}{V} & 0 \\ 0 & 2f_o\frac{s^2}{V} \end{bmatrix}; \quad [K_F] = \begin{bmatrix} 0 & 2f_o \\ -2f_o\frac{s\gamma}{R} & 0 \end{bmatrix}; \tag{5.167b}$$

By examining the matrices $[R_F]$ and $[K_F]$, we can judge the stability of the system, at least in qualitative terms. The equivalent stiffness matrix $[K_F]$ is non-symmetric, indicating the non-conservative nature of the force field, furthermore, the extra-diagonal terms are opposite in sign. Remembering what was said about fields of purely positional forces, we can conclude that the introduction of energy and the establishment of flutter instability are possible. The equivalent damping matrix (diagonal) $[R_F]$ is definitely symmetric and positive definite: this indicates that it is certainly dissipative, but its contribution decreases as forward speed V increases. Flutter instability will, therefore, arise if the positional part of the force field introduces energy and if the equivalent damping of the force field (matrix $[R_F]$), together with structural damping (matrix $[R_s]$) is not sufficient to dissipate the energy introduced. For a quantitative analysis of the problem of instability, we can calculate the values of $\lambda_i = \alpha_i \pm i\omega_i$ which solve the homogeneous algebraic equation that is obtained from (5.167) imposing the solution

$$\underline{z} = \underline{Z}e^{\lambda t} \tag{5.168}$$

i.e.:

$$[\lambda^2[M] + \lambda[[R_s] + [R_F(V)]] + [[K_s] + [K_F(V)]]]\underline{Z} = \underline{0} \tag{5.169}$$

So we can examine the real part of the generic eigensolution $\lambda_i(V) = \alpha_i(V) \pm i\omega_i(V)$: if one of these presents a positive real part $\alpha_i(V) > 0$, then the system is unstable. Given that the force field is also a function of the forward speed

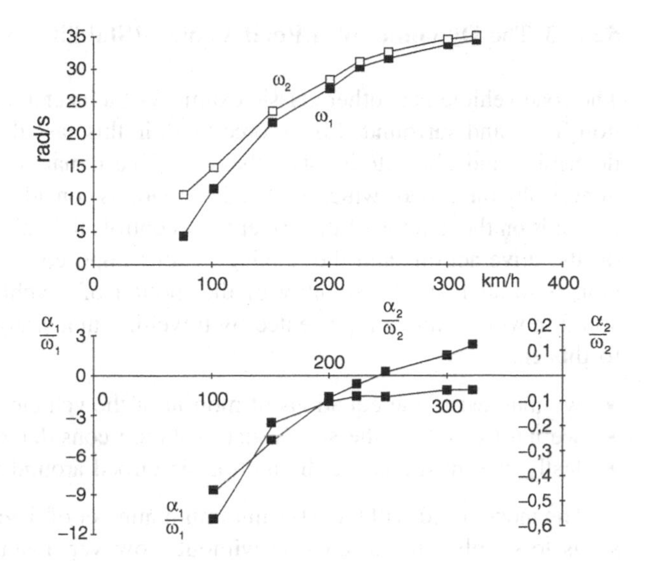

Fig. 5.68 Trend of the natural pulsations of the wheelset as a function of the forward speed V

V and so the actual matrices equivalent to the contact force field are functions of V, by repeating this analysis for different velocities (Fig. 5.68) it is possible to obtain the value of the flutter velocity V_{crit}, corresponding to which the sign of the real part $\alpha_i(V)$ of one of the eigenvalues $\lambda_i(V)$ changes (from negative to positive).

In conditions of instability, the resulting motion is a combination of lateral displacement y and of yaw σ. In the absence of the force field these motion occur at different frequencies:

$$\omega_y = \sqrt{\frac{2k_y}{m}}$$
$$\omega_\sigma = \sqrt{\frac{2k_x b^2}{J}} \tag{5.170}$$

and which the force field brings closer together, coupling them via the extra-diagonal terms of the equivalent stiffness matrix of the force field $[K_F]$. As mentioned previously, studying the instability of the railway wheelset using a two-degree-of-freedom model only qualitatively justifies the behaviour observed experimentally: to quantitatively study the stability of a railway vehicle it is necessary to resort to more sophisticated n-degree-of freedom models, which take into account not just the wheelsets but also the presence of the bogies and the carbody (Fig. 5.45).

5.5.2.3 The Dynamics of a Road Vehicle (Stability in a Straight Line)

The road vehicle is another classic example of a mechanical system that is strongly non-linear and surrounded by a force field, in this case these are contact forces that determine and characterise its behaviour. The dynamics of a vehicle that moves generically on a road (which in this discussion is considered flat) is complex; it also depends on the actions of the driver who controls the vehicle by steering, as well as on the drive actions and the braking actions imposed. The purpose of this discussion is to analysis the stability of the motion of a vehicle around a steady-state solution which can be represented by travelling in a straight line or around a bend; to this end:

- we must write the equations of motion of the vehicle in the most general form;
- we must calculate the steady-state solution considered (bend or straight line);
- lastly, we must analyse the motion perturbed around this steady-state solution.

The model used will have the minimum number of degree-of-freedom necessary, so as to simplify the discussion without, however, losing the essential characteristics of the problem: again, for simplicity's sake, we will also be considering stability of motion in a straight line only, i.e. the response of the vehicle subjected to perturbation around rectilinear motion, assuming that the forward speed V is constant and set. This velocity will be determined by conditions of equilibrium between aerodynamic resultant forces and driving forces reduced to the wheels, resulting from the torques applied to them. This analysis is interesting in its own right, and because it is possible in this way to highlight some characteristics of the behaviour of the vehicle by resorting to simple matrix equations. For a more detailed discussion of the problem, we refer the reader to specialised texts (see the specific bibliography on the subject) where, amongst other things, there are descriptions of more sophisticated mathematical models of the *multi-body* type (linear and non-linear with several degree-of-freedom) to reproduce, not just qualitatively but also quantitatively, all the problems regarding the dynamics of the actual vehicle. The first simplification that we will introduce is that of neglecting the following:

- relative vertical motions between chassis and wheels;
- roll (rotation around the forward movement axis of the vehicle);
- pitch (rotation around the transversal axis of the vehicle).

We also neglect, as a consequence of the previous assumptions, the dynamic variations of the vertical load N acting on the wheels (due to vertical motions, of pitch and roll). So it is legitimate to assume that the behaviour of the left wheel and the right is equal, so that we can reduce to a simplified model of the type shown in Fig. 5.69, i.e. a rigid beam (with mass m and moment of inertia with respect to the vertical axis equal to J) which reproduces the chassis and just the two front and rear wheels. These wheels have radius R and are considered to be without mass (and therefore with moment of inertia equal to zero) and are constrained integrally with the chassis: only rolling motion Ω is associated with them, but as we will see below

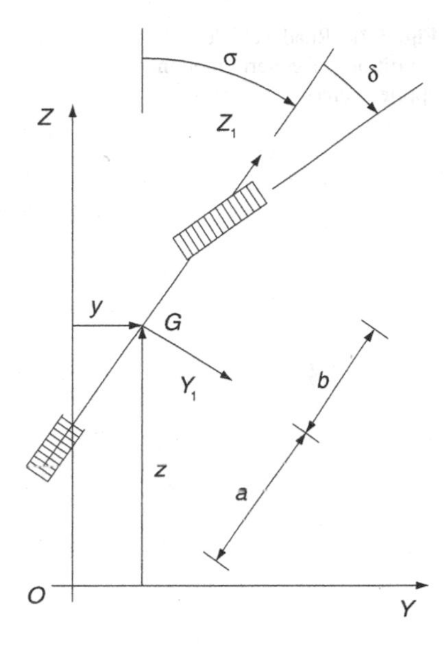

Fig. 5.69 Schematisation of a road vehicle, conditions of generic motion: reference system and independent coordinates

this degree-of-freedom will not be considered in the equations of motion. The simplified model of the vehicle (Fig. 5.69) is, in any case, represented by a system with 3-degree-of-freedom, corresponding to the degree-of-freedom of a rigid body in the plane, plus one that is introduced, to correspond to the steering angle δ of the front wheels. We will be considering an absolute reference system with origin O and axes $X - Y - Z$, (Fig. 5.69) and we will introduce a mobile reference system with origin O_1 coinciding with the centre of gravity of the body G and axes $X_1 - Y_1 - Z_1$ integral with the vehicle. We will consider as degree-of-freedom:

- the coordinates z and y of the vehicle's centre of gravity G which define, with respect to the fixed reference system, the origin of the mobile reference system;
- the angle σ that axis Z_1 forms with axis Z and which corresponds to the yaw rotation of the vehicle.

Lastly, we will use C_A and C_p to describe the torque applied, respectively, on the front and rear wheels of the vehicle. The equations of motion of the system, defined in this way, can be obtained by writing the equations of dynamic equilibrium in longitudinal Z and transversal Y direction and at rotation around the centre of gravity G (Fig. 5.70):

$$\begin{aligned}
&F_{L2}\cos\sigma - F_{T2}\sin\sigma - m\ddot{z} - F_{T1}\sin(\sigma+\delta) + F_{L1}\cos(\sigma+\delta) + F_{aerz} = 0 \\
&F_{L2}\sin\sigma + F_{T2}\cos\sigma - m\ddot{y} + F_{T1}\cos(\sigma+\delta) + F_{L1}\sin(\sigma+\delta) + F_{aery} = 0 \\
&F_{T2}a + J\ddot{\sigma} - F_{T1}b\,\cos\delta - F_{L1}b\,\sin\delta + M_{aer} = 0
\end{aligned} \tag{5.171}$$

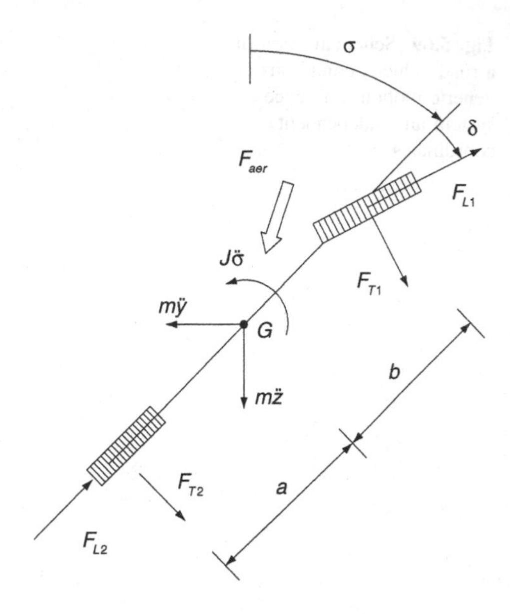

Fig. 5.70 Road vehicle, conditions of generic motion: applied forces

F_{L1} and F_{L2} being the contact forces acting in a parallel direction to the midplane of the generic wheel, F_{T1} and F_{T2} the corresponding transversal contact forces: these forces depend on the longitudinal slip ε_L in the contact patch and on the yaw angles α that the velocity of the wheel centre forms with its midplane. In Eq. (5.171) we have neglected the self-aligning torques M_1 and M_2 on the front and rear wheel, which are also functions of the yaw angles (respectively α_A and α_P). In (5.171) the terms F_{aer} and M_{aer} represent, respectively, the resultant of the aerodynamic force and the aerodynamic moment.

Equation (5.171) make it possible to define the motion of the vehicle by assigning, for example:

- the initial conditions;
- the variation law of the steering $\delta = \delta(t)$;
- the variation laws of the torques (drive and braking) $C_A = C_A(t)$ and $C_P = C_P(t)$.

As far as the longitudinal forces are concerned, assuming that we can neglect the rotary moment of inertia J_{Ry} of the single wheel (the wheels are assumed to be without mass):

$$J_{Ry} = 0 \tag{5.172}$$

from the equilibrium at rotation of the same (Fig. 5.71) it is possible to obtain a link between the generic applied torque C and the corresponding longitudinal component F_L of the contact force which for the front and rear wheel equals:

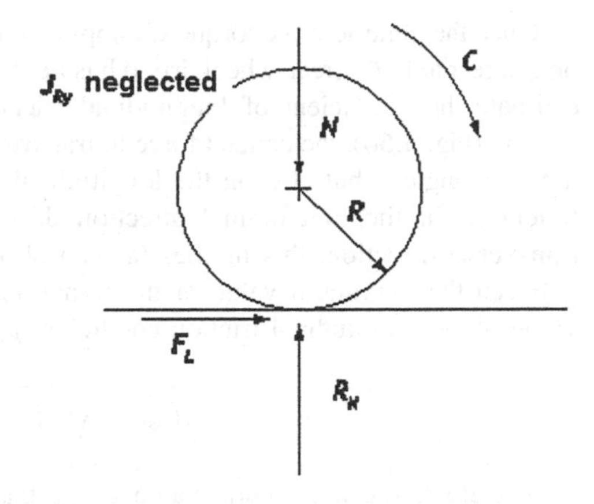

Fig. 5.71 Road vehicle: equations of equilibrium at rotation of the generic wheel

$$F_{L1} = \frac{C_A}{R}; \quad F_{L2} = \frac{C_P}{R} \tag{5.173}$$

As the link between the longitudinal component F_L of the contact force and the normal load acting on the actual wheel N is known:

$$\begin{aligned} F_{L1} &= \mu_{L1} N_A \\ F_{L2} &= \mu_{L2} N_P \end{aligned} \tag{5.174}$$

it is possible to define the coefficient of longitudinal friction on both the wheels, taking into account both (5.173) and (5.174):

$$\begin{aligned} \mu_{L1} &= \frac{C_A}{N_A R} \\ \mu_{L2} &= \frac{C_P}{N_P R} \end{aligned} \tag{5.175}$$

In (5.175) the normal loads N, with a defining the distance of the rear axle from the centre of gravity G and b defining thc distance of the front axle (Fig. 5.69), are distributed as follows[24]:

$$\begin{aligned} N_A &= mg\frac{a}{(a+b)} \\ N_P &= mg\frac{b}{(a+b)} \end{aligned} \tag{5.176}$$

[24]In this discussion, we do not take into account the variations in the normal actions N_A and N_P due to the longitudinal acceleration to which the vehicle is subject.

Once the generic drive torque C_A applied on the front axle (front-wheel drive) or on the rear axle C_p (rear-wheel drive) has been defined, from (5.175) it is possible to estimate the coefficient of longitudinal friction μ_L. As we saw in the previous section (Fig. 5.60), the contact force in transversal direction F_T depends not only on the yaw angle α but also on the longitudinal slip ε_L as the required friction coefficient μ_L in the longitudinal direction decreases the adhesion available in the transversal direction: this implies (as can also be seen in the Footnote 19) a link between the maximum value of the transversal friction coefficient $\mu_{T\max}$ and the value of the longitudinal friction coefficient μ_L (Fig. 5.73):

$$\mu_{\max} = \sqrt{\mu_L^2 + \mu_{T\max}^2} \tag{5.177a}$$

Therefore, the asymptotic value $\mu_{T\max}$ that the transversal friction coefficient $\mu_{T\max}$ can reach, becomes:

$$\begin{aligned} \mu_{T\max\mathrm{A}} &= \sqrt{\mu_{\max}^2 - \mu_{L1}^2}\,\mathrm{e}^{-c\Delta N_A} \\ \mu_{T\max\mathrm{P}} &= \sqrt{\mu_{\max}^2 - \mu_{L2}^2}\,\mathrm{e}^{-c\Delta N_P} \end{aligned} \tag{5.177b}$$

If the longitudinal slip ε_L is such as to completely saturate the constraint, we will obviously have (see, for example, Fig. 5.54b):

$$\begin{aligned} &\text{se}\ \varepsilon_{L1} > 0.25 \Rightarrow \mu_{L1} \Rightarrow \mu_{\max}\text{and}\ \mu_{T\max\mathrm{A}} \Rightarrow 0 \\ &\text{se}\ \varepsilon_{L2} > 0.25 \Rightarrow \mu_{L2} \Rightarrow \mu_{\max}\text{and}\ \mu_{T\max\mathrm{A}} \Rightarrow 0 \end{aligned} \tag{5.178}$$

In (5.177b) the term $\mathrm{e}^{-c\Delta N}$ has been added which, with a simplified empirical formula, takes into account the reduction of the force that can be transmitted as a function of the variation of the normal load compared to the reference load N_{rif}, defined via:

$$\Delta N_A = \frac{N_A - N_{rif}}{N_{rif}}; \quad \Delta N_P = \frac{N_P - N_{rif}}{N_{rif}}; \quad N_{rif} = \frac{N_A + N_P}{2} \tag{5.178a}$$

So the lateral forces become:

$$\begin{aligned} F_{T1} &= \mu_{T1}(\alpha_A)N_A \\ F_{T2} &= \mu_{T2}(\alpha_P)N_P \end{aligned} \tag{5.179}$$

The dependency of the transversal friction coefficient μ_T on the yaw angle α (Figs. 5.54 and 5.72), is subsequently approximated by an analytical expression of the form:

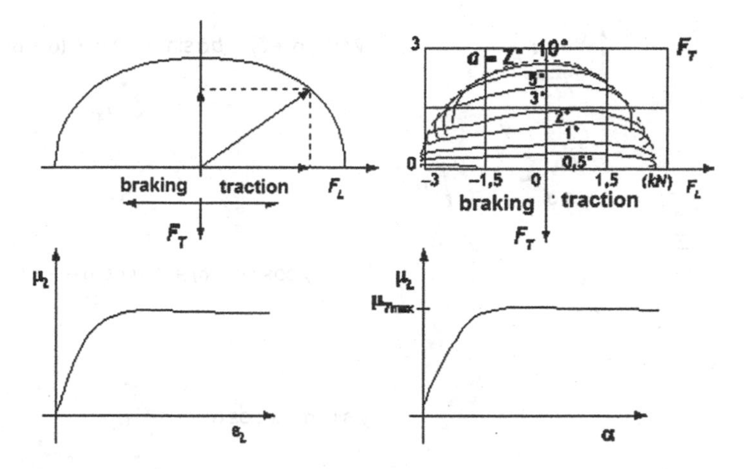

Fig. 5.72 Coefficients of longitudinal and transversal friction

$$
\begin{aligned}
\mu_{T1} &= \mu_{T1}(\alpha_A) = \left(1 - e^{-k|\alpha_A|}\right)\mu_{T\text{maxA}}\, sign(\alpha_A) \\
\mu_{T2} &= \mu_{T2}(\alpha_P) = \left(1 - e^{-k|\alpha_P|}\right)\mu_{T\text{maxP}}\, sign(\alpha_P)
\end{aligned}
\tag{5.180}
$$

where with the terms:

$$
\begin{aligned}
sign(\alpha_A) &= \frac{\alpha_A}{|\alpha_A|} \\
sign(\alpha_P) &= \frac{\alpha_P}{|\alpha_P|}
\end{aligned}
\tag{5.181}
$$

we take into account the fact that the functions μ_{T1} and μ_{T2} are odd compared to the yaw angle α_A α_P. In (5.179) the yaw angles α_A and α_P are functions of the independent variables (Fig. 5.73):

$$
\begin{aligned}
\alpha_A &= \operatorname{atan}\left(\frac{\dot{y}\cos(\sigma+\delta) + \dot{\sigma}b\cos\delta - \dot{z}\sin(\sigma+\delta)}{\dot{z}\cos(\sigma+\delta) + \dot{y}\sin(\sigma+\delta) + \dot{\sigma}b\cos\delta}\right) \\
\alpha_P &= \operatorname{atan}\left(\frac{\dot{y}\cos\sigma + \dot{\sigma}a - \dot{z}\sin\sigma}{\dot{z}\cos\sigma + \dot{y}\sin\sigma}\right)
\end{aligned}
\tag{5.182}
$$

By replacing the expressions of the longitudinal forces of contact F_{L1} and F_{L2} (5.173) and lateral forces F_{T1} and F_{T2} (5.179) in the vehicle's equations of motion (5.171), we obtain equations that are non-linear in the same forces of contact. These equations, integrated numerically, make it possible to calculate the generic motion of the vehicle, once the law of variation of the steering angle $\delta = \delta(t)$ and of the drive or braking torque has been assigned.

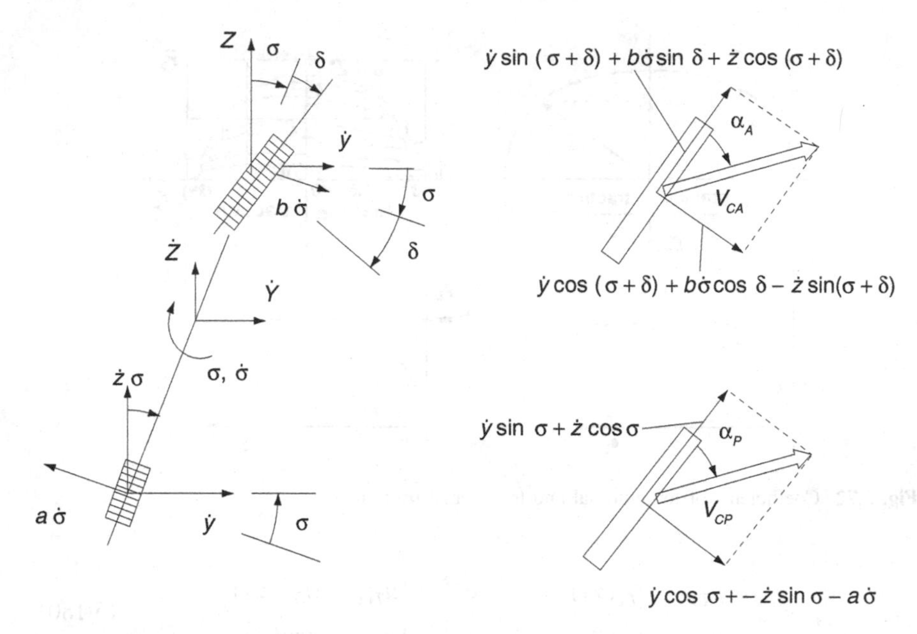

Fig. 5.73 Road vehicle, conditions of generic motion: yaw angles

If we wish to analyse the stability of the vehicle travelling in a straight line, we must linearise these expression, around the steady-state configuration defined by the conditions:

$$\dot{z} = V; \quad y_o = 0; \quad \sigma_o = 0; \quad \delta_o = 0 \tag{5.183}$$

By imposing (5.183) in (5.171), taking into account (5.173), we obtain the equation of steady-state equilibrium:

$$\frac{C_{Po}}{R} + \frac{C_{Ao}}{R} - \frac{1}{2}\rho V^2 S C_x = 0 \tag{5.184}$$

S being the leading surface of the vehicle and C_x the drag coefficient. This equation makes it possible to find the front C_{Ao} or rear C_{Po} drive torque acting on the system to make the vehicle move forward at a certain assigned velocity V. In order to solve the actual equation, in the case of all-wheel drive, we need to put in a system with (5.184), the equation that defines the motion of the transmission and which allows us to correlate the front and rear torque, formally, using a generic relation:

$$C_A = C_A(C_P, \delta, V, \textit{vehicle}, \textit{contact}, \textit{transmission characteristics}) \tag{5.185}$$

It is now possible to analyse the perturbed motion around the steady-state position, defined like so with regard to the transversal motion and the yaw rotation, not considering the perturbation according to Z, thus assuming V = constant and $\delta = \delta_o = 0$. Therefore, we will assume small displacements and rotations around this position while considering the actual terms as infinitesimal. Lastly, in the discussion we will neglect the linearised terms of the aerodynamic forces, i.e. we will not consider the fluid-elastic effects. Now we will impose in (5.171) a change of coordinates considering as independent variables, similarly to the previous section, the perturbation of the motion $\overline{y}$ and $\overline{\sigma}$ around the steady-state, rectilinear configuration:

$$\begin{aligned} y &= y_o + \overline{y} = \overline{y} \\ \sigma &= \sigma_o + \overline{\sigma} = \overline{\sigma} \end{aligned} \tag{5.186}$$

By replacing (5.186) in (5.171) and replacing the cosines and sines of yaw rotation, respectively, with one and with the actual angle, the linearised equations of motion are:

$$\begin{aligned} -\frac{C_{Po}}{R}\overline{\sigma} - F_{T2} + m\ddot{y} - F_{T1} - \frac{C_{Ao}}{R}\overline{\sigma} &= 0 \\ F_{T2}a + J\ddot{\sigma} - F_{T1}b &= 0 \end{aligned} \tag{5.187}$$

$$F_T = F_T(\alpha) \tag{5.188}$$

So the same can be linearised as follows:

$$F_T = F_{To} + \left(\frac{\partial F_T}{\partial \alpha}\right)_o (\alpha - \alpha_o) = \left(\frac{\partial F_T}{\partial \alpha}\right)_o (\alpha - \alpha_o) \tag{5.189}$$

directly expressing the generic term α in a linearised form. Taking into account (5.189), (5.180) and (5.181) the generic term $\left(\frac{\partial F_T}{\partial \alpha}\right)$ becomes:

$$\begin{aligned} \left(\frac{\partial F_{T1}}{\partial \alpha_A}\right) &= -\left(k - \mathrm{e}^{-k|\alpha_A|}\right)\mu_{T\max A} N_A \\ \left(\frac{\partial F_{T2}}{\partial \alpha_P}\right) &= -\left(k - \mathrm{e}^{-k|\alpha_P|}\right)\mu_{T\max P} N_P \end{aligned} \tag{5.190}$$

i.e. remembering (5.181)

$$\begin{aligned} \left(\frac{\partial F_{T1}}{\partial \alpha_A}\right)_o &= -k\mu_{T\max A}N_A = -k\sqrt{f_a^2 - \mu_{L1}^2}\,\mathrm{e}^{-c\Delta N_A}N_A = -\overline{k}_A \\ \left(\frac{\partial F_{T2}}{\partial \alpha_P}\right)_o &= -k\mu_{T\max P}N_P = -k\sqrt{f_a^2 - \mu_{L2}^2}\,\mathrm{e}^{-c\Delta N_P}N_P = -\overline{k}_P \end{aligned} \tag{5.191}$$

The terms present in (5.191) represent the derivative of the transversal force function F_T with respect to the yaw angle α estimated in the origin (Fig. 5.74). The

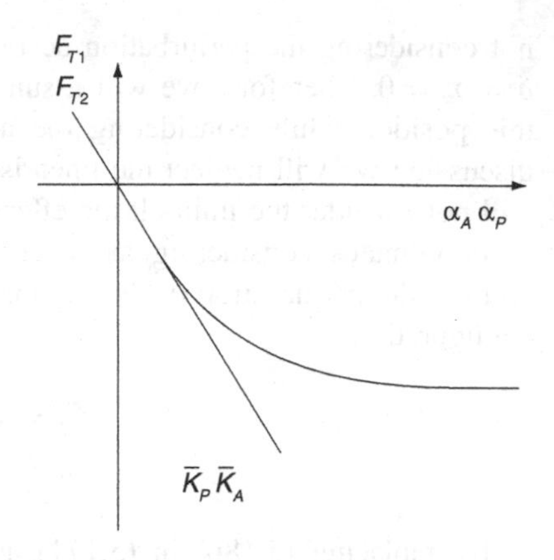

Fig. 5.74 Straight line motion: physical meaning of terms $\overline{k}_A$ and $\overline{k}_P$

linearised expression of the yaw angles, however, (5.182) becomes, being displacements and rotations small (i.e. considered as infinitesimal terms):

$$\begin{aligned}\alpha_A &= \operatorname{atan}\left(\frac{\dot{\overline{y}} + \dot{\overline{\sigma}} b - V\overline{\sigma}}{V}\right) = \left(\frac{\dot{\overline{y}} + \dot{\overline{\sigma}} b - V\overline{\sigma}}{V}\right)\\ \alpha_P &= \operatorname{atan}\left(\frac{\dot{\overline{y}} - \dot{\overline{\sigma}} a - V\overline{\sigma}}{V}\right) = \left(\frac{\dot{\overline{y}} - \dot{\overline{\sigma}} a - V\overline{\sigma}}{V}\right)\end{aligned} \tag{5.192}$$

Taking into account (5.189), (5.191) and (5.192) we directly arrive at the motion equations perturbed around the steady-state rectilinear configuration:

$$\begin{aligned}-\frac{C_{Po}}{R}\overline{\sigma} + \frac{\overline{k}_P}{V}\dot{\overline{y}} - a\frac{\overline{k}_P}{V}\dot{\overline{\sigma}} - \overline{k}_P\overline{\sigma} + m\ddot{\overline{y}} - \frac{C_{Ao}}{R}\overline{\sigma} + \frac{\overline{k}_A}{V}\dot{\overline{y}} + b\frac{\overline{k}_A}{V}\dot{\overline{\sigma}} - \overline{k}_A\overline{\sigma} = 0\\ -a\frac{\overline{k}_P}{V}\dot{\overline{y}} + a^2\frac{\overline{k}_P}{V}\dot{\overline{\sigma}} + a\overline{k}_P\overline{\sigma} + J\ddot{\overline{\sigma}} + b\frac{\overline{k}_A}{V}\dot{\overline{y}} + b^2\frac{\overline{k}_A}{V}\dot{\overline{\sigma}} - b\overline{k}_A\overline{\sigma} = 0\end{aligned} \tag{5.193}$$

Using:

$$\underline{\overline{z}} = \begin{Bmatrix}\overline{y}\\ \overline{\sigma}\end{Bmatrix} \tag{5.194}$$

to define the vector of the new independent variables that describe the perturbed motion around the position of straight line motion at constant velocity V, the equations, which are linearised at this point, become, in matrix form:

$$[M]\ddot{\underline{\overline{z}}} + [R_F]\dot{\underline{\overline{z}}} + [K_F]\underline{\overline{z}} = \underline{0} \tag{5.195a}$$

being in this case (5.194):

$$[M] = \begin{bmatrix} m & 0 \\ 0 & J \end{bmatrix} \quad (5.195b)$$

$$[R_F] = \begin{bmatrix} \left(\frac{\bar{k}_A}{V} + \frac{\bar{k}_P}{V}\right) & \left(b\,\frac{\bar{k}_A}{V} - a\frac{\bar{k}_P}{V}\right) \\ \left(b\,\frac{\bar{k}_A}{V} - a\frac{\bar{k}_P}{V}\right) & \left(b^2\,\frac{\bar{k}_A}{V} + a^2\,\frac{\bar{k}_P}{V}\right) \end{bmatrix} \quad (5.195c)$$

$$[K_F] = \begin{bmatrix} 0 & -\bar{k}_A - \bar{k}_P - \frac{C_{Po}}{R} - \frac{C_{Ao}}{R} \\ 0 & a\bar{k}_P - b\bar{k}_A \end{bmatrix} \quad (5.195d)$$

The limited stability of the steady-state motion is estimated by calculating the eigensolutions of the Eqs. (5.196a) and (5.196b); imposing a solution of the form:

$$\underline{x} = \underline{X}e^{\lambda t} \quad (5.196a)$$

where

$$\lambda_i = \alpha_i \pm i\omega_i \quad (5.196b)$$

represents the generic eigensolution which can be found by bringing (5.196a and 5.196b) back to an eigenvector- eigenvalues problem. The generic free motion perturbed around the steady-state configuration will be stable if all the real parts α_i of all the eigenvalues are negative and unstable if at least one of them is positive. The damping matrix $[R_F]$ is symmetric and positive definite [the terms $\bar{k}_P$ and $\bar{k}_A$ certainly being positive, (5.191)] and therefore, its effect, being inversely proportional to the forward speed V, is purely dissipative. The stiffness matrix $[K_F]$ is, on the other hand, non- symmetric: it may present a term on the main diagonal that could make the system statically unstable. This happens, for example, $\bar{k}_P$ and $\bar{k}_A$ being equal, when the distance b of the front axle from the centre of gravity is greater than the distance a of the rear axle from G, i.e. when the centre of gravity is moved backwards.

Figure 5.75 shows an example of the trend of the eigensolutions $\lambda_i = \alpha_i \pm i\omega_i$ (5.196a) upon variation of the forward speed V (between 50 and 250 km/h) in the case of a car with rear-wheel drive.

More specifically, the configuration with forward centre of gravity (indicated with the "diamond" symbol) is always more stable, since the real parts α_i of the solutions are always zero or negative. Two complex conjugate solutions $\omega_2 = -\omega_4 \neq 0$ correspond to a damped oscillating motion; the solutions $\lambda_1 = \lambda_3 = 0$ are associated with the instability of the system, a fact that is also confirmed by the presence of a column of zero terms in the stiffness matrix $[K_F]$ (Eq. (5.195d).

In the case of a rearward centre of gravity, however ("dot" symbol), for speeds of over 130 km/h, one eigenvalue has a positive real part $\alpha_3 > 0$, indicating, being $\omega_3 = 0$, static instability (or divergence).

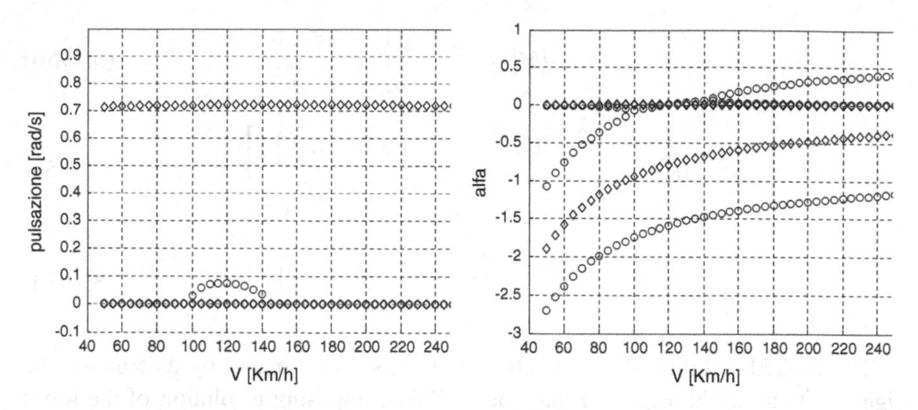

Fig. 5.75 Road vehicle with rear traction in a straight line: trend of the natural frequencies $\lambda_i(V) = \alpha_i(V) + i\omega_i(V)$ upon variation of the forward speed with: forward centre of gravity (*diamond*), rearward centre of gravity (*dot*)

In order to describe the dynamics of a road vehicle more accurately, we must use more complex numerical simulation models: see the bibliography about this, in particular the section regarding the *Dynamics of road and rail vehicles*.

The schematisations that can be used often require 14-degree-of-freedom models (Fig. 5.76, [69, 70]) which take into account the motions of the carbody, and the shaking and rolling of the wheels, or more complex models which adopt the multi-body techniques ([5, 75–78, 106], Fig. 5.76). However, this is not the place for a detailed discussion on the dynamics of the motor vehicle and, for this purpose, we refer the reader to the bibliography.

As an example, Figs. 5.78 and 5.79, show the behaviour of a vehicle travelling on a bend, in steady-state conditions, estimated using a 14-degree-of-freedom model [69]. In the case in question, we assume that the forward speed is constant (V = 100 km/h) and impose rotation (gradually increasing) on the front wheels δ (a manoeuvre defined as *steering pad constant velocity*).

The two figures show, respectively:

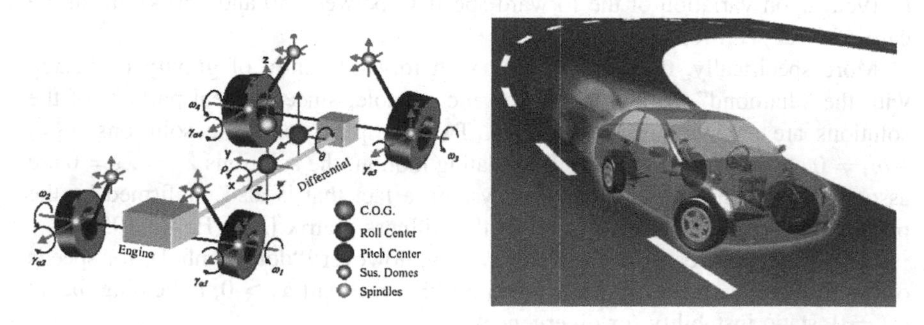

Fig. 5.76 Road vehicle: model with 14 d.o.f.. and vehicle schematised using the multibody system (ADAMS)

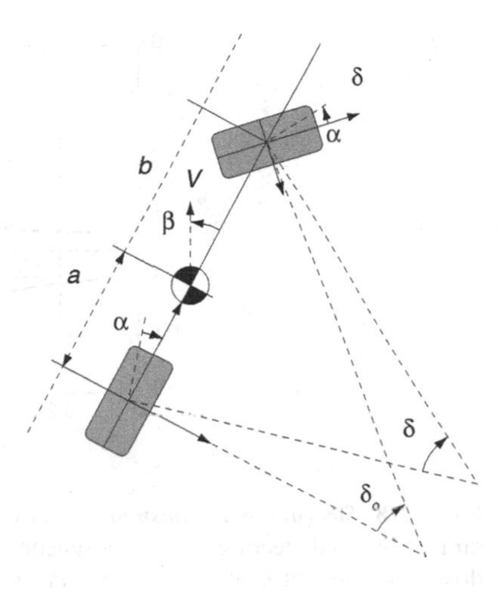

Fig. 5.77 Definition of the angle of Ackermann δ_o and angle of vehicle balance β

- the trend of the actual steering angle δ compared to the angle of Ackermann δ_o [97, 102, 107], that is, to what we would have in the ideal case of kinematic steering, i.e. with yaw angles equalling zero (see Fig. 5.77);
- the trend of the radius of curvature ρ of the trajectory reached in steady-state conditions, compared to that which we would obtain in conditions of kinematic steering ρ_o.

Both these quantities are represented as functions of non-dimensional lateral acceleration compared to gravitational acceleration. The diagrams show the case of a vehicle with rear-wheel drive (dotted line), of a vehicle with front-wheel drive and forward centre of gravity (dashed line) and of a vehicle with rear-wheel drive but with a rearward centre of gravity (continuous line).

As we can see, the vehicle with front-wheel drive and forward centre of gravity is *understeering*, that is, it enters a steady-state trajectory with a greater radius (Fig. 5.79) and, with the same transversal acceleration, requires a greater steering angle (Fig. 5.78). On the contrary, the vehicle with rear-wheel drive and rearward centre of gravity is *oversteering*, that is, it enters a trajectory that is narrower than that defined by the kinematic trajectory (Fig. 5.79) and, to develop the same transversal acceleration, requires a lower steering angle than the kinematic one: this vehicle, above a certain value, is subject to static instability (*yaw*).

When estimating the dynamic behaviour of a vehicle it often becomes important to introduce the active control introduced by the driver into the simulation model [69–71]. The driver represents a complex control system that introduces actions both in feed forward and in feed back (see, for example [55], ADAMS/DRIVER and ADAMS/ANDROID).

Again, as an example, Figs. 5.80 and 5.81 show some simulations, developed with a 14-degree-of-freedom model [69–71], introducing a simplified model of the driver, described as a simple *trajectory tracker* [86]. The simulated manoeuvre is a double change of lanes at constant speed (V = 100 km/h): Fig. 5.80 shows the

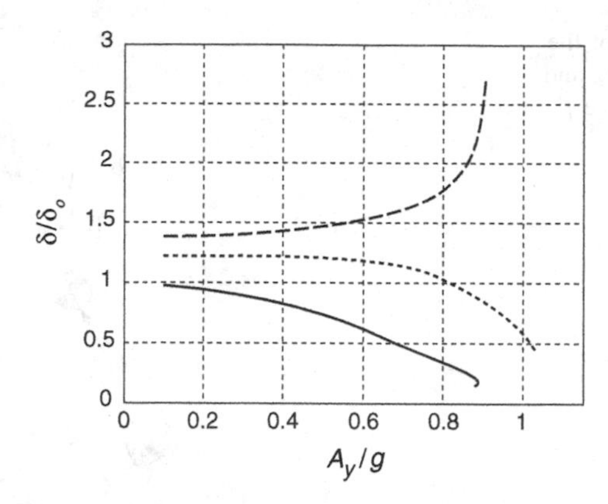

Fig. 5.78 *Steering pad constant velocity* manoeuvre (V = 100 km/h). Trend of the non-dimensionalised steering angle as a function of transversal acceleration: vehicle with front-wheel drive and forward centre of gravity (*dashed line*), vehicle with rear-wheel drive (*dotted line*), vehicle with rear-wheel drive and rearward centre of gravity (*continuous line*)

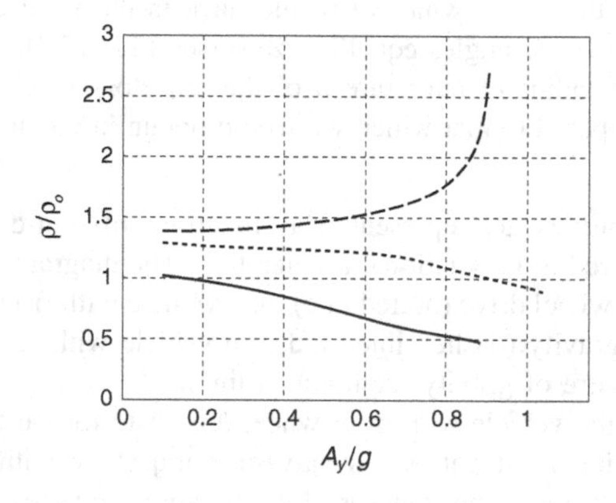

Fig. 5.79 *Steering pad constant velocity* manoeuvre (V = 100 km/h). Trend of the non-dimensionalised radius of the trajectory as a function of transversal acceleration: vehicle with front-wheel drive and forward centre of gravity (*dashed line*), vehicle with rear-wheel drive (*dotted line*), vehicle with rear-wheel drive and rearward centre of gravity (*continuous line*)

trajectory travelled by the vehicle: more specifically, the figure shows the reference trajectory (continuous line), the trajectory followed by the vehicle using a model of it and of the driver (dashed line) and lastly, the trajectory travelled by the vehicle where the steering wheel angle has been provided in an open loop (dotted line). Figure 5.81, however, shows the history of the steering wheel angle imposed on input in the open loop simulation and that imposed by the driver (dashed line) to follow the assigned reference trajectory. The results obtained show, even using a simplified driver model, the effect of whether this "component of the system" is present or not in defining the dynamic behaviour of the actual vehicle.

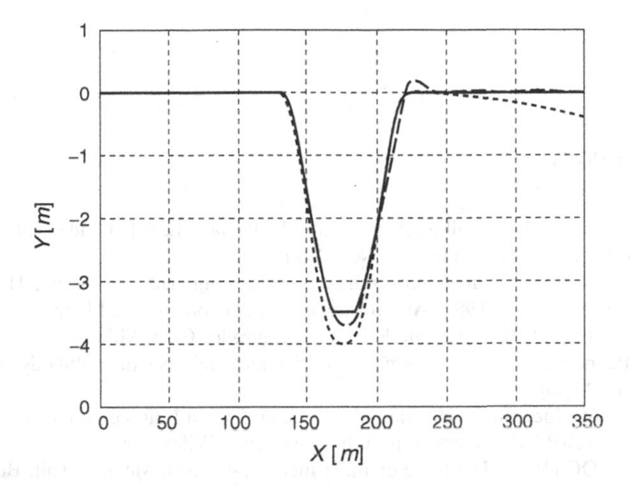

Fig. 5.80 Manoeuvre: double lane change at constant speed (V = 100 km/h): reference trajectory (*continuous line*), trajectory followed by the model vehicle + driver (*dashed line*), trajectory followed by the vehicle with steering wheel angle provided in open loop (*dotted line*)

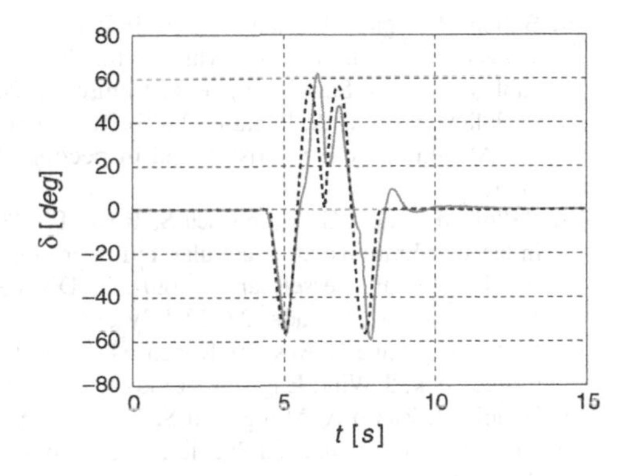

Fig. 5.81 Manoeuvre: double lane change at constant speed (V = 100 km/h) steering angle set by the driver (*dashed line*), steering line provided in open loop (*dotted line*)

Summary This chapter introduces the mechanical systems subjected to force fields, where the excitation depends on the state of thc systcm. As known, these force fields can change the stability characteristics of the mechanical system. The analyses is conducted by introducing the force field characteristics (positional or velocity dependent forces) obtained by linearizing the equations of motion (for 1, 2 and "n" degree-of-freedom systems) in the neighborhood of the static equilibrium position. Stability is analyzed by applying the eigenvalues-eigenvector approach. Some real applications are presented: aerodynamic forces on airfoils, contact force effects in cutting tools, hydrodynamic lubrication in journal bearings and contact force effects in rail and road vehicles. Some real applications are presented: aerodynamic forces on airfoils, contact force effects in cutting tools, hydrodynamic lubrication in journal bearings and contact force effects in rail and road vehicles.

References

General Theory

1. Bachschmid N, Bruni S, Collina A, Pizzigoni B, Resta F (2004) Fondamenti di Meccanica teorica ed Applicata. Mc Graw Hill, New York
2. Bathe KJ (1982) Finite element procedures in engineering analysis. Prentice-Hall, New York
3. Bittanti S, Schiavoni N (1982) Automazione e regolazione. Clup, Milan
4. Bittanti S, Schiavoni N (1982) Modellistica e controllo. Clup, Milan
5. Cheli F, Pennestrì E (2006) Cinematica e dinamica dei sistemi multibody. Casa Editrice Ambrosiana, Milan
6. Sesini O (1965) Meccanica applicata alle macchine. Casa Editrice Ambrosiana, Milan
7. Shabana AA (1989) Dynamics of multibody systems. Wiley, New York
8. Zienkiewicz OC (1987) The finite element method, 4th edn. McGraw-Hill, Berkshire

Aerodynamic Forces Acting on Structures

9. Belloli M, Cigada A, Diana G, Rocchi D (2003) Wind-tunnel investigation on vortex induced vibration of a long flexible cylinder. In: Proceedings of the 5th international symposium on cable dynamics, Santa Margherita Ligure, pp 247–254, 15–18 September 2003
10. Belloli M, Collina A, Resta F (2006) Rope vibrations under wind action. In: Proceedings of OITAF seminar characteristics and inspection of ropes, Grenoble, France, pp 1–6, 27 April 2006
11. Belloli M, Diana G, Muggiasca S, Resta F (2006) A numerical model to reproduce vortex induced vibrations of a circular cylinder. In: Proceedings of PVP2006-ICPVT-11 2006 ASME pressure vessels and Pijournalg Division conference, PVP2006-ICPVT11- 93971, Vancouver, BC, Canada, 23–27 July 2006
12. Belloli M, Diana G, Resta F, Rocchi D (2006) On the vortex shedding forcing on suspension bridge deck. J Wind Eng Ind Aerodyn 94(5):341–363
13. Belloli M, Zasso A, Muggiasca S, Giappino S (2005) Pressure field analysis on oscillating circular cylinder. In: Asia Pacific conference on wind engineering APCWE VI, Seoul Korea, Sept 2005
14. Cheli F, Cigada A, Diana G, Falco M (2003) The aeroelastic study of the Messina Straits Bridge, vol 30. Kluwer Academic Publishers, London, pp 79–105
15. Davenport AG (1961) The application of statistical concepts to the wind, loading of structures. In: Proceedings of Institution of Civil Engineers
16. Diana G et al (1990) Oscillation of bundle conductors in overhead lines due to turbulent wind. In: IEEE winter meeting, New York
17. Diana G, Cheli F et al (1992) Aeolian vibrations of overhead transmission lines: computation in turbulence conditions. In: I interno symposium on computational wind engineering
18. Diana G, Cheli F et al (1992) Suspension bridge parameter identification in full scale test. J Wind Eng Ind Aerodyn
19. Diana G, Gasparetto M (1980) The equivalent oscillator. simulation of vortex shedding on cylindrical bodies by means of an equivalent oscillator. L'Energia Elettrica 8
20. Diana G, Bruni S, Rocchi D (2005) Experimental and numerical methodologies to define the response of a long span bridge to wind. In: Proceedings of the sixth Asia-Pacific conference on wind engineering (APCWE VI), , Seoul, Korea, Sept 2005

21. Diana G, Resta F, Zasso A, Belloli M, Rocchi D (2003) Wind effects on suspension bridges: the case of the Messina strait bridge. In: Invited lecture at 5th international symposium on cable dynamics, Santa Margherita Ligure, 15–18 Sept 2003
22. Diana G, Resta F, Zasso A, Belloli M, Rocchi D (2004) Forced motion and free motion aeroelastic tests on a new concept dynamometric section model of the Messina suspension bridge. J Wind Eng Ind Aerodyn 92:441–462
23. Diana G, Resta F, Zasso A, Belloli M, Rocchi D (2004) Effects of the yaw angle on the aerodynamic behaviour of the Messina multi-box girder deck section. Wind Struct 7(1): 41–54
24. Diana G, Bruni S, Rocchi D (2005) A numerical and experimental investigation on aerodynamic non linearities in bridge response to turbulent wind. In: Proceedings of EACWE4, Prague, Czech Republic, 11–15 July 2005
25. Diana G, Resta F, Rocchi D (2006) A new numerical approach to reproduce bridge aerodynamic non linearity in time domain. In: Proceedings of CWE2006 4th international symposium on computational wind engineering, Yokohama, Japan
26. Diana G, Resta F, Belloli M, Rocchi D (2006) On the vortex shedding forcing on suspension bridge deck. J Wind Eng Ind Aerodyn 94:341–363
27. Diana G, Manenti A, Cosmai U, Hearnshaw D, Laneville A, Papailiou KO (2006) Aeolian vibration. In: Diana G, Cloutier L, Lilien J, Hardy C, Wang J, Goel A (eds) Electric Power Research Institute, EPRI transmission line reference book: wind-induced conductor motion, pp 2.3–2.157
28. Diana G, Cloutier L, Lilien J-L, Hardy C, Wang J, Goel A (2006) EPRI transmission line reference book—wind-induced conductor motion. Electric Power Research Institute, Palo Alto
29. Diana G, Cosmai U, Laneville A, Manenti A, Hearnshaw D, Papailiou KO (2006) aeolian vibration. in: epri transmission line reference book—wind-Induced Conductor Motion. Electric Power Research Institute, Palo Altopp, pp 2.3–2.150
30. Diana G, Bocciolone M, Manenti A, Cigada A, Cheli F (2005) Large wind induced vibrations on conductor bundles: laboratory scale measurements to reproduce the dynamic behaviour of the spans and suspension sets. IEEE Trans Power Delivery 20(2):1617-1624. ISSN: 0885-8977
31. Diana G, Manenti A et al (2005) Modelling of aeolian vibrations of single conductor plus damper: assessment of the technology. Electra 223
32. Diana G, Manenti A et al (2003) State of the art survey on spacers and spacer dampers: part I —general description. Electra 2003(209):1–6
33. Diana G, Zasso A, Vigevano L, Auteri F, Formaggia L, Nobile F (2006) Flow-structure interaction of the oscillating circular cylinder in the lock-in region: computational versus experimental approach comparison. In: Proceedings of CWE2006 4th international symposium on computational wind engineering, Yokohama, Japan, pp 16–19
34. Diana G, Resta F, Belloli M, Muggiasca S (2005) Experimental analysis on vortex induced vibration of a long flexible cylinder. In: Proceedings of the sixth international symposium on cable dynamics, Charleston, SC, September 2005
35. Dowell EH (ed) (1978) A modern course in aeroelasticity. Kluwer Academic Publishers, London
36. Falco M, Curami A, Zasso A (1991) Non linear effects in sectional model—aeroelastic parameters. In: ICWE conference, London, Ontario. Journal of Wind Engineering and Industrial Aerodynamics, Elsevier Ed. Proceeding of the 7th International Conference on Wind Engineering ICWE, Aachen, West Germany, July, 1985
37. Simiu E, Scanlan RH (1978) Wind effects on structures. Wiley, New York
38. Zasso A, Belloli M, Muggiasca S, Giapjournalo S (2006) On the pressure and force field on a circular cylinder oscillating in the lock-in region at sub-critical reynolds number. In: Proceedings of PVP2006-ICPVT-11 2006 ASME pressure vessels and Pijournalg Division conference, PVP2006 ICPVT11-93971, Vancouver, BC, Canada, 23–27 July 2006

39. Zasso A, Belloli M, Giapjournalo S, Muggiasca S (2005) Pressure field analysis on oscillating circular cylinder. In: Proceedings of the sixth Asia-Pacific conference on wind engineering (APCWE VI), Seoul, Korea, September 2005

Lubrication and Bearings with Hydrodynamic Lubrication (See Also Bibliography Chapter 6)

40. Biraghi B, Falco M, Pascolo P, Solari A (1980) Cuscinetto lubrificato in regime laminare e turbolento: parte ii -lubrificazione mista idrostatica- idrodinamica, L'Energia Elettrica, Vol LVII
41. Cheli F, Manenti A (1991) Appunti di lubrificazione idrodinamica. Edizioni Spiegel, Milan
42. Costantinescu VN (1959) On turbulent lubrication, vol 173(38). In: Proceedings of IMechE, London
43. Diana G et al (1982) A forced vibration method to calculate the oil film instability threshold of rotor-foundation system. Int. Conto on Rotordynamic Problems in Power Plant, Rome
44. Diana G, Cheli F et al (1988) Non linear effects in lubricated bearings. In 4th international conference on vibrations in rotating machinery (C300/88 IMechE 1988), Edinburgh, Sept 1988
45. Diana G, Cheli F, Pettone F, Manenti A (1990) Analytical and experimental research on some typical non-linear effects in the frequency response of a rotor on lubricated bearings. In: 3rd IFToMM rotordynamics conference, Lyon, Sept 1990
46. Diana G, Cheli F, Petrone F, Manenti (1990) A non linear effects in lubricated bearings. Vibr Wear High Speed Rotating Mach 174:567–579
47. Falco M et al (1980) Cuscinetto Lubrificato in Regime Laminare e Turbolento: Parte 1- Analisi Statica, L'Energia Elettrica, VoI LVII, 1980; Parte 111- Analisi Dinamica, L'Energia Elettrica, VoI LVII
48. Gallagher RR, Oden JT, Taylor C, Zienkiewicz OC (1975) Finite elements in fluids. Wiley, New York
49. Gunter EJ Jr (1966) Dynamic stability of rotor-bearing systems, NASA, SP-113
50. Journalkus O, Sternlicht B (1961) Theory of hydrodynamic lubrication. Mc Graw-Hill, New York
51. NATO ASI Series (1990) Applied sciences, vol 174, . Kluwer Academic Publishers, The Netherlands
52. Ng CW, Pan CRT (1965) A linearized turbulent lubrication theory. J Basic Eng
53. Pizzigoni B, Ruggieri G (1978) Sulla determinazione delle caratteristiche statiche e dinamiche dei cuscinetti lubrificati, L'Energia Elettrica n.2, VoI LV
54. Ruggieri G (1976) Un metodo approssimato per la risoluzione dell' equazione di Reynolds, L'Energia Elettrica n.2, VoI LIII

Dynamics of the Road and Rail Vehicle

55. ADAMS Software: http://www.mscsoftware.com/products/adams.cfm
56. Belforte P, Cheli F, Corradi R, Facchinetti A (2003) Software for the numerical simulation of tramcar vehicle dynamics. Int J Heavy Vehicle Syst 10:48–69
57. Braghin F, Bruni S (2004) Parametric analysis of wheelset derailment in both quasi steady—state and transient conditions. In: Proceedings of the 9th MINI conference on vehicle system dynamics, identification and anomalies VSDIA04, Budapest, Hungary, 8–10 Nov 2004
58. Braghin F, Bruni S, Diana G (2006) Experimental and numerical investigation on the derailment of a railway wheelset with solid axle. J Vehicle Syst Dyn 44(4):305–325

59. Braghin F, Lewis R, Dwyer-Joyce RS, Bruni S (2006) A mathematical model to predict railway wheel profile evolution due to wear. Wear 261:1253–1264
60. Braghin F, Bruni S, Resta F (2006) Active yaw damper for the improvement of railway vehicle stability and curving performances: simulations and experimental results. J Vehicle Syst Dyn 44(11):857–869
61. Bruni S, Mastinu G (2013) Dynamics of vehicles on roads and on tracks. In: Proceedings of the XIX IAVSD symposium
62. Bruni S, Cheli F, Diana G, Resta F (2003) Active control of the running behaviour of a railway vehicle: stability and curving performances. Vehicle Syst Dyn 37(Supplement): 157–170
63. Bruni S, Collina A, Corradi R (2005) Train-track-bridge interaction: influence of track topology on structure dynamic performances. In: Proceedings of EURODYN 2005, Paris, France, 4–7 Sept 2005
64. Bruni S, Collina A, Corradi R, Diana G (2004) Numerical simulation of train-track-bridge dynamic interaction. In: Proceedings of the 6th World Congress of computational mechanics —WCCM, Beijing (China), pp 1–10
65. Casini C et al (1985) I profili dei cerchioni nei rotabili ferroviari. Situazione attuale e proposte innovative, Ingegneria Ferroviaria
66. Casini C, Cheli R (1992) La sperimentazione ad Alta Velocità in Italia. In: Proc. Congresso Sviluppo e prospettive dei trasporti elettrificati: ricerca ed innovazione, Genoa
67. Castignani L, Cheli F, Giangiulio E, Mauri M (2004) A HIL approach for ABS performance evaluation. In: Proceedings of the international conference on integrated chassis control, effective systems integration as a first step towards the vehicle-by-wire, Naples (Italy), pp 1–14
68. Cheli F, Porcu C (1992) Traffico stradale su ponti di grande luce: interazione dinamica. Modello analitico e risultati sperimentali, XI Congresso Nazionale AIMETA, Trento, Oct 1992
69. Cheli F, Leo E, Melzi S, Mancosu F (2005) A 14 d.o.f. Model for the evaluation of vehicle's dynamics: numerical-experimental comparison. Int J Mech Control 6(2):19–30
70. Cheli F, Leo E, Melzi S, Arosio D, Giangiulio E, Mancosu F (2006) Validation of A 14Dof model for the prediction of vehicle dynamics and its interaction with active safety control systems. In: Proceedings of the 3rd international colloquium on vehicle-tire- road interaction, Stuttgart, Germany, 8 March 2006
71. Cheli F, Leo E, Melzi S, Arosio D, Giangiulio E, Mancosu F (2006) The VDSIM based vertec vehicle model: simulation package. In: Proceedings of the 3rd international colloquium on vehicle-tire- road interaction, Stuttgart, Germany, 8 March 2006
72. Cheli F, Donadio D, Melzi S, Resta F (2003) Experimental and modelling activity on a truck vehicle, heavy vehicle system. Int J Vehicle Design 10:35–47
73. Cheli F, Belforte P, Melzi S, Sabbioni E, Tomasini G (2006) Evaluating of cross wind aerodynamic effects on heavy vehicles through a numerical-experimental approach. J Vehicle Syst Dyn 44(Suppl):791–804
74. Cheli F, Leo E, Melzi S, Arosio D, Giangiuli E, Mancosu F (2006) Implementation of a 14Dof model for the prediction of vehicle dynamics and its interaction with active safety control systems. In: Proceedings of the 3rd international colloquium on vehicle-tire- road interaction, Stuttgart, Germany, 8 March 2006
75. Cheli F, Costantini A, Porciani N, Resta F, Sabbioni E (2004) Controlled semi-active suspension system for high performance vehicles. In: Proceedings of the international conference on integrated chassis control, effective systems integration as a first step towards the vehicle-by-wire, Naples (Italy), pp 1–18
76. Cheli F, Pedrinelli M, Resta F, Travaglio G, Zanchetta M, Zorzutti A (2006) Development of a new control strategy for a semiactive differential for a high-performance vehicle. J Vehicle Syst Dyn 44(Suppl):202–215

77. Cheli F, Dellachà P, Zorzutti A (2006) Development of the control strategy for an innovative 4WD device. In: Proceedings of ESDA2006 8th Biennial ASME conference on engineering systems design and analysis, ESDA2006-95490, Turin, Italy, pp 1–7, 4–7 July 2006
78. Cheli F, Pedrinelli M, Giaramita M, Sandoni G, Travaglio G (2005) A new control strategy for semi-active differential (part I). In: Proceedings of 16th IFAC World Congress, Prague, Czech Republic, 3–8 July 2005
79. Cheli F, Giangiulio E, Mauri M (2004) A mechatronic test bench for ABS control systems. In: Proceedings of SPEEDAM 2004, symposium on power electronics, electrical drives, automation and motion, Capri (Italy), pp 887– 892
80. Cheli F, Cheli R, Corradi R, Diana G, Roberti R, Tomasini G (2003) Cross-wind aerodynamic forces on rail vehicles: wind tunnel experimental tests and numerical dynamic analysis. In: Proceedings of the World Congress on railway research, Edinburgh, UK, 28 Sept–1 Oct 2003, pp 513–522
81. Cheli F, Desideri R, Diana G, Mancini G, Roberti R, Tomasini G (2006) Cross wind effects on tilting trains. In: Proceedings of WCRR 2006 7th World congress on railway research, Montreal, Canada, pp 1–84, 8 June 2006
82. Cheli F, Corradi R, Diana G, Facchinetti A (2005) Experimental validation of a numerical model for the simulation of tramcar vehicle dynamics. In: Proceedings of IDETC/CIE 2005 ASME 2005 design engineering technical conferences and computers and information in engineering conference, Long Beach, California, USA, 24–28 Sept 2005
83. Cheli F, Corradi R, Diana G, Facchinetti A, Gherardi F (2006) Effects of track geometrical defects on running safety of tramcar vehicles. J Vehicle Syst Dyn 44(Supplement):302–312
84. Cheli F, Corradi R, Diana G, Facchinetti A (2003) Wheel-rail contact phenomena and derailment conditions in light urban vehicles. In: Proceedings of the 6th international conference on contact mechanics and wear of rail/wheel systems (CM2003), Gothenburg, Sweden, pp. 461-468, 10–13 June 2003
85. Cossalter V (2006) Motorcycle dynamics, Second English Edition
86. Diana G, Resta F (2006) Controllo dei sistemi meccanici. Polipress, Milan
87. Diana G, Cheli F, Mariani M (1990) Sulla dinamica di veicoli ferroviari: moto in rettilineo ed in curva di carrozze a semplice e doppio stadio di sospensioni (modelli non lineari a 33 e 35 g.d.l.), Report Interno N.1 del Dipartimento di Meccanica del Politecnico di Milano, Milan, Oct 1990
88. Diana G, Cheli F (1991) Ricerche e sviluppi nelle metodologie di analisi dei sistemi meccanici per veicoli metropolitani, L'Energia Elettrica 718
89. Diana G, Cheli F (1990) Sulla dinamica di un veicolo ferroviario in rettilineo ed in curva, X Congresso AIMETA, Pisa
90. Diana G, Cheti F (1990) Sulla dinamica di veicoli ferroviari: moto in rettilineo di un veicolo articolato a 3 carrelli (modello a 34 g.d.l.), Report Interno N.4 del Dipartimento di Meccanica del Politecnico di Milano, Milan, Oct 1990
91. Diana G, Pizzigoni B, Cheti F (1990) Sulla dinamica di veicoli ferroviari: carrozza gran comfort (modello a 23 g.d.l.), Report Interno N.2 del Dipartimento di Meccanica del Politecnico di Milano, Milan, Oct 1990
92. Diana G (1968) Ricerche sperimentali sul movimento in curva a regime dell'autoveicolo, ATA, XXI, n.6
93. Elia M, Diana G, Bocciolone M, Bruni S, Cheli F, Collina A, Resta F (2006) Condition monitoring of the railway line and overhead equipment through onboard train measurement—an Italian experience. Proceedings of the IEE international conference on railway condition monitoring, Birmingham, UK, 29–30 Nov 2006
94. Genta G (1989) Meccanica dell'autoveicolo, Torino, Levrotto e Bella. Int J Vehicle Mech Mob (Swets & Zeitlinger B. V. Ed)
95. Goodall R, Bruni S, Mei TX (2006) Actively-controlled railway running gear: concepts and prospects. J Vehicle Syst Dyn 44(Suppl):60–70
96. Goodall RM, Bruni S, Mei TX (2006) Concepts and prospects for actively controlled railway running gear. Suppl Vehicle Syst Dyn 44:60–70

97. Guiggiani M (2007) Dinamica del veicolo. Città Studi Edizioni, Turin
98. Johnson KI, Vermeulen PI (1965) Contact of non-spherical bodies transmitting tangential forces. J Appl Mech 338–340
99. Kalker JJ (1967) On the rolling contact of two bodies in the presence of dry friction. Doctoral Thesis, Delft University
100. Kalker JJ (1979) Survey of wheel-rail rolling contact theory. Vehicle Syst Dyn 8:317–358
101. Massa E, Diana G (1967) Sulla stabilità del movimento in curva a regime di un veicolo, ATA, XXX, n.8
102. Pacejka HB, De Pater AD (1983) 1st course on advanced vehicle dynamics. In: ICTS-PFT proceeding series, Rome
103. Panagin R (1990) La dinamica del veicolo ferroviario, Torino, Levrotto e Bella
104. Proceeding of the IA VSD Symposium on Dynamics of Vehicles on roads and tracks
105. Rill G (1983) The influence of correlated random road excitation processes on vehicle vibration. In: 8th IA VD symposium, Massachusetts
106. Resta F, Teuschl G, Zanchetta M, Zorzutti A (2005) A new control strategy for semi-active differential (part II). In: Proceedings of 16th IFAC World Congress, Prague Czech Republic, 3–8 July 2005
107. Schiehlen WO (1982) Dynamics of high-speed vehicles. Springer, New York
108. Tomasini G, Collina A, Leo E, Resta F, Cheli F (2006) Numerical–experimental methodology for runnability analysis and wind–bridge–vehicle interaction study. In: Proceedings of the III European conference on computational mechanics solids, structures and coupled problems in engineering; Lisbon, Portugal, 5–8 June 2006
109. Various Authors. Meccanica del veicolo: appunti dalle lezioni, Dipartimento di Meccanica, Politecnico di Milano (yet to be published)

Chapter 6
Rotordynamics

6.1 Introduction

So far in the discussion we have illustrated the methods used for the dynamic analysis of mechanical systems. The main purpose of analysis has always been both to study forced motion and to study the stability of these systems and, therefore, analysis of perturbed motion around the steady-state or rest condition. In describing the various types of system which, we should remember, has been subdivided into dissipative systems and systems subject to force fields, we have included various examples of applications, with the main aim of better highlighting the problem and to better illustrate the methods of analysis used. The subject of rotordynamics which will be addressed in this chapter is now proposed with a dual purpose:

- to provide basic knowledge to the engineer who has to handle the design and use of machinery, since rotors are essential and vital organs in any machinery;
- to offer this subject, given the multitude of problems that it entails, as a valid example of the use of the techniques illustrated in the previous chapters.

6.2 Description of the System Composed of the Rotor and the Supporting Structure Interacting with It

Below we will describe the basic characteristics of the rotor—supporting—structure —foundation system, while showing the schematisation that generally represents it. The purpose is to define a mathematical model aimed at simulating the real behaviour of rotor and foundation, as faithfully as possible, regardless of the specific problem being considered.

Figure 6.1 shows, as an example, a rotor of a turbogenerator group, while Fig. 6.2 shows the rotor of a turbopump: upon examining both the figures, we can

F. Cheli and G. Diana, *Advanced Dynamics of Mechanical Systems*,
DOI 10.1007/978-3-319-18200-1_6

Fig. 6.1 Turbogenerator group

Fig. 6.2 Turbopump

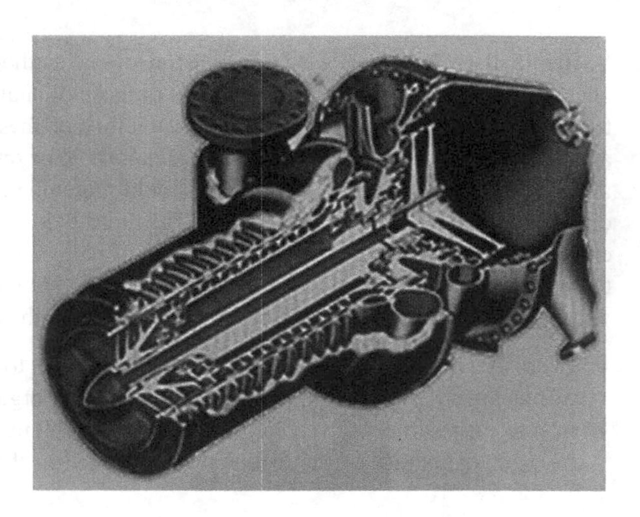

see how every rotor is usually similar to a beam with a circular cross section of variable diameter with fitted elements (disks, blades or the like) that collaborate mainly from an inertial point of view and to a lesser extent from an elastic point of view, and which can be schematised as concentrated weights. The rotor body is constrained to the casing via bearings which, especially in fast and loaded rotors, are lubricated with fluid. The casing can be incorporated into a body that may be cylindrical (see, for example, the pump in Fig. 6.2) or have a more complex form (Fig. 6.1). The casing may, in turn, be constrained to the foundation of the machine by a supporting structure via rigid or flexible elements. Between rotor and casing there can be a fluid that transmits mutual actions to the rotor and the casing or an electromagnetic field: the presence of these fields (fluid or magnetic) can influence the dynamic behaviour of the rotor (fluid elasticity or electro-elasticity phenomena). In the case of turbines, for example, the fluid acts not only on the blade system, but also on the seals, giving rise to a complex series of problems as far as defining the field of forces that interact with the rotor and with the casing is concerned. Regardless of the problem being considered, to be able to adequately simulate the dynamic behaviour of the "rotor-case-foundation" system, with the current state of knowledge and research, the schematisation generally adopted for the structural part (rotor, foundation) uses the finite element method, while the effects caused by bearings and seals are schematised by linearising the force field of the interacting fluid.

6.2.1 Schematising the Rotor

Figure 6.3 shows a photo of an example of a low pressure rotor on a turbogenerator group. As already mentioned, the most suitable way of schematising such a rotor is to consider it as being made up of an adequate number of beam-type finite elements to reproduce the variations in section of the actual rotor. Figure 6.4 shows a beam model that can represent the real rotor in Fig. 6.3: as we know, the greater the number of beams considered in the mathematical model, the more refined the schematisation used will be.

If, as usually occurs, we adopt the finite elements technique (other methods, such as that of transfer matrices, [7, 25, 28] are less common nowadays), these beam elements will generally have 6 degree-of-freedom (d.o.f.) for each node of the actual element (Fig. 6.5) (note that in this example we will use a right-hand reference system, using Z, however, to indicate the longitudinal axis, to the contrary of the method used in Chap. 4, Sect. 4.5.2). Usually the 2 d.o.f. of each node

Fig. 6.3 Low pressure rotor on a turbogenerator group

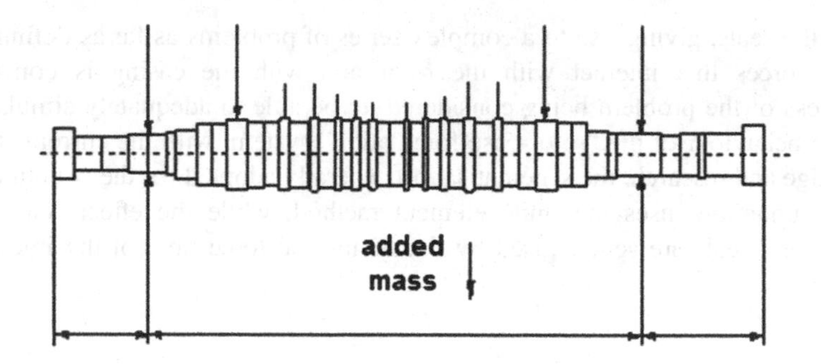

Fig. 6.4 Low pressure rotor on a turbogenerator group: schematisation

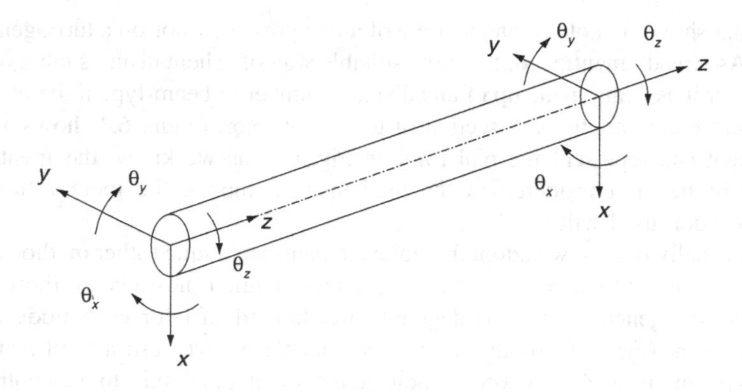

Fig. 6.5 Finite beam element

associated with the axial Z_i and torsional θ_{zi} displacements are considered decoupled as, in reality, axial and torsional vibrations are, in general, actually decoupled from those of bending. If, in this discussion, we first wish to consider the problems associated with bending motions only, it follows that each node of the beam finite element must have 4 d.o.f. associated with it, more precisely:

- two representing the horizontal Y_i and vertical X_i transversal displacements (Fig. 6.6);
- two that represent rotations θ_{yi} and θ_{xi} of the section in correspondence to the node.

Every generic kth finite element thus has 8 d.o.f. which represent the displacements and rotations of the end to the left (indicated in Fig. 6.6 with the subscript "s") and the end to the right (subscript "d"):

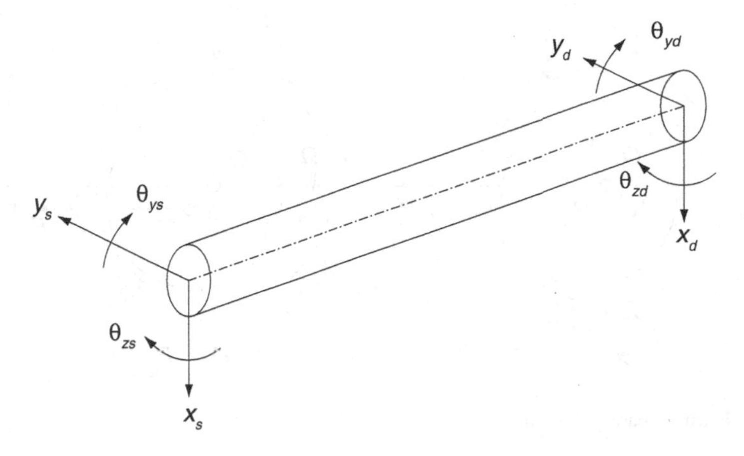

Fig. 6.6 Beam finite element used to schematise the rotor

$$\underline{X}_k = \begin{Bmatrix} \underline{X}_{ks} \\ \underline{X}_{kd} \end{Bmatrix} = \begin{Bmatrix} X_i \\ \theta_{yi} \\ Y_i \\ \theta_{xi} \\ \cdots \\ X_j \\ \theta_{yj} \\ Y_j \\ \theta_{xj} \end{Bmatrix} = \begin{Bmatrix} \underline{X}_i \\ \underline{Y}_i \\ \underline{X}_j \\ \underline{Y}_j \end{Bmatrix} \tag{6.1}$$

according to the conventions used in defining the forces due to oil-film (Sect. 6.2.2, Fig. 6.7). The stiffness matrix $[K_r]$ of the generic kth beam finite element can be obtained from the overall stiffness matrix $[K_{jl}]$ of the beam element with 12 d.o.f. in space (Chap. 4, Sect. 4.5.3, Table 4.3). The lines and columns corresponding to the longitudinal and torsional motion are eliminated and the matrix is reorganised considering (6.1 as Sect. 4.5.2, (4.111)):

$$[K_r] = \begin{bmatrix} [K_{vii}] & 0 & [K_{vij}] & 0 \\ 0 & [K_{wii}] & 0 & [K_{wij}] \\ [K_{vji}] & 0 & [K_{vjj}] & 0 \\ 0 & [K_{wji}] & 0 & [K_{wjj}] \end{bmatrix} \tag{6.2}$$

The matrix $[K_r]$ defines the eight generalised elastic forces at the nodes of the element in relation to the generalised displacements of the same nodes: in (6.2) the single sub matrices are defined as follows:

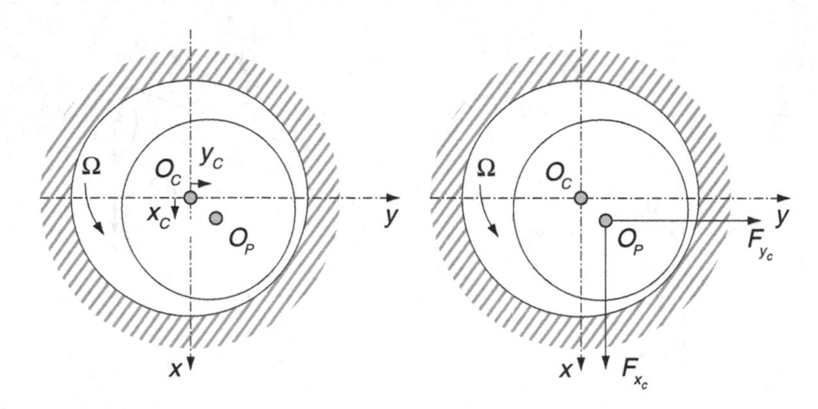

Fig. 6.7 Journal-bearing coupling

$$\begin{bmatrix} [K_{wii}] & [K_{wij}] \\ [K_{wji}] & [K_{wjj}] \end{bmatrix} = \begin{bmatrix} \left(\frac{12EJ_x}{l_k^3}\right) & \left(-\frac{6EJ_x}{l_k^2}\right) & \left(-\frac{12EJ_x}{l_k^3}\right) & \left(-\frac{6EJ_x}{l_k^2}\right) \\ \left(-\frac{6EJ_x}{l_k^2}\right) & \left(\frac{4EJ_x}{l_k}\right) & \left(\frac{6EJ_x}{l_k^2}\right) & \left(\frac{2EJ_x}{l_k}\right) \\ \left(-\frac{12EJ_x}{l_k^3}\right) & \left(\frac{6EJ_x}{l_k^2}\right) & \left(\frac{12EJ_x}{l_k^3}\right) & \left(\frac{6EJ_x}{l_k^2}\right) \\ \left(-\frac{6EJ_x}{l_k^2}\right) & \left(\frac{2EJ_x}{l_k}\right) & \left(\frac{6EJ_x}{l_k^2}\right) & \left(\frac{4EJ_x}{l_k}\right) \end{bmatrix} = [K_w] \tag{6.3a}$$

$$\begin{bmatrix} [K_{vii}] & [K_{vij}] \\ [K_{vji}] & [K_{vjj}] \end{bmatrix} = \begin{bmatrix} \left(\frac{12EJ_y}{l_k^3}\right) & \left(\frac{6EJ_y}{l_k^2}\right) & \left(-\frac{12EJ_y}{l_k^3}\right) & \left(\frac{6EJ_y}{l_k^2}\right) \\ \left(\frac{6EJ_y}{l_k^2}\right) & \left(\frac{4EJ_y}{l_k}\right) & \left(-\frac{6EJ_y}{l_k^2}\right) & \left(\frac{2EJ_y}{l_k}\right) \\ \left(-\frac{12EJ_y}{l_k^3}\right) & \left(-\frac{6EJ_y}{l_k^2}\right) & \left(\frac{12EJ_y}{l_k^3}\right) & \left(-\frac{6EJ_y}{l_k^2}\right) \\ \left(\frac{6EJ_y}{l_k^2}\right) & \left(\frac{2EJ_y}{l_k}\right) & \left(-\frac{6EJ_y}{l_k^2}\right) & \left(\frac{4EJ_y}{l_k}\right) \end{bmatrix} = [K_u] \tag{6.3b}$$

$[K_w]$ and $[K_v]$ being, respectively, the beam's matrices (4 × 4) of stiffness upon analysing the vertical and horizontal plane motion only (Chap. 4, Sect. 4.5.2.2.3).[1]

[1]In the following discussion, it is assumed that the rotor has equal bending stiffness in all directions and, therefore, independent from the angular position of the rotor in directions x and y (in the formulas (6.3a) and (6.3b) $J_x = J_y$).

In a similar manner it is possible to obtain the matrix of mass $[M_r]$ of the beam element with 8 DOF (Fig. 6.6):

$$[M_r] = \begin{bmatrix} [M_{vii}] & 0 & [M_{vij}] & 0 \\ 0 & [M_{wii}] & 0 & [M_{wij}] \\ [M_{vji}] & 0 & [M_{vjj}] & 0 \\ 0 & [M_{wji}] & 0 & [M_{wjj}] \end{bmatrix} \tag{6.4}$$

since [Chap. 4, Sect. 4.5.2.2.3, Table 4.2, (4.69c)]:

$$[M_w] = \begin{bmatrix} [M_{wii}] & [M_{wij}] \\ [M_{wji}] & [M_{wjj}] \end{bmatrix} = m l_k \begin{bmatrix} \frac{13}{35} & -\frac{11}{210} l_k & \frac{9}{70} & \frac{13 l_k}{420} \\ -\frac{11}{210} l_k & \frac{l_k^2}{105} & -\frac{13 l_k}{420} & -\frac{l_k^2}{140} \\ \frac{9}{70} & -\frac{13 l_k}{420} & \frac{13}{35} & \frac{11 l_k}{210} \\ \frac{13 l_k}{420} & -\frac{l_k^2}{140} & \frac{11 l_k}{210} & \frac{l_k^2}{105} \end{bmatrix} \tag{6.5a}$$

$$[M_v] = \begin{bmatrix} [M_{wii}] & [M_{wij}] \\ [M_{wji}] & [M_{wjj}] \end{bmatrix} = m l_k \begin{bmatrix} \frac{13}{35} & \frac{11}{210} l_k & \frac{9}{70} & -\frac{13 l_k}{420} \\ \frac{11}{210} l_k & \frac{l_k^2}{105} & \frac{13 l_k}{420} & -\frac{l_k^2}{140} \\ \frac{9}{70} & \frac{13 l_k}{420} & \frac{13}{35} & -\frac{11 l_k}{210} \\ -\frac{13 l_k}{420} & -\frac{l_k^2}{140} & -\frac{11 l_k}{210} & \frac{l_k^2}{105} \end{bmatrix} \tag{6.5b}$$

6.2.2 Schematising Bearings

The bearings are the elements that connect the rotor to the casing or the external supporting structure. Mutual forces are created between shaft and case, functions of the relative motion between journal and bearing, that depend on the type of bearing being considered. In order to achieve the definition of an adequate mathematical model we must define these forces analytically. Referring to the case of fluid lubricated bearings (see Chap. 5 on this, and more specifically Sect. 5.3.2.2, where it is assumed that the bearing is stationary, i.e., the foundation infinitely rigid), the force field acting between journal and bearing, as a result of the fluid film formed, can be defined via two components, according to the vertical X and horizontal Y direction (respectively, F_{xc} and F_{yc}, see Fig. 6.7) of the resultant action $\mathbf{F_c}$ of the effective distribution of pressures formed in the actual fluid film. The components F_{xc} and F_{yc} are functions of the relative position of the centre of the journal O_p with respect to that of the centre of the bearing O_c, that is:

$$F_{xc} = F_{xc}(x_c, y_c, \dot{x}_c, \dot{y}_c) \tag{6.6a}$$

$$F_{yc} = F_{yc}(x_c, y_c, \dot{x}_c, \dot{y}_c) \tag{6.5b}$$

where x_c and y_c are the relative displacements of the centre of the journal O_p with respect to the centre of the bearing O_c in directions X and Y. These forces can be defined analytically, as mentioned previously in Sect. 5.3.2.2, by integrating the Reynolds equations. For the sake of convenience, below we will show the basic references to the theory of lubrication which are necessary to understand the development of the actual discussion (for further details, see [31, 47, 54]).

6.2.2.1 Analytical Definition of the Static and Dynamic Characteristics of Bearings

For a lubricated bearing, the Reynolds equations [31, 54] have the following expression:

$$\frac{\partial}{\partial\theta}\left(\frac{1}{12\mu C_s}\frac{\partial p}{\partial\theta}h^3\right)+\left(\frac{2R}{b}\right)^2\frac{\partial}{\partial\eta}\left(\frac{1}{12\mu C_z}\frac{\partial p}{\partial\eta}h^3\right)=\frac{\Omega R^2}{2}\frac{\partial h}{\partial\theta}+WR^2 \tag{6.7}$$

where:

- R is the radius of the journal;
- b the width of the bearing;
- η (Fig. 6.8) is the abscissa along the width b of the bearing expressed in dimensionless terms:

$$\eta=\left(\frac{2Z}{b}\right)\quad(-1<\eta<1) \tag{6.8a}$$

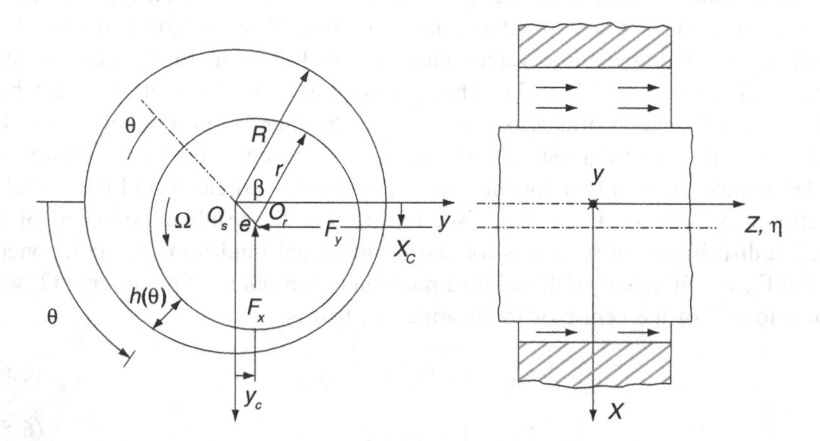

Fig. 6.8 Nomenclature and coordinates in the journal-bearing coupling

- θ the anomaly that defines the angular position;
- p represents the pressure of the fluid inside the meatus, a function, in general, of η and θ:

$$p = p(\theta, \eta) \tag{6.8b}$$

- h represents the thickness of the fluid film, also a function of the position inside the meatus[2]:

$$h = h(\theta, \eta) \Rightarrow h(\theta) = \delta - x_c \sin\theta + y_c \cos\theta \tag{6.8c}$$

- W is the approach velocity between the surfaces of the journal and the bearing:

$$W = \frac{\partial h}{\partial t} = W(\theta, \eta) \Rightarrow W(\theta) = -\dot{x}_c \sin\theta + \dot{y}_c \cos\theta \tag{6.8d}$$

- Lastly C_s and C_z are suitable coefficients, functions of the Reynolds number, which take into account the turbulence that may arise in the meatus [32, 42, 48].

In (6.7) μ represents the viscosity of the lubricant: μ is a function, in turn, of the local temperature in the meatus:

$$\mu = \mu(T) \tag{6.9a}$$

Then as the temperature T is variable along the meatus:

$$T = T(\theta, \eta) \tag{6.9b}$$

viscosity μ is, in turn, a function of θ and η, i.e.:

$$\mu = \mu(\theta, \eta) \tag{6.9c}$$

In order to define function $T(\theta, \eta)$ (6.9b) we must solve the problem relating to the heat exchange that occurs in the bearing (see [52] on this). Given a generic bearing (with radius R of the journal and width b of the bearing) and having assigned the angular velocity of rotation Ω of the actual journal, to define F_{xc} and F_{yc} we proceed in the following manner[3]:

---

[2]In general, h is considered to be constant on the longitudinal direction η, so we assume: $h(\theta, \eta) = h(\theta)$.

[3]More sophisticated methods for defining F_{xc} and F_{yc} also take into account the inertial terms which are, on the contrary, neglected in the Reynolds equations and virtually use the complete form of the Navier–Stokes equations [31], [54], with significant, unwarranted increases in calculation times.

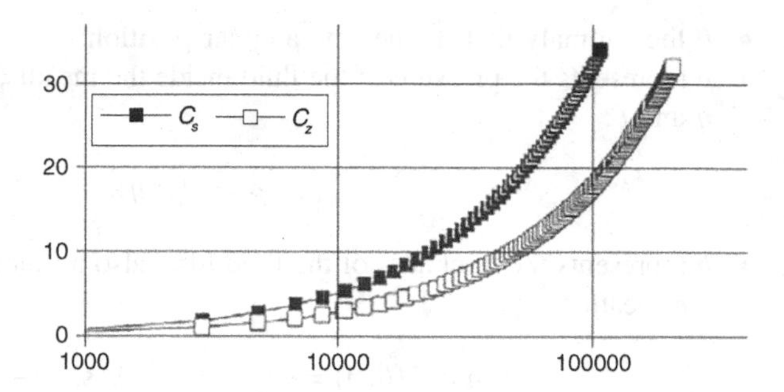

Fig. 6.9 Dependence of coefficients C_s and C_z on Reynolds

- having assigned coordinates x_c and y_c, that is, the positions of the centre of the journal with respect to the bearing;
- the function $h = h(\theta, \eta)$ and its derivate $(\partial h/\partial \theta)$ are univocally defined;
- approach velocity W can also be defined from (6.8d): W is zero if $\dot{x}_c$ and $\dot{y}_c$ are simultaneously zero;
- coefficients C_s and C_z can be brought back [42] to functions of the Reynolds number referring to the height of the meatus, as shown in Fig. 6.9;
- now it is possible to integrate the Reynolds Eq. (6.7) in the only unknown with:
  - finite difference methods [44], [31];
  - finite element methods [46];
  - approximate semi-analytical methods [54];
- in this way we can obtain the trend of the pressure $p = p(\theta, \eta)$ along the meatus, and by integrating this we can calculate the components F_{xc} and F_{yc} of the forces acting on the journal.

Once the forces F_{xc} and F_{yc} corresponding to a given position are known and, obviously, once the various parameters needed to define the bearing's operating conditions have been assigned, i.e. the angular velocity Ω, the radial clearance $\delta = (R - r)$ and the viscosity μ of the fluid used, it is possible to obtain the position occupied by the centre of the journal O_p in the bearing given a certain external load Q.

The locus described by the centre of the journal O_p upon variation of the load Q applied, at constant angular velocity Ω, is defined as the *load locus* (Chap. 5, Sect. 5.3.2.2): Fig. 6.10 shows the trend of the load locus for a cylindrical bearing.

For loads Q tending to zero (or velocity Ω tending to infinity), the position O_p occupied by the centre of the journal tends to coincide with the position O_c of the centre of the bearing. For loads Q tending to infinity (or velocity Ω tending to zero) the centre of the journal O_p tends to the position indicated in Fig. 6.10 with letter A, corresponding to a runout $e_c = \delta$, equal to the radial clearance: the journal in this position touches the bearing along the lower generatrix.

In general, the centre of the journal O_p, for a rotor operating at a given angular velocity Ω, can be found in a generic position along the load locus (for example, point C in Fig. 6.10) defined by a certain runout e_c and by a certain angle β which,

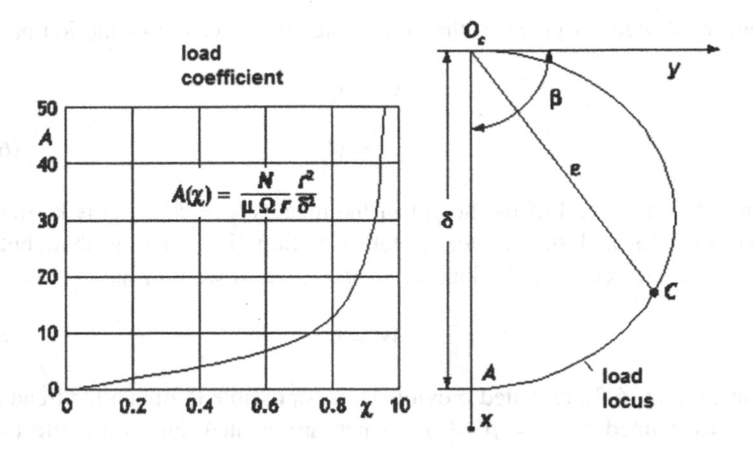

Fig. 6.10 Cylindrical bearing: load locus and load coefficient

for a given type of bearing, is a function of the actual runout, or rather, of the dimensionless ratio $\chi = e_c/\delta$. Typical characteristics of each bearing are the functions:

$$\begin{aligned} A &= A(\chi) \\ \beta &= \beta(\chi) \end{aligned} \tag{6.10}$$

$A(\chi)$ being defined by the following expressions:

$$N = \frac{Q}{b} = A(\chi)\mu\Omega r\frac{r^2}{\delta^2} \rightarrow A(\chi) = \frac{N}{\mu\Omega r}\frac{\delta^2}{r^2} \tag{6.11}$$

where N is the load per unit of width, Q is the applied load, b the width of the bearing, χ the dimensionalised runout and $A(\chi)$ a dimensionless coefficient known as the "load coefficient". The load coefficients $A(\chi)$ (6.11) and the load locus (6.10) for the different types of bearing can be obtained either analytically or experimentally[4]: Fig. 6.10 shows the trend of the load coefficient $A(\chi)$ in relation to the dimensionless runout χ, again in the case of a cylindrical bearing.

Given a generic bearing (with radius R* of the journal and width b* of the bearing), having assigned the angular velocity of rotation Ω^* of the journal, and the applied load Q^* being known, to obtain analytically $\beta(\chi)$ and $A(\chi)$ we must proceed in the following manner:

[4]Once the type of bearing has been defined, a certain type of lubricant has been assigned, the input temperature defined and if the steady-state of the fluid is laminar, $A(\chi)$ and $\beta(\chi)$ are univocally defined. Otherwise, the same quantities are functions of both the input temperature and the Reynolds number.

- a numerical value is given to the coordinates of the centre of the journal:

$$x_c = x_c^* \tag{6.12a}$$

$$y_c = y_c^* \tag{6.12b}$$

- from (6.8c), the trend of the height of the meatus $h^* = h^*(\theta, \eta)$ is known;
- if we wish to analyse the steady-state situation (i.e. rotating shaft but with velocity of the centre of the journal $\dot{x}_c$ and $\dot{y}_c$ zero) we will have:

$$W = 0 \tag{6.12c}$$

- by integrating (6.7), as stated previously, it is possible to obtain the trend of the pressures defined by $p^* = p^*(\theta, \eta)$, which are created due to the effect of the position of the journal x_c^* and y_c^* imposed;
- by integrating along the meatus the function $p^* = p^*(\theta, \eta)$, it is possible to obtain the values F_{xc}^* and F_{yc}^* which, for a generic position, will not generally give a resultant force that is equal and opposite to load Q^* acting on the journal.

The components F_{xc}^* and F_{yc}^* are, therefore, functions of the position x_c^* and y_c^* of the centre of the journal O_p with respect to the centre of the bearing O_c: therefore, we must solve a system of nonlinear equations of the form:

$$\begin{aligned} F_{xc}^*\left(x_c^*, y_c^*\right) &= Q^* \\ F_{yc}^*\left(x_c^*, y_c^*\right) &= 0 \end{aligned} \tag{6.13}$$

The solution (Fig. 6.11) can be obtained by varying the value of x_c^* until F_{yc}^* is zero (using, for example, a bisection method), as load Q^*, according to the conventions used in Fig. 6.7, is acting in direction X: also if $F_{xc}^* = Q^*$ in correspondence to this pair of values x_c^* and y_c^*, then the position defined by the values:

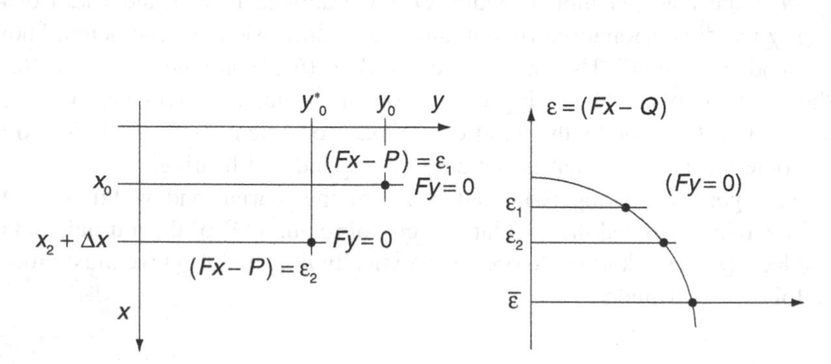

Fig. 6.11 Analytical definition of the load locus and the load coefficient

$$\begin{aligned} x_{c0}^* &= x_c^* \\ y_{c0}^* &= y_c^* \end{aligned} \tag{6.14a}$$

thus obtained, defines the position of equilibrium of the journal under the action of the load Q^* and of the resultant of the pressures that are created in the meatus. Since, in general, $F_{xc}{}^* \neq Q^*$, we vary y_c^* by recalculating, each time, the value of x_c^* which creates the condition $F_{yc}{}^* = 0$, until, for a pair of values $x_c{}^* = x_{co}^*$ and $y_c{}^* = y_{co}{}^*$, we also obtain:

$$F_{xc}^* - Q^* < \varepsilon \tag{6.14b}$$

with a predetermined tolerance ε. Once these values have been obtained, the runout $e_c{}^*$ can be obtained via:

$$e_c{}^* = \sqrt{x_{c0}^{*2} + y_{c0}^{*2}} \tag{6.15a}$$

while the angle $\beta^* = \beta^*(\chi^*)$ is given by:

$$\tan(\beta^*) = \frac{x_{c0}^*}{y_{c0}^*} \tag{6.15b}$$

For the generic bearing with geometric characteristics R*, b*, δ* = (R* − r*) and with a specific lubricant with viscosity μ*, the value of the load Q^* acting on the journal and the steady-state angular velocity Ω^* can be inserted in (6.11) to obtain from the same equation the value of the load coefficient $A^* = A^*(\chi^*)$ (since $\chi^* = (e_c^*/\delta^*)$) for the assigned pair of values Q^* and Ω^*.

By repeating the calculation for different values of Q^* or of Ω^* for the bearing being analysed, it is possible to numerically determine the functions $A(\chi)$ and $\beta(\chi)$ in the entire field of definition $(0 < \chi < 1)$.[5]

These functions, as mentioned previously, are univocally defined if, once a certain ratio is assigned between the width b and the diameter D of the bearing, the temperature T is assumed to be constant in the meatus: on the contrary, when having to define the temperature distribution in the meatus, other independent variables are involved in the problem such as, for example, the input temperature of the lubricant in the meatus, so we obtain, for each pair of values Q^* and Ω^*, families of curves $A(\chi)$ and $\beta(\chi)$ functions, in turn, of the input temperature of the lubricant as well as of the ratio b/D.

Once the force field caused by the presence of the fluid film has been analytically defined, it is possible to linearise the expressions of F_{xc} and F_{yc} around the generic

[5]As mentioned, $A(\chi)$ and $\beta(\chi)$ can also be defined experimentally using proximity sensors to measure the position x_{co} and y_{co} described by the centre of the pin with respect to the bearing, upon variation of the angular velocity Ω and upon variation of the load applied to bearing Q.

position of equilibrium of the journal x_{co} and y_{co}, as already shown in Chap. 5, Sect. 5.3.2.2, regarding systems surrounded by force fields, arriving at an expression of the form:

$$F_{xc} = F_{xc0} + \left(\frac{\partial F_{xc}}{\partial x_c}\right)_0 (x_c - x_{c0}) + \left(\frac{\partial F_{xc}}{\partial y_c}\right)_0 (y_c - y_{c0}) + \left(\frac{\partial F_{xc}}{\partial \dot{x}_c}\right)_0 \dot{x}_c + \left(\frac{\partial F_{xc}}{\partial \dot{y}_c}\right)_0 \dot{y}_c$$
$$F_{yc} = F_{yc0} + \left(\frac{\partial F_{yc}}{\partial x_c}\right)_0 (x_c - x_{c0}) + \left(\frac{\partial F_{yc}}{\partial y_c}\right)_0 (y_c - y_{c0}) + \left(\frac{\partial F_{yc}}{\partial \dot{x}_c}\right)_0 \dot{x}_c + \left(\frac{\partial F_{yc}}{\partial \dot{y}_c}\right)_0 \dot{y}_c \quad (6.16)$$

Using $\overline{x}_c$ and $\overline{y}_c$ to indicate the independent variables that describe the perturbed motion around the position of equilibrium we can make the change of variables:

$$\overline{x}_c = x_c - x_{c0} \quad (6.17a)$$

$$\overline{y}_c = y_c - y_{c0} \quad (6.17b)$$

the expressions (6.16) become:

$$F_{xc} = F_{xc0} + K_{xx}\overline{x}_c + K_{xy}\overline{y}_c + R_{xx}\dot{\overline{x}}_c + R_{xy}\dot{\overline{y}}_c \quad (6.18a)$$

$$F_{yc} = F_{yc0} + K_{yx}\overline{x}_c + K_{yy}\overline{y}_c + R_{yx}\dot{\overline{x}}_c + R_{yy}\dot{\overline{y}}_c \quad (6.18b)$$

The terms:

$$F_{xc} - F_{xc0} = \overline{F}_{xc} \quad (6.19a)$$

$$F_{yc} - F_{yc0} = \overline{F}_{yc} \quad (6.19b)$$

represent the dynamic contributions of the forces caused by the presence of the meatus, since:

$$\overline{F}_{xc} = K_{xx}\overline{x}_c + K_{xy}\overline{y}_c + R_{xx}\dot{\overline{x}}_c + R_{xy}\dot{\overline{y}}_c \quad (6.20a)$$

$$\overline{F}_{yc} = K_{yx}\overline{x}_c + K_{yy}\overline{y}_c + R_{yx}\dot{\overline{x}}_c + R_{yy}\dot{\overline{y}}_c \quad (6.20b)$$

where K_{xx}, K_{xy} etc. are the terms of equivalent damping and stiffness due to the film of lubricant previously defined in Chap. 5, Sect. 5.3.2.2. To define these coefficients analytically we use their actual definition, i.e. (see Chap. 5, Sect. 5.3.2.2 on systems surrounded by force fields):

$$K_{xx} = \left(\frac{\partial F_{xc}}{\partial x_c}\right)_0; \quad K_{yx} = \left(\frac{\partial F_{yc}}{\partial y_c}\right)_0; \ \ldots \ etc. \quad (6.21)$$

As mentioned previously, the derivates are calculated in the position of equilibrium of the journal obtained in the way described previously and defined, for a given load Q^* and for a specific angular velocity Ω^*, from values:

$$x_c = x^*_{c0}; \quad y_c = y^*_{c0}; \quad \dot{x}^*_{c0} = 0; \quad \dot{y}^*_{c0} = 0 \tag{6.22}$$

These derivates cannot be estimated analytically as the functions $F_{xc} = F_{xc}(x_c, y_c, \dot{x}_c, \dot{y}_c)$ and $F_{yc} = F_{yc}(x_c, y_c, \dot{x}_c, \dot{y}_c)$ can only be defined numerically by integrating the Reynolds equation: so the derivates must by estimated by approximating them with the relative difference quotients.

More specifically, defining as described above the position of equilibrium of the journal, starting from that position, variable x^*_{c0}, for example, increases by a value equal to Δx_c and is calculated using the same procedures described previously (i.e. by integrating the Reynolds equation), the value of the forces caused by the meatus F_{xc} and F_{yc} for the new position of the centre of the journal O_p $\left(x^*_{c0} + \Delta x_c,\ y^*_{c0},\ 0,\ 0\right)$:

$$\overline{F}_{xc} = \overline{F}_{xc}\left(x^*_{c0} + \Delta x_c,\ y^*_{c0},\ 0,\ 0\right) \tag{6.22a}$$

$$\overline{F}_{yc} = \overline{F}_{yc}\left(x^*_{c0} + \Delta x_c,\ y^*_{c0},\ 0,\ 0\right) \tag{6.22b}$$

The variations of the forces due to the oil-film corresponding to increase Δx_c of the variable x^*_{c0} only, are, therefore:

$$\Delta\overline{F}_{xc} = \overline{F}_{xc}\left(x^*_{c0} + \Delta x_c,\ y^*_{c0},\ 0,\ 0\right) - \overline{F}_{xc}\left(x^*_{c0},\ y^*_{c0},\ 0,\ 0\right) \tag{6.23a}$$

$$\Delta\overline{F}_{yc} = \overline{F}_{yc}\left(x^*_{c0} + \Delta x_c,\ y^*_{c0},\ 0,\ 0\right) - \overline{F}_{y}\left(x^*_{c0},\ y^*_{c0},\ 0,\ 0\right) \tag{6.23b}$$

By approximating the derivates (6.21) with the difference quotient, we can calculate the quantities:

$$K_{xx} = \left(\frac{\partial F_{xc}}{\partial x_c}\right)_0 \approx \frac{\Delta\overline{F}_{xc}}{\Delta x_c}; \quad K_{yx} = \left(\frac{\partial F_{yc}}{\partial x_c}\right)_0 \approx \frac{\Delta\overline{F}_{yc}}{\Delta x_c} \tag{6.24}$$

By subsequently increasing the variable y^*_{c0} of Δy_c, it is possible to similarly obtain the stiffnesses K_{xy} and K_{yy}. In a similar way it is also possible to determine the value of the equivalent damping terms R_{xx} etc. by starting with the configuration of equilibrium and giving increases in velocity $\Delta\dot{x}_c$ and $\Delta\dot{y}_c$ and defining the equivalent dampings via the respective difference quotients:

$$R_{xx} = \frac{\Delta\overline{F}_{xc}}{\Delta\dot{x}_c} = \frac{\overline{F}_{xc}\left(x^*_{c0},\ y^*_{c0},\ \Delta\dot{x}_c,\ 0\right) - \overline{F}_{xc}\left(x^*_{c0},\ y^*_{c0},\ 0,\ 0\right)}{\Delta\dot{x}_c} \tag{6.25}$$

$$R_{yx} = \frac{\Delta\overline{F}_{yc}}{\Delta\dot{x}_c} = \frac{\overline{F}_{yc}\left(x^*_{c0},\ y^*_{c0},\ \Delta\dot{x}_c,\ 0\right) - \overline{F}_{yc}\left(x^*_{c0},\ y^*_{c0},\ 0,\ 0\right)}{\Delta\dot{x}_c}$$

$\Delta\overline{F}_{xc}$ and $\Delta\overline{F}_{yc}$ being the increases of force caused by the presence of the oil-film due to the effect of an increase in velocity $\Delta\dot{x}_c$. In these calculations we can vary the values of the increases Δx_c, Δy_c, $\Delta\dot{x}_c$ and $\Delta\dot{y}_c$ to confirm that the difference quotient actually converges at the value of the derivates.

6.2.2.2 Experimentally Defining the Dynamic Characteristics of Bearings

The parameters of equivalent damping and stiffness of the oil-film can also be defined experimentally: to do this, the journal or the bearing (or both) is excited to vibrate around the position of equilibrium for a given value of external load Q^* and a given angular velocity of the rotor Ω^*. The excitation is usually sinusoidal with pulsation ω and can be due to either an unbalance (so ω = Ω = angular velocity of the rotor) or to generic external excitations:

$$F = F_0 e^{i\omega t} \tag{6.26}$$

so, not considering the nonlinearity and any disturbances, the response, i.e. the relative vibration between journal and bearing caused by this excitation, is also sinusoidal with equal pulsation. Under these assumptions forces $\overline{F}_{xc}$ and $\overline{F}_{yc}$ and displacements $\overline{x}_{co}$ and $\overline{y}_{co}$ are sinusoidally varying quantities, which can be expressed analytically by expressions such as:

$$\overline{x}_c = \overline{X}_{c0} e^{i\omega t}; \quad \overline{F}_{xc} = \overline{F}_{xc0} e^{i\omega t} \tag{6.27a}$$

$$\overline{y}_c = \overline{Y}_{c0} e^{i\omega t}; \quad \overline{F}_{yc} = \overline{F}_{yc0} e^{i\omega t} \tag{6.27b}$$

where, we should remember, ω is the pulsation of the vibration, which does not necessarily coincide with Ω^*. The Eqs. (6.20a, 6.20b) can thus be expressed as:

$$\overline{F}_{xc0} = K_{xx}\overline{X}_{c0} + K_{xy}\overline{Y}_{c0} + i\omega R_{xx}\overline{X}_{c0} + i\omega R_{xy}\overline{Y}_{c0} \tag{6.28a}$$

$$\overline{F}_{yc0} = K_{yx}\overline{X}_{c0} + K_{yy}\overline{Y}_{c0} + i\omega R_{yx}\overline{X}_{c0} + i\omega R_{yy}\overline{Y}_{c0} \tag{6.28b}$$

where $\overline{F}_{xc0}$, $\overline{F}_{yc0}$, $\overline{X}_{c0}$, $\overline{Y}_{c0}$ are complex quantities.

Using proximity sensors it is possible to measure the relative displacements $\overline{x}_c(t)$ and $\overline{y}_c(t)$ between journal and bearing and, via pressure sensors, we can obtain the distribution of pressure inside the meatus [42, 43] and then obtain forces $\overline{F}_{xc}(t)$ and $\overline{F}_{yc}(t)$ by integrating on the entire bearing; alternatively, these forces can be measured directly by dynamometric supports.

These forces, as well as displacements $\overline{x}_c$ and $\overline{y}_c$, actually contain the contribution to frequency ω and any higher harmonics, as well as the contribution caused by excitation at the angular frequency Ω^* of the rotor and any harmonics of a higher order, not to mention disturbances of an aleatory nature: so we must separate the contribution to the response caused by the forcing element (with frequency ω) from the contribution of the other harmonics caused by unbalances or other causes, by performing harmonic analysis of the different quantities, with fundamental frequency equal to ω, and thus obtain just the synchronous components $\overline{F}_{xc0}$, $\overline{F}_{yc0}$, $\overline{X}_{c0}$, $\overline{Y}_{c0}$ that appear in (6.28a, 6.28b).

From the two vectorial Eqs. (6.28a, 6.28b), corresponding to four scalar equations, it is not possible to obtain the eight unknowns K_{xx}, K_{xy}, …R_{yx} and R_{yy}: to do this, we must write at least another four equations, which can once again be the (6.28a, 6.28b) relating, however, to different values of $\overline{X}_{c0}^{(2)}$ and $\overline{Y}_{c0}^{(2)}$ and, therefore, different values of the forces $\overline{F}_{xc0}^{(2)}$ and $\overline{F}_{yc0}^{(2)}$:

- if we work with the same pulsation of the forcing element ω, we need to change the module and phase of excitation to obtain different values of $\overline{F}_{xc0}^{(2)}$ and $\overline{F}_{yc0}^{(2)}$ and, therefore, of $\overline{X}_{c0}^{(2)}$ and $\overline{Y}_{c0}^{(2)}$, but not linearly dependent on previous $\overline{F}_{xc0}$ and $\overline{F}_{yc0}$ and, therefore, of $\overline{X}_{c0}$ and $\overline{Y}_{c0}$;
- alternatively, we must change the excitation pulsation ω, to equal angular velocity $\Omega*$, in order to obtain different values of $\overline{X}_{c0}^{(2)}$ and $\overline{Y}_{c0}^{(2)}$, again, not linearly dependent on the previous $\overline{X}_{c0}$ and $\overline{Y}_{c0}$.

In both cases it is possible to arrive at a system of eight equations in the eight unknowns K_{xx}, K_{xy}, …R_{yx} and R_{yy} with an approach that is, therefore, deterministic.

Alternatively it is possible to use a procedure of minimisation if the pairs of values $\overline{X}_{co}^{(i)}$, $\overline{Y}_{co}^{(i)}$, $\overline{F}_{yco}^{(i)}$, $\overline{F}_{xco}^{(i)}$, are more than two.

Suppose we have n sets of values $\overline{X}_{co}^{(i)}$, $\overline{Y}_{co}^{(i)}$, $\overline{F}_{yco}^{(i)}$, $\overline{F}_{xco}^{(i)}$ $(i = 1, 2, \ldots n)$ separate from each other and not linearly dependent, obtained by varying, for example, the frequency of excitation ω_i while keeping the angular velocity Ω^* constant: as $n > 2$, (6.28a, 6.28b) will not be strictly satisfied, as the number of equations is higher than the number of unknowns. Therefore, to define the values of stiffnesses K_{xx}, K_{xy}, … and dampings R_{yx}, R_{yy}, …, we must use methods of minimisation such as, for example, the least squares approach: for this purpose we define, as residuals, the differences δ_{ix} and δ_{iy} between the measured value of each component of the force and the corresponding value given by the second member of (6.28a, 6.28b) in correspondence to the generic ith test:

$$\begin{aligned} \delta_{ix} &= \overline{F}_{xc0}^{(i)} - K_{xx}\overline{X}_{c0}^{(i)} - K_{xx}\overline{Y}_{c0}^{(i)} - i\omega R_{xx}\overline{X}_{c0}^{(i)} - i\omega R_{xy}\overline{Y}_{c0}^{(i)} \\ \delta_{iy} &= \overline{F}_{yc0}^{(i)} - K_{yx}\overline{X}_{c0}^{(i)} - K_{yy}\overline{Y}_{c0}^{(i)} - i\omega R_{yx}\overline{X}_{c0}^{(i)} - i\omega R_{yy}\overline{Y}_{c0}^{(i)} \quad (i = 1, \ldots, n) \end{aligned} \tag{6.29a}$$

Using the least squares method, we must minimise the residuals, i.e. impose:

$$f\left(K_{xx}, K_{xy}, K_{yx}, K_{yy}, R_{xx}, R_{xy}, R_{yx}, R_{yy}\right) = \sum_{i=1}^{n} \delta_{ix}^2 + \delta_{iy}^2 = \min \tag{6.29b}$$

at this point, $f\left(K_{xx}, K_{xy}, \ldots\right)$ is a function of the problem's unknowns only, that is, of the equivalent dampings and stiffnesses of the oil-film. To minimise function f we must impose the following 8 relations

$$\begin{aligned} &\frac{\partial f}{\partial K_{xx}} = 0; \quad \frac{\partial f}{\partial K_{xy}} = 0; \quad \frac{\partial f}{\partial K_{yx}} = 0; \quad \frac{\partial f}{\partial K_{yy}} = 0 \\ &\frac{\partial f}{\partial R_{xx}} = 0; \quad \frac{\partial f}{\partial R_{xy}} = 0; \quad \frac{\partial f}{\partial R_{yx}} = 0; \quad \frac{\partial f}{\partial R_{yy}} = 0 \end{aligned} \tag{6.29c}$$

which result in writing 8 complete algebraic equations in the 8 unknowns K_{xx}, K_{xy}, etc. These equations can be rewritten, with a more compact expression, in matrix form: using $\underline{\delta}$ to define the vector containing the scalars δ_{ix} and δ_{iy} (with i = 1, n):

$$\underline{\delta} = \begin{Bmatrix} \delta_{1x} \\ \delta_{1y} \\ \ldots \\ \delta_{nx} \\ \delta_{ny} \end{Bmatrix} = \underline{N} - [C]\underline{\kappa} \tag{6.30a}$$

$\underline{\kappa}$ being the vector containing the problem's unknowns:

$$\underline{\kappa} = \begin{Bmatrix} K_{xx} \\ K_{xy} \\ K_{yx} \\ K_{yy} \\ R_{xx} \\ R_{xy} \\ R_{yx} \\ R_{yy} \end{Bmatrix} \tag{6.30b}$$

[C] the coefficient matrix and, lastly, $\underline{N}$ the vector containing the components of forces $\overline{F}_{xc0}$, $\overline{F}_{yc0}$. The function f to be minimised (6.29b) can then be rewritten as:

$$f = \underline{\delta}^T \underline{\delta} = (\underline{N} - [C]\underline{\kappa})^T (\underline{N} - [C]\underline{\kappa}) = \min \tag{6.31a}$$

or rather, if we wish to weigh each residual according to the accuracy with which we assume it has been calculated, as:

$$f = (\underline{N} - [C]\underline{\kappa})^T [W](\underline{N} - [C]\underline{\kappa}) = \min \tag{6.31b}$$

where [W] is a weighing matrix that can be obtained from the correlation matrix of the measurements, to take into account the greater or lesser reliability of these measurements [64, 68, 70, 72]. So (6.29c) can be rewritten as:

$$\left\{\frac{\partial f}{\partial \underline{\kappa}}\right\}^T = \left\{\frac{\partial(\underline{N} - [C]\underline{\kappa})^T [W](\underline{N} - [C]\underline{\kappa})}{\partial \underline{\kappa}}\right\}^T = \underline{0} \tag{6.31c}$$

i.e.:

$$\begin{aligned} & -2[C]^T[W]\underline{N} + 2[C]^T[W][C]\underline{\kappa} = 0 \\ & \Rightarrow \underline{\kappa} = \left[[C]^T[W][C]\right]^{-1}[C]^T[W]\underline{N} \end{aligned} \tag{6.31d}$$

Thus, similarly to the scalar approach, it is possible to obtain from (6.31d) a system of eight equations, in the eight unknowns $\underline{\kappa}$ that define the equivalent damping and stiffness coefficients of the oil-film.

In this way it is possible, either deterministically or using minimisation approaches, to define the curves that express the trend of the equivalent dampings and stiffnesses of the oil-film, in relation to Ω^* and load Q^* borne by the bearing.

Usually these values, obtained both analytically and experimentally, are shown in relation to the dimensionless runout $\chi = e_c/\delta$ and themselves rendered dimensionless, in respect to load and clearance, using the following relations:

$$C_{xx} = \frac{K_{xx}}{Q/\delta}; \quad C_{xy} = \frac{K_{xy}}{Q/\delta}; \quad C_{yx} = \frac{K_{yx}}{Q/\delta}; \quad C_{yy} = \frac{K_{yy}}{Q/\delta} \tag{6.32a}$$

$$\varepsilon_{xx} = \frac{R_{xx}}{Q/\Omega\delta}; \quad \varepsilon_{xy} = \frac{R_{xy}}{Q/\Omega\delta}; \quad \varepsilon_{yx} = \frac{R_{yx}}{Q/\Omega\delta}; \quad \varepsilon_{yy} = \frac{R_{yy}}{Q/\Omega\delta} \tag{6.32b}$$

Figure 6.12 shows the dimensionless functions (6.32a, 6.32b) calculated using the Reynolds equation, while assuming a constant temperature in the meatus, for a cylindrical bearing. Incidentally, we should remember, as already seen in the chapter on systems surrounded by force fields, that, in general, $K_{xy} \neq K_{yx}$ and therefore the force field created by the presence of the oil film is nonconservative in its positional part, with the exception of tilting pad bearings.

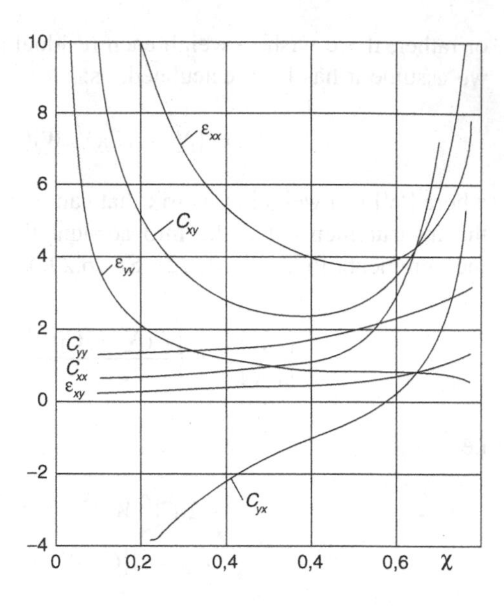

Fig. 6.12 Trend of the dimensionless stiffness and damping coefficients for a cylindrical bearing (T = const)

6.2.3 Defining the Field of Forces in Seals or More in General Between Rotor and Stator

In a compressor, a turbine or similar machinery, the rotor is wet by a fluid that creates a force field which, as stated previously, can significantly influence the dynamic behaviour of the rotor and its supporting structure. As current research stands, in general, this force field is not well defined, either analytically or experimentally, except in specific cases, such as, for example, that of the interstage seals of fluid pumps.

This problem [55–58, 60] can be traced back to the issues already addressed relating to lubricated bearings: the seal can be seen (Fig. 6.13) as a fluid film bearing, fed laterally by the fluid which is at its highest pressure (the difference between the pressures at the two sides of the seal can even be of several dozen atmospheres). To define the force field that is created in the seal it is possible to use the same analytical approaches already described and used for lubricated bearings, as well as the same experimental methods, so we refer the reader to the previous section on this subject. We should remember that while in bearings distribution of pressures, and thus the force field generated by them, is created by the mechanism of hydrodynamic lubrication induced by the circumferential flow, pressure distribution in seals is governed mainly by the axial flow created by the difference in pressure generated between stage and stage (the losses of load that the fluid undergoes at the entrance of the seal are those that mainly define the force field).

Similarly to what we saw for bearings, we can, by linearising the force field, also define the equivalent damping and stiffness coefficients for seals; also for seals, generally, $K_{xy} \neq K_{yx}$ and so there can be problems of instability [55, 59].

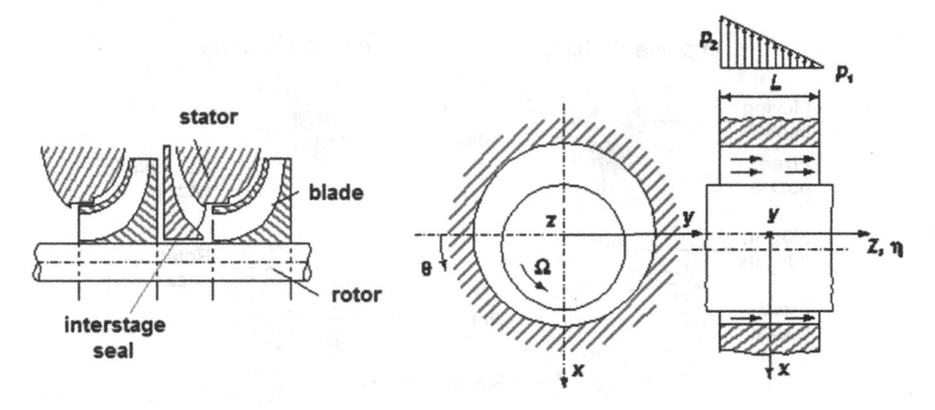

Fig. 6.13 Schematisation of an interstage seal

In the case of seals where the fluid is a gas, like in steam turbines, the problem is similar to that of seals with liquid fluid, but with the added complication associated with the compressible nature of the actual fluid, which makes analytical schematisation of the problem quite difficult, but then again, so is a correct experimental definition of the same force field. As research currently stands, it is possible to confirm that some experimental attempts to define stiffnesses and dampings of this type of seal, although having already been performed, have shown a lack of reliability in the results: in this case too, however, we obtain non-symmetrical equivalent stiffness matrices. Vice versa, as regards the fluid's actions on the entire blade system to define the force field acting on a row of blades (Fig. 6.14) or on the blade disks in relation to the relative shaft-stator displacement, the problem becomes considerably more complicated, as it is affected not only by the fluid-dynamic behaviour of the blade, caused by the displacements relating to the stator, but also by the radial clearance of the row of blades in respect to the casing. In this case, there are currently no approaches, either analytical or experimental, aimed at providing indications on the nature of this force field, so when schematising the rotor this effect is generally ignored.

6.2.4 Schematising the Casing and the Supporting Structure

As stated previously, the casing may, in turn, be similar to either a cylindrical body (Fig. 6.2) or to a body with a more complex shape (Fig. 6.1):

- in the first case it can be easily schematised with beam elements, as shown in Fig. 6.15;
- in the event, however, that the casing has a more complex shape, we can resort to a schematisation of the casing with three-dimensional finite elements.

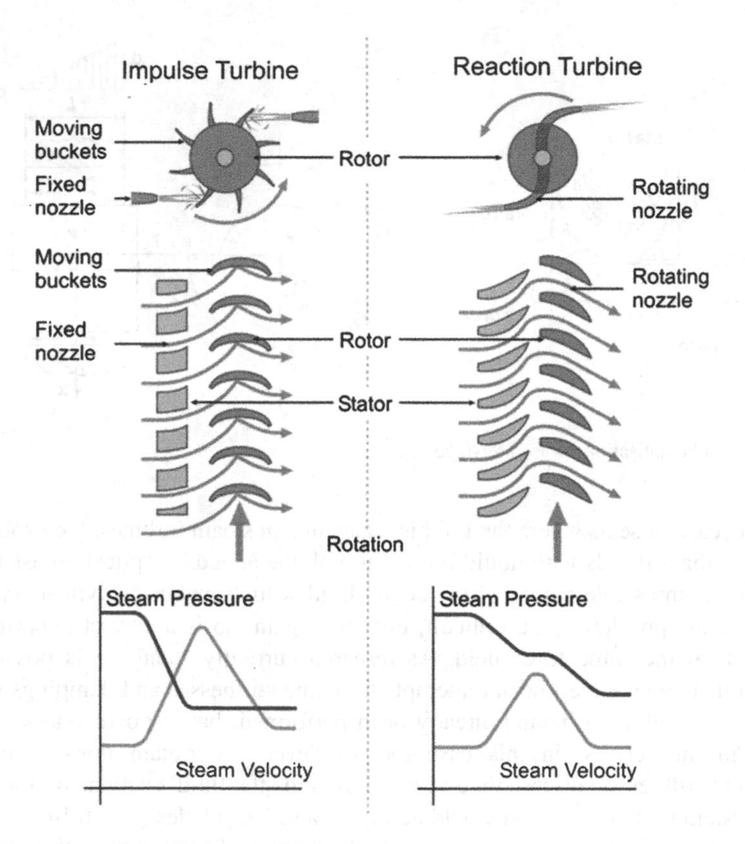

Fig. 6.14 Actions of the fluid on the blade system

However, this approach is very laborious and is also, for reasons that we will explain below, generally discarded: we should also mention that the casing is generally constrained to a foundation, so we must also take into account the "casing + foundation" system. Although it is true that by using a schematisation with finite elements we could take into account the foundation, however complex it is, the difficulties in calculation associated with such a procedure can hardly be justified if we consider the fact that, even using a sophisticated schematisation, it would be difficult to reproduce the problems relating to the connection between casing and foundation plate (elasticity of the connecting elements, more or less accurate coplanarity of the supports, friction in the connections etc.).

So for these reasons, unless the casing can be easily schematised (like the example in Fig. 6.15) it is preferable to reproduce the casing and relevant foundation together, using simple models that reproduce the real forces exchanged between supporting structure and shaft in qualitative and, as far as possible, quantitative terms.

In this regard, we use models, for example, made up of mass-spring-damper type systems with one d.o.f.: in doing so, the model of a shaft, considering the oil-film or

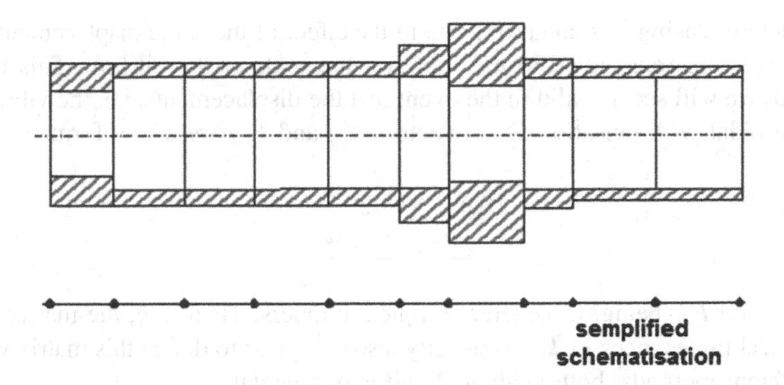

Fig. 6.15 Schematisation of cylindrical shaped casing (Fig. 6.2)

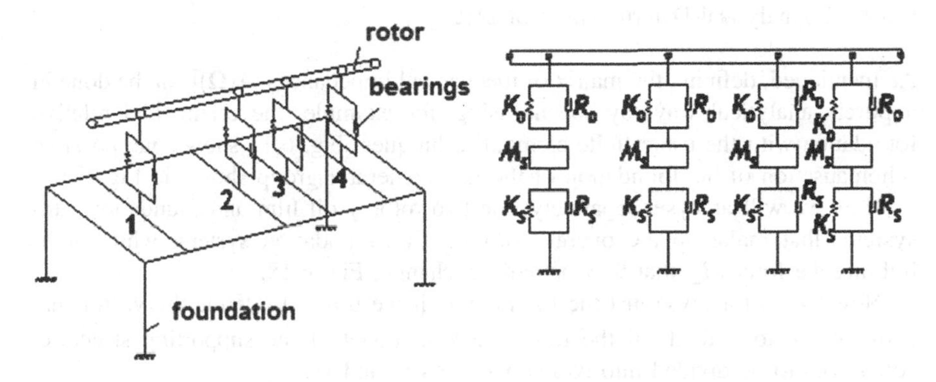

Fig. 6.16 Simplified schematisation of the casing-foundation system in 1 d.o.f. systems

the seals and the foundation, becomes that shown in Fig. 6.16 where K_0, R_0 are used to briefly indicate the stiffnesses and the dampings of the oil-film, while the weight M_s, stiffness K_s and damping R_s simulate the behaviour of the casing and the foundation, at the connection points with the rotor (bearings).

6.2.4.1 Matrix of the Mechanical Impedances $[A(\Omega)]$

A more sophisticated model is that which reproduces the behaviour of the casing and of the foundation using the mechanical impedances of this sub-system. The matrix of the mechanical impedances of the structure supporting the rotor is the matrix $[A(\Omega)]$ defined by the complex relation:

$$[A(\Omega)]\underline{X}_{co} = \underline{F}_{co} \tag{6.33}$$

where $\underline{X}_{co}$ is the vector of the displacements of the supporting structure, in correspondence to the seals or the bearings, $\underline{F}_{co}$ is the vector of the forces transmitted by the structure (casing + foundation) due to the effect of the same displacements $\underline{X}_{co}$: these forces are transmitted to the shaft via the seals or the oil-film. This formulation, as we will see, is valid in the event that the displacements, i.e. the vibrations, are sinusoidal with pulsation Ω, so vectors $\underline{X}_{co}$ and $\underline{F}_{co}$ are of the form:

$$\begin{aligned} \underline{X}_{co} &= \underline{X}_{co0} e^{i\Omega t} \\ \underline{F}_{co} &= \underline{F}_{co0} e^{i\Omega t} \end{aligned} \tag{6.34}$$

with $\underline{X}_{co}$ and $\underline{F}_{co}$ being, in general complex numbers. Therefore, the matrix of the mechanical impedances $[A(\Omega)]$ is usually also complex: to define this matrix we can use different methods, both analytical and experimental.

6.2.4.1.1 Analytical Determination of $[A(\Omega)]$

As mentioned, defining the matrix of mechanical impedances $[A(\Omega)]$ can be done in a purely analytical way, by schematising, for example, the casing and relative foundation with the usual finite element techniques: Fig. 6.17 shows one possible schematisation of the foundation of the turbogenerator group shown in Fig. 6.1.

Now we will analyse, separately, the two rotor + oil film and foundation subsystems that make up the overall rotor-oil film-foundation system, while highlighting the forces $\underline{F}_c$ that they actually exchange, Fig. 6.18.

Now we will analyse just the foundation: if we use $\underline{x}_f$ to define the vector that contains the total d.o.f. of the finite element model of the supporting structure, considered to be divided into two sub-vectors $\underline{x}_c$ and $\underline{x}_{fi}$:

$$\underline{x}_f = \begin{Bmatrix} \underline{x}_c \\ \underline{x}_{fi} \end{Bmatrix} \tag{6.35a}$$

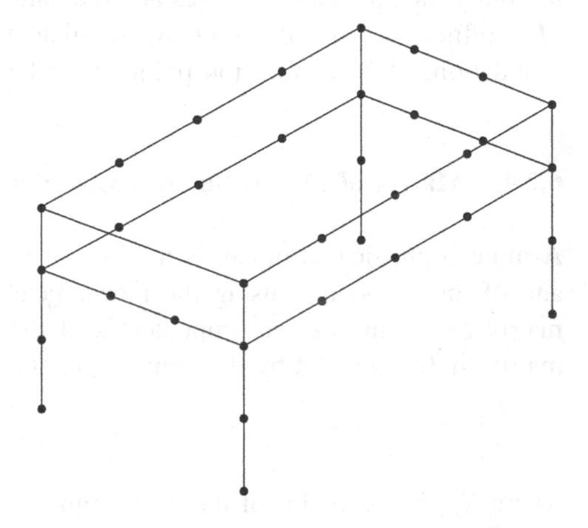

Fig. 6.17 Finite element schematisation of a foundation frame

having used $\underline{x}_c$ to indicate the vector containing the d.o.f. relating to the connection nodes of the shaft-supporting structure and $\underline{x}_{fi}$ for the vector that contains the d.o.f. of the remaining nodes of the foundation.

So the motion equations of just the foundation will be, having divided them as in (6.35a):

$$\begin{bmatrix} [M_{cc}] & [M_{ci}] \\ [M_{ic}] & [M_{ii}] \end{bmatrix} \begin{Bmatrix} \ddot{\underline{x}}_c \\ \ddot{\underline{x}}_{fi} \end{Bmatrix} + \begin{bmatrix} [R_{cc}] & [R_{ci}] \\ [R_{ic}] & [R_{ii}] \end{bmatrix} \begin{Bmatrix} \dot{\underline{x}}_c \\ \dot{\underline{x}}_{fi} \end{Bmatrix} + \begin{bmatrix} [K_{cc}] & [K_{ci}] \\ [K_{ic}] & [K_{ii}] \end{bmatrix} \begin{Bmatrix} \underline{x}_c \\ \underline{x}_{fi} \end{Bmatrix} = \begin{Bmatrix} \underline{F}_c \\ \underline{F}_{fi} \end{Bmatrix} \tag{6.35b}$$

$[M_f]$, $[R_f]$ and $[K_f]$ being, respectively, the matrices of mass, damping and stiffness of just the foundation, assumed to be already constrained[6]:

$$[M_f] = \begin{bmatrix} [M_{cc}] & [M_{ci}] \\ [M_{ic}] & [M_{ii}] \end{bmatrix}; \quad [R_f] = \begin{bmatrix} [R_{cc}] & [R_{ci}] \\ [R_{ic}] & [R_{ii}] \end{bmatrix}; \quad [K_f] = \begin{bmatrix} [K_{cc}] & [K_{ci}] \\ [K_{ic}] & [K_{ii}] \end{bmatrix} \tag{6.35c}$$

In (6.35b) $\underline{F}_c$ and $\underline{F}_{fi}$ represent, respectively, the forces that the rotor discharges on the foundation and the forces (if any) that act on other d.o.f. In the hypothesis of analysing the dynamic response of the overall system at steady-state for sinusoidal forces applied at just the shaft nodes:

$$\underline{F}_r = \underline{F}_{ro} e^{i\Omega t} \tag{6.36a}$$

forces, which are also sinusoidal, will only arrive on the foundation via the bearings:

$$\underline{F}_f = \begin{Bmatrix} \underline{F}_c \\ \underline{0} \end{Bmatrix} = \begin{Bmatrix} \underline{F}_{co} \\ \underline{0} \end{Bmatrix} e^{i\Omega t} \tag{6.36b}$$

so, at steady-state, the solution of (6.35b) will be of the form:

$$\underline{x}_f = \underline{X}_{fo} e^{i\Omega t} = \begin{Bmatrix} \underline{X}_{co} \\ \underline{X}_{fio} \end{Bmatrix} e^{i\Omega t} \tag{6.36c}$$

[6]The matrices defined in this way thus represent just the part relating to the actual free d.o.f. of the finite element model of the foundation: with the nomenclature already defined in Chap. 4, Sect. 4.4.7, we thus have:

$$[M_f] = [M_{fLL}]; [R_f] = [R_{fLL}]; [K_f] = [K_{fLL}]. \tag{6.6.1}$$

which replaced in the same equations leads to:

$$\left[-\Omega^2\begin{bmatrix}[M_{cc}] & [M_{ci}]\\ [M_{ic}] & [M_{ii}]\end{bmatrix} + i\Omega\begin{bmatrix}[R_{cc}] & [R_{ci}]\\ [R_{ic}] & [R_{ii}]\end{bmatrix} + \begin{bmatrix}[K_{cc}] & [K_{ci}]\\ [K_{ic}] & [K_{ii}]\end{bmatrix}\right]\begin{Bmatrix}\underline{X}_c\\ \underline{X}_{fi}\end{Bmatrix} = \begin{Bmatrix}\underline{F}_{co}\\ \underline{0}\end{Bmatrix} \tag{6.37a}$$

Considering the second matrix equation of (6.37a):

$$-\Omega^2[M_{ic}]\underline{X}_{co} - \Omega^2[M_{ii}]\underline{X}_{fio} + i\Omega[R_{ic}]\underline{X}_{co} + i\Omega[R_{ii}]\underline{X}_{fio} + [K_{ic}]\underline{X}_{co} + [K_{ii}]\underline{X}_{fio} = \underline{0} \tag{6.37b}$$

it is possible to express $\underline{X}_{fio}$ as a function of the displacements of the connection nodes $\underline{X}_{co}$:

$$\begin{aligned}\underline{X}_{fio} &= \left[-\Omega^2[M_{ii}] + i\Omega[R_{ii}] + [K_{ii}]\right]^{-1}\left[-\Omega^2[M_{ic}] + i\Omega[R_{ic}] + [K_{ic}]\right]\underline{X}_{co}\\ &= [C_{ic}(\Omega)]\underline{X}_{co}\end{aligned} \tag{6.37c}$$

Replacing (6.37c) in the first matrix equation of (6.37a):

$$\begin{aligned}&-\Omega^2[M_{cc}]\underline{X}_{co} - \Omega^2[M_{ci}]\underline{X}_{fio} + i\Omega[R_{cc}]\underline{X}_{co}\\ &+ i\Omega[R_{ci}]\underline{X}_{fio} + [K_{cc}]\underline{X}_{co} + [K_{ci}]\underline{X}_{fio} = \underline{F}_{co}\end{aligned} \tag{6.38a}$$

we arrive at:

$$\left[-\Omega^2[M_{cc}] + i\Omega[R_{cc}] + [K_{cc}]\right]\underline{X}_{co}\left[-\Omega^2[M_{ci}] + i\Omega[R_{ci}] + [K_{ci}]\right]\underline{X}_{fio} = \underline{F}_{co} \tag{6.38b}$$

i.e.:

$$\begin{aligned}&\left[\left[-\Omega^2[M_{cc}] + i\Omega[R_{cc}] + [K_{cc}]\right] + \left[-\Omega^2[M_{ci}] + i\Omega[R_{ci}] + [K_{ci}]\right][C_{ic}(\Omega)]\right]\underline{X}_{co} = \underline{F}_{co}\\ &\Rightarrow [A(\Omega)]\underline{X}_{co} = \underline{F}_{co}\end{aligned} \tag{6.38c}$$

It is then possible to define the matrix $[A(\Omega)]$ of the Eq. (6.33) as:

$$[A(\Omega)] = \left[\left[-\Omega^2[M_{cc}] + i\Omega[R_{cc}] + [K_{cc}]\right] + \left[-\Omega^2[M_{ci}] + i\Omega[R_{ci}] + [K_{ci}]\right][C_{ic}(\Omega)]\right] \tag{6.39}$$

As we can see, this operation, known as dynamic condensation, makes it possible to formally eliminate the d.o.f. $\underline{x}_{fi}$, relating to nodes that are not connected, from the equations of motion, without introducing any simplification: we should also remember that this operation is only possible in the frequency domain and not in the time domain. Now we will analyse the rotor + oil film sub-system in Fig. 6.18: we use $\underline{x}$ to define the vector that contains the total d.o.f. of the rotor model $\underline{x}_r$ and the connection nodes $\underline{x}_c$:

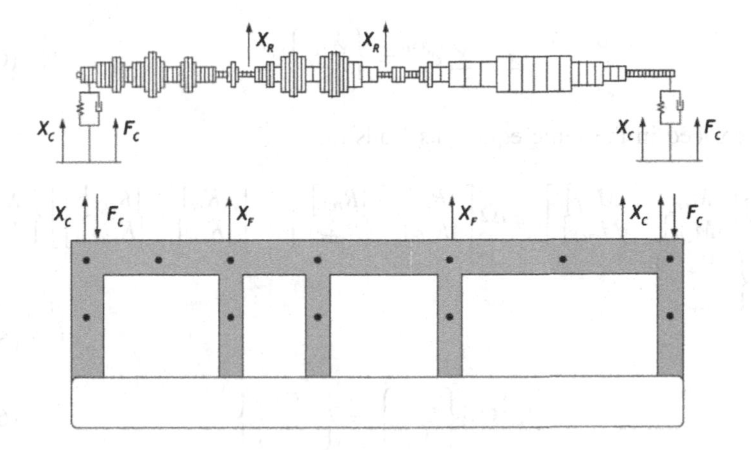

Fig. 6.18 The overall model of the rotor-oil-film-foundation system

$$\underline{x} = \begin{Bmatrix} \underline{x}_r \\ \underline{x}_c \end{Bmatrix} \tag{6.40a}$$

The motion equations of just the rotor, including the terms relating to the equivalent stiffnesses and the dampings of the oil-film, will be, in divided form, as follows:

$$\begin{aligned} &\begin{bmatrix} [M_{rr}] & [M_{rcf}] \\ [M_{cfr}] & [M_{cfcf}] \end{bmatrix} \begin{Bmatrix} \ddot{\underline{x}}_r \\ \ddot{\underline{x}}_c \end{Bmatrix} + \begin{bmatrix} [R_{rr}] & [R_{rcf}] \\ [R_{cfr}] & [R_{cfcf}] \end{bmatrix} \begin{Bmatrix} \dot{\underline{x}}_r \\ \dot{\underline{x}}_c \end{Bmatrix} \\ &+ \begin{bmatrix} [K_{rr}] & [K_{rcf}] \\ [K_{cfr}] & [K_{cfcf}] \end{bmatrix} \begin{Bmatrix} \underline{x}_r \\ \underline{x}_{cf} \end{Bmatrix} = \begin{Bmatrix} \underline{F}_r \\ -\underline{F}_c \end{Bmatrix} \end{aligned} \tag{6.40b}$$

[M], [R] and [K] being, respectively, the matrices of mass, damping and stiffness of just the rotor and the oil-film:

$$[M] = \begin{bmatrix} [M_{rr}] & [M_{rcf}] \\ [M_{cfr}] & [M_{cfcf}] \end{bmatrix}; \quad [R] = \begin{bmatrix} [R_{rr}] & [R_{rcf}] \\ [R_{cfr}] & [R_{cfcf}] \end{bmatrix}; \quad [K] = \begin{bmatrix} [K_{rr}] & [K_{rcf}] \\ [K_{cfr}] & [K_{cfcf}] \end{bmatrix} \tag{6.40c}$$

In (6.40b) $\underline{F}_r$ and $\underline{F}_c$ represent, respectively, the forces applied to the rotor and the forces exchanged with the foundation (the negative sign satisfies the principle of action and reaction, Fig. 6.18).[7] Assuming we analyse the dynamic response of the overall system at steady-state for sinusoidal forcing elements at just the shaft nodes (6.36a), the solution of (6.40b) will be as follows:

[7]It is assumed positive the direction of $\underline{F}_c$ as shown in Fig. 6.18.

$$\underline{x} = \underline{X}_o e^{i\Omega t} = \left\{ \begin{matrix} \underline{X}_{ro} \\ \underline{X}_{co} \end{matrix} \right\} e^{i\Omega t} \tag{6.41a}$$

which replaced in the same equations leads to:

$$\left[-\Omega^2 \begin{bmatrix} [M_{rr}] & [M_{rcf}] \\ [M_{cfr}] & [M_{cfcf}] \end{bmatrix} + i\Omega \begin{bmatrix} [R_{rr}] & [R_{rcf}] \\ [R_{cfr}] & [R_{cfcf}] \end{bmatrix} + \begin{bmatrix} [K_{rr}] & [K_{rcf}] \\ [K_{cfr}] & [K_{cfcf}] \end{bmatrix} \right] \left\{ \begin{matrix} \underline{X}_{ro} \\ \underline{X}_{co} \end{matrix} \right\} = \left\{ \begin{matrix} \underline{F}_{ro} \\ -\underline{F}_{co} \end{matrix} \right\} \tag{6.41b}$$

$$[E(\Omega)] \left\{ \begin{matrix} \underline{X}_{ro} \\ \underline{X}_{co} \end{matrix} \right\} = \left\{ \begin{matrix} \underline{F}_{ro} \\ -\underline{F}_{co} \end{matrix} \right\} \tag{6.41c}$$

[E] being the elastodynamic matrix of the rotor + oil film sub-system:

$$[E] = \left[-\Omega^2 \begin{bmatrix} [M_{rr}] & [M_{rcf}] \\ [M_{cfr}] & [M_{cfcf}] \end{bmatrix} + i\Omega \begin{bmatrix} [R_{rr}] & [R_{rcf}] \\ [R_{cfr}] & [R_{cfcf}] \end{bmatrix} + \begin{bmatrix} [K_{rr}] & [K_{rcf}] \\ [K_{cfr}] & [K_{cfcf}] \end{bmatrix} \right] \tag{6.41d}$$

At this point it is possible to rewrite (6.38c) as:

$$\begin{bmatrix} [0] & [0] \\ [0] & [A(\Omega)] \end{bmatrix} \left\{ \begin{matrix} \underline{X}_{ro} \\ \underline{X}_{co} \end{matrix} \right\} = \left\{ \begin{matrix} \underline{0} \\ \underline{F}_{co} \end{matrix} \right\} \tag{6.42a}$$

By replacing (6.42a) in (6.41b) we then obtain:

$$\left[[E(\Omega)] + \begin{bmatrix} [0] & [0] \\ [0] & [A(\Omega)] \end{bmatrix} \right] \left\{ \begin{matrix} \underline{X}_{ro} \\ \underline{X}_{co} \end{matrix} \right\} = \left\{ \begin{matrix} \underline{F}_{ro} \\ \underline{0} \end{matrix} \right\} \tag{6.42b}$$

In this way, it is possible to rigorously define the dynamic response of the complete rotor-oil film-foundation system, considering as independent variables just the rotor nodes and nodes of connection with the foundation: the mechanical impedance matrix $[A(\Omega)]$ is added algebraically to the elastodynamic matrix $[E(\Omega)]$ of the rotor and the oil film. Once (6.42b) has been solved it is also possible to obtain, via (6.37c), the displacements of the interior nodes of the foundation $\underline{X}_{fio}$.

Using impedances to reproduce the behaviour of the supporting structure is possible if we study the response of the complete system to a sinusoidal forcing action: below we will see how most problems relating to the dynamics of rotors can be analysed via the frequency response of the complete system, so, in most cases of practical interest, it is possible to introduce the effect of the foundation via its mechanical impedances.

The mechanical impedance method is also advantageous because it is possible to estimate the matrix $[A(\Omega)]$ experimentally as shown in the following section.

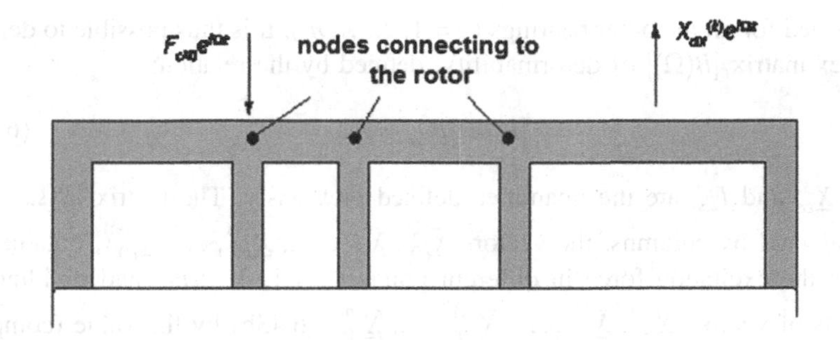

Fig. 6.19 Experimental determination of mechanical impedances

6.2.4.1.2 Experimentally Determining $[A(\Omega)]$

To experimentally evaluate the mechanical impedance matrix $[A(\Omega)]$, we must use a sinusoidal force to excite just the foundation: in other words, the rotor must be removed and in correspondence to the bearings we use suitable actuators to apply harmonic forces (6.43a) and measure the corresponding displacements.[8] Thus applying, in a generic point "k" corresponding to a bearing, as shown in Fig. 6.19:

$$F_{ck} = F_{cko}e^{i\Omega t} \tag{6.43a}$$

a harmonic force, we must simultaneously measure the displacements $\underline{X}_{c0}^{(k)}$ at all the connection points between structure and shaft (i.e. the n_c connection points of the bearings and seals)[9]:

$$\underline{x}_c^{(k)} = \underline{X}_{c0}^{(k)} e^{i\Omega t} = \begin{Bmatrix} X_{c01}^{(k)} \\ X_{c02}^{(k)} \\ \cdots \\ X_{c0k}^{(k)} \\ \cdots \\ X_{c0nc}^{(k)} \end{Bmatrix} e^{i\Omega t} \tag{6.43b}$$

The exciter must transmit a sinusoidal force at frequency Ω and this frequency is made to vary continuously in the frequency range concerned, so as to obtain a vector $\underline{X}_{c0}^{(k)}(\Omega)$ that provides the displacements in the supports (bearings and seals), function of Ω, for a forcing element applied in the kth bearing. This procedure must

[8]This operation involves considerable difficulty during installation of the group and this consideration makes the approach practically useless.

[9]In fact, although the excitation force is harmonic, the effect of noise is added to it: so for this reason the output is, in part, random. To obtain the part that is coherent with the excitation imposed, it is necessary to carry out a harmonic analysis of the output that is synchronous with the excitation, so as to eliminate any possible external disturbances.

be repeated for all the other bearings ($k = 1, 2, \ldots, n_c$). It is thus possible to define a complex matrix $[B(\Omega)]$ of deformability, defined by the relation:

$$[B(\Omega)]\underline{F}_{co} = \underline{X}_{co} \tag{6.43c}$$

where $\underline{X}_{co}$ and $\underline{F}_{co}$ are the quantities defined previously. The matrix $[B(\Omega)]$ contains, aligned by columns, the vectors $\overline{\underline{X}}_{c0}^{(1)}, \overline{\underline{X}}_{c0}^{(2)}, \ldots, \overline{\underline{X}}_{c0}^{(k)}, \ldots, \overline{\underline{X}}_{c0}^{(nc)}$, obtained by shifting the excitation force in different points ($k = 1, 2, \ldots nc$) and dividing the elements of vectors $\overline{\underline{X}}_{c0}^{(1)}, \overline{\underline{X}}_{c0}^{(2)}, \ldots, \overline{\underline{X}}_{c0}^{(k)}, \ldots, \overline{\underline{X}}_{c0}^{(nc)}$ (6.43b) by the value (complex) of the applied force F_{cko} (6.43a): the deformability matrix $[B(\Omega)]$ is, in fact, defined by the displacements $\underline{X}_{co}$ due to unitary force:

$$[B(\Omega)] = \begin{bmatrix} \overline{\underline{X}}_{co}^{(1)} & \overline{\underline{X}}_{co}^{(2)} & \ldots & \overline{\underline{X}}_{co}^{(k)} & \ldots & \overline{\underline{X}}_{co}^{(nc)} \end{bmatrix} \tag{6.44a}$$

Considering relations (6.33) and (6.43b) the following relation is obvious:

$$[A(\Omega)] = [B(\Omega)]^{-1} \tag{6.44b}$$

i.e. it is possible to obtain the mechanical impedance matrix (complex) by inverting the deformability matrix. The experimental approach has the advantage of not relying on any analytical schematisation, but on the other hand is quite laborious as it is necessary to carry out as many tests as there are bearings or shaft-supporting structure connection points while having, amongst other things, to apply the excitation force at each bearing according to two orthogonal directions (generally horizontal and vertical). To make this procedure less laborious we can use analytical methods based on modal identification techniques (Chap. 8): with these methods it is possible, by knowing just one column of the deformability matrix $[B(\Omega)]$, necessarily obtained in experimental form, to define the vibration modes and the relative deformations (i.e. eigenvalues and eigenvectors) of the structure and, via a modal approach (Chap. 2, Sect. 2.5), i.e. using the principal coordinates, define the other columns without, however, repeating the tests regarding the other bearings [29–31, 33, 41, 45, 49–51, 53, 87, 90].

6.2.5 The Overall Model (an Example of Application)

Now we will see how it is possible to arrive at writing the overall equations of the entire system, made up of shaft, bearings and relative foundation, by referring to a specific example. For ease of discussion we will consider the simplified model in Fig. 6.20, where the foundation has been simulated by means of vibrating one d.o.f. systems.

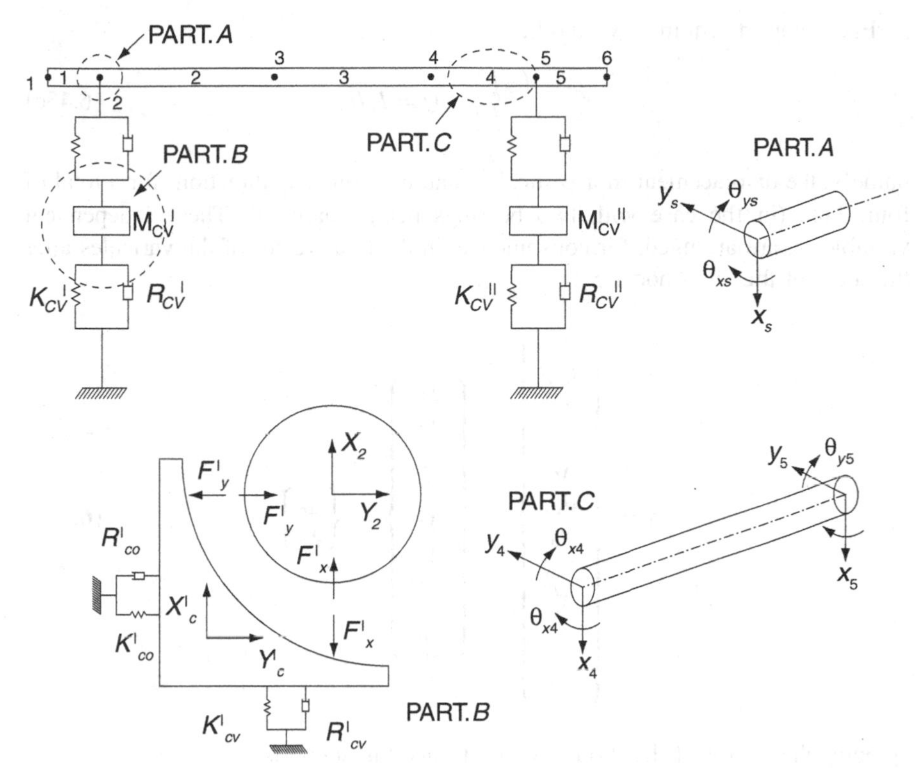

Fig. 6.20 The rotor + oil film + foundation sub-systems for calculating the matrix of mechanical impedances $[A(\Omega)]$

We define with:

$$\underline{x} = \left\{ \begin{array}{c} \underline{x}_r \\ \underline{x}_c \end{array} \right\} \tag{6.45a}$$

the vector containing the displacements of the rotor $\underline{x}_r$ and supports $\underline{x}_c$. We assume, again for simplicity, that the rotor is schematised with five beam elements: schematisation will obviously vary depending on the rotor being considered and the range of frequencies concerned. The following four coordinates will be associated with the generic ith node [Fig. 6.6, (6.1)]:

$$\underline{x}_i = \left\{ \begin{array}{c} x_i \\ \theta_{yi} \\ y_i \\ \theta_{xi} \end{array} \right\} \quad (i = 1, 6) \tag{6.45b}$$

Each support requires two d.o.f.:

$$\underline{x}_c^j = \begin{Bmatrix} x_c^j \\ y_c^j \end{Bmatrix} \quad (j = I, II) \tag{6.45c}$$

namely, the displacement in a vertical x_c^j and horizontal y_c^j direction, for a total of four d.o.f. (in the case with two bearings being analysed). These independent variables $\underline{x}_c$ are arranged, for convenience, in the total vector of the variables after the d.o.f. of the shaft nodes $\underline{x}_r$:

$$\underline{x} = \begin{Bmatrix} x_1 \\ \theta_{y1} \\ y_1 \\ \theta_{x1} \\ \cdots \\ x_6 \\ \theta_{y6} \\ y_6 \\ \theta_{x6} \\ x_c^I \\ y_c^I \\ x_c^{II} \\ y_c^{II} \end{Bmatrix} = \begin{Bmatrix} \underline{x}_1 \\ \underline{x}_2 \\ \underline{x}_3 \\ \underline{x}_4 \\ \underline{x}_5 \\ \underline{x}_6 \\ -- \\ \underline{x}_c^I \\ \underline{x}_c^{II} \end{Bmatrix} = \begin{Bmatrix} \underline{x}_r \\ \underline{x}_c \end{Bmatrix} \tag{6.46a}$$

$\underline{x}_c$ being the vector of the d.o.f. relating to just the supports:

$$\underline{x}_c = \begin{Bmatrix} x_c^I \\ y_c^I \\ x_c^{II} \\ y_c^{II} \end{Bmatrix} \tag{6.46b}$$

and $\underline{x}_r$ that relating to the nodes of just the shaft.

More specifically, we will analyse the conditions of perturbed motion around the "steady-state condition" (i.e. that by which the journal rotates in the bearing with constant angular velocity Ω, without performing oscillations within the bearing): for this reason, the coordinates $\underline{x}$ used (6.46a) represent just the perturbation around this condition, due to dynamic forces applied to the rotor. If we wish to write the motion equations of the complete system using the Lagrange equations, we must define the various forms of energy associated with the system being analysed. The total potential energy V is given by the sum of terms V_r associated with the rotor (schematised, as mentioned, with finite beam elements, see Sect. 6.2.1) and with the terms correlated to the foundation V_c:

$$V = V_r + V_c = \frac{1}{2}\sum_{k=1}^{5} \underline{x}_k^T [K_{rk}] \underline{x}_k + \frac{1}{2} \underline{x}_c^T [K_c] \underline{x}_c \tag{6.47a}$$

(6.45b) being:

$$\underline{x}_k = \begin{Bmatrix} \underline{x}_{ks} \\ \underline{x}_{kd} \end{Bmatrix} = \begin{Bmatrix} \underline{x}_i \\ \underline{x}_{i+1} \end{Bmatrix} \tag{6.47b}$$

the vector relating to the generic kth finite beam element relating to the rotor and $[K_{rk}]$ its stiffness matrix (6.2). In (6.47a) $[K_c]$ is the elastic matrix relating to the elastic support elements (6.46b):

$$[K_c] = \begin{bmatrix} K_{cv}^I & 0 & 0 & 0 \\ 0 & K_{co}^I & 0 & 0 \\ 0 & 0 & K_{cv}^{II} & 0 \\ 0 & 0 & 0 & K_{co}^{II} \end{bmatrix} \tag{6.47c}$$

The kinetic energy E_c is similarly made up of two terms:

$$E_c = E_{cr} + E_{cc} = \frac{1}{2}\sum_{k=1}^{5} \underline{\dot{x}}_k^T [M_{rk}]\underline{\dot{x}}_k + \frac{1}{2}\underline{\dot{x}}_c^T [M_c]\underline{\dot{x}}_c \tag{6.48a}$$

$[M_{rk}]$ being the mass matrix of the finite generic element (6.4) and $[M_c]$ the mass matrix relating to the supports:

$$[M_c] = \begin{bmatrix} M_{cv}^I & 0 & 0 & 0 \\ 0 & M_{co}^I & 0 & 0 \\ 0 & 0 & M_{cv}^{II} & 0 \\ 0 & 0 & 0 & M_{co}^{II} \end{bmatrix} \tag{6.48b}$$

The dissipative function D is also given by the sum of terms D_r associated with the rotor and the terms correlated to the foundation D_c:

$$D_c = D_r + D_c = \frac{1}{2}\sum_{k=1}^{5} \underline{\dot{x}}_k^T [R_{rk}]\underline{\dot{x}}_k + \frac{1}{2}\underline{\dot{x}}_c^T [R_c]\underline{\dot{x}}_c \tag{6.49a}$$

$[R_{rk}]$ being the damping matrix of the generic finite element assumed as a linear combination of the mass and stiffness matrices (Chap. 4, Sect. 4.4.5.1):

$$[R_{rk}] = \alpha[M_{rk}] + \beta[K_{rk}] \tag{6.49b}$$

In (6.49a) $[R_c]$ is the damping matrix relating to the supports (6.46b):

$$[R_c] = \begin{bmatrix} R_{cv}^I & 0 & 0 & 0 \\ 0 & R_{co}^I & 0 & 0 \\ 0 & 0 & R_{cv}^{II} & 0 \\ 0 & 0 & 0 & R_{co}^{II} \end{bmatrix} \tag{6.49c}$$

The virtual work $\delta^* L$ of the external forces $\underline{F}_r$ applied directly to the nodes of the rotor (due to a generically distributed unbalance, Sect. 6.4) and of the forces $\underline{F}_o$ due to the oil-film (Sect. 5.3.2.2 and 6.2.2) can be defined as

$$\delta^* L = \delta^* L_r + \delta^* L_o = \underline{F}_r^T \delta^* \underline{x}_r + \underline{F}_0^T \delta^* \underline{x} \tag{6.50}$$

Since we are analysing the conditions of perturbed motion around the *steady-state condition*, the forces due to the oil film, broken down into two components F_x and F_y, can be defined, with the conventions shown in Fig. 6.20, as:

$$\begin{aligned}
F_x^I &= -K_{xx}^I(x_1 - x_c^I) - K_{xy}^I(y_1 - y_c^I) - R_{xx}^I(\dot{x}_1 - \dot{x}_c^I) - R_{xy}^I(\dot{y}_1 - \dot{y}_c^I) \\
F_y^I &= -K_{yx}^I(x_1 - x_c^I) - K_{yy}^I(y_1 - y_c^I) - R_{yx}^I(\dot{x}_1 - \dot{x}_c^I) - R_{yy}^I(\dot{y}_1 - \dot{y}_c^I) \\
F_x^{II} &= -K_{xx}^{II}(x_6 - x_c^{II}) - K_{xy}^{II}(y_6 - y_c^{II}) - R_{xx}^{II}(\dot{x}_6 - \dot{x}_c^{II}) - R_{yy}^{II}(\dot{y}_6 - \dot{y}_c^{II}) \\
F_y^{II} &= -K_{yx}^{II}(x_6 - x_c^{II}) - K_{yy}^{II}(y_6 - y_c^{II}) - R_{yx}^{II}(\dot{x}_6 - \dot{x}_c^{II}) - R_{yy}^{II}(\dot{y}_6 - \dot{y}_c^{II})
\end{aligned} \tag{6.51}$$

having indicated with:

- x_1, y_1, x_6, y_6 the vertical and horizontal displacements of the journal, in correspondence to the rotor nodes where the bearings are attached (nodes 1 and 6 in Fig. 6.20);
- $x_c^I, y_c^I, x_c^{II}, y_c^{II}$ the displacements of the foundation nodes;
- $K_{ij}^I, R_{ij}^I, K_{ij}^{II}, R_{ij}^{II}$ are the equivalent stiffness and damping coefficients of the oil film relating to the first and second bearing estimated as described extensively in Sect. 6.3.1.

These forces are applied to both the shaft nodes and the support nodes, in a vertical and horizontal direction: for this reason, the virtual work $\delta^* L_o$ due to the forces of the oil-film is[10]:

$$\begin{aligned}
\delta^* L_o = {} & F_x^I \delta^* x_1 + F_y^I \delta^* y_1 - F_x^I \delta^* x_c^I - F_x^I \delta^* y_c^I \\
& + F_x^{II} \delta^* x_6 + F_y^{II} \delta^* y_6 - F_x^{II} \delta^* x_c^{II} - F_x^{II} \delta^* y_c^{II}
\end{aligned} \tag{6.52}$$

[10]The contribution due to the forces of the oil-film must be introduced via the virtual work performed by them as they represent a nonconservative force field.

From the expression of virtual work it is possible to identify the Lagrangian components according to the different d.o.f.:

$$
\begin{aligned}
Q_x &= \frac{\delta^* L_o}{\delta^* x_1} = F_x^I = -K_{xx}^I(x_1 - x_c^I) - K_{xy}^I(y_1 - y_c^I) - R_{xx}^I(\dot{x}_1 - \dot{x}_c^I) - R_{xy}^I(\dot{y}_1 - \dot{y}_c^I) \\
Q_y^I &= \frac{\delta^* L_o}{\delta^* y_1} = F_y^I = -K_{yx}^I(x_1 - x_c^I) - K_{yy}^I(y_1 - y_c^I) - R_{yx}^I(\dot{x}_1 - \dot{x}_c^I) - R_{yy}^I(\dot{y}_1 - \dot{y}_c^I) \\
Q_{xc}^I &= -F_x^I = K_{xx}^I(x_1 - x_c^I) + K_{xy}^I(y_1 - y_c^I) + R_{xx}^I(\dot{x}_1 - \dot{x}_c^I) + R_{xy}^I(\dot{y}_1 - \dot{y}_c^I) \\
Q_{yc}^I &= -F_y^I = K_{yx}^I(x_1 - x_c^I) + K_{yy}^I(y_1 - y_c^I) + R_{yx}^I(\dot{x}_1 - \dot{x}_c^I) + R_{yy}^I(\dot{y}_1 - \dot{y}_c^I) \\
&\ldots\ldots
\end{aligned}
\qquad (6.53)
$$

Hence, by generalising the discussion, the forces of the oil film $\underline{F}_o$ can be rewritten as:

$$\underline{F}_O = -[R_o]\underline{\dot{x}} - [K_o]\underline{x} \qquad (6.54)$$

the overall matrices of damping and of stiffness $[R_o]$ and $[K_o]$ being defined as:

$[K_0] =$ (6.55a)

		$x_2\theta_{y2}y_2\theta_{x2}$			$x_5\theta_{y5}y_5\theta_{x5}$		$x_c^I\ y_c^I$	$x_c^{II}\ y_c^{II}$
x_2 θ_{y2} y_2 θ_{x2}		$K_{xx}\ K_{xy}$ $K_{yx}\ K_{yy}$					$-K_{xx}-K_{xy}$ $-K_{yx}-K_{yy}$	
x_5 θ_{y5} y_5 θ_{x5}					$K_{xx}\ K_{xy}$ $K_{yx}\ K_{yy}$			$-K_{xx}-K_{xy}$ $-K_{yx}-K_{yy}$
x_c^I y_c^I		$-K_{xx}\ -K_{xy}$ $-K_{yx}\ -K_{yy}$					$K_{xx}\ K_{xy}$ $K_{yx}\ K_{yy}$	
x_c^{II} y_c^{II}					$-K_{xx}\ -K_{xy}$ $-K_{yx}\ -K_{yy}$			$K_{xx}\ K_{xy}$ $K_{yx}\ K_{yy}$

$[R_0] =$ (6.55b)

		$x_2\theta_{y2}y_2\theta_{x2}$			$x_5\theta_{y5}y_5\theta_{x5}$		$x_c^{I}\ y_c^{I}$	$x_c^{II}\ y_c^{II}$
$x_2\ \theta_{y2}\ y_2\ \theta_{x2}$		$\begin{matrix} R_{xx} & R_{xy} \\ R_{yx} & R_{yy} \end{matrix}$					$\begin{matrix} -R_{xx} & -R_{xy} \\ -R_{yx} & -R_{yy} \end{matrix}$	
$x_5\ \theta_{y5}\ y_5\ \theta_{x5}$					$\begin{matrix} R_{xx} & R_{yx} \\ R_{yx} & R_{yy} \end{matrix}$			$\begin{matrix} -R_{xx} & -R_{xy} \\ -R_{yx} & -R_{yy} \end{matrix}$
$x_c^{I}\ y_c^{I}$		$\begin{matrix} -R_{xx} & -R_{xy} \\ -R_{yx} & -R_{yy} \end{matrix}$					$\begin{matrix} R_{xx} & R_{xy} \\ R_{yx} & R_{yy} \end{matrix}$	
$x_c^{II}\ y_c^{II}$					$\begin{matrix} -R_{xx} & -R_{xy} \\ -R_{yx} & -R_{yy} \end{matrix}$			$\begin{matrix} R_{xx} & R_{xy} \\ R_{yx} & R_{yy} \end{matrix}$

Now we can apply the Lagrange equations: thus the motion equations become of the form:

$$[M]\,\underline{\ddot{x}} + [R]\underline{\dot{x}} + [K]\underline{x} = \underline{F} \tag{6.56}$$

where $[M]$ is the mass matrix of the complete system (rotor + foundation) which can be obtained from the expression of the kinetic energy E_c shown in (6.48a, 6.48b):

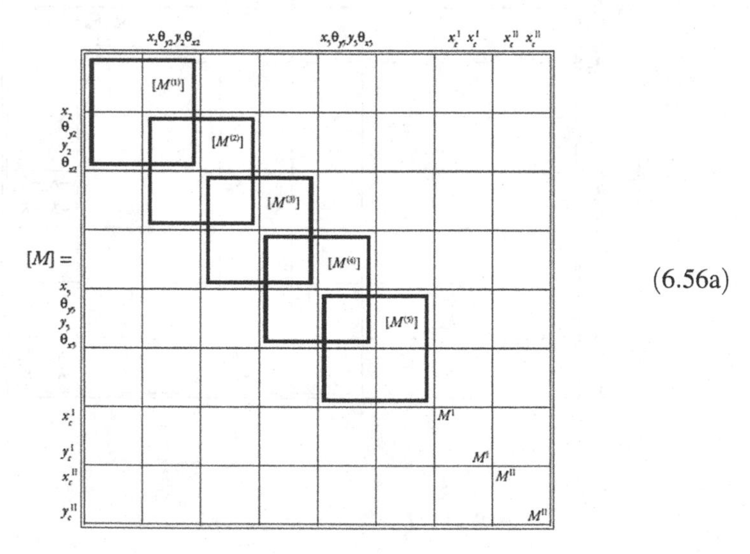

(6.56a)

$[R]$ is the damping matrix (rotor + supports + oil-film):

$$[R] = [R_s] + [R_o] \tag{6.56b}$$

which can be obtained, for the structural terms $[R_s]$, from the expression of the dissipative function D shown in (6.49a, 6.49b) and, for the terms due to the force field of the oil-film $[R_o]$, from (6.55a):

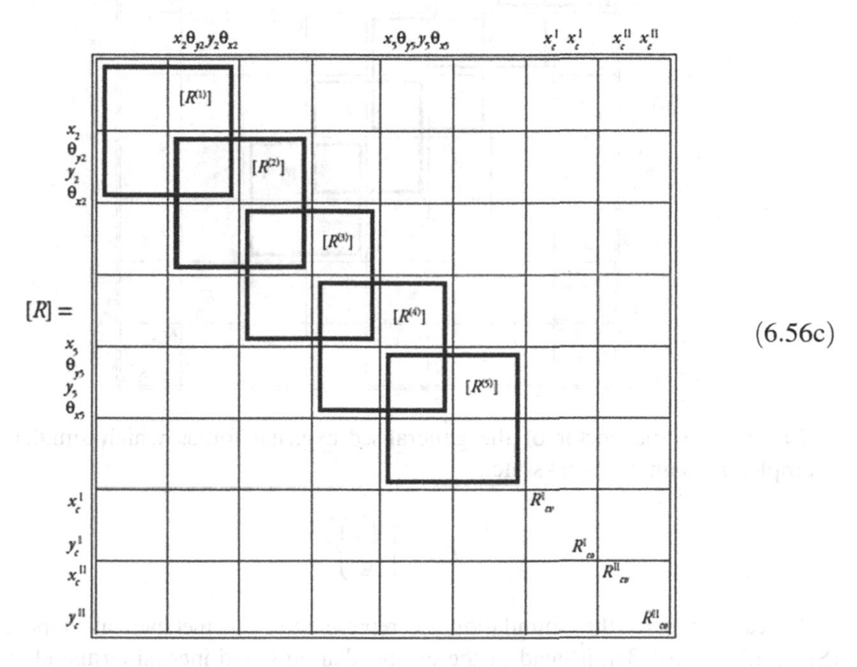

(6.56c)

$[K]$ is the total stiffness matrix (rotor + supports + oil-film):

$$[K] = [K_s] + [K_o] \tag{6.56d}$$

which can be obtained, for the structural terms $[K_s]$, from the expression of the potential energy V shown in (6.47a, 6.47b, 6.47c), and, for the terms due to the force field of the oil-film $[K_o]$, from (6.55b):

$$[K] = \quad (6.56e)$$

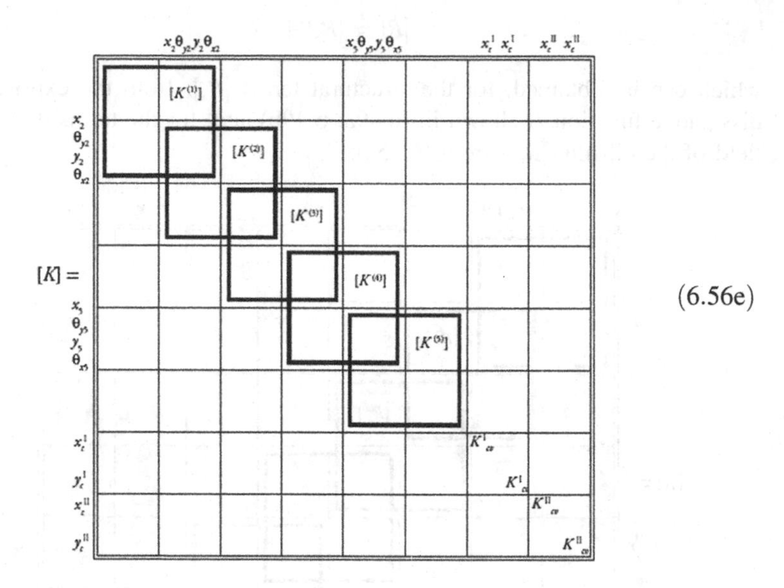

Lastly, $\underline{F}$ is the vector of the generalised external forces which simulate, for example, imbalances, cracks etc.:

$$\underline{F} = \begin{Bmatrix} \underline{F}_r \\ \underline{0} \end{Bmatrix} \tag{6.56f}$$

In cases where the foundation is represented via mechanical impedances [Sect. 6.2.4.1, (6.33)], instead of the elastic, damping and inertial terms relating to the supports schematised with one d.o.f. systems, the terms of the mechanical impedance matrix appear $[A(\Omega)]$: this matrix, see (6.41a, 6.41b, 6.41c, 6.41d) and (6.42a, 6.42b), pairs together the various d.o.f. of the supports, which in the weight-spring model are, however, independent from each other. Lastly, we should remember that the foundation can only be introduced via the mechanical impedance matrix if we are analysing the system in the frequency domain, that is, by means of a frequency response.

6.3 Analysing the Different Vibration Problems Encountered in Rotordynamics

In this section we will review the main issues that can be seen when studying rotordynamics, considering those that are more frequently encountered in machinery and receive more attention in this sector: each topic will be covered in detail in the following discussion.

(a) Response of the rotor to imbalance: critical speeds (Sect. 6.4)
All rotors, whether real, stiff or deformable, are unbalanced. This unbalance generates a force field that rotates with the shaft creating vibrations that are synchronous with the angular frequency of the machine: in general, these vibrations are defined as "the rotor's response to unbalance". The oscillations intensify when the angular frequency Ω, and thus that of the force caused by unbalance, coincides with the rotor's natural bending frequencies ω_i: the corresponding velocities are called "critical speeds".

(a1) Warping in rotors
Rotors may be subject to warping caused by non-polar symmetry temperature distribution which causes bending and, therefore, warping. These effects can appear for different reasons, such as:

- in gas or steam turbines, associated with seals modulated flow;
- in alternators, where asymmetries of electrical impedance in the rotor windings can create different currents and thus different temperatures;
- in rotors where there is a cooling fluid (air or hydrogen cooling) with non-symmetrical flows;
- in cases of friction, where the rotation of the rotor and its synchronous oscillation combine to create repeated contact in the same point of the rotor with consequent local increase in temperature (the mechanism can be self-excited, [12, 13]).

(a2) Misalignments of rotors
Rotors on several bearings are generally coupled with stiff or deformable joints. Small radial or angular misalignments between the joints can cause [1, 3–5, 9–11, 14–16] harmonic excitation forces, similar to the forces caused by an unbalance, with frequency equal to that of rotation. Greater misalignments cause the journals to oscillate inside the bearings and this leads to the onset of excitations, even at frequencies that are multiple that of rotation associated with the nonlinear effects of the forces caused by the oil-film.

(b) Balancing rotors (Sect. 6.5)
This subject is closely linked to the previous: the term "balancing" is used to define those operations where eccentric weights are added to the rotor to minimise vibrations caused by unbalance in the rotor.

(c) Torsional critical speeds (Sect. 6.10)
As shafts also have torsional inertia and elasticity they can be subject to torsional vibrations. The natural torsional frequencies, and the corresponding vibration modes, can be excited by periodic torsional torques applied to the actual shaft. When one of the harmonics of pulsation $\Omega_n = n\Omega$ of the periodic

torque, the frequency of which is linked to that of rotation Ω, coincides with a natural torsional frequency ω_{it} the corresponding angular velocity is defined as "torsional critical speed".

(d) Vibrations excited on rotors for different bending stiffnesses in the two inflection planes (Sect. 6.6)
Many rotors have a bending stiffness that varies as the angular position varies: this is caused by the presence of flaws, shaft keys, polar expansions and, at times, due to the effect of cracks that are generated in the actual rotor. In these shafts, the central ellipse of inertia, in the various sections along the longitudinal axis of the rotor, is not a circumference. It follows that, if the rotor, for example, is horizontal and, therefore, subject transversally to the weight force, the position of static equilibrium will vary as the angular position varies: variation of the static deflection occurs, as will be explained below, with a frequency $\Omega_2 = 2\Omega$ that is double that of rotation Ω. This mechanism intensifies when these frequencies Ω_2 coincide with one of the natural bending frequencies ω_i: in these conditions we describe this as "second harmonic bending critical speeds" due to the different stiffness of the rotor. In addition to this mechanism, the presence of a crack creates vibrations, one per revolution and multiple, due to the nonlinearity of the opening and closing mechanism of the actual crack.

(e) "N" per revolution vibrations
"N" per rev vibrations, i.e. vibrations that have a multiple pulsation of the fundamental pulsation Ω (equal to angular velocity), can often occur in rotors. The causes that trigger these oscillations can be of different natures but, in general, they all depend on the polar non-symmetry of the actual rotor. In a turbine the N per rev vibrations are linked to the number of blades of the rotor and of the stator and to their product. In rotors with gears the forms of excitations are linked to the number of teeth: vibrations 2 or 3 per rev can also be seen when there are problems of friction between the stator part and the rotor part.

(f) Rotor instability
Whilst the problems of vibrations, mentioned earlier, can be classified as vibrations forced on rotors due to different excitation causes, there is also a series of problems where vibrations become unstable due to the presence of nonconservative force fields. Below we will show the most common forms of such instabilities:

(f.1) Oil-film instability (Sect. 6.9): this form of instability, previously discussed in Chap. 5, Sect. 5.3.2.2 on *dynamic systems subject to force fields*, is caused by the nonconservative force field that is generated between journal and bearing when there is a fluid (liquid or gas) forming a supporting meatus. This instability arises, in general, when the runout ε_c is limited compared to the radial clearance δ, and the angular velocity Ω of the shaft is more than double the initial critical speeds ω_1.

(f.2) Instability caused by the flow in the seals: the seals, for example in the inter-stages of steam turbines or in turbomachinery in general, are the sites of nonconservative force fields (Sect. 6.2.3) and, therefore, are responsible for possible phenomena of instability similar to that which occurs in fluid lubricated journal bearings. These forms of instability arise, in particular, when the bearings do not have sufficient damping power.

(f.3) Instability caused by elastic hysteresis (Sect. 6.7): as rotating shafts can be subject to bending oscillation, they may give rise to energy dissipation, of a mainly hysteretic nature, or caused by the friction of elements in relative motion (for example, electric motor windings, teeth of coupling joints, etc.). These dissipation forces give rise to nonconservative force fields and, therefore, to problems of stability.

(f.4) Instability due to the different bending stiffness (Sect. 6.6): the different bending stiffness of a rotor in the two inflection planes, in addition to causing phenomena of vibrations with double frequency compared to that of rotation, as already seen in point d), can also create conditions of instability if the angular velocity Ω falls between the two natural bending frequencies of the rotor, which are of different values in the case where the bending stiffness of the actual rotor is different.

6.4 Critical Speed, Response of the Rotor to Unbalance

A rotor that has angular velocity Ω is the site of a field of centrifugal forces. To define these forces we assume (Fig. 6.21) an absolute Cartesian coordinate system (*X-Y-Z*) and a Cartesian coordinate system (ξ, η, ζ) that is integral with the shaft, i.e. rotating with it. The centrifugal force per unit of length $\vec{F}(\zeta)$ acting on a generic section of the rotor (with abscissa ζ) can be defined, using a complex formulation, by the following expression:

$$\vec{F}(\zeta) = \vec{F_0}(\zeta)e^{i\Omega t} = m(\zeta)\vec{\varepsilon}(\zeta)\Omega^2 e^{i\Omega t} = \left(m(\zeta)\varepsilon(\zeta)\Omega^2\right)e^{i\gamma(\zeta)}e^{i\Omega t} \tag{6.57a}$$

where $\vec{\varepsilon}(\zeta)$ represents the geometric vector defining the runout of the centre of gravity in correspondence to the single elementary sections, defined in the rotating coordinate system as:

$$\vec{\varepsilon}(\zeta) = \left|\vec{\varepsilon}(\zeta)\right|e^{i\gamma(\zeta)} = \varepsilon(\zeta)e^{i\gamma(\zeta)} \tag{6.57b}$$

$\varepsilon(\zeta)$ being the module and $\gamma(\zeta)$ the anomaly of the actual runout and $m = m(\zeta)$ the mass per unit of length, also, generally, a function of the abscissa ζ.

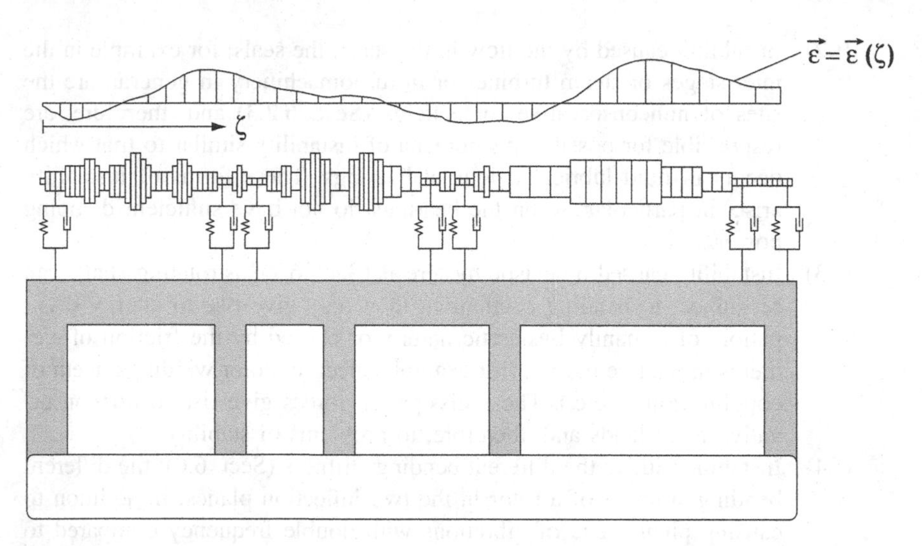

Fig. 6.21 Distribution of unbalance on a real rotor

The projection of this vector $\vec{F}(\zeta)$ on a fixed direction of an absolute reference system produces sinusoidal excitation forces with frequency equal to that of rotation Ω: so the vibrations caused by these unbalances, throughout the entire operating range, have a frequency equal to that of rotation. The effect of these destabilising forces is, therefore, detrimental for different reasons:

- when the angular velocity Ω equals one of the bending natural frequencies ω_i of the rotor, this finds itself in conditions of resonance, with a significant increase in the amplitudes of vibration (in this case, as mentioned previously, this is described as *bending critical speed*);
- throughout the entire operating field, and with amplified phenomena in proximity of the bending critical speeds, the destabilising forces determine alternate reactions to the supports: as a consequence the supporting structure (casing and foundation) can be subject to considerable vibrations.

6.4.1 Two Degree-of-Freedom Model Without Damping

To introduce the problem regarding bending critical speeds, in simplified form, first of all we will refer to a two d.o.f. model, known as the Laval rotor or Jeffcott rotor (Fig. 6.22). This model foresees a homogeneous, elastically deformable shaft, considered to be weight-free, and constrained to rotate around its own elastic curve

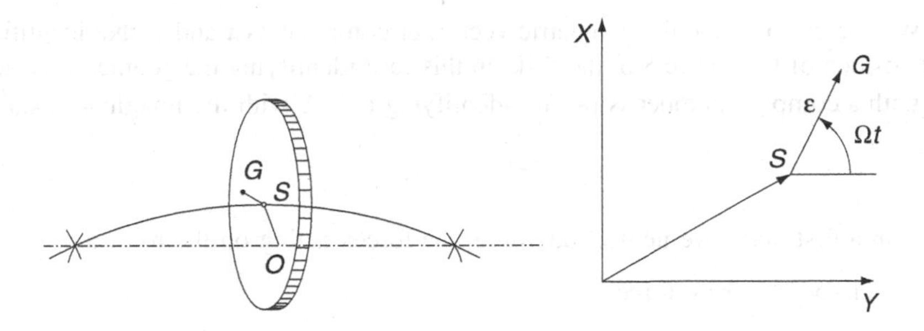

Fig. 6.22 The Laval or Jeffcott rotor (two degree-of-freedom model)

(due to its own weight), with angular velocity Ω. A thin rigid disk is mounted on the centre line of the rotor, it has mass M and is perfectly perpendicular to the axis of the rotor: it is assumed that the disk moves in a direction that is perpendicular to the rotation axis so that the gyroscopic effect can be neglected. The elasticity of the shaft is defined via its bending stiffness k, assumed constant in all directions and defined by the ratio between the constant force F applied to the centre line of the shaft and the corresponding deflection f:

$$k = \frac{F}{f} \tag{6.58}$$

As a coordinate system we assume an absolute right-hand coordinate system with axis Z coinciding with the supports axis, a vertical axis X and, lastly, with O (Fig. 6.22) defining the origin of the coordinate system used, trace of the axis Z on the plane X-Y. We use S to define the position of the geometric centre of the disk and ε for the runout of the centre of gravity G in respect to its geometric centre S, constant in respect to a suitable coordinate system (ξ, η, ζ) rotating integrally with the shaft. For the simplifying assumptions made, the motion of the disk can be considered as plane motion: x and y are the coordinates that define, in the absolute coordinate system (X-Y-Z), the generic position assumed by the geometric centre S of the disk.

The runout of the centre of gravity $\overrightarrow{\varepsilon_a}(\zeta)$ is, in respect to the absolute coordinate system chosen, a rotating vector with angular velocity Ω, thus equal to the angular velocity of the shaft:

$$\overrightarrow{\varepsilon}_a(\zeta) = \overrightarrow{\varepsilon} e^{i\Omega t} = \varepsilon e^{i\Omega t} \tag{6.59}$$

having used ε to indicate the module of the runout and having assumed the vector $\overrightarrow{\varepsilon_a}(\zeta)$ on the real axis at time $t = 0$.

We use $\vec{z}$ to define the geometric vector, of components x and y, that identifies the position of the centre S of the disk: in this case identifying the geometric vector $\vec{z}$ with a complex number is useful, identifying axis X with the imaginary axis[11]:

$$\vec{z} = y + ix \tag{6.60}$$

If in a first stage we neglect damping, the forces acting on the rotor are:

- the elastic restoring force:

$$\vec{F}_e = -k\vec{z} \tag{6.61a}$$

- the weight force (assumed to be directed as axis X):

$$\vec{P} = M\vec{g} \tag{6.61b}$$

- the force of inertia:

$$\vec{F}_i = -M\vec{\ddot{z}}_G \tag{6.61c}$$

where $\vec{\ddot{z}}_G$ is the absolute acceleration of the centre of gravity of mass M which can be defined via the following vector relations:

$$\vec{z}_G = \vec{z} + \vec{\varepsilon} = \vec{z} + \varepsilon e^{i\Omega t} \tag{6.62a}$$

$$\vec{\ddot{z}}_G = \vec{\ddot{z}} - \varepsilon\Omega^2 e^{i\Omega t} \tag{6.62b}$$

Therefore, the equation of dynamic equilibrium in vector form is:

$$M\vec{\ddot{z}} + k\vec{z} = M\vec{g} + M\varepsilon\Omega^2 e^{i\Omega t} \tag{6.63a}$$

and in scalar form:

$$M\ddot{y} + ky = M\varepsilon\Omega^2 \cos\Omega t \tag{6.63b}$$

$$M\ddot{x} + kx = -Mg + M\varepsilon\Omega^2 \sin\Omega t \tag{6.63c}$$

[11]In analysing the dynamic of rotors, and in particular when studying bending vibrations, using the algorithm of complex numbers is useful, associating the complex plane to the plane X-Y of a Cartesian coordinate system with axis Z coinciding with the longitudinal axis of the rotor: conventionally we will assume that axis X coincides with the imaginary axis. With this approach, the displacement vector (which can normally be represented in the two components x and y) becomes defined by just one complex variable z of the kind shown in (6.60). With this approach, as we will see, the 2 scalar equations that define the bending motion in space of a generic point of the rotor can be rewritten with just one complex equation, often more convenient, both as representation and as solution.

As we can see, the two equations obtained are characteristics of a forced vibrating two d.o.f. system, which is thus subject to phenomena of resonance if the pulsation Ω of the excitation force coincides with a natural frequency of the system: in the specific case of the unbalanced rotor, the forcing terms have pulsation Ω equal to the angular velocity of the shaft and intensity that depends linearly on the runout of the rotor and on term Ω^2. In the case being analysed the two natural pulsations of the shaft in the two inflection planes coincide:

$$\omega_x = \omega_y = \omega_0 = \sqrt{\frac{k}{M}} \tag{6.64}$$

Now we will define the particular integrals of (6.63a) while remembering that these integrals, as we know, define the steady-state vibration of the system. The contribution of the weight force leads to:

$$\vec{z} = \vec{z}_G = \frac{M\vec{g}}{k} = \frac{\vec{P}}{k} = \vec{f}_p \tag{6.65}$$

where $\vec{f}_p$ is the static deflection assumed by the disk due to its own weight and is fixed compared to the absolute coordinate system considered. The effect of the centrifugal forces leads, on the other hand, to another particular integral of (6.63a):

$$\vec{z} = \vec{z}_\varepsilon = Y_\varepsilon + iX_\varepsilon = \frac{\varepsilon}{\left[\left(\frac{\omega}{\Omega}\right)^2 - 1\right]} e^{i\Omega t} \tag{6.66a}$$

from which:

$$\frac{\vec{z}_\varepsilon}{\varepsilon} = \frac{1}{\left[\left(\frac{\omega}{\Omega}\right)^2 - 1\right]} e^{i\Omega t} = A(\Omega)e^{i\Omega t} \tag{6.66b}$$

i.e. in scalar form:

$$\begin{aligned} Y_\varepsilon &= A(\Omega)\varepsilon\cos\Omega t \\ X_\varepsilon &= A(\Omega)\varepsilon\sin\Omega t \end{aligned} \tag{6.66c}$$

The vector $\vec{z}_\varepsilon$, therefore, rotates with the same angular velocity Ω of the rotor:

- below resonance ($\Omega < \omega_0$), the excitation force, being $A(\Omega) > 0$, is in phase with the vibration (Fig. 6.23), i.e. $\vec{z}_\varepsilon$ is aligned and in the same direction as $\vec{\varepsilon}$;
- in conditions of resonance $\Omega = \omega_o$, $\vec{z}_\varepsilon$ has a delay of 90° compared to the vector $\vec{\varepsilon}$, with module $\left|\vec{z}_\varepsilon\right|$ tending, due to the absence of damping, to infinite amplitude, since $A(\Omega) \to \infty$;

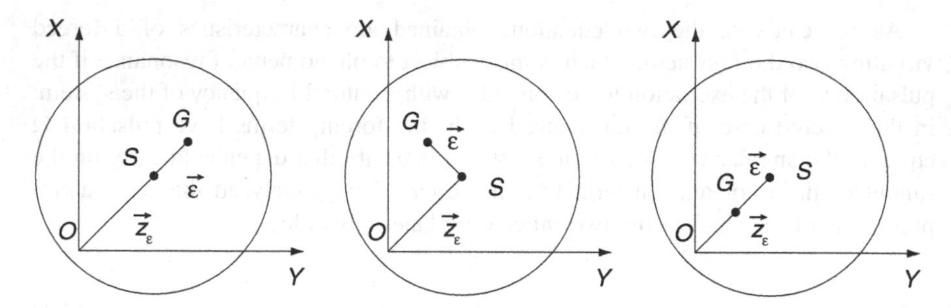

Fig. 6.23 Response of a rotor to unbalance upon variation of the ratio between angular velocity Ω and natural pulsation of the system ω_o

- lastly, above resonance, $\vec{z}_\varepsilon$ returns to alignment with $\vec{\varepsilon}$, but in counter-phase, as in Fig. 6.23: as in this condition $A(\Omega) < 0$ and tends, for $\Omega \to \infty$, to -1, the module of the oscillation amplitude $|\vec{z}_\varepsilon|$ tends to $\vec{\varepsilon} = |\vec{\varepsilon}|$, that is, the shaft tends to self-centre.

As already stated, these results highlight the fact that:

- unbalance causes sinusoidal excitation forces that are synchronous with angular velocity Ω;
- the critical speeds are nothing more than a phenomenon of resonance.

It is, in any case, possible to pass through critical speeds, even in the absence of damping, provided that these passages are carried out quickly, so that the phenomenon of resonance does not have time to amplify.

6.4.2 Two-Degree-of-Freedom Model with Damping

When introducing equivalent viscous damping r into the model shown in Fig. 6.22, the motion equations of the system become of the form:

$$M\vec{\ddot{z}} + r\vec{\dot{z}} + k\vec{z} = M\vec{g} + M\varepsilon\Omega^2 e^{i\Omega t} \tag{6.67}$$

The particular integral $\vec{z} = \vec{z}_\varepsilon e^{i\Omega t}$ which describes the forcing vibrations caused by the unbalance of the disk becomes:

$$\frac{\vec{z}_\varepsilon}{\varepsilon} = \frac{\Omega^2}{\sqrt{\left(\omega_0^2 - \Omega^2\right)^2 + \left(\frac{r\Omega}{M}\right)^2}} e^{i\phi_\varepsilon} e^{i\Omega t} = A_r(\Omega) e^{i\phi_\varepsilon} e^{i\Omega t} \tag{6.68a}$$

thus being:

$$A_r(\Omega) = \frac{\Omega^2}{\sqrt{\left(\omega_0^2 - \Omega^2\right)^2 + \left(\frac{r\Omega}{M}\right)^2}} = \frac{1}{\sqrt{\left(\left(\frac{\omega_0}{\Omega}\right)^2 - 1\right)^2 + \left(\frac{2r\omega_0}{2\Omega M\omega_0}\right)^2}}$$
$$= \frac{1}{\sqrt{\left(\left(\frac{1}{a}\right)^2 - 1\right)^2 + \left(\frac{2h}{a}\right)^2}} \tag{6.68b}$$

$$tg(\phi_\varepsilon(\Omega)) = -\frac{\Omega\frac{r}{M}}{\omega_0^2 - \Omega^2} = -\frac{\frac{2\omega_0\Omega r}{2M\omega_0}}{\omega_0^2 - \Omega^2} = -\frac{2ha}{(1-a^2)} \tag{6.68c}$$

having indicated with:

$$h = \frac{r}{r_c}; \quad a = \frac{\Omega}{\omega_0} \tag{6.68d}$$

In this case, i.e. with the presence of damping, the following boundary conditions may occur:

$$a \ll 1 \Rightarrow \Omega \ll \omega_0 \Rightarrow \phi_\varepsilon(\Omega) \cong 0° \Rightarrow A(\Omega) \cong 0 \Rightarrow \left|\vec{z}_\varepsilon\right| \cong 0 \tag{6.69a}$$

$$a = 1 \Rightarrow \Omega = \omega_0 \Rightarrow \phi_\varepsilon(\Omega) = 90° \Rightarrow A(\Omega) = \frac{1}{2h} \tag{6.69b}$$

$$a \gg 1 \Rightarrow \Omega \gg \omega_0 \Rightarrow \phi_\varepsilon(\Omega) \cong 180° \Rightarrow A(\Omega) \cong 1 \Rightarrow \left|\vec{z}_\varepsilon\right| \cong \left|\vec{\varepsilon}\right| \tag{6.69c}$$

The trend of $A = A(a)$ and of $\phi = \phi_\varepsilon(a)$ is shown in Fig. 6.24 in relation to different values of dimensionless damping $h = \frac{r}{r_c}$.

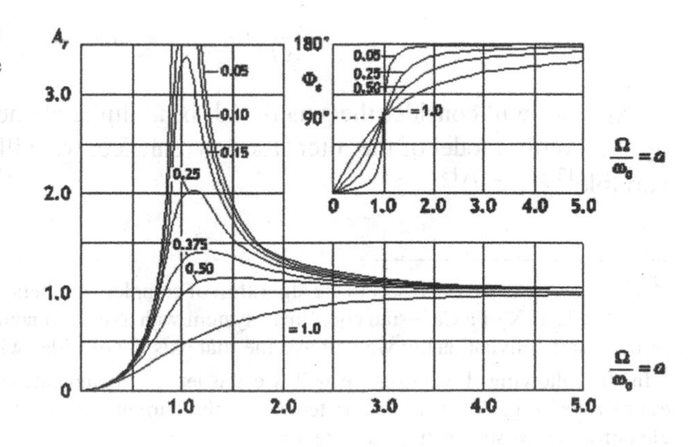

Fig. 6.24 Rotor with 2 DOF with damping: response of the system upon variation of the angular velocity Ω

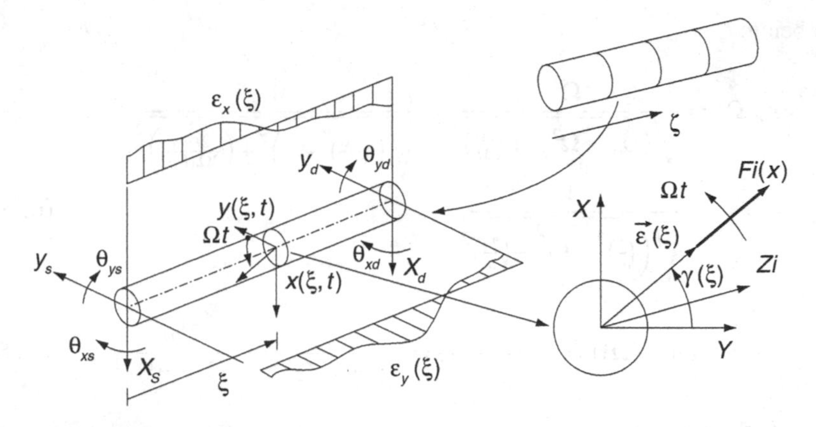

Fig. 6.25 Definition of the generalised forces caused by unbalance on a real rotor schematised with finite elements

6.4.3 Determining the Generalised Forces Acting on a Rotor Due to Unbalance

Now we will see how it is possible to take into account a possible unbalance in a real rotor by adopting the finite element analytical model shown in Fig. 6.20 and described in Sect. 6.2.5: the real rotor (Fig. 6.25) can be considered as being made up of numerous consecutive elementary sections, each of which, as we have seen, has its own runout $\vec{\varepsilon}(\zeta)$ defined in a coordinate system that rotates integrally with the actual rotor:

$$\vec{\varepsilon}(\zeta) = \left|\vec{\varepsilon}(\zeta)\right| e^{i\gamma(\zeta)} = \varepsilon(\zeta) e^{i\gamma(\zeta)} \tag{6.70}$$

This runout is variable along the axis of the rotor from section to section, both in module $\left|\vec{\varepsilon}(\zeta)\right| = \varepsilon(\zeta)$ and in phase $\gamma(\zeta)$. Compared to an absolute coordinate system, however, the runout is represented by a rotating vector:

$$\vec{\varepsilon}(\zeta) e^{i\Omega t} = \varepsilon(\zeta) e^{i\gamma(\zeta)} e^{i\Omega t} \tag{6.71}$$

Now we will consider the generic kth beam finite element (Fig. 6.25) that makes up the overall model of the rotor: for convenience, we will define with the complex variable[12] $\vec{z}(\xi, t)$[13]:

[12]In this case too, we can resort to the algorithm of complex numbers, associating the complex plane to the plane Y-X of a Cartesian coordinate system with axis Z coinciding with the longitudinal axis of the rotor: conventionally we will assume that axis X coincides with the imaginary axis.

[13]In the following discussion we will use the letter ζ to indicate the generic abscissa along the entire rotor (Fig. 6.21), while the letter ξ is the current coordinate in the generic finite "beam" element used to schematise the actual rotor.

$$\vec{z}(\xi,t) = y(\xi,t) + ix(\xi,t) \tag{6.72}$$

the generic displacement of the geometric centre of the generic section of the rotor within the *k*th finite element. So, in (6.72) $y(\xi,t)$ and $x(\xi,t)$ represent, respectively, the two horizontal and vertical components of this displacement in the same absolute coordinate system assumed (*X*-*Y*-*Z*) with axis *Z* coinciding with the rotation axis.

Presuming the trend of the runout $\vec{\varepsilon}(\zeta)$ in correspondence to the segment of rotor schematised from the generic *k*th finite element is known, the generalised forces caused by this unbalance can be obtained from the work $\delta^* L_k$ (scalar product $\cdot$[14]) carried out by the unbalancing force $\vec{F}_i(\xi)$:

[14]In this discussion it is useful to resort to the algorithm of complex numbers. Two generic vectors **V** and **W** in the complex plane can, as we know, be associated with the two complex numbers $\underline{V}$ and $\underline{W}$:

$$\begin{aligned} V \Leftrightarrow \underline{V} = V_y + iV_x = |V|e^{i\phi_V} \\ W \Leftrightarrow \underline{W} = W_y + iW_x = |W|e^{i\phi_W} \end{aligned} \tag{6.14.1}$$

As we know, the scalar product between two geometric vectors equals:

$$V \cdot W = |V||W|\cos(\phi) \tag{6.14.2}$$

and the module of the vector product:

$$|V\Lambda W| = |V||W|\sin\phi \tag{6.14.3}$$

ϕ being the relative phase between the two vectors (see Fig. 6.83).

If we define the complex conjugate of $\underline{W}$ with $\underline{W}^* = W_y - iW_x = |W|e^{-i\phi_W}$ and if we consider the product:

$$\underline{W}^*\,\underline{V} = |W|e^{-i\phi_W}|V|e^{i\phi_V} = |W||V|e^{i(\phi_V-\phi_W)} = |W||V|e^{i\phi} \tag{6.14.4}$$

it follows that the real part $\mathrm{Re}(\underline{W}^*\,\underline{V})$ of the product $\underline{W}^*\,\underline{V}$ represents the scalar product $V \cdot W$ and that the imaginary part $\mathrm{Im}(\underline{W}^*\,\underline{V})$ represents the module of the vector product $|V\Lambda W|$.

In general, it will then be possible to express the work using the algorithm of complex numbers as:

$$L = F \cdot S = Re(\underline{F}^*\underline{S}) \tag{6.14.5}$$

having indicated with:

$$\begin{aligned} F \Leftrightarrow \underline{F} = F_y + iF_x = |F|e^{i\phi_F} \\ S \Leftrightarrow \underline{S} = S_y + iS_x = |S|e^{i\phi_S} \end{aligned} \tag{6.14.6}$$

respectively, the generic vector force and the generic vector displacement, from which:

$$L = \mathrm{Re}\left[\left(F_y - iF_x\right)\left(S_y + iS_x\right)\right] = F_yS_y + F_xS_x. \tag{6.14.7}$$

$$\delta^* L_k = \int_0^{l_k} \vec{F}_i(\xi) \cdot \delta^* \vec{z}\, d\xi = \int_0^{l_k} m\Omega^2 \vec{\varepsilon}(\xi)\, e^{i\Omega t} \cdot \delta^* \vec{z} d\xi \tag{6.73}$$

where $\delta^* \vec{z}$ represents the virtual displacement vector of the position of the geometric centre S of the generic infinitesimal section of the rotor, placed in correspondence to the generic current coordinate ξ (Fig. 6.25) caused by a virtual displacement $\delta^* \underline{x}_k$ of the nodal independent variables, while $\vec{\varepsilon}(\xi)$ represents the relative runout defined with respect to a system rotating integrally with the shaft (Fig. 6.25), which rotates, in turn, with respect to the absolute coordinate system of axes (*X-Y-Z*), with angular velocity Ω. Lastly, in (6.73) m represents the weight per unit of length considered constant within the generic finite element.

By projecting the runout vector in the two directions Y and X the same can be represented as:

$$\vec{\varepsilon}(\xi) e^{i\Omega t} = \varepsilon_y(\xi) + i\varepsilon_x(\xi) \tag{6.74a}$$

since:

$$\begin{aligned} \varepsilon_y(\xi) &= \varepsilon(\xi) \cos(\Omega t + \gamma(\xi)) \\ \varepsilon_x(\xi) &= \varepsilon(\xi) \sin(\Omega t + \gamma(\xi)) \end{aligned} \tag{6.74b}$$

Considering (6.74a), (6.74b) and (6.72), it is possible to rewrite the expression of the virtual work completed by the unbalance within the generic finite element (6.73)

$$\delta^* L_k = m\Omega^2 \int_0^{l_k} \varepsilon_y\, \delta^* y\, d\xi + m\, \Omega^2 \int_0^{l_k} \varepsilon_x\, \delta^* x\, d\xi \tag{6.75}$$

where, for convenience, we no longer indicate the dependence of the runout and the generic displacement on the coordinate ξ.[15]

[15]If we wish to express this work by means of the complex number algorithm, it is possible to define the following relations:

$$\begin{aligned} \varepsilon_y\, \delta^* y &= \varepsilon \cos(\Omega t + \gamma)\delta^* y = \varepsilon(\cos \Omega t \cos \gamma - \sin \Omega t \sin \gamma)\delta^* y \\ \varepsilon_x \delta^* x &= \varepsilon \sin(\Omega t + \gamma)\delta^* x = \varepsilon(\sin \Omega t \cos \gamma + \cos \Omega t \sin \gamma)\delta^* x \end{aligned} \tag{6.15.1}$$

from which:

$$\begin{aligned} \varepsilon_y\, \delta^* y &= \varepsilon(\cos \Omega t \cos \gamma + i \cos \Omega t \sin \gamma)\delta^* y = (\varepsilon(\cos \gamma + i \sin \gamma)\delta^* y) \cos \Omega t \\ \varepsilon_x \delta^* x &= \varepsilon(-i \cos \Omega t \cos \gamma + \cos \Omega t \sin \gamma)\delta^* x = (\varepsilon(\sin \gamma - i \cos \gamma)\delta^* x) \cos \Omega t \end{aligned} \tag{6.15.2}$$

$i = \sqrt{-1}$ being the imaginary unit. Bearing in mind this formalism, the expression (6.75) can be rewritten as:

Now it is possible to correlate (Fig. 6.25) the virtual displacements $\delta^* x = \delta^* x(\xi)$ and $\delta^* y = \delta^* y(\xi)$ of the generic section to the displacements $\underline{x}_k$ (Sect. 6.2, (6.1), Fig. 6.6) of the nodes of the generic finite element:

$$\underline{x}_k = \begin{Bmatrix} \underline{x}_{ks} \\ \underline{x}_{kd} \end{Bmatrix} = \begin{Bmatrix} x_i \\ \theta_{yi} \\ y_i \\ \theta_{xi} \\ x_j \\ \theta_{yj} \\ y_j \\ \theta_{xj} \end{Bmatrix} \tag{6.76}$$

via the following relations:

$$\delta^* x(\xi) = \underline{f}_x^T(\xi)\delta^* \underline{x}_k \tag{6.77a}$$

$$\delta^* y(\xi) = \underline{f}_y^T(\xi)\delta^* \underline{x}_k \tag{6.77b}$$

where (Chap. 4, Sect. 4.5.2.1, Sect. 6.2), the shape functions $\underline{f}_x(\xi)$ and $\underline{f}_y(\xi)$ assume, in this case, the following expressions:

$$\underline{f}_y(\xi) = \begin{Bmatrix} 0 \\ 0 \\ \frac{2}{l_k^3}\xi^3 - \frac{3}{l_k^2}\xi^2 + 1 \\ -\frac{1}{l_k^2}\xi^3 + \frac{2}{l_k}\xi^2 - \xi \\ 0 \\ 0 \\ -\frac{2}{l_k^3}\xi^3 + \frac{3}{l_k^2}\xi^2 \\ \frac{1}{l_k^2}\xi^3 - \frac{1}{l_k}\xi^2 \end{Bmatrix}; \quad \underline{f}_x(\xi) = \begin{Bmatrix} \frac{2}{l_k^3}\xi^3 - \frac{3}{l_k^2}\xi^2 + 1 \\ \frac{1}{l_k^2}\xi^3 - \frac{2}{l_k}\xi^2 + \xi \\ 0 \\ 0 \\ -\frac{2}{l_k^3}\xi^3 + \frac{3}{l_k^2}\xi^2 \\ -\frac{1}{l_k^2}\xi^3 + \frac{1}{l_k}\xi^2 \\ 0 \\ 0 \end{Bmatrix} \tag{6.77c}$$

(Footnote 15 continued)

$$\begin{aligned} \delta^* L_k &= \left(m\Omega^2 \int_0^{l_k} \varepsilon(\cos\gamma + i\sin\gamma)\, \delta^* y\, d\xi + m\,\Omega^2 \int_0^{l_k} \varepsilon(\sin\gamma - i\cos\gamma)\, \delta^* x\, d\xi \right) \cos\Omega t \\ &= m\Omega^2 \left(\int_0^{l_k} \vec{\varepsilon}\, \delta^* y\, d\xi + \int_0^{l_k} i\,\vec{\varepsilon}\, \delta^* x\, d\xi \right) \cos\Omega t. \end{aligned} \tag{6.15.3}$$

(6.75), considering (6.77a)–(6.77c), becomes:

$$\begin{aligned}\delta^* L_k &= m\Omega^2 \int_0^{l_k} \varepsilon_y \underline{f}_y^T(\xi)\delta^* \underline{x}_k \, d\xi + m\Omega^2 \int_0^{l_k} \varepsilon_x \underline{f}_x^T(\xi)\delta^* \underline{x}_k \, d\xi \\ &= m\Omega^2 \left\{ \int_0^{l_k} \varepsilon_y \underline{f}_y^T(\xi)\, d\xi + \int_0^{l_k} \varepsilon_x \underline{f}_x^T(\xi)\, d\xi \right\} \delta^* \underline{x}_k = \underline{Q}_k^T \delta^* \underline{x}_k\end{aligned} \tag{6.78}$$

$\underline{Q}_k$ being the vector that contains the Lagrangian components of the destabilising forces (Chap. 4, Sect. 4.4.6):

$$\underline{Q}_k = m\Omega^2 \left\{ \int_0^{l_k} \varepsilon_y \underline{f}_y(\xi) d\xi + \int_0^{l_k} \varepsilon_x \underline{f}_x(\xi) d\xi \right\} \tag{6.79}$$

If we wish to adopt the complex algorithm (see Eq. (6.14.3) of footnote 14), to calculate the steady-state response of the rotor + bearings + foundation system, (6.79) can be rewritten as:

$$\begin{aligned}\underline{Q}_k &= m\Omega^2 \left\{ \int_0^{l_k} \vec{\varepsilon}\, \underline{f}_y(\xi) d\xi + \int_0^{l_k} i\, \vec{\varepsilon}\, \underline{f}_x(\xi) d\xi \right\} \cos \Omega t = \left\{ \underline{Q}_{ky} + \underline{Q}_{kx} \right\} \cos \Omega t \\ &= \underline{Q}_{ko} \cos \Omega t \Rightarrow \underline{Q}_{ko} e^{i\Omega t}\end{aligned} \tag{6.80}$$

Bearing in mind (6.76) the vector $\underline{Q}_{ko}$ is formed of eight terms representing, respectively, the forces and the generalised torques applied to the end nodes energetically equivalent to the distribution of unbalances on the same segment of rotor:

$$\underline{Q}_{ko}^T = \{ Q_{xio} \quad Q_{\theta yio} \quad Q_{xjo} \quad Q_{\theta yjo} \quad Q_{yio} \quad Q_{\theta xio} \quad Q_{yjo} \quad Q_{\theta xjo} \} \tag{6.81}$$

The vector of the total generalised forces acting on the rotor $\underline{F}_r = \underline{F}_{ro} e^{i\Omega t}$ due to the effect of the unbalance can be obtained with a normal assembly procedure. Writing the equations that govern the vibrations of the unbalanced rotor [Sects. 6.2 and 6.6 and (6.56)], with reference to the model shown in Fig. 6.20, thus becomes:

$$[M]\,\ddot{\underline{x}} + [R]\dot{\underline{x}} + [K]\underline{x} = \underline{F} = \left\{ \begin{matrix} \underline{F}_r \\ \underline{0} \end{matrix} \right\} = \left\{ \begin{matrix} \underline{F}_{ro} \\ \underline{0} \end{matrix} \right\} e^{i\Omega t} = \underline{F}_0 e^{i\Omega t} \tag{6.82}$$

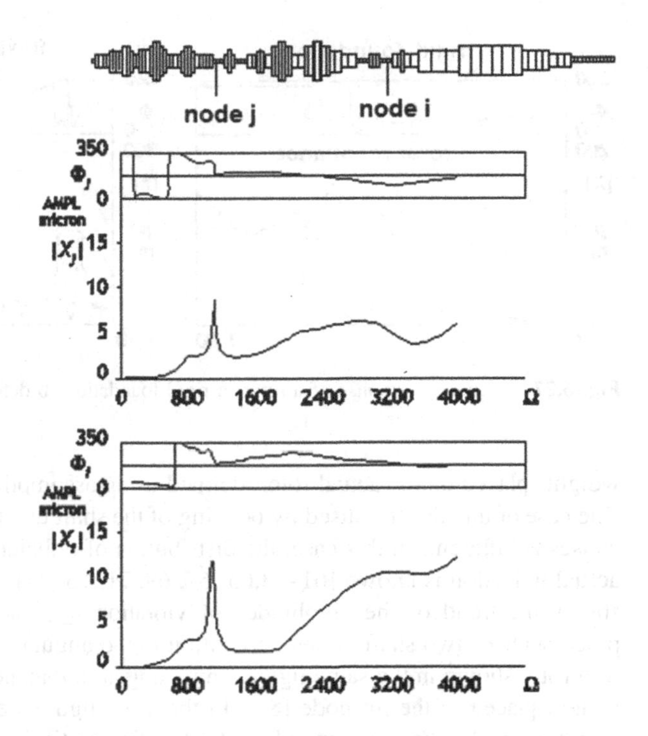

Fig. 6.26 An example of the response of a real rotor to unbalance

where the matrices of mass, damping and stiffness $[M]$, $[R]$ and $[K]$ of the overall rotor + oil-film + foundation system have already been defined in Sect. 6.6,[16] while $\underline{F}$ is the vector of the excitation forces caused by the unbalance defined above [(6.81) and Fig. 6.26]. The solution of (6.82) is given by:

$$\underline{x} = \underline{X}_0 e^{i\Omega t} \tag{6.83a}$$

which, placed in (6.82) leads to:

$$\left[-\Omega^2[M] + i\Omega[R] + [K]\right]\underline{x} = \underline{F}_0 \tag{6.84}$$

The solution $\underline{X}_o$ of (6.84) is estimated with complex algebraic numerical algorithms (Chap. 2, Sect. 2.4.1.4 and Chap. 4, Sect. 4.4.8.2) and provides the steady-state vibrations of the rotor caused by unbalance. In reality, we usually do not know the actual distribution of the unbalances, until now assumed as known; in order to simulate the behaviour of a real rotor we must, therefore, assign destabilising

[16]The matrices $[M]$, $[R]$ and $[K]$ of (6.82) represent, given the particular choice of independent variables $\underline{x}$ (6.40a), the matrices associated with just the free d.o.f., relative to the rotor-oil-film-foundation constrained system, already indicated in Chap. 4, Sect. 4.4.7 as $[M_{LL}]$, $[R_{LL}]$ and $[K_{LL}]$.

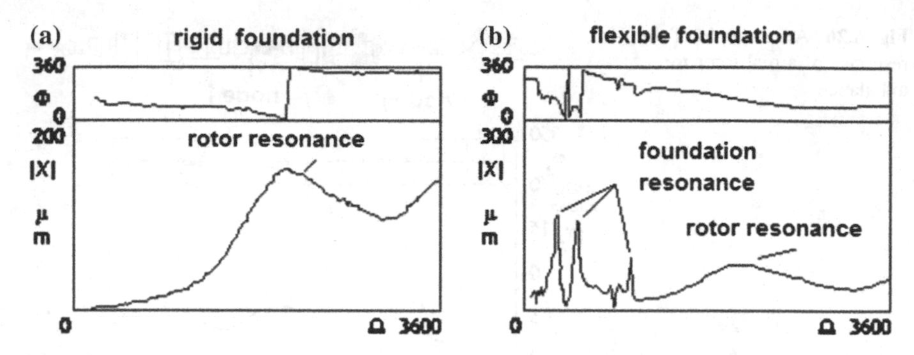

Fig. 6.27 Frequency response of a rotor: **a** rigid foundation; **b** deformable foundation

weights placed on the actual rotor, defined using assumptions of a statistical nature. The case of unbalance caused by bending of the shaft due to thermal effects or other causes is different. In this case, the distribution of unbalances can be defined if the actual inflection is known [61–63,65, 67, 69, 71, 73, 74]. As an example, Fig. 6.26 shows the trend of the amplitudes of vibration $|\underline{X}_o|$ and the phases of the displacements of two shaft nodes, in relation to the angular velocity Ω, for the model of a rotor shown in the same figure simulating an unbalance caused by an eccentric weight placed in the *j*th node [83]. In the same figure we can see the presence of vibration peaks that correspond to the bending critical speeds, and in correspondence to these speeds we can see the typical variation of the phases of the vibrations due to the passage through resonance.

Figure 6.27 shows the frequency response of a rotor presumed to be constrained in one case to a rigid foundation and in the other to a flexible foundation [84]: the peaks of resonance in the first case are due just to the characteristics of the shaft and the oil-film, while in the second case the additional peaks indicate resonance due to the characteristics of the foundation. This figure highlights the importance of introducing a sufficiently sophisticated schematisation of the foundation in order to predict the real behaviour of the rotor (Sect. 6.2.4).

6.5 Balancing Methods

The balancing of rotors is a subject of considerable interest, as the elimination or the attenuation of high vibrations caused by unbalances in the rotors is a crucial element for the good running of a machine. As mentioned, unbalance can be caused by:

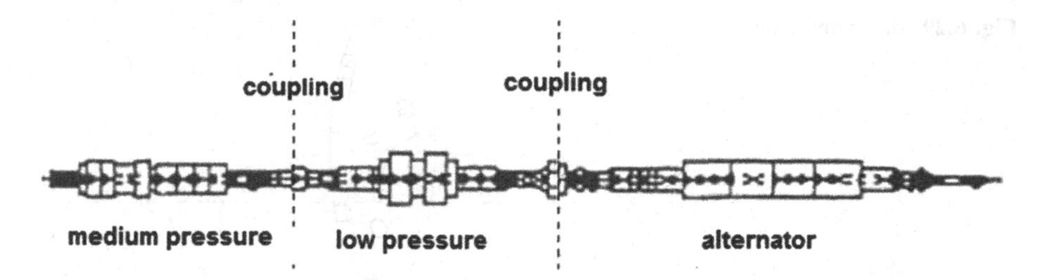

Fig. 6.28 Axis line made up of several rotors

- non-homogeneous distribution of material in the individual sections;
- permanent deformations of the rotor (for example permanent inflection of the rotor causes a runout of the centre of gravity of the individual sections along the rotor with respect to the elastic curve)[17];
- imperfect couplings in the rigid joints, when the axis line is made up of several rotors coupled on the same axis (Fig. 6.28).

The rotating forces that are discharged onto the supports due to the effect of unbalances are a detrimental factor as they can shorten the life of a machine, for various reasons:

- possible fatigue failure of material;
- friction and wear in seals and in bearings;
- difficulty of lubrication in journal bearings with hydrodynamic lubrication;
- problems caused by fatigue stresses in the supporting structure.

All the operations required to reduce to a minimum the forces that the unbalanced rotor discharges on the supports come, as mentioned, under the name of balancing.

[17]Warping in the rotor $\vec{\varepsilon}_g(\zeta)$ causes, as such, centrifugal forces that are proportional both to Ω^2, and to $\vec{\varepsilon}_g$ if the centre of gravity coincides with the geometric centre of the section of the rotor (figure below): this eccentricity $\vec{\varepsilon}_g(\zeta)$ arises even at low revs. This behaviour (characterising a warped rotor) differs from that caused by just eccentricity $\vec{\varepsilon}(\zeta)$ of the centre of gravity: in fact $\vec{\varepsilon}(\zeta)$ does not cause any vibration at low revs as the force generated by it tends to zero.

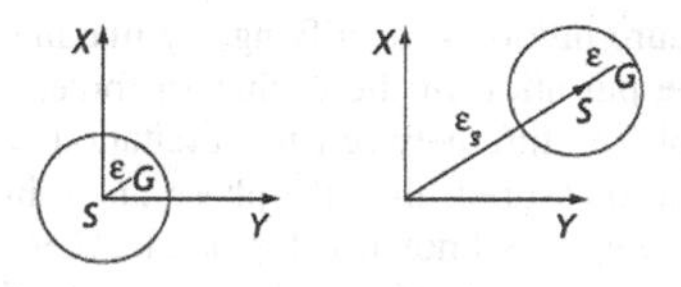

Warping effect

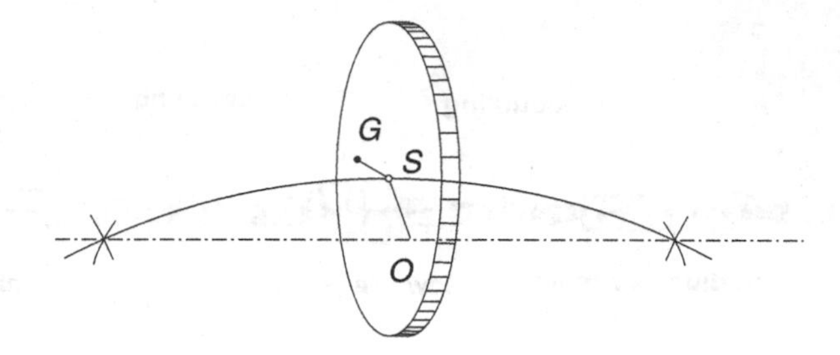

Fig. 6.29 Balancing a disk

6.5.1 Disk Balancing

To introduce the subject of balancing in a simplified manner, we can use the example with two d.o.f. regarding a thin disk (rigid and of mass m) fitted on a massless shaft (Fig. 6.29), already analysed in Sect. 6.4.1:

- conceptually, to balance the disk fitted on the shaft it is enough to place a mass $m^* = m$ positioned in the opposite direction to the position of the centre of gravity G, which thus annuls the effect of the initial unbalance, causing a force that is equal and contrary to that which already exists;
- however, as the position of the centre of gravity of the disk in relation to its geometric centre is not usually known, this can be determined by experimentally measuring the response of the rotor to the unbalance.[18]

6.5.1.1 Measurement Method for Balancing a Thin Disk

A disk (or a generic rotor), if not balanced, vibrates since, as already mentioned, it is stressed by synchronous rotating excitation forces. As it is not possible to define the real distribution of runout, in order to proceed with the balancing operation, we need to experimentally measure the vibrations caused by the actual unbalance. For this purpose, we must fit the shaft to be balanced with the appropriate instrumentation (Fig. 6.30).

[18]One simple balancing procedure involves identifying, by making the angular velocity Ω of the rotor vary, i.e. by changing the pulsation of the excitation force, the peak of the resonance and exploit the condition of 90° phase shift between the excitation force (applied to the centre of gravity) and the actual vibration measured: thus the phase angle in correspondence to which we need to position the balancing weight is known beforehand. Alternatively, it is possible to make the rotor rotate well above resonance (critical speed) and exploit the fact that, in this condition, vibration is about 180° in respect to the excitation force: in this case the balancing weight must be set in phase with the vibration vector measured.

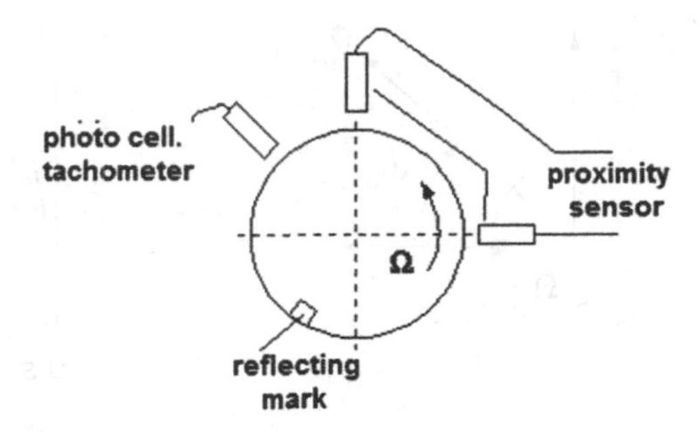

Fig. 6.30 Instrumentation required for balancing

Assuming that we have a sufficiently rigid supporting structure, it is possible to measure the absolute vibrations induced on the disk by mounting appropriate proximity sensors which, constrained rigidly to the supporting structure and facing the rotor, provide an electrical signal that is proportional to the approach of the surface of the rotor, thus measuring the vibrations $V(t)$ along a predetermined plane $n-n$ (Fig. 6.31), without needing to come into contact with the actual rotor[19]:

$$V(t) = \sum_{n=1}^{N} \left|\vec{V}_n\right| \cos(n\Omega t + \Phi_n) + \textit{random vibrations} \tag{6.85}$$

The generic signal $V(t)$ from the vibration sensor is normally periodic, with fundamental frequency equal to the angular frequency Ω,[20] plus random disturbances. Therefore, it is generally necessary to carry out the harmonic analysis of the signals to identify the one-per-rev components associated with the synchronous pulsation Ω, the only one that must be taken into consideration in problems of balancing. In this section, introducing the issues linked to rotor balancing, for simplicity's sake, we presume that the signal is harmonic and synchronous with the angular velocity, thus we will have:

$$V(t) = \left|\vec{V}\right| \cos(\Omega t + \Phi) \tag{6.86a}$$

(refer to Sect. 6.5.2 for the more general discussion). In order to have a coordinate system that is integral with the rotor, in respect to which the individual harmonics are defined, we must place a reflective mark on the rotor or exploit the possible

[19]In fact, these sensors measure both possible surface irregularities and real vibrations: to separate the two synchronous components we need to perform a test at low rotation speed (where the dynamic effect can be neglected) and a test at the desired speed. The signal due just to vibration can be obtained as the vector difference of the two previous signals.

[20]In fact, the output signal from the sensors could, e.g. in the case of flutter instability due to oil-film, (see Chap. 5, Sect. 5.3.2.2), present subharmonics, i.e. pulsating frequencies lower than the fundamental Ω of rotation of the actual rotor.

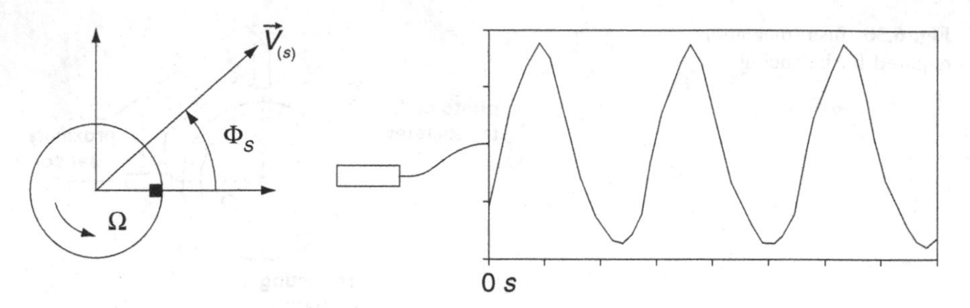

Fig. 6.31 Example of a signal in output from a displacement sensor

presence of a mark: a photoelectric cell, or similar instrument, is then placed facing the rotor to provide an electrical impulse (called the one-per-rev signal) each time that the mark passes in front of it. The signals from the two instruments (proximity sensor and photoelectric cell), can be, for example, displayed on an oscilloscope, where two tracks will appear, as shown, in the case of synchronous vibrations, in Fig. 6.32.

By measuring the amplitude V of the oscillation and the time delay Δt between the one-per-rev signal from the photoelectric cell and the maximum of the signal from the displacement sensor, it is possible to determine the vibration vector $\vec{V}$:

$$\vec{V} = Ve^{i\Phi} \tag{6.86b}$$

defined in respect to a coordinate system that is integral with the rotor[21] (i.e. rotating with it with angular velocity Ω, Fig. 6.32), Φ being the phase defined by the relation:

$$\Phi = \frac{\Delta t}{T_o} 360^\circ \tag{6.86c}$$

[21]In fact, more sophisticated instrumentation is currently used, made up of [102, 109]:

- a data acquisition system that makes it possible to transform the analogical output signal from the sensors into a digital signal stored on a computer, including the filtering operations required to eliminate the disturbances present along with the 1 per rev harmonic components (this operation is carried out with a harmonic analysis synchronised with the rotation of the rotor);
- a part of data processing that makes it possible to carry out (in digital form) the subsequent processing.

This instrumentation directly provides the quantities required for balancing: from an operational point of view, however, the sequence of operations remains that shown in the text and this approach, amongst other things, remains the most significant from a didactic point of view. For further details on the techniques used in automatic experimental data acquisition and in automatic balancing, we refer the reader to specialist texts.

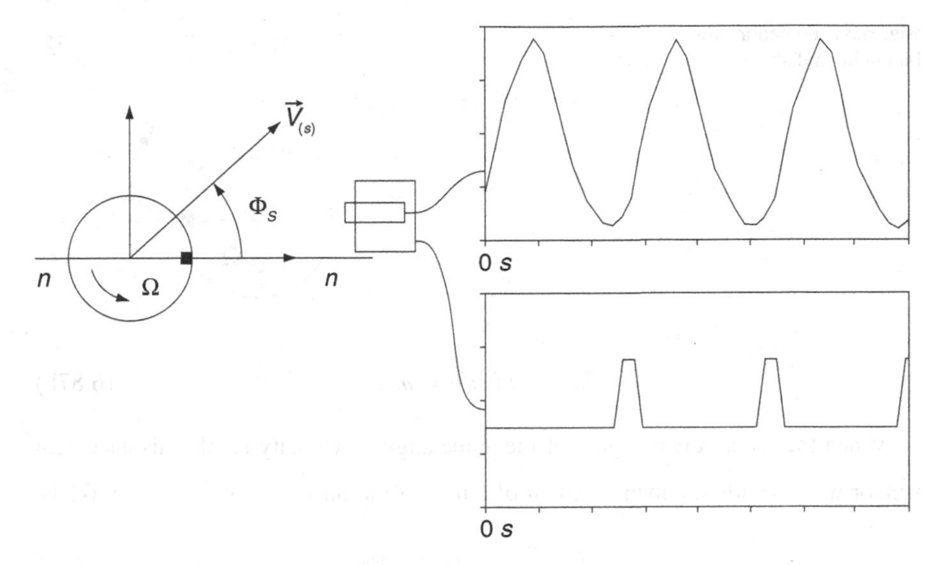

Fig. 6.32 Output signals from the displacement and one-per-rev sensors

where T_o represents the period of vibration (the distance between two consecutive impulses of the one-per-rev) and Δt is the time that lapses between the one-per-rev signal and the maximum of vibration.

6.5.1.2 The Influence Coefficient Method

At this point, there are various methods available for balancing a rotor: the most common approach is that known as *the influence coefficient method* which we will present for the simple case being analysed, but which will also be described below (Sect. 6.5.2) in the case of real rotors. For this purpose (Fig. 6.33) the disk is made to rotate at a given angular velocity Ω and, using the procedure described above, the vibration $\overrightarrow{V}^{(s)}$ caused by just the unknown unbalance is measured:

$$\overrightarrow{V}^{(s)} = V^{(s)} e^{i\Phi_s} \tag{6.87a}$$

Once the rotor has been stopped, an arbitrary mass m_j is positioned on it at a given distance ε_j from the rotation axis (these distances are actually preset, as they are provided by the manufacturer during the design phase) and in a certain angular position γ_j in the coordinate system that is integral with the rotating shaft (as the zero reference angle is conventionally assumed to be the position of the reflective mark). Now we will define, with a single complex number $\overrightarrow{m}_j$, all the characteristics of the test mass (assigned and constant at ε_j) which is thus defined by the vector relation:

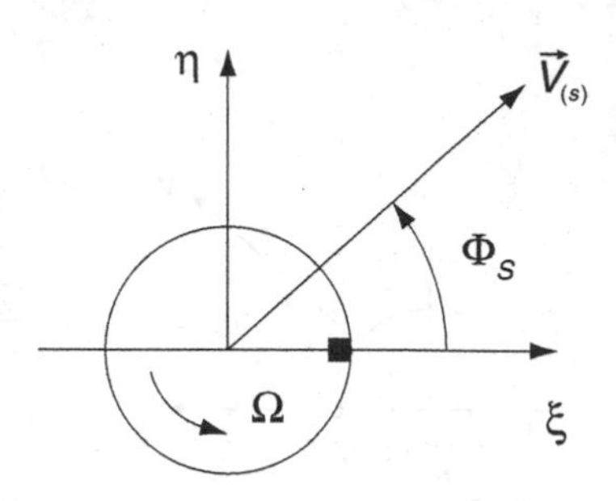

Fig. 6.33 Procedure for balancing a disk

$$\overrightarrow{m}_j = \left|\overrightarrow{m}_j\right| e^{i\gamma_j} = m_j e^{i\gamma_j} \qquad (6.87b)$$

When the rotor rotates again at the same angular velocity Ω, the displacement sensor will provide the measurement of a new vibration vector $\overrightarrow{V}^{(s+m_j)}$ (Fig. 6.34):

$$\overrightarrow{V}^{(s+m_j)} = V^{(s+m_j)} e^{i\Phi_{sj}} \qquad (6.87c)$$

of module $V^{(s+m_j)}$ and phase Φ_{sj} which differ compared to the previous case, now due both to the natural unbalance of the disk and the presence of the test mass $\overrightarrow{m}_j$.

Assuming:

- linear behaviour of the rotor, by which it is possible to apply the superposition principle;
- the repeatability of tests, where all the characteristics of the system remain unchanged between one test and the next;

the vibration induced by just the test mass $\overrightarrow{m}_j$ can be defined by vector $\overrightarrow{W}_j$, which has module $\left|\overrightarrow{W}_j\right| = W_j$ and phase ψ_j, which can be defined by the complex relation:

$$\overrightarrow{W}_j = W_j e^{i\psi_j} = \overrightarrow{V}^{(s+m_j)} - \overrightarrow{V}^{(s)} \qquad (6.88a)$$

At this point it is possible to define as influence coefficient (a quantity that is also complex) the ratio:

$$\overrightarrow{\alpha}_j = \frac{\overrightarrow{W}_j}{\overrightarrow{m}_j} = \alpha_j e^{i\Phi_{\alpha j}} \qquad (6.88b)$$

Fig. 6.34 Meaning of vector $\overrightarrow{V}^{(s+m_j)}$

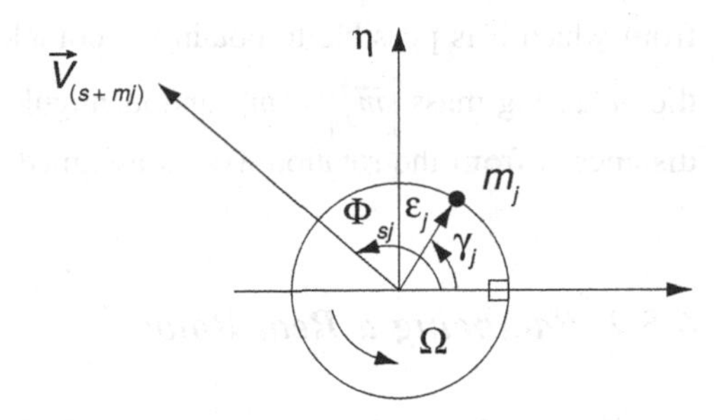

which can be seen as the effect on the disk vibrations of a mass $\overrightarrow{m}_j$ with unitary modulus and zero phase shift compared to the origin[22] (Fig. 6.35).

The disk is balanced when vector $\overrightarrow{V}^{(s)}$, caused by the initial unbalance, is counter-balanced by a vector $\overrightarrow{W}_j^*$ generated by an unknown mass $\overrightarrow{m}_j^*$:

$$\overrightarrow{m}_j^* = \left|\overrightarrow{m}_j^*\right| e^{i\gamma_j^*} = m_j e^{i\gamma_j^*} \tag{6.89a}$$

i.e. when:

$$\overrightarrow{V}^{(s)} + \overrightarrow{W}_j^* = 0 \quad \Rightarrow \quad \overrightarrow{V}^{(s)} + \overrightarrow{\alpha}_j \overrightarrow{m}_j^* = 0 \tag{6.89b}$$

This vector relation corresponds, obviously, to two scalar relations:

$$\begin{aligned} V^{(s)} \cos\Phi_s + \alpha_j m_j^* \cos\left(\Phi_{\alpha j} + \gamma_j^*\right) &= 0 \\ V^{(s)} \sin\Phi_s + \alpha_j m_j^* \sin\left(\Phi_{\alpha j} + \gamma_j^*\right) &= 0 \end{aligned} \tag{6.90}$$

[22]We should remember that the relationship between two complex numbers has as its modulus the ratio of the moduli:

$$|\overrightarrow{\alpha_j}| = \frac{\overrightarrow{W_j}}{\overrightarrow{m_j}} \tag{6.22.1}$$

and as phase the difference of phases:

$$\varphi_{\alpha j} = \Psi_j - \gamma_j \tag{6.22.2}$$

where we should remember that $\varphi_{\alpha j}$ represents the relative angle between the induced vibration vector $\overrightarrow{W}_j$ and the vector $\overrightarrow{m}_j$.

from which it is possible to obtain the complex unknown $\overrightarrow{m}_j^*$, that is, the value of the balancing mass $\left|\overrightarrow{m}_j^*\right| = m_j^*$, and its angular position γ_j^*, since, as mentioned, the distance ε_j from the rotation axis is assigned.[23]

6.5.2 Balancing a Real Rotor

A real rotor (Fig. 6.36) is made up of a solid body, with a certain spatial distribution of mass $m(\zeta)$, rotating around a given rotation axis Z: the unbalance present can be described, as seen, in the various sections, by a runout vector $\overrightarrow{\varepsilon}(\zeta)$ that rotates integrally with the actual rotor:

$$\vec{\varepsilon}(\zeta) = |\vec{\varepsilon}(\zeta)|e^{i(\Omega t+\gamma(\zeta))} = \varepsilon(\zeta)e^{i(\Omega t+\gamma(\zeta))} \tag{6.91a}$$

Due to the effect of rotation Ω only, the distribution of inertial forces is:

$$d\vec{F}_i(\zeta) = \vec{\varepsilon}(\zeta) \cdot m(\zeta) \cdot \Omega^2 d\zeta \tag{6.91b}$$

directed (Fig. 6.36) in the same way as the runout vector $\overrightarrow{\varepsilon}(\zeta)$.

---

[23]Equation (6.90) can be rewritten as:

$$\mathrm{Re}\left(\overrightarrow{V}_s\right) + i\mathrm{Im}\left(\overrightarrow{V}_s\right) + \left[\mathrm{Re}(\overrightarrow{\alpha}_j) + i\mathrm{Im}(\overrightarrow{\alpha}_j)\right]\left[\mathrm{Re}\left(\overrightarrow{m_j^*}\right) + i\mathrm{Im}\left(\overrightarrow{m_j^*}\right)\right] = 0 \tag{6.23.1}$$

from which

$$\begin{aligned} &\mathrm{Re}\left(\overrightarrow{V}_s\right) + \mathrm{Re}(\overrightarrow{\alpha}_j)\mathrm{Re}\left(\overrightarrow{m_j^*}\right) - \mathrm{Im}(\overrightarrow{\alpha}_j)\mathrm{Im}\left(\overrightarrow{m_j^*}\right) = 0 \\ &\mathrm{Im}\left(\overrightarrow{V}_s\right) + \mathrm{Re}(\overrightarrow{\alpha}_j)\mathrm{Im}\left(\overrightarrow{m_j^*}\right) + \mathrm{Im}(\overrightarrow{\alpha}_j)\mathrm{Re}\left(\overrightarrow{m_j^*}\right) = 0 \end{aligned} \tag{6.23.2}$$

the real part and the imaginary part of the unknown appear directly as real unknowns $\mathrm{Re}\left(\overrightarrow{m_j^*}\right)$ and $\mathrm{Im}\left(\overrightarrow{m_j^*}\right)$ from which it is easy to obtain:

$$\begin{aligned} m_j^* &= \sqrt{\mathrm{Re}\left(\overrightarrow{m_j^*}\right)^2 + \mathrm{Im}\left(\overrightarrow{m_j^*}\right)^2} \\ \gamma_j^* &= a\tan\left(\frac{\mathrm{Im}\left(\overrightarrow{m_j^*}\right)}{\mathrm{Re}\left(\overrightarrow{m_j^*}\right)}\right). \end{aligned} \tag{6.23.3}$$

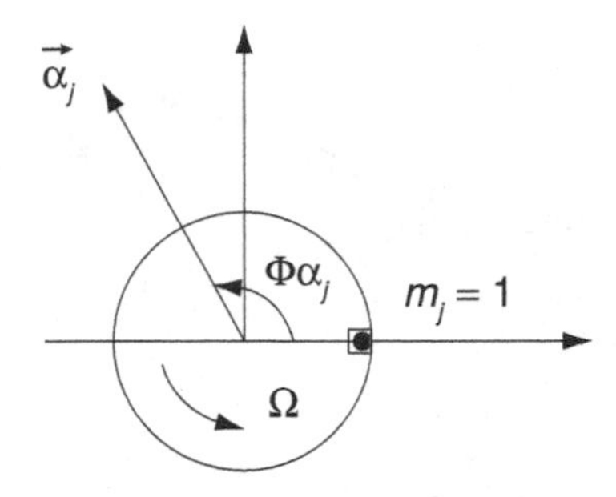

Fig. 6.35 Meaning of influence coefficient $\vec{\alpha}_j$

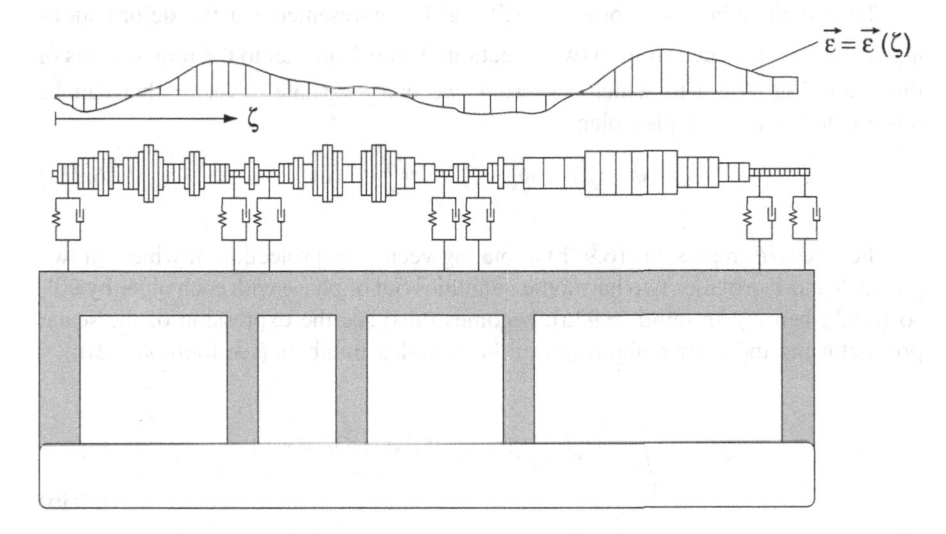

Fig. 6.36 A possible unbalance in a real rotor

The generic vibration mode of the rotor is excited to a greater or lesser extent (and thus the degree of vibration is higher to a greater or lesser extent) depending on the distribution of the destabilising forces $d\vec{F}_i(\zeta)$ along its axis. More precisely, if we use $\vec{\phi}^{(n)}(\zeta)$ to indicate the generic vibration mode, that is, the deformation, generally complex (see Fig. 6.37), function of the position ζ in correspondence to the nth mode of bending vibration, the oscillating motion of the rotor will be excited to a greater or lesser extent depending on the value assumed by the Lagrangian components of the destabilising forces $Q^{(n)}$ (Chap. 2, Sect. 2.5.3, Sect. 6.4.3):

$$Q^{(n)} = \int_0^L d\vec{F}_i(\zeta) \cdot x\, \vec{\phi}^{(n)}(\zeta) \tag{6.92}$$

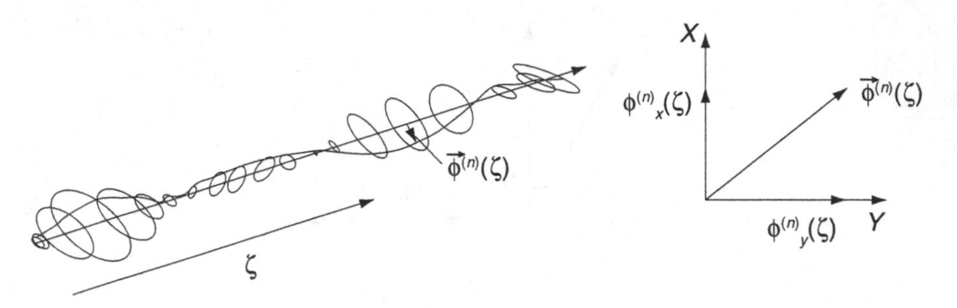

Fig. 6.37 A real rotor: generic vibration mode

The generic vibration mode $\overrightarrow{\phi}^{(n)}(\zeta)$ can be represented via the deformations $\phi_x^{(n)}(\zeta)$ and $\phi_y^{(n)}(\zeta)$ according to two directions *X* and *Y* normal to the rotation axis of the rotor (Fig. 6.37) to which, as usual, we can associate a vector that can be represented in the complex plane:

$$\vec{\phi}^{(n)}(\zeta) = \phi_y^{(n)}(\zeta) + i\phi_{x\ \ y}^{(n)}(\zeta) \tag{6.93a}$$

The force $d\vec{F}_i$ represents (6.91b) a rotating vector the projection of which on two planes *X* and *Y* provides two harmonic quantities out of phase with each other by 90°, so (6.92), bearing in mind (6.93a), becomes (also see the expression of the scalar product using the known algorithm of the complex numbers (see footnote 12)):

$$\begin{aligned} Q^{(n)} = & \int_0^L \phi^{\vec{(n)}}{}_y(\zeta)|\vec{\varepsilon}(\zeta)|m(\zeta)\Omega^2\cos(\Omega t + \gamma(\zeta)) \\ & + \int_0^L \phi^{\vec{(n)}}{}_x(\zeta)|\vec{\varepsilon}(\zeta)|m(\zeta)\Omega^2\sin(\Omega t + \gamma(\zeta)) \end{aligned} \tag{6.93b}$$

This integral, as we know, represents the work of $d\vec{F}_i$ for a deformation corresponding to the *n*th vibration mode $\overrightarrow{\phi}^{(n)}(\zeta)$ of the rotor.

The behaviour of the rotor with a given distribution of unbalance differs depending on the condition in which the rotor is operating: for the same rotor, in fact, the behaviour differs depending on whether it rotates:

- with angular velocity Ω sufficiently lower than the initial critical speed ω_1 calculated on rigid supports:

$$\Omega < \frac{1}{2}\omega_1 : \frac{1}{3}\omega_1 \tag{6.94a}$$

- above this value:

$$\Omega > \frac{1}{2}\omega_1 \tag{6.94b}$$

The balancing methods to be adopted in the two cases, although conceptually identical, are formally different: in the first case (6.94a) we talk of *rigid rotors balancing*, whereas if (6.94b) occurs we talk of *deformable rotors balancing*. We should remember that the spatial distribution of inertial forces $d\vec{F}_i$ always allows a resultant $\overrightarrow{R}_i$ and a moment $\overrightarrow{M}_i$, in whatever way the rotor behaves, that is, both in the case where the rotor is considered to be a *rigid rotor* [condition (6.94a) and in the case of a "deformable rotor" (condition (6.94b)]: a distribution of destabilising forces $d\vec{F}_i$, while having resultant $\overrightarrow{R}_i$ and a moment $\overrightarrow{M}_i$ zero, can, however, have a very high value of generic $Q^{(n)}$, excluding, of course, the rigid modes.

By way of explanation:

- assuming, for simplicity's sake, that the anomaly of the centre of gravity of the individual sections is constant along the rotor ($\gamma(\zeta) = \overline{\gamma} = \text{const}$): in this case the destabilising forces $d\vec{F}_i$ all lie in a single plane;
- assuming that the trend of the $d\vec{F}_i$ is that shown in Fig. 6.38: if the sum of the positive areas equals the sum of the negative ones, the distribution of the forces caused by unbalance has resultant $\overrightarrow{R}_i$ zero and, due to the symmetry of the unbalances, also the moment $\overrightarrow{M}_i$ is zero.

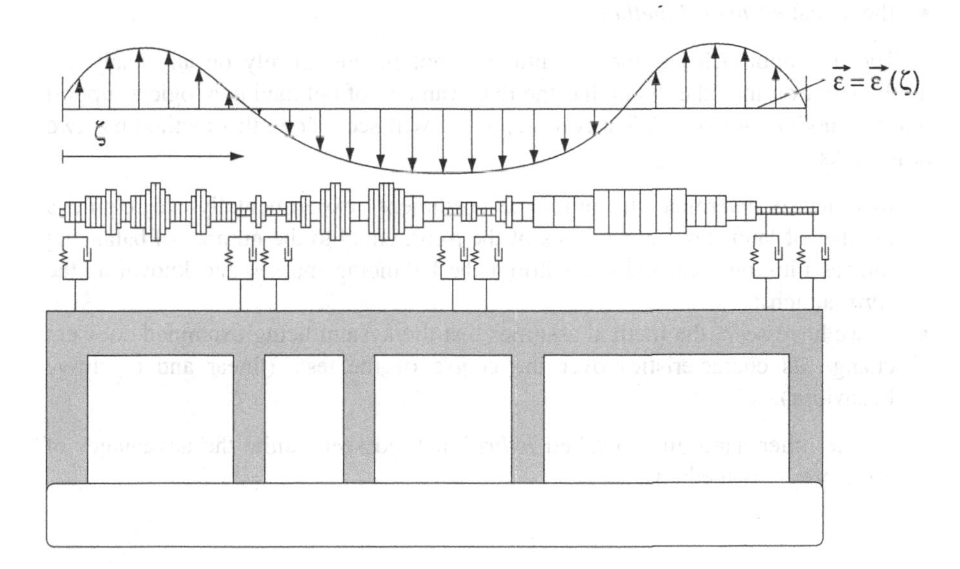

Fig. 6.38 A real rotor: one specific case of runout

In this case, if the rotor rotates in the neighbourhood of its third vibration mode $\Omega = \omega_3$ (Fig. 6.38) the Lagrangian component $Q^{(3)}$ relating to the third vibration mode (see Eq. (6.93b), assuming $n = 3$) is quite high and so the corresponding mode is highly excited. In such conditions, although balanced if considered as a *rigid rotor* (see next section) the rotor, in correspondence to angular velocity Ω close to the frequency ω_3, has high vibration amplitudes which are often intolerable. Considering the corresponding Lagrangian components $Q^{(n)}$ as exciting quantities of a mode, i.e. the work done by the centrifugal forces $d\vec{F}_i$ for a virtual displacement $\vec{\phi}^{(n)}(\zeta)$ that reproduces the form of the mode being considered, is a concept that has already been widely examined in connection with vibrating n d.o.f. systems in the modal approach (Chap. 2, Sect. 2.5). Bearing this in mind, we can say that a rigid or flexible rotor is considered balanced when all the Lagrangian components $Q^{(n)}$ (6.93b) of the destabilising forces $d\vec{F}_i$ relative to the vibration modes that can be excited (i.e. those that give a non-zero contribution to the deformation of the rotor) are zero. This purely modal approach requires knowledge of the vibration modes $\vec{\phi}^{(n)}(\zeta)$ of the complete rotor—bearings—casing—foundation system. The definition of these modes can be done:

- experimentally (but this approach is complicated);
- analytically, with different types of schematisation (the most common is with finite elements, Sect. 6.3).

To balance a real rotor, in addition to the modal method just mentioned, there are other methods that are easier to apply in practice:

- the *method of influence coefficients* (already introduced in Sect. 6.5.1.2);
- the so-called *hybrid method.*

The first approach has the advantage of not having to rely on any analytical method, but on the other hand, has the disadvantage of not having a logical support like the modal approach. What is more, as we will see below, this method has two drawbacks:

- in order to determine the influence coefficients experimentally we require a number of launches, i.e. start-ups of the rotor, equal to the number of balancing masses plus one (carried out without the balancing masses and known as the *zero* launch);
- as we have seen, the method assumes that the system being examined does not change its characteristics over the course of the tests (linear and repetitive behaviour).

On the other hand, the so-called *hybrid* method seeks unite the advantages of both the previous methods.

6.5.2.1 Balancing Rigid Rotors

A rotor characterised by an initial natural pulsation ω_o, calculated on rigid supports, that is sufficiently higher than the angular velocity $\omega_o > 2\Omega$, as already stated, is classified as a rigid rotor.[24] The generic deformation of the rotor z(ζ, t) can always be seen as a linear combination of different modes:

$$\vec{z}(\zeta,t) = y(\zeta,t) + i \cdot x(\zeta,t) = \sum_{n=1}^{\infty} \vec{\phi}^n(\zeta) q_n(t) \tag{6.95}$$

but if (6.94a) are confirmed, the excitation of the deformable modes is negligible, as the frequency of excitation Ω is much lower than the natural frequencies (quasi-static area, see Chap. 1, Sect. 1.3.3 and Sect. 2.3.3.1). So for a rigid rotor the deformation is described by a combination of just the rigid modes: although the balancing operation is always linked to cancellation of the Lagrangian components of the inertial forces for the different excited modes, the fact of having to consider just rigid vibration modes makes it possible to define the balancing conditions in an operationally more convenient way. In fact, we can say that [2, 68]:

1. a rigid rotor is balanced when the resultant of the inertial forces $\vec{R}_i$ and the torque $\vec{M}_i$ due to these forces is zero, since these resultants coincide with the Lagrangian components of the destabilising forces for the rigid modes;
2. a rigid rotor is said to be perfectly balanced when its rotational axis passes through its centre of gravity and is a principal axis of inertia, as it is possible to demonstrate [17, 23] that, for constant angular velocity Ω, the following relations apply:

[24]From this definition we can deduce that a rotor is to be considered rigid or flexible only on the basis of the angular velocity it is subject to: in other words the same rotor can be considered either rigid or flexible depending on the different steady-state angular velocity in different working conditions. Due to the effect of the real deformability of the supports a rotor can have natural pulsations ω_{id} that are lower than the pulsation ω_o defined previously. The rotor in these conditions, that is, with $2\Omega < \omega_o$, remains *rigid* even if Ω is close to or even greater than ω_{id}. The deformation of the rotor in the passage of the four critical speeds ω_{id} is represented by a rigid rotation-translation of the shaft in the two directions X and Y, as shown in the following figure.

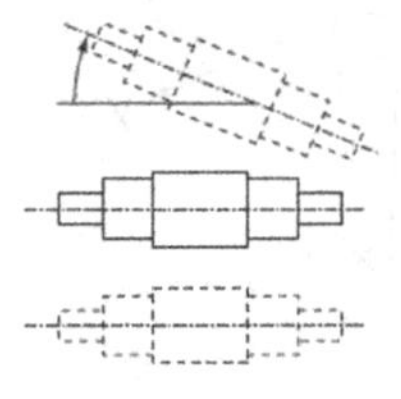

Characteristic deformation of a rigid rotor

$$\vec{R}_i = M\Omega^2\vec{\varepsilon}_G$$
$$\vec{M}_i = \vec{j}_c J_{xz}\Omega^2 - \vec{i}_c J_{yx}\Omega^2 \tag{6.96}$$

having used $\vec{\varepsilon}_G$ to indicate the vector defining the centre of gravity's runout in respect to the rotational axis, with M being the rotor's overall mass and J_{xz} and J_{yz} being the products of inertia defined in respect to the coordinate system rotating integrally with the rotor of versors $\vec{j}_c$ and $\vec{i}_c$; this relation highlights that, in order for the resultant $\vec{R}_i$ of the inertial forces to be zero, it is sufficient for $\vec{\varepsilon}_G$ to be zero, i.e. that the axis of rotation is barycentric; for the torque $\overrightarrow{M}_i$ to be zero, the axis of rotation Z must coincide with a principal axis of inertia (the axis such that $J_{zx} = J_{yz} = 0$).

3 a system of forces equivalent to those generated by the effective distribution of the unbalances is given by the forces caused by two concentrated masses positioned in two arbitrary planes (i.e. to balance a rigid rotor it is enough to position 2 masses in the two balancing planes).

4 so bearing in mind the third point, we can say that in rigid rotors the unbalance, which is actually distributed, can be attributed to two forces $\overrightarrow{F}_1$ and $\overrightarrow{F}_2$ caused by two masses m_1 and m_2, with suitable value and position, placed on two planes π_1 and π_2 and defined by the following relations (see Fig. 6.39):

$$\vec{R}_i = \vec{F}_1 + \vec{F}_2 = m_1\Omega^2\vec{\varepsilon}_1 + m_2\Omega^2\vec{\varepsilon}_2$$
$$\vec{M}_i = \vec{F}_1 \wedge \vec{l}_1 + \vec{F}_2 \wedge \vec{l}_2 = m_1\Omega^2\vec{\varepsilon}_1 \wedge \vec{l}_1 + m_2\Omega^2\vec{\varepsilon}_2 \wedge \vec{l}_2 \tag{6.97}$$

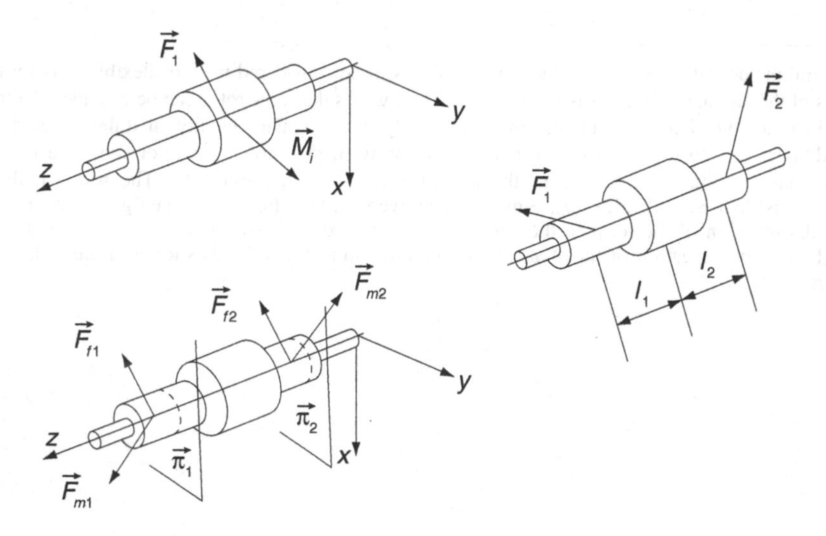

Fig. 6.39 Rigid rotor: equivalent system of forces caused by unbalance

These forces must be such that their vector sum is equal to the resultant of the forces of inertia and that they provide the resultant torque of the forces of inertia. $\vec{F}_1$ and $\vec{F}_2$ can also be defined in the following way: the resultant of the forces of inertia $\vec{F}_i$ can be broken down into two forces $\vec{F}_{f1}$ and $\vec{F}_{f2}$ lying on planes π_1 and π_2, parallel to $\vec{F}_i$. The moment can, in turn, be represented by two forces $\vec{F}_{m1}$ and $\vec{F}_{m2}$ lying on planes π_1 and π_2 equal in module and of opposite direction, so as to give rise to the torque $\vec{M}_i$ (Fig. 6.39). Thus we will have:

$$\begin{aligned} \vec{F}_1 &= \vec{F}_{f1} + \vec{F}_{m1} \\ \vec{F}_2 &= \vec{F}_{f2} + \vec{F}_{m2} \end{aligned} \tag{6.98}$$

The values of $\vec{F}_1$ and $\vec{F}_2$ depend on the arbitrary choice of balancing planes, but, once this position has been chosen, the actual forces are obviously univocally defined. In fact, once planes π_1 and π_2, known as the *balancing planes* have been chosen in an arbitrary manner (i.e. once the values of l_1 and l_2 have been chosen) $\vec{F}_1$ and $\vec{F}_2$ can be expressed as:

$$\begin{aligned} \vec{F}_1 &= m_1|\vec{\varepsilon}_1|e^{i\gamma 1} \\ \vec{F}_2 &= m_2|\vec{\varepsilon}_2|e^{i\gamma 2} \end{aligned} \tag{6.99}$$

where m_1 and m_2 represent the masses which once positioned in the balancing planes π_1 and π_2 at distance $|\varepsilon_1|$ and $|\varepsilon_2|$ from the rotation axis and the angular position γ_1 and γ_2 produce the actual forces due to the initial unbalance of the rotor. This new set of forces represents a system of equivalent forces: it is possible to cancel $\vec{R}_i$ and $\vec{M}_i$ by placing two suitable masses m_1^* and m_2^* on two module balancing planes equal, respectively, to m_1 and m_2 and positioned at 180° in respect to these: to solve the problem of balancing rigid rotors we must define which are the masses m_1 and m_2 that are equivalent, for the purposes of the unbalance, to the real distribution of mass of the rotor.

To do this we must remember, as stated repeatedly, that it is not actually possible to know the actual trend of the runout along the rotor, but we can learn about its effect by measuring the vibrations induced on the rotor. So before describing balancing procedures in real rotors, we will briefly recall the experimental procedures to be adopted for experimentally defining the dynamic behaviour of a rotor (already partly introduced in Sect. 6.2.1 in relation to balancing a thin disk).

6.5.2.1.1 Methods for Measuring Vibrations in Real Rotors

A generic rotor on one or more supports, if not balanced, vibrates because, as mentioned, it is excited by synchronous rotating excitation forces. To proceed with the balancing operation, we must experimentally measure the vibrations $\vec{V}$ at each measurement point: these measurements are made along one or two directions (arbitrary, as long as they are normal to the rotation axis). The sections where these measurements are taken usually correspond to the position of the supports: these supports can be either those of the machine that the rotor actually belongs to, or other suitably built supports, like in the case of balancing machines.

So, in correspondence to the supports we must measure (Fig. 6.40):

- the absolute vibrations $\vec{V}_s$ of the actual supports using accelerometers or velocimeters;
- the relative displacements $\vec{V}_r$ of the shaft with respect to the supports [111] using proximity sensors which, facing the rotor, provide a voltage signal that is proportional to the approach between the surface of the rotor and the relative support, without having to come into contact with the rotor.

Both the absolute vibrations $\vec{V}_s(t)$ of the support and the relative vibrations $\vec{V}_r(t)$ of the shaft with respect to the support, are of a periodic nature, with fundamental frequency equal to the angular frequency Ω: in the same signals, as already mentioned in Sect. 6.5.1.1, we can often see random disturbances caused by external sources of excitation that cannot be defined in advance or generically in presence of noise (6.85). So for these reasons it is usually necessary to carry out harmonic analysis of the signals in order to identify the one-per-rev components, as described in Sect. 6.5.1 notes 17 and 18, and Fig. 6.41 $|V_{s1}|e^{i\phi_{s1}}$ and $|V_{r1}|e^{i\phi_{r1}}$ associated with the fundamental frequency harmonic equal to the angular frequency Ω.

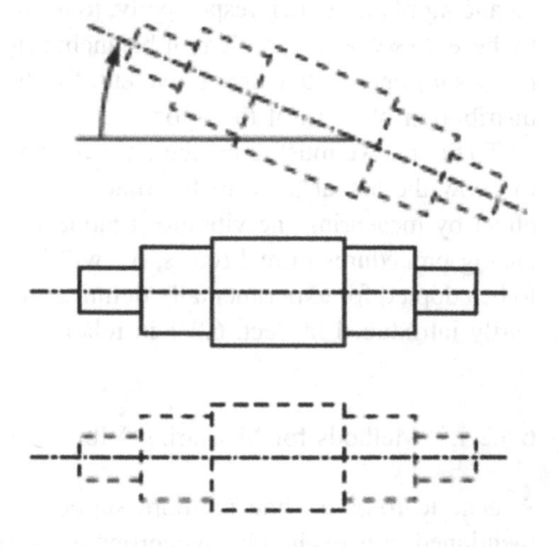

Fig. 6.40 Procedure for measuring the vibrations in a real rotor

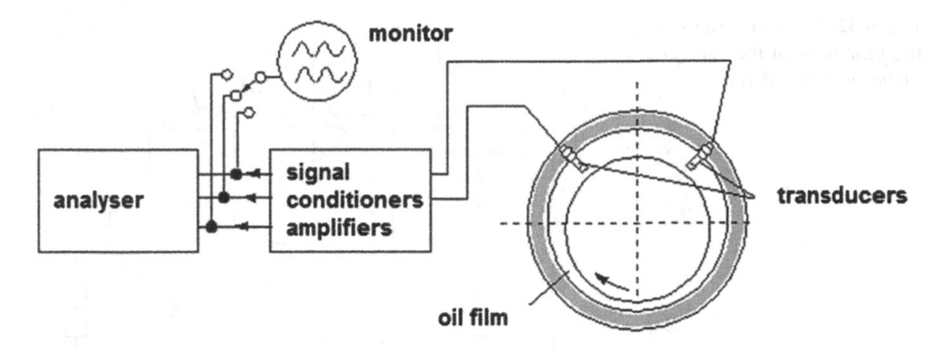

Fig. 6.41 Data acquisition procedure

The absolute vibration $\overrightarrow{V_a}$ with respect to a coordinate system that is integral with the shaft, that is, rotating with angular velocity Ω, is given by:

$$\vec{V}_{a1} = |V_{a1}|e^{i\phi_{a1}} = |V_{s1}|e^{i\phi_{s1}} + |V_{r1}|e^{i\phi_{r1}} \tag{6.100}$$

The values of the vibration amplitudes, defined both in module and in phase, can be conveniently stored and used later as input data for subsequent programs to automatically calculate the balancing masses. A program to prepare data must, therefore, mainly carry out the operations summarised below:

(a) acquisition of the signals from the displacement, speed and acceleration sensors positioned on the rotor and its supports;
(b) analogue-digital conversion of these signals, i.e. transformation of electrical signals into numerical values proportional to the quantities measured, considering scale factors. This operation can be carried out separately from acquisition using magnetic tapes as the support media or it can be done at the same time by an online processor [103–108, 110, 112];
(c) transformation of signals, by means of their harmonic analysis, to calculate the synchronous vibration components;
(d) cleansing of the signals from the sensors of component displacement caused by ovalisation or surface irregularities of the rotor. This operation is usually carried out by subtracting, from the signals acquired at different speeds, the amplitudes of the displacements measured with an angular velocity low enough to be able to exclude vibration phenomena of a dynamic nature;
(e) execution of mathematical operations between the various signals: this step makes it possible to determine the absolute displacements of the shaft as the sum of the signals from the sensor of absolute displacement of the supports and from those from the proximity sensors: more specifically, these signals can be obtained by transforming the signals of the accelerometers into displacements signals by double integration performed using harmonic analysis.

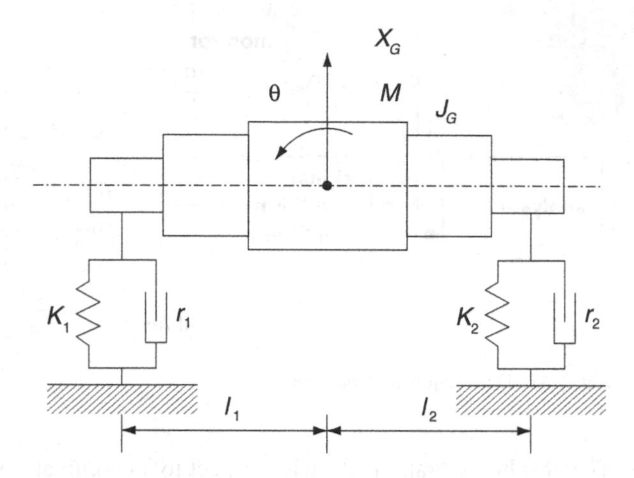

Fig. 6.42 Model for studying the vibrations in the vertical plane of the rigid rotor

6.5.2.1.2 Balancing Machines

The balancing of rigid rotors is usually carried out with special machines known as "balancing machines" which, with suitable procedures, determine the value of the two balancing masses. These machines can be divided into two basic categories:

- balancing machines with measurement of the forces on the supports;
- balancing machines with measurement of the vibrations of the supports.

Figure 6.42 shows the schematisation that can be adopted to reproduce the behaviour of the generic rotor mounted on the balancing machine. The system can be modelled while considering the rigid rotor constrained to two deformable supports with stiffness k, i.e. with a system with 4 d.o.f.

We will now describe the analytical discussion required to reproduce the behaviour of this system, as it is useful in understanding the formulations associated with it. To simplify the discussion, below, we will only consider the motion in the vertical plane, thus reducing the model to a two d.o.f. system. By assuming as independent variables the displacement x_G of the rotor's centre of gravity and the angle θ that the rotor's axis forms with the horizontal plane, the motion equations of the system become,[25,26]:

$$M\underset{G}{\ddot{x}}+(r_1+r_2)\dot{x}_G-(r_1l_1-r_2l_2)\dot{\theta}+(k_1+k_2)x_G-(k_1l_1-k_2l_2)\theta=R_{IV}$$
$$J_G\ddot{\theta}-(r_1l_1+r_2l_2)\dot{x}_G+(r_1l_1^2-r_2c)\dot{\theta}-(k_1l_1+k_2l_2)x_G+(k_1l_1^2-k_2l_2^2)\theta=M_{IV} \tag{6.101a}$$

The forces of inertia caused by the unbalance that arise due to the effect of just rotation, i.e. resultant $\vec{R}_i$ and moment $\vec{M}_i$ which are given by (6.101b), are vectors

[25] In this discussion we neglect the gyroscopic effects.

[26] Similar equations can be written in the horizontal perpendicular plane.

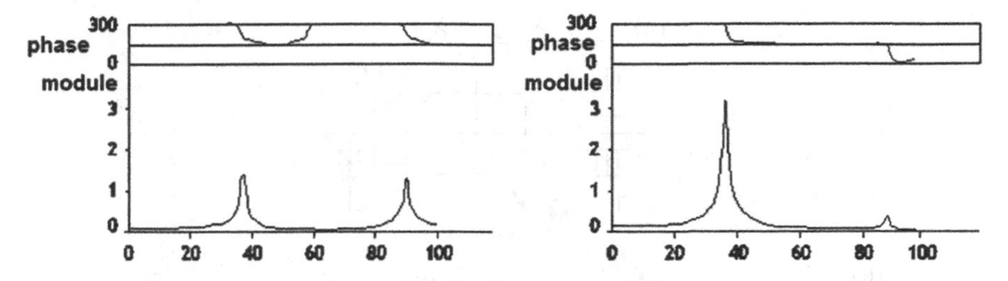

Fig. 6.43 Trend of the modulus of vibration X_G and θ (rigid rotor)

that rotate integrally with the shaft and their projection on the vertical axis provides the force component R_{IV} and the moment component M_{IV}, which are both sinusoidal with frequency that is equal to that of rotation Ω:

$$\begin{aligned} R_{IV} &= \left|\vec{R}_i\right| \sin(\Omega t + \Psi_F) \\ M_{IV} &= \left|\vec{M}_i\right| \sin(\Omega t + \Psi_M) \end{aligned} \tag{6.101b}$$

The steady-state solution to these equations is given by:

$$\begin{aligned} x_G &= X_G \sin(\Omega t + \Psi_1) \\ \theta &= \Theta \sin(\Omega t + \Psi_2) \end{aligned} \tag{6.101c}$$

The values of X_G, ψ_1, θ and ψ_2 depend on the parameters that define the system: in general, the trend of X_G and θ as a function of the angular velocity Ω of the rotor is of the type shown in Fig. 6.43, where ω_1 and ω_2 are the two natural frequencies of the free system.

In force measurement machines, adopting very rigid supports, we must ensure that the angular velocity Ω, i.e. the frequency of the excitation force, is lower than the first natural frequency ω_1 of the system. In this situation the forces that are external to the system are balanced by just the elastic forces and, for this reason, measurement of the forces discharged on the supports makes it possible to directly estimate the resultant and the moment of the unknown inertial forces. In machines for measuring displacement, however, the supports are extremely deformable: measuring the displacements makes it possible, once we know the inertial, elastic and damping characteristics of the system, i.e. the terms to the left of the equals sign in (6.101a), to define the resultant $\overrightarrow{R}_i$ and the moment of the forces of inertia $\overrightarrow{M}_i$ caused by unbalance and so unknown. We will now analyse both types of machine in more detail.

Balancing machines with force measurement

In such machines the rotor is mounted on two supports that are constrained to the machine's structure with dynamometric elements (Fig. 6.44). These dynamometers (either quartz or strain gauges) are chosen so that the deformability of the

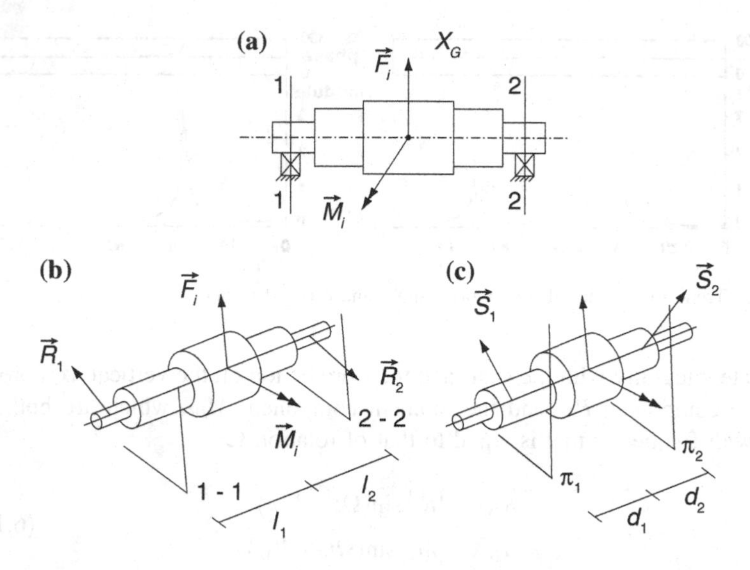

Fig. 6.44 Model of the balancing machine with force measurement

support + dynamometer assembly is low, so that the supports can be considered rigid and the rotor's rotation axis can be considered fixed: these conditions are obtained, as mentioned, by ensuring that the first natural frequency of the rotor ω_1, calculated with these supports, is sufficiently higher than the angular velocity Ω of the actual rotor:

$$\Omega < \frac{1}{2} \div \frac{1}{3} \omega_1 \tag{6.102}$$

The stiffness of the supports k_1 and k_2 is chosen so that it satisfies (6.102). In these conditions the resultant and the moment of the inertial forces (i.e. the excitation forces external to the system) are balanced by the elastic reactions $\overrightarrow{R}_1$ and $\overrightarrow{R}_2$ measured at the supports since the contributions caused by the actions of inertia associated with vibration[27] are negligible: in other words the resultant and the moment of the constraining reactions are exactly equal, except for the direction, to the resultant $\overrightarrow{R}_i$ and the moment $\overrightarrow{M}_i$ of the destabilising inertial forces:

[27]If the axis of rotation, due to the effect of deformability in the supports, should oscillate, the reactions would balance both the forces caused by unbalance (i.e. the unknowns of the problem) and the inertial forces caused by vibration of the axis of rotation: for this reason, in this case measuring the forces acting on the supports would not provide the value of just the unbalance present in the rotor. In this situation (measurement of reactions with oscillating rotational axis) we would not be able to separate the contributions due to vibration of the axis of rotation from those associated with eccentricity as both pulsate with the same frequency Ω as the angular velocity of the rotor.

$$\begin{aligned} \vec{R}_i + \vec{R}_1 + \vec{R}_2 &= 0 \\ \vec{M}_i + \vec{R}_1 \wedge \vec{l}_1 + \vec{R}_2 \wedge \vec{l}_2 &= 0 \end{aligned} \tag{6.103a}$$

where the vectors $\vec{l}_1$ and $\vec{l}_2$ represent the distances of the supports from the centre of gravity. These reactions are defined by measuring, for example, the components $R_1(t)$ and $R_2(t)$ of the reactions in the horizontal plane:

$$\begin{aligned} R_1(t) &= \mathrm{Re}(\vec{R}_1 e^{i\Omega t}) = |R_1| \cos(\Omega t + \Psi_1) \\ R_2(t) &= \mathrm{Re}(\vec{R}_2 e^{i\Omega t}) = |R_2| \cos(\Omega t + \Psi_2) \end{aligned} \tag{6.103b}$$

and obtaining the two modules $\left|\vec{R}_1\right|$ and $\left|\vec{R}_2\right|$ from the amplitudes of the time signal and the corresponding phases ψ_1 and ψ_2, similarly to what was done for the vibrations: these values are obtained via the harmonic analysis of the signal from the dynamometers and by analysing the 1 per rev signal. The real unbalance, as seen in Sect. 6.5.2.1, can be represented with two masses m_1 and m_2, placed at a distance ε_1 and ε_2 from the rotation axis, positioned in two suitable balancing planes π_1 and π_2 distant $\vec{d}_1$ and $\vec{d}_2$ from the centre of gravity (Fig. 6.44c):

$$\begin{aligned} \vec{S}_1 &= \vec{\varepsilon}_1 m_1 \Omega^2 = |\vec{\varepsilon}_1| e^{i\phi_1} m_1 \Omega^2 = \varepsilon_1 e^{i\phi_1} m_1 \Omega^2 \\ \vec{S}_2 &= \vec{\varepsilon}_2 m_2 \Omega^2 = |\vec{\varepsilon}_2| e^{i\phi_2} m_2 \Omega^2 = \varepsilon_2 e^{i\phi_2} m_2 \Omega^2 \end{aligned} \tag{6.104a}$$

where φ_1 and φ_2 are the angles that the radius vector, conducted from the axis of rotation to the mass, form with a reference line that is integral with the actual rotor. Since the system of forces $\vec{S}_1$ and $\vec{S}_2$ (6.104a) represent a system of forces equivalent to resultant $\vec{R}_i$ and moment $\vec{M}_i$ of the forces caused by the unbalance, to cancel the effect of the actual real unbalance the following vector relations must be verified (Fig. 6.44c):

$$\begin{aligned} \vec{S}_1 + \vec{S}_2 &= \vec{R}_i \\ \vec{S}_1 \times \vec{d}_1 + \vec{S}_2 \times \vec{d}_2 &= \vec{M}_i \end{aligned} \tag{6.104b}$$

which, considering (6.104a), become:

$$\begin{aligned} &\vec{S}_1 + \vec{S}_2 = \vec{R}_1 + \vec{R}_2 \quad \Rightarrow \quad \varepsilon_1 e^{i\phi_1} m_1 \Omega^2 + \varepsilon_2 e^{i\phi_2} m_2 \Omega^2 = \vec{R}_1 + \vec{R}_2 \\ &\vec{S}_1 \times \vec{d}_1 + \vec{S}_2 \times \vec{d}_2 = \vec{R}_1 \times \vec{l}_1 + \vec{R}_2 \times \vec{l}_2 \quad \Rightarrow \quad \varepsilon_1 e^{i\phi_1} m_1 \Omega^2 \times \vec{d}_1 \\ &\quad + \varepsilon_2 e^{i\phi_2} m_2 \Omega^2 \times \vec{d}_2 = \vec{R}_1 \times \vec{l}_1 + \vec{R}_2 \times \vec{l}_2 \end{aligned} \tag{6.104c}$$

These vector equations correspond to 4 scalar equations in the 4 unknowns m_1 and m_2, φ_1 and φ_2 that define the value and the position of the balancing masses, as

$\vec{R}_1$ and $\vec{R}_2$ (in module and phase) are known, because measured, as are distances ε_1 and ε_2 since they are imposed.

Balancing machines with vibration measurement

One alternative approach that can be used for balancing rotors is based on evaluating the equivalent system of destabilising forces $\vec{R}_i$ and $\vec{M}_i$ by measuring the displacements x_G and θ (6.101a)–(6.101c). In order to apply this procedure, unlike force measurement balancing machines, we need to know the single parameters of the system beforehand (mass, stiffnesses and dampings) i.e. to know, in analytical form, the transfer matrix of the actual system. In fact, once we know this, as well as the vibrations caused by unbalance x_G and θ, as they are measured, it is possible to trace the forces that generated it (6.101a)–(6.101c). This procedure is not usually followed (given the uncertainties which are, above all, linked to determining the parameters of damping) and we prefer to characterise the system experimentally using *the method of influence coefficients* which, in the case of a linear system, makes it possible to experimentally define the link between excitation forces and vibrations.

6.5.2.1.3 The Influence Coefficient Method

To determine the relationship between excitation forces and vibrations, expressed by (6.101a), it is preferable to proceed with a purely experimental method defined as the *influence coefficient method*, which has already been introduced for balancing a rigid disk in Sect. 6.5.1.1. Let us consider two generic balancing planes π_m and π_n: in correspondence to the supports, we measure the vibrations $\vec{V}_a^{(s)}$ and $\vec{V}_b^{(s)}$ caused by the unbalance, using the procedure already defined in Sect. 6.5.1.1. Then we put a known test mass m_m on the balancing plane π_m at a distance $\vec{\varepsilon}_m$ from the rotational axis and positioned angularly according to angle γ_m in respect to the chosen coordinate system which is integral with the rotor (Fig. 6.45) and represented by the complex number:

$$\vec{m}_m = m_m e^{i\gamma_m} \tag{6.105}$$

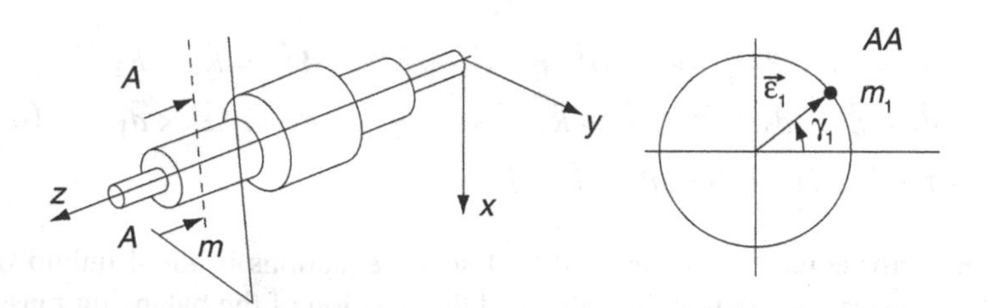

Fig. 6.45 Positioning the generic test mass

and we measure the vibrations on supports, $\vec{V}_a^{(sm)}$ and $\vec{V}_b^{(sm)}$ in correspondence to a given angular velocity Ω. Assuming a linear system and the possibility of repeating testing, by subtracting from these vibrations, the vibrations $\vec{V}_a^{(s)}$ and $\vec{V}_b^{(s)}$ caused only by the existing unbalance, that is, without the test mass, it is possible to estimate vibrations $\vec{W}_{am}$ and $\vec{W}_{bm}$ caused by the presence of just the mass m_m, as:

$$\begin{aligned} W_{am} &= \vec{V}_a^{(sm)} - \vec{V}_a^{(s)} \\ W_{bm} &= \vec{V}_b^{(sm)} - \vec{V}_b^{(s)} \end{aligned} \tag{6.106a}$$

By dividing these vibrations by the mass $\vec{m}_m$ we obtain the two influence coefficients (complex) $\vec{\alpha}_{am}$ and $\vec{\alpha}_{bm}$:

$$\begin{aligned} \vec{\alpha}_{am} &= \frac{\vec{W}_{am}}{\vec{m}_m} = \alpha_{am} e^{i\Psi_{am}} \\ \vec{\alpha}_{bm} &= \frac{\vec{W}_{bm}}{\vec{m}_m} = \alpha_{bm} e^{i\Psi_{bm}} \end{aligned} \tag{6.106b}$$

α_{am} and α_{bm} being the modules of the vibration vectors and ψ_{am} and ψ_{bm} the phases in respect to the coordinate system that is integral with the rotor. These quantities represent the vibrations, to the supports caused by a unitary mass positioned in the balancing plane π_m at an assigned distance $\vec{\varepsilon}_m$ at a zero angle in respect to the coordinate system that is integral with the rotor. In a similar way, by placing a known mass $\vec{m}_n$ on the balancing plane π_n at assigned distance $\vec{\varepsilon}_n$ from the rotational axis:

$$\vec{m}_n = m_n e^{i\gamma_n} \tag{6.107a}$$

by measuring the vibrations on the supports $\vec{V}_a^{(sn)}$ and $\vec{V}_b^{(sn)}$ and subtracting them from the existing ones we obtain the influence coefficients α_{an} and α_{bn} caused by a unitary mass placed on the plane π_n, which are:

$$\begin{aligned} \vec{\alpha}_{an} &= \alpha_{an} e^{i\Psi_{an}} = \frac{\vec{W}_{an}}{\vec{m}_n} = \frac{\vec{V}_a^{(sn)} - \vec{V}_a^{(s)}}{\vec{m}_n} \\ \vec{\alpha}_{bn} &= \alpha_{bn} e^{i\Psi_{bn}} = \frac{\vec{W}_{bn}}{\vec{m}_n} = \frac{\vec{V}_b^{(sn)} - \vec{V}_b^{(s)}}{\vec{m}_n} \end{aligned} \tag{6.107b}$$

Once the influence coefficients have been obtained experimentally, it is possible to set the vibrations caused by unbalance plus those caused by balancing masses to zero, i.e.:

$$\begin{aligned}\vec{V}_a^{(s)} + \vec{m}_m^* \vec{\alpha}_{am} + \vec{m}_n^* \vec{\alpha}_{an} = \vec{V}_a^{(s)} + m_m^* e^{i\Psi_m} \vec{\alpha}_{am} + m_n^* e^{i\Psi_n} \vec{\alpha}_{an} = 0 \\ \vec{V}_b^{(s)} + \vec{m}_m^* \vec{\alpha}_{bm} + \vec{m}_n^* \vec{\alpha}_{bn} = \vec{V}_b^{(s)} + m_m^* e^{i\Psi_m} \vec{\alpha}_{bm} + m_n^* e^{i\Psi_n} \vec{\alpha}_{bn} = 0\end{aligned} \tag{6.108}$$

where ψ_m and ψ_n represent the unknown angles that define the position of the masses in respect to the coordinate system that is integral with the rotor, m_m^* and m_n^* being the masses placed at the same distances $\vec{\varepsilon}_m = \vec{\varepsilon}_n$ from the rotational axis used to determine the influence coefficients $\vec{\alpha}_{ij}$. (6.108) represent two complex equations in the scalar unknowns m_m^* and m_n^*, ψ_m and ψ_n and they make it possible to obtain the balancing masses m_m^* and m_n^* and their position ψ_m and ψ_n once the vibrations $\vec{V}_a^{(s)}$ and $\vec{V}_b^{(s)}$ caused by the unbalance have been measured. The Eq. (6.108) can be solved automatically by a calculator incorporated in the balancing machine; in this case, the calculator:

- receives the information from the vibration sensors;
- calculates the differences between vibration $\vec{V}_i^{(sm)}$ caused by the simultaneous presence of the sample masses and unbalance and vibrations $\vec{V}_i^{(s)}$ caused by unbalance only;
- calculates the influence coefficients $\vec{\alpha}_{ij}$ and places them in (6.108);
- by solving the actual set of equations, provides the value of the balancing masses.

we can see that the influence coefficients $\vec{\alpha}_{ij}$ defined by (6.107b) are a function of the angular velocity Ω in correspondence to which they were actually determined: for this reason, the relation (6.108) is valid for that particular velocity. The relation (6.108) used with the influence coefficient method is essentially equal to (6.101a), used, on the other hand, with the method based on measuring the displacements: the only difference between the two relations is due to the fact that, in the first case, the link between $\vec{V}_a^{(s)}$ and $\vec{V}_b^{(s)}$ and the unbalance is obtained on the basis of an analytical approach to the problem, whereas, in the second case the same relation is obtained directly via experimental testing. With the first approach, the uncertainty in the results obtained is related to the difficulty in determining the characteristics of the vibrating system, while with the experimental approach (influence coefficient method) this difficulty is overcome, even if with this method it is still advisable to verify the repeatability of the results obtained.

6.5.2.2 Balancing Flexible Rotors

As illustrated previously, a generic rotor is considered *deformable* when its steady-state angular velocity Ω is close to or greater than its first natural bending frequency, calculated on rigid supports (Sect. 6.5.2.1). For these rotors the procedures used to balance *rigid rotors*, as seen in the previous sections, cannot be applied since, for flexible rotors, cancelling the resultant $\vec{R}_i$ and moment $\vec{M}_i$ of the inertial forces (Sect. 6.5.2.1) does not represent the necessary and sufficient condition for balancing, which instead, consists of cancelling the Lagrangian components $Q^{(n)}$ caused by unbalance of all the vibration modes that can be excited [66, 72].

6.5.2.2.1 Influence Coefficient Method

This method has already used in balancing thin disks (Sect. 6.5.1) and rigid rotors (Sect. 6.5.2.1): in concept this approach remains unchanged in the case of deformable rotors too. The runout [as seen in Sect. 6.4 (6.57b)] can be represented by a vector $\vec{\varepsilon}(\zeta)$ that rotates with the same angular velocity Ω as the shaft, the module and phase of which are functions of the current coordinate ζ. The steady-state response of the system, in a generic point j, can also be represented with a rotating vector $\vec{V}_j$. If we assume a coordinate system $(\xi - \eta - \zeta)$ that is integral with the rotor (Fig. 6.33) both the runout and vibration vectors are fixed: normally, an axis of this rotating coordinate system is placed in correspondence to the reflective mark applied to the rotor and this axis is taken as the origin for defining the phases of the individual quantities under examination. For this reason, during the balancing operation it is convenient to refer to this coordinate system that rotates integrally with the shaft. Now let us assume that the rotor has n_e balancing planes (planes where the balancing masses can be applied) and n_m measurement planes (sections where the vibrations caused by unbalance can be measured). We can give the following definitions:

- $\vec{V}_i^{(s)}$ is the synchronous harmonic components caused by unbalance (one-per-rev component, measured with the procedure already explained in Sect. 6.5.1.2) at the generic point of measurement $i\,(i = 1, 2, \ldots, n_m)$;
- $\vec{V}_i^{(s+m_j)}$ is the complex amplitude of the vibration measured at the ith measurement point, and caused by the simultaneous presence of the rotor's natural unbalance (unknown) and of a mass $\vec{m}_j$, also assumed to be complex, placed in the generic jth balancing plane (with $j = 1,\ n_e$: so $\vec{m}_j$ represents an arbitrary module mass m_j, placed at a certain distance from the rotation axis ε_j and at a certain angle γ_j with respect to the coordinate system that is integral with the shaft.
- $\vec{\alpha}_{ij}$ is the generic influence coefficient, also complex, estimated via the following relation:

$$\vec{\alpha}_{ij} = \frac{\vec{V}_i^{(s+m_j)} - \vec{V}_i^{(s)}}{\vec{m}_j} \quad (\mathrm{i} = 1,2,\ldots,\mathrm{n_m}),\ (j = 1, 2, \ldots, \mathrm{n_e}) \tag{6.109}$$

This influence coefficient gives the vibration amplitude at the generic measurement point i due to a single arbitrary mass placed with zero phase in respect to the rotational coordinate system assumed in the jth balancing plane. In general, these coefficients depend on the angular velocity Ω: in fact, as this varies the deformation with which the rotor vibrates also varies (the contribution of those modes predominantly excited by the unbalance changes) and so the Lagrangian components of the inertial forces of the rotor caused by unbalance, and the components due to the individual additional masses, also change. By varying the points of measurement $i = 1, n_m$ and the plane where the mass $j = 1, n_e$ is applied, it is possible to define all the coefficients $\overrightarrow{\alpha}_{ij}$. To annul the vibrations caused by unbalance, steps must be taken to ensure that the masses added $\overrightarrow{m}_j^*$, which have now become the unknowns of the problem, satisfy the relation:

$$\begin{aligned} &\vec{V}_1^{(s)} + \vec{\alpha}_{1,1}\vec{m}_1^* + \vec{\alpha}_{1,2}\vec{m}_2^* + \ldots + \vec{\alpha}_{1,n_e}\vec{m}_{n_e}^* = 0 \\ &\vec{V}_2^{(s)} + \vec{\alpha}_{2,1}\vec{m}_1^* + \vec{\alpha}_{2,2}\vec{m}_2^* + \ldots + \vec{\alpha}_{2,n_e}\vec{m}_{n_e}^* = 0 \\ &\ldots\ldots\ldots \\ &\vec{V}_{n_m}^{(s)} + \vec{\alpha}_{n_m,1}\vec{m}_1^* + \vec{\alpha}_{n_m,2}\vec{m}_2^* + \ldots + \vec{\alpha}_{n_m,n_e}\vec{m}_{n_e}^* = 0 \end{aligned} \tag{6.110}$$

With $\underline{\vec{m}}$ defining the vector of the n_m balancing masses placed in the n_e balancing planes:

$$\underline{\vec{m}} = \begin{Bmatrix} \vec{m}_1^* \\ \vec{m}_2^* \\ \ldots \\ \vec{m}_{n_e}^* \end{Bmatrix} \tag{6.111a}$$

and defined with:

$$\underline{\overrightarrow{V}}^{(s)} = \begin{Bmatrix} \vec{V}_1^{(s)} \\ \vec{V}_2^{(s)} \\ \ldots \\ \vec{V}_{n_m}^{(s)} \end{Bmatrix} \tag{6.111b}$$

the vector of the vibrations caused by unbalance on the n_m measurement planes, the link, expressed in scalar form by (6.110), between the vibration vector $\underline{\overrightarrow{V}}^{(s)}$ and that of the balancing masses $\underline{\vec{m}}$ can thus be expressed in matrix form as:

$$\underline{\vec{V}}^{(s)} + [C]\underline{\vec{m}} = 0 \tag{6.112a}$$

where [C] is the complex matrix of the influence coefficients:

$$[C] = \begin{bmatrix} \vec{\alpha}_{1,1} & \vec{\alpha}_{1,2} & \dots & \vec{\alpha}_{1,n_e} \\ \vec{\alpha}_{2,1} & \vec{\alpha}_{2,1} & \dots & \vec{\alpha}_{2,n_e} \\ \dots & & & \\ \vec{\alpha}_{n_m,1} & \vec{\alpha}_{n_m,2} & \dots & \vec{\alpha}_{n_m,n_e} \end{bmatrix} \tag{6.112b}$$

If the number n_e of balancing planes coincides with the number n_m of measurement planes, i.e. the matrix [C] is squared, (6.112a) makes it possible to obtain the values of the masses $\underline{\vec{m}}$ that cancel out the vibrations $\underline{\vec{V}}^{(s)}$, i.e. those that mean that (6.112a) is verified. However, as stated previously, both vector $\underline{\vec{V}}^{(s)}$ and matrix [C] are generally functions of the angular velocity of the rotor Ω. If we wish to stabilise the rotor (i.e. balance it) for different speeds, we must verify the relation (6.112a) for all the n_v speeds chosen, i.e. impose simultaneously:

$$\begin{cases} \vec{V}_s(\Omega_1) + [C(\Omega_1)]\ \underline{\vec{m}} = \underline{0} \\ \vec{V}_s(\Omega_2) + [C(\Omega_2)]\ \underline{\vec{m}} = \underline{0} \\ \dots\dots\dots \\ \vec{V}_s(\Omega_{nv}) + [C(\Omega_{nv})]\ \underline{\vec{m}} = \underline{0} \end{cases} \tag{6.113a}$$

or also, in matrix form:

$$\underline{\vec{W}}_s + [\tilde{C}]\ \underline{\vec{m}} = \underline{0} \tag{6.113b}$$

having indicated $\vec{W}_s$ with:

$$\underline{\vec{W}}_s = \begin{Bmatrix} \underline{\vec{V}}_s(\Omega_1) \\ \underline{\vec{V}}_s(\Omega_2) \\ \dots \\ \underline{\vec{V}}_s(\Omega_{nv}) \end{Bmatrix} \tag{6.113c}$$

a column vector of $n_m \cdot n_v$ terms, made up of the subvectors $\underline{\vec{V}}_s = \underline{\vec{V}}_s(\Omega_1)$ of the vibrations measured for different speeds (made up of $n = n_m \cdot n_v$ dummy measurement points) and with $[\tilde{C}]$ the matrix of the influence coefficients at the various speeds:

$$[\tilde{C}] = \begin{bmatrix} [C(\Omega_1)] \\ C(\Omega_2) \\ \dots \\ C(\Omega_{nv}) \end{bmatrix} \tag{6.113d}$$

Equation (6.113a) cannot generally be satisfied as the number of balancing planes should be equal to the number n_m of measurement planes multiplied by the

n_v speeds ($n = n_m \cdot n_v$). So we must proceed with a process of minimisation. Once the residues of the vibrations have been defined with $\underline{\vec{\delta}}$, these (6.113a) can be rewritten as:

$$\underline{\vec{W}}_s + [\tilde{C}]\,\underline{\vec{m}} = \underline{\vec{\delta}} \tag{6.113e}$$

where $\underline{\vec{\delta}}$ is a vector that contains the residual vibrations in the measurement planes at all the different speeds analysed:

$$\underline{\vec{\delta}} = \begin{Bmatrix} \underline{\vec{V}}_{res}(\Omega_1) \\ \underline{\vec{V}}_{res}(\Omega_2) \\ \cdots \\ \underline{\vec{V}}_{res}(\Omega_{nv}) \end{Bmatrix} \tag{6.113f}$$

In other words, the added masses $\underline{\vec{m}}$ (since $n = n_m \cdot n_v > n_e$) will not be able to cancel the vibrations caused by unbalance of the rotor at all the angular velocities, but they will only be able to limit them: so at the various running speeds Ω_k ($k = 1, n_v$), residual vibrations $\underline{\vec{V}}_{res}(\Omega_k)$ will always exist. Therefore, we must calculate the balancing masses $\underline{\vec{m}}$ so that the residual vibrations are reduced to a minimum, i.e. we must impose that:

$$f(\underline{\vec{m}}) = \underline{\vec{\delta}}^T \underline{\vec{\delta}} = \min \tag{6.114a}$$

Replacing (6.113e) in (6.114a) we obtain:

$$\begin{aligned} f(\underline{\vec{m}}) &= \left(\underline{\vec{W}}_s + [\tilde{C}]\,\underline{\vec{m}}\right)^T \left(\underline{\vec{W}}_s + [\tilde{C}]\,\underline{\vec{m}}\right) \\ &= \underline{\vec{W}}_s^T \underline{\vec{W}}_s + \underline{\vec{W}}_s^T [\tilde{C}]\,\underline{\vec{m}} + \underline{\vec{m}}^T [\tilde{C}]^T \underline{\vec{W}}_s + \underline{\vec{m}}^T [\tilde{C}]^T [\tilde{C}]\,\underline{\vec{m}} \end{aligned} \tag{6.114b}$$

This function takes the minimum value, upon zeroing of the first derivative of the actual function $f(\underline{\vec{m}})$ with respect to vector $\underline{\vec{m}}$[28]:

$$\frac{\partial f(\underline{\vec{m}})}{\partial \underline{\vec{m}}} = 0 \tag{6.115a}$$

By applying (6.115a)–(6.114b) we obtain:

$$\frac{\partial f(\underline{\vec{m}})}{\partial \underline{\vec{m}}} = [\tilde{C}]^T \underline{\vec{W}}_s + [\tilde{C}]^T \underline{\vec{W}}_s + [\tilde{C}]^T [\tilde{C}]\underline{\vec{m}} + [\tilde{C}]^T [\tilde{C}]\underline{\vec{m}} = 0 \tag{6.115b}$$

i.e.:

[28]In the case in question, the minimum condition is guaranteed [61].

$$[\tilde{C}]^T[\tilde{C}]\underline{\vec{m}} = -[\tilde{C}]^T\underline{\vec{W}}_s \tag{6.115c}$$

The system of complex matrix Eq. (6.115c) makes it possible to determine the vector $\underline{\vec{m}}$ of the unknown balancing masses, which minimises the sum of the squares of the vibrations in n_m measurement points and for the n_v speeds being considered. The vibrations $\underline{\vec{V}}^{(s)}(\Omega_k)$ which appear at the second member in (6.115c) in correspondence to a certain value of Ω_k can be given a mass that is proportional to the importance that their minimisation assumes in respect to the vibrations caused by unbalance in correspondence to other speeds. In fact, it is logical that, for example, the behaviour of steady-state vibrations is of greater importance than that relating to other intermediate speeds: to increase the mass to be associated with these vibrations, compared to the others that appear in (6.115c) we can simply introduce a diagonal weight matrix $[P]$, whose elements are greater than one for the speeds that must be more massed and equal to one for the others: when introducing this, matrix (6.114a) becomes[29]:

$$\underline{\vec{\delta}}^T[P]\underline{\vec{\delta}} = \min \tag{6.116}$$

The influence coefficient method is currently the most common method used for balancing high-speed rotors. The advantages and disadvantages of balancing rigid rotors have already been explained (Sect. 6.5.2.1.3): we should remember only the considerable advantage of this method lies in the fact that it does not require the support of any analytical model, unlike the modal balancing method described in the following section.

6.5.2.2.2 Modal Balancing

To balance a flexible rotor we can also use an alternative approach defined as the *modal approach* because it actually uses the concepts of modal analysis seen in Chap. 2, Sect. 2.5. The first step in applying this method is to define the principal vibration modes of the shaft: this can be done either experimentally or analytically. In the case of an industrial plant, the difficulty in defining vibration modes experimentally is linked to the difficulty in obtaining a sufficient number of measurement points to describe the deformation of the rotor in correspondence to the individual critical speeds. On the other hand, if we address the problem in an analytical way, the uncertainties are linked to the schematisation adopted. If we use an analytical approach and adopt the model of the rotor described in Sect. 6.2, the system's equations of motion are:

[29]The matrix [P] can also be introduced to take into account the greater or lesser reliability of some measures compared to others: in this case this can be the matrix of covariance [19–22, 24, 26, 27].

$$[M]\,\ddot{\underline{x}}+[R]\,\dot{\underline{x}}+[K]\,\underline{x}=\underline{F} \tag{6.117a}$$

where $\underline{x}$ is the vector of the generalised displacements relating to both the shaft nodes and the bearing nodes. If we wish to obtain the real and orthagonal vibration modes we must eliminate both the term of the generalised forces $\underline{F}$ and the damping, thus arriving at a matrix equation of the form:

$$[M]\,\ddot{\underline{x}}+[K]\,\underline{x}=\underline{0} \tag{6.117b}$$

In general, due to the presence of journal bearings with hydrodynamic lubrication (Sect. 6.2.2), the matrix $[K]$ is not symmetrical and so the solution of (6.117b) would be of the form:

$$\underline{x}=\underline{X}e^{\lambda t} \tag{6.118}$$

with complex λ and vibration modes $\underline{X}$ that are also complex. To overcome the problem, the overall stiffness matrix of the rotor-oil-film-foundation system can be expressed as the sum of a symmetric matrix $[K_s]$ and a hemisymmetric one $[K_{es}]$:

$$[K]=[K_s]+[K_{es}] \tag{6.119a}$$

The generic term K_{ijs} of $[K_s]$ is given by the relation:

$$K_{ijs}=\frac{K_{ij}+K_{ij}}{2} \tag{6.119b}$$

where K_{ij} and K_{ij} represent the extra-diagonal terms of the system's matrix $[K]$, while the generic term K_{ijes} of $[K_{es}]$ is defined as:

$$K_{ijes}=\frac{K_{ij}-K_{ij}}{2} \tag{6.119c}$$

In real rotors the non-symmetric terms of stiffness K_{ijes} have a negligible effect on calculating natural pulsations, so it is possible to determine them, while neglecting the contribution of $[K_{es}]$, i.e. by solving:

$$[M]\,\ddot{\underline{x}}+[K_s]\,\underline{x}=\underline{0} \tag{6.120a}$$

the solutions are of the form:

$$\underline{x}=\underline{X}e^{i\omega t} \tag{6.120b}$$

with vibration modes $\underline{X}^{(i)}$ characterised by real relationships between the elements. In order to decouple the equations of motion we must calculate the natural frequencies ω_i and the vibration modes $\underline{X}^{(i)}$ of (6.120a) and then impose the transformation of coordinates (Chap. 2, Sect. 2.5.2):

$$\underline{x} = [\Phi]\,\underline{q} \tag{6.121a}$$

where $[\Phi]$ is the real modal matrix defined as:

$$[\Phi] = \left[\underline{X}^{(1)} \quad \underline{X}^{(2)} \quad \ldots\right]. \tag{6.121b}$$

By placing the transformation (6.121a) in Eq. (6.117a) and using $\underline{F}$ to name the vector of the forces introduced by unbalance, we obtain:

$$[m]\,\ddot{\underline{q}} + [\Phi]^T[R][\Phi]\,\dot{\underline{q}} + [k]\,\underline{q} + [\Phi]^T[K_{es}][\Phi]\,\underline{q} = [\Phi]^T\underline{F} = \underline{Q} \tag{6.122a}$$

where $[m]$ is the diagonal mass matrix, in modal coordinates:

$$[m] = [\Phi]^T[M][\Phi] \tag{6.122b}$$

and $[k]$ is also diagonal:

$$[k] = [\Phi]^T[K][\Phi] \tag{6.122c}$$

In general, however, the product $[\Phi]^T[R][\Phi]$ represents a non-diagonal matrix as, above all, due to the presence of hydrodinamically lubricated bearings, $[R]$ is not attributable to a matrix linear combination of the mass matrix $[M]$ and the stiffness matrix $[K_s]$. Similarly, the matrix $[\Phi]^T[K_{es}][\Phi]$ is full. In order to decouple the equations of motion obtained, and then apply the modal approach, we must:

- diagonalise the damping matrix $[\Phi]^T[R][\Phi]\,\dot{\underline{q}}$: one simple way of doing this may be to neglect the extra-diagonal terms;
- neglect the terms of the matrix $[K_{es}]$: this simplification is acceptable when calculating the response to unbalance, if the modes are sufficiently separate.[30]

With these simplifications we can arrive at a system of n decoupled equations in the n principal independent coordinates q_i of the form:

$$m_i\,\ddot{q}_i + r_i\,\dot{q}_i + k_i q_i = Q_i \quad (i = 1, 2, \ldots n) \tag{6.123}$$

where the term Q_i represents the virtual work carried out by the unbalances for the generic vibration mode q_i. For each natural pulsation ω_i concerned the corresponding decoupled equation is then defined and the corresponding deformation is expressed by means of vector $\underline{X}^{(i)}$, whose components are the vibration amplitudes of the generic ith vibration mode. The generic vibration mode $\underline{X}^{(i)}$ is thus defined by

[30]However this assumption is not acceptable when we analyse the free motion of the system, or when we wish to define the conditions of stability, which are linked to the non-symmetrical terms of the stiffness matrix.

a spatial deformation that can be projected onto two orthogonal planes containing the axis of the rotor. Assuming a complex plain with the real axis corresponding to the axis Y and the imaginary axis corresponding to the vertical axis X, the rotor's generic vibration deformation can be represented as a complex vector defined as:

$$\underline{X}^{(i)} = \underline{X}_y^{(i)} + i\underline{X}_x^{(i)} \tag{6.124}$$

having used $\underline{X}_x^{(i)}$ and $\underline{X}_y^{(i)}$ to indicate the vertical and transversal components of the displacements of the rotor nodes. If we assume that the real spatial distribution of the unbalances along the rotor is known (distribution which is actually unknown):

$$d\overrightarrow{F}_i = \overrightarrow{\varepsilon}\,\Omega^2 m(\zeta)\,e^{i\Omega t}d\zeta \tag{6.125a}$$

it is possible to define the work carried out by this unbalance for the ith vibration mode as follows. Once the components of the vector unbalance in vertical and lateral direction have been defined with dF_{xi} and dF_{yi}, the work carried out by the same unbalance becomes (see note 12 about this):

$$Q_s^{(i)} = \int_o^l X_x^{(i)} dF_{xi} + \int_o^l X_y^{(i)} dF_{yi} \tag{6.125b}$$

By introducing the shape function, which binds the physical coordinates to the free coordinates, the work can be rewritten in the form:

$$Q_s^{(i)} = \underline{\overrightarrow{F}}_i^T \cdot \underline{\overrightarrow{X}}^{(i)} \tag{6.125c}$$

where $\underline{\overrightarrow{F}}_i$ is the vector of the generalised forces (vector quantities) at the nodes of the finite element model due to the natural unbalance of the rotor. To ensure that the rotor is balanced, we must add enough n_e balancing masses $\overrightarrow{m}_i$ to ensure that the work $Q_m^{(i)}$ carried out by them for the ith vibration mode cancels out the work performed by the unbalance:

$$Q^{(i)} = Q_s^{(i)} + Q_m^{(i)} \tag{6.126}$$

In fact, in this situation the generic vibration mode is not forced and therefore, for the angular velocity around this pulsation, the rotor is balanced since this mode is not excited by the unbalances and balancing masses, since the Lagrangian components of the sum of the two systems of forces are zero for that vibration mode. As mentioned previously, it is not possible to know the real distribution of the unbalances along the rotor and, for this reason the modal approach is practically useless. However, it is possible to use the so-called *hybrid method*, described in the following section, which uses the method of the modal approach but coupled with the influence coefficients, defined previously.

6.5.2.2.3 Balancing with the Hybrid Method

To be able to actually proceed with balancing, first of all we must define the value of the $Q_s^{(i)}$ unknowns in the problem, to then calculate the balancing masses. To do this we need to solve the inverse problem, that is, once the vibration component $q_s^{(i)}$ has been estimated for each mode obtained from the vibration measurements and knowledge of the vibration modes (6.121a), we need to obtain $Q_s^{(i)}$. In fact, this route is not generally taken but, as mentioned, the *hybrid method* is preferred as an alternative. The method's name stems from the fact that this approach simultaneously uses the techniques of the modal approach and the influence coefficients method. Once the rotor deformation has been measured or calculated, the method involves installing systems of masses that have a high Lagrangian component for the mode that is to be balanced. For example, for a rotor on two supports, to balance the first mode a mass is placed at the centre, for the second mode two masses are placed at one quarter and three quarters, in phase opposition, for the third mode three masses are, again, placed in the dips of the deformation of the mode in question and so on. The value of the masses is obtained with the influence coefficient method described previously, by applying the chosen system of masses directly to the rotor, instead of applying one mass at a time, as will be described below. The mass systems must be orthogonal with the deformations of the vibration modes that are different to the one that needs to be balanced, so as to avoid affecting previous balancing. For example we assume that the operating speed is between the first and second critical speeds. First of all we can use the influence coefficient method to balance the first critical speed $\Omega = \omega_1$, by positioning some masses $\underline{\vec{m}}_1$ in phase along the rotor. Once the masses m_1 that result from this first phase of balancing have been positioned, we can balance the rotor at the steady-state speed $\Omega = \Omega_r$ again with the influence coefficient method while, however, imposing the orthogonality between the new masses $\underline{\vec{m}}_r$ and the first vibration mode $\underline{\vec{X}}^{(1)}$ (already balanced):

$$\underline{\vec{X}}^{(1)T}\,\underline{\vec{m}}_r = 0 \tag{6.127a}$$

where $\underline{\vec{X}}^{(1)}$ is the vector that contains the vibration amplitudes in the first vibration mode (defined up to a constant and derived from the measurements taken at the first critical speed) in correspondence to the n_m balancing planes on which the masses $\underline{\vec{m}}_r$ that balance the rotor at steady-state speed are placed. If we remember that the Lagrangian component $Q_r^{(1)}$ of the forces due to the masses $\underline{\vec{m}}_r$ for the first vibration mode can be expressed as:

$$Q_r^{(1)} = \sum_{j=1}^{nc} m_{rj}\,\vec{\varepsilon}_j\Omega^2 \cdot \underline{\vec{X}}^{(1)} \tag{6.127b}$$

and that this scalar product, using the complex number algorithm (see note 12), is equal to the real part of the product between the two complex numbers $m_{rj}\,\vec{\varepsilon}_j\Omega^2 \cdot \vec{X}^{(1)}$, it is easy to see how the condition (6.127a) is also equivalent to requiring that the Lagrangian component $Q_r^{(1)}$ of the forces due to the steady-state balancing masses $\vec{\underline{m}}_r$ is zero for the first vibration mode, i.e. that:

$$Q_r^{(1)} = 0 \tag{6.127c}$$

In other words, the distribution of masses $\vec{\underline{m}}_r$ that results when we impose the condition of orthogonality (6.127a) is such as to not excite, and therefore unbalance, the first vibration mode. If the rotor does not have a running speed Ω_r that is higher than the second critical, it is possible to extend the method in the following way:

- balancing at the first critical $\Omega = \omega_1$ with the influence coefficient method;
- balancing at the second critical $\Omega = \omega_2$ as for the first, but with the added condition that the distribution of the balancing masses $\vec{\underline{m}}_2$ of the second critical does not excite the first vibration mode;
- balancing at steady-state $\Omega = \omega_r$ as for the previous cases, but with two additional conditions: the distribution of masses $\vec{\underline{m}}_r$ must not excite either the first or the second vibration mode.

One practical application of the aforementioned method involves using mass systems that are applied to the rotor so that each system gives a non-zero Lagrangian component $Q^{(i)}$ for the ith vibration mode, and zero for all the other modes. So each system of masses is applied to balance just one mode: the choice of the value to be attributed to the balancing masses of each system that maintain between them a ratio and a fixed relative position, is defined using the influence coefficient approach. Suppose, for example, that the rotor in Fig. 6.46a has an angular velocity Ω that is higher than both the first critical speed ω_1 and the second critical speed ω_2 and that the vibration shapes are those shown in Fig. 6.46. So to balance the first critical speed it is possible to place two masses $\vec{\underline{m}}_m^{(1)}$ and $\vec{\underline{m}}_n^{(1)}$, in phase with each other, as shown in Fig. 6.47. If the rotor is symmetrical with respect to the centreline of the two supports, the first mode will have an antinode at midspan, while the second mode will always have the node at midspan. If the masses are placed in symmetrical positions with respect to the centreline, then the condition of orthogonality of the first mass system, for the second mode, is that the masses are in phase, with equal modulus. After having measured the vibrations $\vec{U}_1$ caused by the initial unbalance of the rotor for an angular velocity close to, or if possible, equal to the critical speed, two test masses $\vec{\underline{m}}_m^{(1)}$ and $\vec{\underline{m}}_n^{(1)}$ are applied in the aforementioned positions and the vibrations $\vec{U}_{m1}$ caused by the natural unbalance of the rotor and the presence of the two test masses are measured:

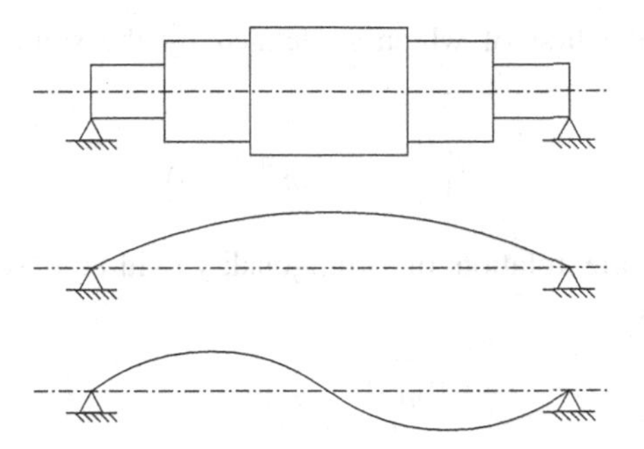

Fig. 6.46 Generic flexible rotor: vibration modes

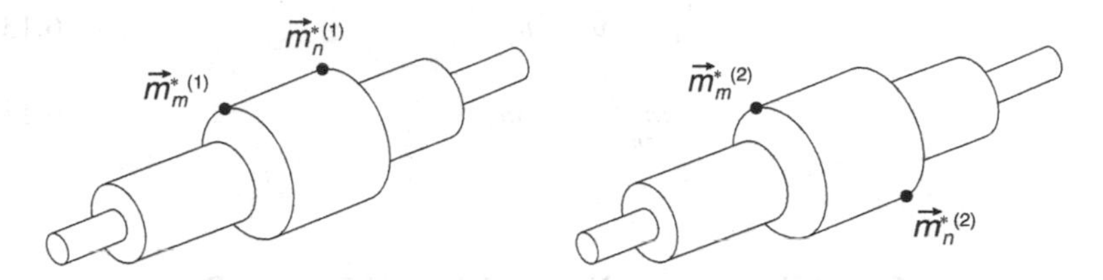

Fig. 6.47 Hybrid balancing: system of balancing masses

$$\underline{\vec{m}}^{(1)} = \left\{ \begin{array}{c} \vec{m}_m^{(1)} \\ \vec{m}_n^{(1)} \end{array} \right\} \tag{6.128}$$

Therefore, the only unknown is the angular position of the mass system and the value of one of them, since due to the constraints imposed by orthogonality, the relative angular position between the two masses (in this case, in phase) and their ratio (in this case, equal to unity) have already been established. As there are only two unknowns, then just one measurement point is necessary to calculate the influence coefficient defined by the complex relation:

$$\vec{\alpha}^{(1)} = \frac{\vec{U}_{m1} - \vec{U}_1}{\vec{m}_m^{(1)}} \tag{6.129}$$

where the difference of the vibrations measured with and without the test mass system, is divided by the value of one of the two masses of the system, given that the relative angular position and the value of the second mass in respect to the first have already been defined. The first critical speed will be balanced by placing a system of masses, the first of which is defined by the solution of the complex equation:

$$\vec{U}_1 + \vec{\alpha}^{(1)} \vec{m}_m^{*(1)} = 0 \tag{6.130}$$

and the second by the relation of orthogonality with the second mode which requires, in this case:

$$m^{*(1)}_{\to m} = m^{*(1)}_{\to n} \tag{6.131}$$

Similar considerations can be made and steps carried out (once the first has already been balanced) for the second critical, see Fig. 6.47:

$$\vec{U}_1 + \vec{\alpha}^{(2)} \vec{m}_m^{*(2)} = 0 \tag{6.132}$$

$$m^{*(2)}_{\to m} = -m^{*(2)}_{\to n} \tag{6.133}$$

6.6 Two-Per-Rev Vibrations Excited by Different Rotor Stiffnesses, in Horizontal Shafts

A horizontal shaft, being subject to the weight force, has a static deflection whose value depends on the elastic and inertial properties of the rotor and which, for example, in large turbogenerator groups, is in the region of several millimetres. If we refer to a shaft resting on two rigid supports, with evenly distributed mass and stiffness (Fig. 6.48), the static deflection assumes a value that depends on the position along the axis of the rotor:

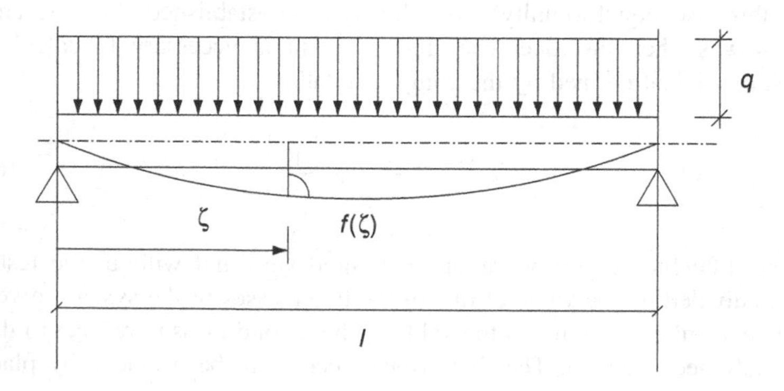

Fig. 6.48 Rotor with uniform section subject to its own mass

$$f(\zeta) = \frac{ql^4}{24EJ}\left(\left(\frac{\zeta}{l}\right)^4 - 2\left(\frac{\zeta}{l}\right)^3 + \left(\frac{\zeta}{l}\right)\right) \tag{6.134}$$

where q is the mass per unit of length of the rotor and EJ its bending stiffness.

If the rotor has a circular cross-section, all the axes lying on this section are principal axes of inertia: in this case, in fact, the inertia ellipsoid degenerates into a circumference. The deflection assumed by the shaft, considering it rotating at a low speed to eliminate dynamic effects on the response, is constant in modulus and direction, so independently of the angular position of the rotor with respect to a fixed coordinate system. However, if we consider a rotor where there is no polar symmetry of the section: a typical case is that of a shaft of a generator or electric motor with two polar expansions, in which there are grooves, needed to accommodate windings, but which contribute little to the overall stiffness of the section (Fig. 6.49).

In order to better understand the problem, without introducing any restrictive hypotheses, we can assume that the section of the rotor is rectangular, as shown in Fig. 6.50, and that the principal axes of inertia ξ and η are those shown in the same figure. If the rotor is in a generic angular position, Fig. 6.50, there is biaxial bending as the plane of stress *s-s* caused by the natural weight does not coincide with a principal axis of inertia: this leads to deflection with components both in a vertical and horizontal direction. The component of deflection in a horizontal direction f_o will be zero whenever the axis of stress, during rotation of the shaft, coincides with a principal axis of inertia; the vertical component f_v will have a maximum in correspondence to the passage of the axis of stress that is responsible for lower moments of inertia and minimum value in correspondence to the passage for the other principal axis.

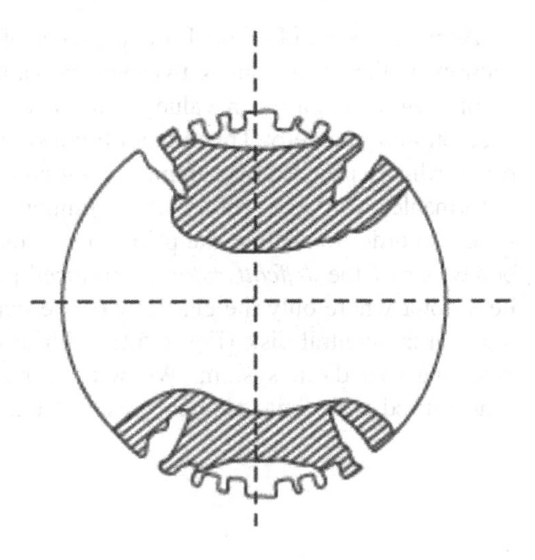

Fig. 6.49 Schematisation of the section of a rotor with non-polar symmetry

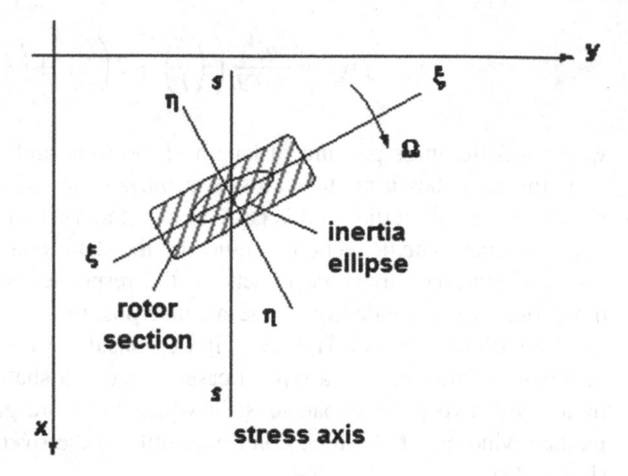

Fig. 6.50 Non-uniform section of an alternator rotor

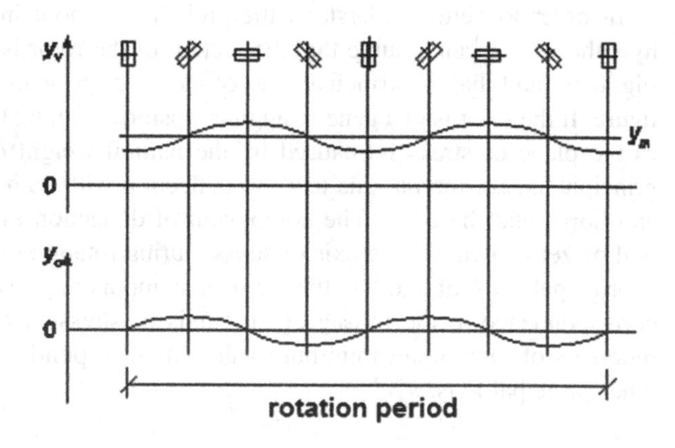

Fig. 6.51 Vertical f_v and horizontal f_o deflection for different angular positions of the rotor (rectangular cross-section)

As we can see in Fig. 6.51, the presence of different stiffnesses of the transversal section of the rotor causes two-per-rev variations of the deflection: the vertical component f_v has a mean value f_m that is different to zero, while the horizontal f_o oscillates around zero. This effect obviously has an impact on the dynamics of the rotor, which is forced by parametric variation of stiffness that concerns, above all, deformable rotors, i.e. rotors with running speeds Ω higher than the first critical speed. In order to analyse the phenomenon qualitatively we can define the dynamic behaviour of the *Jeffcott rotor* (introduced previously, see Sect. 6.6.1, Fig. 6.52), i.e. a rotor where only the elasticity of the shaft is considered, by concentrating the mass in the central disk (Fig. 6.52): with this hypothesis the rotor is reduced to a vibrating two d.o.f. system. We will later extend the discussion to a real rotor schematised with finite elements, Sect. 6.6.2.

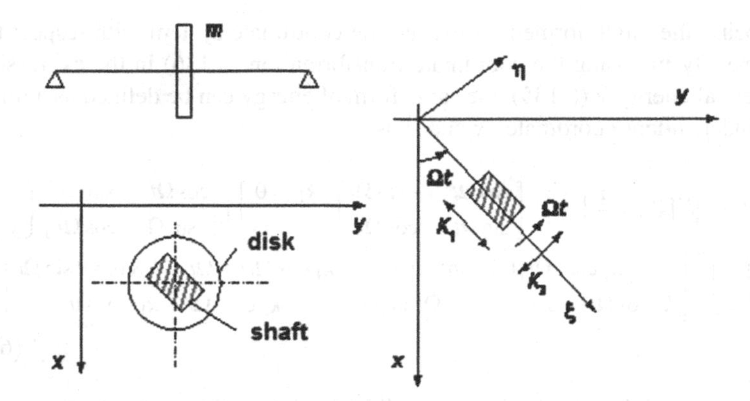

Fig. 6.52 Jeffcott rotor: model with two DOF for studying the two-per-rev vibrations caused by different stiffnesses of the rotor section

6.6.1 Two-Degree-of-Freedom Model

We will now consider a massless shaft with a rectangular section, with a mass m fitted in the centreline (Fig. 6.52) and runout zero (centre of gravity coinciding with the rotation axis). As usual, X-Y-Z is a right-handed Cartesian coordinate system, with axis Z coinciding with the rotor's rotation axis and $\xi - \eta - \zeta$ is a rotating coordinate system that is integral with the rotor and with axes ξ and η coincident with principal axes of inertia of the cross section of the actual shaft. k_1 and k_2 are the stiffnesses of the shaft in the centreline according to directions ξ and η: we can define the kinetic energy E_c, the elastic potential energy V and the virtual work $\delta * L$ performed by the weight force as:

$$\begin{aligned}
E_c &= \frac{1}{2}\underline{\dot{z}}^T[m]\underline{\dot{z}} = \frac{1}{2}\begin{Bmatrix}\dot{x}\\ \dot{y}\end{Bmatrix}^T\begin{bmatrix}m & 0\\ 0 & m\end{bmatrix}\begin{Bmatrix}\dot{x}\\ \dot{y}\end{Bmatrix}\\
V &= \frac{1}{2}\underline{\gamma}^T[k]\underline{\gamma} = \frac{1}{2}\begin{Bmatrix}\xi\\ \eta\end{Bmatrix}^T\begin{bmatrix}k_1 & 0\\ 0 & k_2\end{bmatrix}\begin{Bmatrix}\xi\\ \eta\end{Bmatrix}\\
\delta * L &= \underline{P}^T\delta * \underline{z} = \begin{Bmatrix}mg\\ 0\end{Bmatrix}^T\begin{Bmatrix}\delta * x\\ \delta * y\end{Bmatrix}
\end{aligned} \quad (6.135)$$

having used, to define the generic position of the geometric centre of the rotor with x and y and ξ and η the position of the mass m, respectively, in the absolute and relative Cartesian coordinate system (gathered, respectively, in vectors $\underline{z}$ and $\underline{\gamma}$). Between the two coordinate systems, the following relations apply:

$$\underline{\gamma} = [\lambda]\underline{z} \Rightarrow \begin{Bmatrix}\xi\\ \eta\end{Bmatrix} = \begin{bmatrix}\cos\Omega t & \sin\Omega t\\ -\sin\Omega t & \cos\Omega t\end{bmatrix}\begin{Bmatrix}x\\ y\end{Bmatrix} \quad (6.136)$$

Ωt being the angle formed by the mobile coordinate system with respect to the fixed one. By imposing the coordinate transformation (6.136) in the expression of the potential energy V (6.135), the same form of energy can be defined as a function of the independent coordinates x and y as:

$$V = \frac{1}{2}\underline{z}^T[\lambda]^T[k][\lambda]\underline{z} = \frac{1}{2}\begin{Bmatrix} x \\ y \end{Bmatrix}^T \begin{bmatrix} \cos\Omega t & -\sin\Omega t \\ \sin\Omega t & \cos\Omega t \end{bmatrix} \begin{bmatrix} k_1 & 0 \\ 0 & k_2 \end{bmatrix} \begin{bmatrix} \cos\Omega t & \sin\Omega t \\ -\sin\Omega t & \cos\Omega t \end{bmatrix} \begin{Bmatrix} x \\ y \end{Bmatrix}$$
$$= \frac{1}{2}\begin{Bmatrix} x \\ y \end{Bmatrix}^T \begin{bmatrix} k_1\cos^2\Omega t + k_2\sin^2\Omega t & k_1\cos\Omega t\sin\Omega t - k_2\cos\Omega t\sin\Omega t \\ k_1\cos\Omega t\sin\Omega t - k_2\cos\Omega t\sin\Omega t & k_2\cos^2\Omega t + k_1\sin^2\Omega t \end{bmatrix} \begin{Bmatrix} x \\ y \end{Bmatrix} \tag{6.137}$$

Now by applying Lagrange is possible to define the system's equations of motion, which, in matrix form become:

$$\begin{aligned} m\ddot{x} + k_1x\cos^2\Omega t + k_1y\cos\Omega t\sin\Omega t + k_2x\sin^2\Omega t - k_2y\cos\Omega t\sin\Omega t &= mg \\ m\ddot{y} + k_1y\sin^2\Omega t + k_1x\cos\Omega t\sin\Omega t + k_2y\cos^2\Omega t - k_2x\cos\Omega t\sin\Omega t &= 0 \end{aligned} \tag{6.138a}$$

By ordering the terms according to the d.o.f. x and y we obtain:

$$\begin{aligned} m\ddot{x} + \left(k_1\cos^2\Omega t + k_2\sin^2\Omega t\right)x + (k_1\cos\Omega t\sin\Omega t - k_2\cos\Omega t\sin\Omega t)y &= mg \\ m\ddot{y} + \left(k_1\sin^2\Omega t + k_2\cos^2\Omega t\right)y + (k_1\cos\Omega t\sin\Omega t - k_2\cos\Omega t\sin\Omega t)x &= 0 \end{aligned} \tag{6.138b}$$

Now we can introduce, with k_m, the mean value of stiffness:

$$k_m = \frac{k_1 + k_2}{2} \tag{6.139a}$$

and, with Δk, the variation of stiffness around the mean value:

$$\Delta k = \frac{k_1 - k_2}{2} \quad (con\ (k_1 - k_2) > 0) \tag{6.139b}$$

Considering definitions (6.139a) and (6.139b) the equations of motion (6.138b) can be rewritten as:

$$\begin{aligned} m\,\ddot{x} + ((k_m + \Delta k)\cos^2\Omega t + (k_m - \Delta k)\sin^2\Omega t)x + ((k_m + \Delta k)\cos\Omega t\sin\Omega t - (k_m - \Delta k)\cos\Omega t\sin\Omega t)y &= mg \\ m\,\ddot{y} + ((k_m + \Delta k)\sin^2\Omega t + (k_m - \Delta k)\cos^2\Omega t)y + ((k_m + \Delta k)\cos\Omega t\sin\Omega t - (k_m - \Delta k)\cos\Omega t\sin\Omega t)x &= 0 \end{aligned} \tag{6.140a}$$

i.e.:

$$\begin{aligned} m\ddot{x} + k_m x + \Delta k\left(\cos^2\Omega t - \sin^2\Omega t\right)x + \Delta k\, y \sin 2\Omega t = mg \\ m\ddot{y} + k_m y + \Delta k\left(\cos^2\Omega t - \sin^2\Omega t\right)y + \Delta k\, x \sin 2\Omega t = 0 \end{aligned} \quad (6.140b)$$

or also:

$$\begin{aligned} m\ddot{x} + k_m x + \Delta k\, x\, \cos 2\Omega t + \Delta k\, y \sin 2\Omega t = mg \\ m\ddot{y} + k_m y + \Delta k\, y\, \cos 2\Omega t + \Delta k\, x \sin 2\Omega t = 0 \end{aligned} \quad (6.140c)$$

In matrix terms it is possible to rewrite (6.140c) as:

$$[M]\underline{\ddot{x}} + [K]\underline{x} = \underline{F} \quad (6.141a)$$

having used $[M]$ to indicate the mass matrix:

$$[M] = \begin{bmatrix} m & 0 \\ 0 & m \end{bmatrix} \quad (6.141b)$$

and $[K]$ to indicate the stiffness matrix, sum of two terms:

$$[K] = [K_m] + [\Delta K] = \begin{bmatrix} k_m & 0 \\ 0 & k_m \end{bmatrix} + \begin{bmatrix} \Delta k \cos 2\Omega t & \Delta k \sin 2\Omega t \\ \Delta k \sin 2\Omega t & \Delta k \cos 2\Omega t \end{bmatrix} \quad (6.141c)$$

The system to be solved is made up of two linear differential equations with coefficients that vary periodically over time (Mathieu equations [18]): these equations are difficult to solve analytically. In general, however, we should bear in mind that in real cases, the ratio $(\Delta k/k_m)$ is in the region of a few percent: this consideration makes it possible to introduce suitable simplifications in (6.140a)–(6.140c) aimed at making these equations easier to solve. So, if we assume a change in variables, such as:

$$\begin{aligned} x = x_s + x_d \\ y = y_s + y_d = y_d \end{aligned} \quad (6.142a)$$

where x_d is used to indicate the variation of the deflection assumed by the rotor in a vertical direction with respect to the mean deflection x_s:

$$x_s = \frac{mg}{k_m} \quad (6.142b)$$

and y_d is the variation with respect to the horizontal component, whose mean value y_s is zero. If we consider the small oscillations around the mean position of equilibrium defined by (6.142b), i.e. we consider that:

$$x_d \ll x_s; \quad y_d \, small \tag{6.142c}$$

by introducing the transformation (6.142a) into (6.140c) the same equations of motion can be rewritten as:

$$\begin{aligned} m\ddot{x}_d + k_m x_d + k_m x_s + x_d \Delta k \cos 2\Omega t + x_s \Delta k \cos 2\Omega t + y_d \Delta k \sin 2\Omega t = mg \\ m\ddot{y}_d + k_m y_d + y_d \Delta k \ \cos 2\Omega t + x_s \Delta k \ \sin 2\Omega t + x_d \Delta k \ \sin 2\Omega t = 0 \end{aligned} \tag{6.143a}$$

Considering the hypotheses (6.142b) adopted and remembering that Δk is small, the products $x_d \Delta k$ and $y_d \Delta k$ can be considered negligible, thus (6.143a) becomes:

$$\begin{aligned} m\ddot{x}_d + k_m x_d &= -x_s \Delta k \cos 2\Omega t \\ m\ddot{y}_d + k_m y_d &= -x_s \Delta k \ \sin 2\Omega t \end{aligned} \tag{6.143b}$$

Due to the presence of the term Δk caused by the non-polar symmetry of the section of the rotor, the system can be attributed to a system that is forced by an excitation with natural pulsation equal to 2Ω, i.e. equal to double the angular velocity Ω. The solutions of (6.143b) are of the form:

$$\begin{aligned} x_d = X_d \cos 2\Omega t &= \frac{-x_s \Delta k}{k_m - 4\Omega^2 m} \cos 2\Omega t \\ y_d = Y_d \sin 2\Omega t &= \frac{-x_s \Delta k}{k_m - 4\Omega^2 m} \sin 2\Omega t \end{aligned} \tag{6.144a}$$

Resonance conditions will occur when:

$$k_m - 4\Omega^2 m = 0 \quad \Rightarrow \quad 2\Omega = \sqrt{\frac{k_m}{m}} \quad \Rightarrow \quad \Omega = \frac{1}{2}\omega_o \tag{6.144b}$$

i.e. when the angular velocity Ω of the asymmetric rotor is half of the natural pulsation ω_o of the actual rotor (estimated with mean stiffness).

6.6.1.1 Studying Stability

In the previous section, we saw how the effect of weight in a horizontal rotor with different stiffnesses causes a forcing element with double frequency compared to that of rotation Ω. Now we will investigate the possible conditions of instability, again caused by the presence of different stiffnesses of the rotor. For this purpose we will once again be considering the Jeffcott rotor described in the previous section (Fig. 6.52): the equation of motion obtained (6.140c), with parameters that vary over time, can be defined as a Mathieu equation and characterises the small oscillations not around a condition of static equilibrium, but around a situation of motion: so even if the system is conservative it is possible that conditions of instability may arise. So the periodic motion that the rotor is subject to will be stable if all the solutions of the equation of perturbed motion around the steady-state solution tend to zero as time increases, otherwise it will be unstable. To verify the stability of the system it is easier to rewrite the equations of motion with respect to the rotating coordinate system $\xi - \eta - \zeta$, that is integral with the rotor, introducing, in the various forms of energy associated with the system in question (6.135), the inverse transformation of coordinates compared to that used in the previous section (6.136):

$$\underline{z} = [\lambda]^{-1}\underline{\gamma} = [\lambda]^T\underline{\gamma} \;\Rightarrow\; \underline{\dot{z}} = \left[\dot{\lambda}\right]^T\underline{\gamma} + [\lambda]^T\underline{\dot{\gamma}} \tag{6.145a}$$

i.e. in scalar form:

$$\begin{aligned} \dot{x} &= \dot{\xi}\cos\Omega t - \Omega\xi\sin\Omega t - \dot{\eta}\sin\Omega t - \Omega\eta\cos\Omega t \\ \dot{y} &= \dot{\xi}\sin\Omega t + \Omega\xi\cos\Omega t + \dot{\eta}\cos\Omega t - \Omega\eta\sin\Omega t \end{aligned} \tag{6.145b}$$

The kinetic energy can be expressed in matrix form as:

$$E_c = \frac{1}{2}\underline{\dot{z}}^T[m]\underline{\dot{z}} = \frac{1}{2}\left\{\underline{\dot{\gamma}}^T[\lambda] + \underline{\gamma}^T\left[\dot{\lambda}\right]\right\}[m]\left\{\left[\dot{\lambda}\right]^T\underline{\gamma} + [\lambda]^T\underline{\dot{\gamma}}\right\} \tag{6.145c}$$

and in scalar form, by developing the matrix product (6.145c):

$$E_c = \frac{1}{2}m\left(\dot{\xi}^2 + \dot{\eta}^2 + (\Omega\xi)^2 + (\Omega\eta)^2 - 2\Omega\dot{\xi}\eta - 2\Omega\dot{\eta}\xi\right) \tag{6.145d}$$

By applying the Lagrange equations to expressions (6.135) and (6.145b), this time assuming, as independent variables, the coordinates ξ and η, the equations of motion of the system, expressed in the rotating coordinate system, become:

$$\begin{aligned} m\ddot{\xi} - 2m\Omega\dot{\eta} + \left(k_1 - m\Omega^2\right)\xi &= 0 \\ m\ddot{\eta} + 2m\Omega\dot{\xi} + \left(k_2 - m\Omega^2\right)\eta &= 0 \end{aligned} \tag{6.146a}$$

By dividing both equations by the mass of the disk m and gathering the variables in the only vector $\underline{\gamma}$ [see Eq. (6.135)], (6.146a) become, in matrix form:

$$\begin{bmatrix} 1 & 0 \\ 0 & 1 \end{bmatrix}\ddot{\underline{\gamma}} + \begin{bmatrix} 0 & -2\Omega \\ 2\Omega & 0 \end{bmatrix}\dot{\underline{\gamma}} + \begin{bmatrix} \omega_{1o}^2 - \Omega^2 & 0 \\ 0 & \omega_{2o}^2 - \Omega^2 \end{bmatrix}\underline{\gamma} = \underline{0} \tag{6.146b}$$

i.e.:

$$[M]\ddot{\underline{\gamma}} + [R]\dot{\underline{\gamma}} + [K]\underline{\gamma} = \underline{0} \tag{6.146c}$$

since:

$$\omega_{10} = \sqrt{\frac{k_1}{m}}; \quad \omega_{20} = \sqrt{\frac{k_2}{m}} \tag{6.146d}$$

As we can see, the matrices that define the free motion of the rotor, defined with respect to the coordinate system that rotates integrally with the shaft, do not depend on time. If we wish to analyse the possible conditions of instability, we must first consider just the positional terms: matrix $[K]$ (6.146c) is symmetrical and positive definite if all the principal minors are positive, that is, in particular, if:

$$\left(\omega_{1o}^2 - \Omega^2\right)\left(\omega_{2o}^2 - \Omega^2\right) > 0; \quad \left(\omega_{1o}^2 - \Omega^2\right) > 0; \quad \left(\omega_{2o}^2 - \Omega^2\right) > 0 \tag{6.147}$$

All the conditions (6.147) are satisfied for $\Omega < \omega_{1o}$; on the other hand, for $\Omega > \omega_{1o}$ static divergence type instability arises, i.e. corresponding to a solution that is purely expansive in the rotating coordinate system. Considering the link between the coordinates x and y in the absolute coordinate system and the coordinates ξ and η in the rotating coordinate system (6.145a), this diverging expansive solution in the rotating coordinate system corresponds to an oscillating expansive solution in the absolute coordinate system (Fig. 6.54c). This solution is purely hypothetical as the terms of velocity have been ignored: now by analysing (6.146c) in its complete form and imposing:

$$\underline{\gamma} = \underline{\Gamma} e^{\lambda t} \tag{6.148a}$$

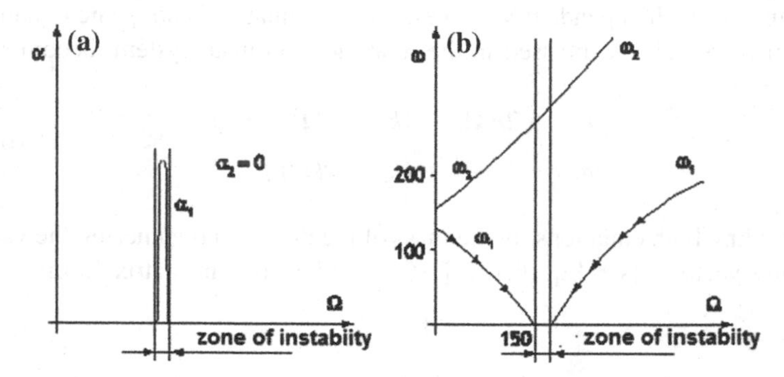

Fig. 6.53 Conditions of stability in a rotor with different stiffnesses

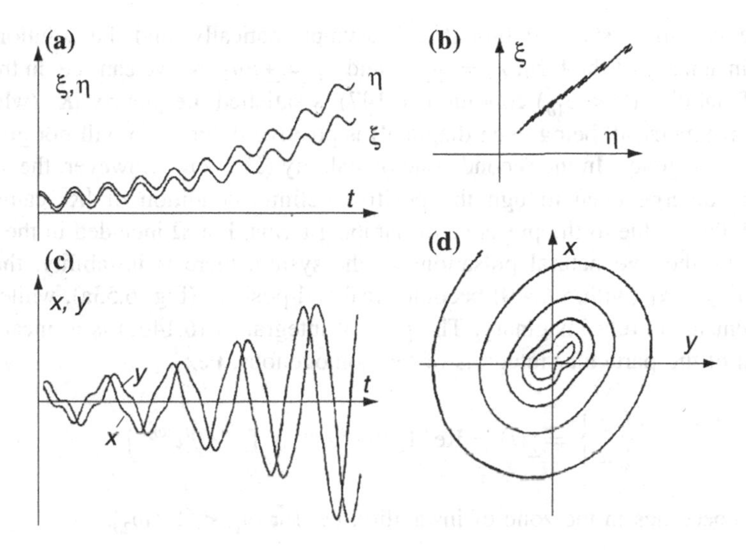

Fig. 6.54 Rotor with different stiffnesses: time history of the rotor's horizontal and vertical displacements in conditions of instability (**a** rotating and **c** absolute coordinates); trajectories in conditions of instability (**b** rotating and **d** absolute coordinates)

we obtain a complex, homogenous algebraic system in unknown $\underline{\Gamma}$:

$$\left[\lambda^2[M] + \lambda[R] + [K]\right]\underline{\Gamma} = \underline{0} \tag{6.148b}$$

Non-trivial solutions of (6.148b) are those that zero the determinant of the coefficient matrix:

$$\begin{vmatrix} \lambda^2 + \omega_{1o}^2 - \Omega^2 & -2\Omega\lambda \\ 2\Omega\lambda & \lambda^2 + \omega_{2o}^2 - \Omega^2 \end{vmatrix} = \lambda^4 + \lambda^2\left(\omega_{2o}^2 + \omega_{1o}^2 + 2\Omega^2\right) + \omega_{1o}^2\omega_{2o}^2 - \left(\omega_{2o}^2 + \omega_{1o}^2\right)\Omega^2 + \Omega^4 = 0 \tag{6.148c}$$

The solutions of free motion, which can be calculated using (6.148c), are of the form:

$$\begin{aligned} \lambda_1 &= \alpha_1 \pm i\omega_1 \\ \lambda_2 &= \alpha_2 \pm i\omega_2 \end{aligned} \tag{6.148d}$$

By way of example, Fig. 6.53 shows the results of a numerical application having considered $\omega_{1_o} = 140.85$ rad/s and $\omega_{2_o} = 165.10$ rad/s: the same figure shows the trend of the natural pulsations ω_1 and ω_2 and exponential terms α_1 and α_2 as a function of the angular velocity of the rotor Ω. For $\Omega < \omega_{1_o}$ and $\Omega > \omega_{2o}$, i.e. in an operational field outside of the range between the two natural pulsations ω_{1_o} and

ω_{2_o} the system is stable, although not asymptomatically, and the solutions are purely imaginary of the form $\lambda_1 = \pm i\omega_1$ and $\lambda_2 = \pm i\omega_2$. As we can see, in the first zone of stability ($\Omega < \omega_{1_o}$) condition (6.147) is satisfied, i.e. matrix $[K]$ (which is always symmetrical, being even diagonal) is positive definite, so will not give rise to static divergence. In the second zone of stability ($\Omega > \omega_{2_o}$) however, the system does not diverge even though the positive definite condition of $[K]$ cannot be verified: this is due to the presence of damping terms. For Ω included in the range defined by the two natural pulsations of the system there is instability: the first solution $\lambda_1 = \alpha_1$ (with $\alpha_1 > 0$) becomes real and positive (Fig. 6.53a), while $\lambda_2 = \pm i\omega_2$ remains purely imaginary. The general integral of (6.146c) is a linear combination of the particular integrals of the homogenous, i.e.:

$$\begin{Bmatrix} \xi \\ \pi \end{Bmatrix} = \underline{\gamma}(t) = \mathrm{Re}\left(\underline{\Gamma}^{(1)} e^{\alpha_1 t} e^{i\omega_1 t} + \underline{\Gamma}^{(2)} e^{\alpha_2 t} e^{i\omega_2 t}\right) \tag{6.149a}$$

(6.153a) becomes in the zone of instability, i.e. for $\omega_{1_o} < \Omega < \omega_{2_o}$:

$$\underline{\zeta}(t) = \underline{\gamma}(t) = \mathrm{Re}\left(\underline{\Gamma}^{(1)} e^{\alpha_1 t} + \underline{\Gamma}^{(2)} e^{i\omega_2 t}\right) \tag{6.149b}$$

that is, an oscillating motion overlying an expansive exponential motion, as shown in Fig. 6.54 which shows both the time history ξ and η of the displacement of the centre of the journal, and its trajectory compared to the rotating coordinate system assumed. If, in these conditions, we wish to define the motion with respect to the absolute coordinate, since:

$$\underline{z}(t) = \begin{Bmatrix} x \\ y \end{Bmatrix} = [\lambda]\underline{\gamma}(t) = \begin{bmatrix} \cos\Omega t & -\sin\Omega t \\ \sin\Omega t & \cos\Omega t \end{bmatrix} \underline{\gamma}(t) \tag{6.149c}$$

(6.149a) becomes:

$$\begin{aligned} x(t) &= \left(\xi^{(1)} e^{\alpha_1 t} + \xi^{(2)} \cos(\omega_2 t + \phi_2)\right) \cos\Omega t - \left(\mu^{(1)} e^{\alpha_1 t} + \eta^{(2)} \cos(\omega_2 t + \phi_2)\right) \sin\Omega t \\ y(t) &= \left(\xi^{(1)} e^{\alpha_1 t} + \xi^{(2)} \cos(\omega_2 t + \phi_2)\right) \sin\Omega t + \left(\mu^{(1)} e^{\alpha_1 t} + \eta^{(2)} \cos(\omega_2 t + \phi_2)\right) \cos\Omega t \end{aligned} \tag{6.149d}$$

i.e. an expanding oscillating motion, represented in Fig. 6.54c and 6.54d.

6.6.2 Schematisation of the Problem on a Real Rotor

We will now consider a rotor that has two supports, and which has a portion of shaft where there are different stiffnesses due, for example, to hollows or stiffening (Fig. 6.55). The finite element model that could be adopted to reproduce the

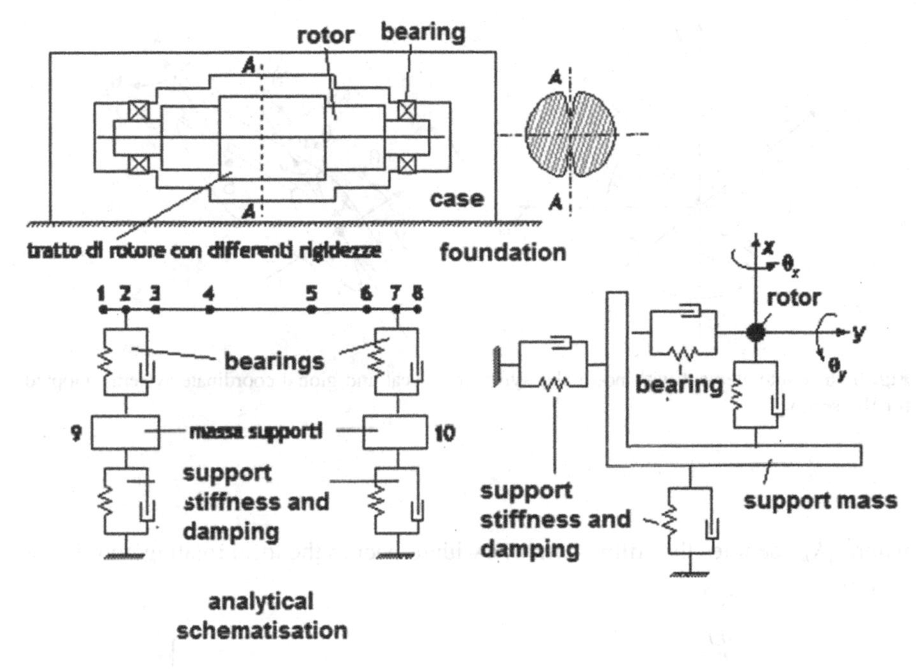

Fig. 6.55 Schematisation of a rotor with a non-circular section

behaviour of the rotor is that shown in the same figure. The definition of the matrices of mass $[M_j]$ and stiffness $[K_j]$ of all the finite elements that make up the mathematical model, have already been fully explained in Sect. 6.2.1. With reference to the discussion in Sect. 6.6.1, Eqs. (6.141a)–(6.141c), we must define the matrix $[\Delta K_j]$ of the element that simulates the section of rotor with different stiffnesses only (element with end nodes 4 and 5 in the Fig. 6.55, [6].

We will now consider this element and define a rotating coordinate system that is integral with the shaft (axes ξ and η): we will assume as independent variables the nodal displacements $\underline{x}_{jl}$ of the end nodes of the element considered in the local rotating coordinate system (Fig. 6.56). These variables are chosen so that we have directions that are parallel to the principal directions of inertia of the cross section. It is then possible to define the potential energy V_j associated with this element:

$$V_j = \frac{1}{2}\underline{x}_{jl}^T [K_{jl}] \underline{x}_{jl} \tag{6.150a}$$

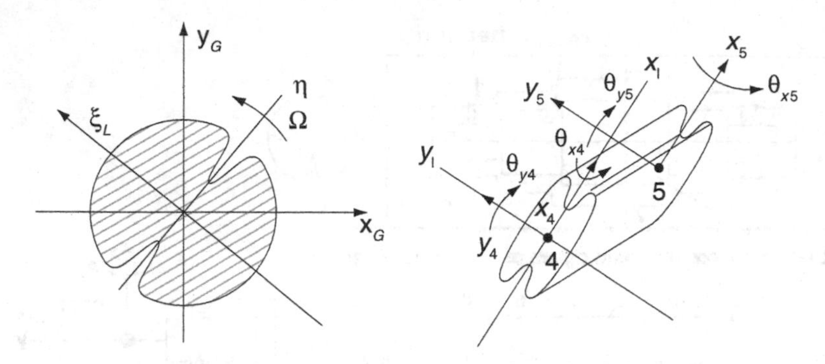

Fig. 6.56 Finite element with non-polar symmetry: local and global coordinate systems adopted for the section

where $\left[K_{jl}\right]$ denotes the stiffness matrix with respect to the local rotating coordinate system:

$$\left[K_{jl}\right] = \begin{bmatrix} \frac{12EJ_y}{l^3} & & & & & & & \\ \frac{6EJ_y}{l^2} & \frac{4EJ_y}{l} & & & & sym & & \\ 0 & 0 & \frac{12EJ_x}{l^3} & & & & & \\ 0 & 0 & -\frac{6EJ_x}{l^2} & \frac{4EJ_x}{l} & & & & \\ -\frac{12EJ_y}{l^3} & \frac{6EJ_y}{l^2} & 0 & 0 & \frac{12EJ_y}{l^3} & & & \\ \frac{6EJ_y}{l^2} & \frac{2EJ_y}{l} & 0 & 0 & -\frac{6EJ_y}{l^2} & \frac{4EJ_y}{l} & & \\ 0 & 0 & \frac{12EJ_x}{l^3} & \frac{6EJ_x}{l^2} & 0 & 0 & \frac{12EJ_x}{l^3} & \\ 0 & 0 & -\frac{6EJ_x}{l^2} & \frac{2EJ_x}{l} & 0 & 0 & \frac{6EJ_x}{l^2} & \frac{4EJ_x}{l} \end{bmatrix} \quad (6.150b)$$

In order to define the equations of motion of the overall system, we need to express all the various forms of energy, and in particular the potential energy V_j (6.150b) of the element with non-polar symmetry, in a single global coordinate system, assumed to be fixed, as shown in Fig. 6.54. For this purpose, we can express the link between the coordinates $\underline{x}_j$ referring to the absolute coordinate system and those relating to the local coordinate system $\underline{x}_{jl}$. The link between the two sets of coordinates can be expressed, in matrix form, as:

$$\underline{x}_{jl} = [\Lambda_{\Omega t}]\underline{x}_j \quad (6.151a)$$

where (see (6.136)) the matrix $[\Lambda_{\Omega t}]$ is of the form:

$$[\Lambda_{\Omega t}] = \begin{bmatrix} \cos\Omega t & 0 & \sin\Omega t & 0 & 0 & 0 & 0 & 0 \\ 0 & \cos\Omega t & 0 & -\sin\Omega t & 0 & 0 & 0 & 0 \\ -\sin\Omega t & 0 & \cos\Omega t & 0 & 0 & 0 & 0 & 0 \\ 0 & \sin\Omega t & 0 & \cos\Omega t & 0 & 0 & 0 & 0 \\ 0 & 0 & 0 & 0 & \cos\Omega t & 0 & \sin\Omega t & 0 \\ 0 & 0 & 0 & 0 & 0 & \cos\Omega t & 0 & -\sin\Omega t \\ 0 & 0 & 0 & 0 & -\sin\Omega t & 0 & \cos\Omega t & 0 \\ 0 & 0 & 0 & 0 & 0 & \sin\Omega t & 0 & \cos\Omega t \end{bmatrix} \tag{6.151b}$$

Therefore, it is possible to express the potential energy V_j in the global coordinates $\underline{x}_j$, to obtain:

$$V_j = \frac{1}{2}\underline{x}_j^T[\Lambda_{\Omega t}]^T[K_{jl}][\Lambda_{\Omega t}]\underline{x}_j = \frac{1}{2}\underline{x}_j^T[K_j(t)]\underline{x}_j \tag{6.152a}$$

The stiffness matrix of the finite element with non-polar symmetry $[K_j(t)]$ defined in the absolute coordinate system will, therefore, be given by the product:

$$[K_j(t)] = [\Lambda_{\Omega t}]^T[K_{jl}][\Lambda_{\Omega t}] \tag{6.152b}$$

Therefore, the terms of this matrix are a function, via sine and cosine terms, of:

- time t;
- different moments of inertia J_1 and J_2 of the section;
- angular velocity Ω.

The term $K_j(t)_{1,1}$, for example, will be equal to:

$$K_j(t)_{1,1} = \frac{12EJ_1}{l^3}\cos^2\Omega t + \frac{12EJ_2}{l^3}\sin^2\Omega t \tag{6.153a}$$

Similarly to what was done for the 2 d.o.f. system, Sect. 6.6, using J_m to define the mean value of the moment of inertia of the section:

$$J_m = \frac{J_1 + J_2}{2} \tag{6.153b}$$

and ΔJ for the semi-difference of the moments of inertia:

$$\Delta J = \frac{J_1 - J_2}{2} \tag{6.153c}$$

the generic term (6.153a) can, therefore, be rewritten as the sum of a constant term and one that is variable over time with pulsation equal to double the angular velocity Ω of the rotor:

$$
\begin{aligned}
K_j(t)_{1,1} &= \frac{12E}{l^3}(J_m + \Delta J)\cos^2 \Omega t + \frac{12E}{l^3}(J_m - \Delta J)\sin^2 \Omega t \\
&= \frac{12EJ_m}{l^3} + \frac{12E\Delta J}{l^3}\cos 2\Omega t
\end{aligned}
\tag{6.153d}
$$

The same transformation can be performed on all the other terms of the stiffness matrix $\left[K_j(t)\right]$, which can then be broken down into the sum of two matrices:

$$
\left[K_j(t)\right] = \left[K_{jm}\right] + \left[\Delta K_j\right] \tag{6.154a}
$$

where $\left[K_{jm}\right]$ is symmetrical, independently of the time t and the angular velocity of the rotor:

$$
\left[K_{jm}\right] = \begin{bmatrix}
\frac{12EJ_m}{l^3} & & & & & & & \\
\frac{6EJ_m}{l^2} & \frac{4EJ_m}{l} & & & & sym & & \\
0 & 0 & \frac{12EJ_m}{l^3} & & & & & \\
0 & 0 & -\frac{6EJ_m}{l^2} & \frac{4EJ_m}{l} & & & & \\
-\frac{12EJ_m}{l^3} & \frac{6EJ_m}{12} & 0 & 0 & \frac{12EJ_m}{l^3} & & & \\
\frac{6EJ_m}{12} & \frac{2EJ_m}{l} & 0 & 0 & -\frac{6EJ_m}{l^2} & \frac{4EJ_m}{l} & & \\
0 & 0 & \frac{12EJ_m}{l^3} & \frac{6EJ_m}{l^2} & 0 & 0 & \frac{12EJ_m}{l^3} & \\
0 & 0 & -\frac{6EJ_m}{l^2} & \frac{2EJ_m}{l} & 0 & 0 & \frac{6EJ_m}{l^2} & \frac{4EJ_m}{l}
\end{bmatrix} \tag{6.154b}
$$

while matrix $\left[\Delta K_j\right]$ is symmetrical but dependent on the time and the angular velocity Ω of the rotor:

$$
\left[\Delta K_j\right] = \begin{bmatrix}
\frac{12}{l^3}\cos 2\Omega t & & & & & & & \\
\frac{6}{l^2}\cos 2\Omega t & \frac{4}{l}\cos 2\Omega t & & & & sym & & \\
\frac{12}{l^3}\sin 2\Omega t & \frac{6}{l^2}\sin 2\Omega t & \frac{12}{l^3}\cos 2\Omega t & & & & & \\
-\frac{6}{l^2}\sin 2\Omega t & -\frac{4}{l}\sin 2\Omega t & -\frac{6}{l^2}\sin 2\Omega t & \frac{4}{l}\cos 2\Omega t & & & & \\
-\frac{12}{l^3}\cos 2\Omega t & -\frac{6}{l^2}\cos 2\Omega t & -\frac{12}{l^3}\sin 2\Omega t & -\frac{6}{l^2}\sin 2\Omega t & \frac{12}{l^3}\cos 2\Omega t & & & \\
\frac{6}{l^2}\cos 2\Omega t & \frac{2}{l}\cos 2\Omega t & \frac{6}{l^2}\sin 2\Omega t & -\frac{2}{l}\sin 2\Omega t & -\frac{6}{l^2}\sin 2\Omega t & \frac{4}{l}\cos 2\Omega t & & \\
-\frac{12}{l^3}\sin 2\Omega t & \frac{6}{l^2}\sin 2\Omega t & -\frac{12}{l^3}\cos 2\Omega t & \frac{6}{l^2}\cos 2\Omega t & \frac{12}{l^3}\sin 2\Omega t & -\frac{6}{l^2}\sin 2\Omega t & \frac{12}{l^3}\cos 2\Omega t & \\
\frac{12}{l^3}\sin 2\Omega t & \frac{12}{l^3}\sin 2\Omega t & -\frac{6}{l^2}\cos 2\Omega t & \frac{2}{l}\cos 2\Omega t & \frac{6}{l^2}\cos 2\Omega t & -\frac{4}{l}\sin 2\Omega t & \frac{6}{l^2}\sin 2\Omega t & \frac{4}{l}\cos 2\Omega t
\end{bmatrix} \tag{6.154c}
$$

By assembling [98–101]:

- the matrices of mean stiffness $\left[K_{jm}\right]$ (6.154b) of the element with non-circular cross-section (Fig. 6.57) in the overall matrix of the system $[K_m]$ of the entire system (rotor + oil-film + foundation) as shown in Fig. 6.56;
- the matrix $\left[\Delta K_j\right]$ (6.154c) in the matrix $[\Delta K]$ as shown in Fig. 6.58;
- the mass matrix $\left[M_j\right]$ in the overall mass matrix $[M]$;

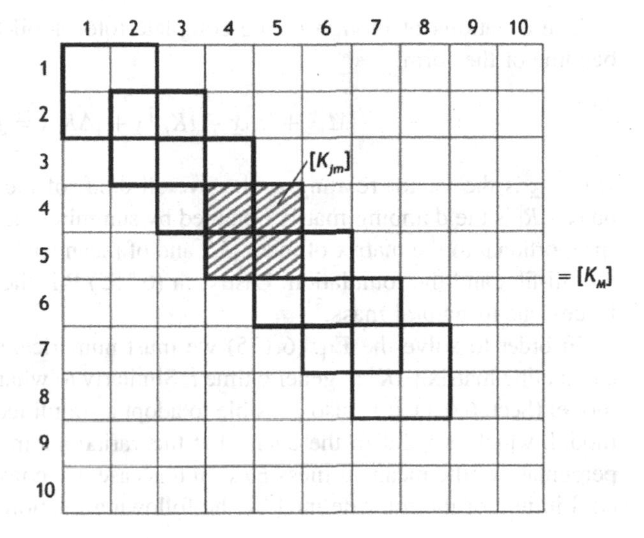

Fig. 6.57 Assembly of the mean stiffness matrix $[K_{jm}]$ in the overall matrix $[K_m]$ of the rotor + foundation + oil-film system

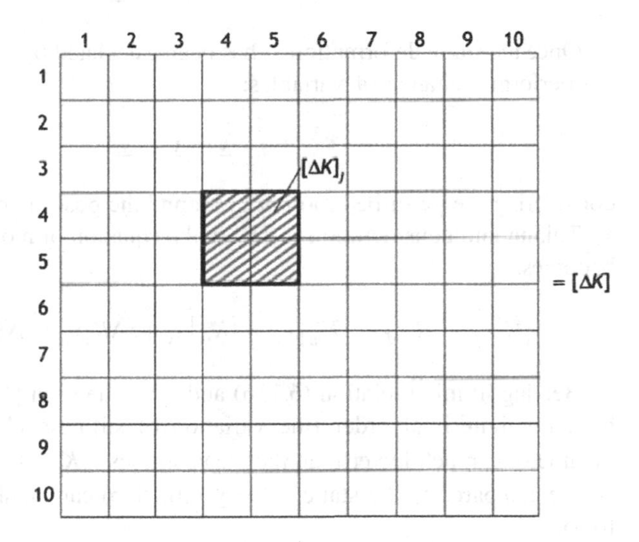

Fig. 6.58 Assembly of the stiffness matrix $[\Delta K_j]$ due to the different stiffnesses in the overall matrix $[\Delta K]$ of the rotor + foundation + oil-film system

and considering:

- the mass of the rotor as an external force;
- the effect of the oil-film (Sect. 6.2.2.1);
- the foundation (Sect. 6.2.4);

The equations of motion of the complete rotor + oil-film + foundation system become of the form:

$$[M]\ddot{\underline{x}} + [R]\dot{\underline{x}} + [K_m]\underline{x} + [\Delta K]\underline{x} = \underline{P} \tag{6.155}$$

where $\underline{x}$ is the vector relating to the overall d.o.f. of the system in global coordinates, $[R]$ is the damping matrix obtained by summing the contributions of the rotor (proportional to the matrix of mass $[M]$ and of mean stiffness $[K_m]$) and those due to the oil-film and the foundation. Lastly, in (6.155) P is the vector of the generalised forces due to natural mass.[31]

In order to solve the Eq. (6.155) we must numerically integrate the system by calculating matrix $[\Delta K]$ at generic time t. Similarly to what we saw for the two d.o.f. model (Sect. 6.6.1), it is also possible to adopt a simplified approach for the n d.o.f. model, which is valid in the event that the variations in stiffness $E\Delta J$ are a small percentage of the mean stiffness EJ_m. In this case, we consider the equation of static equilibrium of the rotor defined by the following relation [derivable from (6.155)]:

$$[K_m]\underline{x} = \underline{P} \tag{6.156}$$

Once the static deformation $\underline{x}_s$ has been calculated by solving the system (6.156), we perform a change of variables:

$$\underline{x} = \underline{x}_s + \underline{x}_d \tag{6.157a}$$

considering the perturbed motion $\underline{x}_d$ around the position of mean static equilibrium x_s. Taking into consideration (6.157a), the equation of motion of the system (6.155) becomes:

$$[M]\ddot{\underline{x}}_d + [R]\dot{\underline{x}}_d + [K_m]\underline{x}_d + [K_m]\underline{x}_s + [\Delta K]\underline{x}_s + [\Delta K]\underline{x}_d = \underline{P} \tag{6.157b}$$

Bearing in mind relation (6.156) and since the term $[\Delta K]\underline{x}_d$ can be considered a higher infinitesimal order (the variation of stiffness $[\Delta K]$ due to the non-polar symmetry is much lower than the mean stiffness $[K_m]$ as well as the displacements xd are compared to the static value $\underline{x}_s$) (6.157b) can be simplified in the following form:

$$[M]\ddot{\underline{x}}_d + [R]\dot{\underline{x}}_d + [K_m]\underline{x}_d = -[\Delta K]\underline{x}_s = \underline{F}_o e^{i2\Omega t} \tag{6.157c}$$

Therefore, for each time t and angular velocity of the rotor Ω it is possible to solve system (6.157c) to obtain the value of variation of displacement $\underline{x}_d$ around the position of static equilibrium $\underline{x}_s$ for each position of the rotor (univocally defined by

[31]In this discussion it has deliberately not considered the effect of other possible excitations such as those induced by the umbalance.

variables t and Ω). This solution reproduces the forced motion of a horizontal shaft caused by the different stiffnesses of the rotor. By solving the homogeneous equation associated with (6.155) we can, similarly to what was done in the two d.o.f. model [6], obtain the conditions of instability that can arise for Ω that falls between the two pulsations associated with the various rotor stiffnesses. In fact, the dissipative terms due to the oil-film cancel out this type of instability. The solution of the homogenous equation corresponding to (6.155) can be obtained using either numerical techniques or approximate methods (perturbation, Flochet, see the bibliography on this subject).

6.7 The Hysteretic Damping Effect

The passive resistances that act on a rotating shaft can be classified into two groups: external resistance and internal resistance. As previously discussed in the introduction (Sect. 6.2.2) and in the chapter on systems surrounded by force fields (Chap. 5), external resistances are mainly caused by resistance to air, to the effects of the oil-film in hydrodinamically lubricated bearings or the actions of fluid in seals. As we saw previously, the effects introduced by these bearings can be traced back to nonconservative force fields: the positional terms of the oil-film (Chap. 5, Sect. 5.3.2.2 and Sect. 6.2.2) can give rise to instability, whereas the corresponding terms of velocity are generally dissipative. The effect associated with the dissipation of energy caused by material deformation or internal slippage between the elements that make up the rotor (slippage between flanged or bolted parts or slippage in the windings of electric motors and alternators) is different: these phenomena are normally defined as *elastic hysteresis*. The characteristic of this hysteresis is that it does actually have a damping effect below the first critical speed ($\Omega < \omega_1$), whereas at higher speeds ($\Omega > \omega_1$) it becomes destabilising. The internal frictions caused by elastic hysteresis are generally less than the external frictions and so this form of instability does not generally arise. Exceptions can be seen in cases where there are other phenomena that give rise to instability, since then the energy supplied to the shaft by elastic hysteresis, for $\Omega > \omega_1$, adds to that of the other concurrent causes of instability.

6.7.1 Two-Degree-of-Freedom Model

We will now begin to examine the problems on the effect of passive hysteretic resistances. We will refer to the simple model of a rotor, assumed to rotate at a constant angular velocity Ω, shown in Fig. 6.59. In the mathematical model the supports are considered to be rigid, we can assume the mass m positioned symmetrically with respect to the supports and we can neglect the gyroscopic effects. O is the intersection between the axis that connects the two supports with the plane

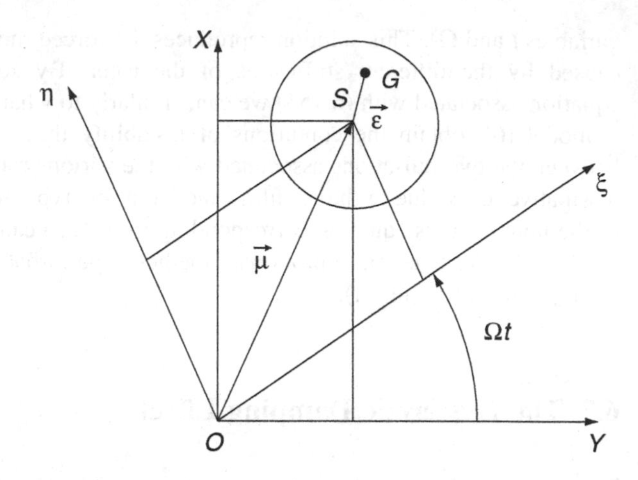

Fig. 6.59 Fixed (*X-Y-Z*) and rotating ($\xi - \mu - \zeta$) coordinate system

that contains the disk, *S* is the geometric centre of the disk and, lastly, *G* is its centre of gravity, assumed to be shifted by ε with respect to *S*. If we neglect hysteretic damping inside the rotor for the moment, then the equations of motion of the system, using the complex notion, are again those obtained in Sect. 6.3, Fig. 6.59:

$$m\ddot{\vec{z}} + r\dot{\vec{z}} + k\vec{z} = m\vec{\varepsilon}\,\Omega^2 e^{i\Omega t} + m\vec{g} \tag{6.158}$$

having indicated with:

- $\vec{z}$ the geometric vector that defines the position of the centre of the disk *S* with respect to a Cartesian coordinate system (*X-Y-Z*) that has axis *Z* coinciding with the rotation axis;
- *k* the stiffness of the shaft;
- *r* the viscous damping due to external damping.

The hysteretic damping, associated with the deformation of the shaft, in turn rotating with respect to the absolute coordinate system, can easily be introduced into a coordinate system ($\xi - \mu - \zeta$) that rotates with angular velocity Ω and axis ζ coinciding with axis *Z*: compared to this rotating coordinate system the geometric centre of the disk S may be subject to displacements defined by a geometric vector $\vec{\mu}$, since $\vec{\mu}$ is correlated to the vector $\vec{z}$ by means of the following relations:

$$\begin{aligned}
\vec{z} &= \vec{\mu}\, e^{i\Omega t} \\
\dot{\vec{z}} &= \dot{\vec{\mu}}\, e^{i\Omega t} + i\Omega\vec{\mu}\, e^{i\Omega t} \\
\ddot{\vec{z}} &= \ddot{\vec{\mu}}\, e^{i\Omega t} + 2i\Omega\,\dot{\vec{\mu}}\, e^{i\Omega t} - \Omega^2\vec{\mu}\, e^{i\Omega t}
\end{aligned} \tag{6.159}$$

By replacing the links between the two independent coordinates defined in relations (6.158) in the motion Eq. (6.159), we obtain:

$$\left(m\left(\ddot{\vec{\mu}} + 2i\Omega\dot{\vec{\mu}} - \Omega^2\vec{\mu}\right) + r\left(\dot{\vec{\mu}} + i\Omega\vec{\mu}\right) + k\vec{\mu}\right)e^{i\Omega t} = m\,\vec{\varepsilon}\,\Omega^2 e^{i\Omega t} + m\vec{g} \tag{6.160}$$

If we wish to define these equations in respect to the rotating coordinate system $(\xi - \mu - \zeta)$, we will need to pre-multiply both members of the term $e^{-i\Omega t}$, to obtain:

$$\left(m\left(\ddot{\vec{\mu}} + 2i\Omega\dot{\vec{\mu}} - \Omega^2\vec{\mu}\right) + r\left(\dot{\vec{\mu}} + i\Omega\vec{\mu}\right) + k\vec{\mu}\right) = m\,\vec{\varepsilon}\,\Omega^2 + m\vec{g}\,e^{-i\Omega t} \tag{6.161}$$

As we can see, compared to an observer rotating integrally with the system $(\xi - \mu - \zeta)$, the unbalance $m\,\vec{\varepsilon}\,\Omega^2$ becomes a geometric vector that is constant in modulus and direction, while the weight is represented by a vector that counter-rotates with speed $-\Omega$. In these equations it is possible to easily introduce the hysteretic damping, associated, as mentioned, with the deformation of the shaft, with an equivalent viscous damping with constant r_I[32]:

$$r_I\dot{\vec{\mu}} \tag{6.162}$$

thus obtaining:

$$\left(m\left(\ddot{\vec{\mu}} + 2i\Omega\dot{\vec{\mu}} - \Omega^2\vec{\mu}\right) + r\left(\dot{\vec{\mu}} + i\Omega\vec{\mu}\right) + k\vec{\mu} + r_I\dot{\vec{\mu}}\right) = m\,\vec{\varepsilon}\,\Omega^2 + m\vec{g}\,e^{-i\Omega t} \tag{6.163}$$

It is simpler, for the following discussion, to bring this equation back to a dimensionless form: by dividing (6.163) by the mass m and defining the different constants as follows:

$$\begin{aligned} \alpha &= \frac{r}{m} \\ \beta &= \frac{r_I}{m} \\ \omega_o^2 &= \frac{k}{m} \end{aligned} \tag{6.164a}$$

[32]This model takes into account the fact that the energy is dissipated only if there is actually a deflection of the rotor, bending associated with a motion in respect to the rotating coordinate: in the case where the rotor rotates rigidly inflected, i.e. for $\dot{\vec{\mu}} = 0$, there is no energy dissipation.

we will obtain:

$$\ddot{\vec{\mu}} + 2i\Omega\dot{\vec{\mu}} - \Omega^2\vec{\mu} + \alpha\dot{\vec{\mu}} + i\Omega\alpha\vec{\mu} + \omega_o^2\vec{\mu} + \beta\dot{\vec{\mu}} = \vec{\varepsilon}\,\Omega^2 + \vec{g}\,e^{-i\Omega t} \tag{6.164b}$$

i.e.:

$$\ddot{\vec{\mu}} + \dot{\vec{\mu}}(2i\Omega + \alpha + \beta) + \vec{\mu}\left(-\Omega^2 + i\Omega\alpha + \omega_o^2\right) = \vec{\varepsilon}\,\Omega^2 + \vec{g}\,e^{-i\Omega t} \tag{6.164c}$$

Now using the inverse relations of (6.159), i.e. by expressing the link between the vector m, which defines the position of the centre of the disk S in the rotating coordinate system $(\xi - \mu - \zeta)$, with the vector $\vec{z}$, that defines the position in the absolute coordinate system (X-Y-Z):

$$\begin{aligned} \vec{\mu} &= \vec{z}\,e^{-i\Omega t} \\ \dot{\vec{\mu}} &= \dot{\vec{z}}\,e^{-i\Omega t} - i\Omega\,\vec{z}\,e^{-i\Omega t} \\ \ddot{\vec{\mu}} &= \ddot{\vec{z}}\,e^{-i\Omega t} - 2i\Omega\,\dot{\vec{z}}\,e^{i\Omega t} - \Omega^2\vec{z}\,e^{-i\Omega t} \end{aligned} \tag{6.165}$$

and by replacing (6.165) in the motion equation of the rotor expressed in the rotating coordinate system (6.164c), we arrive at:

$$\begin{aligned} &\ddot{\vec{z}}\,e^{-i\Omega t} - 2i\Omega\,\dot{\vec{z}}\,e^{-i\Omega t} - \Omega^2\dot{\vec{z}}\,e^{-i\Omega t} + (2i\Omega + \alpha + \beta)\left(\dot{\vec{z}}\,e^{-i\Omega t} - i\Omega\,\vec{z}\,e^{-i\Omega t}\right) \\ &\quad - \Omega^2\vec{z}\,e^{-i\Omega t} + \omega_o^2\vec{z}\,e^{-i\Omega t} = \vec{\varepsilon}\,\Omega^2 + \vec{g}\,e^{-i\Omega t} \end{aligned} \tag{6.166a}$$

After the appropriate simplifications and by multiplying all the terms by $e^{i\Omega t}$ so as to return to the absolute coordinate system, (6.166a) becomes:

$$\ddot{\vec{z}} + (\alpha + \beta)\dot{\vec{z}} + \left(\omega_o^2 - i\Omega\beta\right)\vec{z} = \vec{g} + \vec{\varepsilon}\,\Omega^2 e^{i\Omega t} \tag{6.166b}$$

The presence of hysteretic damping, defined by the dimensionless term β (6.166b), produces two effects: on the one hand it increases the equivalent structural damping, to add to the external damping α (6.164a), on the other, it introduces an imaginary equivalent stiffness $i\Omega\beta$ proportional to the angular velocity of the shaft (6.164a). Now we will examine the effects of these terms both on the steady-state response of the system, and in its actual stability.

6.7.1.1 Studying the Steady-State Solution (Particular Integrals)

First of all we will analyse the particular integral of (6.166b) caused solely by runout of the centre of gravity of the disk (an unbalancing effect):

$$\ddot{\vec{z}} + (\alpha + \beta)\dot{\vec{z}} + (\omega_o^2 - i\Omega\beta)\vec{z} = \vec{\varepsilon}\,\Omega^2 e^{i\Omega t} \tag{6.167a}$$

By imposing the steady-state solution in (6.167a):

$$\vec{z}_\varepsilon = \vec{A}\,e^{i\Omega t} \tag{6.167b}$$

we obtain a complex algebraic equation in the unknown (complex) $\vec{A}$:

$$\left(-\Omega^2 + i\Omega(\alpha + \beta) + (\omega_o^2 - i\Omega\beta)\right)\vec{A} = \vec{\varepsilon}\,\Omega^2 \tag{6.167c}$$

So the solution of (6.167c) becomes:

$$\vec{A} = \frac{\vec{\varepsilon}\,\Omega^2}{\left(-\Omega^2 + i\Omega(\alpha + \beta) + (\omega_o^2 - i\Omega\beta)\right)} = \frac{\vec{\varepsilon}\,\Omega^2}{\omega_o^2 - \Omega^2 + i\Omega\alpha} = \left|\vec{A}\right| e^{i\phi} \tag{6.167d}$$

of which modulus and phase equal, respectively:

$$\left|\vec{A}\right| = \frac{\left|\vec{\varepsilon}\right|}{\sqrt{\left(\left(\frac{\omega_o}{\Omega}\right)^2 - 1\right)^2 + \left(2h\frac{\omega_o}{\Omega}\right)^2}}$$
$$\phi = a\tan\left(-\frac{2h\frac{\omega_o}{\Omega}}{\left(\frac{\omega_o}{\Omega}\right)^2 - 1}\right) \tag{6.167e}$$

having used $h = \frac{r}{r_c}$ to indicate the dimensionless damping ratio. As we can see, by comparing (6.167c), (6.167d) and (6.167e) with the solution of the same damped system, not considering hysteretic damping [(6.68b), Sect. 6.4.2], the two rotors show the same particular integral $\vec{z}_\varepsilon$ regardless of whether any hysteretic damping is considered or not. This result can be attributed to the fact that, in steady-state motion, the forces caused by hysteretic damping are not able to dissipate energy as the shaft assumes a constant deflection with respect to the rotating coordinate system $(\xi - \mu - \zeta)$ and so hysteretic damping, introduced by the term (6.162), cannot dissipate energy. Now we will estimate the particular integral due solely to the rotor's weight, solution of:

$$\ddot{\vec{z}} + (\alpha + \beta)\dot{\vec{z}} + (\omega_o^2 - i\Omega\beta)\vec{z} = \vec{g} \tag{6.168a}$$

The steady-state response, in this case, is defined by:

$$\vec{z}_p = \frac{\vec{g}}{(\omega_o^2 - i\Omega\beta)} = \left|\vec{z}_p\right| e^{i\psi} \tag{6.168b}$$

where the modulus and the phase equal:

$$\left|\vec{z}_p\right| = \frac{g}{\left(\omega_o^2 - i\Omega\,\beta\right)}$$
$$\psi = \operatorname{atan}\left(\frac{\Omega\,\beta}{\omega_o^2}\right) \tag{6.168c}$$

Under the action of the natural mass, due to the effect of the hysteretic damping, the centre of the disk S lowers by a quantity $\left|\vec{z}_p\right|$ that is less than static deflection. Furthermore, the deformation does not lie on the vertical, but is rotated by angle ψ in the direction of rotation of the rotor.[33] Figure 6.60 shows the trend of the modulus of static deflection, compared to the value of deflection in the absence of hysteretic damping, as a function of the ratio $\left(\frac{\Omega}{\omega_o}\right)$, upon variation of parameter β. As we can see, for high values of β or of Ω, in the presence of elastic hysteresis the modulus $\left|\vec{z}_p\right|$ tends to zero and the phase tends to 90°, i.e. the disk tends to lie horizontally.

6.7.1.2 Analysing the Stability of the System

Now we will analyse the stability of the motion of the rotor under consideration: to do this, we must study the solutions of the homogenous equation:

$$\ddot{\vec{z}} + (\alpha + \beta)\dot{\vec{z}} + \left(\omega_o^2 - i\Omega\,\beta\right)\vec{z} = 0 \tag{6.169a}$$

Placing a possible solution in (6.169a):

$$\vec{z} = x + iy \tag{6.169b}$$

the same motion equation becomes:

$$(\ddot{x} + i\ddot{y}) + (\alpha + \beta)(\dot{x} + i\dot{y}) + \left(\omega_o^2 - i\Omega\,\beta\right)(x + iy) = 0 \tag{6.169c}$$

which can be brought back to two scalar equations of the form:

$$\begin{aligned}\ddot{x} + (\alpha + \beta)\dot{x} + \left(\omega_o^2 - i\Omega\,\beta\right)x = 0\\ \ddot{y} + (\alpha + \beta)\dot{y} + \left(\omega_o^2 - i\Omega\,\beta\right)y = 0\end{aligned} \tag{6.170}$$

[33] In particular, we can see that we can obtain parameter β from the experimental measurement of angle ψ.

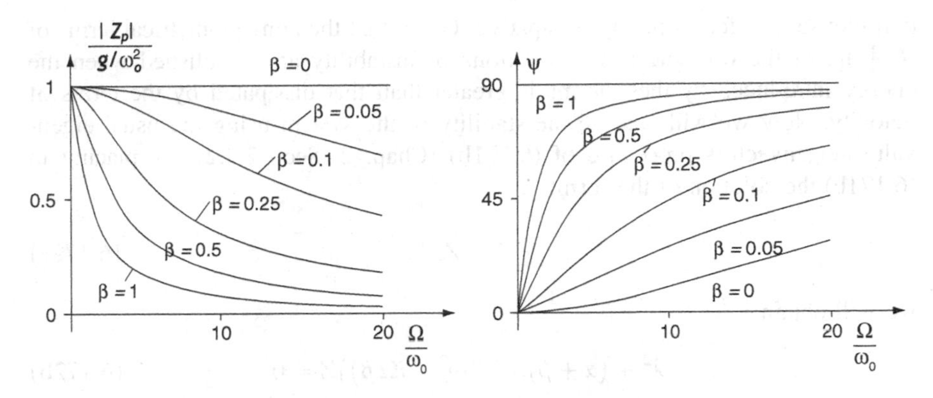

Fig. 6.60 Trend of the static deflection in the presence of hysteretic damping

By grouping the independent variables x and y in vector $\underline{z}$:

$$\underline{z} = \left\{ \begin{array}{c} x \\ y \end{array} \right\} \tag{6.171a}$$

the equations of motion in scalar form (6.170) can be rewritten in matrix form as:

$$[M]\ddot{\underline{z}} + [R]\dot{\underline{z}} + [K]\underline{z} = \underline{0} \tag{6.171b}$$

where $[M]$ is the mass matrix:

$$[M] = \begin{bmatrix} 1 & 0 \\ 0 & 1 \end{bmatrix} \tag{6.171c}$$

$[R]$ is the damping matrix:

$$[R] = \begin{bmatrix} r_{xx} & r_{xy} \\ r_{yx} & r_{yy} \end{bmatrix} = \begin{bmatrix} \alpha + \beta & 0 \\ 0 & \alpha + \beta \end{bmatrix} \tag{6.171d}$$

and $[K]$ is that of stiffness, as the sum of two terms:

$$[K] = \begin{bmatrix} k_{xx} & k_{xy} \\ k_{yx} & k_{yy} \end{bmatrix} = [K_s] + [K_F] = \begin{bmatrix} \omega_o^2 & 0 \\ 0 & \omega_o^2 \end{bmatrix} + \begin{bmatrix} 0 & \Omega\beta \\ -\Omega\beta & 0 \end{bmatrix} = \begin{bmatrix} \omega_o^2 & \Omega\beta \\ -\Omega\beta & \omega_o^2 \end{bmatrix} \tag{6.171e}$$

As we can see, the matrix $[K]$, which takes into account both the elastic terms $[K_s]$ and those $[K_F]$ due to the positional terms of the force field that is generated by the effect of elastic hysteresis, is, therefore, non-symmetrical with extra-diagonal terms of opposite sign: so the system may become unstable. However, due to the terms of velocity of the hysteretic force field, matrix $[R]$ is symmetrical and positive definite: so its effect is purely dissipative. Given that the non-symmetrical terms of $[K_F]$ are of the opposite sign, conditions of instability are established when the energy introduced by these terms is greater than that dissipated by the terms of velocity. Now we will analyse the stability of the system using the usual eigen-values-eigenvectors procedure of (6.171b) (Chap. 2, Sect. 2.4.2): by placing in (6.171b) the solution of the form:

$$\vec{z} = Ze^{\lambda t} \tag{6.172a}$$

we will obtain:

$$\left(\lambda^2 + (\alpha + \beta)\lambda + \left(\omega_o^2 - i\Omega\beta\right)\right)Z = 0 \tag{6.172b}$$

an equation that allows non-trivial solutions when:

$$\left(\lambda^2 + (\alpha + \beta)\lambda + \left(\omega_o^2 - i\Omega\beta\right)\right) = 0 \tag{6.173a}$$

Solutions of the Eq. (6.173a) are:

$$\lambda_{1,2} = -\frac{(\alpha + \beta)}{2} \pm \sqrt{\left(\frac{(\alpha + \beta)}{2}\right)^2 - \omega_o^2 + i\Omega\beta} \tag{6.173b}$$

which, once the natural pulsation of the damped system has been defined, as:

$$\omega_d = \omega_o^2 - \frac{(\alpha + \beta)}{2} \tag{6.173c}$$

become:

$$\lambda_{1,2} = -\frac{(\alpha + \beta)}{2} \pm i\sqrt{\omega_d^2 - i\Omega\beta} \tag{6.173d}$$

By developing the square roots that appear in the expression (6.173d) in series, β being small:

$$\sqrt{\omega_d^2 - i\Omega\beta} = \omega_d - \frac{1}{2} i \frac{\Omega\beta}{\omega_d} \tag{6.174a}$$

we obtain:

$$\begin{aligned} \lambda_1 &= \left(-\frac{(\alpha + \beta)}{2} + \frac{\Omega\beta}{2\omega_d}\right) + i\omega_d = \alpha_1 + i\omega_d \\ \lambda_2 &= \left(-\frac{(\alpha + \beta)}{2} - \frac{\Omega\beta}{2\omega_d}\right) - i\omega_d = \alpha_2 - i\omega_d \end{aligned} \tag{6.174b}$$

As α and β are definitely positive (6.164a), the following is always true:

$$\alpha_2 = -\frac{(\alpha+\beta)}{2} - \frac{\Omega\beta}{2\omega_d} < 0 \tag{6.175a}$$

and the solution λ_2, therefore, represents a stable motion, i.e. decreasing over time, oscillating with pulsation ω_d. The solution:

$$\lambda_1 = \alpha_1 + i\,\omega_d \tag{6.175b}$$

could, however, give an unstable expansive solution if:

$$\alpha_1 = -\frac{(\alpha+\beta)}{2} + \frac{\Omega\beta}{2\omega_d} > 0 \tag{6.175c}$$

i.e. for:

$$\Omega > \omega_d \frac{(\alpha+\beta)}{\beta} \tag{6.175d}$$

Therefore, the relation (6.175d) defines the range of stability of the rotor being analysed while taking into account the presence of internal hysteresis: if the hysteretic damping is zero $\beta = 0$ there is no kind of instability. The presence of hysteretic damping introduces destabilising effects at angular speeds Ω that are higher than $\omega_d \frac{(\alpha+\beta)}{\beta}$ i.e. for speeds higher than the first critical speed, whereas at lower values it leads to a dissipation of energy. To analyse the behaviour of the system it is also possible to use the following energetic procedure, which is approximate but closer to the physical problem, based on the hypothesis of perturbing the shaft around the steady-state position and forcing it to oscillate freely with its natural pulsation ω_o[34]; lastly, we assume that the trajectory of the centre of gravity of the mass disk m is circular, that is, that the perturbed motion is defined by the coordinates:

$$\begin{aligned} x &= A\cos\omega_o t \\ y &= A\sin\omega_o t \end{aligned} \tag{6.176}$$

Now we will consider the work performed by the nonconservative positional force field, that generates matrix $[K_F]$ and that performed by the dissipative terms, linked to matrix $[R]$ (the work performed by the forces of inertia in a cycle is, as we know, zero): by equalising these two expressions we will obtain the boundary condition of instability. L_f being the work carried out by the positional forces:

[34]This perturbation is that which, for example, would occur due to the effect of the unbalance at the passage of the critical speed, i.e. for $\Omega = \omega_o$.

$$F_x = k_{xx}x + k_{xy}y$$
$$F_y = k_{yx}x + k_{yy}y \tag{6.177a}$$

in a cycle:

$$L_f = \int_0^X F_x dx + \int_0^Y F_y dy = \int_0^T F_x \dot{x} dt + \int_0^T F_y \dot{y} dt \tag{6.177b}$$

Taking into account (6.171e), (6.177b) becomes:

$$L_f = -2\Omega A^2 \pi \beta \tag{6.177c}$$

The work L_d of the dissipative forces is, however:

$$L_d = \int_0^X F_{\dot{x}} dx + \int_0^Y F_{\dot{y}} dy = \int_0^T F_{\dot{x}} \dot{x} dt + \int_0^T F_{\dot{y}} \dot{y} dt \tag{6.178a}$$

since:

$$F_{\dot{x}} = r_{xx}\dot{x} + r_{xy}\dot{y}$$
$$F_{\dot{y}} = r_{yx}\dot{x} + r_{yy}\dot{y} \tag{6.178b}$$

which, again, taking into account (6.171d), becomes:

$$L_d = 2(\alpha + \beta) A^2 \pi \omega_o \tag{6.178c}$$

By establishing $L_f > L_d$, from (6.177c) and (6.178c) we obtain:

$$\Omega > \omega_d \frac{(\alpha + \beta)}{\beta} \tag{6.179}$$

With this energetic approach we also arrive at the same expression (6.175d) already obtained previously with the usual eigenvalue-eigenvector procedure.

6.8 The Gyroscopic Effect

In the previous sections we neglected the moments of inertia caused by the rotation and deflection of the axis of the rotor: if we wish to take into account the moments of inertia, we must now define the geometry of the weights.

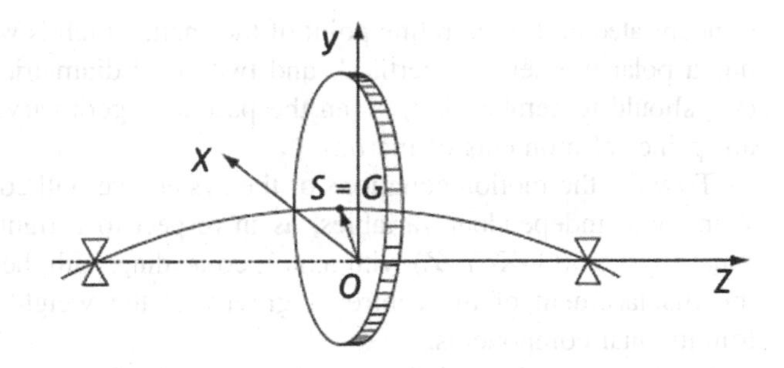

Fig. 6.61 The Jeffcott rotor

The most common case that is of practical interest, is that of a rotor which can usually be considered a solid of revolution with the axis coinciding with the longitudinal axis of the shaft: so this rotor has the main axis of inertia coinciding with the axis of the undeformed shaft and moments of inertia that are equal in respect to the other two main axes. To examine the problem, we can consider a simplified model as shown in Fig. 6.61, as already mentioned: in literature this model is often referred to as the Jeffcott rotor. In this case, if we wish to highlight the gyroscopic effects, we now assume that the disk, in addition to moving perpendicularly to the axis of rotation, can also rotate as shown in Fig. 6.62.

In this case, in addition to the rotation component Ω, the absolute angular velocity of the disk has another two components around axes that are perpendicular to axis Z.

We will now consider the rotor mounted on two rigid supports with a constant angular velocity Ω imposed by the motor. We assume that weight M (with the centre of gravity centred with respect to the geometric centre of the disk) is

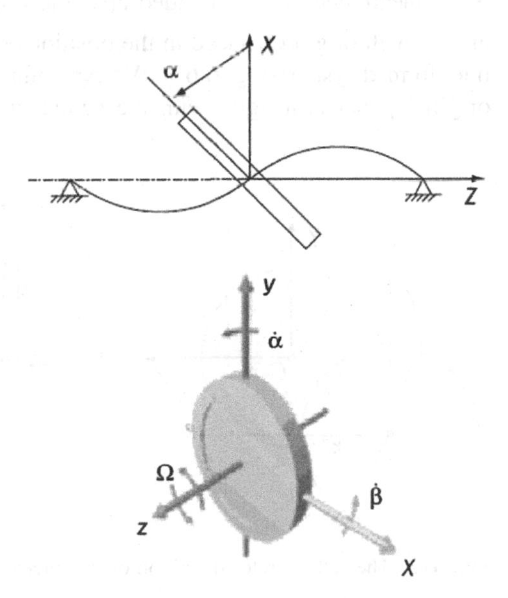

Fig. 6.62 The Jeffcott rotor - gyroscopic effect: independent variables adopted

concentrated at the centreline point of the shaft, which is without weight, and that it has a polar moment of inertia J_1 and two equal diametrical moments of inertia J_2 (we should remember that, given the particular geometry, all the diametrical axes are principal moments of inertia).

To write the motion equations of the system we will consider the displacements x and y as independent variables, as in respect to a right-handed Cartesian coordinate system (O–X–Y–Z) with axis Z coinciding with the rotation axis they define the displacement of the centre of gravity of the weight M while neglecting the longitudinal components.

Furthermore, the disk has another three d.o.f. associated with the three components of rotation:

- the rotation of the disk around axis Z, with assigned angular velocity Ω assumed to be constant, represents a constrained d.o.f.;
- the other two components α and β represent, together with x and y, the independent variables needed to describe the motion of the system in question.

More in particular, the coordinates x and y define the perturbed motion around the system's position of static equilibrium: thus the weight of the rotor will not appear in the motion equations. (Chap. 1, Sects. 1.4 and 1.5).

To define the motion equations we will use the Lagrange equation, adopting several triplets of reference in order to write, directly and conveniently, the various forms of energy of the system. In this example it is useful to introduce the matrix of multi-body systems (Chap. 1, Sects. 1.6 and 1.7) to define the kinematics of the system and, consequently, to define its dynamics.

To describe the kinematics of the system in the discussion that follows, it is easier to introduce several reference triplets: the first reference triplet (O–X–Y–Z) is, as mentioned, a right-handed absolute Cartesian system of vector units $\vec{i}$, $\vec{j}$ and $\vec{k}$ with origin O placed in the position occupied by the centre of gravity in the undeformed system (Fig. 6.63). We can define a second triplet (O_1–X_1–Y_1–Z_1) with origin O_1 that is integral with the centre of gravity G of the weight M and unit

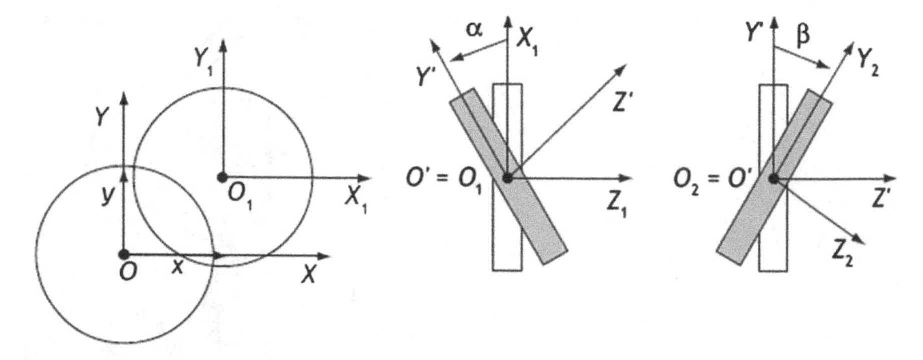

Fig. 6.63 The Jeffcott rotor definition of the reference triplets adopted

vectors $\vec{i}_1 = \vec{i}$, $\vec{j}_1 = \vec{j}$ and $\vec{k}_1 = \vec{k}$ that translate parallel to the absolute triplet: we assume the components x and y of the displacement of the origin O_1 with respect to the absolute triplet as the first two independent variables. Lastly, we consider a third reference triplet (O_2–X_2–Y_2–Z_2) that is integral with the disk for all the motions except for where the rotation defined by the vector Ω is concerned: the reason for this choice is linked to the fact that all the axes perpendicular to that axis are principal moments of inertia. The triplet (O_2–X_2–Y_2–Z_2) has origin O_2 that coincides with O_1 and axes defined by unit vectors $\vec{i}_2$, $\vec{j}_2$ and $\vec{k}_2$.

In order to fully define the motion of the weight M, we first need to define the angular position assumed by the triplet (O_2–X_2–Y_2–Z_2) with respect to the triplet (O_1–X_1–Y_1–Z_1) that is, with respect to the absolute triplet (O–X–Y–Z). The angular position of the triplet (O_2–X_2–Y_2–Z_2) will be defined by means of Cardan angles (Chap. 1, Sects. 1.6.4 and 1.6.5) which we indicate with the independent variables α and β, which define the angular position of the disk.

In the discussion, which is necessary in defining the kinematics of the disk, it is convenient to introduce an additional reference triplet of convenience $(O'-Z'-Y'-X')$ with origin $O' = O_1 = O_2$ rotated by an angle α with respect to the translating triplet $(O_1-Z_1-Y_1-X_1)$ (Fig. 6.63) and axis $Y' = Y\left(\vec{j}' = \vec{j}_1\right)$.

A subsequent rotation β around the axis $X' = X_2\left(\vec{k} = \vec{k}_2\right)$ makes it possible to define the angular position of the triplet $(O_2-Z_2-Y_2-X_2)$ that is integral with weight M, with respect to the absolute triplet.

We will now write the motion equations for the Jeffcott rotor, while taking into account the angular deflection that may affect the disk, using the Lagrange equations.

The kinetic energy of the system (see Chap. 1, Sect. 1.7) expressed in matrix form, can be defined as

$$E_c = \frac{1}{2}\underline{V}^T[M]\underline{V} + \frac{1}{2}\underline{\omega}_2^T[J]\underline{\omega}_2 \tag{6.180}$$

$\underline{V}$ being the geometric vector that defines the absolute speed of the centre of gravity of the weight M:

$$\underline{V} = \underline{\mathrm{h}}^{\mathrm{T}}\,\dot{\underline{y}} = \begin{Bmatrix} \vec{i} \\ \vec{j} \\ \vec{k} \end{Bmatrix}^{\mathrm{T}} \begin{Bmatrix} 0 \\ \dot{x} \\ \dot{y} \end{Bmatrix} \tag{6.181}$$

and $[M]$ being the mass matrix of the system:

$$[M] = \begin{bmatrix} M & 0 & 0 \\ 0 & M & 0 \\ 0 & 0 & M \end{bmatrix} \tag{6.182}$$

In (6.180) $\underline{\omega}_2$

$$\underline{\omega}_2 = \underline{h}_2^T \underline{\omega}_2 \tag{6.183}$$

represents the vector angular velocity of the disk defined via the components ω_{2x}, ω_{2y} and ω_{2z} projected on triplet (O_2–X_2–Y_2–Z_2) which is integral with the actual weight and axes parallel to the principal axes of inertia of the body:

$$\underline{\omega}_2 = \begin{Bmatrix} \omega_{2x} \\ \omega_{2y} \\ \omega_{2z} \end{Bmatrix} \tag{6.184}$$

and $[J]$ being the inertia tensor which, given the choice of the physical coordinates adopted, becomes constant and diagonal (Sect. 1.7.5):

$$[J] = \begin{bmatrix} J_1 & 0 & 0 \\ 0 & J_2 & 0 \\ 0 & 0 & J_2 \end{bmatrix} \tag{6.185}$$

We must now define the physical variables $\underline{\omega}_2$ as a function of the independent variables assumed, i.e. of the two components x and y of the displacement of the centre of gravity G with respect to the absolute triplet (O –X – Y – Z) and the two Cardan angles α and β, defined previously: for this purpose, we will now define the absolute geometric vector angular velocity $\vec{\omega}$ as the vector sum of the three components of angular velocity[35]:

$$\vec{\omega} = \vec{j}' \dot{\alpha} + \vec{k}_2 \dot{\beta} + \vec{i}_2 \Omega \tag{6.186}$$

which can be expressed in matrix form, using the algorithm just introduced, as:

$$\vec{\omega} = \begin{Bmatrix} \vec{i}' \\ \vec{j}' \\ \vec{k}' \end{Bmatrix}^{\mathrm{T}} \begin{Bmatrix} 0 \\ \dot{\alpha} \\ 0 \end{Bmatrix} + \begin{Bmatrix} \vec{i}_2 \\ \vec{j}_2 \\ \vec{k}_2 \end{Bmatrix}^{\mathrm{T}} \begin{Bmatrix} \Omega \\ 0 \\ \dot{\beta} \end{Bmatrix} = \underline{h}'^T \underline{\omega}' + \underline{h}_2^T \underline{\omega}_{2rel} \tag{6.187}$$

The vector angular velocity is thus expressed as the vector sum of two different terms expressed in two different bases; when analysing expression (6.187) it is necessary to define the components of this speed with respect to the triplet that is

[35]Versus $\vec{j}'$, $\vec{k}_2$ and $\vec{i}_2$ attributed to components $\dot{\alpha}$, $\dot{\beta}$ and Ω derives from the definition of the reference triplets assumed and the definition of the Cardan angles assumed (see Fig. 6.63).

integral with the rotor (O_2–X_2–Y_2–Z_2). For this purpose it is necessary to use the coordinate transformation matrix (see Chap. 1, Sect. 1.6) by rewriting (6.187) as:

$$\begin{aligned}\vec{\omega} &= \underline{h}_2^T[\Lambda_{'2}]\underline{\omega}' + \underline{h}_2^T\underline{\omega}_{2rel} = \underline{h}_2^T[\Lambda_{2'}]^T\underline{\omega}' \\ &+ \underline{h}_2^T\underline{\omega}_{2rel} = \underline{h}_2^T\left\{[\Lambda_{2'}]^T\underline{\omega}' + \underline{\omega}_{2rel}\right\} = \underline{h}_2^T\underline{\omega}_2\end{aligned} \tag{6.188}$$

To estimate the components $\underline{\omega}_2$ (physical variables adopted in the expression of the kinematic energy (6.188) as a function of the independent variables) we must, therefore, define the coordinate transformation matrix $[\Lambda_{2'}]$ (see Chap. 1, Sects. 1.2.4 and 1.2.5):

$$[\Lambda_2'] = \begin{bmatrix} \cos\beta & -\sin\beta & 0 \\ \sin\beta & \cos\beta & 0 \\ 0 & 0 & 1 \end{bmatrix} \tag{6.189}$$

Considering the transformation matrix of the coordinates, the components projected $\underline{\omega}_2$ on the triplet that is integral with the body (6.188) become:

$$\begin{aligned}\underline{\omega}_2 &= \begin{Bmatrix} \omega_{z2} \\ \omega_{y2} \\ \omega_{x2} \end{Bmatrix} = \left\{[\Lambda_{2'}]^T\underline{\omega}' + \underline{\omega}_{2rel}\right\} \\ &= \left\{\begin{bmatrix} \cos\beta & -\sin\beta & 0 \\ \sin\beta & \cos\beta & 0 \\ 0 & 0 & 1 \end{bmatrix}\begin{Bmatrix} 0 \\ \dot{\alpha} \\ 0 \end{Bmatrix} + \begin{Bmatrix} \Omega \\ 0 \\ \dot{\beta} \end{Bmatrix}\right\} = \begin{Bmatrix} -\sin\beta\,\dot{\alpha} + \Omega \\ \cos\beta\,\dot{\alpha} \\ \dot{\beta} \end{Bmatrix}\end{aligned} \tag{6.190}$$

This expression makes it possible to clarify the physical velocity variables $\underline{\omega}_2$ (see also Chap. 1, Sect. 1.6.5) as a function of the independent variables organised in the vector:

$$\underline{q}_\theta = \begin{Bmatrix} \theta_r \\ \alpha \\ \beta \end{Bmatrix} \tag{6.191}$$

since $\theta_r = \Omega t$, as:

$$\underline{\omega}_2 = \begin{Bmatrix} -\sin\beta\,\dot{\alpha} + \Omega \\ \cos\beta\,\dot{\alpha} \\ \dot{\beta} \end{Bmatrix} = \begin{bmatrix} 1 & -\sin\beta & 0 \\ 0 & \cos\beta & 0 \\ 0 & 0 & 1 \end{bmatrix}\begin{Bmatrix} \Omega \\ \dot{\alpha} \\ \dot{\beta} \end{Bmatrix} = \left[A\left(\underline{q}_\theta\right)\right]\underline{\dot{q}}_\theta \tag{6.192}$$

$\left[A\left(\underline{q}_\theta\right)\right]$ being the Jacobian matrix, function of the same independent variables, which correlates the physical variables $\underline{\omega}_2$ to the independent variables $\underline{\dot{q}}_\theta$. The expression of the kinematic energy as a function of the independent variables becomes:

$$E_c = \frac{1}{2}\,\underline{\dot{x}}^T\,[M]\,\underline{\dot{x}} + \frac{1}{2}\,\underline{\dot{q}}_\theta^T\left[A\left(\underline{q}_\theta\right)\right]^T[J]\,\left[A\left(\underline{q}_\theta\right)\right]\underline{\dot{q}}_\theta \tag{6.193}$$

By applying Lagrange equations directly to this energy, the Jacobian matrix $\left[A\left(\underline{q}_\theta\right)\right]$ being a function, in turn, of the same independent variables, we will arrive at nonlinear expressions of the generalised forces of inertia. When developing the expressions (6.193) in scalar form, we will have:

$$E_c = \frac{1}{2}\,M\,\dot{x}^2 + \frac{1}{2}\,M\,\dot{y}^2 + \frac{1}{2}\,J_1(\Omega - \sin\beta\dot{\alpha})^2 + J_2(\cos\beta\dot{\alpha})^2 + J_2\dot{\beta}^2 \tag{6.194}$$

If we wish to study the small oscillations around the position of steady-state equilibrium, i.e. estimate the stability of perturbed motion, it is possible, by assuming small Cardan angles, to confuse the sine with the angle and the cosine with unit, to obtain:

$$E_c = \frac{1}{2}\,M\,\dot{x}^2 + \frac{1}{2}\,M\,\dot{y}^2 + \frac{1}{2}\,J_1(\Omega - \beta\dot{\alpha})^2 + J_2(\dot{\alpha})^2 + J_2\dot{\beta}^2 \tag{6.195}$$

In this case it is not possible to neglect the term $\beta\dot{\alpha}$ since it is summed with a finite term, i.e. the angular velocity of the rotor Ω: upon performing squaring $(\Omega - \beta\dot{\alpha})^2$ in fact, the product $-2\Omega\beta\dot{\alpha}$ remains squared. By applying Lagrange in a scalar manner, i.e. by deriving the kinetic energy with respect to the independent variables and their derivatives, we obtain the generalised forces of inertia, while neglecting the infinitesimals of higher order:

$$\begin{cases} Q_{ix} = M\,\ddot{x} \\ Q_{iy} = M\,\ddot{y} \\ Q_{i\alpha} = -J_1\Omega\dot{\beta} + J_2\ddot{\alpha} \\ Q_{i\beta} = J_1\Omega\dot{\alpha} + J_2\ddot{\beta} \end{cases} \tag{6.196}$$

The potential energy of the system V can be subsequently defined as:

$$V = \frac{1}{2}\begin{Bmatrix} x \\ y \\ \Theta_y \\ \Theta_x \end{Bmatrix}^T \begin{bmatrix} k_{11} & & k_{12} & \\ & k_{11} & & k_{12} \\ k_{21} & & k_{22} & \\ & k_{21} & & k_{21} \end{bmatrix} \begin{Bmatrix} x \\ y \\ \Theta_y \\ \Theta_x \end{Bmatrix} \tag{6.196a}$$

having used Θ_y and Θ_x to indicate the rotations around the axes X and Y (unit vectors $\vec{i}$ and $\vec{k}$) of the absolute reference system. In (6.196a) the generic term k_{ij} represents the stiffness of the shaft, i.e.:

- k_{11} the force created on the shaft due to the effect of a unitary displacement imposed on it;
- $k_{12} = k_{21}$ the force created on the shaft due to the effect of an imposed unitary rotation[36];
- k_{22} the torque created on the shaft due to the effect of a unitary rotation.

Rotations Θ_y and Θ_x do not correspond, for the more general case of large displacements, to the independent variables α and β assumed to define the rotation of the body M (which instead define, respectively, the rotations around the axes defined by unit vectors $\vec{j} = \vec{j}_1 = \vec{j}'$ and $\vec{k}_2$). However, in the case of small rotations, which can therefore be compared to infinitesimals, the actual rotations Θ_y and Θ_x become strictly:

$$\begin{aligned} \Theta_y &= \alpha \\ \Theta_x &= \beta \end{aligned} \tag{6.196b}$$

This physically coincides with the fact that, if the rotations are small, the axes around which the rotations connected with the Cardan variables occur can be confused with the initial axes. The potential energy V of (6.216a) can then be expressed solely as a function of the independent variables as:

$$V = \frac{1}{2}\begin{Bmatrix} x \\ y \\ \alpha \\ \beta \end{Bmatrix}^T \begin{bmatrix} k_{11} & 0 & k_{12} & 0 \\ 0 & k_{11} & 0 & k_{12} \\ k_{21} & & k_{22} & 0 \\ 0 & k_{21} & 0 & k_{21} \end{bmatrix} \begin{Bmatrix} x \\ y \\ \alpha \\ \beta \end{Bmatrix} \tag{6.196c}$$

and the corresponding generalised elastic forces become:

$$\begin{cases} Q_{ex} = k_{11}x + k_{12}\alpha \\ Q_{ey} = k_{11}y + k_{12}\beta \\ Q_{e\alpha} = k_{21}x + k_{22}\alpha \\ Q_{e\beta} = k_{21}y + k_{22}\beta \end{cases} \tag{6.196d}$$

Considering (6.195) and (6.196d) it is possible to write the equations of motion of the system in scalar form:

$$\begin{cases} M\ddot{x} + k_{11}x + k_{12}\alpha = 0 \\ M\ddot{y} + k_{11}y + k_{12}\beta = 0 \\ -J_1\Omega\dot{\beta} + J_2\ddot{\alpha} + k_{21}x + k_{22}\alpha = 0 \\ J_1\Omega\dot{\alpha} + J_2\ddot{\beta} + k_{21}y + k_{22}\beta = 0 \end{cases} \tag{6.197}$$

[36]If the disk (Fig. 6.59) is positioned on the centreline of a symmetrical shaft, we have $k_{21} = k_{12} = 0$.

which in matrix form, having gathered the independent variables of the system in vector $\underline{x}$:

$$\underline{x} = \begin{Bmatrix} x \\ y \\ \alpha \\ \beta \end{Bmatrix} \tag{6.198a}$$

become:

$$[M]\underline{\ddot{x}} + [R]\underline{\dot{x}} + [K]\underline{x} = \underline{0} \tag{6.198b}$$

having indicated, with the usual nomenclature, with $[M]$:

$$[M] = \begin{bmatrix} M & 0 & 0 & 0 \\ 0 & M & 0 & 0 \\ 0 & 0 & J_2 & 0 \\ 0 & 0 & 0 & J_2 \end{bmatrix} \tag{6.198c}$$

the mass matrix of the system, with $[R]$:

$$[R] = \begin{bmatrix} 0 & 0 & 0 & 0 \\ 0 & 0 & 0 & 0 \\ 0 & 0 & 0 & -\Omega J_1 \\ 0 & 0 & \Omega J_1 & 0 \end{bmatrix} \tag{6.198d}$$

the damping matrix and, lastly, with $[K]$

$$[K] = \begin{bmatrix} k_{11} & 0 & k_{12} & 0 \\ 0 & k_{11} & 0 & k_{12} \\ k_{21} & & k_{22} & 0 \\ 0 & k_{21} & 0 & k_{21} \end{bmatrix} \tag{6.198e}$$

the stiffness matrix. As we can see, the matrices of mass and of stiffness are obviously symmetrical and positive definite, while the damping matrix, due to the gyroscopic terms, is non-symmetrical and a function of the angular velocity of the rotor Ω. To estimate the natural frequencies of the system it is possible to attribute, as usual, this calculation to evaluation of the eigenvalues and eigenvectors. However, be exploiting the polar symmetry of the rotor, a more compact discussion can be made by resorting to the algorithm of complex numbers [7]. As done previously in Sect. 6.3, it is possible to associate a Guassian plane with plane Y-X where the centre of gravity G of the mass moves (integrally with the origins O_1 and O_2 of the mobile reference systems), by attributing (quite arbitrarily) axis X with the significance of real axis and Y that of imaginary axis: in this way, the displacement of the centre of gravity $G = O_1 = O_2$ instead of being described by the two components according to the axes:

$$\begin{aligned} x &= x(t) \\ y &= y(t) \end{aligned} \tag{6.199a}$$

will be defined by vector z (complex number):

$$z = x + iy \tag{6.199b}$$

In the same way, as far as the rotations α and β are concerned, it is possible to use just one complex number γ:

$$\gamma = \alpha + i\beta \tag{6.199c}$$

the module of which represents the angle formed between the local axis of the disk Z_2 and axis Z_1 (parallel to axis Z). Now by multiplying the second and fourth equation by the imaginary unit $i = \sqrt{-1}$ and summing, respectively, the first and third equation, by grouping the terms we obtain, in complex terms:

$$\begin{cases} M\ddot{z} + k_{11}z + k_{12}\gamma = 0 \\ -iJ_1\Omega\dot{\gamma} + J_2\ddot{\gamma} + k_{21}z + k_{22}\gamma = 0 \end{cases} \tag{6.200}$$

In this way, the two complex equations are coupled just in the elastic terms $k_{21} = k_{12}$. The solution of (6.196) then becomes of the form:

$$\begin{aligned} z &= Ze^{i\lambda t} \\ \gamma &= \Gamma^{i\lambda t} \end{aligned} \tag{6.201a}$$

which replaced in (6.200) lead to the following homogenous complex algebraic system:

$$\begin{cases} (k_{11} - \lambda^2 M)Z + k_{12}\Gamma = 0 \\ k_{21}Z + (k_{22} + J_1\Omega\lambda - \lambda^2 J_2)\Gamma = 0 \end{cases} \tag{6.201b}$$

The non-trivial solutions of (6.201b) are those that annul the determinant of the coefficient matrix, i.e.:

$$|\Delta(\lambda, \Omega)| = \det\begin{vmatrix} (k_{11} - \lambda^2 M) & k_{12} \\ k_{21} & (k_{22} + J_1\Omega\lambda - \lambda^2 J_2) \end{vmatrix} = 0 \tag{6.201c}$$

i.e.:

$$|\Delta(\lambda, \Omega)| = \left(k_{11} - \lambda^2 M\right)\left(k_{22} + J_1\Omega\lambda - \lambda^2 J_2\right) - k_{21}^2 \tag{6.201d}$$

(6.201d) provides four values of λ, as a function of the angular velocity Ω: these values are purely imaginary and can be obtained by placing (6.201d) in the form:

$$\Omega = \frac{\left(k_{11} - k_{12}\right)\left(k_{22} - \lambda^2 J_2\right) - k_{21}^2}{J_1\lambda(k_{11} - Mv)} \tag{6.202a}$$

and plotting the curve $\Omega = \Omega(\lambda)$ (Fig. 6.64): the 4 curve branches start from the values:

$$\lambda_{1o},\ \lambda_{2o} = -\lambda_{1o},\ \lambda_{3o},\ \lambda_{4o} = -\lambda_{3o} \tag{6.202b}$$

which are obtained by setting the numerator of (6.202a) [7] to zero:

$$\lambda_{1,2,3,4} = \pm\sqrt{\frac{1}{2}\left(\frac{k_{11}}{M} + \frac{k_{22}}{J_2}\right) \pm \sqrt{\frac{1}{4}\left(\frac{k_{11}}{M} - \frac{k_{22}}{J_2}\right)^2 + \left(\frac{k_{12}^2}{MJ_2}\right)}} \tag{6.202c}$$

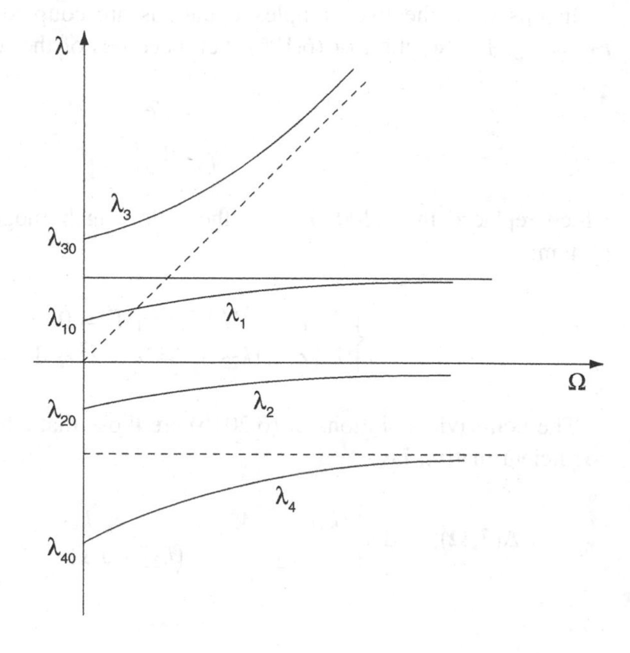

Fig. 6.64 Trend of the 4 solutions in presence of the gyroscopic effect

These values, with two by two coincidents, represent the natural pulsations of the free vibratory motion of the rotor at $\Omega = 0$. As the angular velocity of the rotor increase:

- λ_1 grows, tending to a straight line $\lambda = \sqrt{\frac{k_{11}}{M}}$;
- λ_2 lessens, tending to 0;
- λ_3 grows, tending to a straight line $\lambda = \frac{J_1}{J_2}\Omega$;
- λ_4 grows, tending to a straight line $\lambda = -\sqrt{\frac{k_{11}}{M}}$.

Therefore, the general integral will be given by:

$$\begin{aligned} z &= Z_1 e^{i\lambda_1 t} + Z_2 e^{i\lambda_2 t} + Z_3 e^{i\lambda_3 t} + Z_4 e^{i\lambda_4 t} \\ \gamma &= \Gamma_1 e^{i\lambda_1 t} + \Gamma_2 e^{i\lambda_2 t} + \Gamma_3 e^{i\lambda_3 t} + \Gamma_4 e^{i\lambda_4 t} \end{aligned} \tag{6.203}$$

The solutions λ_1 and λ_3 represent circular motions that comply with the angular velocity Ω and for this reason they will be defined as forward precession motions, while the solutions λ_2 and λ_4 represent backward precession motions: the resultant motion is, therefore, in general elliptic in the plane X–Y. Associated with the four natural pulsations of the rotor we can define 4 vibration modes. Because of the variability of the solutions with the angular velocity of the rotor, the phenomenon of the bending critical speeds arises in a more complex form, similar to that defined in presence of hydrodinamically lubricated bearings (Sect. 6.2). Upon analysing the system forced with an excitation force $F_o e^{i\Omega t}$ the conditions of resonance can be defined by intersecting the curves of Fig. 6.65 with the curve $\lambda = \Omega$: the corresponding values of Ω represent the critical speeds of the rotor. As we can see from Fig. 6.65, the critical speeds can be one or two depending on whether the asymptote $\lambda = {}^{J_1}/_{J_2}\Omega$ is more or less inclined than the bisectrix $\lambda = \Omega$: more in particular, we will have just 1 critical speed for $J_1 > J_2$ and 2 critical speeds for $J_1 < J_2$.

6.9 Oil-Film Instability

Studying the dynamic behaviour of rotating machines includes, as an essential stage of design, verifying the dynamic stability of the complete shaft line. One of the phenomena that can give rise to conditions of instability is, as already seen in Sect. 5.3.2.2, that due to the presence of hydrodinamically lubricated bearings which is known as *oil-film instability*. In Sect. 6.2.2.1 the forces that the oil-film transmits both to the rotor and to the stator, in *static* conditions (rotating shaft with no transversal vibrations) and in *dynamic* conditions (rotating shaft subject to transversal vibrations) have already been analysed. Knowing these forces makes it possible to study the conditions of stability, which can be carried out on a simplified 2 d.o.f. model, in the case of a shaft with two identical supports (as shown in

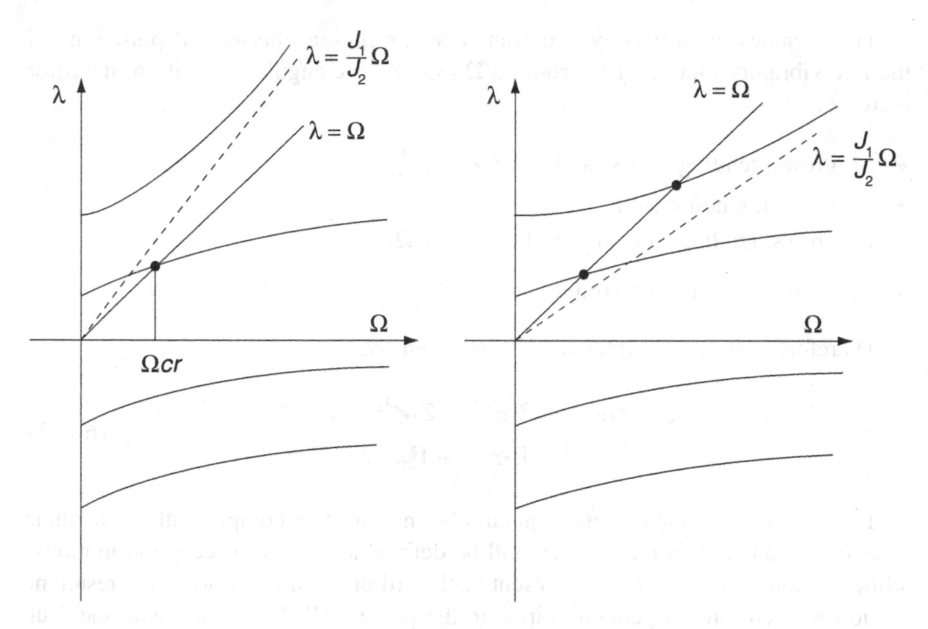

Fig. 6.65 Critical speeds in a rotor considering the gyroscopic terms

Chap. 5, Sect. 5.3.2.2). Wherever the rotor is equipped with different bearings with different characteristics, it is possible, with a more meticulous approach, to study the system using the model with several d.o.f., described in Sect. 6.2.5. The forces that the shaft and support exchange via the oil-film assume, as seen, nonlinear expressions of the variables that define the relative journal-bearing position (Sect. 6.2.2). To analytically describe the nonconservative force field due to the oil-film, we linearised the expression of these forces around the position of static equilibrium defined for a given angular velocity Ω, to obtain for each bearing, as a consequence, a non-symmetrical equivalent elastic matrix and a symmetrical damping matrix, positive definite, the effect of which is purely dissipative (although it decreases as the angular velocity Ω increases). As described in Sect. 6.2.5, by introducing the stiffnesses K_{xx}, K_{xy}, K_{yx} and K_{yy} and the equivalent dampings R_{xx}, R_{xy}, R_{yx} and R_{yy}, we obtain a motion equation of the entire shaft + casing + foundation + oil-film system of the form:

$$[M]\ddot{\underline{x}} + [R]\dot{\underline{x}} + [K]\underline{x} = \underline{0} \tag{6.204}$$

Adding the contribution of the oil-film causes not just the non-symmetry of the overall stiffness matrix $[K]$, but also the presence of terms of coupling between the vertical and horizontal displacements of the rotor: these facts do not allow for analysis of the vibrations separately in the two planes (vertical and horizontal), as the vibration modes are coupled due to the presence of the bearings: for this reason, the study should be carried out simultaneously in the two planes of vibration.

6.9.1 Estimating Instability Using the Eigenvalue and Eigenvector Solution

Equation (6.204) can thus be attributed to a problem of eigenvalues-eigenvectors, defined (Sect. 2.4.2.1, Chap. 2) by the equation:

$$[B]\underline{\dot{z}} + [C]\underline{z} = \underline{0} \Rightarrow \underline{\dot{z}} = [A]\underline{z} \tag{6.205a}$$

since:

$$\underline{z} = \begin{Bmatrix} \underline{\dot{x}} \\ \underline{x} \end{Bmatrix}$$
$$[B] = \begin{bmatrix} [M] & [0] \\ [0] & [M] \end{bmatrix}; \quad [C] = \begin{bmatrix} [R] & [K] \\ -[M] & [0] \end{bmatrix}; \quad [A] = [B]^{-1}[C] \tag{6.205b}$$

The solution (particular integral) of the homogenous Eq. (6.205a) is given by:

$$\underline{z} = \underline{Z}e^{\lambda t} \tag{6.205c}$$

and the solutions (complex conjugates $\lambda_i = \alpha_i \pm i\omega_i$) are obtained (Sect. 2.3.1) as eigenvalues of the matrix $[A]$ and the vibration modes $\underline{Z}^{(i)}$ are estimated as associated eigenvectors. Analysis of the sign of the real part α_i of the generic eigensolution makes it possible to highlight any conditions of instability of the entire vibrating system (with $\alpha_i > 0$). The non-symmetrical terms are, however, generally small compared to the symmetrical terms of the conservative system: this fact often creates difficulties of a numerical nature in calculating the solutions, difficulties that may affect the results obtained and, therefore, call into question the reliability of the conditions of stability calculated.

6.9.2 Estimating Instability with the Modal Method

There are various alternative methods to this procedure: first of all we should recall the modal approach which rewrites (6.204) as:

$$[M]\underline{\ddot{x}} + [R]\underline{\dot{x}} + [K_s]\underline{x} + [K_{es}]\underline{x} = \underline{0} \tag{6.206a}$$

$[K_s]$ being the symmetric part of the stiffness matrix, the individual terms of which are defined by the following relation:

$$K_s^{(i,j)} = \frac{K^{(i,j)} + K^{(j,i)}}{2} \tag{6.2026b}$$

while $[K_{es}]$ represents the antisymmetric part defined by the following relation:

$$K_{es}^{(i,j)} = \frac{K^{(i,j)} - K^{(j,i)}}{2} \tag{6.206c}$$

By neglecting, in a first approximation, both damping $[R]$, and the skew-symmetric stiffness matrix $[K_{es}]$, it is possible to calculate eigenvalues ω_i and eigenvectors $\underline{X}^{(i)}$ of the reduced equation:

$$[M]\ddot{\underline{x}} + [K_s]\underline{x} = \underline{0} \tag{6.206d}$$

By using, for example, the first mainly vertical vibration mode $\underline{X}^{(1)}$ and the first mainly horizontal mode $\underline{X}^{(2)}$, it is possible to impose the coordinate transformation:

$$\underline{x} = [\Phi]\underline{q} = \left[\underline{X}^{(1)}\underline{X}^{(2)}\right]\begin{Bmatrix} q_1 \\ q_2 \end{Bmatrix} \tag{6.207a}$$

where $[\Phi]$ is the modal matrix (Chap. 2, Sect. 2.5.1), made up, in this case, of just the first two vibration modes of the undamped system. This relation, placed in (6.206a), leads to:

$$[\Phi]^T[M][\Phi]\ddot{\underline{q}} + [\Phi]^T[R][\Phi]\dot{\underline{q}} + [\Phi]^T[K_s][\Phi]\underline{q} + [\Phi]^T[K_{es}][\Phi]\underline{q} = \underline{0} \tag{6.207b}$$

Considering the known property of orthogonality (Chap. 2, Sect. 2.5.2.1) (6. 207b) can be rewritten as:

$$[m]\ddot{\underline{q}} + [\Phi]^T[R][\Phi]\dot{\underline{q}} + \left[[k] + [\Phi]^T[K_{es}][\Phi]\right]\underline{q} = \underline{0} \tag{6.207c}$$

with $[m]$ and $[k]$ diagonal matrices:

$$\begin{aligned} [k] &= [\Phi]^T[M][\Phi] \\ [m] &= [\Phi]^T[K_s][\Phi] \end{aligned} \tag{6.203d}$$

In (6.207c) the damping matrix in principal coordinates $[\Phi]^T[R][\Phi]$ is not diagonal as, due to the presence of the oil-film, the matrix $[R]$ is not a linear combination of $[M]$ and $[K_s]$: similarly the stiffness matrix in principal coordinates of skew-symmetric terms $[\Phi]^T[K_{es}][\Phi]$ is not symmetric or diagonal. So now it is possible to calculate the solutions of (6.207c) by imposing the relation:

$$\underline{q} = \underline{Q}e^{\lambda t} \tag{6.207e}$$

to bring the same equation back to a problem of eigenvalues-eigenvectors. Sometimes, due to the presence of terms $[\Phi]^T[K_{es}][\Phi]$, though working on reduced matrices (*mxm* if *m* are the vibration modes considered), the problem of determining the threshold of instability can entail numerical uncertainties.

6.9.3 Estimating Instability with the Forced Method

To overcome these drawbacks, it is possible to use an alternative approach defined with the name of *forced method*. The method is essentially based on calculating the energy introduced by the harmonic excitation forces applied on the global rotor-oil-film-casing-foundation system. The ratio between the energy introduced by these forces and the maximum kinetic energy of the system makes it possible to obtain[37] the dimensionless damping ratio r/r_c and by estimating this it is then possible to estimate any threshold of instability. The motion equation of the entire forced system then becomes:

[37]To demonstrate this property it is easier to consider a forced 1 d.o.f. system, the equation of which is (Chap. 2, Sect. 2.3.1):

$$m\ddot{x} + r\dot{x} + kx = F_o e^{i\Omega t} \tag{6.37.1}$$

and the steady-state solution of which is given by:

$$x = Xe^{i\Omega t} \tag{6.37.2}$$

since

$$X = |X|e^{i\phi} \Rightarrow \phi = a\tan\left(\frac{i\Omega r}{k - \Omega^2 m}\right) \Rightarrow |X| = \frac{1}{k - \Omega^2 m + \Omega r} \tag{6.37.3}$$

The energy introduced in a cycle by the excitation force is:

$$E_F = \int_0^{T_o} xF(t)dt = -|X|F_o\pi \sin\phi = -\mathrm{Im}|X|F_o\pi \tag{6.37.4}$$

Hence, the maximum kinetic energy is:

$$E_{c\max} = \frac{1}{2} m\Omega^2 |X|^2 \tag{6.37.5}$$

As the energy introduced by the excitation force E_F is equal to the energy dissipated E_d:

$$\frac{E_F}{E_{c\max}} = \frac{E_d}{E_{c\max}} = \frac{|X|^2 r\Omega\pi}{\frac{1}{2} m\Omega^2 |X|^2} = \frac{2r\pi}{m\Omega} = 4\pi\frac{r}{r_c} \tag{6.37.6}$$

hence:

$$\frac{r}{r_c} = \frac{E_F}{4\pi E_{c\max}} = \frac{E_d}{4\pi E_{c\max}} = \frac{1}{4\pi}\frac{\mathrm{Im}(X)F_o\pi}{\frac{1}{2} m\Omega^2 |X|^2}. \tag{6.37.7}$$

$$[M]\underline{\ddot{x}} + [R]\underline{\dot{x}} + [K_s]\underline{x} + [K_{es}]\underline{x} = \underline{F}_o e^{i\Omega t} \tag{6.208a}$$

where vector $\underline{F}_o e^{i\Omega t}$ represents the external forces applied to the system for which we wish to verify the stability. Solutions of (6.208a) are of the form $\underline{x} = \underline{X}e^{i\Omega t}$ which placed in (6.208a) lead to:

$$\underline{X} = \underline{X}(\Omega) = \left[-\Omega^2[M] + i\Omega[R] + [[K_s] + [K_{es}]]\right]^{-1}\underline{F}_o \tag{6.208b}$$

The energy introduced by the excitation force for $\Omega = \omega_i$, where ω_i is one of the natural frequencies of the system, is, assuming $\underline{F}_o$ is real, that is, forcing in the phase or opposition of phase between them:

$$E_F^{(i)} = \pi \ \mathrm{Im}\left(\underline{X}^T(\omega_i)\right) \ \underline{F}_o \tag{6.208c}$$

having used $\mathrm{Im}(\underline{X}^T(\omega_i))$ to indicate the vector that contains the imaginary part of the solution $\underline{X}(\Omega)$ obtained from (6.208b) for $\Omega = \omega_i$. The maximum kinetic energy is in approximate form:

$$E_{c\max}^{(i)} = \frac{1}{2}\,\omega_i^2\,\underline{X}^T(\omega_{\mathrm{i}})[M]\underline{X}(\omega_{\mathrm{i}}) \tag{6.208d}$$

having used $\underline{X}(\omega_{\mathrm{i}})$ to indicate the vector that contains the module of the solution $\underline{X}(\Omega)$ obtained from (6.208b) for $\Omega = \omega_i$. Considering (6.208c) and (6.208d), it is possible to obtain the value of dimensionless damping h_i:

$$h_i = -\left(\frac{r}{r_c}\right)^{(i)} = \frac{1}{4\pi}\frac{E_F^{(i)}}{E_{c\max}^{(i)}} = -\frac{\alpha_i}{\omega_i} \tag{6.208e}$$

Thus it is possible to estimate coefficient α_i: if the latter assumes negative or zero value the motion of the system is stable, or if not it is unstable. The forced method illustrated is approximated [75–78, 80–82, 85, 86, 88, 89, 91–97]: in the practical problem of rotors, the approximations introduced can, however, be neglected and, on the other hand, there are numerous advantages offered by this approach with respect to calculating eigenvalues and eigenvectors:

- the method is numerically more stable;
- it is possible to use the same procedure adopted for calculating the frequency response;
- it is possible to introduce the mechanical impedances relating to the foundation into the model, which would be impossible with the normal approach in the time domain (eigenvalues-eigenvectors).

6.9.4 Effect of Load Variations on Supports on the Conditions of Instability

At the end of this section we show, as an example, the results obtained on a real rotor of a 300 MW steam turbogenerator (taken from [79]. Figure 6.66 shows the model of the complete line (made up of 4 rotors: high pressure, medium pressure, low pressure rotors and alternator) and Table 6.1 shows the main properties of the bearings. In these rotors (hyperstatic), changing the running conditions, can vary the alignment of the supports and, as a consequence, the distribution of loads on the actual supports. A variation in load causes a variation in the position of the static equilibrium of the journal in each bearing, with a consequent variation of the runout $\chi = e_c/d$ (see Chap. 5, Sect. 5.3.2.2 and Sect. 6.3.1). In correspondence to the new position reached, there is variation in the stiffnesses and the equivalent dampings due to the linearised force field related to the oil-film and, as a consequence, the conditions of stability of the overall rotor + oil-film + foundation system change. The same effect can be caused by a change in thermal working conditions or temperature of the lubricant. An increase in temperature of the lubricant on input determines a decrease in the viscosity of the lubricant and an increase, all other conditions being equal, of the runout χ of the journal inside the bearing; this leads to an increase in the stability of the overall system (for $\chi < \chi_{\text{lim}}$ the bearing is always stable, Sect. 5.3.2.2, Sect. 6.3.1). In the real case shown, due to the effect of a deformation of the foundation caused by thermal effects, there is a change in alignment that causes a decrease in the load borne by the bearing of the turbo-

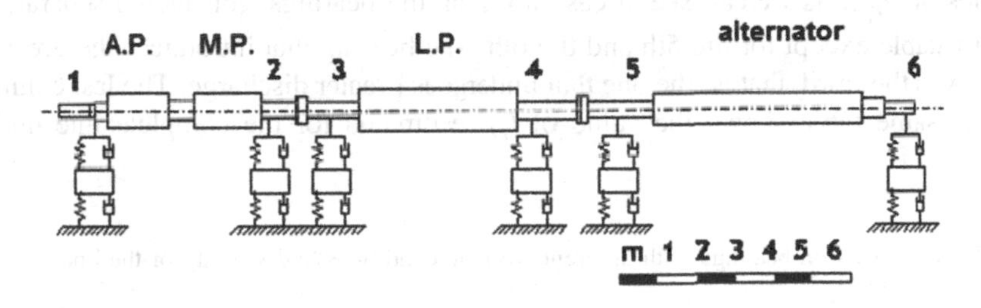

Fig. 6.66 Model of the axis line analysed

Table 6.1 Principal properties of the bearings of the axis line in Fig. 6.66

Bear. no.	1	2	3	4	5	6
Type	Elliptic					
R(mm)	152.5	195.0	240.0	255.0	267.5	212.5
Width./R	1.50	1.28	1.46	1.58	1.17	1.38
Clear./R	0.266 %					
h/R	0.133 %					

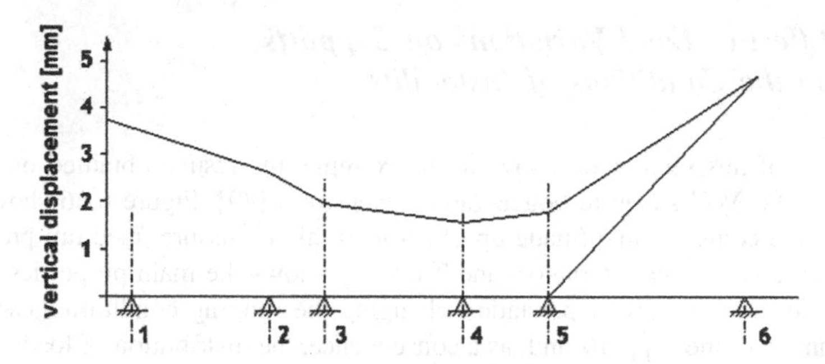

Fig. 6.67 Different conditions of alignment upon variation of the temperature

generator: Fig. 6.67 shows the conditions of alignment in correspondence to different temperature conditions.

The part of the shaft line that proves to be responsible for instability is that of low pressure in correspondence to supports 3 and 4. Table 6.2 shows the loads on the single bearings corresponding to cold machinery (test no. 1), with hot machinery (lubricant input temperature at 42°, test no. 2) and again with hot machinery but with a temperature of 49° (test no. 3). On the three different conditions, the first and last bearing do not undergo any significant load variations, whereas bearing 3 (between medium and low pressure) discharges considerably (from 29.8 tons to 19.2): the same occurs, although less markedly, for bearing 4. The same table also shows the value of dimensionless damping $^{r}/_{r_c}$ relating to the single bearings analysed separately by estimating the energy introduced or dissipated by the forces of the oil-film (the generic bearing dissipates energy for positive values of $^{r}/_{r_c}$); as we can see in case no. 2 all the bearings, considered separately, are unstable except for the 5th and the 6th. The bearing that introduces the greatest energy is the third, that is, the one that undergoes greater discharge. The last column of the same table shows the value of $^{r}/_{r_c}$ estimated for the complete line under

Table 6.2 Loads on bearings with different working conditions and stability of the line

	r/r_c						
Test no./ bear. No.	1 load [kg]	2 load [kg]	3 load [kg]	4 load [kg]	5 load [kg]	6 load [kg]	
1	10841 −11E-5	6628 −63E-5	29795 −10E-3	32462 −68E-4	16525 −10E-4	23039 +12E-5	1.3E-3 stable
2	8803 −43E-6	18481 −59E-5	19256 −18E-3	25496 −22E-3	25756 +65E-5	21498 +26E-7	−4.2E-4 unstable
3	8803 −38E-6	18481 −41E-5	19256 −17E-3	25496 −19E-3	25756 +92E-5	21498 −72E-6	6.8E-5 threshold of instability

different operating conditions: the machinery is stable when cold (test 1, $^r/_{r_c} = 1.3\,\%$) or when the oil temperature is above 49° (test 3, $^r/_{r_c} = 0.07\,\%$), whereas it is unstable at 42° (test 2, $^r/_{r_c} = -0.4\,\%$). The stability of the entire line with the simultaneous presence of local instabilities of the single bearings, is due to the effect of damping of the shaft and of the foundation on the overall behaviour of the entire system.

6.10 Torsional Vibrations

In crankshafts or straight shafts that have several disks fitted to them (for example shafts with several gears or pulleys, turbine or pump couplings) fatigue failures were experimentally noted for prolonged running at specific angular velocities Ω without observing particularly high bending vibrations either on the shaft or the support. These failures can be attributed to the torsional vibrations that shafts can be subject to. To better understand the mechanism associated with these vibrations, we can imagine the generic mechanical system shown in Fig. 6.68, which is part of the transmission of a motor vehicle. The torque applied by the side of the motor varies, in general, with law that depends on the characteristics of the motor used, while from the side of the wheels it is possible to hypothesise the application of a resistant torque which, in this case, for simplicity's sake, is assumed to be constant.

The torque M_m that stresses the shaft is provided by the contribution of actions produced by the pressure of gas in the cylinders, by the forces of inertia, by the forces of gravity and by frictional forces. Figure 6.69 shows, as an example, the trend, as a function of the crank angle φ, of this torque in the case of a two or four-stroke internal combustion engine. As we can see, this torque is periodic and, therefore, can be broken down in a Fourier series:

$$\begin{aligned} M_m &= M_o + M_1\cos(\Omega t + \psi_1) + M_2\cos(\Omega t + \psi_2) + \dots \\ &= M_o + \sum_{i=1}^{n} M_i\cos(\Omega_i t + \psi_i) \end{aligned} \tag{6.209a}$$

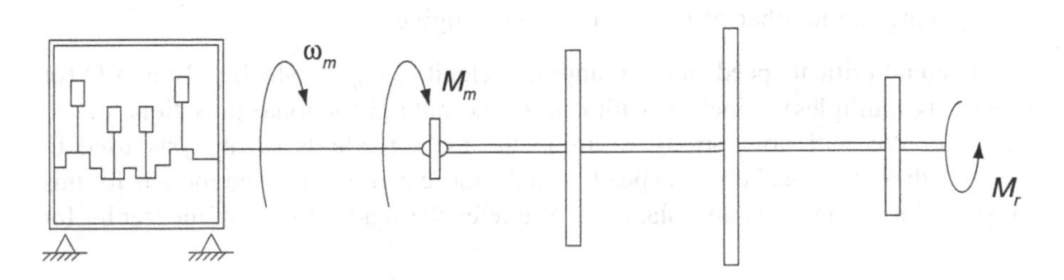

Fig. 6.68 Schematisation of a generic power transmission

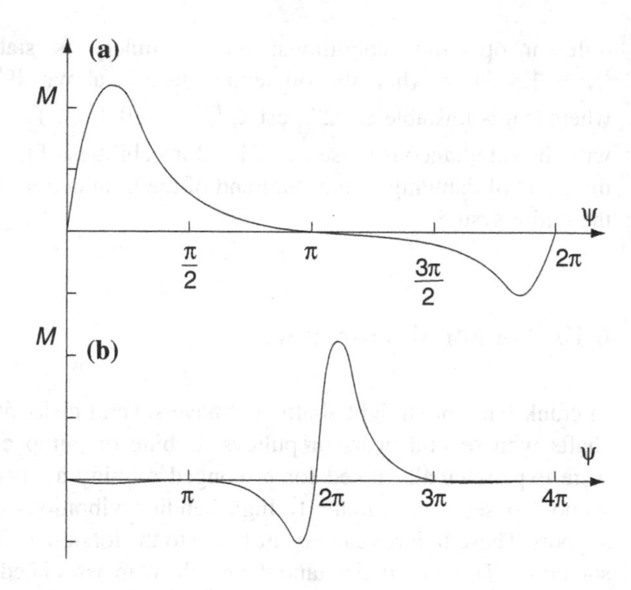

Fig. 6.69 Torque transmitted by an internal combustion engine: two stroke (**a**) and four stroke (**b**)

Ω being the pulsation of the fundamental harmonic. In general the fundamental harmonic Ω is linked to the angular velocity of rotation ω_m; this link depends on the number of cylinders and on the type of engine, for example:

- in the case of a 2-stroke single cylinder engine (one combustion per rev):

$$\Omega = \omega_m = \frac{2\pi}{60} n_m \ [\text{rad}/s] \tag{6.209b}$$

- in the case of a 4-stroke single cylinder engine (one combustion every 2 revs):

$$\Omega = \frac{\omega_m}{2} = \frac{\pi}{60} n_m \ [\text{rad}/s] \tag{6.209c}$$

- in the case of a 4-stroke, 4 cylinder engine (4 combustions every 2 revs):

$$\Omega = 2\omega_m = \frac{4\pi}{60} n_m \ [\text{rad}/s] \tag{6.209d}$$

n_m being the number of revs of the actual engine.

Torsional critical speeds are the angular velocities ω_m at which pulsation Ω (or one of its multiples) coincides with one of the natural torsional pulsations of the shaft ω_i. We will now briefly examine the main methods of analysis used to estimate these torsional critical speeds and the more commonly used models for this purpose during analysis and planning. We refer the reader to the bibliography for

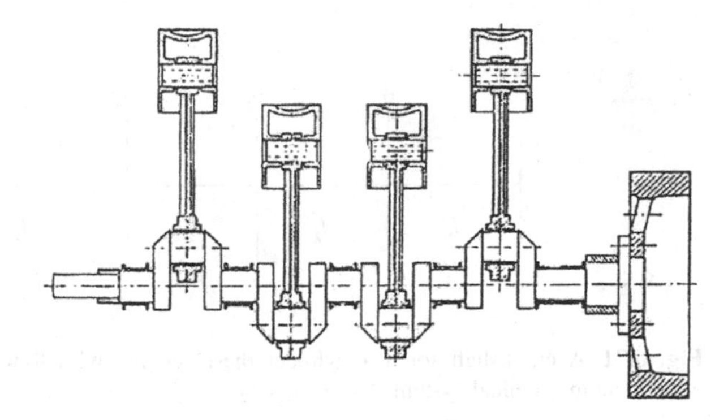

Fig. 6.70 A crankshaft for a 4 cylinder diesel engine with flywheel

more detailed discussion. The generic torsional elastic system of the engine and the associated vehicle (Fig. 6.70) represents quite a complicated mechanical system due to the real distribution of mass and of elasticity.

In order to make studying these systems easier, they are traditionally schematised with much simpler dynamically equivalent models, which consist of (Fig. 6.71) a straight massless shaft which supports a certain number of thin disks that reproduce the inertias of the different elements connected to the shaft.[38] These disks are considered as rigid bodies with equivalent moments of inertia, connected to each other by equivalent torsional elastic elements. The disks are usually placed in correspondence to the centre of each crank and in correspondence to propellers, engines and gears. More refined models even make it possible to analytically reproduce the real distribution of mass and of stiffness. Obviously, schematisation must include all of the shaft line, including the engine, transmission and user. So with this approach, analysis of the torsional critical speeds must be preceded by preliminary calculations based on static analysis of the complete system (crankshaft, disks, flywheels, clutches etc.) to define the equivalent stiffnesses of the single segments of the rotor, and on dynamic analysis to determine the equivalent moments of inertia of the individual rotating parts and the parts that move with them (connecting rods and pistons).

Once the dynamically equivalent model has been defined (Fig. 6.71), it is then possible to estimate both the natural frequencies and the relative vibration modes of the free system and the response of the system forced by the torques and resistant torques acting on it, using the normal analytical and numerical techniques already widely described in the previous chapters. In the following sections we will address:

- the problem of reduction of the real mechanical system to a simplified dynamically equivalent model (Sect. 6.10.1);

[38]Schematisation can be done using the usually three-dimensional finite element method, or we can use it to define the parameters to be introduced into the simpler model with concentrated parameters.

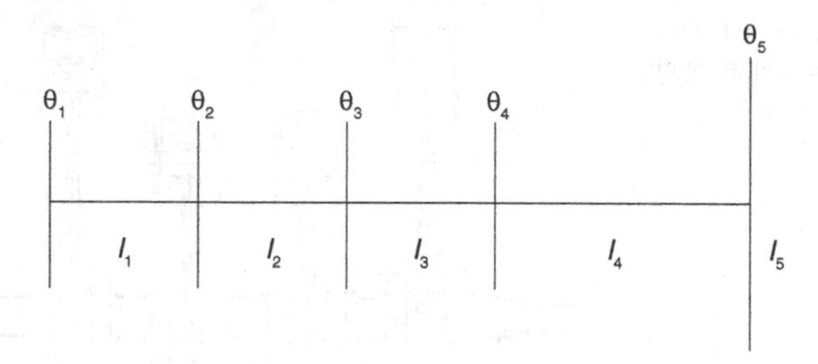

Fig. 6.71 A crankshaft for a 4 cylinder diesel engine with flywheel: schematisation with an equivalent mechanical system

- the schematisation of the problem with discrete systems with 1, 2, …, n d.o.f. (Sect. 6.10.2);
- the schematisation of the problem using the typical continuum approach (Sect. 6.10.3);
- the schematisation of the problem with the finite element approach (Sect. 6.10.4) each time using real examples to simplify and lighten the discussion.

Lastly, in Sect. 6.10.5 we will mention some mechanisms that are useful in reducing these torsional vibrations.

6.10.1 Methods for Reducing to an Equivalent System

For didactic purposes, it is easier to understand the following explanation if we analyse the simple vibrating system shown in Fig. 6.72, which is made up of a vertical shaft fixed at one end and which has a rigid disk fitted onto it. In the hypothesis that the mass of the shaft is negligible compared to the added mass, this system has just one d.o.f.: in this case it is easy to obtain the equivalent dynamic model, which in this case, is similar to the real structure.

Assuming the rotation of the disk θ as an independent variable, the equations that describe the free motion simply become:

$$J\ddot{\theta} + k_t\theta = 0 \tag{6.210a}$$

J being the mass moment of inertia of the disk[39] and k_t the torsional stiffness of the shaft. To estimate this stiffness it is useful to remember that a generic shaft

[39]With the assumption of a circular disk of weight W, with even thickness and diameter D:

$$J = \frac{WD^2}{8g}. \tag{6.38.1}$$

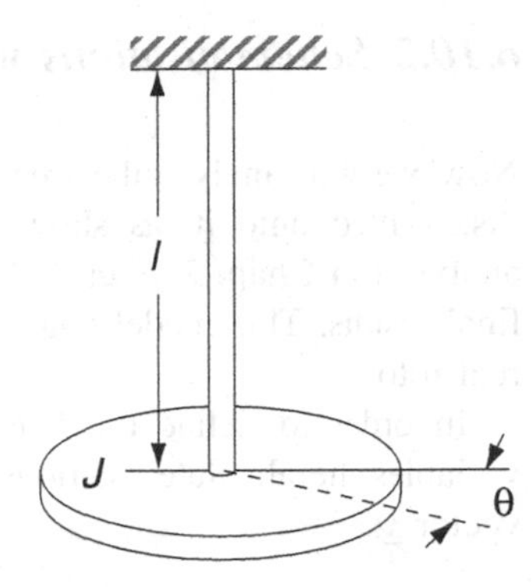

Fig. 6.72 Torsional vibrations: 1 d.o.f. system

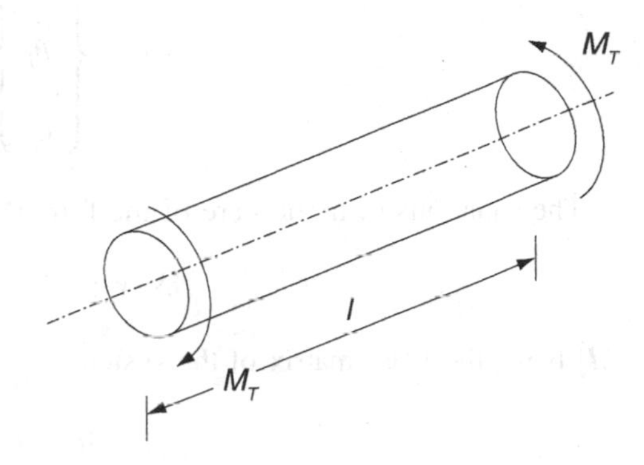

Fig. 6.73 Effect of torsion

subject to torque M_t due to the effect of its elasticity (Fig. 6.73) torsionally deforms and its relative rotation between the ends applies:

$$\Delta\Theta = \frac{M_t l}{GJ_p} \tag{6.210b}$$

where:

- G is the shear modulus;
- J_p is the polar moment of inertia of the section;
- l is the length.

Similarly to what is normally done for linear springs, we will call k_t the torsional elastic constant, i.e. the torque to be applied to the ends to obtain their unitary rotation, that is $\Delta\Theta = 1$, hence, from (6.210b) we have:

$$k_t = \frac{GJ_p}{l} \tag{6.211}$$

6.10.2 Schematisations with N-Degree-of-Freedom Systems

Now we will analyse the free torsional vibrations of a generic shaft with n rigid disks fitted onto it, as shown in Fig. 6.74: this model has already been fully analysed in Chap. 2, Sect. 2.3.5, below we will show, for ease of discussion, the final results. This model could simulate, for example, the dynamic behaviour of a real rotor.

In order to define the behaviour of the system we consider as independent variables the absolute rotations of the single disks θ_i ($i = 1, 2, \ldots, n$) clustered in vector $\underline{x}$:

$$\underline{x} = \begin{Bmatrix} \theta_1 \\ \cdots \\ \theta_i \\ \cdots \\ \theta_n \end{Bmatrix} \quad (6.212a)$$

The equations of motion are of the form (Sect. 2.4):

$$[M]\ddot{\underline{x}} + [K]\underline{x} = \underline{0} \quad (6.212b)$$

$[M]$ being the mass matrix of the system:

$$[M] = \begin{bmatrix} J_1 & 0 & 0 & 0 & 0 \\ 0 & \cdots & 0 & 0 & 0 \\ 0 & 0 & J_i & 0 & 0 \\ 0 & 0 & 0 & \cdots & 0 \\ 0 & 0 & 0 & 0 & J_n \end{bmatrix} \quad (6.212c)$$

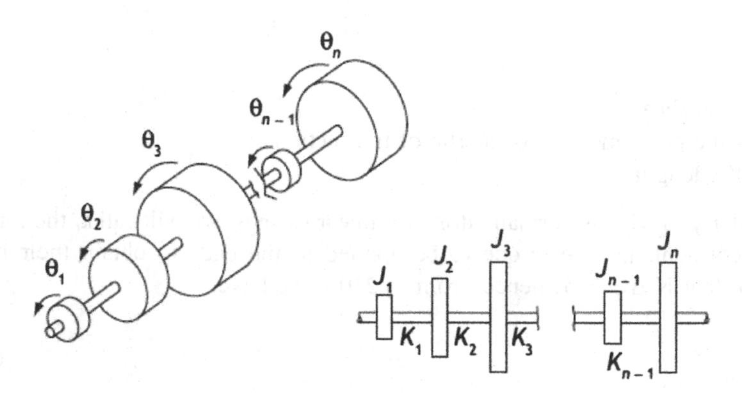

Fig. 6.74 Model with n d.o.f. for studying torsional vibrations

and $[K]$ the corresponding stiffness matrix:

$$[K]=\begin{bmatrix} k_1 & -k_1 & 0 & 0 & 0 \\ -k_1 & k_1+k_2 & -k_2 & 0 & 0 \\ 0 & -k_2 & \dots & \dots & \dots \\ 0 & 0 & \dots & \dots & -k_n \\ 0 & 0 & 0 & -k_n & k_n \end{bmatrix} \tag{6.212d}$$

The solutions of (6.212b) are of the form:

$$\underline{x}=\underline{X}e^{i\omega t} \tag{6.213a}$$

which assume non-trivial values when the determinant is zeroed:

$$\det\left|-\omega^2[M]+[K]\right|=\underline{0} \tag{6.213b}$$

Each value of ω_i has a certain corresponding eigenvector, that describes the vibration mode of the system being analysed in resonance. As mentioned, the torques applied M_m to the system are generally periodic (Fig. 6.69) and they can be represented as reported in (6.209a), where M_i is the generic complex harmonic component and $\Omega_i = i\Omega$ is the generic pulsation, multiple of the fundamental harmonic Ω (rad/s) equal to the angular velocity. The dynamic response to several excitation forces is, therefore, equal to the sum of the responses to each single forcing element. Once the forcing element M_m has been broken down into the single harmonic components M_i (6.209a), it is possible to calculate the response of the system to each component of the forcing element. If the system is forced by a generic harmonic component:

$$\underline{F}=\begin{Bmatrix} 0 \\ \dots \\ M_i \\ \dots \\ 0 \end{Bmatrix} e^{i\Omega_i t}=\underline{F}_{oi}e^{i\Omega_i t} \tag{6.214a}$$

applied to the generic ith disk, the equations of motion become:

$$[M]\ddot{\underline{x}}+[R]\dot{\underline{x}}+[K]\underline{x}=\underline{F}_{oi}e^{i\Omega_i t} \tag{6.214b}$$

The solution of (6.214b) is given by a particular integral of the form:

$$\underline{x}=\underline{X}e^{i\Omega_i t} \tag{6.214c}$$

which can be calculated with the method described in detail in the Chap. 2 regarding 2-n d.o.f. systems. As we have seen several times, whenever the pulsation of the forcing element Ω_i coincides with one of the natural pulsations of the system

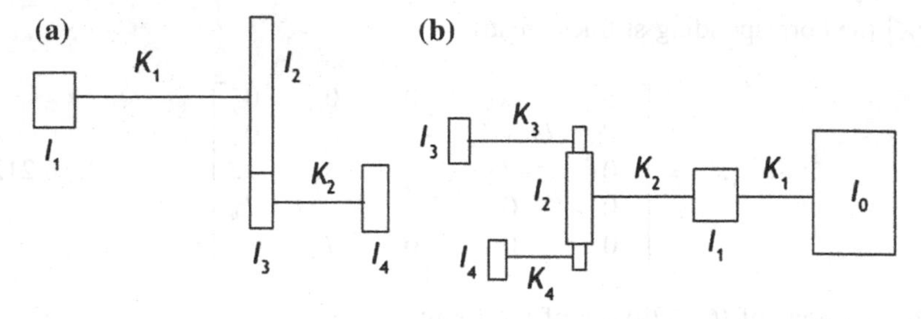

Fig. 6.75 Shafts linked with gears

ω_i there are conditions of resonance and the amplitudes $\theta_1, \ldots, \theta_n$ of the vibration tend to infinity (actually, the vibration reaches only high amplitudes due to the presence of damping). In the case of systems with shafts connected with gears (Fig. 6.75), in the hypothesis of neglecting the clearance and deformability of the teeth and in the hypothesis of considering the wheels to be rigid, the d.o.f. decrease as the rotations of the coupled wheels are linked to the known kinematic relations defined by the transmission ratio:

$$\frac{\theta_i}{\theta_j} = -\tau_{ij} \tag{6.215}$$

As an example, we will analyse the model shown in Fig. 6.75a: the absolute rotations of the four disks being assumed as physical variables $\underline{Y}$:

$$\underline{Y} = \begin{Bmatrix} \theta_1 \\ \theta_2 \\ \theta_3 \\ \theta_4 \end{Bmatrix} \tag{6.216a}$$

kinetic energy E_c and potential V are:

$$\begin{aligned} E_c &= \frac{1}{2}\underline{\dot{Y}}^T[M_Y]\underline{\dot{Y}} \\ V &= \frac{1}{2}\underline{Y}^T[K_Y]\underline{Y} \end{aligned} \tag{6.216b}$$

$[M_Y]$ being the matrix of mass of the system in physical coordinates:

$$[M_Y] = \begin{bmatrix} J_1 & 0 & 0 & 0 \\ 0 & J_2 & 0 & 0 \\ 0 & 0 & J_3 & 0 \\ 0 & 0 & 0 & J_4 \end{bmatrix} \tag{6.216c}$$

and $[K_Y]$ the corresponding stiffness matrix:

$$[K_Y] = \begin{bmatrix} k_1 & -k_1 & 0 & 0 \\ -k_1 & k_1 & 0 & 0 \\ 0 & 0 & k_2 & -k_2 \\ 0 & 0 & -k_2 & k_2 \end{bmatrix} \tag{6.216d}$$

Assumed as independent variables $\underline{x}$ the absolute rotations of three disks:

$$\underline{x} = \begin{Bmatrix} \theta_1 \\ \theta_2 \\ \theta_4 \end{Bmatrix} \tag{6.217a}$$

and imposing the link between physical variables and independent variables:

$$\underline{Y} = \begin{Bmatrix} \theta_1 \\ \theta_2 \\ \theta_3 \\ \theta_4 \end{Bmatrix} = [\Lambda]\,\underline{x} = \begin{bmatrix} 1 & 0 & 0 \\ 0 & 1 & 0 \\ 0 & -\tau_{32} & 0 \\ 0 & 0 & 1 \end{bmatrix} \begin{Bmatrix} \theta_1 \\ \theta_2 \\ \theta_4 \end{Bmatrix} \tag{6.217b}$$

τ_{32} being the gear ratio between the two wheels in contact (the minus sign depending on the conventions assumed for the rotations, all assumed equally oriented positive). Considering the transformation of coordinates (6.217b), the (6.216b) become:

$$\begin{aligned} E_c &= \frac{1}{2}\dot{\underline{Y}}^T[M_Y]\dot{\underline{Y}} = \frac{1}{2}\dot{\underline{x}}^T[\Lambda]^T[M_Y][\Lambda]\dot{\underline{x}} = \frac{1}{2}\dot{\underline{x}}^T[M]\dot{\underline{x}} \\ V &= \frac{1}{2}\underline{Y}^T[K_Y]\underline{Y} = \frac{1}{2}\underline{x}^T[\Lambda]^T[K_Y][\Lambda]\underline{x} = \frac{1}{2}\underline{x}^T[K]\underline{x} \end{aligned} \tag{6.217c}$$

$[M]$ and $[K]$ being the matrices of mass and of stiffness of the system in independent coordinates:

$$\begin{aligned} [M] &= [\Lambda]^T[M_Y][\Lambda] = \begin{bmatrix} J_1 & 0 & 0 \\ 0 & J_2 + \tau_{32}^2 J_3 & 0 \\ 0 & 0 & J_4 \end{bmatrix} \\ [K] &= [\Lambda]^T[K_Y][\Lambda] = \begin{bmatrix} k_1 & -k_1 & 0 \\ -k_1 & k_1 + \tau_{32}^2 k_2 & \tau_{32}k_2 \\ 0 & \tau_{32}k_2 & k_2 \end{bmatrix} \end{aligned} \tag{6.217d}$$

and the equations of motion of the form shown in (6.31b).

6.10.3 Schematisation with Continuous Bodies

In the case where the mass of the shaft is not negligible in respect to the concentrated masses, the shaft cannot be schematised with a purely elastic element, we must take into account the actual distribution of mass. To solve this type of problem we can use the typical continuum approach (Chap. 3, Sect. 3.5) or the finite element approach (Chap. 4, Sect. 4.2.2). As an example, we will analyse the rotor in Fig. 6.72 where, however, we take into account the mass of the shaft (ρ is its density). The equation that defines the torsional motion of a generic homogenous section of shaft with no irregularities is given by the following relation (Sect. 3.5, Chap. 3):

$$\theta_x = \left(A\cos\frac{\omega}{C}\xi + B\sin\frac{\omega}{C}\xi\right)e^{i\omega t} \tag{6.218a}$$

C being the wave propagation speed:

$$C = \frac{G}{\rho} \tag{6.218b}$$

The constants A and B can be defined, together with the natural pulsation of the system ω, by imposing the boundary conditions which, in the case under examination shown in Fig. 6.72, are:

$$\begin{aligned} \theta_x(\xi)|_{\xi=0} &= 0 \\ GJ_p\frac{\partial\theta_x}{\partial\xi}\bigg|_{\xi=l} &= J\frac{\partial^2\theta_x}{\partial t^2}\bigg|_{\xi=l} \end{aligned} \tag{6.218c}$$

from here we can obtain:

$$\begin{aligned} &A = 0 \\ &B\left(GJ_p\frac{\omega}{C}\cos\frac{\omega}{C}l + J\omega^2\sin\frac{\omega}{C}\xi\right) = 0 \end{aligned} \tag{6.218d}$$

Upon finding the zeroes of the function (coefficient matrix):

$$GJ_p\frac{\omega}{C}\cos\frac{\omega}{C}l + J\omega^2\sin\frac{\omega}{C}\xi = 0 \tag{6.218e}$$

it is possible to calculate the natural frequencies ω_i of the system under examination: hence the generic vibration mode becomes:

$$\Phi^{(i)}(\xi) = \sin\left(\frac{\omega_i}{C}\xi\right). \tag{6.218f}$$

6.10.4 Finite Element Schematisation

Similarly to what we observed with regard to bending vibrations, it is also possible to analyse the torsional vibrations of a shaft by means of a finite element approach. In this regard, we consider a beam of which only the torsional motions are taken into account (these motions are actually, with excellent approximation, uncoupled from the other motions). To describe the motion of this system we assume the rotations of the end nodes of the generic finite element θ_{xi} and θ_{xj} (Fig. 6.76). Assuming small oscillations around the steady-state position (shaft that rotates at a certain constant angular velocity with no oscillations) it is possible to approximate the deformation within the generic *j*th finite element via a linear shape function that coincides with the static deformation and obtain (Chap. 4, Sect. 4.2.2) the stiffness and mass matrix, which we show here for convenience:

$$[K_{\theta x}] = \frac{GJ_p}{l_j}\begin{bmatrix} 1 & -\frac{1}{2} \\ -\frac{1}{2} & 1 \end{bmatrix}; \quad [M_{\theta x}] = \rho J l_j \begin{bmatrix} \frac{1}{3} & \frac{1}{6} \\ \frac{1}{6} & \frac{1}{3} \end{bmatrix} \tag{6.219}$$

If we wish to study the free torsional vibrations of a generic rotor, for ease of discussion we will use the specific rotor shown in Fig. 6.77: we will schematise it using 5 beam elements (the corresponding matrices of mass $[M_{\theta x}]$ and of stiffness $[K_{\theta x}]$ are shown in (6.219). The complete matrices of stiffness $[K]$ and mass $[M]$ are assembled as shown in Fig. 6.78. Thus the usual techniques can be used to calculate the natural frequencies and the relative vibration modes. It should be highlighted that as there are no external constraints, a structure made in this way has a possibility of free motion (rigid rotary motion). So for this model there is a zero pulsation $\omega_1 = 0$ with deformation of the form:

$$\underline{X}^{(1)} = \begin{Bmatrix} \bar{\theta} \\ \dots \\ \bar{\theta} \end{Bmatrix} \tag{6.220}$$

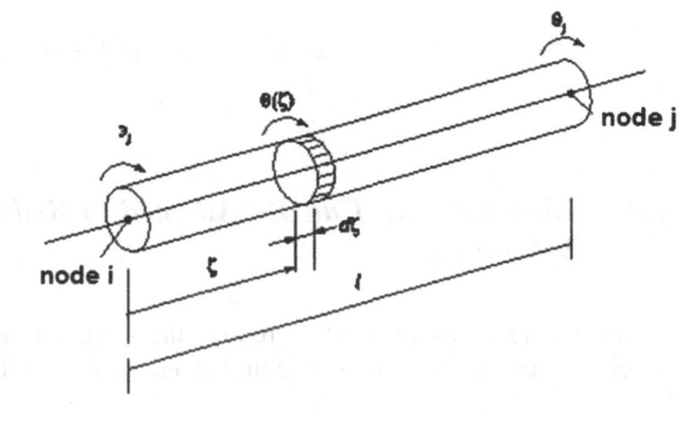

Fig. 6.76 Torsional vibrations in the beams: d.o.f. relating to the generic finite element

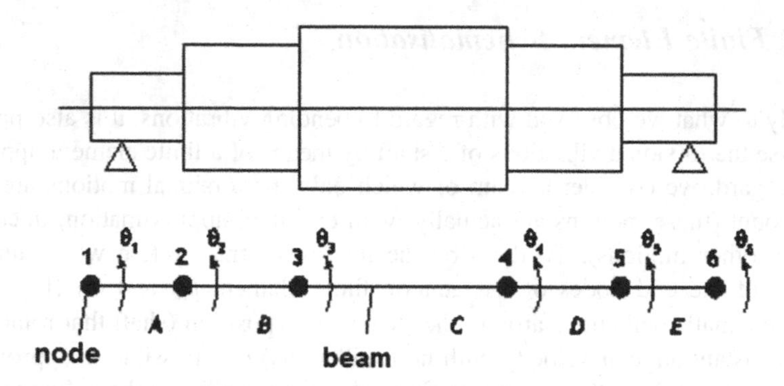

Fig. 6.77 Torsional vibrations: finite element schematisation of a rotor

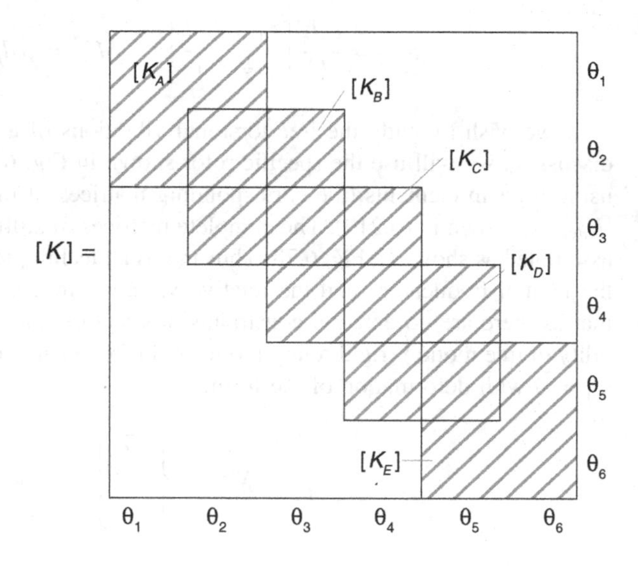

Fig. 6.78 Torsional vibrations: assembly of mass and stiffness matrices relating to the finite element schematisation of a rotor in Fig. 6.74

thus corresponding to a rigid motion

$$\theta_1^{(1)} = \cdots = \theta_i^{(1)} = \cdots = \theta_n^{(1)} = \overline{\theta} = const. \tag{6.221}$$

6.10.5 Elements that Can Be Adopted to Reduce Torsional Vibrations

Several methods can be used to reduce the torsional vibrations based on a free wheel or free inertial element coupled elastically or by means of viscous or

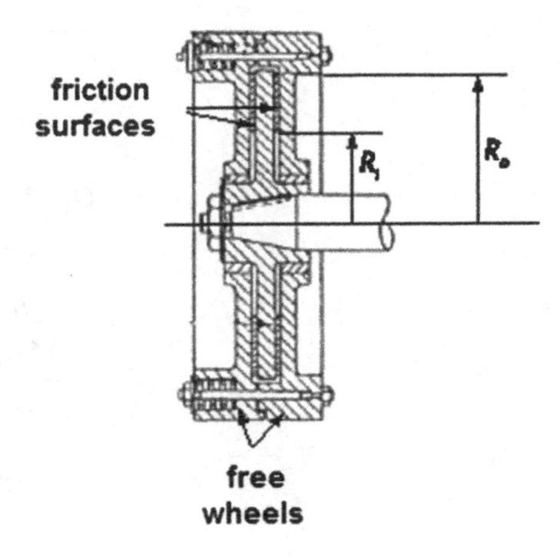

Fig. 6.79 Lanchester damper

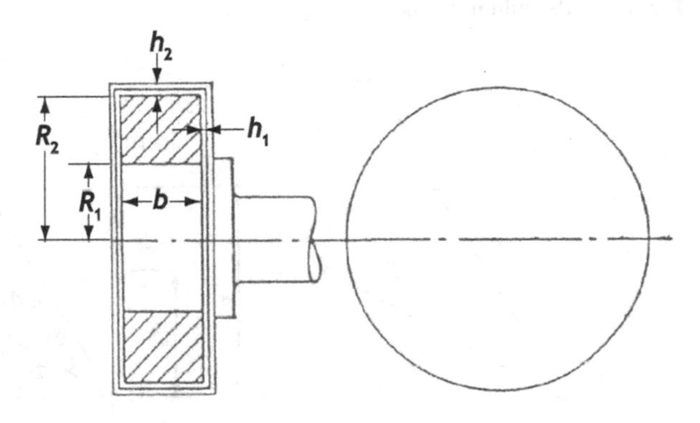

Fig. 6.80 Houdaille damper

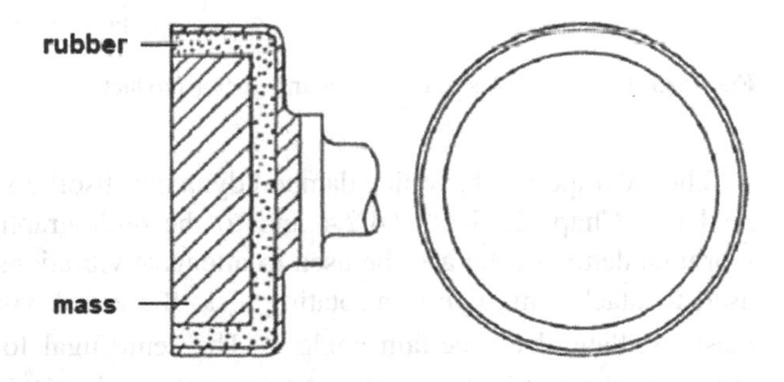

Fig. 6.81 Damper with rubber element

Coulomb friction elements. Figure 6.79 shows the Lanchester damper (based on dry friction). Figure 6.80 shows a model of the Houdaille damper where a free wheel is mounted in a casing with reduced clearances and filled with silicone fluid. Figure 6.81 shows a damper made of rubber which acts both as an elastic element and as a dissipating element.

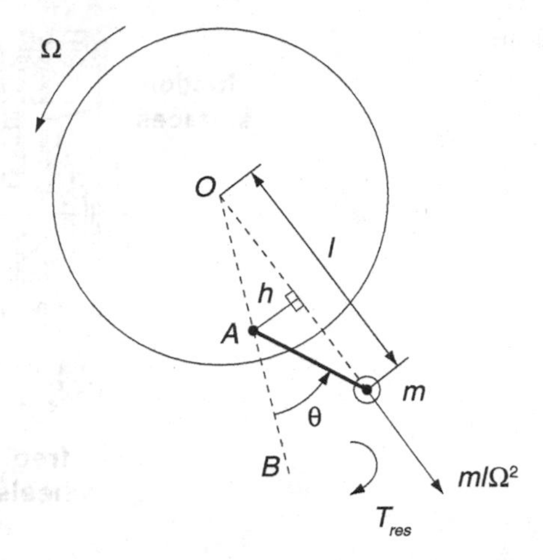

Fig. 6.82 Pendulum detuners

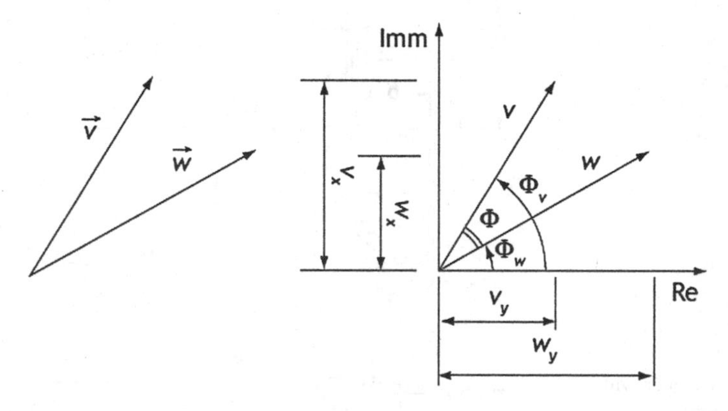

Fig. 6.83 Definition of scalar product and vector product

These dampers behave like damped dynamic absorbers: for their use we refer the reader to Chap. 2, Sect. 2.4.2.4 and to the bibliography. Centrifugal pendulum vibration detuners can also be used to attenuate vibrations (Fig. 6.82): a thin rod is used to attach a mass m to a rotating disk. If the disk vibrates during rotation, the mass oscillates by a certain angle θ. The centrifugal forces $F_c = m\Omega^2 L$ tend to straighten the rod by generating a torque $C_{res} = h\,m\,\Omega^2 L$, thus bringing it back to the OAB position.

Summary This chapter deals specifically with rotor dynamics, introduced as a complete application example of a mechanical system subjected to force fields. The text defines the most common layouts used in these machines and the numerical models adopted to simulate their dynamic behaviour. Vibration and stability problems are analysed, also with reference to critical bending and torsional speeds. Other problems associated with "n-per-rev" vibrations, material hysteresis and gyroscopic effects are also analysed. Some methods used for rotor balancing are described at the end of the chapter.

References

Rotordynamics: general theory

1. AA VV Proceedings of International Conference on Rotordynamics
2. Bachschmid N, Bruni S, Collina A, Pizzigoni B, Resta F (2004) Fondamenti di meccanica teorica ed applicata. Mc Graw Hill, New York
3. Bachschmid N et al (1982) A theoretical and experimental investigation on the dynamic behaviour of an axial pump. In: IFToMM Conference on rotordynamic problems in power plants, Rome (Italy)
4. Bachschmid N, Curami A, Del Vescovo D (1992) Non-linear vibrations due to roller-bearing clearances. In: IMechE, international conference on vibrations in rotating machinery, pp 591–596
5. Bachschmid N, Curami A, Petrone F (1992) Vibrational behaviour of rotors with gear couplings in case of insufficient coupling lubrication. In: Proceedings of rotating machine dynamics conference. Springer, London, pp 232–239
6. Bachschmid N, Pizzigoni B, Passerini R (1983) Rilievi sperimentali del comportamento di una pompa ad asse verticale. L'Energia Elettrica, no 3
7. Bianchi G, Ruggieri G (1973) Dinamica dei rotori di centrifughe. Report Interno della Sezione di Meccanica delle Macchine del Politecnico di Milano, Milan
8. Bishop RED, Johnson DC (1960) The mechanics of vibrations. Cambridge University Press, London
9. Cheli F, Pennestrì E (2006) Cinematica e dinamica dei sistemi multibody. Casa Editrice Ambrosiana, Milan
10. Curami A et al (1982) An on-line computer system as an aid to solving rotor dynamic problems. In: 1FT0MM Conference on Rotordynamic Problems in Power Plants, Rome (Italy)
11. Curami A, Diana G (1981) Un programma di calcolo automatico per l'analisi statica di una Linea di Alberi. L'Energia Elettrica, no 12, Vol LVIII
12. Curami A, Pizzigoni B, Vania A (1986) On the rubbing phenomena in turbomachine. In: 1FT0MM international conference on rotordynamics, Tokyo
13. Curami A, Vania A, Gadda R, Pagani A (1985) Sui fenomeni di friction in turbomacchine. Rilievi sperimentali e loro simulazione. LXXXVI Riunione AEI, Pavia
14. Den Hartog JP (1956) Mechanical vibrations. Mc Graw-Hill Company, New York
15. Diana G, Curami A, Pizzigoni B (1988) Computer analysis of rotor bearing systems. P.A.L. L.A.: a package to analyze the dynamic behaviour of a rotor- supporting structure. Interntional Centre For Mechanical Sciences, Rotordynamics 2 -problems in turbomachinery. Springer, Wien

16. Falco M, Petrone F (1998) Analisi di vibrazioni su pompe di alimento di caldaie. Rapporto Interno del Dipartimento di Meccanica del Politecnico di Milano
17. Finzi B (1978) Meccanica razionale. Zanichelli, Bologna
18. Harris CM, Crede CE (1980) Shock and vibration handbook. McGraw-Hill Company, New York
19. Jacazio G, Piombo B (1977) Meccanica applicata alle macchine. Levrotto and Bella, Turin
20. Meirovitch L (1986) Elements of vibration analysis. McGraw-Hill Company, New York
21. Nestorides EJ (1958) A handbook on torsional vibration. Cambridge at the University Press, Cambridge
22. Sansò F (1991) Il trattamento statistico dei dati. Clup, Milan
23. Sesini O (1964) Meccanica applicata alle macchine. Casa Editrice Ambrosiana, Milan
24. Shabana AA (1989) Dynamics of multibody systems. Wiley, New York
25. Thomson WT (1974) Vibrazioni meccaniche—teoria e applicazioni. Tamburini Editore, Milan
26. Timoshenko S, Young DH (1968) Vibration problems in engineering. D. Van Nostrand Company Inc, Princeton
27. Timoshenko SP, Young DH, Weaver W (1974) Vibration problems in enginnering, 4th edn. Wiley, New York
28. Tondl A (1965) Some problems of rotor dynamic. Chapmann & Hall, London

Rotordynamics: Hydrodinamically Lubricated Bearings (see Also the Bibliography of Chap. 5)

29. Biraghi B, Falco M, Pascolo P, Solari A (1980) Cuscinetto lubrificato in regime laminare e turbolento: parte II—lubrificazione mista idrostatica- idrodinamica. L'Energia Elettrica LVII
30. Cheli F, Diana G, Vania A (1987) Identificazione dei parametri modali delle fondazioni di macchine rotanti. L'Energia Elettrica no 6
31. Cheli F, Manenti A (1991) Appunti di lubrificazione idrodinamica. Edizioni Spiegel, Milan
32. Costantinescu VN (1959) On turbulent lubrication. In: Proceedings of the IMechE London, vol 173, no 38
33. Diana G, Borgese D, Doufur A (1980) Experimental and analytical research on a full scale turbine journal bearing. In: Second international conference vibrations in rotating machinery 1980. The Institution of Mechanical Engineers L Birdcage Walk Westminster London
34. Diana G, Cheli F et al (1988) Non linear effects in lubricated bearings. In: 4-th international conference on vibrations in rotating machinery (C300/88 IMechE 1988), Edinburgh
35. Diana G, Cheli F, Manenti A, Petrone F (1990) Analytical and experimental research on some typical non-linear effects in the frequency response of a rotor on lubricated bearings. In: 3rd international conto on rotordynamics, Lyon (France)
36. Diana G, Cheli F, Manenti A, Petrone F (1988) Non linear effects in lubricated bearings. In: V international conto on vibrations in rotating machinery, Edinburgh
37. Diana G, Cheli F, Petrone F, Manenti A (1990) Analytical and experimental research on some typical non-linear effects in the frequency response of a rotor on lubricated bearings. In: 3-rd lFToMM rotordynamics conference, Lyon
38. Diana G, Cheli F, Petrone F, Manenti A (1990) Non linear effects in lubricated bearings, vibration and wear in high speed rotating machinery. NATO ASI Series, Applied Sciences, vol 174. Kluwer Academic Publishers, The Netherlands, pp 567–579
39. Diana G, Dufour A, Falco M, Serravalli W, Trebbi G (1977) Ricerche sperimentali sul cuscinetto di un turbogeneratore da 320 MW, Confronto con i risultati analitici. 20° Congresso di Tribologia, Turin, October 1976. Relazione Studio e Ricerca ENEL no 329

40. Diana G, Ruggieri G (1974) Sull'instabilità da film d'olio; metodo analitico per cuscinetti a lobi. L'Energia Elettrica, vol LI, no 4
41. Diana G, Ruggieri G (1974) Sull'instabilità da film d'olio; metodo analitico per cuscinetti cilindrici. L'Energia Elettrica, vol LV, no 3
42. Falco M et al (1980) Cuscinetto lubrificato in regime laminare e turbolento: parte I -analisi statica. L'Energia Elettrica, vol LVII, Parte III- analysis dinamica
43. Falco M, Macchi A, Vallarino G (1980) Cuscinetto lubrificato in regime laminare e turbolento, Parte II: Lubrificazione mista idrostatica-idrodinamica. L'Energia Elettrica, vol L VII, no 2
44. Falco M, Pascolo P (1979) Lubrificazione idrodinamica in regime turbolento. L'Energia Elettrica, vol LIV, no 3, 4
45. Frigeri C, Gasparetto M, Vacca M (1980) Cuscinetto lubrificato in regime laminare e turbolento. Parte I -analisi statica. L'Energia Elettrica, no 2
46. Gallagher RH et al (1975) Finite elements in fluids. Wiley, New York
47. Journalkus O, Stemlicht B (1961) Theory of hydrodynamic lubrification. Mc Graw-Hill, New York
48. Ng CW, Pan CHT (1965) A linearized turbulent lubrication theory. J Basic Eng 675–688
49. Pizzigoni B (1987) Effetti non lineari nei cuscinetti lubrificati. Tesi finale Dottorato di Ricerca in Meccanica Applicata Curriculum Vibrations Meccaniche, Milan
50. Pizzigoni B, Ruggieri G (1978) Sulla determinazione delle caratteristiche statiche e dinamiche dei cuscinetti lubrificati. L'Energia Elettrica, vol LV, no 2
51. Pizzigoni B, Ruggieri G (1978) Un procedimento approssimato per l'analisi dei cuscinetti in regime turbolento. L'Energia Elettrica vol LV, no 3
52. Pizzigoni B, Ruggieri G (1980) Caratteristiche di funzionamento dei cuscinetti lubrificati in regime turbolento. V congresso AIMETA, Palermo
53. Pizzigoni B (1983) Static and dynamic properties of tilting pad journal bearings: turbulence and thermal field effects. In: Proceedings of the sixth world congress on theory of machines and mechanisms, New Delhi
54. Ruggieri G (1976) Un metodo approssimato per la risoluzione dell'equazione di Reynolds. L'Energia Elettrica, vol LIII, no 2

Rotordynamics: seals

55. Allaire PE, Lee CC, Gunter EI (1978) Dynamic analysis of SSME plain and plain stepped turbulent seals. School of Engineering and Applied Science, University of Virginia Charlottesville, Report no UVN643092/MAE 81/145
56. Childs SB, Childs DW, Dresden I (1980) Estimation of seal bearing stiffness and damping parameters from experimental data. II conference vibration in rotating machinery, Cambridge
57. Falco M et al (1984) Plain seal dynamic behaviour, experimental and analytical results. C303/84 IMechE -III conto vibrations in rotating machinery, York
58. Falco M, Diana G, Marenco G, Mimmi G (1982) Experimental research on the behaviour of hydrodynamical plain seals by means of a specific testing device (A.P.S.A.L). In: IfToMM conference on rotordynamic problerns in power plants, Rome (Ita1y)
59. Falco M, Mimmi G, Marenco G (1986) Effect of seals on rotor dynamic. In: IfToMM international conference on rotordynamics, Tokyo (Iapan)
60. Marenco G, Falco M, Gasparetto M (1980) On the influence of the dynamic response of the supports on the criticai speed of a boiler feed pump. Convention Fluid Machinery Failures Prediction, Analysis and Prevention

Rotordynamics: Balancing

61. Bachschmid N, Diana G, Pizzigoni B (1984) The effect of unbalance on cracked rotors. In: 11° congress on vibrations in rotating machinery, York
62. Bachschmid N, Frigeri C (1982) Some results on the twice-per-revolution balancing of generators. In: IfToMM conference on rotordynamic problems in power plants, Rome (Italy)
63. Balde H (1975) Computer aided balancing of flexible rotors. IFToMM Newcastle
64. Bigret R, Curami A, Frigeri C, Mecchi A (1977) Use of in-field computer for balancing high power turbomachinery, Chicago
65. Bragley RH (1976) Modern influence coefficient techniques for multiplane rotor balancing in the factory. Test Cali and Field, Cambrige
66. Drechsler J (1975) A combination of modal balancing and the influence coefficient method. IfToMM Newcastle
67. Drechsler J (1976) Systematic combination of experiments and data processing in balancing of flexible rotors. Cambrige
68. Giers A (1976) Pratice of flexible rotor balancing. Cambrige
69. Godlewsky (1975) Self adjustable disturbance filters in balancing techniques. IFToMM Newcastle
70. Goodmann TP (1963) A least-square method for computing balance correction. ASME Trans
71. Gosch D (1978) Modales Answrichen Elastischer Laufer Ohm Testgemicht Setzungen, Karlsruhe
72. Larsen LO (1976) On the determination of the lnfluence coefficients in rotor balancing using linear regression analysis, Cambrige
73. Lund JW (1976) A method for using the free shaft modes in rotor balancing, Cambrige
74. Worwzecky J (1975) A method of site balancing of machine systems with damping included. lfToMM NewcastIe

Rotordynamics: Phenomena of Instability

75. Bachschmid N, Cavalca KL, Cheli F (1992) Relevance of the dynamc behaviour of the supporting structure in calculating critical speeds of multistage centrifugai pumps with lnterstage seals. In: Proceedings of the international conference on rotating machine dynamics, Venice
76. Bianchi G (1983) Dynamic problems in flexible rotors. In: Proceedings Sixth lfToMM congress, New Delhi
77. Diana G, Falco M, Gasparetto M (1977) On the flutter lnstability of a suspension bridge using the finite elements method. ASME 77-DET-140
78. Diana G, Massa E, Pizzigoni B, Di Pasquantonio F (1975) A finite element method for computing oil whirl lnstability of a rotating shaft supported by elastic bearings. In: Proceedings of the IFToMM 4th congress
79. Diana G, Massa E, Pizzigoni B, Taddei C (1982) A forced vibration method to calculate the oil film instability threshold of rotor-foundation systems. In: lfToMM conference on rotordynamic problems in power plants, Rome (Italy)
80. Diana G, Ruggieri G (1974) Sull'instabilità da film d'olio; metodo analitico per cuscinetti a lobi. L'Energia Elettrica, vol LI, no 4
81. Diana G, Ruggieri G (1974) Sull'instabilità da film d'olio; metodo analitico per cuscinetti cilindrici. L'Energia Elettrica, vol LI, no 3
82. Gunter Jr EJ (1966) Dynamic stability of rotor-bearing systems, NASA, SP-113

Rotordynamics: Effects of the Foundation on the Rotor

83. Bachschmid N, Bernante R, Frigeri C (1982) Dynamic analysis of a 660 MW turbogenerator foundation. In: IFToMM conference on rotordynamic problems in power plants, Rome (Italy)
84. Bachschmid N, Cavalca KL, Cheli F (1992) Relevance of the dynamc behaviour of the supporting structure in calculating criticai speeds of multistage centrifugai pumps with lnterstage seals. In: Proceedings of the international conference on rotating machine dynamics, Venice
85. Bachschmid N, Lucchesi K, Cheli F (1992) The relevance of the dynamic behaviour of the supporting structure in calculating criticai speeds of multistage centrifugai pumps with lnterstage seals. In: Proceedings di rotating machine dynamics conference. Springer, London, pp 158–165
86. Bachschmid N, Pizzigoni B, Di Pasquantonio F (1977) A method for lnvestigating the dynamic behaviour of a turbomachinery shaft on a foundation. In: International conference, Chicago, ASME paper 77-DET-16
87. Cheli F, Curami A, Diana G, Vania A (1985) On the use of modal analysis to define the mechanical impedance of a foundation. In: ASME–mechanical vibration and noise conference, Cincinnati
88. Cheli F, Vania A, Cavalca KL, Dedini FG (1991) Dynamical behaviour analysis of rotor structure system by modal truncation method. In: 9-th International modal analysis conference, Florence
89. Cheli F, Vania A, Cavalca KL, Dedini FG (1992) Supporting structure effects on rotating machinery vibrations. In: Proceedings of the 5-th international conference on rotating machine dynamics, Bath, 1992, C432/119, IMechE
90. Diana G, Cheli F, Vania A (1988) A method to identify the foundation modal parameters through measurements of the rotor vibrations. In: 4-th international conference on vibrations in rotating machinery (C300/88 IMechE 1988), Edinburgh
91. Diana G, Cheli F, Vania A (1988) A method to define the foundation modal parameters through the measurements of the rotor vibrations. In: IV international conto on vibrations in rotating machinery, Edinburgh
92. Diana G, Cheli F, Vania A (1987) Identificazione dei parametri modali delle fondazioni di macchine rotanti. L'Energia Elettrica, no 6
93. Diana G (1983) Foundation effects in rotor dynamic behaviour. In: Session proceeding: technical committee on rotordynamics—sixth world congress on theory of machines and mechanisms, New Delhi
94. Diana G, Ruggieri G (1974) Sulla determinazione delle frequenze proprie di alberi su più supporti elastici. L'Energia Elettrica, vol LI, no 2
95. Diana G, Bachschimd N (1978) Influenza della struttura portante sulle velocità critiche flessionali di alberi rotanti. L'Energia Elettrica, no 9
96. Lucchesi Cavalca K, Dedini GF, Vania A (1991) Dynamical behaviour analysis of rotor structure system by modal truncation method. In: 9th international modal analysis conference, Florence
97. Vania A, Lucchesi Cavalca K, Dedini FG (1992) Supporting structure effects on rotating machinery vibrations. In: Proceedings of the 5th international conference on rotating machine dynamics, Bath, C432/119, IMechE

Rotordynamics: 2 x rev Vibrations, Effect Due to Cracks

98. Bachschimd N, Diana G (1980) Reduction of twice per revolution vibration levels due to mass effect in large turbogenerators. Vibration in rotating machinery congress, Cambrige 1980, pp 203–208, paper C 281/80
99. Bachschmid N (1983) A method for calculating the dynamic behaviour of cracked shafts. In: Proceedings of the sixth world congress on theory of machines and mechanisms, New Delhi
100. Bachschmid N, Del Vescovo D, Pezzuti E (1990) Sforzi e deformazioni nell'intorno di una discontinuità in un albero assial-simmetrico in differenti condizioni di carico. X congresso nazionale AIMETA, Pisa 3–5 Oct 1990
101. Bachschmid N, Del Vescovo D (1990) Thermal bow in an axi-symmetrical shaft—containing transverse mark and related vibrational behaviour. In: 3rd international congress on rotordynamics, Lyon (France), l0–12 Sept 1990
102. Diana G, Bachschmid N, Angeli F (1986) An on-line crack detection method for turbogenerator rotors. In: IFToMM international conference on rotordynamics, Tokyo (Japan)

Rotordynamics: Diagnostics

103. Bachschmid N et al (1985) Possibilità e limiti delle misure di vibrazione per la diagnosi dei mal funzionamenti meccanici del macchinario rotante Risultati di una indagine su una turbina da 320 MW. LXXXVI Riunione AEI, Pavia
104. Bachschmid N, Pizzigoni B, Vania A (1987) Some results in the diagnosis of malfunctions in rotating machinery. In: International conference on mechanical dynamics, Shenyang (China)
105. Capelli M, Diana G, Lajournali GL, Vallini A (1990) Diagnosis of malfunctions in large turbine-generators: some case histories. In: 3rd international conto on rotordynamics, Lyon (France), 10–12 Sept 1990
106. Curami A, Gasparetto M, Vania A (1990) An on line application of a turbomachinery supervisory system. In: 3rd international conto on rotordynamics, Lyon (France), l0–12 Sept 1990
107. Curami A, Vania A (1988) Un'applicazione di metodologie statistiche nella diagnostica del macchinario rotante. Rapporto Interno del Dipartimento di Meccanica del Politecnico di Milano
108. Diana G et al (1986) A diagnostic system to define malfunctioning conditions of a turbogenerator. In: International conference on rotordynamics, JSME, IFToMM, Tokyo
109. Diana G, Gasparetto M, Cheli F, Pizzigoni B, Vania A (1986) A diagnostic system to define malfunctioning conditions of a turbogenerator. In: IFToMM international conference on rotordynamics, Tokyo (Japan)
110. Diana G, Gasparetto M, Vania A (1987) Machinery diagnostics: examples of application to thermoelectric plants. In: 7° IFToMM congress 'theory of machines and mechanisms', Seville
111. Frigeri C, Gasparetto M, Pizzigoni B (1980) Metodologia di misura e di elaborazione dati per la definizione delle vibrazioni assolute di un albero rotante. In: 5 congresso nazionale di meccanica teorica ed applicata, Palermo
112. Gasparetto M, Diana G, Vania A (1989) Diagnostica del macchinario: esempi di applicazione agli impianti termoelettrici. La Manutenzione, Anno 1°, n O

Chapter 7
Random Vibrations

7.1 Introduction

In the previous chapters we analysed the behaviour of linear systems (Chaps. 1 and 2) and non-linear systems (Chap. 5) that are subject to deterministic forces, that is, forces whose performance can be defined univocally over time (constant, harmonic, periodic, non-periodic, see Table 7.1 and Fig. 7.1) the value of which can be known in advance for each time t.

In f, excitations may also be of an aleatory or random nature (Fig. 7.2; Table 7.2): these excitations are not deterministic, that is, the value assumed by them, at a generic time t, cannot be defined in advance, except in probabilistic terms [2, 4, 5, 9, 11, 16, 23]. Such sources of excitation cause the structure to behave in a non-deterministic manner, examples include:

- the response of structures to wind turbulence (see the relative bibliography) [26–34];
- the response of fluid machinery (turbine, pump, etc.) to excitation resulting from the turbulent motion of the actual fluid;
- the response of structures to an earthquake;
- the response of offshore structures to wave motion (where the fluid-dynamic forces are not periodic) (See the relative bibliography);
- the behaviour of land vehicles travelling along an uneven route (see the relative bibliography);
- the response of an aircraft to atmospheric turbulence;
- the dynamic response of a missile to the noise emissions generated by the engine.

These sources of excitation cause a random response in the system, in terms of displacements, accelerations, and the corresponding state of stress, which can be analysed using specific techniques to define the behaviour of the system, as is done for deterministic type excitations. A typical example is the fatigue analysis of a system subject to random type stress. In fact, for this type of system it is not

F. Cheli and G. Diana, *Advanced Dynamics of Mechanical Systems*,
DOI 10.1007/978-3-319-18200-1_7

Table 7.1 Classification of deterministic forces

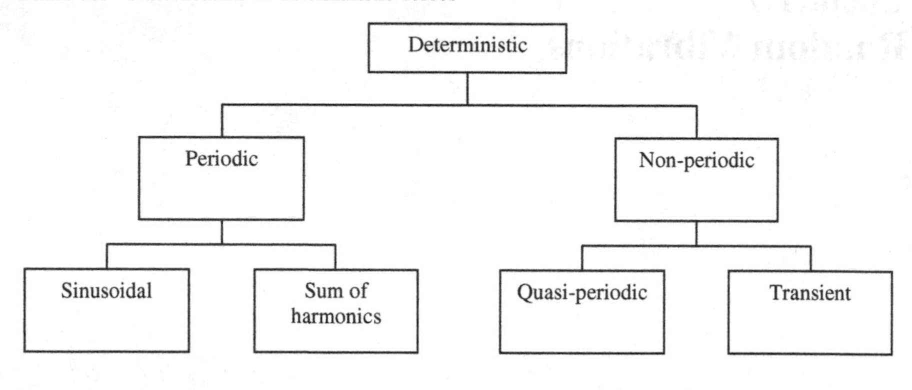

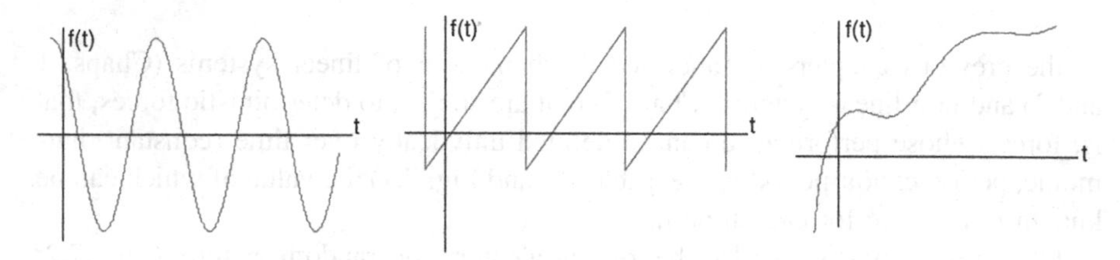

Fig. 7.1 Examples of deterministic forces

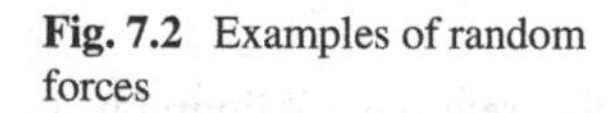

Fig. 7.2 Examples of random forces

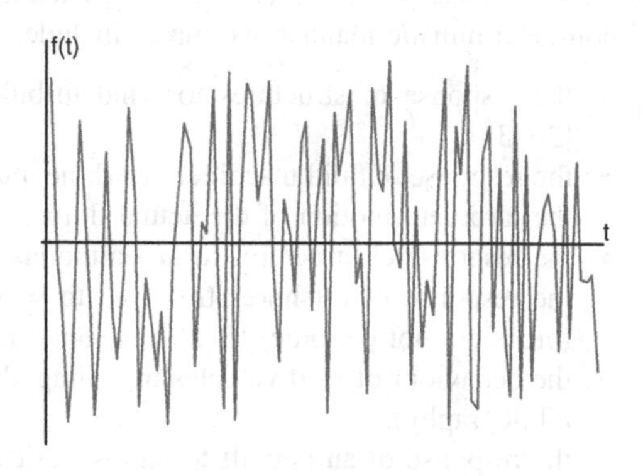

possible to estimate the fatigue accumulation of the material, as we do for harmonic excitation, by simply using the material's Wohler Curve. For systems subject to random forces other methods should be considered, e.g. the *rain-flow* method [14] amongst others. Similar problems can be seen, for example, when judging the comfort of passengers of vehicles subject to random excitation, such as that caused by uneven roads.

Table 7.2 Classification of random forces

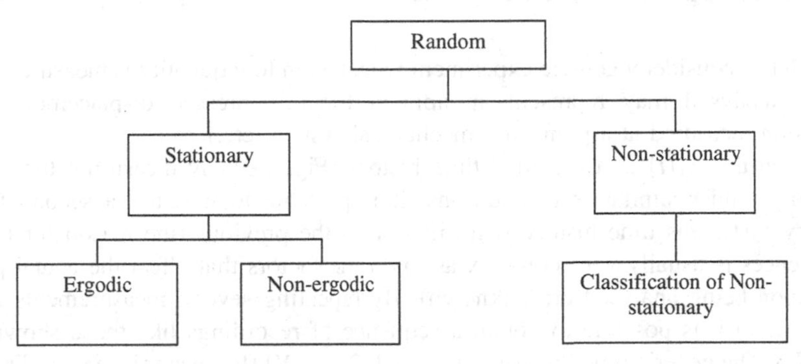

The study of random vibrations, using the theory of stochastic processes, is a new branch in engineering and has developed significantly over recent decades. Clearly the issues linked to random excitation cannot be analysed on a deterministic basis, as discussed up until now in the previous chapters, a statistical or rather, a probabilistic approach must be taken, where both the excitation, and the response of the mechanical system to this excitation, can be defined in relation to several statistical parameters.

Many random phenomena show characteristic statistical regularity in that, although the time histories differ from one another, some average quantities remain constant. In this case, the excitation and the response are considered random processes (or stochastic processes) that can be described using a few statistical parameters and appropriate statistical functions, such as the power spectral density function.

This chapter merely seeks to offer an introduction to the complex issues associated with random vibrations: for a more thorough analysis refer to the bibliography at the end of the chapter. In particular, the following discussion will introduce:

- the concept of the stochastic process;
- descriptions of the characteristic quantities (mean, variance, autocorrelation function, power spectral density etc.) associated with it, both in the time and the frequency domain;
- methods required to evaluate the response of vibrating mechanical systems (linear and non-linear) that are subject to random excitations;
- some examples of applications.

7.2 Defining a Random Process

Now let us consider a generic experiment where a random quantity is measured (the signal analysed may represent an applied force, impressed displacement, the vibration measured at a point in a mechanical system, etc.).

Given that $x_1(t)$ is a recorded time history (Fig. 7.3); by measuring the same quantity, under similar test conditions, it is possible to measure a second time history $x_2(t)$: this time history will differ from the previous (the reason for these differences is usually very complex as often the factors that affect the actual phenomenon being analysed are unknown). By repeating several measurements ($i = 1, 2, \ldots, N$) it is possible to obtain a sequence of recordings like those shown in Fig. 7.3. The generic time history $x_i(t)$ ($i = 1, 2, \ldots, N$) (for example, a recording of wind speed made with an anemometer) is referred to as random variable, whereas, the entire set of time histories (a set of recordings made under the same conditions, for example, same site and same mean value and average wind direction) represents the so-called random process or stochastic process.

Each observation (aperiodic) has an overall trend that is similar to the others, but differs in the detail: it is possible to define some of the constant statistical parameters of these quantities, known as statistically normal random quantities.

In general, a stochastic process can be a function of both time and of space $x(t, \xi)$: in this chapter we will be discussing time-dependent processes only. In general, the generic random variable $x_i(t)$ does not represent the entire process, except in those random processes defined as stationary and ergodic where, as we will discuss in more detail, it is possible to obtain the statistical information for an entire random process by analysing just one of its variables.

As mentioned, to analyse the random process we must define some significant quantities (statistics) that characterise the actual process and the value of which tends to a constant value for N (number of observations) tending to infinity. When this occurs, the stochastic quantity is defined as being statistically normal.

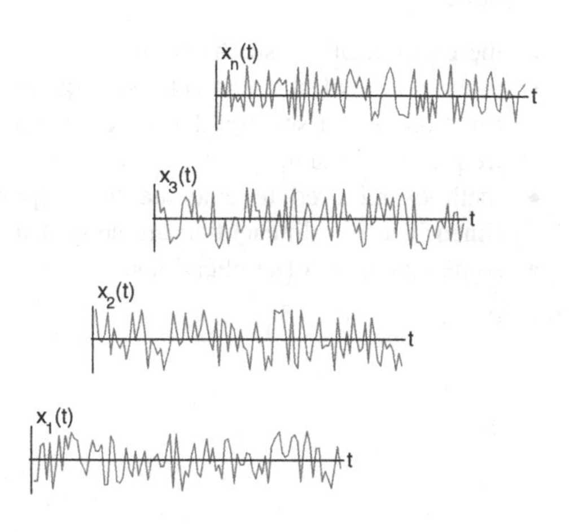

Fig. 7.3 Random process defined by several random variables

7.3 Parameters Defining the Statistical Characteristics of a Random Process

Before performing an in depth analysis of aleatory phenomena, we will briefly introduce the significant quantities to statistically define their characteristics. Given a generic time history $x(t)$ (Fig. 7.2) is defined as:

- mean value (or static component)[1]:

[1]We should remember that the symbol E[g] is used to define the expected value of the generic quantity g(x) [2, 9] defined as

$$E[g] = \int_{-\infty}^{+\infty} g(x)p(x)dx \tag{7.1.1}$$

p(x) being the probability density function, defined as:

$$p(x) = \frac{dP(x)}{dx} \tag{7.1.2}$$

where P(x) is the probability distribution function which represents the probability that x(t) is lower than a given value x (figure below). Let us briefly recall the properties of this function:

$$\begin{aligned} &0 < P(x) < 1 \\ &P(-\infty) = 0 \\ &P(\infty) = 1 \\ &P(x) = \int_{-\infty}^{x} p(\xi)d\xi \end{aligned} \tag{7.1.3}$$

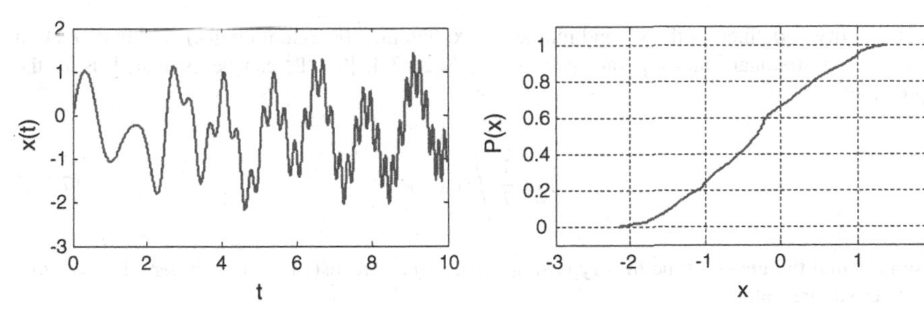

Meaning of the probability distribution function

The mean value μ_x is calculated, in general, as the first moment in relation to the probability density by assuming that $g = x$ in (7.1.1): in the case of a stationary and ergodic process (see Sect. 7.4) the mean value can be estimated using:

$$\mu_x = E[x] = \lim_{T\to\infty}\frac{1}{T}\int\limits_0^T x(t)dt \quad (7.1)$$

- mean square value (i.e. dynamic component)[2]:

$$\psi_x^2 = E[x^2] = \lim_{T\to\infty}\frac{1}{T}\int\limits_0^T x^2(t)dt \quad (7.2)$$

- *RMS* (root mean square), the square root of the mean square value:

$$RMS = \sqrt{\psi_x^2} \quad (7.3)$$

- variance:

$$\sigma_x^2 = \lim_{T\to\infty}\frac{1}{T}\int\limits_0^T (x(t) - \mu_x)^2 dt = \psi_x^2 - \mu_x^2 \quad (7.4)$$

(Footnote 1 continued)

$$\lim_{T\to\infty}\frac{1}{T}\int\limits_0^T x(t)dt. \quad (7.1.4)$$

[2]This quantity is defined as the second moment of x, obtained by assuming $g(x) = x^2$ in (7.1.1): in the case of a stationary and ergodic process (see Sect. 7.4, [9]) Ψ_x^2 can be estimated using the expression:

$$\lim_{T\to\infty}\frac{1}{T}\int\limits_0^T x^2(t)dt \quad (7.2.1)$$

Assuming that the generic time history is sinusoidal $x(t) = Arccos(\omega t + \phi)$, with period $T_o = 2\pi/\omega$, the mean square value is

$$\psi_x^2 = \frac{1}{T_0}A^2\int\limits_0^T \cos^2(\omega t + \phi)dt = \frac{A^2}{2}. \quad (7.2.2)$$

- standard deviation (quantity that defines the fluctuations of the quantity in question around the mean value μ_x[3]):

$$\sigma_x = \sqrt{\sigma_x^2} \tag{7.5}$$

Furthermore, the following is also defined as an autocorrelation function[4]:

$$R_{xx}(\tau) = \lim_{T\to\infty} \frac{1}{T} \int_0^T x(t)x(t+\tau)dt \tag{7.6}$$

---

[3]If, for example, the time history is characterised by an x Gaussian (or normal) distribution, the probability density function p(x) becomes:

$$p(x) = \frac{1}{2\pi\sigma_x} e^{-\left(\frac{(x-\mu_x)^2}{2\sigma_x^2}\right)} \tag{7.3.1}$$

That is, it is defined univocally by the mean value and by the standard deviation: the following table shows the probability $p(\bar{x})$ that an event $\bar{x}$ falls within the range of values around the mean value μ_x and of a size that is 1, 2 and 3 times the standard deviation σ_x:

$$\begin{aligned} \bar{x} &= \mu_x \pm \sigma_x \Rightarrow p(\bar{x}) = 68.26\,\% \\ \bar{x} &= \mu_x \pm 2\sigma_x \Rightarrow p(\bar{x}) = 95.44\,\% \\ \bar{x} &= \mu_x \pm 3\sigma_x \Rightarrow p(\bar{x}) = 97.74\,\%. \end{aligned} \tag{7.3.2}$$

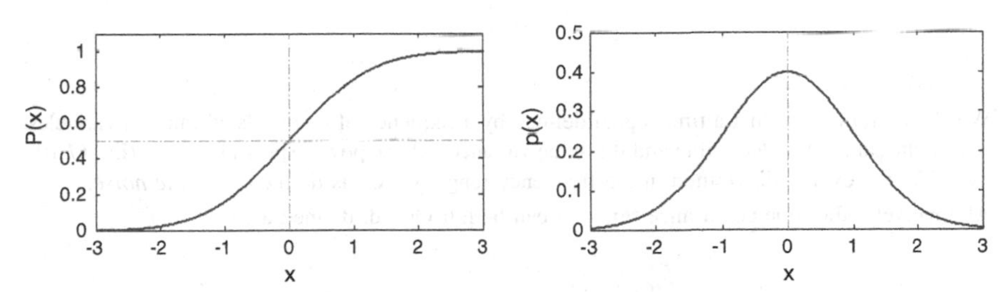

Gaussian (or normal) function of probability distribution P(x) and probability density p(x), for $\mu_x = 0$ and $\sigma_x = 1$

[4]Alternatively, the time covariance function can be introduced, defined as:

$$C_{xx}(\tau) = \lim_{T\to\infty} \frac{1}{T} \int_0^T (x(t) - \mu_x)(x(t+\tau) - \mu_x)dt. \tag{7.4.1}$$

This quantity has the following properties [1, 12, 13, 18]:

$$R_{xx}(0) = \lim_{T\to\infty} \frac{1}{T} \int_0^T x(t)x(t+0)dt = \psi_x^2 \tag{7.7a}$$

$$R_{xx}(\tau) = R_{xx}(-\tau) \tag{7.7b}$$

$$R_{xx}(0) \geq |R_{xx}(\tau)| \tag{7.7c}$$

The autocorrelation function is often defined in non-dimensional form:

$$\bar{R}_{xx}(\tau) = \frac{R_{xx}(\tau)}{R_{xx}(0)} = \frac{R_{xx}(\tau)}{\psi_x^2} \Rightarrow -1 < \bar{R}_{xx}(\tau) < 1 \tag{7.8a}$$

For:

$$\begin{array}{rl} \bar{R}_{xx}(\tau) = 1 & \text{correlation is maximum} \\ \bar{R}_{xx}(\tau) = 0 & \text{correlation is null} \\ \bar{R}_{xx}(\tau) = -1 & \text{correlation is maximum,} \\ & \text{but variations are in counterphase} \end{array} \tag{7.8b}$$

Figure 7.4 shows the trend of the autocorrelation function $R_{xx}(\tau)$ for some characteristic functions.[5]

Let us consider the case where there are two different histories, $x(t)$ and $y(t)$, defining the function of cross-correlation[6]:

$$R_{xy}(\tau) = \lim_{T\to\infty} \frac{1}{T} \int_0^T x(t)y(t+\tau)dt = -R_{yx}(\tau) \tag{7.9}$$

---

[5]We should remember that a time signal defined by a sequence of uncorrelated random variables, all with the same expected value and the same variance, whose power spectral density (Eq. (7.10), Sect. 7.3.1) is evenly distributed in the frequency range $[0, \infty]$, is defined as *white noise*.

[6]Alternatively, the time covariance function can be introduced, defined as:

$$C_{xy}(\tau) = \lim_{T\to\infty} \frac{1}{T} \int_0^T (x(t) - \mu_x)(y(t+\tau) - \mu_y)dt = -C_{yx}(\tau) \tag{7.6.1}$$

which, in non-dimensional form, is given by the correlation coefficient function $|\rho_{xy}(\tau)| \leq 1$:

$$\rho_{xy} = \frac{C_{xy}}{\sigma_x \sigma_y} \tag{7.6.2}$$

For $\tau = 0$, $C_{xy}(0)$ is the covariance and $\rho_{xy}(0)$ is the correlation coefficient.

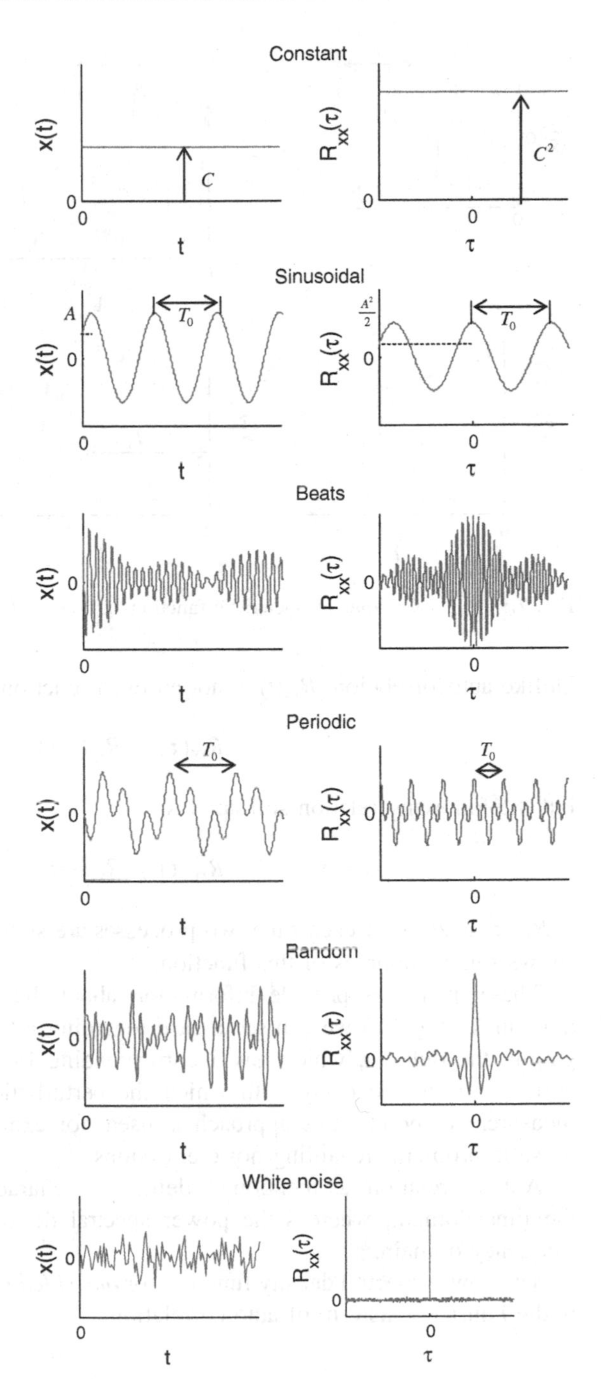

Fig. 7.4 $R_{xx}(\tau)$ for some characteristic functions

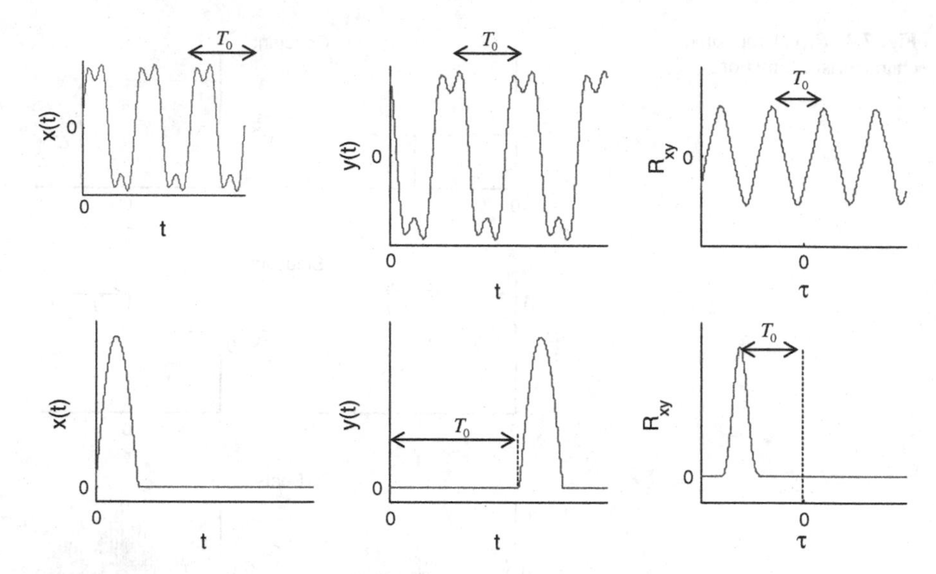

Fig. 7.5 $R_{xy}(\tau)$ for some characteristic functions x(t) and y(t)

Unlike autocorrelation, $R_{xy}(t)$ is not an even function, i.e.:

$$R_{xy}(\tau) \neq R_{xy}(-\tau) \tag{7.9a}$$

but the following relation still applies:

$$R_{xy}(\tau) \neq R_{yx}(-\tau) \tag{7.9b}$$

If $R_{xy}(\tau)$ is zero for each t the two processes are statistically uncorrelated: Fig. 7.5 shows some examples of this function.

These quantities provide information about the random process in the time domain. In Fig. 7.5 we can see a case regarding a typical phenomenon of propagation: the delay τ, which has a corresponding high value of cross-correlation, corresponds to the delay with which the perturbation reaches the two different measurement points: this approach is used, for example, to locate the source of noise in problems regarding noise emissions.

Auto-correlation, as mentioned, defines the characteristics of the signal $x(t)$ in the time domain, whereas the power spectral density function defines it in the frequency domain.

The power spectral density function (or *one sided auto-spectral density function*) is the Fourier transform of auto-correlation:

$$G_{xx}(f) = 2\int_{-\infty}^{\infty} R_{xx}(\tau)e^{-i2\pi f\tau}d\tau$$
$$G_{xx}(\omega) = 2\int_{-\infty}^{\infty} R_{xx}(\tau)e^{-i\omega\tau}d\tau, \quad \text{con } 2\pi f = \omega \tag{7.10}$$

For this definition, the power spectral density provides the same information about a random process that is given by the defined auto-correlation function, but in the frequency domain.[7] This quantity has the following properties:

$$\int_0^{\infty} G_{xx}(f)df = \int_0^{\infty} G_{xx}(\omega)d\omega = \psi_x^2 = \sigma_x^2 + \mu_x^2 \tag{7.11}$$

i.e. the area under the curve defined by the function $G_{xx}(f)$ represents the mean square value. This property immediately gives the following relation:

$$G_{xx}(f)\Delta f = \psi_x^2(f)$$
$$G_{xx}(\omega)\Delta\omega = \psi_x^2(\omega) \tag{7.12}$$

where $\psi_x^2(f)$, by sufficiently small Δf or $\Delta\omega$, represents the average mean square value associated with just the generic frequency f, i.e. $\psi_x^2(f)$ represents the portion of the mean square value filtered at the frequency f of band Δf.

Now we will examine other properties of the power spectral density function [9]:

- $G_{xx}(\omega)$ is a real quantity[8]

[7] The presence of the term 2 in the definition of (7.10) is due to the fact that the one-sided auto-spectral density function [9] is defined only in the positive frequency field $[0, \infty]$.

[8] The relation:

$$G_{xx}(\omega) = 2\int_{-\infty}^{\infty} R_{xx}(\tau)e^{-i\omega\tau}d\tau \tag{7.8.1}$$

can, in fact, be developed as follows:

$$G_{xx}(\omega) = 2\int_{-\infty}^{\infty} R_{xx}(\tau)(\cos(\omega\tau) - i\sin(\omega\tau))d\tau \tag{7.8.2}$$

By analysing the integral:

$$\int_{-\infty}^{\infty} R_{xx}(\tau)\sin(\omega\tau)d\tau = \int_{-\infty}^{0} R_{xx}(\tau)\sin(\omega\tau)d\tau + \int_{0}^{\infty} R_{xx}(\tau)\sin(\omega\tau)d\tau \tag{7.8.3}$$

- $G_{xx}(\omega) \geq 0$
- the inverse transform of $G_{xx}(\omega)$ is given by[9] [2, 9]:

$$R_{xx}(\tau) = \int_0^\infty G_{xx}(\omega)\cos(\omega\tau)d\omega = \Re e\left(\int_0^\infty G_{xx}(\omega)e^{i\omega\tau}d\omega\right) \tag{7.13}$$

If a generic function $x(t)$ is periodic, we can define its complex line spectrum, where the generic harmonic is constant in magnitude and phase; for a random function $x(t)$, however, the power spectral density (real quantity) only defines the mean square value of the amplitudes of the single harmonics.

Figure 7.6 shows some significant examples of *power spectral density function* for different periodic or random signals. Those random processes characterised, respectively, by a narrow peak form (centred around the frequency corresponding to the peak) of the power spectral density or where the actual quantity is characterised by significant values of the power density in a wide frequency camp, are defined as narrow band processes or wide band processes.

In extreme cases where there is only one harmonic (with random phase) the power spectral density presents a function δ (Dirac function), whereas a constant power spectral density value defines the so-called white noise.

In parallel, see Fig. 7.4, narrow band processes show a trend in the sinusoidal auto-correlation function $R_{xx}(\tau)$ modulated by a decreasing exponential, whereas in wide band processes $R_{xx}(\tau)$ decreases rapidly. In the case of white noise we have $R_{xx}(\tau) \neq 0$ only for $\tau = 0$.

If we analyse two different random processes $x(t)$ and $y(t)$ is possible to define:

- the cross-spectrum, a complex quantity (also known as *cross-spectral density function*), defined as the Fourier transform of the cross-correlation:

$$G_{xy}(f) = G_{yx}(f) = 2\int_{-\infty}^{\infty} R_{xy}(\tau)e^{-i2\pi f\tau}d\tau \tag{7.14}$$

(Footnote 8 continued)

Taking into account (7.7a–7.7c) whereby $R_{xx}(\tau) = R_{xx}(-\tau)$ and the fact that the sine function is an odd function so $\sin(\omega\tau) = -\sin(-\omega\tau)$, this integral cancels itself out, therefore (7.8.2) is reduced to:

$$G_{xx}(\omega) = 2\int_{-\infty}^{\infty} R_{xx}(\tau)\cos(\omega\tau)d\tau = 4\int_0^\infty R_{xx}(\tau)\cos(\omega\tau)d\tau \tag{7.8.4}$$

hence it is possible to affirm that, as $R_{xx}(\tau)$ is real, then $G_{xx}(\omega)$ is also real.

[9] Equations (7.8.4) and (7.13) go under the name of Wiener-Khintchine equations.

Fig. 7.6 $G_{xx}(\omega)$ for some characteristic functions $x(t)$

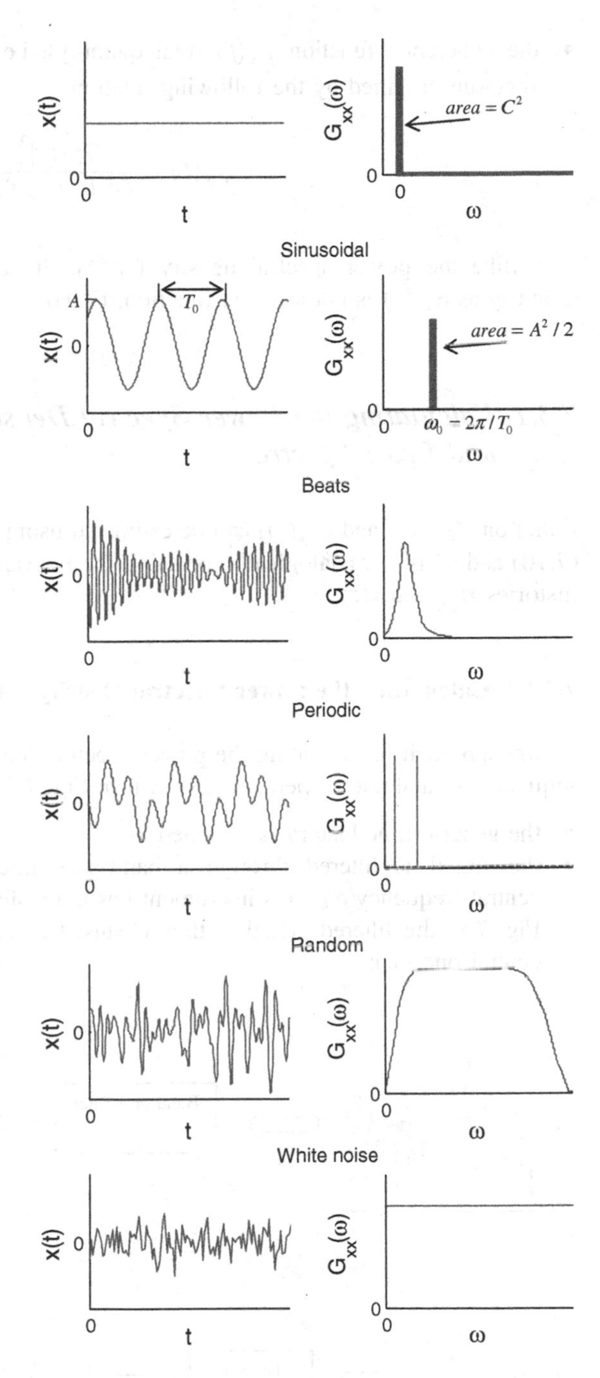

- the coherence function $\gamma_{xy}^2(f)$ (real quantity), i.e. the non-dimensional cross-spectrum, defined by the following relation:

$$\gamma_{xy}^2(f) = \frac{|G_{xy}(f)|^2}{G_{xx}(f)G_{yy}(f)} \tag{7.15}$$

Unlike the power spectral density $G_{xx}(f)$, the cross-spectrum is a complex quantity as $R_{xy}(\tau)$ is not an even function, (7.9a).

7.3.1 Calculating the Power Spectral Density Function and Cross-Spectra

Functions $G_{xx}(\omega)$ and $G_{xy}(\omega)$ can be estimated using the same definition shown in (7.10) and (7.14) or analogically, or using the Fourier transform of the actual time histories $x(t)$ and $y(t)$.

7.3.1.1 Calculating the Power Spectral Density Function Analogically

A first approach in estimating the power spectral density consists of the following sequence of analogical operations, shown in Fig. 7.7:

- the generic time history is obtained $x(t)$;
- the signal is filtered through a band-pass filter with bandwidth $\Delta\omega$ and central frequency ω_k (this instrument has a transfer function like that shown in Fig. 7.8: the filtered signal is thus cleansed of all the frequencies except the central one ω_k);

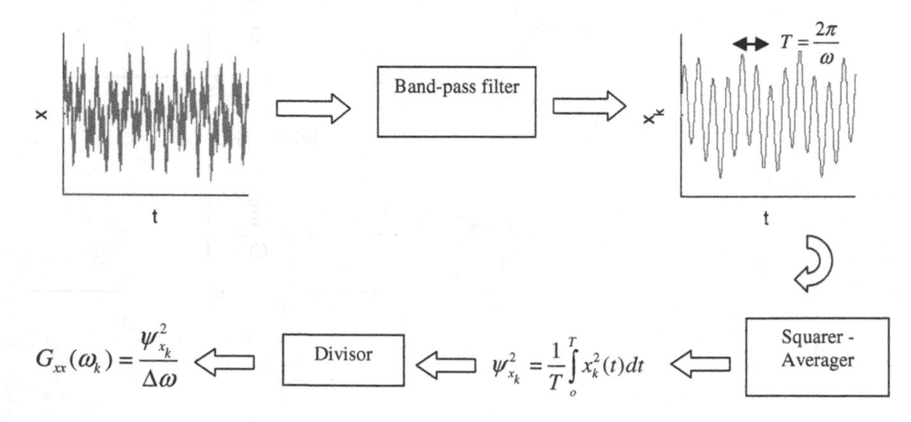

Fig. 7.7 Analogue definition of power spectral density

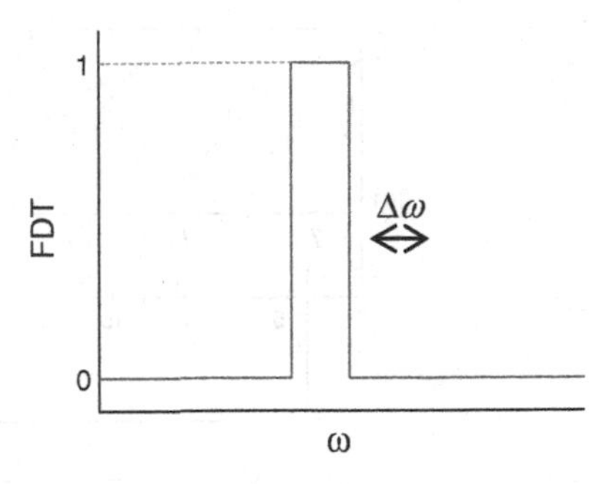

Fig. 7.8 Transfer functions of a band-pass filter

- the filtered output signal $x_k(t)$ is almost harmonic, given the randomness of the input signal;
- the instantaneous value of the signal $x_k(t)$ is squared;
- and it is averaged in a period T using:

$$\psi_{x_k}^2 = \frac{1}{T}\int_o^T x_k^2(t)dt \tag{7.16}$$

- thus obtaining the mean square value of the filtered signal $\psi_{x_k}^2$;
- this value is divided by the bandwidth $\Delta\omega$ to obtain, by the definition shown in (7.12), an estimation of the power spectral density function in correspondence to the central frequency of the filter ω_k:

$$G_{xx}(\omega_k) = \frac{\psi_{x_k}^2}{\Delta\omega} \tag{7.17}$$

- by subsequently changing the frequency of the filter ω_k, it is possible to define the function $G_{xx} = G_{xx}(\omega)$ in the entire field of frequency concerned.

At present, this approach is usually replaced by an approach based on the Fourier transform of the signal $x(t)$, as described in the following section.

7.3.1.2 Calculating the Power Spectral Density Function Using Fourier Series Development

At present the most common method used to define the power spectral density is numeric and is based on Fourier series development:

- The generic time history $x(t)$ (Fig. 7.9) is divided into equal periods $T_{01} = T_{02} = \cdots = T_{0N} = T_0$; each individual history $x_i(t)\ (i = 1, 2, \ldots, N)$ is assumed periodic: to this end, *windowing* is carried out on it [3, 13, 18], i.e. $x_i(t)$ is multiplied by a specific function, for example, the Hanning window, making

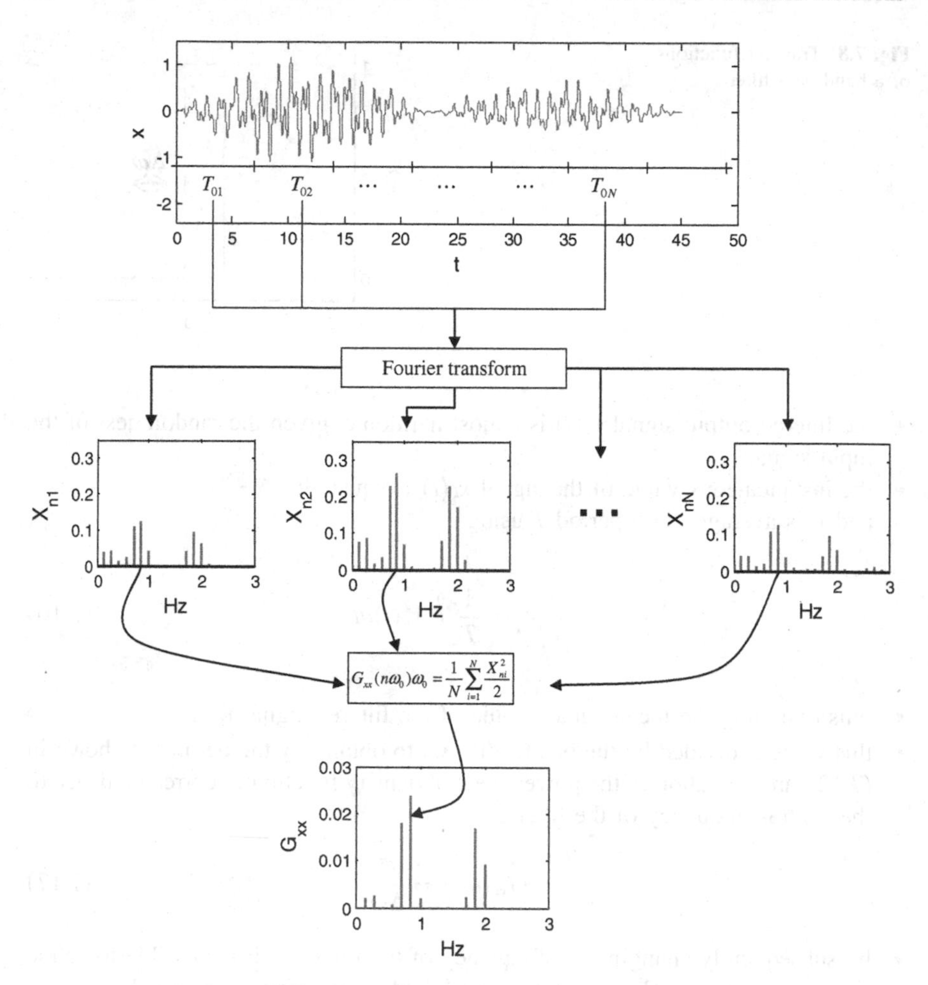

Fig. 7.9 Defining power spectral density using Fourier series development

it possible to eliminate the discontinuities at the beginning and at the end of the history (Fig. 7.10);

- once the signal has been made periodic, the Fourier series development is performed:

$$x_i(t) = \sum_{1}^{N} |X_{ni}| \cos(n\omega_0 t + \phi_{ni}) \quad i = 1, 2, \ldots, N \tag{7.18a}$$

being:

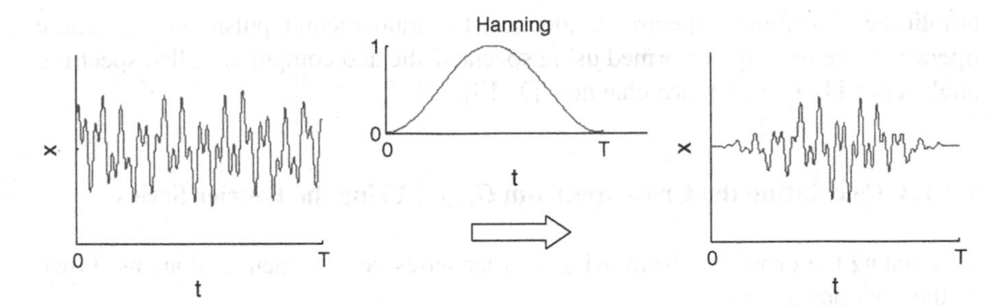

Fig. 7.10 Effect of the Hanning window

$$X_{ni} = X_i(n\omega_0) = X_i(\omega_n) = \frac{2}{T_0}\int_0^{T_0} x_i(t)\mathrm{e}^{-in\omega_0 t}dt \quad i = 1, 2, \ldots, N \tag{7.18b}$$

having indicated with:

$$\omega_0 = \frac{2\pi}{T_0} \tag{7.18c}$$

the pulsation of the fundamental harmonic associated with the period T_0 of the chosen window and X_{ni} the amplitude of the generic harmonic corresponding to the pulsation $\omega_n = n\omega_0$. X_{ni} is a complex number defined as:

$$X_{ni} = |X_{ni}|\mathrm{e}^{i\phi_{ni}} \tag{7.18d}$$

- in correspondence to the generic recording $x_i(t)$ it is possible to define a complex spectrum $X_{ni}(\omega)$, as shown in Fig. 7.9: each spectrum will be different both in magnitude $|X_{ni}|$ and in phase ϕ_{ni}, given the randomness of the process being analysed.

It is possible to show [1–3, 17] that:

$$G_{xx}(\omega_n) = E\left[\frac{X_{ni}X_{ni}^*}{2\omega_0}\right] = E\left[\frac{|X_{ni}|^2}{2\omega_0}\right] \tag{7.19}$$

having indicated the expected value with $E[\,]$ and the complex conjugate generic quantity with the superscript*.[10] In other words, the power spectral density is estimated as the expected value, i.e. the mean value, of the square mean value of the

[10] Equation (7.19) simply derives from the function $G_{xx}(\omega) = \Psi_x^2/\Delta\omega$ (see Eq. (7.12)) where $\Delta\omega$ in this case equals ω_0 and from the fact that Ψ_x^2 of a harmonic signal of amplitude $|X_{ni}|$ is equal to $|X_{ni}|^2/2$.

amplitude of different spectra divided by the fundamental pulsation ω_0. These operations are usually performed using special dedicated computers called spectrum analysers with 1, 2 or more channels [3, 13].

7.3.1.3 Calculating the Cross-spectrum $G_{xy}(\omega)$ Using the Fourier Series

Calculating the cross-spectrum using Fourier series development is done as shown in the previous section.

- The two different time histories $x(t)$ and $y(t)$ (Fig. 7.11) are divided into equal periods $T_{01} = T_{02} = \cdots = T_{0N} = T_0$;
- each single history $x_i(t)\,(i = 1, 2, \ldots, N)$ and $y_i(t)\,(i = 1, 2, \ldots, N)$ is made periodic with the windowing operation;
- Fourier series development is performed on these signals that have been made periodic (Fig. 7.11):

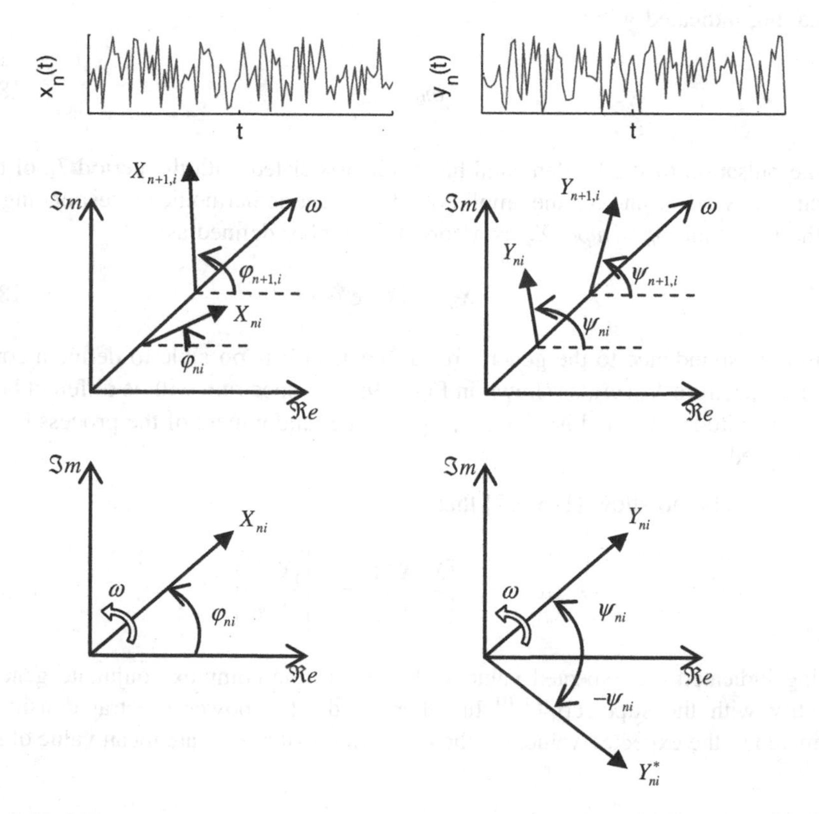

Fig. 7.11 Definition of the generic harmonic of the cross-spectrum

$$X_{ni} = X_i(n\omega_0) = X_i(\omega_n) = \frac{2}{T_0}\int_0^{T_0} x_i(t)e^{-in\omega_0 t}dt$$
$$Y_{ni} = Y_i(n\omega_0) = Y_i(\omega_n) = \frac{2}{T_0}\int_0^{T_0} y_i(t)e^{-in\omega_0 t}dt \quad (7.20)$$

X_{ni} and Y_{ni} being the amplitudes of the generic harmonics corresponding to the frequency $\omega_n = n\omega_0$:

$$X_{ni} = |X_{ni}|e^{i\phi_{ni}}$$
$$Y_{ni} = |Y_{ni}|e^{i\psi_{ni}} \quad (7.20a)$$

- in correspondence to the generic recording $x_i(t)$ and $y_i(t)$ it is possible to define a complex spectrum $X_{ni}(\omega)$ and $Y_{ni}(\omega)$, as shown in Fig. 7.12: each spectrum will be different both in magnitude and in phase, given the randomness of the process being analysed.

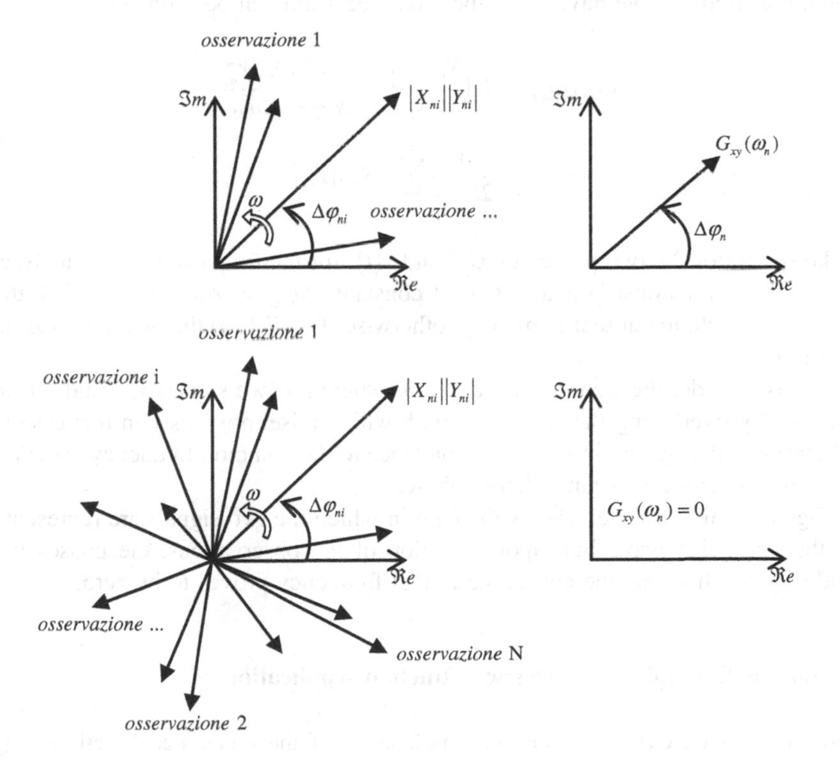

Fig. 7.12 Trend of the product upon variation of the observation performed (i = 1, 2, …, N): **a** case of high coherence, **b** case of low coherence

It is possible to show [1–3, 17] that the cross-spectrum can be estimated using:

$$G_{xy}(n\omega_0)\omega_0 = G_{yx}(n\omega_0)\omega_0 = \frac{1}{N}\sum_{i=1}^{N}\frac{X_{ni}Y_{ni}^*}{2} = E\left[\frac{X_{ni}Y_{ni}^*}{2\omega_0}\right] \tag{7.20b}$$

Now let us try to understand the physical significance of this operation: the generic product defined in (7.20b) can be developed, taking into account (7.20a) as:

$$\begin{aligned} X_{ni}Y_{ni}^* &= |X_{ni}||Y_{ni}|e^{i\phi_{ni}}e^{-i\psi_{ni}} \\ &= |X_{ni}||Y_{ni}|e^{i(\phi_{ni}-\psi_{ni})} = |X_{ni}||Y_{ni}|e^{i(\Delta\phi_{ni})} \end{aligned} \tag{7.21}$$

Therefore, the product $X_{ni}Y_{ni}^*$ represents a complex number whose magnitude $|X_{ni}||Y_{ni}|$ and phase $\Delta\phi_{ni}$ change as the frequency varies f_n $(n = 0, 1, \ldots, nfreq_{max})$ and, given the same frequency, change as the observation performed varies $(i = 1, 2, \ldots, N)$.

Figure 7.12 shows a possible trend for this product, at the same frequency $f = f_n$, upon variation of the observation, in the case (a) of correlation and in the case (b) of non-correlation. As we have seen, the cross-spectrum can be estimated as:

$$\begin{aligned} G_{xy}(\omega_n) &= E\left[\frac{X_{ni}Y_{ni}^*}{2\omega_0}\right] = \frac{1}{N}\sum_{i=1}^{N}\frac{X_{ni}Y_{ni}^*}{2\omega_0} \\ &= \frac{1}{2\omega_0 N}\sum_{i=1}^{N}|X_{ni}||Y_{ni}|e^{i(\Delta\phi_{ni})} \end{aligned} \tag{7.22}$$

If the spectra of the two processes $x(t)$ and $y(t)$, for the frequency $f = f_n$ analysed, have a phase-relationship that is almost constant, $\Delta\phi_{ni} = cost.$, (Fig. 7.12a), then there is a correlation at that frequency; otherwise (Fig. 7.12b) the two processes are uncorrelated.

Let us consider the example in Fig. 7.13a where the two signals $x(t)$ and $y(t)$ are obtained by overlaying the sine wave with white noise: obviously in this case the coherence will only be high in correspondence to the common frequency present in both signals with a constant relative phase.

Figure 7.13b, however, shows the case in which the two signals are represented by the same sine wave but, upon variation of the observations, the phases vary randomly: in this case the coherence at that frequency proves to be zero.

7.3.1.4 An Example of Coherence Function Application

Now we will describe one possible application of the coherence function $\gamma_{xy}^2(f)$ (7.15), i.e. of the non-cross-spectrum $G_{xy}(f)$. We will analyse a linear system

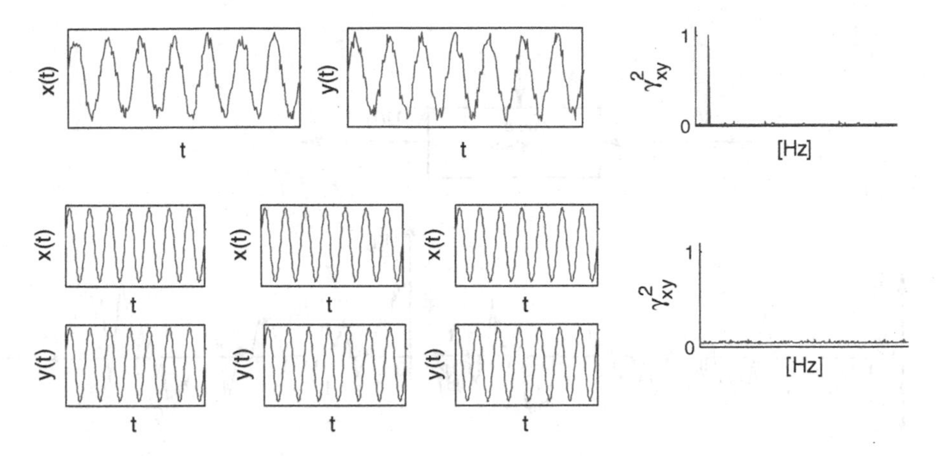

Fig. 7.13 Examples of coherence forces

(Fig. 7.14) subject to a known random input $I(t)$ with $U(t)$ being the corresponding output: the input could represent a force, whereas the output a displacement or acceleration at a generic point of a structure. The same figure shows a possible trend of the spectral density function of the input $G_{II}(\omega)$ and the output $G_{UU}(\omega)$.

If $I(t)$ is the only signal present on input, i.e. assuming there is no noise or other form of excitation present simultaneously at $I(t)$, between the power spectral density function on input $G_{II}(\omega)$ and the power spectral density function at output $G_{UU}(\omega)$ the following relation applies:

$$G_{UU}(\omega) = |H(\omega)|^2 G_{II}(\omega) \tag{7.23}$$

where $H(\omega)$ is the transfer function of the system in question (Chap. 1, Sect. 1.3.5): this relationship will be demonstrated subsequently in Sect. 7.5, Eqs. (7.34a, 7.34b).

In this case, when we know the power spectral density on input $G_{II}(\omega)$ and measure the power density on output $G_{UU}(\omega)$, (7.23) makes it possible to obtain the magnitude $|H(\omega)|$ of the transfer function of the system in question.

However, in general, $I(t)$ is not the only input as the system is not isolated from the outside world but is also excited by noise $N(t)$ introduced by other disturbances which often cannot be eliminated, are not easy to identify or are even impossible to estimate. If we wish to use these measurements to obtain the magnitude of the transfer function of the system $|H(\omega)|$, assumed in this analysis to be unknown (this procedure goes under the name of *identification process*, see Chap. 8 on this subject), it is possible to exploit the fact that the coherence function γ^2_{UI} of the two signals $I(t)$ and $U(t)$, i.e. the cross-spectrum $G_{UI}(\omega)$ provides the part of $U(t)$ that is coherent with $I(t)$, i.e., due only to the known input $I(t)$, i.e. where the following relation applies (as demonstrated in [1, 2], Chap. 4):

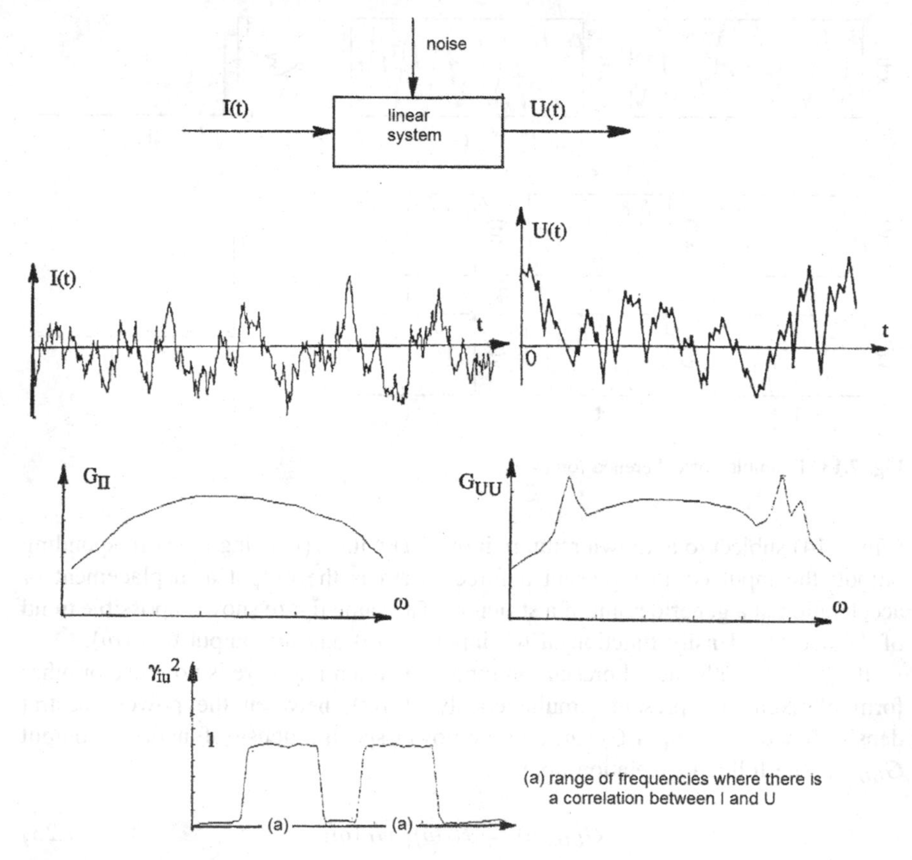

Fig. 7.14 Using the coherence function in identification processes

$$G_{UI}(\omega) = H(\omega)G_{II}(\omega) \tag{7.24a}$$

The system being analysed, which is assumed to be linear, transforms the input $I(t)$ by modulating and phase shifting it, at the same frequency, in a constant manner: the output $U(t)$ effectively caused by the input is, therefore, univocally coherent, whereas the portion of the output due to noise $N(t)$ is not. If coherence γ_{UI}^2 is high in the entire field of frequency being analysed, the output can be virtually attributed to the input only $I(t)$, while, on the contrary, the data will show little correlation with the actual input: under these conditions it is impossible to define an estimation of the magnitude of the system's transfer function $|H(\omega)|$ from the measurements. From these considerations the transfer function $H(\omega)$ (complex quantity) of a generic linear system subject to a random input $I(t)$ and, simultaneously, to noise $N(t)$ can be obtained from the following relation:

$$H(\omega) = \frac{G_{UI}(\omega)}{G_{II}(\omega)} \tag{7.24b}$$

7.4 Defining the Random Stationary and Ergodic Process

Often the complex mechanism that generates a stochastic process does not significantly change its characteristics with longer time periods than the fluctuation time scale: for example, if we analyse different recordings of wind speed at a given point, the various time histories are obviously different, but the statistical characteristic remain almost constant (since the actual *mean* quantities are linked to the orography of the terrain). For these processes it is legitimate to assume the random phenomenon, as a stationary phenomenon.

The term stationary phenomenon (in the strict sense) is used to define those phenomena where all the statistical quantities are not affected by a translation of the time axis, i.e. they do not change over time. Normally, it is not possible to estimate all the statistical quantities, so it is never actually possible to verify whether the process is stationary in the strict sense: by analysing a generic random process (defined by a sequence of random variables $x_i(t)$), the process can be defined as stationary in a broad sense if (Fig. 7.15):

- the mean value of the random process at a given generic time t_1 (assuming statistical regularity):

$$\mu_x(t_1) = \lim_{N\to\infty} \frac{1}{N} \sum_0^N x_i(t_1) \tag{7.25a}$$

- the mean square value of the random process at a given time t_1:

$$\psi_x^2(t_1) = \lim_{N\to\infty} \frac{1}{N} \sum_0^N x_i^2(t_1) \tag{7.25b}$$

- the autocorrelation function:

$$R_{xx}(t_1, \tau) = \lim_{N\to\infty} \frac{1}{N} \sum_0^N x_i(t_1) x_i(t_1 + \tau) \tag{7.25c}$$

prove to be independent of the time t_1 considered, i.e. when the process being analysed maintains constant the statistical characteristics that define it.

When analysing just one random variable (i.e. a single recording), as is often the case, we only have one recording $x_i(t)$ of a stationary process available: in this case it is natural to estimate the statistical quantities as the mean μ_x and the correlation

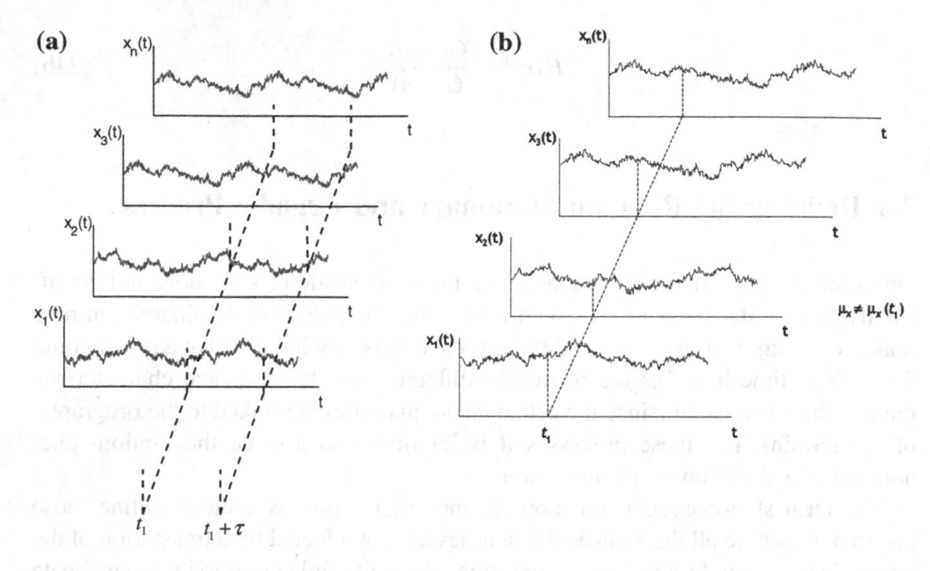

Fig. 7.15 **a** Definition of a random stationary and ergodic process. **b** Example of a stationary but not ergodic process

function R_{xx} with averaging on the time history $x_i(t)$ (Fig. 7.15). One possible estimation of the mean μ_x for a recording of duration T may be given by:

- mean time value:

$$\mu_x = \lim_{T \to \infty} \frac{1}{T} \int_0^T x(t)dt \tag{7.26a}$$

- square mean time value:

$$\psi_x^2 = \lim_{T \to \infty} \frac{1}{T} \int_0^T x^2(t)dt \tag{7.26b}$$

- time autocorrelation function:

$$R_{xx} = \lim_{T \to \infty} \frac{1}{T} \sum_0^T x(t)x(t+\tau) \tag{7.26c}$$

The process is also assumed to be ergodic if the mean times defined by (7.26a–7.26c) coincide with:

$$\begin{aligned} \mu_x &= \mu_x(t_1) = \cos t. \\ \psi_x^2 &= \psi_x^2(t_1) = \cos t. \\ R_{xx} &= R_{xx}(t_1) = \cos t. \end{aligned} \tag{7.27}$$

In this case just one recording is sufficient to obtain the statistical quantities that define its characteristics (in (7.15) we can see an example of a stationary, but not ergodic, process).

In actual fact, random quantities (such as, for example, wind turbulence) are not strictly stationary and ergodic, but they are assumed to be so by defining the statistically significant quantities with a limited number (not infinite) of observations: in other words, by assuming that the phenomenon of turbulence is a stationary and ergodic phenomenon, it is possible to obtain the statistical quantities from a limited, and not infinite, number of time recordings.

In practical applications, when the process is stationary it is commonly assumed to be ergodic too: thus it is possible to define the necessary statistical characteristics (mean value and correlation) from one, or just a few, recordings.

7.5 The Response of a Vibrating System to Random Excitation

Before beginning to study the response of a vibrating system subject to random excitation, we should recall some of the concepts used for studying the response of a system to periodic forces.

As we saw in Chap. 1, the periodic function has a waveform that constantly repeats itself after a period T_0:

$$x(t) = x(t + T_0) \tag{7.28}$$

This quantity can thus be transformed in the frequency domain via Fourier series development:

$$x(t) = \sum_n |X_n| \cos(n\omega_0 t + \phi_n) \tag{7.29a}$$

being:

$$X_n = |X_n| e^{i\phi_n} = \frac{2}{T_0} \int_{-\infty}^{\infty} x(t) e^{-in\omega_0 t} dt \tag{7.29b}$$

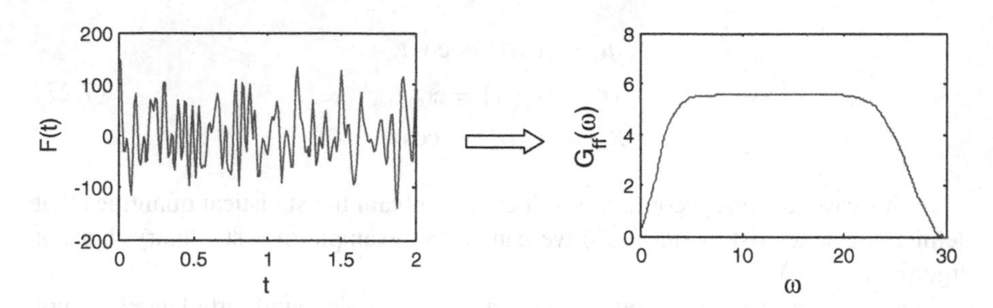

Fig. 7.16 The generic random force

The time history can then be represented by the line spectrum which shows in magnitude and phase the amplitude and the phase of the individual harmonics that make up the signal.

By repeating this operation in different periods, as it is periodic, we always obtain the same complex spectrum.

We will now analyse a generic periodic force (Fig. 7.16) which can be represented as:

$$F(t) = \sum_n |F_n| \cos(n\omega_0 t + \phi_n) \tag{7.30}$$

being:

$$F_n = |F_n| \mathrm{e}^{i\phi_n} = F(n\omega_0) = \frac{2}{T_0} \int_{-\infty}^{\infty} x(t) \mathrm{e}^{-in\omega_0 t} dt \tag{7.31}$$

The steady-state response of a vibrating system can be performed using the transfer function $H(\omega) = |H(\omega)| \mathrm{e}^{i\alpha(\omega)}$ (Chap. 1, Sect. 1.3.5) as:

$$X_n = |X_n| \mathrm{e}^{i\psi_n} = X(n\omega_0) = H(n\omega_0) F_n(n\omega_0) \tag{7.32}$$

from which it is possible to reconstruct the response over time $x(t)$ as:

$$x(t) = \sum_n |X_n| \cos(n\omega_0 t + \psi_n) \tag{7.33}$$

being:

$$\psi_n = \phi_n + \alpha_n \tag{7.33a}$$

Now we will analyse the response of a linear system with an aleatory force and with no disturbances. $G_{ff}(\omega)$ is the power spectral density of the aleatory force. The power spectral density of the response $G_{xx}(\omega)$ [1, 3, 13] is:

$$G_{xx}(\omega) = H^2(\omega)G_{ff}(\omega) \tag{7.34a}$$

while the cross-spectrum is defined by the relation:

$$G_{xf}(\omega) = H(\omega)G_{ff}(\omega) \tag{7.34b}$$

being:

$$H^2(\omega) = H(\omega)H^*(\omega) \tag{7.35}$$

a real quantity. Figure 7.17 illustrates the sequence of operations needed to define the system's response.

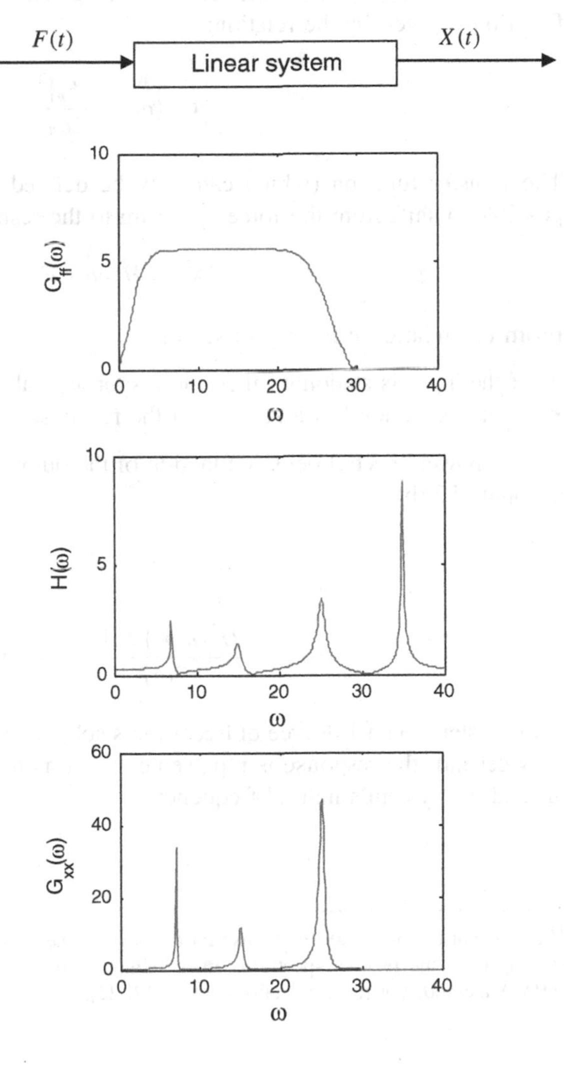

Fig. 7.17 Definition of the power spectral density function of the output

If we wish to reconstruct the time history of the response $x(t)$ of the system, we can use (7.29a)[11]:

$$x(t) = \sum_n |X_n| \cos(n\omega_0 t + \theta_n) \tag{7.36}$$

θ_n being a random phase chosen from 0 and 2π, while $|X_n|$ is the amplitude of the generic harmonic obtained from the power spectral density function (Eq. 7.34a) from:

$$G_{xx}(\omega) = \frac{|X_n|^2}{2\omega_0} \tag{7.37}$$

We will now demonstrate the relation (7.34a, 7.34b). The power spectral density function is given by the relation:

$$G_{ff}(\omega) = \frac{|F_n|^2}{2\omega_0} \tag{7.38a}$$

The transfer function (which can only be defined for a linear system) makes it possible to shift from the force spectrum to the response spectrum as:

$$X_n = H(n\omega_0)F_n \tag{7.38b}$$

From this relation it is easy to see that:

- if the input is stationary then the response is also stationary;
- if the excitation has a zero mean the response will also have a zero mean.

The power spectral density function of the output G_{xx} thus becomes, taking into account (7.38b):

$$\begin{aligned} G_{xx}(\omega) &= \frac{|X_n|^2}{2\omega_0} \\ &= \frac{H^2(n\omega_0)|F_n|^2}{2\omega_0} = H^2(n\omega_0)G_{ff}(\omega) \end{aligned} \tag{7.39}$$

If the system, with 1 degree of freedom, is subject to low damping and if excitation is wideband, the response is represented by a narrow band process with the peak around the system's natural frequency.

---

[11] This is not the only way to reconstruct a possible time history of the aleatory quantity, defined starting from the relative spectrum: other methods are discussed in literature, for example, the ARMA method. For further details see [15, 17, 22].

If the excitation is Gaussian and the system is linear then the response is also Gaussian and so the probability distribution of the response is univocally defined by the mean and by the variance.

If the system is non-linear, superposition of the effects cannot be applied so it is not possible to work in the frequency domain. In this case, once the power spectral density function of the force has been assigned G_{ff}, for example, we must reconstruct a possible time history for the force using, for example, development in series:

$$F(t) = \sum_{n} |F_n| \cos(n\omega_0 t + \theta_n) \tag{7.40}$$

θ_n being a random phase between 0 and 2π and from (7.39):

$$G_{ff}(\omega) = \frac{|F_n|^2}{2\omega_0} \tag{7.40a}$$

This force can be assigned on input in the system's equations of motion which, in the case of a linear system, are of the form:

$$[M]\ddot{\underline{x}} + [R]\dot{\underline{x}} + [K]\underline{x} = \underline{F}(t) \tag{7.41a}$$

while generically, in the case of a non-linear system:

$$F_i(x, x, x) + F_s(x, x) + F_e(x) = F(t) \tag{7.41b}$$

from which it is possible to obtain the response $x(t)$ of the system, using numerical integration.

However, (7.40) represents just one of the possible time histories of the random process; in order to fully reproduce the process we must generate several time histories with different random phase generations.

When generating the time history (7.40) we must also ensure that the period T_0 assumed to correspond to the fundamental harmonic, with which the force was developed in series, is greater than the period of integration: otherwise a fictitious periodicity is introduced into the response of the system, a periodicity that is not present in the real random phenomenon. One alternative approach [10] is to use linear filters: with this approach a sequence of random numbers, uniformly distributed and with zero mean value (white noise), are introduced in a numerical filter defined with an appropriate transfer function. In this way, the random process is generated by the filter whose transfer function is defined in such a way as to obtain a signal upon output with the desired power spectral density. In this case, by using, for example, an ARMA method, i.e. autoregressive-moving-average model [6–8, 15, 25], we can generate the history in the time domain starting from our knowledge of the autocorrelation function, or of its power spectral density transfer function.

7.5.1 *Analysis with Several Correlated Processes*

Up until now the discussion has been limited to analysing just one random process: if the random processes are more than one, we must define the correlation between them. In this regard, there are several different methods [19–22] of reproducing correlated processes:

- the superposition of effects method (WSWA methods, [24]) using development in series of harmonics, obtained from the matrix of cross-spectra (similar to the covariance matrix);
- the ARMA method with several processes, already described briefly for the generation of a single process, which again uses the covariance matrix [10].

These procedures are remarkably complex; please refer to the bibliography at the end of the chapter for further reading: in the next section (Sect. 7.6) we will introduce a simplified method for reproducing several correlated random processes (Sects. 7.6.1 and 7.6.3).

7.6 Some Examples of Application

7.6.1 *Response of a Structure to Turbulent Wind*

Real wind is turbulent, that is, it is variable in time and in space in an aleatory manner: Fig. 7.18 shows a generic space-time history of real wind speed (horizontal component). However, the characteristics of this phenomenon are such that it may be considered as a stationary and ergodic process. The action of turbulent wind on some types of structures (such as, for example, tall buildings, towers, chimneys, bridges, electrical power transmission lines) causes vibrations of significant amplitudes and therefore high static and dynamic stresses caused both by force phenomena and problems of instability. In Sect. 5.3.2, the issues regarding the definition of aerodynamic forces acting on structures, having assumed a known and constant confined flow, were introduced: in this section we will analyse the main characteristics of real wind (see the bibliography in this chapter, section *Effects of wind on structures*).

7.6.1.1 Characterisation of Turbulent Wind

Wind is caused by differences in atmospheric pressure. At a great height the motion of the air is independent from the roughness of the surface of the earth, while in the area below a certain height, described as the gradient height δ (boundary layer), the flow is modified by the surface friction, thus generating turbulent motion of the

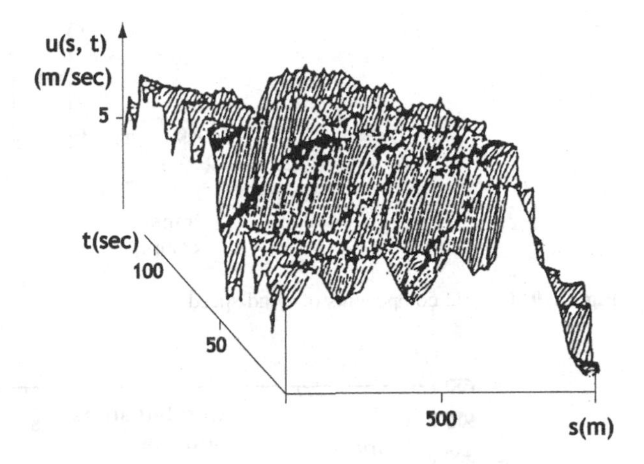

Fig. 7.18 A generic space-time history of real wind

confined flow. The gradient height δ varies depending on the type of terrain. As an indication, we can, for example, assume[12]:

The time history of the longitudinal wind speed can be expressed as:

$$U(t) = \bar{U} + u(t) \tag{7.42}$$

where $\bar{U}$ is the mean value of the speed (averaged on a period that ranges from 10 min to 1 h) and $u(t)$ is the dynamic or fluctuating part. Real wind, in addition to having a fluctuating component in a horizontal direction, also has a vertical and transverse component with zero mean values[13]: Fig. 7.19 shows the generic wind speed vector and the relative components.

The mean value of wind speed $\bar{U}$ depends on the height z at which the actual speed is measured as shown, for different types of terrain, in Fig. 7.20.

The variance in the mean value of the speed, shown in Fig. 7.20, can be expressed in relation to the height by an exponential law of the form:

$$\bar{U}(z) = \bar{U}_g \left(\frac{z}{z_g} \right)^{\alpha} \tag{7.43}$$

$\bar{U}_g$ being the mean speed defined in correspondence to the gradient height $z_g = \delta$: the coefficient α is a function of the type of terrain and assumes the following values:

[12]The orography of the site modifies the flow lines even at much greater heights than the defined gradient height, in the case of orography with the presence of tall obstacles like hills, headlands and mountains.

[13]This is true by definition as the mean value is usually attributed to the so-called horizontal component which, in fact, is itself a vector that has three components of which the orthogonal component on the surface of the terrain is small, but not null.

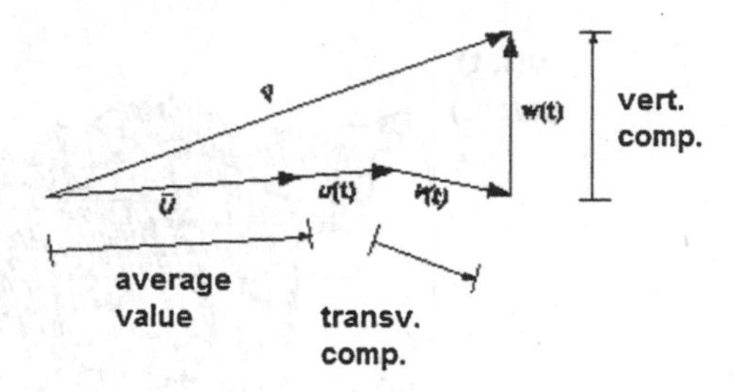

Fig. 7.19 Generic components of wind speed

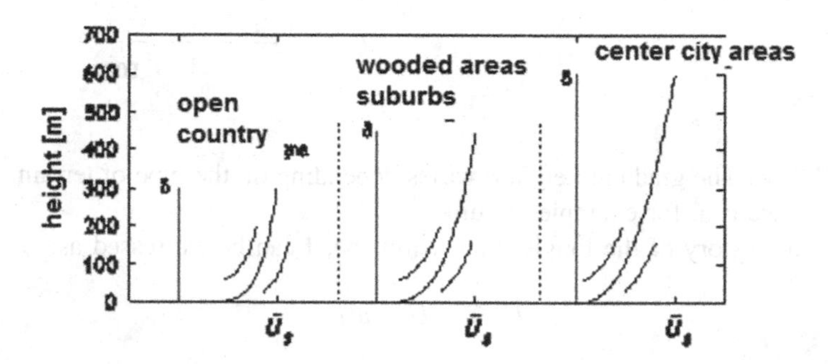

Fig. 7.20 Vertical profiles of mean speed $\bar{U}$ (*continuous line*) and range of fluctuations (*dotted line*)

The fluctuating part $u(t)$ of the wind is less sensitive to variation with height and is a random variable quantity both in space and in time. The main statistical quantities that characterise fluctuations in the confined flow are:

- the intensity of turbulence I_t;
- the power spectral density $G_v(f)$;
- the cross-correlation between measured speed at different points in space $R_{xy}(\xi)$;
- the speed probability distribution.

The intensity of turbulence I_t is defined as:

$$I_t = \frac{\sigma}{\bar{U}} \tag{7.44}$$

σ being the standard deviation of the flow speed.

The index I_t is usually made to depend on the drag coefficient of the terrain k using the relation:

$$I_t = \beta\sqrt{k} \tag{7.45}$$

where k applies, for different terrains:

The coefficient β for the horizontal component is normally equal to $\beta = 2.45$. The index of real wind turbulence varies between 5 and 25 %. The energy associated with the fluctuations in the flow is distributed over a wide field of frequencies: this distribution in relation to the frequency is described using the power spectral density function $G_v(f)$ which, we should remember, is correlated with the variance via the relation:

$$\int_0^\infty G_v(f)df = \sigma^2 \tag{7.46}$$

Often this function is plotted in non-dimensional form, as shown in Fig. 7.21 where it is shown in terms of experimental data (blue line) and in terms of analytical interpolation formulation (Von Karman formula (7.48), red line), where:

- f is the frequency in Hz;
- σ^2 the variance, obtained from the index of turbulence using (7.44), i.e. the type of terrain (Table 7.4 and 7.5);
- δ is the gradient height (a function of the type of terrain, Table 7.3).

The relationship $(f/\bar{U})$ is defined as "inverse of wavelength" and is associated with the sizes of atmospheric vortices. This length can then be compared to a characteristic dimension, known as an integral scale L, defined as the wavelength of vortices corresponding to the peak of the spectrum (this quantity is also defined as

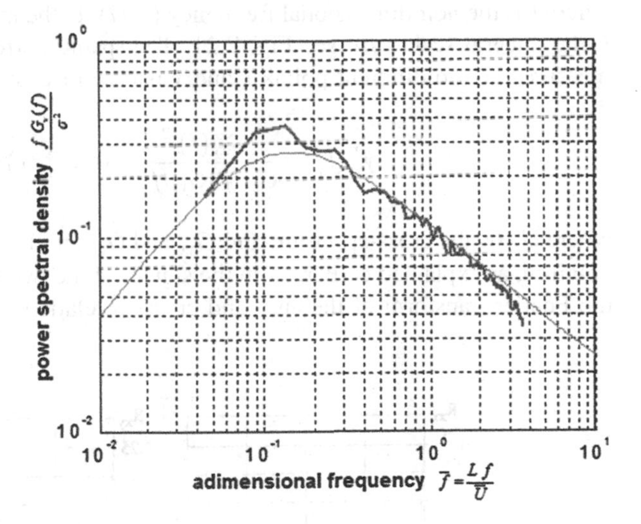

Fig. 7.21 The power spectral density of wind

Table 7.3 Gradient height δ depending on the type of terrain

Open sea terrain	$\delta = 300$ m
Wooded areas and city suburbs	$\delta = 450$ m
Central areas of large cities	$\delta = 600$ m

Table 7.4 Coefficient α as a function of the type of terrain

Open sea terrain	$\alpha = 0.16$
Wooded areas and city suburbs	$\alpha = 0.28$
Central areas of large cities	$\alpha = 0.40$

Table 7.5 Drag coefficient k in relation to the type of terrain

Open sea terrain	$k = 0.005$
Wooded areas and city suburbs	$k = 0.015$
Central areas of large cities	$k = 0.050$

the autocorrelation barycentre): normally L is in the region of a hundred metres. This makes it possible to define a non-dimensional frequency $\bar{f}$ (known as the Strouhal number or reduced frequency):

$$\bar{f} = \frac{Lf}{\bar{U}} \tag{7.47}$$

The power spectral density function $G_v(f) = G_v(\bar{f})$ is obtained, by measuring the wind, with the methods defined in the previous section and it is possible to interpolate this, as mentioned, using the so-called Von Karman formula:

$$G(\bar{f}) = 4k\bar{U}\frac{L/\bar{U}}{(2+\bar{f}^2)^{5/6}} \tag{7.48}$$

where $\bar{f}$ is the non-dimensional frequency (7.47), $\bar{U}$ the mean speed value (m/s) and L the integral scale (m), see Fig. 7.21. The spatial correlation of the wind speed measured at two different points x and y is described by the coherence function:

$$\gamma_{xy}^2(f) = \frac{|G_{xy}(f)|^2}{G_{xx}(f)G_{yy}(f)} \quad 0 \leq \gamma_{xy}^2(f) \leq 1 \tag{7.49}$$

where G_{xy} is the cross-spectrum and G_{xx} and G_{yy} are the power spectral density G_v obtained, as explained earlier, from the signals of the two measurement points using the Fourier transform of the auto and cross-correlation.

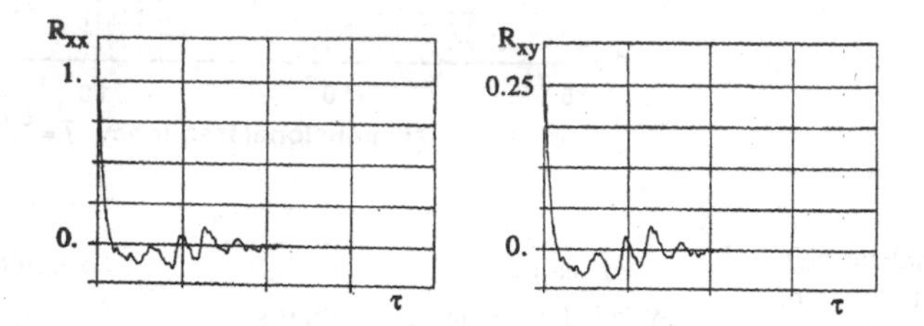

Fig. 7.22 Auto and cross-correlation functions ($\Delta x = 150\,\text{m}$)

For example, Fig. 7.22a shows the autocorrelation function $R_{xx}(\tau)$ estimated from a generic time history of the wind speed, measured using an anemometer. The characteristic autocorrelation trend $R_{xx}(\tau)$ clearly shows (for clarification see the examples shown in Sect. 7.3, Fig. 7.4) how the phenomenon of wind turbulence is characterised by a wide band of frequencies. Figure 7.22b, on the other hand, shows the cross-correlation function $R_{xy}(\tau)$ relating to the wind speed measured at two points located at such a distance that the two time histories prove to have little correlation.

The coherence function depends on the parameter $(f\Delta x/\bar{U})$, Δx being the distance between the two points and $\bar{U}$ the average speed: an appropriate approximation for representing this function could be the following:

$$\gamma_{xy}^2(f) = \mathrm{e}^{-C(f\Delta x/\bar{U})} \tag{7.50}$$

C being a constant ($C = 7$ for the vertical component of the turbulence, $C = 15$ for the horizontal component). Figure 7.23 shows the trend of the coherence function in relation to the non-dimensional frequency $(f\Delta x/\bar{U})$: as we can see, at the same frequency f and mean value $\bar{U}$, coherence decreases as the distance Δx increases between two points of measurement; $\gamma_{xy}^2(f)$, on the other hand, increases as the mean value $\bar{U}$ increases, while it decreases as the frequency f increases (i.e. upon decrease of the dimensions of the vortex associated with that frequency).

7.6.1.2 Calculating the Dynamic Response of a Structure to Turbulent Wind

In Chap. 5, Sect. 5.3.2, we saw how it is possible, once the time history of the wind speed affecting a structure is known, to calculate the forces (drag, lift and torque) exerted on it by the fluid.

The response of the system, considering the fluid-elastic coupling, must be estimated by numerically integrating the non-linear equations of motion of the form:

$$[M_s]\,\underline{\ddot{x}} + [R_s]\underline{\dot{x}} + [K_s]\underline{x} = \underline{F}(\underline{\dot{x}}, \underline{x}, U(t, \xi)) \tag{7.51}$$

$\underline{x}$ being the vector that contains the displacements of the mathematical model of the structure (for example, a model with finite elements or a model using the principal coordinates), $[M_s]$, $[R_s]$ and $[K_s]$ the structural stiffness, damping and mass matrices (diagonal if working with principal coordinates) and finally $\underline{F}$ the vector of the generalised forces.

The generalised forces in (7.51) depend, in non-linear form, as seen in Sect. 5.2, on the displacements and the velocities of the structure's nodes and on the space-time history of the turbulent wind $U(t, \xi)$.

The generic time history in a point of the structure can be obtained from the power spectral density function $G_v(f)$:

- assuming a certain type of terrain, and therefore, a certain value of gradient height δ (Table 7.3), a certain value of average speed.
- it is possible to obtain the power spectral density function $G_v(f)$ from the non-dimensional expression shown in (7.48), (7.21);
- assuming a certain fundamental frequency $\omega_0 = 2\pi f_0$ and a certain number of frequencies n_{max}.
- from the known relation:

$$G_v(f_n) = G_v(nf_0) = \frac{|C_n|^2}{2\omega_0} \tag{7.52}$$

- it is possible the obtain the amplitudes of the individual harmonics $|C_n|$;
- one possible time history can be estimated using the following expression:

$$U(t) = \bar{U} + \sum_{n=1}^{n_{max}} |C_n| \cos(n\omega_0 t + \phi_n) \tag{7.53}$$

where ϕ_n is a random phase between 0 and 2π.

If we wish to define the wind history at several points of the structure, i.e. define the space-time trend of the wind, we can proceed in several ways:

- using methods of harmonic superposition (Wave superposition Methods) that are based on the use of cross-spectrum matrices [10];
- using ARMA methods (autoregressive-moving-average models, [24];
- other simplified approaches.

For an introduction to the complex issues linked to the space-time generation of wind, we will briefly describe a more didactically suitable method, similar to the harmonic superposition method.

We will consider two generic points i and j, distant from each other Δx_{ij}, where we wish to reconstruct the wind time history. Given the characteristics of the aleatory phenomenon analysed, we will have:

$$G_{vi}(f) = G_{vj}(f) \tag{7.54}$$

Furthermore, the coherence function is known $\gamma^2_{nxy}(f)$, Fig. 7.23, (7.15), upon variation of the frequency $f_n = nf_0$ (Δx_{ij} and $\bar{U}$ being set and constant).

The time history at generic point i can be defined using (7.53):

$$U_i(t) = \bar{U} + \sum_{n=1}^{n_{max}} |C_n| \cos(n\omega_0 t + \theta_n) \tag{7.55}$$

The time history of the speed $U_j(t)$, estimated at point j, can be seen as the sum of a coherent term and U_{coer} a non-coherent term U_{uncoer}:

$$U_j(t) = U_{coer} + U_{uncoer} \tag{7.56}$$

The coherent part [see Sect. 7.3.1.4 and the definitions shown in (7.35) and (7.15)], will be expressed as:

$$U_{coer} = \bar{U} + \sum_{n=1}^{n_{max}} \gamma_{nij}|C_n| \cos(n\omega_0 t + \theta_n) \tag{7.57}$$

that is, a history reconstructed with the same phase θ_n used to generate $U_i(t)$ (Eq. (7.55)). The non-coherent part is generated with different random phases ϕ_n and by requiring that the two spatial histories have the same spectrum (7.54):

$$U_{uncoer} = \sum_{n=1}^{n_{max}} \sqrt{1-\gamma_{nij}^2}|C_n| \cos(n\omega_0 t + \phi_n) \tag{7.58}$$

Ultimately we will have:

$$U_j(t) = \bar{U} + \sum_{n=1}^{n_{max}} \gamma_{nij}|C_n| \cos(n\omega_0 t + \theta_n) + \sum_{n=1}^{n_{max}} \sqrt{1-\gamma_{nij}^2}|C_n| \cos(n\omega_0 t + \phi_n) \tag{7.59}$$

By analysing (7.59) we can see that, regardless of the frequency, if there was perfect coherence $\left(\gamma_{nij}^2 = 1\right)$ we would have $(U_i(t) = U_j(t))$, whereas in the hypothesis of complete incoherence $\left(\gamma_{nij}^2 = 0\right)$ the time history $U_i(t)$ would be completely different to the speed history at j, $U_j(t)$, since the phases of reconstruction are uncorrelated.

Subsequently, if we wish to reconstruct the time history of the speed $U_k(t)$ at a further generic point k removed Δx_{jk} from point j, it is possible to rewrite (7.59) as:

$$U_j(t) = \bar{U} + \sum_{n=1}^{n_{max}} |D_n| \cos(n\omega_0 t + \psi_n) \tag{7.60}$$

being:

$$|D_n|e^{i\psi_n} = \gamma_{nij}|C_n|e^{i\theta_n} + \sqrt{1-\gamma_{nij}^2}|C_n|e^{i\phi_n} \tag{7.61}$$

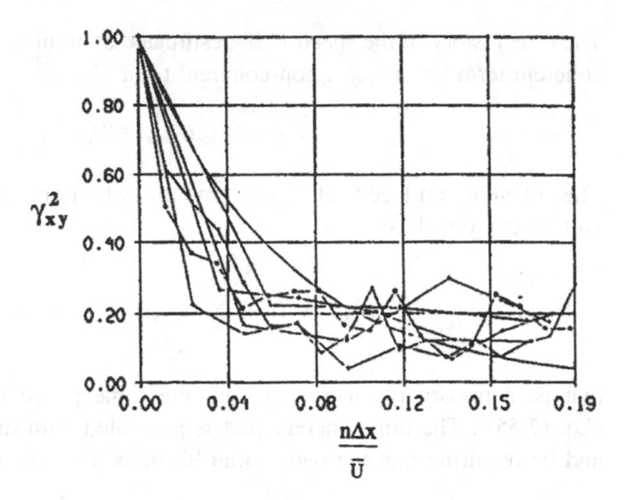

Fig. 7.23 Coherence function in relation to the non-dimensional frequency

Thus, in the same way as for point j (7.56), the speed at point k is estimated using the following expression:

$$U_k(t) = \bar{U} + \sum_{n=1}^{n_{max}} \gamma_{njk}|D_n|\cos(n\omega_0 t + \psi_n) + \sum_{n=1}^{n_{max}} \sqrt{1 - \gamma_{njk}^2}|D_n|\cos(n\omega_0 t + \alpha_n) \tag{7.62}$$

where γ_{njk} is the coherence function between points j and k. This procedure can be repeated for all the other necessary points.

Returning to the resolution of the equations of motion (7.51), in cases where we can ignore the fluid-elastic coupling, the response of the system can be estimated by integrating the equations of motion, this time linear, in the time domain, of the form:

$$[M_s]\underline{\ddot{x}} + [R_s]\underline{\dot{x}} + [K_s]\underline{x} = \underline{F}(U(t,\xi)) \tag{7.63}$$

where F is the vector of the generalised forces, in this case a known explicit function of time and space. The resolution of (7.63) can, in this case, also be performed in the frequency domain. To simplify the matter (the discussion can also be extended to systems with several degrees of freedom, see the bibliography at the end of the chapter) we will consider the vibrations of a rigid cylinder restrained in such a way as to represent an oscillating system with a single degree of freedom, free to move in the direction of the average speed of the fluid: thus (7.63) are reduced to:

$$M_s\ddot{x} + R_s\dot{x} + K_s x = F(U(t)) \tag{7.64}$$

Assuming that the dimensions of the body are negligible in relation to the wavelength of the influential flow and considering that the only resistance (drag) is the force exerted by the wind, this becomes:

$$F = F(U(t)) = \frac{1}{2}\rho C_r A U^2(t) \tag{7.65a}$$

C_r being the drag coefficient, generally a function of the Reynolds number only. Taking into account (7.42), then (7.65a) can be expressed as:

$$F(U(t)) = \frac{1}{2}\rho C_r A \bar{U}^2 \left(1 + 2\frac{u(t)}{\bar{U}} + \frac{u^2(t)}{\bar{U}^2}\right) \tag{7.65b}$$

The term $\bar{D}$ in (7.65b):

$$\bar{D} = \frac{1}{2}\rho C_r A \bar{U}^2 \tag{7.66a}$$

represents the static thrust acting on the structure due to the mean value, whereas the terms:

$$F_d(U(t)) = \bar{D}\left(2\frac{u(t)}{\bar{U}} + \frac{u^2(t)}{\bar{U}^2}\right) \tag{7.66b}$$

represent the fluctuating part: if, as is usually the case, $u(t) < \bar{U}$, the quadratic term of (7.66b) can be neglected, so:

$$F_d(U(t)) = \bar{D}\left(2\frac{u(t)}{\bar{U}}\right) \tag{7.66c}$$

The power spectral density of the wind $G_v(f)$ ((7.46) and (7.48), Fig. 7.21) can, in this way, be correlated to the power spectral density of the force $G_d(f)$ using relation [9]:

$$\frac{G_d(f)}{\bar{D}^2} = 4\frac{G_v(f)}{\bar{U}^2} \tag{7.66d}$$

If the body being analysed is large (Fig. 7.24), or rather, if the generic wavelength can be compared to its dimensions, the forces acting on the cylinder $F_d(U(t))$ change. In any case, if we wish to use the expression (7.66d) we must modify the spectrum of the aerodynamic drag force using an appropriate transfer function X_{aero},

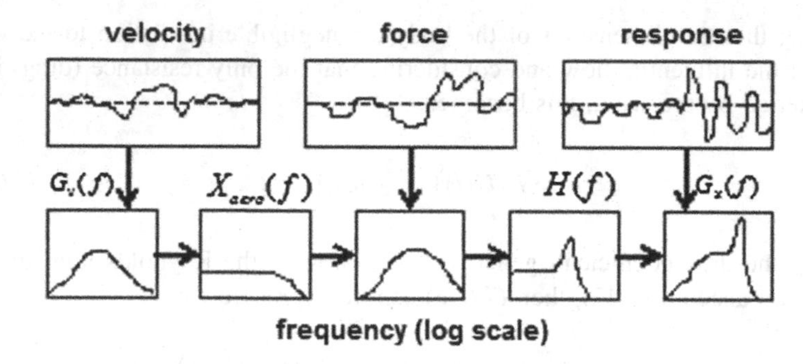

Fig. 7.24 Passage from the power spectral density function of the wind to the power spectral density of the response

known as the aerodynamic admittance function, to obtain the following relation, which replaces (7.66d):

$$G_d(f) = 4|X_{aero}|^2 \frac{G_v(f)}{\bar{U}^2} \bar{D}^2 \tag{7.67a}$$

This function allows us to take into account the variations in the drag coefficient and the decrease in the correlation of the vortices as the wavelengths approach the diameter of the cylinder (in correspondence to wavelengths much greater than the dimensions of the cylinder, the aerodynamic admittance function X_{aero} assumes values that are close to unity).

The power spectral density of the system's response $G_x(f)$ (Sect. 7.5, (7.34a, 7.34b)) is defined by the product of the square of the system's transfer function magnitude $H(f)$ by the power spectral density of the force, i.e.:

$$\frac{G_x(f)}{\bar{x}_s^2} = |H_f|^2 \frac{G_d(f)}{\bar{D}^2} = 4|H_f|^2 |X_{aero}|^2 \frac{G_v(f)}{\bar{U}^2} \tag{7.67b}$$

$\bar{x}_s$ being the static deflection:

$$\bar{x}_s = \frac{\bar{D}}{K_s} \tag{7.67c}$$

and $H(f)$ the transfer function of the system, defined in this case as:

$$H(f) = \frac{1}{-\Omega^2 M_s + i\Omega R_s + K_s}, \qquad \Omega = 2\pi f \tag{7.67d}$$

Recalling the link (7.11) between the variance σ_x and the power spectral density function $G_x(f)$ it is possible to immediately obtain [see (7.46) and (7.2)] the maximum and minimum values of the response given by:

$$x_{max}(t) = \bar{x}_s \pm 3\sigma_x \tag{7.68}$$

Furthermore, remembering the known relation:

$$G_x(f_n) = G_x(nf_0) = \frac{|X_n|^2}{2\omega_0} \tag{7.69a}$$

it is possible to reconstruct possible time histories of the response using:

$$x(t) = \bar{x}_s + \sum_{n=1}^{n_{max}} |X_n| \cos(n\omega_0 t + \theta_n) \tag{7.69b}$$

θ_n being a random phase between 0 and 2π.

7.6.2 Response of a Structure to Wave Motion

When a generic disturbance perturbs the state of rest of the free surface of a fluid, a motion is triggered that tends to bring the surface back to its state of rest: in this way, waves of different shapes and sizes are created (Fig. 7.25) due to the presence of gravity.

This surface motion corresponds to a speed distribution of the single particles throughout the entire volume of fluid. The formation of this wave motion generates forces, generically variable over time, that can dynamically excite both submerged structures (such as, for example, off-shore structures, Fig. 7.26) and vessels (see the bibliography shown in this chapter, section *The Effects of wave motion on structures*).

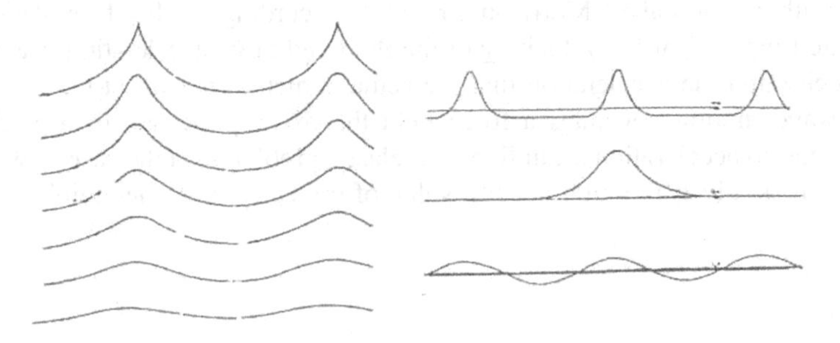

Fig. 7.25 Various wave forms (sea state)

Fig. 7.26 Typical off-shore structure

Knowing the characteristics of these waves, and the relative motion induced in fluid, is, therefore, a key element in planning any marine work or for fully optimising the characteristics of a vessel.

The complex phenomenon of wave motion can be classified as a stationary and ergodic random phenomenon and can, therefore, be characterised with the quantities illustrated in the previous sections (see the bibliography on this subject).

In this section, intended to be an introduction to the complex issues linked with wave motion and its effects, we will analyse the forces $\underline{F}_m$ transmitted by wave motion on a submerged structure assuming that the wavelength associated with the wave motion is much greater than the transverse dimension of the actual structure.

In this case, the induced forces can be estimated, as we will see in more detail below, with the so-called Morrison formulas: according to this formulation, the forces are functions of the velocity $\underline{v}$ of the fluid and of its acceleration $\dot{\underline{v}}$ as well as the velocity and the acceleration that the actual structure is subject to.

So, wave motion generates a force field that is proportional to displacement, velocity and to acceleration as, unlike aero-elastic problems, in the case of water the inertial forces, given the high density value of fluid ρ, are not negligible.

Therefore, the equations of motion of a structure submerged in water and subject to wave motion, and schematised, for example, with finite elements, become of the form:

$$[M_s]\,\ddot{\underline{x}} + [R_s]\dot{\underline{x}} + [K_s]\underline{x} = \underline{F}_m(\ddot{\underline{x}}, \dot{\underline{x}}, \underline{x}, \dot{\underline{v}}, \underline{v}, t) \qquad (7.70)$$

where $\underline{x}$ is the vector of the nodal displacements; $[M_s]$, $[R_s]$ and $[K_s]$ are the system's matrices of mass, damping and stiffness and $\underline{F}_m$ is the vector of the generalised forces caused by the wave motion.

Wave motion or *sea state* can always be seen as a superposition of *elementary states* and is, therefore, of an aleatory nature: each single elementary state is defined by a certain wave period, length and height (monochromatic wave).

The velocity $\underline{v}$ and the acceleration $\dot{\underline{v}}$ of the fluid (i.e. of the generic fluid particle) are, therefore, random quantities, generally time and space functions: these quantities, if referring to a single *elementary state* that makes up the wave motion, depend on the height of the wave on the surface, on the wavelength, on the depth of the ocean floor and, lastly, the distance of the point being considered from the free surface.

Now considering the generic *elementary state*, there are different theories that define the wave shape and that can be used to calculate the velocities and the accelerations of fluid particles. We should remember, in particular (see the specific bibliography at the end of the text):

- the linear theory, a theory that we will refer to in this discussion;
- the Stokes formulas (where motion is assumed to be the sum of small perturbations):
- the Dean stream function solution, etc.

Figure 7.27 shows a graph illustrating the fields of validity of these theories.

The different wave theories have different fields of application depending on the assumed values of parameters:

- depth of the ocean floor d;
- wave period T;
- wave height H;
- gravity acceleration g.

These theories make it possible to obtain the flow velocity since the trend of the wave motion on the surface is known: for this reason, it is clear that in order to properly estimate the forces that fluid exerts on a submerged structure, we need to know the characteristics of the wave motion at the construction site.

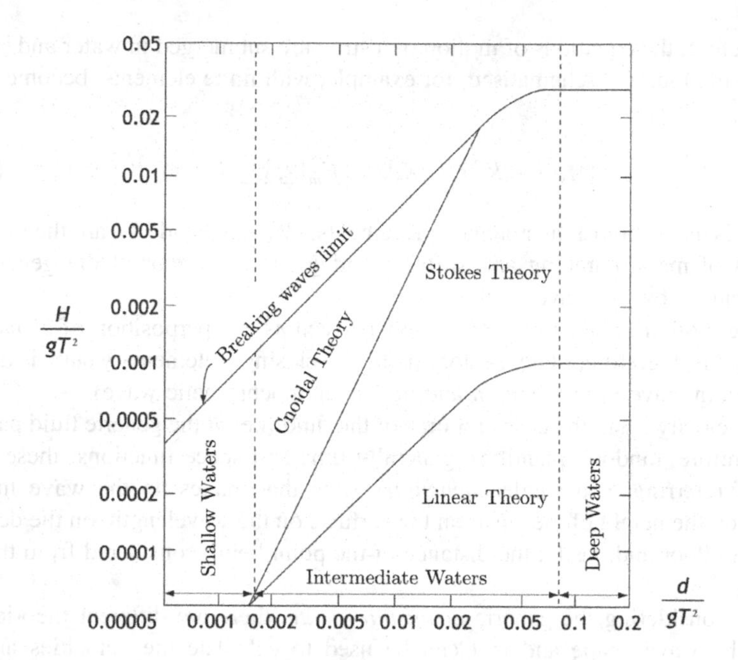

Fig. 7.27 Fields of application of the different wave theories

7.6.2.1 Characterisation of Wave Motion

Characterisation of wave motion can be carried out in three different ways, based on:

- visual observation of the principal statistical characteristics of the wave motion (significant height, period, direction of the wave motion and recurrences);
- experimental readings that make it possible to also provide information on correlations and spectra of the phenomenon;
- processing of meteorological data in the area in question.

Visual observations (historically the first to be used) are done from vessels so the indications obtained are often unreliable; above all on routes that are used less frequently and where there are severe conditions (conditions that ships tend to avoid).

Experimental measurements, which provide more complete information about sea states, can be performed using:

- instruments located above the surface (ships, platforms etc.) based on the emission of laser, electromagnetic or acoustic waves and their reflections;
- instruments located on the surface, such as, for example, accelerometers mounted on buoys (wave rider buoys);
- resistive, capacitive or inductive displacement sensors;
- instruments located under the surface, such as inverted echo sounders, pressure transducers etc.

Data are generally acquired analogically and stored on magnetic tape or on computer in digital form, to be processed subsequently. On the basis of the experimentally obtained data, subsequently we must (see specific bibliography):

- analyse the wave motion data and determine the fundamental statistical parameters of the sea state;
- define the frequency of occurrence;
- define the functions of auto and cross-correlation and the power spectral density functions;
- extrapolate these results to define the possible boundary conditions (extreme waves) with both deterministic and probabilistic methods.

Normally, these indications are obtained from 15–20 min long recordings, taken every three hours (approx. 2900 acquisitions a year). The fundamental quantities that characterise the phenomenon are:

- the significant wave height H_s, i.e. the average height of the wave being observed;
- the significant wave period T;
- the wave direction (it is generally accepted that wave motion is unidirectional);
- the occurrence value.

These quantities would be enough to define the phenomenon if it was harmonic: in actual fact, the time history of the wave motion height is, as mentioned, of a stationary and ergodic aleatory nature, so the following must also be defined:

- its auto and cross-correlation function
- its power spectral density functions, i.e. the mean square value in relation to each frequency in the wave.

We will now illustrate one possible method of investigation used to analyse the behaviour of a generic submerged structure subject to wave motion (an exhaustive discussion of the subject is given in the texts recommended in the bibliography).

7.6.2.2 Generation of a Possible Time History of Wave Motion and of Flow Velocity

As a hypothesis, the linear wave theory assumes an irrotational flow (corresponding to a conservative field): the result being [39–41] that the flow velocity can be represented by a potential function $\Phi(x, y, z, t)$. Taking into account the law of conservation of mass (continuity equations) and assuming incompressible flow, the motion of the single particles must satisfy the Laplace equation (Fig. 7.28):

$$\frac{\partial^2 \Phi}{\partial x^2} + \frac{\partial^2 \Phi}{\partial y^2} = 0 \tag{7.71}$$

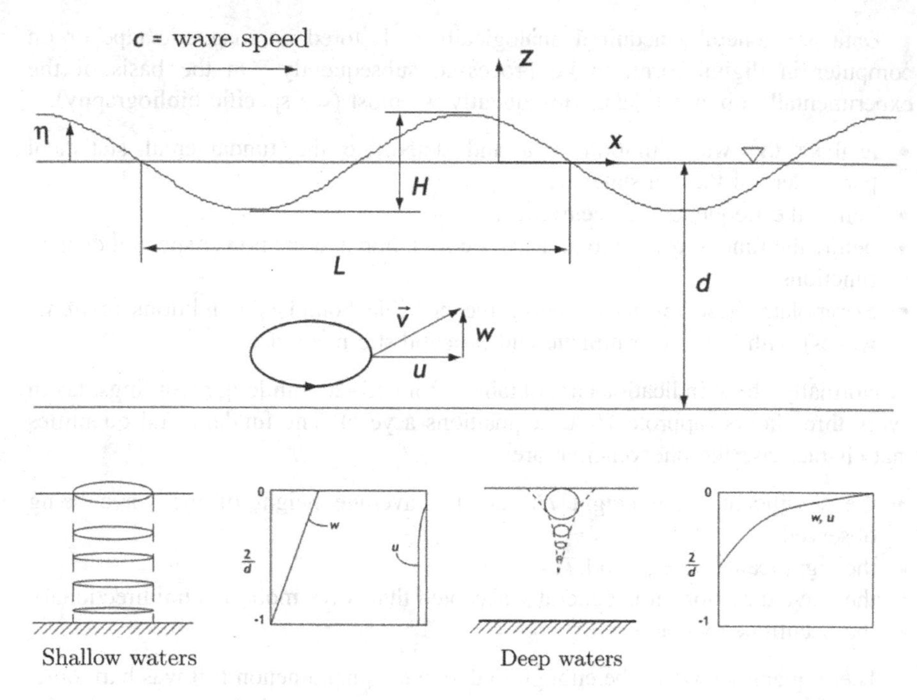

Fig. 7.28 Definition of the motion of the generic particle

The problem is solved by finding a solution to this equation that satisfies the following boundary conditions:

- the velocity of the particles on the surface must be equal to the velocity of the free surface;
- the particle on the bottom must have a normal velocity, null to the actual ocean floor;
- the pressure on the surface must be equal to the atmospheric pressure.

In the hypothesis of a sine wave, the motion of the single particle of fluid, at a certain depth z, is of an elliptical nature (Fig. 7.28), made up of two harmonic components [horizontal $u(x,z,t)$ and vertical $w(x,z,t)$] the magnitude of which decreases exponentially towards the bottom (i.e. as the z magnitude increases). More specifically, we can define with:

- $X - Y - Z$ a Cartesian coordinate system, defining the height of the free surface with $z = 0$;
- $\eta(t,x)$ the rise of the free surface in correspondence to the generic x-coordinate: this hypothesis assumes a unidirectional wave and the problem is, therefore, reduced to a plane problem;
- H the height of the wave (crest-crest), see Fig. 7.28;
- k the wave number $k = \frac{2\pi}{L}$;

- L the wave length;
- ω the wave pulsation (or frequency) $\omega = \frac{2\pi}{T}$;
- T the period;
- d the depth of the ocean floor;
- $C = \frac{L}{T}$ the wave speed;

The variation in the height of the free surface $\eta(t,x)$ becomes, again in the case of a sine wave, of the form:

$$\eta(t,x) = \frac{H}{2}\cos(kx - \omega t) = \frac{H}{2}\cos\left(2\pi\left(\frac{x}{L} - \frac{t}{T}\right)\right) \tag{7.72}$$

The linear wave theory [35–38, 40] makes it possible, by linearising the boundary conditions, to obtain the distribution of the velocity components $u(x,z,t)$ and $w(x,z,t)$ and the accelerations of the generic particle of fluid using the following relations:

$$\begin{aligned} u(t,x,z) &= \frac{g\pi H}{\omega L}\left(\frac{\cosh\left(\frac{2\pi(z+d)}{L}\right)}{\sinh\frac{2\pi d}{L}}\right)\cos\left(\frac{2\pi x}{L} - \omega t\right) \\ w(t,x,z) &= \frac{g\pi H}{\omega L}\left(\frac{\sinh\left(\frac{2\pi(z+d)}{L}\right)}{\sinh\frac{2\pi d}{L}}\right)\sin\left(\frac{2\pi x}{L} - \omega t\right) \end{aligned} \tag{7.73a}$$

$$\begin{aligned} \dot{u}(t,x,z) &= \frac{g\pi H}{L}\left(\frac{\cosh\left(\frac{2\pi(z+d)}{L}\right)}{\sinh\frac{2\pi d}{L}}\right)\sin\left(\frac{2\pi x}{L} - \omega t\right) \\ \dot{w}(t,x,z) &= \frac{g\pi H}{L}\left(\frac{\sinh\left(\frac{2\pi(z+d)}{L}\right)}{\sinh\frac{2\pi d}{L}}\right)\cos\left(\frac{2\pi x}{L} - \omega t\right) \end{aligned} \tag{7.73b}$$

being

- g gravity acceleration;
- z the depth from the free surface;
- L the wave length, which can be defined as:

$$\begin{aligned} L &= \frac{g2\pi}{\omega^2}\tanh\left(\frac{2\pi d}{L}\right) && \text{per bassi fondali} \quad \frac{d}{L} < 0.04 \\ L &= \frac{g2\pi}{\omega^2} && \text{per bassi fondali} \quad \frac{d}{L} < 0.50 \end{aligned} \tag{7.73c}$$

In actual fact, the oscillation of the free surface is not sinusoidal, but is a random phenomenon (which can be considered as stationary and ergodic with a certain sequence of wave heights): therefore, as such, the wave motion must be defined in a statistical manner, starting from the unidirectional power spectral density function of the wave height (for example the Pierson-Moskwitz spectrum). The approach used is based on:

- deterministic theories relating to waves (which univocally establish the link between the shape of the free surface and the corresponding particle velocity, for example, see the linear theory mentioned previously);
- determination of the statistical properties of the real wave motion;
- methods of spectral analysis.

More specifically, the real height of the wave is modelled as a sum of harmonics with different magnitudes H_n and phases:

$$\eta(t,x) = \sum_n \frac{H_n}{2} \cos(k_n x - \omega_n t + \phi_n) = \sum_n \frac{H_n}{2} \cos\left(2\pi\left(\frac{x}{L_n} - \frac{t}{T_n}\right)\right) \tag{7.74}$$

being:

- H_n the height of the wave (crest-crest) in relation to the generic harmonic;
- k_n the generic wave number $k_n = \frac{2\pi}{L_n}$;
- $L_n = \frac{L_0}{n}$ the generic wavelength (sub-multiple of the fundamental wavelength L_0);
- ω_n the wave pulsation (or frequency) $\omega_n = \frac{2\pi}{T_n} = n\omega_0$;
- T_n the period associated with the generic harmonic.

In (7.74) ϕ_n is a generic random phase chosen from 0 and 2π, while H_n, as mentioned, is the amplitude of the generic harmonic determined by the power spectral density function $G_0(\omega_n)$ of the sea state defined by the usual relation:

$$G_0(\omega_n) = G_0(n\omega_0) = \frac{|H_n|^2}{2\omega_0} \tag{7.75}$$

Figure 7.29 shows, as an example, the Pierson-Moskwitz spectrum $G_{0PM}(f_n)$ compared to that of Jonswap $G_{0J}(f_n)$ [40, 41][14]: the diagram shows the power spectral density $G_0(f_n)$ in relation to the frequency $f_n = \frac{\omega_n}{2pi}$ non-dimensionalised compared to the reference frequency f_0 (i.e. in correspondence to which the power spectral density function assumes maximum value). These spectra $G_0(\omega_n)$ can be defined using a function of the form:

$$G_0(\omega_n) = H_s^2 \alpha \left(\frac{\omega_p}{\omega_n}\right)^5 \exp\left(-\frac{5}{4}\left(\frac{\omega_p}{\omega_n}\right)^4\right)\gamma^\beta \tag{7.76}$$

[14] Only an in depth knowledge of the real characteristics of the wave motion at the construction site of the structure can indicate the most suitable spectral formulation.

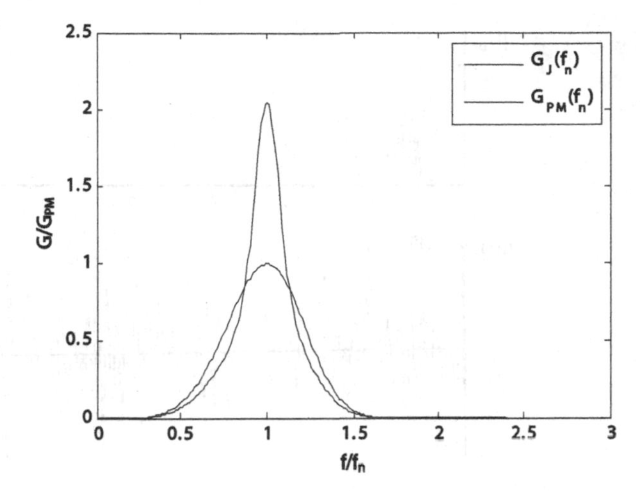

Fig. 7.29 Wave spectrum of Pierson G_{PM} (f) and Moskwitz G_J (f)

where:

- H_s represents the significant wave height, i.e. the average height of the measured wave;
- γ is a coefficient that depends on the type of spectrum considered:

$$\gamma = 1 \text{ Pierson}; \quad \gamma = 3 \text{ Jonswap} \tag{7.76a}$$

- ω_n is the generic pulsation;
- ω_p is the pulsation in correspondence to the peak of the power spectral density function;
- α and β are two functions:

$$\alpha = \frac{0.0624}{0.23 + 0.0334\gamma - \frac{0.185}{1.9+\gamma}}; \quad \beta = \exp\left(-\frac{\left(\frac{\omega_n}{\omega_p} \quad 1\right)^2}{2\sigma^2}\right) \tag{7.76b}$$

being:

$$\sigma = 0.07 \text{ per } \omega_n < \omega_p; \quad \sigma = 0.09 \text{ per } \omega_n > \omega_p \tag{7.76c}$$

Once we know the construction site of the structure, the significant wave height H_s, the wavelength L_n and the corresponding period T_n, it is possible to define from (7.76) the spectrum $G_0(\omega_n)$, from which it is then possible to obtain the magnitude of the single harmonics $|H_n|$ using (7.75), thus making it possible to reconstruct the time history of the wave motion on the surface $\eta(t, x)$ (7.74).

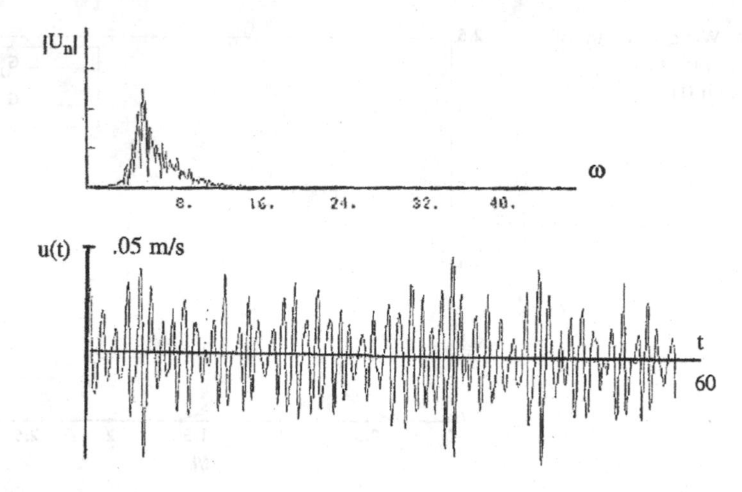

Fig. 7.30 Example of time reconstruction of the flow velocity

The generic space-time history of the flow speed $\underline{v} = \mathbf{i}u + \mathbf{j}w$, at a certain depth z, can be obtained using a wave theory, e.g. the linear theory, by applying the superposition principle, like so:

$$u(x,t,z) = \sum_n u_n(x,t,z); \quad \dot{u}(x,t,z) = \sum_n \dot{u}_n(x,t,z); \tag{7.77}$$

$$w(x,t,z) = \sum_n w_n(x,t,z); \quad \dot{w}(x,t,z) = \sum_n \dot{w}_n(x,t,z); \tag{7.78}$$

where u_n and ω_n are obtained by assuming in (7.73a, 7.73b) $H = H_n$ [obtained from (7.75)] and $\omega = \omega_n$. Equations (7.77) and (7.78) analytically represent the state of the sea, obtained as a superposition of the individual elementary states. Figure 7.30 shows, as an example, the reconstructed trend of the component $u(t,z)$ at a certain point in the flow velocity and the corresponding spectrum.

7.6.2.3 Definition of the Fluid Dynamic Forces Acting on a Cylinder

Assuming that the dimensions of the structure do not alter the influential fluid dynamic field, and once the trend of the two velocity components of the stream is known, using the Morrison equations it is possible to estimate the forces that the fluid exerts on a segment of unitary length of cylinder placed crossways to the direction of the flow: these forces are the sum of a drag component and an inertial component:

$$\begin{aligned} F_x &= \frac{1}{2}\rho D C_d |\underline{v} - \underline{v}_c|^2 \cos\theta + (C_m - 1)\frac{\pi D^2}{4}\rho(\dot{u} - \ddot{x}) + \rho\frac{\pi D^2}{4}\dot{u} \\ F_z &= \frac{1}{2}\rho D C_d |\underline{v} - \underline{v}_c|^2 \sin\theta + (C_m - 1)\frac{\pi D^2}{4}\rho(\dot{w} - \ddot{z}) + \rho\frac{\pi D^2}{4}\dot{w} \end{aligned} \quad (7.79)$$

being:

- ρ the density of the fluid;
- D the diameter of the cylinder;
- C_d the drag coefficient (Fig. 7.31a);

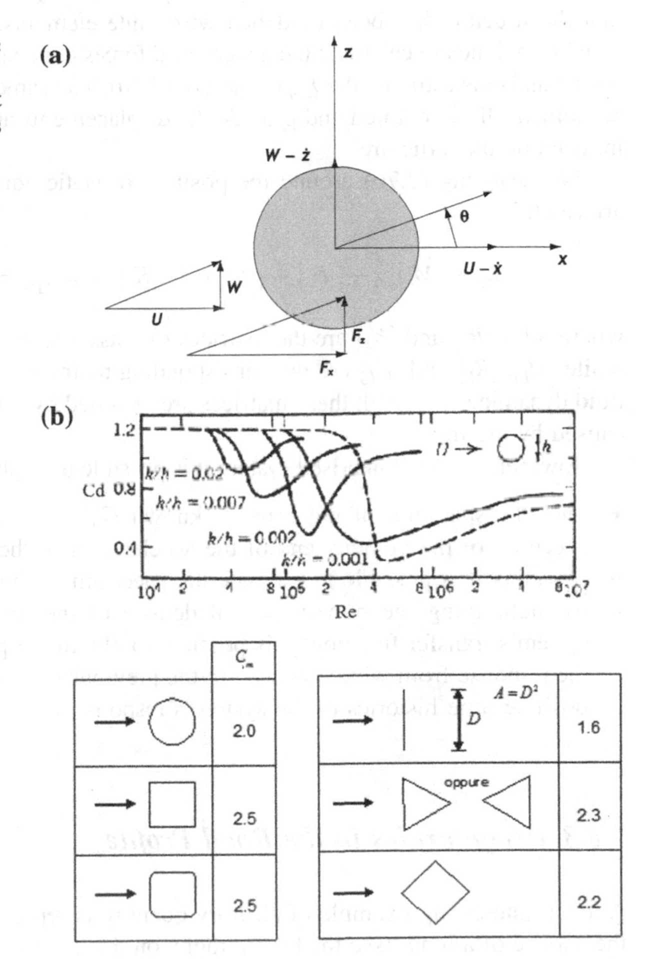

Fig. 7.31 **a** The aerodynamic forces acting on a cylinder. **b** Trend of the drag coefficient C_d in relation to the Reynolds number for cylinders upon variation of the surface roughness (smooth surface, *dotted line*); mass coefficient value C_{mf} for some characteristic sections

- $\underline{v}$ the absolute velocity of the fluid particle:

$$\underline{v} = \mathbf{i}u + \mathbf{j}w \tag{7.80a}$$

- $\underline{v}_c$ the velocity of the cylinder:

$$\underline{v}_c = \mathbf{i}\dot{x} + \mathbf{j}\dot{z} \tag{7.80b}$$

- C_m the added mass coefficient (Fig. 7.31b);

$$\sin\theta = \frac{u - \dot{x}}{|\underline{v} - \underline{v}_c|}; \quad \cos\theta = \frac{w - \dot{z}}{|\underline{v} - \underline{v}_c|} \tag{7.80c}$$

Once the forces exerted on the single length of the cylinder are known, assuming that the structure has been modelled with finite elements, it is possible to use the usual techniques to calculate the generalised forces concentrated on the nodes of the model and thus estimate the $\underline{F}_m$ of the Eq. (7.70). The same equations of motion can be numerically integrated and provide the displacements and, therefore, the stresses induced on the structure.

By linearising (7.70) around the position of static equilibrium it is possible to arrive at:

$$[[M_s] + [M_s]]\,\ddot{\underline{x}} + [[R_s] + [R_f]]\,\dot{\underline{x}} + [[K_s] + [K_f]]\,\underline{x} = \underline{F}_m(\underline{v}, \dot{\underline{v}}) \tag{7.81}$$

where $[M_s]$, $[R_s]$ and $[K_s]$ are the matrices of mass, damping and structural stiffness while $[M_f]$, $[R_f]$ and $[K_f]$ are the corresponding matrices due to the presence of the fluid dynamic force field: these matrices are obtained by linearising the actual forces caused by the fluid.

However, for this linearised system it is possible to apply the spectrum approach:

- once the spectrum of the wave is known $G_0(\omega_n)$ it is possible to obtain the spectrum of the velocity and of the acceleration of the wave motion;
- from this it is possible to estimate the spectrum of the force;
- by multiplying the power spectral density of the force by the square of the system's transfer function, it is possible to obtain the power spectral density of the response from which, as seen in the previous section, it is possible to obtain possible time histories of the system's response.

7.6.3 Irregularities in the Road Profile

Another interesting example of aleatory quantity is represented by irregularities in the surface of a road: (see the bibliography on this subject provided in this chapter, in the section *Effects of irregularities on the dynamic behaviour of road and rail*

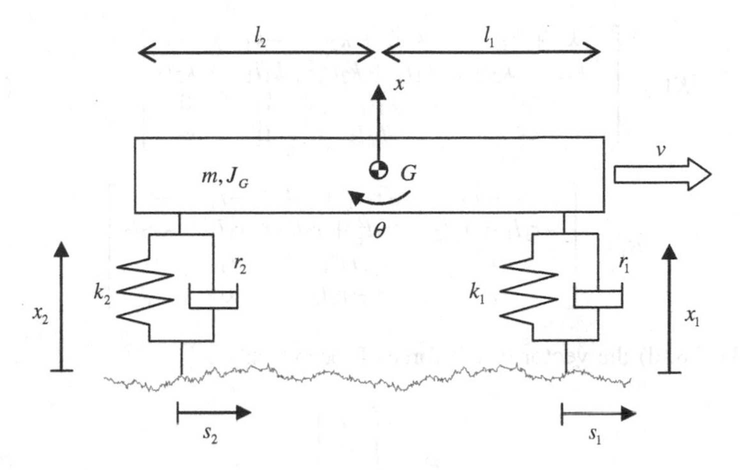

Fig. 7.32 Model of the vehicle for estimating response to irregularities in roads

vehicles): to analyse the effect of these irregularities on a generic vehicle, in a first approximation, we must schematise and analyse the surface using a rigid body connected to the ground with purely elastic and dissipative elements that represent the elasticity and overall damping of the suspensions and tyres (Fig. 7.32).

More specifically we will, first of all, analyse only movement along a straight line and on a level, while assuming a set and constant speed v. Having assumed the vertical displacement of the carriage x, the rotation θ and displacements x_1 and x_2 impressed at the two connection points as independent variables:

$$\underline{x} = \begin{Bmatrix} x \\ \theta \\ x_1 \\ x_2 \end{Bmatrix} \tag{7.82}$$

In this case the equations of motion are of the form:

$$[M]\,\underline{\ddot{x}} + [R]\underline{\dot{x}} + [K]\underline{x} = \underline{F}(t) \tag{7.83}$$

being:

$$[M] = \begin{bmatrix} M & 0 & 0 & 0 \\ 0 & J_G & 0 & 0 \\ 0 & 0 & 0 & 0 \\ 0 & 0 & 0 & 0 \end{bmatrix} \tag{7.83a}$$

$$[K] = \begin{bmatrix} k_1 + k_2 & -k_1 l_1 + k_2 l_2 & -k_1 & -k_2 \\ -k_1 l_1 + k_2 l_2 & k_1 l_1^2 + k_2 l_2^2 & k_1 l_1 & -k_2 l_2 \\ -k_1 & k_1 l_1 & k_1 & 0 \\ -k_2 & -k_2 l_2 & 0 & k_2 \end{bmatrix} \quad (7.83b)$$

$$[R] = \begin{bmatrix} r_1 + r_2 & -r_1 l_1 + r_2 l_2 & -r_1 & -r_2 \\ -r_1 l_1 + r_2 l_2 & r_1 l_1^2 + r_2 l_2^2 & r_1 l_1 & -r_2 l_2 \\ -r_1 & r_1 l_1 & r_1 & 0 \\ -r_2 & -r_2 l_2 & 0 & r_2 \end{bmatrix} \quad (7.83c)$$

In (7.83a–7.83d) the vector of the forces F becomes:

$$\underline{F} = \begin{Bmatrix} 0 \\ 0 \\ R_1 \\ R_2 \end{Bmatrix} \quad (7.83d)$$

R_1 and R_2 being the dynamic reactions transmitted from the terrain to the vehicle in correspondence to the DOFs x_1 and x_2.

By partitioning the same equations of motion (7.83a–7.83d) and using $\underline{x}_V$ to define the set degrees of freedom and $\underline{x}_L$ for those that are free:

$$\underline{x} = \begin{Bmatrix} x \\ \theta \\ \ldots \\ x_1 \\ x_2 \end{Bmatrix} = \begin{Bmatrix} \underline{x}_L \\ \underline{x}_V \end{Bmatrix} \quad (7.84a)$$

it is possible to rewrite (7.83a–7.83d) as two subsystems of matrix equations, the first of which:

$$[M_{LL}]\ddot{\underline{x}}_L + [R_{LL}]\dot{\underline{x}}_L + [K_{LL}]\underline{x}_L = -[M_{LV}]\ddot{\underline{x}}_V - [R_{LV}]\dot{\underline{x}}_V - [K_{LV}]\underline{x}_V = \underline{F}(t) \quad (7.84b)$$

makes it possible, once the irregularity of the runway is known, i.e. the value of $\underline{x}_V$, $\dot{\underline{x}}_V$ and $\ddot{\underline{x}}_V$, to define the response $\underline{x}_L$ of the system. Once this is resolved, the second equation:

$$[M_{VL}]\ddot{\underline{x}}_L + [R_{VL}]\dot{\underline{x}}_L + [K_{VL}]\underline{x}_L + [M_{VV}]\ddot{\underline{x}}_V + [R_{VV}]\dot{\underline{x}}_V + [K_{VV}]\underline{x}_V = \underline{R}(t) \quad (7.84c)$$

makes it possible to obtain the reactions $\underline{R}(t)$ exchanged with the terrain.

Assuming that the runway is considered to be rigid (an assumption that is not acceptable if we wish to simulate a vehicle travelling on a deformable structure, for example, a viaduct), the displacements of the wheels $\underline{x}_V$ are to be attributed solely to irregularities, so:

$$\underline{x}_V = \begin{Bmatrix} i_1 \\ i_2 \end{Bmatrix} \tag{7.85}$$

Irregularity, as mentioned, is a random phenomenon (Fig. 7.33), which can be classified as stationary and ergodic: its characteristics can be defined by the power spectral density function $G_i(1/\lambda)$ and the spatial cross-correlation function $R_{xy}(\xi)$, ξ being the spatial coordinate and λ the generic wavelength. These functions can be obtained using the procedures described previously in Sect. 7.3.1. Figure 7.34 shows a typical trend of the power spectral density function of the irregularity.

Once the power spectral density function is known, it is possible to reconstruct a possible spatial history of the irregularity for the front wheel starting from the actual definition of PSD:

$$G_i(1/\lambda_n) = \frac{|C_n|^2}{2(1/\lambda_0)} \tag{7.86}$$

as:

$$i_1(s_1) = \sum_n |C_n| \cos\left(\frac{n2\pi}{\lambda_0} s_1 + \phi_n\right) \tag{7.87}$$

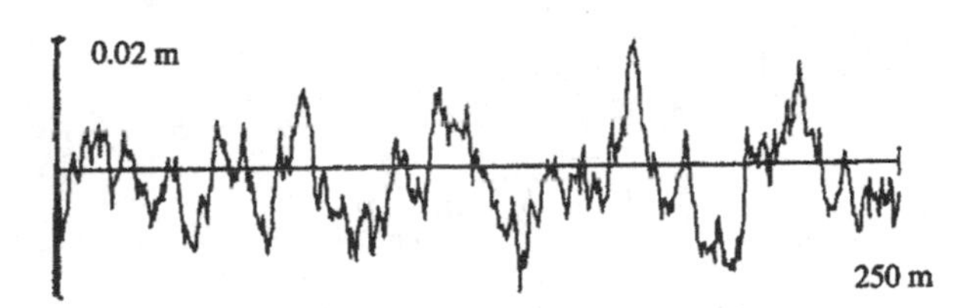

Fig. 7.33 An example of road surface irregularity

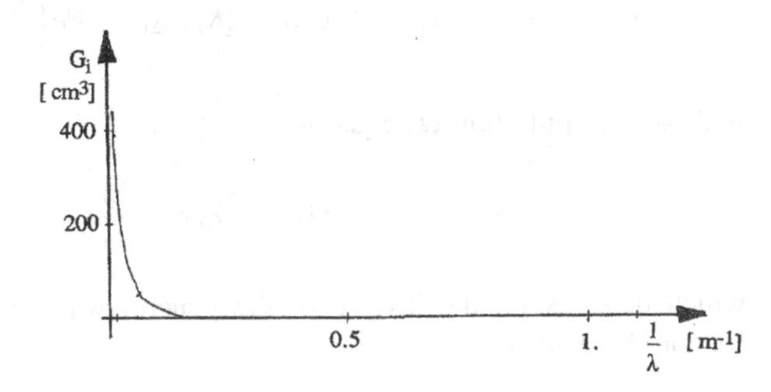

Fig. 7.34 Power spectral density of the irregularity

s_1 being the position reached by the front wheel, $|C_n|$ the generic amplitude, $1/\lambda_0$ the fundamental spatial frequency, λ_0 the fundamental wavelength (usually assumed to be 1 km) and lastly ϕ_n a random phase chosen from 0 and 2π.

Taking into account that the position reached by the front wheel equals:

$$s_1 = vt \tag{7.88}$$

(7.88) becomes:

$$i_1(s_1) = i_1(t) = \sum_n |C_n| \cos\left(\frac{n2\pi}{\lambda_0} vt + \phi_n\right) \tag{7.89}$$

and assuming

$$\Omega_0 = \frac{2\pi v}{\lambda_0} \tag{7.89a}$$

$$i_1(s_1) = i_1(t) = \sum_n |C_n| \cos(n\Omega_0 t + \phi_n) = \sum_n |C_n| \cos(\Omega_n t + \phi_n) \tag{7.89b}$$

The rear wheel "sees" the same irregularity of the pitch p out of phase, therefore, as $s_2 = vt + p$ is the x-coordinate reached by the rear wheel, we will have:

$$\begin{aligned} i_2(t) &= \sum_n |C_n| \cos\left(\frac{n2\pi}{\lambda_0} s_2 + \phi_n\right) = \sum_n |C_n| \cos\left(\frac{n2\pi}{\lambda_0}(vt + p) + \phi_n\right) \\ &= \sum_n |C_n| \cos(n\Omega_0 t + \psi_n) = \sum_n |C_n| \cos(\Omega_n t + \psi_n) \end{aligned} \tag{7.90a}$$

being:

$$\psi_n = \phi_n + n\frac{2\pi}{\lambda_0} p \tag{7.90b}$$

Thus the system is excited by sinusoidal forces:

$$\underline{F}(t) = -[M_{LV}]\underline{\ddot{x}}_V - [R_{LV}]\underline{\dot{x}}_V - [K_{LV}]\underline{x}_V = \Re e\left(\sum_n \underline{F}_{0n} e^{i\Omega_n t}\right) \tag{7.91}$$

with generic pulsation Ω_n equal to:

$$\Omega_n = n\Omega_0 = n\frac{2\pi}{\lambda_0} v \tag{7.91a}$$

which depends on the inverse of the generic wavelength $\lambda_n = (n/\lambda_0)$ and the forward velocity v.

Therefore, there is a possibility that one or more of these pulsations, for particular wavelengths $\lambda_n = (n/\lambda_0)$ and particular forward velocities v, excite the system into resonance (a fact which often occurs when passing over a viaduct where there are expansion joints at regular intervals).

The generalised forces on the front and rear axle are out of phase with each other as a function of the pitch p and the harmonic being considered.

In this case the equations of motion (7.84b), taking into account (7.91), become:

$$[M_{LL}]\ddot{\underline{x}}_L + [R_{LL}]\dot{\underline{x}}_L + [K_{LL}]\underline{x}_L = \Re e\left(\sum_n \underline{F}_{0n} e^{i\Omega_n t}\right) \tag{7.92}$$

and can be numerically integrated with split-step methods.

The resolution of (7.92) can also be carried out in the frequency domain: to simplify the issue, we will consider the system with a single degree of freedom in Fig. 7.35 (the discussion can then be extended to systems with more degrees of freedom, see the specific bibliography at the end of the text), so (7.92) are reduced to:

$$m_s\ddot{x} + r_s\dot{x} + k_s x = k_s x_1(t) + r_s\dot{x}_1(t) = F(t) \tag{7.93}$$

Once a specific forward velocity v has been assigned to the vehicle, it is possible to define the spectrum of irregularities as a function of Ω instead of $(1/\lambda)$ as:

$$\Omega = \frac{2\pi}{\lambda} v \tag{7.94a}$$

thus moving from function $G_i(1/\lambda)$ to function $G_i(\Omega)$.

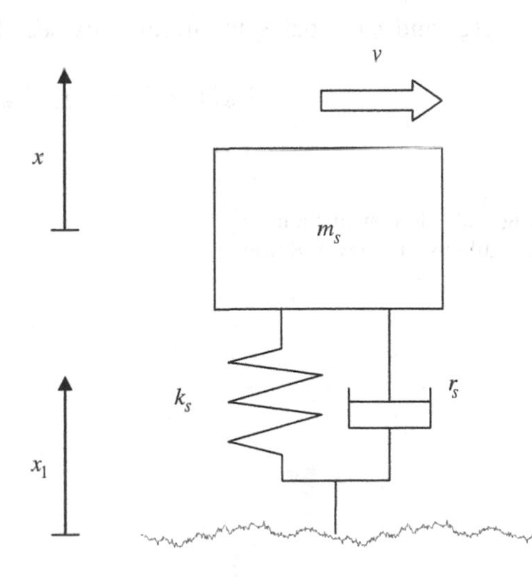

Fig. 7.35 Vehicle as a system with 1 DOF

The power spectral density of the force, similarly to what we saw for the aerodynamic forces, is equal to:

$$G_f(\Omega) = |X_{if}(\Omega)|^2 G_i(\Omega) \tag{7.94b}$$

X_{if} being the transfer function that makes it possible to obtain the spectrum of the force as a function of the spectrum of the irregularities:

$$X_{if} = i\Omega r_s + k_s \tag{7.95}$$

The power spectral density of the system's response $G_x(\Omega)$ is thus defined by:

$$G_x(\Omega) = |H(\Omega)|^2 G_f(\Omega) = |H(\Omega)|^2 |X_{if}(\Omega)|^2 G_i(\Omega) \tag{7.96}$$

$H(\Omega)$ being the transfer function of the actual system:

$$H(\Omega) = \frac{1}{-\Omega^2 m_s + i\Omega r_s + k_s} \tag{7.97}$$

Recalling the link (7.11) between the variance σ_x and the power spectral density function $G_x(\Omega_n)$ it is possible to immediately obtain the maximum and minimum values of the response given by:

$$X_{min-max}(t) = \pm 3\sigma_x \tag{7.98}$$

In the event that we should wish to create an irregularity, taking into account the three-dimensional nature of the problem, we must also consider the coherence between the irregularity seen by the left and right-hand wheels of the same vehicle (Fig. 7.36).

$i_s(\xi)$ and $i_d(\xi)$ being the irregularity seen by the two wheels and (from (7.87)):

$$G_{is}(1/\lambda_n) = G_{id}(1/\lambda_n) = G_i(1/\lambda_n) \tag{7.99}$$

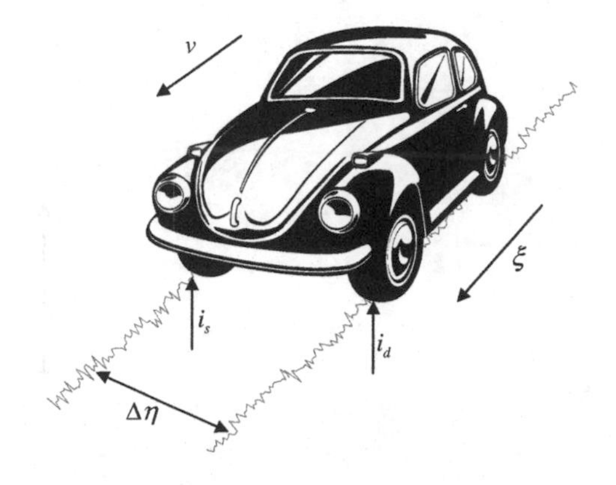

Fig. 7.36 Reconstruction of irregularity on a level runway

being the power spectral density function (quantity which, as seen is defined experimentally and known, Fig. 7.34).

The two spatial histories are obviously correlated: we will use γ_{sd}^2 to define the coherence function:

$$\gamma_{sd}^2 = \frac{|G_{sd}|^2}{G_{is} G_{id}} \tag{7.100}$$

G_{sd} being the cross-spectrum between the irregularity measured on the left-hand wheel and the irregularity measured on the right-hand wheel. Figure 7.37 shows the trend of the coherence function in relation to the non-dimensional frequency $\Delta\eta/\lambda_n$, where $\Delta\eta$ is the distance between the two wheels (Fig. 7.36).

Once $G_i(1/\lambda_n)$ is known, as seen previously, it is possible to generate a possible history of the irregularity as a function of the generic x-coordinate ξ:

$$i_s(\xi) = \sum_n |C_n| \cos\left(n\frac{2\pi}{\lambda_0}\xi + \theta_n\right) \tag{7.101}$$

The history of irregularity of the other wheel can be written as:

$$i_d(\xi) = i_{d\,coer} + i_{d\,uncoer} \tag{7.102}$$

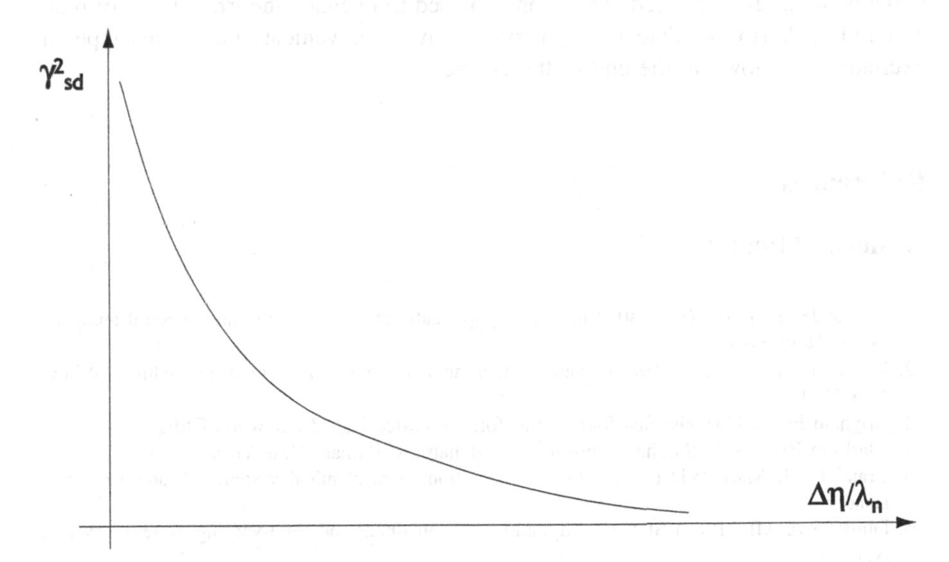

Fig. 7.37 Road surface irregularity: coherence function γ_{sd}^2

i.e. as the sum of one part $i_{d\,coer}$ coherent with $i_s(\xi)$, and of one non-coherent part $i_{d\,uncoer}$. The coherent part, by definition (7.100), will be expressed as:

$$i_{d\,coer}(\xi) = \sum_n \gamma_{sd}|C_n| \cos\left(n\frac{2\pi}{\lambda_0}\xi + \theta_n\right) \quad (7.103)$$

that is, a history reconstructed with the same phase θ_n used to generate $i_s(\xi)$ (7.101). The non-coherent part is generated with different random phase ϕ_n and by requiring that the two spatial histories have the same spectrum (7.99) [42–48]:

$$i_{d\,uncoer}(\xi) = \sum_n \sqrt{1-\gamma_{sd}^2}|C_n| \cos\left(n\frac{2\pi}{\lambda_0}\xi + \phi_n\right) \quad (7.104)$$

Ultimately we will have:

$$i_d(\xi) = \sum_n \gamma_{sd}|C_n| \cos\left(n\frac{2\pi}{\lambda_0}\xi + \theta_n\right) + \sum_n \sqrt{1-\gamma_{sd}^2}|C_n| \cos\left(n\frac{2\pi}{\lambda_0}\xi + \phi_n\right) \quad (7.105)$$

Summary This chapter introduces the basic concepts related to random vibrations. After a brief description of the methods that can be used to analyse such processes, three application examples, related to structure response to turbulent wind and to wave motion, are presented. The approach used to evaluate the irregularity of road and rail-track (to simulate the dynamic behavior of vehicles under this type of excitation) is shown at the end of the chapter.

References

General Theory

1. Bendat JS, Piersol AG (1980) Engineering applications of correlation and spectral analysis. Wiley, New York
2. Bendat JS, Piersol AG (1986) Random data: analysis and measurement procedures. Wiley, New York
3. Brigham EO (1974) The fast fourier transform. Prentice-Hall, Englewood Cliffs
4. Clarkson BL (1977) Stochastic problems in dynamics. Pitman, New York
5. Crandal SH, Mark WD (1963) Random vibration in mechanical systems. Academic Press, London
6. Dimentberg MF (1988) Statistical dynamics of non linear and time-varying systems. Wiley, New York
7. Gawronski W, Natke HG (1986) On ARMA models for vibrating systems. Probab Eng Mech 1(3):150–156
8. Guicking D (1988) Active noise and vibration control: reference bibliography, 3rd edn. University of Gottingen

9. Harris CM, Crede CE (1980) Shock and vibration handbook. McGraw-Hill, New York
10. Jannuzzi A, Spinelli P (1987) Artificial wind generation and structural response. Dip. Ing. Civ., Università di Firenze
11. Laning JH, Battin RH (1956) Random processes in automatic control. McGraw-Hill, New York
12. Loeve MM (1977) Probability theory, 4th edn. Springer, Berlin
13. Marple SL (1987) Digital spectral analysis with applications. Prentice-Hall, Englewood Cliffs
14. Murakami Y (1991) The rainflow method in fatigue: the Tatsuo Endo Memorial Volume. In: International symposium on fatigue damage measurement and evaluation under complex loadings, Fukouka, Japan. Elsevier, Amsterdam
15. Natke HG (1988) Application of system identification in engineering. Springer, Berlin
16. Newland DE (1975) An introduction to random vibration and spectral analysis. Longmans, London
17. Otnes RK, Enochson L (1978) Applied time series analysis. Wiley, New York
18. Pandit SM, Wu SM (1983) Time series and system analysis with applications. Wiley, New York
19. Proceeding of the 9th IFAC/IFORS symposium on identification and systems parameter estimation, Budapest, Hungary, July 1991
20. Proceeding of the IFAC symposium on design methods of control systems, Zurich, Switzerland, Sept 1991
21. Proceeding of the IMACS world congress, Dublin, Ireland, July 1991
22. Proceedings of the 10th international conference on modal analysis, Union College, S.E.M., San Diego, CA 1992
23. Roberts JB, Spanos PD (1990) Random vibration and statistical linearization. Wiley, New York
24. Shinozuka M, Jan CM (1972) Digital simulation of random processes and its applications. J Sound Vib 25(1):111–128
25. Tomlinson GR (1987) Vibration analysis and identification of non linear systems, short course. Herriot-Watt University, Edinburgh

Effects of Wind on Structures

26. Bocciolone M, Cheli F et al (1991) Wind measurements on the Humber bridge and numerical simulations. In: 8th international conference on wind engineering ICWE, London, Canada, June 1991
27. Davenport AG (1961) The application of statistical concepts to the wind loading of structures. In: Proceedings, Institution of Civil Engineers, London
28. Diana G et al (1990) Oscillation of bundle conductors in overhead lines due to turbulent wind. In: IEEE winter meeting, New York
29. Diana G, Cheli F et al (1992) Aeolian vibrations of overhead transmission lines: computation in turbulence conditions. J Wind Eng 52. In: First international symposium on computational wind engineering, Tokyo, CWE 92
30. Dowell EH (1978) A modern course in aeroelasticity. Kluwer Academic Publishers, London
31. Irwin HP, Shuyler GD (1977) Experiments on the full aeroelastic model of the Lion's Gate Bridge in smooth and turbulent flow. N.R.C. report LTR-LA-206
32. J Wind Eng Ind Aerodyn. Elsevier, Amsterdam
33. Proceeding of the 7th international conference on wind engineering ICWE, Aachen, West Germany, July 1987
34. Simiu E, Scanlan RH (1978) Wind effects on structures. Wiley, New York

Effects of Wave Motion on Structures

35. Broglio S, Cheli F et al (1991) Dynamic response of an undersea cable under current and waves. In: V international conference on flow induced vibrations, Bristol, May 1991
36. Chen SS (1987) Flow-induced vibration of circular cylindrical structures. Springer, Berlin
37. Kaliski S (1992) Vibrations and waves, studies in applied mechanics. Elsevier, New York
38. Politecnico di Milano, Cenni sulla valutazione delle onde estreme - Fonti di informazione sul moto ondoso e metodi di elaborazione, Corso su Problemi attuali di strutture marine, Milan, 1979
39. Proceedings of the international conference on hydrodynamics in ocean engineering, Trondheim, 1981
40. Sarpkaya T, Isaacson M (1976) Mechanics of wave forces on offshore structures. Van Nostrand Reinhold Company, New York
41. Zienkiewicz N (1978) Numerical methods in offshore engineering. Wiley, New York

Effects of Irregularity on the Dynamic Behaviour of Road and Rail Vehicles

42. Cebon D, Newland DE (1972) The artificial generation of road surface topography by inverse FFT method. J Sound Vib 25(1)
43. Cheli F, Porcu C (1992) Traffico stradale su ponti di grande luce: interazione dinamica. Modello analitico e risultati sperimentali. XI Congresso Nazionale AIMETA, Trento
44. Diana G, Cheli F (1990) Sulla dinamica di veicoli ferroviari: carrozza gran comfort (modello a 23 g.d.l.). Report Interno N.2 del Dipartimento di Meccanica del Politecnico di Milano, Milan
45. Int J Veh Mech Mobility, Swets Zeitlinger B. V
46. Palamas I et al (1985) Effects of surface irregularities upon the dynamic response of bridges under suspended moving loads. J Sound Vib 99(2):235–145
47. Proceeding of the IA VSD symposium on dynamics of vehicles on roads and tracks
48. Rill G (1983) The Influence of correlated random road excitation processes on vehicle vibration. In: 8th IA VD symposium, Massachusetts

Chapter 8
Techniques of Identification

8.1 Introduction

The need to precisely and accurately predict the dynamic behaviour of increasingly complex machines or, more generally, of installations and structures, and the need to predict their reliability and life span, have led to the need to develop increasingly refined and complex calculation methods which have become indispensable tools for design and operation. Such methods, based on the use of suitable mathematical models, such as multi-body methods and finite element methods for mechanical systems, seek to reproduce the behaviour of the real system and make it possible not only to predict the system's dynamic response to the action of different forms of excitation but also its stability in different working conditions. During the design phase, the advantage of having a sophisticated mathematical model offers the possibility of avoiding partly, if not entirely, any experimental investigations on prototypes or individual components, which would otherwise be necessary and which is usually quite costly and requires often prohibitive implementation times. The mathematical model associated with the real system is usually defined by:

- assuming that the system in question has certain properties (related to the laws of physics or to purely mathematical hypothesis);
- including consolidated partial mathematical models.

So far in the discussion we have always assumed that the elastic, inertial and damping characteristics of the mechanical system to be analysed are known (in terms of stiffness, mass and concentrated and/or distributed damping), as is the type of excitation that the system is subject to. With these assumptions, once a certain mathematical model had been defined that was able to reproduce the true behaviour of the system, the relevant equations of motion were defined and then they were solved. Often, however, it is not possible to accurately define the value of some of

F. Cheli and G. Diana, *Advanced Dynamics of Mechanical Systems*,
DOI 10.1007/978-3-319-18200-1_8

the system's typical parameters (such as, for example, the stiffness and damping coefficients of some components). In this case the mathematical models are subject to uncertainties and the comparison between analytical and experimental results may show significant discrepancies, thus making the results obtained with the mathematical model unreliable. Sometimes, given the complexity of the system, it may not even be possible to define a mathematical model a priori whose equations make it possible to adequately reproduce the true behaviour of the system. To overcome these difficulties several methods have been developed over recent decades, known as "identification techniques", (Proceeding IFAC/IFOR, Proceeding on Modal Analysis, [20]) that use experimental measurements on real systems, making it possible to obtain:

- the characteristics of several parameters of a pre-established mathematical model of the considered system (techniques of parameter identification);
- the equations relating to a mathematical model not corresponding to a physical scheme, but reproducing the system as a "black box" (system identification techniques);
- the characteristics of the sources of excitation that it is subject to.

So the term "identification techniques" means the set of methodologies which, starting from experimental measurements, aims to define the characteristics of a generic system: these methodologies were developed mainly as part of controls and were only later extended to the study of systems of a mechanical, electrical, electronic, economic, biological nature etc., as well as to the study of generic processes. In the specific issues of identification particular nomenclature and symbology are traditionally used, usually by experts in the field of controls, where these techniques, as already mentioned, have undergone important developments. The generic term "system" is used to define a set of several components (material objects) which, being physically interconnected, act as a whole. In general, a mechanical system is made up of a combination of interconnected rigid or flexible bodies with elastic and damping elements interfaced with actuators or motors controlled in an open or closed loop (e.g. a road or rail vehicle, a robot, a rotating machine, including a rotor, bearings, case and foundation etc.). Anything that is not part of the system is usually defined as the "outside world" separated from the system by a physical (or ideal) surface. The objects that make up the system are known as "components" or "subsystems": the second term refers to the fact that, depending on the type of investigation being carried out, a certain combination of objects may be considered to be a stand-alone system or a subsystem (e.g. an articulated rod and crank mechanism can itself be a system, while it can also become a subsystem of a complete machine). A system that is subject to certain sources of excitation shows certain behaviour defined as the system response: an unbalanced rotor shows bending vibrations, a structure that is hit by turbulent wind oscillates around its position of equilibrium etc. The dynamic behaviour of a generic system, or the dynamic process that system is subject to, can often be described by relations between:

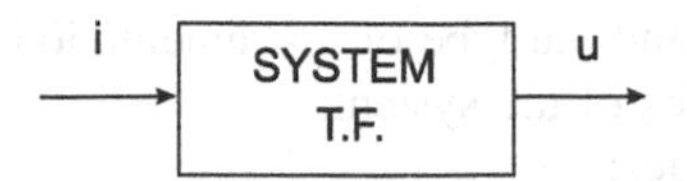

Fig. 8.1 Definition of a system: inputs and outputs

- input i (for a mechanical system inputs are represented by external forces or by displacements impressed on the constraints)
- output u (the motion of the system, the stresses transmitted to the individual components etc.).

These relations can be defined by the equations of a suitable mathematical model which can generally be expressed both in the time domain and in the frequency domain. The relations that link inputs and outputs can be displayed using appropriate block diagrams, such as the one shown in Fig. 8.1. These methodologies are based on an analysis (statistical) of the response of the system subject to inputs affected by disturbances, as well as deterministic inputs: with this assumption the output also becomes a statistical variable. Definition of the parameters is, therefore, performed in probabilistic terms, that is, the most likely values of the parameters are defined and an estimation of them is made. In this context, filtering techniques, meaning techniques aimed at separating the true signal from disturbances, are very important [17, 28]. As mentioned, the mathematical model to be used in these identification procedures can be classified into two main families:

- *black box* model;
- parametric model.

The *black box* model does not require a predefined structured mathematical model (for example, the transfer function of a vibrating system can be considered as such). The structured model, on the contrary, is obtained from the description of the individual components of the given system, through laws that reproduce the behaviour of the parts and to which a specific physical meaning is associated. Therefore, identifying the characteristics of parametric models requires the use of structured mathematical models, i.e., those whose equations are defined in advance and the problem of identification can be traced back, in this case, to a problem in estimating the parameters of the actual model. When discussing identification, the term *estimation* is often adopted, and not *evaluation*, since the same techniques are based, as already mentioned, on measurements carried out using experimental tests and these are always physiologically affected by random errors: for this reason, we need to introduce, as mentioned, appropriate statistical methodologies to reduce these errors and obtain valid information in order to estimate, with sufficient accuracy, the system's unknown characteristics. A problem of identification, regardless of the type of system or process to be analysed, thus involves both theoretical considerations and experimental tests. For the experimental tests, the following must be defined:

- the testing conditions and the type of instrumentation;
- the excitation modalities of the system;
- the measuring techniques;
- the methods of obtaining and processing data.

Whereas the required theoretical analyses involve:

- choosing the most suitable mathematical model for defining the behaviour of the system;
- choosing appropriate methods for estimating parameters;
- the numerical methods for carrying out identification.

The methods for identifying parameters can be applied to any type of system (mechanical, electrical, electronic, etc.): in the following discussion we will obviously be referring to mechanical systems.

8.1.1 Identifying the Parameters of a Mechanical System in the Time and Frequency Domain

The methods of identification, regardless of the type of model of the system adopted, can be developed both in the time domain and in the frequency domain. The equations of motion of a generic mechanical system, assuming it is schematised with a reasonable mathematical model, are generally of the form shown below (see Chaps. 1 and 2):

$$\underline{F}_i(\underline{x},\dot{\underline{x}},\ddot{\underline{x}}) + \underline{F}_s(\underline{x},\dot{\underline{x}}) + \underline{F}_e(\underline{x}) = \underline{F}(\underline{x},\dot{\underline{x}},t) \tag{8.1a}$$

where $\underline{x}$ is the vector of the independent variables, $\underline{F}_i$, $\underline{F}_s$, and $\underline{F}_e$ are the forces of inertia, damping and elastic (in general, non-linear functions of the independent variables $\underline{x}$). Lastly, $\underline{F}$ can group together all the forces not included to the left of the equals sign, associated with field actions, with the presence of a control system or, more generally, with external forces that are explicit functions of time. In the case where the system is linear, or it is linearized (see Chap. 5, regarding systems surrounded by force fields), then (8.1a) becomes:

$$[M]\ddot{\underline{x}} + [R]\dot{\underline{x}} + [K]\underline{x} = \underline{F}(t) \tag{8.1b}$$

where $[M]$, $[R]$ and $[K]$ are, respectively, the equivalent matrices of mass, damping and stiffness of the actual system. As already illustrated in Sect. 2.4.2.2, (8.1b) defined with $\underline{y}$ the vector of velocity:

$$\underline{y} = \dot{\underline{x}} \tag{8.2a}$$

Equation (8.1b) can be redefined on the basis of the so-called state variables $\underline{z}$:

$$\underline{z} = \left\{ \begin{array}{c} \underline{x} \\ \underline{y} \end{array} \right\} \tag{8.2b}$$

i.e. they can be rewritten in the generic form, also known as the state transition function:

$$\dot{\underline{z}} = \underline{f}(\underline{z}, \underline{F}(t)) \tag{8.2c}$$

and for the linear or linearized system:

$$\dot{\underline{z}} = [A]\underline{z} + \underline{u}(t) \tag{8.2d}$$

$[A]$ being the so-called transition matrix (Sect. 2.4.2.2):

$$[A] = \begin{bmatrix} -[M]^{-1}[R] & -[M]^{-1}[K] \\ -[I] & 0 \end{bmatrix} \tag{8.2e}$$

and $\underline{u}(t)$ the vector of known input terms:

$$\underline{u}(t) = \left\{ \begin{array}{c} \underline{F}(t) \\ \underline{0} \end{array} \right\} \tag{8.2f}$$

(8.2c) or (8.2d) make it possible to determine *the state of the system.* In (8.2c) or (8.2d) the vector $\underline{F}(t)$ or $\underline{u}(t)$ represents the input of the system $\underline{i}(t)$, while usually, in mechanical systems, the output is identified with one or more of the state quantities $\underline{z}$ (8.2d). More generally, the system outputs $\underline{y}_u(t)$ are correlated to the state variables $\underline{z}$ by a relation, also defined as the output transformation function, of the form:

$$\underline{y}_u(t) = \underline{y}_u(\underline{z}(t), \underline{i}(t)) \tag{8.3}$$

which defines the output $\underline{y}_u(t)$ of the system in relation to the existing input $\underline{i}(t)$ and of the state of the system $\underline{z}$. In the case of a generic vibrating mechanical n-degree-of-freedom (d.o.f.) system, for example, the state variables may be represented by displacement $\underline{x}$ and by velocity $\dot{\underline{x}}$ of the d.o.f. associated with the motion of the individual bodies making up the actual system. One possible output transformation can be represented by the stresses induced on a component of the system, a function, in general, both of the inputs and the state of the actual system. Identification of the frequency domain generally aims to identify the harmonic transfer function of the system. The techniques used to identify the modal parameters of a mechanical system that is considered linear can be traced back to this category. A non-structured model, i.e. a black box model can, in actual fact, be represented in the frequency domain (Sect. 2.4.1.4.1) by a relation of the form:

$$\underline{X}(\Omega) = [H(\Omega)]\underline{F}(\Omega) \tag{8.4}$$

where $\underline{X}(\Omega)$ represents the vector of complex amplitudes of the system's response, $\underline{F}(\Omega)$ is the vector of input harmonic forces and lastly $[H(\Omega)]$ is the matrix of transfer functions (unknown): the terms of this matrix can be estimated regardless of whether a mathematical model of the actual system has been defined a priori or not. The basic principle of the identification techniques is that of minimising the difference between the analytical response, assessed in the time domain using (8.1a, 8.1b) or (8.2d) and in the frequency domain using (8.4), and the measured response of the real system subjected to the same excitation. Various methods may be used to minimise this difference (see the bibliography attached to this chapter), though they can all be conceptually linked to a least squares approach. For this reason, for mainly educational purposes, the following section will explain the basic concepts of this method. In the discussion that follows (Sect. 8.2) we will give, as an example, the description of some typical methods of modal identification in the frequency domain; these methods, which are based on defining the transfer function, are currently the most commonly used in the field of mechanical systems. Lastly, in Sect. 8.3 we will refer to, as an example, a method of identifying modal parameters in the time domain.

8.1.2 The Least Squares Method

Introduced by Gauss in 1795 [1, 17] in relation to studies on the orbital motions of the planets, the least squares method can be used as an identification technique. Gauss realised that estimation of unknown parameters using experimental observations was conditioned by the number of measurements carried out and by the errors that affect them. The studies by Gauss also led to the conclusion that estimation was also influenced by the degree of approximation of the dynamic model. The basic idea of the method is to minimise, in probabilistic terms, the difference between the value predicted by the mathematical model and the observed value; in other words it defines the best estimation of the most likely value of the unknown parameters, assuming errors in the measurements that are independent of each other. Let us assume that we wish to describe the behaviour of a certain real system using the dynamic equation of motion (8.2a)–(8.2f) where we consider as unknowns the masses, stiffness and damping ranked in a vector $\underline{p}$:

$$\underline{p}^T = \{ m_{11} \quad m_{12} \quad \ldots \quad k_{11} \quad k_{12} \quad \ldots \} \tag{8.5a}$$

Suppose that we have n measured quantities $\underline{x}_k^{(s)} = \underline{x}^{(s)}(t_k)$ at m discrete moments of time $t_1, t_2, \ldots, t_m$.[1] We use $\underline{x}_k = \underline{x}(\underline{p}, t_k)$ to indicate the value of displacements $\underline{x}$ obtained at the generic moment of time $t = t_k$ to integrate the equation of motion (8.1a, 8.1b) having introduced a set of trial parameters $\underline{p}$. The least squares method makes it possible to obtain the most likely values of the parameters $\underline{p}$ by minimising the sum of the squared residuals which can be formally expressed as:

$$J = \frac{1}{2}\sum_{k=1}^{m}\left(\underline{x}_k^{(s)} - \underline{x}_k\left(\underline{p}\right)\right)^T [W_k]\left(\underline{x}_k^{(s)} - \underline{x}_k\left(\underline{p}\right)\right) \tag{8.5b}$$

where $[W_k]$ is a matrix of weights [28]. This least squares approach can also be used in the frequency domain. For example, when referring to a 1 d.o.f. system, the transfer function $h(\Omega)$ is (Sect. 2.4.1.4.1):

$$h(\Omega) = \frac{1}{-\Omega^2 m + i\Omega r + k} \tag{8.6a}$$

It is assumed that we have a set of experimental measurements with different frequencies Ω_k:

$$h^{(s)} = h^{(s)}(\Omega_k) \tag{8.6b}$$

The values of unknown parameters m, r and k, organised in vector $\underline{p}$:

$$\underline{p} = \begin{Bmatrix} m \\ k \\ r \end{Bmatrix} \tag{8.6c}$$

can be estimated by minimising the function:

$$J = \frac{1}{2}\sum_{k=1}^{m}\left(h^{(s)} - h_k\left(\underline{p}\right)\right)^T W_k\left(h^{(s)} - h_k\left(\underline{p}\right)\right) \tag{8.6d}$$

$h_k\left(\underline{p}\right)$ being the expression (8.6a) estimated for the different values of Ω corresponding to those of the experimental measurements $h^{(s)}$. Often (8.6d) is written in matrix form and the weights W_k organised in a matrix $[W]$ which is normally assumed to be diagonal, if the measurements are mutually independent and uncorrelated, where the terms of the diagonal are assumed to be inversely proportional to the elements of the matrix of covariance. The unknown parameters m, r and k can be estimated by minimising the function J, i.e. by cancelling the

[1]In the following discussion we will use the superscript(s) to indicate the experimentally measured generic quantity.

derivatives with respect to the actual parameters; this corresponds to solve a system of non-linear algebraic equations.

8.2 Modal Identification Techniques

8.2.1 Introduction

The term modal identification refers to the set of methods used to describe the dynamic properties (characterised by modal parameters) of a structure or machine based on experimental data obtained by testing the actual structure.

Modal identification uses the modal approach as analytical support, meaning that it reconstructs the analytical transfer function $h_{jk}(\Omega)$ of the system considering it as if it were made up of many one d.o.f. systems (Sect. 2.5), since the various d.o.f. are defined by the modal variables. $h_{jk}(\Omega)$ represents the harmonic transfer function for an input at the generic point k and the output at the generic point j of the n d.o.f. system considered. So this transfer function will have modal parameters as unknown quantities:

$$h_{jk}(\Omega) = h_{jk}\left(\Omega, \omega_i, m_i, k_i, r_i, \underline{X}^{(i)}\right) \tag{8.7}$$

being:

- ω_i the natural frequencies;
- m_i the generalised masses;
- k_i the generalised stiffness;
- r_i the modal damping;
- $\underline{X}^{(i)}$ the modes of vibration.

With this approach the dynamic response of the structure subjected to a set known excitation is measured at several points (the relative quantities will be shown below with the subscript j); this dynamic response, expressed in terms of experimental transfer functions $h_{jk}^{(s)}(\Omega)$, is then compared to the analytical response $h_{jk}(\Omega)$ defined beforehand by minimising the difference between the analytical values and the experimental ones. It is thus possible to determine the set of modal parameters needed to characterise the static and dynamic behaviour of the system being analysed.

8.2.2 An Outline of the Basic Equations

Now, for ease of discussion, we will outline the main results obtained by analysing a discrete generic n d.o.f. system using the modal approach (the full discussion of

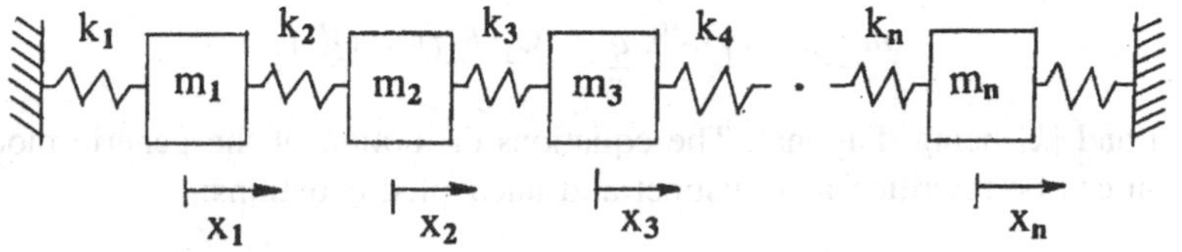

Fig. 8.2 Generic discrete vibrating n d.o.f. system

this issue has already been dealt with in Sect. 2.5.2). So, we will consider a generic linear vibrating system, discretized with a n d.o.f. model (Fig. 8.2): the vibrating system can be described, regardless of the method of discretization (with concentrated parameters, with finite elements etc.) by a system of second-order linear differential equations with constant coefficients. In matrix form we have:

$$[M]\ddot{\underline{x}} + [R]\dot{\underline{x}} + [K]\underline{x} = \underline{F}(t) \tag{8.8a}$$

where $\underline{x}$ represents the vector of the independent coordinates that represent the displacements of some points of the structure:

$$\underline{x}^T = \{x_1 \quad x_2 \quad \ldots \quad x_j \quad \ldots \quad x_n\} \tag{8.8b}$$

$[M]$, $[R]$ and $[K]$ are, respectively, the matrices of mass, damping and stiffness and lastly, $\underline{F}$ is the vector of external forces. We can use transformation into principal coordinates, see Sect. 2.5:

$$\underline{x} = [\Phi]\underline{q} \tag{8.8c}$$

where the matrix $[\Phi]$ is the modal matrix (generally rectangular $(n \cdot p)$ obtained by arranging in columns the first p modes of vibration $\underline{X}^{(i)}$ of the free, undamped structure:

$$[\Phi] = \left[\underline{X}^{(1)}; \underline{X}^{(2)}; \ldots; \underline{X}^{(i)}; \ldots; \underline{X}^{(p)}\right] = \begin{bmatrix} X_1^{(1)} & \ldots & X_1^{(i)} & \ldots & X_1^{(p)} \\ \ldots & \ldots & \ldots & \ldots & \ldots \\ X_j^{(1)} & \ldots & X_j^{(i)} & \ldots & X_j^{(p)} \\ \ldots & \ldots & \ldots & \ldots & \ldots \\ X_n^{(1)} & \ldots & X_n^{(i)} & \ldots & X_n^{(p)} \end{bmatrix} \tag{8.8d}$$

$X_j^{(i)}$ is the generic jth component relative to the ith mode of vibration. So by applying the transformation of coordinates (8.2c) to the Eq. (8.8a), the same equation of motion of the system can be rewritten, if the damping matrix $[R]$ is a linear combination of those of mass and stiffness, such as:

$$[m]\underline{\ddot{q}} + [r]\underline{\dot{q}} + [k]\underline{q} = [\Phi]^T \underline{F}(t) = \underline{Q}(t) \tag{8.9}$$

$[m]$, $[r]$ and $[k]$ being diagonal. The equations of motion of the generic mode of vibration can be rewritten as n distinct and uncoupled equations:

$$m_i\ddot{q}_i + r_i\dot{q}_i + k_i q_i = Q_i(t) = \underline{X}^{(i)T}\underline{F}(t) = \sum_{k=1}^{n} X_k^{(i)} F_k(t) \quad (i = 1, p) \tag{8.10}$$

$F_k(t)$ being the generic k-th generalised component of the forces vector $\underline{F}$ according to the kth d.o.f. X_k. So to fully describe the dynamic behaviour of the discretized system with a n d.o.f. model it is enough to know:

- the natural pulsation of the system $\omega_i^2 = \frac{k_i}{m_i}$;
- the generalised mass m_i and, from the previous definition, the generalised stiffness k_i;
- the generalised damping r_i;
- the modes of vibration $\underline{X}^{(i)}$.

These modal parameters may be determined:

- analytically, when the characteristics of the elements making up the structure in question are known a priori, i.e. the matrices $[M]$, $[R]$ and $[K]$ of the Eq. (8.8a), and by subsequently applying the transformation of coordinates (8.8c) to obtain (8.10);
- experimentally, by measuring the response of the real structure and assessing the same modal parameters at a later stage.

The modal identification techniques described in this section are, as already mentioned, based on the experimental definition of the dynamic response of the system using a harmonic transfer function $h_{jk}^{(s)}(\Omega)$ (in this regard, see Sect. 2.4.1.4.1), which can be obtained, for a linear system, by exciting the system with one or more harmonic forces like $F_k(t) = F_{ko}e^{i\Omega t}$ or with one or more assigned spectrum forces (8.3).

Basically, there are two different approaches used to identify the modal parameters of a mechanical system (as will be further explained and illustrated below):

- using a single external excitation (single-input; multi-output);
- using different excitations simultaneously (multi-input; multi-output).

Here, for educational purposes, we will analyse several methods that require just one input: so when adopting such an approach the Lagrangian component $Q_i(t)$ relative to the generic ith mode of vibration, caused solely by the harmonic force applied $F_k = F_{ko}e^{i\Omega t}$ becomes:

$$Q_i(t) = F_{ko}e^{i\Omega t}X_k^{(i)} = Q_{io}e^{i\Omega t} \tag{8.11}$$

The equation relating to the generic mode of vibration (8.10) thus becomes (steady state motion for a single harmonic force $F_{ko}e^{i\Omega t}$):

$$m_i\ddot{q}_i + r_i\dot{q}_i + k_iq_i = F_{ko}e^{i\Omega t}X_k^{(i)} \tag{8.12}$$

Establishing in (8.12) the steady state solution as:

$$q_i = q_{io}e^{i\Omega t} \tag{8.12a}$$

we have:

$$q_{io} = \frac{F_{ko}e^{i\Omega t}X_k^{(i)}}{-\Omega^2 m_i + i\Omega r_i + k_i} \tag{8.12b}$$

The response of the system, in terms of independent variables and nodal displacements $\underline{x}$, will be of the form:

$$\underline{x} = \underline{X}_o e^{i\Omega t} \tag{8.13}$$

where, taking into account the coordinate transformation (8.8c),:

$$\underline{X}_o = \begin{Bmatrix} X_{o1k} \\ X_{o2k} \\ \cdots \\ \cdots \\ X_{ojk} \\ \cdots \\ \cdots \end{Bmatrix} = \underline{X}^{(1)}q_{1o} + \underline{X}^{(2)}q_{2o} + \cdots + \underline{X}^{(i)}q_{io} + \cdots + \underline{X}^{(p)}q_{po} \tag{8.14}$$

The response $X_{ojk}(\Omega)$ of the system at the generic point j estimated analytically for an applied force $F_{ko}e^{i\Omega t}$ at generic node k is, therefore, taking into account (8.8c) and (8.12b):

$$X_{ojk}(\Omega) = \sum_{i=1}^{n} X_j^{(i)}q_{io} = \sum_{i=1}^{n} X_j^{(i)} \frac{F_{ko}X_k^{(i)}}{-\Omega^2 m_i + i\Omega r_i + k_i} \tag{8.15}$$

We should remember the definition of harmonic experimental transfer function $h_{jk}^{(s)}(\Omega)$, defined (see Fig. 8.3) as the ratio between the response $X_{ojk}^{(s)}(\Omega)e^{i\Omega t}$ of the system at generic point j to a harmonic force $F_k = F_{ko}(\Omega)e^{i\Omega t}$ of variable pulsation Ω, applied on generic node k and the actual force:

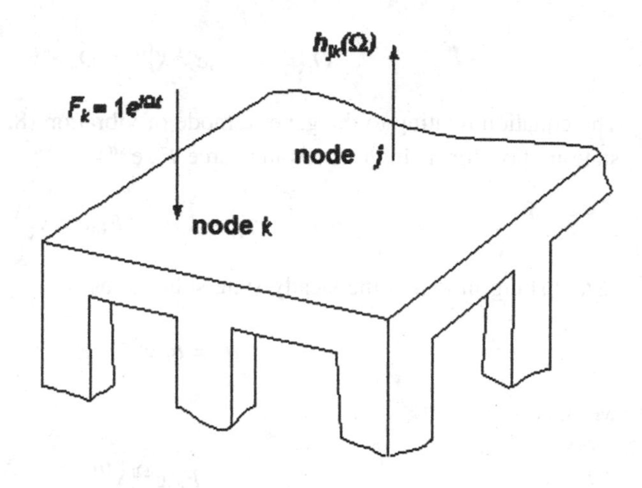

Fig. 8.3 Definition of transfer function $h_{jk}(\Omega)$

$$h_{jk}^{(s)}(\Omega) = \frac{X_{ojk}^{(s)}(\Omega)}{F_{ko}} \tag{8.16}$$

which, therefore, represents the response of the system for a unit force applied at generic point k. So, similarly, the analytically estimated transfer function $h_{jk}(\Omega)$ is defined as:

$$h_{jk}(\Omega) = \frac{X_{ojk}^{(s)}(\Omega)}{F_{ko}} = \sum_{i=1}^{n} \frac{X_j^{(i)} X_k^{(i)}}{-\Omega^2 m_i + i\Omega r_i + k_i} \tag{8.17}$$

In conditions of resonance, in correspondence to the generic ith mode of vibration ($\Omega = \omega_i$), if the contribution of the adjacent modes is negligible (true condition if damping is small and the modes are distanced), the analytical transfer function $h_{jk}(\Omega)|_{\Omega=\omega_i}$ can be rewritten as:

$$h_{jk}(\Omega)\big|_{\Omega=\omega_i} = \frac{X_j^{(i)} X_k^{(i)}}{i\omega_i r_i} = -\frac{i X_j^{(i)} X_k^{(i)}}{\omega_i r_i} \tag{8.18a}$$

The same quantity may be expressed, by dividing numerator and denominator by the value of critical damping $r_{ci} = 2m_i\omega_i$ as:

$$h_{jk}(\Omega)\big|_{\Omega=\omega_i} = -\frac{i X_j^{(i)} X_k^{(i)}}{2m_i \left(\frac{r}{r_c}\right)^{(i)} \omega_i^2} \tag{8.18b}$$

and defining as damping factor σ_i the product:

$$\sigma_i = \left(\frac{r}{r_c}\right)^{(i)} \omega_i \tag{8.19}$$

lastly (8.18b) may be rewritten as:

$$h_{jk}(\Omega)\big|_{\Omega=\omega_i} = -\frac{i X_j^{(i)} X_k^{(i)}}{2 m_i \sigma_i \omega_i} \tag{8.20}$$

(8.17) and (8.20) represent, in the case of pulsation of the generic force Ω, and in the case of $\Omega = \omega_i$, the response of the vibrating system in terms of harmonic transfer function, i.e. the response of the jth generic d.o.f. for a harmonic force applied at the kth point. As we can see, this function is uniquely defined as the modal parameters of the system are known. Using the identification techniques to minimise the difference between the measured experimental transfer function $h_{jk}^{(s)}(\Omega)$ (8.16) and the analytical transfer function $h_{jk}(\Omega)$, defined in relation to the function of modal parameters (unknown) (8.17) and (8.20), it is possible to estimate the actual modal parameters, i.e. natural pulsation ω_i, generalised mass m_i, generalised damping σ_i and mode of vibration $\underline{X}^{(i)}$. In the definition of the analytical transfer function $h_{jk}(\Omega)$, it is possible, about ω_i, to use the expression (8.17), while neglecting the contribution of non-resonant modes in the summation: in this case a defined modal approach with one d.o.f. is used. The use of (8.17) in complete form, on the other hand, considers the contribution of the other modes and so, in this case, a several d.o.f. approach is used (in this regard, see [8, 11, 22]).

8.2.3 Graphic Representations of the Transfer Function

The generic transfer function $h_{jk}(\Omega)$ (both analytical and experimental) can be plotted in different ways taking into account the difficulty in representing, in a plane, a complex quantity fully defined by 3 quantities:

- frequency Ω;
- two scalar quantities [magnitude $|h_{jk}(\Omega)|$ and phase $\phi_{jk}(\Omega)$];
- real part $\mathrm{Re}(h_{jk}(\Omega))$ and imaginary part $\mathrm{Im}(h_{jk}(\Omega))$.

The three most common graphic representations are summarised below:

- Bode plot, which consists of two graphs that show, respectively, magnitude $|h_{jk}(\Omega)|$ and phase $\phi_{jk}(\Omega)$ of the transfer function in relation to frequency Ω (Fig. 8.4 shows this quantity for a 1 d.o.f. system);

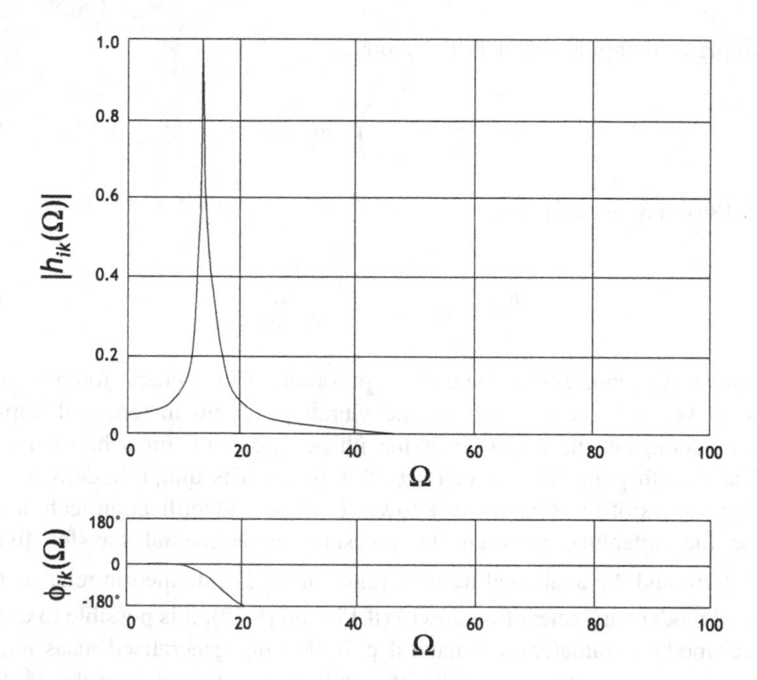

Fig. 8.4 Bode plot for a 1 d.o.f. system (linear scale)

- the graph of the real part $\mathrm{Re}(h_{jk}(\Omega))$ and the imaginary part $\mathrm{Im}(h_{jk}(\Omega))$ of the transfer function in relation to the frequency Ω (Fig. 8.5);
- the graph of the imaginary part $\mathrm{Im}(h_{jk}(\Omega))$ in relation to the real part $\mathrm{Re}(h_{jk}(\Omega))$ upon transformation of Ω: this graph, known as the Nyquist plot or plot in the Argand plane, does not explicitly contain information about the frequency Ω (Fig. 8.6).

This graph (in the Argand plane) represents a circumference: the vector $(P - O)$ defines the magnitude and phase of the transfer function $h_{jk}(\Omega)$ in correspondence to a generic value of Ω. The vector $(B - O)$ represents the transfer function in conditions of resonance with phase $\phi_{jk}(\Omega)$ equal to 90°.

Often the Bode plot or the graph of the real and imaginary parts is represented in logarithmic scale to avoid the issues linked with the graphic representation of quantities, such as, in fact, the transfer function, which can attain values that are very different to each other.

These different representations of the same quantity $h_{jk}(\Omega)$ are used during the identification phase in relation to the algorithm adopted. For example, Figs. 8.7, 8.8 and 8.9 show, respectively, the Bode plots of the real and imaginary part and the Nyquist plots for a vibrating 2 d.o.f. system.

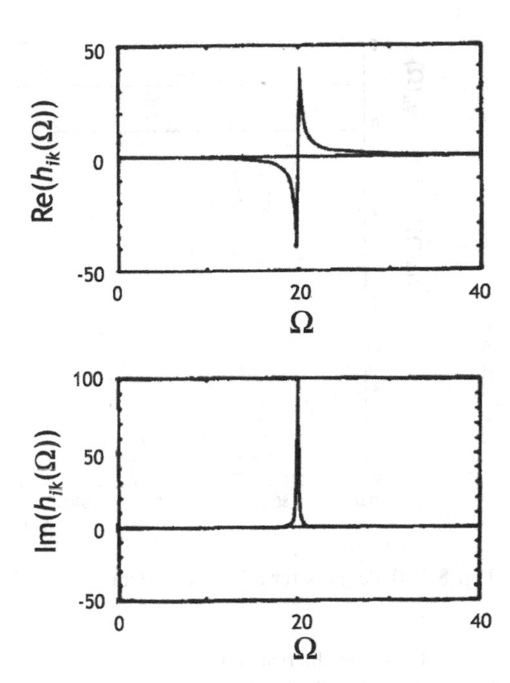

Fig. 8.5 Graph of the real part and the imaginary part of the transfer function for a 1 d.o.f. system (linear scale)

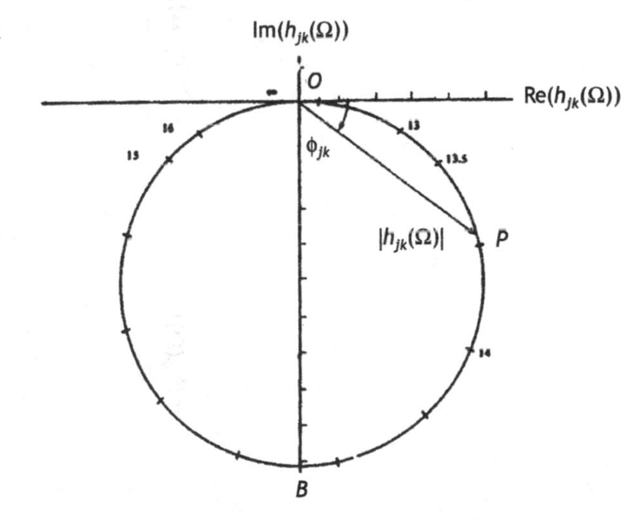

Fig. 8.6 Nyquist plot for a 1 d.o.f. system

8.2.4 Defining the Experimental Transfer Function

In this section we will briefly deal with the main methods that can be used to experimentally measure the transfer function $h_{jk}^{(s)}(\Omega)$ in (8.16), while referring to specialised texts (listed in the bibliography) and to other courses (e.g. courses on

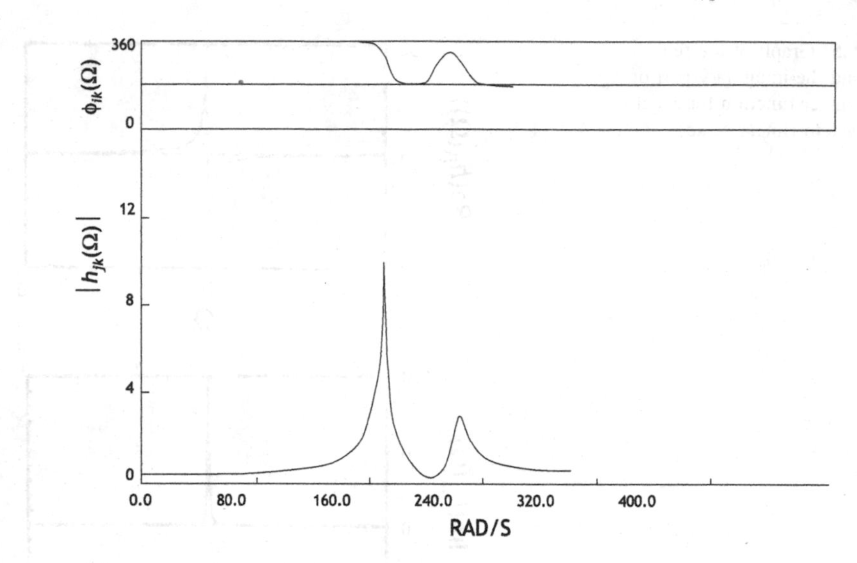

Fig. 8.7 Bode plots for a 2 d.o.f. system

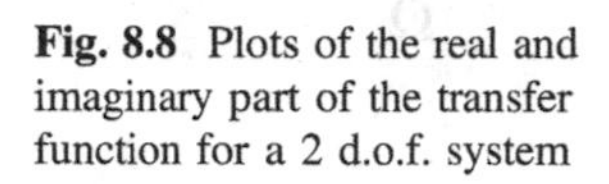

Fig. 8.8 Plots of the real and imaginary part of the transfer function for a 2 d.o.f. system

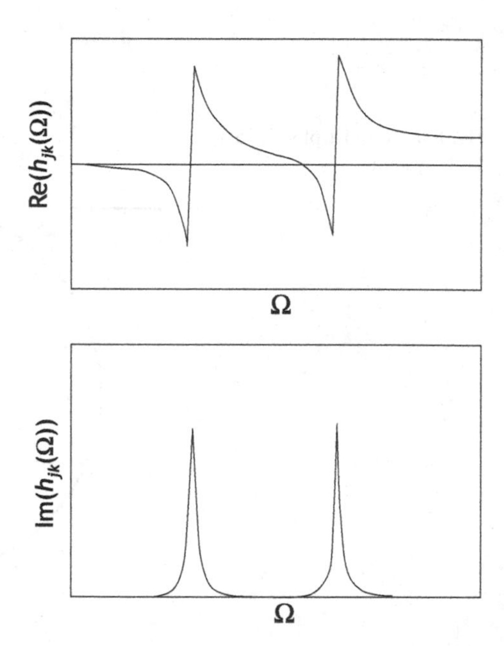

Mechanical and Thermal Measurements, *Industrial Diagnostics*, etc.) for a more detailed discussion. As we have already mentioned, there are two categories of excitations:

- the method known as *single-point excitation*, which consists of exciting the structure with a single source of excitation;

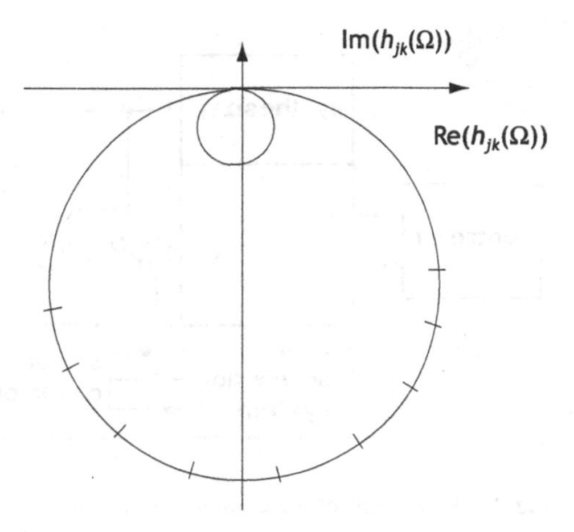

Fig. 8.9 Nyquist plot for a 2 d.o.f. system

- the method known as *multi-points excitation*, which consists of simultaneously exciting the stricture being analysed with several forces placed at different points of the structure.

In this section we will refer solely to techniques of excitation with just one force. The system for measuring the experimental transfer function is very simple, even if there are many variants. The system must include:

- an excitation mechanism;
- a setup of transducers to measure the various quantities;
- a system to acquire and process data which makes it possible to extract the necessary information from the signals, i.e. allowing estimation of the transfer function.

Figure 8.10 shows a diagram of a typical measurement system: an additional component that has become a requirement in all chains of measurement is the *controller*; this role is performed by a mini or micro-computer which automates all the operations relating to measurement and may also be used to analyse and display the results obtained.

8.2.4.1 Preparing the Structure

The structure may be analysed either when *free* or *constrained*. If the structure is analysed while free (i.e. free from constraints with the outside world) it is possible to define its stiffness modes and therefore the overall characteristics of mass and moments of inertia. In reality this condition cannot be achieved and, in practice, to reproduce the behaviour of the free structure the actual structure must be constrained with interconnecting elastic elements (springs) that are extremely *soft* so that the initial modes of the structure are rigid with frequencies close to zero and, in any case, much lower than those relating to the vibration modes that cause the

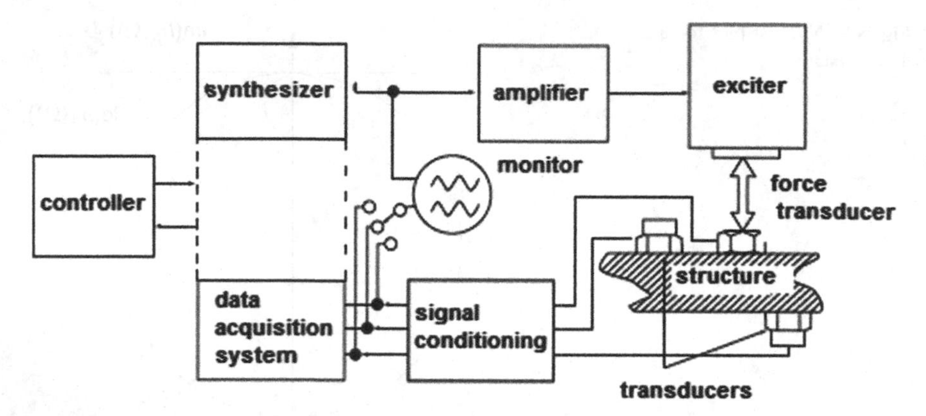

Fig. 8.10 Diagram of a measurement system

structure to begin to deform. When analysing a constrained structure we must verify that the supporting structure is sufficiently rigid so as not to introduce effects of interaction between the structure to be analysed and that of support.

8.2.4.2 Exciting the Structure

Different forms of excitation can be transmitted to the structure: exciters can be classified into two main families:

- contact exciters;
- non-contact exciters.

Contact exciters are connected to the structure during all the operations of excitation, regardless of the type of excitation generated. Non-contact exciters do not have any connection with the structure (for example, electromagnetic exciters) or only remain in contact with the structure being analysed during the period that excitation is applied (for example, impact hammers). The first family includes:

- mechanical exciters (unbalanced rotating masses);
- electromagnetic exciters (or tables) (a magnet moving within a magnetic field);
- electro-hydraulic exciters.

As already mentioned, system excitations can be of different types:

- simple sinusoidal excitation;
- sinusoidal excitation with slow variable frequency (*slow sine sweep*);
- periodic excitation;
- random excitation;
- impulse (or transient) excitation.

Before examining the different types of excitation in more detail, we should remember that a linear behaviour of the system is assumed and so the transfer function is univocally defined. We should also recall that the system may be subject to non-definable or uncontrollable external excitations, i.e. noise: the response of the system, as widely discussed in Chap. 7, will of course be affected by such excitations. The quantity that makes it possible to define the reliability of the measurements performed is the function of coherence between the response of the system and the applied force: when this quantity is close to the unit, the response of the system is, in fact, caused solely by the force applied, when it approaches zero the response of the system is due mainly to noise. As we have already seen in Chap. 7, in the case of random excitation of the system, the transfer function of the actual system can be defined using the cross-spectrum (7.24b) between force and response.

We will now examine the main types of excitation adopted:

- *Simple sinusoidal excitation* This method is used to excite the system with a harmonic force $F_k = F_{ko}\mathrm{e}^{i\Omega t}$ with a fixed amplitude F_{ko} and assigned frequency Ω. Variation in the pulsation Ω of the force is controlled in discrete steps $\Delta\Omega$, either manually or automatically. Between one measurement of the dynamic response $X_{ojk}^{(s)}(\Omega)$ and the next $X_{ojk}^{(s)}(\Omega + \Delta\Omega)$, with different force pulsations, we must wait, however, until the transient, triggered by the discrete change of frequency $\Delta\Omega$, has disappeared. The experimental transfer function $h_{jk}^{(s)}(\Omega)$ is defined for each pulsation of the force Ω and for each point of measurement j on the structure by:

$$h_{jk}^{(s)}(\Omega) = \frac{X_{ojk}^{(s)}(\Omega)}{F_{ko}} \tag{8.21}$$

One advantage of this methodology is that it offers the possibility to set the pitch $\Delta\Omega$ and to therefore densify the points of analysis in proximity of the natural frequencies ω_i and disperse them in areas further away from these.

- *Sinusoidal excitation with slow variable pulsation* This methodology requires a variable frequency oscillator $\Omega = \Omega(t)$ that sends an analogue control signal to the exciter: the frequency of excitation caused by the force $F_k = F_{ko}\mathrm{e}^{i\Omega t}$ is thus made to vary slowly but constantly within the *range* of relevant frequencies. Variation in pulsation (up or down) must take place quite slowly so that the system can sweep constantly from one steady state to another: we should remember that the harmonic transfer function is defined as being the relationship between the response of the system in a steady state $X_{ojk}^{(s)}(\Omega)$ and the harmonic force introduced F_{ko}. To verify the validity of the *sweep* speed a *run-up* is usually performed (i.e. a sweep with increasing pulsation frequency Ω) and a *run-down* (with decreasing frequency Ω), to check that the two curves obtained are coincident.

In this case too, the experimental transfer function $h_{jk}^{(s)}(\Omega)$ is defined for each pulsation of the force Ω and for each point of measurement j on the structure by (8.28).

- *Periodic excitation* An extension of the previous method is to set periodic excitation, i.e. a force that contains all the frequencies in the field of analysis:

$$F_k(t) = \sum_n F_{ko}\, e^{in\Omega_o t} \tag{8.22}$$

Ω_o being the pulsation of the fundamental frequency. Unlike the method that uses a harmonic excitation, in this case the force is obtained as an overlap of different harmonics in the frequency range concerned. The definition of the experimental harmonic transfer $h_{jk}^{(s)}(\Omega)$ using a periodic excitation is based on the fFourier transform: more specifically, harmonic analysis is performed (i.e. the Fourier transform) of both the force signal $F_k(t)$ and the dynamic response of the system measured at generic point $X_{ojk}^{(s)}(n\Omega_o)$:

$$F_{kon}(n\Omega_o) = \frac{2}{T}\int_0^T F_k(t)\, e^{-in\Omega_o t} dt \tag{8.23a}$$

$$X_{ojk}^{(s)}(n\Omega_o) = \frac{2}{T}\int_0^T X_{jk}^{(s)}(t)\, e^{-in\Omega_o t} dt \tag{8.23b}$$

This procedure makes it possible to simultaneously extract each harmonic component both of the strength $F_{kon} = F_{ko}(n\Omega_o)$ and the response $X_{ojk}^{(s)}(n\Omega_o)$ for each frequency $n\Omega_o$. The experimental transfer function $h_{jk}^{(s)}(\Omega)$ is defined simultaneously (for each point of measurement j on the structure) in correspondence to all the pulsations of the force $\Omega_n = n\,\Omega_o$ using:

$$h_{jk}^{(s)}(\Omega) = \frac{X_{ojk}^{(s)}(\Omega)}{F_{kn}} \tag{8.24}$$

The experimental transfer function $h_{jk}^{(s)}(\Omega_n)$ can be defined by simply dividing the spectrum of the response $X_{ojk}^{(s)}(n\Omega_o)$ (8.23b) by the spectrum of the force applied $F_{kon}(n\Omega_o)$, Eq. (8.23a).

- *Random excitation* With this type of excitation the structure is forced with a white noise (corresponding, as seen in Sect. 7.3, to a spectral power density function of constant amplitude). Thus the spectrum of the response corresponds to the transfer function of the system.
- *Impulse excitation* When using this methodology, the system is subjected to an assigned impulse (for example, using an impact hammer). The Fourier transform of the ideal impulse is characterised by a constant amplitude throughout the field of frequencies. However, considering that the impulse is not actually perfect, transformation of the force signal and the displacement signal is, in any case, performed at the different points of measurement: the ratio of the corresponding harmonics at various frequencies Ω_n defines the transfer function $h_{jk}^{(s)}(\Omega)$:

$$h_{jk}^{(s)}(\Omega_n) = \frac{X_{ojk}^{(s)}(\Omega_n)}{F_{kn}(\Omega_n)} \tag{8.25}$$

8.2.5 Determining Modal Parameters

Having defined in Sects. 8.2.2 and 8.2.3 the basic equations needed to define the link between analytic transfer function $h_{jk}^{(a)}(\Omega)$ and unknown model parameters to be used in the modal identification techniques and, having experimentally obtained the transfer function $h_{jk}^{(s)}(\Omega)$ by measuring the dynamic response of the system with assigned forces (Sect. 8.2.4), we will now proceed to describe some of the most basic algorithms used to define the actual modal parameters, i.e.:

- natural frequencies ω_i;
- non-dimensional damping $\left(\frac{r}{r_c}\right)_i$;
- vibration modes $\underline{X}^{(i)}$;
- generalised mass m_i.

8.2.5.1 Determining the Natural Frequencies of the System Ω_I

8.2.5.1.1 Determining the Peaks of the Transfer Function

A simple method used to define the natural pulsations of the system ω_i is based on the consideration that the amplitudes of vibration $X_{ojk}^{(s)}(\Omega)$, and, therefore, the corresponding transfer functions $h_{jk}^{(s)}(\Omega)$ of a resonance forced system, are amplified dynamically: in the summation given in (8.17) the contributions made by non-

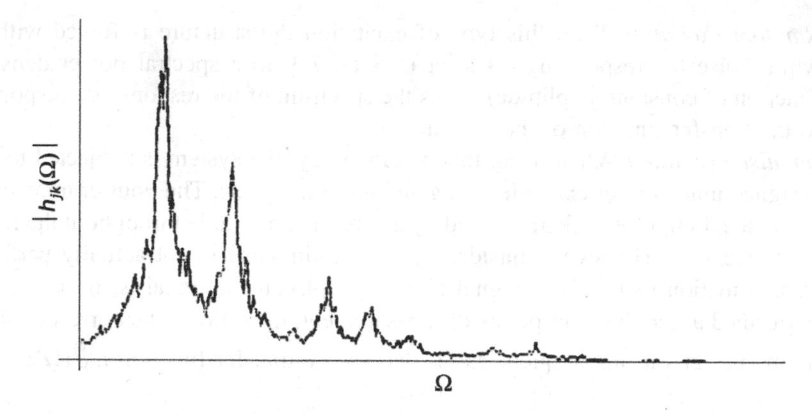

Fig. 8.11 Trend of the experimental transfer function magnitude $h_{jk}^{(s)}(\Omega)$ in relation to the frequency Ω of the force

resonant modes are minimal, while that of the resonant mode is higher, being $\Omega = \omega_i$. After having plotted the trend of the experimental transfer function magnitude $\left|h_{jk}^{(s)}(\Omega)\right|$ in relation to the frequency Ω of the force (Fig. 8.11), the method allows us to define the natural frequencies as those frequencies to which there is a corresponding maximum dynamic amplification, that is, a peak in the magnitude of the harmonic transfer function $\left|h_{jk}^{(s)}(\Omega)\right|$.

The lesser the damping of the system the more accurate the results obtained will be, as the contribution of the non-resonant vibration modes is less significant. The method may cause errors in determining the natural frequencies if the experimental transfer function analysed is that of a vibration node for the generic mode of vibration: in actual fact, in this case even in conditions of resonance, displacement is very limited so a peak in the plot of the $\left|h_{jk}^{(s)}(\Omega)\right|$ is unlikely to be highlighted. To overcome this drawback, the transfer functions $\left|h_{jk}^{(s)}(\Omega)\right|$ at different points of the structure should be analysed: correlation of the various experimental harmonic transfer functions also makes it possible to eliminate random errors in measurement and to exclude the effect of any other external sources of excitation. The point of application of the force could be in a nodal point for one or more modes: in this case, these modes would not be excited. In these situations the point of application of the force must be moved to a different position.

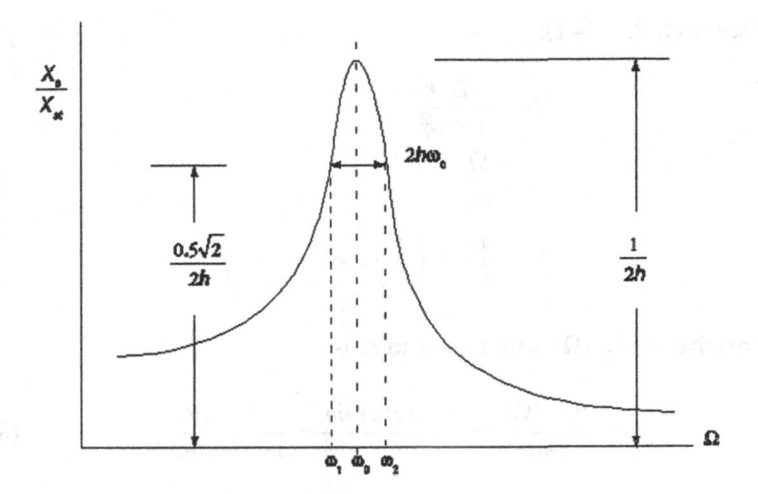

Fig. 8.12 Determining damping with the half power peaks method

8.2.5.2 Calculating Generalised Damping $\left(\frac{r}{rc}\right)_i$

8.2.5.2.1 Derivative of the Transfer Function Phase Method

This method makes it possible to obtain the value of $\left(\frac{r}{r_c}\right)_i$ starting with analysis of the phase $\phi_{jk}^{(s)}(\Omega)$ of the experimental transfer function $h_{jk}^{(s)}(\Omega)$ being:

$$h_{jk}^{(s)}(\Omega) = \left|h_{jk}^{(s)}(\Omega)\right| e^{i\varphi_{jk}^{(s)}(\Omega)} \tag{8.26}$$

The justification for such an approach is as follows: the response of the system, in terms of analytical transfer function, has been defined in (8.17): obviously this response is complex. Under conditions of resonance $\Omega = \omega_i$, if we consider negligible the contribution of non-resonant modes, we have:

$$h_{jk}(\Omega) = \frac{X_j^{(i)} X_k^{(i)}}{-\Omega^2 m_i + i\Omega r_i + k_i} = X_j^{(i)} X_k^{(i)} \frac{(-\Omega^2 m_i + k_i) - i\Omega r_i}{(-\Omega^2 m_i + k_i)^2 + (\Omega r_i)^2} \tag{8.27}$$

while the phase $\phi_{jk}^{(s)}(\Omega)$ is defined as:

$$\phi_{jk}(\Omega) = arctg\frac{-\Omega r}{(-\Omega^2 m_i + k_i)} = arctg\frac{-2a_i h_i}{1 - a_i^2} = arctg(b) \tag{8.28}$$

being (see Sect. 2.4.1.4.1):

$$b = \frac{-2a_i h_i}{1 - a_i^2}$$
$$a_i = \frac{\Omega}{\omega_i} \tag{8.28a}$$
$$h_i = \frac{r_i}{r_{ci}} = \left(\frac{r}{r_c}\right)^{(i)} = \left(\frac{r_i}{2m_i\omega_i}\right)$$

The derivative of $\phi_{jk}^{(s)}(\Omega)$ with regard to a_i is:

$$\frac{\partial \phi_{jk}(\Omega)}{\partial a_i} = \frac{\partial arctg(b)}{\partial b}\frac{\partial b}{\partial a_i} = \frac{1}{1 + b^2}\frac{\partial b}{\partial a_i} \tag{8.29a}$$

that is:

$$\frac{\partial \phi_{jk}(\Omega)}{\partial a_i} = \frac{1}{1 + \left(\frac{2a_i h_i}{1-a_i^2}\right)^2}\frac{-2h_i}{1 - a_i^2} - \frac{-2a_i h_i}{\left(1 - a_i^2\right)^2}(-2a_i) = \frac{-2h_i\left(1 - a_i^2\right) - 4a_i h_i}{\left(1 - a_i^2\right)^2 + 4a_i^2 h_i^2} \tag{8.29b}$$

Being in resonance conditions $\Omega = \omega_i \Rightarrow\ a_i = 1$ (8.36b) simply becomes:

$$\left.\frac{\partial \phi_{jk}(\Omega)}{\partial a_i}\right|_{a_i=1} = -\frac{1}{h_i} = -\frac{1}{r_i/r_{ci}} \tag{8.30a}$$

hence:

$$\left|\left.\frac{\partial \phi_{jk}(\Omega)}{\partial a_i}\right|_{a_i=1}\right| = \frac{1}{h_i} = \frac{1}{r_i/r_{ci}} \tag{8.30b}$$

The most commonly used expression, considering (8.30b) and (8.28), is the following:

$$\left|\left.\frac{\partial \phi_{jk}(\Omega)}{\partial \Omega}\right|_{\Omega=\omega_i}\right| = \left|\left.\frac{\partial \phi_{jk}(\Omega)}{\partial a_i}\right|_{a_i=1}\right|\frac{\partial a_i}{\partial \Omega} = \frac{1}{r_i/r_{ci}\,\omega_i} = \frac{1}{\sigma_i} \tag{8.31}$$

the so-called damping factor being:

$$\sigma_i = h_i\omega_i \tag{8.31a}$$

The non-dimensional damping h of the i-th mode of vibration can thus be obtained by plotting the progress of the phase $\phi_{jk}^{(s)}(\Omega)$ of the experimental transfer function

$h_{jk}^{(s)}(\Omega)$ in relation to the pulsation Ω of the force and by estimating the derivative of the phase compared to the actual pulsation Ω in resonance, that is, for $\Omega = \omega_i$.

8.2.5.3 The Half Power Point Method

This method can be used in vibrating systems with several d.o.f. where the natural frequencies are sufficiently differentiated and damping is sufficiently low, so that, under resonance conditions the predominant contribution to the response of the actual system is mainly due to the resonant mode, making it behave like a 1 d.o.f. system. For a 1 d.o.f. system we have seen how in resonance the coefficient of dynamic amplification (Sect. 2.4.1.4.1) is:

$$\frac{X_o}{X_{st}} = \frac{1}{2h} \quad \text{per} \quad \frac{\Omega}{\omega_o} = 1 \tag{8.32}$$

$h = \frac{r}{r_c}$ being the non-dimensional damping coefficient. The two frequencies ω_1 and ω_2 are defined as half power points (Fig. 8.12), about the resonance (i.e. in the sidebands to the peak) which correspond to:

$$\frac{X_0}{X_{st}} = \frac{1}{2}\sqrt{2}\frac{1}{2h} \quad \text{for} \quad \Omega = \omega_1 \quad \text{and} \quad \Omega = \omega_2 \tag{8.32a}$$

It can be shown[2] that the non-dimensional damping h is linked to the pulsations ω_1 and ω_2 and to the natural pulsation of the system ω_0, with the relation:

---

[2]The response of the system in correspondence to a generic pulsation of the force, in non-dimensional form (Sect. 2.4.1.4.1) can be expressed as:

$$\frac{X_o}{X_{st}} = \frac{1}{\sqrt{\left(1 - \left(\frac{\Omega}{\omega_o}\right)^2\right)^2 + \left(\frac{2h\Omega}{\omega_o}\right)^2}} \tag{8.2.1}$$

So assuming:

$$\frac{1}{\left(1 - \left(\frac{\Omega}{\omega_o}\right)^2\right)^2 + \left(\frac{2h\Omega}{\omega_o}\right)^2} = \frac{1}{2}\left(\frac{1}{2h}\right)^2 \tag{8.2.2}$$

we obtain

$$\left(\frac{\Omega}{\omega_o}\right)^4 - 2(1 - 2h^2)\left(\frac{\Omega}{\omega_o}\right)^2 + (1 - 8h^2) = 0 \tag{8.2.3}$$

hence

$$4h = \frac{\omega_2^2 - \omega_1^2}{\omega_0^2} = 2\frac{\omega_2 - \omega_1}{\omega_0} \tag{8.32b}$$

Therefore, to estimate the non-dimensional damping h it is possible to draw a horizontal line in the dynamic amplification graph in correspondence to the value $0.5\sqrt{2}$ of the maximum peak value: the intersections between the two curves (Fig. 8.12) define the two pulsations ω_1 and ω_2, corresponding to the half power peaks from which it is possible to obtain the value of damping, using (8.39b).

8.2.5.4 Determining Modes of Vibration $\underline{X}^{(i)}$

8.2.5.4.1 Determining the Imaginary Part of the Transfer Function

Several methods can be used to determine the vibration mode. One simple approach can be that which uses the imaginary part of the generic transfer function. As mentioned, the unknown modal parameters are estimated to minimise the difference between the experimental $h_{jk}^{(s)}(\Omega)$ and the analytical $h_{jk}(\Omega)$ transfer function (8.15) expressed as a function of the same unknown parameters.

Both the analytical $h_{jk}(\Omega)$ and the experimental $h_{jk}^{(s)}(\Omega)$ transfer function are, as already mentioned, complex and therefore, using Re(...) to indicate the real part of the generic quantity and Im(...) we have

$$h_{jk}(\Omega) = \sum X_j^{(i)} q_{io} = \sum X_j^{(i)} \mathrm{Re}(q_{io}) + i \sum X_j^{(i)} \mathrm{Im}(q_{io}) \tag{8.33}$$

(Footnote 2 continued)

$$\left(\frac{\Omega}{\omega_o}\right)^2 = (1 - 2h^2) \pm 2h\sqrt{1 + h^2} \tag{8.2.4}$$

Assuming that $h \ll 1$ it is possible to neglect the higher order terms in (8.2.4), to obtain:

$$\left(\frac{\Omega}{\omega_o}\right)^2 = 1 \pm 2h \tag{8.2.5}$$

By assuming that the two corresponding frequencies at the roots of this equation are equal to ω_1 and ω_2 we have:

$$4h = \frac{\omega_2^2 - \omega_1^2}{\omega_0^2} = 2\frac{\omega_2 - \omega_1}{\omega_0} \tag{8.2.6}$$

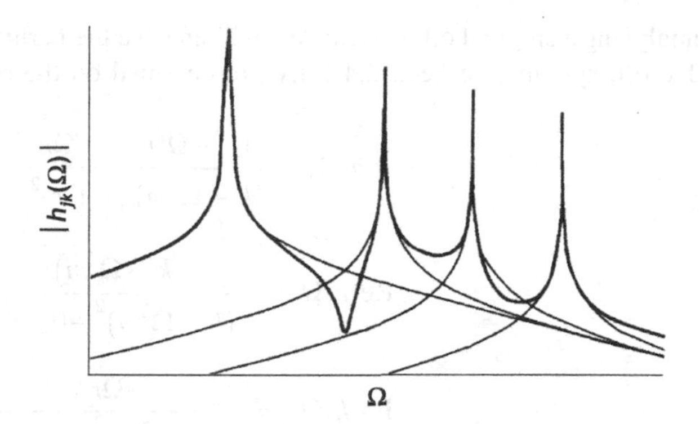

Fig. 8.13 Frequency response of a vibrating system with several d.o.f.: contribution of the various vibration modes

and

$$h_{jk}^{(s)}(\Omega) = \mathrm{Re}\left(h_{jk}^{(s)}(\Omega)\right) + i\mathrm{Im}\left(h_{jk}^{(s)}(\Omega)\right) \tag{8.34a}$$

Under conditions of resonance $\Omega = \omega_i$, ignoring the contribution of non-resonant modes and equating $h_{jk}^{(s)}(\Omega) = h_{jk}(\Omega)$, we find that the contribution of the resonant mode to the experimental transfer function $h_{jk}^{(s)}(\Omega)$ is purely imaginary:

$$h_{jk}^{(s)}(\Omega) = \mathrm{Re}\left(h_{jk}^{(s)}(\Omega)\right) + i\mathrm{Im}\left(h_{jk}^{(s)}(\Omega)\right) = -i\,\frac{X_j^{(i)}X_k^{(i)}}{\Omega r_i} \tag{8.34b}$$

In actual fact, under conditions of resonance $\Omega = \omega_i$, if we ignore the other non-resonant modes, the vibration caused solely by the ith resonant mode is out of phase by 90° compared to the force and is, therefore, purely imaginary. Even if the other modes are not ignored, if the corresponding natural frequencies are sufficiently distant from that of the resonant mode and if the dampings are low (see Fig. 8.13), the non-resonant modes have phases compared to the force that are close to 0° or to 180° and so with a contribution that is in quadrature and thus, in any case, negligible. Up to a constant value $\frac{X_k^{(i)}}{\omega_i r_i}$ (dependent, as can be seen, only on the mode and on the position of the force) the component relative to the jth mode of the eigenvector (mode of vibration) of the generic mode of vibration $X_j^{(i)}$ is definable in this way like the imaginary part of the function of $\mathrm{Im}\left(h_{jk}^{(s)}(\Omega)\right)$ estimated in resonance.

8.2.5.4.2 The Nyquist Method

This method makes it possible to simultaneously determine the natural frequencies, damping factors and modes of vibration. For the sake of simplicity we will begin by analysing a single d.o.f. system. We will analyse the harmonic transfer function of a 1 d.o.f. system (see Sect. 2.4.1.4.1), represented on the Nyquist plot in Fig. 8.6:

$$h(\Omega) = \frac{\left(k - \Omega^2 m\right) - i\Omega r}{\left(k - \Omega^2 m\right)^2 + (\Omega r)^2} \tag{8.35a}$$

$$\mathrm{Re}(h(\Omega)) = \frac{\left(k - \Omega^2 m\right)}{\left(k - \Omega^2 m\right)^2 + (\Omega r)^2} \tag{8.35b}$$

$$\mathrm{Im}(h(\Omega)) = \frac{-\Omega r}{\left(k - \Omega^2 m\right)^2 + (\Omega r)^2} \tag{8.35c}$$

If we introduce the concept of mobility, defined as the relationship between velocity and force, the same one d.o.f. system can be represented by:

$$\alpha(\Omega) = i\Omega h(\Omega) \tag{8.36a}$$

We can show that:

$$\left(\mathrm{Re}(\alpha(\Omega)) - \left(\frac{1}{2r}\right)\right)^2 + \mathrm{Im}(\alpha(\Omega))^2 = \left(\frac{1}{2r}\right)^2 \tag{8.36b}$$

This function is always shown in the Argand plane as in Fig. 8.14. The plot of the imaginary part $\mathrm{Im}(\alpha(\Omega))$ as a function of the real part $\mathrm{Re}(\alpha(\Omega))$ by Ω which ranges from zero to infinity is thus represented by a circle of radius R equal to $\mathrm{R} = {}^1\!/_{2r}$ and its centre has coordinates:

$$\mathrm{Re}(\alpha(\Omega)) = \frac{1}{2r} \tag{8.37a}$$

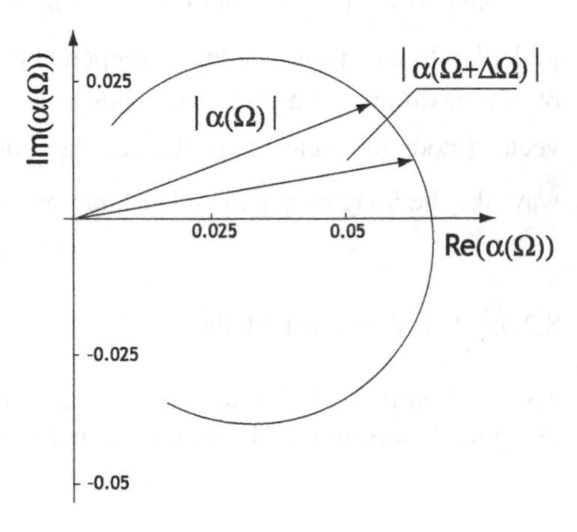

Fig. 8.14 Graph of mobility in the Argand plane

$$\mathrm{Im}(\alpha(\Omega)) = 0 \tag{8.37b}$$

It can also be shown that at the same pitch $\Delta\Omega$ of the definition of the transfer function, the arc of the circle defined by two successive positions of the vector $\alpha(\Omega)$ and $\alpha(\Omega + \Delta\Omega)$ in the Argand plane (see Fig. 8.14) is greater in the vicinity of the natural frequency. Considering this property it is, therefore, easy to identify the system's natural frequencies.

Taking into account (8.37a) it is also possible to obtain the damping of the system. For this purpose it is possible (proceeding in a deterministic way or using a process of minimization) to draw a circumference through the points in the Argand plane, corresponding to $\alpha(\Omega)$ about the natural frequency identified. Once the centre and radius of the interpolating circumference has been defined, it is possible to estimate the damping r of the system, using (8.44). By expanding to a n d.o.f. system with a damping matrix $[R]$ that typically may not even be a linear combination of mass and of stiffness matrices (thus *non-proportional* damping), in the frequency range where the mode is predominant we have:

$$\left(\mathrm{Re}\left(h_{jk}^{(s)}(\Omega)\right) + \frac{U_j}{2\sigma_i}\right)^2 + \left(\mathrm{Im}\left(h_{jk}^{(s)}(\Omega)\right) + \frac{V_j}{2\sigma_i}\right)^2 = \frac{U_j^2 + V_j^2}{4\sigma_i^2} \tag{8.38}$$

having used ω_i to indicate the generic natural pulsation of the system, $\sigma_i = \left(\frac{r}{r_c}\right)^{(i)} \omega_i$ the damping factor associated with it and also $U_j + iV_j$ to define the so-called residue from which it is possible to obtain the generic mode of vibration. Equation (8.38) represents the equation of a circumference passing through the origin of the axes, with centre of coordinates:

$$\begin{aligned} x_c &= -\frac{U_j}{2\sigma_i} \\ y_c &= -\frac{V_j}{2\sigma_i} \end{aligned} \tag{8.39a}$$

while the radius equals:

$$R = \frac{\sqrt{U_j^2 + V_j^2}}{2\sigma_i} \tag{8.39b}$$

The residue $U_j + iV_j$ associated with the ith vibration mode and relative to the jth point of measurement is proportional to the diameter of the circumference. Determining these circumferences allows us to define the vibration modes of the structure when the values of the system's natural frequencies have been defined beforehand. So the residues may be defined, in magnitude and phase (or real part and imaginary part), by imposing the passage of the circumference being sought through a suitable number of points of the transfer function close to the considered resonance, then by applying a minimisation between the analytical function being sought and the experimental points available. The accuracy of the results obtained using this method is closely linked to the resolution with which the transfer functions have been defined and to the number of points through which the passage of the circumference sought is imposed.

8.2.5.5 Defining the Generalised Mass m_i

Generalised mass m_i can be obtained by minimising, in correspondence to the generic natural pulsation ω_i i.e. by $\Omega = \omega_i$, the difference between the analytical $h_{jk}(\Omega)$ and experimental $h_{jk}^{(s)}(\Omega)$ transfer functions at all the system's points of measurement, while ignoring the contribution of non-resonant modes: to this end, the quadratic error function $\sum_{j=1}^{N}\left(\varepsilon_j(\omega_i)\right)^2$ is minimised:

$$\sum_{j=1}^{N}\left(\varepsilon_j(\omega_i)\right)^2 = \sum_{j=1}^{N}\left(h_{jk}^{(s)}(\omega_i) - h_{jk}(\omega_i)\right)^2 = \sum_{j=1}^{N}\left(h_{jk}^{(s)}(\omega_i) - \frac{1}{m_i}\frac{X_j^{(i)}X_k^{(i)}}{i2\omega_i^2 h_i}\right)^2$$
$$= f(m_i) \tag{8.40}$$

In (8.47) $h_{jk}^{(s)}(\Omega)$ is the experimental transfer function measured in correspondence to the resonance, $X_j^{(i)}$ and $X_k^{(i)}$ are the eigenvectors of the ith mode of vibration measured using the method shown above (and therefore known) in correspondence to the generic point j considered and point k where the force has been placed: the generalised non-dimensional damping $h^{(i)}$ and natural pulsation ω_i have already been defined previously. The only unknown in (8.40) is the generalised mass m_i which can be obtained by deriving $f(m_i)$ with respect to m_i, to obtain:

$$m_i = \frac{\sum_{j=1}^{N} X_j^{(i)}X_k^{(i)}}{\sum_{j=1}^{N} i2\omega_i^2 h_i h_{jk}^{(s)}(\omega_i)} \tag{8.41}$$

In (8.47) the transfer function $h_{jk}^{(s)}(\Omega)$ calculated in resonance is purely imaginary (if we neglect the contributions of the non-resonant adjacent modes) and thus the mass m_i is real.

8.2.5.6 Defining the Generalised Stiffness $\mathbf{k_i}$

Remembering the definition of the natural frequency of the system ω_i, the generalised stiffness can be easily defined using:

$$k_i = m_i \omega_i^2 \tag{8.42a}$$

8.2.5.7 Defining Generalised Damping $\mathbf{r_i}$

Once we know the natural pulsation ω_i and the non-dimensional damping $\left(\frac{r}{r_c}\right)_i$ the generalised damping r_i can be estimated by simply using:

$$r_i = 2m_i\omega_i^2\left(\frac{r}{r_c}\right)_i \tag{8.42b}$$

8.2.6 Applications and Examples

We will now consider one possible application of the results obtained using the illustrated techniques of modal identification. We will consider the system in Fig. 8.15 which outlines the foundation of a rotor.

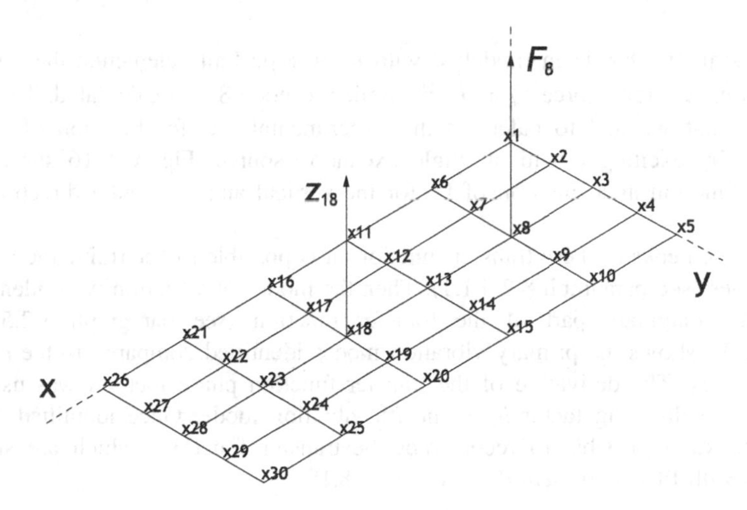

Fig. 8.15 Outline of the real structure

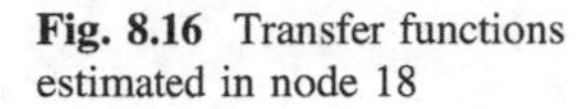

Fig. 8.16 Transfer functions estimated in node 18

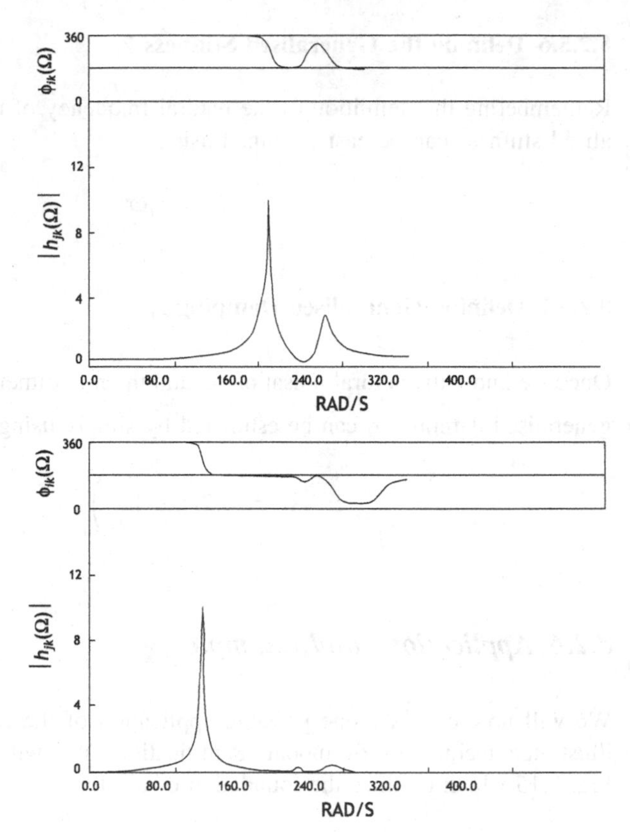

This structure has been modelled with *beam* type finite elements; the transfer function $h_{jk}(\Omega)$ for a force $F_k = 1\,\mathrm{e}^{i\Omega t}$ applied to node 8 was calculated. Then we assumed that we had to calculate the experimental transfer function of a real structure by exciting it with a single excitation source. Figure 8.16 shows the transfer function as a function of Ω, for the vertical and horizontal directions of node 18.

From the peaks of these transfer functions it is possible to determine the natural frequencies (see paragraph 8.2.5.1.1). Then the modes of vibration were identified from the imaginary part of the transfer function (see paragraph 8.2.5.3.1). Figure 8.17 shows the primary vibration modes identified compared to the *experimental* ones. The derivative of the transfer function phase method was used to estimate the damping factor h_i of the ith vibration mode. Once identified, these values make it possible to reconstruct the transfer functions which are shown together with the *experimental* ones in Fig. 8.18.

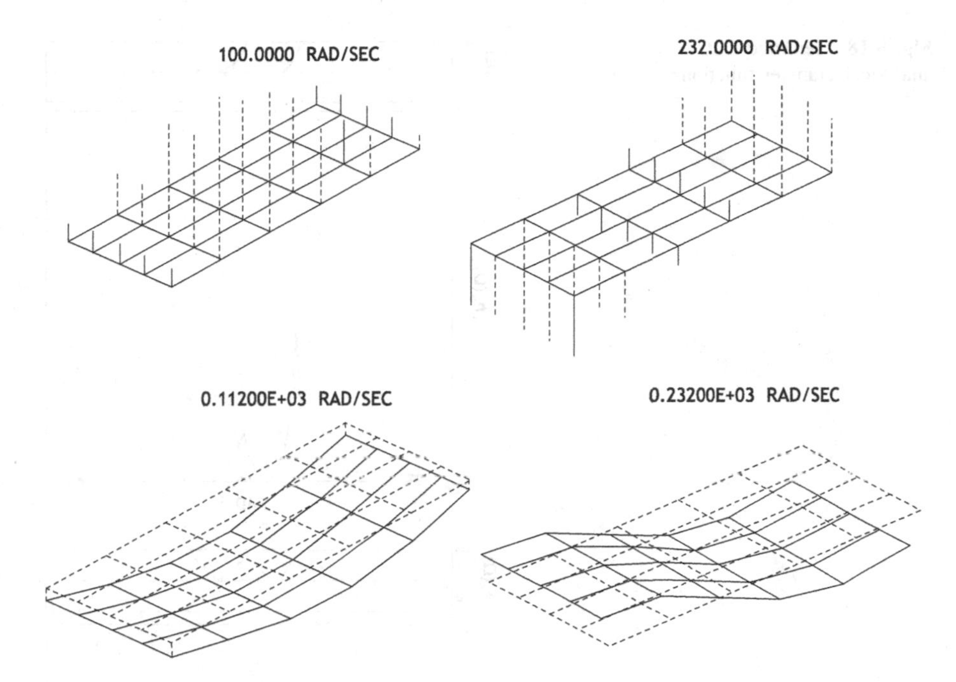

Fig. 8.17 Identified and analytical vibration modes

8.3 Identification in the Time Domain

Techniques for identifying physical or modal parameters (structured or *black box*), either in the frequency or in the time domain, have become diversified both from a point of view of the calculation algorithm, and the basic idea (objective function) that supports them.

So we can refer to filter methods in the frequency domain [17, 19, 28], methods for identifying the parameters of ARMA models [3, 4], identification using neural networks [13, 18], etc.: as we are not able to explore all these topics in this discussion, we refer the reader to the relative bibliography [5–7, 16, 20, 22–25, 28].

As an example we have included an algorithm that can be used to identify the modal parameters of a mechanical system in the time domain [14, 15]. As we have already seen, in general the experimental identification of the modal parameters of a structure is performed by measuring the forces applied to it during the test and by measuring the dynamic response resulting from these inputs. To simplify the identification procedure, some testing techniques use the free motion of the defined structure, when possible, by assigning it appropriate initial conditions or by analysing the dynamic response downstream of removal of the forces applied to the system. In other cases it is possible to exploit environmental excitation, such as turbulent wind on the structure, the passage of heavy vehicles over infrastructures, excitation associated with the irregularities in the pavements or tracks of road and

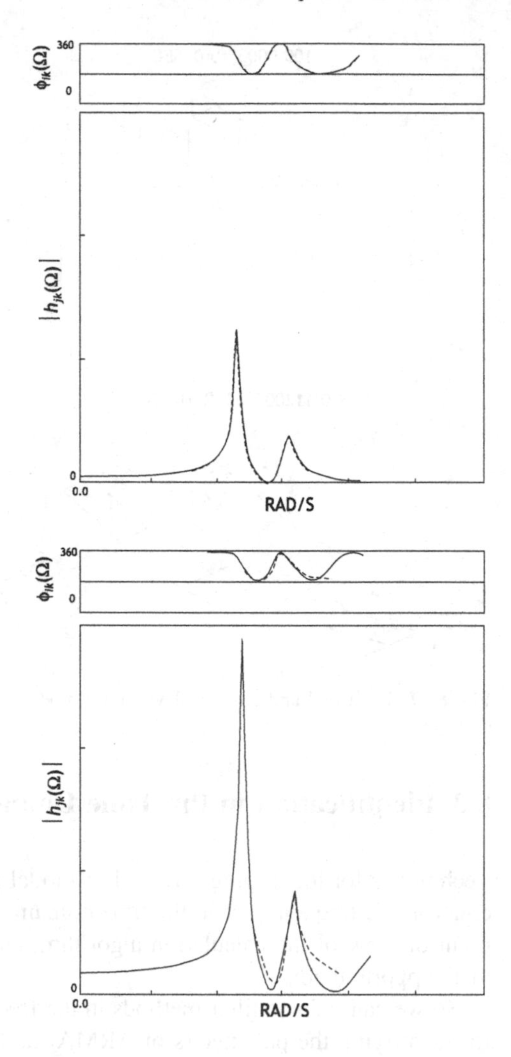

Fig. 8.18 Identified and analytical Transfer functions

rail vehicles: in which case it is possible to obtain the response of the free system using a specific technique known as the *Random Decrement Technique* [14].

This technique is based on the idea that, by extracting and averaging suitable segments of measured time history $Y_1(t)$, associated with the response of a generic mechanical system subjected to stochastic loads, it is possible to describe the intrinsic properties of just the mechanical system being considered, thus eliminating the effects caused by the actual random load applied.

The method is based on the concept that the response of the system at different moments in time is made up of two components: the response to initial conditions (general integral) and the response to forcing (particular integral). When operating with numerous means on several segments $Y_1(t)$ characterised by the same initial conditions, the part associated with random forcing which, as such, tends to disappear, and the result of this averaging operation converges to the response

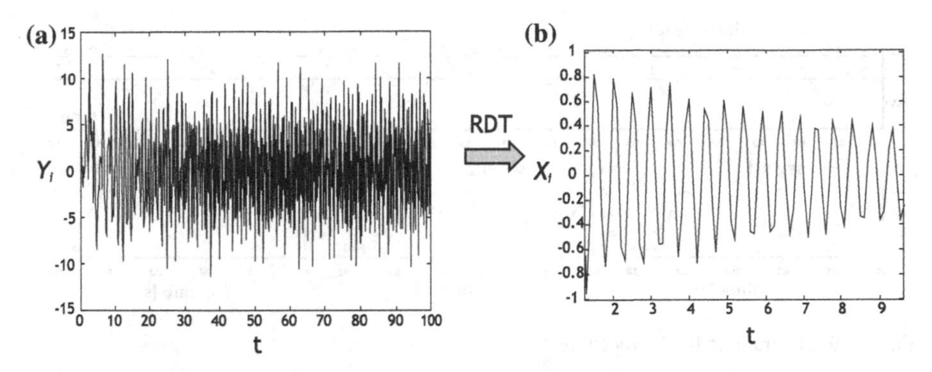

Fig. 8.19 Time histories **(a)** response of mechanical system to stochastic excitation; **(b)** free motion of the system identified with R.D.T

of the system just at the initial conditions (Fig. 8.19). Another effect associated with this averaging operation introduced by the R.D.T. is that of reducing the presence of noise in the actual signal acquired.

Effectively [14] $Y_1(t)$ segments that satisfy a certain trigger condition that determines the start T_1 of each segment considered are extracted (in literature there are several proposals for the choice of such triggers: the most commonly used include those that consider, as a trigger, the passage of the signal measured by zero with positive derivative, exceeding a given threshold value, peak values, etc.)

The R.D.T. technique subsequently foresees a time averaging operation on these segments: since the time histories of the analysed signals are usually sampled at discrete time with intervals of constant time Δt, this operation is described by the following relation:

$$X_i(m\Delta t) = \frac{1}{N}\sum_{i=1}^{N} Y_i(m\Delta t + T_i) \tag{8.43}$$

where N *is the number of means* adopted, $m = 1, 2, \ldots, M$, $T = M\,\Delta t$ being the period of observation of the generic time segment being analysed.

Figure 8.20 shows a diagram of the R.D.T. procedure.

8.3.1 The Ibrahim Method

The method of identification in the time domain described here (suggested by Ibrahim [15]) is a structured approach with modal parameters that analyses the free motion of the system (obtained using one of the methods described previously) and which aims to identify its natural frequencies and vibration modes.

The method is based on reconstructing the free response of a mechanical system obtained as a sum of the contribution of the individual vibration modes expressed in

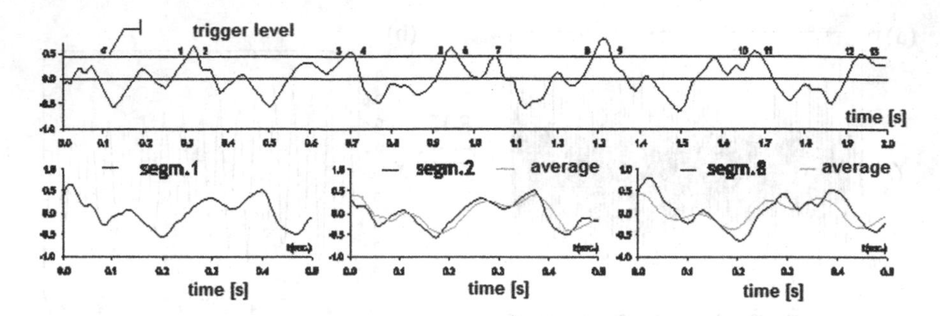

Fig. 8.20 Diagram of RDT procedure

complex terms. The approach assumes a certain number of d.o.f. of the system (number of modes considered in the reconstruction of the response) and considers a certain number of measured experimental time histories, appropriately shifted over time. After a specific set of operations has been carried out on the actual measurements, it is possible to extract the system's unknown modal parameters by tracing the problem back to a problem of eigenvalues-eigenvectors of a matrix containing the same measurements. The method involves identifying the modes that are actually present in the structure using an index of synthesis defined as the *modal confidence factor*.

So, in order to describe and justify Ibrahim's method, we will consider the equation of free motion of a n d.o.f. system:

$$[M]\underline{\ddot{x}} + [R]\underline{\dot{x}} + [K]\underline{x} = 0 \tag{8.44}$$

which, as we know, accepts as general integral:

$$\underline{x} = \sum_{i=1}^{2n} \underline{X}^{(i)} e^{\lambda_i t} \tag{8.45}$$

i.e. a linear combination of vibration modes $\underline{X}^{(i)}$, being λ_i the eigensolutions of (8.44). We assume to have a system to which p sensors have been associated and which provide a signal with frequency $1/\Delta t$, Δt being the sampling interval; $\underline{x}^{(s)}$ is the vector that contains these time signals. We excite the system with a pulse, so that the response is the free motion of the system. The experimental response is discrete both in time and space, as it is only estimated in the p measurement points at certain instants of time $t_k = k\Delta t$ for a total time of $2r\Delta t$ (where $2r$ is the total number of points acquired). The method involves calculating the natural frequencies and vibration modes by reconstructing the free motion of the system, requiring that, at any given moment, the following relation applies:

$$\underline{x}^{(s)}(t_k) = \underline{x}(t_k) \tag{8.46}$$

having used $\underline{x}^{(s)}(t_k)$ to indicate the vector containing the experimental responses at the p measurement points and with $\underline{x}(t_k)$:

$$\underline{x}(t_k) = \sum_{i=1}^{2n} \underline{X}^{(i)} e^{\lambda_i t_k} \quad (k = 1\ldots 2r) \tag{8.47}$$

i.e. the response obtained analytically in terms of natural frequencies and vibration modes. Assuming we have p measurement points equal to n d.o.f. and considering the matrix:

$$\left[X^{(s)}\right] = \left[\underline{X}^{(s)}(t_1)\underline{X}^{(s)}(t_2)\ldots\underline{X}^{(s)}(t_{2r})\right] \tag{8.48}$$

formed by placing in columns the $2r$ vectors $\underline{x}^{(s)}(t_k)$, it is possible to arrange (8.47) into an appropriate matrix form:

$$\begin{aligned} \sum_{i=1}^{2n} \underline{X}^{(i)} e^{\lambda_i t_k} &= \left[\sum_{i=1}^{2n} \underline{X}^{(i)} e^{\lambda_i t_1} \sum_{i=1}^{2n} \underline{X}^{(i)} e^{\lambda_i t_2} \ldots \sum_{i=1}^{2n} \underline{X}^{(i)} e^{\lambda_i t_{2r}} \right] \\ &= \left[\underline{X}^{(1)}\underline{X}^{(2)}\ldots\underline{X}^{(2n)}\right] \begin{bmatrix} e^{\lambda_1 t_1} e^{\lambda_1 t_2} \ldots e^{\lambda_1 t_{2r}} \\ \ldots\ldots\ldots\ldots\ldots\ldots \\ e^{\lambda_{2N} t_1} e^{\lambda_{2N} t_2} \ldots e^{\lambda_{2N} t_{2r}} \end{bmatrix} \end{aligned} \tag{8.49}$$

Thus (8.46) can be conveniently written in the form:

$$\left[X^{(s)}\right] = [X][\Lambda] \tag{8.50}$$

having arranged in the matrix:

$$[X] = \left[\underline{X}^{(1)}\underline{X}^{(2)}\ldots\underline{X}^{(2n)}\right] \tag{8.51}$$

the $2n$ eigenvectors, estimated in p measurement points and being:

$$[\Lambda] = \begin{bmatrix} e^{\lambda_1 t_1} & e^{\lambda_1 t_2} & \ldots & e^{\lambda_1 t_{2r}} \\ \ldots & \ldots & \ldots & \ldots \\ e^{\lambda_{2N} t_1} & e^{\lambda_{2N} t_2} & \ldots & e^{\lambda_{2N} t_{2r}} \end{bmatrix} \tag{8.52}$$

With this equation parity is established between the experimental and analytical responses of the system's free motion. The n d.o.f. system has $2n$ eigenvalues λ_i and just as many vibration modes; as we only have, for the initial hypothesis, $p = n$, measurement points, duplication is carried out over time. The individual experimental time histories are organised in the usual matrix form with the instruction to phase shift them in time by one delay Δt_3. By rewriting (8.46) for a $t_k + \Delta t_3$, we obtain:

$$\underline{x}^{(s)}(t_k + \Delta t_3) = \underline{x}(t_k + \Delta t_3) = \sum_{i=1}^{2n} \underline{X}^{(i)} e^{\lambda_i (t_k + \Delta t_3)} = \sum_{i=1}^{2n} \underline{X}^{(i)} e^{\lambda_i t_k} e^{\lambda_i \Delta t_3} = \sum_{i=1}^{2n} \underline{\overline{X}}^{(i)} e^{\lambda_i t_k} \tag{8.53a}$$

having indicated with:

$$\underline{\overline{X}}^{(s)} = \underline{X}^{(i)} e^{\lambda_i \Delta t_3} \tag{8.53b}$$

This leads to a second system of equations using the same operations described above and which has the form:

$$\left[\overline{X}^{(s)}\right] = [\overline{X}][\Lambda] \tag{8.54}$$

where in $\left[\overline{X}^{(s)}\right]$ the experimental time histories have been organised, shifted by a time interval Δt_3 compared to those shown in (8.48), in $[\overline{X}]$ the products $\underline{X}^{(i)} e^{\lambda_i \Delta t_3}$ have been organised as in (8.51) and the matrix $[\Lambda]$ has the form (8.52) (the symbols are used to identify the quantity measured at time $+\Delta t_3$). By ordering (8.50) and (8.54) in a single matrix equation, we obtain:

$$\left[\Phi^{(s)}\right] \begin{bmatrix} \left[X^{(s)}\right] \\ \left[\overline{X}^{(s)}\right] \end{bmatrix} = \begin{bmatrix} [X] \\ [\overline{X}] \end{bmatrix} [\Lambda] = [X_d][\Lambda] \tag{8.55}$$

The method now involves accompanying (8.55) with a second equation, built in a similar way, with the instruction to time shift the experimental time signals and the analytical response by a time Δt_1. By rewriting the Eqs. (8.47) and (8.53a) for one time $t_k + \Delta t_1$ and $t_k + \Delta t_3 + \Delta t_1$, we obtain:

$$\begin{aligned} \underline{x}^{(s)}(t_k + \Delta t_1) &= \underline{x}(t_k + \Delta t_1) = \sum_{i=1}^{2n} \underline{X}^{(i)} e^{\lambda_i (t_k + \Delta t_1)} \\ &= \sum_{i=1}^{2n} \underline{X}^{(i)} e^{\lambda_i t_k} e^{\lambda_i \Delta t_1} \underline{x}^{(s)}(t_k + \Delta t_3 + \Delta t_1) \\ &= \sum_{i=1}^{2n} \underline{X}^{(i)} e^{\lambda_i (t_k + \Delta t_3 + \Delta t_1)} = \sum_{i=1}^{2n} \underline{X}^{(i)} e^{\lambda_i t_k} e^{\lambda_i \Delta t_1} \end{aligned} \tag{8.56}$$

By organising in a single equation, as already done in (8.55), and identifying with $\hat{X}$ the terms relative to the phase shift of Δt_1, we obtain:

$$\left[\hat{\Phi}^{(s)}\right] = \begin{bmatrix} \left[\hat{X}^{(s)}\right] \\ \left[\hat{\bar{X}}^{(s)}\right] \end{bmatrix} = \begin{bmatrix} [X] \\ [\bar{X}] \end{bmatrix} [\alpha][\Lambda] = [X_d][\alpha][\Lambda] \tag{8.57}$$

where the appropriately time shifted experimental time histories estimated at the measurement points were organised in $\left[\hat{X}^{(s)}\right]$ and $\left[\hat{\bar{X}}^{(s)}\right]$, the modes of vibration in $[X]$ (8.51) and the same multiplied by the term $e^{\lambda_i \Delta t_3}$ in $\left[\overline{X}\right]$ (8.53b). Lastly, time shift Δt_1 is taken into account in the diagonal matrix $[\alpha]$:

$$[\alpha] = \begin{bmatrix} e^{\lambda_1 \Delta t_1} & 0\ldots0 \\ \ldots & \\ 00 & \ldots e^{\lambda_{2N} \Delta t_1} \end{bmatrix} \tag{8.58}$$

By putting (8.61) and (8.63a, 8.63b) in a system we obtain:

$$\begin{aligned} \left[\Phi^{(s)}\right] &= [X_d][\Lambda] \\ \left[\hat{\Phi}^{(s)}\right] &= [X_d][\alpha][\Lambda] \end{aligned} \tag{8.59}$$

Assuming that $2p = 2r$ demonstrates [14, 15] that the matrix $\left[\Phi^{(s)}\right]$ is square and invertible: obtaining the matrix $[\Lambda]$ from the first equation of (8.59) and replacing it in the second we obtain:

$$\left[\hat{\Phi}^{(s)}\right]\left[\hat{\Phi}^{(s)}\right]^{-1}[X_d] = [X_d][\alpha] \tag{8.60a}$$

where the known experimental histories are organised in matrices $\left[\Phi^{(s)}\right]$ and $\left[\hat{\Phi}^{(s)}\right]$, and matrices $[X_d]$ and $[\alpha]$ contain the unknown eigenvectors and eigenvalues of the system. $[A]$ indicates the matrix containing the appropriately memorised measurements:

$$[A]\left[\hat{\Phi}^{(s)}\right]\left[\hat{\Phi}^{(s)}\right]^{-1}. \tag{8.60b}$$

(8.60a) becomes:

$$[A][X_d] = [X_d][\alpha] \tag{8.60c}$$

(8.60c) can be written as:

$$\left[[A]\underline{X}_d^{(1)}[A]\underline{X}_d^{(2)}\ldots[A]\underline{X}_d^{(2n)}\right] = \left[\underline{X}_d^{(1)}\underline{X}_d^{(2)}\ldots\underline{X}_d^{(2n)}\right]\begin{bmatrix} e^{\lambda_1\Delta t_1} & \ldots & 0 \\ \ldots & \ldots & \ldots \\ 0 & \ldots & e^{\lambda_{2N}\Delta t_1} \end{bmatrix} \tag{8.61}$$

where $\underline{X}_d^{(i)}$ is the generic column of the matrix $[X_d]$ (8.57):

$$\underline{X}_d^{(i)} = \begin{Bmatrix} \underline{X}^{(i)} \\ \underline{\dot{X}}^{(i)} \end{Bmatrix} \tag{8.62}$$

We now develop the matrix product to the second member of (8.61):

$$\left[[A]\underline{X}_d^{(1)}[A]\underline{X}_d^{(2)}\ldots[A]\underline{X}_d^{(2N)}\right] = \left[\underline{X}_d^{(1)}e^{\lambda_1\Delta t_1}\underline{X}_d^{(2)}e^{\lambda_2\Delta t_1}\ldots\underline{X}_d^{(2N)}e^{\lambda_{2N}\Delta t_1}\right] \tag{8.63a}$$

Considering the single column of (8.63a) we obtain:

$$[A]\underline{X}_d^{(i)} = \underline{X}_d^{(i)}e^{\lambda_i\Delta t_1} = \alpha_i\underline{X}_d^{(i)} \tag{8.63b}$$

where α_i is the ith element of the diagonal of the matrix $[\alpha]$. Equation (8.63b) is the equation of an eigenvalue problem whose solution allows us to obtain the eigensolutions α_i and eigenvectors $\underline{X}_d^{(i)}$. The eigensolutions α_i are linked to the eigenvalues λ_i of the original motion Eq. (8.44) by the relation:

$$\alpha_i = e^{\lambda_i\Delta t_1} \tag{8.64a}$$

Therefore, the generic eigenvalue λ_i will be:

$$\lambda_i = a_i + ib_i \tag{8.64b}$$

with:

$$b_i = \frac{1}{\Delta t_1}\left[\tan^{-1}\left(\frac{\gamma_i}{\beta_i}\right)\right] \tag{8.64c}$$

$$a_i = \frac{1}{2\Delta t_1}\ln(\gamma_i^2 + \beta_i^2) \tag{8.64d}$$

b_i and γ_i being the real and imaginary part of α_i (8.64a). The approach described is of a deterministic nature as it is assumed that we have $2r$ acquired points equal to p points of measurement making it possible to have the square and invertible matrix

$[\Phi]$ (8.60a). In general this does not occur and it is shown [14, 15], using a method of minimisation, that the solution can be obtained from an eigenvalue problem of a new matrix $[A]$ defined as:

$$[A] = \left[\left[\hat{\Phi}^{(s)}\right]\left[\hat{\Phi}^{(s)}\right]^T\right]\left[\left[\Phi^{(s)}\right]\left[\Phi^{(s)}\right]^T\right]^{-1} \tag{8.65}$$

The initial hypothesis, of p points of measurement equal to the number n of system modes is, in the case of a real system, unacceptable. If p is less than n, a new set of pseudo-measurements is created by time shifting the experimental time signals by an arbitraryΔt_2. Equation (8.46) becomes:

$$\begin{aligned} \underline{x}^{(s)}(t_k + \Delta t_2) = \underline{x}(t_k + \Delta t_2) &= \sum_{i=1}^{2n} \underline{X}^{(i)} e^{\lambda_i (t_k + \Delta t_2)} \\ &= \sum_{i=1}^{2n} \underline{X}^{(i)} e^{\lambda_i t_k} e^{\lambda_i \Delta t_2} = \sum_{i=1}^{2n} \underline{X}'^{(i)} e^{\lambda_i t_k} \end{aligned} \tag{8.66}$$

(the superscript will be used to indicate the quantities relating to the latter time delay Δt_2). (8.66) is written in the form (see (8.50)):

$$\left[X'^{(s)}\right] = [X'][\Lambda] \tag{8.67a}$$

having indicated with:

$$[X'] = \left[\underline{X}'^{(1)} \underline{X}'^{(2)} \ldots \underline{X}'^{(2N)}\right] \tag{8.67b}$$

$$[\Lambda] = \begin{bmatrix} e^{\lambda_1 t_1} & e^{\lambda_1 t_2} & \ldots & e^{\lambda_1 t_{2r}} \\ \ldots & \ldots & \ldots & \ldots \\ e^{\lambda_{2N} t_1} & e^{\lambda_{2N} t_2} & \ldots & e^{\lambda_{2N} t_{2r}} \end{bmatrix} \tag{8.67c}$$

The time shifts Δt_2 are repeated until a number of measurements (real or pseudo-measurements) greater or equal to the number of d.o.f. of the system are achieved. The procedure is strictly that already described:

- with the time shift Δt_2 we can obtain a real or fictitious number of sensors that is greater than n and write the Eq. (8.67a) in the form:

$$\begin{bmatrix} \left[X^{(s)}\right] \\ \left[X'^{(s)}\right] \end{bmatrix} = \begin{bmatrix} [X] \\ [X'] \end{bmatrix} [\Lambda] \tag{8.68}$$

- The time histories obtained in this manner are out of phase by an arbitrary Δt_3 and organised with (8.55) in a single system:

$$\left[\hat{\Phi}'^{(s)}\right] = \begin{bmatrix} \left[X^{(s)}\right] \\ \left[X'^{(s)}\right] \\ \left[\bar{X}^{(s)}\right] \\ \left[\bar{X}'^{(s)}\right] \end{bmatrix} = \begin{bmatrix} [X] \\ [X'] \\ [\bar{X}] \\ [\bar{X}] \end{bmatrix} [\Lambda] = [X_{d'}][\Lambda] \tag{8.69}$$

- A second equation, estimated in a time increased by Δt_1 is added (8.57):

$$\left[\hat{\Phi}'^{(s)}\right] = [X_{d'}][\alpha][\Lambda] \tag{8.70}$$

- The solution is reduced again to an eigenvalue problem with a matrix $[A]$ built as in (8.66) for a deterministic approach or in (8.65) using a procedure of minimisation depending on the number of points acquired.

The method must reconstruct a signal via the summation (8.45): if this signal is particularly disturbed it is not possible to establish equality (8.46) by taking into account just the n presumed vibration modes. Therefore, the meaning of n loses importance as more ways are needed to take into account the presence of noise and, furthermore, in general the mechanical system is a continuous system. So, the eigenvalues and eigenvectors, real and related to noise (this describes the modes that are indispensable for reconstructing the signal due to the presence of noise) become $2m$ (of which only $2n$ will be those associated with the actual vibration modes). The method does not distinguish between real and noise related modes and proceeds by calculating $2m$ eigenvalues and eigenvectors exactly as described above. In order to identify the real vibration modes of the system, it is necessary to have an index of goodness of fit. The model used is the *Modal Confidence Factor* (M.C.F.). Remembering (8.62):

$$\underline{X}_d^{(i)} = \begin{Bmatrix} \underline{X}^{(i)} \\ \underline{\bar{X}}^{(i)} \end{Bmatrix} \tag{8.71}$$

the method calculates the eigenvectors $\underline{X}_d^{(i)}$ that contain, in the first m positions, the vibration modes (real or noise) of the system in question and in the subsequent positions the same modes multiplied by the factor $e^{\lambda_i \Delta t_3}$. The control is performed by exploiting this very characteristic: we calculate the complex ratio (M.C.F.) between the identified eigenvector $\underline{X}^{(i)}$ (first m positions of the vector $\underline{X}_d^{(i)}$) multiplied by $e^{\lambda_i \Delta t_3}$ and that identified in the second m positions.

$$(M.C.F.)_j^{(i)} = \frac{X_j^{(i)} e^{\lambda_i \Delta t 3}}{\bar{X}_j^{(i)}} 100 \qquad (8.72)$$

having indicated X_j the jth element of the vector $\underline{X}^{(i)}$. The magnitude and phase of the ratio $(M.C.F.)_j$ (for each component of each eigenvector) are calculated: thus the vibration modes for which the $M.C.F.$ has magnitude equal to 100 and zero phase will be taken as the real vibration modes of the system.

8.3.1.1 An Example

We analysed the 4 d.o.f. system illustrated in Fig. 8.19, of which the inertial, elastic and viscous characteristics are known.

While giving very general initial conditions, we estimated the displacement of the four masses shown in Fig. 8.21 and this was taken as the experimental response (Fig. 8.22).

Fig. 8.21 4 d.o.f. system

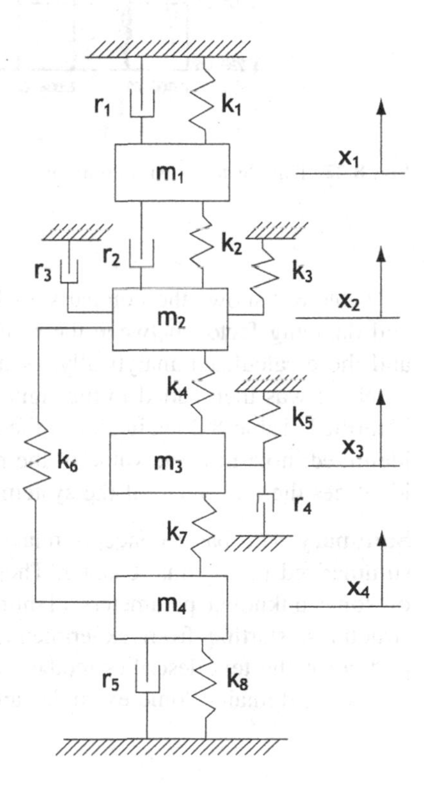

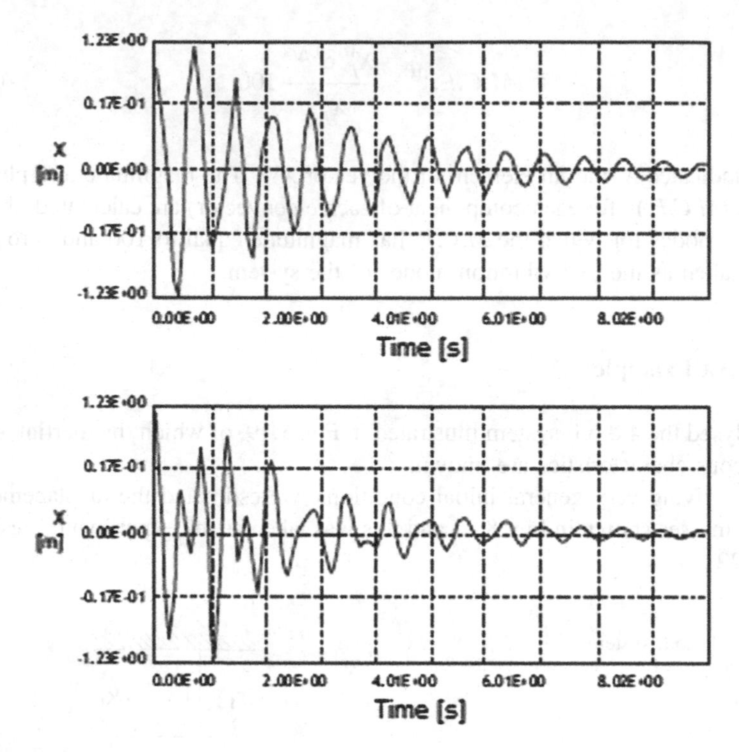

Fig. 8.22 Time histories with no noise

Table 8.1 shows the comparisons between natural pulsations, vibration modes and damping factor, between the results of the analyses with the Ibrahim method and those calculated analytically, assumed to be *accurate*.

Noise was then added to the signal (Fig. 8.23) and the modal parameters were identified. Table 8.2 again shows the accurate vibration modes compared to those identified: note how the value of the magnitude and phase of the *M.C.F.* uniquely identifies the real mode of the system.

Summary The basic concepts related to parameter identification techniques are summarized in this final chapter. These methodologies allow for the identification of some unknown parameters of numerical simulation models of machines and structures, starting from experimental tests on prototypes or real structures. In particular, the text describes modal identification techniques, both in the time and frequency domain. Some examples are given.

Table 8.1 Natural frequencies and vibration modes—calculated (**a**) and identified (**b**)

Mode I		**Mode II**	**(a)**
Alpha	Omega	Alpha	Omega
−0.30	8.87	−0.56	14.6
Eigenvector		Eigenvector	
Magnitude	Magnitude	Magnitude	Phase
1.00	−80°	0.44	120°
0.42	−80°	1.00	−60°
0.04	−80°	0.31	−60°
0.02	−80°	0.14	−60°
Mode I			**(b)**
Alpha	Omega		
−0.30	8.87		
Eigenvector			
Magnitude	Phase	MCF magnitude	MCF phase
1.00	−90°	99.9	1.40E−5
0.42	−90°	100.0	4.45E−5
0.04	−90°	99.9	−1.40E−4
0.02	−90°	100.0	−2.70E−4
Mode II			**(b)**
Alpha	Omega		
−0.56	14.6		
Eigenvector			
Magnitude	Phase	Magnitude	Phase
0.48	90°	96.1	1.39
1.0	−90°	100.8	−0.21
0.32	−90°	100.3	−0.06
0.14	−90°	100.3	−0.07

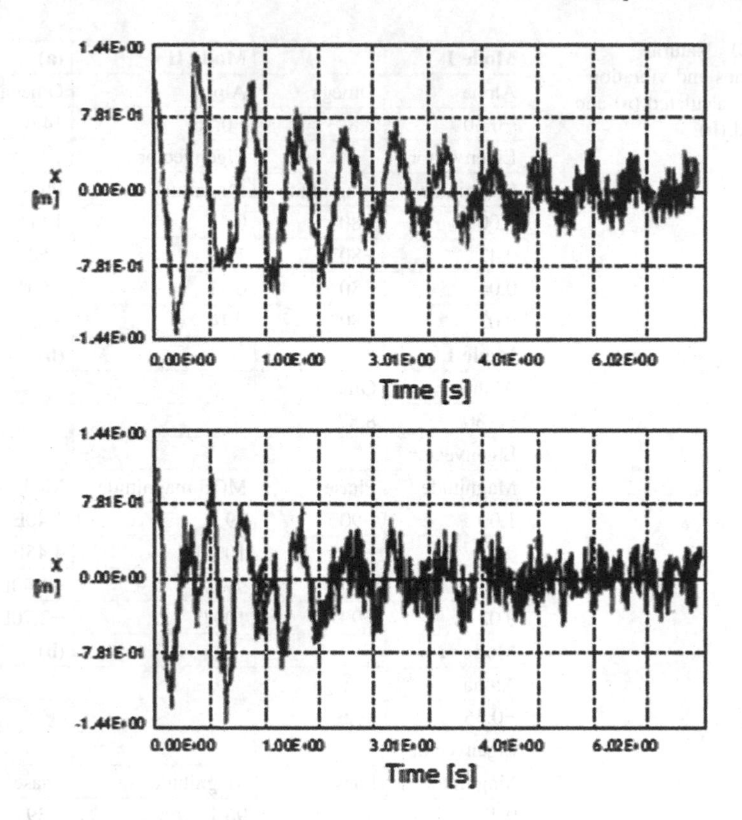

Fig. 8.23 Time histories with noise

Table 8.2 Natural frequencies and vibration modes—calculated **(a)**, identified **(b)**

Mode I		**Mode II**	**(a)**
Alpha	Omega	Alpha	Omega
−0.30	8.87	−0.56	14.6
Eigenvector		Eigenvector	
Magnitude	Phase	Magnitude	Phase
1.00	−80°	0.44	120°
0.42	−80°	1.00	−60°
0.04	−80°	0.31	−60°
0.02	−80°	0.14	−60°
Mode I			**(b)**
Alpha	Omega		
−0.35	8.84		
Eigenvector			
Magnitude	Phase	MCF magnitude	MCF phase
1.00	−10°	100.7	−1.00E−1

(continued)

Table 8.2 (continued)

Mode I		**Mode II**	**(a)**
0.41	−10°	101.5	1.20E−1
0.06	0°	117.2	-8.90E−0
Mode II			**(b)**
Alpha	Omega		
−0.56	14.6		
Eigenvector			
Magnitude	Phase	MCF magnitude	MCF phase
0.42	140°	99.3	0.63
1.0	−40°	102.0	0.17
0.27	−20°	89.6	−3.40
0.11	20°	102.8	−0.68
Mode III			
Alpha	Omega		
−4.7	10.4		
Eigenvector			
Magnitude	Phase	MCF magnitude	MCF phase
0.42	140°	4.8	163.0
0.38	−40°	54.5	104.4
0.93	20°	118.8	−158.9
0.81	−140°	72.5	85.57

References

1. Various Authors (1992) Identificazione Strutturale: Metodi Dinamici e Diagnostica. Seminario ISMES, Bergamo
2. Bendat JS, Piersol AG (1966) Measurement and analysis of random data. Wiley, New York
3. Billings SA, Chen S (1991) Modelling and analysis of non-linear time series. Int J Control
4. Billings SA (1985) Input-output parametric models for non-linear systems. Int J Control
5. Bittanti S, Schiavoni N (1982) Modellistica e controllo. Clup, Milano
6. Bittanti S, Guardabassi G (1988) Sistemi incerti. Clup, Milano
7. Bittanti S (1988) Identificazione parametrica. Clup, Milano
8. Curami A, Vania A (1991) An application of modal analysis techniques. Report del Dipartimento di Meccanica del Politecnico di Milano
9. Diana G, Cheli F, Vania A (1988) A method to define the foundation modal parameters through the measurements of the rotor vibrations. In: IV international conference on vibrations in rotating machinery, Edinburgh
10. Ewins DJ (2000) Modal testing theory and practice. Research Studies Press
11. Ewins DJ (1986) Modal Testing Theory and Practice. Bruel and Kjaer
12. Fritzen CP (1985) Identification of mass, damping and stiffness matrices of mechanical system. In: 10th ASME conference on mechanical vibration and noise, Cincinnati OH, paper 85 DET 91
13. Garibaldi L, Fasana A, Giorcelli E, Riva R (1991) Reti neurali per l'identificazione di sistemi meccanici. ISMES

14. Ibrahim SR, Pappa R (1982) Large modal survey testing using the ibrahim time domain identification technique. J Spacecr
15. Ibrahim SR, Mikulcik EC (1973) A time domain modal vibration test technique. Shock Vibrat Bull
16. Jazwinski AH (1970) Stochastic process and filtering theory. Academic Press, London
17. Kalman RE (1960) A new approach to linear filtering and prediction problems. J Basic Eng 35–43
18. Masri SF, Chassiokos AG, Chaughey TK (1988) Identification of non-linear dynamic systems using neural networks. J Appl Mech
19. Mottershead JE, Stanway R (1986) A comparison of time and frequency domain filters for parameter identification of vibrating systems. In: Proceedings of the international conference on vibration problems in engineering Xi'an China, pp 580–585
20. Natke HG (1988) A brief review on the identification on non linear systems. In: Proceedings of the 6th international modal analysis conference, Kissimmee Fl, Schenectady. pp 1569–1574
21. Natke HG (1988) Application of system identification in engineering. Springer, Berlin
22. Proceeding of the 9th IFAC/IFORS symposium on identification and systems parameter estimation, Budapest, Hungarian, July 1991
23. Proceeding of the IFAC symposium on design methods of control systems, Zurich, Switzerland, Sept 1991
24. Proceeding of the IMACS world congress, Dublin, Ireland, July 1991
25. Proceedings of the 10th international conference on modal analysis, Union College, S.E.M., San Diego, CA, 1992
26. Otnes RK, Enochson L (1982) Digital time series analysis. Wiley, New York
27. Randall RB (1988) Application of Bruel Kjaer equipment to frequency analysis. Research Studies Press LTD, Bruel Kjaer, Letchworth, Hertfordshire
28. Sorenson HW (1985) Kalman filtering theory and applications. IEEE Press, New York
29. Tomlinson GR (1987) Vibration analysis and identification of non linear systems, short course. Herriot-Watt University, Edinburgh
30. Young P (1984) Recursive estimation and time series analysis. Springer, Berlin

Printed in the United States
By Bookmasters